International Conference on Computer Applications in Shipbuilding (ICCAS 2017)

AA005474

Singapore
26 - 28 September 2017

Volume 1 of 3

ISBN: 978-1-5108-8303-1

Printed from e-media with permission by:

Curran Associates, Inc.
57 Morehouse Lane
Red Hook, NY 12571

Some format issues inherent in the e-media version may also appear in this print version.

Copyright© (2017) by The Royal Institution of Naval Architects
All rights reserved.

Printed by Curran Associates, Inc. (2019)

For permission requests, please contact The Royal Institution of Naval Architects
at the address below.

The Royal Institution of Naval Architects
8-9 Northumberland Street
London, WC2N 5DA
United Kingdom

Phone: 44 207 235 4622
Fax: 44 207 259 5912

publications@rina.org.uk

Additional copies of this publication are available from:

Curran Associates, Inc.
57 Morehouse Lane
Red Hook, NY 12571 USA
Phone: 845-758-0400
Fax: 845-758-2633
Email: curran@proceedings.com
Web: www.proceedings.com

CONTENTS

© 2017: The Royal Institution of Naval Architects

© 2017: The Royal Institution of Naval Architects

Vehicle) Based On ROS (Robot Operating System)

H W Lee, M I Roh, L Zhao, S H Ham and N Kim, Seoul National University, Republic of Korea
C W Yu, Agency for Defense Development, Republic of Korea

© 2017: The Royal Institution of Naval Architects

© 2017: The Royal Institution of Naval Architects

International Conference on Computer Applications in Shipbuilding (ICCAS 2017)

26-28 September 2017, Singapore

Volume 1

SHIPBUILDING
ICCAS
SINGAPORE | 2017

Sponsored by:

THE ROYAL INSTITUTION OF NAVAL ARCHITECTS
8-9 Northumberland Street, London, WC2N 5DA, Telephone: +44 (0)20 7235 4622

AUTHOR'S CONTACT DETAILS

Hogyun Park
Korean Register
hgpark@krs.co.kr

David Card
DNV GL
dnncard60@gmail.com

Aldo Zini
CETENA S.p.A.
aldo.zini@cetena.it

Bernard FERRIER
Hoffman Engineering
bferrier@hoffmanengineering.com

Anish S
Cochin Shipyard Ltd.
anish.s@cochinshipyard.com

Gordan Sikic
USCS d.o.o.
gordan.sikic@uscs.hr

Alexandre TEW KAI
Dassault Systemes
Stephanie.FOURNIER@3ds.com

Jan D. Wilgenhof
MecDes
jdwilg@mecdes.nl

Michael J. Sypniewski
The University of Michigan
mjsyp@umich.edu

Kota Okada
the University of Tokyo
kokada@s.h.k.u-tokyo.ac.jp

Rodrigo Perez
SENER
rodrigo.fernandez@sener.es

John L Martin
SAMOSC Ltd.
samosc@btinternet.com

Qingsong Zeng
Delft University of Technology
Q.Zeng@tudelft.nl

Alfonso Cebollero
SENER
alfonso.cebollero@sener.es

Jin-Young Park
Korean Register
parkjy@krs.co.kr

Jacqueline Christmas
University of Exeter
J.T.Christmas@exeter.ac.uk

Mustafa Al-Ani
University of Exeter
J.T.Christmas@exeter.ac.uk

Sangeet Sankaramangalam Ulhas
Cochin University of Science and Technology
sangeetsu@gmail.com

D Morais
SSI
denis.morais@ssi-corporate.com

Kohei Matsuo
National Maritime Research Institute
kohei@nmri.go.jp

© 2017: The Royal Institution of Naval Architects

Seong-Hoon Kim
Seoul National University
ansls102@snu.ac.kr

Seung-Min Lee
Seoul National University
digimin100@snu.ac.kr

Mohammad Danil Arifin ST.MT
Hiroshima University
d145238@hiroshima-u.ac.jp

Hajime Kimura
Kyushu University
kimura@nams.kyushu-u.ac.jp

Taiga Mitsuyuki
University of Tokyo
mitsuyuki@sys.t.u-tokyo.ac.jp

Jinsup Cheong
Samsung Heavy Industries
michael.abbruscato@intergraph.com

Conner Goodrum
The University of Michigan
cgoodrum@umich.edu

Kazuhiro Aoyama
The University of Tokyo
aoyama@sys.t.u-tokyo.ac.jp

Kodungallur Sivaprasad
Cochin University of Science and Technology
yashwa36@yahoo.co.in

Hongyu Cui
Dalian University of Technology
cuihongyu@dlut.edu.cn

Robert Bronsart
University of Rostock
robert.bronsart@uni-rostock.de

Jesús A. Muñoz
SENER
jesus.munoz@sener.es

Matthias Grau
PROSTEP
Matthias.Grau@prostep.com

Hye-Won Lee
Seoul National University
gpdnjs0215@snu.ac.kr

Stalin Ybiernas
Keppel Singmarine
michael.abbruscato@intergraph.com

Kimmo Salmi
Intergraph PP&M
michael.abbruscato@intergraph.com

Myeong-jo Son
Korean Register
mjson@krs.co.kr

Jose Esteve
Bureau Veritas Marine & Offshore
JOSE.ESTEVE@BUREAUVERITAS.COM

Kunihiro Hamada
Hiroshima University
hamada@naoe.hiroshima-u.ac.jp

Wang Ji
Dalian University of Technology
wangji@dlut.edu.cn

© 2017: The Royal Institution of Naval Architects

Christophe CHAUVIERE
Bureau Veritas
christophe.chauviere@bureauveritas.com

Hannes Lindner
University of Rostock
hannes.lindner@uni-rostock.de

Mikko Yllikäinen
CADMATIC
sander.kuik@cadmatic.com

Sebastian H. Greshake1
University of Rostock
sebastian.greshake@uni-rostock.de

Rogero Fernando
Patel Micro Data
rogerofdo@gmail.com

Taisuke Kunisada
Sanoyas Shipbuilding Corporation
t-kunisada@sanoyas.co.jp

Nikolaos Kouriampalis
University College London
n.kouriampalis.11@ucl.ac.uk

Mingyu Kim
SAMSUNG Heavy industries
mgyu.kim@samsung.com

Fairuz Aledroos
Nakilat
faledroos@qgtc.com.qa

Sri Harish Kalidass
University Of Plymouth
srihari.kalidass@bbsr.co.uk

Reinhard Ahlers
BALance Technology Consulting GmbH
reinhard.ahlers@bal.eu

Yujiro Wada
Hiroshima University
wada-y@nmri.go.jp

Ludmila Seppälä
CADMATIC
sander.kuik@cadmatic.com

Peiyuan Feng
Marine Design & Research Institute of China (MARIC)
pyfeng23@163.com

Joo Hock ANG
Sembcorp Marine Ltd
joohock.ang@sembmarine.com

ChunWee Ng
Newcastle University
C.W.Ng2@newcastle.ac.uk

Zulfah Zikrina
Universitas Indonesia
zulfah.zikrina@ui.ac.id

Minseok Kang
KAIST
gnk2000@kaist.ac.kr

Roope Kotiranta
Surma Ltd
roope.kotiranta@surma.ltd

Antonios V. Lalechos
LePlan
lalechos@leplan.gr

© 2017: The Royal Institution of Naval Architects

Winston Pynn
Marine Institute of Memorial University
Winston.Pynn@mi.mun.ca

Dimitrios Konovessis
Singapore Institute of Technology
Dimitrios.Konovessis@SingaporeTech.edu.sg

Tapio Hulkkonen
NAPA ltd.
tapio.hulkkonen@napa.fi

Torben-H. Stachowski
Digitread
torben.stachowski@digitread.com

LR Wang
China Classification Society
lrwang@ccs.org.cn

Molly Heskitt
TSUNEISHI SHIPBUIDING CO., Ltd.
Lmh5@altair.com

S W Ferguson
Siemens Industry Software
christine.goh@siemens.com

Jesús A. Muñoz
SENER
jesus.munoz@sener.es

Ole Christian Astrup
DNV GL
ole.christian.astrup@dnvgl.com

Mohamed Zeid
Caterpillar Propulsion
zeid_mohamed@cat.com

Prabjot Singh Chopra
SeaTech Solutions International (s) Pte. Ltd.
Marketing@seatechsolutions.com

Dorian C. Brefort
University of Michigan
doriancb@umich.edu

RAHUL KAMILLA
ITT Kharagpur
rahulkam93@gmail.com

Jiangchao Wang
Huazhong University of Science and Technology
WangJiangchaocn@gmail.com

Joseph Baumer
Intergraph PP&M
michael.abbruscato@intergraph.com

Marcel Veldhuizen
Intergraph PP&M
michael.abbruscato@intergraph.com

Ken Goh
Knud E Hansen Australia
keg@knudehansen.com

Damir J. Kolich
University of Rijeka
dkolic@riteh.hr

Zhu Shengqing
GLB
sinozsq@gmail.com

Jiayi He
Marine Design and Research Institute of China
hejiayifans@gmail.com

© 2017: The Royal Institution of Naval Architects

A FRAMEWORK FOR ASSESSING THE EFFECTS OF BIAS IN EARLY-STAGE SHIP DESIGN

M J Sypniewski and **D J Singer**, University of Michigan, USA

SUMMARY

Bias can influence the design process by impacting the design decisions that are made. While in some cases this bias is beneficial, there are many instances where bias negatively affects the design process and causes complete design failure. Designers require leading or real-time indicators to respond quickly and prevent bias from propagating. This paper presents a framework for studying the effects of bias by investigating the structure of the temporal design process. To provide a comparison, biased and nonbiased datasets are generated using an adapted genetic algorithm. The design process datasets are then represented in network space and analysed by out-degree and closeness centrality metrics. A case study is presented using the Sen Bulker ship synthesis model to demonstrate the framework and measures of bias. In conclusion, the network metrics were found to successfully differentiate between the biased and nonbiased datasets within the scope of the case study.

1. INTRODUCTION

The effects of bias can impact design decisions and thereby influence the design process [1, 2]. While in some cases this bias is beneficial, there are other instances in which it causes complete design failure [3, 4]. Even when failure is not experienced, bias can cause inefficiencies that consume valuable design resources [5, 6]. For these reasons, it is important to be able to identify and address bias within the design process.

Engineering design continues to utilize an increasing number of simulation and synthesis methods. Due to embedded assumptions and estimations, each of these methods presents an opportunity for bias to influence the design process. In Watson and Gilfillan's seminal paper on ship design, a transcribed discussion critiqued the method in following way:

> *"At the technical level the paper has main sections devoted to dimensions, displacement, form, powering, and so on, but nowhere is there even a sub-heading for ship strength, much less structural materials."* [7]

In [8], Parker analyses the design structure of Watson and Gilfillan's method and identifies that the ship's structural disciplines have minimal influence on the design space. This aspect of Watson and Gilfillan's method provides an example of how bias can affect the decisions and activities within a design. A method is needed to identify these biases and assess their impacts on a design.

This paper presents a framework for studying the effects of bias by investigating the structure of temporal design data. The paper is organized as follows: Section 2 presents a method for generating biased and nonbiased temporal design data. Following this, Section 3 details a design process representation that enables an analysis of design relationships. Section 4 proposes two network metrics for investigating the design datasets. Then Section 5 presents a case study using the Sen Bulker ship synthesis model to demonstrate the framework and measures of bias. Finally, conclusions and opportunities for future work are discussed.

2. GENERATING TEMPORAL DESIGN DATA

This research aims to investigate bias by analysing the structure of design relationships through time. A genetic algorithm (GA) is modified to imitate a design activity and generate the design data to be analysed. GAs have been used for this purpose within the engineering design research community [9, 10].

The data generation approach is independent of a specific design problem, and therefore the created datasets share the same design space. By this property, the GA may be modified without affecting the design space of the created dataset. This preservation of the design space enables comparisons of datasets that are created using the same design problem.

Modifying the GA provides a method for introducing bias into the design process. Given a design problem, both biased and nonbiased datasets can be created. Since the datasets share the same design space, they are comparable. This allows bias to be controlled for, and its effects on the process can be investigated.

This section of the paper details our approach for generating these design datasets. A brief overview of GAs is presented followed by an explanation of the mechanisms used to induce bias. For additional information on GAs, see [11].

2.1 GENETIC ALGORITHMS

Genetic algorithms are a family of biomimetic, stochastic search algorithms. Until its stopping condition is met, a GA iteratively applies operators to a set of individuals, called the *population*. These individuals are called *chromosomes* and are composed of *genes* which represent a single dimension of the design space. The

© 2017: The Royal Institution of Naval Architects

chromosome's list of genes allows it to represent a fully defined point within the design space [11, 12].

The operators defined in a traditional GA are crossover, mutation, and selection. Crossover operates on two chromosomes within the population and generates offspring by combining both chromosomes' features. In this process, the reproducing chromosomes and the newly created chromosomes are referred to as parents and children, respectively. The probability that a chromosome undergoes crossover is a hyperparameter of the GA called the *crossover probability*. In design, crossover represents the way designers combine elements of different alternatives to generate new ideas. After crossover, the children are added to the current population. Mutation operates on a single chromosome and randomly alters one of its genes. Like in crossover, the probability that a chromosome mutates is another hyperparameter of the GA called the *mutation probability*. In design, mutation represents innovation and revolutionary ideas. Finally, selection is the process by which the population of chromosomes is reduced to its original size. Selection occurs at the end of every iteration, or *epoch*. In design, selection represents the way designers choose which set of alternatives should be analysed further and which are rejected.

Many implementations of these operators exist, but only the methods used in the case study are discussed in this paper. The case study does not implement a mutation operator, but its crossover and selection processes are explained below:

2.1 (a) Two-point Crossover

The case study uses two-point crossover as its crossover operator. When two chromosomes are chosen for crossover using two-point crossover, two random cut-points are made in the parents' gene list. From these cut-points, two children are created by combining the middle segment of genes from one parent with the end segments from the other.

2.1 (b) Tournament Selection

The case study uses tournament selection as its selection operator. Tournament selection holds a tournament among a few randomly chosen chromosomes from the population. The winner of this tournament is determined by a ranking function and added to the new population. This process continues until the population's original size has been restored [12].

The *tournament size* refers to the number of chromosomes that engage in each tournament. As the tournament size increases, the ranking function gains more influence on the selection process. A tournament size of 1 defines a random search; therefore, a tournament size of 2 defines a guided search having the least amount of influence from the ranking function.

2.2 GENERATING DATASETS

One of the goals of this research is providing a method to differentiate between biased and nonbiased processes. To facilitate this goal, the dataset generation needs to be able to produce both datasets. This process is detailed below.

2.2 (a) Inducing Bias

The crossover and selection operators of a GA provide an opportunity to introduce bias within the process. In the case study, bias is induced by altering the ranking function within the tournament selection process. The ranking function represents the way which a subset of designs is compared. In a traditional GA, this ranking function is defined by the problem's objective function, and the nonbiased datasets are generated using this approach. The data generator introduces bias by providing an alternative ranking function. In the case study, length to beam ratio is used to define the ranking function for creating the biased datasets.

2.2 (b) Tracking Unique Identifiers

In addition to any bias-driven modifications, the GA alters its chromosomes by enabling them to store more information about the design process. Chromosomes are given unique identifiers, and each child chromosome records its parents' identifiers. This is done so that the structure of design data can be constructed using this information.

This study frequently references two sets of chromosomes within a dataset: the set of all created chromosome and the set of Pareto front chromosomes. For convenience, the set of all created chromosomes will be referred to as the *superset*, and the set of Pareto front chromosomes will be referred to as the *Pareto-set*.

3. REPRESENTING TEMPORAL DESIGN RELATIONSHIPS

After creating a dataset, its temporal design relationships need to be represented. Network theory provides a platform for representing these relationships, and it defines metrics for analysing this structure. This section reviews basic network concepts and presents two networks to represent the temporal relationships found in design.

In its simplest form, a network is a collection of points joined together by lines. These points are commonly referred to as *nodes* or *vertices*, and the lines are referred to as *edges*. Networks can be further defined as being either directed or undirected. In directed networks, each edge has a direction: pointing from one node and to another. In undirected graphs, no edge directionality is specified and edges just signify a bi-directional relationship between the nodes it connects. A *path* is a route across the network that runs from nod to node

© 2017: The Royal Institution of Naval Architects

along the edges of the network. A *component* is a subset of nodes within a network such that there exists at least one path from each member of that subset to each other member of that subset [13]. Using the basic network elements of nodes and edges, the relationships within temporal design data can be represented.

This study defines and investigates two major networks. First is a directed network of chromosomal parent-child (P-C) relationships. In this network, each chromosome in the superset is represented as a node, and an edge pointing toward itself from each of its parents represents the relationship. Second is an undirected network of chromosomal parent-parent (P-P) relationships. For each chromosome in the superset, its parents are represented as nodes, and an edge is created between them.

4. IDENTIFYING BIAS

Using a GA to represent the design process, a source for generating design data has been established. Bias is introduced into the design process by altering the tournament selection's ranking function within the GA. This section proposes two metrics to investigate the effects of bias while using the proposed representations of design process.

Network theory provides a number of metrics for analysing the structure of a network. These metrics can provide insight to the network when applied appropriately. However, the applicability of these metrics is dependent on the network being considered, as is the interpretation of results. To properly define an area of interest, both a metric and a network are necessary.

This study focuses on two metrics: out-degree and closeness centrality. The remainder of this section provides an overview of these metrics, their associated networks, and their interpretations.

4.1 OUT-DEGREE

In undirected networks, degree is a single number; in directed networks, nodes have two different degrees: in-degree and out-degree, corresponding to the number of edges point inward to and outward from those nodes [13]. As out-degree is only applicable to directed networks, this metric will be applied to the P-C network.

Due to the setup of the GA, nodes have an in-degree of either zero or two. This follows consequently from the mechanics of crossover: each child has two parents. Therefore, the initial chromosomes are the only ones with an in-degree of zero since they don't have any parents.

A chromosome's out-degree corresponds to the number of children it has, and by necessity, the amount of times it was chosen for crossover. Chromosomes with a higher relative out-degree can be interpreted as having greater influence on other chromosomes.

4.2 CLOSENESS CENTRALITY

More complex than out-degree, closeness centrality is the inverse of the mean length of shortest paths from a node to any other node [13]. A closeness score of zero is defined when two nodes are not connected to each other along any path. As closeness centrality increases, the network becomes tighter and more connected. In other words, high closeness centrality indicates that a node has short paths to all other nodes, while low closeness centrality indicates that a node has long paths to other nodes. Since the sum of distances depends on the number of nodes in the graph, closeness is normalized by the sum of minimum possible distances [14].

Closeness centrality is commonly used in social networks as a measure of influence [15]. The P-C network consists of parents that influence children, and this relationship is not bi-directional. As such, closeness centrality is not an appropriate metric for analysing this network. However, the P-P network consists of bi-directional relationships, and analysing this network using closeness centrality may produce valuable indicators of bias-dependent relationships.

5. SEN BULKER CASE STUDY

To apply the proposed framework and metrics, the following case study was developed. This section covers a review of the problem, the specifics of the GA, and an analysis of the results.

5.1 SEN BULKER SYTHESIS MODEL

The Sen Bulker ship model outlined in [16] is used to define the case study's design space. Although constructed for optimization applications, this model has been used to study design dynamics [17]. The complete details of the model can be found in the original paper. A summary is presented below.

The Sen Bulker model defines a set of input variables, intermediate functions, objectives, and constraints. This model does not contain any stochastic elements, and therefore objective values follow deterministically from the inputs. The inputs to the model are length, beam, draft, depth, block coefficient, and speed. There are nine constraints relating to displacement, powering, and stability criteria. The multi-objective synthesis aims to minimize transportation cost and light ship mass while maximizing annual cargo.

5.2 DATASET CREATION

The GA was implemented on the design space defined by the Sen Bulker model. The parameters used by the GA are: a population size of 50 chromosomes, a stopping

© 2017: The Royal Institution of Naval Architects

condition of 30 epochs, and a probability of crossover of 0.5. A tournament size of 2 was used to reduce the influence of the bias as much as possible. The case study uses two-point crossover and tournament selection, as detailed in Sections 2.1 (a-b).

A Sen Bulker chromosome consists of six genes: one to represent each of the input variables. To initialize the population, 50 chromosomes are created from a uniformly distributed dataset of 50,000 feasible design points. Feasibility is defined by the constraints: if any constraint is violated, the chromosomes becomes infeasible. Infeasible chromosomes are not removed from the population, although they cannot be represented in the pareto front. Instead, infeasible chromosomes are weighted proportionally by an external penalty function.

5.3 RESULTS AND ANALYSIS

Each of the datasets was generated 50 times using different seed values. For each of these trials, network representations of the datasets were constructed and analysed for the P-C network and the P-P network discussed in Section 3. The section presents the results of this analysis, averaged over the 50 trials. Results are compared by investigating the superset and Pareto-set chromosomes within each dataset.

5.3 (a) Out-degree

This subsection investigates the out-degree analysis on the P-C network. Figures 1 and 2 display the logarithmic relationship of out-degree to its frequency within each dataset. The x-intercepts of these trend lines provide a measure of the corresponding dataset's upper bound of its out-degree distribution. Figure 1 displays these results across the superset, and Figure 2 displays those across the Pareto-set.

Shown in Figure 1, the superset trends for each dataset are similar although the nonbiased dataset is slightly higher. This indicates that when compared to the average biased chromosome, the average nonbiased chromosome is more connected to other chromosomes in its network. Since connections represent parent-child relationships, higher connectivity corresponds to creating more children and therefore having greater influence on the design space. When considering the Pareto-set trends shown in Figure 2, the difference between datasets increases significantly. The biased dataset experiences a much lower out-degree distribution than the nonbiased case. This difference suggests that the bias is causing an insufficient weighting of the chromosomes that are deemed to be good.

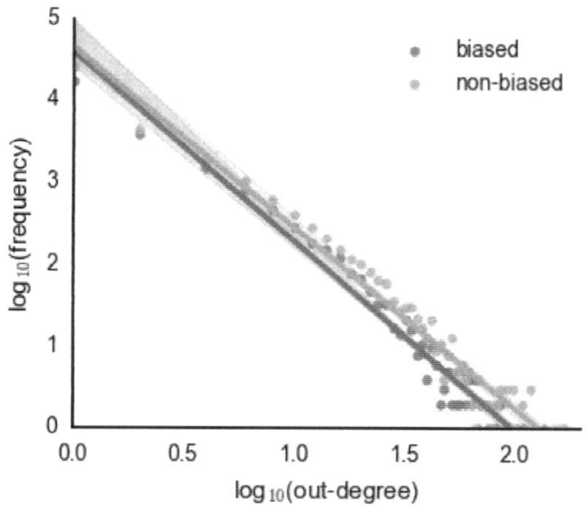

Figure 1: The logarithmic relationship results between out-degree and frequency within the P-C network for the supersets of the biased and nonbiased datasets. Their trend lines are not significantly different from each other.

Comparing the results in Figures 1 and 2 reveals more interesting observations. The x-intercepts of the nonbiased trends are similar, while those of the biased trends vary considerably. Within the biased dataset, the Pareto-set's x-intercept is significantly lower than that of its superset. This difference indicates that an average Pareto-set chromosome is less likely to be involved in crossover and is less influential than an average superset chromosome. These results indicate a misalignment of the evaluation and selection methods.

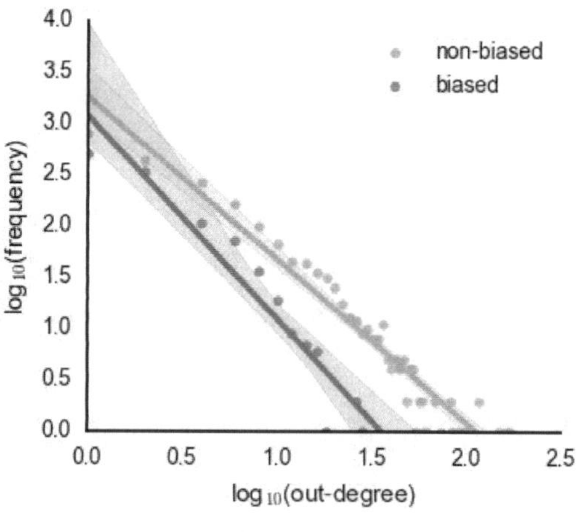

Figure 2: The logarithmic relationship results between out-degree and frequency within the P-C network for the Pareto-sets of both datasets. In contrast to Figure 1, a significant difference is experienced between the two trends. This difference indicates that nonbiased Pareto-set chromosomes have more connection and are therefore more influential than chromosomes in the biased Pareto-set.

© 2017: The Royal Institution of Naval Architects

5.3 (b) Closeness Centrality

This subsection investigates the closeness centrality metric applied to the P-P network. The closeness results are represented by value distributions in Figures 3 and 4. Figure 3 displays the results across the superset while Figure 4 displays those across the Pareto-set. Each distribution exhibits two regions of positive values, one of which is located near zero centrality. This region is referred to as the zero region, and the other as the collective region. The zero region can indicate either a disproportionate number of edges existing within certain components or lack of relational diversity within the network.

Displayed in Figure 3, the closeness distributions of the supersets are similar but not identical. In the collective region, the nonbiased set spans a larger range of scores than the biased set. Investigating the zero region, the nonbiased set has two-thirds the amount of representation that the biased set has. High values in the zero region is generally unwanted, indicating either a random or misguided search.

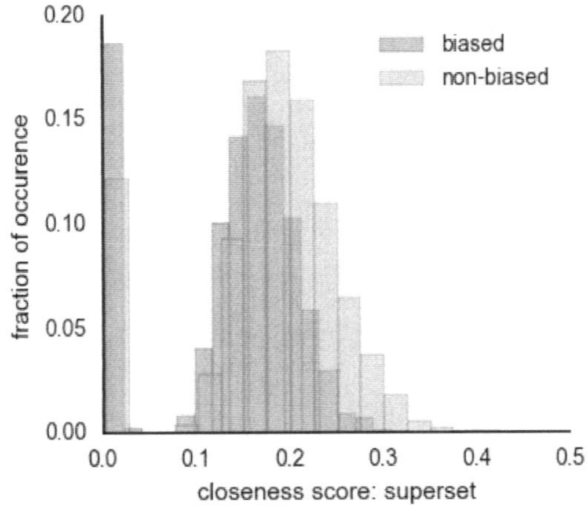

Figure 3: A normalized histogram representing the value distribution of closeness centrality within the P-P network for the superset. The higher closeness scores of the nonbiased tests indicate that this instantiation of bias causes the chromosomes to become less connected to each other.

In Figure 4, the closeness distributions of the Pareto-sets share fewer similarities than they did on the superset. In the collective region, the nonbiased distribution appears to be representative of its superset. The same cannot be said for the biased distribution. In the zero region of the biased distribution, the fraction of representation increases by a factor of nearly 2. This would indicate that the biased process is either populating this Pareto-set by coincidence or focusing on a specific region of Pareto front.

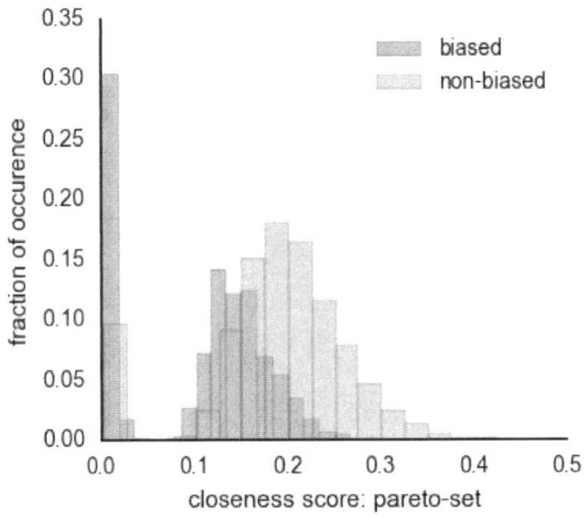

Figure 4: A normalized histogram representing the value distribution of closeness centrality within the P-P network for the Pareto-set. The nonbiased scores are distributed in a representative fashion to that of the superset. The biased scores do not reflect this behaviour.

5.3 (c) Effects of Bias on the Design Space

The analysis thus far has only considered the structure of relationships within design process datasets. This section of results investigates if and how the physical design space was affected by bias.

Since the chromosomes represent inputs to the design problem, the input space is analysed. Rather than attempting to represent a six-dimensional space, this analysis investigates two of the inputs: length and beam. To visualize the effect of bias within the length-beam space, bivariate distributions of the superset and Pareto-set were calculated using a gaussian kernel density estimator (KDE). KDEs are used to estimate an underlying distribution given a set of observations. In this case, the observations refer to the chromosomes within each of the sets. The KDE distribution for the superset signifies where the design effort is being allocated. The KDE distribution of the Pareto-set represents the primary region of interest within the design space.

Figures 5 and 6 display these two distributions for the nonbiased dataset and biased dataset, respectively; the Pareto-set is shaded blue in both figures. In Figure 5, the nonbiased superset (in green) overlaps much of the Pareto-set. This indicates that the design effort is being effectively appropriated and the directionality of the design is correctly aligned. In Figure 6, the biased superset (in red) overlaps much of the Pareto-set. In contrast to Figure 5, much of the design effort is conducted away from the Pareto-set.

© 2017: The Royal Institution of Naval Architects

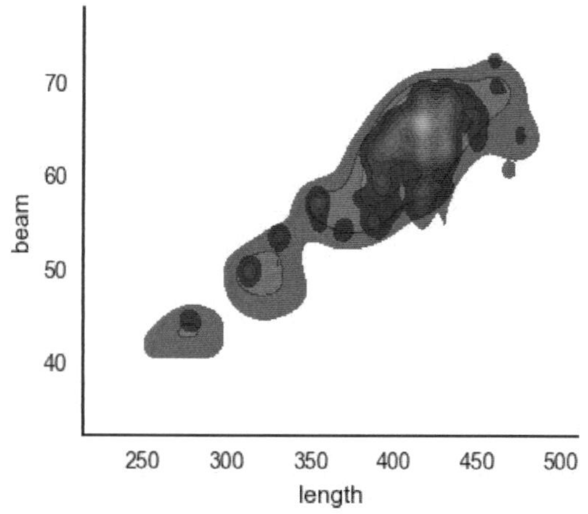

Figure 5: KDE distributions of the length-beam space occupied by the nonbiased Pareto-set (in blue) and superset (in green). The overlap of the distributions suggests that a majority of the design effort was spent progressing the Pareto-set.

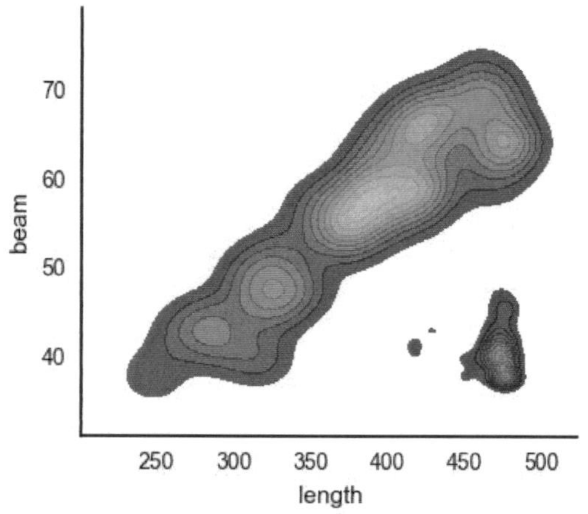

Figure 6: KDE distributions of the length-beam space occupied by the biased Pareto-set (in blue) and superset (in red). Unlike the nonbiased case, the two distributions only slightly overlap, indicating that a majority of the design effort is wasted.

Considering only the Pareto-sets (in blue) from Figures 5 and 6, differences between their contour plots can be observed. The nonbiased Pareto-set is more defined than the other. This definition can represent the effectiveness of resource allocation within the design. Since both processes receive the same amount of resources, the Pareto-set definition can be compared between process. Averaging over the 50 trials, the sizes of the biased and nonbiased Pareto-sets are 27.4 and 52.5, respectively, with standard deviations of 3.5 and 7.9. In the nonbiased dataset, the more efficient search effort provides a larger set of non-dominated alternatives than the biased case.

In addition to the sizes of the Pareto-sets, their relative quality is also investigated. The Pareto-sets are compared by calculating the number of chromosomes from one set that were dominated by the other set. Averaging over the 50 trials, the biased Pareto-set dominated 3.2% of the nonbiased Pareto-set, compared to the nonbiased Pareto-set dominated 27.0% of the biased Pareto-set.

This analysis illustrates the impact that bias can have on the design process. In this case study, the network metrics are able to use the network representation to differentiate between the bias and unbiased datasets by analysing the structure of temporal design process relationships.

6. CONCLUSIONS & FUTURE WORK

Since the design process is vulnerable to bias-related failure, it is important to be able to address bias according.

To facilitate the investigation of bias, this paper presented a framework for representing temporal design data. Two network metrics were proposed and found to be successful in differentiating the biased and nonbiased datasets within the context of the case study. While the case study was limited, the findings present grounds for further investigation and moves towards establishing a framework that may help with identifying biases that negatively impact acquisitions and design. Opportunities for future work include:

- Investigating the component structure within the two networks for additional indications of bias and its behaviour in design;
- Investigating the effects of beneficial biases and comparing the temporal design structure results. These beneficial biases may present as expert opinion or as a matter of coincidence;
- Investigating additional mechanisms and biases for the framework;
- Applying different GA operators and comparing these results to the ones presented here;
- Adapting the framework to consider mutation within the simulated design process; and,
- Testing the framework on more complex synthesis models and on layers of models.

7. ACKNOWLEDGEMENTS

We would like to thank Kelly Cooper from the Office of Naval Research for providing support for this project. This work was funded under grant number N00014-15-1-2752. Additionally, this work received Government support under contract FA9550-11-C0028 awarded by the Department of Defense, Air Force Office of Scientific Research, National Defense Science and Engineering Graduate (NDSEG) Fellowship, 32 CFR 168a.

© 2017: The Royal Institution of Naval Architects

8. REFERENCES

1. HALLIHAN, G., 'Confirmation and Cognitive Bias in Design Cognition', *ASME Proceedings*, 2012

2. NICKERSON, R., 'Confirmation Bias: A Ubiquitous Phenomenon in Many Guises', *Review of General Psychology*, 1998

3. PARK, S., ROTHROCK, L., 'Systematic analysis of framing bias in missile defense', *European Journal of Operational Research*, 2007

4. PIASZCZYK, C., 'Model Based Systems Engineering with Department of Defense Architectural Framework', *Systems Engineering*, 2011

5. JARRATT, T., ECKERT, C., JOHN, P., 'Providing an overview during the design of complex products', *Design Computing and Cognition*, 2004

6. SAVOIE, T., FREY, D., 'Detecting mistakes in engineering models: the effects of experimental design', *Research in Engineering Design*, 2012

7. WATSON, D., GILFILLAN, A., 'Some Ship Design Methods', *The Royal Institution of Naval Architects*, 1977

8. PARKER, M., 'A Contextual Multipartite Network Approach to Comprehending the Structure of Naval Design', *University of Michigan Doctoral Dissertation*, 2014

9. AUSTIN-BRENEMAN, J., YU, B., YANG, M., 'Biased information passing between subsystems over time in complex system design.' *Journal of Mechanical Design*, 2016

10. SOBIESZCZANSKI-SOBIESKI, J. HAFTAK, T., 'Multidisciplinary aerospace design optimization: survey of recent developments', *Structural optimization*, 1997

11. HOLLAND, J., 'Genetic Algorithms', *Scientific American*, 1992

12. GEN, M., CHENG, R., LIN, L., 'Network Models and Optimization', ISBN: 9781848001800, 2008

13. NEWMAN, M., 'Networks: An Introduction', *Oxford University Press*, 2010

14. HAGBERG, A., SCHULT, D., SWART, P., 'Exploring network structure, dynamics, and function using NetworkX', *Proceedings of SciPy*, 2008.

15. FREEMAN, L., 'Centrality in Social Networks Conceptual Clarification', *Social Networks*, 1978

16. YANG. J., SEN., P. 'Interactive trade-off analysis and preference modelling for preliminary multi-objective ship design', *Systems Analysis, Modelling and Simulation*, 1996

17. PARKER, M., SINGER, D., 'Analyzing the dynamic behavior of marine design tools using network theory', *Ocean Engineering*, 2015

9. AUTHORS BIOGRAPHY

Michael J. Sypniewski, is a doctoral graduate student in the Naval Architecture and Marine Engineering department at the University of Michigan College of Engineering, where he earned B.S.E and M.S.E degrees. He specializes in artificial intelligence and machine learning applications to design.

David J. Singer, Ph.D., is an Associate Professor in the Naval Architecture and Marine Engineering department at the University of Michigan College of Engineering. He earned his B.S.E., M.Eng., and Ph.D. degrees in NA&ME and a M.S.E. in Industrial and Operations Engineering from the University of Michigan. He conducts research in the areas of naval design theory, design optimization, and ship production.

© 2017: The Royal Institution of Naval Architects

© 2017: The Royal Institution of Naval Architects

Optimal Arrangement Method for Naval Surface Ship Considering the Stability, Operability, and Survivability

S K Jung, M I Roh, S M Lee, K S Kim, Seoul National University, Republic of Korea

SUMMARY

Several ship performances, such as stability, operability, and survivability have to be considered at the same time. However, there is no way to evaluate the arrangement quantitatively. Due to these reasons, the probability of design change increases as design progresses. Therefore, in this study, the optimal arrangement method of a naval surface ship considering stability, operability, and survivability is proposed. At first, 'ship template model' which is a data structure for representing naval ship arrangement information was configured. Second, a method that can quantitatively evaluate the performances at preliminary design stage was proposed. Third, the optimum arrangement can be obtained using the design variables, objective functions, and constraints of the problem. Finally, user interface which visualize of the optimum arrangement in 3-D is developed. The method was applied to the example of a 7,000-ton class destroyer, and it is confirmed that the method and program can be an efficient tool.

1. INTRODUCTION

Many compartments and equipment are placed in a limited space of naval ship, many design alternatives can occur. There are so many things to be considered when arrangement design is ongoing, but among them, the most important factors could be a stability, operability, and survivability. At first, stability is an ability to restore the ship when the ship inclines due to several reasons. Second, the operability is an ability that represents effectiveness of the flow of the supplies and crew in the ship. The survivability can be regarded as the most important performance of a naval ship in terms of battle mission, and also important part of the recent naval ship design. It is important to carefully consider the performances that were described above in the earlier design stage. At present, there is no way to quantitatively evaluate required performances of the draft arrangement. Therefore, the optimal arrangement method is needed because to evaluate the arrangement and also to find an arrangement that has desired performances. Also, the method can reduce design changes because the expected values for the desired performance can be checked in advance.

A lot of studies for the optimal naval ship arrangement have been conducted by many scholars. Han et al. [1] proposed an algorithm in order to solve the compartment layout problem involving watertight transverse bulkheads and passages by using the genetic algorithm (GA). Boulougouris, E., and A. Papanikolaou. [2] analyzed hull forms of a modern naval combatant using a multi-objective genetic algorithm (MOGA). Nick [3] divided entire arrangement problem into two stages, allocation and arrangement. Roh et al. [4] proposed naval surface ship bulkhead arrangement program when the hull shape had already chosen. Shin [5] proposed the method of spatial arrangement of naval ships considering survivability. Kim [6] proposed optimum arrangement design method of submarine based on the expert system. Jung [7] proposed spatial arrangement method of naval ship using layer-based integrated survivability evaluation.

2. OPTIMAL ARRANGMENT METHOD

The optimal arrangement problem is formulated by two stages. The first stage is a bulkhead arrangement and the second stage is a room arrangement. After the second stage, the stability of the solution for second stage will be checked.
This arrangement problem has a lot of design variables, objective functions, and constraints for each arrangement stage. In this study, entire arrangement problem was divided into two stages.

2.1 OPTIMAL BULKHEAD ARRANGEMENT

The purpose of the first stage, optimal bulkhead arrangement is finding optimum positions of the bulkheads and the decks. In this study, since it is assumed that the shape of the hull has been determined, a hull shape mesh model is necessary. And the number of bulkhead and deck has to be determined before optimization. Tank information such as utility, required space, and position and, warhead information for calculation of survivability are also needed.
Design variables in an optimization problem are set of variables that describes the system. It is allowed to change during the design optimization. Design variables of this stage are positions of the bulkhead in x, y direction and positions of the deck in z direction.
There are three objective functions in the stage. The objective functions include space utilization efficiency, structural stability, and survivability as shown in Table 1.

First, volume for liquid cargo tank have to be minimized in order to maximize the space area for armament and crew.

© 2017: The Royal Institution of Naval Architects

Second, the maximum bending moment have to be minimized to ensure structural stability at the intact state of the ship.

Table 1 Objective functions of the 1st stage optimization problem

Objective functions	
Minimize. Volume of liquid cargo tank	$f_1 = V_{FOT} + V_{FWT} + V_{WBT} + V_{LOT}$
Minimize. Maximum bending moment	$f_2 = maximum\ BM$
Minimize. Bulkhead damage vulnerability due to the attack	$f_3 = \sum f(P_r)$

Lastly, bulkhead damage vulnerability due to attack have to be minimized. According to Said [8], ship survivability has three major elements that express it well and they are susceptibility, recoverability, and vulnerability. Susceptibility and recoverability has weak relation with arrangement design of the ship. This study handles bulkhead and room arrangement, so the vulnerability of the ship was concerned only and that represents the survivability of the ship in this study. If the pressure acting on the bulkheads due to the explosion can be minimized, we can say that the vulnerability of the ship is reduced. Explosive pressure acting on every bulkhead with certain explosive has to be minimized as an objective function.

There are nineteen different constraints used in this study for bulkhead arrangement stage. The constraints are shown as in Table 2.

Table 2 Constraints of the 1st stage optimization problem

Constraints	
Intact stability criteria	$g_1 = \frac{GZ_0}{GZ_{max}} - 0.6 \le 0$
	$g_2 = 1.4 - A_2/A_1 \le 0$
	$g_3 = \phi_0 - 15 \le 0$
Damage stability criteria	$g_4 = \phi_0 - 15^\circ \le 0$
	$g_5 = 1.4 - A_2/A_1 \le 0$
	$g_6 = 0.076 - (GZ_{max} - HA) \le 0$
	$g_7 = 0.020(m - rad) - A_1 \le 0$
	$g_8 = -D\_marginline \le 0.0(m)$
Deck height	$g_9 = 1{,}950(mm) - h_{deck} \le 0$
Engine room area	$g_{10} = 12 - \lvert x_5 - x_6 \rvert \le 0$
	$g_{11} = 12 - \lvert x_7 - x_8 \rvert \le 0$
Required space of liquid tanks	$g_{12} = V_{FOT}^{min} - V_{FOT} \le 0$
	$g_{13} = V_{FOT} - V_{FOT}^{max} \le 0$
	$g_{14} = V_{FWT}^{min} - V_{FWT} \le 0$
	$g_{15} = V_{FWT} - V_{FWT}^{max} \le 0$
	$g_{16} = V_{WBT}^{min} - V_{WBT} \le 0$
	$g_{17} = V_{WBT} - V_{WBT}^{max} \le 0$
	$g_{18} = V_{LOT}^{min} - V_{LOT} \le 0$
	$g_{19} = V_{LOT} - V_{LOT}^{max} \le 0$

2.2 OPTIMAL ROOM ARRANGEMENT

The purpose of the second stage, optimal room arrangement, is finding optimum positions of the rooms and passages for a superstructure area. To arrange rooms, room information and deck information are needed.
Design variables of 2nd stage of optimization consist of two kinds of a variable set: a sequence of the room, positions of the longitudinal and transverse passages. Firstly, the sequence of the room means place the rooms with its application in the proper position. Second variable set means when the shape of the passage for each deck is decided, determine the position of the longitudinal and transverse passages.
The objective function of this optimization stage contains adjacency between rooms and room damage vulnerability due to the attack. The objective functions are as shown in Table 3.

Table 3 Objective functions of the 2nd stage optimization problem

Objective functions	
Minimize. Adjacency index	$f_1 = \sum\sum f_{adj}$
Minimize. Room damage vulnerability due to the attack	$f_2 = \sum f(a_i * p_i)$

First, adjacent index has to be minimized. The adjacency index is an index for checking how much each room is placed with a suitable adjacent degree by each usage. To obtain adjacency index, the adjacency matrix and the distance matrix have to be defined.

The adjacency matrix is a matrix that has adjacency between each room in its element. Adjacency matrix is defined as a subtraction of antagonism and affinities matrix. Antagonisms matrix represents a quantitative value of the risk when two rooms are located in adjacent zones close. Affinities matrix represents a quantitative value of the advantage when two modules have frequent movement of staff each other in the aspect of manning and engineering affinities.

The distance matrix contains distances between every compartment of the ship. To find a distance between two rooms, Dijkstra's algorithm was used.

In order to maximize adjacency of the whole ship, objective function can be defined as minimization of summation of the multiplication of distance matrix and adjacency matrix of the entire room.

As second objective function, room vulnerability due to the attack have to be minimized. Room damage vulnerability is defined as a summation of multiplication of the probability being damaged and room importance factor. If a certain room is placed in the certain position of the ship, the room has its own probability of damage according to its x, y, and z positions. Importance factor

© 2017: The Royal Institution of Naval Architects

of the room is defined by summation of a multiplication of the equipment's important and quantity in the room. The objective function of room damage vulnerability due to the attack can be defined as minimization of summation of multiplication with room damage probability and importance factor of the room.
The constraints of this optimization stage contains required minimum area of each room and required deck of rooms as shown in Table 4.

Table 4 Constraints of the 2nd stage optimization problem

Constraints	
Required minimum area of each room	$g_k = a_k^{min} - a_k \leq 0$
Required deck of rooms	$g_j = H_{deck_{required}} - H_{deck_{room}} \leq 0$

3. APPLICATION

In this study, a prototype program was developed based on the proposed method, which contains user interface such as menu and tool bar, tree view which can represent arrangement template model, and 3D visualization view, etc.

With this program, the method was applied to a 7,000-ton class destroyer, US navy's DDG-51.At first, optimal bulkhead arrangement was conducted. As input information, the utility of the tank and compartment have to be determined. This study considered two loading conditions of the naval ship: minimum operation condition and full loading condition. As a result, the positions of the bulkhead and deck are changed and improve the objective function.

Secondly, the 2nd stage, optimal room arrangement was conducted. For the room arrangement, another 12 kinds of 33 rooms were considered. As a result, the arrangement of the room was improved. The room of similar utilities are arranged closer while grouping some kinds of the zone, and it is considered that this grouping improved adjacency index.

4. CONCLUSIONS

In this study, optimal arrangement method was proposed to solve the problem of complicated naval surface ship arrangement efficiently. The entire optimization problem was divided into two parts, optimal bulkhead arrangement and optimal room arrangement. Subsequently, stability verification was performed.

To confirm the effectiveness and usefulness of the proposed method, the prototype program was developed. Also, the 7000-ton destroyer was applied to this program.

Through these methods, it is possible to support the ship design so that the compartment arrangement that takes into consideration all the stability, operability and survivability can be done from the preliminary design stage.

For future work, the optimal arrangement prototype program will be refined so that the user can input more information on the program. Also, Moreover, in addition to the standards and rules applied in this research, we will apply an expert system that systematically reflects the knowledge of naval ship experts and uses it for evaluation of arrangement alternatives.

5. ACKNOWLEDGEMENTS

This work was partially supported by:

(a) BK21 Plus

(b) Research Institute of Marine Systems Engineering of Seoul National University.

6. REFERENCES

1. Han, S.N., 'A Study on the Optimum Compartment Layout Design for the Naval ship', *Master's Thesis, Seoul National University*, 2001.
2. Boulougouris, E., and A. Papanikolaou., 'Optimization of the survivability of naval ships by genetic algorithms', *Proc. 3rd Int. Conference on Computer and IT Applications in the Maritime Industries (COMPIT'04), Siguënza, Spain*, 2004.
3. Nick, E.K., 'Fuzzy Optimal Allocation and Arrangement of Spaces in Naval Surface Ship Design', *Ph.D. Thesis, University of Michigan*, 2008.
4. M.I Roh et al., 'Development of an optimal compartment design system of naval ships using compartment modeling and ship calculation modules', *Journal of the Society of Naval Architects of Korea 46.4: 424-434*, 2009.
5. Shin, J.H., 'A Study on the Spatial Arrangement of Naval Ships Considering Survivability', *Master's Thesis, Seoul National University*, 2013.
6. K.S Kim, 'Study on the application of an expert system to arrangement design of submarine', *Master's Thesis, Seoul National University*, 2014.
7. J.W Jung, 'Optimization of Spatial Arrangement at an Early Design Stage of Naval Ship through Layer-Based Integrated Survivability Evaluation', *Master's Thesis, Seoul National University*, 2016
8. SAID, M.O., 'Theory and Practice of Total Ship Survivability for Ship Design', *Naval Engineers Journal, 107.4: 191-203*, 1995.

© 2017: The Royal Institution of Naval Architects

7. AUTHORS BIOGRAPHY

Sun-Kyung Jung is a M.Sc. of department of naval architecture and ocean engineering in Seoul National University, Republic of Korea. Her main area of research interests includes optimization of arrangement design of naval ship and submarine, and so on.

Myung-Il Roh is a professor of department of naval architecture and ocean engineering in Seoul National University, Republic of Korea. He holds a B.S., M.Sc., and Ph.D. in naval architecture and ocean engineering from Seoul National University. His main area of teaching and research interests includes ship and offshore structure design, simulation-based design, optimum design, CAD, CAM, CAE, CAPP, integrated design, and so on.

Seung-Min Lee is a M.Sc. candidate of department of naval architecture and ocean engineering in Seoul National University, Republic of Korea. His main area of research interests includes flooding analysis with mesh-free particle methods, and so on.

Ki-Su Kim is a Ph.D. candidate of department of naval architecture and ocean engineering in Seoul National University, Republic of Korea. His main area of research interests includes optimization of arrangement design of offshore plants and ships, expert system for optimization, statistical & optimization method for weight estimation of ship and offshore plant topsides, flooding analysis with mesh-free particle methods, optimal route planning for ship using various algorithm including A*, Dijkstra and GA, and so on.

© 2017: The Royal Institution of Naval Architects

A CASE STUDY IN OPTIMIZING 3D EARLY DESIGN FOR SHIPBUILDING

JS Cheong, Samsung Heavy Industries, Korea
J Baumer, Intergraph, USA
HW Lee, Intergraph, Korea
SH Ha, JM Hong, Samsung Heavy Industries, Korea

SUMMARY

The use of 3D models provides proven benefit throughout the ship design lifecycle and particular focus is put on extraction of high-quality production output. But due to the unique challenges of the early design stage of the shipbuilding process, a multi-purpose 3D model can also be leveraged to significantly reduce design hours and improve accuracy of analysis. The key to optimizing the process is the ability to re-use the same model data from early design through production design, for both outfitting and structure. This requires an ability to import (and modify) data from outside sources, a drawings extraction process that transitions from classification to production using the same model data, flexible analysis interfacing capability, and an ability to seamlessly carry the model forward to detail design and production.

This paper presents a case study of the re-engineering of the early design CAD modelling process at Samsung Heavy Industries. Improvements focused on optimization of the 2D to 3D modelling workflow, smoothly transitioning from early design to detail and production design and interfacing 3D model data to analysis software, resulting in significant productivity and quality improvements. Lessons learned from this project suggest further optimization is possible based on emerging technologies.

1. INTRODUCTION

Since 2011, Samsung Heavy Industries (SHI) has been working towards the implementation of a 3D design process for shipbuilding early design in order to improve their overall design process.

A number of alternatives have been explored to transition from a drawing-based early design process to a 3D design process. Early efforts looked at employing specialized early design systems. [1] While these systems provided benefits over the 2D, drawing-based process, there were still inefficiencies in the overall design process and duplication in 3D modelling activities when transitioning to detail and production design.

This paper will present a case study of the design process innovations enabled by using a 3D model earlier in the design cycle and the benefits to the overall design process. SHI calls this new design process their Advanced Hull Design or AHD process. A key feature of the AHD process is the creation of a 3D model during early design and the evolution of that same 3D model through detail design and into production design. By this process, duplicate modelling operations are eliminated, better, more consistent design information is available earlier and the entire design cycle is streamlined saving both man-hours and design cycle time.

Most important of all, it is possible to prepare the compartment model at the beginning of the design stage, all of the disciplines of the design can start at the same time concurrently including Outfitting as well as hull production design, therefore, it can be reduced the design schedule from 11 months to 8 months.

For the purposes of this case study, early design phase, sometimes referred to as basic design, is defined as the portion of the overall ship design cycle focused on the engineering and design of the ship's primary structural scantlings and the development and approval of classification drawings. Concept design, hull form and naval architecture calculations like powering, hydrostatics, and stability are upstream and outside the scope of this case study.

2. BACKGROUND

2.1 SAMSUNG HEAVY INDUSTRIES

Samsung Heavy Industries is one of the three largest shipbuilders in the world. Since its inception, SHI has delivered more than 1100 ships. SHI specializes in complex, high-technology and high-value specialty ships and offshore projects, including LNG FPSOs, arctic shuttle tankers, LNG carriers, ultra-large container ships and drill ships.[2] See Figure 1.

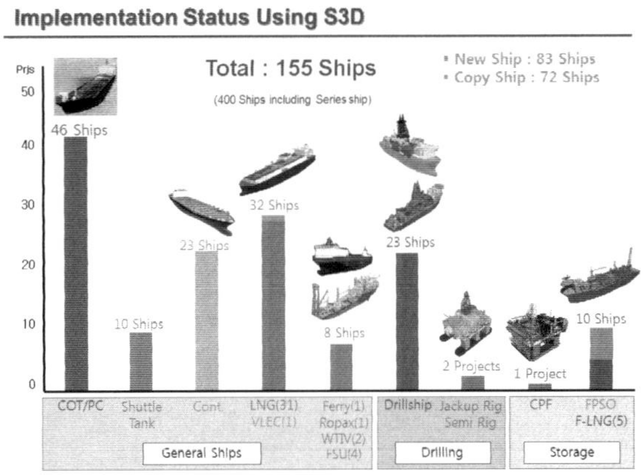

Figure 1: Recent Product Deliveries at SHI

© 2017: The Royal Institution of Naval Architects

2.2 PREVIOUS EARLY DESIGN PROCESS

Prior to the AHD project, SHI utilized a 2D, drawing-based, design process to produce early design deliverables, key plan drawings and classification drawings for approval. This drawing-based process had remained in place because it was generally accepted that drawings were quicker and easier to modify than a 3D model in support of the rapid design changes occurring during early design and because of strong institutional resistance to changing the process. This 2D, drawing-based, design process carried into detail design and the creation of detailed structure scantling drawings.

It was not until midway through the detailed design process that a 3D model was started for production design. SHI uses Intergraph's Smart® 3D (S3D) for detail and production design. Since 2006, SHI has produced more than 400 ships with S3D. The view that 3D modelling is too labor intensive to support the early design phase is widely prevalent in the marine industry, however, this study and the results achieved at SHI should go a long way towards changing this perception.

There were a number of problems and inefficiencies inherent in SHI's legacy early design process. These include:

- The 2D, drawing-based design process was error prone, entirely dependent on designer skill to ensure design consistency between drawing views.
- Interference checking between structure and outfitting was difficult.
- It was required to manually create separate models for class rule checking and finite element analysis. Creating these models resulted in duplicate modelling efforts and additional opportunities for design errors and inconsistencies.
- Weight calculations were performed manually with drawings and spreadsheets, another inefficient, error prone process.
- The overall duration of the design schedule, 11 months, was too long due to the "Bottom-Up" design process that required Detail design not be started until after Basic design finished.

In order to reduce the overall design duration and provide the ability to better support short-term projects, SHI initiated their AHD project two years ago. SHI and INTERGRAPH co-developed the several key functions, which will be introduced in the section 4, to accomplish the ultimate this goal.

3. SMART® 3D

Smart 3D (S3D) is the Intergraph 3D design product for shipbuilding, offshore and plant design projects. Smart 3D offers a range of features that make it well suited to support the entire shipbuilding design cycle and allow a shipyard to innovate and evolve their design process. The most important of these features are:

- S3D is based on an object oriented, data centric architecture. Geometry, attribute data and relationships are all properties of the data object.
- A single model database supports concurrent design by all disciplines. Designers see in real-time what other disciplines are doing. Interference detection can be run interactively against distributed systems (pipe, duct, cableways) as they are routed or run in the background against the entire database.
- Smart 3D supports intelligent design through topological modeling and associative relationships between objects. The object relationships allow modifications to propagate through the model such that if one design object is modified, the system automatically updates related objects. The propagation of change ensures consistency between the related objects both within and across disciplines.
- Smart 3D supports a top-down system based modelling approach. Structure systems are initially defined at a high level that can span the entire length of the ship. Data in the form of seams, openings, stiffeners, structure detailing features, etc. are added to the structure system until manufacturable parts can be extracted from the system in the end. The system based structural modelling workflow is described in general terms below:
  1. The plate system molded surface and boundaries are defined for the root plate system. The root plate is the top level structural system object and controls overall system properties and permissions.
  2. Major openings and design seams are defined on the root plate system.
  3. The root plate system is split by the seams and major openings into leaf plate systems, each with its own part. Plate thickness and material grade properties can be set at the leaf level or can be inherited from the root level.
  4. Stiffeners are defined on a root plate system in a similar manner. The stiffener landing curve, cross section and boundaries are first defined.

© 2017: The Royal Institution of Naval Architects

Root stiffeners systems can run the entire length of the root plate system and design seams are placed and split where stiffener cross section, size or material change. Stiffener systems are children of the root plate system.

5. Minor openings, brackets and miscellaneous plates and stiffeners can then added to the existing major structure systems.
6. All of the above can be modified as needed over the course of the design cycle.

- Smart 3D's rule-based and specification-driven design automation improves both design quality and productivity.
  - Rules *automate* repetitive tasks.
  - Rules *enforce* design standards.
  - Rules *ensure* design integrity.
  - Rules *protect* design consistency.
- Smart 3D's planning functionality is integrated into the design functionality. This allows work breakdown and construction planning decisions to be made and construction planning data generated early in the design cycle where it can be used to influence the design and improve construction efficiency.
- Smart3D's integrated structural and outfitting modelling capability allows the early design user to work with a more complete view of the whole ship and the arrangement and interaction of and between the various systems and disciplines.
- An open, Microsoft COM-based, architecture and commercial relational database, SQL Server or Oracle, enables shipyards to develop their own custom applications and interfaces to both create and exploit their 3D model data. Smart 3D also provides Microsoft .NET application interfaces to further automate design operations. [2]

4. SMART 3D FOR EARLY DESIGN

Smart 3D was developed with the intent to support the full shipbuilding design cycle, from early design, detail design to production design within a single model database. The features defined above all support that objective. However, there were some practical considerations that needed to be addressed to achieve the vision of full early design through production design in a single database. Intergraph and SHI worked together to address the gaps and enhance Smart 3D features in conjunction with SHI developing business processes and workflows. This section describes those considerations and how they were addressed.

4.1 SPLIT ROOT SYSTEMS

Throughout the time S3D has been used at SHI for detail and production design, the standard practice has been to have one structure designer entirely responsible for any given block, a block being the smallest unit of work of structure production work. In order to ensure any given designer was completely and solely responsible for the structure within his block, root plate systems were modelled by block and connections (relationships) were added at block boundaries to drive the generation of welds, bevels and other features at block boundaries. The result of this modelling approach is that a large deck running the length of the ship would be modelled as dozens of root plate systems.

While this modelling approach worked very well within SHI's detail/production design processes and work practices, this block based modelling approach is not practical for early design because 1) it's very inefficient for early design where a handful of designers need to quickly model and later change large areas of ship structure and 2) block boundaries will not be known during early design. Early design calls for large root plate systems, modelled by zone or functional region such that the entire ship's primary structure can be quickly and easily modelled and then changed by a relatively few designers.

Thus there was a need to be able to transition from large root plate systems being worked by a handful of designers during early design to block-based root plate systems being worked simultaneously by up to over 100 designers during detail and production design.

The solution was the Split Root System command (SRS). SRS allowed the large root plate systems needed for early design to be split into block sized root plate systems for detail/production design and thus maintain the single designer ownership and accountability for the structure within a block. SRS splits both root plates and root profile systems into smaller roots and supports the use of previously defined and split design seams, planes and complex surfaces as splitters. Prior to SRS, design splits within a root system created leaf systems, and while it is possible for multiple designers to work concurrently within a single root system, modelling operations by one user in one area of a plate system can cause objects in another area, created by another user, to re-compute (remember changes propagate automatically through associative relationships) and thus blur the clear identification, control and accountability with regards to which designer made a given model change.

4.2 PARTIAL DETAILING

Structure Detailing is a design task that probably makes the most extensive use of S3D's rule based automation. For shipyards with well-defined design standards, it is possible to correctly generate greater than 90% of structure details (slots, clips, collars, end cuts and welds) automatically. The structure detailing operation is also where the structure parts' 3D solid geometry is trimmed by bounding and intersecting structure to obtain the

© 2017: The Royal Institution of Naval Architects

accurate 3D geometry with bevels needed for manufacturing. Prior to detailing, parts have a lightweight, approximate geometry that is thickened but not trimmed. The trimming of the 3D part solids and creation of the detailing feature objects (slots, collars, end cuts, welds) significantly increases the model data size and can reduce the performance of structure modifications. Once structure is detailed, it becomes more difficult to quickly make large scale design changes typical of early design.

The practical consideration missed in earlier versions of S3D is that classification drawings generated in early design need to show slots, collars, end cuts and other details. The previous approach was to generate these details symbolically on the class drawings without actually creating the objects in the model and thus maintain the lightweight, easily modified early design model. The same detailing rules that would place the actual slots, collars and end cuts objects in the model, were used to place the 2D feature symbols on the drawing.

The problem with this approach was designers wanted to see in the model the slot or end cut that would show on the drawing. They insisted on WYSIWYG or what you see is what you get.

The solution was to implement a Partial Detailing capability. With Partial Detailing, the slots and end cuts needed for class drawings can be generated without the full structure detailing operations and the computationally expensive part trimming operation and the generation of welds in the model. See Figure 2.

With Partial Detailing, the features necessary for class drawings can be selectively generated without significantly increasing model size. The model remains lightweight with the modification performance required for early design. Partial detailing utilizes the same design rules as full detailing which will be performed later, after SRS operations have been performed to split the model by block for detail and production design. At that time, structure that is partially detailed can be taken to full detailing by essentially flipping a switch and re-executing the Structure Detailing rules.

4.3 EFFICIENT DATA REUSE

S3D has a number of structural data reuse features to promote fast and efficient modelling. These include Copy / Paste and Copy-By-Family commands. Copy / Paste allows structure systems to be copied from one location and pasted in a different location and the topological dependencies to be re-specified at the new location.

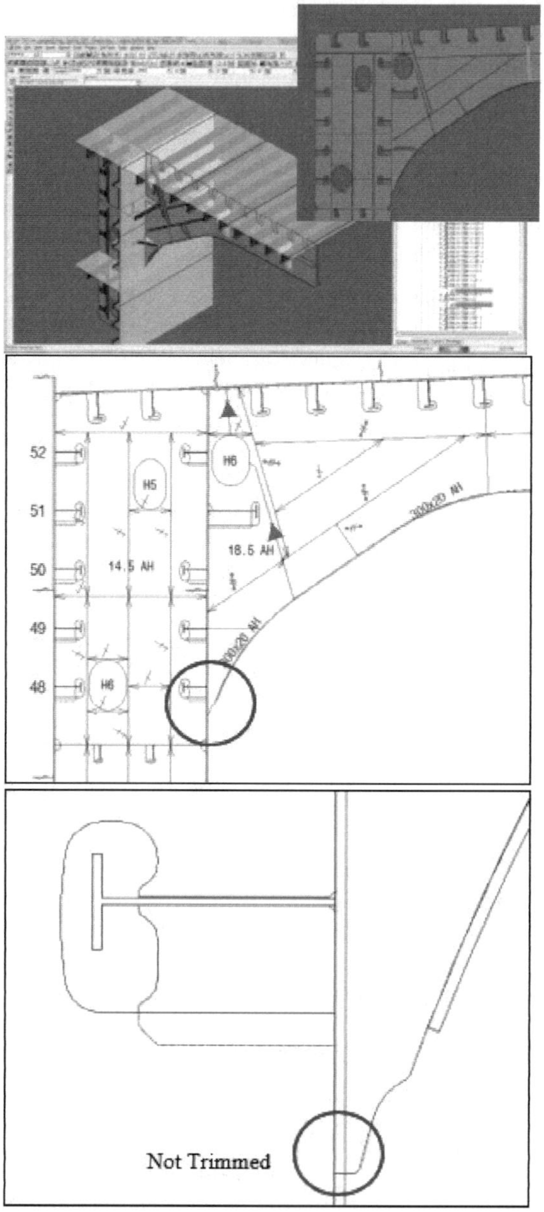

Figure 2: Untrimmed Structure with Partial Detailing

The Copy-by-Family copies groups of structural systems from one location to a similar or symmetrical location and automatically finds and reconnects the topological boundaries. An example of how Copy-by-Family is used would be to model the plate and profile systems of a complete transverse frame in the cargo hold and then copy this structure to the other frame locations within the cargo bay. Then, copy all of the structure in that entire cargo bay to the next bay. Because Copy-By-Family automatically reconnects topological boundaries, the copied frames can adjust to changing hull shape or

© 2017: The Royal Institution of Naval Architects

longitudinal bulkhead locations. In addition, a change made to any one of the members of the family can be propagated to all the other copies quickly, accurately, and completely. [3]

What's new in the latest version of S3D is the ability to disconnect the topological boundaries and relationships when structure systems are copied from the source model, and then reconnect the relationships when the structure systems are pasted in a different target model or location. This allows parametric webs (plate systems with associated stiffeners, openings and face plates) to be copied from a design library model into the early design model being worked. See Figure 3. Topological relationships are automatically reconnected and parameters can be adjusted to quickly model webs and other common structure.

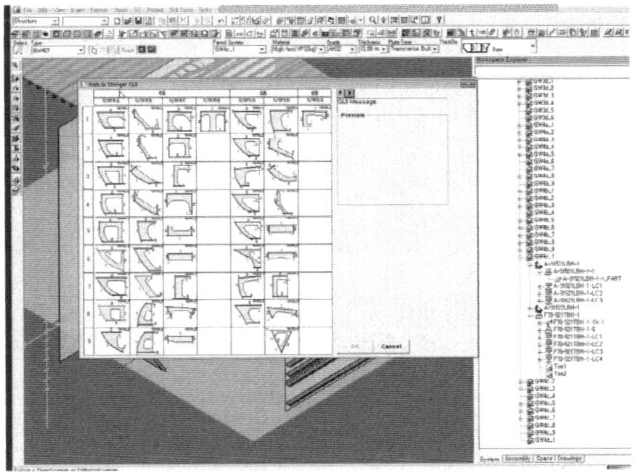

Figure 3: Parametric Web Browser

4.4 NEUTRAL FORMAT XML EXPORT

Structural analysis and class rule checking are mandatory tasks during early design. Efficient interfaces between the 3D model and rule checking and analysis applications are key to avoiding costly remodelling work.

S3D provides a neutral format XML interface to export early design structural scantling data. The neutral format XML is intended to support the exchange of geometric and property data of S3D structure objects with other 3rd party applications. Plate system geometry is exported as zero-thickness molded surfaces with stiffeners exported as landing curves with cross section type and size. Structure detailing features, if they exist in the model, are omitted as the intent is to export early design scantling data suitable for analysis. The geometric representation includes the outer contour defining the plates and any inner contours (i.e. openings) representing the removal of material affecting the early design. Property data, such as material, grade, thickness, cross-section, etc. is also exported along with topological information such as boundary connectivity expressed as relationships between the structural objects. In addition to the typical structural objects, the neutral format also exports spatial geometry and properties representing physical compartments and their cargo and functional zones and their purpose. The spatial object geometry is expressed using industry standard file formats such as IGES or .sat and boundary topology is expressed as relationships to the structural plate or grid objects.

The objective of the neutral format XML is to provide a neutral schema that can be mapped to other application schema. The schema is easily extensible to add optional data not required in earlier versions. [3] Utilizing the neutral format XML with a minor extension, Korean Registry (KR) was able to map the S3D neutral schema to their HullScan application in a matter of weeks and then import S3D structure data. See Figure 4.

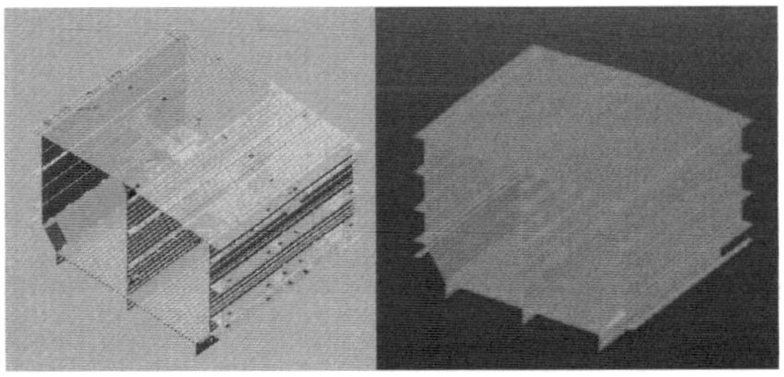

Figure 4: S3D Model and FEA Model via Neutral Format XML.

4.5 AUTOMATED CLASS DRAWINGS

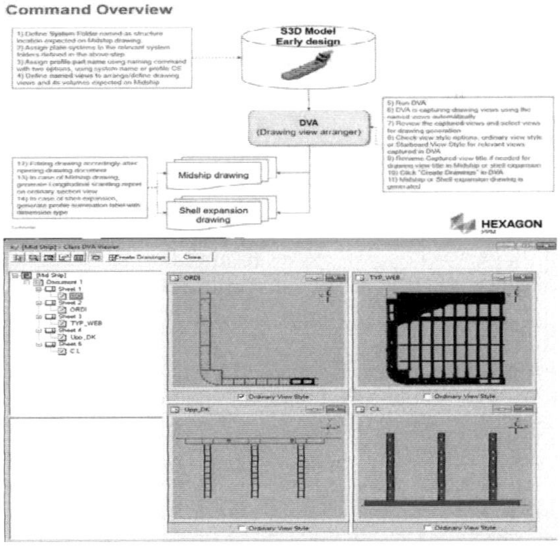

Figure 5: S3D DVA Command for Class Drawing

S3D has long had the capability to generate automated drawings from the 3D model. However, a new drawing command was added to improve the efficiency of the definition of the classification and detail scantling

© 2017: The Royal Institution of Naval Architects

drawings. The new Drawing View Arranger (DVA) command allows users to quickly define the structure to be shown in a given drawing view and then arrange the views on sheets in the drawing. See Figure 5. The drawing can then be generated with automated labels, dimensioning and reports. After initial generation, the drawings can be updated on-command to reflect any changes to the target structure. See Figure 6.

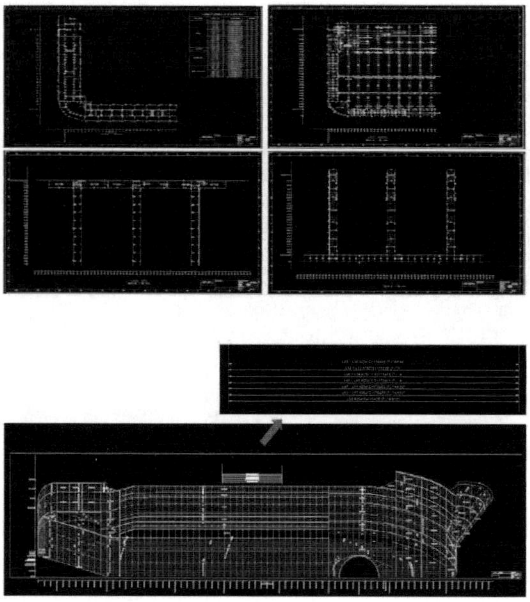

Figure 6: S3D Automated Class Drawings

4.6 BLOCK DIVISION

With a 3D model available early in the design cycle, before block boundaries have been determined, it only makes sense to utilize this model to help optimize block boundary definition. Working together, SHI and Intergraph developed a Block Division System. The Block Division System utilizes historic data from similar type ships to create block volumes and then analyse the model objects, both structure and outfitting, within the block to compute or estimate weights, pipe lengths and weld lengths for the resulting blocks. Block volumes and boundaries can then be adjusted, optimized and then transmitted back into the S3D early design model for the SRS operation. See Figure 7.

SHI expects the following benefits from the 3D Block Division System when it is in production.

- Paperless communication between production engineers and designers.
- Reduced block erection costs from optimized block definitions.

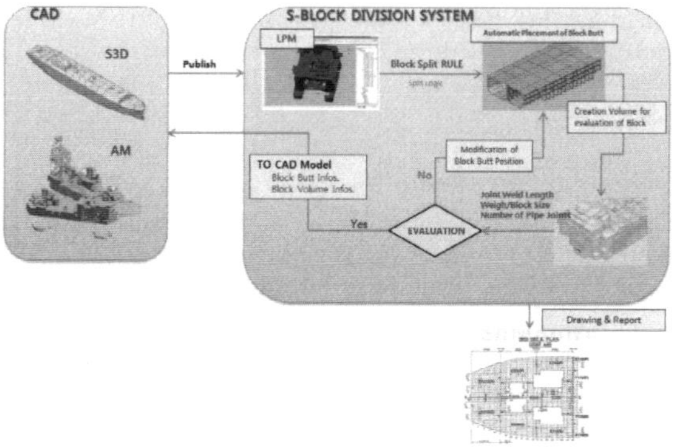

Figure 7: Block Division System Workflow

5. SHI CASE STUDY – ADVANCED HULL DESIGN (AHD) PROCESS

5.1 AHD PROCESS DEVELOPMENT BACKGROUND

SHI had been using the traditional 2D work methods in the generation of the early design drawings and detail (classification) drawings since it is easy to reflect the frequent changes during the early stage design. But SHI faced the issues to be solved as following, which cannot be solved with the traditional 2D method.

- Impossible to perform the short-term, quick turnaround project with old 2D, drawing-based methods.
- Inconsistent quality: Dependent on designer's skill and knowledge.
- Requiring numerous man-hours from 2D manual works.
- Accurate weight calculation is needed for the prediction of the production in early stage.

SHI decided that the only solution for these issues was to use a 3D model from the early design stage and hand over that 3D model to the production design. These requirements have led to the creation of the AHD process, enabling a truly unified single-database ship design process covering the entire design cycle from early design through production design for both structure and outfitting.

Figure 8 shows the SHI legacy process (top) with entirely manual 2D processes within the grey early design box and the 3D model created from drawings for detail and production design. The AHD process, shown in the bottom half of Figure 8, is a much more automated 3D model-based process.

© 2017: The Royal Institution of Naval Architects

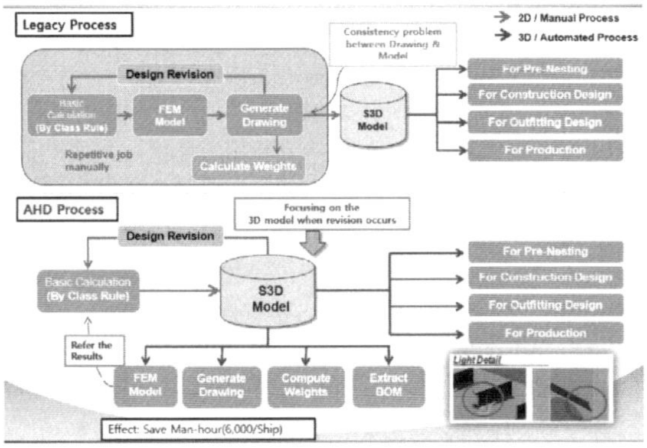

Figure 8: SHI Design Process, Before and After AHD

5.2 AHD PROCESS DESIGN WORKFLOW

The SHI legacy process used a Bottom-Up design method with S3D for production design, which is being used in many shipyards traditionally. The Bottom-Up design method is very useful in the block based production design but there are some limitations in expanding the 3D design to the early design area as follows.

- The final block division information cannot be provided by the production engineers at the early design stage.
- The early designers need the whole ship design model for their analysis work.
- It was impossible to design ships at the early stage considering the production (planning) information.

The best solution to overcome the above limitation is to create the bigger structure systems and then split them when design seams and block boundaries are known. We call that "Top-down" design method. This design methodology will allow both the early designers and production designers to share one 3D model database and this one unified 3D model database can be used more extensively throughout the shipyard for design, planning and estimating activities. This new adopted design work flow is as shown in Figure 9 below.

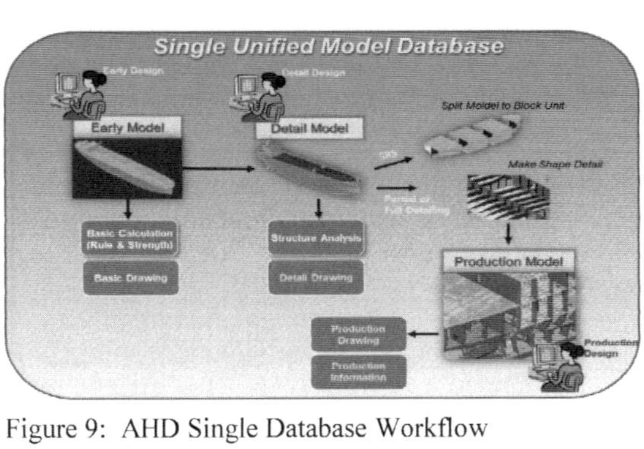

Figure 9: AHD Single Database Workflow

5.3 AHD PROCESS SAVINGS

SHI achieved design man-hour savings with this new workflow in the following areas:

S3D Modelling – 15% Man-Hour Increase
The AHD process created additional modelling activities in early design, primarily working through all of the early design modifications, thus the total modelling man-hours increased.

Drawings – 52% Man-Hour Savings
Since the user can generate all drawings from the S3D model with automatically placed labels, SHI can reduce inconsistency between the plan, elevation and section views, which were drawn independently in the old workflow. Also since the designers are communicating through the single unified model, SHI can reduce the number of drawings needed for the discussion.

Weight Calculation – 89% Man-Hour Savings
Since the designers can get the weight for the production prediction periodically from the S3D model using automation, they could save man-hours compared to the old manual calculation based on 2D drawings and Excel spread sheet.

FE Modelling for Structural Analysis – 75% Man-Hour Savings
Since the early designed S3D model is interfaced directly to KR HullScan software using the neutral format XML, the designers could save the man-hours by eliminating the creation of a separate model for FE analysis in the each FE software.

Class Rule & Strength Calculation – 83% Man-Hour Savings
Since the S3D model is interfaced to rule calculation software with the neutral XML, the designers could save the time to spend section modelling (about 30 sections/ship).

Man-hour savings are summarized in Figure 10.

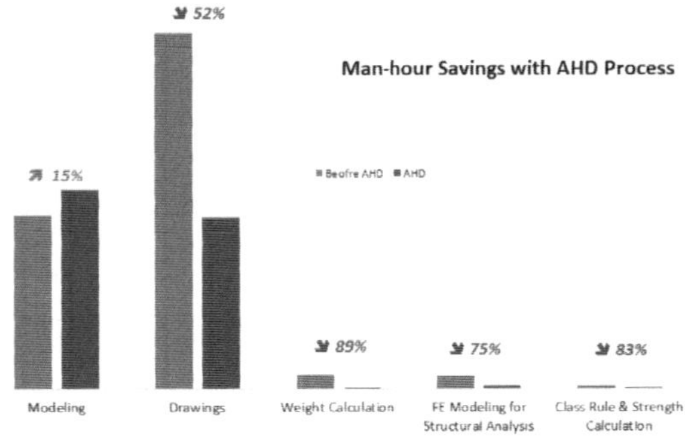

Figure 10: AHD Process Man-Hour Savings

© 2017: The Royal Institution of Naval Architects

Design Cycle Time Reduction
In addition to man-hour savings, the most important result of the AHD project is to reduce the total design duration from 11 months to 8 months.

6. ADVANCED HULL DESIGN RESULTS

6.1 BENEFITS

Combined, the design man-hour savings of the AHD process represent a 30% overall savings over the legacy design process. These man-hour savings are a direct result of starting the 3D model during early design and building upon the same model for both structure and outfitting through detail and production design. As part of the AHD process, both classification and detail construction drawings are generated directly from the 3D model. In addition to drawings, model geometry and attribute data is extracted via the neutral format XML interface for design rule checking and finite element analysis. This interface is providing both man-hour savings and quicker turnaround in these areas.

Weight and center of gravity data can be reported from the 3D model and managed in SHI's in-house weight management system. This provides accurate weights early in the design process that can be easily updated, tracked and managed throughout the design cycle. Fast, accurate material and work processes estimates like welding lengths and paint areas can also be made early in the design process directly from the 3D model.

More important than the man-hour savings is the compression of the overall design cycle. The AHD process has reduced what was previously an 11 month design cycle by 3 months. This allows the 3D modelling of outfitting to begin over 3 months earlier in the design cycle because 1) the structure model starts earlier in the design cycle and 2) it reaches a sufficient level of maturity much more quickly.

6.2 LESSONS LEARNED

The benefits derived from the AHD process, reduced man-hours and cycle times were largely the expected results of the project. However, some of the lessons learned were surprising and unexpected.

Moving the 3D modelling process earlier into the design cycle changed the profile of the 3D design modeller. In the SHI legacy process, production design S3D modellers worked from detail design drawings created by the 2D drawing-based legacy design process. These designers could be lower skilled resources with less understanding of ship structure engineering. In the AHD process, the initial 3D model is being created by highly skilled engineers who do not have the benefit of drawings to work from. They are creating a 3D model from which drawings will be extracted.

Thus, the AHD process required a 3D model that can be easily and productively created and modified because 1) expensive, highly skilled engineers were creating the structure model and 2) early design is a period of rapid and sometimes extensive change in the ship's structural design. S3D features enabling this rapid and productive early stage design changes include:

- Associative, topological modelling architecture that allows changes to propagate through related objects in the model.
- Partial detailing that allows necessary details to be viewed in the model during early design while maintaining a lightweight, easily modified data structure.
- Efficient data reuse capabilities supporting the quick, easy copy of large sets of structure both within a model and from outside of the model.
- Top-down structure modelling capabilities including the ability to split root systems which enable an efficient transition from early design to detail and production design.
- Interfaces that allow 3D geometry and attribute data to be share between the 3D model and 3rd party analysis and rule checking systems.

Finally, the AHD process both reduced the number of drawings required during the overall ship design process and it simplified the structure drawings developed using the process. The overall number of drawings was reduced because many downstream design activities that previously worked from structure drawings, now go directly to the 3D model for their structure background information. Structure drawings were simplified because, during the process re-engineering, obsolete information no longer needed was identified and eliminated. This step-change moves SHI closer to the vision of a paper-less design and build process based on digital product models.

7. FUTURE CONSIDERATIONS

While good progress has been made with SHI's AHD process, work remains. On the structure side, there is the industrywide initiative for paperless or plan-less ship classification approval. Both SHI and Intergraph are working with multiple class societies to both enhance the S3D neutral format XML interface and to expand its use to additional analysis and rule checking applications. The ultimate goal here is an industry standard interface that will both move design data from ship design systems to class analysis systems and bring scantling and other design modifications back from analysis.

On the outfitting side, SHI is working on an Advanced Outfitting Design (AOD) process that will provide similar savings in man-hours and cycle times for outfitting design. The AOD effort is focused on the

© 2017: The Royal Institution of Naval Architects

ability to extract an entire space's worth of multi-discipline design data from an experienced ship design (existing model) and "catalogue" that design data in a S3D design library database from which it can be copied into a new ship design model. Once this set of multi-discipline data is copied into the new ship model, it can be modified to fit and reconnected to integrate into the new design. Work is ongoing to optimize the sets of data within the design library database with regards to:

- Granularity of the data with regards to discipline, level of detail and topographic extent.
- Volume of data that can be effectively managed.
- Tools to efficiently identify, catalogue and manage data within the design library database.
- Tools to automatically modify rule based design data upon copy such as update a piping system to a new piping specification.

8. CONCLUSION

Today, many shipyards still rely on 2D drawing-based workflows for all or portions of their ship design process. This is true even for large, technologically advanced shipyards. Advances in ship design software and computer processing power continue to make ship design using 3D models faster and easier than ever.

Opportunities for real savings, in costs and project times, exist for shipyards willing to break with tradition and adopt or expand the use of 3D design in their ship design process.

Design software features like top-down design, efficient data reuse, automated drawings and interfaces to other applications like analysis and rule checking make it easier and more effective than ever to expand the use of a single design system and database into more phases of the ship design process eliminating duplicate work, improving data consistency and moving closer to the goal of the 3D model being the single source of truth for ship design.

9. REFERENCES

1. HULKKONEN, T., KANG, B.S., CABOS, C., TIETGEN, B., KIDOGAWA, M., *Process Innovation for Hull Design at Samsung Heavy Industry*, International Conference on Computer Applications in Shipbuilding (ICCAS), 2015.

2. BAUMER, J., KIM, Y.G., *True Benefits of 3D Design,* International Conference on Computer Applications in Shipbuilding, (ICCAS), 2011

3. LIMON DUPARCMEUR, Y., BUTLER, J., PATIENCE, R., *Experienced Marine Design and Data Use,* International Conference on Computer Applications in Shipbuilding (ICCAS), 2007

4. POLINI, M., *A Neutral XML Schema for Basic Design Stage Interface to Class Societies*, International Conference on Computer Applications in Shipbuilding (ICCAS), 2011

10. AUTHOR'S BIOGRAPHIES

Joseph Baumer holds the current position of Executive Technical Director at Intergraph and is based in Hampton, Virginia, USA. There, he is responsible for Product Management of 3D Marine and Structural applications. Joe has BS and ME degrees in Engineering with 15 years of shipbuilding industry experience. He has worked with Smart 3D marine applications since their inception.

Jinsup Cheong holds the current position of Senior Manager at Samsung Heavy Industries. There, he is responsible for the customization of S3D structural application. He has 20 years of experience in S3D customization from 1997 to now at Samsung Heavy Industries.

Sungho Ha holds the current position of Vice President at Samsung Heavy Industries. There, he is responsible for Structure Design Team. He has 30 years' experience at the hull structure design and involvement in Process Innovation of GS-CAD (Prior name of S3D) since 1997.

Jinmoo Hong holds the current position of Senior Manager at Samsung Heavy Industries. There, he is responsible for the structure basic design. He has 15 years' experience at the hull structure design from 2003 to now at Samsung Heavy Industries.

Heewon Lee is Intergraph's Marine Implementation Manager based in Goeje Island, Korea. Mr Lee is responsible for pre-sales activities and leading deployments of Intergraph products at Marine customers around the world. Mr. Lee has been with Intergraph for 5 years. Prior to joining Intergraph, Mr. Lee worked in the shipbuilding industry for 15 years.

© 2017: The Royal Institution of Naval Architects

© 2017: The Royal Institution of Naval Architects

INTENT-DRIVEN CAD VS. MECHANICAL CAD IN SHIPBUILDING – A REVIEW AND SOLUTION OUTLINE

T Dusch, B Franke, M Grau and **C Zerbst,** PROSTEP AG, Germany

SUMMARY

Ships are complex, often unique one-of-a-kind products. Even sister ships can significantly differ from their prototype. They are made of hundreds of thousands to millions of single parts and components which are created and positioned as part of the design process, managed in bills of materials for part manufacturing or purchasing and assigned to assembly stages for production planning and logistics. Being able to carry out highly integrated and concurrent cross-discipline design is essential for shipyards to be competitive.

The toolsets used to design and manage these complex products are either dedicated "intent-driven" ship building systems or based on more general mechanical CAD systems. The differences between these two approaches are significant. We will show the ways parts are created and stored, organized in structures and BOMs as well as look into topological relationships, space-based navigation and how they are realized in those systems. Semantics is important if it finally comes to exchanging ship design information or feeding downstream processes. Intent-driven systems carry semantically rich primary information that can be used to generate 3D geometry, 2D drawings, and topological relationships as required and when needed. This can even be used to feed mechanical CAD–based systems. The reverse direction however is the supreme discipline and is harder to achieve due to weaker semantics of information on the source side.

The authors of this paper have successfully realized exchanges in both directions. We will report about this challenging endeavor and show the preconditions and limitations of CAD conversion in shipbuilding. This paper was triggered by a converter development project carried out to transfer both shipbuilding steel and outfitting for a complete cruise liner ship between two shipyards from CATIA to AVEVA Marine Hull and CADMATIC Outfitting.

NOMENCLATURE

API	Application Programming Interface
CAD	Computer Aided Design
COTS	Commercial of the shelf, software without customization
HVAC	Heating, Ventilation and Air Conditioning
JT	Binary format used to transfer product structure and geometry, ISO-14306
MDL	Model Description Language
PLM	Product Lifecycle Management
STEP	Standard for the exchange of product model data, ISO-10303.

1. INTRODUCTION

Looking at the CAD tools currently used in the shipbuilding industry two distinct basic approaches can be found. In this paper, we are naming these two different approaches

- intent-driven
- mechanical CAD-based

The idea for the intent-driven approach goes back to the time when numerical controlled cutting machines in shipyards needed to be fed with geometric information to make their operation more efficient. At this time, both processor power and memory space in computers were limited resources. Therefore, ways had to be found to store the plate geometry in a very efficient way.

The trick that did the job was to store the recipe how to create that geometry as a little "script" – and thereby lean ASCII text – and recreate the geometry every time it was needed instead of storing the – more expensive – explicit geometry information. Over time, the "script language" evolved into a semantically rich set of commands including parameters (such as plate thickness and material offset) and topological references to other structural elements thereby implicitly defining plate boundary or more general part body geometry.

It quickly turned out that not only memory consumption was reasonably low, but also the way to create and maintain the CAD model was more like programming than drawing. This enabled structural designers to maintain really large models of entire ships without a need for expensive graphical interaction making them very efficient. The intent-driven CAD system approach is therefore defined as not to persist explicit part geometry but rather the recipe and the ingredients to "cook" – i.e. create – the part whenever needed.

The mechanical CAD based approach on the other side, has its origin in the application of numerical methods to describe geometry such as Bézier splines which eventually evolved into full support of all kinds of bodies and surfaces featured by todays 3D CAD tools. This approach was mainly driven by industries like automotive, where the ability to work with complex free form surfaces allowed much better designs in a much shorter time and without building physical mockups for car body shapes. The central entity in a mechanical CAD system is the geometric model per part, where "part" can be any-

© 2017: The Royal Institution of Naval Architects

thing, from e.g. a complete car body down to nuts and bolts.

The widespread application of mechanical CAD systems in many industries created the need for additional capabilities such as the concept of "solid" (bound by a free form surface or described using analytical math and then called "primitive"). It became the basis for the concept of "constructive solid geometry" (CSG) which allowed to combine several solids to a new more complex one using Boolean operations. This eventually evolved into the concept of "feature" which basically is a negative solid such as a hole or a chamfer applied to a solid.

Current mechanical CAD systems allow defining geometrical bodies and surfaces while keeping the controlling parameters of the geometry changeable. Nevertheless, the major aspect of mechanical CAD models has persisted until today: It is centered on an explicitly stored geometry model. The mechanical CAD system approach is therefore defined as to hold explicit part geometry.

Since mechanical CAD system vendors started to enter the shipbuilding industry two decades ago, both approaches compete for the same customers and more and more converge, offering similar functionality. The intent-driven CAD tools today enable the designer to interact with geometry in a 2D and 3D environment, replacing the script editor in the daily work. Nevertheless, the rich model intent is still recorded from the interactions and stored in a script-like manner. Although the mechanical CAD system approach supports parametric features, it did not know about domain specific terminology like plates or stiffeners in the beginning. By adding this information on top of the geometry over time, today systems support definition of plate thickness and pipe diameter.

This is not at the latest a result of the "architecture, engineering and construction" (AEC) approach invented to support the requirements of the civil engineering industry.

Today's shipbuilding CAD tools do not follow exactly one of the two approaches anymore, but rather tend to combine the advantages of both. Nevertheless, the differences, even if not visible to a normal designer, still exist in the internal data model. Those tools originating from the intent-driven approach still tend to have rich semantics, whereas some of the shipbuilding semantics in the CAD based tools is contained in the geometry. This plays a role, when converting data from one approach to the other.

Drivers for such conversions are collaboration scenarios between shipyards or the need to integrate the models of different tools for shipbuilding steel and outfitting towards a common ship data model. In an ideal world, the CAD tools would offer data exchange based on protocols like STEP AP218 (Ship Structure) or AP227 (plant spatial configuration) as export or import format to enable a loss-less transfer within design tools or for calculation or classification perspective [1].

In reality those interfaces are not available in any of the COTS tools. At best, there are vendor specific formats offering an intent-based export or import. In most cases the only option is a pure geometry export and import. Size restriction on the import side partially cause severe limitations.

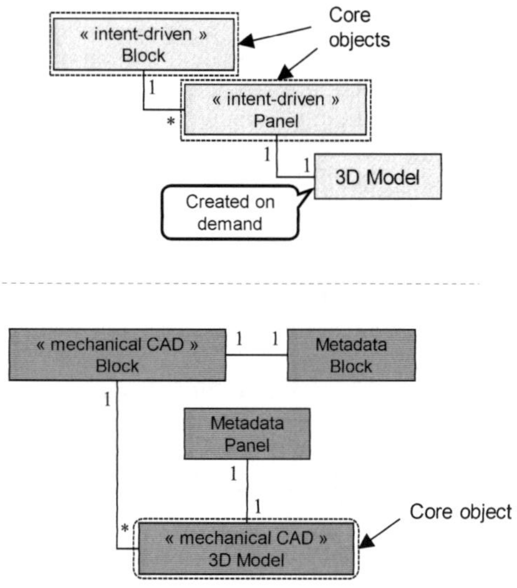

Figure 1: Intent-driven and mechanical CAD

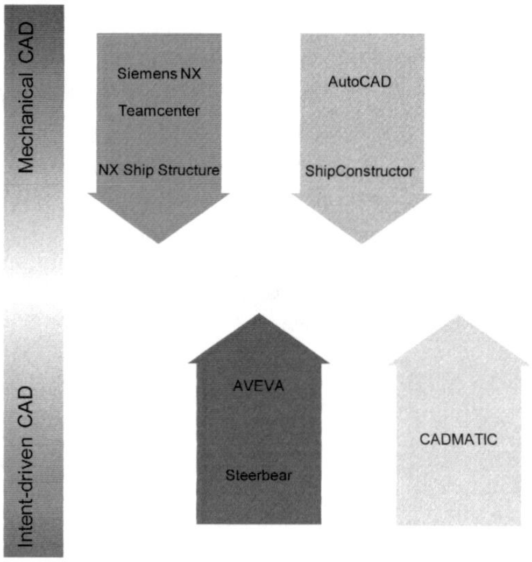

Figure 2: Historical convergence of the CAD toolsets

© 2017: The Royal Institution of Naval Architects

Some experience about conversion from intent-driven to mechanical CAD based tools can be found in [2], [4]. Developing a converter for the opposite direction, i.e. mechanical CAD based to intent-driven for ship steel structure and outfitting meant a new challenge which was successfully carried out in 2016 for a complete model of a large cruise liner. This article highlights the results of the conversion and some of the steps needed to get there.

2. CAD TOOL MARKET ANALYSIS

The principle of the internal structure of the CAD tools following the "intent-driven" or "mechanical CAD-based" approach is visualized in Figure 1. In the first approach, shipbuilding objects representing the intent are the core objects. The resulting 3D geometry is nowadays cached in some CAD tools, but still is no bearer of additional information. In the second approach, the 3D models are the core objects, they are referenced by additional metadata to carry shipbuilding specific information.

To give an overview of the state-of-the-art for data models, semantics and import/export possibilities, four CAD- / PLM vendors have provided support for this publication, which are displayed in Table 1. Among the participating software tools, some systems have their origin in the mechanical CAD area (Siemens PLM / NX & Teamcenter and SSI / ShipConstructor) and others in the intent-driven area (AVEVA / AVEVA Marine Hull and CADMATIC / CADMATIC).

Regarding the evolution of the tools and the shipbuilding sector, separated semantics and data models were needed to fulfil the requirements in its entirety. Therefore, it is not easy anymore to determine if the tool is purely either one or the other type of CAD system. As told by the vendors themselves, SSI (ShipConstructor) and Siemens PLM (NX/Teamcenter and NX Ship Structure) have a mixed environment with partially mechanical and intent-driven CAD based approaches. On the other hand, AVEVA and CADMATIC have also moved in the direction to give the user the capabilities mechanical CAD systems have in some parts of the toolsets. In general terms, this means that the vendors are moving in the same direction of mixed environments as it is schematically displayed in Figure 2.

Table 1: Participating vendors and relevant tools

Company	Shipbuilding Steel	Ship Outfitting
AVEVA	AVEVA Marine Hull	AVEVA Marine Outfitting
CADMATIC	CADMATIC Hull	CADMATIC Outfitting
Siemens PLM	NX, Teamcenter	NX, Teamcenter
SSI	ShipConstructor	ShipConstructor

2.1 REFERENCE CAD MODEL

To investigate the different tools more deeply, all participating vendors were asked to replicate a reference CAD model and provide information about used data models, topological information and applied constraints.

The neutral 3D model which has been distributed to all vendor participants is containing a variety of objects from shipbuilding steel to outfitting (e.g. piping, air condition, raceway). The motivation for this task is to gain necessary insights into the toolsets and to understand how objects are stored, data is handled and references are generated. In detail, the reference model contains objects from the disciplines as specified below.

- Steel: panel, frames, flanges, cutouts, stiffeners
- Piping: pipes, elbows, flanges, valves, hangers
- HVAC: ducts, elbows
- Electric: raceways, bends

For all objects within the reference model, dedicated locations, design parameters and topological relationships have been provided to guarantee that all toolsets are replicating the scenario in the same manner. To help the vendors in creating the documented model, a STEP-file containing the pure geometry was provided. This also allowed to compare the new model with the distributed one. Some excerpts of the provided model specification are given in Table 2. These examples provide explanations for steel objects (stiffeners) and outfitting (pipes and raceways). The locations and topological relations are given with general settlement information (e.g. deck) and distances to other objects, whereby the design parameters describe the geometry of the items.

The results of the reference model are then displayed in Figure 3, where all four native vendor models are presented. At first sight, it is obvious that the models do not differ considerably from each other.

Table 2: Overview of location, design and topological parameters allocated for the reference CAD model

Object	Location	Design Parameters	Topological Relationships
Stiffeners	Deck 2, Section 1	Type: HP 100x6. Straight endcut. Stiffeners length: 6320mm.	Connected on panel plate upper side.
Pipes	Parallel to shorter frame.	Standard pipes with nominal size of DN50.	Distance between parallel pipes: 850mm. Distance between shorter frame and both pipes: 200mm. Shortest available distance between panel plate and pipes: 350mm.
Raceways	Parallel to shorter frame.	U-shaped trays: Height: 84mm, Width: 200mm. One straight raceway and one raceway with elbows.	Distance to steel panel: 200mm. Shortest available distance between trays: 750mm.

© 2017: The Royal Institution of Naval Architects

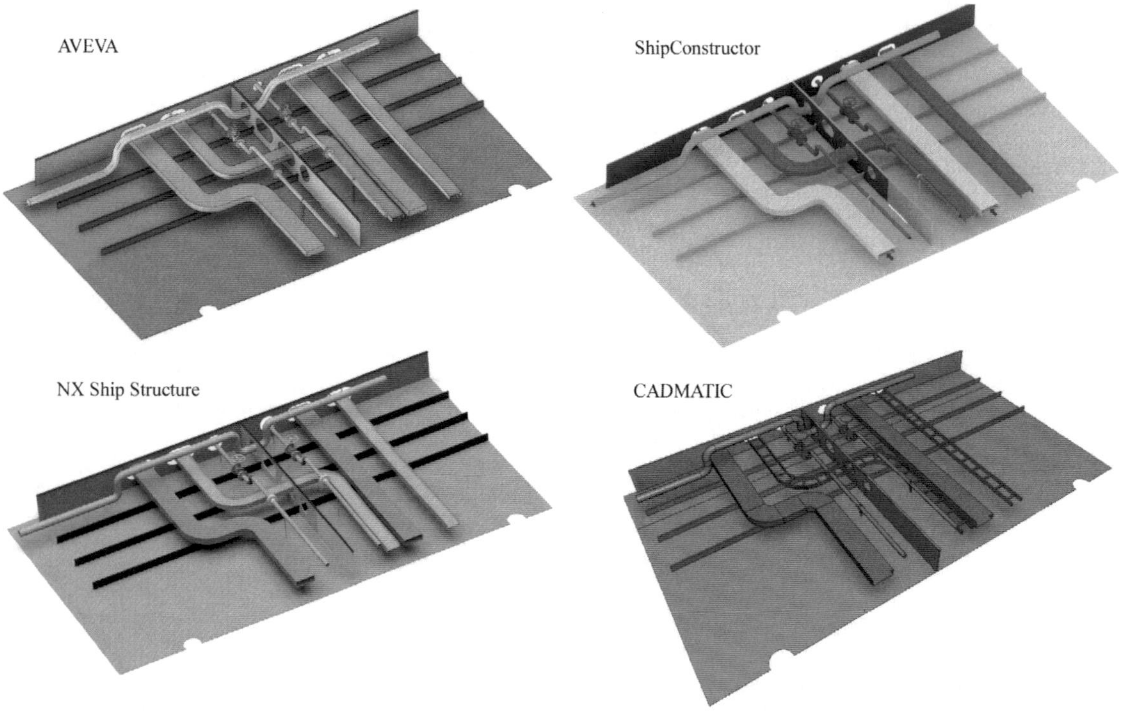

Figure 3: Reference model with shipbuilding steel and outfitting objects, created from all vendors

With this common basis, it is possible to investigate a variety of aspects within the CAD/PLM systems, using a detailed and corresponding questionnaire: information regarding the tool type, data model, steel or outfitting specific information and possibilities for the input and output formats.

2.2 DESCRIPTION OF CAD AND PLM DATA

As explained above, the basis for the data model is the major differentiator between the analyzed toolsets. Nevertheless, the questionnaire has shown that there are not only differences between all participants and the originally mechanical or intent-driven CAD systems, but also similarities. These will be summarized within the next paragraphs and separated into three main topics: shipbuilding steel, ship outfitting and general aspects.

2.2 (a) Shipbuilding Steel

All tools examined support ship basic design – e.g. the definition of panels or blocks without details like endcuts or brackets – as well as ship detailed design – e.g. fully defined steel structure with all details like clips or notches. However, different methods are applied for creating panel boundary definitions between the more mechanical and the intent-driven systems: On the one hand, the utilized/preferred method is purely geometric; on the other, using references to the shell/panels is the way of working.

With respect to the question if geometry is automatically recalculated when shifting a referenced part (e.g. shell, deck or panel), the approaches are not homogeneous: It is feasible for some, but not for all systems. There are no differences when it comes to the definition of customer-specific profile types (for example, special profiles used for railings, balconies or windows) and endcut formats. This is possible regardless of the used approach in all investigated systems.

Within the design process, panels can be developed as planar, curved, knuckled or even corrugated objects. The handling of these objects is not uniform in the analyzed tools. In some, processing and management is identical through the whole design phases, in others, differences appear while editing the panels.

All toolsets provide the capability to automatically split and merge panels, not only in basic design, but also in detail design. The differences arise when having a deeper look at the effort which is necessary for these actions: for almost all, merging is more laborious than splitting.

2.2 (b) Ship Outfitting

All analyzed tools include capabilities to design and evaluate the necessary aspects of the different disciplines of piping, HVAC, cableways, cable routing, outfitting structures and equipment. Furthermore, not only the

© 2017: The Royal Institution of Naval Architects

disciplines are uniformly supported, the integration between schematics (2D) and 3D is also realized. The only difference is that not all systems offer a consistent solution for all kind of reviews, like comparison checks (back and forth) or update commands for synchronization.

If the user wants to generate pipeline or HVAC networks, it is mandatory in all analyzed tools to associate it to dedicated systems, i.e. cooling seawater system or potable water system. In addition to this, all the systems are designed as specification-driven tools. At least one specification is associated to a certain system and can be understood as a filter on the general catalogue with all parts inside.

The specification-driven attempt gives administrators the authority to provide only permitted parts inside the systems. The modelling of piping, HVAC and raceway supplementing elements (insulation and hangers) is obtainable in all tools with 3D data and further, mostly realized with additional metadata.

2.2 (c) Miscellaneous

Besides the given discipline specific descriptions, general topics are of great interest. Cross domain references, between shipbuilding steel and outfitting, are fully supported by some systems, and not at all or only partially feasible by others. On the contrary, it is possible in almost all systems to create and maintain references (e.g. positioning or dimensioning constraints) between objects belonging to the same discipline.

Both system types, mechanical CAD-based and intent-driven, provide capabilities to associate spaces to rooms or volumes. However, the tools are offering different methods to 'ad-hoc' collect and display a subset of objects belonging to a specific location (e.g. 'deck 8' and 'section 6'), using volume search, attribute queries or grouping functionalities. The metadata which is needed to grant capabilities for space allocation can be assigned to the objects either as user-defined attributes or within an automatic calculation at run time. Here again we have different behaviors among the toolsets: some offer an explicit working method (stored on the object), others automatically associate them at run time.

In terms of input and output formats, all systems offer a variety of different options. Pure geometry export as DXF, AP242 or JT is available in all systems, the same applies for geometry import. The support for high-level formats is less then optimal. PCF format could be used for outfitting export or import, but PCF import may again result in pure geometry. None of the systems supports neutral high-level formats like STEP AP218 (shipbuilding steel) and STEP AP227 (outfitting) for export or import.

3. CONVERSION BETWEEN CAD SYSTEMS

Ship detailed design to define panels or pipes is normally created in one single system. This is then also used to create the manufacturing information for developed plates or pipe bending information. Applying different tools for the same domain is uncommon in the industry. This is caused at least partly due to the poor export / import capabilities. In an ideal world, the CAD tools would offer a high-level protocol like STEP AP218 (Ship Structure) or AP227 (Plant Spatial Configuration) to enable a lossless transfer. Though some academic efforts took place in late '90 to early 2000', no high-level protocol is supported by COTS tools. The available formats are geometry-oriented and do not support exchange of shipbuilding semantics.

As result, multiple CAD tools are only exceptionally used during detailed design. A typical scenario are engineering offices providing complete work packages for a yard with a different CAD tool. In those cases the engineering office typically provides the complete manufacturing data like drawings or nesting sketches and the other resulting geometry is loaded as pure geometry model for coordination purpose in the yards system. The geometry is generated using either the capabilities of the

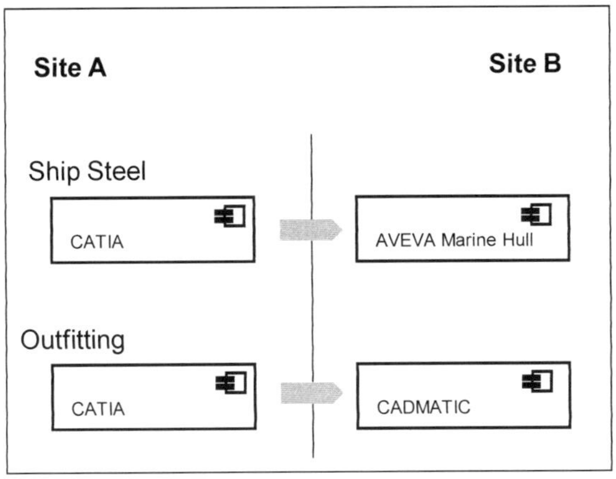

Figure 4: Tool used on the sites

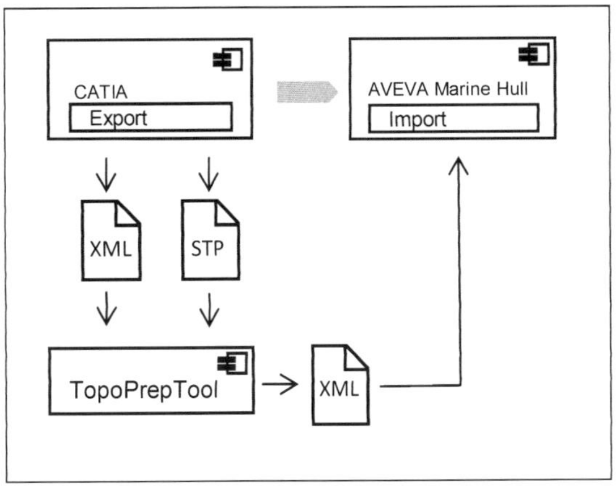

Figure 5: Steel Conversion Chain

© 2017: The Royal Institution of Naval Architects

COTS tool or by 3rd party tools as implemented in the past [4]. Similar scenarios are found if multiple yards with different tools share navel projects. Coordination work is then based on pure geometry, whereas manufacturing information is directly created from the source systems.

This was the state-of-the-art when PROSTEP was asked to convert a complete cruise ship from a mechanical CAD based tool to an intent-driven CAD tool. The customer's motivation behind was to build a series of similar ships, but to design and manufacture them at two different sites using different CAD tools. His idea was to reuse the design work done on the first prototype as starting point for a ship to be built on the second yard. The tools in question were CATIA on the source site and AVEVA Marine Hull for ship steel and CADMATIC for outfitting on the target site as seen in Figure 4.

Two viable options to transfer the design were worth considering:

- Redesign everything in the target CAD tool based on the drawings of the prototype design
- Create a conversion toolchain to create a native model in the target CAD tool

To evaluate the second option, a feasibility study was performed with a small subset of steel items and outfitting types. The study allowed to estimate the effort for the implementation of the needed toolchain and to compare it with the estimated effort for manual re-design. As a result, the conversion of steel from planar panels and all outfitting types was identified as viable solution both from the commercial as well as from the technical point of view. Only for curved panels, manual conversion was favored for two reasons:

- Different sizes of shell plates used on the yards make 1:1 transfer useless on the target site.
- AVEVA Marine Hull uses a different way to describe planar and curved panels. Adding support for curved panels nearly doubles the implementation effort

3.1 SHIPBUILDING STEEL

The shipbuilding steel of the cruise ship consists of 30 blocks with approximately 400.000 single parts. The prototype to transfer from site A to B is designed in CATIA for the basic and detailed design, whereas the target site uses AVEVA Marine Hull as CAD tool. As said before, neither the source tool nor the target tool

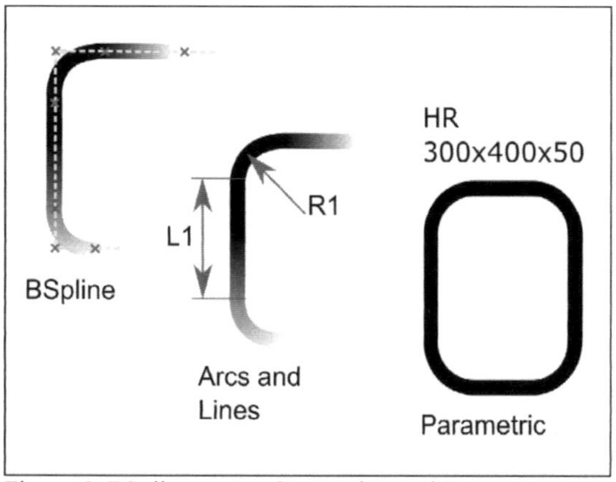

Figure 6: BSpline to Arc Conversion and Feature Recognition

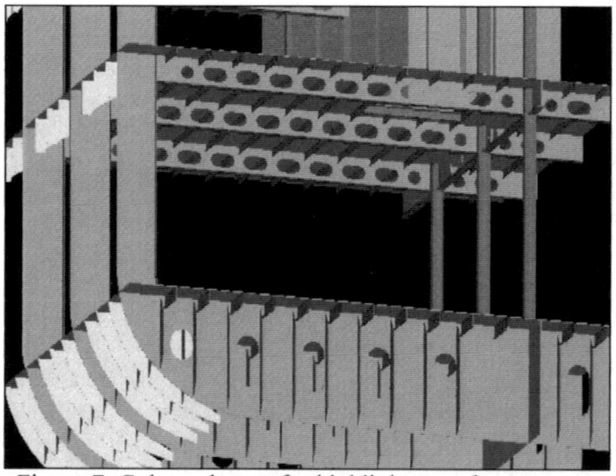

Figure 7: Coloured parts for highlight transfer status

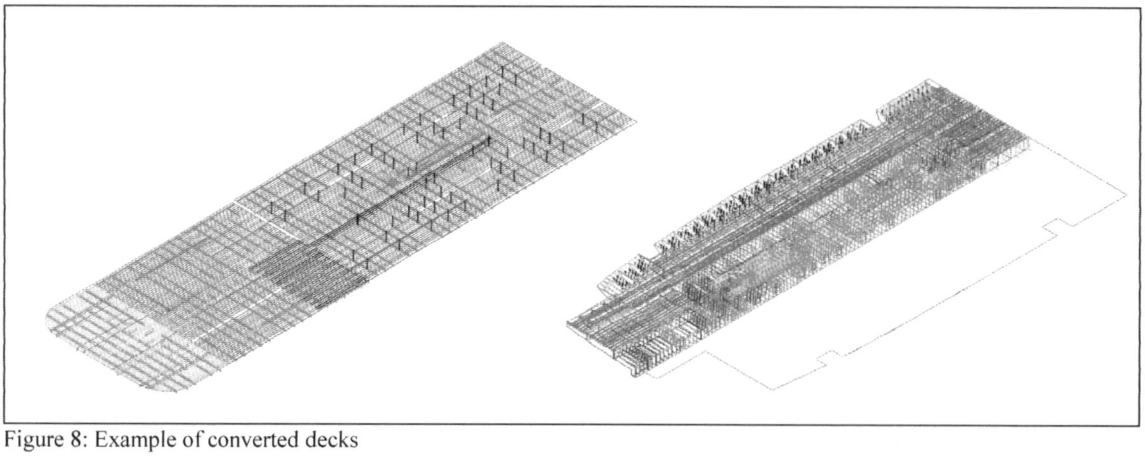

Figure 8: Example of converted decks

© 2017: The Royal Institution of Naval Architects

provides a high-level interface to support a lossless transfer out of the box. Lossless means to have the shipbuilding steel available in the target CAD tool as native data in detailed design quality with entities like panels, plates, stiffeners, brackets, parametric and non-parametric holes, pillars, etc.

Native data is a must have for this project, to be able to apply the typical sister-ship changes and to support manufacturing on site B. In addition to the pure IT-related mapping from one CAD system to the other, some business level mapping was required right from the start. As the naming rules for parts and standard items like holes and end-cuts differ on the yards, additional name mapping work was required in the conversion.

The overall architecture of the implemented conversion chain as shown in Figure 5 was driven by available file interfaces and APIs in the CAD tools. An existing, customer-specific plugin exports the steel data on the source side in a combination of design-intent as XML. Additionally, standard geometry export in STEP AP203 format is used as not all the necessary design-intent information is available for the export. The design goal for this component is to simply dump available information without applying additional logic to it.

The component "TopoPrepTool" is used to evaluate the geometrical and topological information, with the objective to create a dataset containing a topologically correct model and geometry in AVEVA compatible format. The last component is then the import plugin deployed within AVEVA Marine Hull containing the needed import logic. It reads the result file from the "TopoPrepTool" and feeds it into the CAD tool using the available APIs.

Some experience we gained when implementing the toolchain could be valuable for similar conversion projects in the future. Depending on the source CAD tool, the full intent information for an entity is not always available in the desired quality. This can be explained with the example of a hole. On the project, the kind of hole (parametric round, rectangular or user defined contour) was not specified. That's why the "TopoPrepTool" performs three steps on each hole found in the geometry given as a STEP file to resolve the issue (see Figure 6):

- Find the plate geometry in the STEP file and associate it to the plate in the design intent
- Find all hole contours in the plate geometry
- Convert the BSpline hole contour to a minimum of arcs and straight lines
- Check if hole is one of the standard hole types and – if so – extract their parametric description

In case of a non-standard hole, performing these steps allows us to import at least a hole using the arc-based geometry description normally used in AVEVA Marine Hull. Whenever possible, the imported hole has a full parametric description based on type and parameters, although these parameters where not available from the source system. Implementing such a feature recognition greatly increases model quality compared to a pure geometry transfer.

A common problem when implementing conversions is handling of unexpected or incorrect input data. A 100% conversion rate is a dream. In real life, there will always be some entities which cannot be converted, because they are incorrectly modelled, because their conversion was not foreseen when implementing the converter, because it is technically impossible due to restrictions on the export or import side, or simply because it would be too expensive to implement. Stopping the complete transfer on those parts is not a viable solution, but simply ignoring them is neither.

That's why the import tool uses a multi-step import with decreasing quality. An example for this is the import of panels into AVEVA Marine Hull. The first import attempt uses the full topological description with references to the grid or other panels. If the first attempt is not successful due to missing or circular references, the missing topological references are replaced by a geometrical description. If even this fails, the complete panel is defined by its pure geometrical contour. These steps are implemented using the available APIs, making the resulting import process much more robust than simply writing input schemes and trying to import them.

The designer on the target tool somehow needs to know, what parts are correctly transferred, what parts have issues and what parts are completely missing. As they normally refuse to read reports and log files, the parts are color coded during import. The brackets and floor panels seen in Figure 7 are either partly (free edge of floor panels) or completely (brackets) imported using their contour. The designer could now easily identify the parts which may need some manual re-work after import using the color code.

The complete conversion and import process takes about 2 hours for a deck of 750 panels. The resulting panels have native quality which allows manual changes to adapt to the sister-ship with the same functionality as if the panels would have been designed directly in AVEVA Marine Hull. Features omitted in the conversion could be added by detailed designer in the normal design work. Some examples of the converted ship model are contained in Figure 8.

3.2 SHIP OUTFITTING

The second conversion to be done was ship outfitting, containing several hundred systems with approximately one million parts to transfer. The prototype on site A is designed in CATIA, whereas the target site uses CADMATIC Outfitting as CAD tool. The task differs from that in the steel conversion where a new intent-driven model had to be created based on only few features like panels, plates or stiffeners. Outfitting has a

© 2017: The Royal Institution of Naval Architects

much richer data model covering multiple domains like piping, HVAC, cable trays, support structures made from profiles and plates etc. A second difference is that in outfitting at least half of the data is created by placing existing catalogue items in their correct position instead of describing them from scratch. That's why an outfitting conversion has three aspects:

- Mapping the items from the source catalogue to their counter parts in the target catalogue
- Creating the right placement of fixed size catalogue items (valves, flanges) in the target system
- Creating the right placement of flexible sized items described by intent + dimensions (plates, pipes, ducts) based on catalogue material (sheet material, pipe material)

A mandatory precondition of any conversion in outfitting is harmonization of the catalogues on both sides. At first look, the task seems quite simple, as e.g. the number of flanges or sheet material found in the catalogues seems to be manageable. In reality, a project may contain 10 to 15 thousand different catalogue items, some of them used only once, others several thousand times. Identifying equivalent catalogues and adding missing items is a manual task and is handled in parallel to the conversion process. To drive this catalogue setup, some dedicated report tools help the catalogue team to find missing catalogue entries and get an overview on their usage counts.

The architecture of the conversion chain is implemented as found in Figure 9. On the source side, an export plugin deployed in CATIA exports the outfitting data, combining design-intent plus some 2D geometry for plate parts in a text file. There is no need for a 3D model, as the input data contains enough intent information. The idea behind the component is to simply dump available information without applying additional logic to it.

The second component "MDLConverter" is used to convert the design-intent description from CATIA into an MDL file. This tool was designed with the objective to apply it to any conversions needed to create an equivalent intent description for CADMATIC. This includes mapping of the intent description and used geometrical descriptions. The geometrical aspect is mostly concerned with two aspects. One is the different reference points used to describe the same physical product. E.g. branches in pipes are described by the intersection points of the two centerlines in the one system, the intersection point of the centerline and the cylinder in the other system. A second aspect is moving of reference points to bypass omitted objects. There is no third component needed to import the MDL files into CADMATIC, as it is part of the COTS tool. The import capability was enhanced by the vendor to apply a predefined catalogue mapping. This has the benefit, that the catalogue definition already managed in the target tool is available when defining the mapping of standard components.

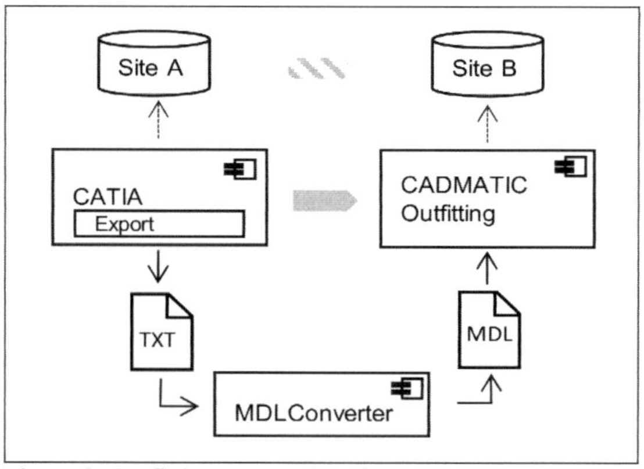

Figure 9: Outfitting conversion chain

One of the difficulties when implementing outfitting mapping is to identify errors as there are two possible sources: catalogue setup and conversion chain. An odd-looking pipeline may have its root either in writing the wrong MDL, using the wrong catalogue mapping or both. Sometimes tiny differences like different slip on length for flanges causes disastrous results. These errors required quite extensive testing to gain the needed quali-

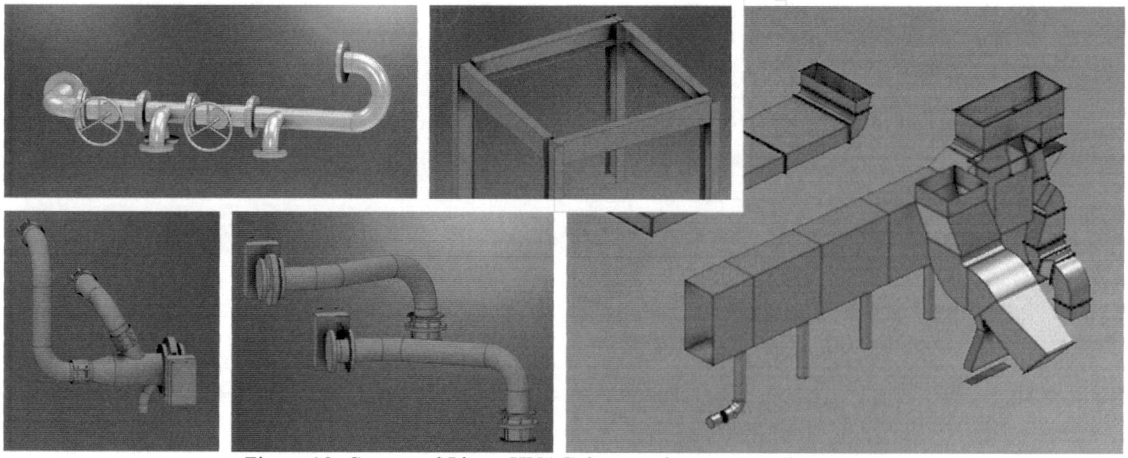

Figure 10: Converted Pipes, HVAC ducts and support steel structure

© 2017: The Royal Institution of Naval Architects

ty. To quickly find the underlying problem, traceability is a major concern. Especially when the catalogue setup was not 100% completed, it was very important to quickly identify the source part to a faulty looking item. By adding all needed information about the source part to dedicated attributes configured in CADMATIC, designers could access this information without searching in huge mapping reports. This helps to visualize the source item, check on correct mapping in either catalogue setup or automated conversion process and trigger a fix.

Another major concern for the conversion is the necessary time. A very time-consuming task is the creation of pipe runs, which are represented as continuous sequences of parts in CADMATIC, whereas CATIA manages individual parts. Finding adjacent parts out of a cloud containing ten thousand of pipes could be massively speeded up by using octree algorithms [3].

4. CONCLUSIONS

The realization, implementation and related safeguarding of the converter have been presented. For both, shipbuilding steel and outfitting, we were able to transfer more than 90% of the parts from the source CAD tool (CATIA) to the target CAD tool (AVEVA Marine Hull, CADMATIC Outfitting). The objective to spend less money on the conversion as compared to manual transfer was fully met.

This study reveals that the gap between intent-driven CAD tools and generic mechanical CAD systems is getting smaller over the time. To drive this, both categories of CAD systems are adding features like access to a pure geometric description on the one side, more and more intent information on the other side. We were able to close the remaining gap by applying technologies as described in this paper enabling even projects such as a full conversion.

In conclusion, to achieve this goal, some efforts are necessary on the IT and on the business side. To implement the conversion chain, some technology from our side is required to ensure a real high-level transfer, even if some intent is not readily available from the source side. By providing this feedback about the achieved quality, designers were enabled to easily find parts that needed manual intervention due to missing model quality. On the other side, the catalogue setup proved to be challenging. Developed report tools helped to provide the needed insights into the used catalogue items.

In total, converting a complete cruise liner as a high-level model from one CAD tool to another proved to be a manageable challenge.

5. ACKNOWLEDGEMENTS

We would like to acknowledge the support from software vendors AVEVA, CADMATIC, Siemens PLM and SSI when compiling the CAD tool survey. The following persons provided great help and input: Andrew Gordon and Jan Bitomsky (AVEVA), Ludmila Seppälä and Teemu Valtonen (CADMATIC), Bert Geisler and Markus Mende (Siemens PLM), Jason Paterson, Denis Morais and Tracy Stevens (SSI).

6. REFERENCES

1. WILKEN, M., CABOS, C., GRAFE, W., Using CAD Data for Computation and Assessment of Global Ship Structures, COMPIT'08

2. BAUCH, J., GRAU.M, ZERBST.C, Providing the Missing Link, Paper for ECTPS, April 2010

3. MEAGHER, D, Octree Encoding: A New Technique for the Representation, Manipulation and Display of Arbitrary 3-D Objects by Computer. Rensselaer Polytechnic Institute (Technical Report IPL-TR-80-111).

4. GRAU, M., SACHERS, M., ZERBST, C, Building the Business Process View into Shipbuilding Data Exchange, COMPIT'08

7. AUTHORS BIOGRAPHY

Thomas Dusch holds the position as "Senior Consultant" at PROSTEP. His professional background is mechanical engineering with a strong focus on both 3D Product Creation and PLM. Previous experience includes projects in the fields of CAD, CAD Data Exchange and CAD/PLM Integration for the aerospace, automotive and shipbuilding industry.

Benjamin Franke holds the position as "PLM Consultant" at PROSTEP. His professional background is production planning in mechanical engineering. At present, he is responsible for CAD/PLM integration, systems engineering and process assessment for shipbuilding customers.

Matthias Grau holds the position as "Principle Consultant" and "Account Manager Shipbuilding" at PROSTEP. His professional background is naval architecture. Previous experience includes PLM, interdisciplinary engineering data management between design disciplines and at the interface from design to manufacturing from various projects in shipbuilding industry and beyond.

Carsten Zerbst holds the position as "Manager Systems Integration" at PROSTEP. Carstens team provides consultancy and software implementation to customers in the shipbuilding, automotive and aeronautic industries. His previous experience includes CAD and PLM integration, conversion and migration projects in the design and manufacturing domains.

© 2017: The Royal Institution of Naval Architects

© 2017: The Royal Institution of Naval Architects

BOW SHAPE OPTIMIZATION OF A CRUISE SHIP FOR MINIMUM ADDED RESISTANCE BASED ON ACTUAL OPERATIONAL PROFILE

P Y Feng, J Y He, S M Fan, J B Wang, Marine Design & Research Institute of China (MARIC), China

SUMMARY

Prediction of added resistance in waves is getting more and more attention in ship design, because of its high economical interest and environmental influence. This study focuses on the design optimization of a cruise ship for minimum wave added resistance based on her actual operational profile. A state of the art numerical code for the estimation of wave added resistance is developed based on a fast and reliable semi-empirical formula and is further extended to incorporate the estimation of ship's mean wave added resistance along her route based on long-term wave statistics. The optimization process based on a fully parametric model is formulated within the CAESES software. Extensive computations are performed to investigate the influence of hull geometrical parameters on the ship's wave added resistance. The long-term mean wave added resistance of the cruise ship can be reduced through the proposed optimization process.

1. INTRODUCTION

With the rapid advancement of fluid hydrodynamics and computer-aided design (CAD) technologies, it is nowadays common practice to perform design optimization for the ship hydrodynamic performance in order to obtain the hull form with minimum calm water resistance. However, according to statistics, there is slightly more than 11% probability worldwide that an open sea ship will encounter waves with height less than 0.5m [1]. In other words, the actual performance of the ship in waves is a better reflection of her economical competitiveness and should be chosen as the design objective.

In the presence of waves, ships will experience an additional resistance known as the added resistance. Added resistance affects ship performance in terms of fuel consumption and engine/propulsion system selection. Moreover, it is a significant factor when determining ship Energy Efficiency Design Index (EEDI), which has become mandatory for all new ships built after January 1, 2013. Therefore, the prediction of added resistance in seaways is nowadays of high scientific and practical interest. A reduction of the ship's added resistance in waves can lead to more efficient and eco-friendly designs.

Hence, the objective of this study is to investigate how the ship's performance is affected in various sea states due to added resistance, and carry out hull form optimization based on the obtained information.

In order to achieve the automatic optimization process, three main ingredients are required to be integrated into a design optimization framework:

- An optimization scheme to minimize the objective functions under given constraints.
- Hull geometry modelling and shape deformation according to the variation of design variables.
- Computational solvers as analysis and evaluation tools to obtain the value of the objective function.

A number of researchers have utilized the computer aided optimization techniques to minimize the user defined objective functions, e.g. calm water resistance, seakeeping, stability, structure, etc. [1-5]

In this study, a fast and reliable semi-empirical formula developed by National Technical University of Athens—Ship Design Laboratory (NTUA—SDL) is adopted for the estimation of wave added resistance [6]. Different from conventional multi-objective optimization schemes that optimize the calm water resistance and wave added resistance as two separate objectives, this study proposes a single objective optimization scheme for both calm water resistance and wave added resistance by introducing a mean wave added resistance estimation model based on the ship's actual operational profile.

The optimization process is then formulated within the CAESES software. A fully parametric model of a medium size cruise ship is created and used for the bow shape optimization investigation. Extensive computations are performed to investigate the influence of hull geometrical parameters on the ship's wave added resistance. The long-term mean wave added resistance of the cruise ship can be reduced through the proposed optimization process.

2. ADDED RESISTANCE IN WAVES

It is well established that the wave added resistance can be decomposed into two major components:

$$R_{AW} = R_{AWM} + R_{AWR} \quad (1)$$

- R_{AWM}—radiation component due to ship motions;
- R_{AWR}—reflection component due to diffraction effect (reflection effect).

Motions of the ship caused by incoming waves consume energy in the form of radiated waves, which corresponds to the R_{AWM} component. Furthermore, the incoming

© 2017: The Royal Institution of Naval Architects

waves are partly reflected and partly diffracted, causing an additional resistance, namely R_{AWR}.

Wave added resistance is primarily related to ship's bow region [7]. Therefore, hull form optimisation for wave added resistance should focus on the ship bow. For large ships, such as the medium size cruise ship in this study, R_{AWR} contributes the most to the wave added resistance. It is particularly important to calculate accurately this component in short waves. Potential theory based near-field and far-field methods are commonly not satisfactory for this task, since they neglect viscous effects. Thus, a validated semi-empirical formula, which takes into account the effect of all relevant design variables on wave added resistance, should serve best the purpose of the envisaged optimisation problem [1].

Consequently, the formulae derived by NTUA-SDL in reference [6] are adopted in this study, as shown in the equations below. The correctness and accuracy of the formulae have been proved by extensive sample ship validations and verifications, including tanker, bulk carrier, container ship and cruise ship types.

For the R_{AWR} component, the following empirical formulae are used:

$$R_{AWR}(\omega)=\frac{1}{2}\rho g\zeta_a^2\left(1+5\sqrt{\frac{L}{\lambda}}Fn\right)\left(\frac{0.87}{C_B}\right)^{1+4\sqrt{Fn}} \times\int_C \alpha_T \sec\alpha_{WL}\sin^2\theta\sin\theta \mathrm{d}l, \quad (2)$$

$$\alpha_T=\frac{\pi^2 I_1^2(k_eT)}{\pi^2 I_1^2(k_eT)+K_1^2(k_eT)}, \quad (3)$$

where ρ is water density; g is gravitational acceleration; ζ_a is the regular incoming wave amplitude; L is the ship length; λ is the wave length; Fn is the Froude Number; C_B is the ship's block coefficient; α_T is the draft correction coefficient. α_{WL} and θ are the flare angles and slope angles along the waterline, respectively, which are defined according to Fig.1.

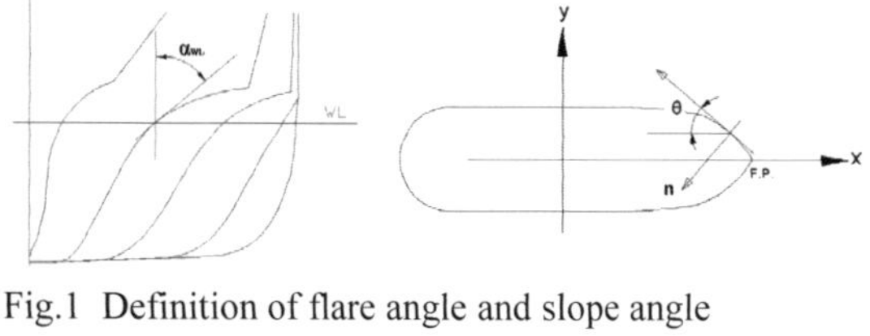

Fig.1 Definition of flare angle and slope angle

As can be seen from the equations, for the R_{AWR} component, the shape of the waterline and the flare angle distribution are the most relevant design variables.

For the R_{AWM} component, the following empirical formulae are used, which only involves the main particulars of the ship, thus no design variable is chosen for the R_{AWM} component optimization.

$$R_{AWM}(\omega)=4\rho g\zeta_a^2B^2/L_{pp}\bar{\omega}^{b_1}\exp\left[\frac{b_1}{d_1}\left(1-\bar{\omega}^{d_1}\right)\right]a_1a_2 \quad (4)$$

$$a_1=60.3C_B^{1.34}\left(0.87/C_B\right)^{1+Fn} \quad (5)$$

$$a_2=\begin{cases}0.0072+0.1676Fn & Fn<0.12\\ Fn^{1.5}\exp(-3.5Fn) & Fn\geq 0.12\end{cases} \quad (6)$$

$$\bar{\omega}=\begin{cases}\sqrt{L_{pp}/g}\sqrt[3]{k_{yy}/L_{pp}}\,0.05^{0.143}\omega/1.17, & Fn<0.05\\ \sqrt{L_{pp}/g}\sqrt[3]{k_{yy}/L_{pp}}\,Fn^{0.143}\omega/1.17, & Fn\geq 0.05\end{cases} \quad (7)$$

$$b_1=\begin{cases}11.0 & \bar{\omega}<1\\ -8.5 & \bar{\omega}\geq 1\end{cases} \quad (8)$$

$$d_1=\begin{cases}C_B\leq 0.75 & \begin{cases}14.0 & \bar{\omega}<1\\ -556\left(L_{pp}/B\right)^{-2.66}*6 & \bar{\omega}\geq 1\end{cases}\\ C_B>0.75 & \begin{cases}556\left(L_{pp}/B\right)^{-2.66} & \bar{\omega}<1\\ -556\left(L_{pp}/B\right)^{-2.66}*6 & \bar{\omega}\geq 1\end{cases}\end{cases} \quad (9)$$

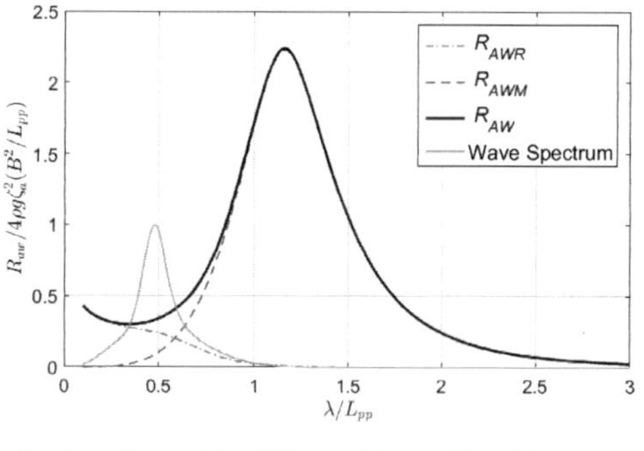

Fig.2 Non-dimensional R_{AW} of the cruise ship

The obtained wave added resistance response amplitude operators (RAOs) of the target cruise ship at Fn=0.2 are shown in Fig.2, along with the unitized ITTC wave spectrum corresponding to the EEDI sea state, i.e. wave height 3m and wave period 6.7s. It is clearly demonstrated that the R_{AWR} component is the main contributor to wave added resistance in this case.

The short-term mean wave added resistance under a given sea state can be calculated from the RAOs:

$$R_{AW}^{MS}=\int_0^\infty[R_{AWR}(\omega)+R_{AWM}(\omega)]\frac{S(\omega;H_S,T_Z)}{\zeta_\alpha^2}\mathrm{d}\omega, \quad (10)$$

© 2017: The Royal Institution of Naval Architects

Where ω is the wave frequency; S is the wave spectrum, which is a function of significant wave height H_S and wave period T_Z.

However, the real long-term performance of the ship at actual seas cannot be represented by certain short-term sea conditions of fixed wave heights and wave periods. This is the reason why the short-term wave added resistance cannot be directly added to the calm water resistance to merge into one single optimization objective, i.e. total resistance at actual seas. To put it in simple words, optimized hull form under one sea state will no longer be optimal under other sea states.

Moreover, a multi-objective optimization process of both calm water resistance and short-term wave added resistance will give no specific optimal design but an ambiguous Pareto Front. In this study, a long-term mean wave added resistance model based on the ship's actual operational profile is proposed, in order to reflect the ship's wave added resistance performance in a more comprehensive way.

The proposed long-term mean wave added resistance to be estimated is:

$$R_T^{ML} = \sum_{j=1}^{N_V} w_j \left[R_T\left(V_j\right) + R_{AW}^{ML}\left(V_j\right) \right], \quad (11)$$

where w_j is the weight of ship speed, which is determined according to the ship's operational profile; R_T^{ML} is the weighted total resistance that can be used as the final optimization objective; R_T is the calm water resistance; R_{AW}^{ML} is the estimated long-term mean wave added resistance, which is related to long-term wave statistics, and is estimated by the following equations:

$$R_{AW}^{ML} = \sum_{i=1}^{N_Z} w_i \cdot R_{AW,i}^{ML}, \quad (12)$$

$$R_{AW,i}^{ML} = \sum_{m=1}^{N_H} \sum_{n=1}^{N_T} w_{mn,i} R_{AW,mn}^{MS}\left(H_{S,m}, T_{Z,n}\right), \quad (13)$$

where $R_{AW,i}^{ML}$ is the long-term mean total resistance in sea zone i; w_i is the weight determined by the percentage of ship's operation time in each sea zone; $w_{mn,i}$ is the occurrence of each sea state in sea zone i, determined according to long-term wave statistics.

With this model, most of the relevant influencing factors on wave added resistance are properly considered. Hence, a single-objective optimization framework can be formulated to yield an optimal design from the long-term perspective.

3. FULLY PARAMETRIC MODEL

It was shown that the wave added resistance for large size ships is owed mainly to the diffraction component. Thus, based on the formula for R_{AWR}, suitable design variables for the parametric model were selected and implemented with the CAESES/Friendship Framework.

By using the parametric modelling technique, the entire hull surface is defined by parameters. It is much easier to make changes to a parametric model than a non-parametric model, and the smoothness of the hull surface can be better guaranteed. The parametric model in this study is achieved by the so-called Fspline, which allows the user to set the starting and terminating positions with their tangents, as well as settings for the curve's area and centroid values.

For example, by setting the area and end-point tangents of the design waterline curve, the waterline coefficient C_{WP} and entrance angle can be controlled. Similarly, the flare angle distribution is controlled by the end-point tangents and the maximum flare angle of the distribution control curve.

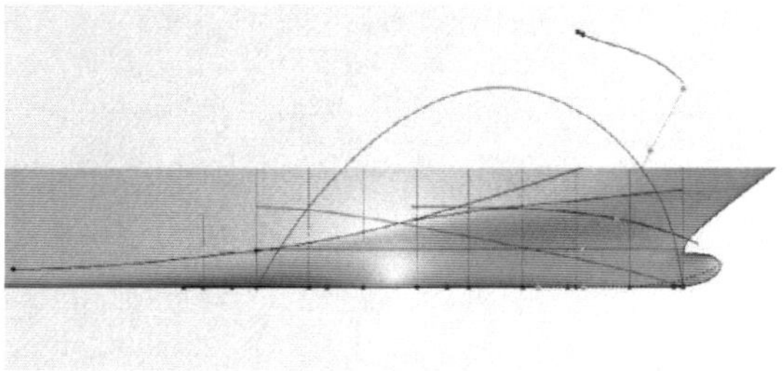

Fig.3 Parametric model of the cruise ship

4. OPTIMIZATION SETUP

The main particulars of the target cruise ship are listed in Table 1. Optimization of the ship's bow region is carried out in order to obtain the design with minimal total resistance. Due to the limitation of computation resources, CFD evaluations of the calm water resistance are not performed. The trade-off between calm water resistance and wave added resistance due to hull form modifications will be studied subsequently. For the current purpose, only the long-term mean wave added resistance is considered.

Table 1 Cruise ship main particulars

Name	Symbol	Unit	Value
Length	L_{PP}	m	239.0
Breadth	B	m	32.25
Main Deck Height	D	m	10.6
Draught	d	m	7.4
Block Coefficient	C_B	[-]	0.658
Waterplane Coefficient	C_{WP}	[-]	0.813

© 2017: The Royal Institution of Naval Architects

The speed profile of the cruise ship is shown in Table 2. Long-term wave statistics of three sea zones (shown in Fig.4) are used for this study. These are the potential areas the cruise ship is going to operate. According to the planned routes, the weights for each sea zone are w_1=0.6 (Z01), w_2=0.2 (Z02) and w_3=0.2 (Z03), respectively.

Table 2 Operational profile of the cruise ship

Condition	Design Speed	Cruise Speed	Slow Steaming
Speed (kn)	22.0	19.5	14.5
Percentage	10	60	30

As listed in Table 3, three design variables are studied: fore waterplane coefficient C_{WPF}, waterline entrance angle θ_E and maximum flare angle α_{max}.

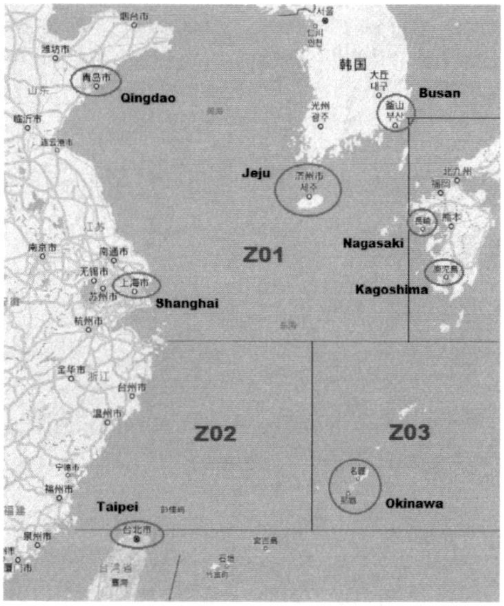

Fig.4 Cruise ship operation sea zones

Table 4 Setting of design variables

Design Variable	Lower	Initial	Upper
C_{WPF}	0.6	0.637	0.7
θ_E	10	14.66	20
α_{max}	40	47	60

Finally, the Non-dominated Sorting Genetic Algorithm II (NSGA-II) is used for searching the optimal design, with 8 generations and 12 populations for each generation.

5. OPTIMIZATION RESULT

As shown from Figs.5~8, all the three design variables and the objective converge at the end of the optimization. The resulting optimal waterline shape and flare angle distribution are shown in Figs. 9 and 10, respectively. An obvious trend can be noticed that the smaller the C_{WPF} and α_{max}, the smaller the resulting wave added resistance. This strong positive correlation is depicted in Figs. 11 and 12. The correlation between θ_E and wave added resistance is not very obvious, as shown in Fig. 13.

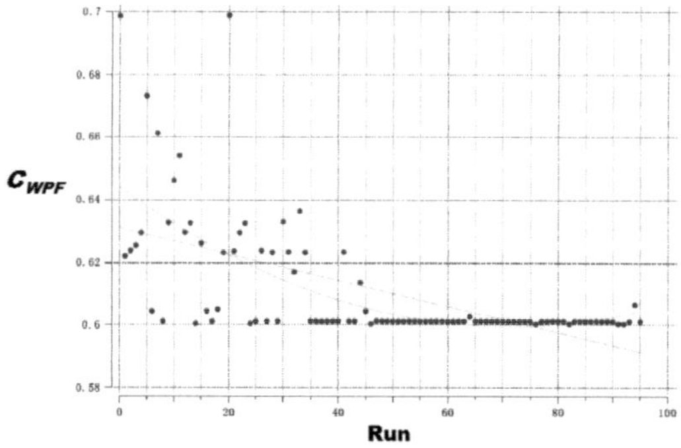

Fig.5 Convergence of C_{WPF}

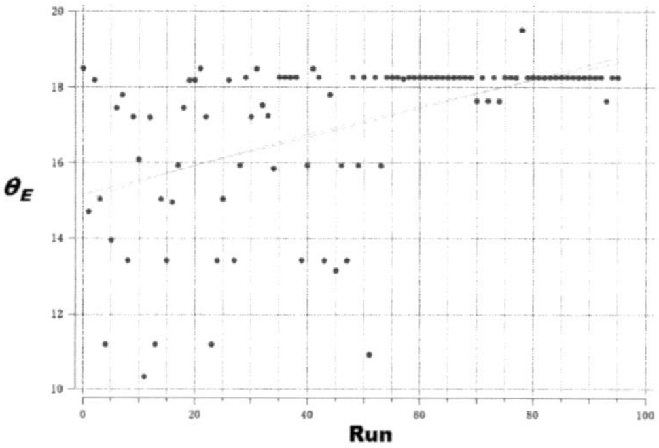

Fig.6 Convergence of θ_E

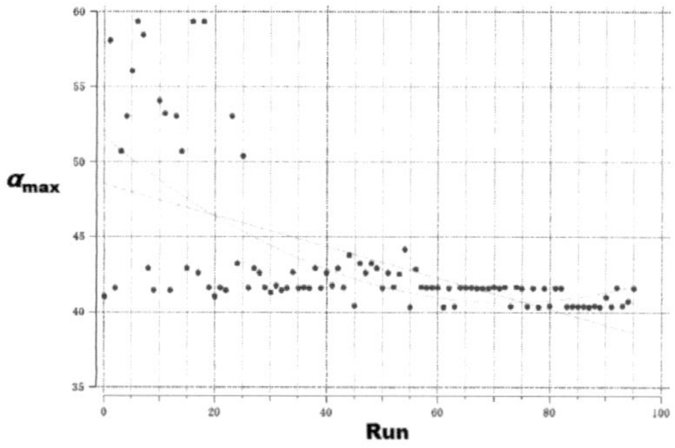

Fig.7 Convergence of α_{max}

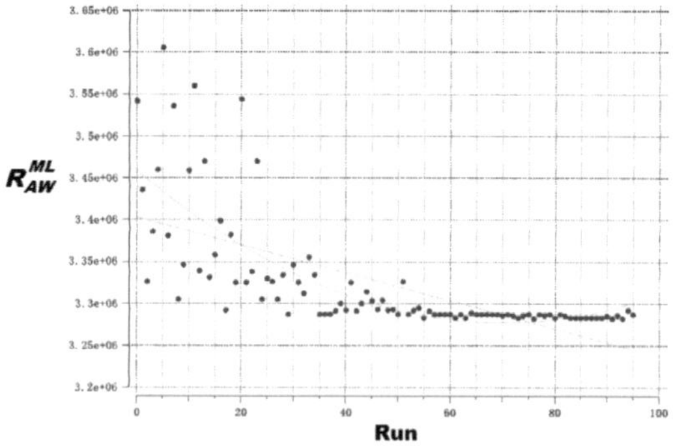

Fig.8 Convergence of objective function

© 2017: The Royal Institution of Naval Architects

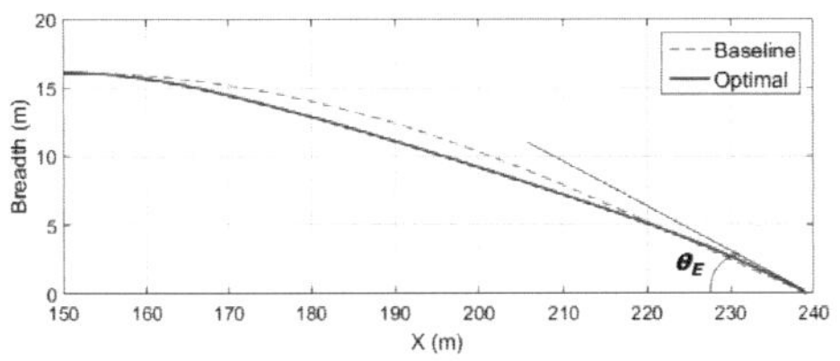

Fig.9 Comparison of waterline shape

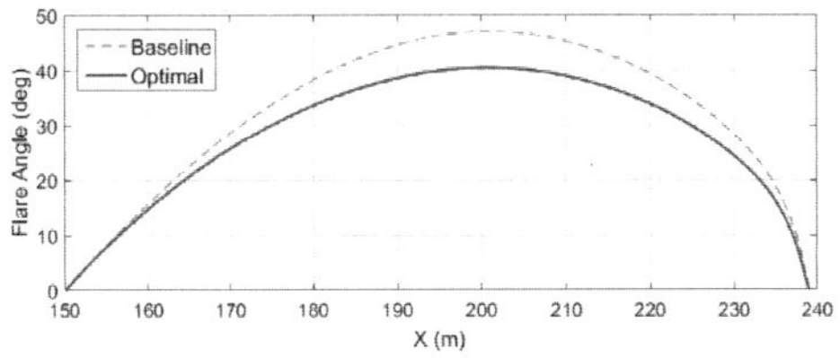

Fig.10 Comparison of flare angle distribution

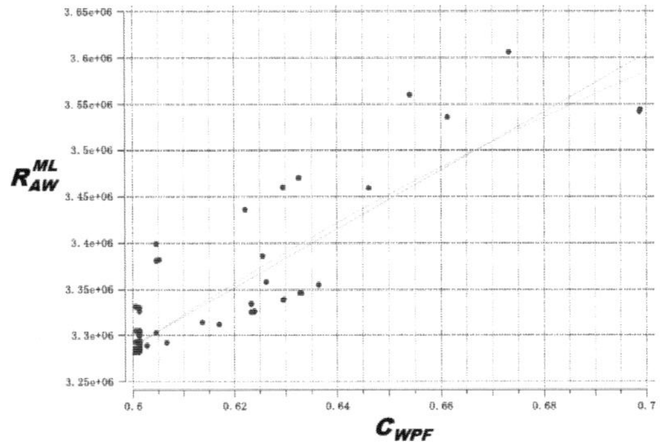

Fig.11 Correlation between C_{WPF} and wave added resistance

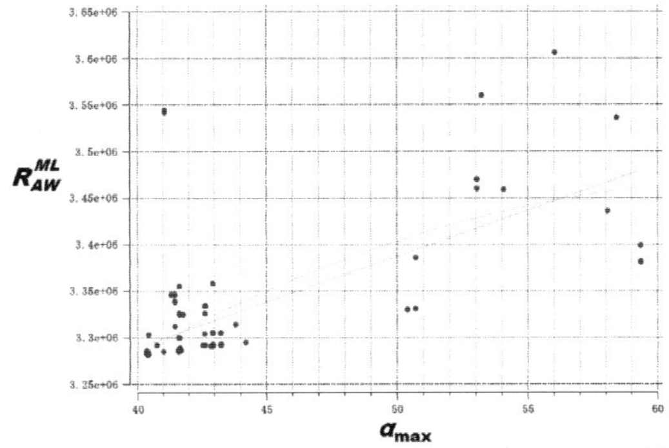

Fig.12 Correlation between α_{max} and wave added resistance

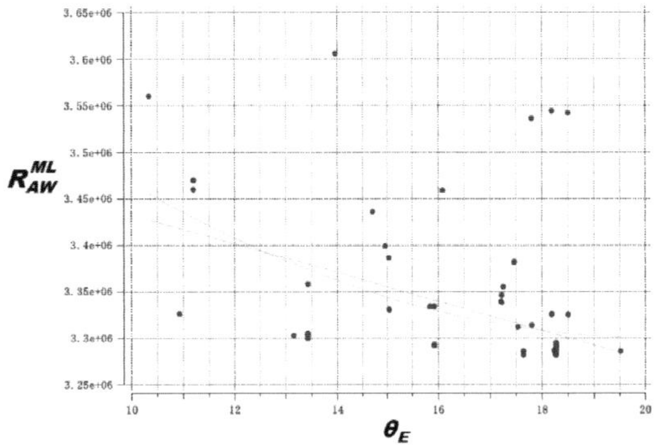

Fig.13 Correlation between θ_E and wave added resistance

The outcome of the optimization is shown in Table 5. As can be seen, the average reduction of the mean long-term wave added resistance is about 3.4%. This degree of reduction can also be achieved in each of the three sea zones. The comparison between the baseline and the optimal hull sectional lines is shown in Fig. 14.

Table 5 Optimization result

Zone	Baseline (kN)	Optimal (kN)	Reduction (%)
Z01	3670	3549	3.3
Z02	3260	3146	3.5
Z03	2714	2621	3.4
Weighted Sum	3397	3283	3.4

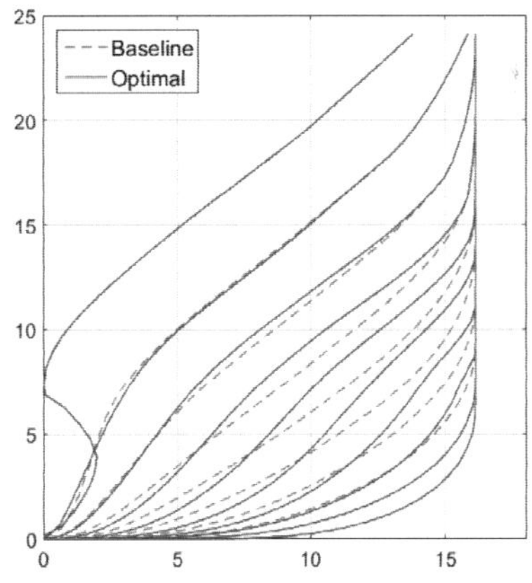

Fig.14 Comparison of hull sectional lines

© 2017: The Royal Institution of Naval Architects

6. CONCLUSIONS

This work presents a long-term wave added resistance estimation model that can capture most of the relevant influencing factors on ship's wave added resistance during actual operation. Based on this model, a single-objective hull form optimization framework can be formulated, considering both performances in calm water and at actual seas, to yield an optimal design from the long-term perspective.

Through this study, it is demonstrated that there is strong positive correlation between C_{WPF}, $\alpha_{\max}$ and the resulting wave added resistance. A smaller C_{WPF} (or $\alpha_{\max}$) will lead to reduced wave added resistance.

The trade-off/interaction between calm water resistance and wave added resistance will be investigated in future studies.

7. ACKNOWLEDGEMENTS

This study is supported by the High-tech Ship Research Project funded by the Ministry of Industry and Information Technology of China. The authors would like to express their gratitude towards Prof. Papanikolaou and Dr. Shukui Liu for the valuable advice and technical discussions.

8. REFERENCES

1. BOLBOT, V. and PAPANIKOLAOU, A., 'Optimization of ship's bow form for the added resistance in waves', *Ship Technology Research*, 2016.

2. KIM, H., YANG C., LOHNER R., NOBLESSE F., 'A practical hydrodynamic optimization tool for the design of a monohull ship', *ISOPE*, 2008.

3. GRIGOROPOULOS, G., and CHALKIAS D., 'Hull-form optimization in calm and rough water', *Journal of Computer-Aided Design*, 2010.

4. GAMMON, M., 'Optimization of fishing vessels using a multi-objective genetic algorithm', *Ocean Engineering*, 2011.

5. ZHANG, B., 'Research on optimization of hull lines for minimum resistance based on rankine source method', *Journal of Marine Science and Technology*, 2012.

6. LIU, S. and PAPANIKOLAOU, A., 'Fast approach to the estimation of added resistance of ships in head waves', *Ocean Engineering*, 2016.

7. GUO, B. and STEEN, S., 'Evaluation of added resistance of KVLCC2 in short waves', *Journal of Hydrodynamics*, 2011.

9. AUTHORS BIOGRAPHY

Peiyuan FENG holds the current position of senior engineer at Marine Design & Research Institute of China (MARIC). He is responsible for design optimization and ship hydrodynamics related researches.

Jiayi HE holds the current position of engineer at Marine Design & Research Institute of China (MARIC). He is responsible for parametric hull form modelling and multi-objective design optimization.

Sheming FAN holds the current position of Chief Engineer at Marine Design & Research Institute of China (MARIC). He is responsible for research project managing and coordinating.

Jinbao WANG holds the current position of Department Director at Marine Design & Research Institute of China (MARIC). He is responsible for research project managing and coordinating.

© 2017: The Royal Institution of Naval Architects

A STUDY ON THE IMPROVEMENT AND APPLICATION OF SYSTEM DYNAMICS MODEL FOR DEMAND FORECASTING OF SHIPS

Y Wada, Knowledge and Data System Department, National Maritime Research Institute, Japan
K Hamada and N Hirata, Graduate School of Engineering, Hiroshima University, Japan

SUMMARY

Shipbuilding is an industry in which the change in demand has been extremely drastic. Therefore, it is important to develop a method of demand forecasting for new ships in order to realize sustainable development of the shipbuilding industry. Recently, sufficient actual data on the shipbuilding market has become available, and it is expected to define an accurate demand-forecasting model by using these data. In this study, the system dynamics model of previous study is improved by using latest market data, and optimal measures planning system for shipbuilding industry is developed based on the proposed model. Thereby, some simulations using the proposed demand-forecasting model and system are conducted, and the effectiveness of proposed model and system are shown.

1. INTRODUCTION

In the shipbuilding industry, the change in demand, i.e., orders for newly built ships, has been extremely drastic. From 2005 to 2008, orders increased rapidly because of the growth of sea cargo traffic, and the shipbuilding industry received the highest level of orders in its history. Therefore, the construction capacity of shipyards, particularly in China and Korea, increased rapidly and a large number of new ships were built. On the other hand, orders for newly built ships decreased rapidly after 2011, and the shipbuilding industry currently faces intense global competition. In consideration of the above characteristics of the shipbuilding market, it is important to develop a method of demand forecasting for new ships and to examine the influence of construction strategy on future demand in order to realize sustainable development of the shipbuilding industry.

Incidentally, orders for newly built ships are influenced by various factors. Although sea cargo movement is influenced essentially by the world economy, orders for new ships are not only affected by sea cargo movement but also by other factors such as ship price, the order books of shipyards, and demolition of ships. In addition, when some shipyards receive orders from shipping companies, the order books and the ship price change in conjunction. From the above, shipbuilding market change by several factors.

Based on the above considerations, we have been developing system dynamics models for demand forecasting of shipbuilding in consideration of shipbuilding market characteristics. In previous studies [1, 2], we have developed system dynamics models for demand forecasting of shipbuilding, and confirmed the validity of the proposed models. However, detail forecasting in shipbuilding market is difficult. Moreover, effective application of the proposed demand-forecasting model is not sufficient to show.

In this study, the system dynamics model of previous study is improved by using latest market data. Particularly, ship price prediction model and detailed demand-forecasting model of each size of ship are defined. Accurate and detailed demand forecasting is realized by the proposed demand-forecasting model. Additionally, optimal measures planning system for shipbuilding industry is developed based on the proposed demand-forecasting model. Sustainable development scenario of shipbuilding industry is discussed using the proposed system. Thereby, some simulations using the proposed demand-forecasting model and system are conducted, and the effectiveness of proposed model and system are shown.

2. OVERVIEW OF PREVIOUS STUDY

2.1 SYSTEM DYNAMICS

System dynamics, which was developed at MIT in 1956, is a well-known numerical simulation technique to analyse complex and dynamic systems [3]. The fundamental concept of system dynamics is modelling causal relations by considering time delays between the elements of the system mathematically and executing a simulation using the developed model. By using this technique, we can analyze complex systems based on logical reasoning, which helps to understand the characteristics and the dynamic behaviors such as shipbuilding market.

2.2 DEMAND-FORECASTING MODEL

The target ship type is the bulk carrier. The target cargo commodities include iron ore, coal, and grain. Demand-forecasting model in previous studies [1, 2] is shown in Figure 1. As shown in the figure, the system dynamics model in previous studies consists of following four models:

(1) Cargo transportation prediction model: This model forecasts the amount of sea cargo movement based on world GDP and cargo transportation distance.
(2) Order prediction model: This model forecasts the number of orders for ships based on sea cargo movement, ship bottoms, ship price, and order books at shipyards.

© 2017: The Royal Institution of Naval Architects

(3) Construction model: This model determines the amount of ship construction. The ordered ships are placed on the shipping market after several years. The construction period is influenced by construction capacity and the order books at shipyards. This model controls these amounts and timings.
(4) Scrap model: This model determines the monthly amount of scrapped ships based on the condition of the shipping market.

World GDP, cargo transportation distance, and ship price are difficult to forecast. Therefore, these variables are treated as exogenous variables and are set by system users. By inputting these variables and initial values at the start of the simulation (bottoms, order books, construction capacity, and amounts of ships under construction), the number of orders, amount of construction, scrap, etc., are forecasted.

Simulation results for orders, construction, scrap, and bottoms are shown in Figure 2. In the figure, black solid lines show the simulation results, grey dotted lines show actual results, and grey solid lines show the actual result by taking the moving average. Simulation results agree well with actual results. However, the simulation result did not reproduce the short-term change with the proposed model. Accordingly, the simulation result was compared with the actual result by taking the moving average.

Compared to the actual result by taking the moving average, the simulation results agree well with actual results taking moving average than actual results without a moving average. The correlation coefficient for orders and construction were a maximum (0.90 (-) and 0.99 (-)) when the moving average period spanned 3 months and 6 months. Additionally, the correlation coefficient for scrap and bottoms were a maximum (0.87 (-) and 1.00 (-)) when the moving average period spanned 9 months and 1 month. It should be noted that the moving average period is determined with the correlation analysis. This is the evaluation index of the proposed model; when this period becomes short, the proposed model reproduces the short-term change seen in the actual results. Additionally, the case when the moving average period is set to 1 month and the case without a moving average period are the same.

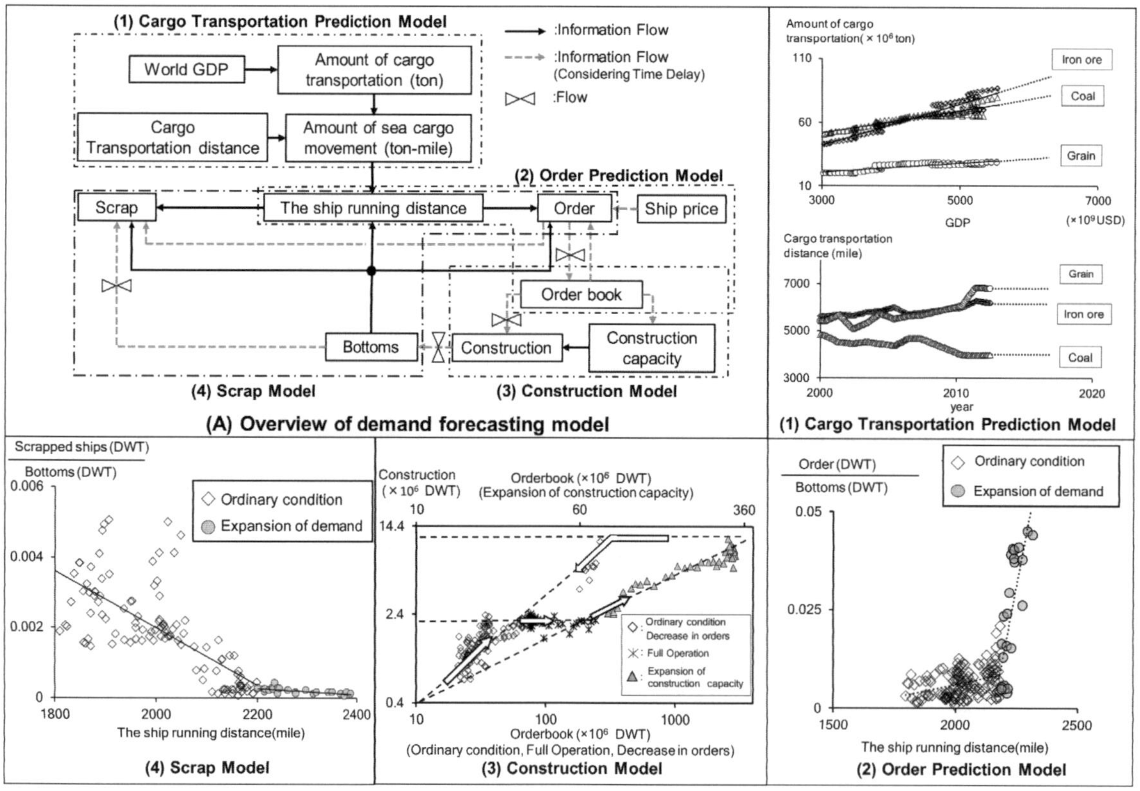

Figure 1: Overview of demand-forecasting model in previous study

© 2017: The Royal Institution of Naval Architects

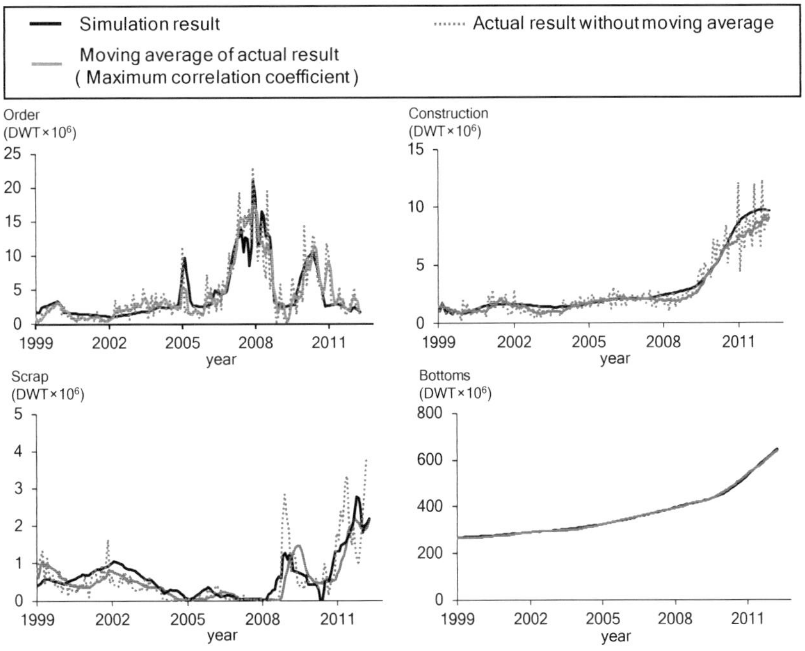

Figure 2: Simulation results in previous study

3. TARGETS OF THIS STUDY

The purposes of previous studies [1, 2] are a long-term demand forecasting in shipbuilding market and forecasting the total number of orders, total amounts of construction, etc., and so on. Ship price and each size of ship orders, etc., are not targets in these studies. Therefore, the system dynamics model in previous studies is difficult to forecast ship price, each size of ship orders, and so on. Additionally, the consideration methods of some internal and external factors which influence on shipyards and shipbuilding industry are not sufficient to show. Thereby, it is difficult to discuss such construction strategy planning and sustainable development scenario for shipbuilding industry.

In order to discuss above, it is necessary to forecast ship price, and the number of orders for each size of ships which shipyards should focus. Additionally, the consideration methods for some factors influence on shipbuilding industry are examined. Moreover, optimal measure planning system supporting sustainable development for shipbuilding industry is developed by using the consideration method of some factors and proposed demand-forecasting model. From the above, we examined the followings in this study:

- In order to forecast ship price, causal relations between order books in shipyards, construction capacity, and ship price are examined, ship price prediction model is newly defined. Additionally, the ship price prediction model is integrated in the previous demand-forecasting model.
- By using the latest data for each size of ship, it is possible to define detail model forecasting each size of ships. The basic concept and detailed model to forecast the demand of each size of ship are shown.
- We focus on construction capacity and modification of international rules, and the consideration method of impact on shipbuilding market were shown. Additionally, in order to realize sustainable development of shipbuilding industry, optimal measures planning system is developed using the consideration method, an optimization method, and the proposed demand-forecasting model. Moreover, overview of the system and formulation as optimization problem are shown.

4. SHIP PRICE PREDICTION MODEL

4.1 MODEL DEVLOPMENT

In general, ship price fluctuates by internal influences in shipbuilding market and external influences such as modification of international rules. It is difficult to forecast the ship price fluctuation by the external influence, therefore, we focus on relation between ship price and the internal influences in the shipbuilding market. Nagatsuka [4, 5] indicates there is a strong correlation between the order books at shipyards and the ship price. We refer on these studies [4, 5], we analyse the relation between order books at shipyards and the ship price.

We define the backlog of the shipyards Bl, which means the stock rate of the order books. The backlog Bl is calculated by using the order books, construction capacity, and the period of facility expansion. Calculation equation of backlog Bl is shown in equation

© 2017: The Royal Institution of Naval Architects

(1). Ship price is predicted by using this backlog and equation (2).

$$Bl_t = \frac{Ob_t}{f_1(Cp_t, ep)} \cdots\cdots (1)$$

$$Sp_t = f_2(Bl_t) \cdots\cdots (2)$$

where Bl is backlog at shipyards (year), Ob is order books in ship yards (DWT), f_1 represents the function to calculate annual construction capacity (-), Cp is the construction capacity at current month (DWT), ep is the period of facility expansion (month), Sp is ship price (index), f_2 represents the function to calculate ship price (-), and t is internal time in the simulation (months).

The relation between the backlog and ship price is shown in Figure 3. Ship price is forecasted by using the following characteristics. It should be noted "Ship price" denotes the average price for each ship size. The unit of ship price is the index value (index), which is the definition used in the Clarkson Shipping Intelligence Network [6].

- Rising period: Ship price increases by the backlog increases. Ship price increases rapidly when the backlog is between 1 year and 3 years. However, the rate of rising speed becomes gentle when the backlog is over 3 years.
- Falling period: Ship price decreases by the backlog decreases. This is causes by rapid expansion of construction capacity in shipyards. Ship price decrease rapidly when the backlog is over 3.5 years. However, ship price becomes constant when the backlog is between 3.5 years and 2.5 years. Additionally, the backlog is less than 2.5 years, the ship price starts to fall down.
- Forcibility falling period: The backlog at shipyards becomes nearly 1 year, ship price is forcibility falling by shipyards in order to keep the amounts of order books considering future recession. Additionally, when ship price reaches at the bottom, ship price enter the rising period (1).

These relations are defined by using following data:

(1) The amounts of order books (DWT) from 1996 to 2011;
(2) The amounts of construction capacity (DWT) from 1996 to 2011;
(3) Ship price (Index) from 1996 to 2011.

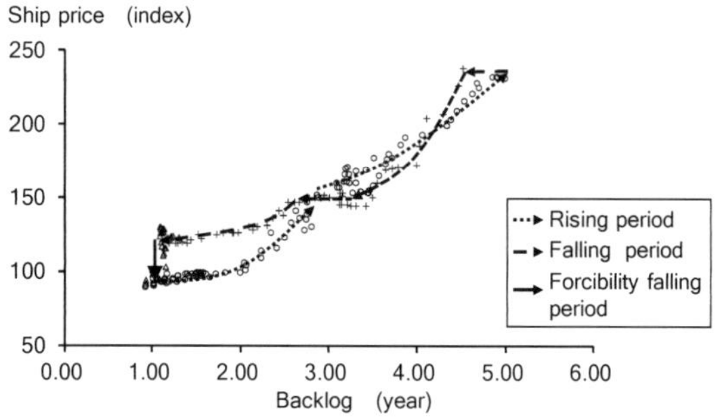

Figure 3: Relation between the backlog and ship price

4.2 MODEL VALIDATION

To confirm the validity of the proposed entire model included the ship price prediction model, a demand forecasting simulation from January 1999 to April 2012 was conducted. Input data for the simulation were as follows:

- Input scenario: from January 1999 to April 2012

(1) World GDP (actual data)
(2) Cargo transportation distance (actual data)

- Initial value: January 1999

(1) Bottoms: 2.64×10^8 (DWT)
(2) Order books: 2.41×10^7 (DWT)
(3) Construction capacity: 2.09×10^6 (DWT)
(4) Ship amount under construction: 1.11×10^7 (DWT)

Simulation results for orders and ship price are shown in Figure 4. In the figure, black solid lines show the simulation results while grey dotted lines show actual

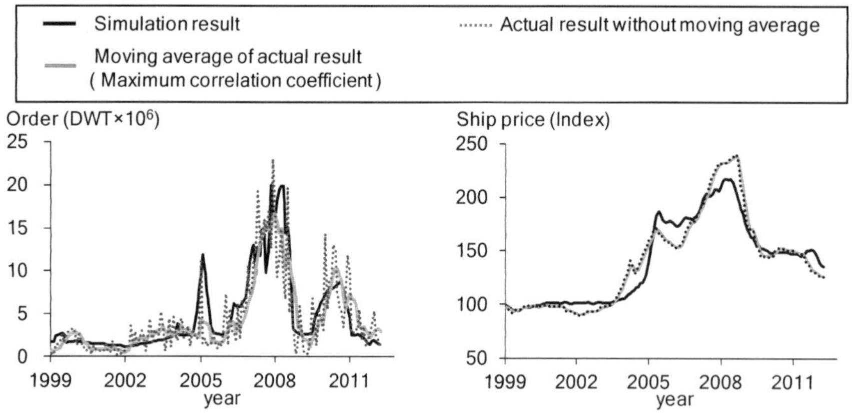

Figure 4: Simulation results for orders and ship price (1999-2012)

© 2017: The Royal Institution of Naval Architects

results. Simulation results for orders agree well with actual results (Figure 4 left). Additionally, the trend of ship price can be simulated (Figure 4 right). However, the simulation result, especially simulation results for orders, did not reproduce the short-term change with the proposed model. In order to confirm the number of months for the proposed model reproduced the actual trend for orders, correlation analysis between the simulation and actual results was conducted. Grey solid lines show actual results by taking the moving average. The correlation coefficient for orders was a maximum (0.89 (-)) when the moving average period spanned 5 months. On the other hand, the correlation coefficient for ship price was a maximum (0.96 (-)) when the moving average period spanned 3 months. Both of correlation coefficient for orders and ship price show a high value. From these results, the proposed model can forecast the average trend in orders for 5 months and the average trend in ship price for 3 months. Moreover, it is reduced the input scenario of ship price by developing the ship price prediction model, therefore, sophisticated simulation is possible compared with previous studies because the ship price scenario is not need to set by system users.

5. DEMAND-FORECASTING MODEL OF EACH SIZE OF SHIPS

5.1 BASIC CONCEPT

By using the latest data for each size of ship [6], it is possible to define detailed model. The basic concept used in forecasting the orders for each size of ship is shown in Figure 5. The orders for each size of ship are calculated as follows:

(1) Calculate the amount of sea cargo movement (tons × miles) by using the cargo transportation prediction model (Figure 5 (A)(1)).
(2) Calculate the total bottoms by summing each size of ship. Thereafter, calculate the ship running distance by using the total bottoms (Figure 5 (A)(2)).
(3) Calculate the total number of orders is calculated by using the ship running distance, order prediction model, and ship price prediction model. After that, calculate the number of orders for each size of ship using order allocation model (Figure 5 (A)(3)).
(4) The amount of construction and scrap for each size of ship are calculated by using each size of model (Figure 5(A)(4)).

By comparing the detailed model (A) and the total model (B) in Figure 5, it can be seen that the calculation flow is the same from the amount of sea cargo movement (tons × miles) to ship running distance. However, the order allocation model is newly defined in the detailed model. Additionally, construction model, and scrap model are defined independently. This is a point of difference between the detailed model (A) and the total model (B). Each size of construction model, and scrap model can be defined following previous study [2] procedure. Therefore, we focus on the order allocation model, and detailed of this model is shown in this paper.

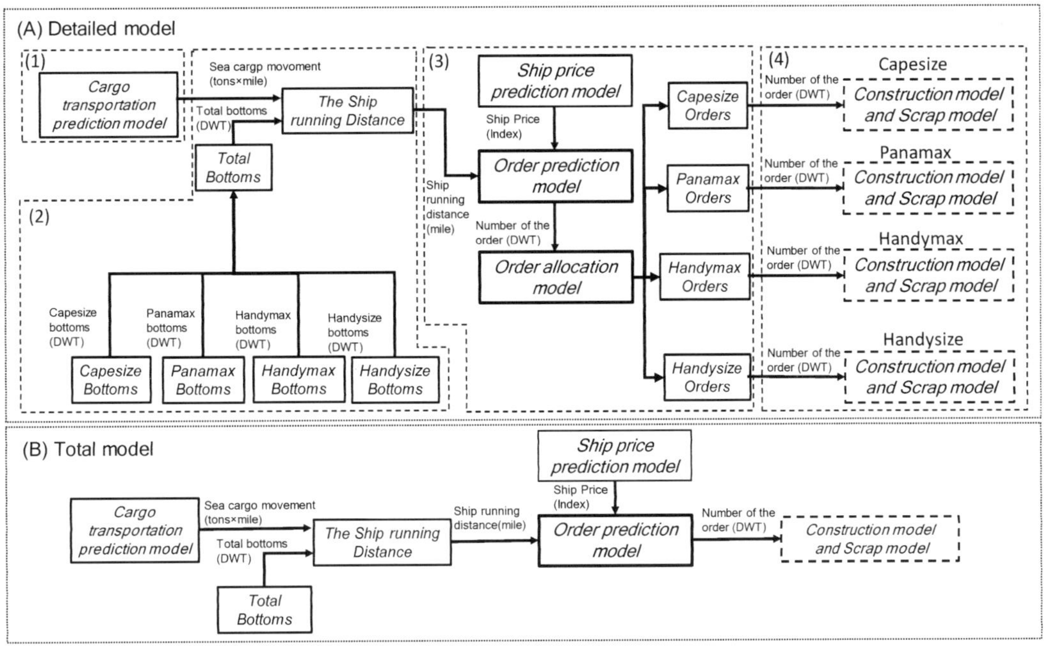

Figure 5: Calculation flow of the detailed model and the total model

The definition of the ship size is based on the Clarkson Shipping Intelligence Network [6] as follows:

Capesize: 100,000 DWT and over;
Panamax: 65,000–100,000 DWT;
Handymax: 40,000–65,000 DWT;
Handysize: 10,000–40,000 DWT.
(Definition on September 12, 2015.)

5.2 ORDER ALLOCATION MODEL

In general, the number of each size orders fluctuates with the state of shipping market. The number of each size orders is calculated by using equation (3).

$$(Oc_t, Op_t, Ohm_t, Oh_t) = Or_t \times f_3(E_t) \cdots (3)$$

where *Oc is* the number of orders for Capesize (DWT), *Op is* the number of orders for Panamax (DWT), *Ohm is* the number of orders for Handymax (DWT), *Oh is* the number of orders for Handysize (DWT), *Or is* the total number of orders (DWT), *E* is the ship running distance (miles), f_3 is calculate function of each size of ships (-), and *t* is internal time in the simulation (months).

The relation between the ship running distance and each size of order is shown in Figure 6. Following relations can be confirmed. It should be noted the ship running distance shows the sailing distance of ships in one month and indicates the condition of the shipping market.

(1) The order rate of Capesize increases by the ship running distance increases.
(2) The order rate of Panamax increases by the ship running distance decreases.
(3) The order rates of Handysize and Handymax are almost constant regardless of the ship running distance.

In general, when shipping market is an active market, the ship running distance increases and the bottoms of ships become insufficient. Shipping companies order the Capesize ships which can transport a lot of cargo in single transport, and the sailing distance of ships becomes short as possible. On the other hand, when shipping market is a recession, the ship running distance decrease and the bottoms of ships is excess. Shipping companies order Panamax ships which can be used in various situations, and respond to the request of various shipper needs. Using this detailed model, demand forecasting simulations were conducted. In reality, the ship running distances for each size of ship are different because the operation for each size is different in actual maritime transportation. By using the average ship running distance for each size of ships, demand forecasting in consideration of ship size was conducted in this study. Relations between the ship running distance and each size of order rate are defined using the following data:

(1) Ship running distance (miles): a result calculated by using actual data for the amount of sea cargo movement and bottoms from 1999 to 2012;
(2) The total number of orders (DWT) from 1999 to 2012;
(3) The number of orders for each size of ships (DWT) from 1999 to 2012.

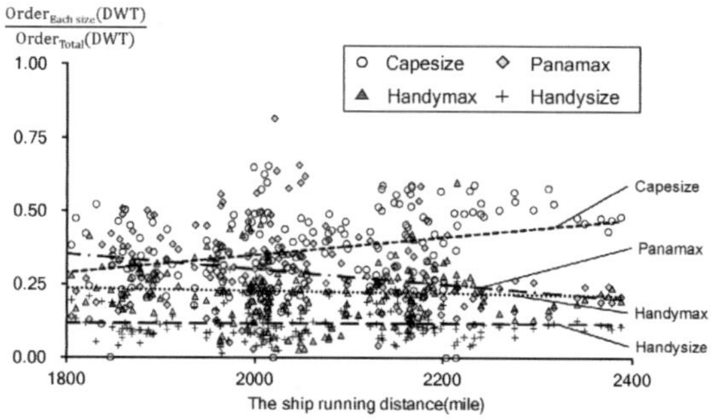

Figure 6: Relation between the ship running distance and order rate

© 2017: The Royal Institution of Naval Architects

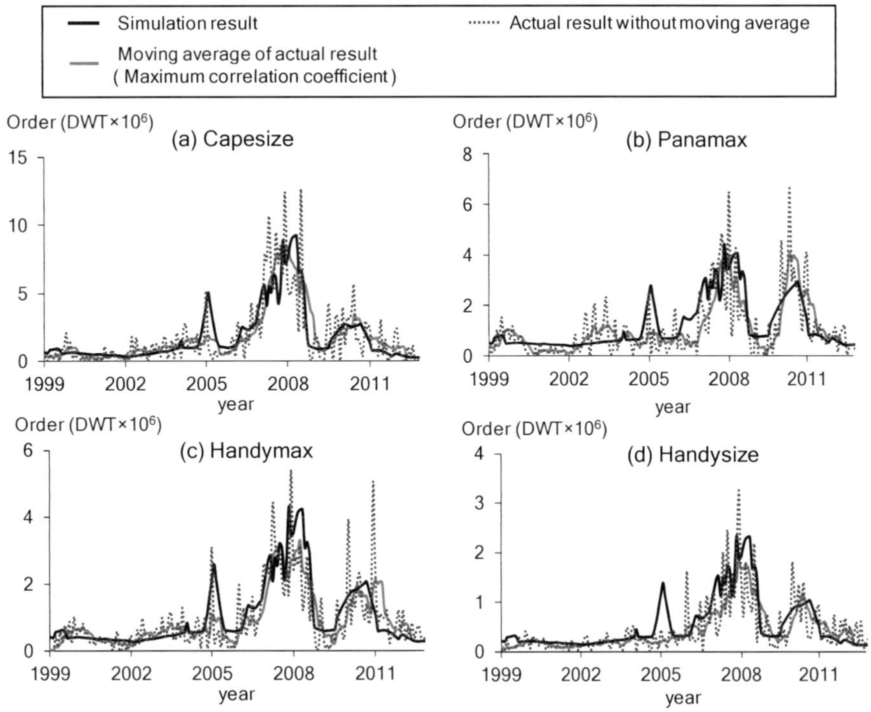

Figure 7: Simulation results for order of each size of ships (1999-2012)
(1999-2012)

5.3 VALIDATING OF DEMAND-FORECASTING MODEL OF EACH SIZE OF SHIPS

To confirm the validity of proposed detailed forecasting model, a demand forecasting simulation from January 1999 to April 2012 was conducted. Input scenario and all of the initial values were the same as in sub-section 4.2.

Simulation results for orders of each size of ship are shown in Figure 7. The simulation results agree well with each set of actual results. The correlation coefficients for Capesize and Handysize were a maximum (0.90 (-) and 0.87 (-)) when the moving average period spanned 6 months. Additionally, the correlation coefficients for Panamax and Handymax were a maximum (0.78 (-) and 0.86 (-)) when the moving average period spanned 5 months. The moving average period for each size of orders was almost same, and the correlation coefficients show a high value of over 0.75. From these results, the proposed model of each size of ships can forecast each size of ship orders.

6. OPTIMAL MEASURES PLANNING SYSTEM

6.1 DEFINITION OF OPTIMAL MEASURES PLANNING

It is necessary to control the number of orders and adjusts the ship price for sustainable development in shipbuilding industry. The number of orders, ship price, etc., are influenced not only internal factors in shipping and shipbuilding market but also external factors such as modifications of international rules. We focus on these factors which influence to shipbuilding market, and discuss sustainable development scenario for shipbuilding market by controlling these factors. This problem defines as the measures planning problem. Particularly, we focus on the following problems.

(1) Construction capacity control problem

Shipbuilding market becomes a long-term recession when total construction capacity at shipyards all over the world become an excess against the shipbuilding market. Therefore, it is necessary to control total construction capacity entire world, and the number of orders and ship price are controlled by the construction capacity control for sustainable development of shipbuilding industry. This problem is defined as an internal factor problem in shipbuilding market in this study.

(2) International rules modification problem

International rules modifications influence on shipbuilding market. Actually, amounts of ship scraps increase for a while because of the influence of double hull and double bottom requirement for oil tanker. Additionally, in order to reduce CO_2 emissions from ships, the Energy Efficiency Design Index (EEDI) has been discussed and reduction of the operating speed of

© 2017: The Royal Institution of Naval Architects

ships is considered for the future. The purpose of these rules is the reduction of marine traffic accident, environment protection of ocean and air, etc., and so on. It is important to consider the influence of international rules modification on shipbuilding market as well as these factors. This problem is defined as an external factor problem in shipbuilding market. In this study, the system which supports to consider the influence on shipbuilding market by the modification of the rules is developed. It should be noted we focus on mainly (1) forcibility scrap of ships, (2) suspension of ship use for a certain period, and (3) reduction of the operating speed of ships.

6.2 OVERVIEW OF THE OPTIMAL MEASURES PLANNING SYSTEM

In construction capacity control problem in measures planning problem, it is necessary to consider the change of prediction results which are calculated by construction capacity assuming several cases. It is similar to the modification of international rules. It is possible to consider the change of prediction results by inputting such as construction capacity and international rules modification scenario as exogenous variables into the proposed demand-forecasting model. By combining the proposed demand-forecasting model and optimization method, optimal measures planning system is developed.

Basic concept of the optimal measures planning system is shown in Figure 8. As shown in the figure, design plan such as the construction capacity control and international rules modification scenario is created in upper problem. Then, the design plan input to the demand-forecasting model and demand forecasting simulation is conducted in lower problem. Prediction results are calculated by the simulation. Fitness value for upper problem is calculated using the prediction results and optimization calculation is executed in upper problem. Thereby, the optimal applied timing and quantity of measures are planned. Simulated annealing [7] is applied in optimization calculation. In a word, construction capacity and modification of international rules are treated as design variable, optimal measures for shipbuilding industry are planned. The calculation flow of construction capacity control and international rules modification is same. Therefore, we focus on construction control problem, overview of the system and simulation results are shown in following sub-section.

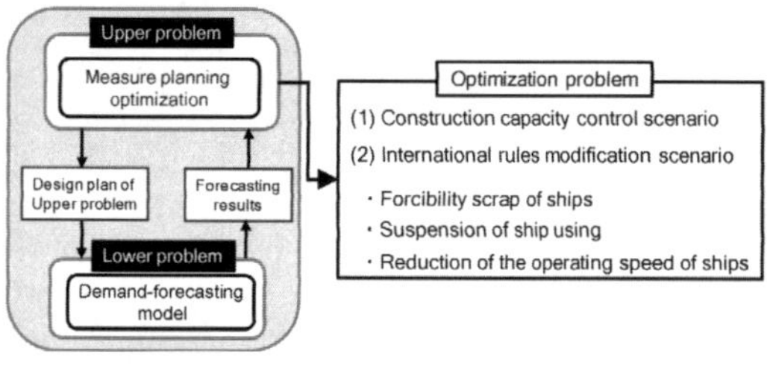

Figure 8: Optimization problem in this study

6.3 FORMULATION OF THE UPPER PROBLEM

The optimal timing and quantity of construction capacity expansion is planned in upper problem. Design variable, objective function, and constraint condition are as follows:

- Design variable: Monthly change of construction capacity is treated as design variable. This construction capacity updates monthly and is set based on each size of ships in equation (4).
- Objective function: Objective function is set as maximum of the shipbuilding industry profit for simulation span in equation (5). Income and cost are referred to these documents [8, 9, 10, 11].
- Constraint condition: Minimum annual profit *Mp* for shipbuilding industry is set in equation (6). This is inputted by system users.

$$Ep = [ep_t^{\,j}] \quad \cdots\cdots \quad (4)$$

$$max(Profit) = \sum_t (Income_t - Cost_t) \quad \cdots\cdots \quad (5)$$

$$Aprofit \geq Mp \quad \cdots\cdots \quad (6)$$

where *Ep* is design plan of construction capacity, *ep* is construction capacity in a month (DWT), *Profit* is the total profit of shipbuilding industry (USD), *Income* is the monthly income of shipbuilding industry (USD), *Cost* is the monthly total cost of shipbuilding industry (USD), *Aprofit* is the annual profit of shipbuilding industry (USD), *Mp* represents minimum annual profit of shipbuilding industry (USD), *j* is respective ship size (1: Capsize, 2: Panamax, 3: Handymax, 4: Handysize), and *t* is internal time in the simulation (months).

6.4 FORMULATION OF THE LOWER PROBLEM

In the lower problem, demand forecasting simulation is conducted by using the proposed demand-forecasting model. Based on simulation span, simulation scenarios (world GDP and cargo transportation distance scenarios), and initial values at the start of the simulation (bottoms, order books, construction capacity, and amounts of ships under construction), the number of orders, amount of construction, ship price, etc., are forecasted. Object functions for upper problems are calculated and constraint condition is checked using the prediction results. These exogenous variables, simulation span, and initial values at the start of the simulation are set by system users in advance. It should be noted it is possible to forecast construction capacity using construction model. However, demand forecasting simulation is conducted using design plan of construction capacity in the case of construction capacity control problem. In other words, internal calculation of construction capacity is not conducted.

© 2017: The Royal Institution of Naval Architects

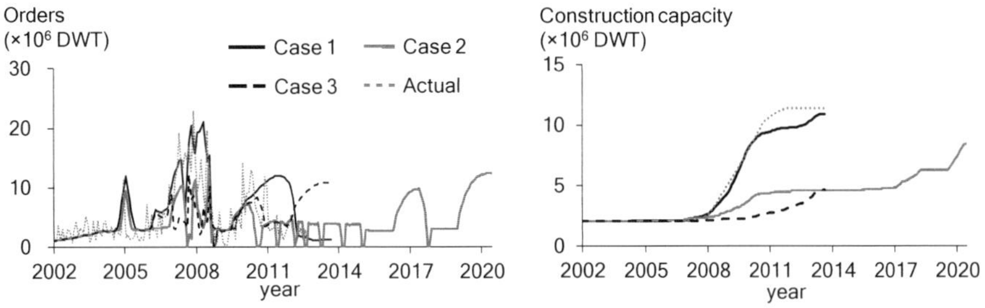

Figure 9: Optimization results for orders and construction capacity

6.5 CASE STUDIES

To consider the optimal timing and quantity of construction capacity expansion, optimization simulations were conducted. Input scenario, initial value, and simulation case are follows:

- Input scenario: from January 2002 to January 2015
(1) World GDP (actual data)
(2) Cargo transportation distance (actual data)
- Initial value: January 2002
(1) Bottoms: 2.86×10^{8} (DWT)
(2) Order books: 2.39×10^{7} (DWT)
(3) Construction capacity: 2.09×10^{6} (DWT)
(4) Ship amount under construction: 1.23×10^{7} (DWT)
- Simulation case

Case 1: Constraint condition is not set. Simulation span is set from 2002 to 2013.
Case 2: Constraint condition is not set. Simulation span is set from 2002 to 2020. World GDP growth rate is set 3.5% and cargo transportation distance assumed constant after 2015.
Case 3: Constraint condition *Mp* (equation (6)) is set 1.0×10^{10} (USD). Simulation span is set from 2002 to 2013.

Results of the simulation for orders and design plan of construction capacity are shown in Figure 9. In the case of Case 1, construction capacity is expanded in conjunction with a rapid increase in orders from 2007 to 2008. Compared with the actual results, the trends of orders and construction capacity agree well. However, in the Case 2 and Case 3, the speed of construction capacity becomes slower and quantity of construction capacity expansion is decreased compared with actual result. Simulation results for orders are changed significantly by this influence.

Optimization results for the profit of shipbuilding industry are shown in Figure 10. In the case of actual result and Case 1, the profit of shipbuilding industry increases from 2007 to 2008 in conjunction with the orders. However, it cannot be acquired profit around 2011 in whole shipbuilding industry. As shown in the figure, shipbuilding industry is in deficit around 2011 because of excessive supply. On the other hand, whole shipbuilding industry can acquire sustainable profit in the case of Case 2 and Case 3. In particular, profit trend of Case 3 is always satisfied constraint condition. From these results, it is considered a recession in current situation is not occurred by properly control of construction capacity expansion. From this result, we can consider the optimal timing and quantity of construction capacity influence on shipbuilding market by using proposed system.

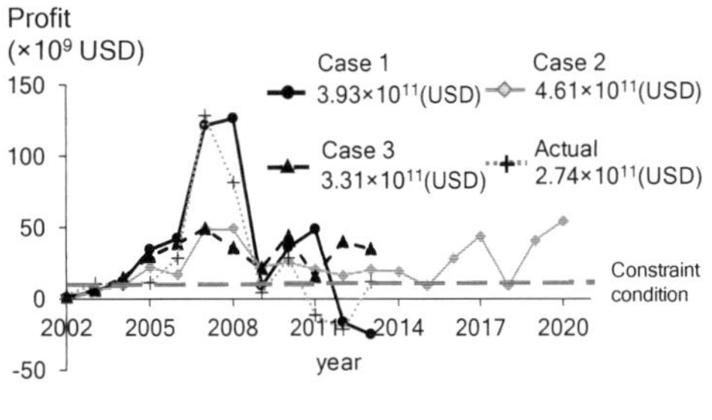

Figure 10: Optimization results of profit

7. CONCLUSIONS

In this study, the system dynamics model of previous studies was improved by using latest market data, and optimal measures planning system for shipbuilding industry was developed based on the proposed model. The key findings were as follows:

(1) In order to predict ship price, causal relations between order books in shipyards, construction capacity, and ship price were examined, and ship price prediction model was newly defined. Additionally, the ship price prediction model was integrated in the previous demand-forecasting model. Additionally, to confirm the validity of the proposed entire demand-forecasting model, demand forecasting simulations were conducted.
(2) By using the latest data for each size of ship, it was possible to define detailed model to forecast each size of ships. The basic concept and order allocation model were shown. Additionally, to confirm the validity of the proposed model, demand forecasting simulations were conducted.
(3) We focus on construction capacity and modification of international rules, and the consideration method of impact on shipbuilding market were shown.

© 2017: The Royal Institution of Naval Architects

Additionally, in order to realize sustainable development of shipbuilding industry, optimal measures planning system was developed using the consideration method, an optimization method, and the proposed demand-forecasting model. Moreover, overview of the system and formulation as optimization problem were shown.

(4) In order to consider the optimal timing and quantity of construction capacity expansion, optimization simulations were conducted. Shipbuilding industry continuous to develop by properly control of construction capacity expansion, and the effectiveness of proposed system was shown by the simulation.

The target ship type in this study was limited to the bulk carrier market. Therefore, in future work, we will focus on tanker market, and container ship market. Additionally, the target industry of optimal measures planning system were only shipbuilding industry, therefore, it was difficult to consider the influence on some stake holders such as shipping industry, ship equipment industry. In future work, we will develop newly optimal measures planning system considering the influence of these stake holders.

8. REFERENCES

1. Wada, Y., Hamada, K., Hirata, N., Miura, S., Ishihara, Y., Arifin, MD., Seki, K., 'A Study on System Dynamics Model for the Demand Forecasting of Shipbuilding',12th International Marine Design Conference (IMDC2015), Vol.2, pp.167–182, 2015

2. Wada, Y., Hamada, K., Hirata, N., Seki, K., Yamada, S., 'A system dynamics model for shipbuilding demand forecasting ', Journal of marine science and Technology, 2017. (in press)

3. Meadows, DH., Meadows, DL., Randers, J., Behrens, WWIII., 'The Limits to Growth' Universe Books, 1972.

4. Nagatsuka, S., 'Analysis on order books at shipyards in japan and system dynamics model', Japan Maritime Research Institute Bulletin, 93-154, pp.1-73,1993. (in Japanese)

5. Nagatsuka. S., 'Analysis and transition on the change of newbuilding price of ships', Japan Maritime Research Institute Bulletin, 2003-236, pp.1-55, 2003. (in Japanese)

6. Clarksons, Clarksons Shipping Intelligence Network, Available at: <http://www.clarksons.net> [Accessed 4 October 2013]

7. Kirkpatrick, S., C. Daniel G. and Mario PV, 'Optimization by simulated annealing', science 220.4598, pp.671-680, 1983.

8. Ministry of Land, Infrastructure Transport and Tourism, 'Foreign exchange tolerance of Japanese shipbuilding Industry-Analysis of shipbuilding cost structure', 2016. (in Japanese)

9. Ministry of Economy, Trade and Industry, 'Current status and problems of steel industry (mainly on blast furnaces)', 2015. (in Japanese)

10. The Shipbuilders' Association of Japan, 'SAJ news', Vol.112, 2008 (in Japanese).

11. The Shipbuilders' Association of Japan, 'The Shipbuilders' Association of Japan news', Vol.124, 2009. (in Japanese)

9. AUTHORS BIOGRAPHY

Yujiro Wada holds the current position of researcher at national maritime research institute (At the time of this research: Hiroshima University). He is responsible for maritime economics analysis.

Kunihiro Hamada holds the current position of professor at graduate school of engineering Hiroshima University, Japan. He is responsible for design support system for shipbuilding.

Noritaka Hirata holds the current position of assistant professor at graduate school of engineering Hiroshima University, Japan. He is responsible for ship manoeuvrability and image processing.

© 2017: The Royal Institution of Naval Architects

A STUDY ON THE SUPPORT SYSTEM OF SHIP BASIC PLANNING BY USING MARINE LOGISTICS BIG DATA

Mohammad Danil Arifin, Kunihiro Hamada, Noritaka Hirata, and **Kai Ihara**, Hiroshima University, Japan
Yuki Koide, Churyo Engineering Co., Ltd.

SUMMARY

In last decade, a large amount of data is growing exponentially which forms as Big Data. In maritime industries, marine logistics Big Data i.e. port data, ship data, route data, international trade data, and data provided by AIS increased rapidly. If these data are effectively utilized, a great innovation might be achieved. The objective of this study is to develop a support system of ship basic planning which can examine the demand and the principal particulars of bulk carrier by utilizing the marine logistics Big Data. In order to realize this, the authors develop ship allocation model consists of three distinct models i.e. shipper model, shipowner model, and operator model that developed by using the marine logistics Big Data. In this paper, a bulk carrier which is operating between Australia and Japan is taken as an example. The details and the effectiveness of proposed method are discussed in the paper.

1. INTRODUCTION

Big Data is clearly more than a buzz word and a business benefit of utilizing Big Data is widely known. A study by MIT found that data-driven organizations perform 5% to 6% better per year [1]. Big Data is already playing a larger role in shaping the future of the maritime industry. By embracing analytics and turning data into actionable insights, marine logistics players have an opportunity to drive improved efficiency and quality [2]. Moreover, it is widely believed that Big Data can aid in improving forecasts, and Big Data can be effective for forecasting demand and planning process [3][4].

There are many potentials and highly useful values hidden in the huge volume of Big Data which is used in various fields [5]. In maritime industries, marine logistics Big Data i.e. port data, ship data, route data, international trade data, and data provided by AIS (Automatic Identification System) can be used in the present situation. These data has a potential to innovate the maritime industries.

By the way, the global marine logistics industry has changed significantly which is influenced by the global goods movement [7]. In this situation, it is important to develop the ships that meet the market requirements. Actually, research of ship demand prediction has been conducted [8]. However, the detailed prediction, on the route which the demand will increase, is difficult to be executed at the present time.

The objective of this study is to develop the support system of ship basic planning by using the marine logistics Big Data which can predict the demand of bulk carrier and examine the effective ship principal particulars for cargo transportation. As a marine logistics Big Data, we utilize the Marine Logistics Database (MLDB) which was developed in our previous study (see section 2). The bulk carriers which operate between Australia and Japan are taken as an example. The details and effectiveness of proposed model is discussed in the paper.

2. OVERVIEW OF PREVIOUS STUDY

2.1 OBJECTIVES OF THE MLDB

In the previous study, the authors developed the MLDB using AIS data and statistical data [6]. The MLDB consists of the latest information of marine logistics Big Data, i.e. operation information from AIS data, ship, port, route and international trading information as shown in Figure 1. The data is integrated to find the estimation of cargo type and cargo volume. The objectives of developing MLDB are shown in the followings;

- To manage the unstructured marine logistics Big Data into structured data.
- To insight the valuable information which is buried in marine logistic Big Data.
- To extract valuable information by developing MLDB.

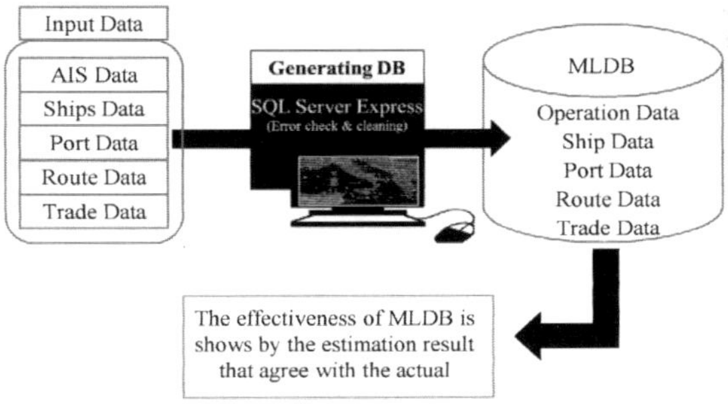

Figure1: Basic concept of previous study

2.2 INPUT DATA OF MLDB

Input data of the MLDB is described as follows:

- AIS Data [11]; e.g., indicated speed, draft, position, and time (arrival and departure date, arrival and departure port from MINT).
- Port Data [12]; e.g., port name, port number, longitude, latitude, port dimension, cargo handling.
- Ship Data [13]; e.g., ship name, DWT, IMO number, ship dimension, operator, shipbuilder, etc.

© 2017: The Royal Institution of Naval Architects

- Route Data [12] [14]; e.g., distance, route, via, etc.
- Trade Data [15]; e.g., commodity trade, a period of trade between country, code, trade value ($), trade quantity, reporter, and partner etc.

2.3 DATA STRUCTURE

In order to extract valuable information easily, the structure of the MLDB was defined and the unstructured data was changed into a relational database. For example, by integrating ship and port data with operation data, some information related to ships operational states can be analyzed (e.g., berthing, anchoring, or sailing).

2.4 ERROR CLEANING

In high-density shipping area where thousands of ships may transmit AIS messages, it is a challenge for the AIS system to efficiently collect, process, and download all the messages. It results that many messages are lost and sometimes error data collection occurred. The samples of errors from AIS data are shown as follows:

- The draft value (d) is zero (0).
- Unrealistic ships movements.
- Duplication information.
- Null information or blank space.

Therefore, in order to ensure the quality of the data, the duplicate and NULL data was deleted, and the draft data between arrival and departure were checked.

2.5 GENERATING CARGO INFORMATION

Cargo information on an operating ship is important for demand forecasting and basic planning. However, such information does not exist in AIS data. Therefore, we estimated the cargo type and volume of each operation as follows:

2.5 (a) Checking the reliability of the data

The confirmation of the reliability of data is required in order to get a good result during estimating cargo volume. In our study, the reliability of data is evaluated by checking the draft rate d_i (%) of the ship during the operation.

2.5 (b) Estimating the cargo type using port data

The port data has an information of the cargo type which is handled at the port. By using this data, cargo of each operation is estimated [6]. Estimation of cargo type is conducted by checking the combination of cargo from arrival port and departure port. By checking the combination of cargo from arrival port and departure port, cargo types of 75% operations are fixed.

2.5 (c) Estimating the cargo type using ship size

If two or more common cargo type exists by executing the estimation using port data, the cargo types are estimated by using ship size [6]. The threshold of each country is decided by using the operation data between the countries.

2.5 (d) Estimating the cargo volume

Ship data has information on deadweight and max draft of the target ship, while AIS data has information in the sailing draft. The cargo volume basically estimated by using the following equation:

$$V_i[ton] = DWT_i[ton] \times \frac{(d_i - 0{,}2)}{(1 - 0.2)} \quad (1)$$

Where;
V_i (ton) is cargo volume, DWT_i (ton) is deadweight and d_i (%) is the draft rate.

2.6 CONFIRMATION OF THE CARGO ESTIMATION

In order to verify the result from proposed methods, the estimation result is evaluated and compared with the actual value of trade data. Target ships are bulk carriers between Japan and Austraria. The estimation results covered 90% of iron ore, 94% of coal and 97% of grain and others compared with the result from trade data.

2.7 EXTRACTING DATA FOR BASIC SHIP PLANNING USING MLDB

The structure of the relational database in MLDB allows a user to get some valuable knowledge. The operation information and another important information e.g. DWT (ton), LOA (m), B (m) d (m), design speed (knot) can be extracted easily. By identifying the extracted data from MLDB, the characteristics of the bulk carrier from Australia to Japan which is important for basic ship planning could be identified [6]. Moreover, the important information for predicting demand of ship in the future can be obtained.

3. BASIC CONCEPT OF THIS STUDY

3.1 BASIC PLANNING SUPPORT USING SHIP ALLOCATION MODEL

The core of this study is a ship allocation model which is developed by using the information from MLDB. Ship allocation model predicts the ship allocation when user inputs the trade volume, economical situation and so on. Therefore, when user inputs the future scenario e.g., the state of the world economy, fluctuation of fuel price, expansion of canal and port, the new ship allocation is generated and what types of ships are effectively used can be estimated based on the simulation. Moreover, by inputting the new ship in the simulation, we can estimate whether new ship will be used or not.

The objectives of the ship allocation model can be summarised as follows:

© 2017: The Royal Institution of Naval Architects

- To realize the actual ship allocation based on the inputting data (e.g., cargo trade between two ports or countries, fuel prices, etc.).
- If some conditions are changed, the allocation model estimates the new state. Therefore by inputting future scenario and new types of ship, we can forecast the ship demand and evaluate the new ship is effective or not.

3.2 CONFIGURATION OF SHIP ALLOCATION MODEL

The ship allocation model in this study consists of the following three models:

- The shipper model issues a request for cargo transportation between two or more ports.
- The operator model requests all shipowner models to estimate the cost, cargo volume, and transport time based on shipper requests. Then, based on the answer from the shipowner model, the operator model decides a ship for the allocation.
- The shipowner model performs estimations (cargo volume, costs and so on) in response to requests from the operator model.
-

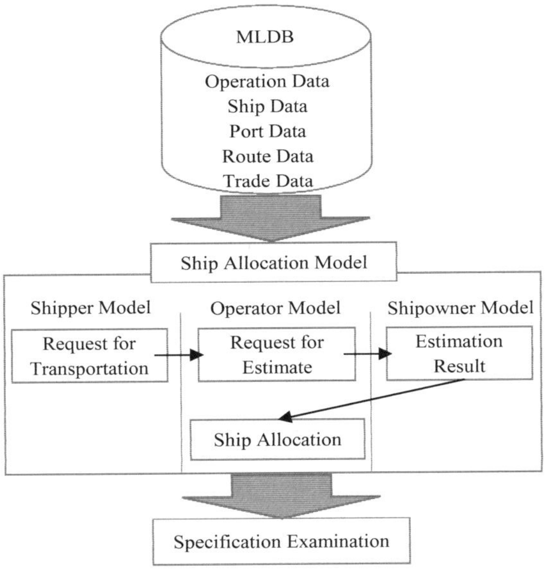

Figure 2: Basic concept of this study

3.3 REQUIREMENTS TO DEVELOP SHIP ALLOCATION MODEL

The ship allocation model should be developed by considering the followings requirements;

- To realize actual ship allocation conditions, three distinct models, the shipper model, shipowner model, and operator model should be developed.
- All data that is important for developing the three models should be extractable from the MLDB.
- The reproducibility of the allocation model should be evaluated and confirmed.

4. DEVELOPING THE SHIP ALLOCATION MODEL

The information extracted from the MLDB is used to develop the ship allocation model. In this study, cargo shipments of iron ore from Australia to Japan in 2014 are taken as an example. As shown in the previous section, the ship allocation model consists of three models: shipper model, shipowner model, and operator model. These models are detailed as follows:

4.1 SHIPPER MODEL

The shipper model issues a request for cargo transportation between two or more ports from Australia to Japan. Herein, the shipper model is generated using cluster analysis. Cluster analysis is a method of defining the similarity in data, grouping similar things together, and classifying them into several groups (clusters). To generate the shipper model in this study, the following steps were taken:

4.1 (a) Extracting Operation Data from the MLDB

All data from 2014 that were important for defining shippers from Australia to Japan were extracted from the MLDB. The information from the MLDB includes operation data, port data, and ship data. By utilizing this data, we can easily analyze the number of port callings from Australia to Japan.

4.1 (b) Defining the Shipper using Cluster Analysis

To define a shipper between Australia and Japan, we must identify the number of port calling in 2014 using cluster analysis. The following steps were taken:

(1) Extracting the number of Port Callings
The number of port callings is extracted from MLDB. Extracted data are managed by using a matrix between the ports (P1...Pn) and ships (S1... Sn) as shown in Table 1(1).

(2) Measuring the Euclidean Distance
Data standardization is shown in Table 1(2) and performed by using the following equation.

$$z = \frac{X-\mu}{\sigma} \tag{2}$$

Where, X; the observation, μ; average number of calls at each port, and σ; standard deviation.

Then, the Euclidean distance is calculated using the standardized data by using the equation (3). The calculation of Euclidean distance is shown in Table 1(3).

© 2017: The Royal Institution of Naval Architects

$$d(x,y) = \sqrt{(x_1 - y_1)^2 + (x_2 - y_2)^2 + \cdots + (x_n - y_n)^2} \quad (3)$$

Where; x_i; the number of calls after standardization of ship i at port x, and y_i; the number of calls after standardization of ship i at port y.

Table 1: Cluster analysis process

(1) Port Calling Calculation

Port \ Ship	S1	S2	S3	S4	S5	S6
P1	0	0	1	0	2	0
P2	0	0	1	0	3	2
P3	0	0	0	1	0	0
P4	5	1	0	0	0	0
P5	2	0	0	0	0	0

(2) Standardization

Port \ Ship	S1	S2	S3	S4	S5	S6
P1	-0.7	-0.7	0.65	-0.7	1.96	-0.7
P2	-0.9	-0.9	0	-0.9	1.73	0.87
P3	-0.4	-0.4	-0.4	2.24	-0.4	-0.4
P4	2.19	0	-0.5	-0.5	-0.5	-0.5
P5	2.24	-0.4	-0.4	-0.4	-0.4	-0.4

(3) Calculation of Euclidean distances

Port \ Port	P1	P2	P3	P4	P5
P1		1.71	3.94	4.04	3.94
P2	1.71		4.08	4.21	4.08
P3	3.94	4.08		3.87	3.79
P4	4.04	4.21	3.87		0.49
P5	3.94	4.08	3.79	0.49	

(3) Clustering using Hierarchical Cluster Analysis
In this study, to measure the distance between two clusters, the average linkage method has been applied. The equation is shown below;

$$d(C_1, C_2) = \frac{1}{|C_1||C_2|} \sum_{x_1 \in C_1} \sum_{x_2 \in C_2} d(x_1, x_2) \quad (4)$$

Where;
C_n; cluster, and x_n; port, and $d(C_1, C_2)$: Euclidean distances between C_1 and C_2.

The goal of this method for this study is to group heterogeneous of the port into homogeneous clusters of the port. As shown in Figure 3, the ports are grouped into four Clusters (Shipper A - D).

The ports in Shipper A match with the JFE Steel locations, and the ports in Shipper B and Shipper C match the Nippon Steel Sumikin locations. That is, the clusters match actual conditions.

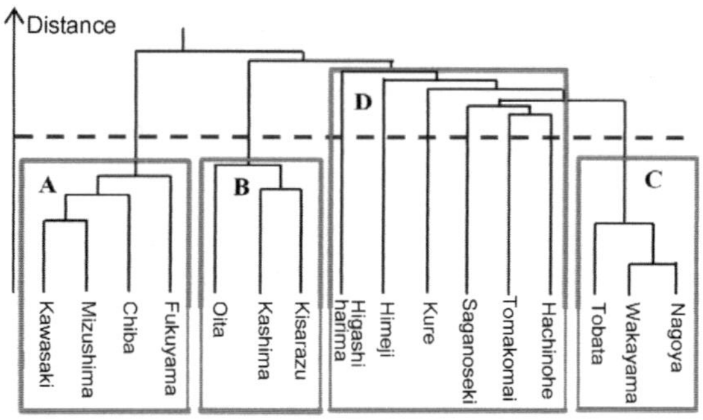

Figure 3: Defined clusters using dendrogram

4.2 SHIPOWNER MODEL

The shipowner model estimates draft rate, average speed in loading and ballast conditions, and days in port due to loading and ballast conditions in response to cargo transportation request from an operator. In this study, the shipowner model is defined using data extracted from the MLDB. To perform these estimations, we employ deep learning. In this study, the estimations are executed based on the followings steps;

4.2 (a) Training Data

In this study, all of the ship data (from the world to Japan, 2014-2015) which is extracted from MLDB is used for training data.

4.2 (b) Generating Learning Model

To predict the following items as mentioned in the previous explanation, the input layer and output layer are set as the followings;

- Input Layer; DWT (ton), length (m), breadth (m), depth (m), draft (m), service speed (knot), horsepower (HP), years of built, new construction of new shipbuilding price index, etc.
- Output Layer; the expected output to be obtained from this generation e.g., draft rate, average voyage speed, and arrival time at the port for loading and unloading condition.

4.2 (c) Calculating the Shipment Time, Amount of Cargo, and Cost

Shipment days are calculated by considering the route distance, navigation speed, and number of days in port. The cargo transport volume and shipment cost are calculated based on the method of Kigure et al. [16] and Aoyama et al. [17] by using the generated data shown in above.

© 2017: The Royal Institution of Naval Architects

(1) Calculation of the total cost and cargo volume

Shipper	Route	Cargo Volume (t)	Ship A ($/t)	Ship B ($/t)	Ship C ($/t)	Ship D ($/t)
A	A1	3.5×10^6	14.8	14.1	16.9	19.9
	A2	2.0×10^6	14.7	13.9	16.4	19.4
B	B1	4.7×10^6	13.6	13	15.1	18.3
	B2	6.0×10^6	13.1	12.6	14.5	18.2

(2) Calculation of standard deviation

Shipper	Route	Cargo Volume (t)	Deviation Value			
			Ship A	Ship B	Ship C	Ship D
A	A1	3.5×10^6	57.2	60.31	47.9	34.59
	A2	2.0×10^6	56.64	**60.43**	45.58	34.35
B	B1	4.7×10^6	56.81	59.74	49.51	33.92
	B2	6.0×10^6	56.84	59.12	50.45	33.58

(3) Ship assignment

Shipper	Route	Cargo Volume (t)	Ship A ($/t)	Ship B ($/t)	Ship C ($/t)	Ship D ($/t)
A	A1	3.5×10^6	14.8	—	16.9	19.9
	A2	**0.6×10^6**	14.7	—	16.4	19.4
B	B1	4.7×10^6	13.6	—	15.1	18.3
	B2	6.0×10^6	13.1	—	14.5	18.2

The average error by using the deep learning analysis shows that the draft rate error is 3.4%, the service speed error is 0.2 knots, and the error of staying time in port is 0.9 days. These errors are smaller than the response surface methodology.

4.3 OPERATOR MODEL

The operator model selects the best ship regularly based on the estimation results from shipowner models. The procedure to determine the ship allocation is as follows:

4.3 (a) Calculating the Total Cost and Cargo Volume

As shown in Table 2(1), shipowner are bidding for all shipment requests. The cost per unit transports volume is calculated by considering the total operation cost ($) and the total amount of transportation volume (t).

4.3 (b) Calculating the Standard Deviation

The deviation values of some ships are calculated for each route as shown in Table 2(2). The deviation value is an index for judging which ship is good to what kind cargo transportation of a certain route.

4.3 (c) Ship Assignment

Ship with the highest standard deviation value is assigned to a shipment of the selected route. For example, as shown in Table 2(3), Ship B is assigned to route A2.

4.3 (d) Recalculating the Amount of Cargo Shipment Request

When the shipment is assigned to the selected route as shown in step 4.3 (c) the amount of cargo shipment is updated and the process is repeated from steps 4.3 (a) – 4.3 (c) until all of the cargo successfully transported.

Table 2: Ships allocation process

4.4 CONFIRMATION OF PROPOSED MODEL

To evaluate the effectiveness of the proposed model, we simulate the ship allocation between Australia and Japan. The result of the simulation was compared with the actual result as shown in Figure 4.

As shown in Figure 4, the result of ship allocation (shows by the number of the operations) from the simulation for all Shippers i.e., Shipper A, Shipper B, Shipper C, and Shipper D generally agrees with the actual result. From this point of view, it can be concluded that the effectiveness of the developed model in this research is confirmed.

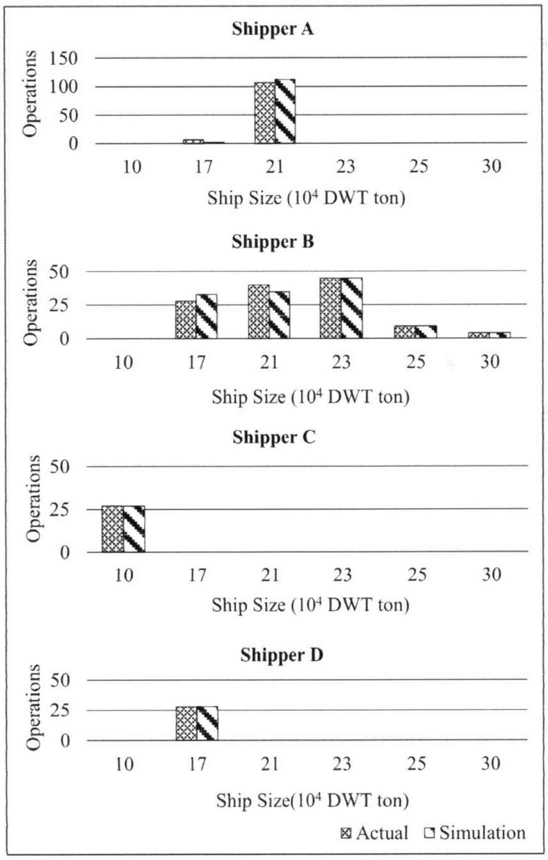

Figure 4: Comparison of the actual and simulation result

5. CASE STUDIES

We conducted a simulation aimed to examine the ship supply-demand balance and examine the influence of

© 2017: The Royal Institution of Naval Architects

ship size and performance to the ship allocation as described below:

5.1 EXAMINATION OF SHIP SUPPLY-DEMAND BALANCE

We define ship supply-demand balance to examine the present condition of ship supply-demand operating between Australia and Japan in 2014, where the cargo types transported was iron ore. Firstly, we carry out a ship simulation when there is no restriction considered, it means there is no limited number of cargo shipment by ship for one-year operation. Secondly, we define supply and demand. In the case using restriction (using the actual number and ship types which are used in 2014 between Australia and Japan), the simulation result is defined as supply. In the case without using restriction, the simulation result is defined as demand.
As shown in Figure 5, we conducted ship supply-demand estimates. Without restriction, the allocation of ships in the 210,000, 250,000, and 300,000 DWT ton clusters increased. However, ship allocation in the 170,000 and 230,000 DWT ton clusters decreased. Therefore, 170,000 and 230,000 DWT ton ships are not very competitive for shipments from Australia to Japan. Meanwhile, there is an insufficient supply of 210,000, 250,000, and 300,000 DWT ton ships. Hence, these ships are competitive for shipments from Australia to Japan and are expected to be in demand in the future.

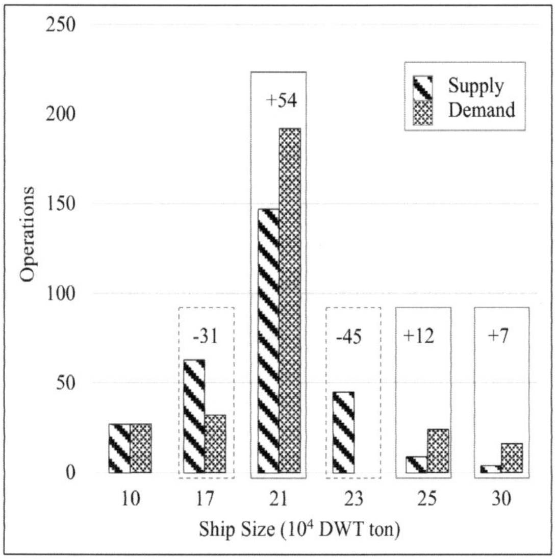

Figure 5: Supply-Demand Balance of Ships

5.2 EXAMINATION OF SHIP ALLOCATION BY THE SHIP SIZE

To examine the influence of ship size for the ship allocation and examine which size is the most competitive and could be operated on multiple routes, the ship allocation simulation is conducted. The simulation is conducted by considering the result from the previous section. Based on the discussion in the section 5.1, ships in the 210,000, 250,000, and 300,000 DWT ton clusters are expected to be in demand. Therefore, we simulate the ship allocation by changing the size of new ships.

As shown in Figure 6, Ship A (300,000 DWT ton) is in demand on a single route. In contrast, Ship B (250,000 DWT ton), and Ship C (210,000 DWT ton) can be expected to be in demand on multiple routes. In other words, the Ship B (250,000 DWT ton), and Ship C (210,000 DWT ton) is more competitive than Ship A (300,000 DWT ton).

However, when the fuel efficiency improves by 10%, Ship C (210,000 DWT ton) is more competitive than Ship B (250,000 DWT ton). As shown in Figure 7, the number of allocated ship and routes for Ship C (210,000 DWT ton) is increased rapidly (5 additional ships and 3 additional routes).

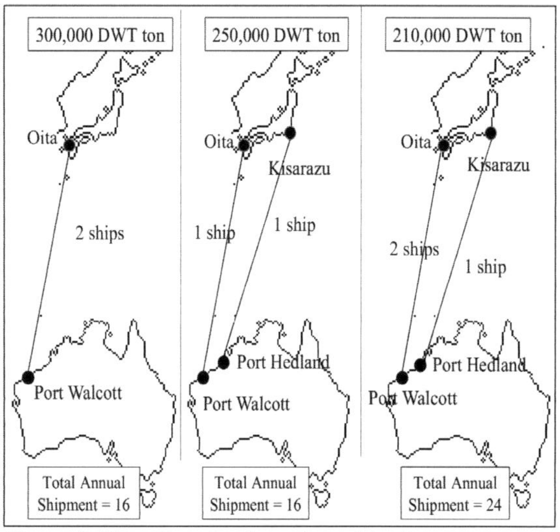

Figure 6: Ship allocation in various ship sizes

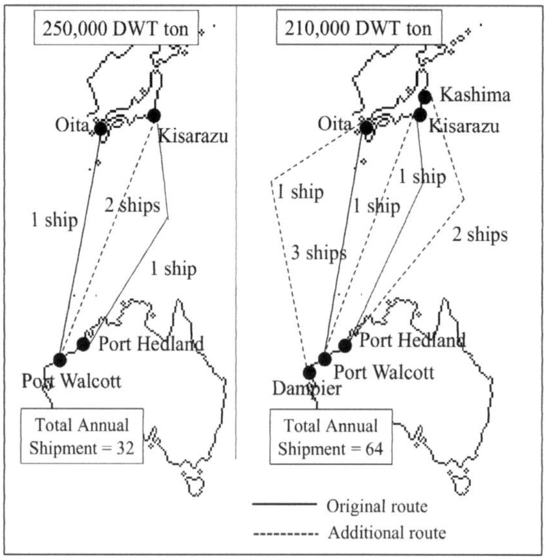

Figure 7: Ship allocation result when the fuel efficiency

© 2017: The Royal Institution of Naval Architects

improved by 10%

5.3 EXAMINATION OF SHIP ALLOCATION BY THE PERFORMANCE

To examine the impact of the performance (fuel efficiency) on ship demand, we simulate by improving fuel efficiency by 5%, 10%, and 15%. A ship with no fuel efficiency change is defined as S_0. Ships with fuel efficiency improves of 5%, 10%, and 15% are denoted by S_1, S_2, and S_3, respectively. Since ships in the 210,000 DWT ton cluster are the most competitive, these ships are simulated. Furthermore, to evaluate these ships effectiveness, the simulation result (replacement using this ship) is compared with the actual ship allocation.

By using the proposed methods, the simulation result of ship allocation in cluster 210,000 DWT ton that is simulated by modifying the fuel efficiency which defined as S_0, S_1, S_2, and S_3 shows as follows;

- S_0, the number of operation is 40 within 3 routes.
- S_1, the number of operation is 47 within 4 routes.
- S_2, the number of operation is 61 within 5 routes.
- S_3, the number of operation is 128 within 10 routes.

Furthermore, the simulation result of replacement from the ship in cluster 210,000 DWT ton which simulated with various fuel efficiency is compared with the actual condition.

As shown in Figure 8, when fuel efficiency improves by 10%, ships in the 170,000 and 230,000 DWT ton clusters show minor demand on that route. Moreover, ships in the 210,000 and 250,000 DWT ton clusters also do not need to be replaced by new ships, since the simulation result shows the same number of the operations compared with actual ship allocation.

When efficiency improves by 15%, replacing ships in the 210,000 DWT ton cluster greatly improves the number of operations for Shipper A. However, when the efficiency improves by 15%, the ships in the 210,000 DWT ton cluster are the same as actual conditions for Shipper B. In contrast, the ship in the 250,000 DWT ton cluster shows that a replacement has occurred that greatly improves the number of operations.

In summary, using the proposed model, we can simulate ship supply and demand. Moreover, the ship allocation that influenced by ship size and ship performance can be simulated. In addition, we can obtain the impact of fuel efficiency on ship demand.

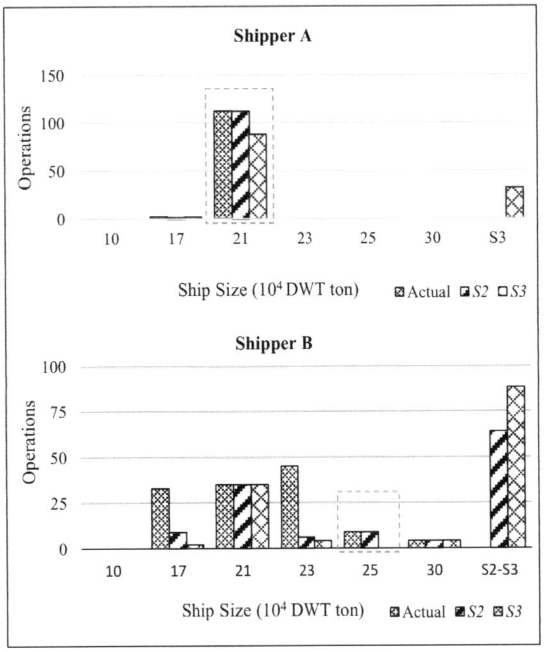

Figure 8: Differences of ship S_2 & S_3 for Shipper A and B

6. CONCLUSIONS

In this study, we have focused on developing a support system for basic ship planning using Marine Logistics Big Data, and have drawn the following conclusions:

- By utilizing the data extracted from MLDB, the ships allocation models which composed of three distinct models; shipper model, shipowner model, and operator model was successfully developed.
- By inputting the future scenario into the simulation, we can examine future ship supply-demand balance, the influence of ship size and fuel efficiency of ship allocation.

7. ACKNOWLEDGNMENT

This work was supported by JSPS KAKENHI Grant Number 16H04602. Authors would like to express special thanks and appreciate their support.

© 2017: The Royal Institution of Naval Architects

8. REFERENCES

1. Esben P., 'Big Data in the Digital Age – The Impact on Ships and Ports', *The International Maritime Transport & Logistics Conference (MARLOG 5).* Pp 1-10, 2016.

2. Martin K et al., 'Ship Insight-The Buzz of Big Data', *the Quarterly Journal*, pp. 35-41, 2015.

3. Hossein H et al., 'Forecasting with Big Data, *Springer, Ann. Data. Sci.* 2(1); 5–19 DOI 10.1007/s40745-015-0029-9, pp.5-19, 2015.

4. Charles W et al., 'Using Big Data to Enhance Demand-Driven Forecasting and Planning', *Journal of Business Forecasting*, Volume 32, Issue 2, pp. 26-32, 2013.

5. Dongmei Huang et al., 'Modeling and Analysis in Marine Big Data', *Mathematical Problems in Engineering*, Article ID 384742, pp. 1-14, 2015.

6. Mohammad Danil Arifin et al., 'Development of marine logistic database AIS and statistical data', *Proceeding of the 30th Asian-Pacific Technical Exchange and Advisory Meeting on Marine Structures (TEAM 2016).* Pp 395-402, 2016.

7. James, J. Corbett et al., 'The Impacts of Globalization on International Maritime Transport Activity', *Global Forum on Transport and Environment in a Globalizing World*, pp. 1-31, 2008.

8. Wada Y. et al., "A Study on System Dynamics Model for the Demand Forecasting of Shipbuilding", *12th International Marine Design Conference 2015, Vol 2*, pp. 167-182, 2015.

9. D. Laney., '3D Data Management: Controlling Data Volume, Velocity and Variety', *Gartner Inc.* 2001

10. D. Laney., 'The Importance of 'Big Data': *A Definition'. Gartner. Retrieved 21 June 2012.*

11. Market Intelligence Network, "Automatic Identification System Data," Available at <https;//www.marketintelligencenetwork.com/app/>, Start using from 2015-present.

12. IHS Maritime & Trade, "Sea-web Ports," Available at <http;//maritime.ihs.com/Areas/Seaweb/authenticated/authenticated_handler.aspx?control=portsearch&bp=1&type=clear>, Start using from 2015-present.

13. IHS Maritime & Trade, "Sea-web Ships," Available at <http;//maritime.ihs.com/Areas/Seaweb/authenticated/authenticated_handler.aspx?control=shipsearch&bp=1&type=clear>, Start using from 2015-present.

14. IHS Fairplay Ports & Terminals Guide, CD-ROOM, 2013

15. http;//www.gtis.com/gta/ Start using from 2016-present.

16. Kigure et al., "Research on basic planning support of ships using Maritime logistics big data", *Proceedings of the Japan Society of Naval Architects and Ocean Engineers*, CD-ROM, 2016. (in Japanese)

17. Aoyama et al., "Assessment Method of CO_2 Reduction Effect of New Technology and International Regulation on Maritime Logistics", *Eco Design 2009 6th International Symposium on Environmentally Conscious Design and Inverse Manufacturing, Sapporo (Japan)* 2009. CD-ROM, 2009.

9. AUTHORS BIOGRAPHY

Mohammad Danil Arifin is a Doctoral course student at Transportation and Environmental System Department, Hiroshima University, under the guidance of Prof. Kunihiro Hamada. He earned his Bachelor and Master degrees from the Department of Marine Engineering, Faculty of Marine Technology, Sepuluh Nopember Institute of Technology (ITS) Surabaya, Indonesia.

Kunihiro Hamada holds the current position of professor at Graduate School of Engineering, Hiroshima University, Japan. His major is design support system for the shipbuilding.

Noritaka Hirata holds the current position of assistant professor at Graduate School of Engineering, Hiroshima University, Japan. His major is ship maneuverability and image processing.

Kai Ihara is a Master course student at Graduate School of Engineering, Hiroshima University under the guidance of Prof. Kunihiro Hamada.

Yuki Koide is an Engineer of Churyo Engineering Co., Ltd.

© 2017: The Royal Institution of Naval Architects

INVESTIGATING THE IMPACT OF DISTRIBUTED SYSTEM ROUTING DENSITIES ON VESSEL OPERABILITY

C. J. Goodrum, C. P. F. Shields, and **D. J. Singer,** University of Michigan – Ann Arbor, USA

SUMMARY

This study presents a distributed system vulnerability analysis that is compatible with the limited information present in early stage design through the use of networks. Distributed system routing densities are evaluated using structural network metrics in an attempt to better understand the effect of damage on a vessel's overall operability in areas with various system densities. This study considers several representative distributed systems which are converted to logical architecture networks. Systems are routed through a physical architecture network, which is analyzed to quantify the routing densities through the physical spaces. The effect of damage on the overall system operability is investigated by simulating damage in the network and measuring the overall system operability using a developed scoring metric. Systems are analyzed both independently and interdependently. The routing density of the damaged nodes is compared to the overall system operability using two correlation coefficients. It is found that system operability is negatively correlated to various structural network metrics, but is not strong enough to capture the complex cascades through system interdependencies, which reinforces the need for survivability research in early stage ship design.

1. INTRODUCTION

Modern multi-mission naval vessels are becoming increasingly complex due to a rise in the number of distributed systems on board. The number of components in each system is also increasing, and these components are becoming more interconnected. This has led to more distribution system routings throughout vessels, which presents new difficulties in systems integration and analysis with the limited information present in early stages of design. The increased interdependencies between systems has led to vessel operability becoming an emergent property of component and system interactions. Increased interconnectedness has also led to the emergence of cascading failures through systems, where a small amount of damage in one system causes widespread failures [1]. The addition of software, automation and more integrated power systems, have resulted in cascading failures being of particular interest to naval designers. In 1998, the USS Yorktown went dead in the water after losing its propulsion system due to a divide by zero error passed into its computers [2]. More recently, the UK Royal Navy's Type 45 destroyer have suffered complete system shutdowns due to overloaded intercoolers in warm climates [3].

The failure modes experienced by the US and UK Navies are both the result of dysfunctional interactions between components, rather than with individual component failures. In simpler systems of the past, analysis and testing allowed all undesired interactions to be understood and accounted for in the design. The increasing complexity and size of modern systems have reduced the ability to enumerate and understand these undesirable interactions [4]. These interactions present unique sources of vulnerabilities which require novel methods of analysis to be understood. This has increased the value of identifying such vulnerabilities earlier in the ship design process [5] when committed costs are lower and design decisions are more flexible.

The analyses presented in this paper focus on identifying the vulnerabilities created by these interdependent and connected routings between system components, using the level of information present in early stages of design. Network analyses are combined with an architectural framework which negates the need for detailed CAD models or simulations that are typically used, and only available in late stage design. The system architecture networks are analyzed to determine if the vessel's emergent operability can be quantified in early stage design using system structural centrality metrics.

2. SYSTEM ARCHITECTURES

To structure the distributed system analysis process presented in this paper, an architectural approach is taken. The architectural approach described in this section reflects the findings of an ongoing research program that include students, researchers, and faculty at The University of Michigan, Virginia Polytechnic Institute and State University, The University College London, and Delft University of Technology.

In an effort to move system design earlier in the design process, significant progress has been made in structuring the approach to shipboard system design [6]. A ship's distributed system design can be discussed in terms of three primary architectures and their interactions with one another:

- *Physical Architecture (PA)*: The spatial environment in which the systems exist. This encapsulates information about locations and interactions of spaces within the vessel.
- *Logical Architecture (LA)*: The interactions of components with a system or group of systems which are required to produce a given system functionality.
- *Operational Architecture (OA)*: The way which the system or systems are used through time to achieve a given system functionality.

© 2017: The Royal Institution of Naval Architects

The analysis conducted in this paper focuses on the physical and logical architectures, the combination of which create the *Physical System Solution (PSS)*. The PSS represents the physical manifestation of logical connections, and encompass the routings of the systems through the vessel, which has a significant effect on a vessel's performance [7]. A qualitative representation of the architectural framework is shown in Figure 1.

Each of the architectures require different amounts of information that need to be modelled, and for operability analyses to be conducted. To model the PA, high-level knowledge of the spaces within the vessel must be known, as well as the way in which the spaces are connected. The LA requires knowledge of the components in each of the systems, and how they are functionally connected to one another. Depending on the survivability analysis conducted, component load and flow demands as well as routing capacities may also be included.

The PSS requires the most amount of information as the knowledge contained within the PA and LA must be merged. The combination of spatial and logical aspects of the PSS adds additional complexity to the analysis, and the creation of the system routings requires significant designer input.

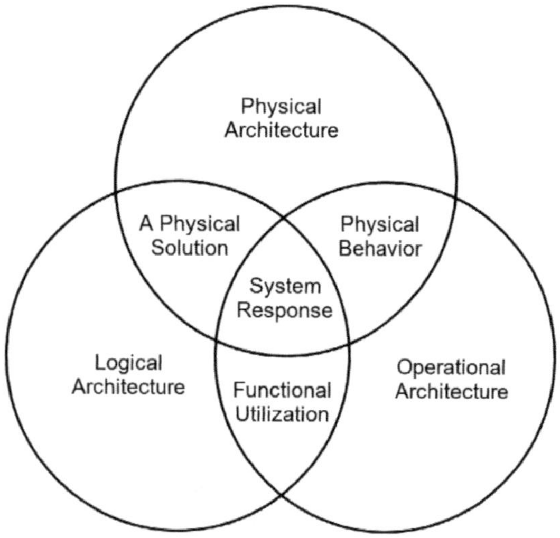

Figure 1: Visual representation of distributed system architectural framework from [6]. This paper considers the Physical and Logical Architectures, as well as their intersection (Physical Solution).

While the information encapsulated in the PA, LA and PSS varies in both magnitude and scope, each provides useful context of operability analyses and are able to be modelled in early design stages. The information of these architectures is modelled and analyzed through the use of networks and network theory.

3. NETWORK APPLICATIONS

Networks are powerful tools which create simplified representations of interactions between aspects of a system using points (nodes) and lines (edges). The versatility of these abstractions enable networks to be widely applied to many social and scientific disciplines. Their ability to represent and utilize information of varying fidelity make networks a powerful design tool. Recently, networks have been applied to the naval design process [7][8] as well as to distributed systems on board [9][10][11][12][13].

The versatility of network representations enables additional insights into system design. By representing the LA, PA, and PSS using networks, emergent system phenomena such as operability can be elicited. By examining the way in which aspects of the network are connected, the *network structure*, additional information can be garnered from the system which may otherwise not have been evident.

The way that information is represented in a network greatly influences the information garnered as well as the quality of the information generated [14]. Thus it is crucial that proper care is taken when converting systems to network space such that the information they provide is of value to the designer, and does not pre-predicate a solution. Table 1 outlines the network representations of each aspect of the architectural framework considered in this paper.

Table 1: Table outlining the network representations of each considered aspect of the architecture framework.

	Nodes	Edges
Physical Architecture	Compartments and physical locations in vessel.	Connections between compartments.
Logical Architecture	System components.	Dependencies between components for system functionality.
Physical System Solution	Compartments where a component or a system exists.	System routings between physical spaces. The logical connections represented spatially.

As the PA contains no information about the distributed systems, it alone cannot be considered for operability analysis. The LA and PSS are those which are considered for the analyses in this paper. Component nodes in these networks are modelled as demand sources or sinks depending on whether they supply or require load from each system, respectively. The edges through these networks are assigned capacities, which limit the amount of demand which can be transmitted. A capacity scaling

© 2017: The Royal Institution of Naval Architects

minimum cost flow algorithm is then applied to the network to determine the flows along each edge to satisfy the demands of all components.

3.1 NETWORK METRICS

A network's structure provides insight to the way nodes influence one another, and can help elucidate hidden behaviors within the overall system. Network structures can provide information about the way information or resources flows through a system, and can thus reveal crucial design parameters [14]. Identifying vital nodes is no trivial task as there is no index that universally categorizes node importance. Many metrics have been developed to quantify node importance through many lenses, the use of which depend on the network in question. The concept of *structural centralities* is widely used to quantify a node's importance in the context of the network structure, however, many such measures exist and 'importance' is widely debated in the network science community [15]. For this reason, a number of structural centralities have been considered in this analysis, in an attempt to determine which, if any, correlate best with system operability. The network metrics used for comparison in this paper are described in Table 2.

Table 2: Brief descriptions of the structural network centrality metrics considered in the operability analysis.

Network Metric	**Description**
Betweenness Centrality (BC)	The extent to which a node lies on shortest paths between other nodes.
Degree (DEG)	The total number of edges connected to a node.
Current Flow Betweenness (CFB)	Extends traditional betweenness to consider non-shortest paths.
Out Degree Centrality (ODC)	The total number of directed edges emanating from a given node.
Katz Centrality (KC)	Considers a node's connectedness to other highly connected nodes.
Closeness Centrality (CC)	The average geodesic path distance from a node to other nodes.
Coreness (C)	'Weights' degree centrality based on a node's location in the overall network structure.
Flow Scaling (FS)	Weights the importance of the node based on the amount of resource flow through it.

The network metrics described above are robust in their ability to be applied across a wide variety of networks, however, they each contain specific limitations. One such limitation applicable to all of the above metrics is that they are unable to place weights on different sources of information contained in the network. For example, a node's structural centrality in a single network does not necessarily represent the systemic variables as well as the greater spatial landscape. For this reason, four hybrid centrality measures were created, which aim to combine spatial and system related data into a single metric. These are labelled Physical System Degree (PSD), Physical System Centrality (PSC), Physical System Hybrid 1 (PSH_1) and Physical System Hybrid 2 (PSH_2).

$$PSD = (DEG[sys]*DEG[PA])/(DEG_{max}[sys]*DEG_{max}[PA])$$

$$PSC = (BC[sys]*BC[PA])/(BC_{max}[sys]*BC_{max}[PA])$$

$$PSH_1 = (BC[sys]*DEG[PA])/(BC_{max}[sys]*DEG_{max}[PA])$$

$$PSH_2 = BC[sys]/BC_{max}[sys] + DEG[PA]/DEG_{max}[PA]$$

These hybrid measures aim to incorporate the logical information of the PSS with the spatial information present in the PA. The metrics aim to merge the connectedness of a compartment with the systems existing within it, to represent the logical and spatial aspects of the system design. To represent the connectedness of the compartments, the degree was used, as it encapsulates local information about the number of other compartments a node is connected to. Betweenness centrality was selected to represent routings due to the directionality and consideration of shortest path routings. Each metric is normalized by the maximum value to ensure equal weighting of each variable. Additional hybrid measures could be investigated, but is out of the scope of this analysis.

3.2 OPERABILITY METRIC

An operability metric was developed to represent the final operability of the vessel in the presence of damage. Damage is inflicted to the network by removing each node sequentially, in addition to all edges in all systems which point to, or emit from, that node. The capacity scaling algorithm is then reapplied to all considered systems, and the demands of all remaining components are checked. If the flow demands of a component are satisfied, the component is considered functioning, and vice versa. The number of functioning components is recorded, and used to calculate the scoring metric.

For the purposes of this analysis, the operability of the vessel is defined as the percentage of components across all systems still functioning after damage has been inflicted to the network.

$$OP(sys) = (Func.\ Components) / (All\ Components)$$

It is important to note that this operability metric solely considers the final state of the vessel, without any time-domain considerations. Additionally, this metric does not distinguish capabilities or operational aspects affected through the loss of certain components (that is to say, for example, the ship's ability to fight back vs. flee). This metric was selected due to its simplicity and its flexibility in comparing different network metrics.

© 2017: The Royal Institution of Naval Architects

Although the different capabilities of each system are not distinguishable, the developed system of systems model within the case study does consider the interdependencies between systems, and allows for component or system failures to cascade throughout all other systems. This enables operability to be an emergent property of the networks which is all that is required for early stage distributed system design when system placement is the primary focus.

4. SHIP CASE STUDY

The operability analysis is applied to a representative ship based on the notional Electric Ship Research and Development Consortium Medium-Voltage DC (ESRDC MVDC) all-electric warship topologies described in [16] and [17].

Using this vessel as a reference, two representative distributed systems are considered: a power system and a cooling system. Each developed system makes use of ring bus topologies for routings, zonal layouts, and includes redundant load sources in each system. A total of 25 components are included, and are all connected to both systems (either as requiring or emitting demand in each system).

The power system consists of two gas turbine generators, two auxiliary generators, and an energy storage magazine used to power two electric motors (prime movers), a pulsed load, four zonal load centers, radar, and the bridge. The load centers represent power sinks within each of the four designed damage zones, each of which has two redundant power control modules (PCM). All of these components feed into a ring bus architecture which spans the length of the vessel, and is port high, starboard low.

The cooling system on board was developed specifically for analysis within this paper's ESRDC MVDC notional ship and is not based on any previous topology. The system consists of a fully connected ring bus along the ship's keel, with two cooling water sources (chillers). These two redundant sources are placed in adjacent damage zones and supply all components with chill water.

4.1 PHYSICAL ARCHITECTURE NETWORK

The notional ESRDC MVDC ship is converted to a physical architecture network containing 135 nodes shown in Figure 2. These nodes are separated into three transverse planes (45 nodes each) representing the port, center, and starboard shipboard locations. The network has 8 nodes longitudinally, which was included to represent a forward and aft section in each of the four damage zones. The ship network contains 5 vertical decks within the hull, and an additional 3 decks in the superstructure.

Nodes are connected using undirected edges between adjacent geometric nodes within each transverse layer, and are connected transversely to those nodes in the same vertical and longitudinal locations in each layer. This forms a perpendicular lattice structure representing the connections between shipboard physical spaces. The developed physical architecture network is depicted in Figure 4.

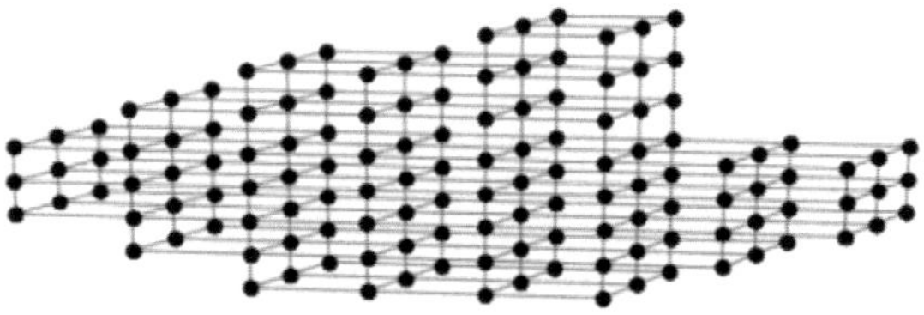

Figure 2: Undirected Physical Architecture network of ESRDC MVDC vessel.

4.2 LOGICAL SYSTEM ARCHITECTURE NETWORKS

The logical system architecture for the ESRDC MVDC notional ship is shown in Figure 3. Logical connections shown in red correspond to the power system, and those shown in blue correspond to the cooling system. Edges within this network are directed, and represent the direction of the flow of resources in each system. Nodes represent the components present in both systems. The node's corresponding component is listed in the accompanying table.

The structure which emerges from the logical connections displays a few interesting features of the interactions between the two systems. For example, there are many more connections in the power system than there are in the cooling system, evident by the increased number of directed edges. Also, it becomes evident that the components are linked in the power system in a much more complicated manner than in the cooling system, where all components are simply fed by the chillers.

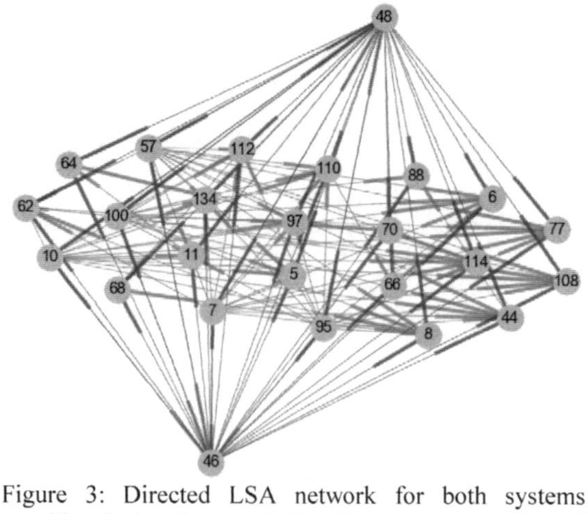

Figure 3: Directed LSA network for both systems considered in the analysis. Power system logical

connections are shown in red, and cooling system in blue. See Table 3 for component keys.

Table 3: Node-component key for nodes in Figure 3.

Node	Component	Node	Component
5	Electric Motor	68	LC2
6	PCM	70	LC1
7	Main Gen.	77	Pulse Load
8	PCM	88	Radar
10	Aux. Gen.	95	Main Gen.
11	PCM	97	Electric Motor
44	Radar Pow. Conv.	100	Energy Storage
46	Chiller	108	PCM
48	Chiller	110	PCM
57	Aux. Gen.	112	PCM
62	PCM	114	PCM
64	LC4	134	Radar Pow. Conv.
66	LC3		

By representing the logical connections between components, the LA network provides a level of detail which could aid designers in preliminary design stages by providing context as to the scopes of systems.

4.3 PHYSICAL SYSTEM SOLUTION NETWORK

Each system's LA network is combined with the PA network to create a single instantiation of a PSS. Figure 4 illustrates the PSS networks for the power, cooling, and combined systems, shown in red, blue, and purple, respectively. The power system solution is based on the routings present in the ESRDC MVDC topologies, and the cooling system routings were developed as a representative case.

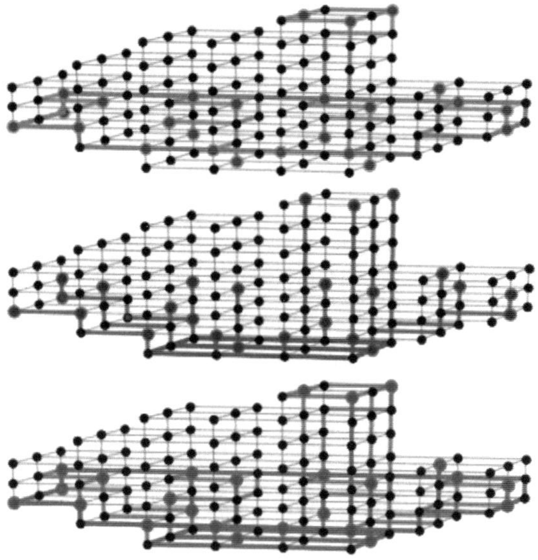

Figure 4: PSS routings for representative power system (red) and cooling system (blue). The total integrated system is shown in purple, and contains the connections of both systems.

5. RESULTS

The operability analysis described in Section 3.2 is first conducted on the LA network, and is extended to consider the PSS. The analysis provides a vessel operability score when each node is removed in the network. A series of network metrics are calculated for each removed node. For each case, two correlation coefficients are calculated between the operability and network metric data sets for each considered network metric. In both cases, the systems are considered independently as well as interdependently.

The first correlation metric used across the operability and network metric data sets is the Pearson correlation [18] (bivariate correlation). The Pearson correlation is employed as a metric of the linear statistical correlation between the two data sets and has the range [-1,1] implying a direct negative or positive linear relationship, respectively. A value of zero implies the data sets are independent.

In addition to the Pearson correlation the Kendall rank correlation is used to measure the ordinal association between the two data sets, which also has the range [-1,1] (with a value of zero implying independence). As operability is an emergent property of the systems and is inherently non-linear, ordinal associations are also of value [19]. As the Kendall coefficient is a rank statistic, it perhaps better accounts for emergent behaviors, which tend to be non-linear.

In comparing the operability metric with the network structural metrics, it is expected to observe a negative correlation. As the connectedness and importance of a node in the network increases, one would expect an observable decrease in vessel operability when that node is removed.

5.1 LOGICAL ARCHITECTURE RESULTS

The closeness centrality, current flow centrality, and degree centralities for each system in the logical architecture (Figure 3) are compared to the operability metric in the presence of damage. Each system is analysed separately, and interdependently, however, the operability metric encompasses all systems. The results from this analysis are displayed in Figure 5.

The Pearson and Kendall correlation values vary greatly depending on the system considered, and the structural metric being used. The interdependent and cooling systems display negative correlations, and the power system displays a weak positive correlation across both metrics. In general, the Pearson correlations are stronger than the Kendall correlations for negative values. This trend does not hold for the power system.

© 2017: The Royal Institution of Naval Architects

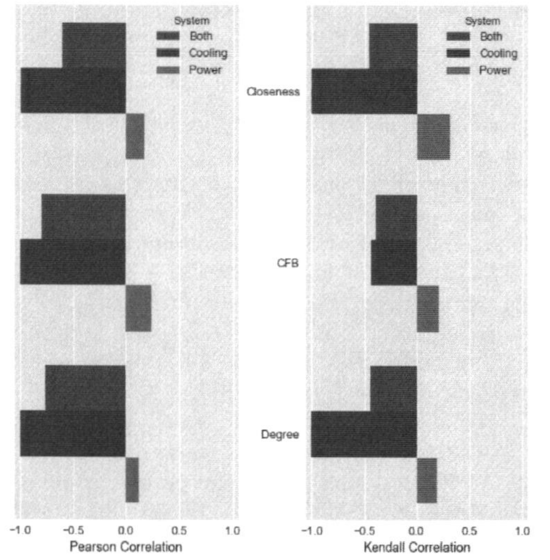

Figure 5: Pearson (left) and Kendall (right) correlation results between three centrality metrics and the developed operability metric when applied to the logical system architecture.

The cooling system exhibits a direct negative correlation (correlation of -1) with the system's operability metric in all metrics but the current flow betweenness centrality. This is due to this metric's requirement for an undirected graph, rather than a directed graph.

The power system displays a weak positive correlation, which suggests the system operability is essentially independent of the operability measure. The power system has a more complicated structure than that of the cooling system, in addition to more redundant sources, which contributes to the disconnect between the measures.

When considering the interdependent system, the resulting correlations are strongly negatively correlated, which is likely a result of the cooling system's perfectly negative correlation.

The large variation across the independent system networks is explained by their respective structures. The results suggest that network structures with a strong hub and spoke structures (where the degree distribution is heavily skewed to a few nodes) will be more highly correlated to operability. This is the case of the cooling system (Figure 3), which contains two source nodes (the chillers) which supply all other components' demand. These nodes have high degrees, and centrality measures, which point to all other nodes in the network. Removing either of these sources causes cascading failures through the networks and the failure of all components. Removing any of the component nodes has no effect on the rest of the system. This illustrates the way vulnerabilities in one system become crucial design considerations in terms of vessel operability. The results also suggest that simple network structures may correlate well to operability, but the correlation decreases as the network grows and increases in complexity.

The conclusions from the LA results is limited in many aspects. The primary limitation of this analysis is that the damage analysis only considers the damage of system components, and is independent of spatial considerations. When a component is removed in the LA network, all edges associated with that node are also removed. This removes all functional connections to that node, but neglects to consider redundancies in system routings. By neglecting to consider the spatial aspect of the connections, the analysis is also bounded in that logical routings which are coupled spatially are not considered.

5.2 PHYSICAL SYSTEM SOLUTION RESULTS

To address the limitations associated with the logical system architecture results, the analysis is expanded to consider the PSS. The results from this analysis is shown in Figure 6.

The results from this analysis illustrate that all tested network structure metrics for all systems are negatively correlated with operability, to varying strengths. The Kendall correlation values are larger than the Pearson coefficient, which implies the relationships are less linearly correlated, and more ordinally related. This contradicts the relationships observed in the logical system analysis, and is due to the increased size of the PSS network compared to the LA networks, in terms of both nodes and number of edges. As vessel operability is an emergent property of the network structures, it is intuitive that it would not be linearly correlated to a structure metric, and that it would become more complicated as the size of the network increases.

In none of the cases is there a perfect negative correlation for either correlation coefficient. The maximum Pearson correlations observed for the cooling, power and combined systems occurred at PSD, PSD, and PSH_1, with values of 0.367, 0.342, 0.562 respectively. Similarly, the maximum observed Kendall correlations for the cooling, power and combined systems occurred at KC, PSD, and DEG, with values of 0.767 0.325, and 0.645 respectively.

The hybrid measures perform well and account for the majority of the highest correlated values. The hybrid measures' success is likely a function of better encapsulating the spatial and logical information of the PSS, which is intuitive as operability emerges as a function of the two. However, the variability across values and across the best performing metric confirms the finding of [15] that there is no single metric to determine node importance, and that the hybrid metric employed must be carefully tailored to the network being evaluated.

© 2017: The Royal Institution of Naval Architects

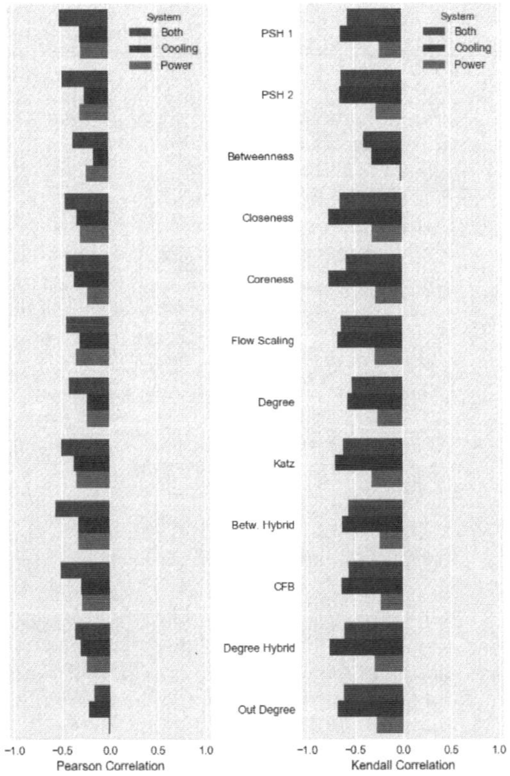

Figure 6: Pearson (left) and Kendall (right) correlation results between centrality and operability metrics applied to the physical system solution.

6. CONCLUSIONS

This paper presents an investigation into the correlation between network routing densities and the resultant operability in the presence of inflicted damage. Network structure metrics are applied to the logical architectures (LA) and physical system solutions (PSS) of a few representative distributed ship systems, and are generally shown to negatively correlate with operability. The strength of the correlation is strongly dependent upon the structure of the network in question.

The LA network analysis reveals that operability information can be elicited from the simple logical connections between components. The results indicate that the simple hub-and-spoke cooling system directly correlated centrality metrics with operability, but this was not the case in either the power or interdependent system cases. The analysis is limited in its ability to include spatial information in the way which damage affects the network structure, and thus how the logical routings between components are affected. Thus, the LA provides valuable information to designers about the context of the system connections, but is insufficient on its own to accurately quantify vessel operability, especially for large and complex system structures.

The correlations elicited in the PSS analysis were lower overall than that of the logical architecture networks due to the increased network sizes, complexity, and edge interactions. The developed hybrid structural centrality measures marginally outperformed the traditional measures, however, no single measure proved best across all networks. The increase in correlation is likely due to better encapsulating the information contained within the physical architecture (PA), LA and PSS. The variance associated with the best performing metric strongly depends on the structure of the solution.

The weak correlation between vessel operability and network routing density is insufficient to definitively use network structure as a means of identifying all failure modes. Complex interactions between components in different systems can often cause cascading failures throughout the vessel, which are often times not captured by structural analysis alone. This means that although network structure can be used to gain insight into the critical areas of a system, it should not be treated as a direct indicator. While the structure of the network alone is insufficient to highlight aspects of a system or systems which have direct impacts on operability the method can still be used to increase the level of knowledge of system layouts and its implications in early stage design. This highlights the importance not only of conducting operability analyses in early stages of design, but also the necessity to include and determine metrics which indicate and quantify vulnerable areas for ship system designers.

This research can be enhanced by considering additional aspects of the analysis. This further work includes:

- Expanding the network analyses to consider time dependent operational architectures.
- Considering the effect of network representations on the solutions generated and the correlation with operability metrics.
- Expanding this analysis to consider ensembles of physical system solutions rather than a single instance, as a means of maximizing correlation.

Additional effort should also be placed in developing network metrics for ship operability which are useful in early design stages, as this will be a requirement to develop the leading warships of the future.

7. ACKNOWLEDGEMENTS

We would like to thank Kelly Cooper from the Office of Naval Research (ONR) for providing support for this project. This work was funded under grant number N00014-15-1-2752. Additionally, this work received Government support under contract FA9550-11-C0028 awarded by the Department of Defense, Air Force Office of Scientific Research, National Defense Science and Engineering Graduate (NDSEG) Fellowship, 32 CFR 168a.

© 2017: The Royal Institution of Naval Architects

8. REFERENCES

1. Buldyrev, S.V. et al. 'Catastrophic cascade of failures in interdependent networks'. *Nature, Vol. 464*, April 2010.

2. Slabodkin, G. 'Software glitches leave Navy smart ship dead in the water'. *Government Computer News*, July 1998.

3. Britton, B., 'Britain's Royal Navy warships are breaking down because sea is too hot', *CNN World*, June 2016.

4. Leveson, N., 'A New Accident Model for Engineering Safer Systems'. *Safety Science (42)*. April 2004.

5. Chalfant, J.S. 'Early-Stage Design for Electric Ship'. *Proceedings of the IEEE 103 (12):2252-2266*. Dec 2015.

6. Brefort, D. et al. 'An architectural framework for distributed naval ship systems'. *Submitted to Ocean Engineering*. 2017.

7. Rigterink, D. et al., 'The use of network theory to model disparate ship design information.' *Proceedings of the 12th Symposium on Practical Design of Ships and Floating Structures*, October 2013.

8. Rigterink, D.T., 'Methods for Analyzing Early Stage Naval Distributed Systems Designs, Employing Simplex, Multislice, and Multiplex Networks.' *(Ph.D. thesis) University of Michigan*. 2014.

9. Trapp, T.A., 'Shipboard Integrated Engineering Plant Survivable Network Optimization'. *(Ph.D. thesis) Massachusetts Institute of Technology*, May 2015.

10. Shields, C.P.F., Rigterink, D.T., Singer, D.J., 'Investigating physical solutions in the architectural design of distributed ship service systems'. *Ocean Engineering 135*. March 2017.

11. Shields, C.P.F., Brefort, D.C., Parker, M.C., Singer, D.J., 2015. 'Adaptation of path influence methodology for network study of iteration in marine design.' *In: Proceedings of the 12th International Maritime Design Conference, vol. 2. Tokyo, Japan, pp. 43-56,* May 2015

12. Shields, C.P.F., Rigterink, D.T., Singer, D.J., 2015. 'The information dual network and its applications to naval architecture.' *In: Proceedings of the 12th International Marine Design Conference, vol. 2, Tokyo, Japan*, pp. 1-8, May 2015.

13. Shields, C.P.F., Sypniewski, M.J., Singer, D.J., 2016. 'Understanding the relationship between naval product complexity and on-board system survivability using network routing and design ensemble analysis.' *In: International Symposium on Practical Design of Ships and Other Floating Structures 2016, Copenhagen, Denmark, pp. 219-225*, September 2016.

14. Newman, M.E.J. 'Networks: An Introduction'. *Oxford University Press*. 2016.

15. Lu, L. et al. 'Vital nodes identification in complex networks'. *Physics Reports*. July 2016.

16. Chryssostomidis, C., Cooke, C.M. 'Space reservation for shipboard electric power distribution systems. *In 2015 IEEE Electric Ship Technologies Symposium (ESTS), 2015 IEEE, pages 187-192. IEEE, 2015.*

17. Chalfant, J.S., Chryssostomidis, C. 'Analysis of various all-electric-ship electrical distribution system topologies.' *In 2011 IEEE Electric Ship Technologies Symposium ESTS 2011, pages 72-77. IEEE 2011.*

18. Sedgwick, P., 'Pearson's correlation coefficient'. *BMJ*, July 2012.

19. Lu. L. et al. 'The H-index of a network node and its relation to degree and coreness'. *Nature Communications*, Nov. 2015.

9. AUTHORS BIOGRAPHY

Conner J. Goodrum is a Ph.D. student at the University of Michigan College of Engineering. He earned his B.S.E. and M.S.E. in Naval Architecture and Marine Engineering, and is currently pursuing a M.S.E. in Industrial and Operations Engineering. He works in the Advanced Naval Concepts Research (ANCR) Lab, focusing on complex systems and network applications in ship design.

© 2017: The Royal Institution of Naval Architects

Colin P. F. Shields, is a Ph.D. student in the Naval Architecture and Marine Engineering department at the University of Michigan College of Engineering. He conducts research in the areas of design theory, the study of complex systems, and network science. He earned a B.S.E, M.S.E., and is pursuing a Ph.D. in NA&ME at the University of Michigan.

Dr. David J. Singer, Ph.D., is an Associate Professor in the Naval Architecture and Marine Engineering department at the University of Michigan College of Engineering. He earned his B.S.E., M.Eng., and Ph.D. degrees in NA&ME and a M.S.E. in Industrial and Operations Engineering from the University of Michigan. He conducts research in the areas of naval design theory, design optimization, and ship production.

© 2017: The Royal Institution of Naval Architects

© 2017: The Royal Institution of Naval Architects

EFFECTIVE UTILIZATION OF DIGITAL DESIGN ASSETS DURING THE POST-DESIGN PHASES IN MODERN SHIPBUILDING

M. Yllikäinen, Cadmatic Ltd, Finland

SUMMARY

The shipbuilding industry has a long history of producing and consuming paper-based documentation for the manufacturing, assembly and verification phases of shipbuilding projects. The wide adoption of modern technologies and advancements in mobility has allowed the use of product models in novel ways, for example in construction and installation work. Traditional documentation cannot fully be omitted in the near future, but for example the adoption of easily accessible interactive documentation as a part of the aforementioned processes, however, streamlines construction workflows, increases speed and reduces errors. Interactive construction documentation also reduces work in the design phase by removing the need to create excessive amounts of static materials.

1. INTRODUCTION

Over the decades, ship design has moved from drawing by hand to current three-dimensional design work, and drawings and other documents produced from that. Major improvements to the design process have been made during the years and efficiency of design work is higher than ever. Traditionally the aim of the design work has been to produce drawings and other documentation, like section drawings, layout drawings, material take-offs and information for automated cutting and bending, that is suitable for example for construction and assembly phases. Ultimately the target of the design work is, of course, to make it possible to actually construct the ship. Thus design assets, for example the intelligent 3D model, which is currently produced as a "side product" of drawing production, can be seen as more than just a way to generate the required documentation.

The design material produced during the design process is still heavily under-utilized during post-design phases like manufacturing, construction, installation and operation phases. Technical advancements for example in mobile technologies make it possible to use the information when-ever, where-ever. This paper discusses through examples existing options to utilize the design data assets in a modern way in order to reduce costs and improve efficiency and quality of the post-design phases.

2. DIGITAL SHIP MODEL

Modern design tool landscape contains dedicated tools for the actual design work, but also for information management. Nowadays in a modern ship design process all necessary information is available at all times at the office, using information management solutions like the one described by Juntunen [1].

The digital ship model created during the design phase contains valuable information for many non-designers: pipe workshop staff, installation teams, and even people responsible for operating the ship. It offers not only a way to produce the production documentation, but also a platform for sharing all the ship-related information.

The basis for the digital ship model is usually an accurate three-dimensional geometric representation of the ship. The digital ship model is usually intelligent, meaning that it contains for example metadata attributes for the items and information about relations between the items. Depending on the tool landscape, the digital ship model can also contain information that can be external from the point of view of the design applications, like status information, workflow related information, instructions, documents, discussion related to the design, purchase information, logistic information, process information, and so forth.

The digital ship model is useful only if the information in it is correct, accurate, up to date and available when needed. To guarantee this, the same digital ship model is usually shared across all disciplines; this way everyone is always seeing the latest version of information from all project parties, and for example problems in data validity can be easily spotted. Traditionally this has been reached through using a replication mechanism in multi-site design environments [5] and single-point data access in less complex design environments. Utilizing the digital ship model outside of the office

© 2017: The Royal Institution of Naval Architects

increases the need for synchronization mechanisms to guarantee that also the offline copies are sufficiently up to date at any given moment.

3. BRINGING THE DIGITAL SHIP MODEL TO THE CONSTRUCTION SITE

The digital ship model has been utilized close to the construction site by using computers, but increased performance of tablets has opened doors for solutions in which the digital ship model is available directly at the construction site. This removes the need to move away from the construction site to view the digital ship model, thus drastically lowering the bar for utilizing it. Modern tablets are performant enough to handle the complete digital ship model, even up to the largest cruise ship models. This means reduced need for preselection – users can select the information they want to see when they need it. They need neither to select the information in beforehand nor to return to the office to get it.

To enable efficient data access, mobile devices, like tablets, need specialized implementation of software tools for browsing the digital ship model. Mobile devices are often used in environments in which no online connection can be arranged. Also, the absence of the traditional input devices might make it hard to use existing applications that have not been designed for mobile use.

Work at the construction site contains for example planning, the construction and installation work itself, verification procedures and reporting. Usually these operations are largely based on drawings, listings and other kind of documentation generated from the digital ship model. Paper-based documentation has several drawbacks. First of all, someone needs to produce it, which causes extra effort for the design and work planning phases. Second, the material is hard to handle, easily out-of-date and non-interactive. It is unlikely that paper-based documentation would completely disappear from the construction site, as their abstract nature makes reading them in some respects quite efficient. But even with that in mind, the presence of the three-dimensional digital ship model can greatly speed up the work and reduce costly errors. The digital ship model can also work as an access point to the traditional drawings in a digital form, combining the good sides of both worlds.

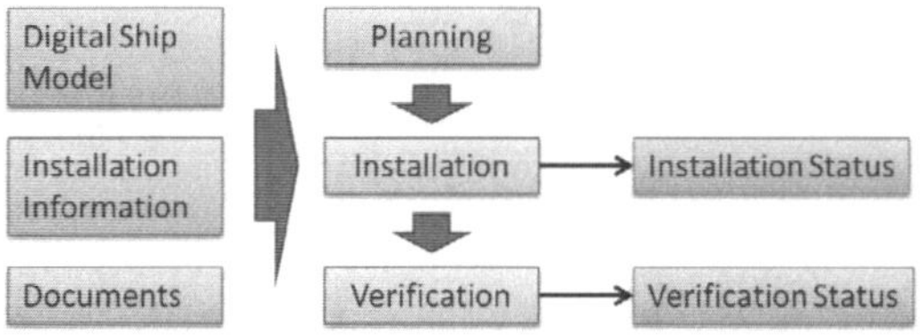

Figure 1: An illustration of the information flow when utilizing digital design assets at the construction site.

4. EASIER INSTALLATION PLANNING

When the installation team is preparing for the installation work, seeing the complete design in an easily perceivable form is more efficient compared to just looking at the drawings that are merely static projections. The 3D model can be, for example, colorized to show both the existing status of the construction and the desired status after the next installation step. The installation team can for example see pipe parts used for fit-up with a glance using the digital ship model. The planned installation order can be visualized, which helps the installation team to decide whether the planned order is optimal or if it still needs to be fine-tuned. The proximity of the items being installed can be checked for items that could cause problems during the installation. The digital ship model can also help in solving the problem of finding the required materials for the installation work [3], due to having easily readable visual representation of the parts.

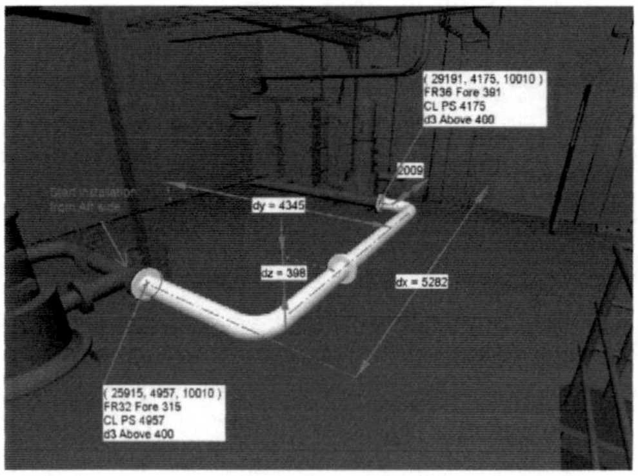

Figure 2: The possibility to check surroundings, add necessary measurements, query item metadata and add position labels makes it possible to utilize the digital ship model when planning the installation.

5. FASTER OUTFITTING INSTALLATIONS

During the actual construction and installation phases the digital ship model lets the construction staff and installation teams do necessary measurements directly on top of the digital ship model. The ability to take measurements as required reduces the need for installation drawings, which benefits the designers as less time is spent creating the documents, and also the team itself as they can choose themselves which measurements are needed.

Having an offline access to the digital ship information allows utilizing it both during the block outfitting phase and also after the block has already been erected and no online access can be arranged. An extra benefit that we get from the interactivity is that the digital ship model can be easily viewed upside-down, which is usually the way the block outfitting phase is performed [4].

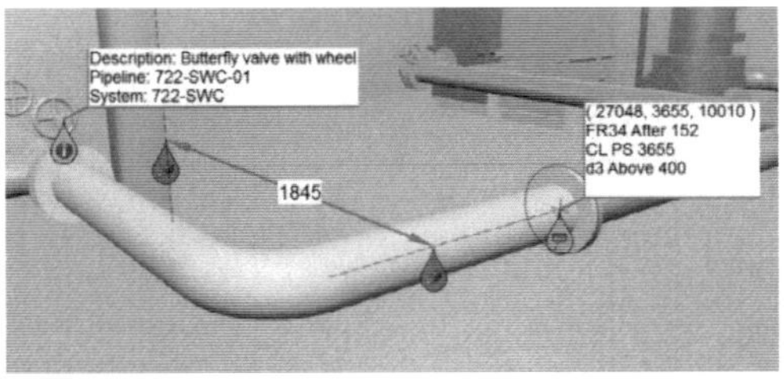

Figure 3: In an interactive drawing the view can be freely rotated, measurements added and removed, and all of the information in the digital ship model is available.

6. MORE ACCURATE REPORTING

In a ship building project it is challenging to keep track of the status of the project. Especially complicated it is if much of the work is sub-contracted. Most steps in the building process have a relation to the other steps – they are either blocking or blocked by the other steps – therefore, successful execution of the project needs accurate and up-to-date information about which tasks have been completed. Successful execution of for example lean principles [2] requires a good view on what others in the project are doing. Several traditional approaches exist to keep track of the progress, but usually they are based on manual work by making use of the drawings or requiring the teams to carry out other paper work.

By having an up-to-date 3D model with the installation team at the site, it is very natural and easy to make a status report immediately after the installation, using the 3D view as the user interface. For example, a visual comparison of the reality and the 3D model which is colorized based on the reported installation status provides an easy way to check that information that was entered is correct and up to date. This information can then be shared and utilized in a digital form, by synchronizing the information back to the shared digital ship model. An additional benefit is that the probability of errors decreases, because manual steps are removed from the process.

Possibility to track the project status in almost real time is hard to achieve using the traditional approach in which the status of the construction is usually recorded on paper to reports or drawings. With digital reporting the project status is up to date and available immediately after the data has been synchronized back to the shared digital ship model.

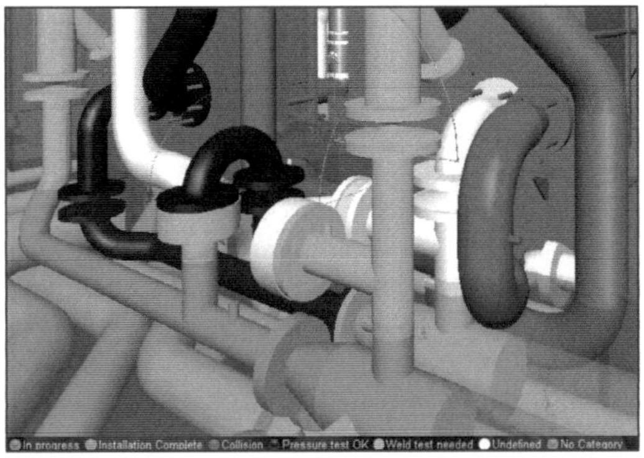

Figure 4: Item statuses are both straight-forward to update and easy to read utilizing the digital ship model.

7. MORE CONTROLLED VERIFICATION

In the verification phase the construction is checked so that what was built is according to what was planned, and that what was built also fulfills the requirements imposed on the construction. Regarding planning verification, the digital ship model can function as an easily readable reference for seeing that the construction is as designed. In requirement verification, the model can assist for example in weld inspections or in pressure testing the piping. The digital ship model can be used as a powerful tool for collecting information about parts

© 2017: The Royal Institution of Naval Architects

that have been verified, thereby also guiding the work by showing which parts still need to be checked. Collecting this information in digital form has the added benefit of being available to other interested parties as soon as the information is synchronized to everyone, which means that other activities can be scheduled more accurately. The digital platform can also record who entered the information, which improves traceability.

8. BENEFITS OF THE DIGITAL SHIP MODEL IN REVAMP PROJECTS

Revamp and retrofit projects can be challenging due to the possible additional tear-down step; the design and work planning processes must take into account the items that are going to be removed to make room for the new items. In addition, the project is usually very fast-paced as the time the ship spends on the shipyard is very limited. This means that project management plays a key role in these projects. Project management can be made easier by improving the communication between the designers, work planning staff, the tear-down crews and the installation teams. If a digital model of the ship does not exist, the ship is often laser-scanned and at least partially remodeled in digital form to get the as-built model. Based on the as-built model, designers add the required new piping, pumps, and so forth. The designers also mark the parts that need to be torn down from the existing construction. Visual information shown in the digital ship model makes the tear-down work faster and less error-prone. By using the digital ship model at the construction site the progress of the tear-down work can be easily reported, allowing more accurate control over when specific new build parts can be installed, thus making it easier to manage activities and also spot tasks that are falling behind the schedule.

9. CADMATIC APPROACH TO INFORMATION MANAGENT

CADMATIC has developed an information management ecosystem that allows efficient utilization of the digital ship model. The model can also be used during the manufacturing and construction phases, all the way to the operation phase.

At the core of the system there is a replicated design tool. Due to the advanced replication mechanism, all designers are working with the most recent design. For information management there is a server based system called eShare, described by Juntunen [1], that is connected to the design application and to other information sources like ERP and DMS systems. Using the shared online data and the shared digital ship model everyone, designers as well as non-designers, have access to the latest design information.

For utilizing the digital ship model at the construction site a tablet application called eGo has been created. It can be connected to the shared digital ship model in eShare. The digital ship model on the tablet can be easily synchronized with the shared digital ship model. The two-way synchronization makes sure that all changes to the shared model become visible on the tablet and all status updates made on the construction site become available for all viewers of the shared model.

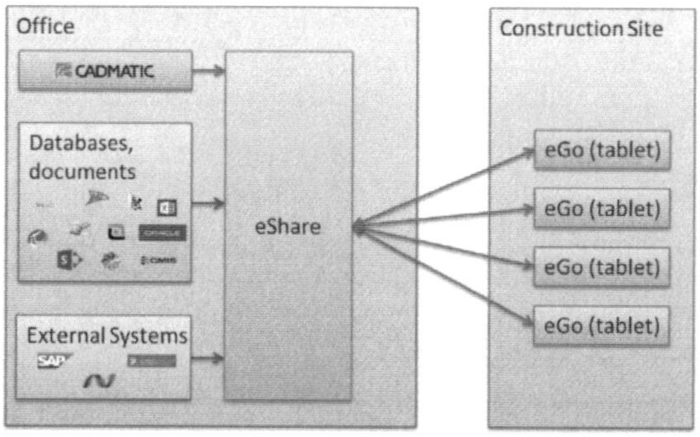

Figure 6: An illustration of the CADMATIC solution for efficient utilization of design assets at the post-design phases. eShare collects information from both the CAD system and the other applications involved, and eGo is used for offline access at the construction site.

10. CONCLUSIONS

The information produced during the design phase can be utilized in many ways after the actual design phase to improve efficiency and reduce errors. Using shared digital ship model information during post-design steps, like piping installation and pressure checks, improves workflows compared to the traditional paper-based approach. Sharing a single digital ship model between all project parties allows faster projects and reduced costs. At the core of the process needs to be the shared model data, which is up to date and accessible from everywhere, including the construction site. Cases shown here serve only as examples as additional use-cases exist. As technologies evolve there will be an increasing number of concrete cases in which the design assets can be utilized to make the post-design phases more efficient.

11. REFERENCES

1. JUNTUNEN M., 'Integrated 3d Ship Design and Information Management', International Conference on Computer Applications in Shipbuilding, 2013

2. LANG S., et al., 'Shipbuilding and Lean Manufacturing - A Case Study', Ship Production Symposium, 2001

3. TOMMELEIN, IRIS D., 'Pull-Driven Scheduling for Pipe-Spool Installation: Simulation of a Lean Construction Technique', ASCE Journal of Construction Engineering and Management, July/August 1998

4. CHAKRABORTY S., 'What Is Advanced Outfitting in Shipbuilding?', Marine Insight, http://www.marineinsight.com/naval-architecture/advanced-outfitting-in-shipbuilding/, 2017

5. JUNTUNEN M., 'How Can The Most Efficient Design Recourses From All Over The World Be Easily Utilized?', International Conference on Computer Applications in Shipbuilding, 2011

12. AUTHOR'S BIOGRAPHY

Mikko Yllikäinen is a Master of Science in Information Technology. Mikko has spent his whole professional career working with ship design information management, and currently he is working as the product owner for information management tools at CADMATIC Ltd. Mikko is an eager adopter of new technologies, but always does so with a practical approach.

© 2017: The Royal Institution of Naval Architects

PERFORMANCE VERIFICATION OF THE AIR-CONDITIONING SYSTEM IN A SUBMARINE

Jan D. Wilgenhof, MSc. , MecDes, The Netherlands,
Jesús Molina Toledo, MSc. , Navantia, Spain

SUMMARY

One of the large energy consumers in a conventional submarine is the air-conditioning system. The system removes a large part of equipment waste heat as well as metabolic heat and water vapor produced by the crew. It also reduces the humidity of intake air during and after snorting in tropical waters. Air temperatures and humidity in the compartments are important in view of crew endurance and equipment limits (e.g. humid air in electronics).

In the first design stage the air conditioning cooling power is estimated. In the detail design stage it is desired to verify the performance capability with the then available data on structure details, insulation, deck covering, accommodation separation walls, etc., as well as ventilation duct arrangement and air flow rates. Besides cooling power verification it is also desired to know resulting temperatures in the compartments in different sailing conditions. A conventional heat balance cannot provide this because it assumes fixed (maximum allowable) space temperatures and it is performed only for steady state.

Therefore a dynamic air energy balance has been developed. It is a software tool that simulates the heat transfer with the environment and the heat transfer between the spaces in the submarine as well as the thermal contribution of the ventilation flow, based on variable space temperature. Equipment and crew are included as heat sources. Humidity is simulated as water vapor concentration.

Several types of air cooling equipment were developed: a dry air cooler model and a cooling coil model in an air conditioning unit which provides the temperature and humidity ratio difference as well as sensible and latent cooling power. A large number of tests of air conditioning units were used to develop the model. Also attention was paid to efficiency. In this way the tool enables calculation of the resulting temperature and humidity in each space in the submarine as well as power consumption.

The model has been applied on the S80 submarine now under construction. It was requested to determine the cooling power margin, power consumption and future growth margin. The installed cooling power will result in lower compartment temperatures than the maximum allowable in a certain sailing condition. By adjusting the cooling power downward the available cooling power margin can be found. Then the average power consumption for this sailing condition can be found.

The new software tool enables faster and more efficient analysis, in particular if a lot of submarine operation conditions have to be analyzed; demonstration of compliance with building specification requirements; optimization of the cooling power distribution over the spaces; finally it provides more accurate energy consumption data for the submarine endurance and range calculations.

1. INTRODUCTION

The waste heat of equipment in the submarine is removed by a number of cooling systems: the fluid cooling systems, such as the fresh water, sea water and the air cooling system. The air cooling also controls humidity, therefore it is called air-conditioning system. For this system different designs exist: a chilled water plant producing low temperature water which is distributed to the several units with a cooling coil located in the compartments; or it can exist of a number of individual air-conditioning units, where the rejected heat is removed by a fresh water circuit [1]. The chain of fluid cooling systems exists for a number of reasons: safety, efficiency, corrosion and fouling, but heat accumulating in the air cannot be avoided and air cooling is almost always required. In this paper we will focus on the air cooling system with separately installed Air Conditioning Units (ACU's), which remove a large part of the heat generated by equipment and all of the heat generated by the crew as well as humidity.

In the design stage the maximum allowable temperatures in all the spaces are known, as well as the operating conditions and maximum sea water temperature, mostly defined in a contract specification or a similar requirements defining document. A conventional heat balance is carried out to determine the required cooling capacity for each space and thus equipment sizes. The heat balance is the equality between the total amount of heat from the heat loads which is accumulated in the air and the heat removal capacity of the air conditioning system. Other definitions for a heat balance are possible e.g. if the fluid cooling systems are included, but the

© *2017: The Royal Institution of Naval Architects*

discussion in this paper will be limited to the energy in the air in the form of heat.

In the following stages when more details become available, such as ventilation ducting and flow rates, bulkheads and deck structure, separation walls in accommodation, coverings, isolation, etc. it is of interest to verify the installed cooling power and to know the actually occurring space temperatures in varying operating conditions. It may be required to demonstrate compliance with requirements before the installation on board is completed.

Also it is desired to have a prognosis of the electric consumption in operating conditions or in general to know the behaviour of the air conditioning system for instance to know how the temperature in a space varies in dependency of equipment power or number of crew present, or ventilation flow rate.

In this paper the aspects involved for the calculation of the occurring temperatures, humidity and the cooling power will be discussed and some example results from the new software tool that is under development.

Waste heat from equipment accumulates in the internal air; is removed by air cooling and is rejected into the environment via a fluid cooling system, see figure 1. A number of air coolers is installed in several spaces. For each space a heat balance is prepared, which in principle contains the terms equipment heat load, human heat load, ventilation heat load, heat transmission through the boundary (walls including pressure hull) and cooling power of the ACU's. Most of the terms in this thermal energy equation are temperature dependent and the software calculates the equilibrium temperature and humidity.

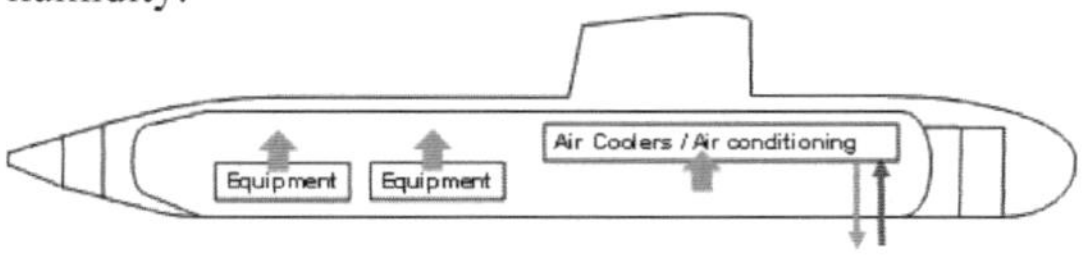

Figure. 1:. Waste heat from equipment accumulates in the internal air; is removed by air cooling and rejected into the outboard seawater via fluid cooling.

2. HEAT SOURCES AND HEAT FLOWS

2.1 OVERVIEW

The dynamic air energy balance contains terms describing the heat sources, heat flows and the cooling power. In the figure 2 the situation for one particular space is schematically illustrated. The space contains heat generating equipment, crew members generating metabolic heat (sensible and latent heat) and air cooling equipment.

Furthermore the ventilation flow contributes to the internal energy of the air in the space. The model assumes a uniform temperature in the space (perfect mixing). The density of the air can be assumed constant without large error, as long as temperatures remain within a certain range. Apart from air temperature it is of course possible to take into account other variations and influences or vary other parameters such as outboard seawater temperature, number of persons in a space, the equipment operation (on/off switching and power variation).

The problem is translated into a set non-linear coupled differential equations describing the temperature changes in all the spaces, which are solved simultaneously by a numerical integration method.

The situation in the figure is somewhat simplified because the air conditioning unit may be located outside the space. In this case air is drawn from the space, mixed with ventilation and cooled down in the airco-unit and then returned into the space. In other cases one airco-unit may distribute cool air to a number of spaces.

In the following the elements of the heat transmission model of one space are concisely discussed, except for the cooling equipment, which will be discussed in the next chapter.

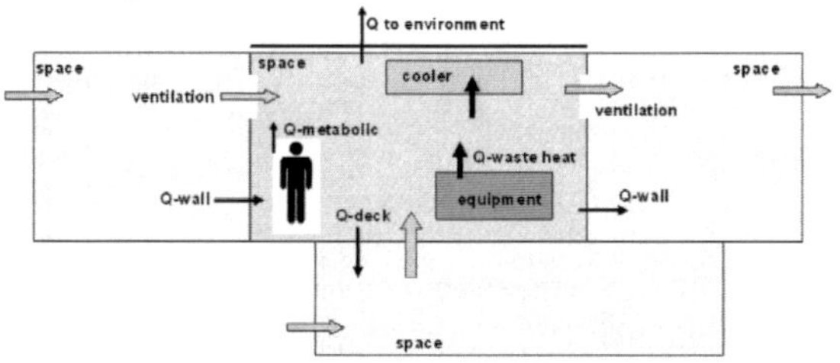

Figure. 2:. Heat sources in one space and heat flows to/from adjacent spaces and environment.

2.2 HEAT SOURCES

Several types of heat sources can be defined. The first, most simple one, has the property of constant power, independent of space air temperature. This type can be used for most equipment as a first approximation. For instance cabinets with electronic equipment will generally emit a constant amount of heat, so that the difference between internal electronics temperature and space temperature will remain approximately the same. Electric motors of permanently running equipment will also emit constant heat for the same reason.

The second type has constant surface temperature, so that the heat release is dependent on the surrounding air temperature and can for instance be used to estimate the heat release of a diesel engine (apart from radiation).
The third type is metabolic heat release of human bodies existing of sensible heat and latent heat. As known, both are dependent on the surrounding air temperature and also the amount of work effort. Mathematical descriptions can be found in literature [2], [3]. The latent heat does not contribute to air temperature increase, so

© 2017: The Royal Institution of Naval Architects

that it may be omitted in the thermal balance however it is of interest for the air enthalpy and thus total cooling power.

2.3 HEAT TRANSFER THROUGH WALLS

The heat transmission over boundaries of spaces involves walls, bulkheads, decks, pressure hulls and other boundaries. The basic equation is probably well-known: $q=k.A.\Delta T$ [Watt]

With k = heat transfer coefficient [W/m2.C], A = area of heat transfer [m2], □T = temperature difference over the wall [C], and q the heat flow rate [Watt]. The direction of the heat flow is dependent on the temperature of the adjacent space, which is unknown on beforehand. In the □T – term both temperatures are variable, which causes mutual interaction between the spaces which has to be solved by the simulation model. The determination of the heat transfer coefficient k may require relatively large effort, see [2], [3] and [4], especially when the wall is built up of several layers. If knowledge from measurements is available this will increase the final accuracy. The model is capable to take into account spaces with different density e.g. fluid filled tanks.

2.4 HEAT TRANSFER TO ENVIRONMENT

The same applies more or less to the heat transfer to the environment, it can be estimated by basic physics as first approximation. The submarine can be seen as one object with a well-defined boundary with the environment, the surrounding sea water. The average heat transfer coefficient at the boundary sea water – submarine hull is not difficult to determine, although it may be of interest to include the dependency of the ships speed.

The heat transmission through the hull occurs over different paths (in the following a single hull submarine is assumed): a direct path from sea water through pressure hull, with or without sound insulation layer at the outer side, and with or without heat insulation layer at the inside; a second path may be heat transmission via the superstructure outer hull, the (almost still) sea water in the superstructure and the pressure hull; a third path may be heat transmission via an internal tank and through the pressure hull.

In this way the submarine hull can be divided into a number of different heat transmitting areas with each their own heat transmission coefficient.

2.5 VENTILATION

A submarine is actually a confined space, divided in a large number of individual spaces, some of which are connected to each other by deck openings, a door, a hatch or passageway, while others are connected by air ducts, which together from a network of ducting and spaces. In the submerged condition the air is circulated throughout this network. The ventilation system circulates the air to maintain a uniform atmosphere, i.e. to equalize carbon dioxide and oxygen concentration; to ventilate the battery spaces (to reduce the hydrogen concentration and to contribute to cooling); and to distribute oxygen supply evenly. In snorting condition the air circulation path is different due to the air intake by the diesel engines. In chapter 4 a schematic example of a ventilation ducting network is shown.

The ventilation air flow contributes to the energy content of the air, in other words the increase or decrease of the temperature. The air flow mixes with the air in the space and in this way the ventilation system works as heat transporter (and may cool spaces without air cooler). The final mixing temperature in one space will vary in a complex way dependent on the air pathway through the spaces. The spaces interact with each other via the ventilation flow dependent on the varying temperatures. This causes mutual interaction between the spaces which has to be solved in the simulation model.

The flow rates in all the ventilated spaces need to be known. This can be determined by means of the design flow rates or by calculation of the actual flows from the fan performance curve and the pressure losses in the duct network.

3. AIR-CONDITIONING UNIT MODEL

3.1 COOLING POWER

Several cooler types may be defined. The simplest model is an air cooler which absorbs constant power (or heat). A better approximation of the air cooler behaviour is offered by the NTU-epsilon method, which is described in [2]. This method provides an exchanger model in which the heat flow depends on the air inlet temperature and temperature of some cooling fluid. Other well-known parameters are heat transmission coefficient and area, and cooling fluid properties and flow rates.

The more important cooling equipment is the air conditioning unit (ACU) model. The air cooler in the ACU is modelled as a heat exchanger with heat transfer dependent on temperature and humidity, while cooling power is based on an expression for the air enthalpy decrease. In figure 3 the orange line represents the ACU cooling line (from warm and high humidity air to cooler air with lower humidity). The relation with the total enthalpy decrease is also shown in the temperature – humidity diagram. The functions for total enthalpy, sensible and latent enthalpy in the calculation model have been derived from cooling power and air flow rate measurements on the ACU's that will be installed in the system.

© 2017: The Royal Institution of Naval Architects

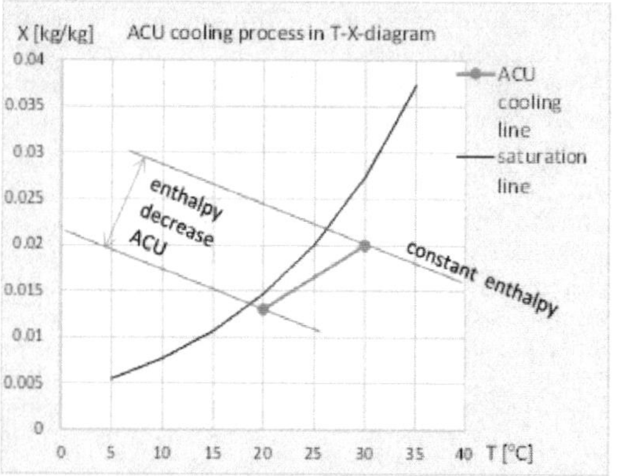

Figure.3: Temperature – humidity ratio diagram

Figure 4 shows examples of the ACU performance: total cooling power dependent on inlet enthalpy and the total enthalpy decrease in the unit dependent on inlet temperature and humidity ratio. The sensible enthalpy is also dependent on inlet temperature. The models of the different ACU's has been developed using a large number of test data of the ACU's that will be installed on board.

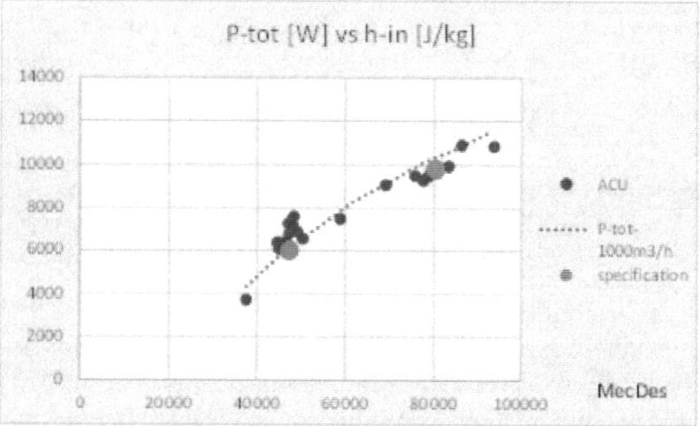

Fig. 4a. Example of cooling power (Watt) versus inlet enthalpy (J/kg), according test data. Dotted line is the model approximation. Blue dots are values acc. specification.

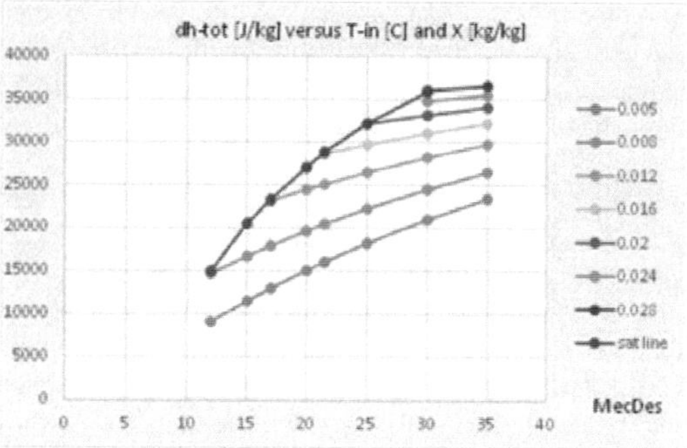

Figure 4a: Example of an ACU model represented by total enthalpy decrease versus inlet temperature at constant humidity. The limiting curve on the left side is the saturation line.

3.2 EFFICIENCY AND ELECTRIC POWER CONSUMPTION

Equipment considered in this paper are Air Conditioning Units (ACUs) used in S80 Submarine series. Those ACUs are direct expansion equipment through a cooling fluid, consisting in one air-liquid heat exchanger (evaporator), one condenser liquid (water)-liquid (cooling fluid), one compressor and one fan. Condenser phase is done through a fresh water circuit, which is cooled by sea water available on board; therefore the whole heat dissipated through the water means an energetic consumption for the submarine.

The electric consumption of ACUs means an important percentage of the whole electrical consumption of the submarine. This has a direct impact on the discretion during the mission, therefore is of critical importance to reduce this consumption by means of high efficiency ACUs designed for the real working condition. A first step to achieve the maximum efficiency of equipment is to know all the performances in the different operation conditions on board. For that purpose a test program of the ACUs has been done. Different measurements have been taken during the tests for different running conditions. All of these tests have focused to obtain efficiency of the ACUs used on board.

This will enable us to adjust the ACUs behaviour during the submarine operation, with accuracy higher than usual for this kind of design. To know the performance of the ACUs during the first stage of submarine design allows us to obtain efficiency measures of the submarine which have been designed.

The several ACU Performances, which have been considered are:

- Total Cooling Power available (TCP)
- Sensible Cooling Power available (SCP)
- Coefficient Of Performance (COP), which is a measure of the efficiency.
- Electrical consumption of compressor (CC)
  Note: Electrical consumption for the internal fan of the ACU is not the matter of the paper, because it depends on others subject different of thermodynamic criteria.
  Variables considered to obtain the above mentioned performances are:
- Air temperature at inlet of ACU
- Relative Humidity (r.h.) of the air at inlet of ACU
- Cooling fresh water temperature (condenser phase)

This paper includes the result for one ACU as example, since those results are representative of the whole tests conclusions. Efficiency values obtained through the COP measured in the test allow the designer to obtain electrical consumption for the different operative conditions on board.

Figure 1 shows relationship between TCP and SCP. Both of those values depend on temperature and r.h at the air inlet. The graphic has been obtained with a fresh cooling water of 30°C; it became evident that differences between TCP and SCP are very similar for different temperatures between 20°C and 40°C. The graphic shows how the difference between TCP and SCP is as higher as the increasing of both, temperature and r. h. at the air inlet. In other words, as higher is enthalpy of air inlet as higher is the difference between TCP and SCP.

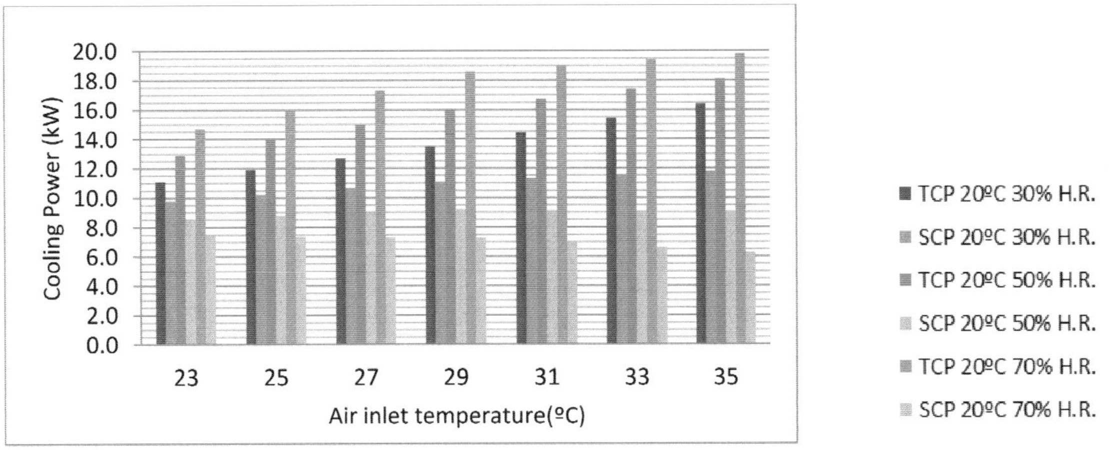

Figure 5: Relationship between TCP & SCP, for fresh cooling water of 30°C.

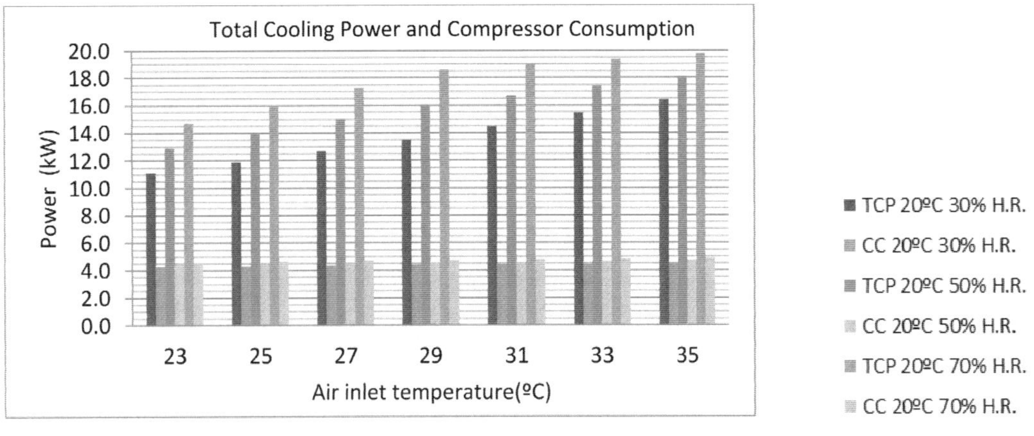

Figure 6: Relationship between TCP & CC, for fresh cooling water of 30°C.

© 2017: The Royal Institution of Naval Architects

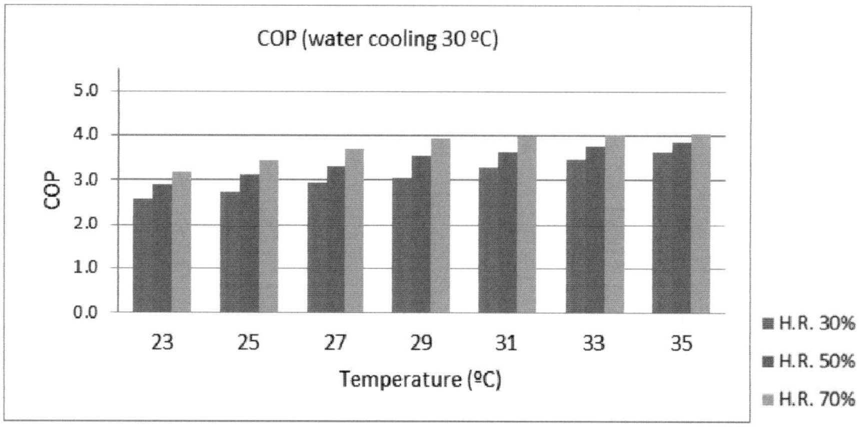

Figure 7: COP values for different temperatures and RH

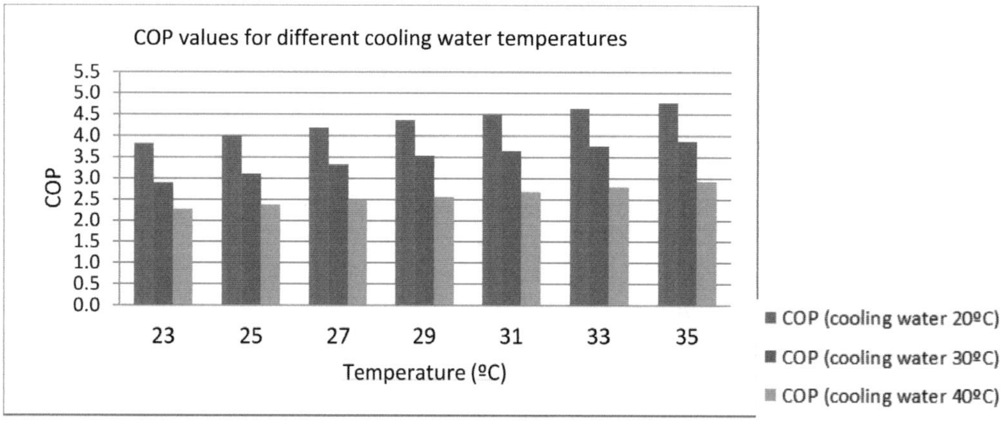

Figure 8: COP values for different cooling water temperatures

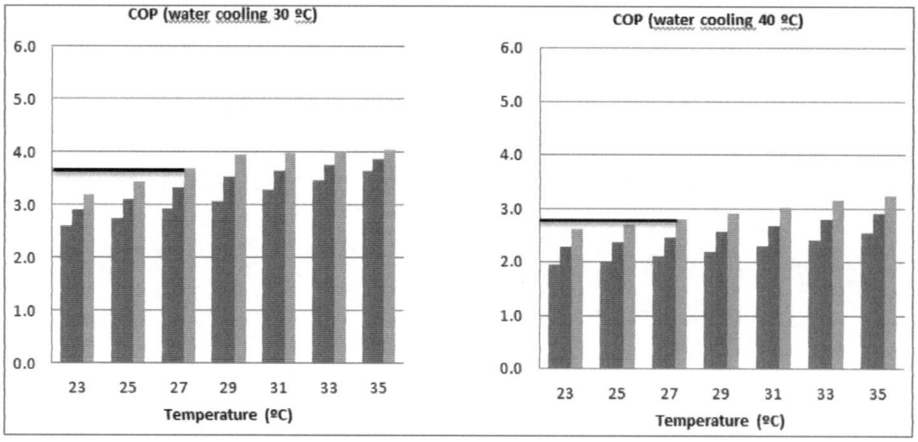

Figure 9: Example. Practical application

© 2017: The Royal Institution of Naval Architects

Figure 5 shows relationship between the TCP and compressor consumption CC for different air inlet temperature and r.h. The graphic has been obtained with a fresh cooling water of 30°C; it became evident that as higher fresh cooling temperature as higher are the differences between TCP and CC. If fresh water cooling temperature is getting lower, also differences between TCP and CC are lower. It is observed that differences observed between TCP and CC are proportional to the fresh water cooling temperature.

Figure 6 shows COP values for different air inlet temperatures in the ACU. Each one of the group of three columns shows COP values for each temperature value and 3 different r.h. values. COP is increasing with both, temperature and r.h. increase, in other words, an increasing of COP (efficiency) is observed with larger air inlet enthalpy.

COP values showed in 7 are for fresh water cooling of 30°C.

A fresh water cooling temperature increase means a reduction of COP values, as showed in Figure 8, where the COP values are for r.h. of 50% for the whole temperature range. Differences observed in the test results for other r.h. values are analogous.

All this information allows the designer to obtain information of the electrical consumption in the earlier stage of submarine design. As shown in the next example.

The example shows a concrete sailing condition of which the designer knows the environmental condition (inside and outside the submarine). It is observed that increasing 10°C the fresh water cooling temperature (in the condenser) the compressor consumption is increased by 1 kW (around 25% increased of each ACU).

Despite the electrical consumption increase observed the evaporator efficiency, measured as delta T, does not change (13°C±0,2). Therefore the TCP of ACU has not significant variation (about 16 kW).

It is also observed that a high r.h. value means that the most of the cooling capacity of ACU is involved in condensing of the water in the evaporator. As far as the contractual requirement allows, this phenomenon should be reduced to the minimum since it is not contributing to the maximum advantage of the available energy.

4. ACU PERFORMANCE RESULTS IN EXAMPLE SUBMARINE

4.1 EXAMPLE SUBMARINE

Almost all input data for the simulation model comes from the S80 Submarine project. The S80 submarine is currently under construction in the Navantia Shipyard in Cartagena, Spain. For reasons of confidentiality some details in the results are withheld or slightly modified. The performance verification was carried out to investigate the margins in capability of the projected ACU's in the submarine and to provide averaged operation time data.

The performance verification includes the determination of actually occurring cooling power and space temperatures in several operating conditions of the submarine, such as submerged transit and snorting, at the maximum sea water temperature.

Further it includes the determination of averaged operation time (duty factors) of individual ACU's. These duty factors depend on the number of ACU's in operation and the heat load.
Finally it includes verification of the capability of the system for the future growth margin on the heat loads.

4.2 VENTILATION NETWORK

The ventilation flow rates for each space are required as input data for the simulation model and were provided from a separate air flow rate analysis. Figure 5 shows an example ventilation network for submerged operation consisting of air ducts and spaces; the blocs indicate spaces and ACU's; the lines indicate the air flows.

4.3 ACU PREFORMANCE – SPACE TEMPERATURES

The analysis was carried out with ACU's performing 100% cooling power continuously in order to verify that they are able to maintain the space temperatures below the maximum allowable. Figure 5 shows the equilibrium temperatures in submerged operation condition at 25 C sea water temperature. As expected in only a few spaces the temperature is close to the maximum, in other words a cooling power margin exists. In practice the ACU's do not operate continuously due to the on/off switching by the temperature control. For instance as indicated in the figure the ACU in the galley is capable to reach 17 C but normally this will not occur.

© 2017: The Royal Institution of Naval Architects

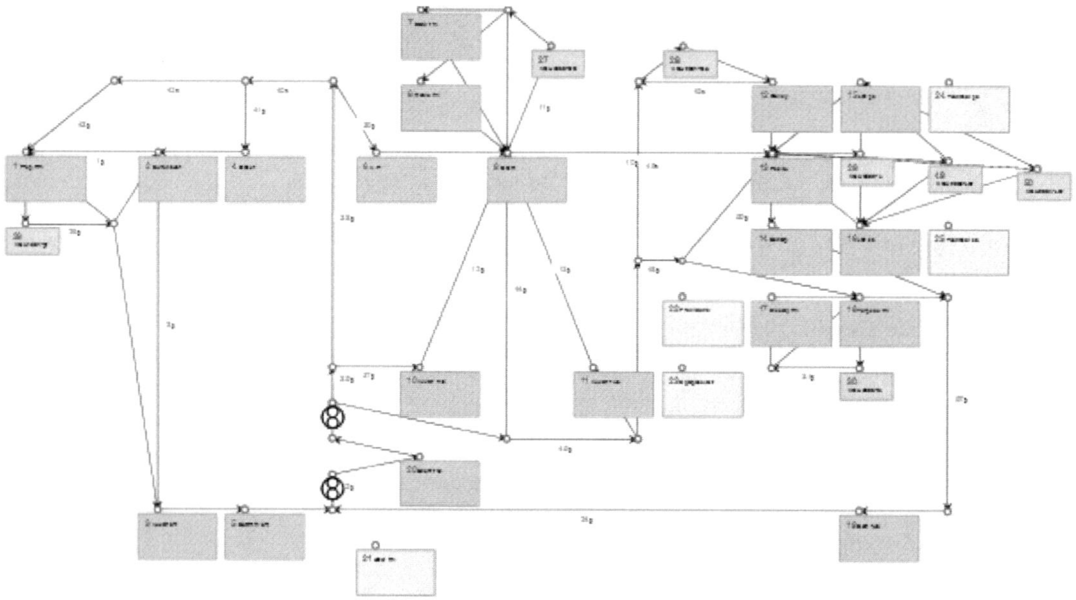

Figure 10: Ventilation network for submerged operation. Green: ventilated spaces; yellow-orange: naturally-ventilated spaces; blue: ACU's serving more than one space (ACU's serving one space are located in their space and not visible).

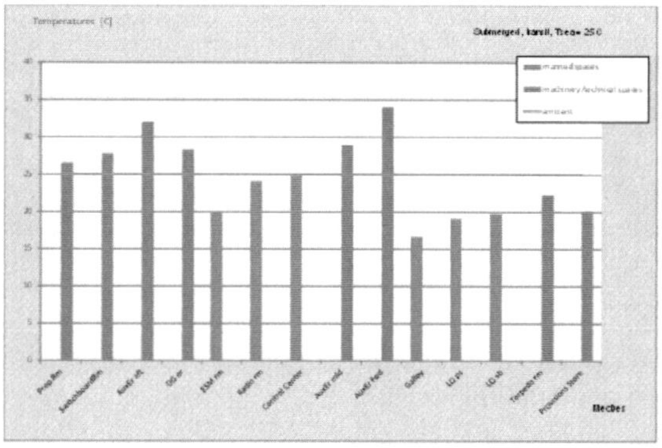

Figure 11: Calculated temperatures in manned spaces and machinery/technical spaces.

4.4 ACAU PERFORMANCE – DUTY FACTORS

ACU performance - Duty Factors

In the previous paragraph the results are presented of the temperature levels that are reached when the air conditioning equipment is in operation continuously, without temperature control. We have also seen that the sensible cooling power of the selected ACU's is sufficient. So, if less than full power is applied the temperature may rise up to the allowable level and this reduced power is a measure for the time that the ACU is in operation. The duty factor is defined as the fraction of time of operation of an ACU (based on on-off control of the ACU). The reduced sensible cooling power has to be found by trial and error because it is dependent on the temperature.

The table with an example of resulting duty factors of a selected number of spaces confirms that some ACU's have a significant cooling power margin, however this is valid for only one operating condition; other operating conditions have to be investigated as well of course. At lower seawater temperatures the duty factors will also be lower. In this example only temperatures were considered, thus sensible cooling power is relevant. If the calculation is extended to include the requirement of the humidity window then the ACU total cooling power has to be taken into account.

Finally, when the duty factor is known the electric power consumption of the ACU compressor and fan can be found for each ACU. It may serve as one of the consumption inputs for the calculation of underwater time.

Table Duty Factor

Room	available sensible power Watt	calculated sensible power Watt	duty factor %
Auxiliary Engine Room mid	13000	7000	54
Central Control Room	20600	19000	92
Galley	3900	1850	47
Accommodation	7900	4400	56
Torpedo Room	13900	12000	86
	59300	44250	

*) Calculated sensible power is valid for the maximum allowable room temperature

© 2017: The Royal Institution of Naval Architects

On the average about 75% of available sensible cooling power is required.

4.5 GROWTH MARGIN ON FUTUR HEAT LOAD

Often the building specification contains a requirement for the air-conditioning system in the form of a future growth margin on heat load in a number of compartments, for instance where electronic equipment is located.
In the study on performance verification it was found that if the heat load is increased, the associated increase in cooling power varies for each ACU and is always lower in terms of percentage of heat load.
A 10% growth of the complete heat load resulted in a cooling power increase of 2 to 6 % for the different ACU's. The growth in cooling power is lower than the heat load increase, because this increase is divided over the wall heat transmission, ventilation air discharge and the cooling power (therefore a margin on the heat load is not the same as a margin on the cooling power).

5. CONCLUDING REMARKS

In the detail design stage it is desired to verify the performance capability of the airconditioning with the available data on structure details, insulation, deck covering, accommodation separation walls, etc., as well as ventilation duct arrangement and air flow rates. Also the actual temperatures and cooling power that will occur in normal conditions are of interest as well as the energy consumption.

The equilibrium temperatures in the spaces depend on each other due to the mutual interaction of wall heat transmission and ventilation contribution. Moreover the ACU cooling power is temperature dependent, so that a simulation model is required to solve for the resulting temperatures in the spaces. The software tool calculates all the terms in the heat balance that are dependent on temperature and determines the temperature at which all terms are in equilibrium as well as sensible and latent cooling power of the ACU's. The software tool also includes a water vapour gas concentration model for the calculation of the relative humidity in every space.

The performance verification of the airconditioning system comprised several subjects:

- The determination of actually occurring cooling power of each ACU in several operating conditions of the submarine, such as submerged transit and snorting.
- The determination of duty factors of individual ACU's. When the duty factor is known the electric power consumption of the ACU compressor and fan can be found for each ACU. It may serve as one of the consumption inputs for the calculation of underwater time.
- The verification that the cooling power is sufficient to cope with a possible larger heat load in the future.

The software tool enables faster and more efficient analysis, in particular if a lot of submarine operation conditions have to be analysed; demonstration of compliance with building specification requirements; optimization of the cooling power distribution over the spaces; finally it provides more accurate energy consumption data for the submarine endurance and range calculations.

6. REFERENCES

1 Optimising the submarine cooling system, J. McLaughlin, 2013, ASC, Australia.
2 2001 ASHRAE Fundamentals Handbook, chapter 8, Thermal comfort.
3 Mechanical Engineering Handbook, Chapter 9 airconditioning and refrigeration, SKWang And Z LAvan, 1999.
4 VDI-Wärmeatlas, achte auflage, 1997.
5 Heat balance S80 submarine [internal project document], Navantia, Spain.
6 Dynamic heat balance of a submarine, paper presented at MAST Asia, Yokohama, May 2015 and UDT Europe, Rotterdam, The Netherlands, June 2015.
7 Evaluation of ACU tests and determination ACU parameters, apr. 2016, [internal project document], MecDes.

7. ACKNOWLEDGEMENTS

The authors are grateful to the shipyard Navantia s.a. (Cartagena, Spain), who has given the opportunity and consent to prepare this paper as well as contributions and support during the carrying out of the work.

7. AUTHORS BIOGRAPHIES

Jan D. Wilgenhof is senior mechanical engineer and technical consultant in MecDes, The Netherlands. He has been working on the S-80 project for several years, on a large number of mechanical systems.
Jesús Molina Toledo is HVAC senior engineer in the mechanical engineering department in Navantia, Spain.

© 2016: The Royal Institution of Naval Architects

© 2017: The Royal Institution of Naval Architects

NAVAL SHIP DESIGN RULES EMBODIMENT IN A CAD TOOL

J.A. Muñoz, SENER Ingeniería y Sistemas, S.A., Spain
R. Pérez, SENER Ingeniería y Sistemas, S.A. – Universidad Politécnica de Madrid, Spain
S. Santana, SENER Ingeniería y Sistemas, S.A., Spain

SUMMARY

The design of a naval ship must be subjected to very restricted design rules that have to be applied during the design stage. There is a very high number of these rules, although most of them can be reduced to different measurements. These measurements have various parameters that give the different locations where the design rules have to be implemented. For any client who designs this type of vessels, having the option to impart the design rules that could help the user while carrying out their daily design task with a CAD tool, is a very high priority due to the great time needed to do the validation. Validation of design rules can take a variety of forms, including, when possible, the capability of preventing the user from endangering a design aspect that does not obey an associated design rule. For the prevention of the user trying to pass on a non-conforming aspect of the design, a pop-up warning that informs the user about the associated rule can be showed. This warning may inform, if applicable, of the design aspect that is breaking a required rule. There is a high level of inferred associativity between components and certain types of these would need to be easily identifiable to allow for the application of the associated rule.

This paper discusses the various situations that can appear and describes the adopted solution to be integrated in a CAD system.

NOMENCLATURE

CAD	Computer Aided Design.
CSR	Common Structural Rules.
DR	Design Rule
EHP	Effective horse power
FEM	Finite Element Model.
HVAC	Heating Ventilation and Air Conditioning
IMO	International Maritime Organization.
NSRP	National Shipbuilding Research Program.
QFD	Quality function deployment
SE	Systems Engineering.
SMAT	Specification of Materials
XML	Extended Mark-up Language.

1. INTRODUCTION

According to Roozenburg and Eekels [1], design methodology is defined as a branch of science that critically studies the structure, methods and rules for designing products. The design of a product is a complex process that needs to take into account many different aspects, sometimes in accordance, other times contradictory. It is difficult to align all of the related aspects. The design of ships and marine artefacts is one of the largest and most complex task in the modern world, because it involves many disciplines and technologies within a critical and many times dangerous environment.

The main goal of a design is to fit the requirements in an optimal manner. But the design is certainly constrained by several limitations. Limitations, restrictions and regulations will be adjusted by the engineer at the time of design. . This set of limitations is known as or can be called the design rules. The nature of the design rules (DR) is very heterogeneous and several classifications can be applied. Following is a classification that can be considered in the shipbuilding design and construction.

One of the first steps of the design process is the requirements definition. The target of the product is to meet all of the requirements, and the types of requirements and their allocation into the design space will clarify the architecture of the system (in our case the vessel or marine artefact). The following types of requirements can be considered:

- Functional requirements define what functions need to be done to accomplish the mission objectives.
- Performance requirements define how well the system needs to perform the functions.
- Constraints are requirements that cannot be traded off with respect to cost, schedule or performance.
- Interface, environmental and other requirements such as human factors, reliability requirements, and safety requirements.

The requirement definition process consist of identifying all the requirements whereas the design process tries to answer to all of these requirements. This process of mapping functions with requirements is governed by the following controls an enablers:

- Applicable Laws and Regulations.
- Industry Standards relevant industry specifications and standards.
- Agreements, terms and conditions of the agreements.

© 2017: The Royal Institution of Naval Architects

- Project Procedures and Standards including project plans.
- Project Directives.
- Organization/Enterprise Policies, Procedures and Standards, including guidelines and reporting mechanisms.
- Organization/Enterprise Infrastructure.
- Project Infrastructure.

All of above mentioned controls are materialized in applicable design rules in the various stages of the project. The application of the several design rules has a heavy impact along the entire process of design, concept, project and finally building the final product. For this reason, several approaches have been taken to try to speed up the process, but the number and variety of the applicable design rules makes almost impossible to have a unique and single tool to solve this problem.

2. STATE OF THE ART

To address this issue, several approaches have been implemented. The most important and efficient is the application of the *Systems Engineering* (SE) to the entire project of shipbuilding: "*Systems Engineering is the art and science of developing an operable system capable of meeting requirements within often opposed constraints. Systems engineering is a holistic, integrative discipline, wherein the contributions of structural engineers, electrical engineers, mechanism designers, power engineers, human factors engineers, and many more disciplines are evaluated and balanced, one against another, to produce a coherent whole that is not dominated by the perspective of a single discipline*"[2].

During the SE process the application of the criteria and design rules are verified in different reviews. When the reviews are successfully passed, having verified that the rules are complied, it is possible to go through the next stage of the process. The advantage of this method is that it is very precise and highly controlled. The advantage of this method is that it is very precise and systematic, so it leaves no gaps in the process. On the contrary, the disadvantage is that it is a very slow process, which requires a lot of staff and has a high economic cost. However this last argument is opposed by the fact that finding errors and making changes in the early stages of a project has a much lower cost than if the same modifications are necessary when the project is already in an advanced stage of development. SE has not been widely implemented in the naval industry and so far only a few military shipyards have partially implemented some of the systems engineering processes.

3. DESIGN RULES

Some time ago (2000-2004), the U.S. National Shipbuilding Research Program (NSRP) promoted and funded a project entitled *WORLD CLASS MATERIAL STANDARDS and PARAMETRIC DESIGN RULES*. The main objective of the project was to develop and implement a rule-based methodology for ship design and material selection. The motivation of the project was to enable associated U.S. shipyards to respond quickly to customer inquiries and develop contract design packages, including cost estimates, which in addition to quick response have minimum risk to the yard in terms of price and schedule. As it was exactly described in the project web page [3], the goal was a 33 percent reduction in cost and cycle time for pre-production processes during the contract, transition and detail design phases. The rule-based methodology, Common Item Database, standards and parametric design tools are applicable to the U.S. shipbuilding industry at large. Among the several tasks covered by this project was the development of a technical approach for the early stage and a parametric ship design tool, which basically consisted of elaborating a kind of integration between several different software devoted to the different aspects of the design.

Another task of the same project was to develop metrics and rules for the *Entire-Ship* design. As anyone can imagine the complexity was of such magnitude that it had to be focused on three generic ship types for a medium sized shipyard: container ship, product/crude tanker and Ro-Ro ship. The catalogue of captured rules had over 500 rules. This gives an idea of the difficulty of managing this into a single tool. Even more difficult when considering the possible changes in regulations. The task number 6 of that project was to develop zone design rules and material templates for each ship zone (Structure, Cargo, Machinery, accommodation), to define generic interim products and to define functional volume design methodology and processes. The conclusion of the project made clear the need and the interest to continue advancing in the aspects related to facilitate the application of design rules to the projects. But the difficulty of the advance is so relevant that no clear advances have been made or applied in real cases.

Depending on the object in question the rules to be applied will have a different treatment. Since the application spectrum is so broad, it is necessary to make a detailed analysis and propose solutions to very specific sets of elements. When searching for one or several solutions, several criteria can be considered: strictly conform to the design criteria, comply with administrative rules, and reduce design time as much as possible. It is also necessary to keep in mind that some of the criteria may conflict with one another. In such a case, it is necessary to choose priorities carefully, bearing in mind that certain rules are binding

In general, the criterion of reducing cost is a correct choice. However, it may be the case that time-saving estimates are not accurate. For example, incorporating tools to control the application of design rules will make the use of design tools, ultimately the *Computer Aided Design* (CAD), more costly in terms of performance of the application, but on the other hand it will result in a design that hopefully does not need modifications. One could also

© 2017: The Royal Institution of Naval Architects

allow the use of unrestricted design tools and incorporate validation tools for design rules later on. This would make depending on the training and knowledge of the rules that have the designers, the modifications of the design, were more or less ample. How could you estimate what is most beneficial?. Certainly it is very difficult to make a prediction and the result will depend very closely on the type of vessel being projected.

Probably, the most suitable approach is to use an intermediate solution. First, to embed design methods that properly conduct the project by adjusting the use of specific materials or standard elements, equipment, fittings, etc. Second, to provide tools that help to know and to apply the design rules established by the different stakeholders. Finally to provide some method or tools to validate the rules in order to assure that the design conforms to them fits the rules.

3.1 DESIGN IN SHIPBUILDING

It is completely out of the question whether to use software to make the design model. The complexity of naval designs makes it necessary to use different software applications for the realization of the designs, obtaining the relevant calculations and validating them. It is also evident that the greater integration between different software, the greater the time savings in project development, basically due to the use of the different models in the different applications.

Each stage of the design project will be subject to its own rules as the variety of the same requires to make a specific use of software applications to have the design adjusted to the requirements. This assortment of functionalities and therefore of rules makes it impossible to combine in a single source all the rules. To make matters worse, the rules are constantly changing. This is due to the fact that the rules have different objectives and with each project it may be necessary to readjust them.

The *Common Structural Rules for Bulk Carriers and Oil Tankers* (CSR) came into force last 2015. The new CSR not only replaced separate rules for bulk carriers and oil tankers with harmonized rules but also introduced some new additional requirements. The application of these new rules has impacted the structural design period, man hours and design cost of shipbuilders, mainly because of the necessity of making *Finite Element Model* (FEM) analysis. As a matter of fact the number of zones to be analysed by FEM has been increased from 1 to 5 or 6.

Mirko Hadjina et al. [4] propose a methodology based on the *Quality function Deployment* (QFD) to take into account during the designs of the naval projects, the experiences obtained with the same by the ship owners and crew. The authors take into account the percentage of remarks obtained after delivery the product to prioritize the set of subjective customer requirements into a set of system level requirements for a future system conceptual design. As a result, they stablish a set of guidelines, among them highlights the improvement of the ship design rules and procedures and finally they suggest the necessity of elaborating a document of guidelines. In their work it is very remarkable that the 26.4% of remarks (second position after defects in assembly) are regarding noncompliance to the good shipbuilding and seamen practice in terms of working and living on board a ship. And in the fourth level with a 9.3% are the Remarks regarding the accessibility and obstruction during the movement of the crew through the ship. These two groups of remarks could have been reduced if during the design of model in the different applications it had been possible to validate or to consult the design rules.

So, currently the ship design process uses CAD as a tool and the application of the design rules relies in the designers and engineers. It is therefore necessary to incorporate new methods and tools to facilitate this application.

3.2 TYPES AND CLASSIFICATION

Focusing in the different stages of the project can give a good classification of the design rules and allow to identify which of the DR deserves to be embed into the CAD or which is the best approach to address this issue.

In the conceptual design, the design rules to be considered are related with naval architecture calculations. The evaluation of the stability criteria can be considered as design rules. These rules are established by international agreements and published by the *International Maritime Organization* (IMO). Having the evaluation of the criteria integrated in the CAD shall have a positive impact in terms of cost savings. Otherwise it will be necessary to export the forms into separate format and to make some rework in the specialized software. Most of the specialized software covers the majority of them, but sometimes it is necessary to add specific calculations made with parametric formula. Not all the software are enabled to address it.

Initial and basic design also requires the validation of the structural design from the point of view of the classification societies. For this purpose, models have to be exported into the specific format for the structural calculation. There is an additional step that adds complexity to this process and it is that the classification societies normally wants to review the models by their own means.

General arrangement in the basic design stage also requires the fulfilment of specific design rules, as those related with the compartmentation, spaces, volumes, etc. During detail design stage the project is expected to be finished from the point of view of the design. However, it is when more in depth observation of the rules has to be done.

© 2017: The Royal Institution of Naval Architects

Definition of the specific structural elements have to fits the classification rules, but also a high amount of equipment have to be located in the ship. The location is in many cases subject to determined rules, for dismounting, for functionality or even for aesthetic reasons. The amount of equipment means that following the design rules expends a large amount of time and hence, it is where the most of time savings can be obtained. So, it is in the detail stage when the validation is more necessary.

4. DESIGN RULES EMBODIMENT

The adoption of intrinsic design rules have been studied in different ways. One of the most interesting approaches was describe by Albers et al. [5]. In that article the authors propose a methodology of integrating into the CAD software, a set of rules for the design of micro parts. This methodology is useful for objects to be manufactured and for which the rules are to be applied in the object itself. Of course, that control could be adopted by the shipbuilding industry, but the rules to apply in the vessels affect not only the parts but also the relationships between them, for the mounting process, the location, etc. However one of the interesting points in the described methodology is the use of the knowledge based engineering and the systematic storage of the design rules in an agnostic format independent of the CAD system, with 6 descriptors:

- A brief verbal description of the rule.
- Eventually a picture.
- A mathematical equation as a condition that refers to the part's geometry.
- A unique key that identifies the rule and it scope.
- Instructions what to do if the condition is true.
- Instructions what to do if the condition is false.

However, when capturing the detail of the process, it is necessary that the mathematical equation can be applied to the dimensioning of the parts.

Validation of design rules within a CAD System can take a variety of forms, including, when possible, the capability of preventing the user from endangering a design aspect that does not obey an associated design rule. For the prevention of the user trying to pass on a non-conforming aspect of the design, a pop-up warning that informs the user about the associated rule can be showed. This warning, may inform, if applicable, about the design aspect that is breaking a required rule. It is noticed that there is a high level of inferred associativity between components and certain type of these, would need to be easily identifiable to allow for the application of the associated rule.

Some rules are conducive to automation but other are not. Furthermore, to automate the validation during the design, might lead to slow down the process. So it is necessary to provide means of validation that don't result in obstruction of the design processes.

FORAN system provides several tools and methodologies for controlling the project model from the point of view of the design rules and for facilitating the transfer of the model for validation purpose to external and specific applications. The different tools are adapted in the different stages of the project and also for the different elements of the ship project and the ship design.

4.1 INITIAL AND BASIC DESIGN

In the initial and basic design subsystem, used in the early stage of design, FORAN system has a complete integration of the stability evaluation methods and it also provides capabilities to create user defined formula with user defined parameters to evaluate and to apply any kind of rule or stability criteria to the preliminary designs. As the model is completely integrated in a single database, if the evaluation is not successful, any changes in the hull's forms, compartment arrangement or loading situation can be directly evaluated without the use of an external and specialized application. The openness of the system facilitates the adoption of new rules.

The use of user mathematical formulae is a powerful tool for customizing the rules, as it can be seen in the figure 1.

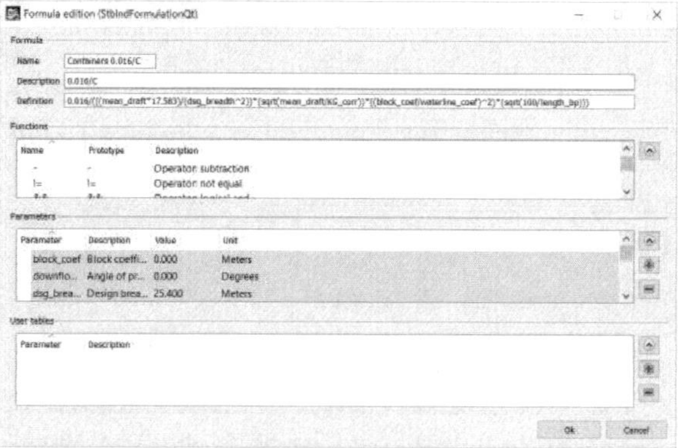

Figure 1: Stability Criteria Formulation

4.2 DETAILED STRUCTURE DESIGN

FORAN system offers several applications for structural detailed design. All the application work with the unique single database, in which the ship model is stored. There are significant aspects that help in the application of the established design rules.

The use of a unified application, called FNORM, where the project manager establishes the standard norms and

© 2017: The Royal Institution of Naval Architects

rules for applying to the entire project, as the following.

- Materials catalogue
- Fabrication methods for plates, profiles, etc.
- Standard shapes
- Fabrication rules
- Plate edge tabulation

Although all of the above items are related to a greater or lesser extent with the design rules, the fabrication rules and the plate tabulation table are the most specific of all of them. With those tools, the manager can set admissible sizes and shapes for the structure objects to be used. So, it is possible to force the use of parts that fits within a range of weights and sizes. See, reference to figure 2. When this control is activated, if a part to create is not within the set limits of the fabrication rules, FORAN will not permit the user to save the part until it is modified to pass the rules. This behaviour ensures that all the parts can be manufactured accordingly.

Figure 2. Fabrication Rules

Another element of embodiment in CAD to help in the application of the design rules is the ability of generating finite element models. This is done through the creation of adapted step files and an XML file to map the concepts and attributes of the plates and profiles. The result can be directly imported in the meshers or in the specific calculators.

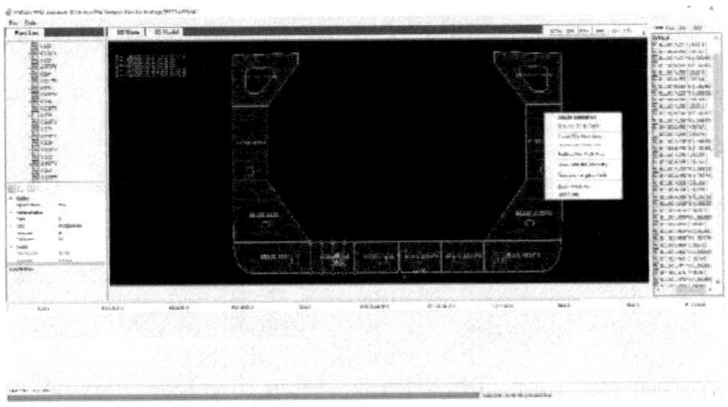

Figure 3. FORAN Fem Assistant

Additionally it is possible to load into the FORAN FEM Assistant to make appropriate idealization of the structural parts. This assistant enables to create idealization, normal or fine mesh, see figure 3. Although this is not exactly and embodiment of the classification rules, it provides a tighter integration with the model, which results in greater design time savings, because the re-design can be done directly in the model which resides in the database.

4.3 MACHINERY DESIGN

For machinery areas there are some aspects covered as design rules embedded in the CAD. The most important is the ability to define the specification of materials to use in the project.

The *Specification of Material* (SMAT) works as a filter for the piping components defined in the project. Typically, a ship shall contain a very large number of different piping components, and all these need to be defined in the project. However, handling such a big amount of components can be tedious when routing a pipe, and could easily lead to mistakes when selecting the correct one. To make this task easier, complying with the design rules and to avoid errors, the material specification limits the type and amount of components available when routing a pipe segment, figure 4. Only the components which comply with the specific values of nominal diameter, material, pressure, and other characteristics of the selected SMAT will be available. So, in this manner it is possible to control what components are available for the project.

© 2017: The Royal Institution of Naval Architects

Figure 4. Specification of materials

4.5 ELECTRICAL DESIGN

FORAN system provides different facilities to integrate the model of the electrical system with user defined rules:

Electrical load analysis assists the user in the dimensioning of the generators depending on the consumption of the equipment of the installation. Equipment may be nested in consumer groups and may be given a percentage of use in the ship's operating situations that the user configures.
The cable dimensioning wizard that allows to dimension a cable with the different criteria that are usually used for such purpose, such as: Maximum allowed current, voltage drop, and short circuit current.

Figure 5. Filling of cable trays

The tool for defining cable tray, see figure 5, also displays the filling of them, helping to anticipate the size of the same, and to control that the chosen size is correct and ina latter instance to make sure that the cables that will be installed comply with the filling rules.

The wizard for the automatic calculation of cable lengths uses the routing information of cables through the network of trays of the 3D model, plus a series of parameters and rules that the user can configure for the project in general or for each particular cable.

Finally a complete set of rules allows to ensure that a cable routed and later installed in the boat will comply with a series of rules such as being installed separately from one or more cables, or otherwise ensure that they will be installed together, or a certain distance away, etc.

4.5 ADDITIONAL TOOLS

However, the number of rules can increase exponentially according to the type of the project, and in some cases, such as naval ships, the amount of rules is so high that it might be impossible to manage all of them, because the application could collapse during the definition of the different elements. For instance, let's think of the following rule, as described:

"*Primary Passageways:*

- *Minimum clear width of 1400 mm except where impinged upon by hatches, inclined ladders; (See Design Rule YYYY for impinged detail);*
- *The 1400 mm minimum clearance to be maintained within 1500 mm of main bulkhead door openings;*
- *The clear width shall be maintained to a height of 2200 mm above the deck covering;*

Exception: The only exception is compartment XXX - where the clear width shall be a minimum of 1100 mm between bulkheads B1 and B2. No bulkhead or deck mounted protrusions shall be fitted in this area.

The clear width should be maximized as far as reasonably practical and the clear route should be as straight as possible; i.e. bulkhead and deck mounted protrusions should be fitted on one side of the passageway only".

Studying a design rule like this one concludes that it could be more interesting to add some facilities:

- To provide methods to validate the application of the rules.
- To enable additional functionalities during the definition, as it could be the live dimensioning.

In this sense, the FVIEWER module of FORAN provides functionality to easy validate design for the project manager. This application easily allows to add design incidences during the project review. As it was described above, the design incidences are stored in the database and their status can be checked in real time.

The FORAN application includes specific tools for the creation and control of design incidences. These are used to store information regarding specific issues, remarks or information that will need to be taken into account later during the ship's design. A design incidence can be easily created whenever necessary, but it also can be

© 2017: The Royal Institution of Naval Architects

automatically created when some design rule is violated, for instance when a clash is detected by the software. It will have all the associated elements assigned, a specified point on the ship with a label and a state (open/analysed/closed/permitted), see figure 6. This makes it very easy to trace the progress of the defined incidences throughout the process of design.

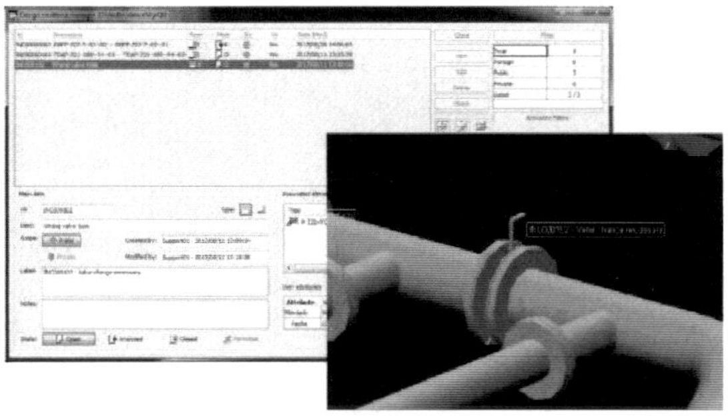

Figure 6 Design incidences manager

For facilitating the detection of the non-complying arrangements, a new tool have been developed to detect the objects that are at a given distance or lower.

The detection is done among any kind of component existing in a determined space or compartment, but located in the two semi-spaces defined by a given polygonal representing a passageways. This utility facilitates the verification of the width of the passageways, which is one of the important tasks at the time of design and verification.

5. CONCLUSIONS AND FUTURE WORK

Software developers have taken their time preparing oriented solutions including those that allow to minimize user errors In recent projects, there has been the need, for ship designers, to use CAD tools to work better, avoiding to introduce mistakes into the 3D model, with all parts information and classification and construction drawings as much accurate and precise as possible. A CAD tool with design rules embedded would allow to reduce time and cost, and produce detailed 3D models and accurate classification drawings, in addition to others. Other advantages for the future of the CAD Systems can be grouped by discipline. During the first stages of the ship design, it is where there are more implications downstream, so more mistakes could imply more losses of productivity. At top of this some of future improvements regarding General Arrangement and Naval Architecture Calculation are:

- The CAD tools should not allow to place compartments in incorrect locations, as they are stated on the Rules of the Classification Societies, and they should also request the inclusion of a cofferdam. As an example, the Systems should avoid locating a fuel tank next to the external hull.
- This Systems should avoid the wrong positioning of an equipment inside the wrong compartment.
- According to power estimation, the software should apply the most convenient method in order to calculate the *Effective Horse Power* (EHP), because the program already has the necessary information about hull forms, as for example Froude number, block and prismatic coefficients.

Also in the hull structure discipline there is plenty of room for improvement. Some of the ideas for the future:

- The system should avoid the location of macro-holes closer than a minimum distance from a watertight bulkhead.
- The CAD tools should avoid piping, *Heating, Ventilation and Air Conditioning* (HVAC) and cable trays penetrations if in the structure there is a hole closer than a minimum distance according to the Classification Society's rules.
- Improvements in the welding management. There are so many areas where the designer could make an error applying welding procedures or tasks.

As mentioned before, there are many areas where the CAD Systems could improve the design rules, and the electrical discipline is not an exception:

- The System should control the minimum and maximum distance between the cable tray and the surface where the supports lay.
- There is no control of the cable tray material to be used based on the zone of the ship, i.e. steel, aluminium, PVC, etc. The CAD tools should control the cable characteristics, as it is isolated against fire, oil, corrosive, low emission in the case of fire
- There is not control regarding connectors/equipment linked with specifications.

Sensors, big data and the cloud are some of the most interesting changes in the way the technology collect and manage data in recent years. These changes have not significantly influenced the common practices in condition monitoring and validating CAD Data. In part this is due to the reduced trust in data security, data ownership issues, lack of technological integration and obscurity of direct benefit. It is necessary to design and prepare a method for incorporating smart sensor techniques and distributed processing in data acquisition for condition monitoring to assist during the design process.

New technologies are appearing in the world that in a very short term shall be linked with the design systems in order to facilitate the design in self. The technologies of treatment of big data and fast searches, shall allow to engage the design of a part or any concept with the applying rules. The integration of the validation of the structural models by the Classification societies shall be

© 2017: The Royal Institution of Naval Architects

done by using the cloud computing or with direct connection with cloud applications. This will result in a faster and more reliable design.

Monitored full-scale performance data should be analysed in the future in order to improve the performance conducted at the design and manufacturing stages. Performance evaluations should be made, employing developed monitoring and performance analysis methods, so full-scale performance can be evaluated with a high degree of confidence, and its results can be effectively utilized in the ship design stage. CAD simulation tools and Artificial Intelligence for user guidance, Big Data using Artificial Intelligence to help ship designer for optimizing the 3D model, so it is easy to predict that this is the future.

There are many advantages of using CAD in shipbuilding: ease of design with the design rules embedded, speed of design, use and reuse of information, etc. It is expected that in the future CAD tools will advance further and allow greater information management through these new improvements. In general, CAD Systems provide tangible benefits while the process is optimized, reducing design time and production, and therefore costs. As a summary, there are several scenarios of improvements, as the design rules explained in this paper, for the next years. Some of these improvements may seem unrealistic in the short term, but reality often exceeds expectations in any field, and probably more in technology

6. ACKNOWLEDGEMENTS

The authors would like to express sincere gratitude to SENER for providing the support and information for elaborating this conference article. They also thanks to Roland Gildea for his help in the review of the written paper.

7. REFERENCES

1. N. J. M. ROOZENBURG AND J. EEKELS. Product Design: Fundamentals and Methods (A Wiley Series in Product Development). John Wiley & Sons Inc, 1995.
2. NASA. Systems engineering handbook: NASA/SP-2007-6105 rev1. Technical report, National Aeronautics and Space Administration, December 2007.
3. NSRP. World class material standards and parametric design rules, 2004.
4. MARKO HADJINA, TIN MATULJA, AND RAJKO RUBESA. Methodology for the ship exploitation feedback inclusion for improving the ship design and production process based on adjusted qfd method. Scientific Journal of Maritime Research, 29:107–114, 2015.
5. ALBERS, A., BURKARDT N., MARZ J., HAUSER S. Computer aided evaluation of design rules for micro parts. INTERNATIONAL CONFERENCE ON ENGINEERING DESIGN. August 2005. Melbourne.

8. AUTHORS BIOGRAPHY

Jesús A. Muñoz, Ph.D. holds the current position of FORAN System Product Manager at SENER Ingeniería y Sistemas S.A. He is responsible for the technical development guidance of the FORAN System. He is also professor of Applied Systems Engineering at the Marine Engineering School of the Technical University of Madrid.

Rodrigo Pérez, Ph.D. is the head of Naval Shipbuilding Business Development in SENER. He is also associate professor imparting several subjects at the Marine Engineering School of the Technical University of Madrid.

Silvia Santana is M.Sc. in Naval and Oceanic Engineering by the Universidad Politécnica de Madrid. She works in the FORAN customer support group at SENER Ingeniería y Sistemas, S.A.

© 2017: The Royal Institution of Naval Architects

TOOLS FOR PREDICTING PARAMETRIC ROLLING AND ITS APPLICATION FOR SECOND GENERATION INTACT STABILITY CRITERIA

J A Muñoz, SENER Ingeniería y Sistemas, S.A.,-Universidad Politécnica de Madrid, SPAIN

SUMMARY

This paper explains the process of developing a computer application by implementing the numerical algorithms necessary for evaluating the appearance of parametric rolling in a vessel. In the analysis of this problem, it is considered of interest the geometrical representation of the hull and other elements, which have decisive influence in these phenomena, such as the sea situation and loading conditions. Ideally, the application would determine the roll angle that occurs when a ship is on waves of different characteristics.

The present work describes the software methodology for addressing the problem of predicting parametric rolling, as this concept is one of the elements involved in the evaluation of the intact stability criteria of second generation. The foreseen come into force of the second generation of intact stability makes necessary getting tools to make easier the evaluation of the behaviour of the vessel subject to hydrodynamics phenomena as the appearance of the parametric rolling. This paper basically describes the results of the doctoral thesis of the author [1], which was defended last year 2016.

NOMENCLATURE

DSA	Direct stability assessment
GM	Metacentric height (m.)
L1	First level of vulnerability criteria
L2	Second level of vulnerability criteria
OG	Operational guidance
OL	Operational limitations
SGISC	Second generation of intact stability criteria
ρ	Density of water (kg m^{-3})

1. INTRODUCTION

The IMO in its 29th session assembly adopted the finalization of the *Second Generation Intact Stability Criteria* (SGISC) by the year 2019 as one of the high level action plan of the organization and priorities for the 2016-2017 biennium. Although it is known this is just an intention, work has advanced significantly, being necessary to consider that sooner or later the criteria will come to force and it is crucial to offer different approaches.

These new criteria take into consideration the operational conditions within five stability failure modes. The considered failure modes are the following:

- Parametric roll
- Surf-riding/broaching
- Pure loss of stability
- Dead ship condition
- Excessive acceleration

For each failure mode, three levels of assessment can be considered. The accomplishment of the different level of assessment will have wider or narrower possibilities of complex designs. The first vulnerability level criteria (L1) are set in order to require simple and approximate evaluations of phenomena related parameters and , therefore, they force to have larger "safety margin", which obviously will reduce the possibilities of the design. A second level of assessment (L2) is based on more precise computations based in statistical measurements of the corresponding phenomena related with the respective failure mode. One of the objectives of the work group is to tune the safety margins of the assessment and to select adequate database of ships to populate the statistical results. The third level of assessment consist of a direct stability assessment (DSA) using numerical simulations that adequately replicates ship motions, and to have a statistical prediction of failure where probability is directly used as safety criterion.

Compliance with any of these assessments is sufficient to fulfil SGISC. Alternatively, ship-specific *Operational Limitations* (OL) or an *Operational Guidance* (OG) can be developed for loading conditions failing to fulfil the criteria. Whereas L1 and L2 are close to finalisation, work on DSA has not been completed yet, mainly due to the intrinsically complexity of the involved calculations.

For any of the assessments levels required for the SGISC, it is necessary the use of specifically developed software, that help not only the evaluation of the criteria , but also to make numerical simulations, validations of the results and verification of the inputs for the calculations. The development of that software must be subject to determined requirements considered from the very beginning in order to avoid problems, misunderstandings and mistakes in further design stages. This paper presents the methodology used for developing a software application oriented to the evaluation of the parametric rolling of a vessel, emphasizing the most important steps to follow in order to provide an adequate method of prediction of this failure mode. The key of this thesis is that it is essential to implement correctly the algorithm, but it is also important to design the

© 2017: The Royal Institution of Naval Architects

application in order to facilitate the validation of the results or to optimize the tests among other things.

2. PARAMETRIC ROLL

Parametric roll is a phenomenon due to the repetition in time of variation of ship restoring moment in waves, when the ship is sailing in head seas. This is due to the variation of the surface of the flotation and as a direct consequence of the variation of the metacentric height (GM), which causes a movement of rolling that, under certain circumstances, it is not damped but accelerates, causing the capsizing of the vessel.

It occurs when the wave encounter frequency is approximatively twice the ship's roll natural frequency. This phenomenon is strongly dependant of the forms of the ship. In that way, a ship with wide forms around the cylindrical body is more likely to the appearance of this problem. For this reason this failure mode is mostly observed on container ships (France, et al., 2001) [2] because the hull shape of these ships may generate a large restoring moment variation.

2.1 PHYSICS OF THE PHENOMENA

The explanation of this phenomenon can be clearly explained in the referenced article *'Criteria for parametric roll of large containerships in longitudinal seas'*. (Shin et al., 2004, [3]) Suppose a ship located in a wave, in the maximum sheer position, i.e., this is with the middle section of the ship located in the bottom of the wave. In this situation, the ends of bow and stern are more submerged than in calm waters, and therefore the average surface of the flotation is significantly higher than in calm waters. On the contrary, when the vessel is situated on the crest of the wave, the bow and stern ends are less submerged and the average surface decreases in relation to the situation of calm waters.

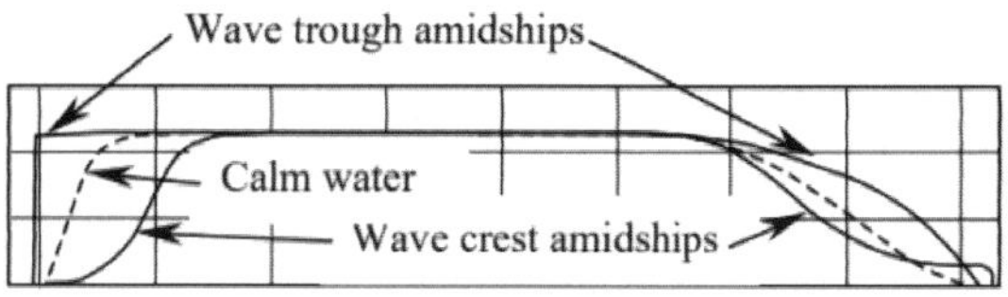

Figure 1: Waterline at calm water, wave crest amidships and wave trough amidships.(Shin et al., 2004, [3])

This fact results in a variation of the metacentric height of the ship (GM) and therefore a variation of what we know as cross stability. This transverse variation will depend on the relative velocity between the ship and the wave, that is, the frequency of encounter.

A transverse disturbance to the static situation of a ship leads to a natural transverse rolling movement, whose period will depend on the properties of the moment of restoration and the mass of the vessel. This period is known as the natural roll period, and its value is:

$$\omega_\phi = \frac{2\pi}{T_\phi} = 2\pi\sqrt{\frac{\Delta GM}{I_{xx} + A_{44}}}$$

Where:

$I_{xx} = Inertia\ moment\ of\ mass\ in\ logitudinal\ axis$

$A_{44} = Added\ masses\ coefficient\ in\ rolling$

When the ship is subject to bow or stern waves, the GM varies in the described sense and the restoring arm varies in the same sense, trying to recover the position of the ship, but also due to inertia of the movement this effect can increase the amplitude of the rolling.

The greatest increase in the parametric balance is observed in the following two circumstances:

- When the frequency of encounter is nearly twice the natural frequency of rolling.
- When the roll disturbance occurs in the time interval between the wave crest and trough amidships position.

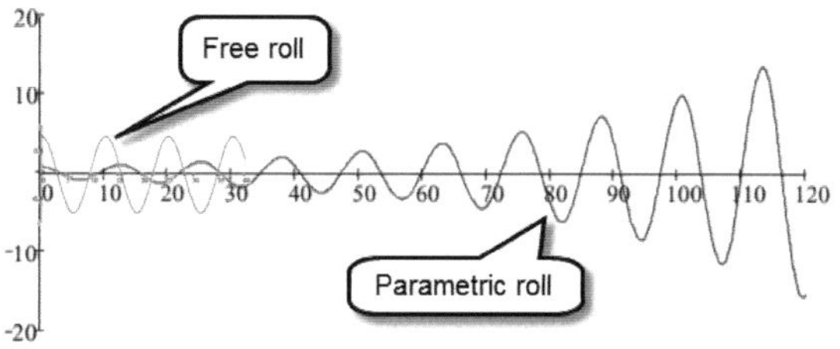

Figure 2: Free roll vs. parametric roll

If there were no other elements participating in the phenomenon, the rolling angle would gradually increase until it caused the ship's capsizing. Fortunately there are other factors that can influence in opposite way to the increase of the rolling angle and these are:

- The viscosity.
- The formation of waves and swirls.

These two factors contribute to the roll damping by reducing its amplitude. If the reduction of the roll angle is greater than the gain by the parametric self-excitation, rolling will be reduced; otherwise the movement will grow, unless conditions change or the ship capsizes.

Two questions arise at this point:

- How long can the vessel remain in this situation, without capsizing?
- Is it possible to predict when capsizing occurs?

There is actually a single answer for both questions. In any case the software application tries to answer it.

© 2017: The Royal Institution of Naval Architects

2.2 MATHEMATICS OF THE PHENOMENA

There are several methods to address the problem of determining the rolling angle of a vessel during the time when the ship sails in following or head seas. So, the first step is to select an affordable mathematical model to solve the physical problem of calculating the roll angle in that situation. This application implements the proposed method of Ovegärd, [4]. The software will implement a numerical algorithm to solve the differential equation of the rolling movement simplified by the author

$$(I_{xx} + A_{44})\ddot{\phi} = M_{rest} + M_{F-K} + M_{damp} + M_{diff}$$

Where:

$\ddot{\phi} = Roll\ angle\ acceleration$
$M_{rest} = Restoring\ moment$
$M_{F-K} = Moment\ due\ to\ Forude - Krylov\ forces$
$M_{damp} = Dampingo\ forces\ moment$
$M_{diff} = Difraction\ forces\ moment$

All the forces and moment are therefore implemented following the work of E. Overgärd.[4]

3. SOFTWARE APPLICATION

There are many ways of doing the application, but the objective is to create one that allows the realization of the necessary calculations to determine the possible appearance of parametric roll. These calculations will be made with very different vessel shapes, and for this, it is necessary that they meet certain requirements.

Many academic programs implementing complex algorithms are console applications, where it is almost impossible to verify if the forms used are properly defined. Also, the evaluation of different alternatives is of high complexity, by using non documented data archives. This is one of the elements that the designed application must supersede.

3.1 APPLICATION REQUIREMENTS

In order to have a versatile application that allows agile analysis while having access to different forms of ships, the following requirements have been considered:

- Interactive application that allows managing different studies for the same vessel or different vessels for the same sea conditions or even different combination of any.
- Ability of reading geometries in different formats (IGES, STEP, DAT, etc.)
- Ability of displaying the curves or surfaces read in order to know if data are correctly read.
- Dialog for easy input of data for identifying loading situations
- Ability of obtaining hydrostatics values for ship condition including sea wave situations.
- Ability of calculating naval architecture values in the time domain: GM, GZ, pressures, forces, moments and rolling angle.
- Numerical and graphic results in order to quickly validate results.

Any software application to be considered as valid for this purpose needs to meet these requirements, and therefore, these allows to select the different components to build the software.

Aligning the items required with the decisions to comply with them, lead to the following assertions:

- To have an interactive application is necessary a powerful library of widgets and general purpose. After evaluating several existing in the state of the art, we select the QT ® Toolkit (Qt is a registered trademark of The Qt Company Ltd. and its subsidiaries). This library offers several levels of licensing and some of them are useful for research and development.
- For reading the different geometric formats, as IGES, STEP or BREP proprietary files, OpenCascade ® library is an optimum alternative. It also provides an extensive set of geometric algorithms and general purpose software.
- There is open source code based in QT ® and OpenCascade ® to easily display geometric forms. This is a simple requirement of high importance because it shall allow the verifying of the goodness of the imported or read geometries.
- By using the basic libraries it is possible to design dialogs forms that allow a quick entering of different data for studying the different cases.
- The different calculations shall be developed using the different elements available in the OpenCascade ® library. These algorithms shall cover traditional and well known naval architecture calculations. The innovation of this application is the implantation of the calculation of the roll angle over the time domain when the ship sailing on a wave.
- The numerical results will be collected in visual widgets of the application. The results shall be also written to file, but they need also to be graphically displayed because it is a faster method to detect if the results are incorrect, and therefore the algorithm has been wrongly implemented.

The selection of this additional software forces the selection of the programming language: C++. This programming language allows a more powerful and efficient application, giving the opportunity of using the standard template library, and the boost library. These

© 2017: The Royal Institution of Naval Architects

two libraries of software are code template based that give flexibility and speed. There is a lot of open source code based in this software, specifically for complex algorithms, such ordinary differential equations. The main advantage of using this is that it is a solid good software covering any need of solution for the problems that we want to solve in this matter. One of the libraries used is ODEINT for integrate the ordinary differential equations set for the solution of this problem.

3.2 ALGORITHMS

The core of these applications is the algorithms. These calculations need several algorithms, which are implemented from the theoretical formulation and therefore they must be also validated against contrasted results.

3.2 (a) Hydrostatics calculation

The hull shape shall be represented by two different types of inputs: lines, which are wide extended format of representation for academic use; and surfaces. The results will be obviously different, but the differences are not relevant for the purpose of the program.

The modelled shapes by lines are normally transversal sections equally spaced along the x-axis. These sections are obtained from the intersection of the hull shape with the flotation. It is possible to use several methods to obtain the sections. Surface geometrical intersection can be obtained with Boolean operations among solids, as represented in the Figure. 3.

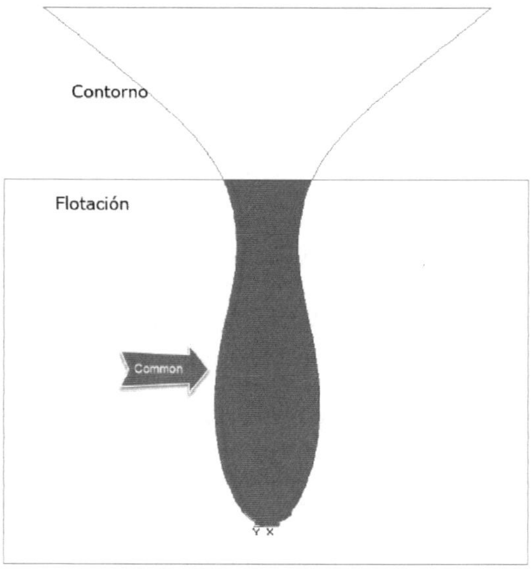

Figure 3. Section obtained as common Boolean operation

When the lines are used, the application will calculate the pressures over the different lines composing the section crossed by the waterline corresponding to the considered wave. The figure 4 explains the method.

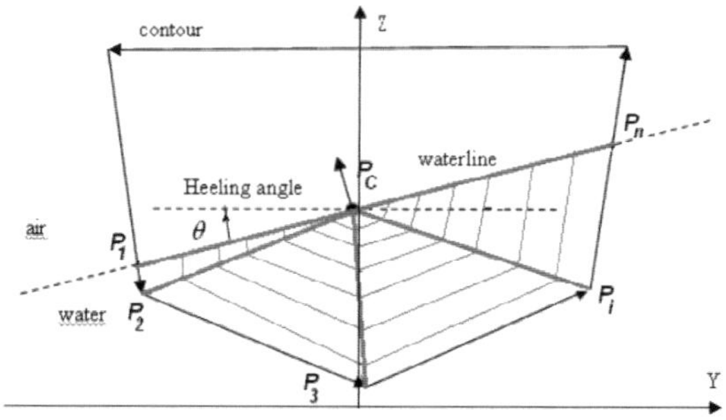

Figure 4. Section calculation

The values calculated for each section are integrated along the x axis. Three different methods of integration are used for validation: Trapezoidal, Simpson and Mean value. After validation just one method is used. If sections are equally spaced, Simpson method can be used, otherwise mean value shall be employed.

Finally a third method to obtain volume properties is used. This method calculates the volume of pyramids formed by the tessellated faces, calculating the submerged part of them, and later on integrated along the surface.

3.2 (b) Wave formulation

The considered wave is the classic formulation for gravitational waves described in the book Principles of Naval Architecture (Beck et al., 1989) [5]. The type of wave is assimilated to sinusoidal one. The application will implement an algorithm to evaluate the position of the wave according to its amplitude and the time, by taking into consideration the relative position of the ship to wave.

As it is implemented in C++, the algorithm is one of the methods of the object wave. The relation between the wave length and the period is the know formulation:

$$T_w = \frac{L_w}{V_c} = \frac{L_w}{\sqrt{\frac{gL_w}{2\pi}}} = \sqrt{\frac{2\pi L_w}{g}}$$

Where:

- T_w Period of the wave (s)
- L_w Wave length (s)
- V_c Wave speed or celerity (m/s)
- g Gravity (m/s^2)

© 2017: The Royal Institution of Naval Architects

All calculated in units of the international system. The object constructed by this class allows us to know other parameters of the wave:

- Period
- Wave number
- Frequency
- Angular Frequency
- Wave amplitude on a given abscissa

3.3 CALCULATIONS

Any application should provide the necessary tools to perfrom all the necessary calculations to obtain the expected results. However, sometimes this is not possible in the early stages of development. In such a case, the application must be designed to incorporate the pending calculations as intermediate parameters to be entered by the user, and at a later stage to be added. This requirement forces the structure of the application in an adequate manner. The interface must be enabled to enter values, and the calculation engine must be prepared to easily incorporate the calculations when the algorithm is developed.

The objective of the developed program is definitely to evaluate the rolling angle over time. The application developed is prepared to calculate:

- Obtainment of hydrostatic properties
- Simulation of roll decay test, for a better estimation of the damping coefficient.
- Calculation of restoration forces and Froude-Krylov.
- Numerical evaluation of the rolling angle over time.

However, it has not been possible to integrate the calculation of the forces of memory and diffraction, so it is necessary to foresee the inclusion in the application of the estimated user parameters or values for these concepts and the calculation engine is prepared to use the values when they are ready for inclusion.

To perform the calculations, the program provides the dialog interface that is displayed in figure 5 and allows the user to enter specific values for the calculation to be performed. Of course, this dialog can only be activated when the forms of ship were previously read.

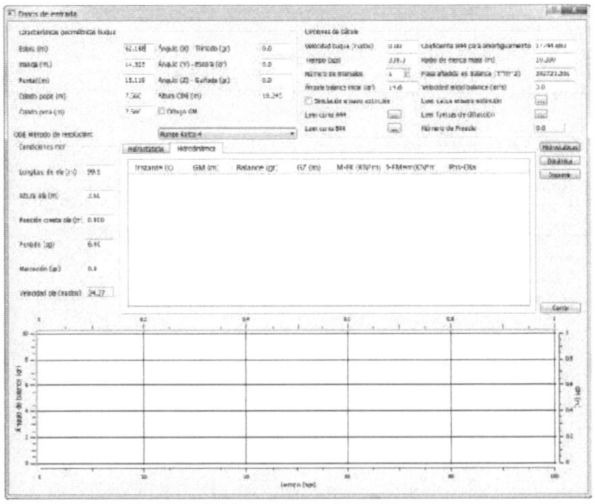

Figure 5 Calculation Form Dialog

The calculation of hydrostatic values is a must in any application program to evaluate the behaviour of the vessel in the water. The program calculates the most relevant hydrostatic values of the ship in calm water or ship on a wave. The location of the ship as regards the wave can be defined by the user or calculated accordingly with the speed of the ship and the speed of the wave when the calculation is made dynamically.

The evaluation of the diffraction forces and the memory forces are respectively calculated from the imported coefficient curves A_{44} and B_{44}. These curves should be calculated from the forms of the vessel, but that is one of the pending parts in the application. This is why a methodology to import that curves is provided. With those, the program is able to calculate the forces, but they definitely need the inclusion of some sort of user estimation.

The damping coefficient and the added mass are also values, whose calculation needs a more developed algorithm. In order to avoid the difficulty of evaluate that parameters for complex shapes, the program includes a simulation of the roll decay essay. With this calculation, if the user gets the real results of a roll decay test, it is possible to import the values of the test. Later, the user can tune the damping coefficient so, the roll angle values obtained for the simulated roll decay matches with the real values. See the figure 6. The application contains embedded natively the results of the roll decay test of the SAFEDOR project [6]

© 2017: The Royal Institution of Naval Architects

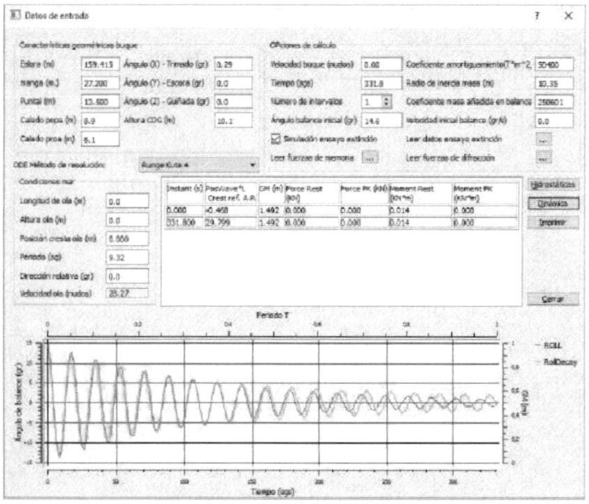

Figure 6. Decay test simulation and comparison

Finally the rolling angle is calculated over time. To evaluate the rolling angle is necessary to solve the differential equation of that motion described in the Equation 1. The numerical solution of same will be solved by means of a simple Runge-Kutta scheme that in principle is considered sufficient. Separating the viscous damping terms and not Viscous the equation can be expressed as follows:

$$\ddot{\phi} = -2\delta\dot{\phi} + \frac{1}{(I_{xx} + A_{44})}(M_{Rest} + M_{F-K} + M_{Mem} + M_{Diff})$$

Three alternatives have been implemented for the value of M_{Rest}:

$M_{Rest} = \Delta \cdot (GM) \cdot \varphi$: Linear approximation

$M_{Rest} = \Delta \cdot (GZ)$: Righting moment

$M_{Rest} = \int(-\rho g \cdot Z)$: Direct calculation

Although, initially a Runge-Kutta schema was used to solve the equation, as the library ODEINT has been used to solve the differential equation, it was easy to implement different solution schemas. The different results were not significantly different.

In addition, an assessment of the situation can be carried out by the method analytical approach proposed by Shin et al. [7] and determine the stability Or instability of the solutions by locating the parameters p and q in the Ince-Strutt diagram, for which this program incorporates an output of the Values p and q of Mathieu. This way it is possible to make an evaluation of the susceptibility criterion proposed in that article.

3.4 RESULTS

The results of simulation calculations of the physics of a phenomenon implemented in computer applications require the appropriate means of validation. Validation should be done using alternative media using the same sources. In the case of this application, it is about three results that it becomes necessary to validate.

For the validation of the hydrostatic results, the FBASIC module of the FORAN System *(© SENER Ingeniería y Sistemas, S.A.)* was used to make these calculations using the same vessel shape. The results were identical for the set of waterlines used, both in calm water and with the ship on a wave. Three different ships were used, being one of them, the model used by the SAFEDOR test [8].

The roll decay test was validated by using the vessel of the SAFEDOR test, as the values of the real test were embedded in the application, so it was possible to compare the results of the simulation with a real situation. The values obtained in the simulation matched each other with no significant difference for the 180 seconds, and after that appeared a divergence.

Finally, the roll angle over time was validated through the comparison of the appearance of the parametric rolling for the vessel of the SAFEDOR project. The simulations done with the application tried to represent the tests of the SAFEDOR project. In the following graphics we can see the results corresponding respectively to the tests numbered as 14 (Figure 6) and 15 (Figure 7).

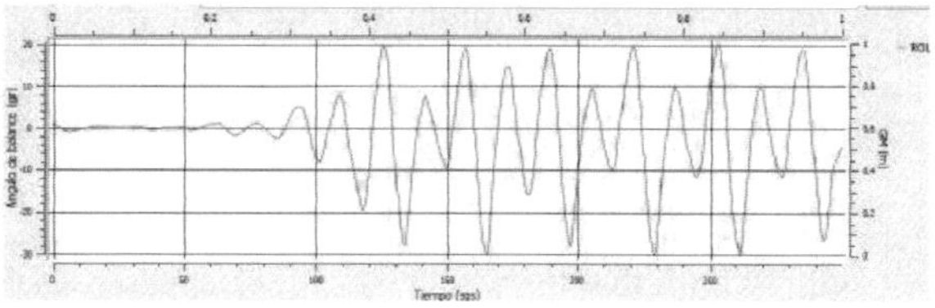

Figure 6. Rolling angle with direct calculation of hydrostatic pressure test T14

Test T14 corresponds to a situation of stability in the rolling, while the test 15 shows the appearance of the parametric rolling, which leads to capsize.

© 2017: The Royal Institution of Naval Architects

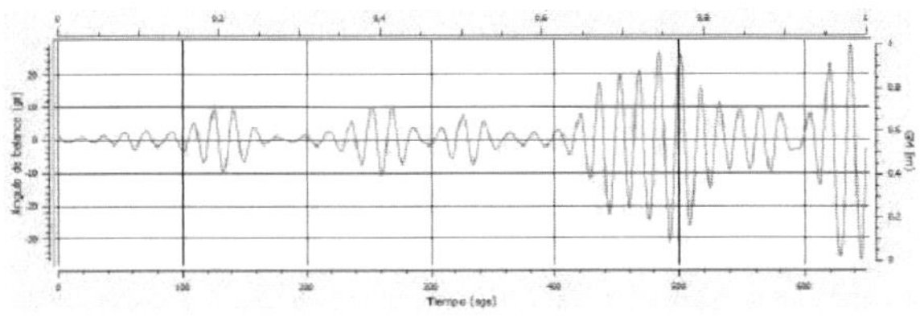

Figure 7. Rolling angle with direct calculation of hydrostatic pressure test T15

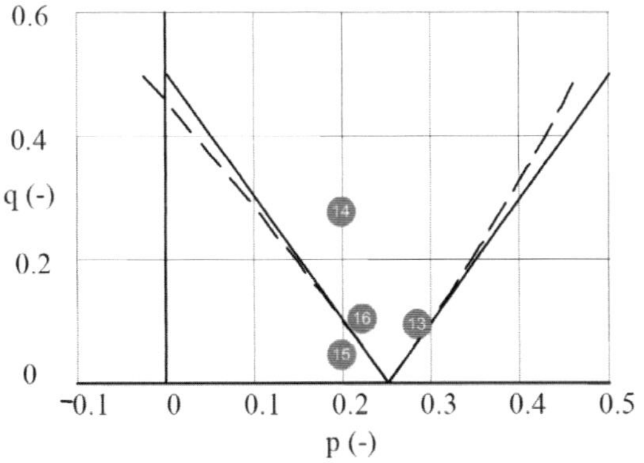

Figure 8. Results of tests in Ince-Strut diagram

Together with other tests simulated by the application, the values of the related parameters p y q for the tests were plotted in the Ince-Strut diagram for susceptibility criteria. The values between the lines correspond to stability situation and out of the lines correspond to parameters p y q that represent instability situation and appearance of parametric rolling.

The comparison of the rest of the tests allowed to conclude that, although the trend was similar, significant differences were observed for the different tests performed. The more difference was obtained when the restoration forces were calculated by the linear approximation, while using the direct calculation the results were close to expected, or at least to those obtained in the SAFEDOR test. However, the differences could be explained in the non-implementation of some necessary calculations, precisely:

- Diffraction forces not implemented.
- Restoration forces modelled with linear approximation using *GM*.
- Damping and added mass coefficients not validated with alternate methodology.

4. CONCLUSIONS AND FUTURE WORK

The second generation intact stability criteria are next to come into force, and it is necessary to provide more realistic calculation for the five failure modes selected. One of them is the parametric rolling, which affects to fine shaped vessels, as container ships. The tool presented is able to simulate the calculation of the rolling angle over time. It also provides additional calculations to estimate some of the parameters and values that currently cannot be calculated. The application allows to predict the appearance of the parametric rolling effect and gives the designers a tool to do a preliminary estimation of the appearance of this effect, so the designer can improve the shape of the vessels.

However, the application does not take into account some specific aspects that need to be included in the program. Due to this issue, the program is prepared to incorporate the missing calculations. Furthermore, the program is provided with auxiliary functionalities to get a proper analysis of the effect. Because of this, it is considered of tremendous interest to continue the work to incorporate some functionalities.

- Integrated coefficient calculation. A calculation of the damping coefficients and added masses. This should be done in a separate tasks associated with the model being read.
- The calculation of diffraction forces has to be incorporated to allow a more accurate prediction of the occurrence of the parametric rolling.
- Combination of harmonics. Several of the tests performed in the benchmark used different harmonics and it would be interesting to enter data for several waves allowing to add the effects of said combination on a specific result.
- Combination of other movements. Although the rolling movement is the most relevant in the appearance of the parametric rolling, it is not the only one that it should be analysed. It would be very interesting and very valuable to combine the movements of pitch and heave as they would give a more precise perception of the behaviour of the vessel subjected to a wave.
- Use of surface meshing. The algorithms used have been developed for models corresponding to cross-sectional areas. The pressure calculation could be done directly on the panels triangular meshes of surfaces. This calculation is expected to give more precision to the results.

5. ACKNOWLEDGEMENTS

The author would like to express sincere gratitude to SENER for providing the support and information for this paper. He also thanks to Cristina Calvo and Alicia Ramirez for their help in the elaboration of this paper.

6. REFERENCES

1. MUÑOZ, J.A., *Estudio de una metodología de software para la predicción y análisis del*

© 2017: The Royal Institution of Naval Architects

balance paramétrico de buques para su aplicación en los criterios de estabilidad de segunda generación. PhD Thesis, Universidad Politécnica de Madrid, 2015. Unpublished thesis.

2. FRANCE, W.N., LEVADOU, M., 'An investigation of head sea parametric rolling and its influence on container lashing systems', *SNAME Annual meeting,* 2001.
3. SHIN, Y., BELENKY, V., PAULLING, J., WEEMS, K. Y LIN, W. 'Criteria for parametric roll of large containerships in longitudinal seas'. *ABS TECHNICAL PAPERS*, 2004.
4. OVEGARD, E., 'Numerical simulations of parametric roll in waves'. *KTH Centre for Naval Architecture,* 2009
5. BECK, R. F., CUMMINS, W.E., DALZELL, J. F., MANDEL, P., WEBSTER, W.C. 'Principles of Naval Architecture, vol. 3'. *The Society of Naval Architects and Marine Engineers*, 1989.
6. PAPANIKOLAOU A., SPANOS, D. SAFEDOR international benchmark study on numerical simulation methods for the prediction of parametric rolling of ships in wave. *NTUA-SDL Report*, 7 2009.
7. SHIN Y.S., BELENKY V.L., PAULLING, J.R. WEEMS K.M., LIN, W.M. Criteria for parametric roll of large container-ships in longitudinal seas. *ABS TECHNICAL PAPERS*, 2004.

8 European Union European Commission. SAFEDOR report summary, 2009. Contract No. FP6-IP-516278.

7. AUTHORS BIOGRAPHY

Jesus A. Muñoz, Ph.D. holds the current position of FORAN System Product Manager at SENER Ingeniería y Sistemas S.A. He is responsible for the technical development guidance of the FORAN System. He is also professor of Applied Systems Engineering at the Marine Engineering School of the Universidad Politécnica de Madrid.

© 2017: The Royal Institution of Naval Architects

DEVELOPMENT OF AN EFFICIENT METHOD FOR EQUIPMENT INSTALLATION OF VERTICAL PRESSURE VESSELS ON OFFSHORE PLANTS WITH 3D SCANNING TECHNOLOGIES

M G Kim, S H Kim, S I Choi, J W Pyo and **J T Lee,** Central Research Institute, Samsung Heavy Industries, KOREA

SUMMARY

On offshore plants, most of the equipment is manufactured by widely well-known specialized suppliers with a view to guaranteeing the product functionality. However, misalignment issues arise in the course of installation equipment due to unavoidable discrepancies between drawings and real objects. In order to resolve this issue, a great of deal of additional rework such as re-adjustment or even re-installation is required. This has had a serious impact on productivity as not only the size of offshore plants but also its complicatedness has grown over the years. In this regard, the aim of this study is to identify ways to reduce potential rework as dramatically as possible by predicting difficult areas to install equipment in advance with application of 3D scanning technologies. Since the 3D scanning technologies enable us to easily obtain the real shape of objects, the installation procedure can be virtually simulated and possible problems detected at an early stage. While utilization of diverse equipment may be considered, vertical pressure vessel is selected. Because its shape is extremely high in the vertical direction, it is difficult to identify and fulfil design tolerances with conventional measurement methods. This study consists of three chapters: pre-processing of 3D points cloud, measurement equipment using 3D scanner and procedure of virtual installation simulation. Consequently, the present paper presents the result of experiments with two vertical pressure vessels installed without any rework and additional facilities usage. This result indicates that installation time can be reduced by 70% using the proposed method.

1. INTRODUCTION

In shipbuilding industries, installation of equipment manufactured by widely well-known specialized suppliers has been considered as a crucial shipbuilding process since it requires huge amount of working time as well as expensive facilities such as goliath crane or floating crane. Thus, an advanced control like geometric dimension and tolerance (GD&T) is necessary to accurately install at the first try.

In case of shipbuilding, the proportion of equipment is relatively low compared of offshore plant, so that all installation of equipment had been dependent on worker's experience rather than technical manners. It was because trial and error installation method were more cost effective despite of additional rework. Development of new measurement system and concepts which requires new investment was not commercially viable. However, with growing demand for construction of large offshore plant, shipyards face a completely different environment. In case of construction of offshore plant, the proportion of installation of equipment among entire production processes is extremely high compared to conventional shipbuilding, since the offshore plant consists of much equipment like pressure vessel and refinement unit located in top sided structures. At the beginning of construction of offshore plant, many additional works occurred on installation stage due to a trial and error method. It resulted in relay of unforeseen delays and the huge amount of additional cost and loss for the shipyards. It is one of primary reasons for difficulties offshore builders face now.

While utilization of variety of equipment may be considered, this study focuses on vertical pressure vessel. Because its shape is extremely high vertically, it is difficult to identify and fulfil design tolerances. For installation of vertical pressure vessel, it requires the cylinder axis of vertical pressure vessel perpendicularly stand on horizontal axis. Thus, orientation of vessels should be determined on basis of in-let nozzle. Since there is no appropriate conventional device to measure geometric information, we were forced to install by trial-and-error way and to partially measure with contactable device after installation.

(a) 3D point cloud for vertical vessel

(b) Foundation of vertical vessels

Figure 1: The pictures of vertical vessel and its foundation

In order to resolve this issue, we introduce new development of measurement system with 3D scanning technologies to shipbuilding industries. 3D scanning is the one of best technologies to capture real-world shape and handle all of data on virtual space. It allows us to check out existence of sub-parts and geometric information. In shipbuilding industries, many researchers have used 3D scanning technology to check out geometric information of specific objects[1]. After Kim et al.[2] paper published, we could see the possibility of virtual simulation for assembly of large structures. Kim et al.[2] presented the virtual clash detection for pre-erection process. In particular, Kim et al.[2] simulated pre-erection stage by virtually locating two registered 3D

© 2017: The Royal Institution of Naval Architects

point cloud data in the target location. However, it did not consider geometric information.

This study is to represent virtual simulation of installation process for offshore plant equipment with geometric information using the registration method by Choi et al.[3] This study consists of 5 chapters: Frameworks of efficient method for installation, Virtual simulation of installation for equipment, Experimental results and Conclusion. We discuss the proposed method in detail as below.

2. FRAMEWORK OF EFFICIENT METHOD FOR INSTLLATION WITH 3D SCANNING TECHNOLOGIES

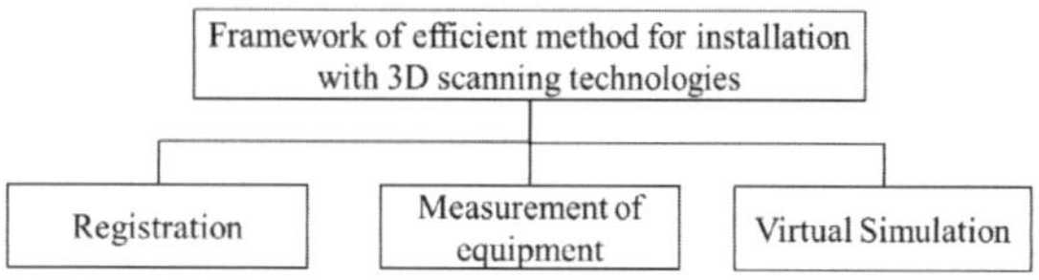

Figure 2: Outline of Framework of installation method with 3D scanning technologies

In order to take advantage of 3D scanning technologies, we create the framework consisting of three parts. At first, registration part should be prepared. Registration means point clouds obtained in different coordinate is aligned as closely as possible with respect to a reference coordinate system. 3D scanning application for large structures is dependent on quality of registration since it is not possible to measure all surfaces of a specific object without occlusion by one try. As mentioned, we introduce proven method from Choi et al.[3] to our solution. Secondly, it is important to use 3D scanner to measure a datum point accurately. Generally, black and white target has been used to extract datum points in 3D scanning technologies. We also adopt black and white target to this study. Thus, we propose a method that considers not only the accurate measurement but also the reflection of the simulation results. The last part is virtual simulation of installation for vertical pressure vessels. It is a key idea of this study to propose the 3D point cloud manipulation flow using geometric information. Since registration method is well explained in Choi et al.[3], we mainly discuss the rest of parts: measurement of equipment and virtual simulation.

3. MESUREMENT OF VERTICAL PRESSURE VESSELS WITH 3D SCANNER

For vertical pressure vessels, the datum is individually cylinder axis and orientation of in-let nozzle as mentioned in chapter 1. Since cylinder axis of vertical pressure vessels can be only defined by means of cylinder fitting after registration, targets are not necessary during measurement. On the other hands, in case of orientation, most equipment are marked with an orientation information at 0°, 90°, 180°, 270° by the manufacturer. In order to identify orientation of

Figure 3: Black and white targets on vertical pressure vessels

in-let nozzle, we are required to attach targets on the vertical pressure vessel according to marked orientation. It can be a datum to indicate current orientation. However, it is with high possibility that all of marks have not been written correctly because they were drawn before welding and heating treatment. If all marks are not correctly written, we can figure out exact orientation of the in-let nozzle from a registered 3D point cloud and calibrate correctly on basis of measured datum. In this study, we suggest that black and white targets be attached on vertical pressure vessels as Figure 3. After attachment of targets, we measure the equipment at various locations with a 3D scanner to obtain data without occlusion.

Next, foundation of the vertical pressure vessel located in a hull structure can be also measured. According to design tolerance, the vertical pressure vessel located onto its foundation should perpendicularly stand on the horizon. Therefore, we should measure a specific object that can provide a reference coordinate relative to the horizon. By measuring the foundation with a device having the coordinate system based on the horizon, reference coordinate system can be constructed. When we measure the foundation of equipment, we should carefully calibrate the coordinate of 3D scanner.

4. VIRTUAL SIMULATION OF INSTALLATION FOR EQIUPMENT

Virtual simulation of installation for equipment consists of manipulation of 3D point cloud and reverse engineering. Reverse engineering is responsible for evaluating geometric information and manipulation of 3D point cloud is responsible for simulation of installation process. Entire flow is demonstrated in Fig 4. Firstly, cylinder fitting for the in-let nozzle is performed

© 2017: The Royal Institution of Naval Architects

to define cylinder axis used for obtaining accurate orientation of the vertical vessel. Secondly, the terminal point of obtained cylinder axis is moved to bottom of the vertical vessel. This task allows us to see exact orientation from the bottom.

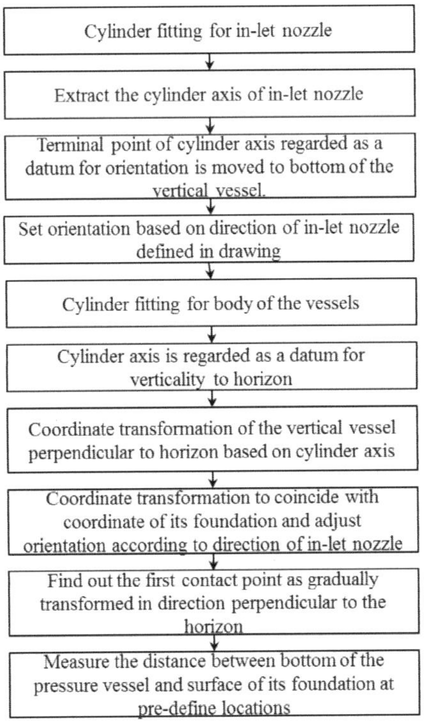

Figure 4: Flow chart of virtual simulation of installation for vertical pressure vessel

Third of all, cylinder fitting for the body of the pressure vessel is conducted to prepare the datum point for vertically positioning the pressure vessel. Next, using obtained cylinder axis at the previous step, the coordinate system of a vertical pressure vessel is transformed to stand perpendicular to the horizon and coincide with the coordinate system of its foundation. After coincide with corresponding both coordinate systems, the coordinate of the vertical pressure vessel is gradually transformed down in direction perpendicular to the horizon until the first contact with its foundation. The course of tasks mentioned above describes the virtual installation for equipment. By means of virtual simulation, we can measure not only all of distances between the vertical pressure vessel and its foundation at pre-determined points but also maintain the required design tolerance. Based on the calculated distances, it is possible to prepare a way to fill the gap between the vertical pressure vessel and foundation before installation.

6. EXPERIMENTAL RESULTS

For experimental results, we applied the proposed method into two vertical pressure vessels. The dimension of each pressure vessel is described in Table 1.

Table 1: Main dimension of two vessels in experiments

	Vessel 1	Vessel 2
Height [mm]	7500	11140
Diameter [mm]	3500	4700

First of all, we present the distance between the vertical pressure vessel and foundation in Figure 5 as experimental results. Through the proposed method, we can figure out the gap between the vessel and foundation using only 3D point clouds before installation.

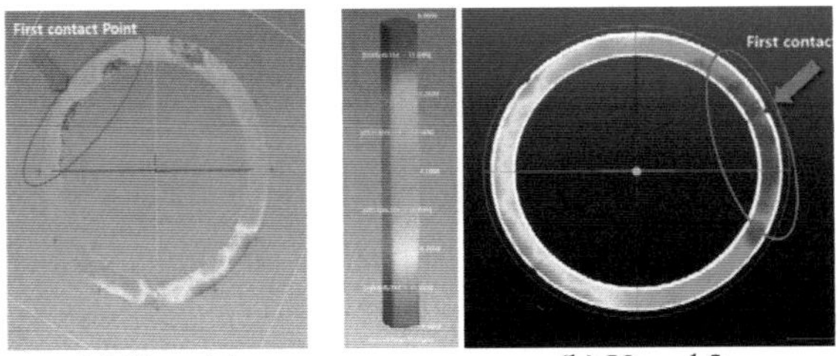

(a) Vessel 1 (b) Vessel 2

Figure 5: The distance between bottom of vessel and foundation

In order to evaluate the quality of the proposed installation method, we present the results of the verticality of each vessel by conventional method. The conventional method is measuring distance from the surface of equipment to a pre-determined specific point at the top and bottom. If the difference between the top and bottom distances is within allowable tolerance, installation process can be completed. For accurate assessment of installation, we individually perform the conventional method at 0°, 90°, 180°, 270° locations and use the average of the four difference as a measure of evaluation. When we evaluate in a conventional way, the installation by the proposed method can be completed within the designed allowable tolerance and is proved to be used immediately for the practical purposes.

Table 2: The results from conventional measurement

	Allowable Tolerance [mm]	Average of difference [mm]
Vessel 1	7.5	5
Vessel 2	10	10

© 2017: The Royal Institution of Naval Architects

Additionally, we can complete the installation of equipment without re-use of crane while satisfying the allowable tolerance and all of pre-works can be prepared in advance as Fig 6.

(a) Pre-works for filling the gap (b) The gap after installation

Figure 6: The result of one time setting for installation of vertical vessel

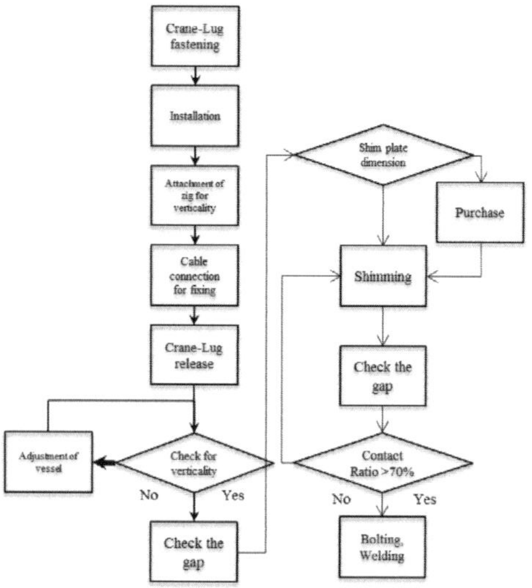

(a) Conventional Installation flow

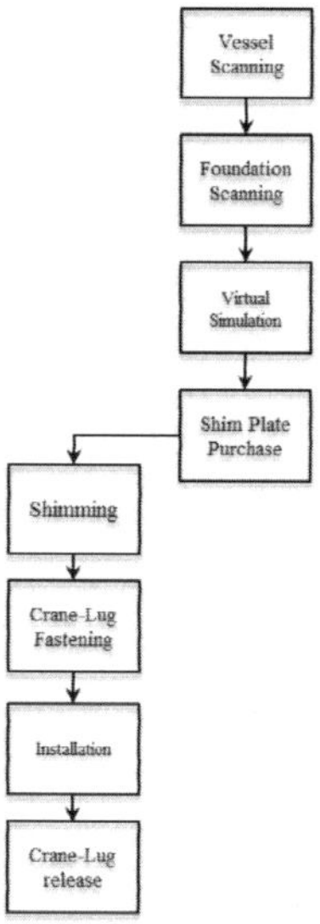

(b) Proposed installation flow

Figure 7: Comparison with the conventional installation flow and the proposed installation flow

It means that the proposed method enables us to make the simple flow of installation as Figure 7 by eliminating repetitive tasks. It also leads to dramatic reduction in working time by as much as 70%

7. COLCLUSION

In this paper, an efficient method for installation of vertical pressure vessels with 3D scanning technologies is presented. The method consists of registration, measurement of equipment with 3D scanner and virtual simulation of install process, each of which is developed for preparation of pre-works and removal of additional works. In particular, virtual simulation is designed for GD&T by using reverse engineering such as cylinder fitting and coordinate transformation.

The proposed method is formulated in conjunction with conventional measurement. Therefore, the proposed method can be efficiently used in the real equipment installation. In terms of management, it can reduce working time 70% by getting rid of repetitive tasks.

However, this study is limited as it only considers the vertical pressure vessel among various type of equipment. We will develop a plan to expand the application introduced in this study to other equipment such as box girder type and horizontal pressure vessel.

Nevertheless, this study can be meaningful find that 3D scanning technologies can used to perform virtual installation considering geometric information.

8. ACKNOWLEDGEMENT

This work was supported by the development of industrial technology program (No. 10063452) of the Korea Evaluation Institute of Industrial Technology grant funded by the Ministry of Trade, Industry and Energy

9. REFERENCES

1. Sun, J., Kiekata, K., Yamato, H., Nakagaki, N. and Sugawara, A. 'Efficient point cloud data processing in shipbuilding: Reformative component extraction method and registration method', *Journal of Computational Design and Engineering*, 1, 202-212(2014).

2. Sunghan Kim, Mingyu Kim, Jeongtae Lee, Jinwhi Pyo, Heeyoung Heo, Dongho Yun and Kwanghee Ko 'Registration of 3D point clouds for ship block measurement', *SNAME*, 2015. 2011.

© *2017: The Royal Institution of Naval Architects*

3. Sungin Choi, Sunghan Kim, Jinwhi Pyo and Mingyu Kim '3D auto registration for large structures in shipbuilding industries by point feature histogram', *30th Workshop on image processing and image understanding*, 2017(In Korean).

10. AUTHORS BIOGRAPHY

Mingyu Kim, Research Engineer, Samsung Heavy Industries.

Sunghan Kim, Principal Research Engineer, Samsung Heavy Industries.

Sungin Choi, Senior Research Engineer, Samsung Heavy Industries.

Jinhwyi Pyo, Assistant Engineer, Samsung Heavy Industries.

Jeongtae Lee, Assistant Engineer, Samsung Heavy Industries.

© 2017: The Royal Institution of Naval Architects

© 2017: The Royal Institution of Naval Architects

MINIMISING THE DESIGNER / END USER KNOWLEDGE GAP USING VIRTUAL REALITY

W Pynn, Marine Institute of Memorial University, Canada

SUMMARY
It is important that the practice of naval architecture continue to evolve as new, cost-effective tools become available. This paper explores how one such tool, virtual reality, may be used to improve the designer's skillset as well as the design process using digital mock-ups and participatory design practice. By incorporating feedback from professional mariners at various points throughout the design spiral, knowledge is transferred and the final design improved. The paper also includes a survey of professional mariners on the quality of bridge and machinery space design and a review of the current tools available for incorporating virtual reality into the design process.

1. INTRODUCTION

Naval architects do not manage the daily operations of ships, nor does the ship's crew design the ship. However, the vessel's design process requires the naval architect to capture the ideal, often with little to no input from the final users of the space. It is of no surprise then, that a ship's crew may perhaps be less than enamored at the layout of its key workspaces. A survey conducted of professional mariners employed by the Marine Institute of Memorial University confirmed that while naval architects are not doing too badly when it comes to meeting mariner's needs, we could be doing better. This of course is not an uncommon design problem. Many of us have used a product, driven a vehicle, or walked through a house or a building and thought "what was that designer thinking?"

I will argue that incorporating participatory design into the ship design workflow at various stages of the design spiral is good practice. And the maturity and availability of relatively inexpensive virtual reality (VR) hardware/software make this process straightforward. Creating a virtual mock-up of key areas of the vessel for evaluation and input earlier in the design process from the end users would benefit the design as well as the designer greatly.

This isn't so much a scientific paper as it is the sharing and discussion of some ideas on the use of a technology that has been around for quite a while, is perfected to the point where it is nearly flawless, the average person or design office can afford and use it, and there is software readily available that is useful and user friendly.

2. THE KNOWLEDGE GAP PROBLEM

Few naval architects have the professional marine background of a deck officer or marine engineer. The ones that do, bring the valuable experience of onboard ship operations to the design process. The remaining majority, although trained experts in the areas of structural design, resistance and propulsion, and stability, unfortunately do not have first-hand, valuable shipboard experience and are therefore left to try and anticipate how shipboard operations work, or rely on ad hoc explanations and regulation.

Mallam, Lundh and MacKinnon (2017) summarize the problem nicely in the paper "Evaluating a digital ship design tool prototype: Designers' perceptions of novel ergonomics software". Noteworthy points include the following:

- Ship owners place more importance on cargo carrying capacity, speed and versatility than on human factors.
- End user considerations are rarely integrated into the planning process.
- A well designed, user centered onboard work environment can contribute to improved safety, as well as improved productivity and economics.

Mallam et al. (2017) go on to say that incorporating participatory design practices is a logical way to "fill the knowledge gaps between seafarers, who hold tacit, domain-specific experience of onboard ship operations, and naval architects, who are experts in engineering and design methods, and create the ships and work environments which seafarers work on/in."

This, of course, is a common problem in many design and engineering disciplines, not solely naval architecture or ship design. So how do other disciplines handle this knowledge gap problem?

3. PARTICIPATORY DESIGN

Participatory design is a design approach in which the input of the stakeholders is incorporated into the design process (Hoffman, 2014). Stakeholders in this case would be professional mariners; in other words, ship's crew. Not necessarily *the* ship's crew, but professionals who can provide the designers with meaningful and actionable feedback. The process is used widely in other disciplines such as urban design, architecture, automotive and product design.

3.1 ARCHITECTURE

The incorporation of user feedback extends back centuries. In the case of ancient Greece, it was citizens having a meaningful say in how their society and communities were formed and conducted. Rather than a

© 2017: The Royal Institution of Naval Architects

select few with perhaps expert, but scope-limited knowledge making decisions for the masses, the masses were invited to participate in the design. Or at least representatives of the masses. We now call that democracy (Sanoff, 2008). Fast forward to more recent times, since about the beginning of the 20th century architects and city planners have had community stakeholders more directly involved in urban planning, and have incorporated that feedback into creating cities that are healthier and more people focused and friendly (Sanoff, 2008).

Susanne Hoffman in her book "Architecture is Participation" argues that users of the spaces architects create are quite capable of having meaningful input in the design process. She goes on to discuss the "Design Methods Movement" that started in the 1960's was another early attempt to involve end users in the architectural design process. While critics argued that the design process is opaque and often requires expert knowledge to understand technical drawings, Hoffman (2014) points out that even with these issues *"Feedback and evaluations of various projects have confirmed the success of (these) method(s): the users' degree of identification with the finished building is high."*

It is true that we as architects, engineers and design experts spend years in technical schools learning the language of engineering. This time is not just dedicated to learning how to solve problems and crunch numbers, but how to communicate design ideas to other design professionals and the builders of our designs. To a non-design professional, these communication methods can be difficult to decipher. Fortunately, we now have tools that compensate for a user's lack of technical "design" expertise and make the design process, or at least the output, less opaque.

3.2 PRODUCT / AUTOMOTIVE

It is not just architecture that has benefited from user feedback. In fact, one might argue that an entire branch of the marketing profession, responsible for focus groups and customer feedback, is designed specifically to capture what product designers have missed along the way. Products in this sense include everything from breakfast cereal to amusement park ride experiences.

In 1933 the automobile manufacturer General Motors created a permanent consumer research department for the single reason of designing cars that customers wanted. Five years later they had amassed more than 185 design and technical improvements driven entirely by customer feedback (Boler-Davis, 2016). More recently, automotive designers for Ford and automotive design studios such as Seymourpowell have incorporated the use of VR into their design process. And because the design is taking place inside a virtual and networked world, it is possible to have designers (and customers) worldwide participate concurrently in the process (Aouf, 2017).

3.3 THOSE AGAINST

Not everyone is in favour of the concepts espoused by Participatory Design. For example, Steve Jobs was famously quoted as saying "A lot of times, people don't know what they want until you show it to them." (Mui, 2011). Jobs doubted market research largely because he believed he was designing products so different from what was already out there, that his customers couldn't really articulate what was or what was going to be outside the box. In his view, he was designing revolutionary products, whereas customers can only give feedback in an incremental, evolutionary way.

Fortunately, revolutionary shipboard workspace design is not the primary role of the naval architect. The main objective is to design to relevant regulations whilst maximizing safety and usability for the end user.

3.4 PARTICIPATORY SHIP DESIGN

While ship designers and builders may not be subject to the same market pressures as mass manufacturing companies such as GM or Ford, how these companies approach end user evaluation of design is something we can learn from and apply. For example, through personal experience in yacht design and construction, the vessel owners tended to be very involved in the design of the final product, in some cases, annoyingly so. Even 50 and 60 metre vessels had significant involvement of the crew, specifically the captain and chief engineer, in the design of the relevant spaces. Often the yacht's captain acted as owner's representative and was involved in the construction of the bridge, including the layout of the console and positioning of the equipment. It was not recognized or labeled as such at the time, but this process was participatory design. The feedback received from these professionals was invaluable for the current design/build as well as improving the author's design skillset.

In addition to the owner's involvement with the design/build yacht process, maritime professionals in the form of owner representatives were present throughout the construction phase, something not particularly common in shipbuilding as a norm. Mallam et al. (2017) posit that "The major challenge for effectively supporting participatory practices in the shipping industry is bridging the geographical, cultural and professional gaps between disciplines involved in ship design and construction." While the methodology espoused in this paper does not solve all the challenges listed above, it does connect end users to designers regardless of their discipline or geographical location.

4. PROFESSIONAL MARINER SURVEY

It is possible that there isn't a knowledge gap problem at all, and that shipboard workspace design that naval architects create are indeed adequate. Interest in the

© 2017: The Royal Institution of Naval Architects

matter led to a survey conducted for this paper of the professional mariners at the Marine Institute of Memorial University (Canada) as well as the local chapters of both the Master Mariners of Canada and the Canadian Institute of Marine Engineers. Two surveys were prepared, one targeted at the Bridge Crew and Bridge Design while the other was designed for Marine Engineers and Machinery Spaces. Although the number of responses were predictably dismal (of more than 100 professional mariners, nineteen responses were received) the results were interesting, particularly the comments the mariners made.

In general, the first half of each survey probed the mariner's thoughts on bridge and machinery space design in general whereas the second half of the survey focused on the least functional bridge and machinery spaces they had experienced in their careers. The surveys used a Likert scale from 1 to 5 with 1 being Strongly Disagree and 5 Strongly Agree. For each question there was an opportunity for the mariners to add comments. The results and their comments have been included in the Appendix.

4.1 SURVEY DISCUSSION

The findings were a little surprising as a more scathing review of the naval architect's collective ability to design shipboard workspaces was expected. The regulations and experience of naval architects seems to be reasonably effective however, it seems that some spaces reported on are less than ideal - so there is still room to improve. The Likert scale was not as revealing as the comments that accompanied each question and while anecdotal, they are useful feedback.

An unexpected response from the survey emphasized the difficulties with lighting and in particular, the moving from light to dark spaces proved problematic. The poor sizing and location of bridge windows was also a reoccurring theme. Other responses highlighted the lack of space and poorly located equipment within machinery spaces, which was expected.

Much of the design issues mentioned throughout the survey responses could have been minimized or eliminated with participatory design practices. Although this survey yielded few responses, the replies gained from end users would have been useful for even an experienced naval architect when laying out a bridge or machinery workspace.

5. A KNOWLEDGE GAP SOLUTION

A tried and tested solution to the verification of workspace (or any space) design is to build a physical mock-up of the proposed space and have the anticipated users of the space experience it. Variations of this method have been used for years in designing and building recreational craft for production boat builders as well as large, one-off yacht manufacturers.

Essentially, these mock-ups are basic skeletons of the space with the intent to verify the spatial layout and its functionality. It enables shipwrights and carpenters to make relatively easy and quick changes as instructed by the user evaluators and when the exercise is complete, the designers are then able to record changes with photos, tape measures and sketches, adjusting the original design as needed.

In my experience, these exercises in participatory design were useful to all involved, and helped to ensure the final space met the requirements of the customers that would be using the space. These types of exercises are particularly important when building high volume vessels, otherwise the potential is to have a less than ideal space replicated potentially hundreds of times.

Figure 1: Simple Physical Mock-up. LymanMorse Boatbuilding

We are now at a place where these "mock-ups" for participatory design can be built digitally rather than physically using VR.

The Virtual Reality Society of the UK provides us with a good working definition of virtual reality: "Virtual reality is the term used to describe a three-dimensional, computer generated environment which can be explored and interacted with by a person. That person becomes part of this virtual world or is immersed within this environment" (The Virtual Reality Society, 2017). We may be entering the era in ship building of Virtual Reality Aided Design or VRAD; or perhaps augmented CAD is the more accurate term.

The digital mock-up has several advantages over the physical. For example:

- The digital mock-up can be built easily in an engineering office on a computer using designers already employed, rather than having

© 2017: The Royal Institution of Naval Architects

to have the physical space in a warehouse for example.

- The digital version requires no building materials or shipwrights or materiel disposal when the project is finished.
- Many design variations can be tested quickly and modifications can also be made to the digital models, almost while you wait.

While completely immersive, a user is still aware that they are in a virtual environment. The virtual mock-ups do however, have the advantage of being able to add as much detail as required along with correct colouring and textures. For example, a radar can actually look like a radar, as the vendor 3D model can be placed inside the mock-up and a graphics texture can be added to simulate the working screen.

From a design workflow perspective, the creation of these virtual mock-ups entails taking the 2D general arrangement drawing and creating a 3D model from it, and it is this model that becomes the virtual world, or in our use case – the virtual mock-up. In my experience, design offices still start with 2D when doing spatial planning or general arrangements, however the tools are beginning to become commercially available where initial design can take place inside a virtual environment and in 3D. Mindesk for example (www.mindeskvr.com) has introduced a set of drivers for Rhinoceros 3D modelling software that facilitates exactly that - actual model creation inside the virtual environment.

Regardless, the CAD tools themselves are not all that important as the VR model viewers accept various file types as demonstrated below and a quick proof of concept model can be created by a skilled modeler in just a few hours. While this quick model may lack detail or texturing, it enables a designer to evaluate the space and make changes to the design before committing to the time required for a more comprehensive "mock-up" model. This effectively closes the time gap between design and construction and allows the designer to experiment with variations and take design risks they may not otherwise take in the conventional design workflow.

5.1 THE VR PROJECT

The Marine Institute acquired the HTC Vive VR hardware in November 2016 for the purpose of investigating where these new tools might fit in the ship design workflow.

Several of the senior naval architecture technology students volunteered their time to model their bridge designs in full to use them as VR mock-ups and have the designs evaluated by professional mariners that are instructors at the Marine Institute. This was for the purpose of testing VR for participatory design as well as improving the student's designs and the design skills of the student architects.

Several of the students that took part in the exercises of bridge modelling reported that a more complete model with detailed parts and textures generally took them in the region of forty hours. This depends to some degree on the readily available vendor furnished object models that can be imported into the bridge model as well as the extent of the library of prebuilt parts such as electronics, helm seats, etc., that the design office already owns or has access to.

By all accounts the project was a tremendously useful exercise, particularly for the students, but also for the mariners that participated. As predicted the students received excellent and practical advice from the experienced ship masters, but perhaps as significant, the mariners recognized the importance of participating in the design process and are committed to be available for evaluating student design mock-ups even earlier in the design process next year.

The project will continue again with the 2017/2018 cohort with the idea of expanding the project to include more student participants and extending the mock-ups from just bridge design to machinery spaces for evaluation by the marine engineers.

Figure 2: Evaluation of Digital Bridge Mock-up

5.2 VR TECHNOLOGY

Currently there are two types of VR devices commercially available. The simple and inexpensive headsets that work with a smartphone (smartphone VR) and the more powerful, but expensive headset and controllers that require a reasonably powerful computer processor and higher end video card (PC Connected Headset VR). Neither is particularly complicated to get working quickly. For example, the smartphone headset was working within five minutes, and the higher end HTC Vive was a little more involved and required approximately one hour to set up the first time through.

© 2017: The Royal Institution of Naval Architects

5.2 (a) Smartphone VR

The smartphone VR headset tested was the Utopia 360 by ReTrak. These were readily available at a local electronics retailer for about $15 USD. While there are simple apps and games available for the smartphone VR headset, the goal was to use it for participatory design. From this point of view, the smartphone VR headset is limited to panoramas of the mock-up. The word limited is used only in comparison to what is possible with the PC connected Vive, the higher end VR system. The panoramas are still useful in terms of immersing the user inside a virtual environment. The view is limited to one single 360° pivot point at a time as the headset tracks head movements but the user cannot "walk" around. The software used for the viewing the shipboard workspace panoramas does, however, allow the user to teleport from predetermined panorama to panorama.

Figure 3: Utopia 360 VR Headset

The basic workflow process used to create the panoramas for viewing in the smartphone VR starts the same for both high and low-end VR product types - the creation of the shipboard workspace model or virtual mock-up. This is the only commonality however. The students that participated in the VR Project created their 2D general arrangements in AutoCAD and their hull models in Maxsurf. Both of which were easily imported into Rhinoceros 3D, which is a general-purpose 3D modelling software by Robert McNeel and Associates (www.mcneel.com).

After the creation of the virtual bridge mock-ups in Rhinoceros 3D (and in the case of the VR Project students - the vessel exterior was also modelled) the models and textures were imported into Lumion 6.5 (www.lumion3D.com) for panorama creation. It should be noted however that V-Ray (www.chaosgroup.com), which works natively in Rhinoceros 3D can also be used in the creation of panoramas that are used in the following workflow. Once the ship workspace model has been imported into Lumion, an environment can be added. Having viewed panoramas with and without environments, the additional minutes that it takes to add the environments in Lumion make it worth it as more detail matters when it comes to the user's immersive experience.
Within Lumion the camera viewpoint is moved around using the keyboard until it is placed in one of the panorama positions and a panorama "snapshot" is taken to tell the software that a 360° panorama is required at this viewpoint. In the student created OSV model used for the VR Project, four different bridge positions were created at key evaluation points while the sample containership model had two panoramas on the bridge, one at the foredeck and one on the mast. Lumion then automatically creates the graphics files for each of the indicated viewpoint positions necessary for the next step. Depending on the quality settings and number of viewpoints, this can take anywhere from two minutes to several hours. The four viewpoints in the OSV model at the highest quality setting rendered in approximately ninety minutes.

Figure 4: One of Four Key Panorama Views for OSV Evaluation

The final step is to import the panoramas into a format that can be used for smartphone VR headset use. IrisVR (www.irisvr.com) is a company that creates software specifically for this purpose. Their Scope app can be downloaded for free to any iOS or Android based smartphone for sharing panoramas for viewing in a headset such as the Utopia 360. The group creating and sharing the virtual mock-up panoramas must create an account at the IrisVR website to upload and share. Once they have done this, a link can be sent to anyone with a smartphone and the free Scope app to give them access to that library of panoramas. In this way, any potential end user/evaluator of the workspace can have access to the mock-ups anywhere in the world.

The panoramas can be viewed in a single frame on a computer or tablet screen using the mouse, stylus or finger to navigate, or on a smartphone placed in the headset for stereo viewing. Again, because the headsets are inexpensive (Google's "cardboard" VR headset retails for about $15 USD for example) the user can likely obtain a headset wherever they are and view the virtual workspace mock-ups as intended. Inside the Scope app, there is the function of bringing up the related key panorama viewpoints in the shared panorama library

© 2017: The Royal Institution of Naval Architects

so the user can move from key viewpoint to key viewpoint with relative ease.

In terms of software cost, and assuming the engineering office already has some sort of 2D CAD software, the current basic cost of the being able to provide this service to clients or users is as below. However, it is dependent largely on the modeling software chosen and one has to consider that the computer workstation has minimum requirements as well. The workstation at the Marine Institute cost about $1500 USD and $300 of that was the video card.

- Rhinoceros 3D – $995 USD
- Lumion Basic – $1700 USD
- Scope (ability to upload panoramas for sharing) - $480 USD annual subscription
- Scope (for viewing) – No Cost

The alternative to Lumion in the workflow would be V-Ray for creating panoramas. As mentioned, V-Ray works natively in Rhinoceros 3D (and other CAD packages) and retails for about $700 USD. It should be noted that the learning curve is much steeper for V-Ray than for Lumion and the ease with which environments are added in Lumion is also advantageous.

As an aside, one of the comments from the professional mariner's survey spoke about the need for the ability to do low light testing of workspaces. Interestingly, because the panoramas are created in Lumion, various lighting scenarios can be captured and tested in the less expensive headsets. The higher end Vive/Prospect combination described below is still not capable of this particular test. An educated guess as to why is also below.

5.2 (b) PC Connected Headset VR

The hardware that started the VR Project at the Marine Institute was the acquisition of the Vive. This hardware by HTC comes with the headset, two controllers, the scanners that are used to track the headset and controllers and all the necessary cables to power and connect it to the workstation. The headset is currently tethered (a wireless version is in the works), and it must be connected to a computer to function. The Vive retails for about $800 USD. The only additional hardware required for the VR Project at the Marine Institute were portable stands to mount the scanners on. The scanners can be permanently installed if the office had a space dedicated to VR; however, the VR Project required the portability of collapsible stands, and these were reasonably priced at about $30 USD. The scanner diagonal spacing determined the size of the available walkabout space, and the current incarnation provided a virtual space of about 3m x 3m. Alternatively, to make the experience even more portable, Alienware gaming laptops seem to be the tool of choice and a VR capable laptop will cost an engineering office about $2000 USD.

In terms of workflow, the process of creation of the model for use with the Vive is identical to that with the smartphone headset and panoramas described earlier. Regardless of what technology the user experiences the modelled workspace on, that workspace must first be modelled. It is post model creation however, where the workflow paths diverge.

Figure 5: HTC Vive and Components

To be immersed in the model with the Vive we again turn to IrisVR for their architectural VR viewing package Prospect. Prospect is designed to work with both the Vive and its competition, the Oculus Rift ($650 USD). Once again, as we modelled the bridges in Rhinoceros 3D, IrisVR has created a plugin for Rhinoceros, which automatically meshes the model for importing into Prospect. Depending on the complexity of the model this can take anywhere from a few minutes to an hour. Technically this just needs to be done once, but practically we found that the model required many tweaks once the user could "walk around" on the model and so the model ended up being converted several times. However, once this process is completed, the same mock-up/model can be viewed as many times as necessary.

After conversion, the Prospect software opens automatically on the workstation and the user is presented with the model on a virtual table, much like one would be looking at a physical architectural model. Now comes the power of what the Vive brings to the table over the simple panoramas. The model can be made larger and rotated using the Vive controllers. One controller has a blue virtual laser pointer coming from it and by directing the laser pointer anywhere on the model, when it meets a horizontal "walkable" surface, a humanoid appears indicating that by pulling the trigger on the controller the user will be taken to that point in the model.

© 2017: The Royal Institution of Naval Architects

Figure 6: Viewing the Tabletop Model

Pull the trigger and the scaled-down model becomes full size, and you are standing at the point at which the humanoid dangled. In terms of the benefit from the ability of being able to increase the size of the model and rotate the model is that it enlarges and orients the walking surfaces, making them easier to choose with the laser pointer.

Figure 7: Headset View of Tabletop OSV Model

The experience of pulling the trigger on the controller and having the model become a full-size ship is awe inspiring the first time. This has been the common experience of most people that have tried out the Vive/Prospect combination. On the first try at evaluation with a student model, several issues were discovered that were initially missed when reviewing the model on a screen versus being immersed in the model full-size. For example, there were areas in the model that were designed as walking spaces just outside the bridge that were actually too small and could not be walked on.

To get around inside the virtual world and explore the mock-up, Prospect gives you the ability to "teleport." By pointing the controller at a horizontal surface, a target appears and when the trigger is pulled on the controller that point becomes the new centre of the workspace. Recall that in the virtual environment for the VR Project there are roughly nine square metres to walk around in. Once the trigger is pulled and the user teleports, there is again nine square metres at the new position.

Figure 8: The Full Size Virtual Ship

Built into the Vive itself are several tools to keep you from wandering outside the workspace. These revolve around variations of a blue mesh that appears in the headset when the user gets too close to edge of the workspace. However, this has nothing to do with the virtual mock-up, just the actual space between the Vive scanners.

As the Prospect application is designed as a tool for reviewing land based architecture, it comes standard with review and markup tools built into the package. For example, there are markup tools for annotating problem areas, taking snaps or grabs from the headset for later review outside of VR, as well as the ability to turn layers on and off inside the VR model to investigate design variations. Again, as Prospect is meant for land based architecture, there are also tools for doing things specific to civil architecture such as solar studies, although these are not useful for shipboard workspace review.

As mentioned previously, there is currently no way to test low level lighting on a bridge mock-up. Perhaps in a future release of Prospect this will be addressed. Although generally speaking, it is the calculation of lighting that requires a tremendous amount of power from the graphics processing unit (GPU) so it may be that we are still several years from having reasonably priced GPUs powerful enough for complex multi-light source calculations.

The experience is convincing. For instance, when the user is high up on a mast, the inherent fear of falling is present. When observing people exploring and walking around in a virtual space, they are avoiding objects as if they were real. For all intents and purposes the user feels like they are experiencing the space in a way much like it will be when constructed. There are disadvantages too, however. For example, some may feel an underlying nausea after wearing the headset for more than five or ten minutes. While small spaces actually feel like small spaces, a user can bump against a bulkhead in physical mock-up, whereas currently, the controllers and headset,

© 2017: The Royal Institution of Naval Architects

(the virtual you) can pass through solid objects without any feedback whatsoever. There are companies working on this (www.axonvr.com), with the idea of providing haptic feedback jackets and gloves so that when the user bumps into a virtual bulkhead, feedback is received telling the user where on his or her torso or hands that they bumped into a constraint.

Some of the other limits of the technology are that currently, the models themselves are not interactive. Objects cannot be picked up or moved, steering wheels or valves cannot be turned; the mock up is really just for spatial planning and object positioning. There is also no physics engine or gravity built into Prospect so objects don't fall, etc. To achieve this one would have to build more of a simulation rather than a simple mock-up using tools such as gaming creation engine like Unreal Engine for example (www.unrealengine.com). There are several such engines that are made to work with the VR but this would require more time and some programming expertise. Depending on why the design office is considering VR, it may be something to investigate.

6. FURTHER AREAS OF STUDY

As VR becomes more of a common tool for ship designers, it is easy to envision many more uses for the technology. For example:

- Evaluating the buildability of complex structure.
- Early verification of the human ergonomics of safety systems before physical prototype.
- Key views on a cruise ship.
- Multiple design variations of a galley layout.
- Simulations for crew training.

The verification of design decisions seems to be the major use of the technology in the short term, however the tools for designing inside a virtual world are developing quickly. Even as this is being written, the software drivers that allowed the Vive to be used with Rhinoceros 3D went from viewing a design to actually being able to create and modify curves and surfaces within a virtual environment. Perhaps the time when Virtual Reality Aided Design will become as common place as the computer screen is now. Creating a workflow to incorporate these new tools into the actual design phase, rather than as just an evaluation tool will be the subject of further study.

There may be opportunity in the future as well, with multiple headsets for a crew with members in disparate geographical areas, to work on a single bridge or machinery control room VR model. In such a scenario, the design itself or the work processes may be evaluated in a more thorough way, perhaps even a multi hour watch simulation with various test situations in order to give a new design a more rigorous evaluation.

7. CONCLUSION

Is it possible to expect someone who has never driven a car to design the ultimate driving experience? Even with an exhaustive list of required features and regulations to work with, the design and feel of the user interface matters. As a natural consequence of a more tightly integrated human machine interface, the experience is more pleasant and safety is improved because the driver can focus on what is happening outside the vehicle rather than be distracted by a clumsy interface.

While the design workflow of the naval architect doesn't follow the same path as that of automotive engineer, engineering disciplines have been borrowing from each other in other ways for decades. For example, the ability to mathematically represent curves and complex surfaces using a computer, pioneered by Pierre Bezier, a French automotive engineer, is essential in the designing of complex hulls and topsides. Naval architects should be looking at other disciplines that are pioneering interesting technological innovations for the purpose of applying it to the practice of ship design.

This paper has explored some of the benefits of the virtual mock-up for end user verification. From the designer's perspective, the ability to experience the space being designed long before it is built will change the way we think about and execute the process. It is typically months at minimum from when a general arrangement is completed to the time of construction, and with that kind of time gap, it is too great a risk to wander away from design ideas that are just status quo. Shorten that time to days or even hours, the time frame of digital construction, and the architect can experiment, tweak, and take design risks in a virtual ship that are not possible with conventional practice. Adding virtual prototyping to the design spiral will allow the designer to ask spatial what-if questions in the same way other CAD tools have provided for stability and hull design, further improving the iterative, optimization process.

Finally, the practice of obtaining critical feedback from end users through the process of participatory design facilitated by VR has been used by other industries for years; it is a readily available and reasonably priced solution for creating better shipboard work spaces for professional mariners. Virtual reality aided design seems like the logical next step in the evolution of ship design practice.

8. ACKNOWLEDGMENTS

The author would like to thank Catherine Dutton, Head of the School of Maritime Studies at the Marine Institute for taking the leap and purchasing the equipment for the VR Project. He would also like to thank Devin Sooley and Braden Howse for volunteering their senior design projects for scrutiny by professional mariners as well as the professional mariners themselves who took the time

© 2017: The Royal Institution of Naval Architects

to evaluate the designs. A thank you is very deserving to the many editors that have reviewed this work, some several times. The author takes responsibility for any errors or issues with the paper. In addition, special thanks to Kaelyn Burke for her interest and help in researching this paper.

9. REFERENCES

1. Aouf, R.S. (2017, May). *Seymourpowell demos VR software for collaboratively designing cars.* Retrieved from https://www.dezeen.com/2017/05/08/seymourpowell-demonstration-virtual-reality-software-collaborative-design-cars-technology

2. Boler-Davis, A. (2016, February). How GM Uses Social Media to Improve Cars and Customer Service. *Harvard Business Review.* Retrieved from https://hbr.org/2016/02/how-gm-uses-social-media-to-improve-cars-and-customer-service

3. Mui, C. (2011, October). *Five Dangerous Lessons to Learn From Steve Jobs.* Retreived from https://www.forbes.com/sites/chunkamui/2011/10/17/five-dangerous-lessons-to-learn-from-steve-jobs/#7177d0373a95

4. Hofmann, S. (2014) *Architecture Is Participation: DieBaapiloten: Methods and Projects.* Berlin: Jovis

5. Mallam, S.C., Lundh, M., MacKinnon, S.N. (2017). Evaluating a digital ship design tool prototype: designers' perceptions of novel ergonomics software. *Applied Ergonomics*, 59, 19-26

6. Sanoff, H. (2008). Multiple Views of Participatory Design. *International Journal of Architectural Research, 2* (1) 57-69

7. The Virtual Reality Society. (2017) *What is Virtual Reality?* Retrieved from https://www.vrs.org.uk/virtual-reality/what-is-virtual-reality.html

10. AUTHORS BIOGRAPHY

Winston Pynn holds the current position of naval architecture instructor at the Marine Institute of Memorial University of Newfoundland. He is responsible for teaching aspiring naval architecture technologists all aspects of ship design. His previous experience includes twenty years of yacht design and management. His research interests include the gamification of learning as well as the application of technology in design. He is an avid musician and holds an MBA from the University of Warwick.

© 2017: The Royal Institution of Naval Architects

APPENDIX

4.1 (A) Shipboard Workspace Design Assessment - Bridge and Wheelhouse

The following 5 question are prefaced with: Thinking of the ship's bridges you have worked on, generally:

1. The layout of the bridge aided in proficient bridge operations. Weighted Average: 4.09

Comments

- *Generally yes. SOLAS Ch V sets minimum requirements for design and placement of particular equipment.*
- *Typically, 20ish year old ships that have poor light/dark space assignment.*
- *All ships have that little deficiency that still needs perfection in its design.*

2. The design allowed me to collect and process the information I needed to make decisions quickly and accurately. Weighted Average: 3.91

Comments

- *Generally yes; however, I have experienced some bridge's where the collection of some important information was remote (i.e. radio room) from the central conning/observation area.*
- *Again, 20ish year old ships usually have equipment just tacked on where ever it fits with little regard for usability.*
- *Same comment as above*

3. The design contributed to the effective communications of the ship's officers and crew. Weighted Average: 4.00

Comments

- *Yes. In my experience the manning of the bridge consisted between: 1 officer, 1 officer + 1 rating, 1 officer + 1 rating + captain, 2 officers + 1 rating (or 2), + captain. In most instances the design of the bridge did not hamper effective communications.*
- *Poor layouts make comms difficult on the bridge.*

4. The design was adequate in terms of bridge size, window placement and heights (for visibility) and door (interior and exterior) placement for access.
Weighted Average: 3.73

Comments

- *Generally yes. However, some bridges do not have 360degree visibility or have other blind sectors created by external structures. One ship I worked on the access to the bridge wing was at the rear of the bridge and not at the bridge wing itself (where it should be).*
- *All bridges have different designs, there are some good and some bad, but they are improving with time.*

5. There is nothing I can think of that would have improved the usability of the bridge design.
Weighted Average: 2.82

Comments

- *I worked on one ship where the radars were positioned aft of where the central control console was located. It is not uncommon for the captain or senior officer to navigate the vessel (hands-on) through congested port areas and this particular bridge design did no contribute to effective situational awareness.*
- *Light/dark spaces are often an afterthought (no point putting the coffee pot in a dark space), work flow is often not considered... and alternate work flows are often difficult (OOW as lookout, larger bridge teams, pilot station, planning stations)*
- *Duplication of navigational and control, communications. better lines of site with least obstructions*
- *Monitor for CCTV cameras covering all ship areas*
- *I would make the chart table central to the bridge and/or have the ECDIS displays (which are increasingly replacing charts) to be multi-angled display for easy accessibility.*

The following 4 question are prefaced with: Thinking of the least functional ship's bridge you have worked on:

6. What design feature(s) made it least functional?

Comments

- *Size, placement and types of windows*
- *great distances between port and starboard radars*
- *Poor light dark space, poor flexibility for atypical manning plans. No dedicated emergency response space. I could go on forever...*
- *Unnecessary equipment in the bridge area i.e. Fax machines with info pertaining to another department. This results in noise distraction and clutter.*
- *Old design with poor ergonomics*
- *Inability to see the bow from the conning position (Supply Vessel)*
- *Visibility and Organization of Equipment*
- *The chart room removed to one side of a bridge, making it difficult to access in a hurry or emergency situations.*
- *Chart Room was a separate room*
- *No access to open air bridge wings. Bridge wings were totally enclosed. Washroom located away from bridge. Basically, had to leave the Bridge to use the washroom.*

© 2017: The Royal Institution of Naval Architects

7. Did the design issues impact the crew's ability to collect and process the information needed to make decisions quickly and accurately? Weighted Average: 4.18

Comments
- *Noisy and distracting and caused restricted movement*

8. Thinking of the least functional ship's bridge you have worked on: Did the design flaw negatively impact the effective communication of the ship's officers and crew? Weighted Average: 3.73

Comments
- *Yes, see comment Q.5. Often had to relay information which should not be necessary in scenario described in Q5.*
- *The noise*

9. What would you have done differently in terms of the design?

Comments
- *To me, effect(ive) design is backward design. In other words, include stakeholders to determine the exact specific requirements and outcome. This process would be thorough and include all stakeholders. Once this is finalized the drafting of plans can begin.*
- *Placed the radars (at least 1) where it can be effectively seen from the central conning console.*
- *moved navigational stuff together*
- *See above*
- *Move the equipment elsewhere*
- *Required modification of the bridge lay out and its equipment. Duplication of controls and communications at various stations and better lines of sight*
- *Move the conning position or change the window design.*
- *Better visibility for both OOW and a Bride Watch rating. Better Bridge Resource Management Organization*
- *In hindsight, most definitely*
- *All around bridge*
- *Have access to open air... the ability to walk outside. Have the washroom on the bridge.*

10. Would you have been willing to participate in or consult on the design of that bridge to improve its design, or would you be in the future? Weighted Average: 3.55

Comments
- *My experience is with small fishing vessels so I'm not sure if my insights would be of use.*
- *Sure.*

3.1 (b) Shipboard Workspace Design Assessment - Machinery Spaces

The following 5 question are prefaced with: Thinking of the machinery spaces you have worked in, generally:

1. The layout of the space aided in proficient operations. Weighted Average: 3.88

Comments
- *Good visibility of equipment is use is essential*
- *Smaller ships (Coast Guard ships) did not have large machinery spaces, space is limited for equipment. Tankers with a larger machinery space generally were pretty good, however, there were a few cases where pumps had been installed in areas where access was limited for maintenance. I don't think maintenance for these units had been considered in the design phase.*

2. The design allowed me to collect and process the information I needed to make decisions quickly and accurately. Weighted Average: 3.88

Comments
- *With automation and close circuit TV, great overview is readily available*
- *Information now collected and fed to control room so no need to collect it where it is taken*

3. The design contributed to the effective communications of the ship's engineers and crew. Weighted Average: 3.71

Comments
- *Sufficient telephones in work areas for direct communication*

4. The design was adequate in terms of size, ladder and stair placement and access. Weighted Average: 3.38

Comments
- *Some vessel there could have been improvements, but when you have restraints due to available space, size of equipment and amount of equipment desired by the owner, it is not always possible. Although some time less can be more.*
- *Adequate yes, but posed some problems with respect to moving equipment into and out of the space.*

5. There is nothing I can think of that would have improved the usability of the space. Weighted Average: 2.38

Comments
- *Initial planning and layout of equipment. Designers knowing all the requirements of the clients so that last minute additions are not needed. Also not using up every inch of available space for equipment as*

© 2017: The Royal Institution of Naval Architects

over time when things need to be changed for maintenance reasons or regulatory purposes, still having space to install new equipment would be beneficial.
- *Many spaces could be better utilized if proper storage shelves and cabinets were available for spare parts storage*

The following 4 question are prefaced with: Thinking of the least functional machinery spaces you have worked in:

6. What design feature(s) made it least functional?

Comments
- *Tracing piping runs and valving.*
- *Overcrowding and insufficient clearances with bulkheads and other items.*
- *The lack of lift points above machinery made maintenance difficult.*
- *Little or no space required for maintenance of the equipment. Accessibility of the valves specially in bilge area.*
- *Crowed space. Too much machinery in too small a space. Piping layout was a mess. Too much confusion below the deck plates. Not enough room to lift machinery. Not enough room to get around equipment to do regular maintenance comfortably.*
- *Not enough room and having machines and devices that can not be moved.*
- *Control Room location on tankers. Would have been better suited for watchkeepers had it been it the middle deck of the Machinery Space instead of the top deck.*

7. Did the design issues impact the crew's ability to collect and process the information needed to make decisions quickly and accurately? Weighted Average: 3.88

Comments
- *Decisions no but responding to them, yes.*
- *In many cases, remote reading instrumentation would have provided safer and easier access to required information*
- *On one ship, unless you were skinny, you were not able to get in position to check the oil level in the gearbox*
- *Least functional yes, current no.*

8. Did the design issues negatively impact the effective communications of the ship's engineers and crew?
Weighted Average: 3.00

Comments
- *Not really as other technologies have aided and compensated for that.*

9. What would you have done differently in terms of the design?

Comments
- *Make an equipment more accessible for maintenance.*
- *Better labeling of piping and valves.*
- *Not getting in to specifics for a certain ship - Better planning in equipment layout and not trying to use up every inch of available space in the initial design process.*
- *Lifting points would have been installed as part of build - not as afterthought to meet maintenance requirements*
- *Re-routing pipes and ventilation ducts.*
- *Wherever possible, re-position equipment and bulkheads, increase head room. Less pipework. The use of manifolds in pumping systems rather than individual pipes. Improve view of machinery space from MCR.*
- *Hard to say, better layout of instrumentation would help*
- *Ensure easy access to equipment to perform maintenance.*

10. Would you have been willing to participate in or consult on the design of that machinery space to improve its design, or would you be willing in the future?
Weighted Average: 3.75

Comments
- *Have been involved in the process and found it very useful to improve operations including maintenance.*

© 2017: The Royal Institution of Naval Architects

The Virtual Ship: from design to training

M. Peverero, D. Tozzi, A. Zini, CETENA S.p.A., Italy

SUMMARY

Simulation has always been used during ship design activities in order to verify and demonstrate the feasibility of the choices made. During the years virtual reality has given a boost to the implementation of what we can call virtual prototype: set of models, algorithms, methodologies and best practices to be implemented in the computer as a realistic model of the ship both from a visual and behavioural perspective. The virtual prototype usually is the only prototype of the ship herself and represents an interactive real-time simulation of the ship under design: the Virtual Ship.

In the perspective of reusing models and simulations in the whole ship life cycle, an important success story is the implementation of training simulators which use, at the best, all models, algorithms and simulations implemented and utilized during the ship design phase. CETENA has always been active in simulation based design and virtual prototyping and has now implemented a simulation system which utilize all the achieved experience and can be used for both training and design assessment purposes. The fact that a shipbuilding company is entering directly the ship training arena provides a new perspective in training simulation.

1. INTRODUCTION

The Ship is a very complex system of systems and the design process is consequently particularly complicated due to the great number of aspects which have to be studied and evaluated to make the right choices. As in any other design process, simulation has always been used in order to investigate the effects of design choices on the whole product. Nevertheless model trials and real tests are often performed for be more confident that the ship and her systems are actually behaving in the way they were designed for. In other words, the designer needs a prototype of the product under design in order to verify, optimize and visualize the work. A major problem in ship design is that the ship usually is the only prototype of herself. Large ships are usually stand alone constructions. Even twin ships are usually slighting different. It is quite difficult and very expensive to modify the ship when built.

These assumptions are the reasons why simulations, physic based analyses, computational calculations have been widely used since long time in ship design process. Unfortunately these tools and methods are usually used separately, in different ways and by different experts. Moreover, the final user is rarely involved in the process and interactions are usually limited among the experts/designers themselves and not the final users or client.

The above aspects can be faced applying new technologies aimed at presenting the product being designed in a common and objective way using a computerized representation of it. In other words a Virtual Prototype able to reproduce in virtual reality all the features expected in a real prototype: visual aspects, physics based behaviours, interactive real-time environment.

2. SHIP AND SIMULATION

Computer simulation has always been used in ship design for analysing and evaluating different design solutions.

The availability of computation power and strong software environment for computer graphics have lead to the availability of virtual reality tools which are able to represent the ship and the environment in a more realistic manner from both visual and behavioural point of view.

In particular, interactive simulations represent the most demanding application of today's available technologies: algorithms, 3D design, real-time environment, ship manoeuvring and sea keeping behaviour, virtual reality and immersive visualization.

CETENA has been using this kind of simulation for several years and continues to invest and implement new SW solutions to improve its products. CETENA has developed competencies and experiences in two particular application areas: Virtual prototyping and Port assessments.

Both of them uses a common simulation and visualization framework in a way that any development and improvement in one area is automatically reused also in the other.

2.1 THE LABORATORY

Virtual reality, interactive and distributed simulations are all items which need a dedicated area to be studied, tested and explored. CETENA developed years ago a dedicated laboratory, the VISLab (Virtual Integrated Ship Laboratory), for this kind of research and development. In this laboratory experts from different

© 2017: The Royal Institution of Naval Architects

ship design fields and IT experts created a technology architecture devoted to incorporate and employ available software tools (commercial or developed in-house) normally used for different purposes, with the objective of building a common environment for studying, use and operate virtual prototypes of existing vessel as well as new designs[4].

The architecture implements a distributed simulation framework where different entities can cooperate in a common environment. The underlying approach is HLA (High Level Architecture – IEE1516) which allows for the development of complex distributed simulations [5]. As the ship is an intrinsic system of systems, the HLA approach results well suited for modeling and building federations which can have different levels of details in the implementations of complex behaviors.

In particular the approach allows to create simulation systems which can easily be updated including new or more detailed physical effects and new interacting entities in the synthetic environment.[10]

First target of the VISLab was the implementation of the manoeuvring simulator (SAND) that CETENA uses for port assessments. SAND is the simulation system used in all activities of the laboratory. It is compliant with the HLA 1516 architecture and includes a ship console and high quality visualization system. A dedicated management station is used to set the simulation conditions, in terms of scenario, environmental conditions, ship characteristics. The controls on the console can easily be set through reconfigurable LCD touch panels in order to reproduce those of the simulated ship. During each simulation, all relevant data are stored for subsequent debriefing sessions.

Figure 1: SAND and 3D visualization system

In order to reuse all existing tools and competences and thanks to more than twenty years of R&D in ship design and consultancy for harbor designers and Port Authorities, SAND has rapidly reached a very high level of confidence in the representation of the ship behavior.

The ship mathematical model included in SAND is based on a Maneuverability and Seakeeping model integration, that allows to calculate the six degrees of freedom (Surge, Sway, Heave, Roll, Pitch, Yaw) and has been validated over the years by Port pilots who has used the simulator during the port assessment activities providing CETENA with valuable input and feedback based on their real life experience and sensitiveness in conducting the vessel. Moreover CETENA has a huge database with real data from the sea trial activities CETENA performs onboard ships.

The simulation model is designed to be "open" and configurable for all types of ships (cruise, bulk carrier, petrol, LNG, container ship) and for multiple environmental scenarios (map, sea state, wind, sea current, rain, fog, time of day).

Different types of vessels can be easily configured using a wide range of characteristics like:

- hull data
- engine type (diesel, diesel-electrical, turbine, or every combination of them)
- propulsion (azimuth, fixed/controllable pitch propeller, POD, water-jet)
- thruster
- rudder (simple type, compound butt, under hung deep horn, shallow horn, spade)

The physical model is completely modular allowing the implementation of add-ons to take into account new or more detailed environmental and interaction effects.
The user can choose the environmental conditions through a set of sea states, wind intensity and sea current and the software computes the interactive ship behavior for very accurate and realistic vessel simulations taking into account several effects like:

- shallow water
- navigation in narrow channel
- collisions
- anchors and chains

SAND can be used either as a stand-alone system or as a federate, hence as part of a more complex HLA federation. This can be useful when other simulation entities needs to be considered as actors in the same simulation environment. For this reason a basic standard HLA federation was designed and developed within the VISLab and is structured as follows:

- *Execution Manager federate* is used to set up the whole federation and to specify all initial parameters.
- *Data Logger* records all data sent through the RTI, in order to be able to replay the federation in offline mode.
- *3D Visualization federate* visualizes the simulation scenario in a 3D environment.
- *Environment federate* sets all parameters concerning

© 2017: The Royal Institution of Naval Architects

the meteo-marine conditions, as sea state, current, wind etc

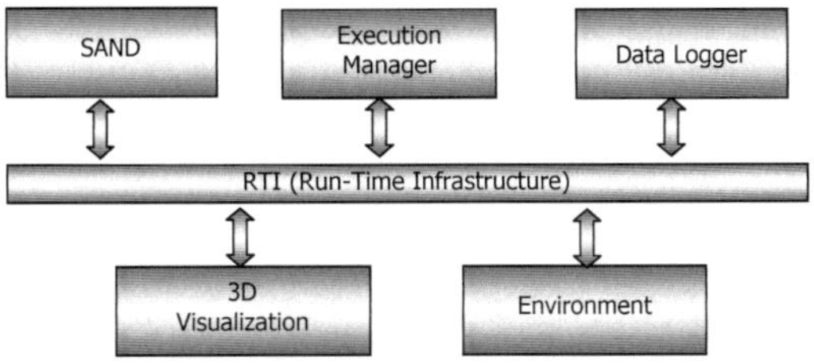

Figure 2: CETENA's basic standard HLA Federation.

In order to be compliant with HLA distributed simulations standards, the above mentioned federation was developed using the VS-FOM (Virtual Ship – Federation Object Model)[9], deriving from the RPR-FOM (Real-Time Platform Reference – FOM).
Multiple entities of SAND can be run in a federation allowing different ships to interact in the same synthetic environment.

3. VIRTUAL PROTOTYPING IN SHIP DESIGN

A prototype is any preliminary working implementation of a product, component or system. A virtual prototype is a computer simulation model of a final product, component or system. A virtual prototype can be used in a design process specifically for: exploring design alternatives, demonstrating design concepts, testing of requirements, satisfaction and/or correctness.[1]
A virtual prototype must have at least the following elements:

- Visualization: virtual reality can be used for navigating inside the model. The user can walk through the complex project viewing exactly how the product will look like. Since the environment in which the user moves is totally virtual and controlled by the computer, the user can easily query physical, topological and geometrical characteristics to any system or component.
- Simulation: a static 3D view or a flythrough can be insufficient to point out problems deriving from the movement of objects on deck or interaction between correlated components. Virtual reality can help in connecting or combining simulation results and real time photo-realistic visualization in synthetic environments.
- Interactivity: the capability to consider the human interaction with the object we are designing, represents a considerable step forward in the ergonomic field or in taking into account stressing operational efficiency during design time. The "man in the loop" represents an important design aspect rarely or hardly taken into account in standard design processes.

Virtual prototyping is a good way to synthesis and visualize all design efforts both for the designer and for the client/final user who can be more directly involved in strategic decisions from the very beginning of the design process since they can view and choose between different solutions [2][3].

Moreover a single simulator can increase its potential and effectiveness if included in a common environment where more simulators interoperate with each other to reproduce the behavior of a more complex system.

Great effort is necessary to combine design expectation (simplicity, integration, etc.) with physical and mathematical modeling and simulation. Some interaction effects which are automatically accounted for in physical models need to be consciously accounted for in numerical modeling. E.g. a problem that we frequently face during simulations is the "hull to hull" interaction which involves interactions between two or more ships and hence ship motion models. Few numerical codes are able to completely model the interaction between two ships, while reality accounts for interactions automatically.

From a strict engineering point of view, a traditional design approach is sufficient and expert people are able to take decisions, but non-expert people, operative people, the final users, are generally not involved in these decisions and cannot bring their experience in the design loop. The Virtual Prototyping approach can facilitate this "man in the loop" approach. The Virtual Ship can be tested by the final user even before cutting metal.

3.1 EXPERIENCES AND APPLICATIONS

With the main target of implementing a virtual prototyping environment able to help the designer and the customer to verify the ship design from an operational point of view, several activities have been put in place to demonstrate that interactive real-time simulation can be used for investigate complex ship operations for verifying performances and receiving feedback for the design phase.

Virtual Ship is the virtual representation of the ship and we use virtual ships in our virtual prototyping environment not only to verify contractual requisites but also extreme conditions scenarios which may be dangerous to verify in reality.

© 2017: The Royal Institution of Naval Architects

The implemented architecture and the open structure of the SAND "federate" are the basic ingredients for implementing complex distributed federation simulating operations involving multiple ships or multiple other entities.

3.1(a) Replenishment at Sea (RAS) Operation

Replenishment At Sea is one of the more complex operations performed by modern navies. The operation is executed in three main phases:

- The vessels approach each other and sail alongside.
- The transfer system is established between the vessels using a cable system and the fuel or other cargo is transferred.
- The transfer system is removed and the ships depart (breakaway).

Basically the operation is focused on the capability of the two ships to navigate parallel for a enough long period of time avoiding collisions and early breakaway.

The two ships are interactively manoeuvrable and the systems reacts in real-time to any change in speed and direction of each ship. In particular the federation can take into account problems deriving from a malfunction in the manoeuvring devices (rudder, propeller...) of one of the two ships in order to accordingly evaluate the capability of the other ship to avoid collision and to continue the parallel course.

In all phases there is a risk of ship collision, especially in high sea states, and when small receiving vessels are involved. A simulation of the RAS operation will allow to determine safe operating conditions and may influence the design of the supply ship, receiving ship and/or transfer rigs.

From an implementation point of view, two SAND federates were involved in the federation in order to simulate both the supply and the receiving ships. In addition to the standard CETENA's federation two federates were developed and inserted in the federation: a *RAS equipment federate* and an *Interaction federate*.

RAS equipment federate was designed for the calculation of the forces applied to the cables used for the fuel and load transferring operation. In case the RAS cables were subjected to a tension exceeding the maximum allowed, federate is able to give a warning representing the unsuccessfully result of the operation.

The *Interaction federate* was designed in order to simulate the forces due to side-by-side navigation; it was developed using mathematical models deriving from CFD calculations and/or towing tank tests and it can be configured to simulate this specific behaviour for different kinds of ships.

Figure 3: Screenshot of the RAS simulation

The implemented simulation system is completely configurable and parameterized so that it can be used to simulate ship-to-ship transfer for:

- Any pair of ships
- Any sea state
- Any type of transfer rig
- Any type of liquid/solid cargo

The used architecture allows the reuse of the federates in other federations.

Some of the federates have been used in international research project having the purpose of validating the simulation results with sea trials data and other towing tanks tests results.

3.1 (b) Landing Craft Unit operations inside amphibious ship

The manoeuvres of a landing craft (LCU) inside the wet dock of an amphibious ship (LPD) and the transfer of a tank from the Landing Craft to the dock of the ship is another example of complex operations involving multiple interactive but independent entities.

The development of the federation takes inspiration and reuse all the already available competences, expertise and software used in the past for studying similar scenarios [8]. In particular stand alone simulation for collision detection of the landing craft with the interior (bottom and sides) of the dock was utilized in the new federation.

All entities in the scenario are completely manoeuvrable interactively: the ship, the landing craft and the tank. This last one was simulated in a very simplified but enough realistic way while the ship and the landing craft are modelled in detail and their behaviours are accurate. Also in this case, the ship is implemented using the SAND federate while the Landing Craft Unit is a completely new mathematical model due to the particular type of hull and propulsion and moreover due to the shallow water effect inside the dock.

© 2017: The Royal Institution of Naval Architects

Particular attention has been paid to hydrodynamics effects:

- the sloshing effect which is very complicated due to the fact that it is generated by the LPD Ship and is characterized by the volume of the water in the wet dock. Experimental model tests have been performed to tune the mathematical model.
- the behavior of the LCU behind the Ship. The LCU travels in a wave field that is the superposition of the LPD's stern generated waves and the sea waves modified by the presence of the LPD.

Figure 4: Screenshot of the Landing Craft simulation

3.1 (c) Vertical Take Off and Landing (VTOL)

Take-off and landing operations involves different issues related to the interaction between the ship and the air vehicle.

One of the problems is associated to the forces during touch down: the ship moves and air vehicle approaches choosing the best moment for touching the deck. The forces involved are the ones coming from the ship motions and the ones coming from the drop of the vehicle. Especially in heavy sea state the sum of the forces can be very high and can damage the air vehicle landing devices [6].

This effect has been studied in one of the first federations developed within an international group studying the best landing period for a helicopter to land on a frigate [7]. Besides impact effects, another important aspect to consider is the turbulence generated by the superstructures in the vicinity of the landing spots. This aspect involves a heavy interaction between fluid dynamic calculations (CFD) for simulating the flows around the structures (Air Wake) and the relative position of the two moving entities: the helicopter and the ship. These two separated effects were already studied in CETENA in the past and we took advantage of this knowledge developing the relative tool. A federation that simulates the interactive scenario of an helicopter landing on a ship dynamically taking into account in real-time of the air-wake effects and the relative movements and forces between the two moving entities was designed and implemented.

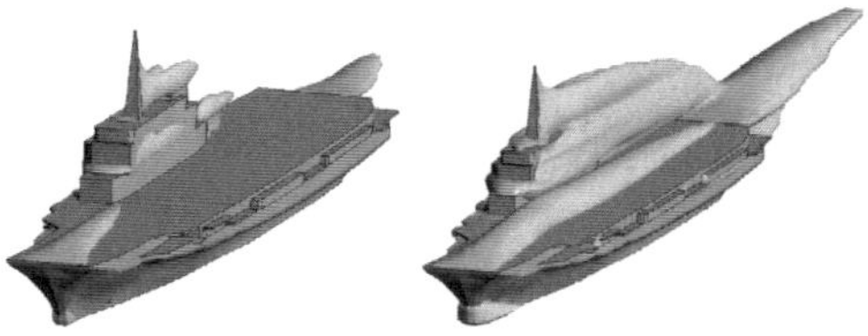

Figure 5: CFD results for VTOL simulation

The developed system was composed by the standard basic federation for the ship with the addition of two new federates: the helicopter and the air wake.

The air wake federate is basically a sophisticated interpolator of pre-calculated values coming from offline CFD processing. Depending on the relative position from the ship, the speed, the wind direction and other physical parameters, the federate determines the flow pressure and speed in a certain point.

Figure 6: Screenshot of the VTOL simulation

Since the federation was developed in compliance with both the VS-FOM and the RPR-FOM, any Helicopter federate compliant with the above mentioned FOMs can be easily integrated.

3.1 (d) Onboard handling of ammunition or spare parts

The technologies and the infrastructure developed can also be used for studying less complicated interactive simulations such as the ones needed for verifying handling operations onboard. In this case the physics of the motion can be less sophisticated but the 3D model and the collision algorithms should be more detailed and performing.
In many cases CETENA used VISLab for verifying handling of loads onboard, checking spaces and manoeuvres, bottlenecks and obstructions adding also ergonomic investigations of field of views, necessary forces and working clearance analysis.

© 2017: The Royal Institution of Naval Architects

This type of investigation benefits from the full immersive environment with head mounted display or from the use of real dimension projection on large screens.

Due to the close cooperation with the design team, this type of analysis should be very quick and need to consider that many changes can be requested during the design phase. This means a great effort in quickly changing/updating the 3D model due to multiple design modifications.

While for internal handling the ship motions may not be important, they become essential for studying movements of heavy entities on the fly deck or in the dry dock. In this case several types of assessments can be performed: from checking available spaces for movements to the evaluation of the forces applied on the suspension of the airplane etc.

3.1 (e) Ergonomic Assessment

The availability of an explorable 3D model immersed in a virtual environment provides the possibility to model the interaction between humans and equipments. This allows for safety and ergonomic issues to be easily examined.

Anthropometrics considerations can be addressed by, for instance, varying the size of the crew models in the simulations. The software will indicate if the human postures, required for a particular sized crew member to operate and maintain equipment, are safe, comfortable and physically realisable without undue strain.

The virtual reality environment permits the user to check fields of view and to estimate the visibility of portions of the ship for particular operations since the outside scenario is also available and the user can see what is happening outside the windows. This is a very important feature which differentiate a physical mock-up from the virtual one.

Lights can be taken into account in order to verify reflexions and internal illumination during the whole day and in different positions.

4. SIMULATION FOR PORT ASSESSMENT

One of the main activities and services CETENA performs with the simulator is the evaluation of ship manoeuvres in port. This risk assessment task is very delicate since it is used to decide strategic statements: port layouts, restrictions on dimensions of ships, how many tug are needed, extreme weather conditions for safe manoeuvres etc.

The decisions derived from the port assessment activities have usually great impact on budgets and investments - public and private. Hence the simulation must be as objective and affordable as possible.

CETENA has a long experience in this type of simulation and the manoeuvring algorithms developed are very stable and detailed, taking into account several physical effects which are essential in port manoeuvres. The mathematical model constantly being developed and updated, adding new effects and optimizing the existing ones.

Since the manoeuvres in the port are performed by experienced pilots or captains, the simulator has been subject to intense and continuous validation activities.

The port assessment activity requires also the calculation and collection of numerous information related to the manoeuvre (fuel consumption, number of changes in the direction…) which can be used for post processing the interactive sessions performed.

5. THE VIRTUAL SHIP IN THE WHOLE LIFECYCLE

The development of simulation applications and models costs time and money. If the return of the investment is not very clear or appreciated by stakeholders, the simulation may not be performed extensively as it should but could be reduced to small aspects with a limited use of software and 3D models implemented. In this case the impression generated is that simulation is too expensive and can be avoided.

This aspect is particularly visible in the ship design phase where usually the trend is to reduce costs and optimizing consolidated procedures. Especially if the customer doesn't require any virtual representation of the ship being designed. It is still complicated to build virtual prototypes primarily because they rarely are reused for different analyses and verifications due to several reasons: the design process is not ready yet, the people don't realize or understand the possible benefits of using virtual prototyping - tradition is more important than innovation. In this type of scenario, it is important to discover other applications of technologies and methodologies investigated for virtual prototyping also in other phases of the whole ship lifecycle.

Algorithms and software developed for building the virtual ship can be efficiently reused for training purposes. In this case the costs and time are more comprehensive and the return of the investment is more clear to the customer. The simulator is used for training the crew and hence optimize the ship procedures reducing costs and increasing safety of the ship in exercise.

© 2017: The Royal Institution of Naval Architects

Using the virtual ship for training purposes brings positive feedbacks also to the virtual prototyping application in the design phase benefiting from all the features which has to be implemented for training.

This produce a proactive effect: all upgrades for training will be available also for virtual prototyping and all details and new implementation in the design phase will be reused in training.

This aspect is particularly important for a company which is both involved in training and in design activities.. In one case indeed the training simulator is always guaranteed to have the top of the mathematical models and physic algorithms because of the continuous algorithms development and the stringent user requirements in the design activities. From the other side, having a more complete, immersive and coherent simulation infrastructure and user interface with real equipment and complete console layout inherited by the training simulator helps the virtual prototyping applications to take into account also that aspects and features which are not essential for the analysis to be performed, but which make the user and the final operator more confident and involved and hence optimising the design

One example : in virtual prototyping and also in port assessment a radar system is not necessary, but its presence in the console used for the analysis activity, makes the user more involved and more concentrated as if being really onboard. This "side effect" can be very important because it means that the actions and the reactions of the person involved in the assessment activities is more coherent with the real ones achieving a more realistic evaluation of the operation investigated with the virtual prototype.

6. VIRTUAL SHIP FOR TRAINING

The idea to reuse Virtual Prototyping experience obtained for design also for training purposes, find the boost with the profitable interaction with a company mainly focused on training simulators. The fact that both approaches (design and training) in terms of architecture, software development assets and application philosophies are close brought to the definition of a shared architecture which actually takes into account all the best practices and experiences from both areas.
In particular for training, other synergies can be found, for instance:

- serious games are very near to the analysis of handling entities onboard and both have common aspects
- noise analyses are a key component for echo sounder
- infrared calculation is the basis for IR vision
- takeoff and landing, air wake and fluid dynamics analysis are very important input to the helicopter simulator.

This dual use complementarities is a key point for encouraging the use of simulations for both training purposes and design activities.

6.1 THE ARCHITECTURE

Using simulation the single person and moreover the entire crew will train themselves in the operations and situations they will or could experience onboard before actually being on board the ship.

In this perspective a comprehensive simulation approach for naval training has been developed originating the framework proposed by CETENA called MANTA (Multipurpose Advanced Naval Training Architecture).

MANTA integrates different training methods and tools in a common scenario to fully cover the most advanced training needs: multi task personnel, team cooperation and coordination, human behaviour analysis in controlled realistic and stressful situations.

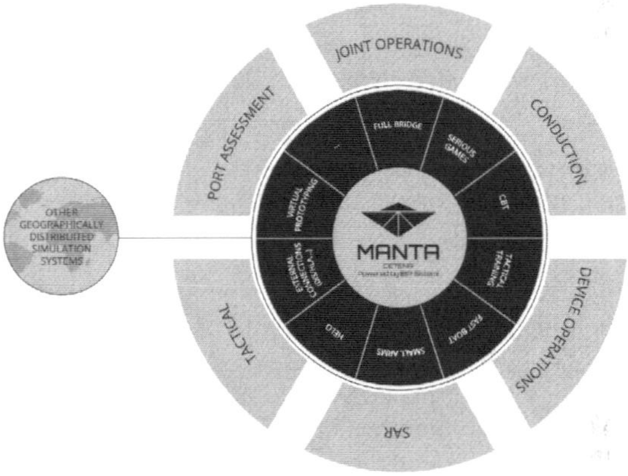

Figure 7: MANTA Architecture

MANTA represents the ultimate training simulation environment which follows the single operator together with the complete crew from the beginning to the end of training activities.
In particular the MANTA architecture covers the following different levels of training:

- Educational level
- System level
- Ship level
- Integrated team level

© 2017: The Royal Institution of Naval Architects

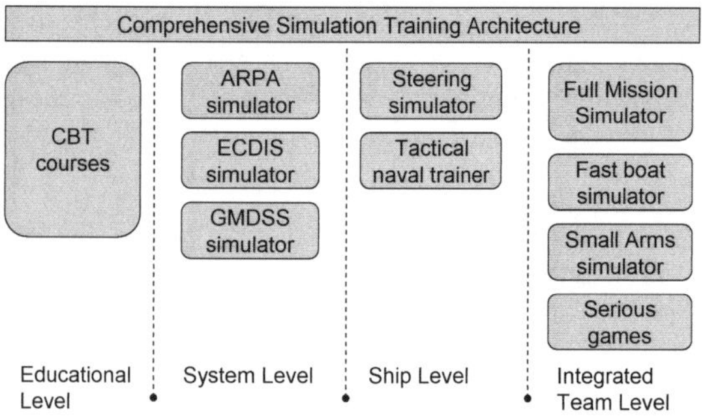

Figure 8: Levels of training covered by MANTA

All the tools, simulation applications and simulators in the MANTA framework are strictly connected to each other in order to benefit as much as possible from all the information, data, scenarios and exercises. In this way the user is always facing a known environment and a standard way to interact with the training systems avoiding misunderstanding and loss of time in understanding how to use the software rather than learning what is necessary.

MANTAs integrated architecture provides the possibility to use different simulation systems in conjunction with others so that many trainees can share the same scenario and can cooperate or oppose in a common exercise. This is achieved by using a ICT infrastructure which favours interactions and integrations of all possible tools in a common platform.

Particular importance should be paid for the Full Mission Simulator which is the most complex component of the MANTA infrastructure since it is on its own a set of simulators and instructor stations strictly interconnected and highly reconfigurable.

The standard Full Mission Simulator infrastructure is usually composed by the following components:

- Primary Bridge,
- primary Combat Operation Centre,
- secondary Bridge,
- secondary Combat Operation Centre
- Instructor Consoles
- Ship Control Centre
- Debriefing room

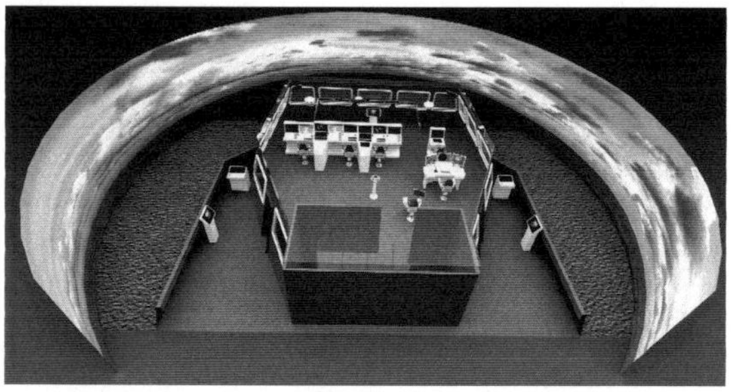

Figure 9: Concept of Full Bridge Simulator

In order to perform training for complex naval operations, other entities can be integrated in the Full Mission Simulator. In particular a simplified Helicopter simulator is foreseen as an important entity to be accounted for. In this perspective the above structure simply has a new simulator which interacts in the common scenario with all the other components of the whole infrastructure.

With this architecture is possible to configure different types of exercises: from the single bridge or single COC operation to a complex scenario with several ships and helicopter cooperating.

As the Full Mission Simulator is part of the MANTA overall architecture, even more complex scenarios can be implemented. For instance the full mission simulator during a RAS operation and be connected with the Serious Games simulator which implement the behaviour of the crew performing the necessary operations while the two ships are moving in parallel. In this case the people at the Serious Games consoles and the people in the bridges can see each other while performing their operations in the same scenario.

7. CONCLUSIONS

The use of a complete "living" virtual prototype even in the conceptual design phase represents a new way of facing the ship design process. Present technologies, software, computers and know-how help the ship designer to test operational requirements other than classic design aspects.

This will lead to a better and more cost efficient ship since several problems and optimizations can be taken into account from the very beginning of the design avoiding expensive modifications in later steps of the process.

In the near future probably more operational tests and procedures will be requested by the client in order to verify ship operational requirements. Virtual Prototyping seems to be an excellent way to respond to the question, especially when the cooperation and interaction of several objects has to be studied and tested in a unstable environment like the sea.

Ship designer and ship owners will appreciate the availability of tools able to display accurately the final product operating in a synthetic environment with detailed physical phenomena simulations in an early design stage. Valuable suggestions from different field experts can be made available so that the final product can take into account more detailed and functional aspects otherwise obtainable only after the ship has been built.

© 2017: The Royal Institution of Naval Architects

Currently all the necessary technologies and competencies for building the Virtual Ship able to take the "missing" place of the real prototype is available. There has been a significant progress in the recent past. However work still has been done to produce a reliable, validated and accredited virtual prototype of the Ship. Two major obstacles remains : costs and culture.

The first one is mainly evident when simulations are seen only as a tool for design purposes only and no other applications are foreseen. In this case, yes, simulation is a cost and the return of the investment seems to be very limited. Making simulations and simulator reusable for other purposes and for other projects is a solution that overcome the problem. In this perspective if most of the models and algorithms developed in the design phase can be reused for instance in a training environment, the ROI is shifted in this second phase which can start before the ship is built. It is a benefit for the customer who can train the crew in advance but also for the shipbuilder who has the possibility to expand business also in the after sales period of the ship lifecycle construction an even more strong relationship with the ship-owner.

The cultural obstacle is more complicated. Why I need to enhance the way I'm doing my job? What is the benefit of making things more sophisticated? Why using a realistic visualization also in real time?

People who usually play computer games have become accustomed to certain visual and interactive performances and recognize the effectiveness. Unfortunately usually these people are not the decision makers of today but they will be the ones of tomorrow.

If visualization and interactivity are important features, what distinguish a Virtual Ship from a ship in a video game is the realism and the reliability of the behaviour which should be extremely physic based and comparable with actual one. In other words it is important that the experts become equally or even more confident in the behaviour of the Virtual Ship compared to the confidence in the results of the detailed analysis tools used today.

An effective and accredited validation of the simulation is one of the most important challenges in Virtual Ship implementation.

In CETENA we are facing all these challenges finding immediate solutions to some of them and caring out long term research for the others.
The mixture between virtual prototype and training simulations is foreseen as a viable solution to these kind of issues.

7. REFERENCES

[1] ANDERT E. P. Jr., Morgan D., "Collaborative Virtual Prototyping and Test", *Naval Engineers Journal, pp. 17 – 23*, November 1998

[2] NILSSON, PER-OLOF, "The digital product model – a valuable tool for shipbuilders and ship-owners", *Scandinavian Yearbook of Maritime Technology, pp. 37-38*, 2000.

[3] KANERVA M., "Virtual Reality – 4D Product Modelling Tool for Efficient Shipbuilding Process."

[4] TOZZI D., ZINI A., NECRISI R., LOMMI A., PERRA F., GUAGNANO A., "Distributed Simulation in Total Ship Design for Naval Vessels", *SIW*, 2004

[5] DEPARTMENT OF DEFENSE, "High Level Architecture", 1988.

[6] FERRIER B., LE BIHAN O., "The use of simulation tools in the calculation of aircraft-ship interface optional limits", *20th Congress on the International Council of the Aeronautical Sciences*, Italy 1996

[7] BUDDE E., "The Multi-National HLA Federation Development Supporting Simulation Based Acquisition", *International Conference on HLA Simulation Methodology and Applications*, 2005

[8] ZINI A., ROCCA A., RAFFA M., COSTA R., "Building a Virtual Ship", *HMS*, 2000

[9] NATO STANAG 4684 "Virtual Ship"

[10] TOZZI D.,VALDENAZZI F., ZINI A., "Use of HLA federation for the Evaluation of Naval operations in ship design", *MESAS*, 2014

© 2017: The Royal Institution of Naval Architects

9. AUTHORS BIOGRAPHY

Aldo Zini holds the current position of head of simulation and virtual prototyping business unit at. He is responsible for developing business and strategic activities in the field of simulation either for design purposes and for training. His previous experience includes the design and development of the VISLab and related experiences in virtual prototyping and simulation based design. He participated to NATO and NIAG groups on the subject.

Massimo Peverero. Senior Researcher, responsible for harbours manoeuvring simulations and ship manoeuvrability models. Graduated in Mechanical Engineering in 1998 at Genoa University, he has been working for CETENA since 2000. He was involved in the Risk Analysis and Simulation Department and later for Simulation and Virtual Reality Department. In particular he was involved in several project dealing with: Virtual Prototyping and Mathematical Models for Ship Manoeuvrability Simulations. In his company, he was a member of the team that developed the new CETENA's Manoeuvring Simulator SAND. Co-inventor and coordinator of the team for Italian Patent N° 0001427579 "Simulatore di navigazione e manovra" – dec. 2014.

Davide Tozzi holds the current position of senior software engineer at CETENA. He is responsible for the development and integration of software tools in simulation based architectures. His previous experience includes the design and development of the VISLab and related experiences in virtual prototyping and simulation based design. He participated to NATO and NIAG groups on the subject.

© 2017: The Royal Institution of Naval Architects

THE IMPLICATIONS OF UNINHABITED VEHICLE TECHNOLOGY ON NAVAL FLEET STRUCTURES AND NAVAL SHIP DESIGN

N. Kouriampalis, **R. J. Pawling**, **D. J. Andrews**, Design Research Centre, Marine Research Group, Department of Mechanical Engineering, University College London, UK

SUMMARY

Uninhabited vehicles (UXVs) are becoming an important tool in naval warfare. Such a technology has provided an entirely new capability for a navy. By projecting power in an affordable way through the use of UXVs, the exposure of human life to military threats can be significantly reduced. The Royal Navy is exploring the employment of UXVs for a variety of applications. However, the concept of a fleet of such vehicles operating together from a mothership that hosts them and supports their overall operations during a mission scenario, is a challenge. This paper describes ongoing work at University College London (UCL) Design Research Centre (DRC) to develop an analysis tool. This is intended to provide ship designers with a framework for early stage assessment of the impact on the capability of a UXV mothership of the various facilities to host and support a fleet of UXVs.

NOMENCLATURE

AAW	Anti- Air Warfare
ASW	Anti-Submarine Warfare
ASuW	Anti-Surface Warfare
CAD	Computer Aided Design
CN3	Communication and Navigation Network Nodes
CONOPs	Concept of Operations
C2	Command and Control
DBB	Design Building Block
DRC	Design Research Centre at UCL
ESSD	Early Stage Ship Design
GA	General Arrangement
GCS	Global Combat Ship
LAR	Launch and Recovery
LARS	Launch and Recovery System
LCS	Littoral Combat Ship (US Navy)
MIW	Mine Warfare
OPV	Offshore Patrol Vessel
OA	Operational Analysis
OR	Operational Research
PSD	Preliminary Ship Design
QT	Queueing Theory
RHIB	Rigid-Hulled Inflatable Boat
UCL	University College London
UXV	Uninhabited (Air, Surface or Underwater) Vehicle
UAV	Uninhabited Air Vehicle
USV	Uninhabited Surface Vehicle
UUV	Uninhabited Underwater Vehicle

1. INTRODUCTION

In recent years, Western navies have been attracted to the concept of deployment of UXVs in naval operations [1] for a wide range of roles, including Communication and Navigation Network Nodes (CN3), detection and identification of threats, Mine Warfare (MIW), Anti-Submarine Warfare (ASW), Anti- Air Warfare (AAW) and Anti-Surface Warfare (ASuW). Western navies are actively exploring the possibility of deploying UXVs from surface ships. The Littoral Combat Ship (LCS) consists of a large class of surface combatant in the U.S. Navy that can accommodate and support a few UXVs, which are deployed for littoral zone operations [2]. The R.N. Type 26 Global Combat Ship (GCS) class of multi-mission warships are being procured for the Royal Navy. They will have an adaptable mission space for some UXV modules, providing the capability to deploy Uninhabited Air (UAV), Surface (USV) and Underwater (UUV) warfare vehicles/modules [3]. In contrast to the limited discrete-type UXVs capability of both LCS and T26 GCS, the concept of hosting, deploying and supporting a large fleet of UXVs from a mother surface vessel can be seen as a new capability. Consequently, there is a need to investigate the ship design implications of this new operational concept.

The deployment of a fleet of UXVs in a theatre of operations requires the presence of a mother vessel (or several vessels) capable of hosting many such assets and supporting their overall operations. This means a design approach needs to be used to explore how a fleet of UXVs at a concept level could be integrated into the design of a mothership. This paper aims to describe how the UCL originated Design Building Block (DBB) approach to Preliminary Ship Design (PSD) can be combined with the development of an analysis tool, which would quantify the implications of a fleet of UXVs in the supporting role of a mothership. This facilitates a more holistic, numerically and architecturally focused investigation of potential UXV motherships at the Early Stage Ship Design (ESSD), while the ship design is still fluid. This is presented through the use of a recently developed ship concept design tool and the application of a proposed method for evaluating the impact of the UXVs and their associated facilities on concept ship designs.

© 2017: The Royal Institution of Naval Architects

2. THE DEPLOYMENT OF UNINHABITTED VEHICLES IN NAVAL WARFARE

2.1 UNINHABITED VEHICLES IN SUPPORT OF NAVAL OPERATIONS

UXVs offer a number of potential benefits to naval operations, as they act as a force multiplier, while reducing the operational risks to which personnel might be exposed [4]. The future vision of naval operations involves the application of autonomous vehicles that could potentially reduce personnel related costs, by reducing the extent of facilities for humans, life support and protection [5]. This would keep humans out of dirty (dealing with hazardous material), dull (long duration) and dangerous (exposure to hostile environment) tasks [6] [7]. Additionally, removing the human from the vehicle can significantly reduce vehicle size, and also lower its signatures [1]. Advances in computer software, speed and processing power, improved sensor technologies, enhanced communications, better image-processing capabilities, efficient and miniaturized propulsion systems giving longer endurance and range, could replace outdated technologies (that previously impeded the wide use of UXVs). This is likely to lead to an even greater deployment of UXVs in a broad range of naval operations [8].

2.2 DEPLOYMENT OF UNINHABITED VEHICLES FROM NAVAL SURFACE SHIPS

A major role of a navy is to be able to provide credible and sustained combat power from the sea, when and where it is required. This vision now can involve naval forces employing autonomous vehicles in sea strike and shield missions [8]. A UXV fleet would be able to carry out tasks, providing an increased capability to naval missions, if supported from a mothership. The mission profile would be determined by the appropriate Concept of Operations (CONOPs). This could define the number and types of UXVs required to operate in the mission theatre, for the mission to be successfully accomplished. Distinct UXV types bring specific capabilities and performance abilities into the theatre of operations, including payload, endurance, range and speed. However, the capabilities the UXVs deliver for a mission would be driven by the needs to meet the relevant mission scenario. Consequently, the mission needs translate into technological requirements for both UXVs and support vessels [5].

The overall operations of UXVs during a naval mission refer to all stages, from the deployment of the vehicles to the mission theatre to their recovery on-board the host ship, including the required support while they are deployed. The off-board tasks involve the support of the UXVs while on-station, i.e. monitoring, identification of potential problems, Command and Control (C2), sensor information interpretation, communications and interoperation, authorisation of weapon systems use or other mission related activities, refuelling/recharging and ordnance on-station. Operations on-board the mothership could include troubleshooting and repair, maintenance, refuelling, rearming, stowage, pre-mission checks and Launch and Recovery (LAR) [1] [9].

A UXV mothership would be the core for a group of daughter vehicles integrated into the overall naval force, supporting not only their stowage, maintenance and LAR operations, but also significantly contributing to C2 and communications to manage them [10]. Although, stowage space demands, on-board handling systems, maintenance and refuel equipment, and ordnance for the UXVs systems have implications on the design of a mothership, the successful overall UXV operations strongly depend on the capabilities of delivering and recovering the vehicles from the operational area [11]. Thus the ability to launch and retrieve a fleet of UXVs would be a critical aspect of such operations [12], with a major impact on the size, configuration and performance requirements of the potential mother vessel.

The development and integration of sensors and control systems that provide data associated with LAR requirements, as well as development of robust Launch and Recovery Systems (LARSs) are necessary for successful and safe LAR operations [13]. There are numerous potential ways of deploying and retrieving UXVs from naval ships, including catapult launcher, crane and stern ramp. Undoubtedly, there are numerous considerations in the selection of the most appropriate LARS employed for with a UXV fleet system from a surface ship. These issues include space requirements, appropriate location to compensate for wave-induced motions in limiting sea state conditions, adequate freeboard, operational availability, deck loading, well-trained personnel, power needs and cost implications [14].

2.3 INTEGRATION OF LAUNCH AND RECOVERY SYSTEMS INTO THE MISSION BAY OF A HOST SHIP

The ship designer is responsible for integrating the capabilities revealed from Operational Analysis (OA) studies, into a balanced design solution, which also has to meet the ship's performance requirements. A dedicated UXV mothership would have the ability to deploy a range of UXVs, creating a network of interconnected UXVs that can be adjusted to the changing operational needs [10]. This could be achieved by providing flexible and functional mission compartments on the host ship, capable of accommodating LARSs, stowage spaces, movement systems for handling the UXVs, support equipment and any associated personnel facilities.

A flexible compartment on-board the host ship, the mission bay, could provide the mission capabilities, including LAR, maintenance and repair, refuelling, ordnance and UXVs stowage. The integration of mission

© 2017: The Royal Institution of Naval Architects

related equipment into a mission bay and subsequently the integration of the mission-oriented spaces into a mothership design is a naval architecture challenge, particularly in meeting the ship's operational and safety criteria [15].

The implications of the UXVs on a host ship design and the resultant interfaces between the ship and the UXVs are of vital importance, and hence need to be considered early in the design of the host ship. To enhance safety, multi-mission capability, mission effectiveness and affordability, the mission-oriented spaces need to be integrated into the host ship's architecture taking into account integration issues. Such mothership integration issues include:

- LARSs;
- Stowage spaces and handling systems, with a lifting allowance required to allow for the vehicles to be moved inside the mission bay [16];
- Maintenance and repair equipment;
- Refuelling and rearming equipment;
- Relevant crew;
- Relevant communications systems [15];
- Capability of a mothership to host and support a UXVs fleet, i.e. ship performance, including speed, range, endurance, seakeeping, manoeuvrability and stability.

A highly functional mission bay, should also be able to provide some enhanced versatility (flexibility) which would allow the mothership to perform a broad range of roles by embarking a wide array of assets, thus allowing the ship to re-role (i.e. using an adaptable mission bay) in order to suit emerging tasks. A flexible mission bay will also have the space provision for future upgrades [17]. However, trade-offs might be necessary in order to come up with a viable solution, as enhanced functionality and adaptability attributes are highly likely to drive up the cost and complexity of a design and could result in non-affordable solutions [15].

The ship designer will need to examine numerous mission bay configurations considering the mission bay occupancy, usage and functionality aspects [15]. It is also essential that the mission-oriented spaces are developed in parallel with a UXVs mothership as a single integrated entity able to deliver the required capabilities. Moreover, more than one mission bay might be integrated into a single mothership, depending on the number and types of UXVs carried on-board, as well as on the extent of LARs requested to handle multiple UXVs being deployed or recovered simultaneously. The designer has to see a mothership as an essential component of the wider naval force/fleet that should be able to continuously provide a certain level of naval capability. Therefore, it is essential to systematically develop and investigate a large number of design alternatives, perform cost estimations and carry out trade-off studies on requirements and affordability.

3. PRELIMINARY NAVAL SHIP DESIGN

3.1 LIMITATIONS OF THE TRADITIONAL PRELIMINARY NAVAL SHIP DESIGN

Brown and Andrews incorporated "Style" as the fifth characteristic of a ship design in their "S^5" term adding it to Speed, Seakeeping, Stability and Strength (S^4) [18]. Style is a cross-cutting concept, where one decision explicitly influences a wide range of solution areas. Consciously addressing style enables the designer to accommodate uncertainty and conceptually connect it, not just in the form of "hard" knowledge (e.g. structural standards) but also "soft" knowledge (e.g. guidance on internal arrangement from a range of perspectives) [19].

An example of a top level (or macro) stylistic issue is the adoption of unconventional ship types. Compared to monohull ships, multihull mothership solutions could offer more dynamically stable options for LAR operations (e.g. catamaran), or increased speed and improved stability (trimaran), as well as providing upper deck flexibility by being able to accommodate a more extensive mission bay to support the operations of a given UXVs fleet. Since style is inherently cross-cutting, such style changes explicitly affect a wide range of solution areas. Thus, the decision on the ship type to be adopted can influence a large range of both overall and detailed design features. For instance, aspects such as LARSs are strongly influenced by the overall ship type. Hence an early evaluation of the UXVs impact on mothership requires the ship architecture to be at the centre of the initial synthesis of naval ship design.

The traditional numerically based approach to naval ship design synthesis delays architectural modelling in focusing on the readily quantified aspects of speed, seakeeping, stability, strength and the combat system demands. Thus, the initial ship sizing ends up only considering the internal arrangement once an initial numerically balanced design has been achieved [20]. The sizing can ignore the disposition of the different space requirements within the overall envelope and hence is bound to limit design variety. The examination of key design-related issues, such as the integration of mission bay systems, are likely to be postponed to later design stages, despite the fact that warships are generally not weight or space limited, but architecturally driven [21]. Therefore, the outcome is a design solution space which is constrained by the numerical-related issues traditionally seen to drive the initial size of a combatant, without sufficient emphasis on those more architecturally driven aspects. Consequently, any design issues that emerge once these aspects are subsequently incorporated could well result in additional cost and schedule delays [22].

© 2017: The Royal Institution of Naval Architects

3.2 THE DESIGN BUILDING BLOCK (DBB) APPROACH AS THE PREFERRED APPROACH TO EARLY STAGE SHIP DESIGN

Andrews' proposal to integrate ship architecture with the traditional numerical ship sizing [23] was followed by the demonstration of "creative synthesis" [24]. From this work, a new approach to ESSD was developed, namely the DBB approach [25]. The basic idea behind the DBB approach is for the designer to identify ship's functions and sub-functions and link those functional elements required to discrete, physically realisable, ship elements (DBBs) required to meet those functions. Each building block contains geometric and technical attributes regarding the functions of that block [25]. The designer positions the DBBs required to achieve primary functions of the vessel, yielding a 3-D block layout around which a hull form could be wrapped with an early configuration of the desired S^5 and combat system features [23].

In addressing the limitations of conventional numerical methods, the DBB approach is holistic, combining the traditional numerical sizing with an initial architectural description to produce more broadly balanced solution. The resultant design description consists of a 3-D representation and the numerical description of the whole ship (i.e. the Master Building Block) that results from the ship designer's arrangement of building blocks. Typical information in the Master Building Block is the overall ship capabilities (e.g. speed, endurance, seakeeping, stability, powering, longitudinal strength, manoeuvring) and ship characteristics (e.g. weight, space, centroids) [26]. The resultant ship design can then be modified, as the DBBs are increasingly subdivided and the necessary level of naval architecture balance is achieved [25]. The DBB approach encourages exploration of novel solutions from the commencement of ship design, as well as enabling the ship designer to reflect many of the user's operational requirements in a initial graphical description that would not be addressable in a purely numerical description [20], hence contributing to requirements elucidation [27]. Figure 1 gives a summary of the DBB approach.

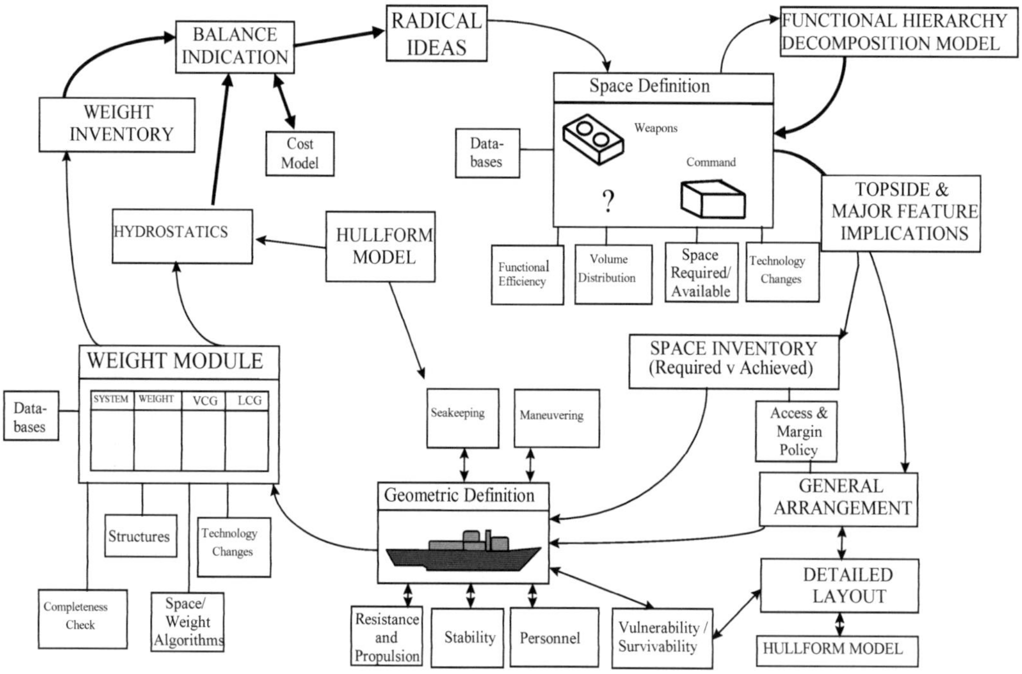

Figure 1: The Design Building Block approach to early stage ship design applied to surface ships [25]

© 2017: The Royal Institution of Naval Architects

4. THE USE OF A NEW SHIP CONCEPT DESIGN SOFTWARE TOOL

4.1 UCL SHIP CONCEPT DESIGN SOFTWARE TOOL

The DBB approach to ESSD was developed and implemented in an industry standard PSD tool following the rapid developments in computer graphics capabilities. This was into the fully developed and working PSD CAD system, PARAMARINE, by the addition of the SURFCON module for surface ships [26], which was developed by QinetiQ - GRC for UCL. This addition gives the ship designer the ability to use other Paramarine modules to assess a resulting DBBs defined design's naval architecture aspects. However, the use of sophisticated 3D Computer Aided Design (CAD) models in ESSD research has several drawbacks. These include: restricted availability, potentially unnecessary high level of detail plus excessive precision, which may be inappropriate for just elucidating the requirement in the early stages of ship design, and additionally a high learning and familiarisation overhead for new users, such as students [28].

In order to address these issues, an online ship design toolset, implementing the DBB approach in ESSD, is being currently developed at UCL for use in ESSD teaching and research [28]. This online tool is being developed in JavaScript, where the ship general arrangement model uses a cellular representation of the arrangeable space on the ship (in 2D, rather than SURFCON 3D approach). JavaScript was chosen because it provides the ability to develop an online tool with interactive elements, high responsiveness and accessible anywhere where there is an internet connection. However, it is not intended that sophisticated naval architecture tools, such as PARAMARINE, will be abandoned, as they offer superior analytical capabilities, such as carrying out damaged stability analysis, seakeeping and longitudinal strength analysis, should these be required in ESSD for specific option studies. Moreover, detailed arrangement drawings, if required, will have to be finalised using specialist CAD tools, such as AutoCAD [28].

4.2 DESCRIPTION OF THE UCL DESIGN TOOL

The ship concept design UCL tool comprises three principal components; namely the "Ship Data", the "DBB Data" and the "General Arrangement" (GA). "Ship Data" and "DBB Data" are Comma Separated Value (CSV) files, whereas "GA" provides a demonstration of the design space solution, as well as information regarding the ship performance:

- Ship Data: contains information with regards to major ship dimensions, watertight bulkhead positions, number and position of hull and superstructure decks, maximum speed and powering margins.

- DBB Data: provides a tabular description of DBBs defining the solution space, including the location in the design area, weight, space and volume requirements. The data for the DBBs are structured according to the UCL functional group breakdown, i.e. Float, Move, Fight/Operations and Infrastructure. An example DBB Data file is presented in Table 1.

Table 1: The UCL Ship Design tool's "DBB Model" component, showing tabular definition of the location, size and weight of the Design Building Blocks [28]

Group	Classification	Name	xFwd	xAft	yCentre	ySize	zHeight	zBase	FixedWeight_lightships	VariableWeight_cond1	VariableWeight_cond2	FluidDensity	AreaRequired	VolRequired
FLOAT	1301	mooring_sp_fwd	102	86.5	0	16	2.7	8.8	21.63	0	0		2.1	583
FLOAT	132	cl	96.5	93.5	0	16	5.2	3.5	12.81	0	0			71.4
FLOAT	1302	mooring_sp_aft	7.5	0	0	16	2.7	8.8	0.746	0	0		7.35	333
FLOAT	16	sturcture	50.5	50.5	0	0	0	6.5	1357.162	0	0			
FLOAT	1011	access_1	82	76.5	0	2	2.5	14	3.421	0	0			
FLOAT	1012	access_2	66.5	56.5	0	3	2.5	11.5	9.329	0	0			
FLOAT	1013	access_3	54.5	22.5	6	2	2.7	6	22.29	0	0			
FLOAT	1014	access_4	54.5	22.5	-6	2	2.7	6	22.29	0	0			
FLOAT	102	paint	51.5	51.5	0	0	0	7.4	27.3	0	0			
FLOAT	112	degaussing	51.5	51.5	0	0	0	7.4	3.385	0	0			
FLOAT	15	workshops	63.5	54.5	4.75	6.5	2.7	6	7.74	0	0		56.921	
FLOAT	231	liferafts	51.5	51.5	0	0	0	9.8	2.31	0	0			
FLOAT	1111	boat_bay_port	66.5	54.5	4.75	6.5	5.2	8.8	7.665	0	0			365.992
FLOAT	1112	boat_bay_stbd	66.5	54.5	-4.75	6.5	5.2	8.8	7.665	0	0			365.992
FLOAT	80	g_b_margin	51	51	0	0	0	7	148.344	0	0			
MOVE	33	fuel_sys	51.5	51.5	0	0	0	1	13.941	0	0			
MOVE	451	mmr_1	66.5	54.5	0	16	5	1	24.075	0	0			795.489

© 2017: The Royal Institution of Naval Architects

The data is fed into the online ship design tool, which then generates the GA of the ship, through the designer's disposition of the DBBs in the design space, as well as displaying the position of the designer determined main transverse watertight bulkheads. The DBBs in the general arrangement are assigned labels and visible outlines for identification, as well as a colour according to the DBB functional breakdown using the DBB functional breakdown colour code (Float - Blue, Move - Yellow, Fight - Red, Infrastructure - Green [26]). An example GA for a UCL ESSD study of a UXV capable Offshore Patrol Vessel (OPV) [16] is reproduced in Figure 3.

Figure 2 shows the simple approach used to represent DBBs in the design space. The cells are placed on a square grid; a group of cells is allocated to each DBB and they are assigned a colour to identify them as belonging to the specific DBB functional group [26]. Each space also has a visible outline and nametag. Figure 2 also displays the deck outlines (blue lines) and the location of the main watertight bulkheads (red lines). The JavaScript-based ship design toolset allows the ship designer to generate a 2-D ship configuration at an appropriate level of detail for the early stages of design process, without the modelling effort of (say) SURFCON.

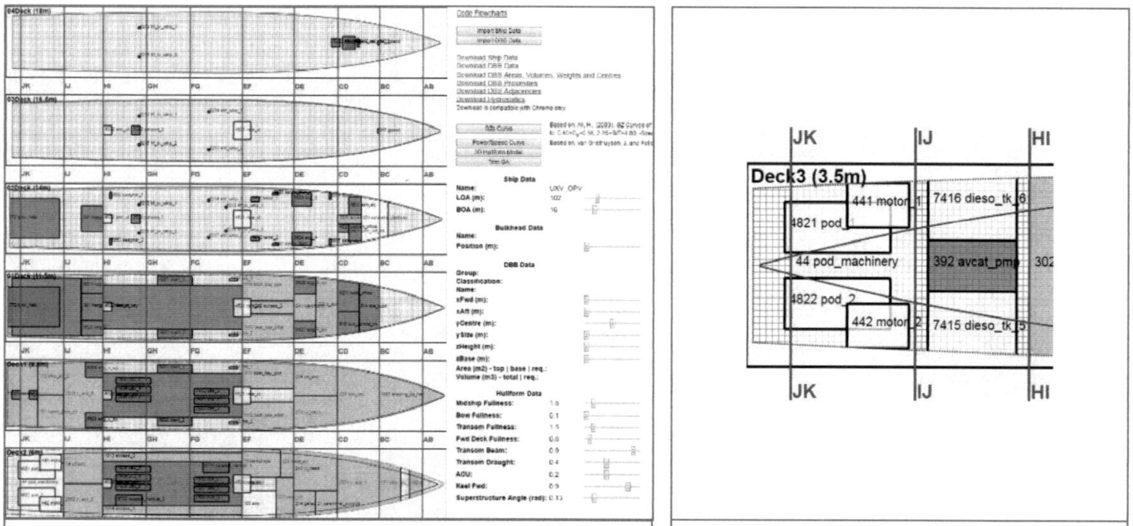

Overall Graphical User Interface (GUI) with controls at right | Close-up showing cellular representation

Figure 2: The UCL Ship Design Tool's "GA" component, showing the layout of a typical early stage design model for a UXV OPV study and a detail of the simplified arrangement [36]

4.3 UCL JAVASCRIPT SHIP DESIGN TOOL CAPABILITIES

The UCL developed ship design toolset has been used to define an initial ship configuration, as well as providing feedback to the ship designer. This data can also be forwarded to external tools, in order to assess the resultant ship performance (e.g. damage stability). The tool essentially allows the naval architect to assess potential ship options with respect to intact stability and powering. Moreover, using network analysis [28] and 2D modelling, the toolset enables the quick extraction of a range of spatial properties, such as block adjacency. These additional geometric metrics may be used for innovative network-based layout analysis [29], or fed into the simulation-based analysis of personnel movement [30]. Thus the UCL tool can provide the basis for investigating additional design aspects early in the design process.

The main scope of concept ship design is not to provide fully worked up design solutions, but to provide insights into requirements through exploring a number of potential design options. For a UXV mothership these could be: UXVs fleet compositions; LARSs; and mission bay arrangements. Such aspects could be examined for their impact on the ship design in achieving believable and reasonably feasible mothership solutions, consistent with ESSD levels of research. Given that the toolset can provide an ESSD architectural model necessary to capture design drivers, it can therefore enable the investigation of the impact of a fleet of UXVs in the early stage design of naval vessels.

© 2017: The Royal Institution of Naval Architects

5. A PROPOSED APPROACH FOR ASSESSING AND EVALUATING THE IMPACT OF A FLEET OF UNINHABITED VEHICLES ON THE DESIGN OF A MOTHERSHIP

5.1 THE BASIS FOR A MOTHERSHIP DESIGN APPROACH

As noted earlier, future naval capabilities may be delivered by a system composed of a variety of conventional, unconventional and uninhabited vessels and vehicles. Whereas traditionally a set of naval capabilities could be provided by a single unit and thus designers could focus on that design, the evaluation of UXV system concepts will require a broader, more holistic approach to designing the complete system. Just as a concept designer must trade-off model resolution for speed and flexibility in the early stages of design, so "system-at-once" design methods and tools will have to be used to make similar trade-offs to inform early stage decision making. The DBB approach can be utilised to assess the performance of individual mothership and daughter craft designs, by addressing the capability of the mothership to accommodate the operations of a fleet of UXVs. The approach can also examine the impact the UXVs fleet and the required facilities, systems and on-board services to support such fleets has on the size, configuration, performance and cost of the resultant mothership solution. The ESSD studies will investigate how certain on-board facilities, systems and resources, required to support potential UXV fleet scenarios (i.e. fleet compositions), as well as how architectural decisions (e.g. integration of mission oriented spaces), could affect the ability of a mothership to host and support such fleets during naval operations.

5.2 FACTORS INFLUENCING THE SELECTION OF THE UXV SYSTEM DESIGN APPROACH

To date considerations of likely ships able to carry and support the operations of a significant number of UXVs has been limited to broad CONOPs studies. A more comprehensive approach could be able to capture the overall operations of UXVs and then integrate a fleet of them into the design of a mothership. CONOPs studies could provide estimates of the numbers and types of UXVs, required to bring distinct capabilities into a theatre of operations. However, an OA restricted just to the study of CONOPs scenarios is not sufficient to explore mothership options. It is necessary to take into account the integration of a fleet of UXVs into the design of a mothership. This must address the required equipment, support systems, space, stowage, complement and on-board services demands on a mothership necessary to host and support a fleet of UXVs.

Therefore, the issue of accommodating the above features in the ship, so it can deliver its primary UXV support function, is the key to its design. Given the immaturity of the UXV fleet concept and LARSs for such a sizeable number of vehicles, the physical demands on the mothership are likely to be speculative. This is because it is not possible to capture technologies, such those of UXVs and LARSs that are under constant evolution. However, the proposed mothership design approach is not aimed at delivering definitive or absolute answers, but to provide the ship designer with an approach that is able to model a wide range of potential UXV systems and the supporting equipment and resources. This is regardless their technological maturity, if insights with regards to ship impact (i.e. size, configuration, performance, cost and ability to support a UXVs fleet) are to be obtained.

Even in concept exploration, significant decisions have to be made and such investigations need to apply the best approach possible to address the problem of integration of a fleet of UXVs into the design of a mothership. Hence, a comprehensive OA approach should couple CONOPs studies with the investigation and identification of integration issues of a fleet of unmanned assets into the design of supporting vessels. This would provide appropriate functional requirements (e.g. extent of LARSs, C2, maintenance, refuelling, rearming, space, etc.) for the mothership to be able to support the overall operations of a fleet of UXVs during a set of mission scenarios.

5.3 PROPOSED QUEUEING THEORY APPROACH

For an indicative UXVs mothership design, the overall UXVs operations that must be carried out have to be examined. Those are likely to reveal necessary physical interfaces between the unmanned assets and the mothership, as well as non-intuitive improvements to the overall operations. The latter can then be taken into consideration in designing the mothership. The likely operations of a fleet of UXVs supported by a mothership can be represented as a number of interconnected nodes, forming a network system (i.e. process represented as a network). This is shown in Figure 3, where each node represents a task from the list of the UXVs overall operations. A number of physical and discrete items, i.e. UXVs, would flow through each node of the sequence of activities to perform appropriate tasks, in the context of the overall UXVs operations. These will include equipment and systems (facilities/resources) for a mothership to be able to host a number of UXVs on-board and support their overall operations during a given mission scenario.

© 2017: The Royal Institution of Naval Architects

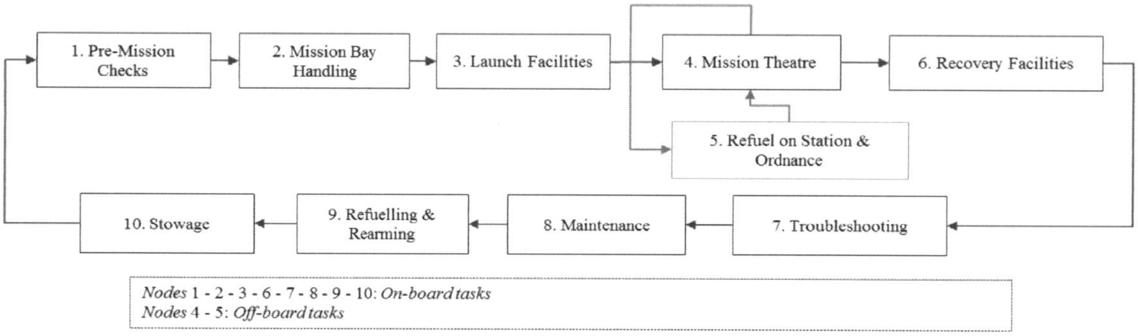

Figure 3: Representation of UXVs overall operations supported by a mothership as a network system

The design based investigations of a UXV mothership demands innovations derived from analyzing, modeling, designing and eventually implementing new systems [31]. Thus a novel quantitative framework for modelling the network of operations of a fleet of UXVs supported by a mother vessel has been proposed. The approach uses Queueing Theory (QT), which is a mathematical representation of waiting lines, i.e. queues. This has previously been adopted in Operations Research (OR), with output used in decision-making on the type and number of resources to be allocated, in order to provide service within certain time limitations [32]. Additionally, QT provides valuable information regarding the design and schedule of queueing systems and is also considered a powerful tool towards evaluating the performance of queueing systems [33]. This metrification of performance provides the ship designer with measures of effectiveness of a potential mothership, in quantifying the ability of the ship to support the operations of the UXVs fleet. The metrics can also be used to compare the capabilities of various potential mothership solutions to support a fleet of UXVs.

QT models are mathematical models of real life systems. They have been employed successfully and proved to be beneficial in both the manufacturing and service domains, including transport, airports, production lines, telecommunications and the internals of computers [33]. In particular, QT can be used to model systems that provide a service and may represent any system where customers arrive looking for service of some kind and depart once the appropriate service has been provided. Such models can be constructed, to predict performance measures of a queuing system, such as the queue length (i.e. number of customers waiting in the queue) and waiting time, which would also contribute to understanding the behaviour of such systems [33]. A simple queuing model can be described as having two distinct elements: the waiting area and the service area, as is schematically shown in Figure 4. Such a model may be used to represent a number of customers (i.e. UXVs) that arrive at a waiting area (i.e. space requirement), where they queue up if all servers are busy and eventually get served from an available server (i.e. facility or resource) and leave after obtaining the required service (i.e. a list of tasks undertaken within UXVs overall operations described in Figure 3).

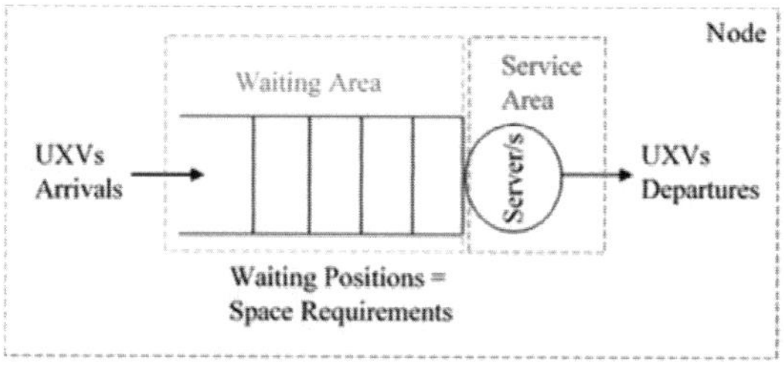

Figure 4: Simple queuing model

It is regarded as quite common for customers to require service from more than one facility. Usually in real-world systems, customers can be served by more than one node and the nodes are arranged in a network structure. This network structure is considered as a collection of service nodes that are interconnected with a path/route [31]. Once a customer has been served at one node, it can then either leave the network or join another service node and queue up for the appropriate service. Therefore, each node of the UXVs overall operations network system, shown in Figure 3, can be described by a simple queueing model that, along with the other nodes, forms a queueing network of interconnected nodes.

One of the major areas of interest in investigating the overall operations of a fleet of UXVs, to ascertain the mothership design impact, is that of LAR. LARSs are seen as providing a significant aspect of mothership design, since they are the necessary means to deploy and retrieve a UXV fleet capability during naval operations effectively. They are likely to have major implications on the design and performance of the resultant mothership (e.g. well dock LAR arrangement mothership).

Different LARSs can achieve significantly different LAR rates and subsequently affect the host ship design. LAR of a number of UXVs could be modelled by the

© 2017: The Royal Institution of Naval Architects

suggested queueing network, where key elements are the UXVs (i.e. customers), as well as the appropriate LARS (i.e. facilities). LARS can be regarded as resources that need to be allocated on a mothership. Moreover, a range of innovative LARSs might be represented using simplified, concept models, and incorporated in queuing network models in order to assess their impact on LAR performance and subsequently on the mothership design.

The information, extracted from QT modelling, seen as most appropriate to make decisions regarding the ship synthesis can be divided into (i) information regarding the pertinent node performance and thus relevant to ship ability to support the operations of a UXV fleet, defined through the node performance obtained from QT, and (ii) ship impact in terms of the necessary equipment, systems and resources for the ship to support such a fleet of unmanned assets (i.e. space requirements, mothership size, layout and performance). The first category can inform ship capability to support the operations of a UXV fleet and can be defined through the node performance obtained from QT: -

Node Performance:

- Total time spent at a node (i.e. facility) by a number of vehicles requesting the appropriate service (e.g. LAR). This is a direct measure of total time required to provide the appropriate service to a number of vehicles (i.e. requests) at the relevant facility (e.g. LARS);
- Node throughput (i.e. number of vehicles processed per unit time). Greater values of throughput show an enhanced (i.e. faster) node performance (e.g. LAR), as more vehicles can be processed through a node in the network system per unit time.

Ship impact can be thought of as:

- Number of vehicles in a queue at a node is a measure of the space (i.e. waiting area) required in front of the facility that provides the appropriate service;
- Number of servers, is a measure of the space required for the facilities (e.g. LARSs) that provide the appropriate service (e.g. LAR). Moreover, a number of servers that might be maintained for redundancy purposes should be taken into account.

The service rate of a node within a queueing network system could be improved by either allocating more servers of identical type, or by employing faster servers (i.e. lower processing time). It is noteworthy that the service time of a facility is determined by the specifications upon which the particular equipment has been designed. Both of these options would enhance the node's performance. The designer must determine the necessity and the extent of the alterations at the underperforming nodes, which subsequently impact on the mothership design. For example, incorporation of a greater number of LAR facilities, in order to enhance the performance of the equivalent nodes (i.e. faster launching and recovering of a fleet of unmanned assets) could be indicated by the metrics extracted from the model. An enhanced LAR capability may result in decreased length of queues, increased throughputs and shorter times spent at these facilities. Such a change in the LAR capability of a mother vessel would then affect the latter's size, configuration and performance.

The queuing network model allows the ship designer to alter the input parameters of the model and assess their impact on the metrics extracted from it. Based on these, network variations would enable the ship designer to assess whether the performance of the resultant network is satisfactory. Thus, the QT tool contributes to the decision-making process with regards to the mothership design in terms of the required facilities for the mothership to deliver its primary functions, see Figure 5. In general, smaller times spent at each node along with higher throughput values and low queue lengths indicate a mothership design of enhanced support capability. The metrics resulting from modelling a network along with the feasibility studies of potential alternative mothership solutions, can inform cost analysis and technical assessment of the various solutions. This would allow the naval architect to obtain a more holistic picture of the design solution space for such an innovative operational concept.

© 2017: The Royal Institution of Naval Architects

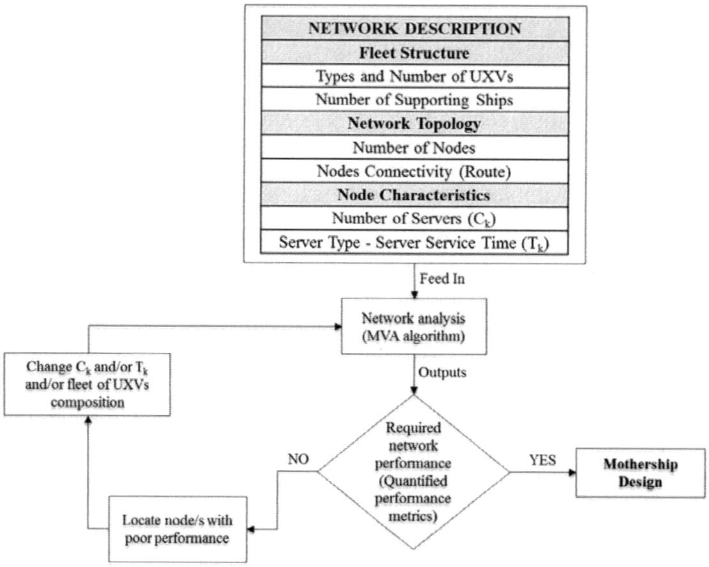

Figure 5: Decision making process through queueing network model

Variations of the network structure and the input parameters, in order to capture different UXV systems and their supporting facilities and resources, would translate into different performance measures of the system. These are likely to lead to distinct mothership designs, as variations of the network nodes might have impacts on the interfaces between the mothership and the UXVs and subsequently affect the mothership size, configuration, performance requirements and cost.

The interfaces between the mothership and UXVs could act as major design drivers for the mothership, given that: -

- They have major implications on the spatial requirements of the mothership, due to the equipment footprint and the necessary clearances for accessibility and operations, e.g. LARSs;
- They would impact the ship layout and size, if the necessary equipment is to be integrated into the mothership configuration (e.g. stern or midships UXVs mission bay(s));
- They would affect the performance of a mothership in terms of the appropriate ship capabilities to support the UXVs' operations, through any as S^5 implications to achieve LAR operations.

The interactions between a mothership and the supported UXVs fleet, as shown in Figure 6, can be modelled as nodes of a queueing network system, allowing the ship designer to determine the supporting capability of a mothership, as well as assessing their ship impacts.

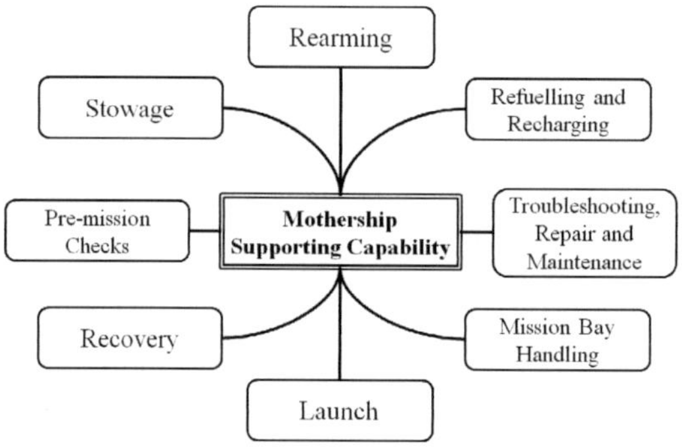

Figure 6: Mothership-UXVs fleet interactions

5.4 DEMONSTRATION OF QUEUEING THEORY TOOL CAPABILITIES AND INTEGRATION TO THE DESIGN OF MISSION ORIENTED SPACES OF A MOTHERSHIP

This section provides a comprehensive example of a network system, representing the operations of a fleet of USVs during a mission. The network has been analysed using QT. The QT tool provides insights regarding the mothership-UXVs fleet interactions. This particular demonstration model involves two different types of USVs, termed Type A and Type B. Vehicles of Type A and B are hypothetical types of USVs: similar size to a RHIB (i.e. 11 m); and smaller non-disposable USV (i.e. ~ 3 m), respectively. Moreover, the USVs fleet composition (i.e. number and types of USVs) brings different capabilities into the theatre of operations to meet the appropriate CONOPs scenario requirements. Two distinct USV types demonstrate that

© 2017: The Royal Institution of Naval Architects

the QT tool is capable of modelling complex UXV systems consisting of different types of uninhabited assets, where different UXV types require distinct service facilities. The example, outlined below, describes the deployment of Types A and B vehicles in the theatre of operations to perform allocated tasks in possible CONOPs scenarios.

Despite the fact that the number and types of vehicles is suggested by the requirement criteria based on a particular CONOPs scenario study, the developed code is able to model any reasonable range of UXV fleet compositions. This is sensible given the concept of UXV fleets is speculative, as future UXV technology, LARSs and plausible mission scenarios cannot be assumed beyond any broad level of detail. Thus for possible scenarios, rough estimates of the number and types of unmanned assets in a fleet of UXVs have been postulated and this information will be fed into the proposed QT tool.

The on-board nodes represent the facilities and resources that must be integrated into the mothership configuration, while the resource levels reflect the maximum number of vehicles that can be simultaneously served by the servers at each node. The network model under study follows the generic structure, shown in Figure 3, and it has been adapted to employ three generic types of LARSs, namely crane and davit systems for side deployment plus stern ramp. A portion of the overall model of the example network used in this demonstration is shown in Figure 7.

Using the QT code developed in FORTRAN, an example queueing network model has been constructed and analysed. The results produced from running the QT tool for the analysis of the network system for a mission scenario that requires a fleet of 60 USVs (i.e. 20 USVs of Type A, 40 USVs of Type B), give the measures of performance of the nodes in the network. The results presented in the current section focus on the performance of the nodes that model the launching process from a mothership, since the employed LARSs were considered to have the most significant impact on the mothership design (as discussed in Section 5.3). This also demonstrates the effect of the QT tool results on the performance of the nodes in the network system and how this information can be employed in the ship design decision-making process. Analogous results can be drawn for the rest of the nodes in the network system, as those nodes modelling on-board services also impact the design of a mothership.

The capabilities of the QT tool and of its results incorporation in the design procedure of a mothership are achieved by displaying the performance of the relevant launching nodes and that implication on the design of a mothership. The following issues have been considered: -

i. The implication of the number of USVs on the design of a mothership, for a given set of LAR facilities on a mothership;
ii. The implication of the **number** of LAR facilities on the design of a mothership, for a given fleet of USVs.
iii. The implication of the **type** of LAR facilities on the design of a mothership, for a given fleet of USVs. Distinct types of LAR equipment provide different LAR/service time per vehicle, using data obtained from the manufacturers. More enhanced and sophisticated LARSs are likely to be quicker processing by LAR per vehicle.

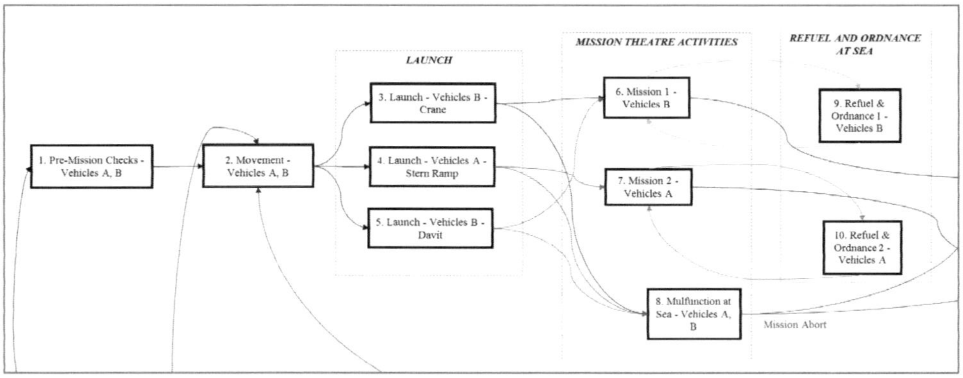

Figure 7: Section of an example network representing USV operations showing the vehicles flows at each node

© 2017: The Royal Institution of Naval Architects

The performance of the launching nodes, and thus the launching capability of the mothership, is quantified via the metrics extracted from the QT. Among these metrics the number of USVs in queue at each of the launching nodes are indicators of the space and on-board service demands on the mothership. These metrics are:

- The total time spent to launch a number of USVs by employing a particular LARS;
- The number of vehicles (i.e. queueing and service activities) processed per unit time at a particular LAR node (throughput);
- The number of USVs waiting in the queue in front of a particular launching facility.

5.4(a) Implication of the number of USVs

For a set of launching facilities (i.e. 1 crane, 1 davit and 1 stern ramp), the QT code was run for four cases: 15, 30, 45 and 60 uninhabited assets present in the network system. For each of those four cases, each launching facility (i.e. crane, davit and stern ramp) has been set to serve/launch 5, 10, 15 and 20 USVs in total, respectively. The time to launch each vehicle from the crane, davit and ramp has been set 8, 7 and 5 minutes, respectively.

The total time to launch the pre-set number of USVs from each launching facility is shown in Figure 8. The time the vehicles spend in each launching node is the summation of the service time plus the queueing delay/time. The queueing delay is the time the vehicles wait in front of the launching equipment in order to be served. As the vehicles flow through the network system they might have to wait in the queue until served, as a consequence of insufficiency of servers.

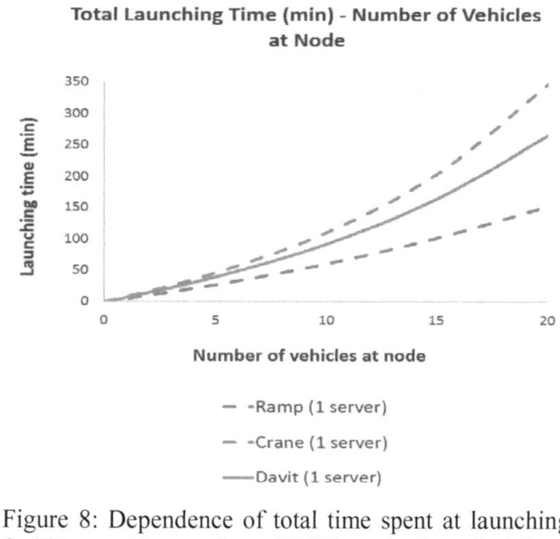

Figure 8: Dependence of total time spent at launching facilities on the number of USVs to be launched from the particular facility for a test mission scenario

For a given number of servers, the time spent at a node increases, due to the incurred queuing delay as more and more vehicles spend time in the queue, since the service time of a particular server type is constant. For the same number of servers at launching nodes (i.e. 1 crane, 1 davit and 1 stern ramp), the total time to launch the same number of USVs from the LARSs differs due to the distinct service time of each launching facility type (i.e. $T_{Crane} > T_{Davit} > T_{Ramp}$).

The throughput data obtained for the launching nodes of the network system show that the number of USVs processed per unit time at each of the relevant nodes (i.e. crane, davit and stern ramp) increases as the number of USVs present in the network increases. However, as the number of vehicles in the network increases, the node throughput increases at a slower and slower rate, due to the fact that (for a given number of servers/facilities available at a node) congestion occurs with the increasing number of vehicles present at the node [34]. Figure 9 shows the dependence of the throughputs for the launching nodes on the number of USVs present in the network.

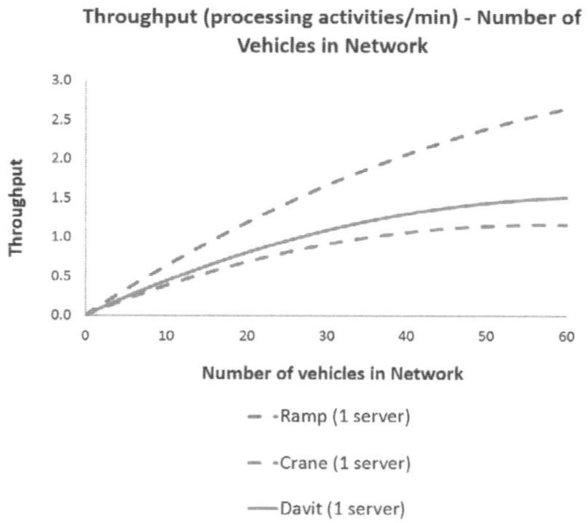

Figure 9: Dependence of the throughput at launching nodes on the number of USVs present in the network system for a test mission scenario

Each node of the network system that represents on-board mothership tasks needs space: - (i) for the facilities providing the appropriate service (node service area - see Figure 4) and (ii) for the vehicles waiting at queue to be served from the particular server (node service area - see Figure 4). Since the space requirement for the node service area is defined from the employed launching facility (i.e. crane, davit and stern ramp) the node space demand for the waiting area in front of the service area, where the vehicles queue and wait to be served from the appropriate facility, is defined by the number of vehicles in the queue at the relevant nodes. The queue length at each launching facility is shown in Figure 10. The number of vehicles in queue, for a given number of servers available at the node, increases with the number of vehicles to be launched from the particular facility.

© 2017: The Royal Institution of Naval Architects

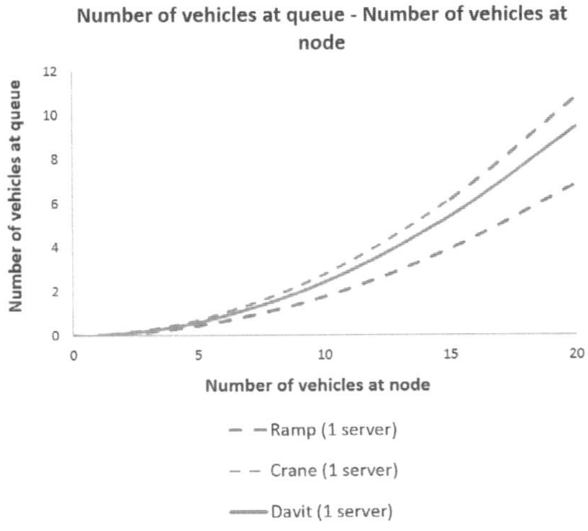

Figure 10: Dependence of the queue length at launching nodes on the number of USVs to be launched from the particular facility for a test mission scenario

5.4(b) Implication of the number of launching facilities

For 60 vehicles present in the network system, where each launching facility, i.e. crane, davit and stern ramp, has been set to serve/launch 20 of the 60 USVs, the QT code has been run for two facilities (i.e. LARSs) available at each launching facility, i.e. 2 cranes, 2 davits and 2 stern ramp. Increasing the number of facilities to provide the appropriate service, i.e. launch, requested by the number of USVs at each of launching nodes in the network, the performance of the pertinent nodes and thus the mothership's supporting ability improved. This was verified by the metrics obtained from running the QT tool for twice the number of servers available at the launching nodes, as shown in Table 2.

In the case of deploying and retrieving a fleet of USVs of 60 assets by employing 2 cranes, 2 davits and 2 stern ramps, the size, configuration and performance of the resultant mothership would be distinctly different from using a single crane, davit and stern ramp. As shown in Table 2, the performance of the mothership (i.e. "Design 2") with respect to the launching capability is enhanced compared to mothership "Design 1", i.e. less time to launch a fleet of 60 USVs, faster processing LARSs and fewer vehicles in queue (i.e. less space required for USVs queueing/piling up until launched). However, the employment of a greater number of LARSs results in greater space requirements on the mothership and likely growth in ship size and cost.

Table 2: Launching nodes performance for twice the number of LARSs

Mothership	Node	LARS Type	Number of LARS	Total Launching Time (min)	Throughput (number of USVs processed/min)	Number of USVs in Queue
Design 1	Ramp	A	1	148	2.7	7
	Davit	A	1	255	1.6	10
	Crane	A	1	329	1.2	11
Design 2	Ramp	A	2	103	3.9	1
	Comparison to Design 1 (%)			-31	44	-86
	Davit	A	2	148	2.7	2
	Comparison to Design 1 (%)			-42	72	-80
	Crane	A	2	172	2.3	2
	Comparison to Design 1 (%)			-48	91	-82

5.4(c) Implication of the type of launching facilities

For 60 vehicles present in the network system, where each launching facility, i.e. crane, davit and stern ramp, has been set to serve/launch 20 of the 60 USVs in total, the QT code has been run for different slower types of single facilities (i.e. LARSs) available at each launching facility, i.e. 1 crane with 9 min service time per USV, 1 davit with 10 min service time per USV and 1 stern ramp with 11 min service time per USV. Employing less sophisticated LARSs of higher service time per vehicle (T_k) to provide the appropriate service (i.e. launch) requested by the number of USVs at each of launching nodes in the network, the performance of the pertinent nodes and thus the mothership's supporting ability decreased, as shown in Table 3.

Table 3: Launching nodes performance for different types of LARSs

Mothership	Node	LARS Type	Number of LARS	Total Launching Time (min)	Throughput (number of USVs processed/min)	Number of USVs in Queue
Design 1	Ramp	A	1	148	2.7	7
	Davit	A	1	255	1.6	10
	Crane	A	1	329	1.2	11
Design 3	Ramp	B	1	405	0.9	12
	Comparison to Design 1 (%)			173	-63	71
	Davit	B	1	516	0.8	13
	Comparison to Design 1 (%)			102	-51	30
	Crane	B	1	665	0.6	14
	Comparison to Design 1 (%)			102	-51	27

© 2017: The Royal Institution of Naval Architects

Different types of cranes, davits or stern ramps would result in different service times initially based on manufacturer specifications. It is likely that more sophisticated and technologically advanced LARSs would offer faster means of deploying and retrieving USVs. In the case of launching and recovering a fleet of USVs consisting of 60 vehicles by employing a slower type of crane, davit and stern ramp, it is likely that the size, configuration and performance of the resultant mothership would be distinctly different, due to the different LARS footprint and on-board service requirements. As presented in Table 3, the performance of the mothership with respect to the launching capability is reduced. Thus when mothership concept "Design 3" is compared to mothership "Design 1", where the former provides more time to launch a fleet of 60 USVs, slower means of processing USVs for deployment and more vehicles in queue, entails further space demands for the USVs queueing/piling up until launched.

The QT tool can be employed to analyse different UXVs fleet operational scenarios to gain an insight into how the design drivers would affect the design of potential UXV motherships. Changes in the performance of a queueing network can be traced through the extracted metrics and are likely to impact the mothership supporting capability. Such changes might result in distinct mothership solutions, due to the implications for the required equipment, systems, spaces, complement and on-board services. Consequently, the QT tool could provide the information, in terms of appropriate metrics and supporting data, to assist in making design decisions. This information would consist of the number and types of facilities that should be allocated in each node of the network system, and thus allow the definition of the appropriate services that a UXV mothership should then provide.

6. CONCLUSIONS AND FURTHER WORK

This paper presents ongoing work at UCL to develop a numerical framework that, along with the DBB ship design approach, would allow the examination of variations of UXV fleet scenarios for assessing mothership drivers for potential future UXV motherships. A more holistic mothership design approach has been proposed, allowing the ship designer to make better informed decisions on the mothership design in ESSD.

The mix of types and total number of servers at a node represent the equipment and crew needed to perform certain tasks, in order to support the USVs throughout a mission. The allocation of servers depend on the number and type of vehicles to be served and must also meet particular performance criteria, such as the maximum time to carry out the appropriate task. Each server within the QT representation of the USV support network will correspond to some combination of physical equipment, crew and procedures. Thus, the network can be used to analyse the performance of a proposed USV support system. Alternatively, if the number and type of servers can be set to be sufficient to meet the relevant performance requirements of the USV system, then the model can be used to derive the requirements in terms of equipment, space, crew and on-board services for the system. This information can then be used as an input to the mothership design process.

Based on the proposed mothership design approach, various fleets of UXVs may be assessed in terms of the equipment, systems, spaces, complement and on-board services, required for a mothership to effectively host and support a fleet of uninhabited assets. Such variations would impact on the mothership supporting capability and might result into distinct mothership design solutions, due to the implications on the size, configuration and performance of the ship. The capability and hence the effectiveness of a mothership to support the operations of a fleet of UXVs can be assessed by the results obtained from the developed QT tool. Since fleet variations reflect distinct QT network structures, that subsequently result in different mothership solutions, the various mothership design options could then be compared on a cost - effectiveness basis.

In conclusion, QT modelling could act as a means for assessing the mothership capability, indicating the operational effectiveness of each mothership design. The integration of QT tool with ship design tool would provide the ship designer with insights into the space demands (e.g. UXVs fleet composition, number and types of on-board servers/facilities). It would also assist in deriving relevant ship's performance requirements (i.e. S^5 plus combat systems capability) and an appropriate configuration for a mothership to successfully carry on-board and support the operations of a given fleet of UXVs.

Further work will include the use of QT modelling as part of the ship design process. As QT tool is capable of assessing the performance of a UXV network system relatively rapidly, the ESSD tools and methods to be used should be capable of exploiting this. Thus, the use of ship concept exploration models and a means of down-selections, as modifications of the DBB approach, will be investigated. This would lead to developing a toolset with an appropriate balance of detail, focusing on the mission areas on-board. Concept exploration requires the generation and assessment of a large number of design variants. To generate these design variants, an approach is being adopted, which is drawing on that previously described by McDonald et al [35], where a large number of variant designs can be generated for a limited number of fixed ship topologies. However, rather than attempt to numerically balance

© 2017: The Royal Institution of Naval Architects

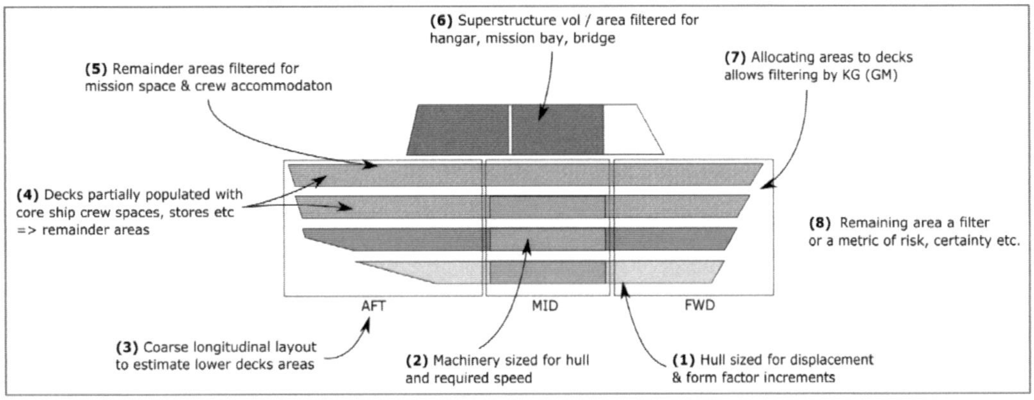

Figure 11: An illustrative example of a topological model [36]

the design at every stage, a search and down-selection process will be used to remove those nominal variants that could never meet the design requirements [36].

It is important to note that what is proposed would not be a purely numerical sizing method, as the configuration of the vessel would be incorporated in the form of topological models, with available areas and volumes being associated with selected locations in the vessel. Figure 11 illustrates a generalised topological ship representation indicating some of the issues relevant to down-selection. The primary reason for avoiding an extensive balanced and iterative model is that such tools are both complicated, complex and time demanding [36].

7. ACKNOWLEDGEMENTS

Financial support for the first author's PhD studentship has been provided through an EPSRC industrial CASE award with BAE Systems sponsorship and is gratefully acknowledged, as is the latter's ongoing input to the research project.

8. REFERENCES

1. CLAPPER, J. R., YOUNG, J. J., CARTWRIGHT, J. E., GRIMES, J. G., "Unmanned Systems Roadmap 2007 - 2032". Department of Defence U.S.A. Washington, DC, U.S.A. December, 2007.

2. MINISTRY OF DEFENCE, DE&S. "Type 26 - The Shape of Things to Come". *Desider. Issue 52.* Bristol, U.K. September, 2012

3. O'ROURKE, R., "Navy Littoral Combat Ship (LCS)/Frigate Program: Background and Issues for Congress". *Congressional Research Service*. June, 2016.

4. SAVITZ, S., BLICKSTEIN, I., BURYK, P., BUTTON, R. W., DELUCA, P., DRYDEN, J., MASTBAUM, J., OSTBURG, J., PADILLA, P., POTTER, A., PRICE, C. C., THRALL, L., WOODWARD, S. K., YARDLEY, R. J., YURCHAK, J. M., "U.S.A. Navy Employment Options for Unmanned Surface Vehicles (USVs)". *National Defence Research Institute*. Santa Monica, CA, U.S.A. RAND Corporation. 2013.

5. CANNING, J. S., "A Definitive Work on Factors Impacting the Arming of Unmanned Vehicles". *Naval Surface Warfare Centre.* Dahlgren, VA, U.S.A. May, 2005.

6. ALKIRE, B., KALLIMANI, J. G., WILSON, P. A., MOORE, L. R., "Applications for Navy Unmanned Aircraft Systems". *National Defence Research Institute.* Santa Monica, CA, U.S.A. RAND Corporation. 2010.

7. PAWLING, R. J., ANDREWS, D. J., "The Ship Design Challenge of Naval Unmanned Aerial Vehicles". *Proceedings of Warship 2009: Air Power at Sea. RINA*. Bath, U.K. June, 2009.

8. COMMITTEE ON AUTONOMOUS VEHICLES IN SUPPORT OF NAVAL OPERATIONS, "Autonomous Vehicles in Support of Naval Operations". *The National Academies Press.* Washington, D.C. 2005.

9. BROADBENT, C., BINNS S., "A Vision for an Affordable Flexible UXV Enabled Versatile Surface Combatant". *Proceedings of Warship 2006: Future Surface Ships. RINA*. Bath, U.K. June, 2006.

10. BRAITHWAITE, C. R., "A Group System Approach to Naval Warfare - The UXV Mothership Concept". *Proceedings of Warship*

© 2017: The Royal Institution of Naval Architects

2013: Minor Warships. RINA. Bath, U.K. June, 2013.

11. THOMSEN, J. E., "The Navy Unmanned Surface Vehicle (USV) Master Plan". *U.S.A. Department of the Navy.* Washington, DC, U.S.A. July, 2007.

12. RAUCH, G. C., PURCELL, J. M., AUSTIN, T., PACKARD, J. G., "Ship of Opportunity Launch and Recovery System for REMUS 600 AUV's". *Proceedings of IEEE Aerospace Conference. Oceans Conference.* Kobe, Japan. April, 2008.

13. BUTT, M. E., "Shipboard Launch and Recovery Procedures Manual". *U.S.A. Department of Homeland Security. United Stated Coast Guard.* Washington, DC, U.S.A. January, 2013.

14. LESTER, G., "REMUS Launch and Recovery Systems". *International Symposium on Unmanned Untethered Submersible Technology.* Durham, NH, U.S.A. August, 2007.

15. EATON, M., AMBROSE, G., PASS, D., "The Integration of Launch and Recovery Systems into the Mission Compartment". *Proceedings of ASNE Launch and Recovery Symposium.* Linthicum, MD, U.S.A. November, 2014.

16. PAWLING, R. J., ANDREWS, D. J., "Large Unmanned Vehicles and the Minor War Vessel". *Proceedings of Warship 2013: Minor Warships. RINA.* Bath, U.K. June, 2013.

17. RANDLES, I., "Designing for the Future: The Type 26 Global Combat Ship Mission Bay". *International Naval Engineering Conference and Exhibition (INEC).* Edinburgh, U.K. May, 2012.

18. BROWN, D.K., ANDREWS, D.J., "The Design of Cheap Warships", *Proceedings of the International Naval Technology Expo 1980*, Rotterdam, Netherlands, February 1980.

19. PAWLING, R.J., ANDREWS, D.J., PIKS, R., SINGER, D., DUCHATEAU, E., HOPMAN, H., "An Integrated Approach to Style Definition in Early Stage Design", *Proceedings of COMPIT 2013*, Cortona, Italy, May 2013.

20. ANDREWS, D.J., "A Creative Approach to Ship Architecture", *Transactions of RINA, Vol. 145,* 2003.

21. BROWN, D.K., "Naval Architecture", *US Naval Engineers Journal, Vol. 105 (1)*, 1993.

22. HEATHER, R.G., "Preliminary Design Considerations for Fast Warships". *Proceedings of Warship 1990: The Future for Surface Warships. RINA.* London, U.K. June, 1990.

23. ANDREWS, D.J., "Creative Ship Design", *Transactions of RINA, Vol. 123,* 1981.

24. ANDREWS, D.J., "An Integrated Approach to Ship Synthesis", *Transactions of RINA, Vol. 128,* 1986.

25. ANDREWS, D.J., DICKS, C.A. "The Building Block Design Methodology Applied to Advanced Naval Ship Design", *Proceedings of IMDC 1997*, Newcastle, UK, May 1997.

26. ANDREWS, D.J., PAWLING, R.J., "SURFCON - A 21ST Century Ship Design Tool". *Proceedings of International Marine Design Conference (IMDC).* Athens, Greece. May, 2003.

27. ANDREWS, D.J., "Marine Requirements Elucidation and the Nature of Preliminary Ship Design", *IJME Vol. 153, Part A1*, pp. 23-29, 2011.

28. PAWLING, R. J., PIPERAKIS, A. S., ANDREWS, D. J., "Developing Architecturally Oriented Concept Ship Design Tools for Research and Education". *Proceedings of International Marine Design Conference (IMDC).* Tokyo, Japan. May, 2015.

29. PAWLING, R.J.; PIPERAKIS, A.; ANDREWS, D.J. (2016), "Applications of Network Science in Ship Design", 15th COMPIT, Lecce.

30. ANDREWS, D.J., CASAROSA, L., PAWLING, R.J., GALEA, E., DEERE, S., LAWRENCE, S., "Integrating Personnel Movement Simulation into Preliminary Ship Design", *IJME Vol. 150 Part A1, Discussion IJME, Vol. 150, Part A3*, 2008.

31. ROBERTAZZI, G. T., "Computer Networks and Systems. Queuing Theory and Performance Evaluation". *Springer-Verlag. Second Edition.* New York, NY, U.S.A. 1994.

32. HILLIER, S. F., LIEBERMAN, J. G., "Introduction to Operations Research". *McGraw-Hill. 7TH Edition.* New York, NY, U.S.A. 2001.

© 2017: The Royal Institution of Naval Architects

33. BHAT, U. N., "An Introduction to Queuing Theory - Modeling and Analysis in Applications". *Birkhauser Boston*. New York, NY, U.S.A. 2008.

34. LAGERSHAUSEN, S., "Performance Analysis of Closed Queueing Networks". *Springer-Verlag*. Berlin. 2013.

35. MCDONALD, T., ANDREWS, D.J., PAWLING, R.J., "A Demonstration of an Advanced Library Based Approach to the Initial Design Exploration of Different Hullform Configurations", *Computer-Aided Design*, Vol. 44, Issue 3, March 2012, pp. 209-223

36. PAWLING, R. J., KOURIAMPALIS, N., ESBATI, S.,BRADBEER, N., ANDREWS, D.J., "Expanding the Scope of Early Stage Computer Aided Ship Design". *International Conference on Computer and IT Applications in the Maritime Industri*

37. *es (COMPIT)*. Cardiff, UK. May, 2017.

9. AUTHORS BIOGRAPHY

Nikolaos Kouriampalis completed his undergraduate engineering studies in NTUA in 2010 followed by MSc in Mechanical Engineering at UCL in 2012. He is currently a PhD student at UCL DRC investigating implications of UXVs on naval ships in the preliminary ship design stage.

Dr Rachel J. Pawling, MEng, PhD, AMRINA, is the BMT DSL Research Fellow for Naval Ship Design at UCL. She completed the MEng in Naval Architecture and Marine Engineering at University College London in 2001 and then joined the UCL DRC as a research student investigating the application of the Design Building Block approach to innovative ship design. Gaining her PhD in 2007, she has continued her research both in the DRC and via a secondment in industry. She is the recipient of the 2008 RINA Samuel Baxter Prize, 2009 RINA WHC Nicholas Prize, and 2012 COMPIT GL Award for papers describing her research.

Professor David J. Andrews, FREng, PhD, FRINA, FIMechE, RCNC, was appointed Professor of Engineering Design at UCL in September 2000 following an extensive career in ship and submarine design and acquisition management in the UK Defence Procurement Agency. He leads UCL's design research in computer aided ship design, design methodology and design practice. He is a Fellow of RINA, Fellow of IMechE and was elected to the Royal Academy of Engineering in 2000 and as a Vice President of RINA in 2006. He is the International Committee Chairman of the IMDC.

© 2017: The Royal Institution of Naval Architects

© 2017: The Royal Institution of Naval Architects

Integrated development environment of autonomic software for USV (Unmanned Surface Vehicle) based on ROS (Robot Operating System)

H W Lee, M I Roh, L Zhao, S H Ham and **N Kim**, Seoul National University, Republic of Korea
C W Yu, Agency for Defense Development, Republic of Korea

SUMMARY

The USV (Unmanned Surface Vehicle) mainly performs mine detection and maritime surveillance and reconnaissance. To carry out various strategies unmanned, the autonomic software mounted on USV takes charge of detection, decision, and command procedure itself. To operate various hardware on USV, the autonomic software should include complicated control algorithms, and the communication between the control algorithms and the hardware is required. In this study, an integrated simulation environment was proposed to verify and improve the control algorithms of the autonomic software, based on a virtual prototype of USV and ROS (Robot Operating System). Then, various tests were performed including mine detection, path following, and entry/departure control scenario. As a result, it can be seen that the environment can be effectively used to develop the autonomic software for USV with less time and cost.

1. INTRODUCTION

1.1 BACKGROUND

The USV (Unmanned Surface Vehicle) performs various missions including mine detection and marine reconnaissance and survey unmanned. As the autonomic software for USV takes charge of detection, decision, and command instead of human, the verification and validation (V&V) of the software is highly required at the early design stage. In this study, the virtual prototype of USV is developed to verify and improve the control algorithms of the autonomic software.

Meanwhile, to operate various hardware such as sensors and devices on USV, the autonomic software should include complicated control algorithms, and the communication between the control algorithms and the hardware is required. Therefore, ROS (Robot Operating System), a robot software platform, is used to construct integrated simulation environment to integrate the hardware and software interfaces through network communication. Finally, three main missions of USV, mine detection, path following, and port entry/departure control scenario are performed.

1.2 RELATED WORKS

There are some related works using ROS for the mission performance of UWS (Unmanned War System). Demarco et al. [1] developed the integrated system of GTRI Yellowfin UUV (Unmanned Underwater Vehicle) including mission planning and performance process using ROS. The MOOS (Mission Oriented Operating Suite), which has been widely used in underwater robot field, was also used together with ROS using MOOS/ROS bridge. Mendnca et al. [2] developed multi robot 3D simulator, Kelpie, based on ROS. Then, the simple path following the mission of USV was simulated using Kelpie.

In this study, the integrated simulation environment is developed to verify the various control algorithms of USV for the realistic mission scenario. ROS was used for the communication of the virtual prototype, control algorithm, and USV hardware.

2. VIRTUAL PROTOTYPE

Due to the difficulty of the verification and validation of control algorithms in a real world, the virtual prototype of USV and ocean space are necessary. The virtual prototype of USV can be divided into several modules depending on the actual hardware it reflects.

2.1 SCENARIO MANAGEMENT / DISPLAY

The scenario management and display module manage the overall information of the USV simulation. In the scenario management module, the virtual USV is created, and the virtual mission and the ocean environment are defined. The display module presents the simulation result on the map.

2.2 MOTION ANALYSIS

The motion analysis module represents the virtual USV. As the USV moves on the sea surface, the motion of the USV is based on the 3 DOF maneuvering equations of motion while the submarine and airplane have 6 DOF motion. Therefore, the 3 DOF motion, surge, sway, and yaw of the USV is calculated using the maneuvering equations of motion suggested by Getler [3].

2.3 SUBSEA IMAGE COMPOSITION

To detect mines, the SSS (Side Scan Sonar) is used to create an image of large areas of the sea floor. The subsea image composition module is the virtual prototype of SSS.
The various technologies to create subsea image have been suggested by some researchers [4, 5, 6]. In this study, the hybrid technique which composes the virtual

© 2017: The Royal Institution of Naval Architects

mine on the real subsea topography image is used. In the subsea image generated by SSS, the height of the sea floor is presented with the brightness. Therefore, the highlight and shadow area due to virtual mine can be generated according to the location of the USV and mine, and composed of the subsea image.

2.4 PORT IMAGE GENERATION

In the port entry/departure process, as the USV reaches docking area, the video of the port is made by the camera on USV for the detailed estimation of the position. The port image generation module represents the virtual camera on the USV.

For this, the real port image set from known viewpoints are first obtained. Then, the key points of each image are extracted for the interpolation of an image from virtual viewpoints. As the virtual USV reaches the docking area, the virtual port images are generated according to the position of the virtual USV.

3. CONTROL ALGORITHM

The USV performs various missions such as mine detection and maritime surveillance and reconnaissance. In this study, the missions for USV is divided into three main scenarios: mine detection, path following, and port entry/departure. For the performance of each scenario, the control algorithms should be developed and mounted in the autonomic software of USV.

3.1 MINE DETECTION

The mine detection module receives the subsea image from SSS and detects mines in the image. In this study, the AdaBoost algorithm suggested by Viola and Jones [7] is used for mine detection. The AdaBoost algorithm extracts the important feature information of the target, and apply it to the given image to find the target. Kim [8] and Sawas [9] have applied this algorithm to mine detection in the subsea image.

3.2 PATH FOLLOWING CONTROL

In the path following scenario, the USV should follow the target path while environmental load such as wave, wind, and current exist. Then, the USV follows the target path minimizing the environmental load effect. In the path following control module, the reinforcement learning is adopted to calculate the target rudder angle. Then, the target rudder angle is sent to the virtual prototype of USV as a command for the control.

3.3 POSITION ESTIMATION

In the port entry/departure scenario, the port image is generated by a camera on USV. The position estimation module receives the port image from the camera, and estimate the relative position of the USV. Firstly, with port image composition module, some image set of the port with known USV positions are generated. Then, the convolution neural network is constructed to estimate the position of the USV with given port image.

3.4 ENTRY/DEPARTURE CONTROL

After the position of the USV is obtained from position estimation module, the entry/departure control module calculates the target rudder angle and the propeller rpm of the USV. For the control, the vector field to the docking point is generated, and the PD controller is used.

4. INTEGRATED SIMULATION ENVIRONMENT

Generally, various communication method such as HLA (High Level Architecture) or TCP/IP is used for the data transmission and reception between modules. In this case, the additional interface or the conversion of the program is required for the communication with hardware. On the other hand, ROS includes hardware platform and provides various type of hardware such as the motor, sensor, and so on. Therefore, if the hardware provided by ROS is used, the modules can be integrated with them easily without a particular interface. In this study, ROS is used for the data communication method.

ROS is communication infrastructure program based on message communication between nodes. ROS is composed of multiple nodes and a master. A node is a minimum execution unit in ROS, and each is a program that is in charge of one function. Master act as a namespace that manages the nodes. The nodes transmit and receive data with TCP/IP communication method.
In this study, the modules of virtual prototype and control algorithm are developed as a node of ROS. Therefore, each module is a particular program which uses the message communication method provided by ROS.

Meanwhile, the operating system that ROS officially supports is Ubuntu. The master of ROS is driven in Ubuntu, but various nodes are often developed in environments other than ROS. In this study, the modules for virtual prototype and control algorithm are developed in C# language and MATLAB. To use the message communication function provided by ROS, the package named Rosbridge is adopted.

5. APPLICATIONS

The overall modules for the virtual prototype and the control algorithms transmit and receive data between each other using communication network provided by ROS. Figure 1 shows the configuration of the modules and the transmitted/received data for mission scenarios.

© 2017: The Royal Institution of Naval Architects

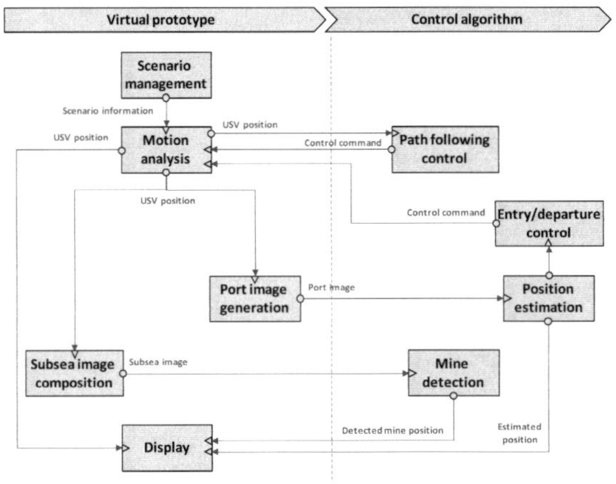

Figure 1:. Configuration of the overall modules for virtual prototype and control algorithm.

5.1 MINE DETECTION

Mine detection is one of the main missions that USV performs. The user inputs initial data of the simulation such as USV path and mine location in scenario management module, and the initial data is transmitted to other modules. As the virtual USV moves along the target path, the motion analysis module transmits the position data to the subsea image composition module, which is the virtual prototype of SSS. Then, the subsea image composition module generates the subsea image according to the position of the USV and mines. Then, the subsea image data is transmitted to the mine detection module, and the mines on the image are detected. The result data such as the position of USV and detected mines are transmitted to the display module.

5.2 PATH FOLLOWING

To follow the target path while environmental loads exist, the rudder angle of the USV should be controlled in real time. The motion analysis module calculates the position of the virtual USV as it moves. Then, the path following control module finds the target rudder angle by reinforcement learning and transmits to the motion analysis module.

5.3 PORT ENTRY/DEPARTURE

As the USV reaches the docking area near the port, the motion analysis module transmits the position of the USV to port image composition module, which is the virtual prototype of the camera on USV. Then, the port image composition module generates the virtual port image by the interpolation of the real image set. With the image, the position estimation module estimates the more accurate position of the USV. Then, the entry/departure control module calculates the target rudder angle and propeller rpm of the USV to get to the port. Then, the control command is transmitted to the motion analysis module.

6. CONCLUSIONS

In this study, the simulation for the verification and validation of the control algorithms in USV is performed using virtual prototype. For the data communication, ROS is used for the integrated simulation environment. Lastly, the main mission scenario of the USV, mine detection, path following, and port entry/departure control is simulated. The result shows that the integrated simulation environment using ROS is properly constructed, and can be used to develop the autonomic software for USV. For the future works, the HILS (Hardware-In-the-Loop Simulation) including simple hardware will be performed.

7. ACKNOWLEDGEMENTS

This work was partially supported by,
(a) Agency for Defense Development (UD160009DD)
(b) BK21 Plus, Education & Research Center for Offshore Plant Engineers (COPE) of Seoul National University,
(c) Research Institute of Marine Systems Engineering of Seoul National University.

8. REFERENCES

1. Demarco, K., West, M.E., Collins, T.R., "An Implementation of ROS on the Yellowfin Autonomous Underwater Vehicle (AUV)", *Proceedings of OCEANS '11 MTS/IEEE*, pp.1-6, 2011.

2. Mendonca, R., Santana, P., Marques, F., "Kelpie: A ROS-based Multi-robot Simulator for Water Surface and Aerial Vehicles", *Proceedings of the IEEE International Conference on Systems, Man, and Cybernetics (SMC)*, pp.3645-3650, 2013.

3. Gertler, M., Hagen, G.R., "Standard Equations of Motion for Submarine Simulation", *Naval Ship Research and Development Center*, 1967.

4. Manning, R., "Small Object Classification Performance of High-resolution Imaging Sonars as a Function of Image Resolution", *Proceedings of OCEANS '02 MTS/IEEE*, pp.2156-2163, 2002.

5. Fawcett, J.A., "Image-based Classification of Side Scan Sonar Detections", *Proceedings of CAD/CAC '01*, 2001.

6. Reed, S., Petillot, Y., Bell, J., "Model-based Approach to the Detection and Classification of Mines in Side Scan Sonar", *Applied Optics*, 43(2), pp.237-246, 2004.

© 2017: The Royal Institution of Naval Architects

7. Viola, P., Jones, M., "Robust Real-Time Face Detection", *International Journal of Computer Vision*, 57(2), pp.137-154, 2004.

8. Kim, J.Y., Shin, C.H., Kim, K.H., "A System Design Method of Mine Warfare Using Information for Sonar and MDV", *The Journal of Korean Institute of Communications and Information Sciences*, 39(12), pp.1243-1249, 2014.

9. Sawas, J., "Automatic Target Recognition in Sonar Imagery Using a Cascade of Boosted Classifiers", *Ph.D. Thesis of Heriot-Watt University*, 2015.

9. AUTHORS BIOGRAPHY

Hye-Won Lee is a Ph.D. candidate in the department of naval architecture and ocean engineering at Seoul National University, Republic of Korea. Her main area of research interests includes simulation based design, multibody dynamics, and so on.

Myung-Il Roh is a professor of the department of naval architecture and ocean engineering at Seoul National University, Republic of Korea. He holds a B.S., M.Sc., and Ph.D. in naval architecture and ocean engineering from Seoul National University. His main area of teaching and research interests includes ship and offshore structure design, simulation-based design, optimization, CAD, CAM, CAE, CAPP, and so on.

Luman Zhao is a Ph.D. candidate in the department of naval architecture and ocean engineering at Seoul National University, Republic of Korea. Her main area of research interests includes Hardware-In-the-Loop Simulation, knuckle boom crane control, and so on.

Seung-Ho Ham is a Ph.D. candidate in the department of naval architecture and ocean engineering at Seoul National University, Republic of Korea. His main area of research interests includes simulation based design, multibody dynamics, virtual reality, and so on.

Nakwan Kim is a senior research engineer at Research Institute of Marine Systems Engineering in Seoul National University, Republic of Korea. He holds a B.S. in the department of mechanical and aerospace engineering at Seoul National University, and M.Sc. and Ph.D. from Georgia Tech. His main area of research interests includes unmanned ship, unmanned underwater vehicle, marine control, and so on.

Chan-Woo Yu is a senior researcher of Agency for Defense Development, Republic of Korea. He holds a B.S. and M.Sc. in the department of mechanical and aerospace engineering from Seoul National University. His research interests are in mission planning of unmanned maritime system, and so on.

© 2017: The Royal Institution of Naval Architects

CYBER PHYSICAL SYSTEM IN SHIPBUILDING TO REALIZE SMART SHIPYARD: INTEGRATION OF MONITORING SYSTEM AND SHIPYARD SIMULATION SYSTEM

Kazuhiro Aoyama, Mayuna Hoshi, Yusei Hiro, Ryo Kitamura and Kazuya Oizumi, Department of Systems Innovation, School of Engineering, The University of Tokyo, JAPAN

SUMMARY

In shipbuilding, creating a smart shipyard is desirable to improve productivity using the concept of a Cyber Physical System (CPS). The key point of a CPS is tightly linking the physical world and cyber world through information technology. In this paper, we propose the integration of a monitoring system and a shipyard simulation system. We developed a shipyard simulator incorporating an optimization method to find the specific improvement points for ship construction. We then integrate this simulation system with a previously developed monitoring system using the monitored data as input data to the simulation. This integration enables a comparison of the real production process with the ideal process and allowed the generation of improvement points for the current process. We implement the CPS concept with some important functions that can characterize the construction progress in real time, detect construction delays in real time, and cope with the delay at an appropriate point in real time.

NOMENCLATURE

CPS:	Cyber Physical System
GA:	Genetic Algorithm
IoT:	Internet of Things
RFID:	Radio Frequency Identification

1. INTRODUCTION

In recent years, the movement towards Industry 4.0 has advanced to optimize the entire production process in order to provide greater value and enhance competitiveness [1]. The objective is to create a Cyber Physical System (CPS) that builds a cyber space reflecting a physical space through the construction of the Internet of Things (IoT) that connects everything related to production with the Internet, digitalizes them in real time, reproduces design, manufacturing and other processes by simulation, and controls and manages real business processes based on the CPS simulation.

Of course, even in the shipbuilding industry, it is desired to implement CPS but acknowledging that is very difficult to do so. Since shipbuilding is a made-to-order production system, the best construction processes are planned for each ship individually. Moreover, due to shipbuilding utilizing a job shop style, both the work targets and the worker's movements for each job in a shipyard are unique so the working environment always changes. Additionally, the large scale of ship construction work requires a lengthy time in a large work area. For these reasons, it is extremely difficult to fully grasp and manage all processes in detail at shipyards.

In order to implement CPS at a shipyard, it is necessary to solve the above-mentioned problem. In the past, research has been conducted that aimed to correctly characterize the work being done at a shipyard and acquire required information. The authors recorded a video of the construction floor of a shipyard with a web camera and applied image processing to the image to characterize workers, blocks, and work [2,3,4]. Additionally, in another research effort, the authors acquired detailed work information from a worker's perspective using a wearable terminal [5]. Through these research efforts, it became possible to acquire characterizing information such as “when, where, who, for what, which work” and “which work corresponds to which plans”. However, the challenges still remain.

Although it is possible to grasp the detailed process, this information has not been able to be utilized to improve the construction process and schedule. It has not been considered how to improve productivity by using the information obtained by monitoring. In order to improve productivity, it is necessary to improve the construction processes and schedule.

A simulation is one of the most effective approaches for improving productivity because complicated and large-scale processes such as ship construction make it very difficult for a human to make improvements. A construction simulation enables visualization and evaluation of the distribution in the shipyard and the behaviour of workers. However, in creating a simulation model, it is necessary to generate an enormous amount of input information, and modelling such a complex job shop system is very complicated, which makes it difficult to create an accurate model.

In this research, we aimed to implement a shipbuilding CPS by integrating a monitoring system and a shipbuilding simulation system. First, we defined and created a construction model that can project the entirety of the shipyard construction process. By using a construction model, we reproduced the construction process performed in the past and searched for an improved process through the simulation. In addition, by using the simulation in combination with current data

© 2017: The Royal Institution of Naval Architects

acquired through monitoring, we realized successive formulation of a quasi-optimal plan. In this paper, we show the feasibility of CPS in shipbuilding by utilizing a monitoring method and simulation in cooperation with a construction model with results showing an improved method for the construction process.

2. PREVIOUS RESEARCH: SHIPBUILDING MONITORING SYSTEM

The authors have previously developed a construction process monitoring system that has two functions. One function is for collecting monitored data (image-processing and sensor data), and the other is for generating and visualizing work data [3,5]. Figure 1 shows an overview of the monitoring system for construction processes.

In the shipyard, there are many types of construction stages. These stages include the fabrication stage, subassembly stage, assembly stage, and erection stage. Of those stages, the subassembly stage is the easiest to record video of the work place. Therefore, this stage is suitable for verifying the effectiveness of the monitoring system and, for this reason, this research focused on the subassembly for monitoring. The following sections provide an overview of this monitoring system from the viewpoints of the two functions.

2.1 COLLECTING MONITORING DATA

In order to collect monitoring data, a device system has been developed to capture video images of the construction process in the subassembly stage. The developed image processing system can detect, track, recognize, and identify objects from video images and sensor data captured by the recording system. Video image processing collects, detects, and identifies work, workers' behaviour, construction modules' working areas, and transfers. In addition, acceleration sensor data is collected to characterize workers' behaviour.

The monitoring device system is composed of a wireless video camera and a video recording server. The video camera, which is placed in a construction shop, monitors the site and transfers video information via a wireless LAN to the recording server. Monitoring data that are meaningful and helpful for effective management can be generated by the image processing system. The processing system framework includes the following main stages: object detection, object tracking, object recognition, and identification.

Welding and grinding operations are commonly performed in shipyard construction shops and are considered to be irreplaceable key operations. Data from welding and grinding operations, such as their locations and durations, can influence or even determine the results of parts manufacturing, assembly, etc. All of the crucial data cannot be manually recorded in the shipyard shops since these operations are performed many times each day. Thus, there is uncertainty involved in each operation and this uncertainty can lead to extra costs. However, the proposed system is a possible low-cost method to effectively solve these problems. The left side of Figure 1 shows the image processing for welding, grinding, gouging, and heating operations.

In a construction shop floor, many construction modules wait to be assembled by workers or moved away by cranes. In addition to the quality of management, the module location, time needed for assembly, and other related information can influence or even determine the working procedure for the next batch of modules.

Video image processing can also detect moving objects, such as workers and forklifts, and analyse their routes. Workers and forklifts are typical moving objects in actual

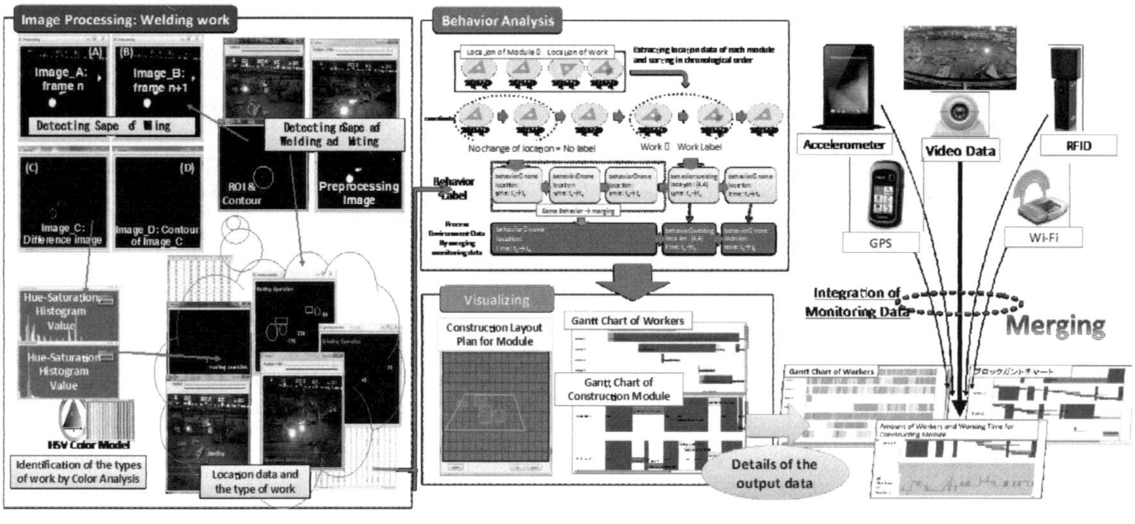

Figure 1: Functions of the Construction Monitoring System

© 2017: The Royal Institution of Naval Architects

shops. A temporal difference–based motion detection method was selected as the main method to detect these two objects in the object detection stage. Contour-based tracking and the motion templates method are used as the main tracking methods during the object-tracking stage. In the recognition and identification stage, the differences between workers and forklifts are recognized in the video data.

Two additional monitoring technologies, Radio Frequency Identification (RFID) and accelerometers, are introduced to collect data for the monitoring system. The collected data are the workers' identification information to match the image-processed data with a worker's identification. RFID technology is utilized to make the worker's unique identification correspond to the "operation" data from the video image-processed data. RFID tags are attached to operation tools, e.g. welding torches, and each worker's RFID reader detects the RFID tags. Thus, the data on how each worker uses each tool can be detected by the RFID reader (RFID data). The system can grasp when and what each worker did because the operation tools and types have a one-to-one correspondence. An acceleration sensor built into smartphones is utilized to estimate the worker's motion state (moving or stationary) by processing its acceleration data.

2.2 GENERATING AND VISUALIZING OPERATION DATA

In the function for generating and visualizing operation data, the data processing system integrates the collected monitoring data to generate operation data (4W1H: when, who, where, what, how). This data processing system analyses, classifies, statistically tallies, and calculates the useful data from the image processing system, such as time and location, for effective management. The output of the monitoring system is displayed by the image and data processing systems. As Figure 2 shows, many types of progress schedule charts (Gantt charts) are generated from the operation data and then displayed.

The role of the operation data generation function is to collect all useful management data, such as times and locations from the image-processing system, and to analyse, classify, statistically tally, and calculate the data for each object to enable feasible and effective management. In construction time management, the subassembly lead-time is an important value that affects the assembly lead-time as the longest activity.

Most operations occur around construction modules with workers completing jobs with the modules. In other words, the module is the centre and main object of subassembly on the construction shop floor of a shipyard. If we want to improve subassembly management, we need to pay more attention to the module so the module's standby state should be chosen as the managed activity. The lead-time for constructing the module has three parts: transportation, operation, and wait time. The transportation time is the time required to transfer the module from one shop floor to another. The operation time is the total time required to perform an operation. In this monitoring system, welding, grinding, and heating operations are considered as monitoring targets. The waiting time is used to describe the time spent by module waiting for an operation or to be moved.

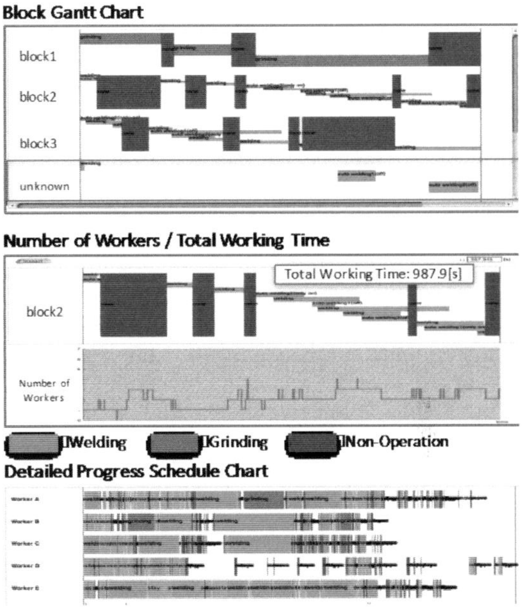

Figure 2: Visualizing Operation Data Generated by the Integration of Monitored Data

After video image processing, the data-processing system receives all data of interest. The data of interest can include information such as the location of a module, operation, or a worker; or the duration of a welding, heating, or grinding operation; or a module's wait time. All these data are analysed statistically to automatically classify and calculate the move time, operation time, and wait time for each activity. The Gantt chart of the modules' construction can be drawn based on the results of the data-processing system. Figure 3 shows one example case of using data processing to generate operation data.

Work result information for workers, operations, and modules is necessary to comprehend the current work situation. Methods for work record pre-processing, integration, and instruction generation are introduced to facilitate the extraction of the collective work result information from the collected work records. However, it is difficult to directly use the raw work records because the data in the work records are isolated due to the different devices and data processing system used to collect them.

In order to integrate the isolated data in the work records, a data-labelling method is introduced. This

© 2017: The Royal Institution of Naval Architects

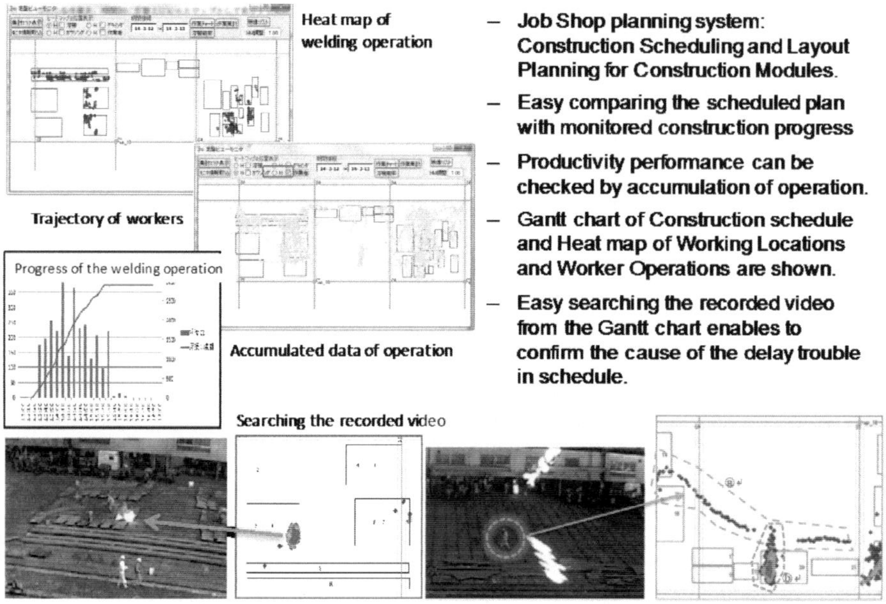

Figure 3: Snapshots of Construction Monitoring Browser

method makes labels for the states of workers, operations, and modules in space and time enabling integration of the data using those labels. A worker's position cannot be specified; however, his RFID data can be specified. Therefore, in order to extract more detailed and accurate operation information, the RFID data and image-processed data are merged by matching corresponding data such as operation time, position, and type (welding, grinding, etc.). The position of each worker can also be characterized by acceleration data. Therefore, in order to extract each worker's position data, image-processed data and acceleration data are correlated.

The Construction Process Monitoring Browser is developed as the function for visualizing operation data. This browser, which is based on the Job Shop Planning system that supports Construction Scheduling and Layout Planning for Construction Modules, provides construction managers with the following functions.

- Compare the scheduled plan with the monitored construction progress.
- Check productivity performance by the accumulation of operations.
- Show a Gantt chart of the construction schedule and a heat map for working points and worker activities.
- Enable a recorded video search from the Gantt chart to investigate the cause of a delay in the schedule.

Figure 3 shows some screenshots of this browser. The screenshots show heat maps of welding operations, trajectories of workers, and accumulated data of operations.

3. SHIPBUILDING SIMULATION FOR IMPROVING PRODUCTIVITY

Our proposed monitoring system that has video image analysis and wearable smart devices enable collection of useful information regarding construction progress and visualization of a detailed process in a job shop. However, we had not considered how to improve productivity applying the information obtained through monitoring. Therefore, so we had to develop utilization methods to support productivity improvement by applying the shipyard CPS.

3.1 SIMULATION BASED IMPROVING PRODUCTIVITY

In order to improve productivity, it is necessary to improve the construction processes and the construction schedule. A construction simulation approach is effective for addressing these two facets, because it is difficult for humans to identify improvement points in complicated and large scale processes such as ship construction. In general, the effectiveness of the simulation for complex systems has been widely known. A simulation enables visualization and evaluation of the distribution in the shipyard and the behaviour of workers.

However, in order to generate a simulation model, it is necessary to collect an enormous amount of input information. In addition, modelling such a complex job shop system is very complicated. It is difficult to generate an accurate model due to these complexities. In this research, we aimed at realizing process improvement by generating an accurate simulation model of the construction process and applying it.

© 2017: The Royal Institution of Naval Architects

As an approach to meet this objective, we proposed the following process.

1) Generate a construction model by integrating information from each element of construction systems such as construction modules and workers to define a master model that can exchange information between several systems.
2) Generate a simulation model from the construction model supplementing data as necessary for execution of the simulation incorporating the monitoring information.
3) Reproduce the current construction process using monitored data.
4) Adjust various conditions in the simulation model and thereby search for process improvements over the current process.

3.2 CONSTRUCTION MODEL: MASTER MODEL OF CONSTRUCTION SYSTEM .

In order to use the data obtained by the current monitoring system for simulation exchanging information between two systems is necessary as shown in Figure 4. This research proposes a construction model

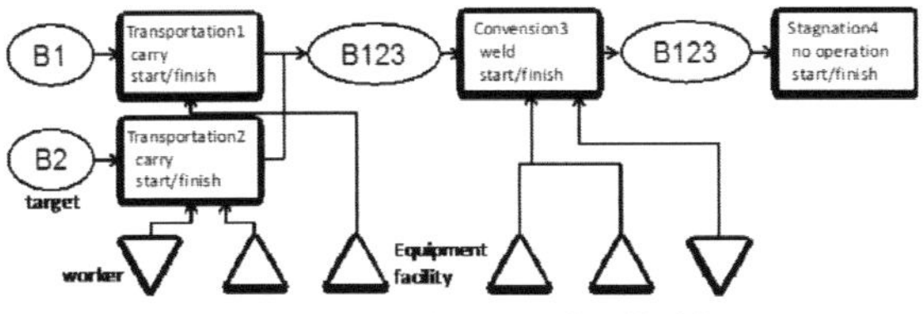

(A) Structure of Construction Model

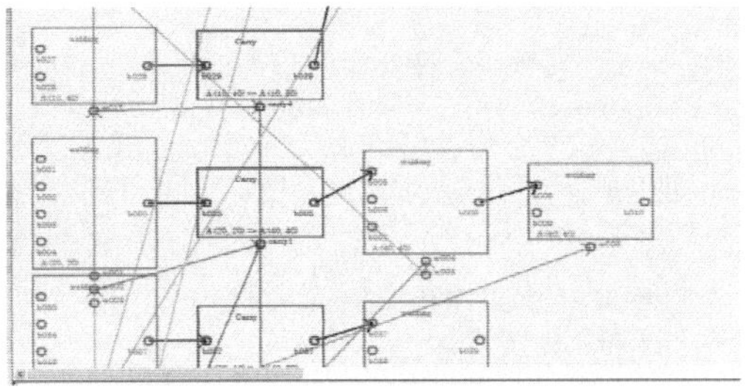

(B) Snapshot of a Construction Model in System

Figure 5: Construction Model

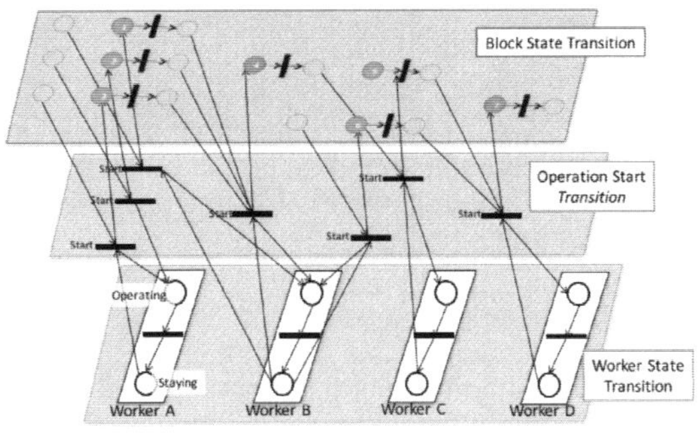

Figure 6: Simulation Model by Petri nets

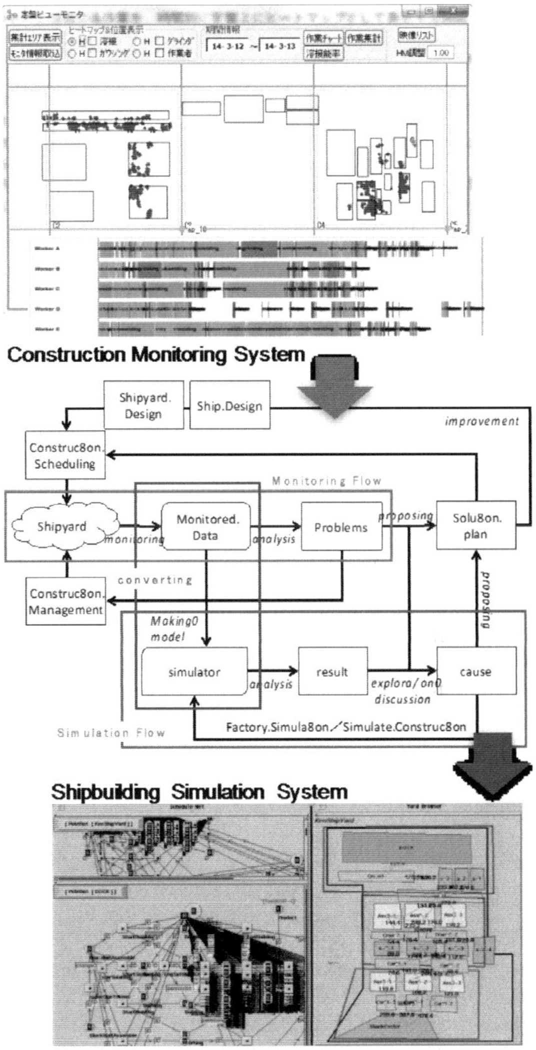

Figure 4: Integrating Monitoring System with Shipyard Simulation System

as a master model for the construction system that can be used along with the simulation system while storing information gathered by the monitoring system.

In order to define a construction model that enables connection of the monitoring system with the simulation system, we have to understand the features of shipbuilding. There are many types of system elements in a shipbuilding production system. Physical elements in the shipbuilding process can be roughly classified into worker or equipment, which are the construction resource, and the construction module, which is a construction target. Information on the production target and production resources should be managed in the construction model. Furthermore, the production target and production resources are associated by an operation. In this research, a process is expressed as a connection of operations with a module and a worker associated with each operation to generate process models. The construction model incorporates all of the process models necessary during construction.

In this research, we define three types of operations: stagnation, conversion, and transportation. Stagnation is an operation making a temporal change to a target. Conversion is an operation making a physical change to a target. In this research, we consider four types of conversions: welding, grinding, gouging, and heating. Transportation is an operation making spatial changes to a target. The operations include information on workers such as how many workers will be allocated to the operation and operation allocation information on who performs which operation. Figure 5(A) shows an association model between production targets and the operation as well as between production resources and the operation. Figure 5(B) shows a snapshot of an operations network in a construction model of the system.

3.3 SIMULATION MODEL FOR CONSTRUCTION IN SHIPYARD

The Petri nets model is introduced to generate a simulation model. Construction activities of a job shop, such as the shipbuilding process, are characterized by parallel and asynchronous activities. Petri nets treat discrete events as a tool and the model is suitable for expressing parallelism and asynchronous activities.

In a construction model, elements of the construction resources and the construction targets are defined and three kinds of operation models are introduced to associate them. As previously mentioned, an operation associates a target with resources. From the view point of the simulation, we can recognize that each of the elements has a state according to the production progress. For instance, the state of a construction module is divided into a stagnation state, conversion state, and transportation state according to associated operations. In the terms of resources, the state of the worker or equipment can be classified into a state of working in such a module or into a dormant state.

A change in the status of each of the elements, i.e. changing between the stagnation state, conversion state, or transportation state, represent a place or a break of the state which is joined together by the transition to indicate the start and end of an event of the work. The condition of the state transition is represented by the Petri nets. The module and the worker's model are integrated by having them share the transition indicating identical operational starting points as shown in Figure 6. This enables the module's and worker's state transition to be expressed at the same time and is simulated by providing the necessary information to the factory model.

3.4 PROCESS REPRODUCTION USING MONITORING INFORMATION

The transportation state of the module, indicated by the simulation model in Figure 6, is the information that has been determined at the design stage. Based on this information, it is possible to carry out the simulation to determine assignment of representative operations for each worker as well as the time required for each operation. This information is expected to be acquired from the monitoring.

As Figure 7 shows, the transition of the model and responsible work of each worker is connected based on information obtained by monitoring. Furthermore, the working time can be obtained by monitoring the factory setting. Therefore, it is possible to reproduce the conditions of the site and to monitor the site using the simulation model. In the simulation, it is possible to reproduce the different steps of the actual module by changing the job assignment and the total number of workers. The model specifies various conditions that can

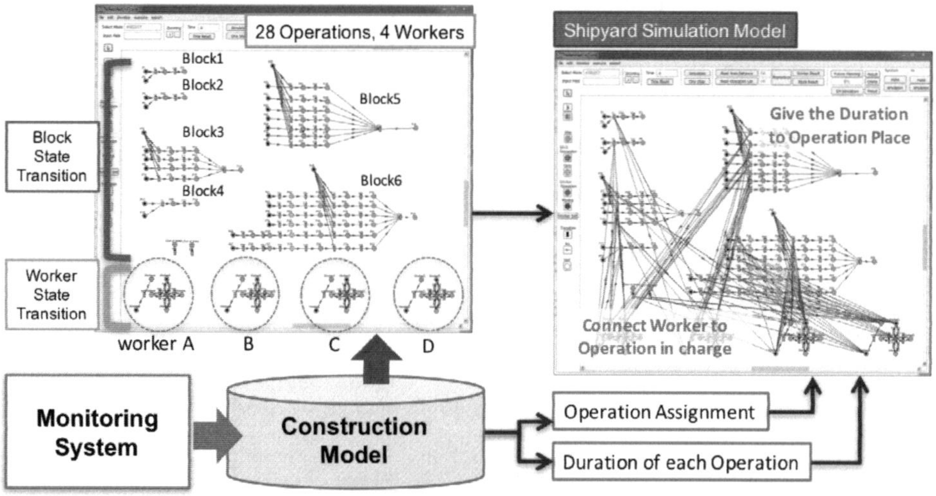

Figure 7: Generating Flow of Shipyard Simulation Model

© 2017: The Royal Institution of Naval Architects

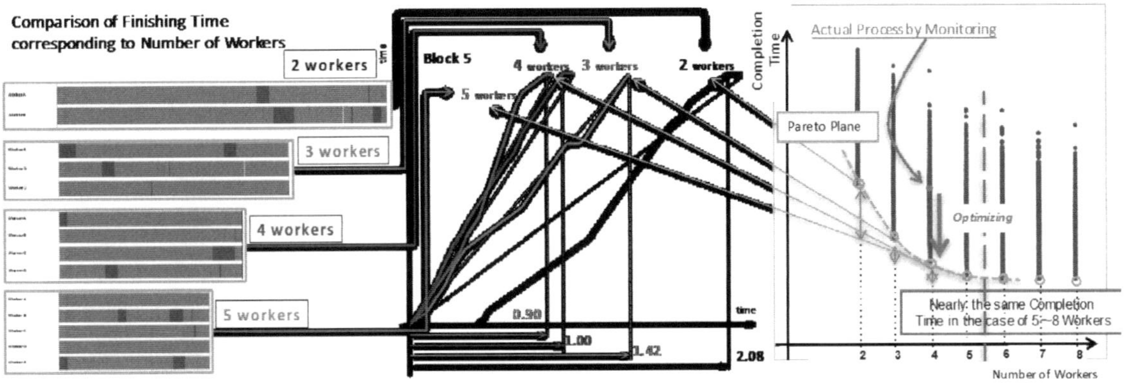

Figure 8: Sample Results from Shipyard Simulation

be used to explore process improvements by running the simulation.

4. FINDING PROCESS IMPROVEMENTS IN THE MONITORED PROCESS

4.1 METHOD OF FINDING IMPROVING PROCESS

During a construction process requiring a large number of operations, such as the shipbuilding process, the number of variable work combinations assigned to the simulation and the number of workers is very large. Therefore, it is very difficult to search for an optimal solution. So, a search for the improvement process was carried out by performing an approximate solution search using a genetic algorithm (GA).

Two evaluation indexes are introduced in the GA: the shortest total work completion time and the time margin of the process. The total work completion time is the end of the simulation time for one of the work assignments. The time margin of the process calculates the amount of the time extension that is possible for each operation in the process. The total amount of possible time extension was determined to be the time margin of the entire manufacturing process.

4.2 SIMULATION FOR PLANNING THE NUMBER OF WORKERS

Using the simulation result, we discuss the effectiveness of the proposed method. The monitoring information collected from three hours at the shipyard is set as the object of adaptation from the simulation. In this case, 28 tasks are executed by four workers. Plans for process improvements are explored for the three axes of minimizing the working time, maximizing the time margin, and minimizing the number of workers by performing simulations using the monitoring information.

4.2(a) The Case of Same Number of Workers

Using the simulation, we identified the shortest possible time to complete the task using the same number of works as used in the actual construction process. The simulation result showed that the total working time was reduced by 12% (Figure 8). In general, a ship is a made-to-order product, so it is rare that exactly the same construction process is reused for another ship. However, similar process may appear at the subassembly stage while building a small size construction module. Therefore, the shortest process identified in this way may be used for similar processes that may occur in the future.

4.2(b) The Case of Different Number of Workers

Comparing the shortest process and the actual process using various numbers of workers, it was shown that the shortest possible time varies according to the number of workers. In this case, as shown in Figure 8, the work completion time got faster until the number of workers reached five but there is no change in results when using more than five workers. In other words, it is meaningless to increase the number of workers to more than five people for this amount of work and it is only necessary to consider the process for two to five workers. It is possible to investigate the optimum value number of required workers by finding a process that satisfies the working time commensurate with the required completion date of the process using two to five workers for the amount of work in this case. As a result, it becomes possible to make proposals concerning the optimum number of workers for productivity improvement while maintaining strict observance of the required delivery date.

Figure 9 shows another sample result of finding the process improvements with three kinds of evaluation axes: "minimize work time", "minimize delay risk", and "minimize the number of workers". The delay risk is

© 2017: The Royal Institution of Naval Architects

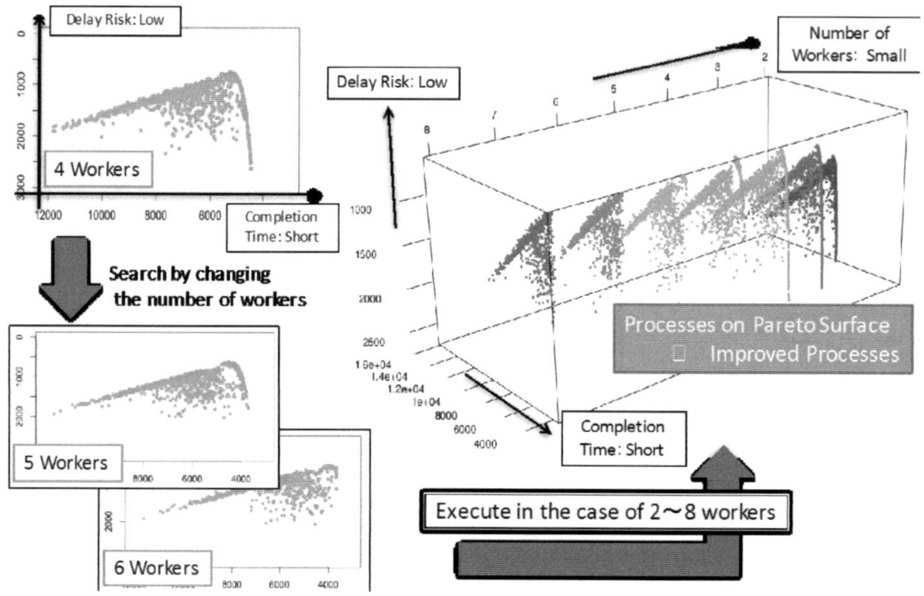

Figure 9: Sample Results from Shipyard Simulation

determined by calculating the estimated extended time possible for each work based on the type of work and the processing time as well as the amount of extended time that causes the entire process to be extended. A process with a smaller delay risk value is evaluated as a better process.

5. FINDING PROCESS IMPROVEMENTS IN THE ONGOING PROCESS

In the previous section, we discussed a method for finding better processes for the already completed process by simulation. In this section, we focus on real-time rescheduling using the proposed simulation method. By managing information using the construction model and monitoring the process, it is possible to compare the initial plan formulated before the start of production with the actual production progress after the start, detect process delays, and continually reschedule and optimize the schedule. Since the criterion for detecting a delay needs to be able to express how much delayed time can be recovered, "the time margin rate" is defined and used. In this research, we deal with re-assignment of workers within the construction module as a countermeasure to the delay as well as the movement of workers between job shops.

5.1 GRASPING PROGRESS AND DELAY OF THE WORK

The progress of the construction process determines how well the production is proceeding according to the plan. It is necessary to deal with a delay before the delay becomes so large that additional production resources are required, that is, while enough production capacity remains for the process. In this research, the production capacity included in the building process is expressed as "margin" and the time at which the margin of the construction process becomes smaller than a certain amount with respect to the incomplete work amount is detected as a delay. The rescheduling process starts with delay detection. In this research, two types of buffers (time margins) corresponding to different types of delays were defined as "non-planned work buffer" and "planned

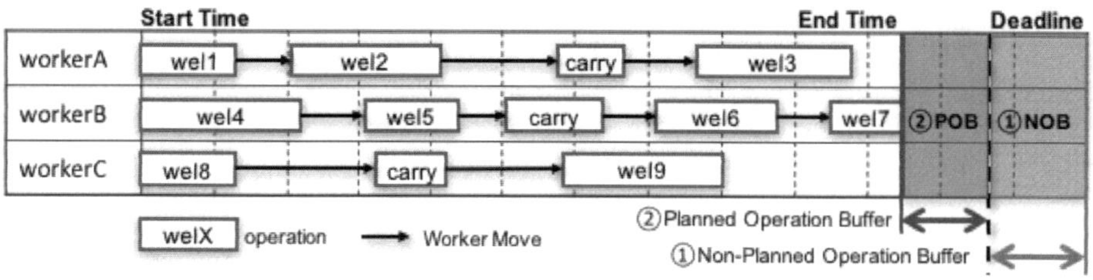

Figure 10: Process Buffer

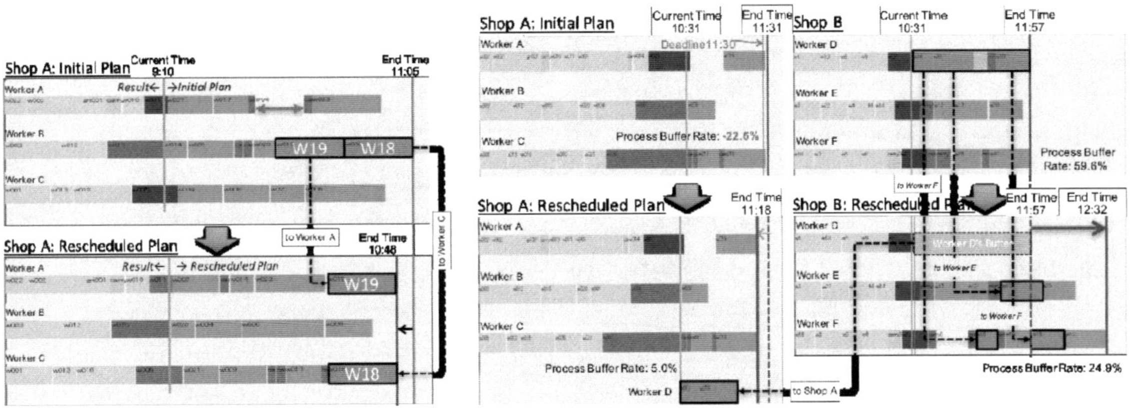

(A) Changing Work Assignment **(B) Movement of workers between Job-Shops**

Figure 11: Sample Results of Rescheduling

work buffer" that can absorb delays (Figure 10) in the construction process.

5.1(a) Non-planned Work Buffer

Non-planned work buffer is defined as margin provided to absorb delays in the construction process due to the occurrence of unplanned work. Non-planned work, such as rework and adjustments, may occur at the production site. Therefore, in preparation for the occurrence of non-planned work, a plan needs to be formulated so that the planned work is completed with a certain time margin available with respect to the required completion time of the block. We define this amount of time intended to be used for unscheduled work as the "non-planned work buffer" and include it in the process at a fixed rate relative to the unfinished work amount of the entire construction process.

5.1(b) Planned Work Buffer

Planned work buffer is defined as margin provided to absorb delays in the construction process due to the extension of the each planned working time. While the non-planned work buffer is a constant rate based on the workload involved in the construction process, the planned work buffer will vary depending on the plan and the progress through production. The time (margin) from the planned end schedule to the completion deadline (the required delivery time of the block) is subtracted from the non-planned work buffer and the remainder is labelled the planned work buffer.

5.1(c) Margin Rate

The "margin rate" is an index used to represent the condition when the margin of the construction process becomes smaller than a certain threshold in proportion to the amount of the unfinished work. Using the planned work buffer, the "margin rate" is determined by the ratio of the planned work buffer to the unfinished work amount of the process. If there is a large amount of work remaining, the risk of prolonging the work time will also increase so it is better to have a large margin for that contingency. If there is only a small amount of work remaining, a small margin time is sufficient to absorb delays. Therefore, the "margin rate" is represented by the ratio of the planned work buffer [person/hour] to the unfinished work amount [person/hour] for the process. The greater the margin rate, the higher probability that the process can meet the completion deadline. If there is a delay in the actual shop relative to the planned timeline, the margin rate of the actual process will decrease. By monitoring the variation of this margin rate, it is possible to grasp the progress in contrast with the plan. Rescheduling is initiated the moment the margin rate falls below a defined threshold.

5.2 RESCHEDULING ON AN APPROPRIATE TIMING

In this research, we deal with two cases as a method of rescheduling: changing work assignment in a job shop and moving of workers between job shops.

5.2(a) Changing Work Assignment in a Job Shop

The assignment of works that have not yet started at the time of rescheduling is changed with the assignments that lead to the shortest completion time is adopted as the optimum plan. Since the number of assignment combinations can be enormous, we searched quasi-optimal assignment using a GA. If there is work being delayed at the time of rescheduling, the optimum plan may change depending on the time at which the work is completed. Therefore, it is necessary to assume a completion time for the work being delayed and prepare a contingency plan with an optimal work assignment based on that assumption. If the work has not yet been finished at the preplanned time, the contingency plan is applied, the supposed completion time is updated, and the optimum plan is recreated and held again.

© 2017: The Royal Institution of Naval Architects

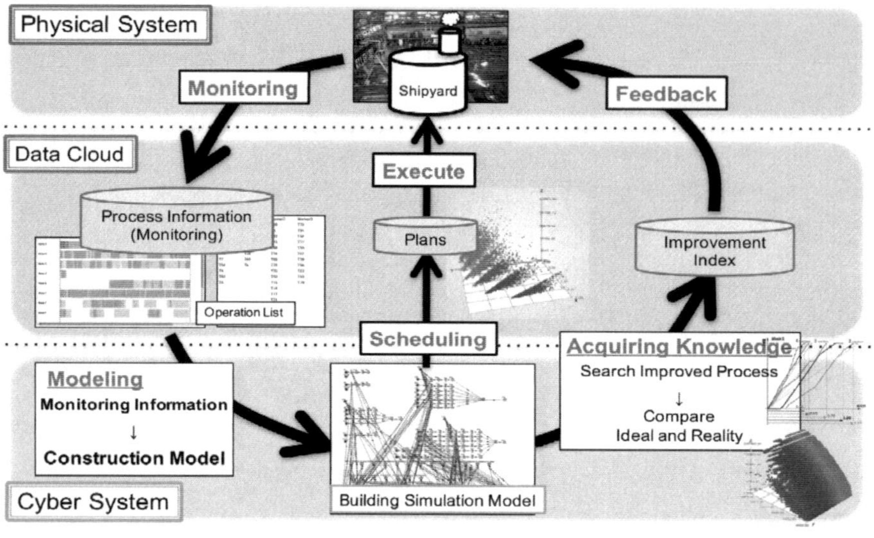

Figure 12: Loops of Improving Productivity by CPS

5.2(b) Movement of workers between Job Shops

In the case that the margin rate is negative even if the work assignment of workers within the construction module is changed, the delay cannot be made up without additional resources so it is necessary to call workers from other job shops. First, the job shop which has negative margin sends out notification that additional personnel are required. Another job shop that has some margin provides information on who can be sent. The simulation is used to verify that this job-shop can complete its process even with reduced number of personnel. If it is confirmed that the movement of workers is possible, the plan after increasing/decreasing the number of workers in both places is applied.

5.3 RESCHEDULING PROCESS VERIFICATION

We assumed a job shop (shop A) with 28 tasks and three workers as an example. In this example, a delay occurred during the construction process due required rework in the grinding task and extension of working time for multiple welding operations. As a result of these variations from the plan, the margin rate dropped and reschedule was performed. Based on the rescheduling, the task assigned to the worker who was the most delayed was distributed amongst the other workers, the lead time was shortened, and the margin rate was recovered. This example shows that it is possible to update the plan so that the entire process can proceed without delay (Figure 12(A)).

As another example, a case was assumed in which shop A is further delayed to the point that it could not meet the required delivery time even if rescheduling within the job shop was performed. In this case, there is job shop B with 29 tasks and three workers progressing as planned that has a high margin rate. Rescheduling between those two job shops, shop A and shop B, is carried out.

As a result of rescheduling, the number of workers increases by one in shop A making it possible for the construction process to be completed in time for delivery. In shop B, even with the number of workers reduced by one, it is confirmed that there is sufficient margin in the construction process to proceed without problems. By interchanging margins effectively between job-shops in this manner, it is possible to recover from delays in the construction process without adding new production resources (Figure 12(B)).

6. CONCLUSION

In order to use the monitoring data to improve productivity in shipbuilding, this paper discusses a method for integrating a shipbuilding simulator with a construction monitoring system (Figure 12).

A construction model was proposed to facilitate this integration. The construction model serves as the master model that can manage all of the system elements of the construction system and is very useful for generating a simulation model. In the construction model, the operation model is a key model for managing data on the construction process. In particular, associations of system elements are very important to the model. The current operation model is not so detailed so we have to check the features of real operations in shipbuilding. Achieving a more detailed operation model necessitates new monitoring methods to collect the detailed operation data.

A method of accomplishing real-time rescheduling is also described. In searching for a more efficient construction schedule, shipbuilding simulations using GA make it possible to explore potentially improved

© 2017: The Royal Institution of Naval Architects

process steps via simulation rather than utilizing actual process steps. Based on the developed simulation, a method for real-time rescheduling to prevent delays in the construction process is proposed. In this method, the margin rate is defined as a method for monitoring a process and a delay can be detected by comparing the margin rates of the planned model with the model produced from monitoring information. Furthermore, a method of dealing with delays is defined which detects a delay using the model and optimizes the plan according to the progress of the process at that time.

Future research into this topic is expected to include analysing the characteristics of the worker from the monitoring information, introducing these results to the model, and, thereby, achieving process analysis for a more accurate simulation. Given the challenges and potential of the proposed system, expansion into a large assembly has been left for further research. Additionally, the total set of available monitoring information not only supports the small assembly process, it can be expected to complement portions that cannot be recorded by a video camera by utilizing wearable technology.

Furthermore, the information available to be analysed includes acquired information necessary to construct an acquisition flow and feedback that can immediately instruct the operator for more efficient operation. Through these methods, cyber physical systems can be expected to reduce the time and resources required for traditional activities in gross business dealings.

7. ACKNOWLEDGEMENTS

This research received financial support from the Nippon Foundation. Additionally, the Japan Ship Technology Research Association (JSTRA) supported us through the research committee for the shipbuilding monitoring system. We would like to express our gratitude to the members of this committee for providing their valuable ideas and comments.

8. REFERENCES

1. Acatech (National Academy of Science and Engineering), Recommendations for implementing the strategic initiative INDUSTRIE 4. 0, 2013.

2. J. Liu and K. Aoyama, "Monitoring system for sub-assembling-shop to manage effective construction," Proceedings of the 15th International Conference on Computer Applications in Shipbuilding (ICCAS 2011), Trieste, Italy, September 2011.

3. J. Liu, J. Hirota, CX. Wu, and K. Aoyama, "Monitoring System for Advanced Construction Management (Integration Methods of Monitoring Data to Generate Virtual Job Shop)", Proceedings of the 16th International Conference on Computer Applications in Shipbuilding (ICCAS 2013), Busan, Korea, September 2013.

4. Y. Hiro, K. Aoyama and J. Liu, Monitoring System for Advanced Shipbuilding Construction Management: Extracting and Utilising Monitoring Data by Considering the Reliability of Monitored Data, 17th International Conference on Computer Applications in Shipbuilding 2015 (ICCAS 2015), Bremen, Germany, September 2015

5. K. Aoyama, T. Mizushima, Y. Hiro and K. Oizumi, Advanced Management of Shipbuilding Construction using Monitoring System and Job Shop Simulation, Proceedings of PRADS2016, Copenhagen, Denmark, September 2016

9. AUTHORS' BIOGRAPHIES

Kazuhiro Aoyama currently holds position of Professor at the University of Tokyo. He is responsible for realizing CPS from the view point of integrated design and manufacturing systems. His previous experience includes topics such as Computer Integrated Manufacturing systems in shipbuilding, information and knowledge management, decision support environment, project management, and process modelling.

Hoshi Mayuna was a graduate student at a Master course of the University of Tokyo. She was responsible for the rescheduling method using the construction simulation system. Currently, she works as a Systems Engineer at Nomura Research Institute.

Yusei Hiro was a graduate student at a Master course of the University of Tokyo. He was responsible for generating the simulation model using monitored data. He currently works as a Systems Engineer at Nomura Research Institute.

Ryo Kitamura is a master course student at the University of Tokyo. He is responsible for realizing CPS of construction system in shipbuilding.

Kazuya Oizumi currently holds position of Assistant Professor at the University of Tokyo. He is responsible for the rescheduling method using construction simulation system. His previous experience includes topics such as system design theory and systems engineering methods.

© 2017: The Royal Institution of Naval Architects

© 2017: The Royal Institution of Naval Architects

POWER WITH SIMPLICITY TO VISUALISE YOUR DIGITAL ASSET

A REVOLUTION FOR DECISION MAKING

G Stonestreet, Marine Products Strategy, AVEVA Solutions Ltd, UK.

SUMMARY

As shipbuilders seek to increase productivity, owners seek more control and visibility across every element of the delivered ship. This paper explores how shipbuilders and shipowners can benefit from developing a unique, centralized, and trusted living source of information to address modern industry trends and to make better-informed decisions.

In the meantime, the industry landscape is evolving, projects are getting more and more challenging and complex from various aspects, workers' behaviours are changing, and so technology needs to adapt to new expectations and new challenges. This paper also presents a modern solution to access, visualise and search efficiently into massive digital assets, avoiding the typical "information chaos" and the time-consuming research of information often implied by complex projects.

Finally, the paper provides a reflection on how Virtual Reality is already taking the interaction with the Digital Asset to the next level and how this applies to the industry.

INTRODUCTION

In the last decades, digital solutions have simply become unavoidable in the shipbuilding industry. Indeed, the use of digital models is driving greater clarity, control and cost-effectiveness in all activities across the ship lifecycle, ranging from planning, design, engineering, to procurement, manufacturing and commissioning.

But the growth of the global offering for digital solutions, led by the continual innovations and the multiplication of independent specialist solutions, combined with an exponential raise of projects complexity, have often driven shipyards into highly complex digital landscapes. Although each bit of information is meant to improve a given part of a given process, the overall efficiency of the full process now tends to be ruined by the difficulty to access, search, verify, update, compare and distribute a "chaotic sea of information". In other words, the effort it takes to look for and to validate information can exceed the benefit we can gain from it.

AVEVA's approach to the Digital Asset is to come with a centralized data repository, focusing mainly on the quality and the flow of information, independently of the originating source of this information.

In this context, the first challenge was to collect information from all sort of sources in a "universal" way, in order to store it all in a generic hub (populating the "Digital Asset").

The second challenge was to establish relevant connections between this massive amount of disconnected bits, in order to consolidate the information, and by this, maximize its intrinsic value.

The third challenge was to find out how the information could be efficiently sorted, filtered and distributed to the end users based on any given context, avoiding disruptive information research and information overloads.

Finally, another important challenge was to find a performant way to physically visualize this information, at the same time in the overall context of the massive digital asset, in order to capture the context of things, and also in the smallest details, to allow an accurate description of things.

THE DIGITAL ASSET DEFINED

Before exploring how a digitalised approach can revolutionise decision making in shipbuilding it is important to establish how 'digital asset' or 'digital twin' may be defined.

The digital storage of all information relating to an asset is the digital asset which, due to its digital nature, can be

© 2017: The Royal Institution of Naval Architects

constantly synchronised with the physical asset; capturing, monitoring and connecting everything in the physical asset.

Shipbuilding creates an enormous amount of information. This is information used not just in the design and construction of the ship but which must then be carried forward to allow ship owners and operators to understand the ship's structure in infinitesimal detail. In short, the digital asset improves predictability and operational reliability:

- **Predictability:** The information is not required for one moment in time alone (the design and build phase), it is essential that it can be accessed and used for all time; the life of the ship, and beyond.

  For the life of the ship it enables confident and informed operations, and preventive maintenance schedules based on a complete view of every component part within the ship.

- **Reliability:** The vast amount of information produced can be harnessed and exploited if captured digitally, enabling it to be treated as the core of every project and facility. Each physical asset within a ship, no matter how small, has information pertaining to it. This may include, but is not limited to, model number, warranty details, maintenance requirements, operating instructions and other operational data, user manuals, inspection history, health status, mechanical integrity, technical information, and isolations. It covers the profile of the asset and effectively summarises the expectations a shipbuilder or an owner can make of it.

- Of course, such information can be kept in files and folders, books, manuals and CDs, as it currently is with many organisations. If we accept that information that is available is all of value, and accept that the amounts of information are therefore enormous, there is a high likelihood that retrieving it from manually curated and often diverse source locations and formats will not only be a complicated and time-consuming process – more so as time elapses between installation and the need to refer again to source information – but that there is a possibility that it may not be retrievable. It may simply have been lost. Alternatively, all this information can be stored digitally, available at the click of a mouse, and visually accessible through digitised fully-rotatable 3D models.

WHY SHOULD DIGITALISATION INTEREST THE INDUSTRY?

OLD WAYS

Inconsistency starts with incompatibility

There are many instances where the digital asset, or the 'digital twin', may be used by one department or specialist team in an organisation, but not by others who feel that their traditional approach has always worked and will always be sufficient for purpose.

This variance may also prevail between partner organisations fulfilling different roles in the development of a project, but needing to share common data. So, while one team works with the technology solutions of today, it may end up handing the project to the next team which then prefers to work with what it believes to be tried and trusted approaches and workflows, not necessarily fully aligned with digitised capabilities.

These incompatible workflows will often be manually focussed, depending and even wholly reliant on individuals' skills, and thus exposed to the risks of human error. Such approaches only employ minimal, if any, elements of process automation; processes which introduce robust and reliable data not subject to compromise which may arise from human intervention and data manipulation.

Seamless efficiency

Partial adoption of a technology solution within an organisation is likely to complicate processes rather than simplify them. Digitalisation is simplification. It is also the seamless transference of information from one user to the next, one department to the next, one stage to the next; a process enabled and accelerated through high performance cloud technologies.

To harness the improved productivity benefits that come from digitalisation, however, requires that the shipbuilding organisation sees itself also as seamless; without barriers between functions in the company caused by the siloed approach – now universally outmoded by the 'joined up' organisation.

Although siloes still exist in the industry, the progressive organisation has removed such barriers to smoother operation and broader collaboration, as will be observed in the partnership between 40 European partners from the maritime industry engaged on the HOLISHIP project.

The partners include HSVA (coordinator), ALS Marine,

© 2017: The Royal Institution of Naval Architects

AVEVA, BALance, Bureau Veritas, Cetena, Center of Maritime Technologies, Consiglio Nazionale delle Ricerche, Damen, Danaos, DCNS, DLR, DNV-GL, Elomatic, Epsilon, Fraunhofer-AGP, Fincantieri, Friendship Systems, Hochschule Bremen, IRT SystemX, Institute of Shipping and Logistics, Lloyd's Register, MARIN, SINTEF, Meyer Werft, Navantia, National Technical University of Athens, Rolls Royce, Sirehna, SMILE FEM, Starbulk, TNO, TRITEC, Uljanik, Univ. Genoa, Univ. Liege, Univ. Strathclyde, van der Velden.

HOLIstic optimisation of SHIP design and operation for life cycle [HOLISHIP]

The HOLISHIP project is funded by the European Commission within the HORIZON 2020 Transport Programme. The project aims to bring together a wide range of software tools to drive a unified Computer Aided Engineering platform to include optimisation and virtual demonstration of vessel performance. The approach emphasises the dependency of systems and components in a ship to ensure that they are considered concurrently.

A ship both is, and is not, a collection of diverse standalone elements.

It is: because every constituent part performs a dedicated function, many of which are not interdependent. Information about these parts and systems is crucial for understanding the ship in operation as well as its maintenance needs into the future.

It is not: because one critical system failure can impact the smooth operation of the ship, viewed as a whole the ship is a cohesive entity with a convoluted anatomy within which every organ has to function efficiently for the greater sustainable performance of the whole.

The HOLISHIP principles have been established thus:

"A system approach to ship design and operation for life cycle considers the ship as a complex system integrating a variety of subsystems and their components, e.g. subsystems for the creation of the hull, generation of energy/power and ship propulsion, for cargo storage and handling, accommodation/comfort/safety of crew/passengers and ship navigation. Inherently coupled with ship design and operation and related decision making is optimisation, *namely the selection of the best solution out of many feasible ones on the basis of a criterion, or rather a set of criteria. All of the above criteria shall be addressed and covered in the new HOLISHIP platform which is currently developed."*

AVEVA is proud to be associated with the HOLISHIP project.

Another example can be found in the collaboration between Rolls-Royce, the Norwegian University of Technology Science (NTNU), SINTEF Ocean and classification society DNV GL who have agreed to work together to create an open source digital platform for use in the development of new ships. The project will allow the creation of digital twins, which the group has said serve as "a virtual test bench to assess the safety and performance of a vessel and its systems, both before its construction and throughout its life". DNV has also stated that a platform of this nature could form the basis for future class services.

REDUNDANT APPROACHES

The current situation in shipbuilding is that digital models are almost exclusively used by specialists, using dedicated software solutions. It is common practice to have isolated and redundant models for each discipline and each purpose.

This 'isolation' provides a potentially fertile breeding ground for errors; both those involved in the recreation of function- or department-specific models and drawings (often involving re-origination of source data, reassessment of data, and judgement-based decisions about information). When unilateral practices are undertaken, difficulties immediately stand in the way of sharing the whole, unified and up-to-date information with all stakeholders.

Not all stakeholders, however, require all the information all the time. By moving towards a digitalised approach the information can be segmented to address the specific needs of specific users, thus avoiding the negative impact of 'information overload'. Filtering information accelerates the collaborative process and focuses specialists' attention directly onto the required task, driving faster responses, in shorter timeframes, and thus improving productivity.

STREAMLINED INFORMATION SHARING – THE DIGITAL ASSET IN ACTION

NEW WAYS

Improving the quality and reliablity of decision making across capital projects and operations

Digitalisation should and will impact and empower every aspect of the shipbuilding process. All the information related to the design, construction and operation – all the

© 2017: The Royal Institution of Naval Architects

way through to the decommissioning of a ship – can be accessible from one single and trusted information platform, enabling collaboration and ensuring consistency. Such is the goal of the HOLISHIP project, scheduled for completion towards the end of 2020. Viewed as a whole, the single source becomes a digital twin of the physical asset; a model in which the entire ship can be explored.

At any point during such an exploration journey, any functioning part of the ship, any fixture and fitting, from the most mission-critical and even through to the cosmetic and non-essential, can itself be further interrogated. From a piston to a propulsion system, from a propeller to a porthole, the ship no longer holds any secrets or conceals any difficult-to-access parts. One can be on board from a keyboard.

All authorised participants or stakeholders in a project can access the digital information easily and intuitively from any device. This is the digital world; the efficient environment in which shipbuilding should now seek to relocate itself and redefine its efficiency and productivity.

To achieve such objectives, the main challenges are:

- Integrating, comparing and consolidating complex and massive information from diverse and disconnected sources;

- Exposing this information in an accurately filtered way, without provoking information overloads.

THE FUTURE OF DECISION SUPPORT

If information sharing can be streamlined, with the end goal of a handover where shipbuilders deliver physical and digital asset at the same time, decisions are more comprehensively supported at every stage. Decisions can be made faster, helping to ensure sticking to construction schedules, and the corollary is that budgets will be less likely to overrun. Once information sharing is streamlined, its practical value becomes amplified. The simpler it becomes to collect, visualise and interact with the data, the greater the power for shipbuilders and owners.

This section looks at those three aspects of improving decisions by learning to work with the digital asset:

- o Streamlining information sharing
- o Live asset and digital twin: the competitive advantage
- o Power with simplicity

- **Streamlining information sharing**

Industrial Revolution 4.0 is upon us. Connected data from the Internet of Things is adding to the world's knowledge repository and can be of significant value to shipbuilders and owners. Other technologies are coming together to create unprecedented choice in how organisations can drive their efficiencies, but this confluence of capability also creates an aura of confusion as organisations consider the limitless options available to them.

Thus shipbuilders are considering not just the digital transformation imperative but associated technologies that can enhance further the benefits it brings. Among these associated technologies may be considered capabilities that make real-time visualisation not just possible, but simple: 3D scanning and models, remote and automated scanning, augmented reality/virtual reality and HoloLens mixed reality.

There will be less confusion as the various new technologies and processes link together to make the industry more efficient without it having to recreate genius every time it seeks to make a leap forward. That leap has largely already been made as real-time visualisation builds on data in the cloud to deliver advanced data visualisation. In a content-rich system important information can be drawn together from predictive analytics, connected IoT devices and other sources and put into context with static near-term information to make better decisions, improving the next generation of design.

Services are available to shipbuilders and owners for visual analysis. Such services give context to understanding the interdependencies in shipbuilding data and give the ability to explore insights. It should be borne in mind that data visualisation works effectively also for retrofits where the 3D environment can be recreated with laser scans to produce a result that looks much as it would if it had originated in a 3D model.

- **Live asset and digital twin: the competitive advantage**

The depth of detail permissible with the digital asset should not be underestimated. This is an

© 2017: The Royal Institution of Naval Architects

aid not just to shipbuilders either at construction or retrofit but also to owners. Objects within a 3D model can be tagged. Every tagged object is selectable and is linked with the full contextual information that resides in the Digital Asset.

Whatever is available in the Digital Asset can be instantly recalled and viewed with a touch of the screen; this includes vendor datasheets, layout drawings, process schematics, purchasing information, and much more – effectively it includes all available information.

If shipbuilders can handover a complete and 'live' digital twin when they handover the physical asset (the ship) they stand to gain a significant competitive advantage. The digital twin is virtually a guarantee that there are no mysteries within the ship's construction; no problems that may emerge as costly issues to the owner at some future time.

Such actions build trust by offering transparency. Owners can themselves utilise the digital twin on an ongoing basis, understanding every component part, mastering every interdependency and controlling every task subsequently involved in ensuring a smooth and cost-effective operation. Adoption of the digital twin approach is now beyond being a preference; it is an imperative. Indeed, many owners are mandating that shipyards delivery a digital twin along with the physical ship. It serves to deliver greater safety levels and operations efficiency. The digital twin enhances the sustainability of functioning and effective ship operations

- **Power with simplicity**

  At AVEVA, we define the wide range of benefits that accrue from adoption of the digital twin as 'power with simplicity'. If views of the asset, once digitised, can become intuitive to work with and dynamic in their continuous relevance, then they do become empowering. This is the value of visual, digital decision support.

  Instant access to asset information improves the speed, accuracy and reliability of decision making throughout the asset life cycle; through 3D visualisation a seamlessly connected view of project and asset information delivers a highly efficient and effective decision support capability for build and retrofit projects and operational environments; from shipbuilder to ship owner.

AVEVA's ROLE IN ADVANCING THE USE OF THE DIGITAL TWIN

AVEVA Engage removes barriers to collaborative decision making, enabling shipbuilders to intuitively understand project and asset data in ways never previously thought possible. It extends the benefits of 3D visualisation and Information Management to decision makers at all levels, in all disciplines. By placing the entire digital asset at the disposal of all project participants, in Ultra High Definition (UHD) detail, finding everything is simplified through the power of touch.

THE WAY AHEAD

VIEW FROM THE LAB: NEW IDEAS IN SHIP DESIGN

Virtual Reality (VR) will improve spatial awareness in ship design, by providing fully-immersive decision support to overlay onto and enhance visual and digital decision support. VR enables the user to engage in walk mode with the 3D model and to navigate the model with a teleport. It is likely that such developments will redefine information sharing even beyond such capabilities, however.

Team members in diverse locations will be able to use Skype for parallel audio communication, supporting multisite collaboration, real-time shared problem solving, and increasingly improved decision making. This shared experience will be enriched by picture-in-picture (PiP) enabling each user to see what the other user is seeing. The real sense of space and distance will put project participants on board, without their having to leave their offices.

CONCLUSIONS

Ship owners and operators are relying more and more on digital information and 3D models in their activity. The need to have one single centralised and up-to-date information hub is becoming crucial to all the actors involved in a ship's life cycle, from design to operation. Ship builders can no longer expect that their processes will stand up to rigorous owner demands and competitive pressures if they do not at least explore the use of the digital twin and at best more to its early adoption.

For several years, AVEVA has been developing tools to link asset documentation by context to create a connected dataset which allows critical and any related information

© 2017: The Royal Institution of Naval Architects

to be found quickly, and expose trends in data providing insight into ways the asset may be changing over time.

Simplicity is becoming simpler. The growth and increased familiarity of mobile devices now allows information to be published and consumed in ways beyond the traditional desktop or laptop, as well as extending how datasets can be continuously updated during operation using sensors on many devices. Such developments are bringing high degrees of agility to the shipbuilding community. Complex ideas can now be presented on tablets on the move. Contributors can collaborate from diverse locations. Decisions can be more fully informed.

AUTHOR'S BIOGRAPHY:

Gauthier Stonestreet holds a Master's degree in mechanical engineering and naval architecture, beginning his professional career as a naval engineer in both design and production departments of several shipyards. He joined AVEVA in 2008, providing customer support to many European shipyards, before becoming Product Strategy Manager for AVEVA's marine software portfolio.

© 2017: The Royal Institution of Naval Architects

International Conference on Computer Applications in Shipbuilding (ICCAS 2017)

Singapore
26 - 28 September 2017

Volume 2 of 3

ISBN: 978-1-5108-8303-1

Printed from e-media with permission by:

Curran Associates, Inc.
57 Morehouse Lane
Red Hook, NY 12571

Some format issues inherent in the e-media version may also appear in this print version.

Copyright© (2017) by The Royal Institution of Naval Architects
All rights reserved.

Printed by Curran Associates, Inc. (2019)

For permission requests, please contact The Royal Institution of Naval Architects at the address below.

The Royal Institution of Naval Architects
8-9 Northumberland Street
London, WC2N 5DA
United Kingdom

Phone: 44 207 235 4622
Fax: 44 207 259 5912

publications@rina.org.uk

Additional copies of this publication are available from:

Curran Associates, Inc.
57 Morehouse Lane
Red Hook, NY 12571 USA
Phone: 845-758-0400
Fax: 845-758-2633
Email: curran@proceedings.com
Web: www.proceedings.com

CONTENTS

© 2017: The Royal Institution of Naval Architects

© 2017: The Royal Institution of Naval Architects

© 2017: The Royal Institution of Naval Architects

© 2017: The Royal Institution of Naval Architects

International Conference on Computer Applications in Shipbuilding (ICCAS 2017)

26-28 September 2017, Singapore

Volume 2

SHIPBUILDING
ICCAS
SINGAPORE | 2017

Sponsored by:

Altair AVEVA BETA SIMULATION SOLUTIONS
CADMATIC SOFTWARE SOLUTIONS INTERGRAPH SENER
SIEMENS Ingenuity for life SSI VIRTALIS

THE ROYAL INSTITUTION OF NAVAL ARCHITECTS
8-9 Northumberland Street, London, WC2N 5DA, Telephone: +44 (0)20 7235 4622

VARIATION OF ASYMMETRIC SIDE HULL OF PENTAMARAN HULL RESISTANCE BY COMPUTATIONAL FLUID DYNAMICS (CFD)

Z Zikrina, Yanuar, and F A Nugroho, Universitas Indonesia, Indonesia

SUMMARY

Fast vessel technology is rapidly developing with the increasing of the demands in maritime field such as military, shipping and tourism. The substantial of designing fast vessel is to create a vessel with highest speed by the lowest fuel consumption. Ship hull optimization is one of methods to reduce fuel consumption in designing the high speed vessel. The advantages of multihull vessel are huge payload, wide deck area, good stability and seakeeping, also decrease the total resistance due to wave making resistance caused by wave interference of the hulls. Multi-hulll vessel are divided by number of hulls which are catamaran, trimaran, quadramaran and pentamaran. In a research by Tuck (1998) shows that the wave resistance of catamaran, trimaran and quadramaran based on Michell's theory (classical thin ship theory) is decreasing with the increasing of number of the hulls and Froude Number at the high speed. Pentamaran hull is hulls form development of catamaran, trimaran and quadramaran which has inner and outer side hulls. This research is aim to show the effective configuration of pentamaran ship hull with inboard and outboard Asymmetric hull at outer and inner position by investigating the characteristic of total resistance at Froude Number 0.1-1 compared to symmetric pentamaran hull. Computational Fluid Dynamics (CFD) method have been performed to predict the resistance of hull-form besides its advantages is economical in time and cost. In this research, the calculation of pentamaran hull resistance is simulated by ANSYS – FLUENT (CFD Software Package). The multiphase region is separated by defining boundary air inlet as the flow of air and water inlet as the flow water in the pentamaran hull ship region. The free surface model is performed in calm water condition by Volume of Fluid (VOF) model and k-epsilon is used to simulate the turbulent viscous model. The validation data obtained from the Computational Fluid Dynamics (CFD) method were found almost similar with latest pentamaran experimental data. The result shows the coefficient resistance characteristic of Asymmetric pentamaran hull ship compare to symmetric pentamaran hull ship.

Keywords: fast vessel; multi-hull; pentamaran; ansys – fluent; volume of fluid

NOMENCLATURE

υ	Kinematic viscosity (N s m^{-2})
ρ	Density of water (kg m^{-3})
P	Pressure (N m^{-2})

1. INTRODUCTION

One of the essential thing in designing fast vessel is to create the hull model with the lowest resistance to reduce fuel consumption. Multihull vessels are categorized by the number of its hulls, those are catamaran, triaran, quadramaran and pentamaran. Pentamaran or five hulls design is new innovation in multihull research to decrease fuel consumption. The pentamaran hull is able to produce very low resistance at high speed for certain configuration of hull which is determined by configuration and form of each hull.

Pentamaran is going to be an interesting research topic to discussed because it able to produce lower resistance than other forms of multi-hull at high speeds by changing the configuration of the side-hulls to main-hull. The hull configuration of the vessel pentamaran varied based on two parameters, namely the ratio of the distance between the hull transversely to the length of the main-hull or so-called staggered and comparison distance between the hull longitudinally to the length of the main-hull or so-called separation. Both of the parameters will be varied to be proper configurations to produce both the maximum and maximum resistance of the wave and total resistance at high speed.

Investigation of pentamaran hull resistance has been performed in 2001 by Peng et al. Peng calculated the pentamaran ship resistances with two various configuration of staggered side hulls by numerical methods. The research has shown the advantages interference of pentamaran ship due to the value of pentamaran resistance coefficient at Froude Number 0.6 [1]. In 2013, Yanuar investigated the resistance of pentamaran ship by experimental methods. Those are nine pentamaran ship configuration varied by staggered and separation side hull arrangement. The result said that separation pentamaran ship configuration with S/L=3/16 has the lowest resistance [2]. Whereas, the staggered pentamaran ship configurations show uncertain value of coefficient resistance. In 1997, a research for asymmetric side hull on trimaran ship shown that it produces the lower resistance than the symmetric [3]. This paper will find out the effect of asymmetric side hull arrangement on pentamaran ship.

The Computational Fluid Dynamics (CFD) method has been succeeded in showing similar result to the experimental method. The simulation of water flow over the ship hull already implemented to determind ship resistance by using RANS Equation [4]. This paper presents variation of pentamaran hull configuration by varying the asymmetric hull arrangement and calculing the resistance by CFD software package, Ansys- Fluent.

2. THEORITICAL APPROACH

2.1 SHIP RESISTANCE

Ship resistance occurs caused by air and water interaction to the hull surface which produce losses wave energy. To analyze the advantages or disadvantages of ship resistance, the coefficient resistance is compared to Froude Number in a Graph. The coefficient resistance is determind by Equation 1.

$$c = \frac{R}{\frac{1}{2}\rho V^2 S} \quad (1)$$

The Froude Number is determind by Equation 2.

$$Fn = \frac{V}{\sqrt{g.L}} \quad (2)$$

Viscous resistance and wave resistance are two main components of ship resistance as well as Molland's explaination (2008) in Figure 1. The viscous resistance are supported by viscous form resistance and skin friction, whereas the wave resistance are effected by pressure resistance [5].

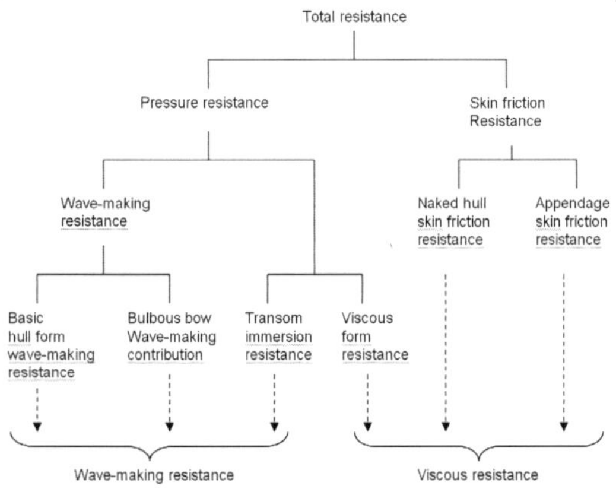

Figure 1: Ship Resistance Components [5]

The calculation result by Tuck (1998) shown the wave resistance has contribute to reduce the total resistance of multihull ship. Based on michell theory, wave resistance decrease caused by wave interaction produced by each hull [6]. In Figure 2 shows the characteristic of wave resistance to the Froude Number due to the increasing of hull number. A research about pentaran hull resistance by Tarafder (2013) conclude that wave resistance coefficient start to decrease at Froude Number 0.8 [7].

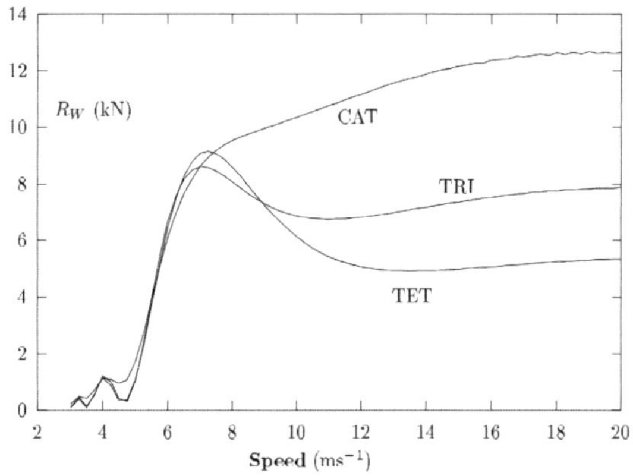

Figure 2: Wave Resistance Vs Speed [6]

2.2 ANSYS - FLUENT

Ansys – Fluent can be used to determind the compressible or incompressible, laminar or tubulent fluid flow and solve the problem based on momentum equation. In this case, the fluid flow is assumed as incompressible, viscous turbulent and multiphase flow. So that, Reynold-Average Navier Stokes (RANS) is used to simulate the case. In a research by Sahoo [2015] about investigating trimaran resistance using CFD, Volume of Fluid (VOF) model is needed for modelling the free surface between air and water [8].

3. CFD SIMULATION

There are three main process of CFD Simulation, those are pre-processing, processing, and post processing.

- Pre-processing
  In this step, the geometry is prepared as well as the real condition and the boundary conditions are defined. Then, meshing process are build by define number and size of grid.
- Processing
  In this step, the computer is trying to solve the problem.
- Post-processing
  In this step, the result is prepared to be shown.

In this research, the CFD simulation started with design the model, defining meshing and configure the CFD setup.

3.1 MODELLING

Before starting the simulation, the essential part is preparing a model. Pentamaran ship has five hulls, the main hull is located in the centre of body and others hull called side hull. Asymmetric hull forms are combined in the pentamaran ship as the side hull. In this research, there are five pentamaran ship designs which varried by configure the asymmetric side hull arrangement as we can see at Figure 3.

- A : Asymmetric Outboard Outer side hull
- B: Asymmetric Inboard Outer side hull
- C: Asymmetric Outboard Inner side hull
- D: Asymmetric Inboard Inner side hull
- E: Symmetric

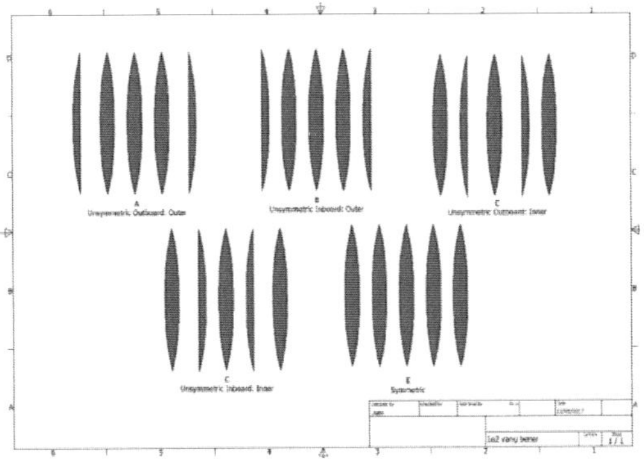

Figure 3: Variation of Pentamaran Ship

Each hull has the same hull form and dimension. The dimension of the hull shown at Table 1.

Tabel 1 Main Dimension

Main Dimension	
Lenght (m)	1.8
Beam (m)	0.18
Height (m)	0.17
Draft (m)	0.07

The wigley hull form has the lowest resistance of monohull. These pentamaran ship geometry are designed by lofted section to other section. The dimensi of domain shown at Table 2.

Tabel 2 Domain

Domain				
Forward	0.5x	Loa	0.9	m
After	4x	Loa	7.2	m
Left Side	1x	Loa	1.8	m
Rigth Side	1x	Loa	1.8	m
Depth	2x	Loa	3.6	m
Upside	0.5x	Loa	0.9	m

The boundary condition are defined as air inlet & water inlet as velocity inlet, outlet as pressure outlet, walls and ship as walls, and top as symmetry as shown in Figure 4.

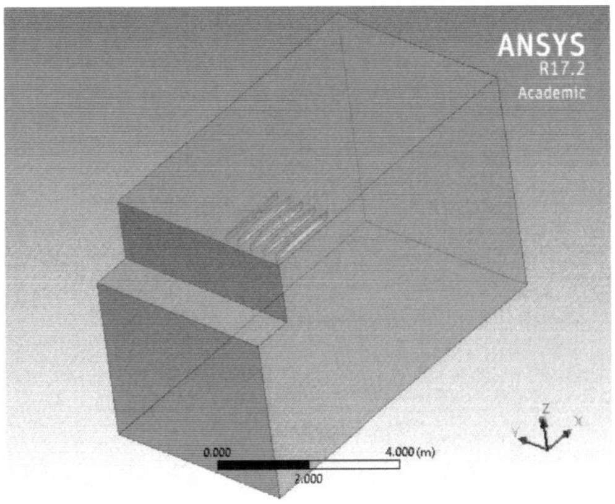

Figure 4: Boundary Condition

3.2 MESHING TECHNIQUE

Mesh quality and skewness are important parameters to create a good quality of meshing. If the value of mesh quality close to one, it describs a good meshing quality, and if the value is close to zero then the quality of mehsing is bad. Contrary to the mesh quality, if the value of skewness close to one, it means a bad skew of meshing and if the value cloe to zero means the skewness is good.

In this section, meshing technique is needed to implemented. There are many things affect mesh quality and skewness, such as size and type of elements. So, the complexity of geometry is considered. Simple geometry is easy to create the structured grid meshing, then possible to decrese time setup. But, for the complex geometry, there will be special treatment to creat a good quality of meshing. There are five various meshing technique already tried to create the best meshing grid. In this case, the geometry are quite complex, so that some special treatment are applied. The size of elements are 0.027, mesh quality is 0.99, and skewness is 0.0000041429.

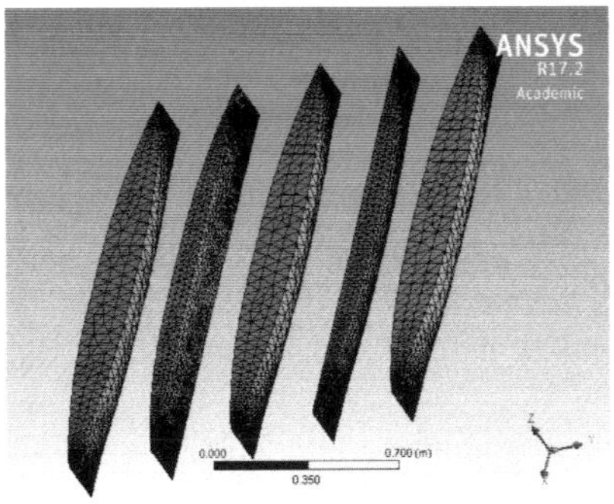

Figure 5: Meshing

3.3 CFD SETUP

To determind the accuration of CFD methods, validaton step is needed by compared the result of CFD with the result of experimental methods. The Figure 6 shows that the CFD methods performed a good agreement for calculating ship resistance by Ansys- Fluent.

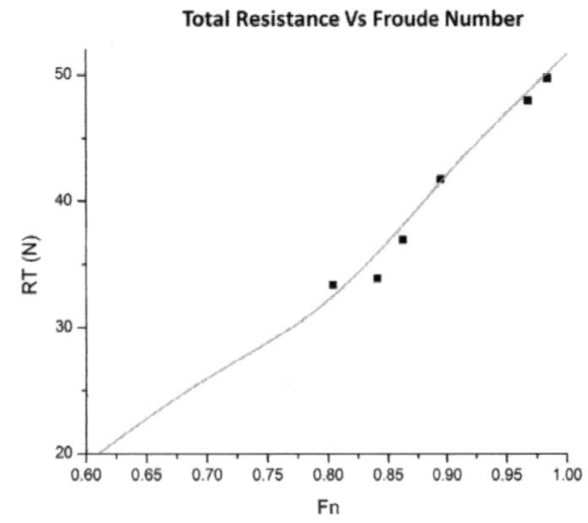

Figure 6: Validation

There are three types of multiphase model, those are Eularian, VOF (Volume of Fluid) and Mixture. In this case, VOF (Volume of Fluid) is used to modelling free surface by defining the value of surface tension between air and water. Tubulence model is important to simulate the fluid flow in turbulence condition. When petamaran ship is moving in a calm water, the fluid flows will become turbulence. Because the flow is assumed to be steady flow and constant velocity of ship, it is predicted that turbulent flow is not too complicated. Hence, k-epsilon turbulent model are applied to describe the turbulence phenomena in the simulation. Also, 2 % are given for the value of turbulent intensity and viscousity as. The volume fraction shown as Figure 7.

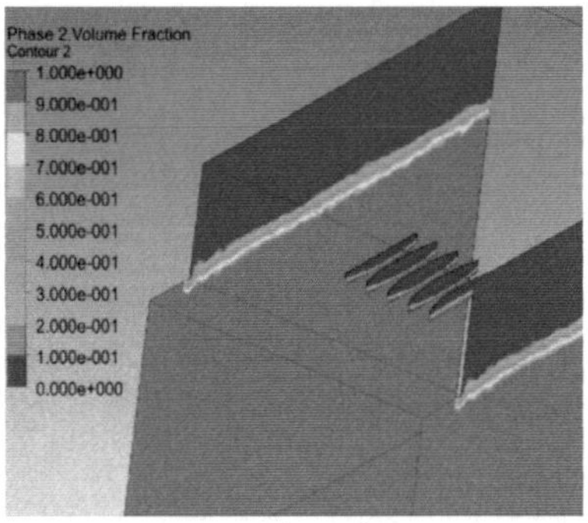

Figure 7:. Volume Fraction

4. RESULT AND DISCUSSION

4.1 COMPONENTS OF PENTAMARAN SHIP RESISTANCE

4.1 (a) Asymmetric Outboard Outer side hull

In Figure 8, the waveform resistance coefficient curve has a very steep decline at low speed precisely in Froude Number 0.1-0.2, then the sloping curve decreases along Froude Number 0.2-06. The coefficient value rises again at Froude Number 0.8 and decreases to Froude Number 1. On the viscous coefficient curve, the viscous coefficient value decreases steeper at n 0.1-0.2 than Froude Number 0.2-0.6. At Froude Number 0.7 there is a significant increase so that next the viscous viscous keofisien curve is above the waveform resistance coefficient curve.

This happens, allegedly the interaction of waves between the hulls. In catamaran ship, the interaction between the hulls is affected by the distance between the hulls and the speed increase [9]. On a pentamaran ship, the wave system will be more complicated than the catamaran vessels, so we suspect that the phenomenon of interaction between hulls waves also occurs on the ship's pantamaran with high complexity system. In general, the coefficient graph for ship A ships illustrates that the viscous coefficient has a more constant value along Froude Number 0.2-0.6 and Froude Number 0.8-1. While the wave resistance coefficient has a steep decline at low velocities and sluggish pengurnan at high speed. This causes the coefficient value of the wave resistance barrier to have a role in decreasing the total coefficient. Suspected this phenomenon occurs due to the interaction of anatar wave hull which causes a decrease in the coefficient of resistance wave pattern at high speed.

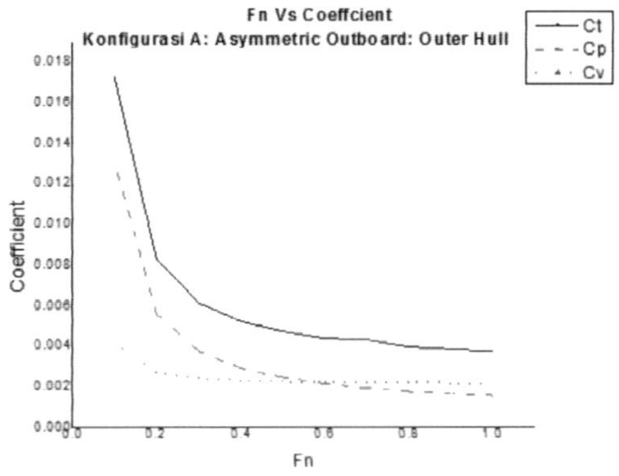

Figure 8: Asymmetric Outboard: Outer Hull

4.1 (b) Asymmetric Inboard Outer side hull

In Figure 9 the shape of the wave resistance coefficient curve in the range 0.1-0.2 is very steep and is above the viscosity coefficient curve. Next to Froude Number 0.2-1 the curve is below the viscous load curve with a downward trend with a non-significant decrease. At Froude Number 0.8 there is an increase in the wave resistance coefficient which affects the total coefficient on the mountain curve. This is because the viscous resistance coefficient in Froude Number 0.2-1 shows a constant curve as the value of Froude Number increases.

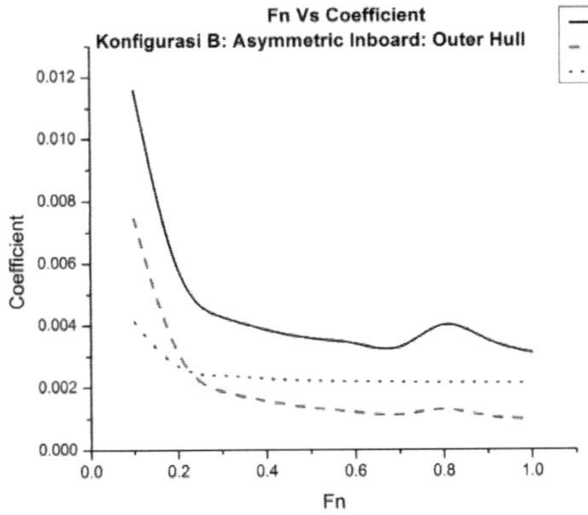

Figure 9: Asymmetric Inboard: Outer Hull

The phenomenon of the elevation form in the total resistance coefficient curve occurs due to the increase in the value of the wave resistance coefficient in Froude Number 0.8. The mountain-shaped or hump curve is due to the dominance of wave resistance at high speeds [2]. In addition, the viscous barrier coefficient does not affect the increase or decrease in total resistance due to the constant-shaped curve. This is because there is no change to the viscos barrier component that is the shape component, the friction resistance component and the appendage barrier or additional components.

4.1 (c) Asymmetric Outboard Inner side hull

On the curve, there are two components of resistance coefficient ie wave resistance coefficient and viscous resistance coefficient. In the wave pattern resistance coefficient curve shows in Froude Number 0.1-0.2 there is a significant decrease coefficient. Furthermore, in Froude Number 0.2-1 there is a decrease in the coefficient of wave resistance. However, the wave resistance curve is still above the viscous barrier curve along Froude Number 0.1-1. On the viscous resistance coefficient curve shows a consistent coefficient value in Froude Number 0.2-1. This occurs because viscous barriers do not contribute significantly to decrease or increase in the multi-hull ship's total coefficients resistance as shown in Figure 10.

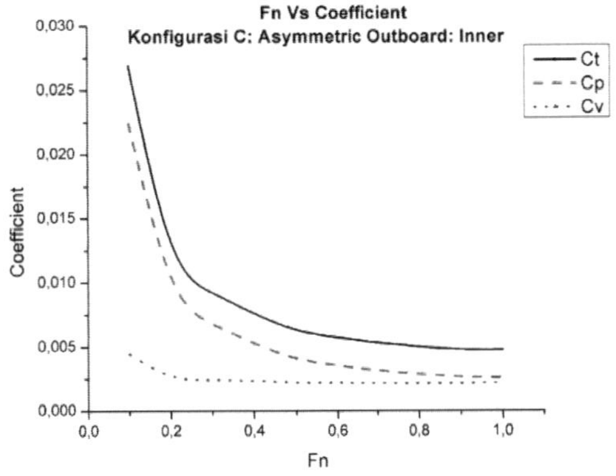

Wave resistance is the most important constraint on multi-hull ship resistance when speed increases, hence wave resistance plays a role in decreasing total resistance value. However, actual wave resistance coefficients may contribute to raising or lowering the total resistance coefficient [10]. In this case, the wave interaction generated by the hulls of the ship configuration which has two asymmetrical hulls on the inner side hull and outward, results in an adverse interaction as it contributes to the rise in the waveform resistance coefficient value.

4.2 (d) Asymmetric Inboard Inner side hull

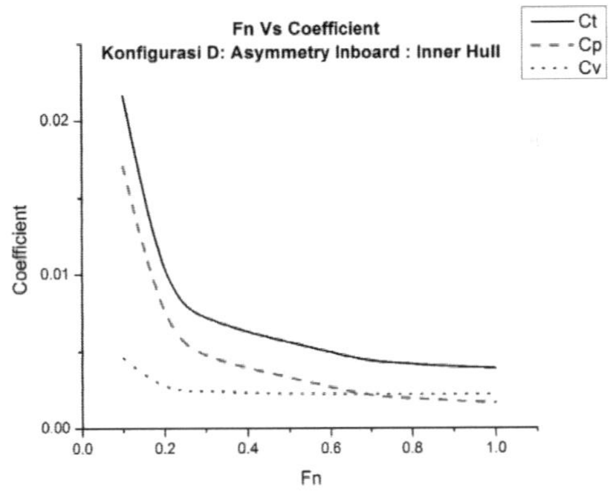

Figure 10: Asymmetric Inboard: Inner Hull

The barrier coefficient is a factor affecting the ship's resistance at any speed. In Figure 10, the wave coefficient curve of the wave pattern shows a drastic decrease in the Froude Number 0.1-0.2 range. Further along with the increase in Froude Number, the value of the wave pattern coefficient decreases with a small decrease to form a sloping curve. In the Froude Number range 0.7 -1, the wave resistance coefficient becomes so small that it is below the viscous viscosity coefficient curve. This is assumed because of the favorable wave interaction between the stomach in the configuration.

4.3 (e) Symmetric

As shown in Figure 11, the coefficient curve of resistance to Froude Number shows the coefficient of wave and viscous pattern resistance to the total coefficient. Curve of wave resistance coefficient has decreased greatly on the vulnerable Froude Number 0.1-0.2, in the range 0.2-0.4 decline also occurs but not too steep and then the decline becomes sloping. Throughout Froude Number 0.1-1 the value of the waveform resistance coefficient is always greater than the viscous resistance coefficient. The viscous resistance coefficient curve becomes constant at Froude Number 0.2-1. It is suspected that the viscous barrier coefficient component does not change as the speed changes.

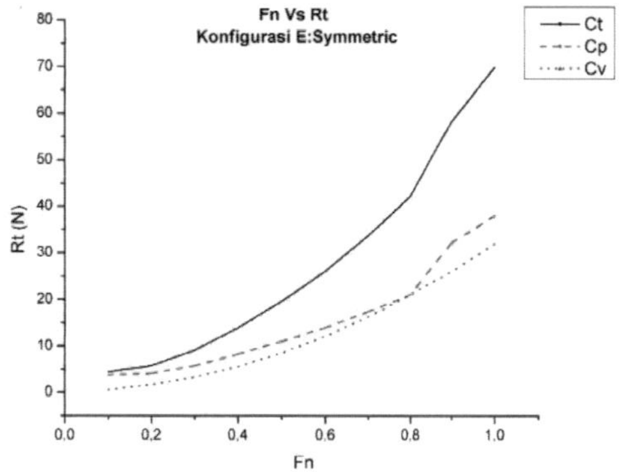

Figure 11: Symmetric

4.3 INTERFERENCE FACTOR

Factors affecting wave interference can be expressed as factor interference. The factor is a comparison between the interference pattern coefficient of the interfering wave resistance and the non-interfering wave resistance coefficient [11]. On the pentamaran ship, factor interference is considered in computed by comparing the coefficient pattern of the ship's wave resistance with the number of coefficients of wave resistance barriers of each hulls

In the graph of the interference factor to Froude Number indicating the negative value means that the wave interference gives the advantage in lowering the resistance. In the graph, the smallest interference factor is in configuration B that is the asymmetric configuration located on the outer hull and inward (Inboard Outer), but at Froude Number 0.9-1 the interference factor value is above the configuration curve C. The interference factor curve at Configurations A and C are always under the E (symmetry) configuration curve. While the D configuration curve intersects the E configuration curve so that it is above the E configuration curve of Froude Number 0.2-0.6 as in Figure 4.35.

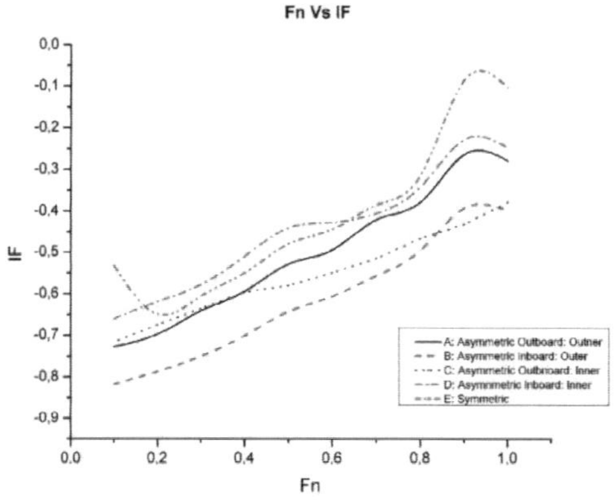

Figure 12: Interference Factor

5. CONCLUSIONS

After investigation, there are variations of advantages and disadvantageous ships. The pentamaran ship with the B configuration is the asymmetric hull inboard: outer hull. This is because the interaction of the wave between the hull is advantageous so that the wave pattern resistance coefficient contributes to decrease the total coefficient value. Based on the calculation of the wave interference factor, configuration B has the lowest value in Froude Number 0.1-9. All configurations have a negative interference factor which means profitable. Configuration B (inboard outer) has the smallest interference factor in Froude Number 0.1-0.9. While the configuration of E (symmetry) and configuration D (inboard inner) in turns have the greatest interference factor than other configurations.

6. ACKNOWLEDGEMENTS

The authors would like to express sincere gratitude to Univeristas Indonesia to support our research. And thank you for company who support in joining International Conference on Computer Application 2017.

7. REFERENCES

1. Peng, Hongxuan, 2001. Numerical computation of multi-hull ship resistance and motion, DALHOUSIE UNIVERSITY Halifax.

2. Yanuar, Gunawan, Talahatu, Indrawati and Jamaluddin. Resistance analysis of unsymmetrical trimaran model with outboard sidehulls configuration. Scopus. Volume 12, Issue 3, September 2013, Pages 293-297

3. Ackers BB, Michael TJ, Tredennick OW, Landen HC, Miller III ER., Sodowsky JP, Hadler JB (1997). An Investigations of the Resistance Characteristics of Powered Trimaran Outriggers Configurations. SNAME Trans. 105, 349-373.

4. Chang Hwan Son. CFD Investigation of Resistance of High-Speed Trimaran Hull Forms. The College of Engineering at Florida Institute of Technology. 2001.

5. Molland, A.F., Turnock, S.R., dan Hudson, D.A. (2011), Ship Resistance and Propulsion: Practical Estimation of Ship Propulsive Power, Cambridge University Press, New York, USA.

6. Tuck EO, Lazauskas L, 1998. Optimum spacing of a family of multi-hull. Ship Technology Research, 45, 180-195.

7. Md. Shahjada Tarafdera, Mir Tareque Alia, Md. Shahriar Nizamb. Numerical prediction of wave-making resistance of pentamaran in unbounded water using a surface panel method. 2013. Bangladesh.

8. ANSYS FLUENT. (2013), "ANSYS FLUENT Theory Guide" ,ANSYS Inc, U.S.A

9. R. Broglia, B. Jacob, S. Zaghi, F. Stern, and A. Olivieri, “Experimental investigation of interference effects for high-speed catamarans,” Ocean Engineering, vol. 76, pp. 75–85, 1/15/2014.

10. Chang Hwan Son. CFD Investigation of Resistance of High-Speed Trimaran Hull Forms. The College of Engineering at Florida Institute of Technology. 2001.

8. AUTHORS BIOGRAPHY

Zulfah Zikrina holds the current position of intern at Ministry of Ocean and Fishery Republic of Infonesia. She is responsible for CFD (Computational Fluid Dynamics) methods.

Firman Ady Nugroho holds the current position of lecturer at Universitas Indonesia. He is responsible for CFD (Computational Fluid Dynamics) methods. His previous experience includes ship structure and welding.

Yanuar holds the current position of professor at Universitas Indonesia. He is responsible for experimental data]. His previous experience includes multihull ship resistance by experimental methods.

EFFICIENT HULL FORM DESIGN OPTIMISATION USING HYBRID EVOLUTIONARY ALGORITHM AND MORPHING APPROACH

J H Ang, Sembcorp Marine Ltd., Singapore, and University of Glasgow, UK, **C Goh,** University of Glasgow, UK, **V P Jirafe,** Sembcorp Marine Ltd., Singapore, **Y Li,** University of Glasgow, UK

SUMMARY

An efficient hull form design can improve the overall efficiency of a marine vessel by reducing drag and therefore lowering carbon dioxide emission and fuel consumption. Traditional methods of hull form design and optimisation process, using trial-and-error approach require many designers' man-hours to produce more efficient hull form designs, which may be only sub-optimal and inefficient. This paper introduces an intelligent hull form design optimisation concept which aims to address the above issues. Combine with Industry 4.0 concept, the objective is to upgrade hull form design into a smart design process. This is accomplished by coupling an intelligent global search method-evolutionary algorithm with an efficient shape manipulation approach known as morphing. By doing so, process of hull form design optimisation can be achieved with minimum user intervention to produce optimal hull form more efficiently. A case study comparing existing proven designs to new hull forms created from the proposed hybrid evolutionary algorithm and morphing approach are presented.

1. INTRODUCTION

Hull form design and optimisation is an important topic in the shipbuilding industry. This is especially so in the face of more stringent environmental regulations and reduction of ship operational cost due to fuel consumption. An efficient hull form will help to reduce overall resistance acting on the vessel and thereby saving fuel and reduce harmful emission. Traditionally, hull form design and optimisation is carried out manually by ship designers using 'trial and error' approach where they will first select the most suitable hull form design from a pool of existing proven designs, improve the shape of hull manually and test the new hull form design using numerical tools or model test. This cycle will continue until the most optimal hull design is obtained. This manual process is very time consuming and only allows a few hull design variation and testing. While latest simulation based design (SBD) methods may help to automate some of these processes, they still require considerable human input and the result often depends heavily on the designer's experience and knowledge.

Hull form design optimisation (HFDO) is one of the SBD methods applied to perform detail investigations of hydrodynamic performance and numerical optimisation of the hull form. Typically, HFDO consists of the following steps; it starts with formulating the problem under (1), followed by design space exploration (2) which carries out the optimisation process. The next process is geometry modification (3) where the shape of hull is modified to produce new designs. Lastly, the hydrodynamic performances are evaluated in (4) using numerical methods or Computational Fluid Dynamics (CFD) analysis. The final output of this process is optimal solutions (5) which consist of hull forms that provide the best performance. The HFDO process is illustrated in Figure 1 as below.

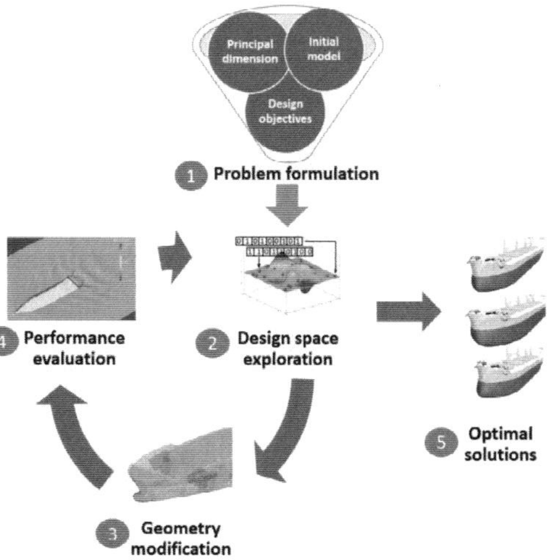

Figure 1: Hull form design and optimisation (HFDO) process [1]

While HFDO has proven to be a very efficient tool for hull form optimisation, it has not been widely adopted-largely due to the lack systematic shape variation and robust optimisation techniques [1]. Research in this area had evolved by using more advance optimisation and evaluation methods. Some recent works related to hull form design and optimisation process include [2] where they used Sequential Quadratic Programming for design space exploration, section area curve for geometry modification and evaluated the performance using potential and the viscous-flow solver. [3] applied Radial Basis Function for geometry modification, optimised using artificial bee colony algorithm and evaluated using CFD. [4] combined Sequential Quadratic Programming and shape modification using geometrical parameterisation and evaluated the performance using vortex line method for sailing yacht foil design.

An emerging trend in design and manufacturing- widely known as industry 4.0 (i4) - integrates cyber and physical systems for smart design and manufacturing and has the potential to transform the future of ship design and manufacturing. With more powerful computers and incorporation of i4 into design and manufacturing, computational intelligence methods such as evolutionary algorithm will play an even more important role in producing more efficient, cost-effective and innovative ships. The objective of this paper is to introduce a hybrid evolutionary algorithm and morphing approach which enables intelligent and automated design optimisation in pursuit of efficient hull form designs. In Section 2, an overview of industry 4.0 and smart design concept is proposed to link up HFDO process and product lifecycle. Section 3 introduces and elaborates the hybrid evolutionary shape manipulation approach for hull form design and optimisation. In Section 4, the preliminary findings obtained are presented and discussed. Section 5 concludes the paper.

2. INDUSTRY 4.0 AND SMART DESIGN

2.1 INDUSTRY 4.0

Industry 4.0 (i4) - often referred to as the fourth industrial revolution- aims to merge the real and virtual space through cyber-physical systems. I4 promises a paradigm shift from traditional segregated manufacturing process to fully connected manufacturing system, which is fast gaining worldwide attention. There are several studies [5, 6] that describe the basic components and enabling technologies of i4, which includes internet of things (IoT), collaborative robots, cyber-security, cloud computing, additive manufacturing, augmented reality and big data analytics. Increasingly, computational intelligence will play more important role in realising the full potential of i4.

2.2 CLOSED-LOOP SHIP PRODUCT LIFECYCLE AND SMART DESIGN

Traditionally, ship design process is carried out manually with ship designers relying on their experience and conventional computer aided design (CAD) or design simulation tools. As computational methods and high-performance computing advances, the use of more high-end optimisation (i.e. evolutionary algorithm) and hydrodynamic performance evaluation procedure such as CFD are becoming more prevalent in the ship design process. However, the adoption of these advance techniques in the marine industry is somewhat limited due to lack of a 'close-loop' approach where the design process can become fully automated with little or no user dependencies. To achieve this goal, a framework that considers the entire product lifecycle of ship into a 2-way closed-loop process is proposed in [7] and presented in Figure 2 below.

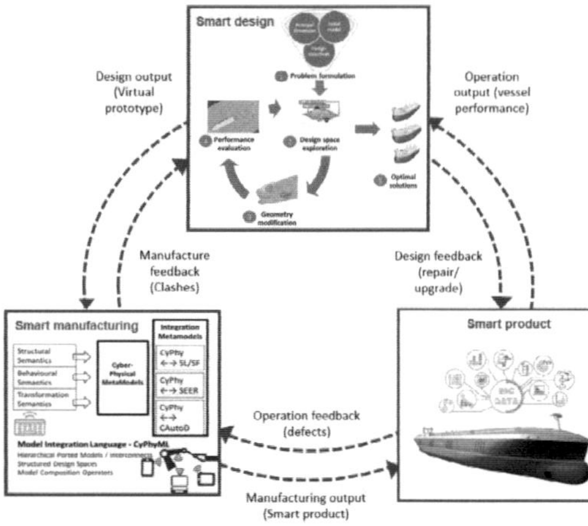

Figure 2: Two way closed-loop framework for smart design, manufacturing and operation [7]

Under this product lifecycle framework, smart design such as HFDO plays an important role to help realise the goal of i4 when integrated with smart manufacturing and smart products. Recently, the digital twin concept has gained popularity as it becomes more cost effective to implement, possibly due to the rise of i4 related technologies such as IoT, cloud and data analytics. Building on these concepts, smart design is hereby proposed as an intelligent and automated design process that collaborates closely with digital manufacturing and digital product throughout the entire product lifecycle. By means of smart design, this design automation tool can i) automatically search from a large database of proven hull designs, modify the hull geometry to create more design variations and evaluate the potential designs using CFD analysis, ii) capture and incorporate designer's knowledge into the design process so as to reduce dependency of experienced designers, iii) interconnect with smart manufacturing and smart product for through-life design, with a final goal of producing more efficient and innovative hull form designs.

3. HYBRID EVOLUTIONARY SHAPE MANIPULATION APPROACH

Considering the various issues and trends highlighted earlier, smart design can be developed by combining design automation process and computation intelligence considering through-life design in i4. This can be achieved by incorporating techniques in evolutionary strategies (genetic algorithm), geometry manipulation technique (surface morphing), advance simulation (CFD) and combine them into an intelligent hull automation design tool. This tool can then be used to automate the hull form design and optimisation process and interface with smart manufacturing and smart product across product lifecycle to produce more innovative hull form designs.

3.1 GENETIC ALGORITHM

Evolutionary computing such as Genetic algorithms (GA) proved to be very useful and become standard techniques in many HFDO process. GA is a nature-inspired search heuristic method based on Darwinian Theory of natural selection and the 'survival of the fittest' principle [8]. Unlike conventional optimisation method, GA has many desirable traits and offer significant advantages to efficiently navigate large and challenging design search space to produce globally optimal non-dominated solutions. Key to the workings of GA is the principle of 'genes' and 'chromosomes'. Through the use of genetic operators namely selection, crossover and mutation, information exchange takes place between these chromosomes over a number of iterations, typically with the fittest solutions replacing weaker ones, eventually leading to a set of optimal solutions. Examples of HFDO process using GA include [9], where the authors developed a hull form design system using GA, successive quadratic programming (SQP) and Rankine-source panel method for minimum wave-making resistance of a container ship. Several other works that applied GA for ship design optimisation includes [10-12].

3.2 GEOMETRY MODIFICATION- MORPHING

In any hull form optimisation, geometry modification plays an important role in ensuring the hull geometry can be easily manipulated to form new shapes in order for the optimiser to investigate and evaluate. This is no trivial task as every new shape generated must be smooth and a feasible design. There are 2 main approaches in hull form modification - direct modification and systematic variation. Direct modification involves manual adjustment of points and curves such as B-splines, which are usually quite localised and time-consuming to modify. Systematic variation includes parametric modelling, free-form modification which allows global modification and allows the hull geometry to be modified more efficiently. More recently, morphing method are being applied for systematic variation of hull forms. Morphing, also known as metamorphosis is a technique that is used widely in the animation industry to generate a sequence of images that smoothly transform a source into a target image. In computer graphic and industrial design, it is also used to compute a continuous transformation from one source shape to another target shape. In ship application, morphing was applied in [13, 134.

3.3 HYBRID EVOLUTIONARY ALGORITHM AND MORPHING APPROACH

Considering the main issues in existing hull form optimisation with respect to the lack of automated shape manipulation and robust optimisation techniques to generate feasible designs, a hybrid evolutionary algorithm and morphing (HEAM) approach based on HFDO concept was first proposed by the authors in [15]. The proposed methodology integrates a Genetic Algorithm (GA) and morphing techniques into a single optimisation platform. By combining the advantages of GA - ability to search for the best global solution - and that of morphing- ability to generate smooth intermittent shapes from the combination of two or more hull form designs, we can potentially create an optimum hull form design with improved efficiency. An overview of the proposed HEAM concept is provided in Figure 3.

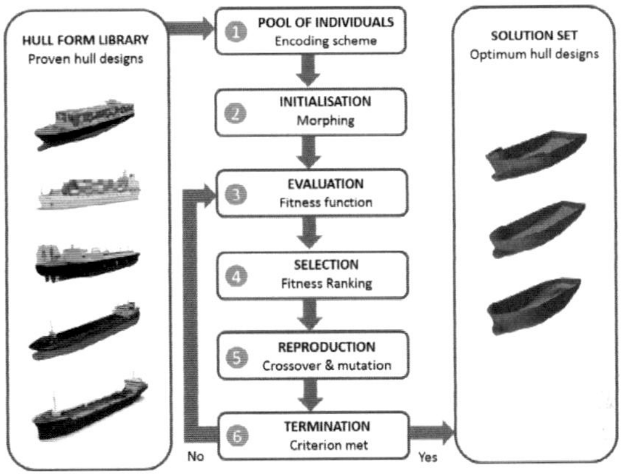

Figure 3: Hybrid evolutionary algorithm and morphing (HEAM) approach

The HEAM proposed in this paper consists of 6 main components namely: (1) pool of individuals- encoding scheme, (2) initialisation- morphing, (3) evaluation- fitness function, (4) selection- fitness ranking, (5) reproduction- crossover and mutation and (6) termination. The following sections details key mechanisms of the HEAM for ship hull form design and optimisation.

3.3 (a) Pool of Individuals- Encoding Scheme

Under this HEAM approach, the first step is to create a pool of 'initial solutions' and 'mapped' into unique encoding scheme. In the context of ship design, this can be drawn from existing hull forms from a design library or created from scratch. The approach developed in this paper is based on the former where existing hull forms from ship design firm or shipyard is used. Similar to existing ship design process, these hull forms are used as reference or parent designs which will be further improved to meet the new design objectives. The advantage of using existing designs is the assurance of their performances, whilst may not be optimal, are validated to meet basic design objectives and could thus potentially shorten the design cycle. For the approach developed in this paper, real-value chromosomes using morphing parameter (t) which captures the ship's geometry in 3D (x,y,z planes) according to their respective location or stations, as illustrated in Figure 4. This provides a simple yet direct representation of the ship geometry and helps to reduce the occurrence of infeasible designs (odd shape, unsmooth surface, etc.) generated during later parts of the optimisation process.

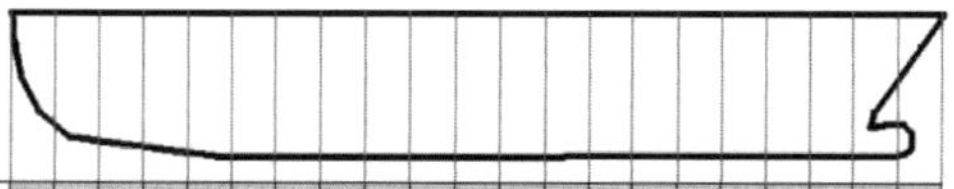

Station no. (stn)	0	0.5	1.0	1.5	2.0	2.5	3.0	3.5	4.0	4.5	5.0	5.5	6.0	6.5	7.0	7.5	8.0	8.5	9.0	9.5	10
Morphing parameter (t)	0	0	0	0	0	0	0	0	0	0	0	0	0	0	0	0	0	0	0	0	0

Figure 4: Encoding scheme using real value chromosome (t=0)

Depending on the number of hull form available in the library, each vessel will be assigned different morphing parameters- e.g. 1st vessel (parent A) will be assigned morphing parameter t=0 and 2nd vessel (parent B) will be assigned t=1, 3rd vessel (parent C) will be assigned t=2 and so on. The advantage of this hybrid approach is we can include as many hull forms in order to increase the variety of shapes and hence increasing the search space to include more novel or optimal designs.

3.3 (b) Initialisation-Morphing

Since the beginning of shipbuilding and subsequent introduction of CAD, two-dimensional (2D) hull lines remains the most fundamental graphical representation of the ship's hull form. This is the starting point where experienced designers model and modify the hull design prior to hydrodynamic calculations. The advantages of using 2D hull lines are it is a simple means to represent the entire shape of the hull and it is relatively easy to modify the hull form by adjusting the lines. It also serves as a primary source of hull form data which are used for subsequent plan approval and construction. In this proposed HEAM approach, we apply morphing to i) provide encoding scheme using morphing parameters (t) to modify the shape of hull, ii) generate intermediate solutions from initial pool (parents) to form initial population and iii) combine 2 or more existing hull forms (parents) to generate new hull designs (child).

At this stage, morphing is applied to transform one hull shape to another, which will generate the 'intermediate' shapes in between the 2 'parents'. Using morphing equation:

$$M(t) = (1-t) \times R0 + t \times R1 \qquad (1)$$

Where M(t) is the morphed shape, t is the morphing parameter, R0 denotes the source shape and R1 the target shape. From above equation, we can see when t =0, M(t) is also equal to 0 and hence the morphed shape is equivalent to source shape R0. Likewise, when t =1, M(t)=R1 which is the target shape.

Using hull lines provided from the body plan of source and target vessels (parents), we can morph and generate large number of intermediate shapes (child) just by changing the morphing parameter (t). Other than using morphing for interpolation, morphing can also be used to 'extrapolate' between 2 hull lines. As an example, we take one hull lines each from sample ship A (source) and sample ship B (target) at station 0.5 in way of stern of both vessels. By applying morphing through interpolation and extrapolation using different morphing parameter (t), we are able to generate both interpolated and extrapolated curves as illustrated in Figure 5.

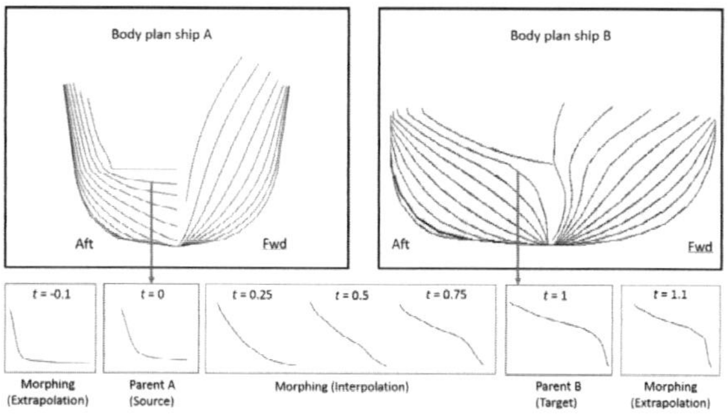

Figure 5: Morphing through interpolation and extrapolation at station 0.5

Using the same principle, we can effectively morph and create many intermittent forms between parent A and parent B by applying constant morphing parameter (t) across all transverse frame or stations. At this stage, we use morphing to create new individuals (child) which will form the initial population. This helps to increase the initial population size in the case of limited hull form in the library and allows the creation of more initial solutions prior the optimisation procedure.

3.3 (c) Evaluation - Fitness Function

The next step is to assign fitness function to each individual by evaluating the performance of each hull form designs based on objective function. In hull form optimisation, this can be reducing of hull resistance or ship's motions. For both objectives, the goal is to minimise the cost functions as follow:

$$\text{Min } f(\chi), \qquad \chi \in X \qquad (2)$$

Where f is the vector of design objectives, χ is vector of design variables and X is the feasible design variable space.

At this stage, we will translate the hull geometry from 2D coordinates to three-dimensional (3-D) surfaces by mapping the offset table into hull lines and surfaces using surface design tool such as NAPA. The 3-D surfaces can then be panelised or meshed and evaluated using CFD analysis. For this HEAM approach, we proposed that the hydrodynamic performance of all candidate design solutions should be assessed using low-fidelity CFD simulation such as potential flow for resistance analysis. This is in consideration of the large number of candidate design solution to be evaluated and potential flow method is usually preferred due to its efficiency and fairly good estimation for early ship design. High-fidelity CFD simulation such as Reynolds Averaged Navier-Strokes Equation (RANSE) can then be applied at final design stage to validate the design.

3.3 (d) Selection - Fitness Ranking

In GA, selection is a process of selecting which solutions will be used in crossover for generating new solutions. The principle is to always select the good solutions in order to increase the chance to obtain better individuals. Generally, there are 3 types of selection strategies- i) rank selection, ii) roulette wheel selection and iii) elitist selection. Rank selection assigns numerical ranking of each individual in the population based on their fitness, roulette wheel mechanism probabilistically selects individuals based on their performance and elitist selection pre-select the best candidates and retain them for next population. For proposed HEAM approach, we use rank selection based on the fitness level of each individual evaluated. This is done by ranking each candidate based on their fitness value and individual that are assessed to be highly ‘fit’ with relate to entire population from the evaluation process are selected for next round of reproduction.

3.3 (e) Reproduction - Crossover

Crossover is a very useful operator in GA where it combines two chromosomes (parents) to form new chromosome (child). Principle of this operator is to create new individuals by mixing the good genes of their parents and subsequently leads to fitter individuals. At this stage, we apply morphing as the main driver within crossover process to combine 2 or more existing hull forms (parents) to generate new hull designs (child). This can be achieved by setting varying morphing parameter (t) at different transverse curve along the stations or frame lines (x-planes), illustrated in Figure 6 below.

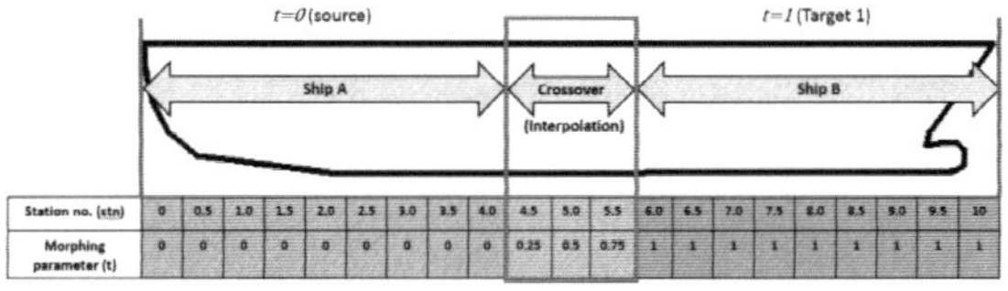

Figure 6: Cross-over through combination of ship A aft body and ship B forward body

The aim here is to create as many variation and possible combination of hull form designs so as to increase the solution space. On top of morphing 2 hull forms together into one hull concept, morphing can also be applied by joining multiple hull forms at different locations by changing the morphing parameters (t) at desired stations.

3.3 (f) Reproduction- Mutation

Mutation is a process in GA where new genes are created in random to produce a new genetic structure, which helps to introduce new elements into the population. In this proposed HEAM approach, we alter the morphing parameter at random stations through interpolation or extrapolation to create a new solutions (chromosome) illustrated in Figure 7 below.

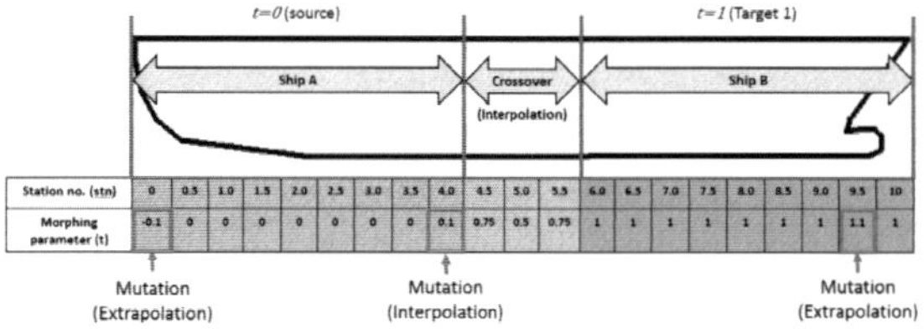

Figure 7: Mutation at random locations through morphing (interpolation or extrapolation)

As demonstrated above, there are many possibilities to ‘inject’ new elements into the chromosome just by changing the morphing parameters (*t*) randomly through interpolation or extrapolation. Other than introducing new elements, another key advantage using this method is the morphed shape are mostly feasible and hence helps to increase the overall shape modification and search efficiency of GA.

3.3 (g) Termination Criterion and Solution Set

Once all solutions are ranked and termination conditions are met, the iteration will stop and provide the results identifying the non-dominated solutions or Pareto optimum designs. It is now up to the designer to choose the final design, which will best meet the customer’s requirement. In this proposed HEAM approach, the termination condition can be set based on total number of iterations or terminate if there are no further improvement after 10 iterations. Ultimately, it depends on the designer requirement as to how much time are available or number of initial hull form in the library.

As highlighted earlier, we also propose to include a high-fidelity validation process such as RANSE method after termination and attainment of a range of Pareto optimum hull designs. This is to ensure the optimum hull form obtained from the optimisation process is truly ‘optimal’ and it also provides an additional reference for the designer when deciding on the final hull form design. It is useful to note while RANSE is a highly accurate CFD method used for ship evaluation within the marine industry, it is highly computation expensive and mostly suited at latter part of design stage where only a few narrowed down designs are required to be evaluated.

4. RESULTS AND FUTURE WORKS

To demonstrate the feasibility of our proposed HEAM approach, we performed a case study to develop an optimised hull form for a container vessel with objective to minimise forward resistance.

4.1 INITIALISATION USING MORPHING TECHNIQUE

Firstly, we selected 2 container vessels and 1 oil tanker vessel to form the 1st pool of individuals. The principle dimensions for 3 vessels are provided as follows:

Table 1: Principle dimensions

	Container vessel A	**Container vessel B**	**Tanker vessel C**
Length between perpendiculars (LBP)	202.1m	185m	314m
Breath (B)	32.2m	32m	58m
Draft (T)	10.5m	9m	9m
Design speed	20 knots	20 knots	16 knots

Using the three parent vessels selected, we performed morphing (interpolation) to generate several new intermediate solutions in order to form the initial population. The generated designs are provided in Figure 8.

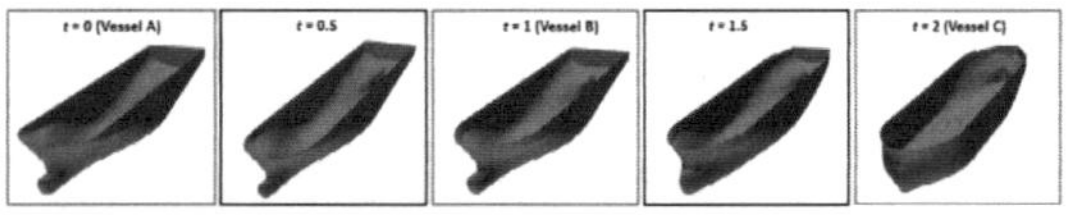

Figure 8: Isometric view of parent A, B and C hull surfaces and intermediate hull designs created using morphing (interpolation) method

With the initial population formed, the designs are evaluated and those solutions which are deemed to be fit are selected for next stage of reproduction. In this study, we evaluate the performance of each candidate design using potential flow method in NAPA program and it takes less than 5 minutes to evaluate one hull form design using standard quad core workstation.

4.2 CROSSOVER AND MUTATION

In this example, we performed crossover and mutation for vessel A and vessel B which provide the best performance before any modification. Firstly, we combined the stern area of vessel A to forward area of vessel B and then vice versa for the 2nd combination. The profile view of the combined vessel and shorten version of crossover encoding scheme are provided in Figure 9 as below.

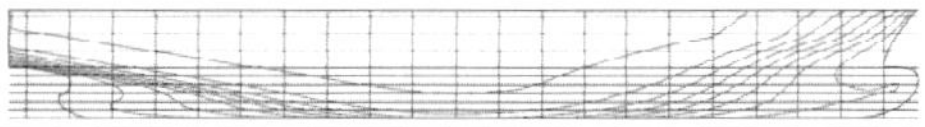

Station no. (stn)	0	0.5	1.0	1.5	2.0	2.5	3.0	3.5	4.0	4.5	5.0	5.5	6.0	6.5	7.0	7.5	8.0	8.5	9.0	9.5	10
Morphing (t)	0	0	0	0	0.05	0.2	0.35	0.5	0.5	0.5	0.5	0.5	0.6	0.65	0.75	0.8	0.9	1	1	1	1

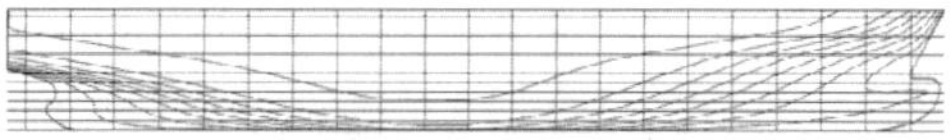

Station no. (stn)	0	0.5	1.0	1.5	2.0	2.5	3.0	3.5	4.0	4.5	5.0	5.5	6.0	6.5	7.0	7.5	8.0	8.5	9.0	9.5	10
Morphing (t)	1	1	1	1	0.95	0.8	0.65	0.5	0.5	0.5	0.5	0.5	0.4	0.35	0.25	0.2	0.1	0	0	0	0

Figure 9: Crossover procedure at midship between (top) vessel A aft- vessel B forward and (bottom) vessel B aft and vessel A forward

For mutation, we selected one of the station lines (at station 10) and performed morphing (extrapolation) to inject new element into the design, illustrated in Figure 10 below.

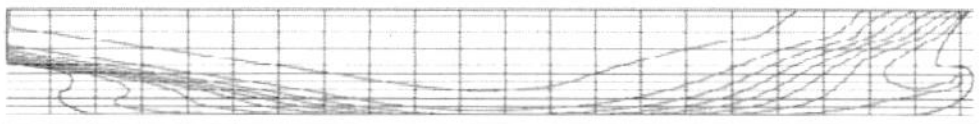

Station no. (stn)	0	0.5	1.0	1.5	2.0	2.5	3.0	3.5	4.0	4.5	5.0	5.5	6.0	6.5	7.0	7.5	8.0	8.5	9.0	9.5	10
Morphing (t)	0	0	0	0	0.05	0.2	0.35	0.5	0.5	0.5	0.5	0.5	0.6	0.65	0.75	0.8	0.9	1	1.1	1.1	1.1

Figure 10: Mutation via morphing (extrapolation) at station 9-10

In actual mutation process, one should note that this modification can be carried randomly within GA to allow more unique combination and increase overall solution space.

4.3 PRELIMINARY RESULTS

From the hull forms generated, we evaluated the fitness level of each solutions and obtained the following results.

Table 2: Principal dimensions and results of candidate solutions obtained from NAPA

ID	**Type**	**Length**	**Breath**	**Draft**	**Total pressure resistance (kN)**	**Forward pressure resistance coefficient (E-3)**
1	Vessel A (t=0)	202.1	32.2	10.5	251.635	0.027
2	Morphed (t=0.25)	197.82	32.15	10.125	179	0.034
3	Morphed (t=0.5)	193.55	32.1	9.75	149.1285	0.021
4	Morphed (t=0.75)	189.27	32.05	9.375	119.257	0.008
5	Vessel B (t=1)	185	32	9	80	0.007
6	Morphed (t=1.5)	249.5	32	9	2222.952	0.281
7	Vessel C (t=2)	314	58	9	3047.202	0.42
8	Crossover (A aft and B fwd)	185	32	9	136.455	0.017
9	Crossover (B aft and A forward)	213	32	9	163.761	0.017
10	Mutation (A aft and B forward)	183.29	32.2	9	76.828	0.003

From above preliminary results, we observed some improvements in performance as compared to the three initial vessels or parents. Taking solution ID 10, where we performed crossover of vessel A (aft) and vessel B (forward) and mutate through extrapolation, there was an overall improvement of 69% for vessel A and 3% for vessel B in terms of total resistance pressure. To further verify the results, we plot out the wave profile for both vessel A and vessel B against solution ID 10 in Figure 11 as below. It is noted the overall wave generated had reduced especially around the forward area of the vessel.

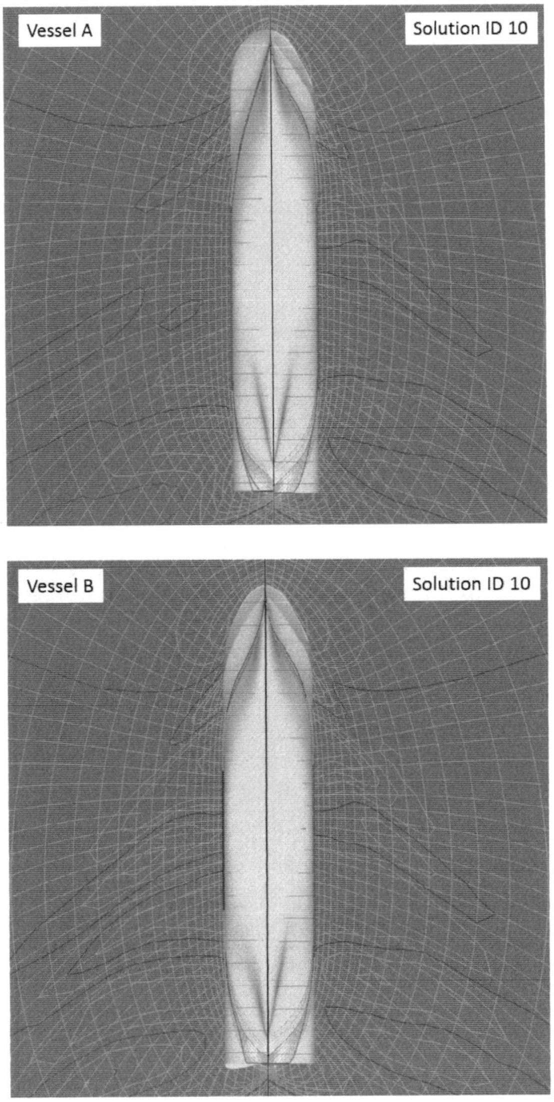

Figure 11: Wave profile comparing (top) vessel A-solution ID 10 and (bottom) vessel B- ID 10

4.4 DISCUSSION AND FUTURE WORKS

By comparing the preliminary results obtained using just a few solutions created using the proposed HEAM approach, we can see the potential of this method- the ability to create large number of feasible solutions in a highly efficient manner. We demonstrated through this case study by coupling up GA and morphing, this creates many possibilities to generate innovative hull forms especially when more parent hull forms are available. As the results shown here are only parts of our ongoing work, the full functionality of proposed HEAM approach and more case studies will be featured in subsequent publications.

The HEAM concept put forth in this paper provides a simple, yet promising solution to hull form design optimisation. Through morphing, it effectively transforms the shape of a hull and can generate a variety of different hull forms. It is also demonstrated through this paper that morphing can be used for both interpolation and extrapolation of the curve just by changing the morphing parameters (t). By coupling morphing parameter (t) into encoding scheme of GA, we utilise GA operators- crossover and mutation function to 'combine' and 'create' new solutions which helps increase the search space and producing more optimal hull forms.

By connecting this approach to smart design under i4, we can potentially link up this intelligent automated design process with smart manufacturing and smart product as part of product life cycle management. Through digital twin and computational intelligence, actual ship performance data or designer's experience can be effectively captured to automatically validate and improve the hull designs.

5. CONCLUSION

Hull form design and optimisation is an important topic in shipbuilding industry due to its ability to reduce emission and fuel consumption of vessels. Traditional method using trial-and-error to improve hull form design is not effective and efficient. While SBD can be used improve the efficiency of the design process, they are not widely adopted due to the lack of systematic shape modification and robust optimisation techniques. It is also highly dependent on designers' experience. With the introduction of industry 4.0 and digital twin concept, we can now link up entire ship design, manufacture and operation process into one connected product lifecycle. The solution put forward in this paper details a hybrid evolutionary algorithm and morphing approach that address the above issues and trends to make design process smarter. This is achieved by combining GA and morphing to modify the shape of the hull and allow more efficient search within the solution space. Morphing through interpolation and extrapolation are applied into crossover and mutation process to create new and more efficient hull form designs. A case study is applied to optimise the hull form of container vessel and the effectiveness of this approach is demonstrated with some preliminary results. We envisioned this proposed HEAM approach can transform the hull form design and optimisation process to create more innovative and efficient hull form designs.

6. ACKNOWLEDGEMENTS

This research is funded under Economic Development Board (EDB) of Singapore and Sembcorp Marine Ltd. (SCM) under Industrial Postgraduate Programme (IPP) grant no: COY-15-IPP/140002

7. REFERENCES

1. ANG, J.H., GOH, C. AND LI, Y., 'Hull form design optimisation for improved efficiency and hydrodynamic performance of ship-shaped offshore vessels', *International Conference on Computer Applications in Shipbuilding*, 2015.

2. PARK, J.H., CHOI J.E. AND CHUN, H.H., 'Hull-form optimization of KSUEZMAX to enhance resistance performance', *Int. J. Nav. Archit. Ocean Eng*, 2015.

3. YANG, C. AND HUANG, F.X., 'An overview of simulation-based hydrodynamic design of ship hull forms', *Journal of Hydrodynamics*, 2016.

4. BERRINI, E., ET AL., 'Geometric model for automated multi-objective optimization of foils', *International Conference on Computational Methods in Marine Engineering*, 2017.

5 KAGERMANN, H., WAHLSTER, W. AND HELBIG, J., 'Recommendations for Implementing the Strategic Initiative INDUSTRIE 4.0', *Acatech National Academy of Science and Engineering*, 2013.

6. MARIO, H., TOBIAS, P. AND BORIS, O., 'Design Principles for Industrie 4.0 Scenarios: A Literature Review', *Technical University of Dortmund*, 2015.

7. ANG, J.H., GOH, C., SALDIVAR, A.A. AND LI, Y., 'Energy-Efficient Through-Life Smart Design, Manufacturing and Operation of Ships in an Industry 4.0 Environment', *Energies*, 2017.

8. BAGHERI, H. AND H. GHASSEMI, 'Optimization of Wigley Hull Form in order to Ensure the Objective Functions of the Seakeeping Performance', *J. Marine Sci. Appl.*, 2014.

9. TAHARA, Y., ET AL., 'Development of CAD_CFD optimizer integrated hull form design system', *J. Kansai Soc*, 2003.

10. JACQUIN, E., ET AL, 'Hull form optimisation using free surface RANSE solver', *Symposium on naval hydrodynamics*, 2004.

11. BAIWEI, F., ET AL., 'Ship resistance performance optimization design based on CAD_CFD', *International Conference on Advanced Computer Control*, 2011.

12. KIM, H., 'Multi-Objective Optimization for Ship Hull Form Design', *George Mason University, Computational and Data Sciences Department*, 2012.

13. KANG, J.Y.; LEE, B.S., 'Mesh-based morphing method for raid hull form generation', *Computer-Aided Design*, 2010.

14. BAIWEI, F. ET AL (2011), 'Ship resistance performance optimization design based on CAD_CFD', *International Conference on Advanced Computer Control*, 2011.

15. ANG, J.H., GOH, C. AND LI, Y., 'Hybrid Evolutionary Shape Manipulation for Efficient Hull Form Design Optimisation', *Conference on Computer Applications and Information Technology in the Maritime Industries*, 2016.

8. AUTHORS BIOGRAPHY

Joo Hock Ang is currently a Manager in Research and Development at Sembcorp Marine Ltd. He joined Sembcorp Marine in 2001 and undertaken various functions including project management, ship production and engineering design before his current appointment. He was involved in wide variety of industry projects including specialised ship building, ship jumbolisation, and Floating-Production-Storage-Offloading (FPSO) vessel conversions. Concurrently, he is also a PhD candidate in University of Glasgow. His research focuses on using computational intelligence for multi-objective optimisation and geometry manipulation of ship hull form design.

Cindy Goh is an Associate Professor and the Director of Research Programmes at the University of Glasgow Singapore. She received her Ph.D. in Electronics Electrical Engineering from the University of Glasgow in 2004. Her research interest is in intelligent optimisation and data analytics for design and decision making to advance the state-of-the-art technology in energy management systems, early detection of diseases and complex engineering systems. Her work has been published in internationally peer reviewed journals and conferences. She is a member of the IEEE.

Vijay is currently Deputy General Manager (Engineering) in Sembcorp Marine, Admiralty Yard, overseeing and guiding the various engineering and design teams. Vijay graduated as Mechanical Engineer and started his career in 1983 with Mazagaon Dock Limited, India, designing naval ships. Since joining Sembcorp Marine in 1995, he held various responsibilities in engineering department. As an engineering project manager, Vijay successfully led the engineering teams for various FPSO, FDPSO, FSO, DP heavy lift vessel, research vessel complex new built and conversion projects engineering.

Yun Li is currently Professor of Systems Engineering at University of Glasgow and has been teaching at Glasgow since 1991. He is Visiting Professor to University of Electronic Science and Technology of China. Previously he was Founding Director of University of Glasgow Singapore, post-doctoral research engineer with Industrial Systems and Control Ltd Glasgow and consultant engineer with UK National Engineering Laboratory East Kilbride. His current research interest is in intelligent design for smart manufacturing and Industry 4.0. He has over 200 publications. Dr. Li is a Chartered Engineer in the UK.

A NUMERICAL AND EXPERIMENTAL STUDY OF RESISTANCE, TRIM AND SINKAGE OF AN INLAND SHIP MODEL IN EXTREMELY SHALLOW WATER

Q Zeng, R Hekkenberg, C Thill and **E Rotteveel,** Delft University of Technology, the Netherlands

SUMMARY

Inland vessels generally experience a resistance increment when the water in which they sail is extremely shallow. In this case, resistance extrapolation from ship model to full scale becomes complicated, and the traditional approaches do not often lead to satisfactory predictions. In this study, both numerical and experimental methods were applied to investigate the ship resistance, trim and sinkage in extremely shallow water. In the numerical calculations, the model initially has a trim and sinkage obtained from the model tests. The overset mesh technique was used to save the meshing effort. A 1/30 scaled model, which is only allowed to pitch and heave, was used in the model tests. It was found that, in extremely shallow water, the ITTC57 correlation line is not sufficient to extrapolate the resistance.

NOMENCLATURE

μ	Molecular dynamic viscosity (m^2 s^{-1})
μ_t	Turbulent viscosity (m^2 s^{-1})
ρ	Density of water (kg m^{-3})
C_f	Frictional resistance coefficient
CFD	Computational Fluid Dynamics
c	Air clearance (m)
EFD	Experimental Fluid Dynamics
Fn	Froude number
Fn_h	Depth Froude number
GCI	The grid convergence index
h	Water depth (m)
I	The turbulent intensity
L	Length of ship model (m)
R_f	Frictional resistance (N)
S	Wetted surface at zero speed (m^2)
U_{ave}	Reynolds averaged velocity (m s^{-1})
u_x	Velocities of flow in x direction (m s^{-1})
u_y	Velocities of flow in y direction (m s^{-1})
u'	The root-mean-square of the turbulent velocity fluctuations (m s^{-1})
ukc	under-keel clearance (m)
V	Velocity of ship (m s^{-1})
W	Width of the towing tank (m)

1. INTRODUCTION

Model tests are the most common way of predicting the resistance of ships. A correlation line [1] with a constant form factor [2] are used to build a relationship between the resistance of the ship and its scaled model. The definition of frictional resistance coefficient (C_f) in model tests is as follows:

$$C_f = \frac{R_f}{0.5 \cdot \rho V^2 S}, \quad (1)$$

where R_f is the drag force, V the ship speed and S the wetted surface when the ship speed is zero.

When the water becomes extremely shallow $(h/T \leq 2.0)$, however, the correlation line may not be valid, and the assumption of a constant form factor becomes false. This is probably due to the backflow and/or a different wetted surface. Therefore, the traditional way might cause avoidable errors in the resistance prediction in shallow water.

To fix this, Toxopeus [3] proposed a modified form factor for shallow water using double-body simulations. Raven [4] suggested considering the scale effect into the viscous resistance calculation in the extrapolation. But the different trim and sinkage, which belongs to the important features of shallow water, are ignored. Likewise, shallow water effects specific on frictional resistance were not well discussed.

Alternatively, ship researchers, like Schlichting [5], Lackenby [6] and Jiang [7], built empirical formulas trying to predict ship resistance in shallow water based on the results from deep water tests. Their methods are physically reasonable and practically useful, but they are often limited by the ship types they used, and the graphs they offered still need to be validated for a different ship type.

Computational Fluid Dynamics (CFD) is another option to do ship research in shallow water [8-10]. Compared with model tests, the physics around the model can be better understood and forces on the ship surface can be easier achieved.

Therefore, this study applied CFD method especially to reveal the changes of friction on the model surface in shallow water. A number of model tests were also used to validate the simulations and investigate ship behaviours in shallow water. It was confirmed that the resistance, trim and sinkage were significantly affected by the water depth and showed nonlinear changes. The discrepancies between the calculated friction and the ITTC57 line were explained, and a piecewise function is probably needed for a shallow water correlation line.

2. CFD METHOD

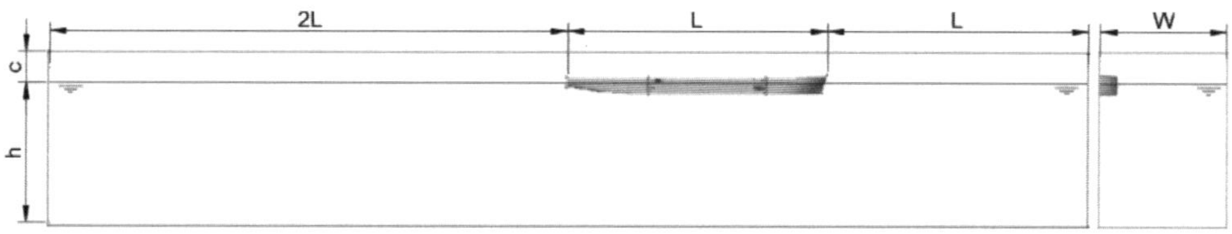

Figure 1: Computational domain

CFD was applied to solve the time-averaged Navier-Stokes equations. All simulations were completed on a commercial package ANSYS Fluent (Version 18.1). In the simulations, ship model had an initial trim and an initial sinkage, and those values were obtained from the model tests. Making the model fixed to its final situation can make it easier to converge and also, to avoid grounding while the algorithm is seeking for the applicable equilibrium situation.

2.1 THE COMPUTATIONAL DOMAIN

The computational domain is shown in **Figure 1**. It stretches L (the length of the ship) in front of the ship and $2L$ in the back. The width (W) is set comparable to the width of the towing tank. The water depth (h) is a variable and depends on the position of the water bottom. The air clearance (c) is set to a constant (10 meters). Due the symmetry of the domain, half of the model was used.

In the simulations, Case 0 simulates a deep water situation for comparison, as shown in **Table 1**. T is the draft of the ship and h the water depth. Two other cases of shallow water ($h/T = 2.01$ and 1.20) were chosen for the shallow water study. The Froude number (Fn) was from 0.0566 to 0.2074, and the corresponding water depth Froude number (Fn_h) was varied from 0.0857 to 0.9386.

Table 1: Cases in the simulations

Cases	*h/T*	*Fn*	Fn_h
		0.0566	0.0857
0(deep)	10.71	0.1320	0.1999
		0.2074	0.3142
		0.0566	0.1983
1	2.01	0.1320	0.4627
		0.2074	0.7271
		0.0566	0.2560
2	1.20	0.1320	0.5973
		0.2074	0.9386

Since shallow water conditions are mostly observed in inland waterways, fresh water at 15°C was used as the upstream flow. For all calculations, the technique of "Open Channel Flow" was implemented to simulate the two-phase flow. The SST k-ω model was chosen as the turbulence model. The scheme of the Pressure-Velocity Coupling was "Coupled", and the accuracy was second order for all items in the spatial discretization.

2.2 MESH

Because the trim and sinkage are different for different cases, it would cost huge effort to build one set of mesh for each case. To avoid this, a function in ANSYS Fluent, overset mesh, was applied. The mesh around the ship and that in the background does not necessarily match, as shown in **Figure 2**.

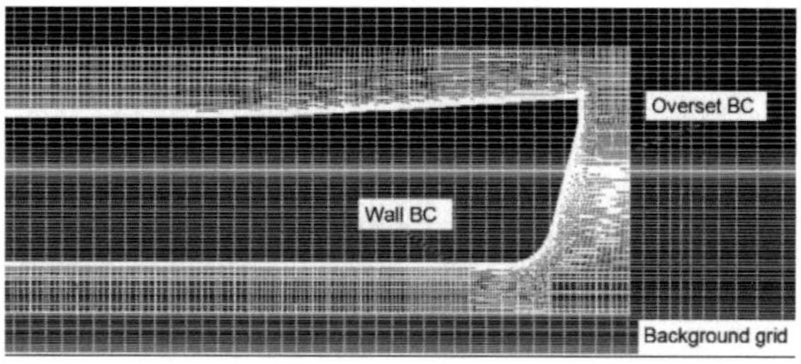

Figure 2: Overset meshing around the ship

Data are interpolated in the overlapping area, which has a larger data-transfer zone than the non-overlapping hybrid mesh. Therefore, a calculation with overset mesh could be faster than a non-overlapping hybrid mesh when they have a similar accuracy. Overset meshing of each part in the domain can be built separately without considering the conformation of each part's grids [11]. Consequently, the trim and sinkage can be realized by rotating and moving the mesh around the ship, and the different water depth can also be easily realized by changing the background mesh only.

2.3 BOUNDARY CONDITIONS

The boundary conditions are illustrated in **Figure 3**.

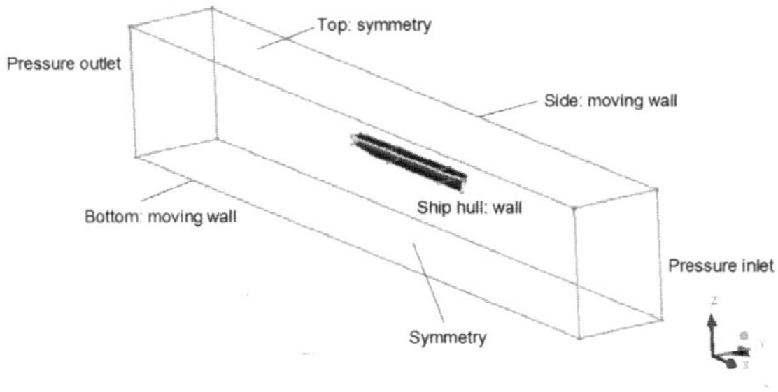

Figure 3: Boundary conditions

2.3 (a) The Inlet And Outlet Boundaries

The flow entering the domain was undisturbed and incompressible. Dirichlet conditions (2) and (3) were applied at the inlet boundary and conditions (3) for the outlet boundary:

$$u_x = U_x, \quad u_y = 0; \qquad (2)$$

$$I \equiv \frac{u'}{U_{ave}} = 5\%, \quad \frac{\mu_t}{\mu} = 10. \quad (3)$$

In equations (2) and (3), u_x and u_y are the velocities of upstream flow in x and y direction, respectively. I is the turbulent intensity, u' the root-mean-square of the turbulent velocity fluctuations, U_{ave} the Reynolds averaged velocity, μ_t the turbulent viscosity, and μ the molecular dynamic viscosity. Free surface level and bottom level were specified at both the inlet and outlet boundaries.

2.3 (b) Other Boundary Conditions

The Side and the Bottom conditions were set to the non-slip "moving wall", which with the same speed as the upcoming flow. The velocity and all derivatives in y direction on these boundaries were set to zero, and the speed in x direction relative to these moving walls was also set to zero.

Dirichlet conditions were set for the ship surface, which was a stationary, non-slip wall. The velocities were zero at both x and y directions. Symmetry conditions were set for the Top and Symmetry boundary. The velocities and all the derivatives of the flow at y direction were set to zero.

2.4 VERIFICATION

In this part, the numerical uncertainties are given after a mesh refinement study. Basically, the numerical error is the difference between a numerical result and its exact solution. According to Roache [12], the numerical error has three components: the round-off error, the iterative error and the discretization error.

The round-off error comes from the limited precision of the computers, but using double precision can usually keep this error negligible[13]. The iterative error results from the non-linearity of the mathematical equations. Using double-precision scheme and sufficient number of iterations can normally keep this error to the level of round-off error. In this study, the convergence criteria of all residuals are set to 10^{-7} (local scaling). However, for a simulation with free surface (like this study), this strict convergence criteria were probably never met. Alternatively, we considered the residuals of some important physical parameters, like resistance coefficients and the position of the centre of gravity of the ship. When these values become stable or change periodically during the calculation, the iterative error is considered negligible.

The discretization error is generated when transforming the partial differential equations into algebraic equations, and usually occupies the majority of the numerical error. Therefore, in this article, only the discretization error is considered.

Roache [12] recommended a grid refinement study for verification, which is commonly accepted as a good way to estimate the numerical uncertainty [14, 15]. In this study, three geometrically similar grids (Grid1 was the finest) were generated for the deep water case. The number of nodes in x, y and z directions had a refinement ratio of 1.25, as shown in **Table 2**.

Table 2: Number of nodes in x, y and z dirctions for the deep water case

	Background part			Ship part		
	Nxb	*Nyb*	*Nzb*	*Nxs*	*Nys*	*Nzs*
Grid 1	327	86	152	444	90	136
Grid 2	263	66	120	344	70	108
Grid 3	207	54	96	276	54	92

The procedure of the numerical uncertainty analysis proposed by Roache [12] is shown as follows:

1) The observed order of accuracy p (1-fine, 2-medium, 3-coarse)

$$p = \frac{\ln\left(\frac{\varphi_2 - \varphi_3}{\varphi_1 - \varphi_2}\right)}{\ln(r_{21})};$$

2) The error estimates

$$e^{21} = \frac{|\varphi_1 - \varphi_2|}{|\varphi_1|};$$

3) The grid convergence index (GCI)

$$GCI_{fine}^{21} = \frac{F_s \cdot e^{21}}{r_{21}^p - 1}.$$

The φ_i is the result from grid i and the r_{21} the grid refinement factor. The F_s is a safety factor. When $0.5 \leq p \leq 2.1$, $F_s = 1.25$; otherwise, $F_s = 3$.

Following the above procedure, the uncertainties of the total and frictional resistance for the finest grid (G1) are shown in **Table 3.**

Table 3 The *GCI* of total and frictional resistance for the finest grid (G1)

	Total resistance (N)	Frictional resistance(N)
Grid 1	5.9630	3.5402
Grid 2	6.0684	3.5552
Grid 3	6.4160	3.6884
p	5.3476	9.7865
GCI(G1)	2.31%	0.16%

The values of GCI are small (less than 3%), which indicates that the results are within the asymptotic region. Since the results of Grid 3 has acceptable accuracy for this study and consumes less computation efforts, this set of grid is used for further calculations.

3. EXPERIMENTS

3.1 SHIP MODEL

A full-loaded inland vessel was used in this study, and its 1/30 scaled model is shown in **Figure 4**. The detailed parameters of the vessel and its model are shown in **Table 4**.

Figure 4: The1/30 scaled model in the towing tank

Table 4: Parameters of the inland ship

	symbol	unit	full-scale	model(1/30)
Length	L	m	86.0	2.867
Beam	B	m	11.4	0.380
Draft	T	m	3.5	0.117
Mass	M	kg	2948118	109.19
Wetted surface	S	m^2	1417.8	1.575
Block coefficient	C_B	-	0.8642	0.864
Horizontal center of gravity	CGx	m	43.726	1.458

All tests were run on the small towing tank of TU Delft which is 85m long and 2.75m wide. Its maximum water depth is 1.25m and the maximum carriage speed is 3m s^{-1}.

3.2 TEST TASKS

Cases with five different water depths were applied in the tests and the smallest depth-to-draught (h/T) ratio was 1.2. Positions of the bottom are qualitatively presented in **Figure 5** and some parameters in these five cases are shown in **Table 5**.

Figure 5: Positions of bottom in each case

Table 5: Parameters of each case in model tests

	Case 0	Case 1	Case 2	Case 3	Case 4
Depth-draft ratio (h/T)	10.71	2.01	1.80	1.50	1.20
water depth (m)	1.25	0.235	0.21	0.175	0.14
under-keel clearance (*ukc*) (m)	1.133	0.118	0.093	0.058	0.023

The case 0 was a deep water condition for comparison. The velocities and the depth Froude numbers in each case are listed in **Table 6**.

Table 6 Velocities and depth Froude number for each case

V (m/s) 1/30 model	V (km/h) full-scale	depth Froude number (Fr_h) Case 0	Case 1	Case 2	Case 3	Case 4
0.3	5.92	0.086	0.198	0.209	0.229	0.256
0.4	7.89	0.114	0.263	0.279	0.305	0.341
0.5	9.86	0.143	0.329	0.348	0.382	0.427
0.6	11.83	0.171	0.395	0.418	0.458	0.512
0.7	13.80	0.200	0.461	0.488	0.534	0.597
0.8	15.77	0.228	0.527	0.557	0.611	0.683
0.9	17.75	0.257	0.593	0.627	0.687	0.768
1.0	19.72	0.286	0.659	0.697	0.763	0.853
1.1	21.69	0.314	0.724	0.766	0.840	0.939

The depth Froude numbers (Fn_h) were mostly within the subcritical range ($Fn_h \leq 1.0$) which was the case for inland vessels.

4. RESULTS AND DISCUSSION

The signs of the drag force, trim and sinkage are defined in **Figure 6**.

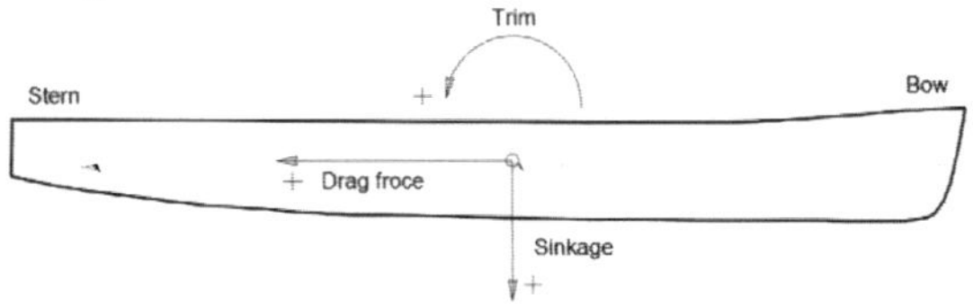

Figure 6: Definition of the signs of the drag force, trim and sinkage

Results of the trim and sinkage in the model tests are shown in **Figure 7** and **Figure 8**. To avoid grounding, some dangerous cases (a high speed but with a small *ukc*) were not involved in the tests.

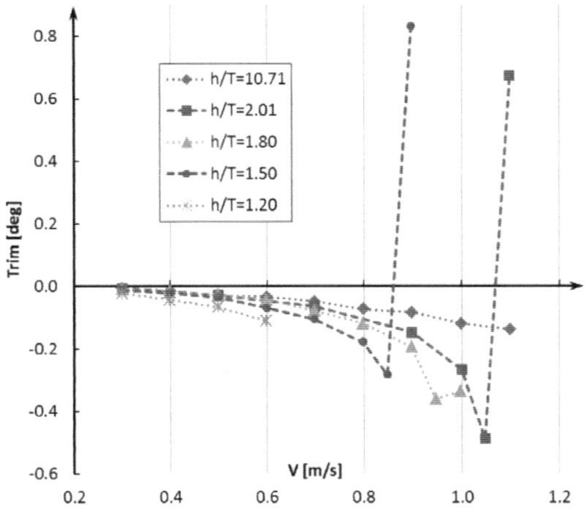

Figure 7: Results of trim with the velocities in the model tests

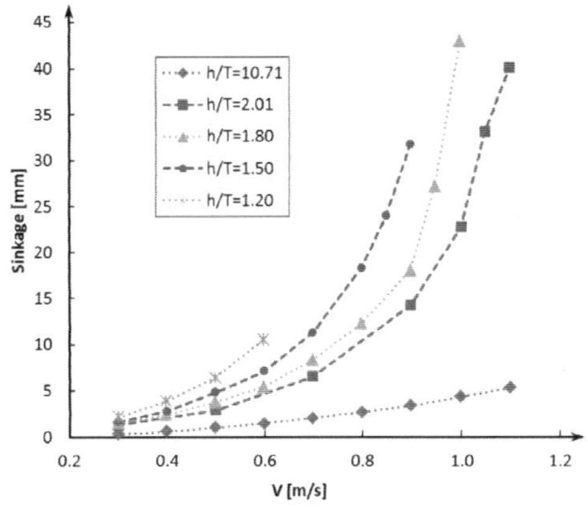

Figure 8: Results of sinkage with the velocities in the model tests

Compared with deep water test (h/T=10.71), the absolute values of trim and sinkage increase significantly in shallow water. It should be remarked from **Figure 7** that the model first trims with the bow down, but when the speed is fast enough, it will (or have a trend to, for h/T=1.80) trim with the bow up.

The results of trim and sinkage were used in CFD simulations, i.e. gave the model in CFD an initial trim and an initial sinkage. Some simulating results of the drag force were validated by the model tests, as shown in **Figure 9**.

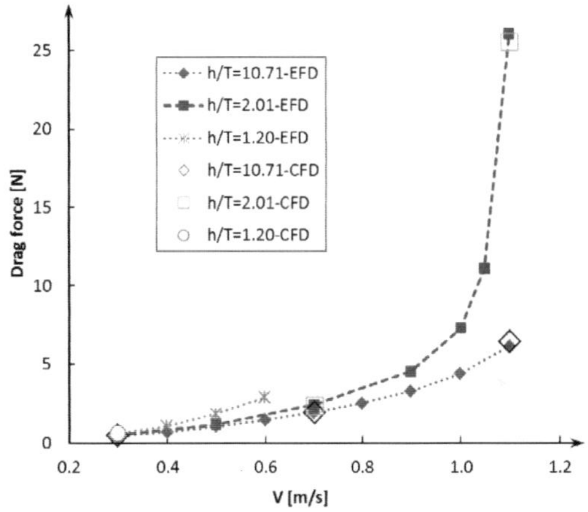

Figure 9: The drag force form CFD and EFD with different velocities

In **Figure 9**, the results of the drag from CFD agree well with the EFD results. After checking the total force in vertical direction and the moment of the model, they were both small values and can be negligible, which means that the values of trim and sinkage were rightly obtained from the tests.

After successfully validating the CFD calculations, in terms of measured and calculated drag, same trim and sinkage and thus same features of the global flow around the ship, the frictional resistance coefficient (C_f), which is hard to get from model tests, can be reasonably achieved from simulations. **Figure 10** compares the results of C_f from CFD with the ITTC57 correlation line, which were both calculated with the equation (1). The case h/T=1.20 with Re =6.44 (V=1.1m/s) is simulated and measured with the model fixed at the zero-speed position (zero trim and sinkage).

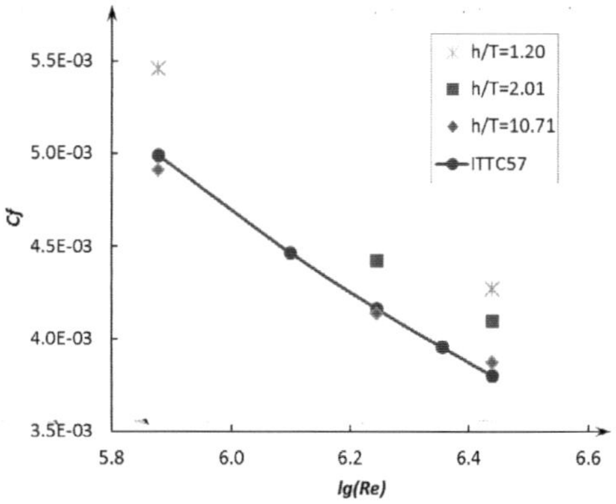

Figure 10: Frictional resistance coefficient (C_f) with Reynolds number in different water depths

This figure confirms that the frictional resistance in shallow water is larger than the ITTC57 correlation line. Even excluding the trim and sinkage for *h/T*=1.20 with *Re* =6.44, the friction still increases. This can be explained by the accelerated flow under the keel, which leads to a thinner boundary layer. And the trim and sinkage will amplify this effect. Seen these results, the ITTC57 correlation line is apparently underestimating the friction in shallow water and cannot be seen sufficient for the extrapolation of shallow water resistance.

One may argue that the ITTC57 line can still be used when the local speed and the exact wetted surface are known. However, these values are difficult to measure in a model test; even so, backflow speed and wetted surface will both become a variable in the extrapolation process, which makes the prediction more complicated and not easy to use. As a result, a better way is to keep the method of using the ship speed and the zero-speed wetted surface unchanged, and reconsider the correlation line together with shallow water effects.

A shallow water correlation line might be a piecewise function. As shown in **Figure 7**, the values of trim are not monotonic at the high speed range, which might have a complex effect on the friction. Additionally, it is hard to affirm that the results in **Figure 10** are suitable for all ship types or for all different waterways. It is the target of ongoing research to find a restricted and easy to determine set of parameters that predicts the shift of the C_f curves independent from individual CFD calculations.

5. CONCLUSIONS

In this study, both numerical and experimental methods were applied to investigate the resistance extrapolation of an inland ship model in shallow water. Due to shallow water effects, the exact frictional resistance increases, and will be further amplified by trim and sinkage. This will eventually lead to an unavoidable discrepancy with the ITTC 57 line. Consequently, the correlation line becomes insufficient in the resistance extrapolation when the water becomes (extremely) shallow.

Therefore, the resistance extrapolation from model to full scale should be improved specific for shallow water. And the first step is updating the correlation line. Ship and waterway geometry, flow acceleration, trim and sinkage are all important factors to be considered.

6. ACKNOWLEDGEMENT

The authors cordially thank for Peter Poot, Hans v/d Hek, Wick Hillege and Jasper den Ouden for their efforts in the experiments in the towing tank of TU Delft.

7. REFERENCES

1. ITTC, in *8th International Towing Tank Conference*. 1957: Madrid, Spanish.
2. Prohaska, C., 'A simple method for the evaluation of the form factor and low speed wave resistance', *Proceedings 11th ITTC*, 1966.
3. Toxopeus, S.L. 'Viscous-flow calculations for KVLCC2 in deep and shallow water', in *MARINE 2011, IV International Conference on Computational Methods in Marine Engineering*. 2011. Lisbon, Portugal.
4. Raven, H. 'A computational study of shallow-water effects on ship viscous resistance', in *29th symposium on naval hydrodynamics, Gothenburg*. 2012.
5. Schlichting, O., 'Ship resistance in water of limited depth-Resistance of sea-going vessels in shallow water', *Jahrbuch Der STG*, 1934. **35**: p. 127-148.
6. Lackenby, H., 'The effect of shallow water on ship speed', *Shipbuilder and Marine Engineer*, 1963. **70**: p. 446-450.
7. Jiang, T. 'A new method for resistance and propulsion prediction of ship performance in shallow water', in *Proceedings of the 8th International Symposium on Practical Design of Ships and Other Floating Structures*. 2001. Shanghai, China.
8. Tabaczek, T., 'Computation of flow around inland waterway vessel in shallow water', *Archives of Civil and Mechanical Engineering*, 2008. **8**(1): p. 97-105.
9. Tezdogan, T., A. Incecik, and O. Turan, 'A numerical investigation of the squat and resistance of ships advancing through a canal using CFD', *Journal of Marine Science and Technology*, 2016. **21**(1): p. 86-101.
10. Caplier, C., et al., 'Energy distribution in shallow water ship wakes from a spectral analysis of the wave field', *Physics of Fluids*, 2016. **28**(10): p. 107104.
11. Fluent, A., '17.2 User's Guide', *Ansys Inc*, 2016.

12. Roache, P.J., 'Verification and validation in computational science and engineering'. 1998, Albuquerque, New Mexico: Hermosa.
13. Eça, L. and M. Hoekstra, 'Evaluation of numerical error estimation based on grid refinement studies with the method of the manufactured solutions', *Computers & Fluids*, 2009. **38**(8): p. 1580-1591.
14. ASME, 'Standard for verification and validation in computational fluid dynamics and heat transfer', *American Society of Mechanical Engineers, New York*, 2009.
15. Eça, L. and M. Hoekstra, 'A procedure for the estimation of the numerical uncertainty of CFD calculations based on grid refinement studies', *Journal of Computational Physics*, 2014. **262**: p. 104-130.

8. AUTHORS BIOGRAPHY

Qingsong Zeng holds the current position of a PhD candidate at Delft University of Technology (DUT). His previous experience includes CFD and ship resistance in shallow water.

Robert Hekkenberg holds the current position of an associate professor and Director of Studies BSc Marine Technology at DUT. He got his PhD at DUT in 2013 on "*Inland Ships for Efficient Transport Chains*".
As researcher at DUT he was partner in many inland shipping related EU projects such as CREATING and MoVe IT! His research fields are design optimisation and manoeuvring of inland ships as well as autonomous vessels.

Cornel Thill holds the position of the manager of the maritime laboratories of DUT and is lecturer at the University of Duisburg-Essen in Germany, where he before joining DUT in 2015 held the position of the head of the hydrodynamic department of the *Development Centre for Ship technology and Transport systems* (DST), one of the most renowned research institute on inland navigating ships. He gained further relevant experience in previous EU projects such as CREATING, SMOOTH (coordinator), STREAMLINE and MoVe IT!

Erik Rotteveel holds the current position of a PhD candidate at Delft University of Technology (DUT). His previous experience includes CFD and inland ship stern optimization in shallow water.

ENABLING A PARADIGM SHIFT IN SHIP STRUCTURAL DESIGN WITH A 3D APPROACH

T Hulkkonen, NAPA Ltd, Finland
H C Shin, Hyundai Heavy Industries, Korea
N H Yi, Hyundai Heavy Industries, Korea
D H Jang, NAPA Ltd, Korea
T G Jang, NAPA Ltd, Korea

SUMMARY

In 2016, Hyundai Heavy Industries (HHI) implemented NAPA Steel's 3D model-based ship structural design tool across its Basic Hull Design Department. One of the main advantages of NAPA's new interactive 3D structural modelling interface is that it allows users to create a single flexible product model for a whole vessel. HHI was able to use these models throughout the design process, including class society rule checks, creating plan approval drawings, Finite Element model generation, weight calculation and other tasks.

Following the introduction of NAPA Steel, HHI expects the design time and man-hours spent in ship initial and basic structural design to be reduced by 30%. By linking structural design databases and drawings to the NAPA 3D model, design quality is improved and the probability of human errors is reduced, as common information is shared and the update process becomes simpler.

NAPA's software has created now a paradigm shift in how HHI designs vessels, and has helped the business become more competitive for the future of shipping. The presentation gives an overview on the process of this paradigm shift at HHI. It highlights the technology used with practical examples, changes to the traditional design process, and the benefits achieved so far.

NOMENCLATURE

BV	A class society, member of IACS.
ClassNK	A class society, member of IACS.
DNVGL	A class society, member of IACS.
FE	Finite Element method
FSRU	Floating storage and regasification unit
HHI	Hyundai Heavy Industries
IACS	International Association of Class Societies
LNGC	Liquefied natural gas carrier
LPGC	Liquefied petroleum gas carrier
LR	A class society, member of IACS.
NAPA	Software company or the software, depending on the case

1. INTRODUCTION

HHI and NAPA have undertaken a long-term development co-operation in developing a 3D based approach for ship initial and basic structural design.

1.1 PROJECT PARTNERS

Hyundai Heavy Industries Shipbuilding Division (HHI) in Korea delivers many different kinds of ships for international shipbuilding market. The records now show more than 2000 ships for about 300 ship owners in 50 countries since 1972. The most common ship types delivered have been:

- Tanker
- Container ship
- LNGC
- LNG FSRU
- LPGC
- Drillship
- Bulk Carrier.

As a part of Shipbuilding Division, the Basic Hull Design department is responsible for project planning, initial design and basic hull design. The typical tasks in Basic Hull Design Department are:

-
- Drawing main plans
- Drawing Mid ship, Construction profile, Shell expansion and Key Sections
- Class Society rule check
- FE analysis
- Estimation of the hull weight
- Development of new structure types and ship concepts

NAPA Ltd is a software developing company providing tools for ship design and operations with the brand name NAPA. The products for design solutions are NAPA Naval Architecture and NAPA Steel, which are used by HHI in achieving the targeted paradigm shift in their process.

1.2 THE DEVELOPMENT PROJECT DONE

This paper presents the process followed and results achieved at HHI in renewing working practices at its Basic Hull Design Department. The paradigm shift was achieved when the traditional, drawing-based 2D approach was changed to a 3D approach by using NAPA software.

The major success factor in achieving the final customer satisfaction was the long term co-operation between HHI and NAPA between 2007 and 2016. This consisted of continuous communication between NAPA and HHI in giving feedback and making development changes accordingly.

The final decision to change the design process was taken at HHI after benchmarking and evaluating several other available software packages during 2015-2016.

2. NEED FOR CHANGE IN DESIGN

Many shipyards have recently found themselves in the situation of having to come up with completely new ship designs, and comply with changing regulations and ship owner's demanding requirements.

2.1 TRADITIONAL 2D PROCESS

Before this paradigm change at HHI, the design process was based on the traditional 2D approach. In this approach, 2D drawings played the central role in sharing information to different stakeholders [Figure 1].

This is a common process adopted by many ship design organizations around the world. It has many benefits, such as:

- 2D drafting tools are well developed and commonly available.
- Using these tools is something which most of the engineers learn during school and college.
- Ship design organizations have long experience in using these tools and it is easy to continue without being afraid of the high risks associated with changing processes.
- After working for several years with these systems, the value of reference documents for older ship projects is increased, as they give good starting points for updated projects, improving productivity.

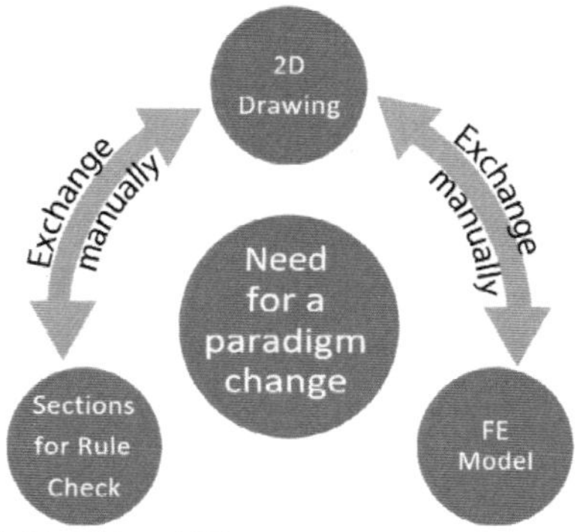

Figure 1: Traditional 2D process

Even though considered as low risk and easy to implement, it can be seen that there are also drawbacks in the 2D approach. These are difficult to overcome without changing the process more radically:

- Possibility for inconsistent information among several documents which are coordinated only by manual inspection.
- Duplicate jobs in regenerating geometry and its properties in several places.
- Heavy workload created by need to modify data in different sources when design changes.

2.2 NEED FOR IMPROVEMENTS

The market environment is getting tougher all the time. Ship building activity in different countries has been at a high level for several years, and several new market players have entered into business.

On the other hand, after high season in shipbuilding in 2008, the world economy has been becoming increasingly unstable due to oil price changes and financial and political crises in many countries. This has led to overcapacity in ship building and shipyards have to compete hard to win the available ship orders.

At the same time the class society rules have become increasingly complex. For example, the CSR rules for tankers and bulk carriers specify increased demands on verifying the hull structure safety [2]. In many other ship types like container ships, offshore and passenger vessels the trends and needs are similar. This forces the shipyards to look for improved design methods and tools.

HHI has also found itself in a situation whereby it has to come up with completely new ship designs and renew its processes to comply with changing regulations and ship owner's demanding requirements.

Based on this background, with advances in technology and competition among shipyards, HHI needed to consider changing their design process.

The requirements for the new process were:

- Come up with completely new ship designs
- Effectively review new types of structures
- Comply with changed regulations and ship owner's requirements
- Share the hull strength verification for more concurrent users
- Shorter design period
- Easy to use and quick implementation process

2.3 CHANGES DONE IN THE PROCESS

In general, 3D CAD tools in ship building were introduced at the same time as commercial computer systems at late 1970's and early 1980's. The benefits of a

3D approach have been evident in production level systems and in general data management.

Since the 1980's the same 3D approach has been proposed for ship basic design many times. The basic 3D approach has been known and it has been clear, but the practical implementation has failed often due to missing functionality for the structural basic design phase.

The basic technical functionality is the ability to automatically create outputs from 3D models. This covers 2D drawings, interfaces to rule check, FE-model, and weight and cost calculations.

The technical functionality is available in many 3D CAD systems, but to achieve the requirement for easy and quick implementation, a major concern has been to create 'easy to learn and easy to use' 3D structural modelling technology.

In this co-operation, User Experience (UX) from customer's point of view [Figure 2] played a central role. UX covers many aspects of software, not only User Interface, but including the user's deeper needs. It is development in UX that has been the leading differentiator from the previous attempts, giving the technology the necessary edge and leading to this practical solution for HHI needs.

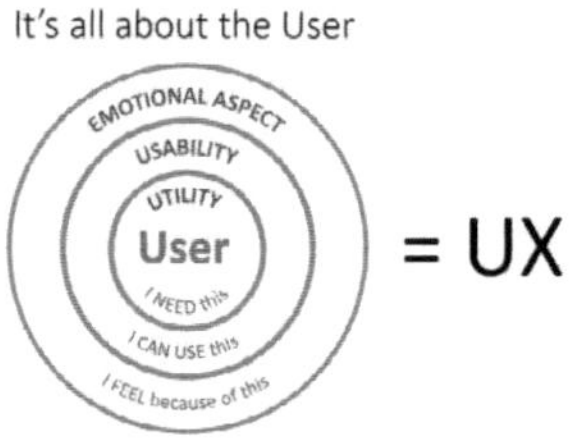

Figure 2: Components of the UX

3. DEVELOPMENT OF THE NEW PROCESS

Close co-operation between NAPA and HHI was a key component in the development of the new process. It has covered both technical, feature-related development and has a strong focus on delivering improved user experience.

3.1 LEARNING PROCESS FOR HHI AND NAPA

The co-operation was based on tight collaboration between NAPA developers and HHI engineers. The long development project has shown the mutual interest at both HHI and NAPA for this subject.

The development process has been iterative and the expectations have changed many times during the process as mutual understanding from both parties has increased.

The result of this is that NAPA has learned how to better understand the user requirements, and HHI has learned how to set requirements for software development to reach practical results. This includes rethinking old processes with new possibilities, rather than attempting to copy all old practices to new tools.

The evaluation process started with the evaluation of functional requirements and coverage of different interfaces from 2007-2015 [Figure 3].

After the development of the functional features it became evident that the system was no longer easy to use and therefore also difficult to implement as a replacement of the old fine-tuned 2D process.

The final breakthrough in the development of the new process was reached during the last few years when the work was concentrated on renewing the NAPA Steel UX requirements. The new user interface was realised through the final, yet vital requirement for an "easy to use and quick implementation process".

Date	Details	Target Ship
2007.11 ~ 2011. 6	• Functions Check - 3D model implementation - 2D drawing extraction - FE model generation	VLOO Tanker FPSO(Top side supports)
~ 2015. 7	- Class program interface - Aveva Marine 3D CAD interface • Performing the ongoing feed back and presenting HHI's opinion	Container ship VLCC LNGC
~ 2016. 7	• The pilot test for new type of NAPA steel(NAPA Designer/Drafting)	Tanker Container
2016. 10	• The Adaption of NAPA Steel	
2016. 10 ~	• Structural design in progress with NAPA steel	

Figure 3: Project phases

3.2 THE UX IMPROVEMENT PROCESS

While developing the new UX and user interface, NAPA adopted a new software development process, which is called the "Design Thinking" method [1], [Figure 4]. This is a guiding philosophy used by many businesses to solve problems by maintaining focus on user experience, and iterating accordingly.

This style of development requires an intensive research phase with users at the centre followed by rapid iterative prototyping. User feedback is gathered as quickly and frequently as possible.

The main stages in the process are:

- Empathize
- Define
- Ideate
- Prototype
- Test and Learn

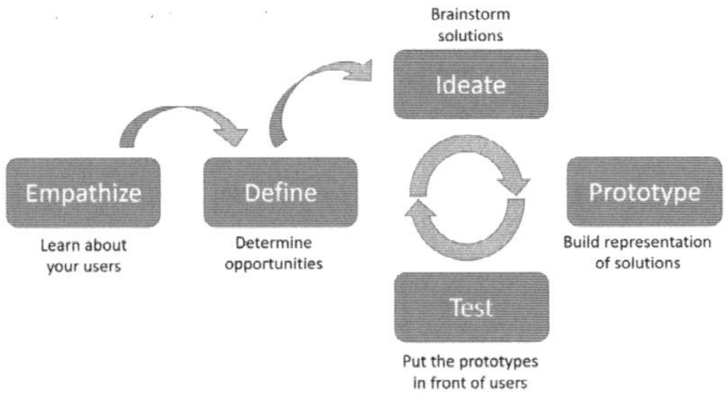

Figure 4: Design Thinking Method

In addition to HHI staff, NAPA has interviewed more than 100 users from 20 different countries to get overall picture of user requirements [Figure 5]. This work is still continuous and it is implemented as everyday practice in NAPA.

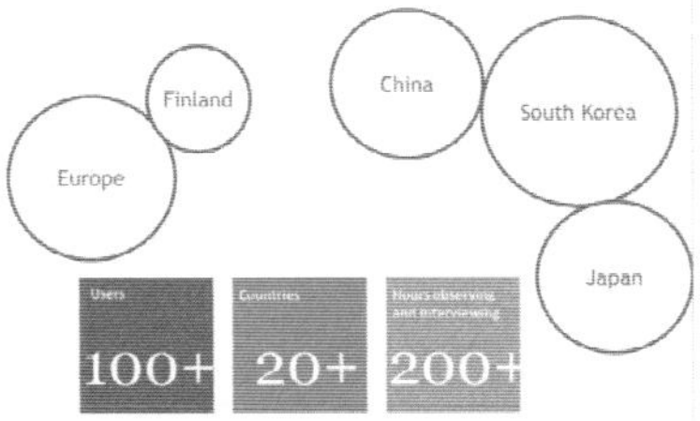

Figure 5: Working with users

Prototypes are developed in the process from ideated sketches to finalised product in an iterative process [Figure 6].

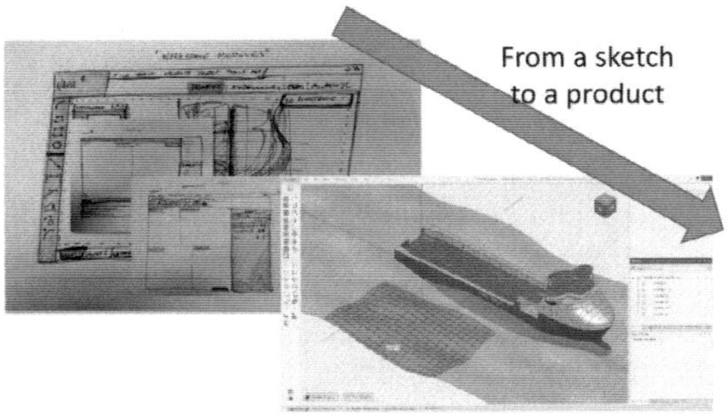

Figure 6: From a sketch to a product

4. CURRENT PROCESS AT HHI

The paradigm shift at HHI has started, and there are already some practical experiences that show the direct and indirect benefits of the new approach.

4.1 PRACTICAL EXPERIENCE

HHI started to work on a large scale with NAPA Steel based design processes from October 2016. Since May 2017, it has been adopted for use on several contracted ship projects [Figure 7].

Ship type	Number of projects
LNGC	3
FSRU	1
VLCC	2
Aframax Tanker	1
Container Ship	1
VLOC	1

Figure 7: Examples of constructed ships.

4.2 THE CHANGE OF DESIGN PARADIGM

The core of the paradigm change in the HHI design process is in taking the 3D model as the source of original information. Starting from naval architecture up to plan approval, the information is shared with all the users, who work on the same 3D model.

On a rough level the process can be presented as a simple information flow from tender stage to later design targeting to class society approval [Figure 8].

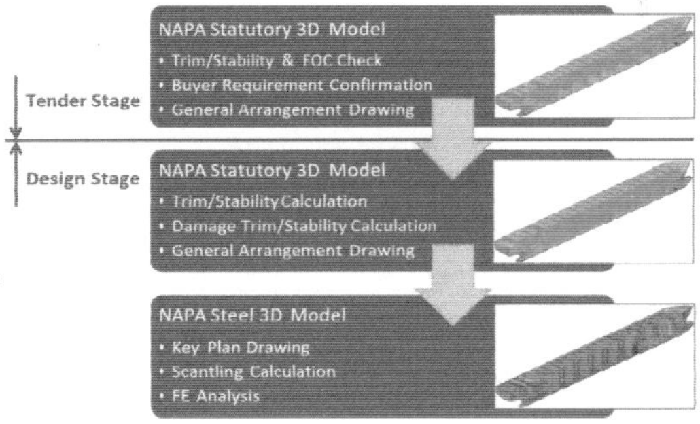

Figure 8: Current Design Process

4.3 COMPARISON OF DESIGN METHODS

There are many benefits in 3D approach, which are not yet fully utilized at HHI and they are currently under consideration for the future development of the process.

The requirement for shorter design period was a key issue, which was achieved during the benchmarking and later verified with contracted ships [Figure 9].

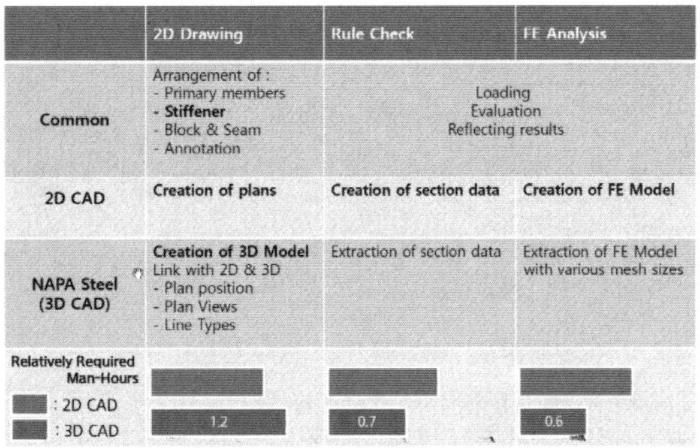

Figure 9: Comparison of 2D and 3D design process

The results show that just for creating 2D drawings, the new 3D process takes 20% more man hours than the old one. This part includes 3D modelling, which does not exist in the 2D process.

The benefit and time saving is realised over the whole process, in which rule checks and FE analyses require only 60-70% of the work needed in 2D approach.

Following the introduction of NAPA Steel, HHI expects the design time and man-hours to be reduced by almost 30% in ship structural design. After further development and customization in partnership with NAPA, HHI expects this to be reduced even further.

4.4 MAIN ADVANTAGES OF NAPA STEEL

As a summary, the main advantages of the new process are:

One Integrated Database

- Includes all information about weight, painting area, welding length, etc.
- Same information available to all users
- Extracts FE model and rule interface

Flexible Model

- Easy modification of design parameters, hull form, frame spacing and primary members
- Quick review of design alternatives

Information Sharing

- Allows other departments to reference 3D data (Detailed Design, Outfitting Arrangements)
- Offers intuitive 3D view to understand complicated structures

5. KEY FEATURES OF NAPA STEEL

An important part of good UX is the technical functionality and proper features. In addition to naval architecture needs, NAPA covers structural design key features for plan approval and interfaces to related activities.

5.1 PARAMETRIC MODELLING

The parametric model features increases the efficiency of the design work. In initial and basic design phases, there are frequent changes which affect the ship project. Parametric concept makes it possible to utilize previous work effectively [Figure 10].

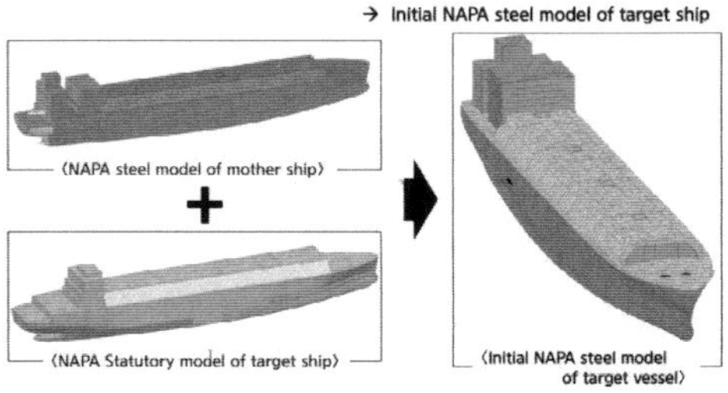

Figure 10: Example of combining ship projects

This parametrization makes it possible to modify design on very a high concept level, like combining information from different ship projects [Figure 10] or sharing information of alternative structural details [Figure 11].

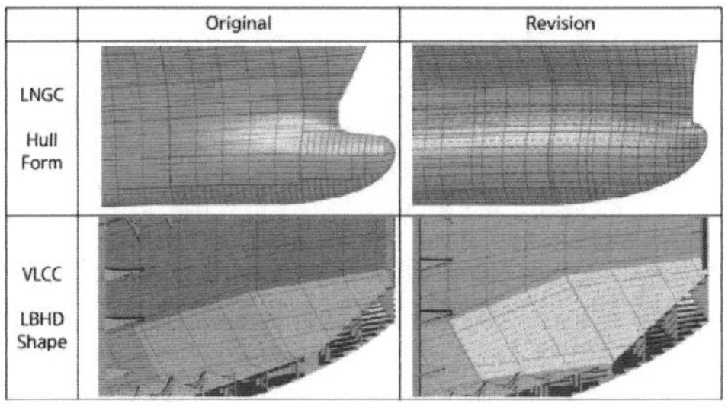

Figure 11: Example of modifying hull form

5.2 CLASS PROGRAM INTERFACE

Prescriptive rule checks are important part of plan approval work in defining local scantlings of structures. This work is most often done in software provided by the classification society involved.

From shipyard's point of view, the class society is selected based on many different aspects. Shipyards have to co-operate with whichever society is selected and prepare to deal with a large variety of different software packages.

The information needed for the rule check software will either be 2D section information or 3D information, depending on the case. Either way, this will require duplicated manual work if original design information is in 2D drawings [Figure 12].

NAPA supports most common class society rule check software, including:

- BV's MARS
- ClassNK PrimeShip-Hull
- DNVGL Nauticus Hull
- DNVGL Poseidon
- LR's RulesCalc
- KR's SeaTrust HullScan

The interface from NAPA to class society's rule check software speeds up the design process and helps to avoid errors common in manual work. The interfaces provide automatic creation of structural geometry including the steel structure attributes.

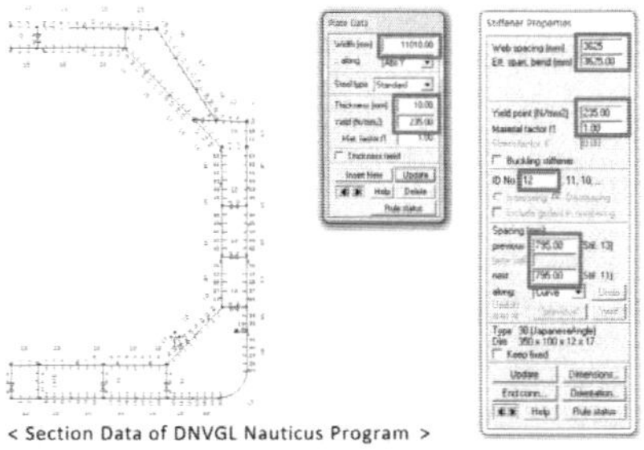

Figure 12: NAPA information in a rule check software

5.3 DRAWING EXTRACTION

Even though the drawings are not the origin of the design information, they are still used [Figure 13] for sharing the design information between different stakeholders, such as:

- Different design departments like machinery and outfitting design departments at the shipyard.
- Different functions like design, production and procurement at the shipyard.
- Ship-owner and shipyard.
- Class society and shipyard.

Completely avoiding drawings in the ship building process would probably bring most benefits and improvements in the process, but at the current state of the process development it was not considered possible.

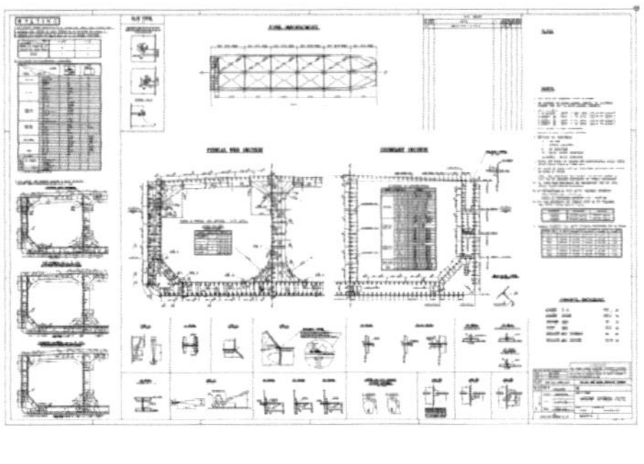

Figure 13: Typical structure drawing

With current NAPA Steel tools, it is possible to create drawings automatically from the 3D model. The drawings can be completed with interactive 2D CAD-like tools with additional information.

This brings benefits especially while updating the drawings when previous work is not lost and the automatic update can handle modifications in 3D model.

5.4 FE MODEL GENERATION

FE analyses are currently an essential and mandatory part of ship structural plan approval. By the nature of the method, a major part of the work is done traditionally in order to produce a good quality FE mesh.

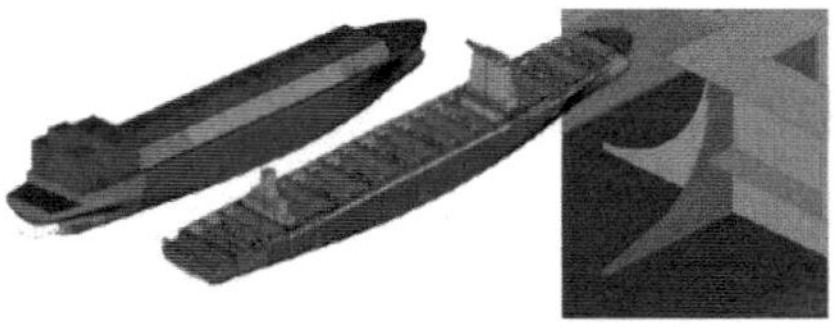

Figure 14: Examples of mesh created from NAPA

This is handled in NAPA with automatic meshing functions [Figure 14], which have the control over the needed parameters like element type and mesh size. The benefits are in saving design work and giving confidence that all other design information is compatible with FE calculations.

5.5 HULL WEIGHT AND COST CALCULATION

With good quality 3D product models, it is quite straight forward to check the weight, welding length and other cost indicators accurately and automatically.

In addition of direct weight calculation from 3D models, NAPA supports several techniques for handling statistical weight information with incomplete 3D information.

The advantage is in keeping the weight estimation up to while the design is evolving and be sure that the most accurate information is always available.

6. FUTURE WORK

The work has so far targeted the most essential key activities in structural basic design. There is still the possibility to improve these functions and this work continues based on "Design Thinking" principles.

The availability of accurate 3D information in initial and basic design opens also new possibilities for supporting other design and production disciplines. A high potential has now been identified for:

- NAPA supporting importing and exporting several different CAD file formats to be used with other CAD and browsing tools. This opens many possibilities for future work.
- Support for early outfitting design in engine room and other complex areas [Figure 15].
- Making simplified viewing and browsing models available to several stakeholders like production planning, ship owners and other design disciplines.
- Using 3D model for internal training at the shipyard. A pure 3D view gives a good visualization on real structures of ships.
- Using NAPA customization tools like scripting, macro language and API for design automation.

Figure 15: NAPA Steel with 3D outfitting

Work is now ongoing to study and evaluate these possibilities.

7. CONCLUSIONS

NAPA and HHI have undertaken a long development project during 2007-2016 making a paradigm shift possible at HHI Basic Structural Design Department.

In 2016, Hyundai Heavy Industries (HHI) implemented NAPA Steel's 3D model-based ship structural design tool across its business in Basic Hull Design Department. NAPA's new interactive 3D structural modelling interface allows users to create a single flexible product model for the whole vessel effectively.

Following the introduction of NAPA Steel, HHI expects the design time and man-hours to be reduced by 30% in ship initial and basic structural design and at the same time the quality of design is improved by having the single source of information for all activities.

NAPA's software has created now a paradigm shift in how HHI designs vessel and has helped the business become more competitive for the future of shipping.

8. ACKNOWLEDGEMENTS

During the many years this project has taken, there has been a large group of people at HHI and NAPA participating in the real development work and making this paper possible. They deserve our gratitude for their efforts.

9. REFERENCES

1. 'March of the digital native', The Naval Architect, February 2017, pp 14-18.

2. T Hulkkonen, C Cabos, M Kidogawa, T Kurki, B Tietgen, 'Utilization of a 3D Product Model to fulfil the new IACS CSR-H Rule Requirements', ICCAS 2013.

10. AUTHORS BIOGRAPHY

Tapio Hulkkonen holds the position of Development Director, NAPA Design Solutions. He is responsible for the development of NAPA Design Solution software and business. His previous experience includes 23 years working at shipyards in Finland and now 12 years in NAPA Group.

Hyung-Cheol Shin holds the position of Senior Engineer, Hyundai Heavy Industries (HHI). He is responsible for the development team leader of Basic Hull Design Department. His previous experience includes 13 years working at Ship Structure Research Department and now 5 years in this Department of HHI in Korea.

Nak-Hoon Yi holds the current position of Senior Engineer, Hyundai Heavy Industries (HHI). He is responsible for the technical strategy of Initial Design Dep't. His previous experience includes 10 years working at Basic Hull Design Department and now 1 years in this Department of HHI in Korea.

Deok-Hoon Jang holds the current position of Sales Director in NAPA. He is responsible for the global sales of NAPA's Shipping Solutions. His previous experience includes 3 years working at Ship Project Planning of HHI, and 7 years at Basic Design of DSME in Korea.

Teuk-Gyu Jang holds the current position of Senior Technical Consultant in NAPA. He is responsible for the technical consultant of NAPA's Design Solutions. His previous experience includes 11 years working at Basic Hull Design Department of HMD and now 2 years in NAPA.

Auto-Fine Mesh Generation for Local Analysis based on the Consistent Finite Element Model

MJ Son, JJ Woo, HG Park and **JY Lee,** Korean Register, Republic of Korea

SUMMARY

Under the circumstances of Common Structure Rules for Bulk Carrier and Oil Tanker (CSR-H; Harmonized Common Structure Rules) which is on effective from July 2015, the modeling area for finite element (FE) has been extended to forward and afterward cargo hold regions of a ship, more design man-hours are required in the FE modeling than that of previous rules. In addition, it is mandatorily required to generate the fine-mesh model for hot spot zones, but the target area has been extended so extensively, more design man-hours are required for fine-mesh modeling than that of coarse mesh modeling. Moreover, the local area for fatigue analysis should be modelled with very-fine mesh. However, these kinds of FE modeling for different level of detail and different purpose of analysis are time consuming and repetitive works. In this research, we would like to introduce the auto fine mesh modeling functionality that can generate fine-mesh and very-fine mesh model from the coarse mesh model, and retrieve and restore them to coarse mesh for certain purpose by re-mesh functionality. Through this process, FE model can be managed and reused with a consistent FE model from the initial design stage to detailed design stage of a ship.

1. INTRODUCTION

1.1 BACKGROUND OF RESEARCH

The new CSR for Bulk Carrier and Oil Tanker (or CSR-H; Harmonized CSR) [1], which was adopted by IACS (International Association of Classification Societies), has been effective on July 1, 2015, and has been applied in newbuilding projects since then. In addition, at the 96th meeting of the Maritime Safety Committee (MSC) of the International Maritime Organization (IMO) in 2016, it was decided that CSR-H would meet the requirements of Goal-based Ship Construction Standards (GBS) so that it has been internationally announced that bulk carriers and tankers of 150 m or above should be designed and constructed in accordance with this rule. There have been many studies and presentations on design changes [2] and weight changes [3], ship structural changes by ship type (158K tanker [4], MR tanker [5], Aframax tanker [6]) after CSR-H. Most of these are affected by enlarged structural analysis scope for fore/after body of a ship and increased mandatory fine-mesh zone [7]. In order to effectively cope with increased design man-hour, we presented in the previous study that 3D modeling has to be performed in the initial structure design phase, and it would be reused in the structural assessment S/W of a ship, which is provided by classification society, using CAD/CAE interface [8]. This has resulted in a significant reduction in design man-hours for generation of structural analysis finite element models (coarse-mesh) and modeling for rule scantling for forward and afterward body of a ship.

In the CSR-H environment, the more influential part of the design man-hours is the fine-mesh modeling rather than the coarse-mesh modeling. For the stress concentration area, it is mandatory to make a fine-mesh model to perform the local structural strength evaluation. Typically, there are connection area between a hopper and inner bottom with transverse web section, an opening such as manhole, and the toe-end of a large bracket. These fine-mesh applications are increased in proportion to the size of a ship, it is necessary to perform fine mesh modeling at 100 to 150 for 50K tanker [5] and over 200 zones for VLCC [9].

In addition, to evaluate the fatigue strength, a very fine-mesh modeling with element size of the thickness of the structural member should be performed, and this part often overlaps with the fine-mesh modeling zone

In this study, we will introduce the development of the finite element modeling automation function for this structure evaluation purpose. Furthermore, the function of automatically changing and restoring the finite element model from the coarse-mesh model to the fine-mesh and very fine-mesh model in the flow of design evaluation will be described. This concept enables the consistent finite element model to be managed within the structural design and evaluation S/W of the classification society, and in conjunction with the consistent use of CAD model, which introduced in the previous study [10], this can lead the innovation in the process of the initial structure design of a ship in the major shipyards.

1.2 OUTLINE OF SEATRUST-HULLSCAN

Korean Register (hereinafter referred to as KR) developed SeaTrust-HullScan (hereinafter referred to as HullScan), a 3D model-based engineering software for evaluating ship design based on CSR-H, and officially launched and distributed it in 2014.

As shown in Fig. 1, it mainly consists of two modes, the rule scantling mode, and direct strength assessment (hereinafter referred to as DSA) mode (Pre/Post system of FEM, except a solver). Various calculations on compliance with ship design rules including CSR-H for bulk carrier and tanker and KR rules for all type of ships can be performed based on the geometry model of a ship.

Furthermore, as shown in Fig. 2, the model paradigm for HullScan has been extended from 2D geometry/ 3D finite element model to 3D geometry/ 3D finite element model to utilize 3D CAD model of a whole ship that has been created and utilized in the initial design phase in the major shipyards in Korea, recently. In addition, we developed the interfaces in HullScan to all initial structure CAD for the shipbuilding that are used in the initial structure design divisions in Korean shipyards.

This allows users of HullScan to create cross-section model for the rule scantling automatically in real time by importing 3D CAD models without the need to create a new model. In result, this can reduce design man-hours buried in the modelling jobs during the initial structure design phase where design changes frequently, so that the design engineers can focus on engineering work. In addition, we developed various functions of 3D CAD model to be automatically modeled as a finite element model. The generated coarse-mesh model can be automatically modeled with a fine-mesh, and very fine-mesh, in accordance with the purpose of design and structural analysis, and vice versa. In this paper, we will introduce these functions in detail.

In addition, the function of this paper has already been applied to HullScan, and it is currently being used in all of the initial structural design divisions in Korean shipyard [11], more than dozens of CSR-H ships, considering only KR register projects, have been utilized these functionalities.

2. COARSE MESH MODEL GENERATION

The finite element model for yield strength evaluation of a ship is based on quad elements with size of the stiffener spacing (700 ~ 900mm). In some cases, triangular elements are used together, but they are limited to at least. The stiffeners are modeled as 1D finite element such as beam, rod, and so on. This coarse-mesh model should be one that shares only the same nodes at absolute coordinates where several elements meet (Singularity), and the elements should be organically connected (Connectivity) so that free-edges do not occur.

These FE models are usually created manually referencing 2D construction drawings using general-purpose FE modelers such as Patran, Hyper Mesh, and FE Gate or specific software provided by the classification society such as SeaTrust-HullScan, SeaTrust-Holdan, DNV-GL Genie and BV VeryStar 3D. Recently, an FE model, automatically generated from the initial structure 3D CAD for shipbuilding such as NAPA Steel and TTM ISD, is also utilized. Furethermore, HullScan can automatically generate FE models after importing shipbuilding 3D CAD models (AVEVA Marine, S3D, TTM ISD, and NAPA Steel) that are used in Korean shipyards [12].

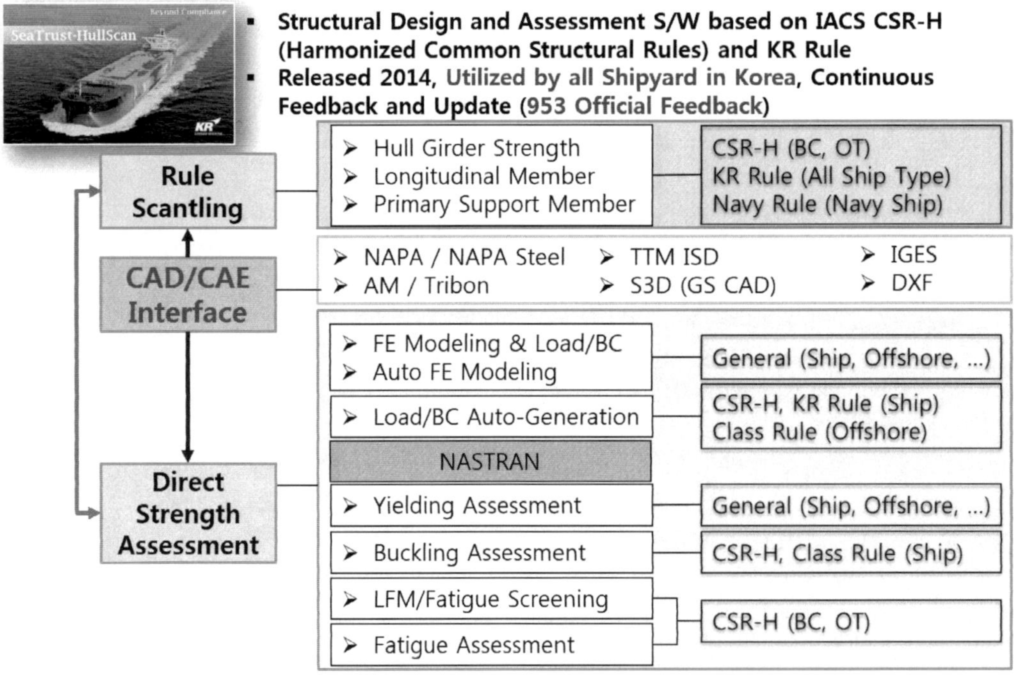

Figure 1: Overview of KR Structural Design and Assessment Software for Ships (SeaTrust-HullScan)

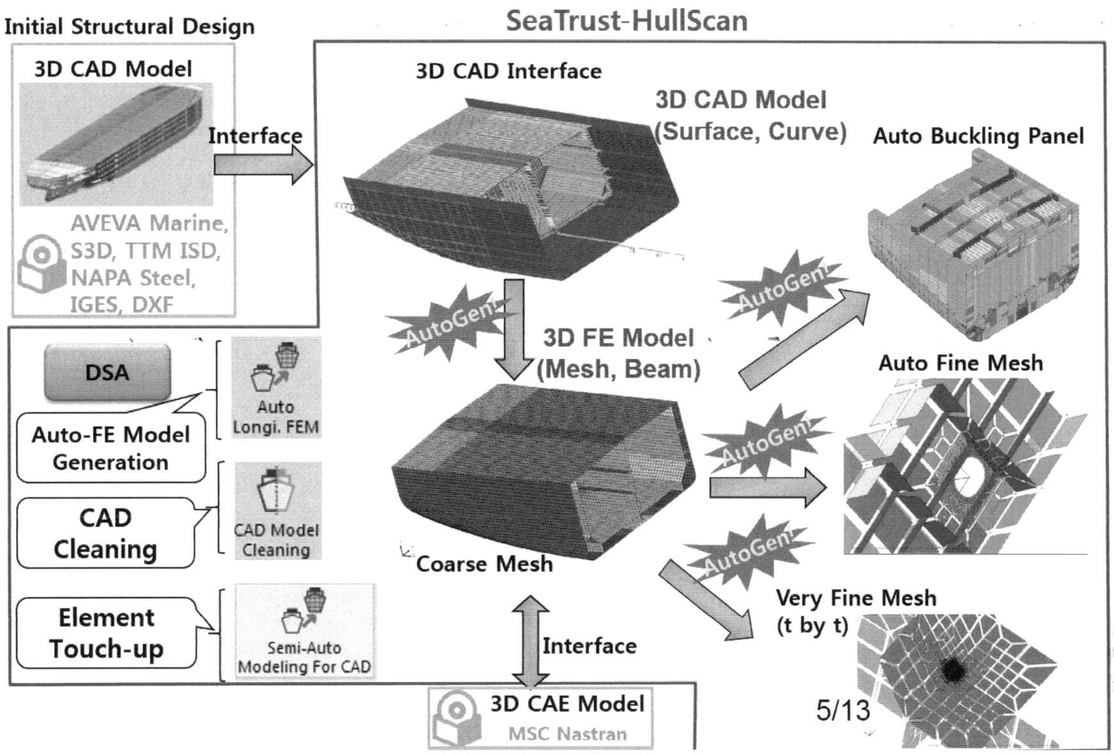

Figure 2: Outline of Finite Element Modeling in the Form of Coarse Mesh

The FE models generated by software other than HullScan are compatible with each other through the BDF file, the input format of MSC Nastran. In other words, although a coarse-mesh model can be generated in various ways depending on the S/W preference of design engineers, and 3D CAD environment of shipyards, in order to set the boundary condition and to set load case automatically in compliance with CSR-H, FE model should be input to the class structure design and strength evaluation S/W to be complete as the analysis model. The FE model, which is prepared for the evaluation of yield strength, is created in BDF format, and the calculation is performed through Nastan. The result is received as F06 file, and it is subjected to be imported again to the class structure design and strength evaluation S/W for the post-processing and analysis of results.

The concept of coarse-mesh modeling, and the linkage with other CAE S/W and CAD systems, and the concept of fine-mesh modeling and automatic generation of very fine-mesh from coarse-mesh to be explained in the next chapter are shown in Fig. 2.

3. AUTO FINE-MESH MODEL

The fine-mesh for the local structural strength evaluation specified in the CSR-H is to model uniformly more than 10 quad elements in a certain size (50 mm x 50 mm) or less in all directions around the stress concentration region. Also, there exists a transition zone where the element size naturally increases for seamless connection with the existing coarse-mesh. The representative zones for fine-mesh modeling with different characteristics are structural member connecting area where element stress is concentrated, manhole and a large opening area and a bracket toe end area. In this paper, we describe the development of the function to complete the fine-mesh model automatically from the coarse-mesh through the parametric input for these three cases.

3.1 FINE-MESH FOR CONNECTION AREA

Representative stress concentration area to which hull structural members are connected is the hopper knuckle area of the double hull type ship. There is also a double knuckle area where the knuckle parts are doubled where the longitudinal member and the transverse member meet at the forward and afterward cargo hold, so that the modeling difficulty is higher in the CSR-H ship.

This double knuckle area is mandatory fine-mesh zone in CSR-H. These parts repeatedly appear on both sides of each area where the transverse member exists in the longitudinal direction of a hull, however it is difficult to repeatedly copy and reuse the same fine-mesh model due to changes in shape of area as it goes to the forward and afterward and changes in thickness and material of the related structural members.

The automatic fine-mesh modeling for this works like as shown in Fig. 3. When user selects the nodes of the stress concentration region and enters the element size of the fine-mesh and how many fine-mesh should be uniformly modeled from the stress concentration region, then, after the function is applied, the meshes of the fine mesh and the transition region are modeled at once, while affecting only the two regions of the entire range from the nodes of the stress concentration region to the coarse-mesh basis.

As shown in Fig. 3, the generated fine mesh elements have different colors, which means that they inherent properties and materials of the existing coarse-meshes, and loading group set and longitudinal member group set for structural strength evaluation are maintained. In addition, if there exist beam elements near the fine-mesh zone, elements are also automatically split while maintaining its properties.

When the fine-mesh model is completed, the transition zone to coarse-mesh is to be modelled. After finding the corresponding coarse-meshes within the transition zone and the edges of each elements are to be extracted. It is possible to find the edges adjacent only to coarse-mesh of the transition zone, except the edges adjacent to fine-mesh zone, then nodes for the pattern mesh, which is mainly used for fine-mesh modeling can be created. According to this, the node within the corresponding coarse-mesh can be defined, so that the automatic mesh algorithm with the node constraint is applied to generate the transition meshes.

3.2 FINE-MESH FOR OPENING

Around area of an opening is a typical stress concentration region so that large size of opening was used to be modelled in fine-meshes. In the CSR-H environment, the scope extends to manholes. As a result, a considerable number of areas around the openings have been subjected to fine-mesh modeling. The feature of the fine-mesh modeling of the opening is that the quad elements of a certain size (50mm x 50mm) or less should be modelled as two layers around the opening.

There are three main methods of the automatic generation for the fine-mesh modeling of the openings as shown in Fig. 4. First method is that if the user selects a node or point of the center position of the opening and inputs the length direction of the parameters of the opening as a vector (①-a) or selects two or more nodes to form a vector (①-b). The second method is to create a local coordinate system at the center of the opening and select it (②). The last one is to draw the shape of the opening directly into a curve and selects it (③).The third method can be used by receiving the shape information from the CAD interface, and is used to apply a specific opening shape that does not correspond to the library provided by this function.

The shapes of the opening, which are typical in Korean shipyards, that can be automatically generated through parameter input in this function is shown in Fig. 5. When user selects the opening type, only the parameter window to be input is activated. If user fills in the corresponding dimensions, the 2-layer model is automatically created for the fine-mesh.

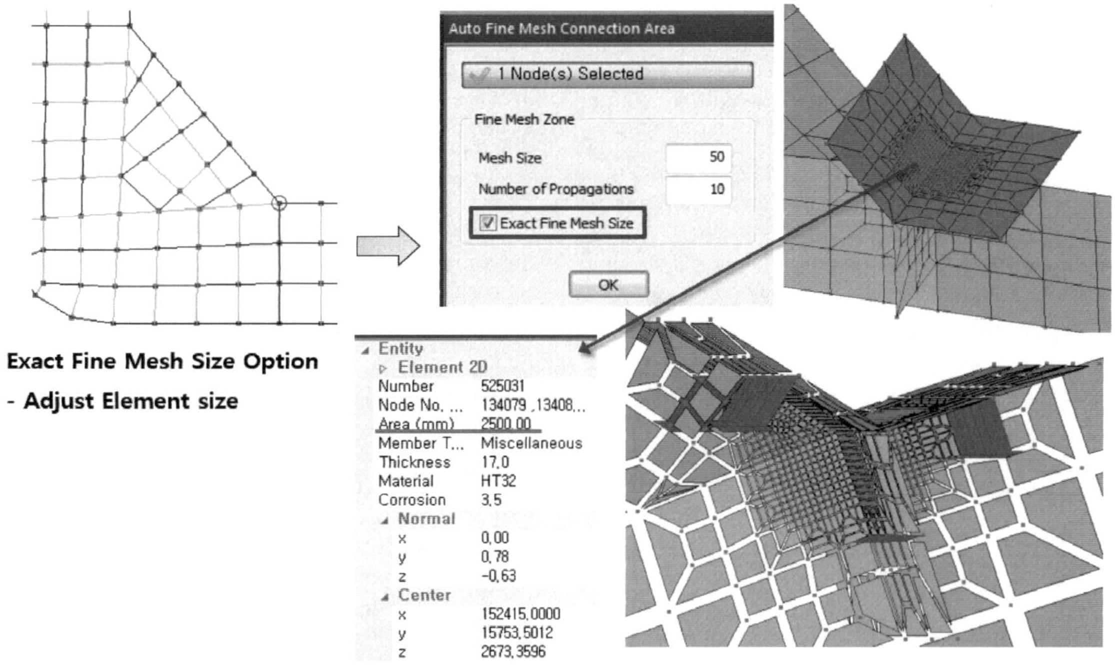

Figure 3: Auto Fine Mesh Generation for Connection Area

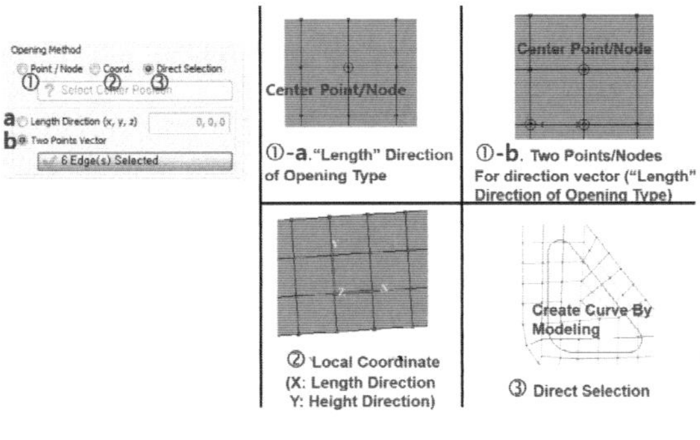

Figure 4: Auto Fine-Mesh (Opening) Creation Method

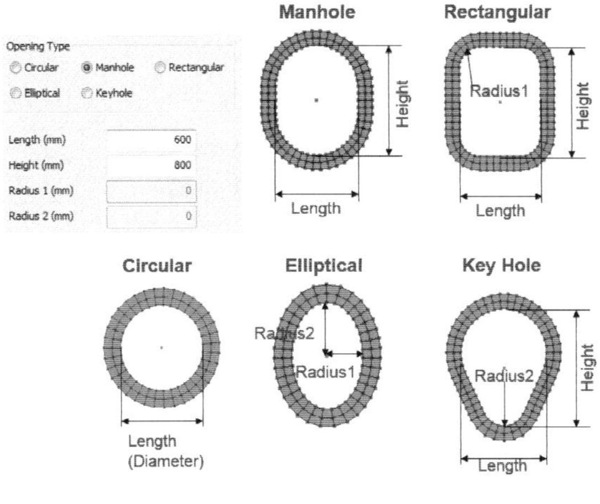

Figure 5: Parameter-based Modeling for Opening

The generation of the transition zone between the fine-mesh area of the opening and the near coarse-mesh area are divided into five methods in consideration of the structural characteristics (Fig. 6).

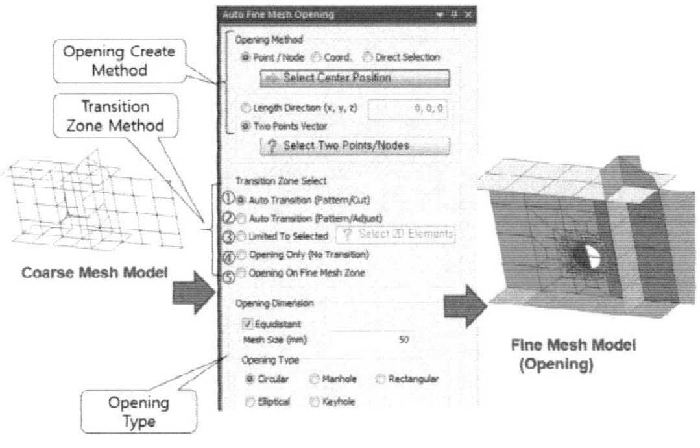

Figure 6: Auto-Fine Mesh (Opening) Overview

It is important to create a model of the transition zone considering the presence of stiffener for reinforcement around the opening. To do this, there exist two different methods. First one is to generate transition meshes first, then to split them with beam elements as a cutter (①).

Second one is to consider beam elements as the constraint of the auto-mesh for the transition zone (②).It is developed by the design engineer to determine according to the structural characteristics of the peripheries depending on whether the opening is located on a double bottom floor or a girder. The results of applying the two options for the same shape are shown in Figure 6.

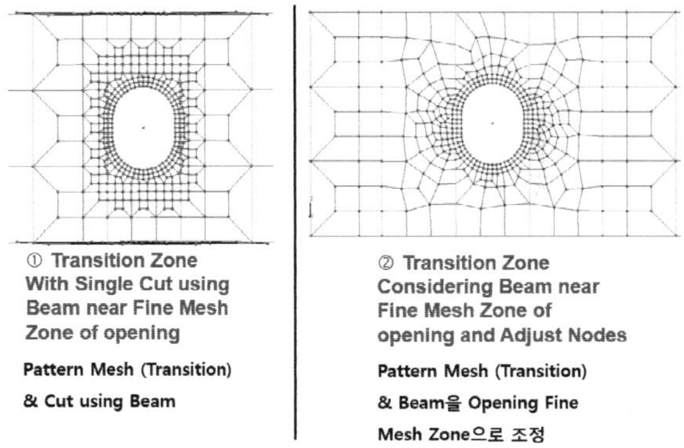

Figure 7: Auto-Fine Mesh (Opening) Transition Zone Modeling Methods

The other three transition zone modeling options are those in which only the elements specified by the design engineer are created and do not affect the unspecified adjacent elements (③).This method is used for the purpose of preventing the automatically generated elements from invading or affecting the range of existing models when adjacent fine-mesh models already exist. Another method is to create a 2-layer fine-mesh model of the openings and to create a transition model so that the design engineer can precisely secure node connectivity without free edges to adjacent fine-meshes that is pre-modeled by the design engineer (④).The last method is to create only the opening 2 layer (⑤), which is used by the design engineer to directly modify the peripheral model to link the fine-mesh model to the coarse-mesh model

3.3 FINE-MESH FOR BRACKET

In a bracket shape including the toe end area of a large transverse member and a hatch corners, the fine-mesh modeling is performed on the toe end area and the free edge area where a stress is concentrated. In the fine mesh model, it is necessary to generate fine mesh models based on the precise geometric shape which is modelled roughly in the coarse-mesh model, thus after deleting the bracket element reflected in the existing coarse-mesh model, the model is created by applying this function.

This function has been developed to generate automatic model based on parameter of bracket shape which is frequently used in Korean shipyard. The method of generating the bracket fine-mesh model is as same as the opening method, the method of selecting the center node and the direction vector by the user, the method using the local coordinate system, and the method of directly inputting the bracket boundary curve (Fig. 8).

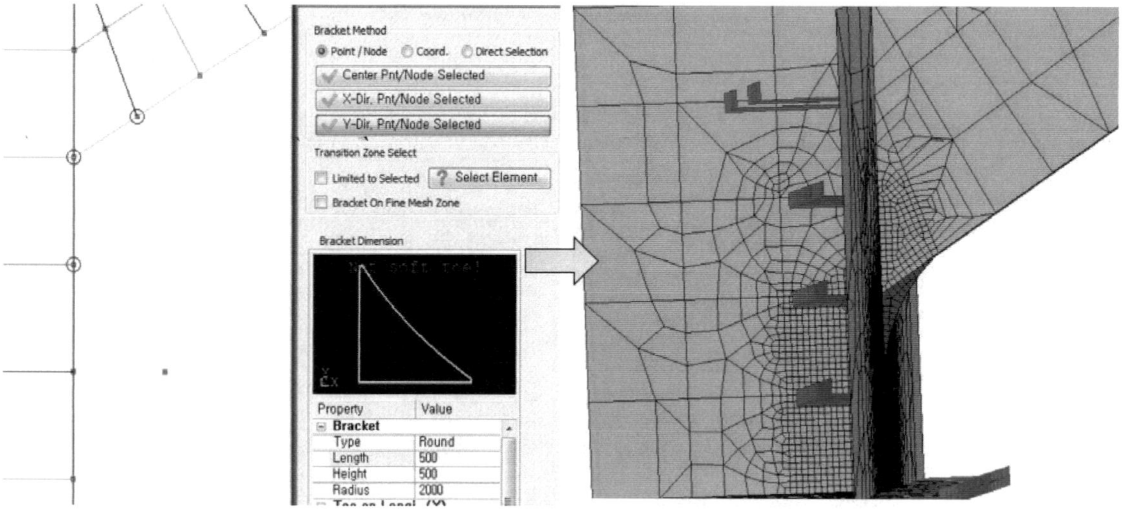

Figure 8: Auto Fine-Mesh (Bracket) Overview

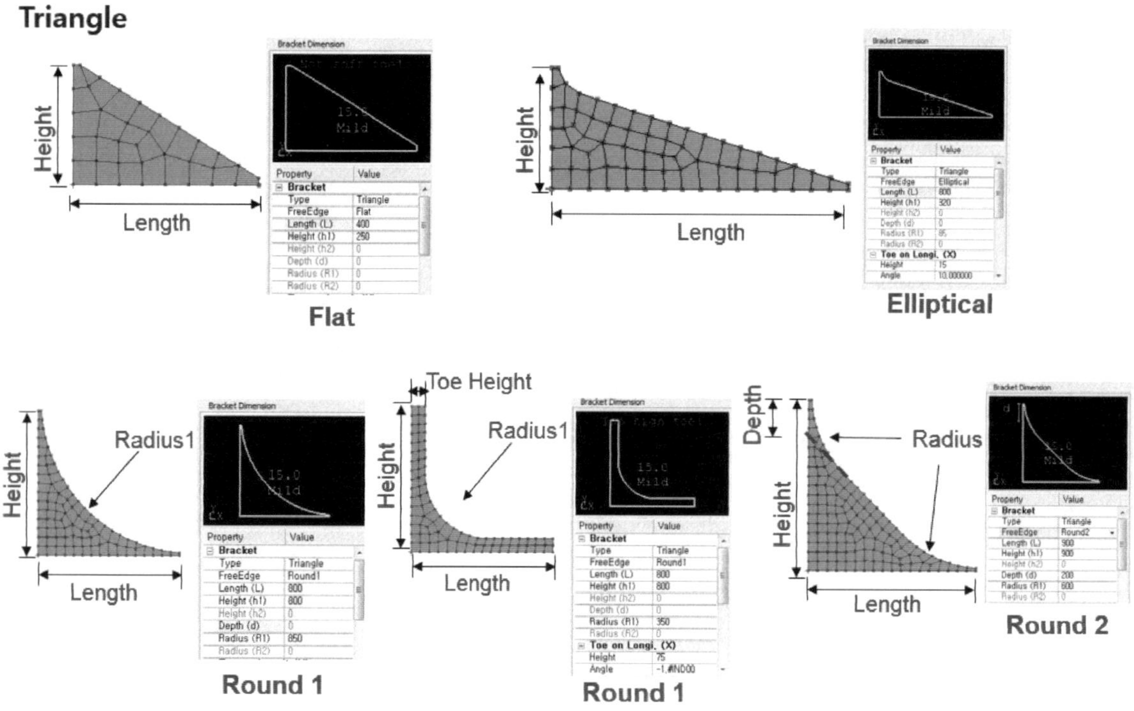

Figure 9: Auto Fine Mesh (Bracket) – Triangle Type

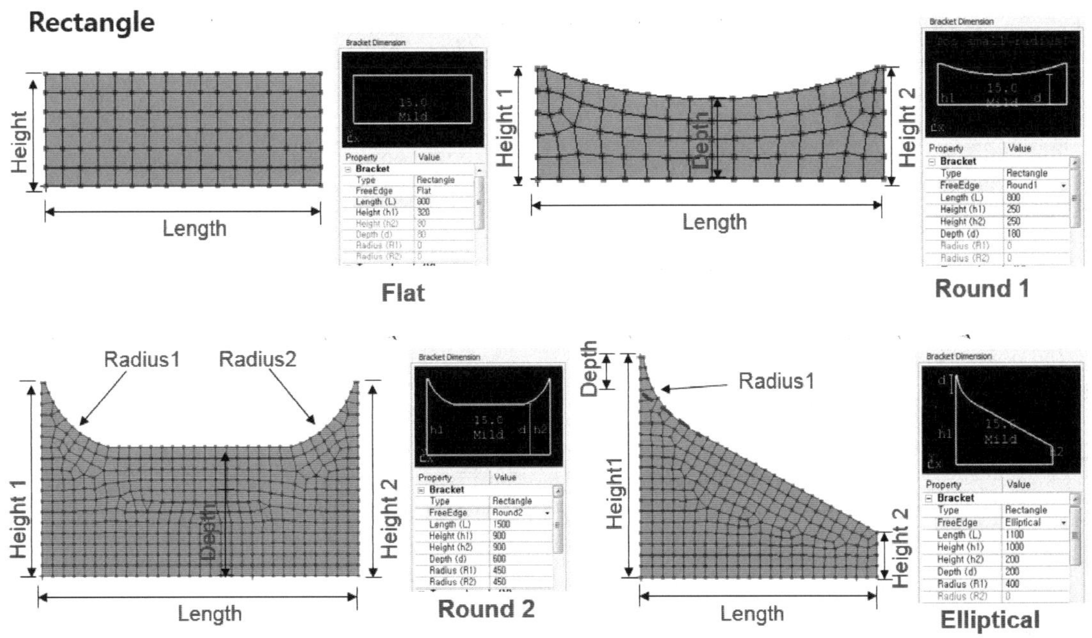

Figure 10: Auto Fine Mesh (Bracket) – Rectangle Type

The models that can generate the bracket shape by parameter-based modeling are divided into three types: triangle type (Fig. 9) and rectangle type (Fig. 10), and the details consists of 'Flat', 'Round1', 'Round2', and 'Elliptical'. When the design engineer enters the dimensions for each input item, the brackets of the 2D geometry are updated in real time. When the auto fine-mesh modeling function is applied, the bracket geometry is converted to fit to the 3D coordinate system to be generated with the fine-mesh element size of user input.

The generation of the transition zone between the generated fine-mesh and the adjacent coarse-mesh model is divided into three method. First method is automatic generation. Second one is the method of automatically correcting the bracket fine-mesh model to fit for the fine-mesh zone created by the user. And last one it to restrict transition modeling within user specified elements.

4. VERY FINE-MESH MODELING AND AUTO RESTORING

For some of the fine-mesh modeling described in Chapter 3, a model for fatigue analysis should be created for fatigue strength evaluation. In this case, very fine-mesh modeling should be performed with a member thickness (t) as the element size (typically 10 to 30 mm), which is smaller than the general element size (50 mm x 50 mm) of the fine-mesh. When creating the "t by t" model, which is commonly referred to, at the fine-mesh model, it is difficult to directly reuse the existing model due to the limitation of the existing fine mesh size. Therefore, in this study, the function is implemented according to the concept of completing the very fine-mesh model by using the automatic fine-mesh function after restoring the existing fine-mesh into the coarse-mesh. That is, if only the input mesh size is changed to the member thickness size in the automatic fine-mesh generation function described in Chapter 3, the corresponding sizes are arranged in ten elements in all directions, and the remaining transition meshes are generated within the range of two coarse-meshes. Since the difference of element size between the very fine-mesh and coarse-mesh is about 30 to 100 times that the size of the transition mesh is gradually increased. However, in order to control the number of elements, the number of transition meshes is minimized by processing with a multi-layer pattern mesh.

When using the automatic fine-mesh generation function of HullScan introduced in the previous chapter, the element group having an absolute coordinate of a corresponding function and a central node as its name is automatically generated and the original coarse-mesh is managed by this group. This element group is stored in HullScan's native format (*.nfm). At this time, not only the newly generated group information but also all the group information to which each element belongs are stored together. When we use this information to find the fine-mesh group where the very fine-mesh model should be created and run the "Restore Coarse Mesh" function, the original coarse mesh model can be retrieved as shown in Fig. 11. In this restored state, the very fine-mesh model can be easily completed by inputting the element size corresponding to the desired "t by t" in the "Auto Fine Mesh" function (Fig. 12).

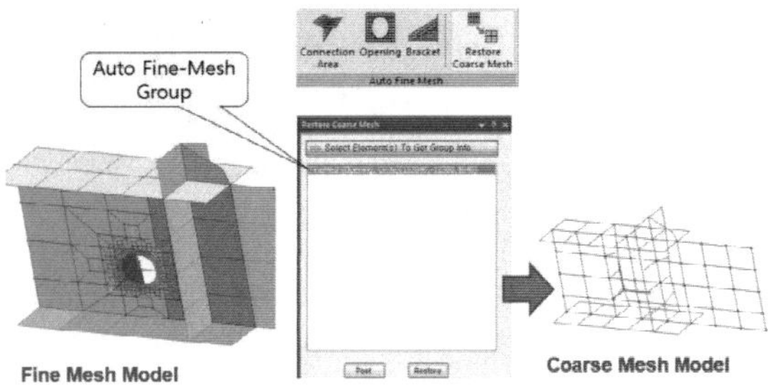

Figure 11: Coarse Mesh Restore

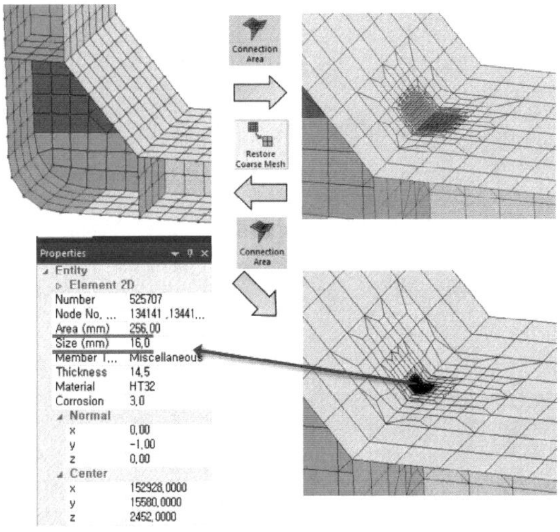

Figure 12: Process for Very Fine-Mesh Modeling

5. CONCLUSION AND FUTURE WORKS

In this paper, we proposed the concept and function development result of various detailed FE models for the purpose, focusing on the consistent FE model, in order to reduce the design man-hours of structural analysis modeling which had been increased more than three times compared with the existing one due to the application of CSR-H.

We presented conventional method and the new method utilizing CAD / CAE interface for generating a coarse-mesh model for yield strength evaluation and introduced automatic fine-mesh modeling functions for the connection area, opening area, and bracket toe end area. In addition, we show the concept and result that the generated fine-mesh model can be restored to the coarse-mesh model or it can be generated to the very fine-mesh model.

The developed functions are installed in the DSA mode of SeaTrust-HullScan, which is the structural design and assessment software of Korean Register, are widely being used in the initial structure design division in Korean shipyards. These functions are already applied to several CSR-H projects, demonstrating its effectiveness by feedback on the modeling time savings of more than 80% [9].

For future works, we will develop the automatic fine-mesh generation function that reflects the presence of a sniped face plate or flange at the end of the bracket, and that reflects bent hopper plates. And the fatigue modeling automation for slots and scallop is also considered to be developed.

6. ACKNOWLEDGEMENTS

This research was performed as a part of the research project below and supported by the organizations indicated. We acknowledge and appreciate the support provided.

- 'Development of a 3D Model-based Consistent System for Ship Design' project funded by Ministry of Trade, Industry and Energy of Korea (No. 10052708).

7. REFERENCES

1. IACS, Common Structural Rules for Bulk Carrier and Oil Tankers, 1st January 2015 Issue: International Association of Classification Societies, 2015.

2. Shibasaki, K., Shipbuilding industry's perspective on the new IACS Common Structural Rules, Proceedings of Tanker Structure Cooperative Forum 2016, Busan, Korea, 2016.

3. Kwon, O.S., Investigations of New(Harmonized) CSR Application, Proceedings of Tripatite 2015, Seoul, Korea, 2015.

4. Choi, U.C., Impact of Harmonized CSR for oil tanker, Proceedings of Tanker Structure Cooperative Forum 2016, Busan, Korea, 2016.

5. So, H.Y., MR. Tanker with Harmonized CSR, Proceedings of Tanker Structure Cooperative Forum 2016, Busan, Korea, 2016.

6. Cai, S., Application of New Common Structural Rules on Aframax Tankers, Proceedings of Tanker Structure Cooperative Forum 2016, Busan, Korea, 2016.

7. Seo, H.D., Study on Fatigue Strength for Tank Structures subject to Harmonized CSR, Proceedings of Tanker Structure Cooperative Forum 2016, Busan, Korea, 2016.

8. Son, M.J., Lee, J.Y., Park, H.G., Kim, J.O., Woo, J.J., and Lee, J.H., Development of 3D CAD/CAE Interface in Initial Structural Design Phase of Shipbuilding

Transactions of the Society of CAD/CAM Engineers, 21(2), pp. 186-195, 2016.

9. Park, K.Y., VLCC DSA by SeaTrust-HullScan, Proceedings of KR Software User Conference 2016, Busan, Korea, 2016.

10. Son, M.J., Lee, J.Y., Park, H.G., Kim, J.O., Lee, J.H., and Woo, J.J., CAE Interface Using Consistent 3D CAD Model, Proccedings of the Annual Autumn Conferece, SNAK, pp.54-58, 2016.

11. Park, H.G., Development of Ship Structural Design Assessment System for Enhacement of Design Productivity and Quality, Proceedings of the 43rd Ship and Offshore Design Workshop 2015, Busan, Korea, 2015.

12. Son, M.J., Park, H.G., Lee, J.Y., Lee, J.H., Kim, J.O., and Woo, J.J., Development of Auto FE Modeling for enhancement of the productivity in modeling based on CSR-H, Proccedings of the 13th International Symposium on Practical Design of Ships and Other Floating Structures, 2016.

8. AUTHORS BIOGRAPHY

[Myeong-Jo SON] holds Ph.D. Degree in naval architecture and ocean engineering of Seoul National University, and holds the current position of senior researcher at Korean Register. He is responsible for development of CAD/CAE Interface, automatic mesh generation, and automatic fine-mesh generation in SeaTrust-HullScan. He is also in charge of development of 3D model-based design approval system and the digital twin of a ship. His previous experience includes shipbuilding CAD/CAM, PLM, modeling and simulation, business process management, and initial/production hull structural design.

[Jeong-Jae WOO] holds M.S. Degree in mechanical engineering of Korean Maritime and Ocean University, and holds senior researcher at Korean Register. He is responsible for rule scantling mode in SeaTrust-HullScan.

[Ho-Gyun PARK] holds M.S. Degree in naval architecture and ocean engineering of University of Ulsan, and holds of senior researcher at Korean Register. He is responsible for DSA mode, especially fatigue analysis of SeaTrust-HullScan.

[Jeong-Youl LEE] holds Ph.D. Degree in naval architecture and ocean engineering of Chungnam National University, and holds the current position of general manager of ICT center at Korean Register. He is responsible for development of all of technical S/W in KR.

AN INTEGRATED PROCEDURE FOR PRELIMINARY TANKER SHIP DESIGN CONSIDERING OIL OUTFLOW RISK FROM SHIP-SHIP COLLISION

X Tan and **J Tao**, Maritime Institute at Nanyang Technological University, Singapore
D Konovessis, Singapore Institute of Technology, Singapore
H E Ang, School of Mechanical and Aerospace Engineering at Nanyang Technological University, Singapore

SUMMARY

This paper addresses the implementation and application of a numerical procedure intended for the preliminary design of oil tankers considering the risk of oil outflow from ship-ship collision accidents. The simulation procedure was established by integrating a probabilistic oil outflow model with common ship design software to evaluate various aspects of performance, e.g. stability, safety in regard to oil spill prevention and mitigation, and economic efficiency. The procedure was implemented in a parametric way, such that by systematically varying the parameters related to the conceptual design of the ship, large amount of design variants can be evaluated. The comparisons of several ship design merits including safety across the different variants are discussed through a case study with a Suezmax tanker to get a preliminary insight into the trade-offs, which is crucial for modern ship design.

NOMENCLATURE

List of Abbreviations:

AP	Aft perpendicular
BN	Bayesian network
IMO	International Maritime Organization
RCO	Risk control options

List of Symbols:

B	Breadth, moulded (m)
C	Oil outflow volume (m^3)
C_B	Block coefficient
D	Depth, moulded (main deck) (m)
Disp.	Ship Displacement (ton)
DWT	Deadweight tonnage (ton)
HDB	Double bottom height (m)
HDS	Width of double skin sides (m)
L_{BP}	Length between perpendiculars (m)
L_{OA}	Length overall (m)
LCB	longitudinal center of buoyancy (m)
p	probability of having a certain amount of oil outflow
R	Risk metric (m^3)
T	Draught (m)
V_{out}	Oil outflow volume (m^3)

1. INTRODUCTION

Marine transportation of oil involves high risks due to the devastating consequences of potential oil spill [1]. Safety in terms of avoiding or mitigating accidental oil spills has thus become the major concern of the ship design and operation industry [2]. In the context of tanker ship design, a number of measures, i.e. risk control options (RCOs), can be taken to reduce the oil spill risks and thus ensure safe operations [2], [3]. The measures within the ship hull structural design discipline may include: employing double hull concepts, adding longitudinal and/or transverse bulkheads, changing the location of transverse bulkheads to avoid having large tanks at parts of the ship that are more likely to be breached, reinforce the local structural elements of the vessel, etc. On the other hand, a ship should be designed to be as safe as practical, as implementation of risk control options can often incur extra costs or lower the profitability and thus economic efficiency. An optimal, "safe enough" design, which is usually chosen from a subset of mathematically optimal designs, is a compromise of costs and benefits to different stakeholders over a multiple of optimization objectives [4], [5]. In order to evaluate and compare the safety level across design variants, quantification of risks of accidental oil outflow has become an important subject.

Risk is in the present work expressed as an expected oil outflow volume following [6] and [7]:

$$R = \sum_{i} p_i \cdot C_i \qquad (1)$$

where,
i is the index of consequences, which in the present paper represent the different magnitudes of oil outflow volume;
C is the outflow volume;
p is the probability that a certain amount of outflow occurs;

This expression is essentially the discrete version of the expectation of the probabilistic distribution of oil outflow volume. It also shows that risk is defined in the present paper as the product of the probability of having a certain consequence and the metric of the consequence.

Ship-ship collision is a result of a chain of failures in a sequence of complicated physical processes where uncertainties are present at every single stages of the interaction between the two ships [8] [9]. The onset of a collision starts as early as when the two ships are on courses that can lead to contact if no evasive maneuver is done properly, i.e., the encounter situation. Commonly adopted method for the frequency of encounter ("geometric probability") is given by e.g: Fujii and

Shiobara [10]. Given an encounter, collision can still be avoided if the evasive maneuver is successful. And the failure probability for this stage is often termed as the "causation probability". Given failure in evasive maneuver, the oil tanker can either be striking or struck, depending on the relative speed and collision location. And finally, if the tanker is the struck vessel, a failure in the structure of the cargo tanks eventually lead to an oil outflow.

In order to obtain a statistically sound damage (outflow) distribution, extensive amount of collision cases have to be evaluated for each design variant. Since even the preliminary design optimization can easily involve hundreds of hull variants, methods such as performing integrated dynamic simulations with the finite element method and rigid body dynamics to collect the damage data to establish the damage distributions directly are considered unsuitable for the purpose of risk-based ship design and optimization. Friis-Hansen [11] proposed a numerical software for risk analysis of two ships colliding in a given sea area, and the use of "super elements" has well balanced computational efficiency and accuracy. However, parametric ship modeling does not seem to be straightforward. Although historical accident statistics are harvested from the real world, it reflects only fragments of reality. Relying solely on accident statistics may lead to biases either because they cannot reflect the future or because they do not take into consideration of specific conditions, e.g. ship structural design, for different cases [12] [13]. This is also part of the reason for the criticism of the IMO oil outflow calculation method.

Since the work by Minorsky [14], the ship-ship collision problem has commonly been simplified by being divided into external dynamics and internal mechanics, coupled by the collision energy, (e.g., [15], [16], [17], [18], [19], [20]). The procedure presented in this paper integrates a model for simulating probabilistic oil outflow given tanker being struck. The advantages of the model can be summarized as follows:

- Variables that describe the collision energy, the damage location and extents are explicit inputs based on respective probabilistic distributions, which means collision cases can be generated by sampling the random variables and aleatory uncertainties in the collision scenario variables can propagate to the damage;
- Main particulars and tank configurations for the struck vessel are input variables too, ensuring the influence from ship (preliminary) design variables being accounted for;
- Correlations between damage extents are accounted for;
- Efficient and thus suitable for risk modeling;

The scope of the present work is as follows:

- Only accidental cargo oil outflow is considered, operational discharge or bunker oil outflow are not the focus this paper;
- Tanker layout is 6 by 2 limited by the damage extent model;
- All oil is spilled in a breached tank, which may lead to a conservative result;
- Safety in the present paper refers to the environmental safety in regard to oil spillage impacts. Safety of human life is not the subject of the paper.
- Since the focus of present paper is on generic methodology for the procedure of preliminary tanker ship design considering oil outflow risks, the "geometric probability", and "causation probability" which are dependent on the traffic conditions for a given sea area, are not addressed.

In the following sections, first the methodology is introduced briefly. Then its intended application is examined by a case study of the preliminary design of a Suezmax crude oil tanker to gain a preliminary insight into the trade-offs among design merits including steel weight, cargo capacity and oil outflow volume.

2. METHODOLOGY

The procedure is a synthesis of existing analytical and statistical models for ship collisions and common ship design software. The basic idea is to establish a procedure that has high computational efficiency and at the same time can still take into consideration of details such as ship's main particulars and randomly distributed collision scenario variables. Parametric ship hull model was setup using the common ship design software NAPA. Traditional ship performance aspects such as the intact and damage stability, calm water resistance, cargo capacity and steel weight were calculated using NAPA too. In parallel, Monte Carlo simulations were conducted to obtain the probabilistic oil outflow for each design variant. The main particulars and inner tank arrangement parameters were fed to the probabilistic oil outflow model together with the randomly generated collision scenario variables. A schematic illustration of the procedure is given in Figure 1. In the following part of this section, the procedure and the adopted models are briefly introduced.

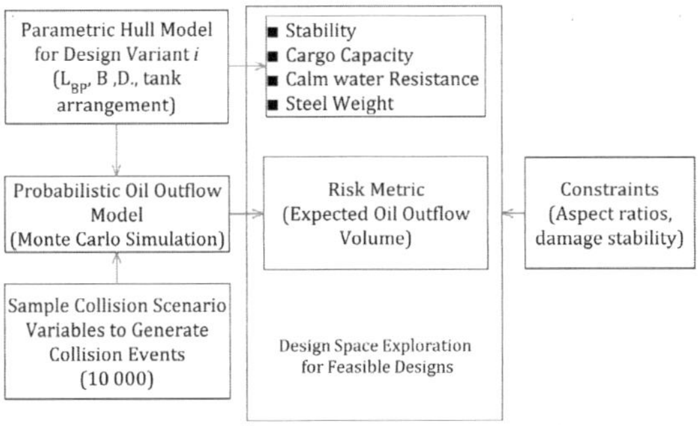

Figure 1: Schematic illustration of the procedure

2.1 STEEL WEIGHT, RESISTANCE AND CARGO VOLUME

In the present stage of the study, simple empirical methods were adopted to calculate ship weight and resistance. The weight component accounted for are the hull steel, outfit and machinery weights [21]. The resistance was calculated using Holtrop and Mennen's formulae [22]. The cargo volume of each tank is given by the geometric model of NAPA.

2.2 PROBABILISTIC COLLISION DAMAGE

The probabilistic ship-ship collision damage model given by [23] was adopted and implemented numerically in the present study to evaluate the probabilistic collision damage and subsequent oil outflow. The model is a regressional meta-model by processing simulation results of 80 000 collision cases for four different tanker designs. The simulations were conducted by using a deterministic numerical procedure of ship-ship collisions developed by [24]. The inputs of the model are: the velocities of the two colliding vessels right before collision, the masses of the two vessels, the collision angle and location and the half bow entrance angle of the striking vessel. The outputs of the model are the damage properties including the length, depth and start and end location along the ship side. 10 000 collision cases were generated based on each input collision scenario variable's probabilistic distribution. The same 10 000 collision cases were applied to each design variant and Monte Carlo simulations were conducted. By knowing the damage properties and inner tank arrangement, the breached tanks can then be identified and oil outflow can be calculated accordingly.

3. CASE STUDY

A case study was conducted to examine the application of the procedure in the preliminary design of a Suezmax tanker ship considering oil outflow risk. Main particulars and parameters that control the inner tank arrangement were varied systematically so that the design space was explored efficiently and effectively. For each feasible design variant, the expected oil outflow volume was calculated using the probabilistic collision damage model introduced in Section 2.2. Intact and damage stability, steel weight and calm water resistance were calculated in NAPA. Finally, the trade-offs among the design merits including risk were discussed.

3.1 REFERENCE SHIP

A representative double hull Suezmax tanker given in [25] was chosen as the reference ship. The main particulars of the ship are given in Table 1.

Table 1: Main particulars of the reference ship

Parameter	Value	Unit
L_{OA}	273.64	m
L_{BP}	264.00	m
B	48.00	m
D	23.1	m
C_B	0.86	/
LCB (to AP)	133.09	m
HDS	2.5	m
HDB	2.8	m
T	17.11	m
Disp.	190243	ton
DWT	158.982	ton
Cargo tank block length	213.14	m

3.2 DESIGN CONSTRAINTS & DESIGN SPACE FILLING

The following design constraints were applied to either setting the bounding limits of the design variables in Table 2 or filtering out the designs that do not fulfill the general requirement of ship design, e.g. stability, to ensure a design space of feasible designs.

- Suezmax are the maximum marine vessels that meet the restrictions of the Suez Canal. The maximum permissible draught of Suezmax is 20.1 m to pass the canal. From 2010, the wetted surface cross sectional area of the ship is limited by 1006 m^2, which means 20.1 meters of draught for ships with the beam no wider than 50.0 m or 12.2 meters of draught for ships with maximum allowed beam of 77.5 meters [26].
- The ratios *L/B*, *B/D* and B/T are limited to between 5.5 and 6.0, 1.9 and 2.1 and 2.25 and 3.75, respectively [27].
- Damage stability of each design is checked against IMO's damage stability code [28].

In addition, the examination is restricted to the feasible designs that have a displacement falls in the range of ±10% the displacement of the reference design, so that the variants are relatively comparable in terms of ship size.

Table 2: Ranges of ship design variables

Variable	Lower bound	Upper bound
L_{BP}	248.7 m	289.5 m
B	45.0 m	50.0 m
LCB	123.1 m	147 m
C_B	0.818	0.896
D	21.5 m	25.2 m
HDB	1.8 m	3.2 m
HDS	1.8 m	3.2 m

In order to explore the design space effectively and efficiently, low variance sequence was used to fill the design space by generating the design variables within the bounds given in Table 2. In the present study, originally 10 000 designs were generated, of which 4 023 are feasible ones that fulfill the constraints listed above. The scatter plots of the generated ship design variables are given in Figure 2 appended at the back of the paper.

3.3 PARAMETRIC SHIP MODEL

A parametric ship hull model (Figure 3) was established with NAPA. As the detailed information as to the lines of the ship is not available, a generic tanker geometry provided by NAPA was used. The design variables of the ship model were varied according to the data points as presented in Figure 2.

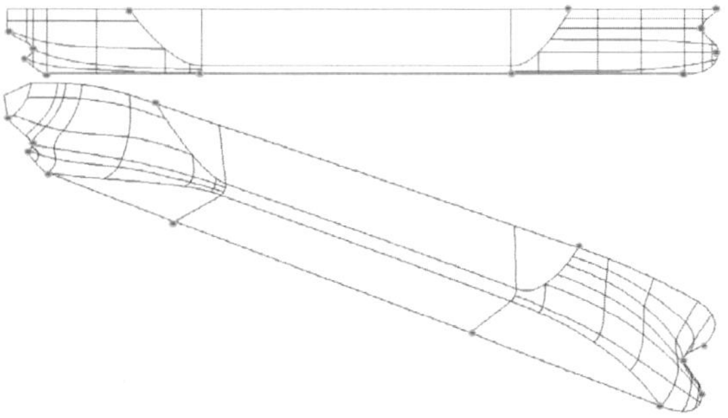

Figure 3: Parametric hull model in NAPA (the points that control the geometry of the hull are highlighted.)

The tank arrangement is illustrated in Figure 4.

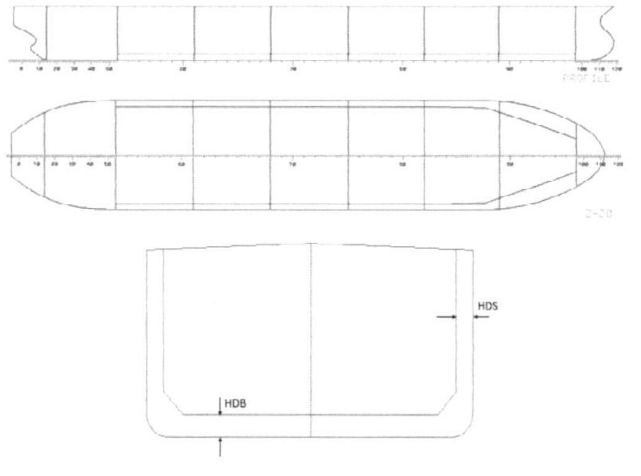

Figure 4: Tank arrangement

3.4 PROBABILISTIC DISTRIBUTIONS OF COLLISION DAMAGES OF THE REFERENCE SHIP

The simulated probabilistic distributions of oil outflow volume and damage extents are exemplified by the results of the reference ship as given in Figure 5 and Figure 6. The probability of zero outflow for the reference ship is 77.69%. The statistics are conditional on tanker being struck. It is clearly shown in Figure 6 that the damage length and depth are correlated.

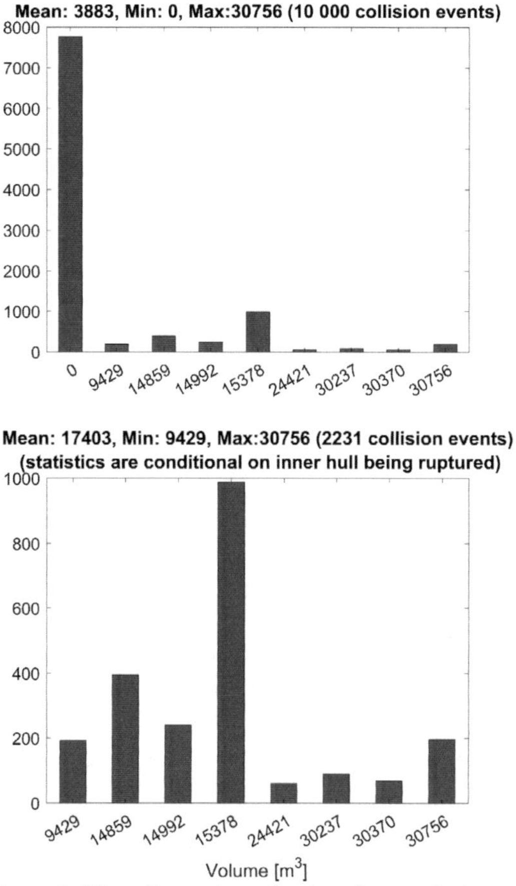

Figure 5: Oil outflow volume for the reference design

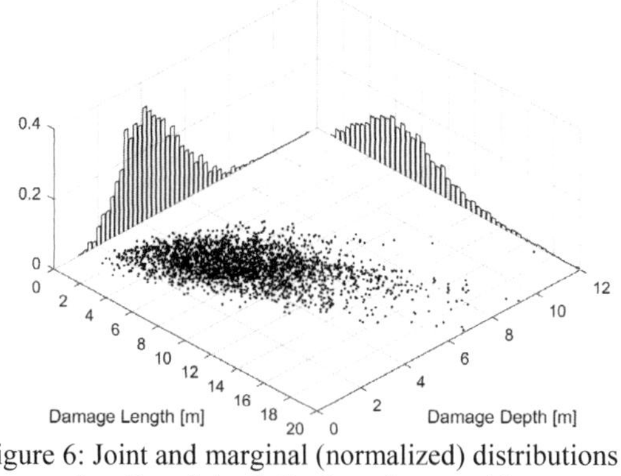

Figure 6: Joint and marginal (normalized) distributions of damage length and depth for the reference ship

3.5 RESULTS

The relationships among the steel weight, calm water resistance, cargo capacity and mean oil outflow as obtained from the procedure are summarized in Figure 7 through Figure 9. The figures indicate the trade-offs between each two conflicting objectives. Also marked in the figure are the reference design, the "Pareto front" of cargo capacity versus oil outflow, and the five most favorable designs in terms of utility levels which combines the effects of all the four objectives with serval sets of predefined weighing factors.

The designs fall on the Pareto front are the "non-dominated" set, which means that the answer to which are better remains open until a specific setting and thus preference is introduced. How a favored design is subsequently to be chosen depends on the specific settings of the intended service profile of the ship, which may include, e.g. the sailing area, design life span etc. For example, it was found that within the examined design space, there did not exist a design that is better than the reference design in terms of the absolute values for all four merits. This is as expected since the reference design is believed to be an optimized one at least for the three conventional merits that do not include oil outflow risk. Nevertheless, depending on the operational profile of the vessel, relatively better designs can still be pursued by putting more weights on the dominant factor(s) and evaluating the overall performance.

In Figure 7, equal weights, i.e., no preference, were given to the four design merits, and the five most favorable designs given by ranking the utility levels of all the designs are in general concentrated in the smaller size segments. This is reasonable because all of the four merits but the cargo capacity favor a smaller ship in general.

In environmentally sensitive areas, more preferences should be given to the designs with relatively low oil outflow level with some sacrifice of economic aspects in order to avoid or mitigate the devastating consequences from oil spill. In contrast to the scenario in Figure 7, when more weights are given to oil outflow and cargo capacity, as shown in Figure 8 and Figure 9, some larger vessels can become competitive too.

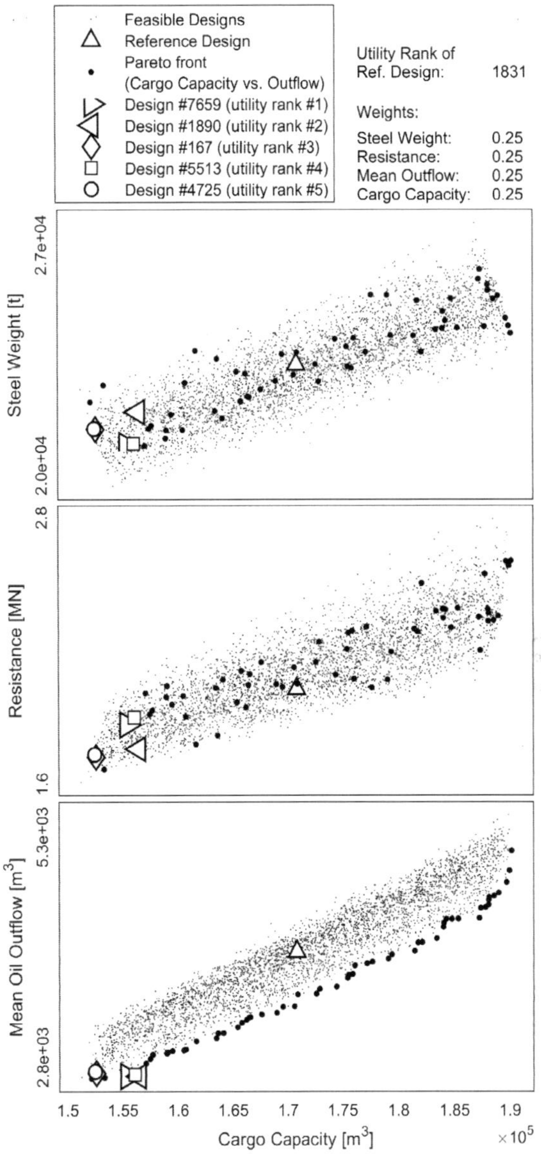

Figure 7: Scatter plots for the design merits (equal weights)

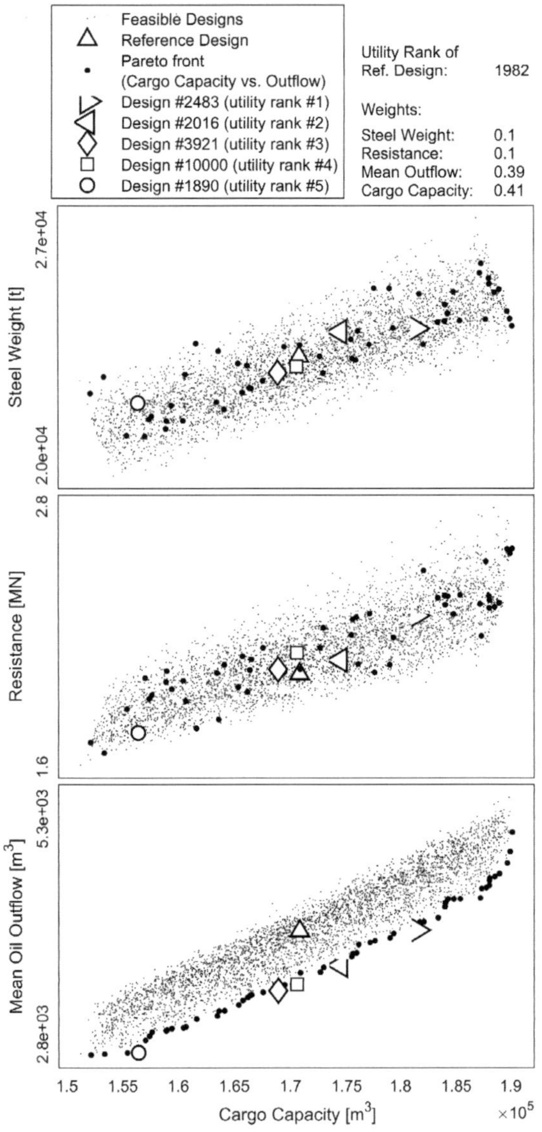

Figure 8: Scatter plots for the design merits (preference is firstly given to cargo capacity and then to oil outflow risk)

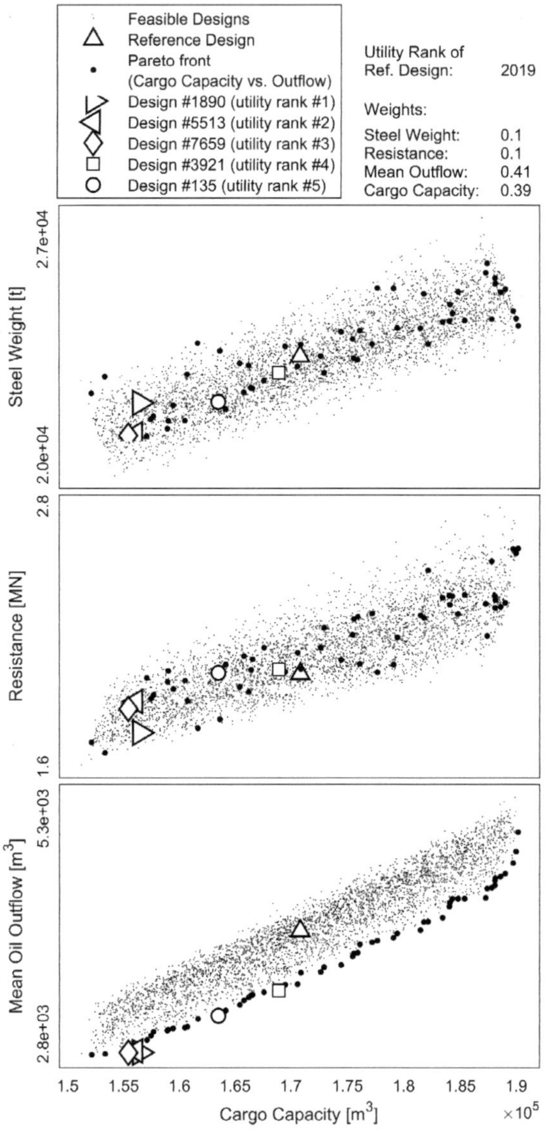

Figure 9: Scatter plots among the design objectives (preference is firstly given to oil outflow and then to cargo capacity)

4. CONCLUSIONS AND DISCUSSION

A procedure for the preliminary design of oil tanker considering oil outflow risk from ship-ship collisions was introduced. The procedure is a synthesis of statistical and analytical models and common ship design software. The application of the procedure was examined by conducting a case study with a representative Suezmax tanker. Risk as measured by the expected oil outflow is considered together with other conventional ship design merits. The utility levels for each of the design variants were compared for different weighing factors.

It was seen that different weighing factors can point to designs with different features. The procedure can facilitate making decisions under various or changing scenarios. For example, in environmentally fragile sea areas, more attention is naturally given to oil outflow risk, at the same time without compromising profitability too much. In this case, similarly high weights can be assigned to these two merits, like design #3921 exemplified in Figure 8 and Figure 9, to achieve a better overall performance. It can also be seen from the study that although there does not exist a design in the explored design space which is superior to the reference ship in regard to all of the four merits, when given a specific scenario, an even better design can be pursued to that direction. For instance, design #3291 rank #3 and #4 for the weighing factors in Figures 8 and 9, as compared to #1982 and #2019 for the reference design.

Future work needs to be carried out to evaluate the gains and losses among the various design by, e.g. converting the respective performance aspect to a monetary value and by considering the overall cost-benefit of the vessel during her design life span. Also, a number of additional issues have to be studied for a "fully probabilistic" oil outflow to be arrived at, namely, encounter number (geometric probability) in the area that the ship under design is intended to sail; and, the "causation probability" for collision due to failure of evasive maneuver given encounter, e.g. by using fault trees or Bayesian networks.

5. ACKNOWLEDGEMENTS

This research is funded by Singapore Maritime Institute.

6. REFERENCES

1. NATIONAL RESEARCH COUNCIL, 'Tanker Spills: Prevention by Design', Washington, DC: The National Academies Press, 1991.

2. IMO MARINE ENVIRONMENT PROTECTION COMMITTEE, 'FSA – Crude Oil Tankers', MEPC 58-INF.2, 2008.

3. PEDERSEN, P. T., 'Review and Application of Ship Collision and Grounding Analysis Procedures', Marine Structures, Volume 23, Issue 3, Pages 241-262, 2010.

4. KLANAC, A., VARSTA, P., 'Design of Marine Structures with Improved Safety for Environment', Reliability Engineering & System Safety, Volume 96, Issue 1, Pages 75-90, 2011.

5. PAPANIKOLAOU, A., ZARAPHONITIS, G., BOULOUGOURIS, E. et al. J Mar Sci Technol 15: 359, 2010.

6. ISO, 'Risk Management Vocabulary, Guidelines for Use in Standards, ISO/IEC Guide. International Standards Organisation, Geneva. ISO Guideline vol 73, 2002.

7. VINNEM, J., 'Offshore Risk Assessment Principles, Modelling and Applications of QRA Studies', Springer-Verlag London, 2014.

8. GOERLANDT, F., MONTEWKA, J., 'A Framework for Risk Analysis of Maritime Transportation Systems: A Case Study for Oil Spill from Tankers in A Ship-Ship Collision', Safety Science, Volume 76, Pages 42-66, 2015.

9. SORMUNEN, O. E., GOERLANDT, F., HÄKKINEN, J., et al., 'Uncertainty in Maritime Risk Analysis: Extended Case Study on Chemical Tanker Collisions', Proceedings of the Institution of Mechanical Engineers, Part M: Journal of Engineering for the Maritime Environment, Volume: 229 issue: 3, page(s): 303-320, 2014.

10. FUJII, Y. & SHIOBARE, R, 'The Analysis of Traffic Accidents'. Journal of Navigation 24(4): 534-543, 1971.

11. FRISS-HANSEN, P., SIMONSEN, B. C., 'GRACAT: Software for Grounding and Collision Risk Analysis', Marine Structures, Volume 15, Issue 4, Pages 383-401, 2002.

12. BROWN, A., TIKKA, K., DAIDOLA, J. C., LÜTZEN, M., and CHOE, I-H, 'Structural Design and Response in Collision and Grounding', In Annual meeting. Vancouver: SNAME, 2000.

13. DNV, 'Risk Assessment – Large Passenger Ships – Navigation', DNV Report 2003-0277, Annex II, 2003.

14. MINORSKY, V.U. 'An Analysis of Ship Collisions with Reference to Protection of Nuclear Power Plants', Journal of Ship Research, Vol. 3, No. 2, pp. 1-4, 1959.

15. BROWN, A.J. and CHEN, D., 'Probabilistic Method for Predicting Ship Collision Damage', Oceanic Engineering International, 6, 54–65, 2002.

16. PEDERSEN, P. T., ZHANG, S., 'On Impact mechanics in ship collisions, Marine Structures', Volume 11, Issue 10, Pages 429-449, 1998.

17. ZHANG, S. 'The Mechanics of Ship Collisions', Ph.D. thesis, Technical University of Denmark, p. 262, 1999.

18. HONG, L., 'Simplified Analysis and Design of Ships Subjected to Collision and Grounding', Norwegian University of Science and Technology, 2009.

19. LÜTZEN, M., "Ship Collision Damage", Ph.D. Thesis, Department of Mechanical Engineering, Technical University of Denmark, Lyngby, 2001.

20. TABRI, K., BROEKHUIJSEN, J., MATUSIAK, J., VARSTA, P., 'Analytical Modelling of Ship Collision Based on Full-Scale Experiments', Marine Structures, Volume 22, Issue 1, Pages 42-61, 2009.

21. WATSON, D.G.M., and GILFILLAN, A.W., 'Some Ship Design Methods', Transactions RINA, Vol. 119, 1977.

22. HOLTROP, J., and MENNEN, G.G.J., 'An Approximate Power Prediction Method', International Ship Building Progress, 1982.

23. VAN DE WIEL, G. and VAN DORP, J.R., 'An Oil Outflow Model for Tanker Collisions and Groundings', Annals of Operations Research, 187, 279–304, 2011.

24. BROWN, A. 'Alternative Tanker Designs, Collision Analysis', NRC Marine Board Committee on Evaluating Double-Hull Tanker Design Alternatives, 2001.

25. IMO MARINE ENVIRONMENT PROTECTION COMMITTEE, 'FSA – Crude Oil Tankers', MEPC 58/INF.2, 2008.

26. WIKIPEDIA, 'Suezmax', https://en.wikipedia.org/wiki/Suezmax

27. LAMB, T, 'Ship Design and Construction'. Jersey City, NJ, Society of Naval Architects and Marine Engineers, 2003.

28. IMO, 'International Convention for the Safety of Life at sea, Chapter II-1, Part B-1, Regulation 7', 1974.

7. AUTHORS BIOGRAPHY

Xiang Tan is Research Fellow at the Maritime Institute at Nanyang Technological University, Singapore. She graduated with MSc in Naval Architecture and Ocean Engineering from Tianjin University and PhD in Marine Technology from the Norwegian University of Science and Technology (NTNU). Her research interests include dynamics and the design of ships and marine structures under ice and wave actions. She is responsible for establishing the procedure, analyzing the results and writing the paper.

Jidong Tao is Research Associate at the Maritime Institute at Nanyang Technological University, Singapore. He is responsible for preparing the tanker model in NAPA and generating results of technical and economic performance of various tanker designs. His previous experience includes ship performance, structural design and analysis of ship and offshore engineering, etc.

Dimitrios Konovessis is Associate Professor at the Singapore Institute of Technology. He graduated with MEng in Naval Architecture and Marine Engineering from the National Technical University of Athens and PhD in Ship and Marine Technology from the University of Strathclyde. His research interests include design methods for ship safety, first-principle and performance-based approaches for risk-based ship design, operation and regulation, marine systems reliability and maintainability, and energy efficiency and management.

Hock Eng Ang graduated with a BSc in Mechanical Engineering from the University of Wales, Swansea and a MSc degree in Mechanical Engineering, with DIC, from University of London, Imperial College. He is Chartered Mechanical Engineer (UK) and Registered Professional Engineer with Practising Certificate (Malaysia). He is the academic focus coordinator for the Naval Architecture & Marine Engineering specialization courses offered by School of Mechanical & Aerospace Engineering, Nanyang Technological University, Singapore.

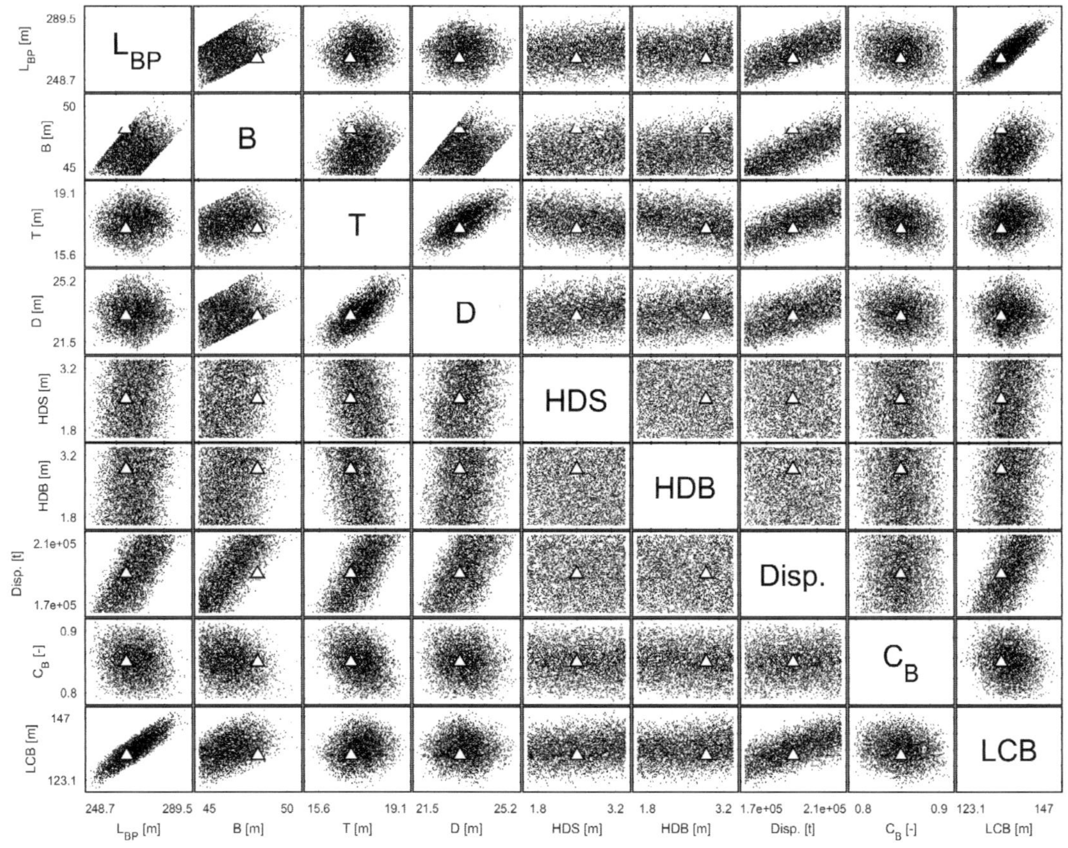

Figure 2: Input variables for feasible designs (the point corresponding to the reference design is highlighted in white triangle; the upper and lower bounds of the data set are labeled along the axes)

DEVELOPMENT FOR EVALUATION SW REGARDING TO CONTAINER SECURING STRENGTH

Jin-young Park, Dong-min Kuem, Jae-hun Jung, Korean Register of Shipping, Republic of Korea

SUMMARY

Recently, As container vessel become larger, the number of loading tiers, lashing bridge tiers and loading patterns have been diversified. And Most of classification society has revised guidelines for securing devices to protect and customer's safety and property.

However, as the various securing device and methods for container cargos are applied, it is very difficult to calculate finding efficient arrangement plan within permissible load. So the Korean Register of Shipping developed SeaTrust-LS(Container securing strength evaluation SW) to evaluate the container stowage and securing in compliance with the KR's guideline, and have continuously revised and upgrade SW.

1. INTRODUCTION

The number of container loading stack is increased because container ship's size is larger than before. And various methods of securing are devised. Therefore, it is hard to find optimum stack arrangement within allowable load according to classification society's guidelines for securing devices and securing methods for ship and crew's safety. Therefore many man-hour, effort and know-how are needed to find proper stack arrangement.
And it is practically impossible to calculate container securing strength manually according to classification society's rule because it serves a lot of load cases considering container ship's route, ship motion and container failure mode.
In addition, the shipping company has been making efforts for efficient operation through cargo boosting by new securing methods, and each class has amended the regulations that safe and efficient loading is possible yearly. And then container securing service providers and shipping company need to apply new rule and regulations after understanding new one fully. So providers need so much time to apply new rule to in-house or their own SW.
Therefor Korean Register of Shipping released container securing strength evaluation SW that new lashing method against ship's enlargement and cargo boosting is applicable and provide a guide to help customers make quick decisions and minimizes the time loss due to changes in classification rules.

2. DEVELOPMENT REQUIREMENTS

2.1 THE INTRODUCTION OF RULE TECHNICAL BACKGROUND

The reaction force considering the lashing method and type, wind power, etc. is calculated. And the allowable limit for this reaction force are presented by each class. In addition, it is also necessary to consider rule because the rule with reduction factor according to operation route are provided in the rules in the domestic and international classification society recently. As shown in the figure below, the ship performs various motions during the operation, so that the container loaded on the ship also performs the same motion. As shown in the following figure, loads are applied to each container by the ship motion and connection between the containers as well as the external force due to the wind acting on the container during the ship operation. As a result, Racking, Shear, Compression, and Lifting Force are generated in the Container, and Tension occurs in the supporting Lashing Rod. And these loads must meet the limits specified by the Society.

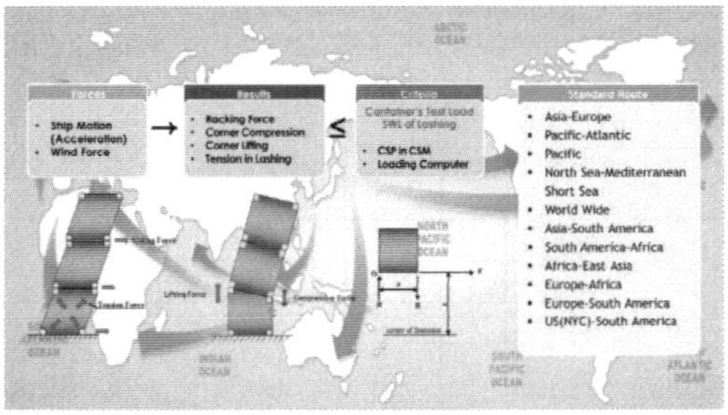

Figure 1: Technical background of rule

2.2 FUNCTIONAL REQUIREMENTS

2.2 (a) the various securing method

As vessels became bigger and the height of the lashing bridge increased, the stack height of containers became higher. This increases the likelihood of failure in case of the compression and lifting of the container. In the past, Internal Lashing alone has been able to find a suitable layout, but in recent years, methods such as External Lashing and Vertical Lashing have also been considered. When external lashing is used, result of compression is better than internal lashing. And it is more advantageous to apply it to the outermost stack which is influenced by wind force.

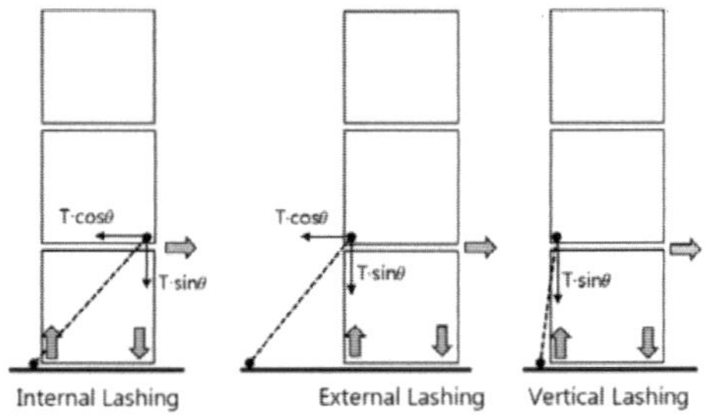

Figure 2: Various lashing type

Generally, when the long lashing is applied to the top of deck or lashing bridge as shown in the first figure below, two lashing rods are applied. However, as the height of the racing bridge increases recently, it is required to secure container at a position other than the top of the lashing bridge as shown in the second figure below. So if long lashing at the top of the lashing bridge would be applied as shown in the second picture below, the number of lashing rods are three. However, it is not easy to calculate the load generated from the secured container using three lasing rods because it is necessary to solve the simultaneous equations with three unknowns. And the regulations only show how to obtain the tension when two lashing rods are applied. Therefore, it is required to develop an algorithm for obtaining the tension when three lasing rods are mounted. Because there is possibility of loading from higher position than the deck height at the after-end or fore-end of the vessel, this should be taken into account.

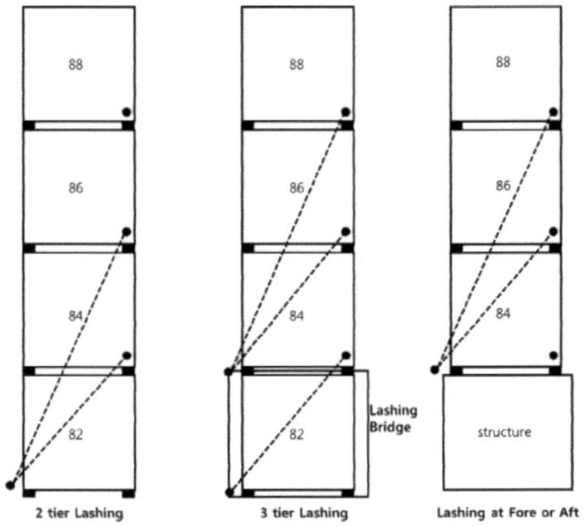

Figure 3: Various cases of securing method

2.2 (b) Optimum stack arrangement

In general, ship owners are trying to find a layout with a maximum vertical center of gravity (VCG) in order to load as many containers as possible within the allowable load. It takes a lot of man-hour to find the optimum arrangement having the maximum VCG. However, as vessels become larger and calculation cases become more numerous with various lashing methods, specialist companies also require one to two weeks of work time. The optimum stack arrangement guide function can maximize the efficiency of the design work of the container ship in the shipyard, the container securing manual (CSM) certification work of the classification society, and the ship operation of the shipping company.

2.2 (c) Software development kit

In recent years, revision of the rules related to container securing has been continuously carried out at domestic and foreign classification of society. In addition, the updated rules were immediately reflected in the software, and domestic and foreign classification of society was updated and distributed. As a result, third-party developers who provide in-house SWs or shipping companies' own SW have been required to develop. However, it takes a great deal of time and expertise to incorporate the rules of Container Securing into SW. In addition, since the safety evaluation of container securing is very difficult even in the shipping companies operating the container, if the software is not ready, there will be a major disruption to container operation. Thus, software development kits (SDKs) are required to save time and money in software development.

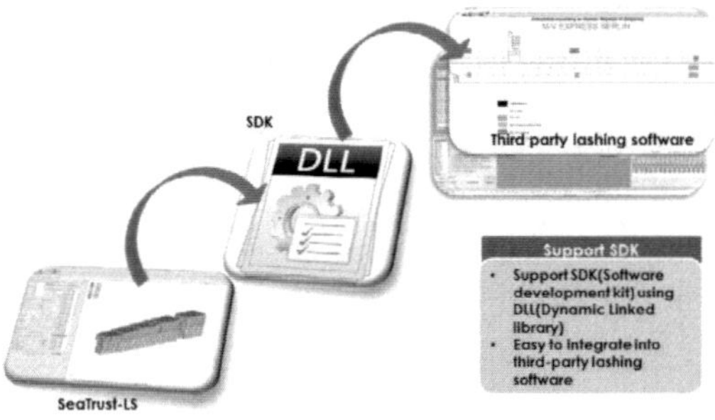

Figure 4: Schematic diagram for SDK

3. DEVELOPMENT RESULTS

3.1 LASHING TENSION CALCULATION ALGORITHM

When three lashings are applied, it is assumed that the container is loaded with lashing bridge on the structure with the height of one stack as shown in the figure to obtain the lashing tension. The reason for assuming this structure is that three lashing rods with the lashing bridge are applied, and container at the after-end and fore-end of vessel can be loaded at the tier other than above the deck.

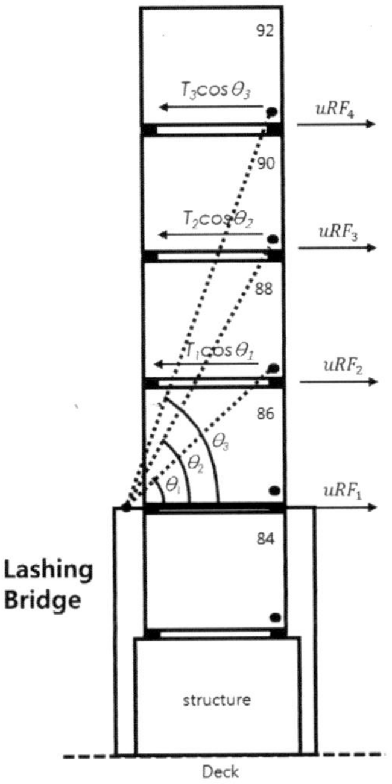

Figure 5: example for securing method using 3 rod

Assuming that the deformation amount of the container is equal to the deformation amount of the lashing rod, the following equilibrium form can be derived.

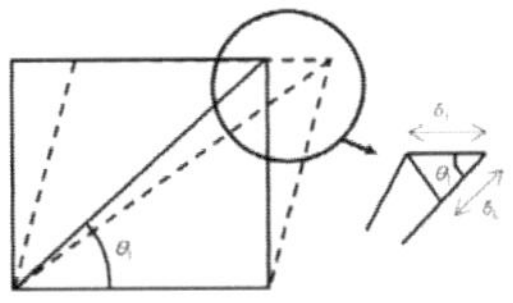

$$\frac{uRF - T\cos\theta}{K_C} = \frac{T\cos\theta}{K_L}$$

Figure 6: the relation between container and lashing rod

Considering the equilibrium of the working load and the lashing rod at each tier, the following equilibrium equation can be derived.

$$(K_C/K_1 + 3 - 1)T_1\cos\theta_1 + (3-1)T_2\cos\theta_2 + (3-1)T_3\cos\theta_3 = uRF_1 + uRF_2$$

$$(3-1)T_1\cos\theta_1 + (K_C/K_2 + (4-1))T_2\cos\theta_2 + (4-1)T_3\cos\theta_3 = uRF_1 + uRF_2 + uRF_3$$

$$(3-1)T_1\cos\theta_1 + (4-1)T_2\cos\theta_2 + (K_C/K_3 + (5-1))T_3\cos\theta_3 = uRF_1 + uRF_2 + uRF_3 + uRF_4$$

uRF_i : Unlashed Racking Force
T_i : Lashing Tension
K_i : Lashing Rod Stiffness
K_C : Container Stiffness

Following matrix formula can be obtained.

$$\begin{bmatrix} (K_C/K_1+2) & 2 & 2 \\ 2 & (K_C/K_2+3) & 3 \\ 2 & 3 & (K_C/K_3+4) \end{bmatrix} \begin{bmatrix} T_1\cos\theta_1 \\ T_2\cos\theta_2 \\ T_3\cos\theta_3 \end{bmatrix}$$

$$= \begin{bmatrix} uRF_1 + uRF_2 \\ uRF_1 + uRF_2 + uRF_3 \\ uRF_1 + uRF_2 + uRF_3 + uRF_4 \end{bmatrix}$$

In order to apply the above determinant to the calculation algorithm, the left side stiffness matrix is derived as follows.

$$\begin{bmatrix} (K_C/K_1+2) & (3-1) & (3-1) \\ (3-1) & (K_C/K_2+(4-1)) & (4-1) \\ (3-1) & (4-1) & (K_C/K_3+(5-1)) \end{bmatrix}$$

$$= \begin{bmatrix} (K_C/K_1+Nl-Ns) & (Nl-Ns) & (Nl-Ns) \\ (Nl-Ns) & (K_C/K_2+Nl+1-Ns) & (Nl+1-Ns) \\ (Nl-Ns) & (Nl+1-Ns) & (K_C/K_3+Nl+2-Ns) \end{bmatrix}$$

Nl : *nth of Lowest Lashed Tier*

Ns : *nth of Structure Tier*

3.2 OPTIMUM STACK ARRANGEMENT

Container arrangement in bay will have the maximum load capacity when the Bay's VCG is high, because the maximum loading capacity of a single container is fixed. Generally, optimum stack arrangement with maximum load within the allowable load specified by the Society has been found by engineer. The algorithm is largely developed in consideration of two cases: Stacking on Deck and in Hold. However, the optimum stack arrangement guide function does not suggest the lasing method. It is a function that finds the optimum stack arrangement according to the lasing method set by the user.

For deck stowage, optimum arrangement is to find the arrangement with the highest VCG in each stack within the allowable load provided by the classification society. In the case of a load in Hold, it is to find the maximum allocated amount properly.

In order to find the optimal placement with the highest VCG, the desired value was found by slightly reducing the load at the top. The constraints for finding the optimal placement are as follows: allowable load, minimum weight desired by the user, general container placement method, and optimum value calculating time.

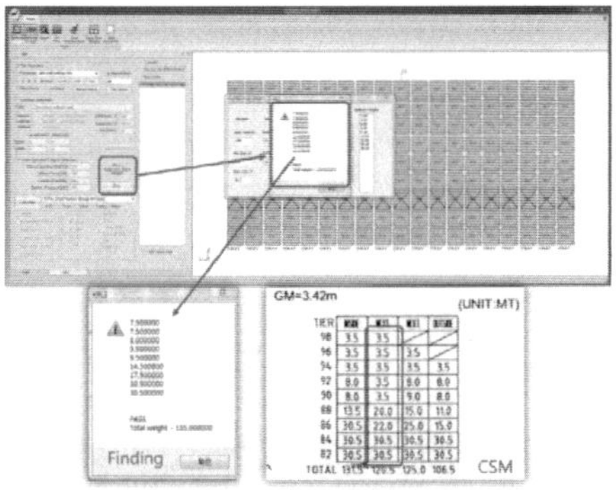

Figure 7: Result of optimum stack arrangement

As a result, results are similar to specialist's CSM(container securing manual) within allowable load. And the above results show higher loading load results than CSM results.

3.3 Software Development Kit

The data structure shown in the figure below allows the user to get the desired output after the basic container information should be inputted. It can be used according to the following procedure.

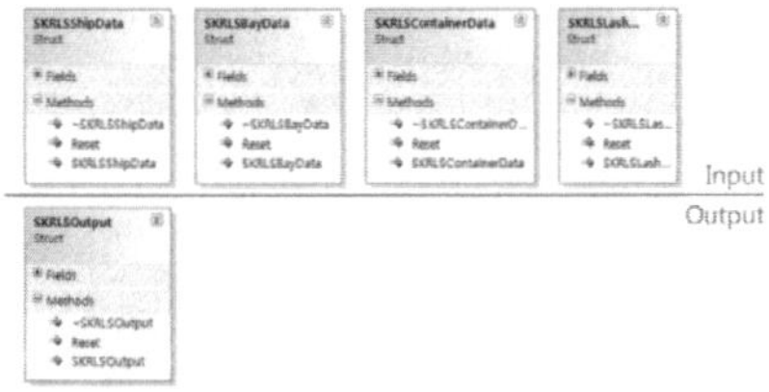

Figure 8 Data structure for SDK

- Reset All Data
- Set SKRLSShipData data(for Ship Principal Dimension : L, B, D, etc)
- Add SKRLSBayData data(for Bay Data : GM, Bay Name, etc)
- Create Stack
- Set SKRLSContainerData data(for Container data : weight, type etc)
- Set SKRLSLashData data(for Lashing information)
- Apply Lashing
- Solve
- Get SKRLSOutput data(for output)

4. DEVELOPMENT RESULT

4.1 CONTAINER SECURING STRENGTH EVALUATION SW

This SW allows intuitive input using the container simplification model for user's convenience. And it distributes the developed software to domestic and overseas related companies and leads to good response. Key features are as follows :

- Applying ship motion combination factor and route specific reduction factor
- Various securing type and methods.
- 3D graphic for container loading
- Result for various loading condition
- Supporting guide for optimum stack arrangement
- Providing software development kit(SDK)

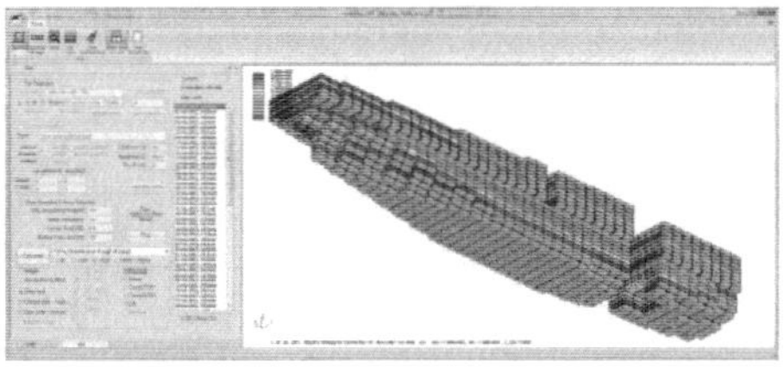

Figure 9: GUI for SeaTrust-LS

4.2 Optimum stack arrangement algorithm

From the stack arrangement results by optimum stack arrangement algorithm, it can be seen that the allowable load of the bottom container is satisfied with 100%. And, it and leads to good response from experts in related industries.

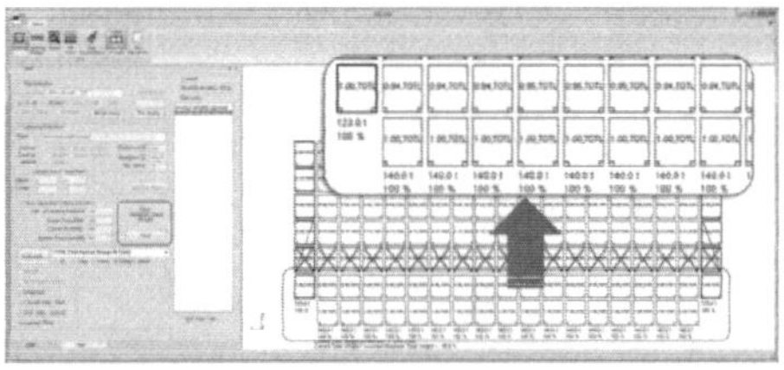

Figure 10 Result for Optimum stack arrangement

4.3 SOFTWARE DEVELOPMENT KIT

We have developed SDK(Software Development Kit) in 2017, it has been tested by related companies, and we received good reviews. To support developers' various development environments and provide easy-use (portability and scalability), Microsoft Visual Studio template files in Unmanaged (C ++ with MFC) and Managed (C #) environments have been provided. SeaTrust-LS SDK can be downloaded directly from the KR Web-Site and require a license file to be provided after a simple agreement procedure for free-use.

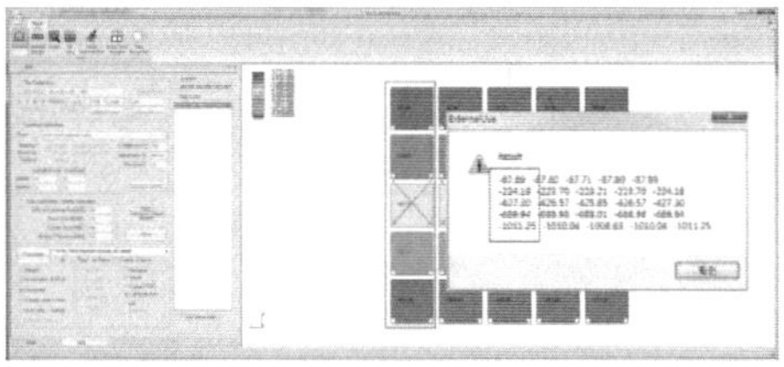

Figure 11 Example for use of SDK

5. CONCLUSIONS

When loading cargoes in container carrier, total weight and stack arrangement for container should be accordance with the guidelines of the classification society which specify permissible loads for securing device and containers. Korean Register has also announced the revision of the guidelines (annex 7-2) for container securing devices. And Container securing strength evaluation SW including optimum stack arrangement algorithm was developed. Also SDK(Software development kit) make ease to develop in-house SW or their own SW for 3rd party company.

6. ACKNOWLEDGEMENTS

This research was performed as a part of the research project below and supported by the organizations indicated. We acknowledge and appreciate the support provided.

- 'Development of a 3D Model-based Consistent System for Ship Design' project funded by Ministry of Trade, Industry and Energy of Korea (No. 10052708).

7. REFERENCES

1. Korean Register of Shipping, 'Part 7 Ships of Special Service(Ch1-4, 7-10)', Guidance Part 7, Annex 7-2 Guidance for the container securing arrangements, 2017.

2. Shin, S. H., Proceedings of the thirteenth International Offshore and Polar Engineering Conference, A Study on Container Securing system for Optimum Arrangement, *2003*.

8. AUTHORS BIOGRAPHY

[Jin-young Park] is presently naval architecture researcher in ICT center of Korean Register of Shipping providing domain expertise and knowledge to customers. With previous experience in ship structure strength assessment, Naval architect software development has continued to meet customer's needs.

MARINE PROPULSION SHAFTING SIMULATION INTERFACES

M. Zeid, Caterpillar Propulsion, Sweden

SUMMARY

Marine propulsion shaft lines are commonly subject to individual simulations in torsional, lateral and axial motion. This types of simulations are basically carried out by the propulsion maker considering the available input data. On the other side, the tools used to perform these shafting simulations are developed significantly along the years. Also, the hydrodynamic simulation tools as well as the structural analysis tools developed significantly. With such high level of sophistication in the analysis tools, simulation process and level of input, it shall be expected a more close to zero failure/vibration observation within the torque transmission equipment. However, there is still existing a significant number of cases on the market. One of the most common reasons for major failures is the missed interfaces among different types of simulations. The paper highlighting a wide range of these simulation interfaces domains and relative practical cases presented as well.

1. INTRODUCTION

The commercial marine vessels are powered floating bodies. The design principles are basically subjected to cover the Naval Architecture and Marine Engineering principles. Respectively there are different types of simulations carried in order to ensure the targeted performance.

The most common way of evaluating the global performance of a marine commercial vessel is to make a certain speed or produce a certain thrust value or a combination of both. Also, the performance of the vessel achieving a certain level of comfort regarding noise and vibration and maintain a healthy working environment for the crew.

In order to maintain the target performance of the vessel, a different types of simulations and analysis are required during design stage.

The paper presenting the most common interfaces between the Marine Rotordynamics simulations, Marine Hydrodynamics simulations and hull structure analysis.

The importance of highlighting such interfaces is to maintain the global output of all simulations performed at high level. This to avoid unexpected conflicts during vessel operation, which are related to that a certain interface is missed during the design stage.

2. MARINE ROTORDYNAMICS SIMULATIONS

Marine Rotordynamics simulations are basically focus on the torque transmission equipment for main propulsion with relative drive train configuration. It consist of Torsional, Lateral and Axial analysis. The aim of the analysis is to identify the vibration behaviour in relative motion.

In addition to the analysis, the selected components for shafting configuration have to be checked in respect to the analysis results. This to verify the expected performance of each relative component (couplings, bearings, etc...).

The system is considered "Rotordynamics optimized" when the shafting configuration is free from excessive vibration in different motions, as well as the shafting components function in a compatible way toward each other.

2.1 TRADITIONAL ANALYSIS.

The traditional analysis is considering the torsional vibration analysis, later analysis and axial analysis. However each motion done individually without checking possible interfaces.

In few cases, when a natural frequency within operating speed range determined (In Axial motion for example), a coupled vibration analysis (Torsional / Axial or Lateral / Axial) could be performed. This is an advanced way to justifying the system, however it is not reflecting the complete picture of system dynamics.

2.2 LOCAL ROTORDYNAMICS INTERFACES.

The principle of studying the interfaces between different shaft motions. The idea is to consider the output from a certain motion analysis (Torsional analysis for example) as an input for another shaft motion (Axial analysis for example). This transmission from one motion to another has to be through the pre-defined interface in the specified configuration, also with the relative dynamics for this interface.

The study can be performed by applying the principles of dynamic response for forced vibration in damped systems. The study can be carried out by applying the input as a time signal, or applying the highest harmonics (orders) of the analysed input signal. This depend on if the input signal dominated by few orders (1 to 5), or a complicated signal where influenced by wide range of orders.

2.2 (a) Torsional / Axial Interface

The torsional dynamics can be reflected in axial direction where the torque transmission process develops axial forces. The gears and cardan shafts are the most common example for torsional / axial interface.

The below example (Figure 1) shows a simple reflection of torsional motion on axial direction.

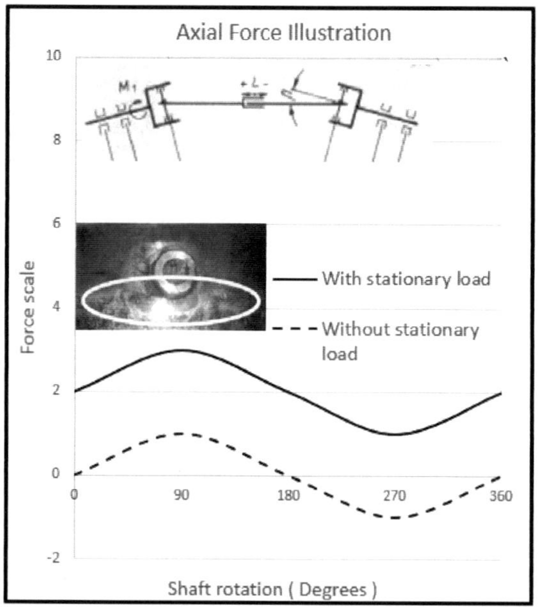

Figure 1: Illustration of torsional / axial interface in a drive line containing cardan shafts.

Due to the axial thermal growth of the shaft line during operation, the axial bearing integrated to the shafting system shall support the developed axial force by the cardan shaft middle part (spline part). In this case the stationary force on the axial bearing is the dominating force component.

When the ambient temperature becomes stable, the thermal axial growth of the shafting line is fully compensated by the sliding of the cardan shaft middle part (spline part). The major axial stationary force on the axial bearing then gets down to a very low level, where it can be neglected. In such case, the major dominating axial force on the axial bearing is the harmonic force component.

Reference to cardan shaft dynamins [6],[7], the axial displacement (L) in the middle part (spline part) of cardan shaft is directly proportional with the transmitted torque (M).

The harmonic axial displacement due to harmonic axial force is a direct reflection of the traditional torsional harmonics on the shafting system. In this example, there was no excessive harmonic torsional amplitudes, however the axial bearing outer race and shear pin started to show serious wear due to metal to metal contact after approximately 100 operating hours. The wear marks are developed due to a combination of a gap between bearing housing and bearing outer race. This gap in combination with the harmonic axial displacement in addition to the absence of the stationary axial load lead to the wear marks. By removing the clearance between the outer race and the bearing housing, the harmonic axial load could load the bearing rollers directly and eliminate the harmonic axial displacement inside the bearing and respectively the problem eliminated.

2.2 (b) Torsional / Lateral Interface.

The torsional dynamics can be reflected in lateral direction as well where the torque transmission process developing lateral force. The gears and cardan shafts are the most common example for torsional / lateral interface.

The below example (Figure 2) shows an experimental test results of torsional motion on lateral direction.

The shafting configuration for the model used for the test considered a prime mover, cardan shaft and shaft line supported by shaft bearings. In this test, the furthest bearing from the cardan shaft was the axial bearing.

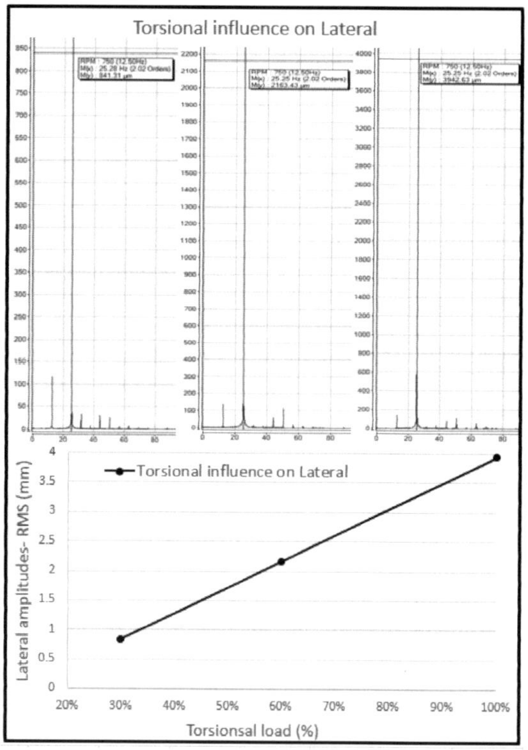

Figure 2: Influence of torsional motion on shaft lateral amplitudes.

The shaft lateral measurements are carried out by proximity transducers, located in the middle of the shaft span between the bearings. The length of the shaft span is modelled in a way to fit the second order natural frequency with the second order radial excitation developed by the cardan shaft dynamics [6], [7].

The shaft lateral measurements carried out at 30%, 60% and 100 % torque load. The results shows (Figure 2) a clear linear relation between the applied torque and the measured lateral amplitudes of the shaft.

2.2 (c) Axial / Lateral Interface.

The shaft lateral behaviour can be influenced by axial motion as well. The shaft in lateral has a certain deflection in the middle between bearings as per natural physics. By applying an axial force at the neutral axis of the shaft, the shaft deflection will increase respectively. The gears and cardan shafts are the most common example for axial/ lateral interface.

The below example (Figure 3) shows an experimental test results of axial motion on lateral direction.

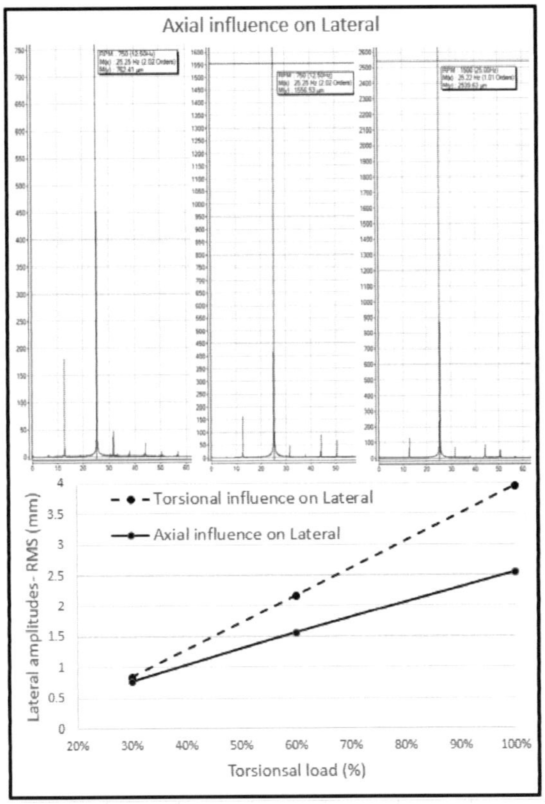

Figure 3: Influence of axial motion on shaft lateral amplitudes.

The model used is the same model used in the previous example, however the axial bearing located the nearest to cardan shaft. This in order to eliminate the axial forces excitation transmission to shaft.

The shaft lateral measurements carried out at 30%, 60% and 100 % torque load (Same load points). The results shows (Figure 3) a clear linear relation between the applied torque and the measured lateral amplitudes of the shaft. However the lateral amplitudes is found to be significantly lower. The reduction of lateral amplitudes by eliminating the axial forces to the shaft line found to be 9%, 28% and 36% at 30%, 60 % and 100 % torque load respectively.

3. MARINE STRUCTURE ANALYSIS.

The structural analysis in ship industry recently has been developed significantly. This has a direct relation with the development of the structural analysis tools and modelling.

In principle, the development of the structural analysis respectively scantling has a big influence on the entire ship building process. A slight change on the traditional designs where can be proven from structural point of view as safe, can lead to hundreds of tons of steel saving. This steel saving value can be a key go / no go for financing possibilities to the new building, shipyard profits, etc…

The approach is positive where it helps in many directions, however it have to be well covered from different point of views and not only structural point of view.

3.1 GLOBAL HULL GIRDER VIBRATION ANALYSIS

The hull girder vibration analysis is essential in order to avoid hull girder natural frequencies at low frequencies where it can be easily excited by main excitation sources in the system (specially the propeller).

In case of hull girder natural frequency at low frequency where it can be excited by the propeller for example, the dynamic response can be excessively high. For example a traditional bending excitation from the propeller where it can lead to a hull deflection of 0.2 mm at the propeller area, the dynamic response of the hull girder can reach more than 30mm deflection amplitudes. This as the hull girder damping is very low.

The hull girder is the physical reference for all installed machinery as well as all structure blocks built relative to hull girder. When the hull girder is suffering from a low natural frequency, it is excited through operating speed range and develops a resonance frequency zone. This will lead to that all machinery equipment (Including propulsion and shafting) as well as relative structure blocks vibrating with same or higher amplitudes at resonance frequency.

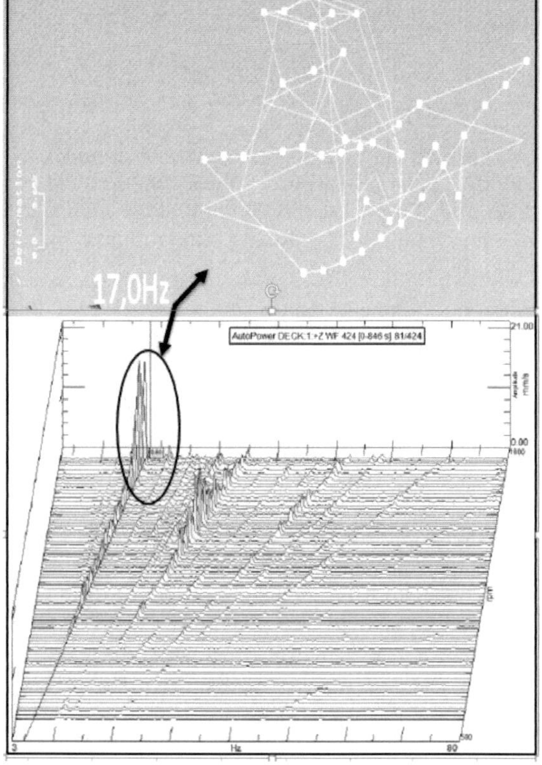

Figure 4: Hull girder amplitudes measurements where shows the hull girder mode shape at the resonance frequency.

When a vessel has been built with such problems (Figure 4), the root solution normally is not applicable as re-scantling a finalized vessel is very costly. Instead the problem is masked by changing the excitation frequency or reducing the excitation amplitudes, however for each similar solution there are relative negative influences where it need further consideration.

This problem can be highlighted during design stage if the vessel design subjected to hull girder vibration analysis. This by considering a high accuracy modelling where it simulate as close as possible the actual vessel condition. In such case, slight adjustment of the scantling to avoid hull girder natural frequency at least at the end of the operating speed range can be the away to avoid such kind of problem.

3.2 SUB-GLOBAL VIBRATION ANALYSIS

The sub-global vibration analysis for the massive structure where built reference to hull girder is important to check. The superstructure toward the hull girder and the bulkhead panels toward hull girder in transverse scantling vessels are the most common examples for this interface.

The below example (Figure 5) shows a practical measurements for a bulkhead panel in a transverse scantling vessel.

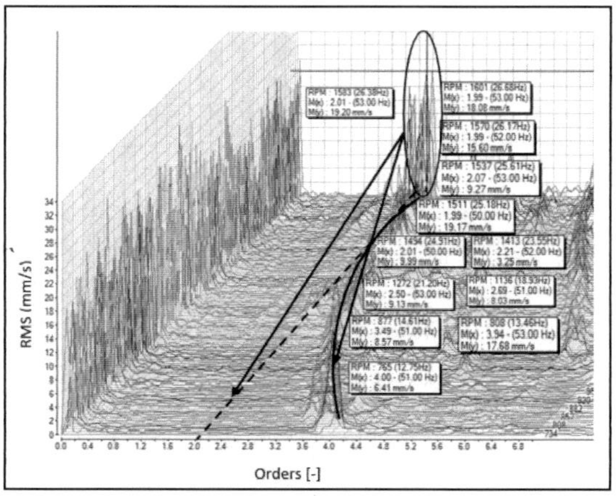

Figure 5: Waterfall diagram for a point on a transverse bulkhead panel on a transverse scantling vessel.

In this example (Figure 5), the vibration levels at a point on the bulkhead panel has been measured. The waterfall diagram shows that the bulkhead at the area of measurements vibrates due to its own natural frequency (50-53 Hz / black curve) through entire operating speed range regardless the speed itself.

On the other hand, when the bulkhead panel excited by a steady excitation (Propeller blades in this case), the dynamic response when the excitation frequency met the bulkhead natural frequency is found to be very high. Looking to the excitation value transmitted from the propeller (dashed line) , is in this case relatively small along the operating speed range, however due to the absence of damping in the bulkhead panel the dynamic response had a high magnification.

For such kind of problems, the influence is more local to the surrounding. However, slight modification in the scantling or changing frequency used to be the way to solve in case of problem.

3.3 BEARINGS SUPPORTS STIFFNESS.

The bearing support stiffness has a major influence on the shafting dynamics in axial and lateral motions. Also, it has a high contribution into the transmitted vibration from propulsion system to the hull.

3.3 (a) In Axial Motion

An axial natural frequency or high axial direct response can be expected in case of low relative stiffness for the thrust bearing support. The risk for axial vibration is coupled to propeller mass, propeller hydrodynamics, shaft diameter and thrust bearing stiffness [4]. Due to relative low propeller RPM, connected to the propeller diameter [2], the risk for having axial vibration problem in real life is limited. However, detailed axial vibration check is beneficial to carry out especially in case of having a non-traditional design for the scantling around the thrust bearing.

3.3 (b) In Lateral Motion

The lateral motion is describing the shaft motion in lateral reference to shaft section neutral axis in the entire radial (360 degrees) of any cross section of the shaft.

The shafting dynamics in lateral is influenced by Propeller mass / hydrodynamics, shafting configuration and shaft lateral support bearings.

The shaft lateral support stiffness is the major contributor into shafting lateral analysis, the high sensitivity is subjected to the high uncertainty of the bearing support stiffness at design stage.

The shaft lateral excitation is subjected to shaft bending due to mass (order 1) and the propeller blades bending fluctuation (blade order) due to propeller thrust fluctuation [3], [4]. The shaft can be also excited by integrated shafting coupling orders so far it is applicable (Cardan shaft excitation for example).

The shafting lateral damping is basically provided by propeller [2]. The propeller damping has its highest significance for the cantilever beam from the propeller to the propeller bearing. The propeller damping influence is reducing as far as going forward of propeller bearing. Propeller damping is insignificant at the first radial flexible connection in the shafting line (radial flexible coupling, gears, etc…).

The shaft lateral support slide surface bearings can provide a secondary damping. Also, the shafting radial flexible coupling can provide a minor lateral damping to the shafting system when applicable.

There is a limited / none damping capability that can be provided by roller bearing to the shafting system in lateral in a traditional configuration.

The resulting of lateral excitation and lateral damping is the shaft lateral amplitudes. The shaft lateral amplitudes shall be minimized as much as possible, however it is always applicable.

In case of shaft lateral natural frequency match with shaft excitation frequency, it will result on resonance zone. The shaft lateral amplitudes in such case expected to be higher than normal. In some cases the response is relatively low due to damping effect, in some other cases the shaft lateral amplitudes are significantly high due to low relative damping (Figure 6).

In the following example (Figure 6), the shafting lateral support stiffness recommended to be 1.0E+009 N/m or higher from Rotordynamics point of view. After few months of operation, the propeller bearing suffered a complete wear down. During inspection, it has been recognized that the shaft suffering a bending fraction (Perpendicular to shaft neutral axis). By further investigation after replacing defected components, It has been found the shafting system suffer a resonance in lateral motion. The resonance occurred when the propeller blade frequency matched with the shaft lateral natural frequency. The only logic reason where could explain the root cause by reverse engineering process is the bearing support stiffness. By reducing the propeller bearing stiffness from 1.0 E+009 N/m to 3.5 E+008 N/m the shaft lateral natural frequency dropped from 323 RPM to 173 RPM. Where the frequency match the determined natural frequency by measurements. The aft part of the vessel has been analysed separately by vessel designer in order to obtain the propeller bearing stiffness. The results showed a very close value to the value obtained by lateral reverse engineering.

Change the scantling in such situation is very difficult, respectively changing shafting configuration was the more practical solution in this case.

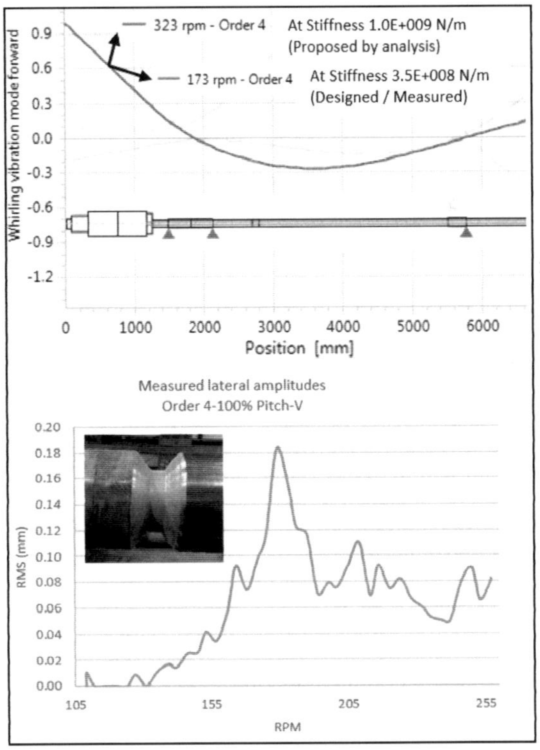

Figure 6: The influence of propeller bearing support lateral stiffness on shafting lateral vibration.

The shafting lateral amplitudes are reflecting a harmonic loading for the lateral supports. The harmonic loads on the lateral supports are the key parameter of propulsion vibration transmission to hull.

In case of well-functioning slide surface bearing and physical damping domain provided by the bearing (enough lubricant film), the harmonic loading will be mostly dominated by bearing damping and respectively a low propulsion vibration will be transmitted to hull (Figure 7).

In the below illustration (Figure 7), the bearing stiffness layout consist of bearing material stiffness, oil film stiffness and bearing foundation stiffness. The bearing material stiffness and oil film stiffness considered same in the two examples. The bearing foundation stiffness considered 2.0E+008 N/m (low stiffness) and 2.0E+009 N/m (high stiffness). The calculation presenting the direct response ratio [1] in the two cases. In the low stiffness example the direct response ratio shows approximately 69% of the applied load transmitted to hull, respectively 31% dominated (damped) by the bearing lubricant. On the other side for the high stiffness example approximately 18% of the applied load transmitted to hull, respectively 82% dominated (damped) by the bearing lubricant.

Propeller Bearing stiffness layout (low)

	Stiffness level	K (10^9 N/m)
	Bearing Material:	256
	Bearing foundation Support:	0.2
	ktot	**0.20**
ASTB	**Oil film**	**0.450**

Propeller Bearing stiffness layout (High)

	Stiffness level	K (10^9 N/m)
	Bearing Material:	256
	Bearing foundation Support:	2
	ktot	**1.98**
ASTB	**Oil film**	**0.450**

Low stiffness propeller bearing	**Transferred vibration levels**	
	White metal	Notes:
Lubricant (Oil):	30.75%	Damping effective.
Bearing Material:	0.05%	No damping effective.
Foundation direct response:	69.19%	No damping effective.
Total transmitted Vibration to hull:	69.25%	No damping effective.
Total possible damping:	30.75%	Applicable only at at good designed bearing.

High stiffness propeller bearing	**Transferred vibration levels**	
	White metal	Notes:
Lubricant (Oil):	81.52%	Damping effective.
Bearing Material:	0.14%	No damping effective.
Foundation direct response:	18.34%	No damping effective.
Total transmitted Vibration to hull:	18.48%	No damping effective.
Total possible damping:	81.52%	Applicable only at at good designed bearing.

Figure 7: Illustration of the influence of bearing stiffness on the value of propulsion vibration transmission to hull.

On the other hand, in case of roller bearing the lubricant film is very thin and can't provide the essential damping, however it is very common to integrate radial flexible couplings with shafting system next to radial bearing in order to obtain as low as possible lateral excitation (harmonic loading) to the bearings, respectively low level vibration transmission to hull.

Practically, the middle size propellers and smaller (3 meter diameter and less) are more sensitive to lateral vibration. On the other hand, middle size propellers and bigger (above 3 meter diameter) are more sensitive to propulsion vibration transmission to hull.

4. MARINE HYDRODYNAMICS SIMULATIONS

The main focus of marine hydrodynamic simulation is to improve the hull / water interaction and hull / propeller interaction. Also, the propeller design hydrodynamic optimization for the propeller itself.

The propeller hull interaction and the propeller design hydrodynamic optimization are the main contributor into marine Rotordynamics.

4.1 THRUST / LATERAL INTERFACE.

The development on hydrodynamics simulations lead to improvement into developed thrust by the propeller, also improved interaction between propeller and hull (less thrust fluctuation). The thrust and thrust fluctuation has a direct impact into the propeller bearing dynamics. This as the thrust and thrust fluctuation are influencing the slope mismatch between the propeller shaft and propeller bearing [5].

In order to maintain a good performance of the propeller bearing, the slope mismatch between propeller shaft and propeller bearing should be well optimized (Figure 8). A slope bored bearing can be proposed in a lot of cases when needed, however more or less slope value (than it should be) for the bearing can lead to a bearing failure [5].

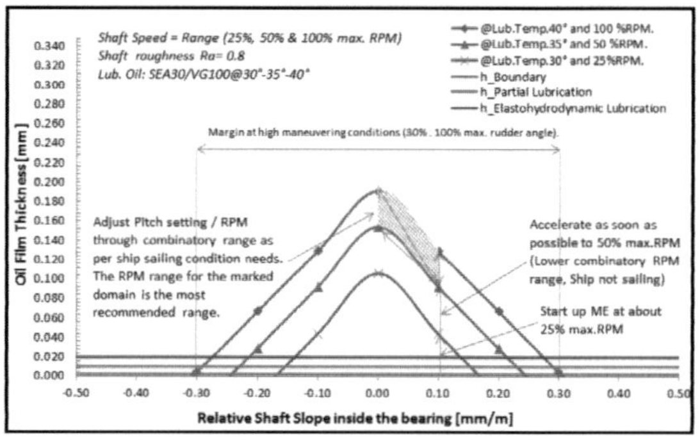

Figure 8: Illustration of the optimum domain for slope mismatch between propeller shaft and propeller bearing.

It is always beneficial to know how much the propeller thrust eccentricity and thrust fluctuation during design stage especially if the design is not a traditional design.

4.2 THRUST / HULL INTERFACE.

The thrust / hull interface where the thrust fluctuation presented by normal propeller function is contributing the dynamic behaviour on the vessel especially on the aft area.

It has a major contribution in dynamic behaviour in case of main propulsion type is a Z-drive propulsion. The excitation in Z-drive configurations is normally higher in compare to conventional propulsion as it is influenced by the shift on the thrust fluctuation axis toward the hull connection. Respectively the Z-drive configuration / type together with the thrust fluctuation presenting the propulsion excitation to hull due to developed bending.

It is important to know what the expected thrust fluctuations are, this improve the possibility to simulate the dynamic response due to thrust fluctuation.

4.3 PROPELLER DESIGN / TORSIONAL INTERFACE.

The Propeller design hydrodynamic optimization has a major contribution into torsional vibration. This as the main objective of the propeller hydrodynamic optimization is to reduce the torque toward the thrust respectively torque coefficient toward thrust coefficient. The propeller is well known is the major contributor on the dynamic behaviour on the vessel, however it is also the main contributor in damping as well especially in torsional motion.

As more improvement in local propeller hydrodynamic optimization, as less torsional damping where can gained from propeller in torsional.

It is important to investigate the applicable damping in torsional where the propeller design is a non-traditional design.

5. INPUT POWER

The input power into the system is the main input for all dynamic behaviours. The most common trend of prime mover development is to develop more power at same RPM rating.

The torque is directly proportional with power at same RPM, this respectively increase the transmitted torque on the system with the same percentage of the input power increase. The system dynamic response is directly proportional with the transmitted torque as well.

In marine ship building, it is very common to repeat a building when the first design succeeded or building a fleet of same type of vessel where commonly known by "Sister Vessel". In such case it is very important to check all components of the shafting configuration considering all updated items including the input power increase on the system so far it is applicable.

6. CONCLUSIONS AND RECOMMENDATIONS

It is important to have a clear understanding about the possible interfaces with relative reference as it helps to have the best gain of simulation process.

A Rotordynamics problem can be created by a direct response of an interface where no concern of this interface been highlighted through traditional simulation.

It is beneficial to have a shared interface value between different parties. For example address a specific amount of power a specific vessel design can handle.
To consider the recommended stiffness addressed by Rotordynamics while designing bearing foundations as another example, etc...

During design evaluation process it is important to look into each component in the shafting system as a local system by itself in different motions, this in addition to look into the entire propulsion system as one system in different motions. This will help to see the different interfaces where it can differ from case to case.

7. ACKNOWLEDGEMENTS

I would like to thank my family for all support I get to do this work. I'm also grateful to Caterpillar Propulsion where supported with all needed resources to finalize this job.

8. REFERENCES

1. Clarence W. de Silva, 'Vibration and Shock Handbook', *CRC Press*, 2005.

2. Carlton, J. S., 'Marine propellers and propulsion', *Elsevier Ltd.*, 2007.

3. Lech Murawski, 'Shaft Line Whirling Vibrations: Effects of Numerical Assumptions on Analysis Results', *Marine Technology*, 2005.

4. Veritec, 'Vibration Control In Ships', *A.S. VERITEC*, 1985

5. M. Zeid, 'Sailing conditions effects into Propeller Bearing for big vessels', *Caterpillar Propulsion internal researches*, 2013

6. Klein Propeller Shaft, 'Product Technical Supplement' Klein website, Internet.

7, Gewes, 'Product catalogue', Gewes website, Internet.

9. AUTHORS BIOGRAPHY

Mohamed Zeid I'm currently holding the position of Rotordynamics specialist at Caterpillar Propulsion. I'm responsible for Rotordynamics group activities within core competence team since 2006. As originally a Naval Architect and Marine Engineer, my previous experience includes intact / damage stability, structure analysis and strength calculations, machinery configuration and arrangements, pipelines design, etc.

RESEARCH ON LARGE-SCALE MARINE PROPULSION SHAFTING LOAD TEST AND ADJUSTMENT TECHNOLOGY

Ji Wang, Collaborative Innovation Center for Advanced Ship and Deep-sea Exploration, School of Ocean Engineering, State Key Laboratory of Structural Analysis for Industrial Equipment, Dalian University of Technology, Dalian, China
Zhongchi Liu and **Dongxin Xue,** School of Power and Energy, Dalian University of Technology, Dalian, China
Shengjun Zhang and **Feixiang Wang,** School of Ocean Engineering, Dalian University of Technology, Dalian, China

SUMMARY

Power shaft is an important component which transmits power from the engine to the propeller in ship power system. Currently, because the sizes of cargo ships become bigger and bigger, the sizes of the power shafts increase. During the installment of shafting, a very important step is making the bearing loads meet the requirements of shafting alignment manual. The Traditional method is just measuring the bearing loads repeatedly and gradually adjusting the bearing displacement by worker's experience. The repeating measurements and adjustments takes a lot of time and the results are not accurate enough, especially for large power shaft. It can lead to high temperature of bearing, tile kilning and ship outage during navigating, which will bring huge economic losses. So a more reliable and accurate adjusting method is very necessary. This paper firstly investigates marine propulsion shafting load test method, improves the jack-up method and strain-gage method respectively, and increases the precision of bearing load measurement. On this basis, this paper studies bearing displacement adjustment calculation method, calculates bearing load influence factors which take bearing stiffness into account, establishes special nonlinear optimization model with absolute values symbol, and establishes bearing displacement adjustment algorithms. Finally, by applying data of several ships, the bearing load measurement method and adjustment techniques promoted by this paper are verified. The result shows that they are able to meet the requirements of actual construction technology.

1. INTRODUCTION

The operation environment of marine propulsion shafting is very complex and they are influenced by many factors that will cause bearing load changes, such as marine draft, main engine rotate speed, surrounding environment, main engine temperature and etc. With the trend of large-scale ship, the factors have constantly bigger influence and the non-uniform of load distribution become more severe. If some shafting bearing load is too big, it can cause serious wear to some bearing and significantly decrease bearing life. On the other hand, if the difference between the largest and smallest bearing loads is too big, which means the total load are not shared by the bearings evenly, it will enhance the vibration significantly and the fatigue life of shafting and surrounding parts will decrease. Therefore, for modern large-scale ship, it is necessary to adopt certain shafting alignment method to ensure every bearing load is close in actual operating situation. The shafting alignment quality has important influence to guarantee normal rotation of shafting and main engine, and decrease hull vibration.

Marine shafting alignment often adopts rational alignment method [1,2]. Under limitations such as bearing load and rotation angle, this method calculates the rational positions of every bearing, installs and adjusts the shaft into certain curve state, so as to reach reasonable load distribution in shaft sections and bearings and suitable rotation angles in allowable range. In this way, there will be some compensation value left to variation in operating process. Shafting rational alignment can mainly divided into two parts. One part is theoretical computing that provide data support for bearing height adjustment. The other part is bearing load measurement. Because of certain error exist in shafting calculation and installment, together with support structure deformation, the actual completed bearing height may not meet the requirement. It is necessary to adjust bearing heights based on current bearing load situation and measure bearing loads after adjustment to check whether the result achieve predicted values.

Currently, shafting installment quality usually reflects by bearing load measurement result. The major applied load measurement methods are jack-up method and strain-gage method that measure shafting bending deformation [3,4]. The jack-up method is often adopted in actual operation because of the simple operating process. However, the precision of this method is very low because it relies on people to read the varying data during the adjustment. Besides that, this method cannot be applied to the stern tube which is a very important part of the shaft. In addition, the actual bearing height adjustments totally depend on human experience, which takes a lot of time.

To increase the precision and efficiency of the bearing load measurement and adjustment, this paper firstly investigates marine propulsion shafting load test method, improves the jack-up method and strain-gage method respectively. Furthermore, by taking bearing stiffness into account, the more accurate bearing load influence factors can be calculated by finite element method, which can contribute to increase the accuracy of the bearing displacement adjustment algorithm. Finally, by applying data of several ships, the bearing load measurement method and adjustment techniques promoted by this paper are verified.

2. RESEARCH ON BEARING LOAD DIGITAL MEASUREMENT TECHNOLOGY

2.1 RESEARCH ON JACK-UP METHOD DIGITAL MEASUREMENT TECHNOLOGY

In the jack-up method, a hydraulic jack is placed near the bearing and the shaft is jacked up by it gradually until the measured bearing is completely separate with the shaft. During the jack up process, dial indicators are used to record the lifting value and record the corresponding load value of jack simultaneously. The jack-up curve can be drawn and the jack-up oil pressure value can be obtained when measured bearing is replaced by jack. Then the data is converted into jack-up force and multiplied by the jack-up coefficient. In this way the actual load of measured bearing can be achieved [5].

The key technologies of jack-up method digital measurement are how to achieve automatic synchronous acquisition of pressure data and displacement data, and how to apply linear regression to pressure data and displacement data accurately, as shown in Fig. 1.

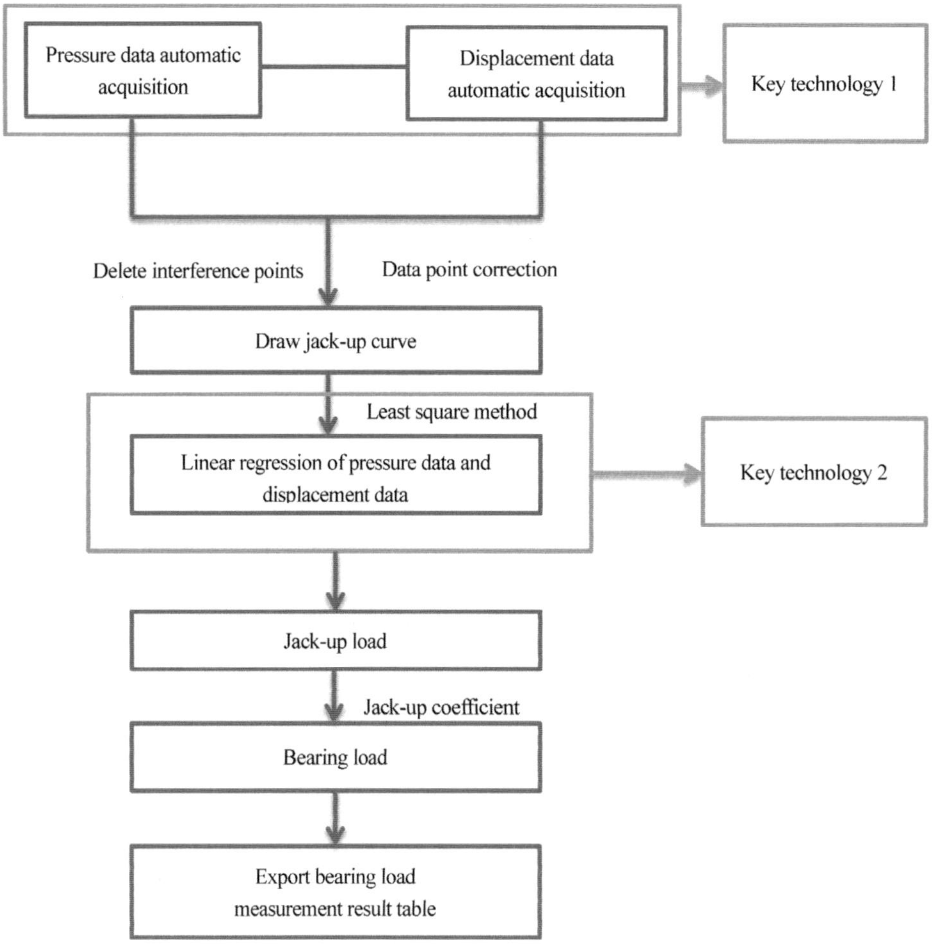

Figure 1. Key technologies of jack-up method digital measurement

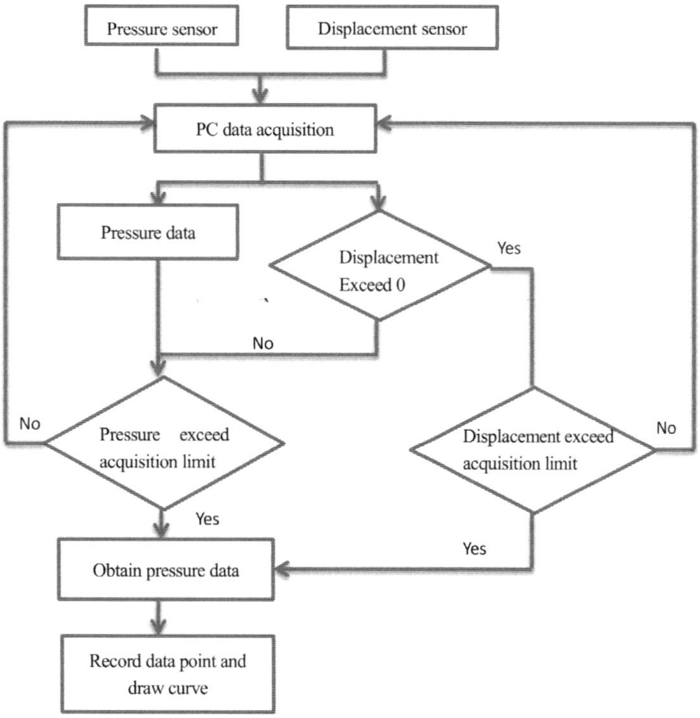

Figure.2. Flow chart of jack-up method automatic synchronous acquisition of pressure data and displacement data

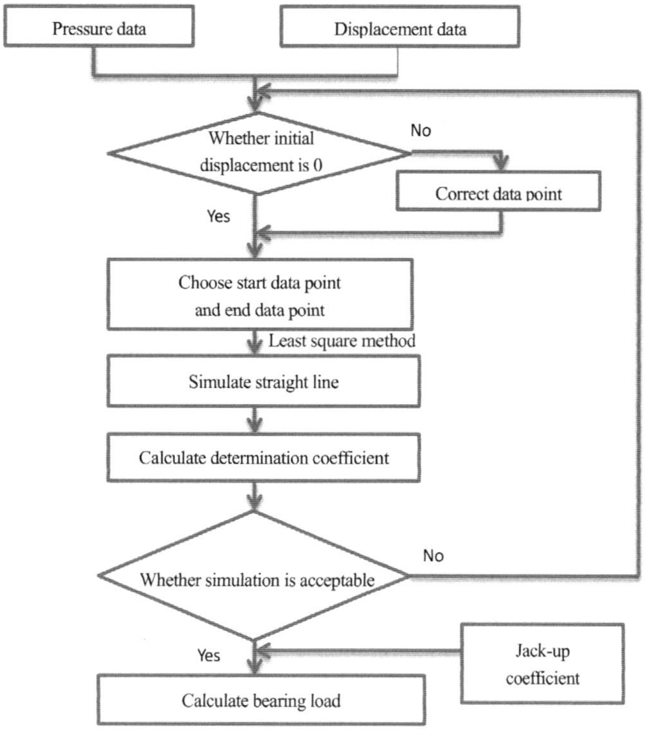

Figure.3. Flow chart of pressure data and displacement data regression analysis

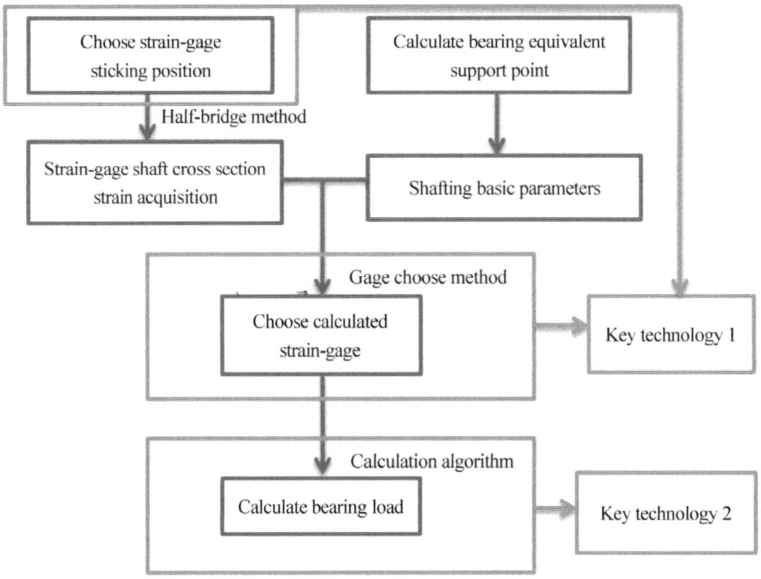

Figure 4. Flow chart of strain-gage bearing load measurement and calculation

2.1(a) Automatic synchronous acquisition of pressure data and displacement data

Based on asynchronous communication protocols, the actualization flow chart of jack-up method automatic synchronous acquisition of pressure data and displacement data is shown in Fig. 2. RS232 output module of pressure sensor realizes the communication between the output module and computer through the three lines in RS232-DB9 port. Displacement sensor uses oscillating circuit to trigger data, realizes transmission through serial bus. When pressure sensor and displacement sensor jointly send measurement data to computer, data with 10Hz of frequency is filtered, which is slightly less than displacement oscillating circuit frequency. When the displacement or pressure exceed the programmed acquisition limit, the recording of the data is stopped. The automatic synchronous acquisition of jack-up data and generates jack-up curve are achieved at the same time.

2.1(b) Pressure data and displacement data regression analysis

Data points formed by pressure and displacement data need to be simulated to jack-up curve and the inflection point of lift-up curve and drop-down curve can be judged. The fitting precision and accuracy of inflection point will directly influence the accuracy of calculated bearing load. The lift-up and drop-down line are simulated based on least square method and distinguish the fitting by determination coefficient. When it meet the precision requirement, the jack-up load can be gotten by corresponding intercept of load shaft and the bearing actual load can be obtained through multiplied by jack-up coefficient of the bearing. The calculation process is shown in Fig. 3.

2.2 STRAIN-GAGE DIGITAL MEASUREMENT TECHNOLOGY

Strain-gage method is sticking strain-gage on certain shaft cross section, calculating the bending moment of shaft cross section by the strain value variation when turning shafting. Combining several cross sections bending moment, the bearing load is calculated. The calculation process is shown in Fig. 4. Details about measurements and calculations of stern tube aft bearing load, stern tube fore bearing load and middle bearing load are given by Wang[6].

However, there are two kinds of error when using strain-gage method to measure bearing load. One is the strain data acquisition error caused by acquisition equipment and acquisition method, the other one is calculation error caused by different sticking position. There are three major factors influence the first error.

- The definite position strain is replaced by average strain, which is the acquisition strain data of whole shaft surface covered by strain-gage;
- The data acquisition error is generated by factors such as glue thickness, insulation between strain-gage and shaft surface, and environment humidity.;

- The transmission due to wire resistance values.

The reduction method of first error are increasing the strain-gage sticking quality, improving protection for strain-gage and sticking it on position with bigger shaft bending moment.

Because of the second kind of error due to sticking position, the calculation error of the stern tube fore, the aft bearing load and the middle bearing load is analysed.

2.2(a) Error analysis of stern tube fore and aft bearing load calculation

Fig. 5 shows the calculation brief diagram of stern tube fore and aft bearing load calculation, the bending moment error is ΔM_i, the stern tube aft bearing load error due to bending moment error is ΔF_α, the stern tube fore bearing load error is ΔF_f, according to the moment balance equation, it can be obtained that:

$$\begin{cases} (F_\alpha + \Delta F_\alpha)l_{\alpha f} - F_0 l_{0f} - G_0 l_{g0f} - G_1 l_{g1f} + G_2 l_{g2f} + (F_{i\#} + \Delta F_{i\#})l_{i\#f} - (M_{i\#} + \Delta M_{i\#}) = 0 \\ F_\alpha l_{\alpha f} - F_0 l_{0f} - G_0 l_{g0f} - G_1 l_{g1f} + G_2 l_{g2f} + F_{i\#} l_{i\#f} - M_{i\#} = 0 \end{cases} \quad (1)$$

Where, $\Delta F_\alpha = \frac{\Delta F_{i\#} l_{i\#f} + \Delta M_{i\#}}{l_{\alpha f}}$, $\Delta F_f = \frac{\Delta F_{i\#}(l_{\alpha f} + l_{i\#f}) + \Delta M_{i\#}}{l_{\alpha f}}$ (2)

F_0-- Actual weight of propeller under immersion state
F_α-- Stern tube aft bearing support force
F_f-- Stern tube fore bearing support force
$F_{i\#}$-- Shearing force of i# strain-gage cross section
G_0-- Actual weight of shaft from end of shafting to stern tube aft bearing equivalent support point
G_1-- Weight of shaft from stern tube aft bearing equivalent support point to stern tube fore bearing equivalent support point
G_2-- Weight of shaft from stern tube fore bearing equivalent support point to i# strain-gage

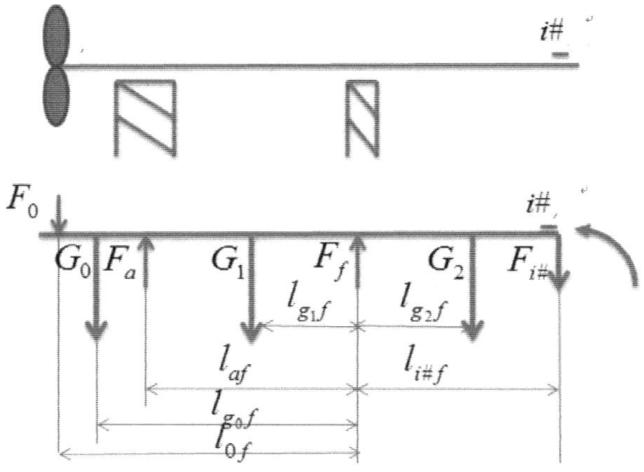

Figure.5. Brief diagram of stern tube fore and aft bearing load calculation

For installed shafting, the distance between stern tube aft bearing equivalent support point and stern tube fore bearing equivalent support point is constant, which is represented by $l_{\alpha f}$. The load calculation error of stern tube fore bearing and stern tube aft bearing is related with calculation error of $i\#$ strain-gage cross section shearing force represented by $\Delta F_{i\#}$, measurement error of bending moment represented by $\Delta M_{i\#}$ and distance between i# strain-gage and stern tube aft bearing equivalent support point represented by $l_{i\#f}$. When shearing force error $\Delta F_{i\#}$, bending moment error $\Delta M_{i\#}$ and distance $l_{i\#f}$ decrease, bearing load calculation error ΔF_α, ΔF_f decrease.

2.2(b) Error analysis of strain-gage shaft cross section shearing force calculation

Fig. 6 shows the brief diagram+ of strain-gage shaft cross section shearing force calculation model, $\Delta M_{j\#}$ and $\Delta F_{j\#}$ are measurement error and shearing force of $j\#$ strain-gage shaft cross section bending moment respectively. According to moment equivalent equation, we obtain:

$$\begin{cases}(F_{i\#}+\Delta F_{i\#})l_{i\#j\#}-G_3 l_{g3j\#}+(M_{i\#}+\Delta M_{i\#})-(M_{j\#}+\Delta M_{j\#})=0\\ F_{i\#}l_{i\#j\#}-G_3 l_{g3j\#}+M_{i\#}-M_{j\#}=0\end{cases} \quad (3)$$

Then,

$$\Delta F_{i\#}=\frac{\Delta M_{i\#}+\Delta M_{j\#}}{l_{i\#j\#}} \quad (4)$$

Where,

$F_{j\#}$ -- j# Strain-gage shaft cross section shearing force;

$M_{j\#}$ -- j# Strain-gage shaft cross section bending moment;

G_3 -- Shaft weight between $i\#$ strain-gage and j#strain-gage;

$l_{i\#j\#}$ -- Distance between $i\#$ strain-gage and j# strain-gage;

$l_{g3j\#}$ -- Distance between shaft weight center from $i\#$strain-gage to j# strain-gage and j# strain-gage.

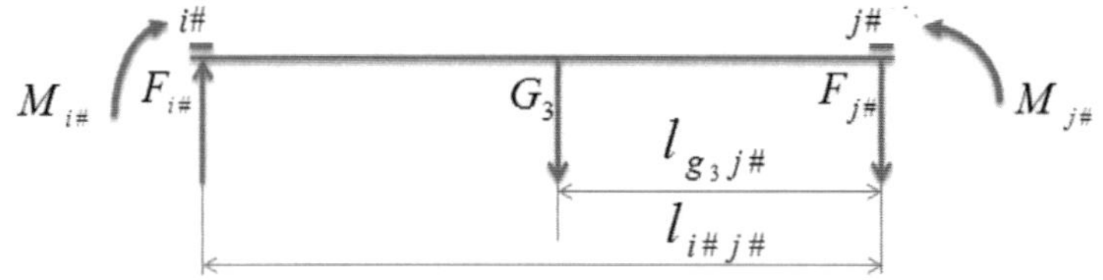

Figure. 6. Brief diagram of strain-gage shaft cross section shearing force calculation model

So to decrease the shearing force calculation error of $i\#$ strain-gage shaft cross section, it is necessary to decrease the bending moment error $\Delta M_{i\#}$, $\Delta M_{j\#}$ and increase the distance $l_{i\#jZ\#}$ between $i\#$ strain-gage and $j\#$ strain-gage.

2.2(c) Error analysis of middle bearing load calculation

Fig. 7 shows brief diagram of middle bearing load calculation error, $\Delta M_{k\#}$ is measurement error of K# strain-gage shaft cross section bending moment, and $\Delta F_{k\#}$ is shearing force calculation error of K# strain-gage shaft cross section. According to moment equivalent equation, it can be obtained that:

$$\begin{cases}(F_i+\Delta F_i)l_{ik\#}+(F_{j\#}+\Delta F_{j\#})l_{j\#k\#}-G_4 l_{g4k\#}+(M_{j\#}+\Delta M_{j\#})-(M_{k\#}+\Delta M_{k\#})=0\\ F_i l_{ik\#}+F_{j\#}l_{j\#k\#}-G_4 l_{g4k\#}+M_{j\#}-M_{k\#}=0\end{cases} \quad (5)$$

Then,

$$\Delta F_i=\frac{\Delta F_{j\#}l_{j\#k\#}+\Delta M_{j\#}+\Delta M_{k\#}}{l_{ik\#}} \quad (6)$$

Where,

F_i-- Middle bearing support force;

$F_{k\#}$-- Shearing force of $k\#$ strain-gage shaft cross section;

$M_{k\#}$-- Bending moment of $k\#$ strain-gage shaft cross section;

G_4-- Shaft weight between $j\#$ strain-gage and $k\#$ strain-gage.

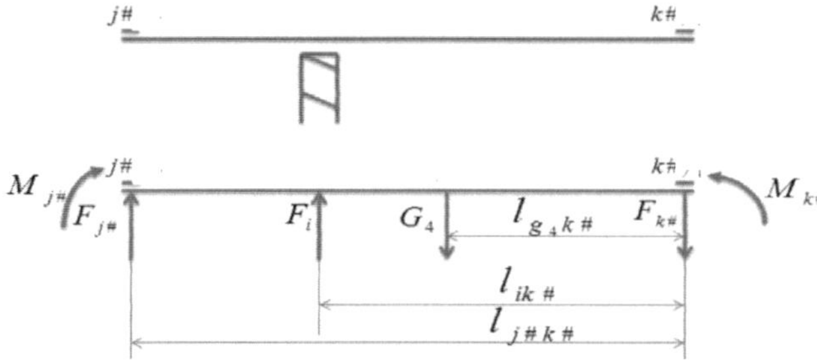

Figure.7. Brief diagram of middle bearing load calculation model

In order to decrease the middle bearing load calculation error, the moment $\Delta M_{j\#}$ and $\Delta M_{k\#}$, the shearing force error $\Delta F_{j\#}$, the distance between two strain-gages $l_{i\#j\#}$ should be decreased while the distance between middle bearing equivalent support point and $k\#$ strain-gage $l_{ik\#}$ should be increased. However, the decrease of $l_{j\#k\#}$ contradicts with the increase of $l_{ik\#}$. How to coordinate the relationship between them is a problem need to be solved.

2.2(d) Rational positions and selection principles of strain-gages

According to the error analysis above, it is known that in order to decrease calculation error of bearing load, it is necessary to decrease the measurement error of bending moment and increase the distance between two strain-gages when there are no bearing between them. The strain gage can obtain a bigger number when it is stuck onto the cross section with a bigger bending moment. At this time The bending moment error ΔM is smaller. Thus, it is better to stick strain-gage on positions with bigger shaft cross section bending moment, which usually near bearing, flange and flywheel. In this way, strain gages should be stuck on position with distance of 50~200mm from bearing, flange and flywheel where there is no mutation of shaft journal size and no defects of surface.

3. RESEARCH ON BEARING DISPLACEMENT OPTIMIZATION ADJUSTMENT TECHNOLOGY

In the aspect of bearing displacement adjustment method theoretical research, Huang Yujian [7] concludes marine main engine bearings displacement adjustment method based on actual construction experience. Xu Li [8] establishes displacement optimization algorithm based on linear optimization to obtain minimum stern tube aft bearing load, but it cannot be used to make displacement adjustment scheme of shafting installed bearing. K ShanKar [9] combined single objective linear optimization algorithm and multi-objective linear optimization algorithm to establish bearing displacement optimization algorithm. With optimization target of minimum load difference between adjusted bearing measured load and expected bearing load, it can only solve the situation of adjusted bearing measured load is bigger than expected bearing load, and it doesn't regard main engine bearings as a whole when setting constraints, which is not in conformity with the actual construction condition.

The bearing displacement optimization adjustment method usually set minimum stern tube aft bearing load as the optimization target, which doesn't conform to actual construction need. In the meantime, the bearing stiffness doesn't be taken into account when calculating bearing load influence factor, the main engine doesn't be regarded as a whole when setting constrains. These don't conform to actual construction technology. So, this paper adopt special nonlinear optimization method with absolute value based on bearing load influence factor that taking bearing stiffness into account. This method sets the minimum absolute value of difference between adjusted bearing load and expected bearing load as optimization target, involves weight coefficient to adjust the optimization proportion of every bearing load, combines actual construction technology and establishes constraints in accordance with actual bearing displacement adjustment. The bearing displacement optimizing adjustment and bearing load forecasting process are shown in Fig. 8.

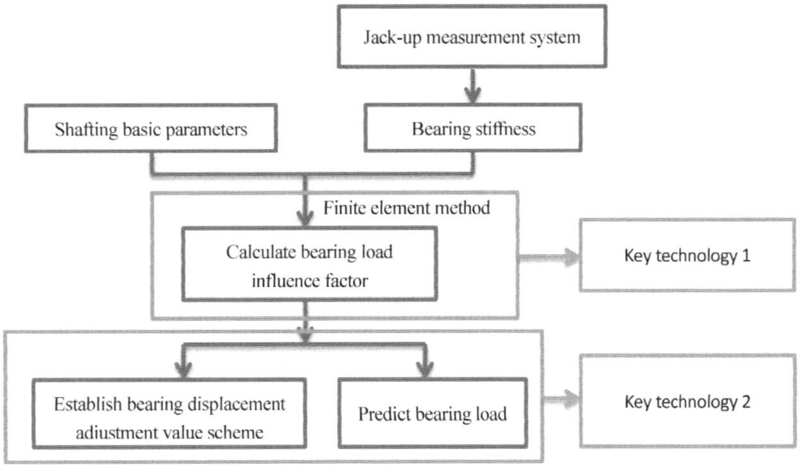

#

Figure.8. Bearing displacement optimizing adjustment and bearing load forecasting process

3.1 BEARING LOAD INFLUENCE FACTOR CONSIDERING BEARING STIFFNESS

Bearing load influence factor is the influence of other bearing load when the *j*th bearing's height varies one unit value, like 1 mm. $R_{i,j}$ represents the variation value of *i*th bearing load due to unit displacement of *j*th bearing[2]. Usually, the bearing load influence factor given by shafting alignment manual doesn't consider the influence of bearing stiffness, which will definitely influence it and then influence the accuracy of bearing displacement adjustment value prediction, especially the calculation of main bearing displacement adjustment value. Thus, this paper establishes bearing displacement adjustment optimization algorithm with especial consideration of bearing stiffness's influence on bearing load influence factor.

Currently, there is no direct measurement or calculation method of marine bearing stiffness. This paper promotes a bearing stiffness calculation method based on jack-up method, which is simple and effective, can calculate the bearing stiffness in conformity with actual construction situation.

3.1(a) Bearing stiffness calculation based on jack-up method

Jack-up method obtains bearing load by measuring hydraulic jack load near measured bearing and then multiplied by jack-up coefficient [10]. After the measured bearing load is obtained, combining with the compression value when bearing is under load, the bearing stiffness K can be calculated. The equation is shown as below:

$$K = \frac{R_B}{H_B} \qquad (7)$$

Where, R_B is measured bearing load, H_B is bearing compression value. By adopting jack-up method to measure bearing load, the measured bearing position doesn't change, jack lift-up continuously. When jack lift up to H_J, the measured bearing is totally free. During the jack-up process, the jack load changes from 0 to R_J, and the measured bearing load changes from R_B to 0. From above analysis, it can be obtained that:

$$\begin{cases} \text{Jack:} & R_J' = 0 + H_J \cdot R_{J,J} \\ \text{Bearing:} & R_B + H_J \cdot R_{B,J} = 0 \end{cases} \qquad (8)$$

Where, $R_{J,J}$ represents load influence factor of jack to itself, $R_{B,J}$ represents load influence factor of jack to measured bearing.

Supposing the jack keeps staying and measured bearing height is lowered, then measured bearing height decrease to H_B, measured bearing will be just free with R_J jack load. During this whole process, jack load changes from 0 to R_J, ensured bearing load changes from R_B to 0. From above analysis, it can be obtained that:

$$\begin{cases} \text{Jack:} & 0 + H_B \cdot R_{J,B} = R_J \\ \text{Bearing:} & R_B + H_B \cdot R_{B,B} = 0 \end{cases} \qquad (9)$$

Where, $R_{B,B}$ represents load influence factor of measured bearing to itself, and $R_{J,B}$ represents load influence factor of measured bearing to jack.

As $\qquad R_{s,J} = R_{J,B}$

Thus, $$R_B = -\frac{R_{B,B}}{R_{J,B}} \cdot R_J = CR_J \qquad (10)$$

Where, C represents jack-up coefficient, which is usually provided by shafting alignment calculation manual, and jack load R_J is obtained by jack-up curve.

Combining above equations, it can be obtained:

$$K = \frac{C^2 \cdot (R_D + R_J)}{2H_J} \qquad (11)$$

Where, R_D, R_J, and H_J are obtained by jack-up curve.

3.1(b) Load influence factor considering bearing load

When marine propulsion shafting is installed completely, adjusting any one or several bearing displacements will cause variation of other bearing load. The new load value by changing bearing displacement can be represented as below:

$$R_i = R_{i,0} + R_{i,1}\Delta y_1 + \cdots R_{i,j}\Delta y_j + \cdots + R_{i,n}\Delta y_n \qquad (12)$$

Where, $R_{i,0}$ represents *i*th bearing load value before bearing displacement adjustment, Δy_j represents *j*th bearing displacement adjustment value, and $R_{i,j}$ represents influence factor of *j*th bearing to *i*th bearing. Building equation as below:

$$\begin{bmatrix} R_1 \\ R_i \\ R_n \end{bmatrix} = \begin{bmatrix} R_{1,0} \\ R_{i,0} \\ R_{n,0} \end{bmatrix} + \begin{bmatrix} R_{1,1} & R_{1,i} & R_{1,n} \\ R_{i,1} & R_{i,i} & R_{i,n} \\ R_{n,1} & R_{n,i} & R_{n,n} \end{bmatrix} \cdot \begin{bmatrix} \Delta y_1 \\ \Delta y_i \\ \Delta y_n \end{bmatrix} \qquad (13)$$

Where,

$$\begin{bmatrix} R_{1,1} & R_{1,i} & R_{1,n} \\ R_{i,1} & R_{i,i} & R_{i,n} \\ R_{n,1} & R_{n,i} & R_{n,n} \end{bmatrix}$$

is the load influence factor matrix.

The bearing stiffness is important to bearing load influence factor, so we need to calculate bearing load considering bearing stiffness, and then calculate bearing load influence factor considering bearing stiffness. Firstly, establish one dimension shafting model through ANSYS finite element software. Then, replace bearing with linear spring, change displacement constraints of model and calculate every bearing load after every bearing change one unit displacement. Finally, the bearing load influence factors which take considering bearing stiffness in account can be obtained by deleting every bearing load before displacement change.

3.2 BEARING DISPLACEMENT ADJUSTMENT OPTIMIZATION ALGORITHM

After actual marine shafting installment, if measured bearing load exceed the allowable range of bearing load in shafting alignment manual, it is necessary to adjust bearing displacement, so as to make adjusted bearing load within allowable range. For shafting with n bearing, solving every bearing displacement value is an optimization problem of n unknown quantities.

3.2(a) Establishment of optimization model

In the real situation, it is necessary to adjust bearing load near shafting alignment prescribed load and within certain deviation range. The optimization target involves optimization of all measured bearing loads that have big deviation with expected load, is able to reflect bigger optimization weight on bearing load with bigger deviation. The optimization conditions involve bearing load range and bearing displacement limit.

Target function: After actual marine measurement, ith bearing load $R_{i\text{-}FwdMea}$ doesn't satisfy shafting alignment requirement. After adjusting bearing height, it is expected that adjusted ith bearing load $R_{i\text{-}AftMea}$ can approach to expected load $R_{i\text{-}Des}$, which is

$$\min \quad \left|R_{i-AftMea} - R_{i-Des}\right| \tag{14}$$

Considering the existing linear relationship between adjusted bearing load value and bearing displacement value, it can be transformed into:

$$\min \quad \left|R_{i-FwdMea} + \sum_{j=1}^{n} R_{i,j}\, \Delta y_j - R_{i-Des}\right| \tag{15}$$

For shafting with n bearing, if there are many bearing loads cannot satisfy alignment requirements, it is necessary to construct target function:

$$\begin{aligned}\min\ & Weight_1\left|R_{1-FwdMea} - R_{1-Des} + \sum_{j=1}^{n} R_{1,j}\, \Delta y_j\right| + \cdots + \\ & Weight_i\left|R_{i-FwdMea} - R_{i-Des} + \sum_{j=1}^{n} R_{i,j}\, \Delta y_j\right| + \cdots + \\ & Weight_n\left|R_{n-FwdMea} - R_{n-Des} + \sum_{j=1}^{n} R_{n,j}\, \Delta y_j\right|\end{aligned} \tag{16}$$

Where,

$$Weight_i = \frac{\left|R_{i-Des} - R_{i-FwdMea}\right|}{\sum_{i=1}^{n}\left|R_{i-Des} - R_{i-FwdMea}\right|}$$

$$\sum_{i=1}^{n} Weight_i = 1$$

Constraint condition: according to shafting alignment requirements, bearing load need to be within certain range, which usually set to ±20% of value given by shafting alignment manual, so it is necessary to set bearing load limit:

$$R_{i-min} \le R_{i-AftMea} \le R_{i-max} \tag{17}$$

Where, $R_{i\text{-}max}$ represents the biggest value of ith bearing load, and $R_{i\text{-}min}$ represents the smallest value of ith bearing load.

So,

$$R_{i-min} \le R_{i-AftMea} + \sum_{j=1}^{n} R_{i,j}\, \Delta y_j \le R_{i-max} \tag{18}$$

Equivalently, the limit of bearing moment can be represented by following equations:

$$M_{i-min} \le M_{i-AftMea} + \sum_{j=1}^{n} m_{i,j}\, \Delta y_j \le M_{i-max} \tag{19}$$

Where, $m_{i,j}$ represents bending moment influence factor of jth bearing to ith bearing; $M_{i\text{-}FwdMea}$ represents ith bearing bending moment value before adjustment; $M_{i\text{-}max}$ represents ith bearing biggest bending moment value; $M_{i\text{-}min}$ represents jth bearing smallest bending moment value.

Considering actual construction situation, bearing displacement cannot exceed certain range, so the following displacement limits are added:

$$\Delta y_{min} \leqslant \Delta y_i \leqslant \Delta y_{max} \tag{20}$$

Considering actual construction technology, it is unable to adjust every bearing height, such as stern tube fore and aft bearing because of construction limit, so the following constraints are added.

$$\Delta y_i = 0 \tag{21}$$

As main engine bearings adhere to main engine pedestal, when adjusting main engine bearings, the total vertical main engine bearing is treated as 0. The following constraints are added.

$$\left[1 - (l_j - l_{fl})/(l_n - l_{fl})\right]\Delta y_{fl} - \Delta y_j + \frac{(l_j - l_{fl})}{(l_n - l_{fl})\Delta y_n} = 0 \tag{22}$$

Where, l_j represents axial position of jth main engine bearing (supposing stern shaft ending point as position starting point); l_{fl} represents axial position of main engine flywheel bearing; l_n represents radial position of main engine bearing near fore end; Δy_{fl} represents main engine bearing displacement near flywheel; Δy_n represents main engine bearing displacement near fore end.

3.2(b) Solution of optimization equations

Through analysis of actual shafting alignment engineering problem, bearing displacement adjustment optimization algorithm based on special nonlinear optimization with absolute value is established. The solution is transforming algorithm into standard form.

$$
\begin{aligned}
&\min Weight_1 \left| R_{1-FwdMea} - R_{1-Des} + \textstyle\sum_{j=1}^{n} R_{1,j}\, \Delta y_j \right| + \cdots + \\
&\quad Weight_i \left| R_{i-FwdMea} - R_{i-Des} + \textstyle\sum_{j=1}^{n} R_{i,j}\, \Delta y_j \right| + \cdots + \\
&\quad Weight_n \left| R_{n-FwdMea} - R_{n-Des} + \textstyle\sum_{j=1}^{n} R_{n,j}\, \Delta y_j \right| \\
&s.t. \textstyle\sum_{j=1}^{n} R_{i,j} \Delta y_j \leq R_{i-max} - R_{1-FwdMea} - \textstyle\sum_{j=1}^{n} R_{i,j} \Delta y_j \leq R_{1-FwdMea} - R_{i-min} \\
&\quad \textstyle\sum_{j=1}^{n} M_{i,j} \Delta y_j \leq M_{i-max} - M_{1-FwdMea} - \textstyle\sum_{j=1}^{n} M_{i,j} \Delta y_j \leq M_{1-FwdMea} - M_{i-min} \\
&\left[1 - \frac{(l_j - l_{fl})}{(l_n - l_{fl})}\right] \Delta y_{fl} - \Delta y_j + \frac{(l_j - l_{fl})}{(l_n - l_{fl}) \Delta y_n} = 0 \\
&\quad \Delta y_i = 0 \\
&\quad \Delta y_{min} \leq \Delta y_i \leq \Delta y_{max} \qquad (23)
\end{aligned}
$$

The general solving algorithms of special nonlinear optimization problem with absolute value are Trust-region-reflective, Active-set, Interior-point-convex, Sequential Quadratic Programming. Trust-region-reflective and Interior-point-convex belong to large-scale optimization algorithm, however Active-set and Sequential Quadratic Programming belong to middle-scale optimization algorithm. Large-scale algorithms suit for situation that only have boundary constraints or linear equation constraints while middle-scale algorithms have larger application range. Thus, Active-set or Sequential Quadratic Programming are applied to solve bearing displacement optimization adjustment problem.

3.2(c) Calculation example

This paper takes actual measured data of C8800TEU container ship and C6600TEU container ship as example, verify and analyze feasibility of bearing displacement adjustment optimization algorithm.

The actual measured data are composed of measured load before adjustment, measured load after adjustment and actual bearing displacement. Actual bearing displacement means bearing adjusted vertical height. This paper regards measured load after adjustment as bearing expected load, obtains bearing displacement predicted value by bearing displacement adjustment optimization algorithm, and then verifies feasibility of optimization algorithm by contrasting bearing displacement actual construction value with bearing displacement theoretical predicted value.

Table 1 gives contrast result of C8800TEU container ship main engine bearing actual construction value and theoretical predicted value. From table data, the target function and load of optimization algorithm include three bearings. As the main engine bearing is adjusted as a whole when actually adjusting bearing vertical height, it is necessary to set displacement adjustment values of middle bearing and stern bearing as 0mm when setting bearing displacement constraints of bearing displacement adjustment algorithm. After calculation, it can be obtained that the main engine fore end should drops 4mm and aft end drops 2.17mm, which have same adjustment direction with actual main engine bearing displacement and the displacement values are close.

Table 1: Contrast between C8800 TEU bearing displacement predicted value and actual construction value

	Bearing name		
	Stern tube fore bearing	2# middle bearing	1# middle bearing
Measured load before adjustment/t	20	30.3	38.3
Measured load after adjustment/t	21.6	28.1	43.6
Actual bearing displacement	Main engine fore end drop 5mm,aft end drop 2mm		
Predicted bearing displacement	Main engine fore end drop 4mm,aft end drop 2.17mm		

Table 2 gives contrast result of C8800TEU container ship main engine bearing actual construction value and theoretical predicted value. The 2# bearing is adjusted when actually adjusting bearing vertical height, it is necessary to set displacement adjustment values of 1# middle bearing, main engine bearing and stern bearing as 0mm when setting bearing displacement constraints of bearing displacement adjustment algorithm. After calculation, it can be obtained that 2# middle bearing should be lifted by 0.56mm, which have same adjustment direction with actual 2# bearing displacement and the displacement values.

Table 2: Contrast between C8800 TEU bearing displacement predicted value and actual construction value

	Bearing name		
	Stern tube fore bearing	2# middle bearing	1# middle bearing
Measured load before adjustment/t	21.5	33	45.8
Measured load after adjustment/t	17.9	37	44.5
Actual bearing displacement	2# middle bearing lift 0.5mm		
Predicted bearing displacement	2# middle bearing lift 0.56mm		

Table 3 is given by an authoritative organization, predicts C6600TEU container ship bearing displacement value based on actual measured bearing load, actual bearing displacement adjustment value and base data with actual

Table 3: Contrast between C8800 TEU bearing displacement predicted value and actual construction value

	Bearing name						
	Stern tube aft bearing	Stern tube fore bearing	2#middle bearing	1# middle bearing	10#main engine bearing	9#main engine bearing	8#main engine bearing
Measured load before adjustment/t	110.1	14.5	33.2	24.3	61.8	27.2	65.8
Measured load after adjustment/t	104.6	14.6	31	29.9	28.5	45.3	68.1
Actual bearing displacement	Main engine fore end drop 1.2mm						
Predicted bearing displacement	Main engine fore end drop 1.24mm						

measured data. When establishing bearing displacement adjustment optimization algorithm of C6600TEU container ship, optimization target and load constraints involve seven bearing shown in table 3. When setting bearing constraint, displacement adjustment values of stern tube bearing and middle bearing are set as 0mm. After calculation, it can be obtained that main engine fore end should drop 1.24mm, which is almost the same with main engine adjustment value.

4. CONCLUSION

With the development trend of large-scale marine, marine propulsion shafting installment quality have important influence on safety and economics of marine operation. Currently, bearing load relies on artificial readings during shafting install construction, and bearing displacement adjustment relies on work's experience, which cause lower measurement precision and longer adjustment time. This paper researches on jack-up method and strain-gage method digital measurement technology respectively, improves bearing load measurement precision and efficiency by solving key technologies like pressure data and displacement data synchronous automatic acquisition of jack-up method, data linear regression, and rational sticking position and choosing method of strain-gage method. Besides, this paper proposes bearing displacement adjustment optimization algorithm based on special nonlinear optimization with absolute value, considers bearing stiffness, takes actual marine engine bearing displacement adjustment technology as reference, and adds marine engine bearing whole adjustment constraints. After test and analysis of several actual ships, it is proved that this research has important realistic meaning to improve marine shafting installment quality and construction efficiency.

5. ACKNOWLEDGEMENTS

This work was financially supported by Innovation Team Project of Liaoning Higher Education (No. LT2014002)

6. REFERENCES

1. Zheng Daizhong, 'Principle and application of rational alignment', Ship Design and Communication, 2010.
2. Zhou Jiliang, Zou Hongju, 'Principle and application of marine shafting alignment',Beijing. China Communication Press,1985.
3. Shen Junru, Zhou Jiliang, 'Experiment research on shafting alignment method quality',Marine Technology, 1983.
4. Hen Junru, Zhou Jiliang, 'Experiment research on shafting alignment method quality(added)', Marine Technology, 1983.
5. Zheng Dingyu, 'Marine shafting jack-up test application',China Water Transport, 2011.
6. Ji Wang, Shengjun Gu, Yu-jun Liu, Zhiyuan Wei, 'Zhi Qu. Development of marine propulsion shafting load measurement and calculation system', Busan, Korea. Royal Institution of Naval Architects, 2013.
7. Huang Yujian, Wen Weizhong, 'The law exploration of marine main engine shafting bearing load adjustment', GSIShipbuilding Technology, 2007.
8. Xu Li., 'Best optimization design of marine

shafting ration alignment', Ship and Ocean Engineering, 2001.

9. BATRA A, SHANKAR K, SWARNAMANI S., 'Propulsion shaft alignment measurements on warships afloat and alignment solution using multi-objective optimization', Journal of Marine Engineering and Technology,2007.

10. SUN J, CHEN D, REN Y. 'Stiffness Measurement Method of Repulsive Passive Magnetic Bearing in SGMSCMG', Instrumentation and Measurement, IEEE Transactions on, 2013.

TECHNOLOGICAL ADVANCEMENTS MADE IN NAKILAT'S FLEET

H.R. Suwaid, Nakilat, Qatar

SUMMARY

Qatari shipping and maritime company, Nakilat, realized the importance of ensuring uninterrupted 24/7 connectivity for its global fleet operations. The company embarked on a *Unified Ship Management Project* to upgrade all its vessels' existing communication systems, utilize new hardware and applications as well as enhance the communications process with reliable back-up systems, not only for operational benefits but also to improve crew welfare. All outdated fleet applications on vessels and fleet office interfaces were replaced to create a futuristic software landscape comprising of a globally-supported maintenance system with source codes belonging to Nakilat, including standard online /offline ship management systems and reporting capabilities. These led to substantial cost savings for the company, leaner manpower requirements, eliminated single point failures and allowed for remote backup of systems, 24/7 connectivity and a centralized database for Nakilat's entire fleet.

NOMENCLATURE

VSAT	Very Small Aperture Terminal (VSAT)
VDR	Voice Data Recorder
SMS	Ship Management System
VOIP	Voice Over Internet Protocol

1. INTRODUCTION

In 2016, a fleet transition agreement was signed for all 25 of wholly-owned LNG carriers managed by another ship operator, to be transitioned in a phased plan to Nakilat. These vessels being transitioned are among the world's largest and most complex LNG carriers in operation. Information Technology (IT) infrastructure and communication systems onboard these gigantic vessels is the critical backbone towards effectively managing fleet operations as well as business continuity preparedness.

Due to the legacy systems and applications onboard these vessels, Nakilat faced enormous challenges and high risks in its day to day business operations. The IT infrastructure, systems and existing technology of each ship was primitive and required major upgrade work. Each ship had a very basic infrastructure, with low availability, no virtualization and no business continuity, which made it difficult to collect information even though we had opened a new high-speed satellite channel. The vessels were exposed to various risks including server failures, no redundancy, unreliable backup and recovery, poor data availability, no cloud integration, no security systems, obsolete software, etc. If a ship goes 'blind', this could result in significant monetary impact to the company. To compound matters, the IT infrastructure and technology on board these vessels were reaching the end of their life cycle.

Nakilat's rapid growth over the past year, especially in ship management, brought about the critical need for significant digital transformation of the IT and communication infrastructure onboard its vessels. An extensive study was conducted and detailed roadmap created, which led to the undertaking of a comprehensive digital transformation initiative to effectively manage its massive fleet. Nakilat embarked on a mission to modernize its fleet based on these qualities – Resilience, Innovation, Agility and Adaptability.

- Resilience - ability to react to new challenges and bounce back from setbacks
- Innovation - sustain growth through tactful solutions
- Agility - seize opportunities, open to venture into vast array or technologies, applications, systems, and socially sound business practices.
- Adaptability – flexibility, compliance and consistency to ensure new environment is relatively customizable and widely acceptable by personnel

2. DIGITAL TRANSFORMATION JOURNEY

The journey began with the upgrade of VSAT and was followed by the IT Infrastructure and Applications upgrade respectively. A summary of the feat is further detailed below.

2.1 VSAT UPGRADE

The demands are growing rapidly for better communication solutions within maritime markets of all types. The demands include much higher broadband speeds, increased reliability and global coverage of the world's oceans and waterways. However, that is not all. Complete, cost-effective VSAT network solutions with value added services, tailored per vessel, for core operations and crew welfare, are essential to be competitive in the maritime markets.
Challenge:

- Slow bandwidth and internet speed with the existing service provider.
- Lack of flexible bandwidth products and modules
- Unavailability of global coverage
- No automatic beam switching
- Excessive costs
- Limited security filters for VSAT

Solution:

Upon extensive research, gap analysis of the present VSAT service and planning, Nakilat narrowed down to select the best service provider for its fleet. The selection process was based on inviting tenders from major companies around the world in this business and the selection criteria were based on the elements of added value, costs, innovation and superior quality and support. Thorough evaluation saw EMC emerge as the service provider of choice for our in-house managed vessels.

Working closely with EMC, Nakilat implemented an integrated, managed maritime VSAT network on all the in-house managed LNG and LPG vessels. The expectation was to meet diverse business needs and budgets with agility. This included global coverage with reliable [1]KU-bandwidth and seamless beam handovers. EMC's flexible "Fleet Bandwidth Bundles" yield continuous internet & intranet connectivity using automatic beam switching with dynamic allocation of bandwidth.

Through its association with EMC, Nakilat achieved double the bandwidth (4 Mbps) line at half the cost of previous service provider – contributing to company's reduction of operational expenditure (OPEX).

Additionally, this VSAT coverage offers free VOIP calls from the vessels to the Head Office and vice versa, free internet usage for the Crew along with low-priced VOIP calling facility for crew to stay in touch with their loved ones. This not only means improved connectivity between vessels on voyage with management at corporate office, it also contributes towards enhancing the lives of our dedicated personnel onboard.

Technology:

One of the major advancements with the new VSAT provider is the usage of KU-band Antenna upgradable to [2]KA-band. This is a major boost and will help in further enhancing global coverage and speed.

Figure 2.1.1 below shows the difference between the two antennas, its advantages as well as outlines the shortcomings.

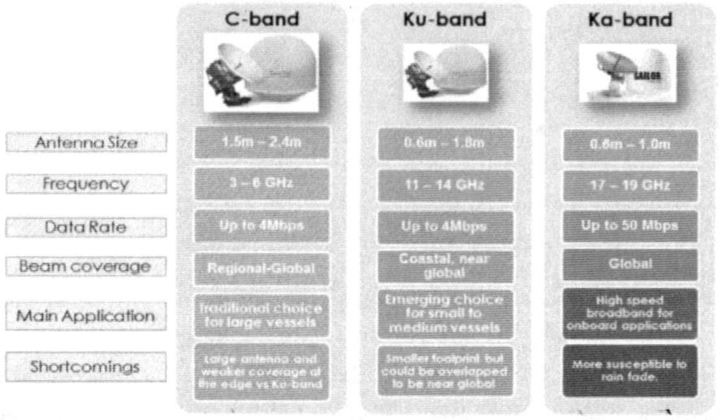

Figure 2.1.1

With the upgradeable KA-band, it provided us the option to go for a data rate of up to a maximum of 50 Mbps. This is very useful during high profile high-definition (HD) video conferencing which requires superior and quality bandwidth. Moreover, with VPN tunneling between the vessel and Head Office for remote data backups and support, this quality of bandwidth is adequate.

Figure 2.1.2 shown below is the Antenna Size and Band Spectrum that is used for the KU / KA Antenna.

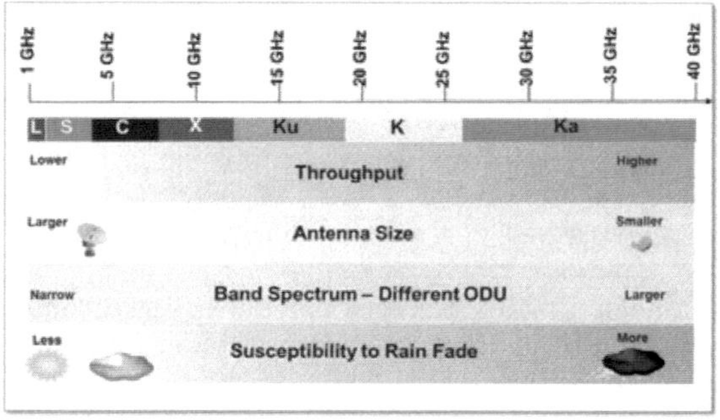

Figure 2.1.2

With EMC VSAT Intellian V100 / 8W Antenna and superior hardware, the VSAT infrastructure offers a secured and reliable platform for the vessels IT & Communication Systems while maintaining a low downtime.

In addition to the security element is the implementation of Fortinet Cloud Firewall to safeguard the network by offering Whitelisting, Proxies, Filtering and such.

Figure 2.1.3 below shows the hardware established on each vessel with the Iridium[3] Failover System

[1] KU band: A portion of the electromagnetic spectrum in the microwave range of frequencies ranging from 11.7 to 12.7GHz. (downlink frequencies) and 14 to 14.5GHz (uplink frequencies)

[2] KA band: A portion of the electromagnetic spectrum covers the radio frequencies of 26.5–40 GHz, i.e. wavelengths from slightly over one centimeter down to 7.5 millimeters. The K_a band is part of the K band of the microwave band of the electromagnetic spectrum.

[3] Iridium: It operates the Iridium satellite constellation, a system of 66 active satellites used for worldwide voice and data communication from hand-held satellite phones and other transceiver units.

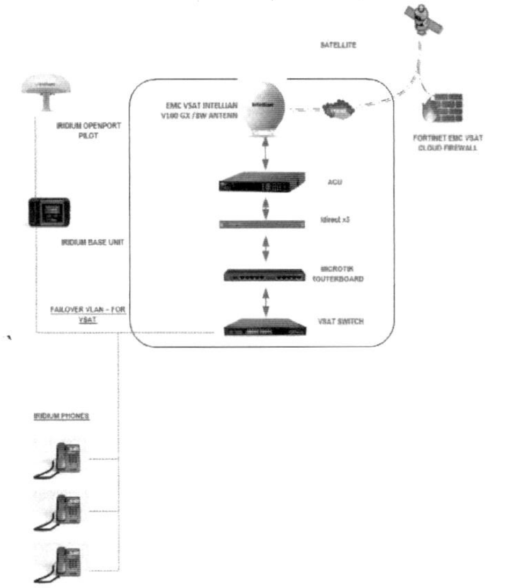

Figure 2.1.3

Benefits:

- Double bandwidth at half the cost of previous Service Provider leading to reduced OPEX.
- Global and reliable VSAT coverage
- Auto beam switching
- Free VOIP calls from the vessel to the Head Office and vice versa
- Secured VSAT network with Cloud Firewall
- Free and fast Internet for Crew with low priced VOIP calling Facility
- KA-band upgradable Antenna
- IRIDIUM Failover as backup system for VSAT Data and Voice
- VLANs & VPN Services
- Bandwidth Management & Ship Tracking System

2.2 FLEET IT INFRASTRUCTURE UPGRADE

Inheriting vessels with primitive IT infrastructure, systems and technology – this called for a significant upgrade of IT infrastructure to guarantee peak performance of vessels during operations. Nakilat required a cost-effective and simple IT platform onboard each ship and thus surveyed the market extensively - to find the right solution.

Challenge:

- Single Server without High Availability (HA) or Disaster Recovery (DR)
- No Virtualized Environment
- No Firewall
- No Auto Backup & Archiving
- Does Not Match the IT Industrial Standards
- Systems were not scalable
- High Maintenance Cost (Manpower, Spares and Travel)

Solution:

The company had already partnered with Fujitsu to deploy FUJITSU Integrated System PRIMEFLEX® for SAP HANA at both its Head Office data centers in 2015. The success of that project, a scalable and flexible SAP private cloud platform based on Fujitsu FlexFrame Orchestrator at half the operating cost, made Fujitsu a trusted partner for this new business-critical IT project. Upon stringent evaluation of the onboard Systems vis-à-vis our requirement's, Nakilat selected Fujitsu PrimeFlex V-shape Solution.

Nakilat chose the pre-configured, plug-and-play Fujitsu PRIMEFLEX vShape solution to provide a robust, virtual on-board environment to modernize its fleet for better connectivity and performance. Combining Prime Flex High Availability Architecture with VM (Virtual Machine)ware Hypervisor and Commvault Backup Solution, Nakilat along with its partner Fujitsu implemented a state-of-the-art system to fully meets the needs of a secure, reliable and efficient IT system in accordance to the highest IT standards.

FUJITSU Integrated System PRIMEFLEX vShape is a virtualization and private cloud reference architecture that integrates the expertise and technologies of leading manufacturers' software. It provides highly scalable industry-standard components, including storage, networking, virtualization software, and incorporates backup to the ETERNUS CS200c with Commvault software.

These components are configured, synchronized and validated as a single solution. Since each vessel is different and therefore has different requirements, the beauty of PRIMEFLEX vShape is that it can be customized to meet each ship's specific needs. Together, Nakilat and Fujitsu created a case study based on the infrastructure of each vessel to ensure that the configuration was correct.

The first four PRIMEFLEX vShape solutions were then put together by Fujitsu, consisting of two FUJITSU Server PRIMERGY RX2540 units and one FUJITSU Storage ETERNUS DX100 device supporting 2TB of data. This platform runs VMware Horizon and connects to both thin clients and traditional PCs on-board, as well FUJITSU Storage ETERNUS CS200c with Commvault software at the head office providing a backup solution.

It took us around three days to equip each ship with the new equipment, from the bridge deck to the engine room. Typically, there are six virtual desktops and six thick clients on each vessel, all connected to a PRIMEFLEX vShape back end that runs numerous essential applications and has in-built resilience in the event of failure. It is also entirely Solid-State Drive (SSD), which is far more suited to life on the waves.

Improved performance with maximum availability and cost efficiency

The new PRIMEFLEX vShape on-board solution is helping Nakilat improve system reliability and guaranteed up-time. Previously, ships needed to dock for an engineer to resolve any IT issues. Now, any incidents can be quickly handled remotely. This has dramatically reduced manpower costs as one engineer can now manage eight ships. Hence travel, administrative and other logistical expenses have reduced dramatically.

Highly reliable FUJITSU Integrated System PRIMEFLEX vShape enabled remote support, eliminates single-point failures, ensures effective monitoring of our fleet and improve responsiveness to business requirements.

The new virtual environment with in-built disaster recovery, ensured that Nakilat could effectively oversee the systems onboard each ship, its cargoes and crew. Furthermore, application performance has improved thus enhancing productivity of crew onboard and enabling the company to introduce new revenue streams, while making smarter business decisions based on real-time shared data.

Ships can now operate as independent assets, collecting information relating to weather or conditions, transmitting that data and become information providers to our charterers, operators and top management. Fujitsu provided the reliable hardware to gather that big data. What this also means, is that we expect to recoup between 70-80% of our initial hardware investment in the first year.

Each datacenter was integrated with:

- 3 Brocade Switches
- Two Fujitsu RX2540 Servers with VMware HV and VDI
- Eternus CS 200 with Commvault Backup and
- Fujitsu PY SX05 S1 with RDX tape
- Fujitsu Eternus DX 100 with 2 TB Storage

The data center is configured to run the VMs and Provide High Availability in case of any hardware failures or contingency. With Commvault Backup System, we have established an auto real-time backup to the Head Office on a Eternus Appliance.

Hence, data is highly secured and can be easily recovered is case of any catastrophes. With the Cisco Meraki Firewall coupled with the Fortinet Cloud Firewall, we have established a two-layer protection for the Ships network leading to enhanced security. Dual Anti-Virus Systems run on VMs and each VDI.

Illustration Figure 2.2.1 below shows the network topology in a standard Nakilat vessel.

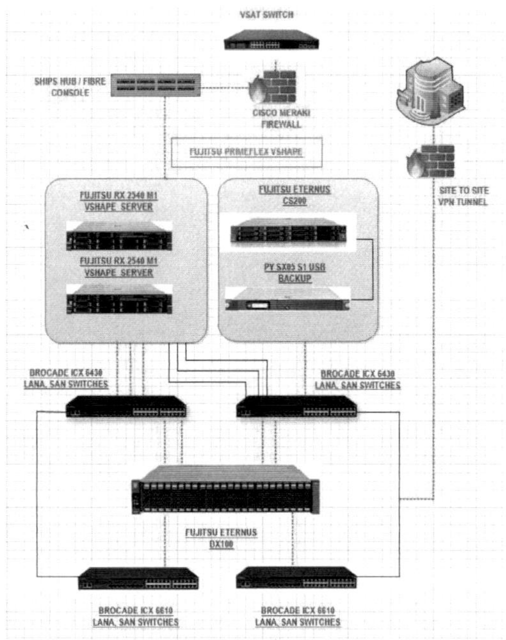

Figure 2.2.1

Benefits:

With the Fujitsu Hyper Converge Systems Prime Flex V-shape on a VMware Hypervisor, the major benefits can be reflected by the enhancement of fleet performance, significant cost reductions and up to 80% Returns on Investment (ROI) on the first year on implementation:

i. Reduced various Points of Failure compared to the old system wherein Hyper converge will provide failover compute and high availability ensuring control and data remain operational even in an event of node failure. It also adds extra redundancy to the system with additional fall over points and dual hardware.
ii. Since it is a scale out platform which suits the scale out workloads and Virtual Desktop Infrastructure (VDIs), it helps deliver consistency for all the Marine Applications especially our database applications in ensuring that the load is equally divided and the database which can grow systematically overtime. Thus, it allows greater mobility between workloads and applications and ensure minimum crashing of application and easier updates to the database.
iii. Unlike legacy systems where we need to perform a hardware refresh every three years (as per policies for shipping vessels), all that would

need to be done for a hyper-converged system is adding additional nodes to the data center cluster thus nullifying the high Operating as well as Capital expenditure that we incur regularly for maintaining and upgrade of the traditional systems. Moreover, with the hardware being in superior quality, the life of the hardware is also extended to a minimum of 5 years.

iv. Integrated Commvault backup system provides snapshot capabilities, data deduplication, inline compression and auto-backup to Disaster Recovery (DR) sites; unlike the traditional method where the back-ups were taken manually by the captain on tape drives and sent to our head office very quarter. This ensures that data can be easily restored and initiated from the DR site to the vessel thus eliminating the burden on the ship crew so that they can focus on business rather than IT system.

v. The switch from physical PCs to virtual desktops, Hyper-converged infrastructure assists us with the processing load because it comes with a lot of flash - which is great for virtual desktop performance. It also improves input/output (I/O), reduces the problems of boot storms, and will let us run virus and other scans in the background without users ever knowing. Additionally, the Hyper-converged systems provide high level of protection and security against viruses and other malware attacks – which occur on shipping vessels frequently and is a highly sensitive area.

vi. For IT resources, the simplicity of this infrastructure on a shipping vessel reduces the variety of skills required to maintain a legacy traditional system such as network, service desk, systems, and etc. due to the smaller footprint and bundled hardware of a Hyper-converged system. With efficient policies to manage Virtual Machines (VM) groups, IT resources can do a better and efficient job of maintaining the datacenter as well as implementation of the same.

vii. Improvement of operational excellence and reliability enhancement through high availability and redundancy which eliminates single point of failures, dedicated bandwidth for corporate office and vessel crew to improve business requirements responsiveness, highest possible uptime thus far with reduction of support calls to IT engineers, improved response time, simplified server management and more.

viii. Efficient and swift management with 24-hours connectivity, centralized administration and automated control enabled subsequently leading to effective analysis and data reporting for commercial and operational needs.

2.3 SHIP MANAGEMENT APPLICATION UPGRADE

As any effective IT system depends on the blend of a robust hardware and software, Nakilat also focused on improving the application side of its IT infrastructure. Nakilat moved to replace the Ship Management Applications on board its managed vessels and at head office, namely:

- Omnisafe[4] / Safety Module
- PROP: Engineering, Finance & Forms
- WRAP: Work Planning & Risk Assessment Module
- Marine Reporting: Shipboard Desktop

Challenge:

a) High cost of support
b) Legacy databases
c) Obsolete software
d) Online / Cloud version not available
e) No future Roadmap or support for latest windows platform
f) Does not match the IT Industrial Standards
g) No source code availability

Solution:

Nakilat invited tenders and proposals from various vendors in the field of Marine Applications to propose a Unified System for Fleet Management which will be interlinked with all modules. Vessels modules was to have the capability of achieving 100% integration with SAP at Head Office and meet all business requirements as detailed below in the key points:

- Fortify the systems needed for taking over ownership and management of LNG and LPG vessels
- Strengthening Management Control
- Globalization of accounting standards & risk standards
- Automated Ship Management System
- Sustainable software with current technologies standards
- Flexibility and scalability for new vessels
- Security of system and data
- Efficiency and Effectiveness of Ship Management
- Integrated process & systems
- Quality of products, services & business processes
- Audit objectives, Risk Management
- Align IT with business requirements
- Security by design
- SAP standard interfaces should be available.

[4] Omnisafe: Utilized as Safety Management System on vessels.

- Ship Staff & Shore Staff Familiarization & Training.

Upon thorough assessment of various proposals, the features vis-à-vis our requirements, meeting with the various Vendors, conducting market research, doing cost analysis, and more, the contract was awarded to SpecTec AMOS.

AMOS primarily included the following main modules of Marine business i.e. Plant Maintenance & Stock Management System (PMS), Health, Safety & Quality Management Module, Procurement Module, Work Planning & Operations Module, Voyage Analysis Module & Staff Management Module.

To effectively plan and execute the project, various workshops were conducted with affected stakeholders and departments to capture the various synergies, needs, ideas and expectations. After numerous meetings, workshops, and gap analysis, AMOS developed the blue print for customizing the application. Next, was to extensively test the application with all the stakeholders and gather their feedback. Integrating with SAP interface was also another challenge that was extensively tested and reworked.[5] After months of tenacious work, Nakilat rolled out the customized AMOS application for its vessels and the corporate office

Benefits:

- Direct Ownership and Control
- Facilitate real-time decision making through Business intelligence advanced reporting and KPI management
- Reduced Total Cost of Ownership (TCO)
- Possibility of future enhancements and integrations with various stakeholders and charterers
- Enabling a maintenance strategy to keep full control of stock levels company-wide, minimizing unexpected machinery downtime. This increases the overall lifetime of most valuable assets whilst remaining in compliance with maritime law. Unexpected downtime is costly for any organization, leading to lost time and revenue. Without a structured maintenance approach, it would not be easy for companies to monitor the life cycle and performance of their machinery, or to maintain control over their available spare parts. With AMOS, the tracking and monitoring became much easier and more efficient. With AMOS, we could better plan the maintenance strategy, monitor the vessel's overall performance, and track required spare parts for any piece of machinery. The key to success is a standardized and centrally controlled approach to planned maintenance, combined with a single view of spare parts company-wide and AMOS ticked all the boxes
- In addition, AMOS helps to prevent problems from happening rather than reacting to a problem. This is achieved through a condition-based approach to ship maintenance
- Integration with SMS
- Standard Technology based Solution
- Optimized, suitable and automated Empowerment of Fleet Management Team
- Fully integrated with Virtualized environment and VDI systems

2.4 CREW WELFARE & ENTERTAINMENT SYSTEM

An increasingly important aspect for Nakilat is with regards to the enhancement of crew welfare management. Nakilat's investments in satellite and VOIP technology onboard our vessels reflects our continuous commitment in providing conducive working environment for all our personnel. With free internet and low cost VOIP calls already provided for its crew onboard vessels, further improvements include installation of IP TV infrastructure directly to the crew's personal devices for video streams.

Challenge:

Having a diverse workforce with various nationalities from different countries, regions and continents also means a wide-range of interests, cultures, languages and personalities. It is understandably challenging to find a common interest in television of entertainment channels. Moreover, having a massive library of CDs or DVDs of movies or films is becoming a thing of the past. There is delay in obtaining latest movies. Often, disks got lost or damaged.

Solution:

Upon extensive research and evaluation of several vendors, Nakilat entered into a contract with a leading IP TV Service Provider for Content Delivery. This unique and economical solution to enhance crew welfare delivers hundreds of gigabytes of content per month. It also enables the crew to access commercially licensed media using personal devices. This addresses several provisions in the Maritime Labour Convention (MLC) 2006which outlines general standards, fundamental principles and guidelines to ensure shipping companies are in compliance with the mandatory regulations pertaining to the welfare of seafarers.[6]

[5] SPETEC, Maintenance,Materials Management, Retrieved from http://www.spectec.net/maintenance-materials-management, 2017

[6] KVH, Maritime Satellite Communications, Retrieved from http://www.kvh.com/Commercial-and-OEM/Maritime-Systems/Communications, 2017

Benefits:

- Watch on any device: Crew have the access to watch any commercially licensed media on their personal devices such as mobile phones, laptops or common room TVs etc. via wireless connection.
- Updated library with the latest releases.
- Unlimited viewing for the crew
- Does not use vessel's data plan and has no effect on onboard data speeds
- Access to different language newspapers and music channels
- File Sharing Services
- Announcement services where Nakilat Shore Management needs to make any announcements to the vessels.
- Training videos prepared by Nakilat can be streamed online.
- Brings maritime training courses onboard easily like VideoTel etc.
- Content is delivered automatically in the background over the top of the mini-VSAT broadband network
- Remote management and control by Service Provider, no burden for IT or crew onboard
- Delivers hundreds of gigabytes per month of content to the vessel using a separate DataStream
- Enables large files to be delivered, such as the entire ECDIS ENC global chart database
- Fast, affordable delivery of charts from leading maritime providers – Jeppesen and Transas*
- Affordably delivers high-resolution weather and ocean data for voyage optimization

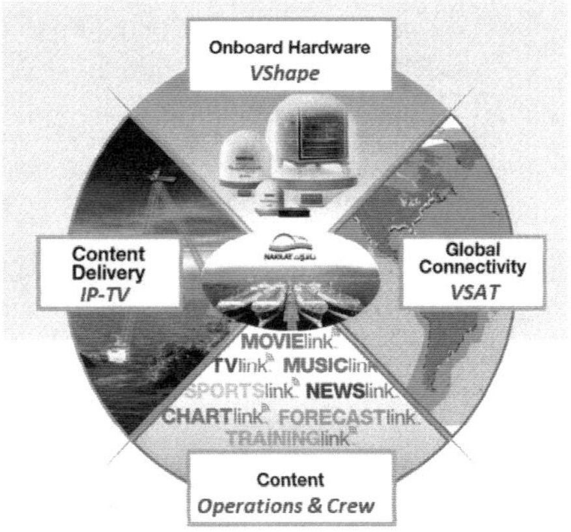

Figure 2.4.1

2.5 IT SECURITY

The traditional method of Information Security was based on the principle to keep threats at bay. Legacy methods used rules and signatures to look for threats that have been pre-identified. While these approaches were helpful in finding known threats, they are not effective against many of today's advanced attackers and insider threats.

Nakilat was driven with the vision to provide an immune system for cybersecurity. This is critical to provide the vessels with evolving internal protection and the ability to fight back against external threats that penetrate the network. Advanced cyber security systems were assessed as necessary by Nakilat.

Challenge:

As the phased vessel transition took place, Nakilat realized the overwhelming exposure and potential target for cyber threats such as ransomware and hacking to its unprotected IT infrastructure. Due to the outdated applications and software, swift actions were required to ensure the safety and security of all our vessels.

Solution:

In alignment with the IT hardware infrastructure upgrade, Nakilat assessed and adopted an advanced security system that would not only nullify existing cyber security risks but also proactively warn us of any impending threats or dangers to the network.

To achieve this, Nakilat implemented a third-party Intelligence Immune System Appliance to the ship network which could analyse network traffic and learn patterns of every device, network and user. Each measure of network behaviour was then monitored in real time to detect anomalous behaviour. [7]

Using cutting-edge visualization techniques, the Threat Visualizer user interface automatically alerted analysts to significant incidents and threats within their environments, enabling analysts to proactively investigate specific areas of the IT infrastructure.

Benefits:

- Using cutting-edge visualization techniques, the Threat Visualizer user interface automatically alerted analysts to significant incidents and threats within their environments, enabling analysts to proactively investigate specific areas of the IT infrastructure
- Fully updated software and applications, with latest patches and updates, ensure that the systems are not in a compromised state
- This Intelligence Immune System coupled with efficient Fortinet Cloud Firewall and Cisco Meraki premise based firewall systems ensure

[7] Darktrace, Threat Detection and classification, Retrieved from https://www.darktrace.com/products/#darktrace, 2017

that Nakilat Vessels has a dual security layer and the IT network is highly protected from the day-to-day evolving cyber security threats

2.6 DUALOG EMAIL SYSTEM

The final frontier in this digital transformation was the changeover to Dualog[8] Email System. Relying on [9]Rydex Email system over the years as a legacy email system for the vessels had a lot of technical issues and problems. Hence, it was time for Nakilat to take the plunge with Dualog Email.

Dualog® Crew Mail is an email system tailored for maritime use. Unlike Rydex, this service provides private email accounts for individual crew members regardless of the communication equipment onboard. Crew details are updated and distributed automatically from the Crew Management System to all vessels – this was a zero-administration solution.

Challenge:

One of the main challenges of the Rydex Email system was that it was restricted to only the Master and Chief Engineer. The rest of the crew had to use third party applications to communicate between themselves. Moreover, the Master had to use another application for forwarding or sharing emails to the Chief Engineer. This entire process was tedious and very time-consuming.

In addition to this, the email system had frequent downtimes and mailbox issues resulting in difficulties of maintaining business continuity and communications by the personnel.

Solution:

Overcoming this challenge, Nakilat strived to ensure an uninterrupted highly versatile email system for all the crew onboard the vessels.

To achieve this, we considered various email solutions and products that would meet our requirements. After evaluating several systems and carrying out a Proof of Concept, we narrowed down to Dualog Email System and AME as a replacement for Rydex.

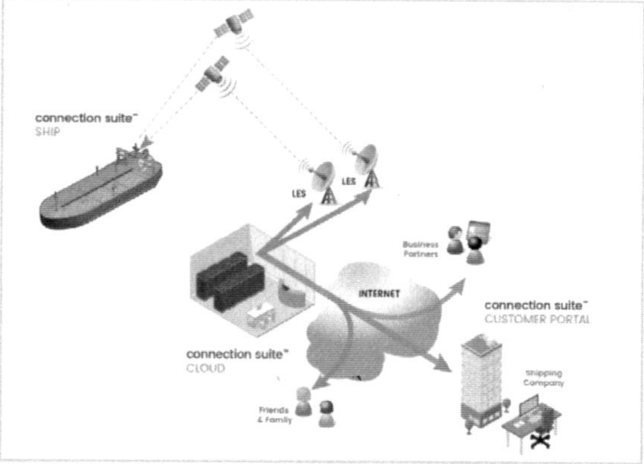

Figure 2.6.1

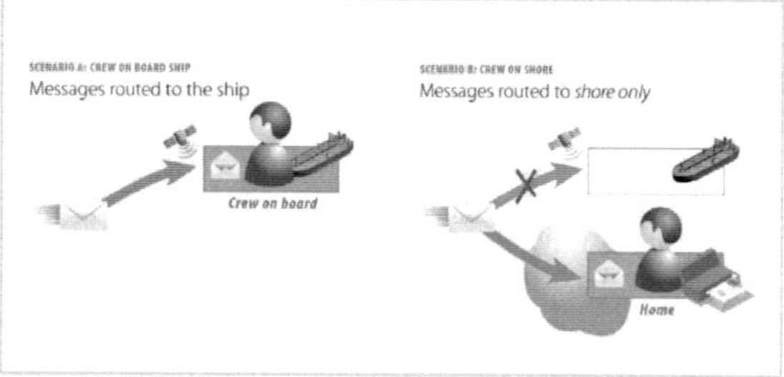

Figure 2.6.2

Benefits:

- Private, zero administration maritime email
- Login with crew credentials
- Includes web (HTTP) e-mail client enabling (if permitted) access from any networked computer on board (if configured)
- Crew can create own message folders
- Reduced burden on the Captain for ensuring email communication and correspondence to all crew
- Send and Forget model wherein the Captain does not need to make a manual connection for Emails or AME exchange. Dualog is auto-configured for this
- Crew keeps the same email address regardless of ship assignment (if configured)

[8] Dualog, Maritime Email for Crew, Retrieved from http://dualog.com/node/69, 2017

[9] Rydex: Auto Email Transfer job and mode of data communication

- The global Dualog Address Book allows each user to store and access his / her contacts' details from anywhere
- Shore-based user administration (if required). Modifications are replicated automatically to ships
- Communication and compression features within Dualog Connection Suite minimize communication cost as it always selects the low cost option available as default
- Cost control and size control (filters, etc.). Users can be exported from the company's Crew Management System and automatically imported to Connection Suite

3. CONCLUSION

The digital transformation journey of Nakilat is far from over. As technological advancements move rapidly, Nakilat continues to look out for more innovative approach and solutions towards a secure and robust IT infrastructure, to further enhance the way we operate.

Nakilat has achieved massive benefits through the implementation of these technological advancements and upgrade from substantial cost savings and OPEX reduction, improved operational excellence, increased vessel reliability, more efficient, swift and centralized flee management as well as enhancement of crew welfare.

In today's progressively digitalized era, connectivity is one of the key elements that enables an organization to be responsive to dynamic global market evolution and is an increasingly important aspect of crew welfare management.

Nakilat's advancement in Information Technology (IT) has seen the company making headway in the industry with multiple awards and invitations to share our expertise and experience at reputable local and regional forums. Nakilat prides itself as being an early adopter of technology in the maritime industry.

4. ACKNOWLEDGEMENTS

I would like to thank the Nakilat Management and colleagues from Fleet Department for their tremendous support and belief in this digital transformation, in assisting me to achieve these outstanding results. I am also grateful to the service providers and technology experts for their dedication and support in overcoming numerous obstacles during the implementation of solutions and making it an enormous success.

I would also like to thank my fellow team members from the Information Technology (IT) Department for their feedback, co-operation and their team work. In addition, I would like to express my gratitude to the staff and crew of Nakilat vessels for being patient with us and supporting me and my team in this remarkable transformation of our Fleet IT Infrastructure.

5. REFERENCES

1. Darktrace, Threat Detection and classification, Retrieved from https://www.darktrace.com/products/#darktrace, 2017

2. Dualog, Maritime Email for Crew, Retrieved from http://dualog.com/node/69, 2017

3. KVH, Maritime Satellite Communications, Retrieved from http://www.kvh.com/Commercial-and-OEM/Maritime-Systems/Communications, 2017

4. SPETEC, Maintenance,Materials Management, Retrieved from http://www.spectec.net/maintenance-materials-management, 2017

6. AUTHORS BIOGRAPHY

Hamad Rashid Suwaid holds the current position of Information Technology Manager at Nakilat. He is responsible for the design and delivery of cost-effective, high-performance technology solutions in support of rapid company growth. He has been bestowed with several recognitions such as the CIO 100 award in 2016 and 2017 and 2017 Catalyst IT Man of the Year award for successful Cloud initiatives in the Middle East region.

Synchronised monitoring of sustainability and life cycle costs with a modular maritime IT-platform

R Ahlers, BALance Technology Consulting GmbH, DE
A Fontana, SUPSI - Scuola Universitaria Professionale della Svizzera Italiana, CH
M Petrucciani, Dassault Systèmes Italia Srl, IT
J Cassina, Holonix S.r.L, IT
D Corti, SUPSI - Scuola Universitaria Professionale della Svizzera Italiana, CH
C Norden, BALance Technology Consulting GmbH, DE

SUMMARY

The reduction on caused environmental impacts getting increasingly important in the maritime industry. At the same time, profitability goals and customer requirements must be matched to remain competitive on the global market. Impact assessment of designs in terms of environmental impact and life cycle costs is becoming crucial. Therefore, intelligent assessment tools are needed to provide decision makers with effective information.

The European project MANUTELLIGENCE (Manufacturing Intelligence Engineering Platform) has developed a modular IT platform to optimise the balancing between life cycle costs and environmental friendly production, operation, and maintenance. The challenges are very similar for different industries. Therefore, end users from the shipbuilding industry but also from construction industry, car manufacturing and rapid prototyping are involved to develop a common approach.

The platform allows to assess the consequences of design changes on sustainability and life cycle costs based on different operational scenarios. It supports data collection from supplier as well as from the product operational phase (sensors implemented in the product) to adapt and improve the next product generation.

NOMENCLATURE

AFP	Aerosol Formation Potential
AP	Acidification Potential
BoM	Bill of Material
CED	Cumulative Energy Demand
DfE	Design for Environment
E-BOM	Electronic Bill of Material
EP	Eutrophication Potential
GWP	Global Warming Potential
IoT	Internet of Things
KPI	Key Performance Indicator
LCA	Life Cycle Assessment
LCC	Life Cycle Costing
LCI	Life Cycle Inventory
LCPA	Life Cycle Performance Assessment
NPV	Net Present Value
PLM	Product Lifecycle Management

1. INTRODUCTION

The consideration of environmental impact while at the same time respecting cost efficiency gets increasingly important in the transport sector. Impact assessment of design changes in terms of environmental impact and life cycle costs is becoming crucial. Intelligent tools are needed to provide decision maker with effective information in real time considering all phases of the vessel´s life cycle from design to end of life. Some software tools are available to support the evaluation using economic and environmental data of the vessel and the subsystems. Nevertheless, the evaluation tools are working as standalone solutions which is a serious disadvantage. Cost efficiency and environmental impact have a high dependency. Additionally, all evaluations lack the feedback of operational data. The European funded project MANUTELLIGENCE has developed an approach for an integrated software-architecture combining Life Cycle Assessment (LCA) and Life Cycle Costing (LCC) as a part of the Life Cycle Performance Assessment (LCPA) with IoT data of the operation phase. Therefore, the interconnection with the 3DEXPERIENCE engineering platform was a big step forward to optimise the input data for the above-mentioned life cycle analysis. Besides the engineering platform 3DEXPERIENCE (Dassault Systèmes), the MaGA - LCA tool (SUPSI), the BAL.LCPA – Life Cycle Performance Assessment tool (BALance TC) and the I-LiKe - sensor platform (HOLONIX) were successfully implemented and tested in a real environment. End users from different industries were involved in the project, since LCA and LCPA is not only a challenge for the shipbuilding industry (Meyer Turku), but also for car manufacturers (Ferrari), building contractors (Lindbäcks) and rapid prototyping centres (FCIM - FundacióCIM).

2. General goal of the Maritime Engineering platform

The maritime platform follows a modular approach which can be used for several activities during the vessel´s life cycle and which guarantees simply analysis with quick results. Therefore, the 3DEXPERIENCE platform was taken as a basis for the overall integration. 3DEXPERIENCE is a PLM (Product Lifecycle Management) platform with a single, easy-to-use interface, based on 3D design and modelling, program, and document management, as well as on engineering change management, FEM simulation and CAD integration. It is available in a public or private cloud. Many data generated by the different modules are the input basis for the LCA/LCPA analysis:

- in terms of energy consumption,
- over an entire life cycle for a product using economic and environmental data,
- by choosing different designs and materials.

Within the design phase the designers can use the LCA and LCPA analysis to optimise the design in respect of economic and environmental impacts based on different design opportunities and materials. The design and engineering data are generated by the different 3DEXPERIENCE software modules and used for the analysis by the specific tools MaGA and BAL.LCPA, which are connected to the engineering platform.

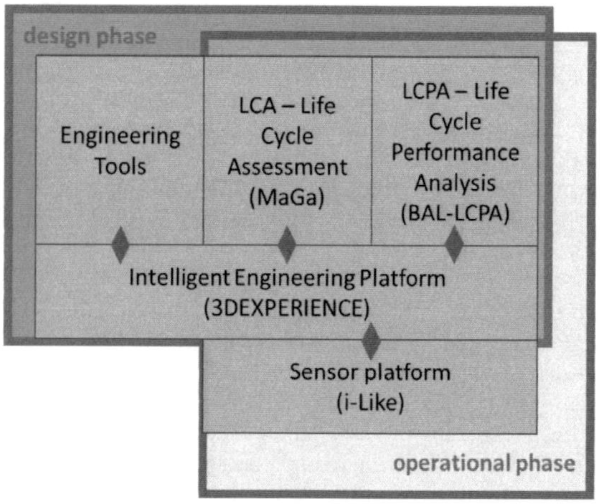

Figure 1: MANUTELLIGENCE Architecture

The platform is not only focusing on the design phase (Figure 1 – red box) but also on the operation phase (Figure 1 – green box) of the platform. The I-LiKe platform connected with 3DEXPERIENCE collects measurement data from sensors installed inside the product (IoT data). Within the MANUTELLIGENCE project, different kind of sensors were installed e.g. in a test car but also in several apartments. The measurement data are used for assessing the environmental impact as well as the life cycle costs to optimize the next generation of products based on real data. As a result, the quality of the LCA and LCPA activities increases based on the real data input.

2.1 Environmental assessment in Maritime Sector

Life Cycle Assessment (LCA) is a methodology that provides quantitative information about the environmental impacts of products and services along their lifecycle, and to evaluate its current application in the maritime sector. The analysis of the available literature on this topic mainly addresses the identification of the needs designers and LCA practitioners may have during the application of the methodology in the environmental assessment of a vessel. In the project particular attention has been paid to understand how tools can effectively support the LCA preparation and providing a basis for the detection of the sustainability assessment software requirements.

2.1.1 LCA Methodology

LCA is a consolidated process that evaluates the environmental burdens of a product, a process, or a service, quantifying the resources used and the emissions released into the ecosphere along the whole lifecycle, and translating this flows and exchanges between the tecnosphere and the ecosphere into impacts [1]. According to ISO 14040 [2] the LCA framework is composed by four phases:

- ***Goal & Scope definition:*** in this phase, the objective of the assessment is identified and declared. Moreover, a set of elements are defined and stated to guide the whole analysis and the interpretation phase: the functional unit, the system boundaries, the assumptions done, the impact assessment categories and the related indicators, and the allocation criteria;
- ***Inventory analysis:*** a list of all inputs and outputs flow in terms of materials and substances related to the studied system is prepared including their quantification. Here, all the resources entered into the system and all the emissions released into the environment along the whole life cycle are collected. This is the most challenging and time/resource consuming phase of LCA;
- ***Impact assessment:*** in this phase, the data gathered in the inventory analysis are classified to the different impact categories selected and translated into impacts and effects to environmental. This evaluation needs to be performed considering a wide range of environmental compartments, in order to cover different problems such as including aspects such as energy use, global warming potential, habitat destruction, acidification, eutrophication, resource depletion, toxicity, etc.;
- ***Interpretation:*** this is the last phase of the LCA, where the results obtained from the impact assessment phase are evaluated and checked, and a set of conclusions and recommendations are prepared.

Nowadays, LCA is exploited to evaluate the environmental impacts of products and services belonging to different industrial sectors: building, food, energy, manufacturing, construction [3]. Since the methodology requires to gather and manage a huge amount of data, a plethora of tools and database are today available to support and automate the information retrieving, the system modelling, the assessment of the indicators and the interpretation of the obtained results.

The scope of LCA is indeed many fold since it enables eco-design, highlighting the system hotspots (e.g. components that cause the highest environmental impact, the most impacting lifecycle phases…), it supports environmental decision making, it avoids problem shifting (moving an impact from a lifecycle phase to another one or from a compartment to another one) and it allows the reliable communication of environmental performances. Within the MANUTELLIGENCE project, LCA is the enabling methodology that ensure designers to evaluate the sustainability performances, providing them with reliable information on how to intervene in order to decrease the global impacts and enabling to assess trade-offs in design (such as those in selecting raw materials, processes, suppliers, etc.).

2.1.2 LCA in the Maritime Sector

The aim of this chapter is indeed twofold: on the one hand, analysing the available literature related to LCA applied in ship-building, it is meant to investigate the use of the methodology and how it has been approached in this particular sector where, because of the characteristics of the product to be evaluated, a huge amount of data need to be collected to perform a reliable environmental evaluation. On the other hand, the literature analysis has the aim to depict the possible problems and open issues to be tackled to support the maritime sector, and its designers, in performing LCA and exploit its potentialities so that the product performances are improved.

Several LCA studies have been conducted in the maritime field evaluating different vessel types and considering different functional units. In 2005, Shama conducted a study [4] emphasizing the importance of using LCA tool to determine the environmental impacts of a ship. The paper describes how to use the LCA methodology and how to exploit it to analyse, propose and evaluate different scenarios to reduce the energy consumption and the environmental impacts of ships during three main phases: i) ship construction, ii) ship operation, and iii) ship scrapping (the end of life of the product). Moreover, Shama suggests that LCA can be a very useful tool to support the Design for Environment (DfE) approach, a methodology meant to reduce environmental impacts during the design phase, emphasising the importance to intervene in this development stage, where there is still time to act. According to this approach, Beizadea et al. [5] present a LCA study of ship, starting from the model and hypothesis developed through SolidWorks software. The study evaluates the environmental impacts of construction, operation (for a period of 20 years on the same route), dismantling and recycling phases stating that the most impacting one is the shipbuilding phase. Chatzinikolaou and Ventikos [6] performed an LCA analysis of a Panamax ship cargo where the main difficulties experimented with the LCA in the maritime sector are identified. The authors highlighted that LCA is a very useful tool to evaluate environmental performances of different products, but in this case, the ship is too complex and its analysis requires to manage a huge amount of data, so they suggested to simplify the model making coherent assumptions. Moreover, this procedure creates another problem related to the reliability of the results considering that some parameters could strongly affect the final environmental impacts. Another stressed problem is related to the lack of specific LCA databases for maritime sector.

A further interesting study has been performed by Fet [7], demonstrating and confirming that LCA-method is applicable for the evaluation of the environmental performance of ship in maritime sector but not without encountering problems especially in relationship to bad data quality, or lack of data that are often taken from computer models, on drawings and assumptions, increasing the probability to also have inconsistent system boundaries. Moreover, Fet also highlighted the feasibility to apply LCA to different ship with different aims (transportation of goods, vessel passengers, etc.).

Eventually, an interesting study has been performed by Kameyama et al. [8] that developed a LCA software for ship sector taking data and information from a predefined database installed together to the software and calculating the impact assessment using the LIME [9] methodologies. The software allows to evaluate the environmental performance considering four life cycle phases: i) shipbuilding, ii) ship operation, iii) ship dismantling, and iv) ship recycling after conducting a detailed inventory analysis about materials, energy consumption and processes [10]. Currently, the software is still a prototype and is not yet available on the market.

The performed literature review shows that LCA methodology can be a useful tool to evaluate the environmental performance of vessels and to support the DfE approach. Nevertheless, problems can be encountered defining functional unit and system boundaries. Despite this, the absence of reliable, effective and user-friendly tools that may rapidly support designer during the conception and the engineering of the ship is registered. Moreover, the absence of specific databases and the huge amount of data to be collected force the LCA designers to carry out important assumptions about the processes, materials and other information that could strongly influence the interpretation phase of the LCA and the accuracy of the final results and of the information obtained.

In Chapter 3.2 a new LCA tool is described. This tool can be useful to overtake the above-mentioned problem performing an LCA for the maritime sector.

2.2 LIFE CYCLE COSTING

Life Cycle Costing (LCC) is an accounting method, which considers every cost flow throughout the lifecycle of a product [11]. LCC is also defined as "an assessment of all costs associated with the life cycle of a product that are directly covered by any one or more of the actors in the product' life cycle (supplier, producer, user/consumer, end-of-life actor), with complimentary inclusion of externalities that are anticipated to be internalized in the decision-relevant future". In the definition, the term "product" should be considered as a reference to a material good, to a service or a product service system: the life cycle concept can be applied without a methodological distinction [12].

In sustainability's terms, the LCC is the economic equivalent of the LCA, it shares with LCA the same life cycle, system boundaries and functional unit concepts [13]. The main objective of the LCC analysis is to maximise the difference between life cycle cost and benefits.

Like the LCA, also the LCC is an interactive process that supports the decision-making analysis. Despite these similarities, there are radical differences in the purpose and approach due to the fact that they arise from different disciplines, namely the environmental and the economic ones [14].

2.2.1 Main Characteristics of LCC

Life cycle costs can be divided into three categories [15]: development costs, utility/service costs and recycling/reprocessing costs. According with the definition provided by the National Institute of Standards and Technology, which defines the LCC as the sum of all discounted total costs of the design, implementation, maintenance and end of the project over a given period, the LCC is characterized by three components within the Life Cycle [13]:

- cost responsibility of the work/investment to be undertaken;
- period of time within the costs occur;
- discounted rate used to discount the future costs at time t0.

Another classification of costs presented by Norris includes [16]:

- direct costs: costs of capital investment, labour, raw material, and waste disposal;
- indirect costs: this kind of costs are not allocable directly to product or process (overhead);
- contingent costs: this category includes cost such as fines and penalties;
- external costs: those costs borne by parties other than the company.

2.2.2 LCC result interpretation

The results of the life cycle cost analysis are also used to optimise the design within an improvement cycle. Possible uses have identified [15]:

- Calculation of total costs for products;
- Identification of cost and revenue drivers;
- Impact outsourcing decisions;
- Cash flow analysis, return on investment (ROI);
- Analysis of what if scenario;
- Optimal due-date for machine replacement;
- Holistic investment budgeting;
- Analysis of trade-offs;
- Analysis of customer lifetime value.

Gluch and Baumann [17] discuss about the usefulness of the LCC approach in making environmentally responsible investment decisions. They state that its practical usefulness is constrained by its oversimplification to a monetary unit and the lack of reliable data on the future. In fact, the environmental consequences of a decision often occur long after the decision was made. There are numerous techniques available that attempt to decrease the uncertainty of future consequences for example scenario forecasting, sensitivity analysis, probability analysis, decision tree and Monte Carlo simulation.

3. MODULAR MANUTELLIGENCE PLATFORM TO REALISE COMPREHENSIVE LCA (LIFE CYCLE ANALYSIS)

3.1 3D-EXPERIENCE – PLATFORM

3D-EXPERIENCE is a PLM (Product Lifecycle Management) Platform, which provides support for business process applications. These applications contain pre-defined schema and processes to support many business industry sectors. A subset of processes that were used for the MANUTELLIGENCE platform are program and document management, engineering change, 3D modelling and FEM simulation and CAD integration.

3.2 MAGA - LCA TOOL

Trying to tackle the open problems of LCA in maritime sector identified in chapter 2.1, the LCA software MaGA (MANUTELLIGENCE Green Application) has been developed by SUPSI within the MANUTELLIGENCE project, that, integrated into the MANUTELLIGENCE Platform presented in chapter 3.1, is meant to allow users to perform LCA during the design of products belonging to different industrial sectors. In this section, the application is presented highlighting the main functionalities that could support designers in the evaluation of the environmental performance of ships. The MaGA functionalities can be mainly grouped in two groups: i) elements meant to facilitate lifecycle modelling and the gathering of the data needed for the assessment, and ii) elements meant to advise the designers and guide their decisions from the sustainability point of view.

Modelling and data gathering elements

In the following the software features that enable the representation of the product model lifecycle and those easing the collection of Life Cycle Inventory (LCI) data are presented. As shown in the literature review, the detailed description of the ship lifecycle with the complete list of the involved processes and the preparation of LCI phase are the most impacting tasks to be faced by practitioners during LCA preparation.

Company specific data management. During the Life Cycle Assessment, LCA experts can collect and define information at different levels (e.g. process, company, country, global, etc.) that are not necessarily related to the single product studied, but that could be inserted into the LCA platform once and then exploited every time is needed to characterize processes or supply chain members. This avoids repeating, for each new project, the data-entry process of company-specific information. For example, a milling operation performed in Switzerland could have different impacts respect to the same operation performed for construction operations in Sweden. Through the MaGA software, and more specifically the *Global editor mode* (Figure 2), the user can insert detailed or general data about the materials, processes, companies, countries, etc. defining the supply chain actors and their related environmental impacts. These data can be reused in several projects and periodically updated (e.g. once a year).

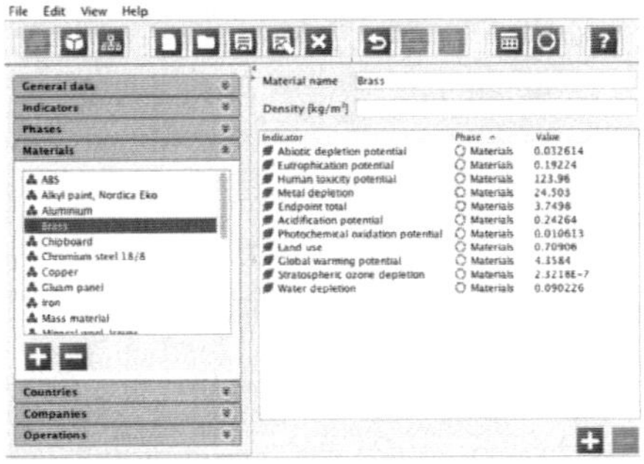

Figure 2: MaGA - Global editor mode

BOM-oriented approach. According to Sorlini et al. [18], the "language" used by MaGA is familiar to designer since it is oriented to the description of every single component, a Bill of Material (BoM) oriented approach. Conversely, most of the LCA software available on the market are based on processes for the description of the product lifecycle, thus a process-oriented approach is used. BOM-oriented approach specifically supports ship designers since they have the possibility to model more easily and rapidly the ship components as they already shaped them through a CAD software. The above-described approach can be implemented through the *Project editor* mode (Figure 3) of the MaGA software. Here, the user can model the products inserting assemblies and components in a hierarchical manner (see the window part indicated with 1 in Figure 3).

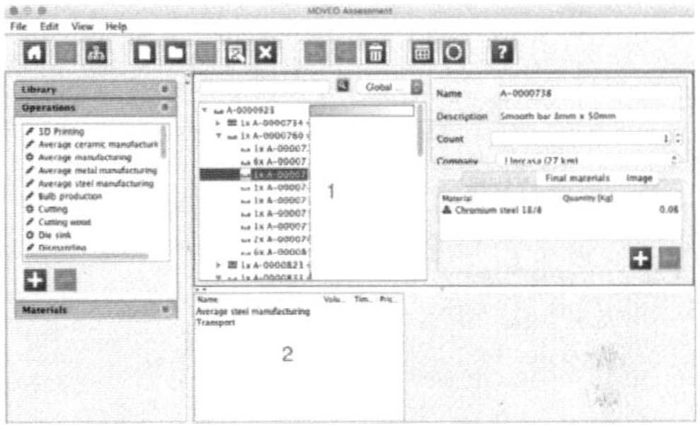

Figure 3: MaGA - Project editor mode

Moreover, the operations accomplished on the single item during its whole lifecycle are inserted in a chronological way through the *Timeline* section (see the window part indicated with 2 in Figure 3). Thus, in the *Timeline* it is possible to describe also the lifecycle phases occurring after the shipyard such as the operation, the maintenance, and the end of life phases.

Interoperability. MaGA has been designed to facilitate its integration with existing platforms and software in order to ease the collection of the data needed for the assessment, share the calculation results to other users or people potentially interested in environmental impacts evaluation (project managers, marketing, company middle and top management) and store them for future utilizations. Through appropriate protocols, the application can be easily connected to external software in order to both receive and transmit data. For instance, thanks to the integration with the MANUTELLIGENCE Platform, the application allows the users to automatically retrieve the E-BOM (the Bill of Material generated inside the MANUTELLIGENCE Platform through the included CAD application) structure and other possible information such as the part quantities, their descriptions, the materials constituting them. This integration supports the drastic reduction of the time needed in the data input process, especially addressing the ship product case since it is composed by many subsystems that in turn are constituted by millions of assemblies and parts. Moreover, the interoperability of the software could be exploited in the near future to further expand the integration with other components of the MANUTELLIGENCE Platform that may provide processes and supply chain information, data that currently are not automatically available.

Data importing. MaGA application allows to import background data from different LCI and Life Cycle Assessment (LCA) databases, such as EcoInvent, one of the principal databases in term of diffusion and data availability and quality. Moreover, the software has been designed to allow the possibility to import information concerning material and processes impacts and company specific data using Excel templates.

LCI and LCA calculation. The last useful element for data gathering is linked to the previous one. In fact, MaGA software allows to create or customize processes from the environmental point of view through their LCI and LCA related data. This element is available in the *Operation editor mode* (Figure 4).

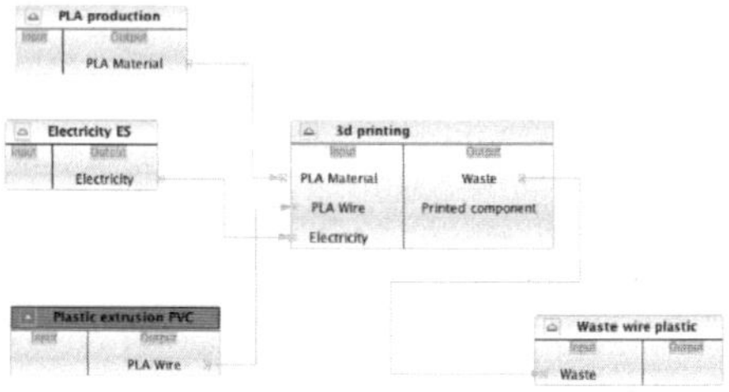

Figure 4: MaGA - Operation editor mode

This mode has been developed to support the creation of custom operations of the product and characterize them from the environmental point of view, starting from zero or modifying imported data. Through this mode, the user can create the desired processes starting from the definition of all its input and outputs, including their environmental impacts, and characterizing the core process with the right amount of input the process requires and the right quantity of output that the process produces. This function is also very useful to model possible service processes that can be hardly found in the current database.

Advisory elements

Hereinafter the functionalities of MaGA related to support the interpretation of the assessment results are presented.

Real-time assessment. MaGA allows the user to view and analyse in real-time the environmental impacts results. This means that the user can instantaneously view how the environmental impacts change modifying material and other information on the parts of the modelled ship. This could be done into a couple of visualization modes: i) directly in the BOM section of the product (Figure 5), and ii) in the two dedicated assessment windows (Figure 6). Concerning the first one, it is possible to select an environmental indicator (upper right part of the window reported in Figure 5) and, alongside each component and assembly, analysing through a bar the environmental contributions of the specific element respecting to the whole product. This mode allows the user to identify parts quickly and easily with high environmental impacts in order to prioritize the effective interventions.

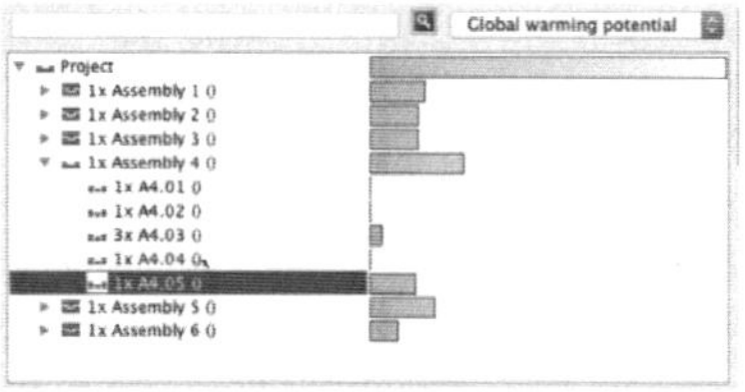

Figure 5: MaGA - BOM Assessment results

On the other hand, the dedicated assessment windows are composed by two visualizations: a tabular and a graphical one (Figure 6). Both provides a detailed representation of the environmental results showing the indicator value and the contributions to that value coming from the different lifecycle phases with a numerical representation (in the tabular view) and a percentage representation (in the donuts-chart view).

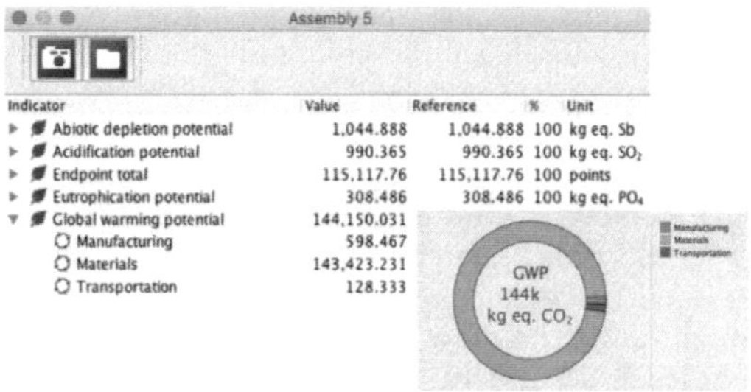

Figure 6: MaGA - Assessment windows

Finally, the MaGA software, besides the possibility to view the environmental impacts related to the whole product, allows the user to analyse and compare the performance at different BOM levels: the entire product, an assembly, a single component or a selection of different parts.

Benchmarking. A further feature related to the interpretation phase of the LCA is the possibility to carry out rapid comparison between different products or components. In fact, through the assessment window (Figure 6) the user can save the current situation of the product through the camera icon visible in the upper left side of Figure 6, in the meanwhile modify materials, process or other elements of the product, come back to the assessment window and view the percentage difference generated between the previous product version and the new one (Figure 7).

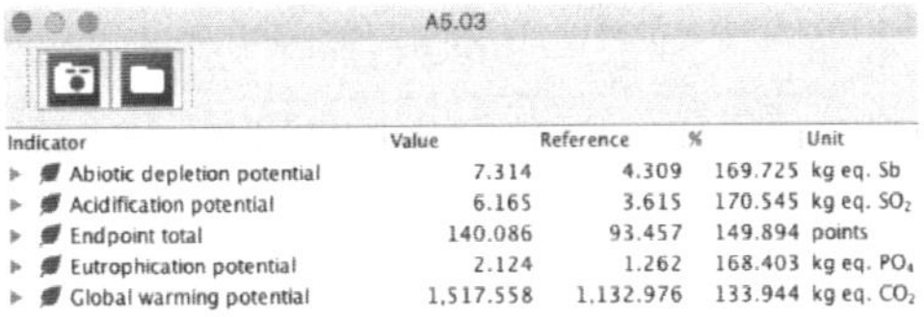

Figure 7: Benchmark representation

Results sharing. As already mentioned during the presentation of interoperability, it is possible to share the calculation of the environmental indicators with other users that are interested in the sustainability theme, ranging from other designers to company management. This is currently performed thanks to the integration with the 3D Experience Platform, where the assessment results are directly uploaded from MaGA allowing the comparison of different versions of the product or different products belonging to the same product category.

Summarizing, the MaGA software can be a very useful tool in order to support LCA designers of vessels. *Data importing* and *interoperability* are relevant functionalities to ease the data inserting process considering the huge amount of parts that are composing a ship. For example, E-BOM can allow to directly import the entire ship structure designed inside the MANUTELLIGENCE Platform and the possibility to plug with multiple LCA database could facilitate the retrieval of impact data. Moreover, when LCA information are not already available, the *Operation editor mode* enables the possibility to create new operation and to characterize them from the environmental point of view, completing the information retrieved from database.

The advisory elements, eventually, can facilitate the environmental-aware design phase of the ship. Thanks to the different ways the assessment results can be displayed, designers can easily and quickly evaluate the environmental impacts of different ship variants considering different levels of details (from the single material, to the single component, to the whole product), whilst supporting and implementing the Design for Environment (DfE) approach.

3.3 BAL.LCPA – Life Cycle Performance Assessment

The development of BAL.LCPA has been started in the course of the EU-funded research project BESST (Grant Agreement No. 233980) (Breakthrough in European Ship and Shipbuilding Technologies, 2013) with the goal to assess the life cycle performance of innovative technologies in European Shipbuilding. Further development with respect to the assessment of the environmental impact of energy and energy converters in shipping has been conducted in the course of the EU-funded research project JOULES (Grant Agreement No. 605190). The overall developed approach investigates several economic and environmental Key Performance Indicators (KPI) like Net Present Value (NPV), Global Warming Potential (GWP), Cumulative Energy Demand (CED), Eutrophication Potential (EP) Acidification Potential (AP) and Aerosol Formation Potential (AFP) in a comprehensive approach that allows to compare, numerically and visually, multiple objects.

The assessment of the life cycle performance starts with the definition of the required life cycle phases, of which the user can determine the number, name, length and accuracy level according to the needs of his assessment. In this regard, the option of operation profiles allows to summarise defined phases and repeat them a determined number of times in order to represent recurring processes, or operation profiles. Life cycle phases can be interrupted by events, which eases the consideration of accidents and repair cases. In the next step, components must be defined and attached to the corresponding life cycle phase, before the associated life cycle data has to be added. Thereby the user can choose from various data types, ranging from the mass of material, to investment costs or annual operational costs to specific energy converters, like diesel engine or generator. Thereby BAL.LCPA considers constant values as well as value timelines in order to account for price or effort increases, like an increase of maintenance costs throughout a product's life cycle or to calculate the life cycle costs considering alternative fuel price forecasts (Figure 8). In addition, BAL.LCPA supports sensitivity analysis to identify the impact of varying input parameters on the NPV. In this regard, the robustness of designs can be evaluated in potential future scenarios. Multiple components can be summarised to a super-component to structure the assessment and to represent systems. Moreover, the flexible approach allows to carry out assessment of the life cycle performance of ships, as well as for airplanes or railway-wagon as proven in the EU-funded research project FIRE-RESIST (Grant agreement No. 246037).

Figure 8: Structure of an exemplary life cycle model in BAL.LCPA

In the same way, processes or services can be considered as components by naming and adding the corresponding life cycle data. The environmental impact is assessed by applying the screening-LCA method, which uses a limited set of data and indicators in order to balance the relation between effort and benefit and to support a quick decision-making process. As a result, the environmental assessment of BAL.LCPA focuses on the KPIs GWP, CED, EP, AP and AFP.

The economic evaluation of the comparative LCPA focuses on the net-present value, the amortisation, and external costs. This flexible approach of the net-present value allows to assess the profitability of an investment at any demanded interest rate. Thereby, each cash flow of the life cycle under investigation is discounted in order to reflect the value of the cash flow considering the point in time of its occurrence. The approach also allows to calculate the undiscounted, aggregated life cycle cost, by setting the discount rate to 0. The duration of the amortisation as an indicator of economic risk can be determined and analysed as a second economic KPI. Moreover, BAL.LCPA calculates the operative external costs based on the emission identified by the screening-LCA or the results of the MaGA tool.

BAL. LCPA supports the result analysis by providing a wide range of result visualisation for each of the introduced KPIs for any specific set of components, KPIs or life cycle phases (see Figure 9). Besides the visual result analysis, BAL.LCPA also generates detailed numeric results, which can be exported as PDF-report.

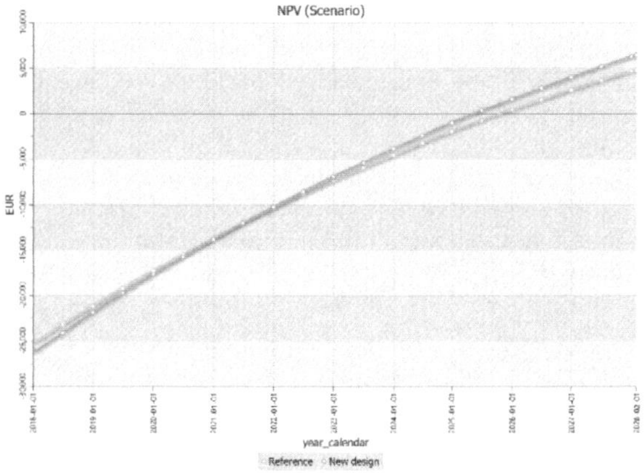

Figure 9: Visual net-present value comparison in an exemplary case study

LCPA is a software for Life Cycle Costing evaluation developed by BALance. The assessment calculation is based on the availability of a Bill of Material (BOM) of the product to be analysed. The MANUTELLIGENCE platform is designed in order to let LCPA import the BOM from 3DEXPERIENCE platform. Once the calculations are completed, the final results can be uploaded in 3DEXPERIENCE platform, making it available to all the users involved.

3.4 Sensor-supported lifecycle analysis I-LIKE - SENSOR PLATFORM

The Holonix I-LiKe is a web platform aimed at retrieving, organising, and visualising all the IoT data that are relevant to know the history and the current status of a machine or product. The core of the solution consists of a cloud platform supporting the following four main tasks:

- Storing of the relevant information collected during the machine or product lifecycle,
- Maintaining a complete representation, at any moment, of the machine or product status,
- Keeping track of the machine or product status history,
- Detecting machines or product alarms and anomalous conditions and notify the users

Within the MANUTELLIGENCE project the I-LiKe platform has been interfaced with 3D-EXPERIENCE. The implementation of an energy monitoring system in the production machinery like a 3D printer has been carried out as well as the collection of temperature and energy data from apartments. Data are sent from the IoT gateway (Raspberry Pi) via Internet connection to the Holonix I-LiKe platform and can be visualized through a simple user interface. The data that represent current and historical status of the product are stored in Holonix I-LiKe platform.

MANUTELLIGENCE explores how embedded systems can be connected to lifecycle analysis. For this purpose, the BAL.LCPA tool is extended with a web service that connects it to the I-Like platform. The I-Like platform contains lifecycle data about the operating conditions of products (including energy consumption data). Operating data, such as energy consumption, are integrated into the LCPA to replace assumptions with more accurate inputs from sensors.

4. MANUTELLIGENCE PROJECT APPROACH

MANUTELLIGENCE intends to improve the product development through cross-disciplinary feedback loops based on modular collaborative secure intelligence engineering platforms. It has been proven that these platforms enable manufacturers to design and develop new innovative products and to improve existing products based on product usage information and customer's wishes. After the implementation of the described components (3DEXPERIENCE, MaGA - LCA tool, BAL.LCPA – Life Cycle Performance Assessment tool and the I-LiKe - sensor platform) several engineering processes have been supported by the platform in different industries.

4.1 DESK LAMP AND ROBOT PROTOTYPE

The design processes as well as the LCA and LCC evaluation using the MANUTELLIGENCE platform have been started with simple products. A desk lamp was designed using 3DEXPERIENCE design tools and optimised in respect of ecological and economical parameters. The MaGA tool used the BoM available on the 3DEXPERIENCE platform for the ecological analysis. Material related LCA data were collected from existing Life Cycle Assessment (LCA) databases. After the assessment, the LCA results were stored on the 3DEXPERIENCE platform. As a next step, the BoM was downloaded into the BAL.LCPA tool and combined with life cycle information to perform the LCC. Material, production, expected service and repair costs were considered within different usage scenarios. The results were uploaded to the platform. Since LCC and LCA have a high degree of dependencies, the optimum is currently only to be determined manually by trial and error. Therefore, the loop of design/material changes, LCA and LCC must be repeated several times. For the desk lamp prototype the design phase and the selection of the best manufacturing process has been supported by the MANUTELLIGENCE platform tools.

The second prototype was more complex. A robot was designed and produced by additive manufacturing using the same processes than for the lamp before. But this time the energy consumption during the production phase has been measured and transferred via the I-LiKe sensor platform to 3DEXPERIENCE. The LCA and LCC for the robot considers the BoM but also took sensor data into account. LCA bases no longer on estimations for e.g. energy consumption but also on real energy measurements. The BAL.LCPA tool is using the sensor data via the 3DEXPERIENCE platform and can clarify the life cycle costs calculation in real time. The use of this approach for the production phase might be limited. But the same application can be used during the operation phase (energy consumption during life time). Both products have been produced by applying additive manufacturing technologies. The MANUTELLIGENCE platform has been used to optimise and assess the design as well as the production phase.

4.2 BATHROOM MODULES DEMONSTRATOR

The construction company Lindbäcks is producing 3-D apartment modules in a flow production process. One of their suppliers deliver complete installed 3D bathroom modules (Figure 10) to the apartment assembly line. The apartment modules as well as the bathrooms are completely produced in a factory. The engineering and production approach is very similar, respectively equal to the production of cruise liner cabins. These cabins are also usually manufactured in a factory as turn-key products and finally installed onboard a vessel. Therefore, it is a good example for further innovations in the maritime sector. The advantage of the apartment bathrooms compared to cruise liner cabins is the fact that the apartments are always at one place and installation adjustments can be done very easily during the test phase.

Figure 10: Production line for batch room modules

The apartment and bathrooms demonstrator is used to gain experiences in a real environment. During the design processes, the 3DEXPERIENCE engineering platform and the related tools MaGA and BAL.LCPA have been used as described for the lamp and the robot prototype. But for this demonstrator the operation phase was more in focus.

The apartment designers usually make assumptions about the conditions, because they lack the information how the product will be operated. In some cases, the assumptions of the designer about the use conditions might be sufficient to plan a product. In other cases, it is favourable to have more accurate data, for instance to avoid over-engineering or to specify the product more user specific. Low cost measurement systems capable of sensing the environment and the user's behaviour can complement the designer's assumptions through measurements. The data can be used to customize the product in-situ or support the design of future versions of the product. The information can also be used to adapt the LCA and LCC assessments by real world measurements. Sensors for temperature, energy consumption and humidity have been installed in apartments and specially in bathrooms. Energy consumption is influencing the cost and environmental impacts of a product very much. The estimation of energy consumption is, in many cases, a matter of assuming the conditions of the products expected use. Conditions include, for instance, environmental conditions and the way users interact with the product.

All sensor data from the apartments were made available on the 3DEXPERIENCE platform via I-LiKe. LCA and LCC assessments are using these measurements to optimise the next apartment generation in respect of ecological and economical parameters.

5. CONCLUSIONS

The MANUTELLIGENCE architecture has proven an efficient way to consider the environmental impact while at the same time respecting cost efficiency for the whole product life time. While using an engineering platform the assessment effort has been reduced and the product optimisation has been improved. Since LCC and LCA have a high degree of dependencies, the optimum (between ecological and economic criteria) is currently only to be assessing manually in several loops. Especially the apartment demonstrator has shown the potential of the assessment in respect of:

- product weaknesses considering over-engineering,
- ecological improvement potentials in the direction of greener products,
- visibility of real production and maintenance costs,
- new material compositions focusing on LCA and LCC improvements,
- weaknesses monitoring over the life cycle (spare parts, unplanned services).

Even if the data on the platform are used for the LCA and LCC assessment, the approach is different depending on the application and the optimization goal. Therefore, the assessment focus must be defined like e.g.:

- environmental impact reduction,
- payback time optimisation by using e.g. alternative materials,
- weight reduction by using e.g. composite materials,
- energy consumption reduction in operation,
- recognising the phases with the highest life cycle impact (production, operation, etc.).

As one example, weight reduction might result in higher material investment and higher emission during the production, but reduces production by pre-production in a factory instead of installation on site and much less emission during the operational phase. All these complex assessments are only possible by using an integrated architecture like developed in MANUTELLIGENCE project. Beside the bathroom calculations also cabin assessments have been done within other projects using the same tools. It is intended to use the approach for the assessment of the whole ship in the future.

6. ACKNOWLEDGEMENTS

The research leading to these results has received funding from the European Union's Horizon 2020 research and innovation programme under grant agreement no. 636951.

7. REFERENCES

1. Guidelines for Life-Cycle Assessment: A 'Code of Practice'; 1st ed.; Consoli, F., Allen, D., Boustead, I., Oude, N. de Fava, J., Franklin, W., Quay, B., Parrish, R., Perriman, R., Postlethwaite, D., Seguin, J., Vigon, B., Eds.; SETAC-Europe: Brussels, Belgium, 1993.

2. ISO14040:2006 – Environmental management – Life Cycle Assessment – Principles and framework https://www.iso.org/standard/37456.html

3. Guinée, J., Heijungs, R., Hupples, G., 2012, Life Cycle Assessment: Past, Present, and Future," Environ. Sci. Technol., 45, 90–96

4. Shama M.A. (2005) "Life cycle assessment of ships", Maritime Transportation and Exploitation of Ocean and Coastal Resources

5. Beizada, E.H., Popa, C., Nicolae, F. (2014) "Applications of Life cycle assessment (LCA) in shipping industry", 14th SGEM GeoConference on Energy and Clean Technologies – Proceedings

6. Chatzinikolaou, S. D., & Ventikos, N. P. (2014, October). Applications of Life Cycle Assessment in Shipping. In Proceedings of the 2nd International Symposium of Naval Architecture and Maritime (pp. 23-24)

7. Fet, A. M. (2002). Environmental reporting in marine transport based on LCA. Proceedings of the Institute of Marine Engineering, Science and Technology. Journal of Marine Design and Operations B (B1), 1476-1556.

8. Kameyama, M., Hiraoka, K., & Tauchi, H. (2007). Study on life cycle impact assessment for ships. Tokyo, Japan.

9. https://www.aist-riss.jp/old/lca/cie/activity/project/lime/index.html

10. Kameyama, M., Hiraoka, K., Sakurai, A., Naruse, T., & Tauchi, H. (2005, October). Development of LCA software for ships and LCI analysis based on actual shipbuilding and operation. In Proc. 6th Int. Conf. on Ecobalance (Vol. 40).

11. International Organization for Standardization, "ISO 15686-5:2008 - Buildings and constructed assets -- Service-life planning -- Part 5: Life-cycle costing." International Organization for Standardization, 2008.

12. Rebitzer, G., Hunkeler, D., Jolliet, O. (2003). LCC – The Economic Pillar of Sustainability: Methodology and Application to Wastewater Treatment. Environmental Progress, 22(4), 241-249.

13. Testa, F., Iraldo, F., Frey, M., O'Connor, R. (2011). Life Cycle Costing, a View of Potential Applications: from Cost Management Tool to Eco-Efficiency Measurement. *Supply Chain Management*, 569-590.

14. Chong, N., Dreijer, L. S., Howard, T. J., Birkved, M., Kreye, M. E., Bey N., McAloone, T. C. (2014). Comparing LCC with LCA to assess PSS sustainability: the case of the eco-box. *International design conference*, Design 2014 Dubrovnik - Croatia, May 19 - 22, 783-790.

15. Niemann, J., Tichkiewitch, S., Westkämper, E. (2009). *Life Cycle Evaluation. Design of Sustainable Product Life Cycle*, 58-59.

16. Norris, G.A., Integrating Life Cycle Cost Analysis and LCA (2001). *International Journal of Life Cycle Assessment 6 (2)*, 118 – 120.

17. Gluch, P., Baumann, H. (2003). The life cycle costing (LCC) approach: a conceptual discussion of its usefulness for environmental decision making. *Building and Environment*, 39(5), 571-580.

18. Sorlini, M., Fontana, A., Alge, M., Diviani, L., Canetta, L., & Specht, I. (2014, September). Sustainability Assessment and Advisory in Mould&Die: Implementation Challenges and Solutions. In IFIP International Conference on Advances in Production Management Systems (pp. 423-430). Springer, Berlin, Heidelberg.

8. AUTHORS BIOGRAPHY

Reinhard Ahlers is Managing Partner at BALance Technology Consulting GmbH and is graduated Electrical Engineering and received his PhD in Production Technology. He is responsible for innovation consulting focusing on technical implementations, re-organization of business processes digitalisation, Life-cycle cost calculation and creation of business plans.

Alessandro Fontana is lecturer and researcher at SUPSI-ISTEPS. Graduated in Material Engineering from Politecnico di Milano, he had a six-year industrial experience in Sustainability implementation, Life Cycle Assessment and Ecodesign of products and manufacturing processes within BTicino S.p.A. (Legrand Group). Currently he is a researcher at Sustainable Production Systems (SPS) Lab of SUPSI where his main research domain is the development of sustainability assessment methodologies in mass production and mass customization context. He teaches Industrial Sustainability issues within MSc. in Engineering and the BSc. in Industrial and Mechanical Engineering.

Maurizio PETRUCCIANI is Senior Project Manager at Dassault Systèmes Italia Srl. and is responsible for the management of the MANUTELLIGENCE project with 12 partners from six different European countries. He received his PhD in Aeronautical Engineering at University of Pisa and manages complex Product Lifecycle Management (PLM) projects in different business areas (aerospace, consumer goods, apparel, etc.) to improve product development processes by reducing time to market, better addressing the target costing, enhance the collaboration, and reducing the inefficiencies costs.

Jacopo Cassina is CEO at Holonix S.r.L., has been graduated Mechanical Engineering and received his PhD in Management and Industrial Engineering at Politecnico di Milano. He is responsible for PLM (Product Lifecycle Management) applied to the product/service, predictive maintenance, and Serious Games. He has expertise in PLM sustainable management and on the assessment of environmental impact of a product/service in its life cycle (LCA).

Donatella Corti graduated in Management Engineering (with distinction) and was awarded a PhD in Management, Economics and Industrial engineering from Politecnico di Milano, (Italy) where she worked as researcher till 2012. Currently she holds the position of lecturer/researcher senior at SUPSI - Scuola Universitaria Professionale della Svizzera Italiana in the Institute of Systems and Technologies for Sustainable Production of the Department of Innovative Technologies. Her main research interests are in the Operations & Service Management field, with a particular focus on global operations management, manufacturing sustainability and servitization of manufacturing. She authored and co-authored a number of scientific publications, including papers in international journal, around 50 papers presented at international conferences, and books chapters. She is responsible for the SUPSI team working on MANUTELLIGENCE project.

Christian Norden is consultant at BALance Technology Consulting GmbH. He is an Economist and worked in a number of international research projects (among others: JOULES, BESST, FIRE-RESIST) responsible for life cycle performance assessment of innovative product designs in the transport area and new materials.

THE CONCEPT OF SPEEDS (SMART PLATFORM OF ENHANCED ENGINEERING DATA FOR SHIPPING AND SHIPBUILDING) AND INNOVATIVE USE OF SHIP 3D DATA

K Hamada, Hiroshima University, JP, **T Hiraki,** Mitsubishi Heavy Industries, Ltd., JP, **M Nagano,** Japan Marine United Corporation, JP and **M Ozaki,** NTT DATA ENGINEERING SYSTEMS Corporation, JP

SUMMARY

Although 3D-CAD has been used in the shipbuilding industry for a long time, effective use of 3D data in the production and new initial design based on the 3D model are not advanced compared with the expectation. Moreover, it is difficult for a ship owner and a classification society to use the 3D model effectively because CADs and the security policies are different according to the shipyard.

Considering the above discussions, Japan society of naval architects and ocean engineers set the project research committee and has examined the information platform named SPEEDS (Smart Platform of Enhanced Engineering Data for Shipping and Shipbuilding). SPEEDS aims to realize the sharing and exchange of various information on ships among the all maritime industries including shipyards, classification society, ship owner, operator and the marine equipment manufacturers. A simplified 3D geometry information and a variety of attribute information are described in the SPEEDS and the player in maritime industries can utilize these information promptly and effectively based on the various use cases of ship lifecycle.

In this paper, overview of SPEEDS is discussed with some examples intended for the use cases in shipyards.

1. INTRODUCTION

Shipbuilding design adopted the use of 3D CAD a long time ago; however, even within shipyards, there can be problems involving the practical use of 3D information in production or difficulties converting upstream designs to 3D. Even in the ship owner and classification society, each shipyard has its own CAD and security policies, which makes the application of 3D information difficult.

With the above background, the Japan Society of Naval Architects and Ocean Engineers set the Project Research Committee P-40 "Committee to examine the common format of ship 3D information to realize the intelligence sharing and advanced use of ship 3D information". This committee has been conducting a study to formulate a standard for maritime industry internal information exchange so that each player in the maritime industry—including marine equipment manufacturers, shipyards, ship owners and classification societies and research institutes—can quickly and efficiently exchange various items of 3D information created throughout a ship's life cycle, from building to operation[1]-[4].

In this presentation, we give an overview of the information sharing platform "Smart Platform of Enhanced Engineering Data for Shipping and Shipbuilding" (SPEEDS) and discuss practical examples of its application to use cases.

2. BASIC CONCEPT OF SPEEDS

2.1 FOR EFFECTIVE USE OF DATA IN THE MARITIME INDUSTRIES

For effective use of the big data associated with a ship's life cycle, it is necessary to evaluate information exchange among equipment manufacturers, shipyards, ship owners and the classification societies. In addition, industry/business-level information sharing becomes necessary. To identify the effectiveness of and issues involved in information sharing, we carried out a questionnaire survey of shipyards across Japan. As a result, it was identified that although an expectation for information sharing exists, providing 3D information may result in leakage of confidential information such as the ship model, and resolving this issue is a necessary condition for the realization of information sharing.

Therefore, this research committee discussed the sharing of ship 3D information with shipyards, ship classification societies, shipping companies, and public institutions. As a result, the following information was obtained:

- The 3D information that shipping companies need is not so detail as the geometry information in 3D CAD format that the shipyards had assumed.
- Shipping companies need various items of attribute information more than they need detailed 3D information.

- The 3D information offered to ship classification societies and related companies may be somewhat detailed.
- By managing the level of detail of the 3D information offered by shipyards, it is possible to prevent information leakage.
- In order to share information and use it effectively, it is important to offer attribute information other than 3D information.

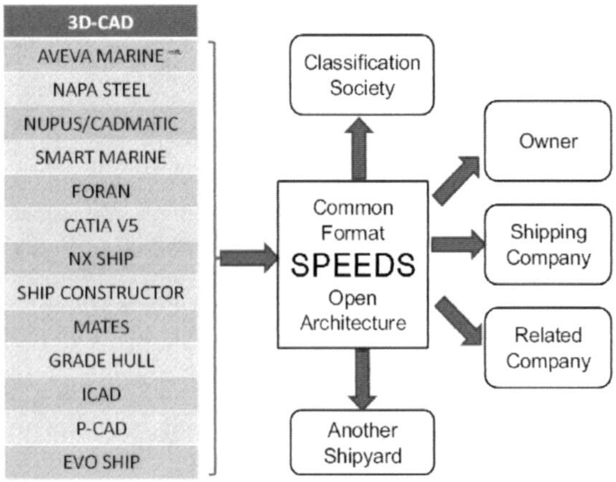

Figure 1: Basic Concept of SPEEDS

2.2 OVERVIEW OF SPEEDS

In order to make effective use of big data in the maritime industry, it is necessary to implement a structure for the circulation and accumulation of data, which will be the core of information sharing at the industry level. Therefore, this research committee proposes Smart Platform of Enhanced Engineering Data for Shipping and Shipbuilding (SPEEDS), an information sharing platform in the maritime industry.

A conceptual diagram of SPEEDS is shown in Figure 1. SPEEDS is a ship product information sharing platform with simplified ship 3D geometry information and various items of attribute information, output in CAD by each company. At the time of this output, the output items and functions are limited according to the use case (ship classification approval, provision of information to related companies, for inspector's use, 3D model for shipboard installation, etc.). Information and functions are restricted, but compatibility with the CAD / viewer of each company is maintained. Additionally, by making it an open architecture that reveals its basic structure, each company related to the maritime industry will be able to freely develop analysis software based on SPEEDS and stimulate competition.

2.3 FOR THE IMPLEMENTATION OF SPEEDS

In order to implement SPEEDS, it is necessary to make an objective evaluation of the expected use cases and the level of detail of functions and information required for each use case. Regarding geometry information, various kinds of viewers have been implemented in recent years, and these can be used for information structure. Thus, in this research committee, we established the following two working groups to study the possible use cases and their corresponding necessary functions:

- Working Group for the Advanced Use of 3D Information within Shipyards; Studies the expected use cases and information needed for the implementation of information sharing and advanced use among shipyards and related companies.
- Working Group for Integrated Utilization of 3D Information in Maritime Industries; Studies the expected use cases, necessary information, and feedback to the shipyard needed for the implementation of information sharing and advanced use between shipyards and classification societies / shipping companies.

3. FOR THE INFORMATION SHAREING WITHIN THE SHIPYARDS

3.1 BASIC CONCEPT

"WITHIN SHIPYARDS" means our own shipyards and companies directly related to design and construction. In the study of SPEEDS' information requirements, the following hull and outfitting elements were considered:

- Hull
- Outfitting: Piping, pipe parts, equipment, iron fittings, ductwork, cable tray, maintenance space, pipe support

The operation considered was the design and manufacturing operation process carried out in the shipyard, from basic design to construction. For each item, we extracted the information item considered necessary and evaluated its content. Because each company uses a different 3D CAD, even the same information item may have different details in content. Therefore, in regulating content details of information items, it is necessary to exclude content for specific 3D CADs and information exclusive to shipyards (e.g., information exclusive to a specific piece of production equipment) and make industry-standard content that has a common denominator among all companies and 3D CADs.

3.2 USE CASES

A 3D CAD contains various pieces of information, but this is not to say that every piece of information is essential for practical use in actual situations. Each expected scenario requires a different set of information. Therefore, regulating usage scenarios—that is, use cases—is extremely important for developing it into a practical application based on the information requirements stipulated in this study.

In the first part of the evaluation, the use cases were divided into the four major groups of basic design, detail design, production design, and production department. Then, for each group, 14 cases were examined as more detailed and concrete usage scenarios focused on specific objectives. For this paper, we selected eight cases—the four major cases and four usage scenario cases—and we present an overview of these. In addition, Table 1 shows excerpts of the correspondence table between regulated information items and use cases. In the table, for each use case, filled circles denote essential information items, and empty circles denote desired information items.

3.2 (a) Basic Design

The hull makes it possible to view and display the geometric shape and attributes of the main structural components described in the Key Plan. The attributes allow the display of plate thickness (in the case of plate part), material properties, scantling (in the case of beam part), and weight and center of gravity. Production information such as plate thickness direction and block name / assembly name does not need to be displayed. The outfitting makes it possible to view the spatial arrangement status of large-diameter pipes, large equipment, the main cable tray, etc., which are subject to arrangement evaluation at the initial planning stage.

3.2 (b) Detail Design

The hull makes it possible to view and display the geometric shape, and attributes of each structural component described in the detailed drawing.

The attributes make it possible to display the plate thickness (the case of plate part), material properties, scantling (in the case of beam part), weight and centre of gravity, and plate thickness direction. Production information such as block and assembly name does not need to be displayed. The outfitting makes it possible to view all pipes, equipment, cable tray, and ducts subject to an integral arrangement, with or without display by type or system.

Table 1: Correspondence between Information Items and Use Cases (within Shipyards)

Part	Information Item	Description	Use Case							
			Major Group				Specific Case			
			Basic Design	Detailed Design	Production Design	Production Department	Connection with Analysis Tools	Hull Assembly	Outfitting Installation	Totaling Painted Area
Hull	Geometry	・Polygon ・Mesh(Surface) ・Primitive ・Box ・Cylinder ・Cone ・Torus, etc	◎	◎	◎	◎	◎	◎	◎	◎
	Comments	Comments	○	○	○	○	○	○		○
	Hull Parts Type	・Plate Part (Planar) ・Plate Part (Curved) ・Beam Part ・Longitudinal Frame ・Face ・Bracket ・Pillar	○	○	○	○	◎	◎		○
	Thickness (Plate Part)		○	◎	◎	○	◎	◎		○
	Plate Thickness Direction	Direction (F,A,P,S,U,L) Vector		○	◎	○		○	◎	◎
	Scantling (Beam Part)	・Web Height ・Web Thickness ・Flange Width ・Flange Thickness	○	◎	◎	○	◎	◎		
	Material	AH36, etc	○	◎	◎	○	○	○		○
	Block Name	Block Name			◎	◎		◎	◎	○
	Part Name	Part Name			◎	◎		◎	○	○
	Assembly Name	Assembly Name			◎	◎		◎		○
	Weight	Weight		○	○	○		◎		○
	Center of Gravity	Center of Gravity		○	○	○		◎		○
	Free Edge Treatment	・1C ・2C ,etc			○	○		○		

3.2 (c) Production Design

The hull makes it possible to view and display the geometric shape and attributes of each structural component displayed in the detailed drawing (working plan). The attributes make it possible to display structural attributes such as plate thickness (in the case of plate part), material properties, scantling (in the case of beam part), weight and center of gravity, plate thickness direction, etc., as well as production information such as block and assembly name, etc.

The outfitting makes it possible to view all the pipes, equipment, cable trays, ducts, and supports subject to production design, with or without display by type, system, or pallet basis after pipe splitting.

3.2 (d) Production Department

The hull makes it possible to view and display the geometric shape and attributes of each structural component described in the detailed drawing (working plan). The attributes make it possible to display structural attributes such as plate thickness (in the case of plate part), material properties, scantling (in the case of beam part), weight and center of gravity, plate thickness direction, etc., as well as production information such as block and assembly name. It also enables the display of construction confirmation considering the construction stage such as large assembly and mid- assembly, as well as the weight and center of gravity.

Outfitting makes it possible to view all the pipes, equipment, cable tray, ducts, and support that are subject to production design, with or without display on a pallet basis after pipe splitting.

3.2 (e) Connection with Analysis Tools

Information on shape and attributes required in the analysis is retrieved from 3D Viewer and then goes through edition and addition of data before being used in tools for the analysis of strength calculation and propulsion performance. It is preferable that the data retrieved from 3D Viewer are in the same format as the input to the analysis tools or in a format that is easily editable.

Then, the analyzed result is returned to 3D Viewer to be used for improvement and discussions by sharing it with the design, quality control, and manufacturing departments. For example, data for structural analysis in hull design are shared with downstream production design and are used for early strength evaluation of block loading and reversing operations.

3.2 (f) Hull Assembly

"Hull assembly" means assembly → Ground assembly → In-dock assembly. The assembly procedure, weights, work posture, etc. for each process is checked with 3D Viewer beforehand. If an error occurs, it is possible to use 3D Viewer to ascertain whether it was a design error or an on-site error. Initially, 3D Viewer will be used alongside the drawing or as its auxiliary tool, but in the future, we will try to achieve a drawing-less workflow (using only 3D Viewer information) and 3D model visualization.

3.2 (g) Outfitting Installation

"Outfitting installation" means the internal and external work as well as the entire process of installing outfitting items in the dock. 3D Viewer will be used mainly to confirm the instructions for the installation of pipes and pipe support. We will try to achieve a drawing-less process such as that for the hull.

3.2 (h) Totaling Painted Area

"Totaling painted area" means the task of summing the areas painted with different specifications for the purpose of estimating and evaluating the painting process. This totaling of an area by paint specification is carried out with 3D Viewer. The geometric shape and area of the hull and outfitting components to be painted can be viewed with the basic function of 3D Viewer. Using the 3D Viewer function that sorts the component surfaces by paint specification (for plate material, both sides are included), as well as the function that sums the areas of each paint specification, it is possible to sort and sum while performing a visual check.

4. FOR THE INTEGRATED UTILIZATION OF 3D INFORMATION IN MARITIME INDUSTRIES

4.1 BASIC CONSEPT

Through the SPEEDS platform for the implementation of information sharing, each player in the maritime industry will be able to use various items of information over the life cycle of a ship. This will not only accelerate shipbuilding design, manufacturing, operation, and maintenance management but also stimulate new designs based on feedback on manufacturing, operation, and maintenance results as well as new services derived from analysis of big data collected in a common data format, thereby leading to revitalizing and innovation in many fields of the maritime industry.

Based on the circumstances outlined above, the Working Group for Integrated Utilization of 3D Information in Maritime Industries carried out an evaluation to identify the new forms of using 3D information (referred to as use cases from this point on) brought about by the implementation of SPEEDS—especially between shipyards and classification society / shipping companies—as well as the information (including 3D information) and functions required in each use case. In addition, we organized the information requirements using the same format as the chart used in Section 3, in a way that does not compromise the commonality.

4.2 USE CASES

Five use cases were selected from the keywords mentioned from the perspective of the ship classification society, shipping companies, and shipyards, and a discussion was held on the organization and functions of their information requirements. Introduced below is the objective and an overview of each use case. Table 2 shows excerpts of the correspondence table between information items and use cases.

4.2 (a) Plan Approval

If plan approval becomes possible with 3D information, the upstream design process of shipyards, which is mainly carried out with the 2D drawing, may suffer significant changes. As large-scale analyses are required by H-CSR and others, if it becomes possible to create 3D models concurrently with the 2D drawing even in upstream design, and if the approval process can be carried out on the 3D model as is, it may contribute to a considerable reduction in design lead time.

4.2 (b) Viewer for Construction Supervisor

A viewer for the construction supervisor could be used to record daily inspection activities and to report work progress to the head office. In terms of functional requirements, it would have the standard functions that a general 3D viewer is equipped with.

4.2 (c) Periodic Inspection Records

Once the ship is operational, a single periodic inspection results in the accumulation of an enormous quantity of data, which involves up to tens of thousands of plate thickness measurements, pictures, and paper records of repair drawings and repaint drawings. Converting these into a database from which users can freely extract information will be of considerable benefit to shipping companies for maintenance management and inspection preparation. In addition, if the 3D model is used as an interface to access data such as Google Maps, an even wider data application, such as for connection with structural analysis, may be expected. A prerequisite for using the application is to input the enormous database of plate thickness measurements into the database and link it to the 3D information, which currently is a time-consuming task. However, with the recent development of Internet of things (IoT) technology, the use of monitoring technology with sensors and image analysis technology may drastically reduce the input load in the coming years.

Normally, shipyards are not involved in periodic inspections, but if shipping companies cooperate to build an environment in which they can obtain the age deterioration status and operation records of their own ships from the SPEEDS common platform, we believe it will result in new perspectives on construction specification, design, and quality, which will lead to further development of shipbuilding technologies and allow shipping companies to develop better ships.

4.2 (d) Repair Records

The repair records application is similar to that of periodic inspection records 4.2(c), but by the addition of a link to a damage database, the range of analysis widens even further. Because of factors such as replacement of components, which demands plan approval-level information, repair records will require a higher information density than periodic inspection records.

In addition, with regard to the addition of new equipment mainly used in ballast water management systems and for replacement of machinery, if marine equipment manufacturers can offer 3D information of equipment in a data format that conforms to SPEEDS specifications, it will be possible to use it together with the ship 3D information to adjust the equipment specifications, study its layout, and plan its installation in advance.

4.2 (e) Crew Training

Ship crews periodically switch to working on a different ship, and since each ship has its own layout, specifications, and operating mode according to the ship type and the shipyard where it was constructed, they require some adaptation time before they can execute the tasks smoothly. If operation manuals, safety guidance, and maintenance records using 3D information can be used, it will serve as support for the adaptation process. On the other hand, there is an issue of high data creation load, since every ship will require its own content to be created.

Table 2: Correspondence between Information Items and Use Cases (Integrated Utilization)

Part	Information Item	Description	Use Cases				
			Plan Approval	Viewer for construction supervisor	Inspection Records	Repair Records	Crew Training
Hull	Geometry	Polygon Mesh (Surfece) Primitive (Box, Cylinder, Cone, Torus, etc)	◎	◎	◎	◎	◎
	Comments	Comments	◎	◎	◎	◎	◎
	Hull Parts Type	Plate Part (Planar) Plate Part (Curved) Beam Part Londitudinal Frame Face Bracket Pillar	◎	◎	◎	◎	○
	Thickness (Plate Part)		◎	◎	◎	◎	◎
	Plate Thickness Direction	Direction (F,A,P,S,U,L) Vector	○	◎	○	○	
	Scantling (Beam Part)	Web Height Web Thickness Frange Width Frange Thickness	◎	◎	◎	◎	○
	Material	AH36 etc.	◎	◎	◎	◎	◎
	Block Name	Block Name		◎	○	○	
	Part Name	Part Name		◎	◎	◎	○
	Assembly Name	Assembly Name		◎	○		
	Weight	Weight					
	Center of Gravity	Center of Gravity					
	Free Edge Treatment	1C 2C etc.	◎	◎	◎	○	

5. DEVELOPMENT OF PROTOTYPE SYSTEM

5.1 SPEEDS BASED APPLICATION DEVELOPMENT

Currently, most shipbuilding companies use CAD systems that support 3D data. However, there are so many different products that even in the same shipyard, upstream and downstream designs may differ, and in many cases, the hull and outfitting were designed with different CAD systems. One of the major issues in this kind of operation is the data linkage and compatibility between CAD systems, in particular, the development of individual interfaces for peripheral applications that use CAD system data.

In order to resolve this issue, SPEEDS selects the most typical use cases for purposes other than design work and evaluates the intermediate files common to all shipyards, and then discloses this intermediate file with the SPEEDS XML Schema specification. The CAD system of each company then outputs its files into this SPEEDS XML Schema file. On the application side, as long as compatibility with the SPEEDS XML Schema file is maintained, the information from all CAD systems can be used by a single application. Figure 2 shows the differences between conventional development as performed to date and development based on SPEEDS.

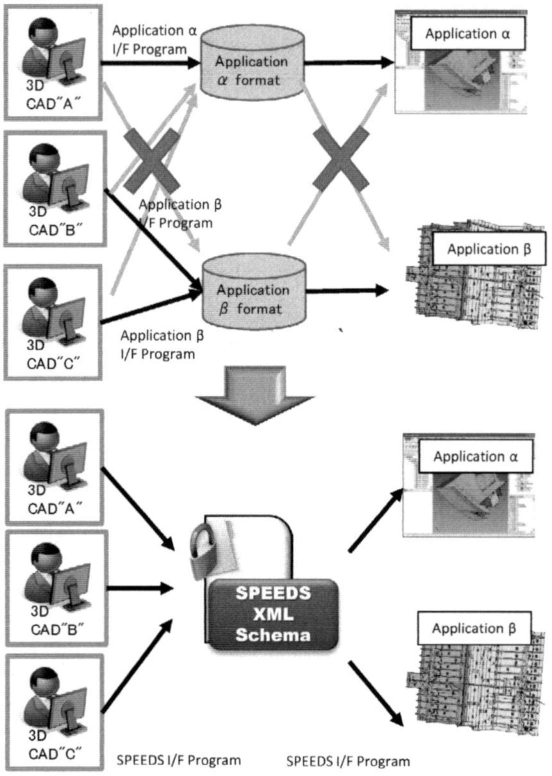

Figure 2: Application Development based on SPEEDS

5.2 DEVELOPMENT OF PROTOTYPE FOR PAINTING STAGE

5.2 (a) Expected Use Cases

Here, we consider a system that covers painting processes and that functions as an application system even if the attribute information retrieved from a CAD system is minimal. Using this application system as a base, we will develop a prototype of an intuitive system with which it is possible to calculate areas painted with different specifications and give on-site work instructions through the use of 3D models. The required items are as follows:

- A function to allocate components to paint sections through the use of 3D geometric information of hull components and thereby to automatically calculate the painted area of each block by paint section.
- A system that can be operated intuitively on site with the information above and that understands the painting range. The system must also be capable of digitizing field work instructions using the visualization of PSPC paint thickness measurement points.
- Compatibility of the prototype system with the CAD of each company and capability of being commonly used in various shipyards.

5.2 (b) SPEEDS XML Schema

We developed the SPEEDS XML Schema, the core of the information sharing, for paint planning and the paint work support process. The three characteristics of the SPEEDS XML Schema developed are as follows:

- The information structure of the schema as a whole is based on the Ship 3D data format developed by ClassNK[5]. The use of the Ship 3D data format may lead to the following results:
  - Instead of using a new, unique format, information will conform to a widely known data format that makes it easier to communicate with the world.
  - It will be possible to externalize the maintenance of a long-term data format, including expandability.
  - The data structure remains neutral rather than being based on the perspective of shipyards and the vendor in charge.
- Geometry information uses polygons with consideration of lightness and ease of handling rather than of increasing the accuracy of geometry information more than necessary.
- Attribute information is based on that specified in the SPEEDS requirements definition document.

Figure 3 shows part of the information structure and attributes of the SPEEDS XML Schema that was developed.

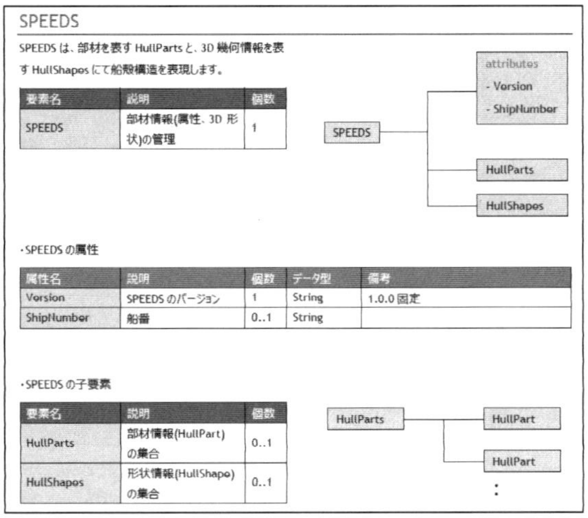

Figure 3: SPEEDS XML Schema for Painting Stage

5.2 (c) Development of Interface with the CAD System

To verify the possibility of common use among shipbuilding companies, we selected three types of CAD system: MATES, used by Mitsubishi Heavy Industries (MHI); Grade Hull, used by Japan Marine United (JMU); and AVEVA MARINE, used by Kawasaki Heavy Industries (KHI). We developed an interface program that converts these CAD data to intermediate files that conform to the SPEEDS XML Schema. An interface program is developed individually by each company for its respective CAD system.

5.2 (d) Development of Interface with Application

Using a commercial software product called Beagle Paint, sold by NTT Data Engineering System Corporation , we developed a prototype of a SPEEDS-version paint support system (referred to as SPEEDS Paint from this point on). It has a direct interface with Beagle Paint and Grade Hull but not with MATES or AVEVA MARINE. By building an interface with SPEEDS instead of a direct interface with MATES and AVEVA MARINE, we developed a SPEEDS Paint that supports MATES and AVEVA MARINE as well.

5.3 EXAMPLE OF APPLICATION OF THE PROTOTYPE SYSTEM

Using the SPEEDS concept, in this section we verify that the geometry information and various items of attribute information outputted from the CAD system of each company can be handled in the same way on SPEEDS Paint, which uses the SPEEDS data format as a base. We also confirm that it works effectively as an application system with the data format adopted.

First, MHI, JMU, and KHI created geometry and attribute information using their respective CADs. Then, using the above-mentioned interface programs, we created SPEEDS intermediate files, which were uploaded to SPEEDS Paint. The result is shown in Figure 4. In Figure 4, different CAD geometry and attribute information created by MHI, JMU and KHI are shown displayed on SPEEDS Paint simultaneously. In other words, using the SPEEDS XML Schema file, it is possible to handle CAD data from each company using the same format. Figure 5 shows the visualization of the painting process using information from AVEVA MARINE, and Figure 6 shows the visualization of PSPC paint thickness measurement points using information from MATES. This confirms that the CAD data of each company can be used in common and that the calculated total area and paint thickness point measurement position are sufficiently accurate for actual operations; in other words, it shows that the data structure of the SPEEDS XML Schema is a practical standard.

In addition, although it was not the main objective of this prototype development, we verified that using the component weight and center of gravity data saved in the SPEEDS XML Schema file, it is possible to verify the block weight and center of gravity, as shown in Figure 7, and that it is sufficiently accurate for practical applications.

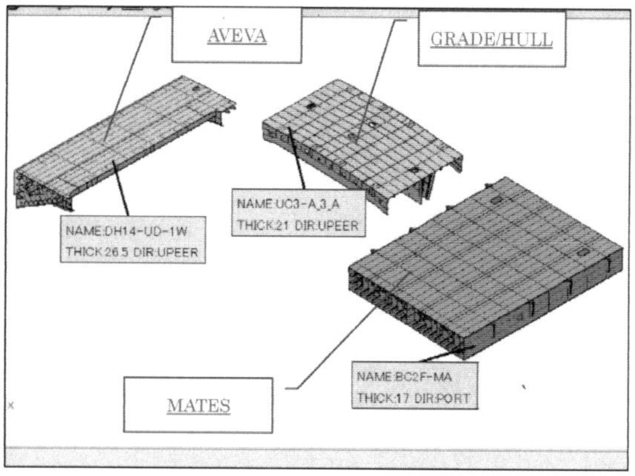

Figure 4: Three different CAD Data in SPEEDS Viewer

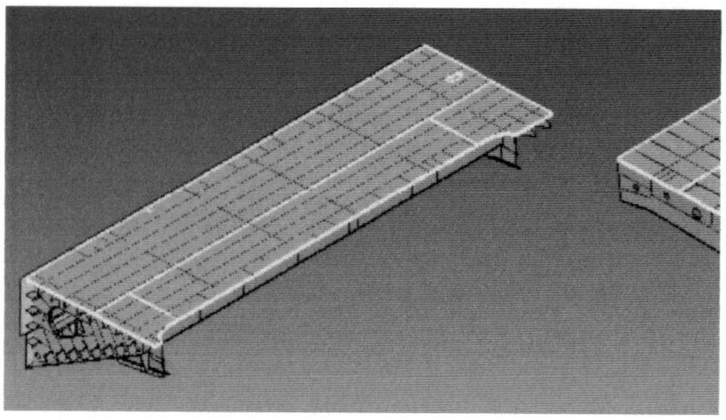

Figure 5: Visualization of the Painting Process using Information from AVEVA MARINE

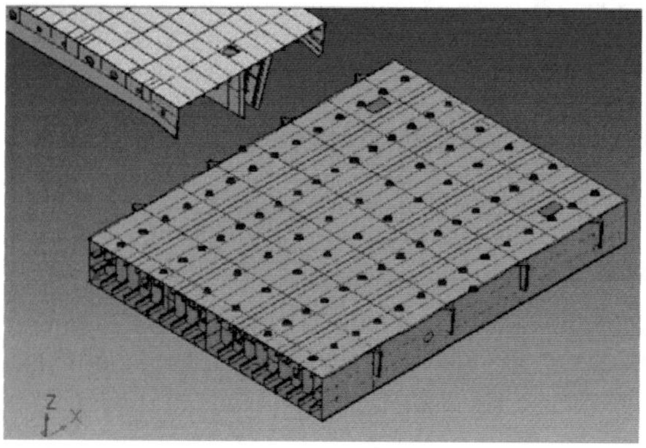

Figure 6: Visualization of PSPC Paint Thickness Measurement Points using Information from MATES

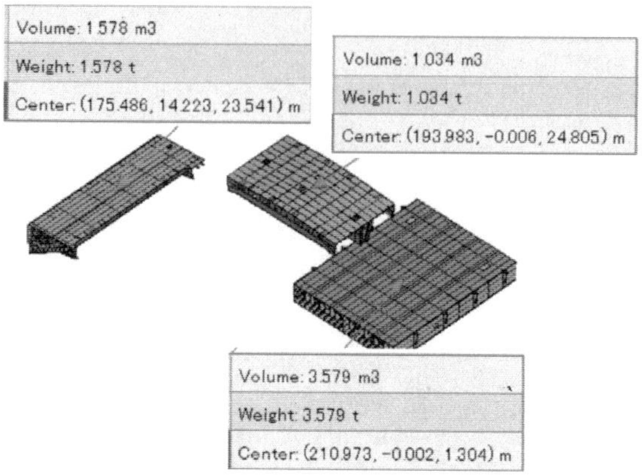

Figure 7: Block Weights and Center of Gravity

6. CONCLUSIONS

Using the SPEEDS concept, by building an interface with a common intermediate file, it is possible to develop a shipbuilding application system targeting multiple CADs. From the vendors' perspective, this means not only higher development speed but also that a single development could reach a greater number of shipyards. Therefore, the shipbuilding software market becomes more appealing. From the shipyard side, there will be more types of software that can be used, which may stimulate the shipbuilding software industry and contribute to information sharing that does not disclose the intellectual property of shipyards to other companies unnecessarily.

However, since this development was limited to the painting process, expansion to other use cases and mitigation of security problems will require further studies in the future. In addition, implementing SPEEDS involves non-technical issues such as standardization, maintenance, and management systems, which we will study in the future.

7. ACKNOWLEDGEMENTS

This research is supported by i-shipping Project (Ministry of Land, Infrastructure, Transport and Tourism, Japan) and the Society of the Japan Society of Naval Architects and Ocean Engineers. Also, this work is the collaborative work of all members of P-40 research committee. Authors would like to appreciate them.

8. REFERENCES

1. K. Hamada, 'Activity report on the Project Research Committee P-40: Committee to examine the common format of ship 3D information to realize the intelligence sharing and advanced use of ship 3D information', *Conference proceedings of JASNAOE*, 2015(in Japanese).

2. T Hiraki, 'Result of the Project Research Committee P-40 Working Group for the Advanced Use of Ship 3D Information within Shipyards', *Conference proceedings of JASNAOE*, 2015(in Japanese).

3. M. Nagano, 'Result on the Project Research Committee P-40 Working Group for Integrated Utilization of 3D Information in Maritime Industries', *Conference proceedings of JASNAOE*, 2015(in Japanese).

4. M. Ozaki, 'An Outline and Conceptual Image of 3D Viewer for Sharing Ship's3D Information', *Conference proceedings of JASNAOE*, 2015(in Japanese).

5. ClassNK, 'Research and development on common format and interface program of CAD data for hull structures', http://report. classnkrx.com/, 2014 (in Japanese).

9. AUTHORS BIOGRAPHY

K Hamada holds the current position of professor at Hiroshima University, Japan. His major is design and manufacturing support system for the shipbuilding.

T Hiraki holds the current position of Team Manager at Ship & Ocean Engineering Department, Shipbuilding & Ocean Development Division, Mitsubishi Heavy Industries, LTD. He is responsible for MATES system development.

M Nagano holds the current position of senior staff at Information Systems Group, Japan Marine United Corporation. He is responsible for maintenance and developing of 3D-CAD and Viewer systems, and some production support systems for outfittings.

M Ozaki holds the current position of Department Manager at System Development Department 2,NTT DATA ENGINEERING SYSTEMS Co. Ltd. He is responsible for Marine systems & solutions development.

SMARTER PRODUCTION, PANEL LINE OPTIMIZATION

K C Salmi, Hexagon PPM, Singapore

SUMMARY

This document discusses panel line optimization and the importance of panel line for fabrication yards. There are several requirements and constraints before any optimization can be performed. Different panel line optimization criteria affect the optimization results. Results can be further analysed and used to execute scheduling for the panel line. Analysed data provides information about the panel line performance, bottlenecks, and other areas of yard that can be improved.

1. INTRODUCTION

Panel line plays a key role at any given fabrication yard. While competition between shipyards grows bigger and processes are being optimized there are still further steps that many fabricators tend to overlook. Optimizing panel line can tremendously increase the throughput, improve overall quality and save money. There are few key performance indexes that define the optimization rules. Multiple requirements must be considered to optimize panel line so that it can support other functionalities of the shipyard.

2. REQUIREMENTS FOR OPTIMIZATION

Before any panel optimization can be carried out there are important things that needs to be known. Panel line is the most important factor that will determine many aspects of the optimization. Design system needs to provide information about the panels that are optimized. Knowledge of plate and profile part prefabrication gives additional parameters for the optimization that can improve the overall efficiency of panel line. Internal logistics of the fabrication shop is the last aspect for ensuring that all parts will be available at the panel line before panel assembly begins.

2.1 STUDY OF PANEL LINE

Panel lines consist of several work stations such as tack welding, final welding, stiffener mounting, bevel cutting and marking. Every work station has a work process, and study of work processes are finally translated into mathematical equations. Each equation needs to return a value of work time in minutes.

Work tasks at each work station at the panel line are either manual, semi-automatic, or fully automatic. Automatic processes such as robotic welding time can be estimated very accurately, but when it comes to manual work tasks that are carried by humans there is always a factor of inconsistency. Work execution times for each work station are only best estimates that will be used for the panel line optimization.

Panel lines can have restrictions when it comes to size of the panel in question. Another typical restriction is that panels cannot overtake each other at the panel line, so the process is only flowing into one direction. Amount of buffer space between work stations limit the queue size for panels. Some work stations might be halted when panel buffer space is full between two work stations, and the other work station is not able to take new works under progress.

2.2 KNOWLEDGE OF PANELS

Panels consist mainly of plate parts, stiffeners and T-girders that are joined together at the panel line's different work stations. Figure 1 illustrates a sample 3D panel drawing.

It is required to know what plate parts are in panels. Which plate parts are joined together and what are the type of welds and welding distances. This information enables the estimation of welding time requirements. Finally it is necessary to know the full panel plate geometry together with the offset margins, and how panel is joined together with other panels. From this knowledge design systems can form the final cutting contour for panel, and add the bevelling information.

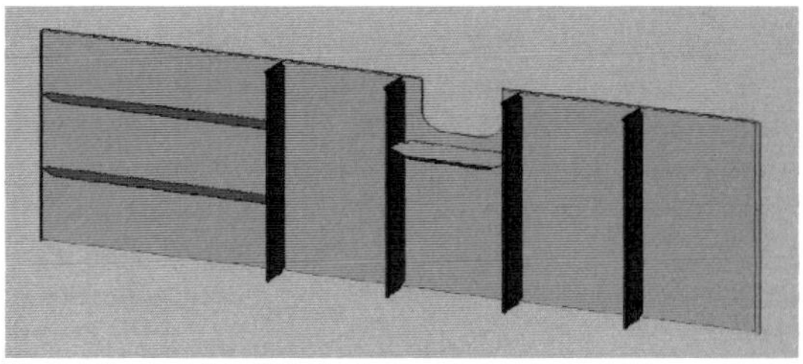

Figure 1: Panel in 3D design system.

Knowing all stiffeners and respective positioning on panel is essential. Stiffener locations are marked on the panel base before they are actually welded. Stiffener information helps to generate time estimations of line marking time and stiffener welding time.

This information should be provided by a digital design solution, which has the capability to manage panels with high details. Further to improve the overall quality the design data needs to be imported into the optimization system via automated interfaces. Figure 2 illustrates a 2D geometry of a panel.

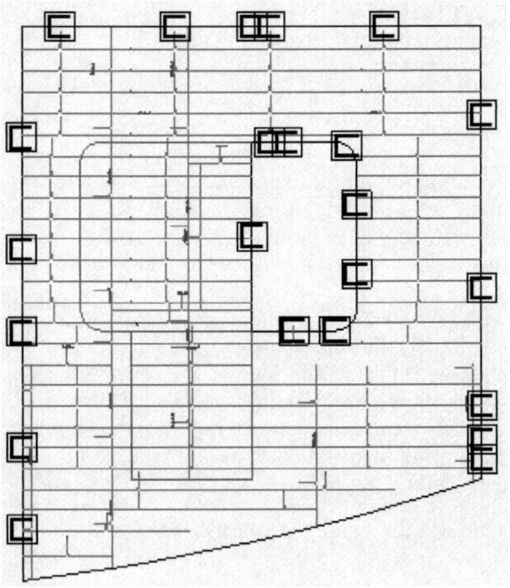

Figure 2: 2D geometry of panel containing cutting contour, stiffener line markings and welding joints.

2.3 PROCESSES BEFORE PANEL LINE

Knowing the plate part cutting process and profile line process can be beneficial for the panel line optimization. Though knowing the prior processes to panel line is not mandatory for optimizing the panel line, but every bit of additional information can be used for creating more accurate and optimal optimization results. More information enables the use of additional optimization parameters for better functionality of the panel line.

It is important to know when required parts are ready for the panel assembly. If some required parts are not ready for the panel assembly, then this can cause the whole panel line optimization result to be void. In the worst-case scenario, panel line could be halted causing delays for the project. Quantity of machines and machine related efficiency gives valuable information for the part production scheduling that can be linked into panel scheduling directly.

2.4 INTERNAL LOGISTICS

Managing materials efficiently enables stable production process flow during the initial cutting stages of materials. Proper material management ensures that preparation of panel parts will be done in timely fashion before panel assembly. Internal logistics management takes care of part location data, so that no cut part is lost and that the parts can be found in front of the correct panel line work station at the right time.

2.5 SCHEDULING SYSTEM

High level scheduling systems or project planning systems provide start and due dates for block assembly. This information outlines the ground restrictions for the optimization algorithm. Having higher level scheduling information on hand enables need dates for panels to be determined.

Once knowing the panel need dates, then it is possible to make scheduling optimization for the panel line. After completing the panel optimization and having the detailed panel assembly schedule, it is possible to do fine scheduling of part production. Managing part production schedule and the status of part production enables providing more information for the panel line optimization.

3. PANEL LINE OPTIMIZATION

3.1 OPTIMIZATION PARAMETERS

There are four main parameters or key optimization criteria: lead time, lateness, earliness and scattering. Each of these parameters try to optimize certain functionalities for the panel line so that the production is most efficient, lean, and just-in-time.

Lead time optimization arranges panels in such order for the panel line so that panels will be manufactured in the shortest time possible. This maximizes the throughput of the panel line.

Lateness optimization sequences panels in order such that no panel would be completed too late. This tries to ensure that panels are manufactured just in time.

Earliness optimization sequences panels in order such that no panel would be completed too early. This parameter tries to minimize the waiting time of panels before the block or sub-block assembly.

Scattering optimizes the waiting time between plate and profile part in stock to panel manufacturing time. In other words, this parameter makes panel line to work more in synchronization with the part manufacturing lines.

It is possible to use two optional parameters in panel line optimization. If panel scheduling is not mastering the part production schedule, then panel line should be optimized with a parameter to ensure that all necessary parts are manufactured before panel fabrication. If shipyard has space restrictions for panels, then it is possible to add in a rule that takes into consideration of the earliest start date of panel assembly. This parameter tries to make sure that panels are not occupying buffer space from block assembly stages.

3.2 PANEL LINE OPTIMIZATION IDEOLOGY

The key ideology for panel line optimization is to maximize the throughput of panels while keeping the other yard restrictions in mind. Sheer amount of possibilities and variables makes panel line optimization very time consuming operation. With elegant algorithms and intelligent data systems this is possible in Smart Production software.

Panel line can be optimized with the parameters explained above, but making the most throughput efficient panel line does not guarantee most functional shipyard. It is still important to keep in mind the block assembly, because having panels waiting for weeks before the assembly itself is not good especially if there are space limitations. Naturally having panels manufactured too late then will risk the project schedule as well, which is again definitely not suitable.

Panel line optimization is combination of all the parameters put into work together. To achieve harmony between the optimization parameters it is best idea to have them as weighted values. Meaning that total sum of optimization parameters should be 1. For every shipyard and for every panel line there is the most benefitting combination of panel optimization parameters. From time to time these parameters might need some adjustment, because panel optimization is based on the panel types itself, and different ship designs favour different set of panels. Figure 3 below illustrates comparison of different optimization results.

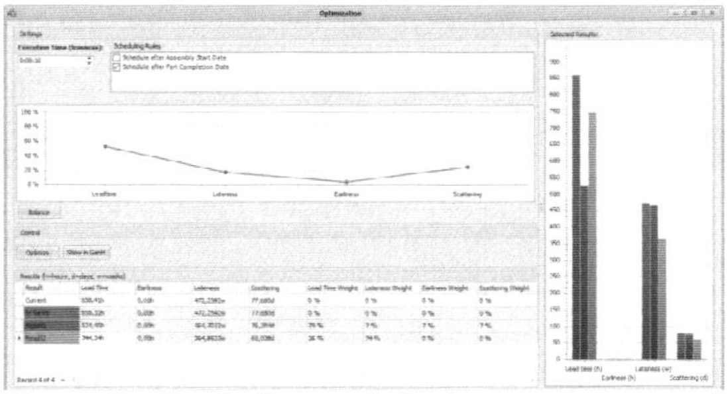

Figure 3: Panel line optimization parameters and comparison of optimization results

3.3 VIEWING OPTIMIZATION RESULTS

After having the optimization results it is time to select the most suitable panel sequence for the panel line. This sequence of panels creates ultimately panel line schedule for the fabrication yard, that works on different shifts and personnel operating the panel line machinery. This schedule shows the panel line loading, and work load for every work station. Figure 4 below shows currently accepted panel line schedule after optimization.

Based on the panel line schedule it is possible to study the overall efficiency of the panel line or performance of any given work station. This process can help to identify possible bottlenecks of the optimized panel line. Collected data can be used to verify whether the part fabrication can keep up with the panel line flow. Another benefit is provided for work stations that require manual work, because if one of these work stations is the bottleneck, then it is possible to increase manpower to overcome the bottleneck rather easily.

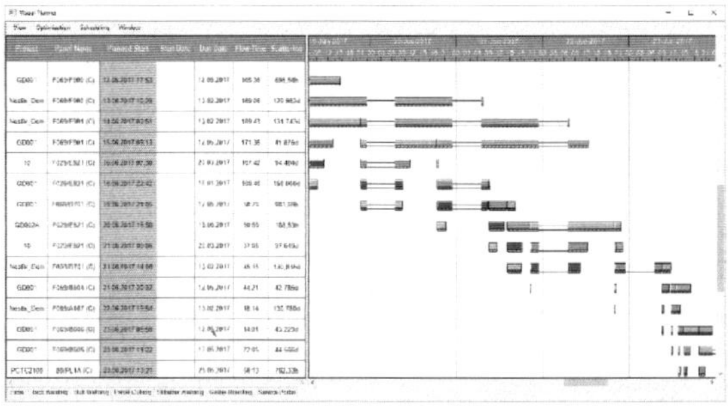

Figure 4: Optimized panel line schedule

4. CONCLUSIONS

Panel line is the heart of any shipyard or offshore rig fabricators yard. It is closely linked to most of the yard's functionalities of part prefabrication and scheduling, while also enabling and driving the block assembly in timely fashion. It is possible to increase productivity and keeping the promised milestones of projects with panel line optimization. Optimization results and the provided panel line schedule can bring valuable information for further improving the panel line itself or to other functionalities at the fabrication yard.

5. ACKNOWLEDGEMENTS

I would like to thank Jani Heiskari for his continuous expert advice and support given for panel line optimization delivery projects.

6. AUTHORS BIOGRAPHY

Kimmo Salmi holds the current position of Business Development Consultant at Hexagon PPM. He is responsible for developing business in APAC region for Smart Production powered by NESTIX solutions. His previous experience includes delivering customized Smart Production solutions for shipyards and steel service centres all over the world.

LEVERAGING ENGINEERING 3D DESIGN MODELS FOR PRODUCTION PLANNING THROUGH THE SHIPBUILDING PROJECT LIFECYCLE

Stalin P. Ybiernas, Keppel Singmarine, Singapore
Najaf Bashir, Hexagon PPM (Previously known as Intergraph PPM), Singapore
Davide Guzzi, Hexagon PPM (Previously known as Intergraph PPM), U.S.A

SUMMARY

Two of the top key factors in efficient shipbuilding are the efficient use of an integrated engineering 3D design solution and efficient production processes. Keppel Singmarine has been exploring new ways to improve the optimization of the data exchange between these two factors through an efficient and early information exchange process. This is done by introducing production planning departments to the same integrated engineering 3D model used in design, which has allowed production planning to start in the earlier stages of the project. The early preparation of production processes has enabled Keppel Singmarine to identify potential bottlenecks very early in the design cycle.

An important aspect of shipbuilding is knowing exactly when certain activities need to be executed in order to support the pre-outfitting of blocks, movement of components, installation of equipment items, etc. Knowing in advance when these activities are to be executed has allowed Keppel Singmarine to further optimize the complete supply chain and provide "just-in-time" support for the complete production process.

This paper will describe how the production planning department at Keppel Singmarine uses an integrated engineering 3D model to provide early feedback for block and assembly boundaries, generate detailed assembly planning drawings, create advanced outfitting block assembly plans, plan equipment loading, and more. This will lead to the realization of Industry 4.0, where direct connection with – and feedback from – the shop floor will be leveraged to optimize production processes and therefore improve overall schedule.

NOMENCLATURE

LPM	Light Weight Production Model
KSM	Keppel Singmarine
S3D	Smart 3D
DAP	Detail Assembly Procedure
PPD	Production Planning Department
DSI	Design System Integrator

1. INTRODUCTION

Keppel Offshore and Marine's (KOM) specialised shipbuilding division, led by Keppel Singmarine (KSM), designs and builds a wide spectrum of highly specialised ships for a global clientele. KSM have the flexibility to customise vessels for every frontier, including some of the world's harshest environments and they design customized vessels and equipment for all operating conditions.

KSM customized solutions

- Multi-Purpose Offshore Support Vessels
- Anchor Handling Tug/Supply vessels
- Tugboats
- Icebreakers & ice-class support vessels
- Floating Storage Offloading vessels
- Pipe lay ships

KSM possesses a strong track record for reliable and versatile new build solutions and have constructed some 400 ships of diverse types and sophistication levels for a global customer base.

KSM Track record includes:

- Anchor handling tug/supply vessels
- Platform supply vessels
- Multipurpose construction vessels
- Ice breakers
- Ice-Class Floating Storage and Offloading vessel
- Pipe lay vessels
- Cable laying vessels
- Harbour and terminal tugs
- Well stimulation vessels

Since 2012 KSM has successfully implemented Hexagon PPM's Smart 3D (S3D) system as their engineering solution for ship detail engineering in 3D environment and auto extraction of production output requirements to support construction of their projects safely, on time, and within budget. Leveraging on Smart 3D's global workshare capabilities KSM can execute and synchronize its detail engineering of multiple projects across its business units across the globe in real time.

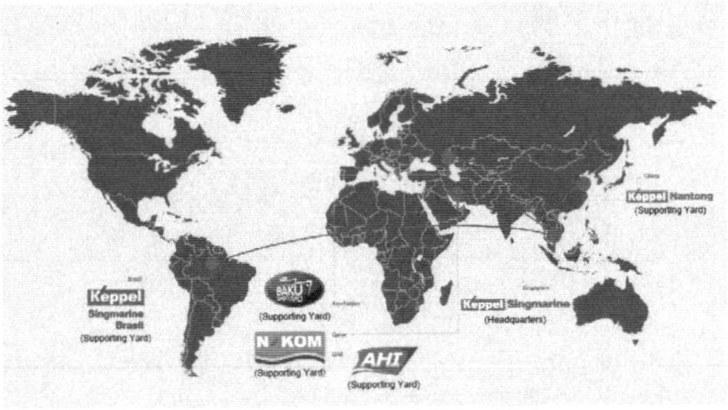

Figure 1: KSM engineering and yards across the globe

As new technology and solutions geared towards a "Smart Yard Integration" have been developed by Hexagon PPM. KSM has intentions to align its engineering and production processes in achieving the same Smart Yard goal.

KSM's vision of a full integrated environment with the support from Hexagon PPM solution will cover 3D detail design (S3D, DSI, SmartPlant Electrical, and SmartPlant PID), shop floor management/spooling (NESTIX), production planning (LPM), fabrication and construction (SmartPlant Construction), and documentation (EDMS light version).

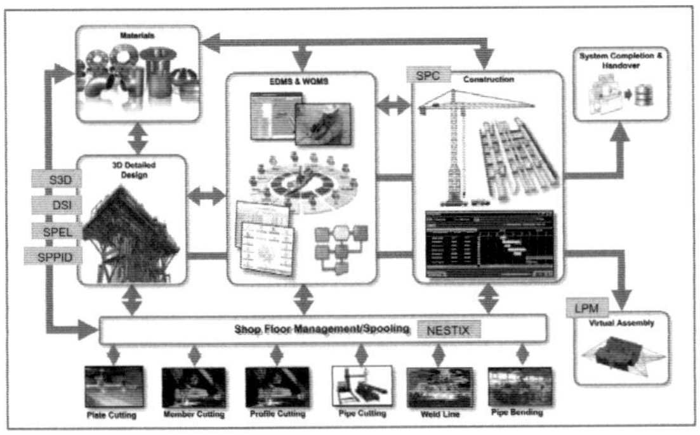

Figure 1a: Hexagon PPM's Smart Yard Integration

The journey in realizing a smart yard for KSM has just started and it is aware of the difficulties and challenges that it needs to address in order to attain the required integration and automation. In partnership with Hexagon PPM KSM is optimistic in fulfilling its smart yard vision in near future.

2. CURRENT PRACTICE

There are several departments under KSM that contributes to the delivery of a project. For this topic, we will discuss more on the engineering and production side on their current workflow and their collaboration with each other.

Detail and production engineering are the main workloads for KSM engineering. Using S3D, engineering team creates a 3D model of the ship in detail containing all relevant objects from all trades (hull, piping, mechanical, outfitting, electrical, HVAC, and accommodation) to be used as reference for drawing generation, clash checking, design review, necessary data estimation and forecast.

The engineering team issues to production team the shop drawings, part lists, and nesting plans generated from S3D.

Production team uses the nesting plans to cut plates and profiles then followed by block fabrication with reference to the shop drawings.

KSM also has a production planning team that works on block division plans, detail assembly procedure (DAP), and planning coordination of project executions. The planning team has access to the live S3D model and provides feedback on the arrangement of the assembly hierarchy for hull structures. They also prepare block DAP drawings using S3D snapshot views from the live model.

Challenges/Issues from production planning team

Currently the KSM planning team is unable to optimize the usage of the live 3D model in S3D to complete their tasks. They are able to access the live S3D model but to the extents of viewing and generating isometric drawings only. They are not allowed to create or modify anything on the live on-going model, which is also a valid point, to avoid model issues. They don't have a separate system where they can work in 3D environment containing the required 3D model and then create their models corresponding to their planning tasks. The absence of this platform makes it quite a challenge for the planners to produce up-to-date, fast, and good quality drawings.

Manual work in the DAP and in the other tasks of planning team are some of the main concerns that require improvement to increase productivity for production. For example, the current S3D does not support automation features for planning team's DAP drawings. The team has to display and orient accordingly each of the S3D block or assemble models and then generate the snapshot of the view. This process consumes a lot of man hours charged to DAP drawing preparation.

Another issue is the planning team can only make use of existing drawings or issued drawings, which may be sometimes outdated, as reference for transport or lifting drawings. The rest of the tasks of planning team are done manually in 2D and following the conventional method which is time consuming and prone to human errors.

With the introduction of Lightweight Production Model (LPM) application, KSM sees the huge benefit it can get in terms of efficiency, quality, and accuracy of drawing output for production planning team. Workflows and process will be changed and improved with the implementation of LPM as the system engages the team in a 3D environment of the published ship from the live projects. Automation and customization in both 3D modelling and drawings are the key features for LPM and the modules it offers covers most of the production planning task. KSM recognizes the impacts and value of LPM as well as the man-hour and cost savings it can contribute.

The LPM solution is part of the integration tools of smart yard and it suites well with KSM's existing S3D system.

3. TYPICAL PRODUCTION PLANNING TASKS

3.1 DETAILED ASSEMBLY PROCEDURE (DAP)

DAP drawings are prepared and issued by planning team if required by projects. The DAP drawings are prepared once the block model in S3D is finalized by engineering hull section. Planning team opens S3D and access the desired block in the live project database.

In S3D assembly tab the blocks will contain assemblies and sub-assemblies which have already been approved by planning team. Planning team will manually use snapshot tool in S3D to generate an isometric drawing view for each assemblies and sub-assemblies. These isometric views are imported to AutoCAD where touch ups and arrangements are done to nicely present the assembly procedure. A sample KSM block DAP drawing flow is shown in Fig.2.

It takes at least three days on the average to complete a block DAP drawing. It would take even more time to finish if the block assembly hierarchy becomes complex and too many, which is usually the case. Manual procedures are associated with the accrued man-hours. Human errors in preparing the drawings may also occur if user may fail to illustrate an assembly.

Having a tool that can generate block DAP drawings automatically from the live S3D projects would be very useful and beneficial to implement. KSM's goal is to decrease the man-hours spent on preparing DAP drawings thus allowing faster response time for production team to proceed with fabrication and erection.

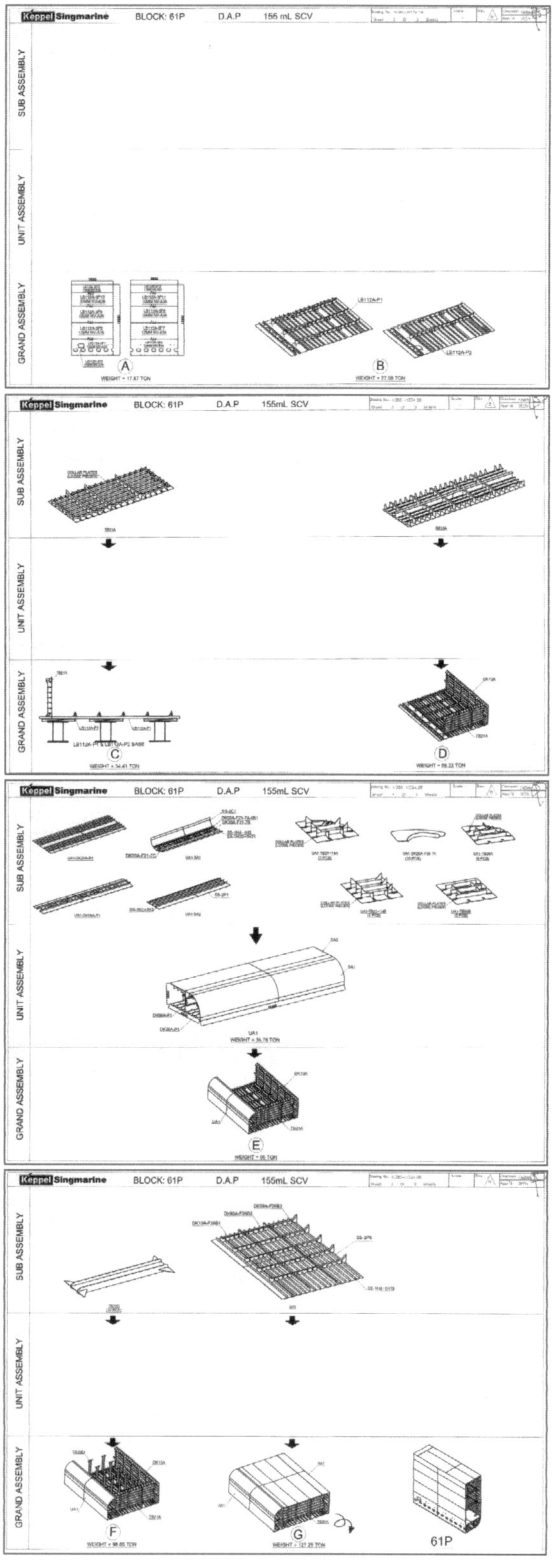

Figure 2: KSM Block 61P DAP drawing sample

3.2 LIFTING CONCEPT/ARRANGEMENT

The preparations to ensure safety and success for lifting operations involve three departments: production planning, engineering, and production.

To comply with the lifting standards and control procedures, the production planning department (PPD) is tasked to prepare and issue the Lifting Concept/Arrangement and the preliminary sling lengths and weight estimate. Fig.3 and Fig.4 shows samples of lifting concept and sling lengths and weight estimate respectively.

Lifting Concept is a document that provides the lifting methodologies in pictorial form, containing essential information like lifting location and destination for unloading. Details on the block transport route is also included in the lifting concept.

The PPD planner prepares these requirements for lifts and unloading of sub-blocks, leg structures, finished blocks or equipment units from one location to another destination. The lifting concept is issued to the related departments and teams for review and approval. The PPD planner addresses all queries and concerns related to the lifting concept and revises the document as applicable.

The lifting concept is all done manually by PPD planner in AutoCAD.

On the other hand, Naval Architect (NA) section will use as reference the approved lifting concept to generate and issue the Lifting Arrangement and Padeye Details. Fig.5 shows a sample for lifting arrangement and Padeye Details. These drawings are also done manually in AutoCAD. The block section views are copied from the issued block shop drawings. The block center of gravity (CG) information, however, is taken from finished S3D block model.

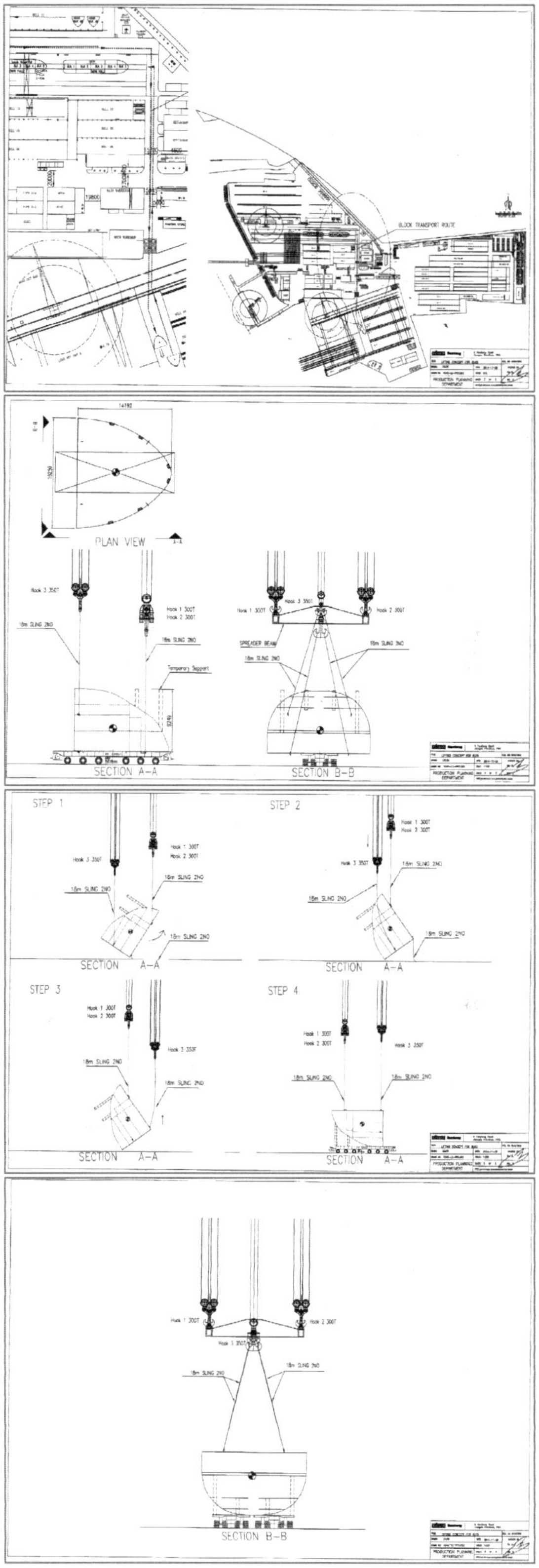

Figure 3: Lifting concept sample

PRELIMINARY WEIGHT ESTIMATE

PROJECT H045/H046
CONCEPT NO. : H045-LC-PPD.003
DATE 28-Nov-14

DESCRIPTION **LIFTING WEIGHT ESTIMATE FOR BLK05**

BLOCK CONTENTS	ASSUMPTION	PPD PRELIMINARY ESTIMATE	ENGNG CONFIRM WEIGHT	REMARKS
STEEL		180		
PIPING		3		
OUTFITS		3		
MACHINERY				
WELDING		4		Factor = 0.02
PAINTING		2		Factor = 0.01
ELECTRICAL				
PAD EYE				
HULL BLOCK		**190**		
SHACKLES		0.8		SWL 35T,0.02T/Pcs
SPREADER		50.0		
STEEL WIRE		5.8		0.97T/Length
TOTAL		**247**		

Figure 4: Preliminary weight estimate sample

1. LIFTING LOADS:
   TOTAL LIFTING LOAD: 267.52 MT. (ESTIMATED)
   BLOCK WEIGHT: 192.14 MT
   (INCLUDE WELDING, PAINTING, PIPING, OUTFITS, EQUIPMENT);
   WEIGHT OF PADEYE, SHACKLE AND SLING: ABOUT 2.84MT.
   SPREADER BEAM: 71.14MT,
   REINFORCEMENT SUPPORT: 0 MT
   TEMPORARY SUPPORT: 1.4 MT

2. TYPES OF SHACKLE:
   2x85 MT SHACKLE . (CONNECTING TO PADEYE)
   2x125 MT SHACKLE . (CONNECTING TO PADEYE)

3. SLINGS:
   2x18M SLING, MAX SLING FORCE 123 MT(MBL-246MT)
   2x18M SLING, MAX SLING FORCE 76.5 MT(MBL=153MT)

NOTE: ALL SLING FORCES SHOWN ARE STATIC AND WITH FACTOR =1.0
ALL SLINGS SHOULD BE SELECTED BY PRODUCTION DEPARTMENT.

4. LIFTING CONCEPT:
   FOR LIFTING CONCEPT & PROCEDURE, REFER TO PRODUCTION PLANNING DEPT'S DOCUMENT.

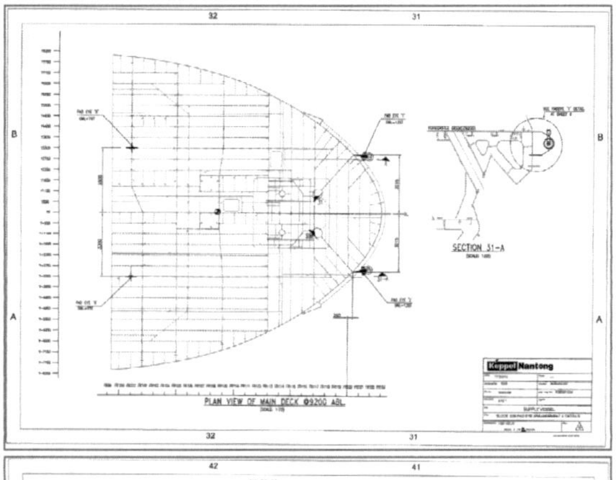

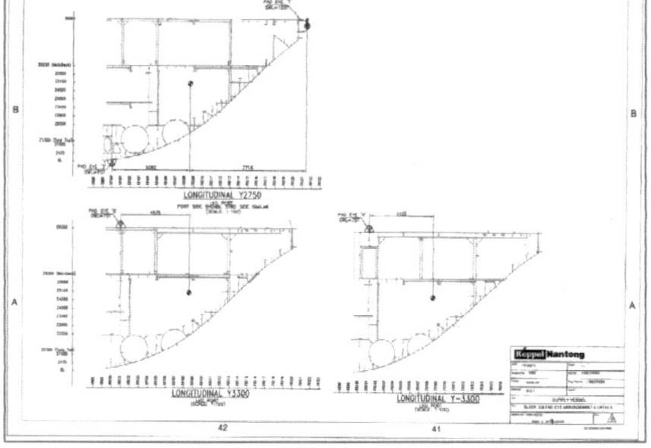

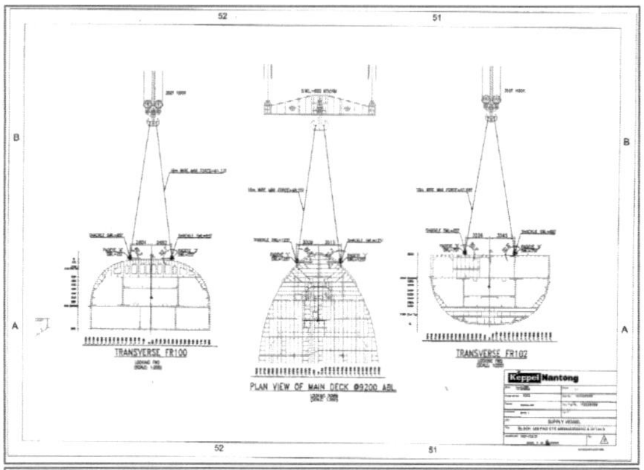

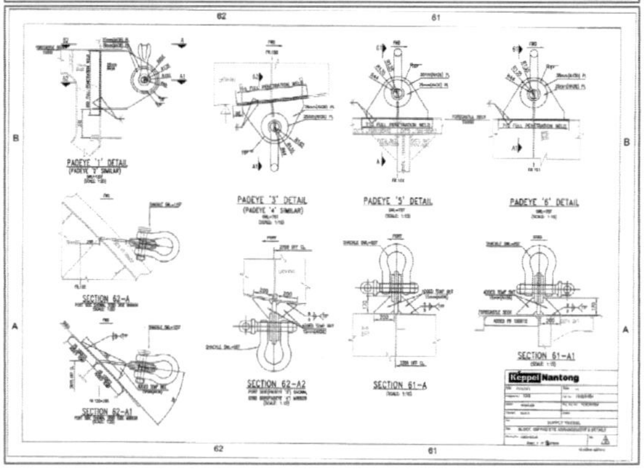

Figure 5: Lifting Arrangement and Padeye Details sample

It takes at least three days on the average to complete a Lifting Concept. It would take even more time to finish if the block lifting procedure is complicated involving multiple lifts, unloading, turns, and transport. Manual procedures are again associated with the accrued man-hours.

Having a tool that can generate similar to the lifting concept automatically from the live S3D projects would be very useful for production planning department. KSM's goal is to decrease the man-hours spent on preparing lifting concept to shorten lead time for lifting operations without compromising safety and accuracy of data.

3.3 PAINT AREA

KSM generates paint area report that includes hull, outfitting, piping, and electrical items subject for painting activity. For hull paint area the information are generated from the live S3D model under Compartmentation task using the out-of-the-box Compartment Painting Area report. For the rest of the trades the engineers estimate manually the painting area with use of excel tables and formulas.

HULL H397_3500 M3 HOPPER SUCTION DREDGER
1. PAINT AREA SUMMARY (BY DEPARTMENT)

PAINT AREA SUMMARY

NO.	ITEAM	AREA (m^2)	REMARK
1	TABLE-1 PAINT AREA SUMMARY (By Painting System)	47017	Page 03 -page 08
2	TABLE-2 SHELL & TANK	42201	Page 09 -page 11
3	TABLE-3 OUTFITING	2818	Page 12 -page 15
4	TABLE-4 & 5 PIPING	1473	Page 16 -page 20
5	TABLE -6 ELECTRIC	524	Page 21

HULL H397_3500 M3 HOPPER SUCTION DREDGER
3. Table -2 Paint Area For Shell & Tank

Table -2 Painting Area For Shell & Tank

NO	NAME	LOCATION	DESCREAPTION	AREA (m^2)	Coating 涂层
1	SHELL		Hull below light waterline	1703.00	5
			Hull above light waterline, including outside of bullwarks	1407.80	3
2	DECK OUTSIDE		All deck outside	1071.22	3
3	DECKHOUSE		Forward accommodation deckhouse and wheelhouse	4754.56	3
			Aft deckhouse	2112.16	3
4	FORWARD MACHINERY SPACE	FR84-FR104(CENTER) (TANK TOP)	Shell, wall and bulkhead above floor level, undersides of deck	1046.62	2
			Steel deck and platforms	191.56	2
5	PAINT STORE	FR90-FR93 (S) (TANK TOP)	Shell, wall and bulkhead above floor level, undersides of deck	95.18	2
			Steel deck and platforms	2023.00	2
6	CHEMICAL STORE	FR93-FR96 (S) (TANK TOP)	Shell, wall and bulkhead above floor level, undersides of deck	93.54	2
			Steel deck and platforms	7.89	2
7	WBD	FR104-FR116(CENTER) (TANK TOP)	All inside	436.87	3
8	VOID 1 (P)	FR92-FR100(P) (TANK TOP)	All inside	118.60	3
9	VOID 1 (CL)	FR92-FR100(CENTER) (TANK TOP)	All inside	353.31	3
10	VOID 1 (S)	FR92-FR100(S) (TANK TOP)	All inside	123.47	3
11	VOID 2(P)	FR84-FR92(P) (TANK TOP)	All inside	158.29	3

HULL H397_3500 M3 HOPPER SUCTION DREDGER
2. Table -1 PAINT AREA SUMMARY (By Painting System)

TABLE-1 PAINT AREA SUMMARY (By Painting System)

SECTION	LOCATION		AREA (M^2)
B30 2 1 Paint system EP	Engine room	Stairways, handrails and ladders	24.5
		Hoisting beams	3.5
		Shell, walls and bulkheads above floor level, undersides of decks	2123.6
		Engines and auxiliaries	
		Pipelines and fittings above floor level	
		Air ducts, outside and inside	
		Cable trays	83.9
		Steel storage racks	
	Thruster room	Stairways, handrails and ladders	10.5
		Hoisting beams	3.5
		Shell, walls and bulkheads above floor level, undersides of decks	1046.6
		Engines and auxiliaries	
		Pipelines and fittings above floor level	
		Air ducts, outside and inside	
		Cable trays	30.2
		Steel storage racks	
	Rudderpropeller room	Stairways, handrails and ladders	
		Hoisting beams	
		Shell, walls and bulkheads above floor level, undersides of decks	318.2
		Engines and auxiliaries	
		Pipelines and fittings above floor level	

Figure 6: Sample Paint Area Summary of a vessel

Hull engineering team generates this compartment painting area report in S3D once request is made. The report is done when hull block modelling has been completed. S3D makes use of volumes in the model to define the locations where painting area information are required. Only hull objects inside the created volume for painting area will be included in the report.

Planning team is not involved in preparing the information for the painting area report however it would be beneficial if they are able to assist on this workload from engineering. The planners are also not well versed in using S3D and they are only limited to viewing and creating snapshot views from the model.

The S3D compartmentation task also has functionality in terms of generating volumes for compartments with irregular shape or curved surfaces. Volumes are constrained to box type only in S3D. Manual work is still required for such cases to get accurate values.

A system that can generate painting area report without the need for manual computation would help boost productivity for this kind of task. It can generate compartment painting area for any kind of compartment shape or volume complexity.

3.4 SCAFFOLDING

Tube and coupler components are commonly used by KSM in its scaffolding works. Staging from the ground up using pipes (for frames, braces, and handrail), clamps, ladders, and boards as platform are used for scaffolding. Hanging scaffolds are also employed if deemed required.

Figure 7: KSM typical scaffoldings of a block

The scaffolding sub contractor builds up the scaffolds base on the standard scaffolding details of KSM. Simple sketches are just used for block scaffolding drawing with reference to the issued block drawings and standard scaffolding details.

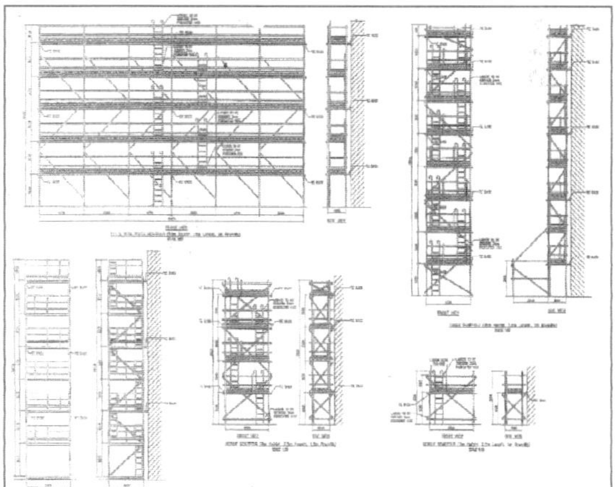

Figure 8: Sample KSM standard scaffolding details

KSM does not have a system where scaffoldings are simulated or modelled into the actual block model for production reference.

4. LPM

At ship and offshore yards, assembling, lifting, and moving around big blocks and modules requires many activities that have to be planned mostly in advance to avoid costly idle time during fabrication and to ensure workers' safety on the shop floor. There are hundreds of documents and drawings that have to be prepared in order to instruct crews on the right operating sequence to accomplish their jobs in the most safe and productive way.

Three years ago, Hexagon PPM started development of a set of applications dedicated to yard production planning in partnership with one of the biggest shipbuilders in the world. What was at the beginning just a customer specific software customization, is now a commercial solution available on the market with the name of LPM, which stands for Lightweight Production Model.

LPM provides a comprehensive set of tools and functionalities to be used on top of the 3D design model to prepare all necessary deliverables to be used at the yard for assemblies and blocks painting, scaffolding installation and removal, lifting lugs positioning, cranes configuration, and much more.

4.1 CAD AGNOSTIC TECHNOLOGY

LPM has been designed to work on top of 3D models created by the most common 3D CAD systems. Through its interoperability technology, third party 3D models are translated in a format that LPM can use and 'consume' to generate drawings, make quantities estimations, design scaffolding, and so forth. This makes LPM a great tool for the yard's production engineering office, independently by the 3D CAD systems that is used by the design office.

In the case of offshore vessels and facilities, where hulls and topsides design are developed by different entities, another challenge that shipbuilders face is the difficulty combine and merge in one single 3D model, the hull and topside modules coming from different design systems. LPM allows the import and merge of 3D models coming from various CAD technologies in order to provide to the shipbuilder an homogenous, single environment where they can perform the most important tasks related to production planning.

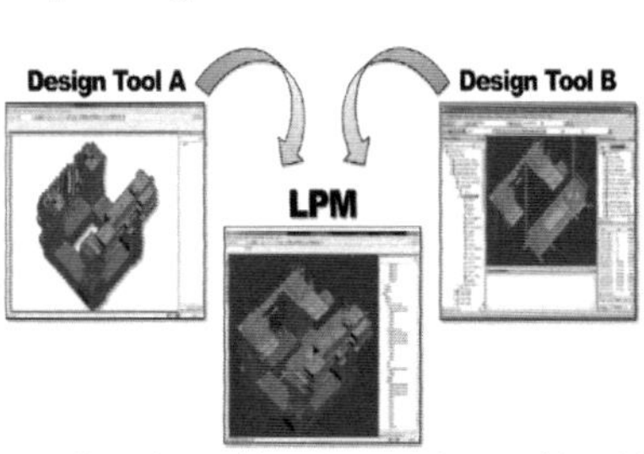

Figure.9: Hull and topside 3D models combined in one

4.2 CORE REVISION MANAGEMENT

Because LPM starts to use the design 3D model in the early stage of the project, new versions of the same model are available almost daily. For this reason it's extremely important to have the right tools in order to include in the production planning model all the updates coming from the 3D design model in the smoothest way possible.

LPM Core Revision Management module ensure the automatic update and refresh of the 3D model new versions, keeping track of all the variations made from one version to the new one. LPM Core Revision Management also allows the scheduling of the design 3D model automatic publishing, making this operation completely transparent to the final user.

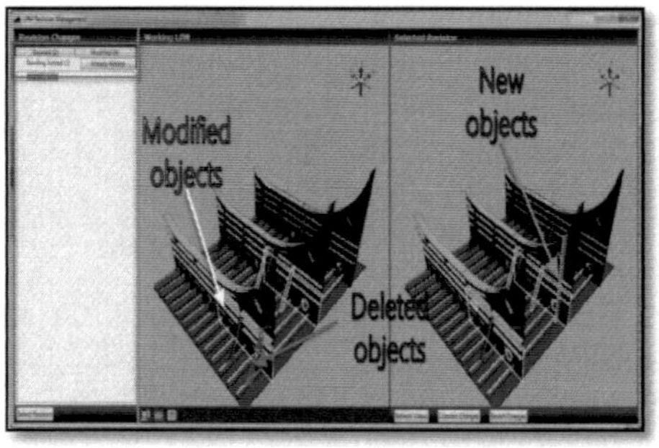

Fig.10 Different versions for the same block or assembly can be compared graphically side by side

4.3 QUANTITIES ESTIMATIONS

The production planning teams at the yard need an easy and quick way to access to the properties and quantities that are available in the 3D model like weights, lengths and volumetric dimensions.

LPM, through its OPM module (Objects Properties Manager), allows a fast and intuitive way to browse the 3D model most relevant data such as weight, center of gravity, welds length, surfaces for painting and insulation, and more. OPM has searching, filtering, and sorting capabilities like in Excel that allow the users to quickly and easily create the list of quantities needed for estimations during the entire phase of the project.

Figure 10: 3D model attributes as they appear in OPM

4.4 TOP-DOWN PLANNING

In order to manage and plan the entire ship fabrication and construction sequence, from the smallest plate part up to the biggest mega-block, it becomes almost mandatory to organize the data associated to the 3D model in a hierarchical way. This can be a very time consuming job and considering also the fact that the production planning team wants to start this process early in the design stage, the planning hierarchy creation needs to be manually updated every time there's a design change.

LPM can quickly group parts to form assemblies and then assemblies to form sub-blocks or blocks and so forth. In this way the production engineering office can create more accurate estimations for each step of the construction phases in the earliest stages of the project, ensuring more efficiency on the shop floor and reducing at minimum idle time. The planning hierarchy grouping operation in LPM is done almost fully automatically with the help of rules defined by the shipyard. Thanks to this automated tool, changes coming from the design model will be quickly integrated in the existing planning hierarchy.

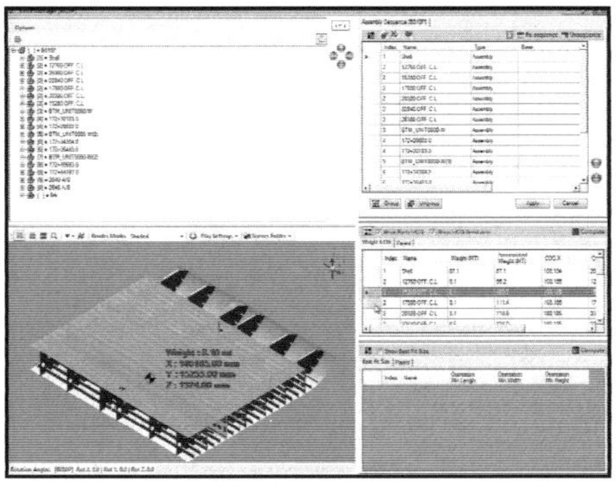

Figure 11: 3D model is grouped and well organized in the planning hierarchy

4.5 DETAILED ASSEMBLY PROCEDURE

It's fundamental for the crew at the shop floor to have the most clear and detailed information about the assembly procedure they have to follow. Can you imagine purchasing an unassembled piece of furniture for your home and trying to put it together without the instructions? At the minimum it will take you twice the time to put it together. For this reason shipbuilders put a big effort to prepare detailed assembly procedure type of drawings.

LPM Detailed Assembly Procedure Command allows production engineers to create in a few clicks and seconds very complex drawings that generally take hours to be done at the yard using traditional CAD systems. From one simple and intuitive window, users can check weight, center of gravity, and orientation for each block, sub-block or assembly and make an effective drawing composition. LPM's outstanding drawing technology will do the rest, creating automatically the final output including labels and material lists when needed.

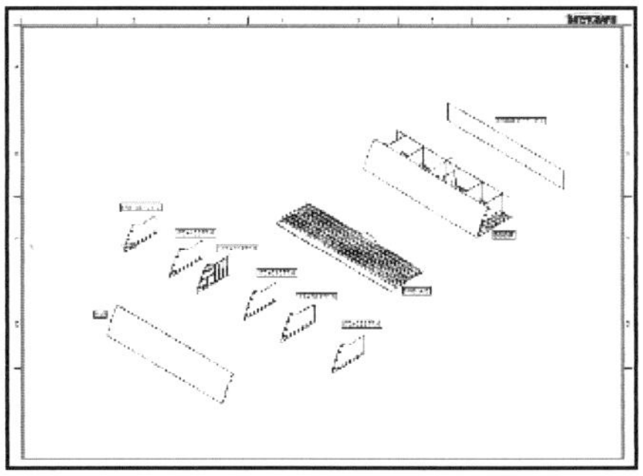

Figure 12: One example of DAP drawing automatically produced by LPM

4.6 PRODUCTION DATA AND DELIVERABLES

There are many activities at the yard that have to be driven by detailed drawings and information. Scaffolding, lifting, transportation, and painting are all tasks that require a lot of documentation to be prepared by the production engineering office at the yard. LPM comes with special features and functionalities that allow a fast and easy placement of construction supporting items such as scaffolding and lifting lugs and also several predefined drawings view styles that can be used to extract all the relevant information from the 3D model in seconds.

4.6.1 Painting Areas

Surfaces that need to be painted, coated or insulated, like plates, pipes, etc... are quickly calculated by compartments or volumes. LPM provides a set of functionalities to quickly create compartments and volumes within the 3D model as shown in Fig.13.

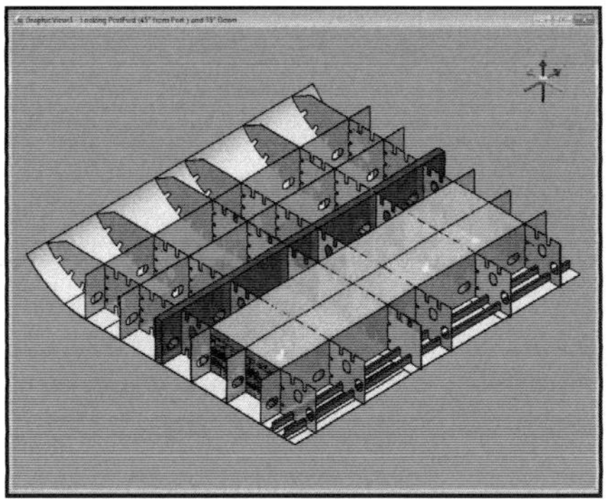

Figure 13: Volumes created for painting and insulation purposes

Once volumes and compartments are available in the 3D model, painting and coating specs are assigned either to the entire volumes or to single parts only. At this point LPM can calculate and produce either surfaces overall quantities by block or by single parts as shown in Fig.14.

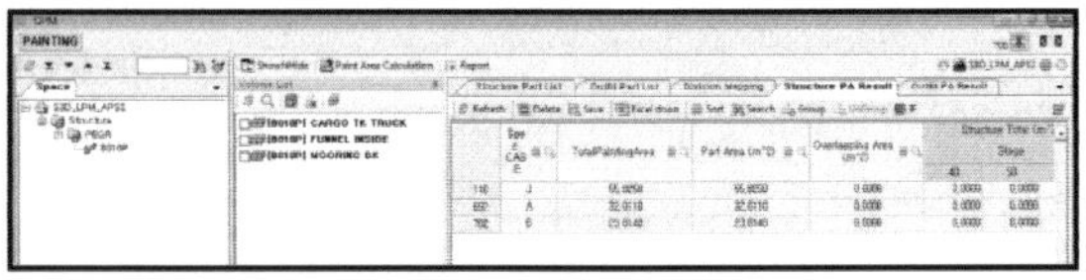

Figure 14: Surface quantities are automatically calculated

Last but not least, orthographic and isometric drawing views showing areas to be painted or coated with a specific product using different hatches styles, are automatically produced by LPM drawings module.

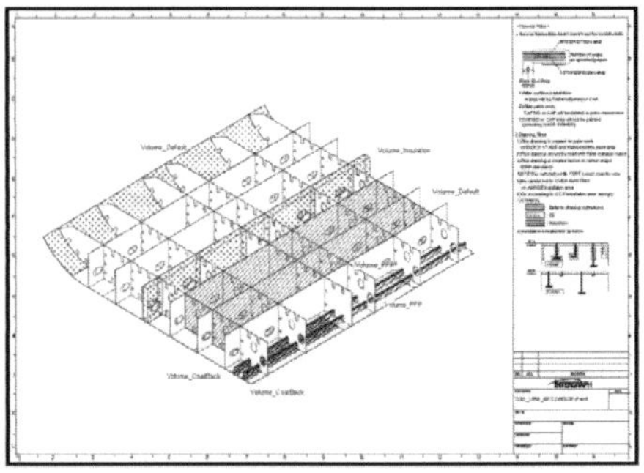

Figure 15: Painting and coating example drawing produced by LPM

4.6.2 Scaffolding

Planning and managing all scaffolding installation and removal operations during the overall project requires a tremendous amount of preparation work.
In LPM scaffolding can be easily placed directly in the 3D model using a series of automation and parametric functionalities.

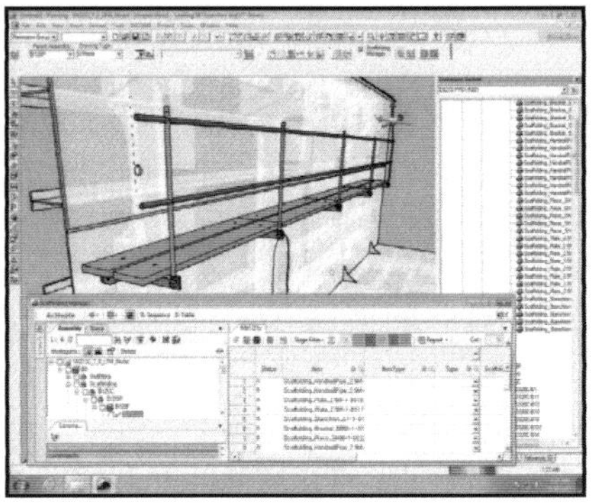

Figure 16: Scaffolding are placed in the 3D model in free space or welded or clamped to existing structures

From the Scaffolding Manager in LPM, scaffolding assemblies are assigned to a specific installation and removal stage along the entire construction sequence and drawings and BOMs can be automatically produced by the system.

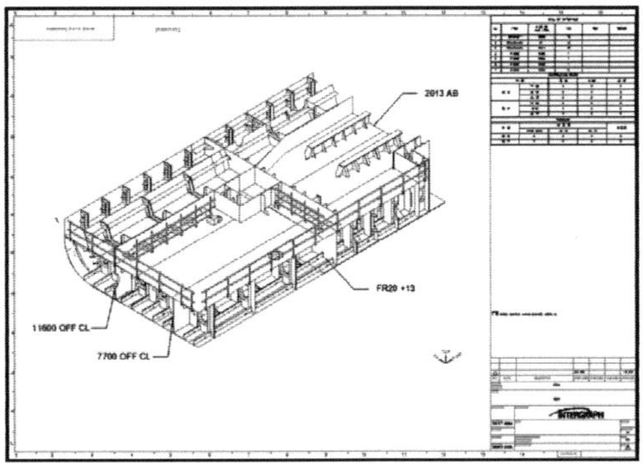

Figure 17: Scaffolding drawing w/BOM example produced by LPM

4.6.3 Lifting Lugs

Blocks and assembly lifting operations at a shipyard are a very frequent task. Depending by the block size and weight, cranes have to be configured and prepared accordingly. Very precise information needs to be provided to the lifting crew in order to make sure every lifting operation can be completed smoothly without any accidents. There are also safety regulations that need to be applied for each lifting operation and for this reason, special drawings have to be prepared at the yard.

LPM provides the functionality to quickly and precisely place lifting lugs along the block using Center of Gravity information embedded in the 3D model. Crane information needs to be defined by the user and this information can also be saved in the LPM catalog for later use.

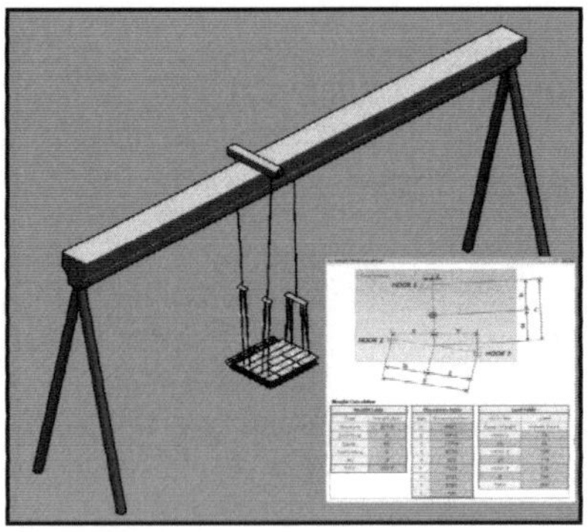

Figure 18: Slings have been placed based upon lugs and hooks position

At the end of the lugs placement and crane configuration

process, drawings can be automatically extracted from LPM. The amount of data and information included in these drawings can be configured to suit the shipyard requirements.

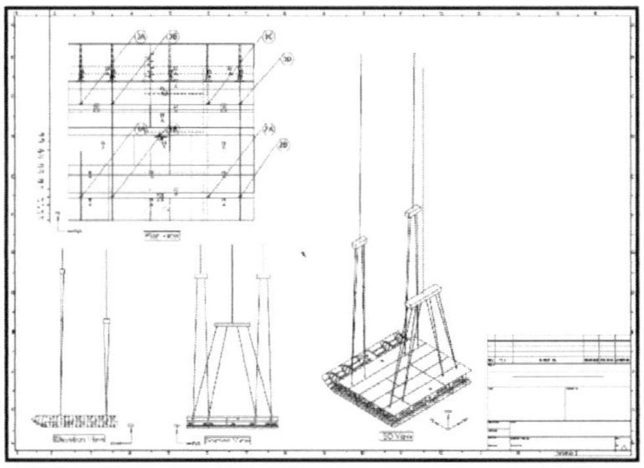

Figure 19: Lifting lugs location and slings drawings example produced by LPM

5. LPM Evaluation by KSM

As a step in-line with their vision of, 'Smart Yard', KSM evaluated Hexagon PPM's LPM solution for their Production planning department use. As part of the evaluation, KSM production engineers were able to work in a lighter version of 3D model directly and were able to produce required deliverables in a fraction of a time compared to their current practise. Also, by working in a 3D model, they were able to produces more accurate outputs compared to their current practise of using 2D drawings.

Based on the feedback from evaluation team and successful results from the evaluation project, KSM is planning to implement LPM solution in their fabrication yards in near future.

Evaluation criteria included training based on selective blocks from one of the KSM model and testing of LPM modules using out of the box configurations. Some of the modules evaluated were DAP, Paint area calculations, lifting and Scaffoldings.

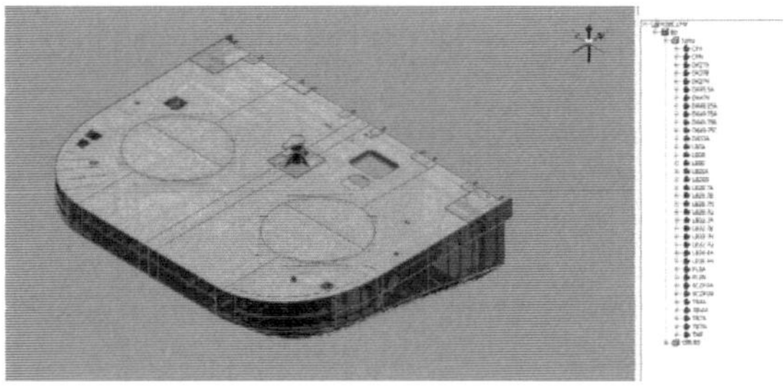

Figure 20: KSM block in LPM used for evaluation and training.

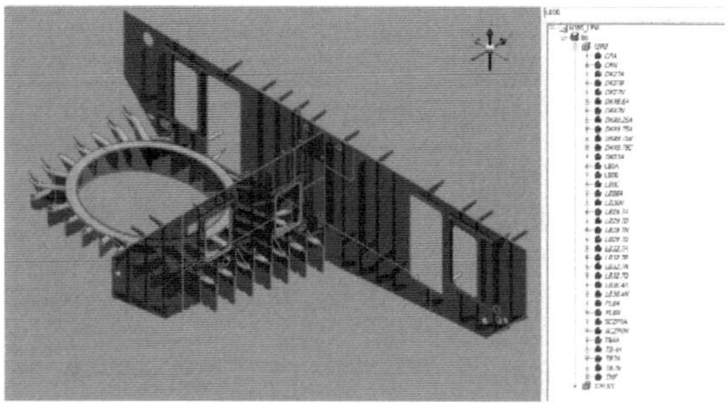

Figure 21: A KSM block in LPM used for evaluation and training.

Below figures shoes some of the outputs generated during evaluation project:

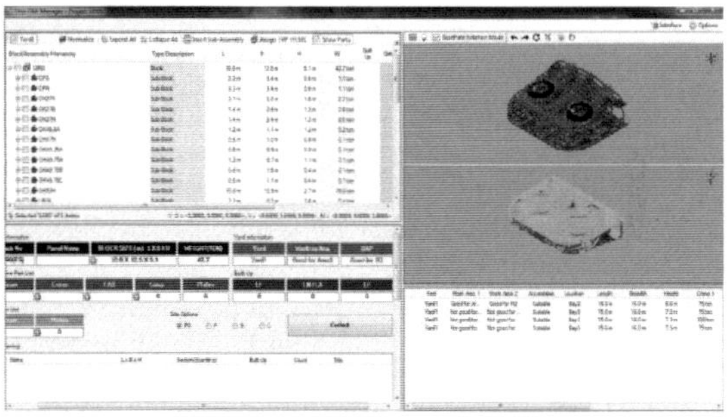

Figure 22: DAP setup from LPM using KSM block

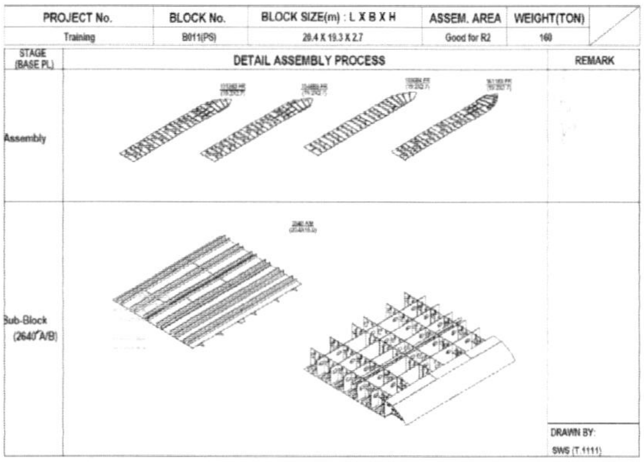

Figure 23: DAP drawing sample from LPM using KSM block produced during evaluation

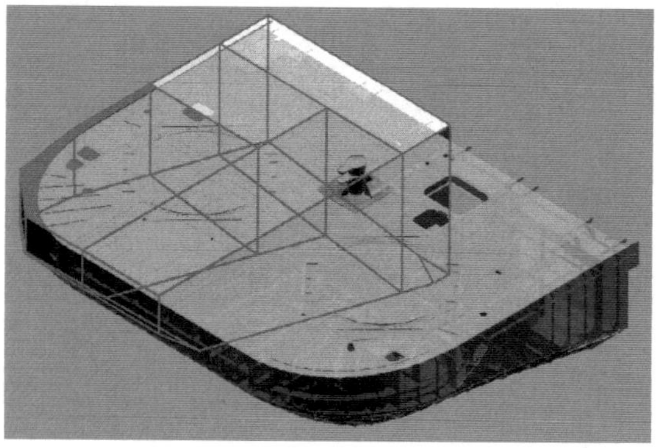

Figure 23: Paint volumes in LPM using KSM blocks

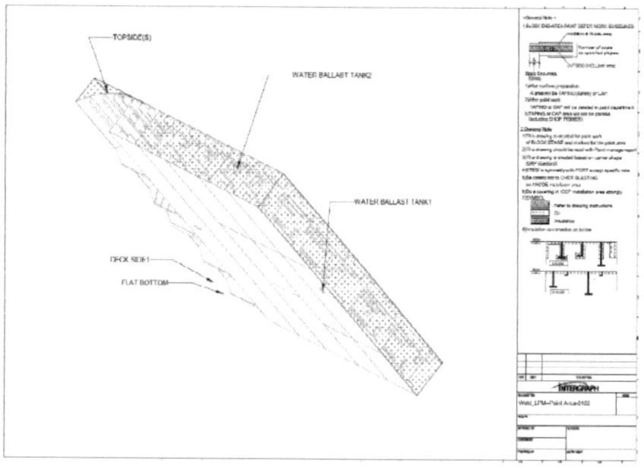

Figure 24: Paint area drawing sample from LPM using KSM block produced during evaluation

Since the evaluation was done using out of the box configuration, as a next step, KSM will work with Hexagon PPM to configure LPM to implement KSM specifications.

6. CONCLUSION

It is becoming more and more prominent that optimization of processes and integration of disciplines are necessary for any yard's survival. Often production planning department is ignored in the optimization process, which should not be the case. To improve production, planning must be improved also. In order to do their job with maximum efficiency, production planning department must be equipped with the right tools. LPM as proven within several customer implementations is capable of handling the light versions of the different engineering models (AM, PDMS, S3D) and hence providing the production planning departments a much earlier insight in what is coming as well as additional time to prepare all the necessary production tasks. This greatly impacts the delivery schedule of the project. It also de-risks the production process since the information is much sooner available.

The design agnostic approach makes it useful to handle any design provided at basic or FEED design stage, while the yard prepares for the heavy detailed and manufacturing models. Using the embedded revision management allows continuous parallel work and utilization of work already done.

Scaffolding, painting, lifting and transportation are always being a major production cost factor can now be much more accurately planned and scheduled using a real 3D model as the basis instead of redrafting 2D or 2.5D information.

8. AUTHORS' BIOGRAPHIES

Stalin P. Ybiernas is a CAD/CAM engineer under engineering department of Keppel Singmarine Pte Ltd, Singapore. He has been with the company for 3 years, providing project hull technical support, training services for S3D, R&D, and system maintenance and support. He has joined Keppel Singmarine Philippines last 2008 working as a hull design engineer and a CADCAM support prior to his posting in Singapore main office last 2014.

Najaf Bashir holds the position of technical Director at Hexagon PPM. he has been with the company for 23 years, providing support, training and consulting services to clients around the globe. For the past 7 years he has been working at Hexagon PPM (Intergraph PPO), Singapore helping Offshore and Marine clients with business development and leading the implementations of integrated solutions.

Davide Guzzi has been with Hexagon PPM for the last 16 years helping customers around the globe deploying and supporting Hexagon PPM technologies. For the past 11 years he has been working together with the Industry Overlay Group with strong focus on shipbuilding and fabrication industries. Since 2010 he holds the position of Business Development Manager for the Americas region working from the Houston office.

THE NEXT GENERATION OF ASSET INTEGRITY MANAGEMENT SYSTEMS

J Esteve, Bureau Veritas, France
A Tew-Kaï, Dassault-Systèmes, France

ABSTRACT

Ship owners, ship managers, and offshore units operators are faced with challenges to manage the lifecycle of their assets as information is often spread across various information systems or software. They need to easily access and consolidate all the data intuitively to quickly figure out the overall situation and thus identify faster the dangers related to the integrity of their assets.

Bureau Veritas and Dassault Systèmes have engaged in a two-year development partnership to address these challenges with a new generation of Asset Integrity Management System powered by the 3DEXPERIENCE platform. At the heart of this solution, a 3DEXPERIENCE twin of the assets enables to connect data coming from the real world to the asset integrity analysis in the digital world through 3D visualization and data analytics at any level (from unitary structural members up to the entire asset) made available to all stakeholders.

This paper will present how this new Bureau Veritas solution can optimize the lifecycle of assets through the following main capabilities: 3D model creation and lifecycle management, asset integrity monitoring, interoperability with information systems (ERP, CMMS, DMS ...), interventions planning, and integration of interventions results and findings in the 3D model. Beyond the software functionalities, both partners are convinced that it will serve as a business digital transformation catalyst.

This solution will enable comprehensive RBI of the hull structure as each individual element is tracked with its current or foreseeable degradation, thus enabling better and faster maintenance and repair decisions, therefore saving time and money.

NOMENCLATURE

AIM Asset Integrity Management
AIMS Asset Integrity Management System
CAD Computer-Aided Design
CMMS Computerized Maintenance Management System
DMS Document Management System
EPC Engineering Procurement Construction
ERP Enterprise Resource Planning
FEM Finite Element Method
FPSO Floating Production Storage and Offloading
FPU Floating Production Unit
IACS International Association of Classification Societies
KPI Key Performance Indicator
PLM Product Lifecycle Management
RBI Risk-Based Inspection
VPM VeriSTAR Project Management

1. INTRODUCTION

Ship owners, ship managers, and offshore units' operators are faced with challenges to manage the lifecycle of their assets as information is often spread across various information systems or software. They need to easily access and consolidate all the data intuitively to quickly figure out the overall situation and thus identify faster the dangers related to the integrity of their assets. In order to help these actors better manage their asset's integrity through their entire lifecycles, Bureau Veritas and Dassault Systèmes decided to leverage their complementarity and their sharing of a similar vision on digital transformation to develop a breakthrough new solution for Asset Integrity Management (AIM), now known as Veristar AIM^{3D}.

1.1 HISTORICAL PATH

Bureau Veritas has had two different development lines that are now converging into the requirements that have led to the birth of Veristar AIM^{3D}.

On one side, the development of tools link to the digitalization of work processes and analyses related to integrity Management. This is illustrated in the Figure 1 below.

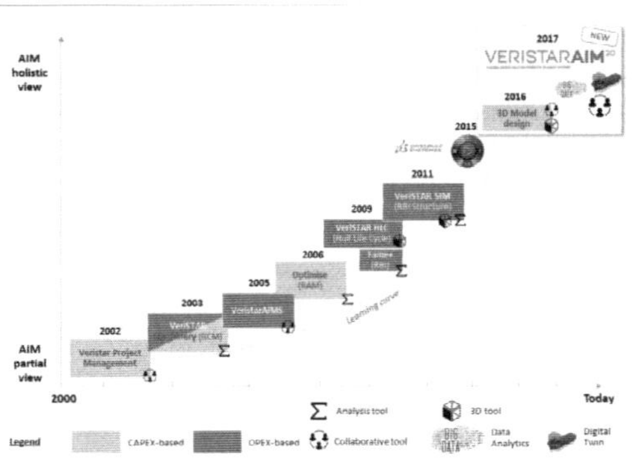

Figure 1: Bureau Veritas tools development related to AIM

It could be said that the main cornerstone was the arrival of Veristar Project Management (known as VPM) in 2002, although it should be mentioned that Veristar Hull, the 3D FEM analysis tool, was already available since 1997. VPM digitalized the plan approval process through a web-based collaborative platform linking Bureau Veritas plan approval offices and the yards and designers engineering departments.

Veristar Machinery followed shortly after, being Bureau Veritas reliability centered maintenance tool for systems installed on board of vessels.

Three years after VPM was launched, Veristar AIM was developed as the operations-focused version of VPM, this time targeting directly the owners and operators as the main users of the platform. Using the same technology as VPM, Veristar AIM was, and still is, a web-based collaborative solution where interventions, issues and corrective actions can be tracked for each individual asset of a client's fleet.

In the next ten years Bureau Veritas will develop:

- its first 3D digital twin software of hull structures, Veristar HLC (Hull Life Cycle) able to track thickness measurements and degradations on structural elements, individually identified.

- its first Reliability, Availability & Maintenability tool, Optimise®

- its first Risk Based Inspection tool for static equipment, cooperation with TOTAL, Fame+, this one based on API RP580 (American Petroleum Institute, 2002)

- its first Risk Based Inspection methodologies and tools for non-hull structures, Veristar SIM and Veristar SIM-topsides, both relying in the 3D representation module developed for Veristar HLC. Veristar SIM was developed in compatibility with the first draft of RP2SIM (American Petroleum Institute, 2009)

Except for Veristar AIM, these were all individual packages, focusing on different disciplines, illustrating the siloed mode of work at the time. Veristar AIM itself could not easily import/export data with these solutions, relying on manual operations to maintain the content in Veristar AIM updated.

On the other side, Bureau Veritas, as Classification Society, has been updating its rules in line with the Marine & Offshore sectors state of the art and requirements. In this sense it could be said that the Veristar AIM^{3D} features have been designed having in mind the requirements as described in the following additional notations (in increasing requirements order):

- **VeriSTAR-HULL,** reflecting that a structural tridimensional analysis has been performed for the hull structures.

- **VeriSTAR-HULL CM** reflecting that in addition to the structural tridimensional analysis a hot spot map has been made available for construction surveys and is kept on board the ship after delivery.

- **VeriSTAR-HULL SIS** reflecting than, in addition to VeriSTAR-HULL CM, the hull structure condition is periodically assessed using the results of the inspections and thickness measurements performed during the survey.

- **STAR -HULL** reflects the fact that Inspection and Maintenance Plan (a procedure including periodical and corrective maintenance, as well as periodical and occasional inspections of hull structures and equipment) exists and is dealt with on board by the crew and at the Owner's offices according to approved procedures.

The additional class notations VeriSTAR-HULL , VeriSTAR-HULL CM and VeriSTAR-HULL SIS are assigned at the design stage or after construction, whereas the notation Veri-STAR -HULL SIS is to be maintained during the service life (Bureau Veritas Marine & Offshsore, 2017).

1.2 OWNERS & OPERATORS CHALLENGES IN ASSET INTEGRITY MANAGEMENT

Only 5.4 out of 10 of Oil & Gas professionals are satisfied with their AIM program (IQPC, 2016), leading 35% of them willing to change their AIM provider. In the following, typical challenges faced by the Marine & Offshore industry are exposed.

First, owners and operators manage a varied fleet with assets of different age. As a consequence they need a global vision on their fleets and not lose time collecting all the relevant information.

Second, at the different stages of the asset lifecycle (design & construction, operation, overhauling, decommissioning) multiple stakeholders are working concurrently on the same single asset (the designers, yard, owners, operators, class societies, inspectors, etc...) A huge amount of data is generated by all of them in different formats, transmitted through different means and stored in several different locations, sometimes in local hard drives. Therefore relevant information gets lost or duplicated with inconsistencies between the different copies and over time.

Without consistent consolidation of the data, owners and operators cannot make data-driven decisions based on actual data and have therefore to make choices with partial information and by isolated individuals with a subjective comprehension on the situation.

Finally, there is an urgent need to reduce costs in a difficult and volatile economic context that lowers actors' financial performances. Finding new efficient solutions to make them save money and avoid costs overruns is critical.

1.3 BUREAU VERITAS AND DASSAULT SYSTEMES PARTNERSHIP

Both Bureau Veritas and Dassault Systèmes are sharing the same vision on Marine & Offshore digital transformation as a business transformation for sustainable profitability of the industry through significant business changes.

One of this changes is to switch from a management based on static documents to one leveraging model-based enterprise approach, switching focus on the data, converging around a centralized and unique 3D model of the asset that carries all the relevant information than can be leveraged to monitor the as-is actual condition and extract on-demand accurate deliverables in an automated fashion.

Digital transformation in Marine & Offshore also implies to break strong silos that are used to work in a disconnected manner in order to have all the stakeholders working concurrently and collaboratively on the same platform through the digitization and automation of business processes.

This shared vision is at the core of the partnership between Bureau Veritas and Dassault Systèmes, materialized in the VeriSTAR AIM^{3D} Solution.

As a leading classification society, Bureau Veritas brings its fifteen years of experience in asset integrity management and its strong expertise in Marine & Offshore. Leader in 3D modelling and PLM, Dassault Systems provide its large range of applications and its expertise in sophisticated software development.

2. VERISTAR AIM^{3D} OVERVIEW

2.1 VERISTAR AIM^{3D} MAIN PRINCIPLES

The objective of Veristar AIM^{3D} is to become an Asset Life Management solution for Industrial Assets combining a collaborative platform through web-access, a digital mock-up continuously updated to represent the condition of its physical homologue and an intelligent indexing system for KPIs reporting. This can be summarized into three solid pillars:

2.1(a) A Detailed 3d Digital Twin Of The Asset

The new VeriSTAR AIM^{3D} solution relies on a single 3D model of the asset that carries all the relevant smart data related to its integrity. This 3D model is the common support in which to:

- capture the integrity data from the real asset throughout the years

- the single source used as input to run the necessary analysis to evaluate the status on the integrity and the associated risks

- the single source from which to prepare the work packs for all physical interventions on the asset and the generation of support drawings, 3D representations and diagrams to help the interpretation of the work packs.

2.1(b) Smart indexation & data analytics

VeriSTAR AIM^{3D} leverages Information Intelligence capabilities of the 3DEXPERIENCE Platform to provide smart data through dedicated dashboarding made available for the various roles involved and 3D visualization of the integrity.

Veristar AIM^{3D}, once connected to legacy IT systems such as a ERP, DMS and CMMS, will be able to consolidate the data coming from each of these systems related to objects in the Digital Twin and bring to the user the knowledge it needs when it needs it.

It is easy to envision that, once a certain number of assets have their Veristar AIM^{3D} representation, enhanced data analytics, trends assessment and degradation forecasts will be a natural request by the integrity management teams, leveraging all the information built over time and now made available to them.

2.1(c) Collaborative Platform

In order to make the most of the first two main pillars of Veristar AIM^{3D} it is necessary that all the information made accesible through it and any new data brought in with time should be made available to all those involved in the asset integrity. Not only should they have the latest information with a simple search but have project management, tasks and corrective actions follow-up in the same platform, connected to the digital twin objects and leveraging the data analytics capabilities.

Bringing new data or identified inconsistencies should be straight forward and easy for all actors collaborating through the platform. This collaboration is to be made possible on different devices from office desktop to mobile devices in connected or disconnected mode.

2.2 TARGETED USERS

VeriSTAR AIM^{3D} is targeting all users that at one point or another will participate in the asset integrity management. Some of these users will not be expected to edit or add the data but find very useful to have access to the information available and updated. Others will be active actors in Veristar AIM3D and will have to edit and add information at one point or another.

Expected "Read-Only" Users of Versitar AIM^{3D} are the following:

- OIM / Superintendent
- Country / Fleet Manager
- QHSE Management team
- Class Surveyors
- Discipline Technical authorities

Expected "Authoring" users with creation and edition rights are the following:

- Inspection team management & members
- Maintenance team management & members
- Asset Integrity Manager
- Engineering team
- Consultants and Contractors
- Chief Engineer
- Mechanical engineer

2.3 TARGETED MODES OF USAGE

The single most common mode of access to Veristar AIM3D is expected to be through the web browser. It user will find its own made dashboard, the projects and tasks he or she has been assigned to and a 3D viewer to inquire directly the 3DEXPERIENCE twin.
The idea behind VeriSTAR AIM3D was to have a platform that would support an asset register through a 3D representation and that could communicate at individual component level with the existing IT systems that will have a part of such asset register, such as CMMS and ERPs. Users will use VeriSTAR AIM3D interface to retrieve/edit/add data independently of where it is stored or what is its custodian (the CMMS, the ERP or the DMS). This connectivity is illustrated by the process at the bottom of Figure 2 below.

Knowing that numerical assessments and simulations are done, not only at the design and construction phase, but also during operations for the assets targeted by Veristar AIM^{3D}, it was necessary to be able to export contents from it (3D models, metadata and component attributes) that could be used as input for such analyses (strength and fatigue 3D assessments, vibration analysis, risk analysis, accidents consequence modeling, hydrodynamics, stability, etc...)

Such analysis and resulting assessments provide important conclusions on the asset integrity and its operational margin. It is then necessary to be able to import back such results and attach them to the corresponding individual components and systems. This cycle of export/import is illustrated with the process on the right in Figure 2 below.

In parallel it is necessary to extract a part of the digital twin with all its richness of information and put it on the hands of the intervention/maintenance and repair teams, in a mobile, disconnected mode. With it, the workpacks can be discussed and prepared in advance using directly an up-to-date 3D representation of the asset, identifying access paths, logistics required and hazards. Reporting directly findings and measurements, attached to the individual components will then allow uploading it back to the Veristar AIM^{3D} that will digest this new information (after being validated) and update the asset register automatically. This cycle is illustrated with the process on the left in Figure 2 below.

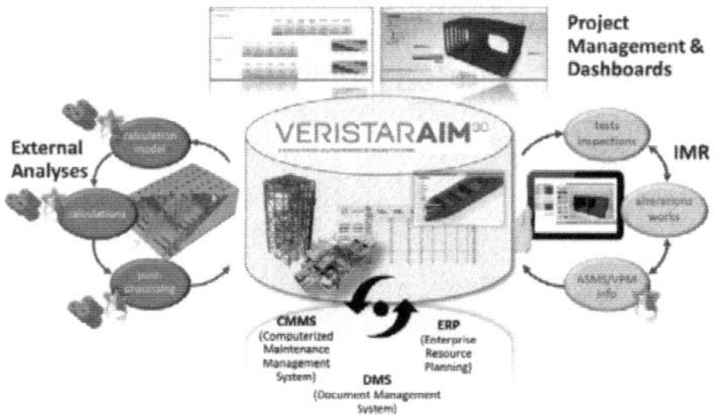

Figure 2 : VeriSTAR AIM^{3D} Solution Principle

2.3(a) SOLUTION HOSTING

VeriSTAR AIM^{3D} supports two kinds of hosting of the solution:

- Full Mode: VeriSTAR AIM^{3D} is installed on premise in customers' offices. In this case complete integration with legacy IT systems can be made.
- Remote Mode: VeriSTAR AIM^{3D} is hosted by by Bureau Veritas and accessible online by customers. In this case connectivity with IT legacy systems will need to be done through secured web-service systems.

2.4(b) Clients & Devices

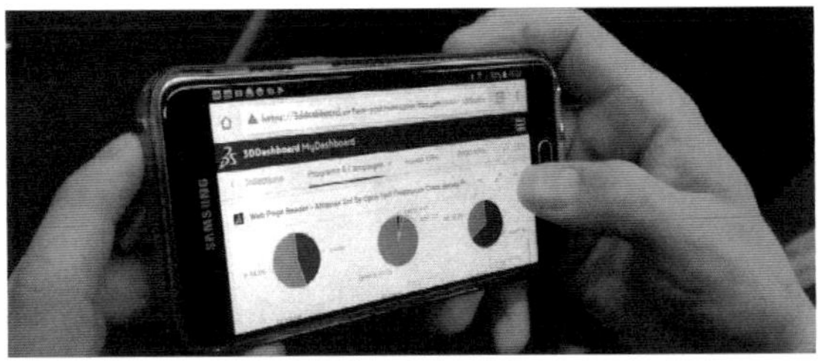

Figure 3 : Mobile access to VeriSTAR AIM^{3D} through

VeriSTAR AIM^{3D} has been conceived to have different modes of user interface, closely correlated to the expected roles:

- Users requiring to update or directly work with the 3D objects, update the digital twin with the latest intervention date and prepare the work package prior to a site entry, will work through a thick client, installed in their PCs
- Users in charge of intervention planning, validation of the inspection reports and coordination activities will need only use a Web Client, accessible through web browsers from any type of devices. Similarly, users simply needing to have an updated dashboard with the main KPIs they want to be aware of will use the web client.
- Finally users directly involved in the actual execution of the intervention will (also) make use of a Light Disconnected Client, installed on a laptop or a Microsoft mobile tablet, on which they will review the work pack requirements for the intervention on an interactive 3D interface as well as to capture the inspections findings. They can retrieve the work pack files through the web client, attached to the corresponding intervention, and they can uploaded in the same location once the intervention is finished, now enriched with the findings and measurements taken.

3. VERISTAR AIM^{3D} SOLUTION DESCRIPTION

3.1 The 3dexperience Twin

It is not only a Digital mock-up representing a physical asset with explicit modeling of actual components; structures, accesses, equipment, piping, instrumentation, electrical, HVAC systems and corrosion protection items., allowing as well the representation of spatial spaces or zones whether physically limited or functionally defined.

The 3DEXPERIENCE twin is also an intelligent asset breakdown arrangement and a dedicated data model that contains attributes and parameters that will allow track the condition of each component and compare it with design values. Thanks to such data model it can be possible to build the digital image of a complex asset even if there were no 3D representations of its components, that is, work as an asset register.
The 3D objects can either be built from scratch or imported from the yard or EPC CAD model.

The VeriSTAR AIM^{3D} 3DEXPERIENCE twin can include the representation of the following:

- Structural tanks (primary, secondary and tertiary structure, hatches, brackets anodes, walkways, coating):
- Hull and deck external (under and above water, sea chests, anodes, rudder and propeller boss)
- Hull piping below main deck (pipes, valves, couplings, pipe supports)
- Rotating machinery and equipment
- Non-hull structures such as topsides structures on Floating Production Units and fixed platforms
- Load bearing structures such as jackets and legs of jack-ups or wind turbines mats
- 3D spaces representing compartments, cargo and ballast tanks, hazardous areas or any other spatial breakdown such as no-fly zones.

3.1(b) Metadata

Metadata carried by the individual asset's elements can include:

- Design factors: Yielding and buckling ratios, fatigue damage factors, as-built thickness, owner addition thickness, pressure, temperature...
- IMR strategy and execution data: Next inspection due, last inspection done.
- Inspection results: degradations, thickness measurements
- Criticality ratings or RBI based factors: Safety Criticality Rating value, Risk rating (pair consequence, probability) resulting from the application of Operator's Criticality assessment or Integrity Management Methodology

3.1(c) Condition Data

VeriSTAR AIM^{3D} can store typical Hull inspection findings linked to the item where the anomaly was detected. Pictures and reports can be attached to each of the Hull Survey findings.

After every intervention (inspection or maintenance requiring a physical visit on the asset) the results are automatically uploaded, after being validated, by the use of a dedicated template (the light disconnected client). Veristar AIM3D identifies on which object of the digital twin have the findings been reported. These include not only thickness measurements but any degradation identified by the inspection team. Veristar AIM3D will then create an intervention results object, referred to the 3D item and dated. Interventions done from different spaces (for example from both sides of a plate boundary of two different hull internal tanks) will each result in different intervention results objects.

It will then be able to know all the findings reported on any individual 3D item chronologically.

In order to simplify the users data entry the template already includes a generic list of degradations, each of which with their specific parameters. Such generic degradations, typical on hull structures include (non-exhaustive):

- Anode Consumption
- Buckling
- Coating condition
- Cracks
- Deformation
- Localized corrosion
- Marine Growth
- Pitting
- Surface Corrosion
- Weld Defects
- ....

3.2 THE WEBTOP INTERFACE

VeriSTAR AIM^{3D} leverages the 3DEXPERIENCE platform to manage the interventions from planning and scheduling to management of interventions results.

3.2(a) INTERVENTIONS PLANNING & STAFFING

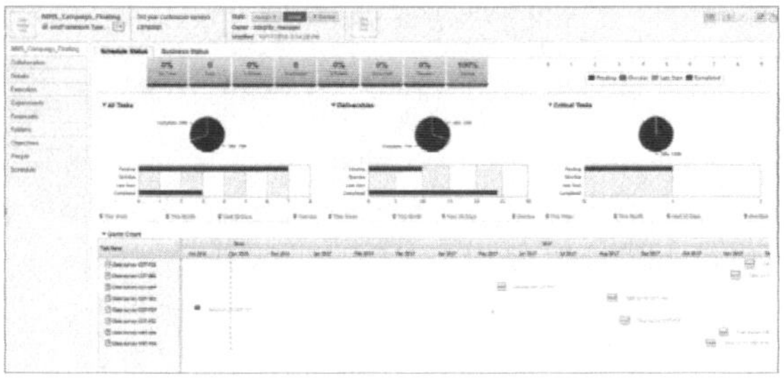

Figure 4: Project Management & Intervention Planning

VERISTAR AIM^{3D} provides the ability to program campaigns of interventions, each intervention being handled as a task within a project

Integrity Managers create the project structure based on templates of intervention tasks and assigned them to the users who access online the project information and the tasks that they are responsible to carry out.

Interventions tasks, also carry all the necessary documentation as well as the interventions procedures to follow.

3.2(b) DASHBOARDING & 3D KPIs

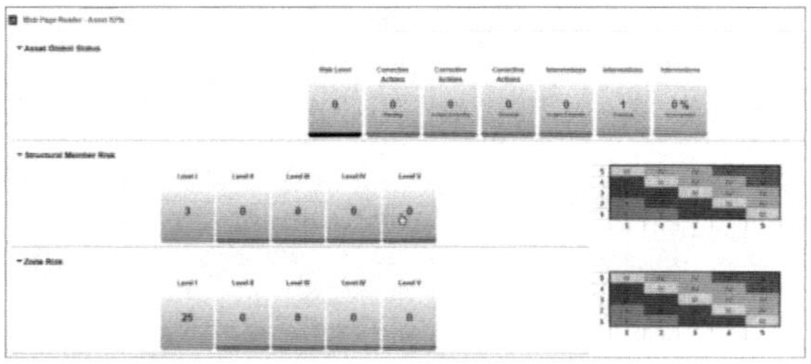

Figure 5: Asset Dashboarding

All users have specific access rights to read and edit objects and information, based on the role assigned to them on the platform. They can also access dedicated dashboards that leverage data from the 3D model to provide a quick access to the actual status on the integrity with comprehensive view on all the issues and anomalies and the ability to drill down in order to better analyse the situation

One of the best means to intuitively identify where the issues are or where the next inspections will be done is to present them directly as a KPI colour code on a 3D representation, as illustrated on Figure 6 below. The 3D viewer is automatically updated with the attributes of the asset components and the results from inspections.

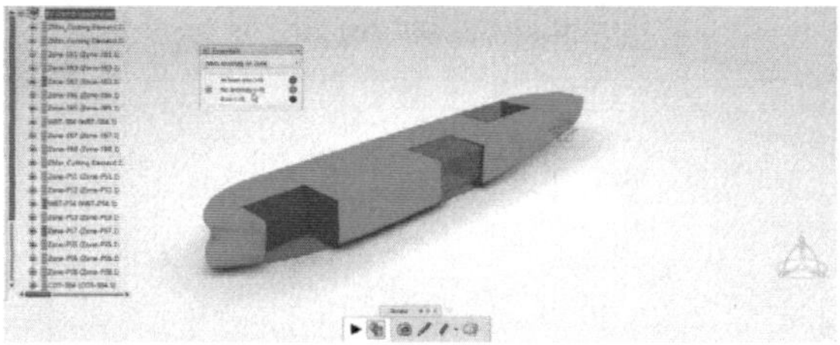

Figure 6: Asset Dashboarding

3.3 DISCONNECTED MODE

In order to prepare an intervention, users need to select in the 3D model the objects that are relevant for this intervention. In order to do so, in the thick client the VeriSTAR AIM^{3D} user queries the data base by zone and by attributes to quickly identify the elements to be inspected and their context.

Then the user will export this intervention context in a light disconnected application in which he can prepare the workplan according to the Intervention tasks defined in the Project Management interface in the webtop. The Figure 7 below illustration the preparation of tasks to be carried out.

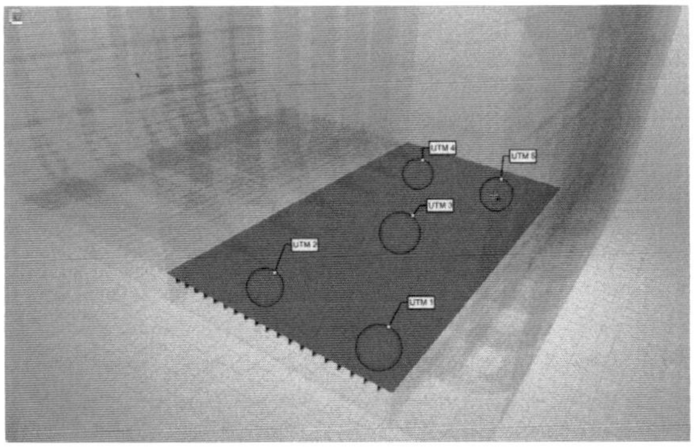

Figure 7: Creation of thickness measurements to be taken on inner bottom plating

This light disconnected application is embedded in a pre-defined Excel template form in which all elements to inspect are automatically listed with the preformatted cells to be filled after execution of the intervention, as illustrated in Figure 8 below.

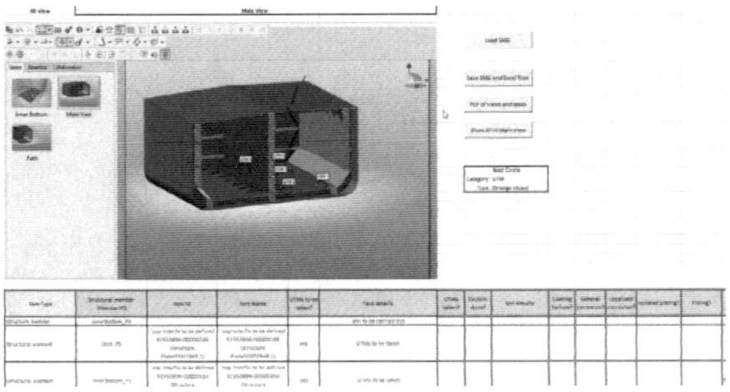

Figure 8: Creation of thickness measurements to be taken on inner bottom plating

After the intervention, the surveyor fills the excel with all the information gathered from the real assets.

The resulting populated Excel file will now act as the intervention report and can be uploaded, through the webtop interface as the final deliverable of the intervention, under the Project Management interface. From there the integrity manager can retrieve it, validate it and have it processed by Veristar AIM^{3D}, meaning that objects attributes are updated and findings created in the 3DEXPERIENCE twin as objects that can be easily retrieved or the information they carry leveraged in the dashboards previously presented.

At any moment the user will be able to export the data introduced in the template according to the IACS reporting formats as defined in UR Z (International Association of Classification Societies, 2017)

4. BENEFITS BROUGHT BY VERISTAR AIM^{3D}

The main objectives targeted by Veristar AIM3D are:

1- Improve the safety of operations and on the personnel involved through a better understanding of the hazards and visualization of the physical barriers

2- Improve the efficiency of the asset as a whole through:

   a. A better comprehension of all the activities to be carried out within the same location, pertaining to the different disciplines
   b. Providing better predictive tools that combined with risk based inspection methodologies and the 3D representation can indicate where an intervention has to be carried out

3- Provide global asset management, both at individual unit and at whole fleet level

4- Improve the reputation and the accountability of the operator as it can use Veristar AIM3D to proof the management of risks to the asset and that it keeps under control the condition of its units

5- Provide to the users an improved experience as well as simplifying their tasks

6- Reducing Inspection, Maintenance and Repair costs through:

   a. Avoiding double entries and keeping a continuous update of the condition of the unit that can be trusted by the users

   b. Reducing the number of exchanges required by different parties, simplifying the decision making and increasing the budget certainty for any intervention on-site

   c. Reducing the need for report preparation, reading and conclusion extraction by different parties. The data is directly made available to the users and the dashboards are automatically updated with the last information introduced

d. Avoiding the need of working on 2D static drawings that need to be manually updated for each intervention.

4. CONCLUSION

Bureau Veritas and Dassault Systèmes have partnered for the development and deployment of what is considered to be the digital transformation of asset integrity management. It follows all the stages of the unit life cycle as:

1- At the design and construction phase (or major refurbishment) it can be used to import and/or build the 3D digital twin, define its operating thresholds directly into the individual components for future use during operations, use it as Project Management tool following the design and fabrication activities and directly track the QA/QC and commissioning activities using the mobile disconnected mode

2- At the operations phase it provides tailor-made Dash Boarding for each user according to his/hers responsibilities and the assets followed, a Collaborative Calendar and Project Management tools tracking all intervention Programmes and the provision of the mobile disconnected application to support the Intervention preparation & reporting

3- At the repairs and modification times as it can be used to carry out modification evaluations directly on the 3D twin, export 3D model to repairs contractors and provide all the required data for decision making

4- At the end-of-life phase as it can provide all the elements to better assess the life extension costs or increase the marginal value of the asset as it can be used to demonstrate to buyers the actual condition of the unit.

5. NEXT STEPS

In addition of the continuous improvement of Veristar AIM3D from new technologies and feedback from users, it is envisioned to make it compatible to data model standards and asset management standards such as

1- ISO 55001 (ISO, 2014) (complemented by ISO 55000 and ISO 55002)

2- ISO 15926 (Internation Organization for Standardisation, 2003)

3- API RP 2SIM (American Petroleum Institute, 2010)

6. REFERENCES

American Petroleum Institute. (2002). *Risk-based Inspection API RECOMMENDED PRACTICE 580.*

American Petroleum Institute. (2009). *Recommended Practice 2SIM.*

American Petroleum Institute. (2010). *API RP 2SIM: Recommended Practice for Structural Integrity Management of Fixed.*

Bureau Veritas Marine & Offshore. (2017). *Rules for Steel Ships Part F (Additional Class Notations), Chapter 1, Section 1 and Section 2.*

Internation Organization for Standardisation. (2003). *ISO.ORG.* Consulté le July 2017, sur https://www.iso.org/standard/29557.html

International Association of Classification Societies. (2017). *Unified Requirements: Survey and Certification (UR Z).* Consulté le JULY 2017, sur International Association of Classification Societies: http://www.iacs.org.uk/

IQPC. (2016, 06). *ASSET INTEGRITY MANAGEMENT ABERDEEN 2016.* Consulté le 06 10, 2016, sur ASSET INTEGRITY MANAGEMENT ABERDEEN 2016: https://assetintegrity.iqpc.co.uk/

ISO. (2014). *Internation Organization of Standardization.* Consulté le July 2017, sur https://www.iso.org/news/2014/01/Ref1813.html

6. AUTHORS BIOGRAPHY

Jose Esteve holds the current position of AIMS Project Director at Bureau Veritas Marine & Offshore. He is responsible for the development and deployment of the Veristar AIM^{3D}. A Spanish Naval Architect (degree obtained at the "Universidad Politécnica de Madrid") with 15 years of experience in the Offshore O&G. Hired in 2001 by Bureau Veritas by the Commercial Department of the Marine Business Line, he started first as project engineer carrying out structural analyses and since 2005 worked as Offshore Projects Manager. In January 2011 he was appointed Manager of the Offshore Centre based in Singapore and from 2014 to 2016 he was the Offshore Business Development Director for the South Asia Zone. Holds a Master Degree on Energy Management delivered by the BI School of Management in Oslo, the ESCP Europe in Paris and the IFP (Institute Francais du Petrole) in Paris.

Alexandre Tew-Kaï holds the current position of Marine & Offshore Senior Business Consultant at Dassault

Systèmes. He holds a degree in n naval engineering from ENSTA Bretagne (ex-ENSIETA) and has over 15 years of experience in the field of PLM for Shipbuilding in Dassault Systèmes. He is currently responsible for the Marine & Offshore solutions and business strategy and in charge of the operations of the Marine & Offshore Industry team at Dassault Systèmes.

ADVANCED REPORTING AS A TOOL IN COMPLEX MULTIDISCIPLINARY ANALYSIS – CASE GENERIC SURVIVABILITY ASSESSMENT

R Kotiranta and K Partiala, Surma Ltd, Finland

SUMMARY

Modern ship design process includes several multidisciplinary analysis which produce significant amount of data. A good example of this kind of problem is a generic survivability assessment, whether used to analyze the safety of a passenger ship, an offshore platform or a naval vessel. The actual result is a complicated network of information, from which it may be difficult to find the key factors affecting the capabilities of the analyzed vessel, or the reasons leading to these. Traditionally, the analysis results have been static documents which do not capture the interaction between the disciplines. The designer may be provided, for example, with the final survivability level but not with the factors which actually lead to that level and how the design could be improved. This means that reporting of the results should also seamlessly support the design process.

This paper describes a novel HTML-based dynamic reporting concept used in SURMA survivability assessment. An introduction to a modern survivability assessment is provided highlighting the goals and key challenges in the process. Then the solution, informative, interactive and user-driven reporting is explained with sufficient use-case examples to show how the challenges mentioned are overcome. Finally some ideas regarding the different possibilities of future enhancements and means to extend the scope of application are discussed.

1. INTRODUCTION

The conventional ship design process is a complex effort. The path from concept design to a sailing ship includes multidisciplinary challenges, which become ever more profound when survivability considerations, a vital aspect from basically any vessel's perspective, are added to the equation. One of the current trends in ship design is aiming for adjustable modular platforms, thus further increasing the complexity and creating a serious challenge for naval architects. From this background it is clear that the design decisions need to be well measured, finely balanced and achievable in timely manner. Survivability considerations are overarching, hence a comprehensive, yet understandably reported survivability assessment is the key to a successful marine design.
Since the question of survivability is old and well known, many means of assessment have been established over the years ranging from rules of thumb to highly sophisticated software systems. In the field of software, basically two different concepts can be found. The first approach contains software dedicated solely for survivability assessments, typically using simplified geometric models and statistical and probabilistic methods to estimate the vessel's condition after encountering an incident. The advantage of this approach is that it doesn't require very detailed information about the vessel or the weapons, and therefore laborious modelling and time consuming calculations are not needed. On the other hand also the results might be very general without accurate consideration of the actual damage.

The other concept is more related to the physical simulation, and various models for numerical calculations (CFD, FEA) are used to predict the damage of a certain kind of physical effect, e.g. fire, collision, explosion, etc., in a specific part of the vessel. This detailed data from different sources is then gathered and the survivability of the vessel is assessed on a more general level. With this approach fairly detailed information can be acquired, yet associated with laborious modeling and computation.

SURMA approach, where the calculation model is fetched from the actual ship design model with all necessary data and the analysis is based on simplified, yet well proven and validated physical methods, is somewhere in between these two. This paper introduces SURMA survivability assessment as an example of a complex multidisciplinary analysis, where the different results might have complex effects on the overall design alternatives. The amount of various kinds of data associated with these results dictates implementation of a non-conventional reporting methodology to be of any use to the designer or any other stakeholder having to form opinion regarding the design. An advanced HTML based reporting utilized currently in SURMA is presented as a solution, with appropriate examples highlighting the usage. Some future development aspects are also discussed, followed by concluding remarks.

2. SURVIVABILITY AND THREAT

Having taken the mentioned general viewpoint, it is clear that the basic aspects of survivability are damage stability, longitudinal strength and the functionality of vital systems. In this respect survivability can be seen as a fundamental measure defining the vessel's performance when subject to any encountered hazard in the designed operating environment and various operating conditions. Thus the scope of survivability should be seen remarkably wide, including beside offshore platforms and navy ships also the merchant and passenger vessels, especially nowadays as the threat of a terrorist attack makes this consideration more significant. Some of the similarities between incidents encountered by offshore or war-fighting vessels and civil ships can be fathomed from Figure 1.

Figure 1: Some well known incidents from the previous history.

However, if focusing for a while to the offshore environment, the most common hazards encountered by the vessels and other platforms operating there can roughly be divided into two categories:

- Explosions and related fires.
- Collisions and related stability and other issues.

The first category follows from the fact that vast quantities of highly volatile materials are constantly being handled in this environment. The second is a consequence of production pressures and tight schedules – the shuttle tankers and supply vessels usually try to keep their timetables despite the sometimes adverse sea conditions around the production fields.

The various sources of both of these threat type are depicted in Figure 2.

Figure 2: Common threats in offshore environment.

As a comprehensive assessment tool, SURMA enables the definition of a threat scenario, which contains the specific type of incident and the parameters defined for it. An explosion scenario could for example consist of the energy content of the assumed gas leakage, and a threat distribution such that all compartments of the platform where gas is processed or through which gas pipes go, are marked as possible explosion locations. Hence even this simple threat definition with a typical sized production platform can lead to tens of explosion cases to be assessed. And the number of cases multiplies by the increasing number of different threat.

3. SURVIVABILITY ANALYSIS AND RESULTS

In a typical SURMA explosion assessment a case analysis starts from the ignition or detonation location, which is defined by the threat scenario. The parameters affecting the event are derived from the previous input provided by the user, thus the analysis itself is automatic. To give a better understanding on what is actually happening, a case where an explosion in air takes place is selected as an example, and the analysis procedure is described in more detail to clarify the different physical phenomena it includes.

3.1 EXPLOSION IN AIR

The analysis of an explosion in air first generates the pressure histories resulting from a detonation like release of equal energy content as the assumed amount of gas based on semi-empirical formulas. If the detonation occurs in a confined space inside the platform, also the quasi-static pressure build-up is calculated and the possible quasi-static pressure spread to adjacent compartments is analysed via a time integration scheme. If there is sufficient knowledge on the energetic properties of a particular gas, these calculations take into account the type and amount of the assumed gas leakage.

The semi-empirical formulas used in SURMA are selected based on their best correlation with performed testing and more complex numerical simulations. An arbitrary number of incidents with different parameters can be defined. The dynamic structural response is then calculated based on a single degree of freedom systems, constructed for every naturally formed structural element. The single-degree-of-freedom (SDOF) analysis includes non-linear structural behaviour such as the membrane forces, material models associated with high strain rates and different strain rates along the cross section of the structural member (higher strain rates in beam outer fibres). The collapse of structural elements is estimated based on equivalent plastic strain which is calculated for every loaded element during the analysis.

The incident type is selected from the alternatives defined in the previous phase, and the location is coming from the scenario definition. Detonation locations can also be defined outside the platform, hence these cases are calculated as non-confined, i.e. free-field explosions. The final result of the analysis part is a case description containing described data about damaged structures and damaged compartments. Thess enables the creation of further damage assessment cases, e.g. damaged stability calculation cases and damaged structural details for finite element analysis, if global structural model is available. Also system functionality is evaluated based on the damaged equipment, and all subsystems out of order are stored to the case description. All the assessed aspect regarding the vessel are shown in Figure 3 below.

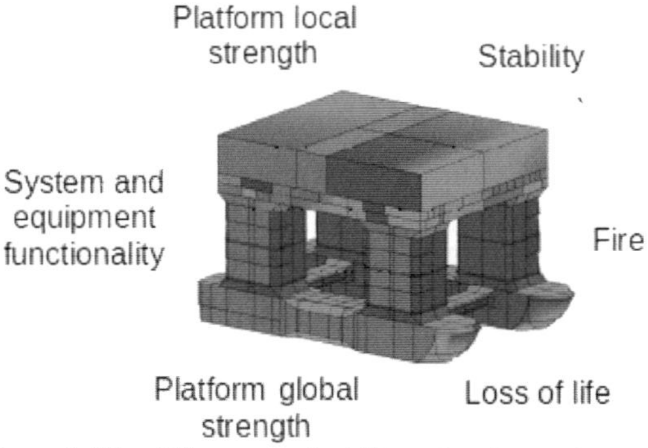

Figure 3: The different survivability related aspects assessed for a marine platform within SURMA.

3.2 SURMA ASSESSMENT PROCESS

In SURMA, survivability assessment process starts with reading in the platform design model e.g. NAPA, Rhinoceros, etc. If need be, additional blast resistant bulkheads can be defined. Also the known explosion related properties for energetic materials may be defined. The actual SURMA calculation model is automatically created within the software based on the platform design model.

A single threat scenario can consist of several hundred of cases. In SURMA threat scenario can be defined several ways, including use of incident grid and even or rule based incident distribution. After scenario definition the user may proceed to actual survivability assessment, which will take 1 to 10 seconds per case depending on the type of assessment. Result listings are produced automatically and are available for review in the user interface, also all assessments results containing the necessary data are stored in the SURMA database for further processing. In addition, an HTML report is automatically produced. The SURMA assessment process is depicted in Figure 4.

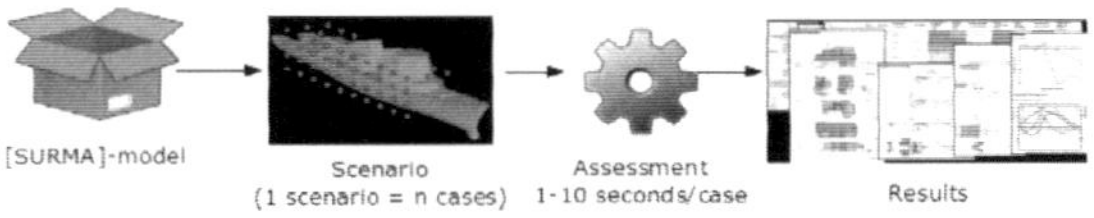

Figure 4: SURMA assessment process.

To interpreter the results, SURMA uses categorization system that consists of five levels. This system is called operational readiness level (ORL) and it takes into consideration the floating position of the platform, system and equipment functionality and the overall damage level. The user is able to set target values for each level and therefore control how the operational readiness is defined. The five levels of operational readiness are presented in Table 1 below.

Table 1: Categorization of operational readiness level. (ORL)

ORL1	Full operational readiness
ORL2	Decreased operational readiness
ORL3	Operational readiness compromised
ORL4	Minimum operational readiness remaining
SUNK	Stability lost or hull girder beam broken

4. SURMA HTML-REPORTING

Perhaps the most challenging part of an assessment software is the presentation of results. The results must be presented in a way that is intuitive, transparent and unambiguous and provides the designer with valuable information that can be used for improving the design. The traditional approach to reporting is to produce countless number of PDF pages, that might be cryptic to understand and therefore fail to actually produce information for the design process.

In SURMA the selected approach to overcome these challenges is the use of HTML format. As most people nowadays are familiar with websites, this is an intuitive format which also enables user interaction with the results. The hyperlinks connecting the different entities to each other, for example a compartment is connected to all the the equipment components it contains, make it easy to track the details within the SURMA model.

The core element of the report is a summary table that presents all the cases belonging to the analyzed scenario and the corresponding operational readiness level. For transparency, and in order to provide useful additional information, the levels comprising the operational readiness are presented with their contribution. An example of the summary table is presented below in Figure 5.

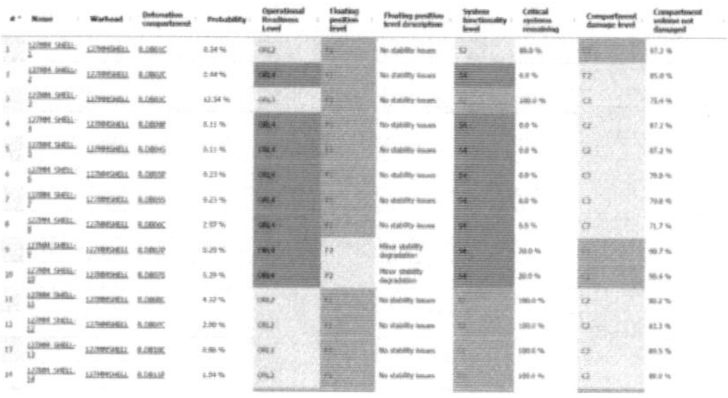

Figure 5: Scenario overview table.

A clear understanding of the big picture is crucial, if the aim is to find improvable issues in the overall design. With this kind of interactive and informative graphics the user is quickly able to get an overview of the assessed scenario.

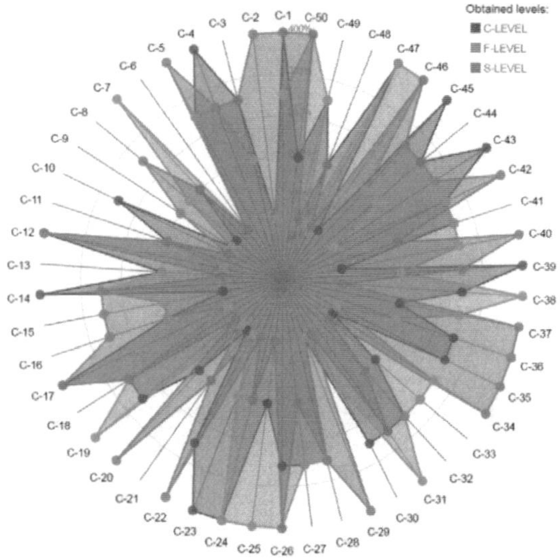

Figure 6: Radar chart presenting the obtained floating position, system functionality and damage levels for 50 cases. Example implementation done using D3 library.

Figure 6 above presents an example scenario that consist of 50 cases. For each case, the levels of floating position, system functionality and overall damage are presented as different coloured dots. Together these levels form the Operational Readiness Level (ORL) which is used to assess how operational the platform is following a certain incident. If all dots are on the centre, the platform retains full operational capability. The further away from the dots are, the worst is the ORL. As can be noted, with this type of representation of the results, it is easy to get an overview of the scenario and to understand why the operational capability is reduced for some cases.

4.1 FINDING THE WEAK SPOTS

In order to enhance the design based on the conducted survivability assessment, the designer should be presented with understandable and applicable information regarding what happens when the platform is influenced by an incident. To do this, SURMA relies heavily on graphical representations and sortable tables and lists regarding the assessment results. An example of the colour indicators denoting the possibility of damage potential is given in Figure 7.

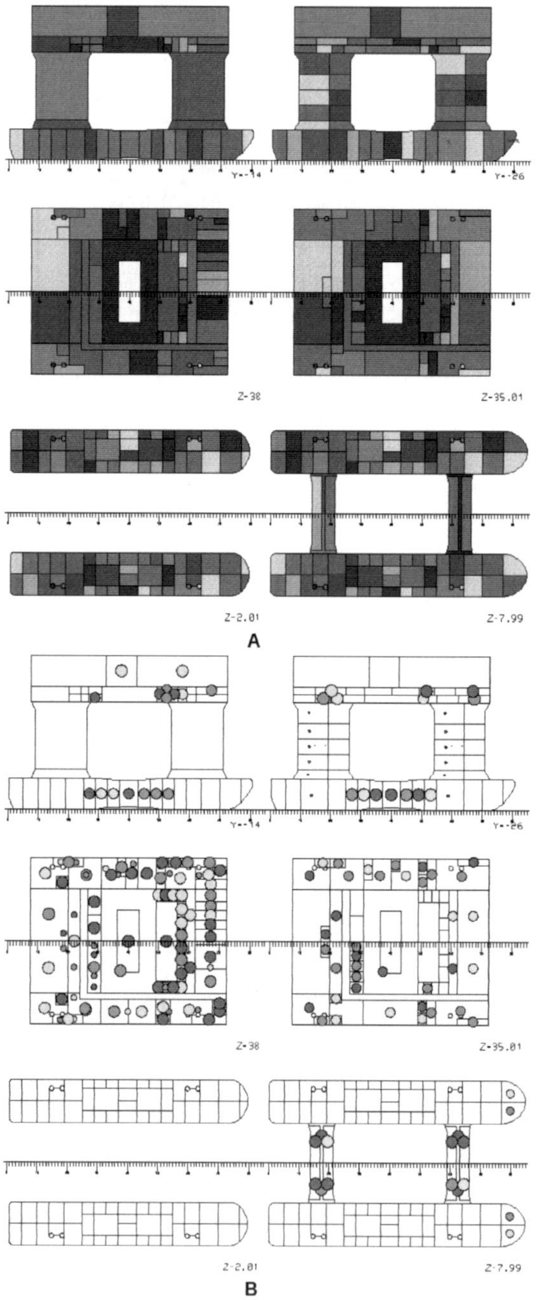

Figure 7: A: Compartment damage probabilities (the darker the color, the more likely the damage to that compartment). B: Explosion points colored according to their damage potential.

Detailed description of the suffered equipment and system damages (Figure 8, A) as well as detailed damage spread (Figure 8, B), aid the designer to understand what are the weak spots in the design.

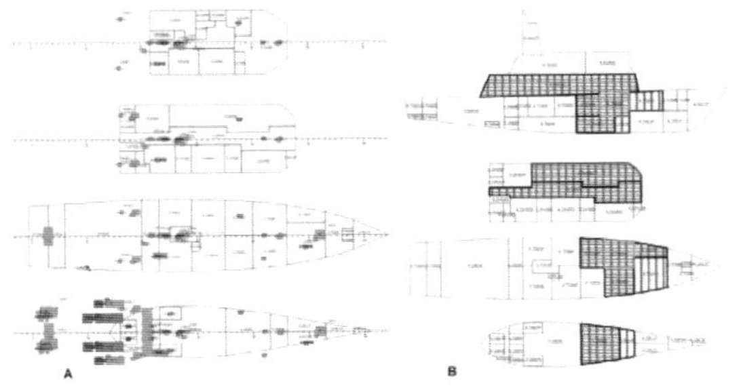

Figure 8: A: Equipment damage. B: Damage spread. Different colors and hatches refer to different types of damage (explosion overpressure, water).

5. LEAVING THE PAPER – 3D

The biggest differentiator between SURMA survivability assessment report and a traditional report is probably the third dimension. The use of HTML format enables the integration of conventional static report and interactive 3D-graphics without requiring any additional software. As mentioned earlier, most people nowadays are used to navigating on web pages with all kinds of fancy graphics, hence this approach shouldn't cause much of a headache. Figure 9 shows an example of 3D graphics used in SURMA reporting.

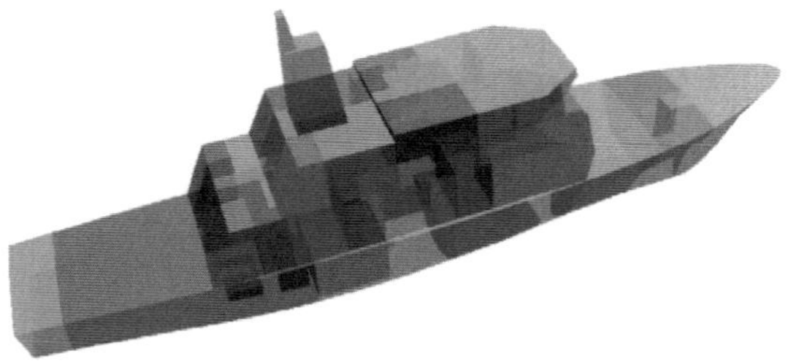

Figure 9: An example of the use of 3D-graphics to show damaged extents. Example implementation done using WebGL.

The use of interactive 3D allows the designer to even walk through the platform and see why, for example, a particular structural design solution might lead to unexpected damage when subject to an explosion. This approach also enables incorporation of 3D animation showing how damage propagates, if deemed appropriate.

6. FUTURE DEVELOPMENT

Regarding the reporting technology, HTML is a constantly developing technology and future enhancements of the format are bound to create new opportunities for improving and further enriching the presented reporting concept.

Future development of the presented reporting concept could include, for example, incorporation of results from other assessment tools using interfaces that enable saving of the results in a common database.

There is also room for improvement considering the methods used for processing the assessment results. As was described, a single threat scenario can consists of hundreds of cases, each of which can be associated with a probability based on statistical values. For these type of scenarios, it might be useful to deploy statistical methods to find the weak spots in the design.

7. CONCLUSIONS

In the current paper, the complexities regarding multidisciplinary analysis and related reporting have been demonstrated. It has been shown that HTML based reporting is able to overcome many of the obstacles the traditional reporting methods have. Being interactive, user-driven and intuitive, HTML technologies offer a flexible and versatile base for creating usable reports for survivability assessment. Being able to pinpoint the problematic areas within the designed concept, the reporting adds value to the design process and survivability assessment process in a way which traditional static reports are never able to achieve.

8. REFERENCES

1. Daley, J. 'Mumbai High North Platform Disaster.', *Coastal and Ocean Engineering Undergraduate Student Forum, COASTAL-2013.* 2013.

2. U.S. Chemical Safety and Hazard Investigation Board. 'Investigation Report Volume 3 – Drilling Rig Explosion and Fire at the Macondo Well.' *Report No. 2010-10-I-OS.* 4/12/2016.

3. http://cimsec.org/jihad-sea-yemen-al-qaedas-new-frontier/9733 Accessed: 29.5.2017

4. http://news.bbc.co.uk/2/hi/in_depth/photo_gallery/3493784.stm Accessed: 29.5.2017

9. AUTHORS BIOGRAPHY

Roope Kotiranta, M.Sc. (Tech.), has been involved in marine industry since year 2000. He has been working as a designer, software developer, consultant and project manager in various projects in several companies. After graduating from Helsinki University of Technology he has studied explosions, their effects to ship structures and equipment components and the use of product models for survivability assessment. Currently he works as the chief technology officer in Surma Ltd.

Kirsi Partiala, M.Sc. (Tech.), has been working in marine industry since 2007. In her previous work, she has been heavily involved in design and model testing of ice going vessels. After joining Surma Ltd in 2012, she has been working as a project manager in various development and research projects, particularly relating to utilization of product model information in numerical simulations. Currently she works as a development manager leading the SURMA software development.

ON A CONCEPT OF A 3D CAD/CAM SYSTEM BASED ON GEOMETRIC THEORY FOR CFRP PLATES MOLDING PROCESS IN SHIPBUILDING

K Matsuo, M Takezawa and **A Sakurai,** National Maritime Research Institute, JAPAN

SUMMARY

The paper describes a concept on a 3D CAD/CAM system for CFRP plates molding process especially for shipbuilding. CFRP should be a new material even for a ship, and we can already see a practical example to use a CFRP as material of a part of a ship. If we consider a manufacturing method of a CFRP molding process for curved plate surfaces, we meet various patterns of arrangements of carbon fibre along the objective free surface. The paper proposes the theoretical method based on geometric theory especially for free surface, and we propose new arrangement patterns of carbon fibre along with lines of curvatures on an objective surface. The paper explains our new method theoretically at first, and we show some examples which is made in our laboratory how our method is effective and reasonable. Finally, we summarize our future concept of a 3D CAD/CAM system especially for using CFRP as material of ships.

1. INTRODUCTION

The use of CFRP for structures has been advanced in various industries. CFRPs can be expensive to produce but are commonly used wherever high strength-to-weight ratio and rigidity are required, such as aerospace, automotive, civil engineering, sports goods and an increasing number of other consumer and technical applications. Also in ships, use for propellers and outfittings has been put to practical use [1]. From now on, with respect to the hull structure, the use of composite materials may be promoted, starting with the upper structure and also for strength members in the future. In the future, material of ship should toward the composite structure of various materials (steel, carbon, and organic material). Furthermore, application of new nanocarbon materials such as carbon nanotubes (CNT) and graphene may be introduced even in shipbuilding industries. Those materials is not only a good material for structure, also becomes functional materials such as semiconductors or the heat exchanger. The way of outfitting to ships also dramatically change, for example, embedding the electrical materials and printed circuit board directly into hull member by printing. Thus, the use of composite materials for ships has the potential to dramatically change not only the weight of the hull but also the shape and form of the ship. In this way, we have conducted research on new design and molding methods, especially for the application of CFRP materials to ships.

The paper consists of 5 chapters. Chapter 1 explains the purpose of this paper and the composition of the entire paper. Chapter 2 explains the theory of a line of curvature in curved geometry. This is because we propose CFRP design method and molding method based on geometric information of lines of curvature in this paper. Chapter 3 is a main topic of this paper, which explains the optimal placement method of CFRP carbon fibre based on the line of curvature on curved surface structure and its molding method. Then, in chapter 4, the concept of CAD / CAM system for CFRP structure design by our proposed method will be explained. Chapter 5 is a conclusion.

2. LINES OF CURVATURE

2.1 THEORY OF LINES OF CURVATURE

The lines of curvature are two pairs of continuous curved lines on the surface, and are set up by following the direction of maximum or minimum normal curvature. The direction is called the principle direction. The pairs of lines of curvature run at right angles to each other and indicate curve direction. In this paper, we set CL1 as the line of curvature that follows the direction of maximum normal curvature, and CL2 as the line of curvature that follows the direction of maximum normal curvature.

If a given arbitrary surface can be represented by a mathematical function, a lattice of lines of curvature can be applied to the surface, such as by using the Runge–Kutta method.

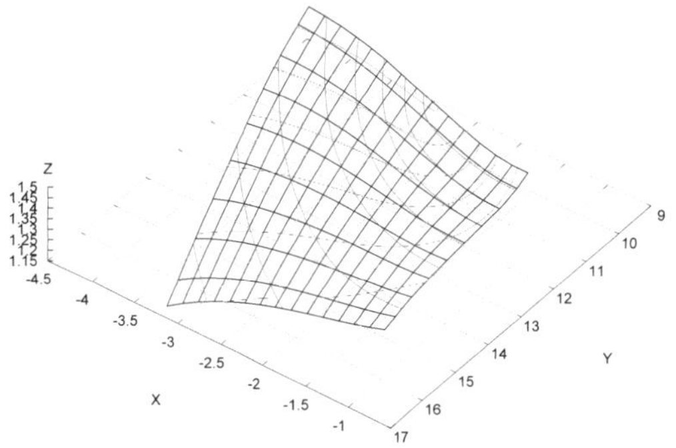

Figure 1: Lines of curvature on a curved surface

2.2 INDUSTRIAL USE OF LINES OF CURVATURE

The authors have already improved the design and construction method in shipbuilding by utilizing information of lines of curvature on the curved surface. As a concrete example, the authors are working on the optimization of shell expansion and sheet metal forming work in shipbuilding [2].

The type of surface that appears in ships is nearly always undevelopable. The Gaussian curvature of the surface is not zero, and the deformation to the objective curved surface requires stretching or shrinking. Previous studies have indicated that the best procedure for sheet metal forming is to first form a rough shape through a cold process, such as pressing or roll bending, and then to utilize a hot process for forming the undevelopable surface. The cold process must be carried out to minimize the amount of adjustment involved in the hot process. Both cold and heating processes should be performed along the set of lines of curvature. The fundamental reason why the set of lines of curvature should be optimum lines for sheet metal forming work is that a line of curvature can be deformed to the corresponding flat line by just removing its normal curvature. This feature eventually achieves the decomposition of the geometrical component on the surface into the outer bending component and the internal stretching or shrinking component. It is also important that there is no interference in the forming during both processes along the set of lines of curvature, because the pairs of lines of curvature run at right angles.

The lattice of lines of curvature for the objective surface is drawn on the developed plane, and this lattice pattern represents the location of press lines or gas heat lines. Press work is first conducted along CL2 to form the rough shape, bending along CL1 in the process. Gas heating is next employed along CL1 to form an undevelopable surface, adjusting shrinkage along CL2. Traditionally, craftsmen have developed their own strategies for forming the plate, based on their experience and intuition. However, it can be an effective measure in increasing efficiency in workers of every skill level to utilize this approach. Figure 2 is an example of actual sheet metal forming work performed along lines of curvature. At present, sheet metal forming work along lines of curvature is implemented by several shipyards in Japan.

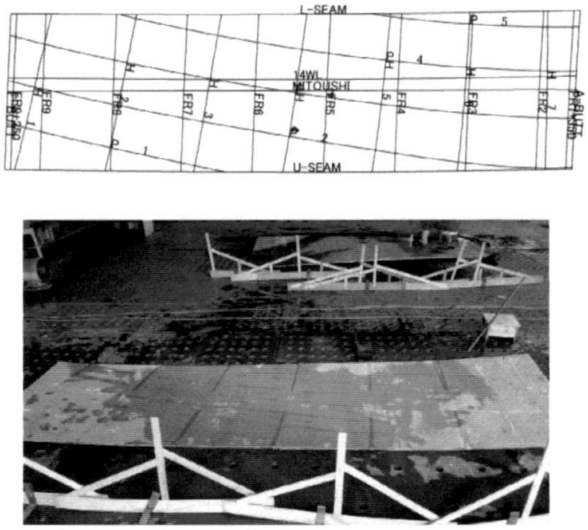

Figure 2: Sheet metal forming work along lines of curvature

Also, optimization of a structure can be cited as an example of utilization of information on a line of curvature for structural design. The vessel supports the load by the membrane and stiffeners on it. Optimization of the structure can be expected by placing a stiffener along the line of curvature of a curved surface. Figure 3 (Upper) is the result of the structural analysis by the conventional structural design in which stiffeners are arranged along frames. On the other hand, the figure 3 (Lower) is the result of the structural analysis by a new structural design in which stiffeners are arranged along lines of curvature with respect to the same membrane. As can be seen from figures, the structure along the line of curvature makes it easier to transmit the load. By placing a stiffener along the curvature line, it can be arranged perpendicular to a curved surface without twisting the stiffener.

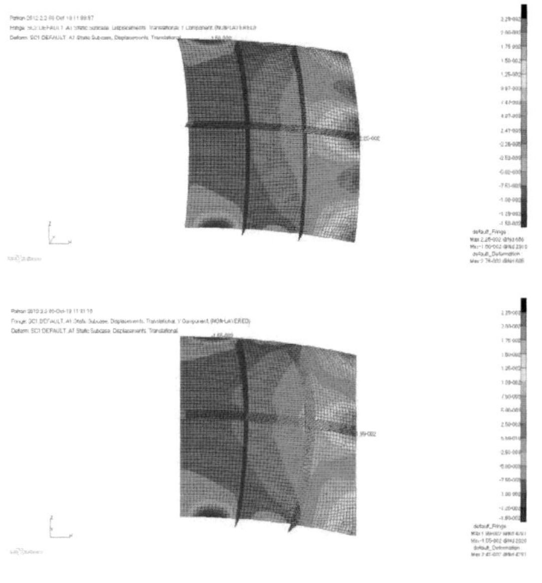

Figure 3: Structural analysis for each arrangement of stiffeners. (Upper: Frames, Lower: Lines of curvature)

In this way, efficiency and rationalization can be realized in various aspects of design and molding process by using information of the line of curvature on a curved surface. In this paper, information of lines of curvature is applied also to the arrangement of carbon fibers of CFRP besides these.

3. ARRANGEMENT OF CARBON FIBERS ALONG LINES OF CURVATURE

3.1 ON GENERAL MOLDING OF CFRP FOR CURVED SURFACE

Fiber-reinforced plastic (FRP) is a composite material made of a polymer matrix reinforced with fibers. The fibers are usually glass, carbon, aramid, or basalt. The polymer is usually an epoxy, vinylester, or polyester thermosetting plastic. Carbon fiber reinforced plastic

(CFRP) is an extremely strong and light fiber-reinforced plastic which contains carbon fibers. Carbon fibers include those in the form of sheets in which carbon fibers are knitted in a cross shape and those in the form of tows in which several thin carbon fiber filament are bundled together. Both of them are arranged according to the shape of a target shape and contain a matrix resin.

Figure 4: Sheet-like carbon fiber (Left) and carbon fiber tow (Right)

Figure 5 shows a scaled propeller molded with a composite material in our laboratory using sheet-like carbon fibers. The more complicated curved surface becomes, and the deeper the curved surface becomes, the more difficult it is to arrange sheet-like carbon fibers, and it is necessary to cut or fold into the sheet. This not only reduces the performance (structural strength, aesthetic appearance of the product) as a product, but also requires a considerable skill in order to precisely arrange the sheet-like fiber along the curved surface. The molding process varies depending on the worker, and the quality of the finished product also varies. In addition, the directional arrangement of the cross-shaped carbon fibers is not usually taken into consideration in many cases. This is also insufficient from the viewpoint of optimizing the performance.

Figure 5: CFRP propeller made of cross-shaped sheets

For this reason, a method of arbitrarily arranging carbon fiber tow on a curved surface, not a sheet-shaped carbon fiber, is adopted for a particularly complicated curved surface structure. Recently, in order to realize this, there are also sewing machines and three-dimensional printers which mechanically arrange carbon fiber tow in arbitrary directions. In this case, there arises a problem as to how to arrange carbon fiber tows, but there is no clear design guideline on this.

3.2 ARRANGEMENT OF CARBON FIBERS ALONG LINES OF CURVATURE

For this reason, the authors propose optimal placement of carbon fibers based on information of lines of curvature. This will optimize the structure of the CFRP curved structure and realize quantitative and objective design and molding process. Also, the arrangement of resin and carbon fiber can be reasonably adjusted in composite materials. This can adjust the material properties (yield stress, Young's modulus, etc.) at the local position on the curved surface and expect a new form of a curved surface structure.

The main point of the optimum arrangement of carbon fibers proposed in the paper is to arrange carbon fiber tow along the line of curvature on a curved surface. This aims to impart the following advantages to conventional CFRP molded products.

- Improvement of structural characteristics of curved surface structures. Reduction of plate thickness, weight and material use.
- Automated design of placement of carbon fibers. Improve molding productivity. Stabilization of product quality.

Regarding the actual molding process, instead of arranging the carbon fiber tow directly on the three-dimensional surface of target curved shape, the carbon fiber tow are temporarily arranged on the two-dimensional developed plane of the target curved surface. The reason for this is for convenience in the molding process. A carbon fiber tow is placed along a corresponding the line of curvature on the two-dimensional developed plane of a target curved surface, and then the three-dimensional shape is formed on the mold of target curved surface. This is basically the same principle as the approach of sheet metal forming work along lines of curvature described in Section 2.2.

We introduce concrete application examples on the arrangement of carbon fibers along the line of curvature. Figure 6 depicts the line of curvature in the shape of the hood of a car. According to our carbon fiber arranging method along a line of curvature, a carbon fiber tow is arranged along this line of curvature in principle. However, it is impossible to place the carbon fiber tow in the path of the geometrically obtained the line of curvature (Figure 7) as it is, and it is necessary to process the raw information on lines of curvature into the arrangement of the carbon fiber tow in the factory. The point here is as follows.

- Since a carbon fiber tow is arranged continuously as one long thread, the arrangement pattern of the carbon fiber tow must be written in a single stroke.
- The width of a carbon fiber tow depends on the type, but about 5 mm is popular. In order to have sufficient strength, arrange the carbon fiber tow as

much as possible without gaps, and cover the entire curved surface as much as possible.

- As shown in Figure 6, in distribution of the raw line of curvature, there is a gap difference between the lines of curvature, and coarse density occurs. Depending on the width, arrangement of carbon fibers should be such that the clearance is even.
- Sewing machines sewing carbon fiber tow in any direction on the fabric have restrictions on their operation. For example, it is the minimum radius of the carbon fiber tow arrangement. Considering these factors, it is necessary to consider the arrangement of carbon fibers.

Considering these, Figure 8 shows the drawing of the tool path of carbon fiber tow from original information of lines of curvature. Also, according to this tool path, Figure 9 is the one in which carbon fiber is sewn into the actual fabric. This is a carbon fiber tow sewn into a glass fiber fabric. Carbon fiber tow sewing in any direction was realized by dedicated sewing machine developed by Tokai Industrial Sewing Machine Co., Ltd.

There are two patterns of arrangement in the CL1 direction and the CL2 direction in lines of curvature. Thereafter, the two are superimposed on each other on a mold, and a resin is poured to form a target curved surface. On the objective three-dimensional curved surface, carbon fiber tows become lines of curvature of a curved surface and cross in the perpendicular direction in principle. From now on, curved surface structure with arrangement along lines of curvature will be actually moulded in our laboratory, and formability and structural strength will be evaluated.

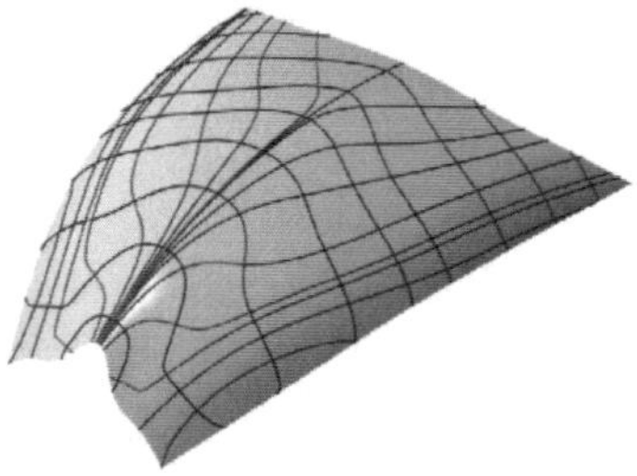

Figure 6: Three-dimensional surface and its line of curvature of automobile hood

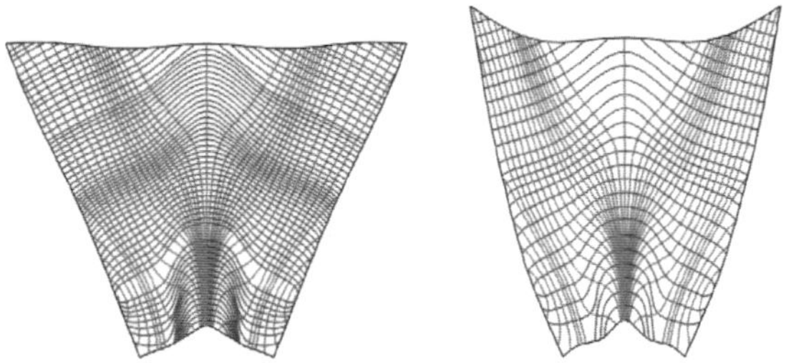

Figure 7: Developed plane and corresponding lines of curvature of automobile hood

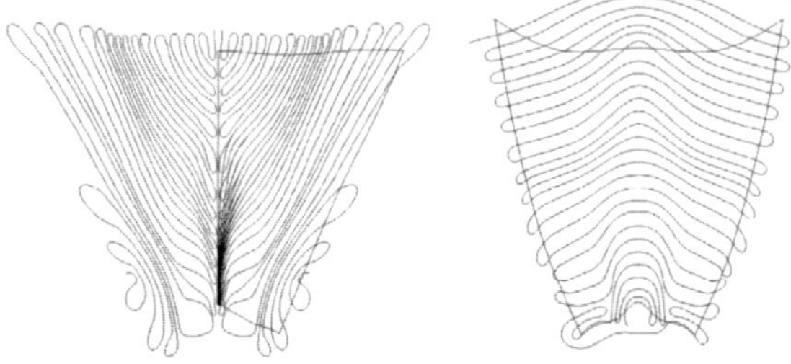

Figure 8: Drawing of the tool path for sewing with actual sewing machine

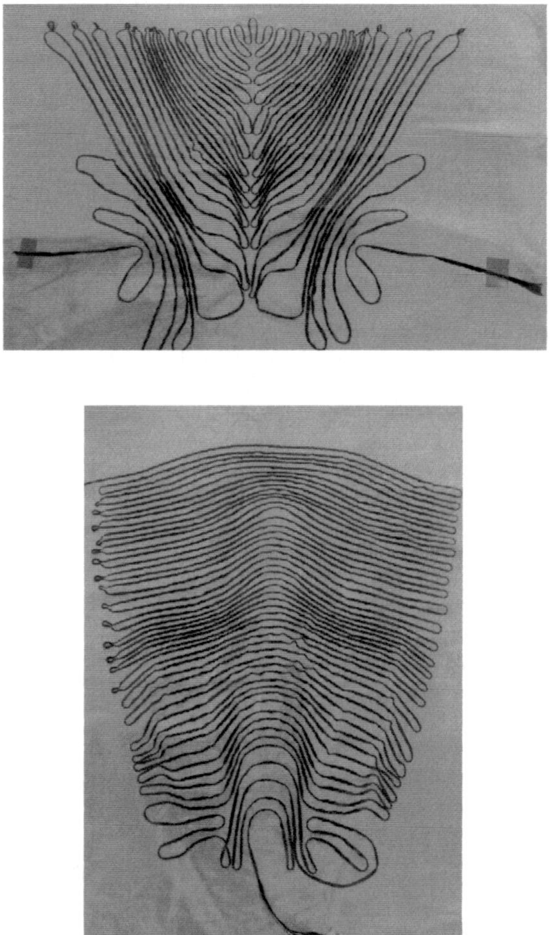

Figure 9: Pictures actually sewn with sewing machines

4. CONCEPT OF CAD / CAM SYSTEM FOR CFRP STRUCTURE

We are developing a CAD / CAM system for ship design and manufacture, assuming that CFRP is used for materials of hull structure and equipment. Here we introduce the whole concept of the CAD / CAM system. The CAD / CAM system provides the following functions when the structural members of hull and equipment are to be CFRP.

- Function to design the necessary structural characteristics for the structure
- Function to determine specification of CFRP to realize necessary structural characteristics
- Function to design tool path of carbon fiber and transfer those information to sewing machine

Figure 10 shows the functional flow of the proposed CAD / CAM system. Details of each function are as follows.

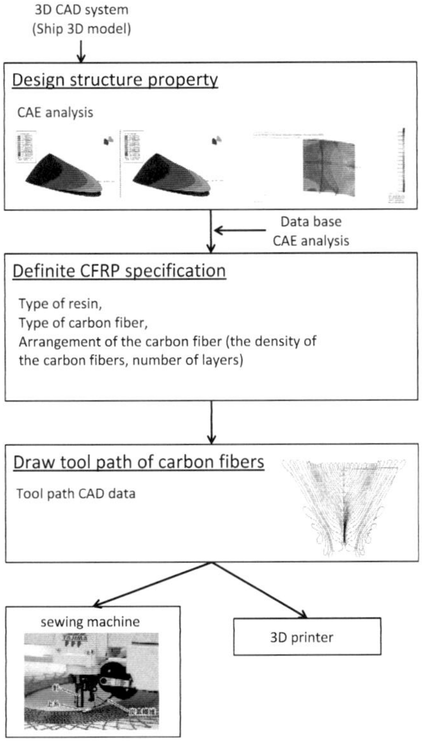

Figure 10: Functional flow of CAD / CAM system of CFRP for ship design and manufacturing

4.1 FUNCTION TO DESIGN STRUCTUAL CHARACTERISTICS

To satisfy the required function of the structure, determine the structural characteristics necessary for the structure. In the case of a hull structure, it generally corresponds to designing the necessary strength (yield stress, etc.) at an assumed load by structural analysis calculation. However, in the case of using a composite material, since the kind and arrangement of the resin and carbon fiber to be used can be flexibly changed, it is possible to realize a structure having material properties which continuously change from place to place. In this respect, it differs from the conventional structural design assuming steel plates.

It also covers functional products that intentionally cause deformation under specific load conditions. For example, for propellers for ships, propeller blades are deformed by water pressure under a specific use environment, and the occurrence of cavitation is reduced. In this case, necessary material properties (Young's modulus) are arranged at various places on the propeller in order to obtain ideal deformation. Elastoplastic simulation of a CFRP propeller is shown in Figure 10 as an example. By adding such CAE simulation to the CAD / CAM system, not only strength design but also various functions will be implemented.

4.2 FUNCTION TO DETERMINE SPECIFICATION OF CFRP

In order to realize the design result of the placement of the material characteristic to the structure designed in 4.1, determine the specification of the CFRP material. Here, the specification of the CFRP material is the type of resin, the type of carbon fiber, and the arrangement along the line of curvature of the carbon fiber (the density of the carbon fibers, number of layers). How to map CFRP specifications to the required material properties (yield stress, modulus, etc.) will be implemented by CAE analysis or calculations from the property database.

4.3 FUNCTION TO DESIGN TOOL PATH OF CARBON FIBER

This function is a function to automatically convert from raw information on lines of curvature to a tool path that actually sews carbon fiber tow with a sewing machine. Implement items organized in 3.2 shall be installed in the CAD / CAM system. The CAD / CAM system will attach the CAM function to transfer the information of the tool path to the sewing machine without burdening the designer.

5. CONCLUSIONS

In this paper, we introduced the concept of CAD / CAM system for CFRP, assuming CFRP usage for hull structure and marine equipment for the future. By realizing the proposed CAD / CAM system, the following effects can be expected.

- CFRP structure can be designed based on rational design philosophy. This can be expected to improve the performance of the structure, such as strengthening the structural strength.
- The design and manufacturing process of CFRP structure can be automated, contributing to improvement of productivity. In addition, the design and manufacturing process can be quantified, which makes it possible to stabilize quality and release from skilled work.

- By utilizing the composite material, the designer can flexibly select the physical properties of the structural member. By doing so, we can expect to create new products with a different concept from previous products.

In the future, we will promote R & D of each component toward practical application of the proposed CAD / CAM system.

6. ACKNOWLEDGEMENTS

This work is supported by the Japan Society for the Promotion of Science, Grants-in-Aid for Scientific Research under grant number 15H03910.
Automotive bonnet data was provided by Professor Maekawa from Yokohama National University. We express our deep appreciation to Prof. Maekawa here.

7. REFERENCES

1. T. Yamatogi, H. Murayama, K. Uzawa, Y. Ishihara, 'Study on Composite Material Marine Propellers', Journal of the JIME (The JAPAN INSTITUTE OF MARINE ENGINEERING) Vol. 46, No.3, 2011.

2. K. Matsuo, S. Fujimoto, 'Development and Practical Introduction of a Support System for a Cold Bending Work in Shipyard', The 12th International Symposium on Practical Design of Ships and Other Floating Structures(PRADS 2013), 2013.

8. AUTHORS BIOGRAPHY

Kohei Matsuo holds the current position of chief researcher at National Maritime Research Institute, Japan. He is responsible for manufacturing engineering of shipbuilding. His experience includes various approaches on ICT development for shipbuilding process.

Masahito Takezawa holds the current position of researcher at National Maritime Research Institute, Japan. He has research skills on CAD development in general and is currently engaged in research on the representation of structures in CAD for shipbuilding.

Akio Sakurai holds the current position of researcher at National Maritime Research Institute, Japan. He is engaged in the study of composite materials and has been proving a lot of work, especially in studies on the use of composite materials for ships. His composite material production technique is also excellent.

VIRTUAL REALITY EMPOWERED DESIGN

A Cebollero and **L Sánchez,** Sener Ingeniería y Sistemas S A, Spain

SUMMARY

In much the same way as other advanced technologies in the past, Virtual Reality has been used until recently only in few professional environments due to its relatively high cost. Remarkably, in engineering projects whose final product is available as a digital mock-up allowing immersive interaction, VR has already shown its great potential for design review, ergonomics and usability studies and simulation of assembly operations, among other practical applications.

At present, the increasing availability of affordable VR gear and amazing portable devices is favouring the development of new solutions in many more fields of application, including home entertainment, which will let us interact with the virtual world like never before. Definitely, VR has burst into our lives.

How these new ways of man-computer interaction are being translated into the shipbuilding business? In particular, how computer aided design and manufacturing systems are incorporating VR technology? How will the new generation of naval architects and marine engineers design?

Starting with an overview of the state of the art of this technology, our paper addresses the above questions and describes the benefits arisen from the immersive experience of true VR in the ship design and production process. Particular attention is paid to the latest VR-related features in Sener's Foran software.

NOMENCLATURE

CAD	Computer Aided Design
CAVE	Cave Automatic Virtual Environment
DLP	Digital Light Processing
GPU	Graphics Processing Unit
HMD	Head-Mounted Display
HVAC	Heating, Ventilation, Air Conditioning
LCD	Liquid Crystal Display
LCoS	Liquid Crystal over Silicon
LED	Light Emitting Diode
MTS	Motion Tracking System
OSVR	Open Source Virtual Reality
PLM	Product Life-Cycle Management
SDK	Software Development Kit
UHD	Ultra-High Definition
VR	Virtual Reality
WQXGA	Wide Quad eXtended Graphics Array
WUXGA	Wide Ultra eXtended Graphics Array

1. INTRODUCTION

The Virtual Reality (VR), which we can define as "a computer-generated digital environment that can be experienced and interacted with as if that environment were real" [1], is rapidly consolidating in response to the industry demand. It is no longer just about having 3D visualisation, moving objects or querying their properties. Now, those objects are expected to have the same behaviour as they would have in the real world, and immersive systems with much more advanced interfaces and a more natural interaction are required.

This paper is intended for ship designers and builders to get an idea of what is required to apply VR. We start with an overview of the most frequent solutions and their key components. Then we introduce the relevant features of Sener's Foran software, outline the main applications of VR in shipbuilding, and describe the new devices that are revolutionising our access to the virtual worlds. Finally, we dare to look into the future of CAD enhanced by VR.

2. VR SOLUTIONS AND CONFIGURATIONS

There are different VR solutions available in the market that are especially suitable for the shipbuilding industry. The decision on which one to implement is not easy, as they require specific hardware components that are in a constant process of improvement and there are a series of factors that may condition the choice, such as space, budget and intended use.

The simplest (and often only viable) way to easily and quickly create a practical VR environment is a transportable, self-contained system. Basically, it consists of a stereoscopic projector with a wide angle lens, a portable tracking system and a laptop computer, all packed in a suitcase, and a screen with foldable aluminium frame. When bright ambient light is a problem, a wide-screen 3D television may replace the screen and projector.

If the purpose is to set up a VR-specific room, there are several large format options. A good compromise between cost and performance is the high-resolution display wall or *powerwall*. Typically, a *powerwall* features a screen with a width larger than 4m and a rear projector to avoid shadows. In front of the screen, the working area is limited by the scope of the tracking system. These large flat screen systems allow a wide field of vision, which provides an excellent feel of immersion and interaction with the full-scale 3D objects shown.

By way of example, Sener provided a *powerwall*-based Virtual Design Center to Navantia Cartagena (pictured) in 2011, which has been widely used in the S80-class submarine programme, and another to Navantia Ferrol in 2012, for surface warship projects.

Figure 1: Cartagena VR Room (*courtesy of Navantia*)

With more space and budget, the solution may include more screens or projection faces, each one displayed by one stereoscopic projector. There are many possible configurations, from the basic two-wall (front and the floor or corner-type), through some additional side walls, to the most complete solution, the Cave Automatic Virtual Environment.

The CAVE consists of four, five or even six projection faces (typically to the front, left and right of the viewer, plus one on the floor, sometimes also on the ceiling and to the back) and delivers the greatest feel of immersion. Several persons can stand and walk freely between the screens and are immersed in the 3D world being displayed.

Figure 2: CAVE (*courtesy of ESI Group*)

In any case, the VR rooms must be dimensioned in accordance with the maximum number of participants to the working sessions, the number and size of the screens required, and the space for rear projection, among other parameters.

When analysing the cost of the solution it is also important to take into account the preparation and setting-up of the room. The performance of the equipment has equally an impact on the final cost.

3. HARDWARE COMPONENTS

The VR simulators require a series of hardware components, the most important being stereoscopic visualisation system, tracking system and computer system. We briefly describe them in the following paragraphs. Usually they also include hi-fi audio system, control system and specific supports. Their perfect integration is what creates the artificial reality where the participants can interact with objects in an immersive mode.

3.1 VISUALIZATION SYSTEM

The visualisation systems used in one-face solutions like *powerwalls* and multiple-face solutions like CAVEs consist of three main elements: screen, stereoscopic 3D projector and glasses.

3.1 (a) Screens

The screens can be for front or rear projection. Front projection screens reflect the light. The surface must therefore be highly reflective, so that the audience captures the image as perfectly as possible. At equal intensity, distance and size of the image, the front projection is brighter than that corresponding to the rear projection.

Rear projection screens diffuse the light through the material. The light source is in front of the observer, behind the screen, where space is needed for equipment and projectors. The surface must transmit an image with true optical clarity and brightness, it is therefore necessary to use a source of higher intensity.

There are screens of different types such as rigid or flexible, fixed or retractable. In terms of composition, there is the possibility of using different tissues and materials specially designed to optimize stereoscopy. Some screens can be used for both front and rear projection.

3.1 (b) Projectors

In addition to the projection size, which will obviously determine the light output, the following characteristics have to be considered when selecting a projector for a VR solution:

Type of stereoscopic projection: active or passive. In both the 3D effect is achieved by "deceiving" the sight: through the use of special glasses right and left eyes see a different image. In the passive stereo, both images are created with different characteristics, such as polarized light or colour. The active stereo is based on image separation either by location or by time.

Resolution: So far, most 3D projectors have provided Full HD (1920x1080), WUXGA (1920x1200) or WQXGA (2560x1600). For large format sizes there are already manufacturers offering projectors with 4K resolution (4096×2160) and it will not take long until projectors with 8K or UHD resolution (7680×4320) are available.

Projection technology: LCD, LCoS or DLP. In addition, DLP has two varieties: one-chip (with colour wheel) and three-chip (one chip for each primary colour).

Image quality: The three-chip DLP technology is widely regarded as the most advanced solution in terms of image quality. However, for many applications it is enough with single-chip DLP or LCD technologies, which within the same range of price, and despite their fundamental differences, offer similar quality and performance, with slight advantage for DLP in VR-related applications. DLP projectors tend to be brighter than LCDs, which lose some light as it passes through liquid crystal displays. As a result, DLP projectors can produce high brightness in small units and use less power to achieve the same brightness levels compared to LCD projectors.

Type of light source: until recently, there was only one type, the lamp. The introduction of metal halide lamps and then ultra-high pressure mercury lamps has greatly reduced the cost of bulbs and improved their life and reliability, but current projectors using LED technology offer significant advantages over lamp systems.

Maintenance costs: the projectors need preventive maintenance. If projection technology is based on lamps, they should be replaced every approximately 2,500-3,000 hours. LED technology projectors guarantee long-term projection, maintaining a stable and operational life with 80% brightness in periods of about 30,000 hours. And new models are already doubling that average life.

3.1 (c) Glasses

The glasses depend on the type of stereoscopic projection, passive or active. In the passive stereo, special glasses filter by polarized light direction (orthogonal polarizing directions for left/right eyes) or by colour (different colour –red/green or red/blue– for left/right eyes). It is light weight and cheap technology. The (cardboard) glasses are more comfortable to wear for longer, reduce eye fatigue, offer more angle of vision, but less resolution of the final image. Examples are large-screen projection systems.

In the active stereo based on image separation by location both stereoscopic images are created at different locations; each image is channeled to the respective eye by a separate optical system. It is heavy and expensive technology (typically two monitors and optics). Examples are head-mounted displays (HMDs).

In the active stereo based on separation by time, both images are separated in rapidly alternating sequence; shutter glasses, synchronized with the rapid display, block out each eye's view accordingly. Examples are CAVE, monitor-based VR and certain large-screen projection systems.

The active glasses offer higher resolution, need a brighter panel (they darken the image), weigh more and can be uncomfortable after a long time. If the refresh rate is not correct, the left and right image channels may leak or bleed into each other, producing a ghosting effect (crosstalk), flickering or blinking.

3.2 TRACKING SYSTEM

The ability to add a motion tracking system (MTS) is the most effective way to improve interaction with complex 3D objects or datasets in an immersive environment. Usually motion capture sensors are placed on the head (attached to the glasses) and the wands for the hands.

There are several tracking technologies. One of the most frequently used is optical technology (visual tracking). Active or passive marks are fixed to the subject to be followed and their position is stored. Passive marks are illuminated by radiation sources (infrared or ultrasound) and the rays are reflected towards a detector. Active marks emit rays into the detector by themselves. Fast and accurate update rates are important.

Other frequently used technologies are electromagnetic, mechanical, ultrasonic and inertial.

For electromagnetic tracking, a stationary transmitter creates electromagnetic fields, within which moving receivers or sensors register an object's position and orientation. It is a popular and reliable method of tracking, but has some drawbacks, such as interferences with metallic objects and limited range. With low-cost devices, it may give problems of accuracy, stability, repeatability and latency.

The mechanical tracking uses a direct mechanical connection between the object to be tracked and a reference point. Typically, this linkage is a mechanical arm with rotating joints that allow for six-degrees-of-freedom control. It features high accuracy and repeatability, and low latency, but also restricted operational volume.

The ultrasonic system utilizes three or more stationary microphones that receive an acoustic signal from a moving emitter. The differences in signal time delay and triangulation yield tracking data. It is a low-cost approach, but has problems with occlusion (line of sight) and echoes, as well as limited accuracy in orientation.

The inertial tracking system uses a couple of miniature gyroscopic sensors to measure yaw, pitch and roll. These are not true six-degrees-of-freedom sensors because they measure only orientation, but they can be effective as head-tracking sensors where position information is not essential.

3.3 COMPUTER SYSTEM

In order to manage in these immersive environment the vast quantity of information contained in ship 3D models, a powerful computer system, tailored to get the most out of processors, memory, graphics cards and software, is indispensable. The hardware is in continuous evolution, nowadays computers with 4-, 6- or 8-core processors are commonly used. Some motion tracking systems require a dedicated computer to process motion calculations.

The main component of the computer system is the graphics card. Most hardware provided by different manufacturers for use in VR solutions features cards with powerful Graphics Processing Units (GPU). These devices allow to incorporate even several graphics cards.

When the VR simulator is equipped with more than one projection face, it is necessary to have computers of similar performance for each of these faces, typically configured in a master-slave architecture.

4. FORAN SOFTWARE

Foran is a computer system specifically developed for design and construction of all types of ship. It has been developed, marketed and maintained by Sener for more than fifty years.

One of the essential characteristics of Foran is the integration of the entire range of ship design disciplines: naval architecture, hull structure, outfitting and electrical. This, combined with the seamless continuity between the basic, detail and production stages of the design and the homogeneous development of all its modules, makes possible the highest level of concurrent engineering. Foran is also fully integrated with PLM systems such as PTC Windchill or Siemens TeamCenter.

Foran employs a comprehensive 3D product model as the single source of all data relating to the design and construction of the ship, which is defined in every detail. Associative or topological relationships between components permit the on-line construction of the model, propagating automatically modifications to all related components. Stored in an Oracle relational database, the 3D model (or digital mock-up) is created once and increases in fidelity as the design matures and progresses from concept through detail.

The design is not oriented to just 3D modelling. The Foran model contains not only each ship component, but also material definition, manufacturing data and production process.

From the outset of the VR technology, Sener incorporated it into Foran through its own and fully integrated development. Around the end of the 90's, the VISUAL3D module for walk-through design review already featured two of the necessary characteristics for creating VR: immersive navigation and stereoscopy. Contrary to the rest of the modules of the System, this program did not read the information from the database, but from proprietary format files, which brought some advantages such as faster reading, simplified geometry and smooth navigation through and around large models.

In the launching of the version 80 of Foran, in 2012, the FVIEWER module superseded VISUAL3D. The new module was developed for 64-bit platform, which allowed using much more RAM and therefore possibility to navigate throughout large 3D models with fluidity and ease, even when the stereoscopy is activated.

Like its predecessor, the use of special 3D files to load the information continued to be the choice in order to get acceptable performance in the rendering of the scene of a whole ship. As to its features, we could highlight the following:

- Native 64-bit / OpenGL
- Immersive navigation
- Access to different product trees
- Consultation of object technological data
- Measurements (distances, coordinates, angles)
- Annotations
- Different clipping planes of the model
- Interference detection
- Simulation of assembly/disassembly of equipment
- Movement paths
- Incorporation of light points and textures
- Multiple camera positions
- Creation of screenshots and videos at 4K resolution
- Virtual humans in different work positions
- Stereoscopy
- Motion tracking system support
- Connection to HMD devices

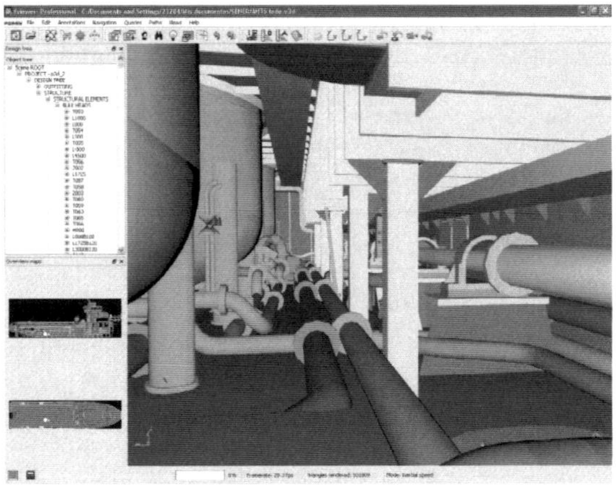

Figure 3: Screenshot of FVIEWER module

This is a complete set of functionalities oriented to design review, evaluation of design alternatives, ergonomic studies, simulation of construction, operation and maintenance tasks, and sales and marketing [2].
Taking the implementation of VR technology one step further, Sener and ESI Group, a leading simulation software company, have teamed up to provide the most advanced solutions for the shipbuilding industry by perfectly integrating Foran with IC.IDO.

IC.IDO is a powerful VR solution by ESI Group that combines high-end visualisation and real-time simulation of product behaviour in its actual size, and allows product operation very close to reality. It offers an excellent compromise between performance, working efficiency, visualisation quality, real time physical simulation of dynamic components and real time user interaction.

The integration of Foran with IC.IDO add some advance features to the above mentioned for FVIEWER, including:

- Advanced immersive user interface
- Advanced rendering of models
- Realistic presentations and animations
- Advanced ergonomics with virtual humans (in any position with total freedom of movements and field of vision)
- Tools operation
- Part and assembly behaviour
- Point cloud management
- Data import from different CAD sources
- Physics simulation (collision, friction, gliding, flexibility, kinematics, ...)
- Multi-channel support (optimized for complex CAVEs or rooms with multiple projection screens)
- Cluster infrastructure
- Multi-site collaboration

5. VR APPLICATIONS IN SHIPBUILDING

VR is one of the technologies that has the greatest future for engineering and manufacturing since it can be used to improve the design and planning in an interactive collaborative environment, providing an unprecedented understanding of the product in a complete 3D immersion.

VR can be used in different phases of the ship project, from the initial stages of engineering to the maintenance phase when the ship is in service.

A constant review of the engineering can avoid unnecessary development costs and delays in the production phase. Designers may make mistakes that can easily be identified in a virtual validation process.

The VR allows, in a much easier way than with a classic 3D CAD environment, to locate faults in the model. In addition, VR users can attach annotations related to problematic elements and share them with designers.

At this stage, it is also vital to consider different design alternatives, especially in complex projects where the information needs to be constantly reviewed, analysed and verified.

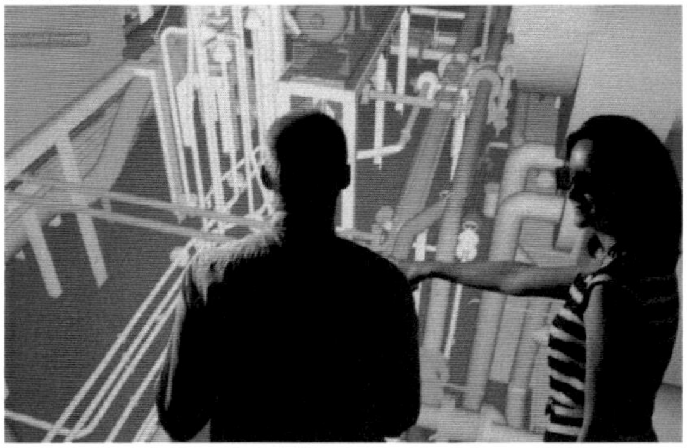

Figure 4: VR-supported design review with Foran

Another important application of VR is the validation of the production information, since any problem not detected will have an impact on the construction cost and schedule. Virtual manufacturing allows to assess if the design can be built and likewise help to define the best sequence of assembly operation.

On the other hand, having a tool that allows to simulate any manufacturing task can help to efficiently train the operators that will execute it and produce documentation, either images or videos, of the whole process.

Once the ship is delivered, regular maintenance should be scheduled. If the design does not consider this fact (which will have to be done in any case) the maintenance operations will be inefficient and will incur increased costs. All service and maintenance operations can be simulated in the immersive system. The goal is not only to learn and practice these operations, but also to optimize them. There are situations in which the crew must make repairs in difficult navigation situations and the optimal training can help. As a complement, creating 3D documentation as videos can help facilitate the work of operators and in many cases overcome barriers such as language.

During and after the development of the project it is necessary to present it to customers, suppliers and other ship-owners. In this regard, VR is a very effective way of communication. The image of the company is reinforced when performing a sales or technical demonstration in which the ship looks real even before it has been built. Definitely, VR can be one more tool for the sales and marketing team.

6. NEW DEVICES

VR has come to our homes. The race to offer new forms of entertainment and open new markets encourages major video game manufacturers to launch new devices and develop games specially designed for them.

For years, the use of HMDs has been reserved to the business world due to their high cost. Segments such as military or aerospace have been the first to adopt this type of devices in different programmes. The picture has changed radically and now there are affordable VR gear and amazing portable devices that allow us interact with the virtual world like never before.

There are from complements to adapt our smartphone to a housing equipped with lenses so that we can experience a stereoscopic visualisation with certain apps, to 3D glasses equipped with displays that provide us a feel of immersion very similar to what we would experience in a sophisticated CAVE.

Here we focus on those portable devices that currently meet the requirements of VR for shipbuilding with acceptable performance, in other words, with graphic processing capability enough for handling the heavy 3D models of a ship. Good examples are Oculus Rift or HTC Vive, in combination with a powerful personal computer equipped with a graphics card suitable for stereoscopic rendering (nVidia GTX 970, AMD Radeon R9 290 or superior) and at least 8GB RAM.

These devices, comfortable and light, offer the possibility of using motion tracking controls as an extension of our body, include different sensors like accelerometers and gyroscopes.

And the manufacturers supply software development kits (SDK) for third parties, for instance CAD programmers, to develop applications, built upon a powerful and complete rendering kernel, that exploit the full capabilities of the new devices.

When developing said applications some fundamental aspects must be considered:

- The design of the motion system within the scene cannot follow the conventional rules, as special care must be taken not to cause excessive dizziness in the user.
- The ability to interact with nearby objects in a more natural and flexible way requires higher quality in textures.
- The interaction with the environment is different: not only through the use of pads, but also by hands tracking.
- The high resolution of the screens (usually 2169x1200) requires graphic processing capability to achieve from 60 to 90 frames per second.

Rendering, image ratio, working resolution, head-tracking operation, point-of-view control, position tracking, scene acceleration control, motion speed, camera position, stereoscopy settings, user interface, ... are just some of the aspects that software developers must be aware of to obtain the desired result in the immersive experience.

It is not easy however to develop a valid application compatible with different devices. That is why the OSVR project presents an open SDK to create universal content for VR that can be targeted by device manufacturers as well as the open development community. The adoption of this SDK by all manufacturers would allow in the future the compatibility of third-party software with a wide variety of glasses and new VR devices that may appear in the market, with few modifications at the code level.

The OSVR project is led by Razer but other leaders in VR devices such as Nod and Leap Motion also participate in it. OSVR also has its own hardware development kit, which includes a device (OSVR HDK2) similar to the Oculus Rift.

Oculus Rift, HTC Vive or OSVR HDK2 are examples of VR gear currently available, but new hardware and software development kits are appearing on the market, for a growing community of developers to design more and better applications. The growth of this market will make prices even cheaper and VR definitely incorporated into our everyday life.

7. THE FUTURE OF SHIPBUILDING CAD

In the industry in general, the shipbuilding field in particular, most VR tools have their application when a 3D product model is available. For uses such as analysis of the production information or operators training, their application is possible as the 3D information is practically the final version given the design maturity.

As previously mentioned, main applications of VR in ship design and engineering are the review of the model aimed at identifying errors and the evaluation of alternatives. This review-and-modification process is realized mainly in projects where there are complex zones, such as the engine room of a ship or a cramped submarine, and it is necessary to iteratively refine the 3D model, moving from the CAD workstation to the VR simulator successively until the project is validated. This process is costly, as it implies generating the 3D models, transferring them to the VR software and reviewing them with some additional information.

The immersive environment, with the designer "inside" the scene, permits much better understanding of the space in a much more natural way. And it would be nice to be able to act directly over the 3D model and make the necessary changes there. This would save time, especially in those of high complexity.

Some typical tasks for which this new way of working would be really feasible are: equipment layout (positioning and displacing items and even affecting surrounding elements), routing of pipes, cable trays and HVAC ducts, insertion of fittings, definition of supports and other outfitting structures, ...
Having a CAD system that allows modifying elements directly within an immersive environment would also be very useful in collaborative sessions in a *powerwall* room or a CAVE where different engineers make decisions. The possibility to make the changes in real time can result in significant time and cost savings.

Another field of application where VR may do the job is production organization. The 3D model is created according to design criteria, but it must also be organised into a product structure based on a build strategy, that is, a breakdown of elements adapted to the different production units. The assignment of objects to the different units could be done easily and intuitively within the virtual scene. This would be applicable to structural assemblies as well as to pipe piece fabrication and mounting.

In our opinion, the CAD systems should evolve in order to ensure that VR technology be used directly whenever necessary or convenient. Thus a new future would open to designers, who could place themselves inside the scene and create or modify elements in a way never experienced before.

The incorporation into ship design of CAD tools compatible with devices such as those described in precedent paragraphs (HTC Vive, Oculus Rift, Razer OSVR HDK2 or similar) would allow designers to use VR more generally and at a less cost, shortening the overall design-review-modification process.

For a CAD system to allow working in an immersive environment, it must evolve in the following aspects:

- Graphics engine: modelling tools must include advanced graphics and rendering engines that allow a fluid immersive navigation with stereoscopy as well as compatibility with VR gear.

- Performance and massive use of data: the system should be prepared for handling huge volumes of information, which may be stored in a database and must be loaded dynamically in the scene during the working session.

- User Interface: the monitor has been replaced with VR glasses; the mouse, with handheld devices such as wands or joysticks. The designer cannot see the traditional keyboard. The interface in traditional CAD does not serve anymore and the commands must be invoked in a much different way (voice, gestures, ...).

Sener software engineers are now working on the development of a new, modern graphics engine, the heart of Foran. This involves not only a visual change, with photorealistic rendering, stunning animations and new visualisation modes (shadows, cell rendering, point clouds, textures, stereoscopy), but above all a functional change to endow the System with new features (object handlers, interactive 2D/3D handling, walk-through navigation mode) to exploit to the full the potential of all the different graphics cards currently available and to connect the visualisation system to new technologies such as CAVEs and HMD devices.

8. REFERENCES

1. JERALD, J., 'The VR Book: Human-Centered Design for Virtual Reality', *Morgan and Claypool Publishers*, 2016.

2. ALONSO, V. et al., 'Advantages of Using a Virtual Reality Tool in Shipbuilding', *Proceedings of COMPIT,* 2012.

9. AUTHORS BIOGRAPHY

Alfonso Cebollero, MSc Naval Architecture & Marine Engineering from Polytechnic University of Madrid and MA Shipping Business from ICADE, currently holds the position of Chief Regional Officer of Sener for China, Japan and Korea. He joined Sener in 1994 after having worked several years for another engineering company, a shipping consultancy and a naval shipyard.

Luis Sánchez, MSc Naval Architecture & Marine Engineering from Polytechnic University of Madrid. He joined Sener in 1997. Currently holds the position of Key Account Manager for Navantia. He has led the implementation of VR *powerwalls* in Navantia Cartagena and Ferrol.

USING VIRTUAL REALITY PARADIGM TO PRESENT SHIP STRUCTURES IN CAD ENVIRONMENT

G Šikić, USCS d.o.o., Croatia

SUMMARY

After implementing Stereo 3D viewing mode in ShipExplorer application, as a one of next steps, it was decided to explore virtual reality paradigm that will allow complete immersion into the ship. Starting point was Stereo 3D module, and for head mounted display, it was decided to use OCULS RIFT hardware. Initially it seemed like well-defined task; everything was already solved, the only new functionality would be additional viewing transformation in connection to head attitude. It turned out that implementation posed completely new set of requirements, much closer to real time visual simulations, compared to classical CAD visualization system like ShipExplorer. Some of demands were expected, but some were completely new, and were connected to virtual reality environment only.

In this article, decomposition of the task of adding virtual reality capabilities to existing software will be presented, starting with theoretical analysis, problems and additional requirements that appeared during implementation will be discussed, along solutions that were used to solve them. At the end, experiences gained during addition of virtual reality extensions into CAD environment will be summarized, including present and future usages of VR in ShipExplorer.

1. INTRODUCTION

After extending ShipExplorer [1] with stereo 3D capabilities, it was decided to take one step further, and implement complete immersion into the ship. New functionality was designed and implemented as new extension into already existing system. Ideally, all major premises of original system should be maintained, namely no special data prepossessing, user experience and easiness of usage. As an output device, Oculus Rift [2] was chosen, because it offered drivers and development toolkit for Windows and Linux.

Starting development point was stereo 3D presentation subsystem [3], that used modified OpenGL rendering pipeline. It was further modified in order to allow changes of body and head attitude.

During implementation phase, some new problems were discovered and resolved, various ways of controlling the position on the ship were tested, and graphics subsystem was redesigned in order to improve drawing speed.

Also, ideas for further improvements and usages that emerged during development were discussed.

2. THEORETICAL BACKGROUND

ShipExplorer rendering system is OpenGL based, and it was natural to adopt it for virtual reality rendering as well. Also, since head orientation had to be tracked, new sensors had to be included into the system.

2.1 OPENGL TRANSFORMATIONS

OpenGL, graphical library that was used for geometry presentation is organized as a state machine with all transformations defined as a matrix multiplications [4]. Matrix concatenation is done as follows:

- First transformation that is applied is so called "ModelView" transformation, that transforms 'world' coordinates into 'eye' coordinates. After it is executed, coordinate system origin is in the position of viewer, and coordinate axes are oriented according to the view direction.
- Second transformation is "Projection" transformation, that defines the way the view of the geometry is projected on a flat surface. Two most commonly used transformations are Orthographic projection and Perspective projection. Perspective based projection is how humans naturally see the world, and it will be used further on. After this transformation all coordinates are in so called "Clip Space", and after perspective division, those are called normalized device coordinates.
- Last transformation is "viewport" transformation, that maps normalized device coordinated onto size and position of viewing window on actual, physical device.

2.2 TRANSFORMATIONS MODIFICATIONS

To achieve complete and natural feeling of immersment in the ship, additional transformations had to be introduced:

- eye offset transformation,
- rotation of head relative to body;

2.2 (a) Eye Offset Transformation

In order to create stereo 3D effect, it is necessary to render two scenes, each one from from point of view of corresponding eye. Rendered results will be similar, with small differences, that actually create stereoscopic sensation. Easiest way to achieve this is to add new transformation, representing eye offset, and it was inserted after "ModelView" transform. Placing transformations this way, we are relieved of taking care of actual orientation of viewer; all transformations are defined as constant offset along currently active 'x' axis, while the orientation of current coordinate system is defined by previous transformations in pipeline.

To keep everything as simple as possible, eye offset transformations were placed into projection transformation. This way, all constant transformations are kept on one side, while transformations that can be changed are dealt with separately.

2.2 (b) Rotation of Head Relative to Body

This transformation represents set of rotations that define head position relative to the body. It was initially included because it physically represents additional degree of freedom that needed to be addressed. Presenting it as additional transformation would allow movement through the ship independent of line of sight, for example, walking while looking to the right.

It was applied after original, "Model View" transformation, and before "Eye Offset" transformation.

2.2 (c) Transformation Pipeline Summarized

Complete transformation pipeline is shown on figure 1, and is stacked as follows:

- Original ModelView transformation, that defines position and orientation of viewer's body relative to absolute, world coordinate system,
- Viewer's head rotation, relative to his body,
- Eye offset transformations. These are two constant translations, one for each eye.
- Projection transformations, that define how is 3D world projected onto flat surface.
- Viewport transformations, that connects normalized device coordinates with physical coordinates of viewing window.

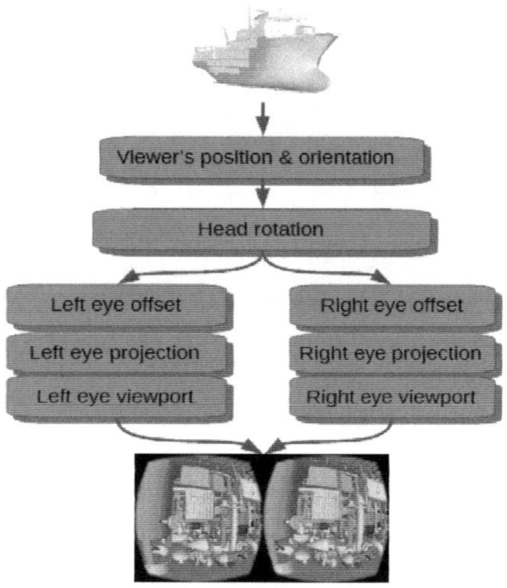

Figure 1: Modified OpenGL rendering pipeline

2.3 HEAD POSITION TRACKING

Data for "Viewer's head rotation" transformation calculation must be sampled from the outside source, using head attitude of the real user in the real world as an input. Main issue here is delay management [5].

Basically, frequency of input data sampling is not correlated to frequency of picture generation. Instead, head position sampling and frame creation might run in parallel, possibly in different threads. For every frame, just before it's drawn, head position is acquired. On the other hand, approximation of real head position is returned, using pre-fetched data, and information about present time. There are two common approaches how to do it:

- interpolation between of last sampled values, that as a consequence introduces delay of at least sampling period,
- extrapolation using last couple of samples, that as a consequence introduces possible large error in predicted data.

There is no definite answer which approach is better, but luckily this problem has to be solved by hardware vendor. For Oculus, here is so called "TimeWarp" functionality that keeps track of exact moments when drawing started, as well as predicted values for head rotation, thus reliving application programmer from dealing with his kind of problems. Nonetheless, problems of data input, head tracking in particular, must be well understood.

3. DESIGN CONSIDERATIONS

Virtual reality capability was designed to be new extension to ShipExplorer system, and it would be ideal to accept all premises of working system. Also, introducing Virtual Reality (VR) environment adds new set of design requirements. All these requirements have their roots in completely different way of user interaction with computer.

3.1 INHERITED DESIGN REQUIREMENTS

Among inherited requirements the most important was that no off line preparation of data is necessary. Anything that can be shown in ShipExplorer, should automatically be available for presentation in virtual reality environment. All necessary data conversion and arrangement should be done in real-time.

The other important requirement was to achieve same level of user experience as in base program. During ShipExplorer development, user sensation was one of most important premises, and easiness of usage, and overall user experience was one of the strongest parts of original program. We wanted to achieve same level of experience in new module.

One of inherited requirements that had to be abandoned was independence of operating system. Original program runs on all relevant windows versions, Linux and Solaris. Unfortunately, due to hardware requirements, namely OS support from hardware vendors, this had to be abandoned, and all development was done on Windows platform.

3.2 NEW DESIGN REQUIREMENTS

3.2 (a) Rendering Pipeline

Since VR extensions are very hardware dependent, one of very first decisions was to choose gadget to be used. At the time of initial development, only viable alternative was to use Oculus DK2 (Development Kit 2). Act of choosing HW immediately imposed new requirements, but it was expected. Transformation pipeline had to be made compatible with API (Application Programming Interface) presented with hardware vendor.

One of very first requirements Oculus hardware presents is required number of frames per second, and for DK2 it is 75 fps. Original ShipExplorer program was built with completely different goals, that were in essence incompatible with requirements of VR. Drawing was connected to refresh rate of monitor, and by definition couldn't exceed it. In practice, it means max refresh rate of 60fps. Also, occasional drop to 15 - 20 fps on older computers was more than acceptable, because input used for drawing (position, attitude, zoom) was filtered in real time by 2nd order differential equations, producing very natural and pleasing sensation even with reduced frame rate.

This approach to graphics generation had to be abandoned, because it would become completely unnatural in VR environment. Simply stated, additional real time filtering that was used in original application would generate movements and/or accelerations that are not naturally present, usually resulting in nausea, which had to be avoided at any cost.

Same applies to position changes; real time filtering was used to implement "slow in, slow out" functionality of speed changing, had to be converted into pure gain, for the same reasons of unnatural movement sensations.

In brief, Oculus hardware added two major changes:

- drawing speed of at least 75 FPS had to be achieved.
- OpenGL transformation pipeline had to be made compatible with API of Oculus SDK.

Switch into Virtual Reality in general added yet another constraint:

- dynamic effects (i.e. real time filtering using 2nd order differential equation) that served us with great results had to be excluded from ShipExplorer's VR extensions.

3.2 (b) Input Considerations

In ordinary desktop applications like ShipExplorer, input is usually performed using keyboard and mouse combination. 6DOF mouse was also used, as a complement to ordinary mouse.

Initial idea was to use ordinary mouse for navigation in a sense that left/right relative position of the mouse would define rotational speed of viewer's body about vertical axis, and vertical position of the mouse relative to the screen center would define speed of forward movement. Immediately, two possible problems were identified:

- position of the mouse is measured relative to the physical size of the screen, i.e. monitor, but output device is something completely unrelated. We were afraid that this fact might create problems for the end user. As it turned out, after solving initial issues, this didn't create any additional obstacles;
- mouse doesn't return to neutral position automatically. For example, if the mouse is moved bit to right in order to initiate right body rotation, it will continue rotating indefinitely. This is acceptable for normal, desktop application, but becomes unacceptable after switch to VR. It is simply unnatural, users do not expect that body attitude is controlled that way. On the other hand, using mouse for forward speed control received quite positive feedback from the users.

Using 6DOF mouse eliminated these problems, but due to hardware independence requirement, inherited from ShipExplorer, support for ordinary mouse was kept. As it will be shown in section 5, this decision allowed us to test various other devices that are seen as mouse, with positive results.

4 IMPLEMENTATION

Virtual Reality extensions are made using C++ programming language and initially, only libraries that were used for core program.

Completely new graphics subsystem was implemented, that was compatible with requirements of Oculus SDK, and with greater drawing speed. Various optimization techniques were tested, and implemented.

During implementation phase, "feeling" regarding input was also checked, until satisfactory results were achieved. Some of initial design decisions had to be changed and it will be explained in chapter 4.2 and 5.

It should be noted that all changes and improved functionality were made in a way to make it compatible with core program, and might be easily re-implemented in base ShipExplorer system.

4.1 GRAPHIC SUBSYSTEM

There were two mutually independent changes to graphics subsystem that had to be implemented:

- connection to Oculus SDK,
- scene drawing speed enhancements;

4.1 (a) Connection to Oculus SDK

Oculus SDK is organized in a way that application programmer gets data form SDK, initializes every frame, draws two scenes, one for each eye, and sends such prepared data back to Oculus for presentation in Head Mounted Display. Complete rendering subsystem works as shown :

```
• initialize Oculus,
• initialize OpenGL,
• for every frame:
    o initialize frame,
    o collect head position,
    o initialize left eye rendering,
        ■ draw scene,
    o initialize right eye rendering,
        ■ draw scene,
    o finalize frame,
    o send frame to HMD,
    o (optional) draw frame on monitor
• shut down Oculus;
```

Scene drawing is organized in the form of self-contained object, also compatible with graphics part in ShipExplorer system.

4.1 (b) Scene Drawing Speed Enhancements

It was clearly visible that some drawing speed improvements required, and great effort has been included to achieve it.

First step was to identify areas of possible improvements, and it was quickly confirmed that original system was bound by geometry transfromations. The bottleneck was in handling huge amount of geometry, sending it to video card, and transforming it.

Also, in original system, due to requirements that program should work on very old computers (i.e. computers produced in 2000.-2005.) everything was organized in orthodox way. For example, no OpenGL extension requiring base version higher than 2.0 was to be used. This design decision inevitably had consequence of only sub optimal drawing on a newer computers, with slower frame rate. It was decided to abandon support for old computers, allowing us to use capabilities of modern OpenGL implementations.

4.1 (c) Octogonal Tree Hierarchical Geometry Organization

Modern OpenGL draws geometry in batches, to up to couple of tens of thousands points in size. Choosing optimal batch size is "open problem", there are some guidelines, but there is no definite answer. It would be ideal if those batches could be constant, and created in initialization. It was done so, and maximum size of every batch was not hard-coded, but was defined as a runtime variable.

Additionally, while creating large drawing batches, structure called adaptive octogonal tree (octree) [6] was implemented. It is a recursive tree structure where every node has either 8 children or none, as shown in figure 2. Every node of octree holds geometry that fits into extents of the node, and exceeds size of node's possible children. Drawing batches are constructed out of geometry of every particular node.

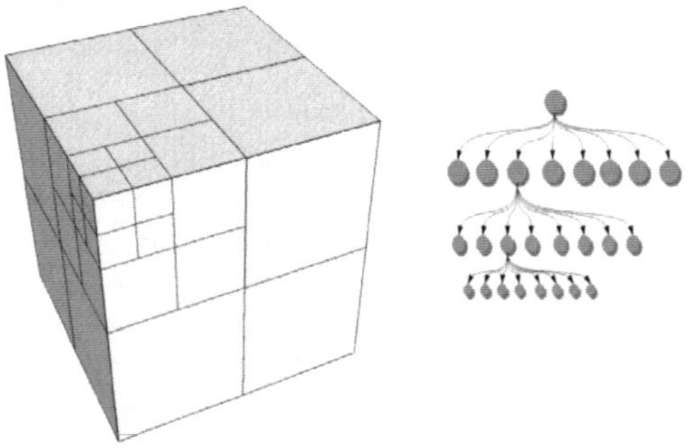

Figure 2: Octree with 4 levels, left: physical appearance, right: corresponding tree structure

Employing octree structure, lowered number of batches from original (model for Uljanik Yard 492) 237507 batches with average of 34 points per batch, to 921 batch with average of 8834 points per batch. Octree containing that 921 batch was up to 9 levels deep. It was possible to lower number of batches even further, but without significant speedup, since in that case it would present geometry with bigger extents, and would be drawn much more frequently. This is explained by the fact that occlusion query was used to discriminate visible geometry, and often only small part of big batch would be visible, thus triggering drawing of complete batch.

Operations on octree were implemented using visitor pattern [7], allowing to split the code that uses octree from the code that defines octree.

Using octree structure allowed for very efficient check whether the geometry is in viewing frustum, and exclude non-visible geometry very early in the process of drawing. For every node, it was necessary to do visibility check of it's bounding volume; if the node was not visible, or it was completely visible, non of it's children needed check, since those are completely contained within original node. Using node's bounding volume visibility check, it was possible to very quickly exclude geometry that was surely not visible, and drawing batches representing this geometry were not drawn, thus lowering the load of hardware graphical system.

4.1 (d) Using Occlusion Query

Still, the remaining geometry that **might** be visible and was drawn, might also be completely obscured by another geometry, and therefore drawn unnecessarily. The classical example of this situation is pipeline, hidden by the wall, as shown on the figure 3. On the left side of the image, final rendering result is shown, and on the right side geometry that is drawn and completely obscured is shown as well.

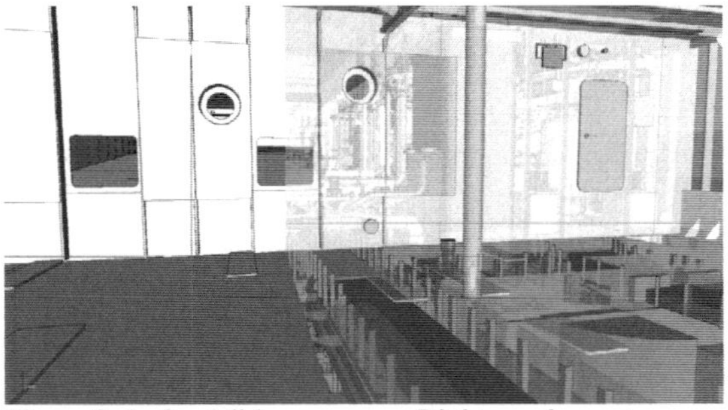

Figure 3: Left: visible geometry, Right: equipment obscured by hull

Classical OpenGL can help solving this problem by performing visibility test (i.e. GL_DEPTH_TEST) and rejecting non-visible fragments early in the rasterization part of OpenGL drawing pipeline. To utilize this capabilities it would be necessary to sort geometry front-to-back before sending it to OpenGL, for actual drawing.

Still, all possibly visible geometry had to be transformed, and at least partially rasterized. It was clear that some other mechanism was needed, and OpenGL also provided it, in the form of GL_ARB_occlusion_query extension [8]. This is a OpenGL extension that allows association of some form of query to all objects sent to draw, and after the drawing is finished results retreival. The basic usage is outlined in next code snippet:

```
•  create query objects and associate
   it with nodes of octree,
•  for every frame:
   o  disable writing to color and
      depth buffers,
   o  for every object occlusion query
      is needed:
      ▪  start query,
      ▪  render object representation
      ▪  end query,
   o  enable writing to color and
      depth buffers,
   o  for each query:
      ▪  get query result,
      ▪  if number of visible points
         is greater than 0:
         •  render geometry of the
            node associated with
            query,
   o  end frame
```

Please note that if used this way, occlusion query requires additional rendering pass. On the other hand, collecting query results doesn't need to be in same frame [9], and results could be collected while rendering the following one. After some tweaking, this is (simplified) representation of algorithm used for occlusion query:

```
•  create query objects and associate
   it with nodes of octree and
   initialize to "visible",
•  for every frame:
   o  clear the list of "objects
      invisible in previous frame",
   o  for each query:
      ▪  get query result
      ▪  if number of visible points
         is greater than 0:
         •  begin query,
         •  render geometry of node
            associated with query
         •  end query,
      ▪  else:
         •  add query into the list of
            "objects invisible in
            previous frame",
   o  disable writing to color and
      depth buffers,
   o  for each query in "objects
      invisible in previous frame":
      ▪  begin query,
      ▪  render object representation,
      ▪  end query,
   o  enable writing to color and
      depth buffers,
   o  end frame.
```

Using approach involving 2 frames, not only eliminated additional rendering pass, but as a consequence, testing is done using real geometry as occluders, and not simplified representation.

4.1 (e) Using Vertex Buffer Objects

One of the rendering bottlenecks that was addressed was transfer of rendering data form computer main memory to the video card memory. For example, model of Uljanik Yard 492 contains almost 400MB (389516224 bytes to be exact) of vertices that has to be transferred from main memory to video subsystem for every single scene draw. For Oculus rendering, amount of data to be transferred is doubled, because the scene is drawn twice for every frame.

Using OpenGL Vertex Buffer Objects (VBO) [10] allows to upload drawing data to video card, and keep it there, thus eliminating need for per-scene-draw transfer.

As the geometry is static, and non-changing, changes to drawing part are relatively simple; every node in octree structure would also be associated with one VBO, and it will be used for drawing:

- `for every node with geometry in octree:`
    - `create vbo`
    - `activate vbo`
    - `associate it with node`
    - `upload vertex data to video card`
- `for every frame:`
    - `for every octree node with geometry:`
        - `activate corresponding vbo`
        - `draw geometry using same api calls as usual`

4.1 (f) Summary of changes to graphics subsystem

During development of VR extensions following changes were implemented:

- geometry is organized using tree structure octree. All operations on octree were implemented using visitor-pattern; drawing data is organized into bigger batches, and every batch corresponds one-to-one to nodes of octree;
- occlusion query is used, and all queries map one-to-one to corresponding node in octree;
- vertex buffer objects are used, and every vbo also corresponds one-to-one to associated node of octree.

4.2 IMPLEMENTATION OF INPUT

Input to Virtual Reality extension of ShipExplorer was implemented twofold:

- head position tracking; it was used to construct transformations that control rotation of the head relative to the body.
- input from mouse it was initially used to control orientation of the body, and movement through the ship.

It should be noted that unlike the original program, input was not filtered in any way, due to additional requirements of VR environment.

4.2 (a) Head Orientation Tracking

Low level part of tracking of orientation of head was done within Oculus driver, and was presented to application programmer via Application Programming Interface (API). Orientation of the head was received in the form of quaternion, or transformation matrix, and it is responsibility of application programmer to interpret it correctly and connect to the rest of program.

4.2 (b) Input from mouse

Input from mouse was initially used to control movement through the ship, and to control rotation of the body, as described in section 2. Mouse position was interpreted relative to the screen size where application was started:

- middle of the screen was interpreted as "neutral point" where no movement was initiated;
- left/right offset of neutral point would initiate rotation to the left/right;
- vertical offset to neutral point was interpreted as a speed of movement, (up→ forward, down→ backwards) along the line defined by body rotation.

During development, it was very quickly realized that body rotation independent of the head rotation, although physically correct, was not producing expected results. Two problems were identified:

- since head attitude was naturally controlled, i.e. with head, it was expected to control position of the body analogously.
- It was very easy to loose track of body orientation, and up/down mouse offset would very frequently initiate movement in unexpected direction. This in turn, would almost immediately result in nausea.

The second problem was very serious one, and was resolved by abandoning possibility to orientate body independently. Instead, angle of head rotation about vertical axis would also be interpreted as body rotation. As a consequence, this also introduced the rule "move in the direction you are looking at". Additionally, placing a user into a rotational chair before initiating jump into VR become common practice. This allowed him/her easy change of real body attitude. While physically not completely correct, movement like this received very positive feedback from the users.

Yet another problem was identified with mouse input; it was very difficult to stop movement at some particular point in the ship. Also, making small adjustments of position was difficult as well; simply stated, mouse was "touchy". It was resolved by introducing non-linearity in a vicinity of neutral point, as shown on figure 4. Instead of direct linear function, cubic function was used for interpretation.

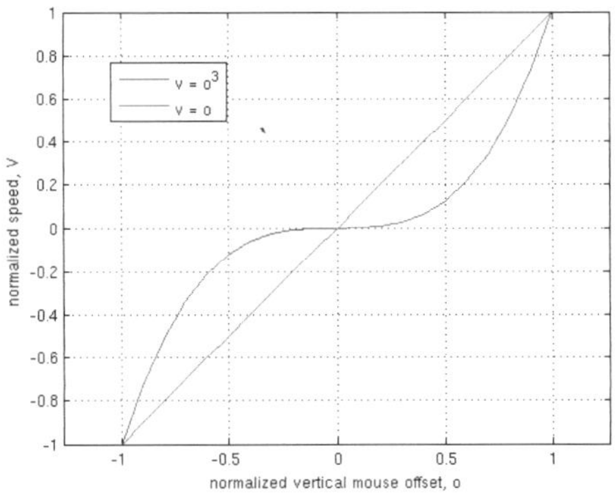

Figure 4: Nonlinearity introduced into vertical mouse position interpretation

Near neutral point, mouse was made less sensitive to the movements, compared to the larger offsets, allowing for very fine control of forward speed and position. This change also received very positive feedback.

5. EXPERIENCE GAINED DURING DEVELOPMENT

During design and development phase of adding VR extensions to ShipExplorer, many of obstacles were anticipated, but there were some issues that were discovered during implementation.

Data agnosticism was preserved, and everything that can be shown in ShipExplorer, can also be viewed in VR environment as well.

After implementing changes to scene generation, requirements of Oculus regarding drawing speed were met in almost all circumstances. On laptop with GTX980 video card, drawing speed was around 80-100 FPS if only hull was loaded; if equipment and hull were loaded, average speed was around 60 FPS, with occasional drop to 45 FPS. However, during testing with users, anyone who didn't know about this fact, was very satisfied with overall drawing speed.

Solutions to the problems of mouse input also proved to more than satisfactory. Eliminating independently controlled rotation of the body although physically wrong, proved to be right decision. Many users stated that the feeling is much more natural, and way input is handled shouldn't be changed. Also, user experience was best if wireless trackball mouse variant was used. The explanation is that user can use one hand only to control position in the ship, leading to more natural control, resulting in better feeling of immersion.

Same applies for rotating chair, that proved to be unavoidable piece of equipment for VR presentations.

After testing, almost all users stated that "experience of the space" within the ship is much more vivid, and the inter relations among different parts of equipment is much better seen. On the other hand, many of the users also stated that presentation of the ship become somehow "artificial", figure 5, and more physically realistic rendering would be desirable. It turns out that move to VR also made a switch in perception of a model, and user expectations radically changed.

Figure 5: Artificial ship in Uljanik shipyard; snapshot from virtual walk through the ship.

6. WHAT NEXT?

There are more than few areas where VR extensions may be improved.

Drawing speed is one of such areas. With relatively small changes, satisfactory results were obtained, and it is certainly possible improve it even further. For example, while creating octree structure, another parameter might be added, to discriminate geometry; elements with relatively small number of points, coupled with big extents, (i.e. hull plates) would be treated differently form geometry containing many points in a small volume. Also, other OpenGL extensions might be used, like GL_ARB_instanced_rendering [11], that would allow geometry with similar characteristics to be sent to video card once, and internally rendered many times.

In case of usage of latest NVIDIA video cards, single pass stereo rendering [12] might also be used, resulting in huge performance gain.

Another area of change is introduction of photo-realistic rendering. Till now, all ship structures were rendered using "non photo realistic" approach, and users were very satisfied with it. After introducing stereo 3D, and especially VR immerse into the ship, many users have asked about adding "physically based rendering" [13]. First step in this direction will be adding soft shadows and ambient occlusion. Ideally, end result bill be similar to rendering shown on figure 6. Some of the improvements are already being implemented; at time of writing, hard shadows are in the final phase of implementation.

One of natural next steps would be to extend VR functionality to Augmented Reality. For example it would make possible to overlay model and elements of real ship. Using appropriate hardware, it would allow on site examination and correlation of build structures with corresponding model.

Figure 6: Results of off-line rendering of ship structure. Used with permission from Sinisa Stambolec, Visualia Festival

7. CONCLUSIONS

After initial skepticism, virtual reality extensions are quickly becoming integral part of ship design process. In figure 7, Uljanik's leading designer Mr Djani Dundara is seen examining hull structure in virtual reality environment.

Figure 7: Uljanik's leading designer examines ship structure in Virtual Reality environment.

While designing and implementing VR extensions, it become more clear that distinct line dividing simulations (with VR) and CAD no longer exists. After seeing VR in action, many users asked additional functionality on top of VR, starting from taking snapshots, attaching notes to parts of model while in VR up to wishes for complete editing in this environment.

Implementing VR extensions was a very nice experience, that resulted in joining the world of simulations to the world of CAD, and future work will certainly involve this combination.

8. REFERENCES

1. Benčić R., Milanović M., Šikić G., Bistričić M., 'Multi-Platform Collaboration Tool For Ship Design And Production', Proceedings ICCAS 2013

2. Oculus Developer Center, https://developer.oculus.com/pc/

3. Šikić, G., Bistricic, M., 'Stereo 3d Presentation Of Ship Structures Using Low Cost Hardware', Proceedings ICCAS, 2015.

4. Mason, W., Neider, J., Davis, T., 'OpenGL Programming Guide', 1997.

5. Šikić, G., 'Simulation of Flight of MiG-21 Aircrft', Master thesis, 2000.

6. Octree, https://en.wikipedia.org/wiki/Octree

7. Gamma, E., Helm R., Johnson, R., Vlissides, J., Design Patterns,2005.

8. Occlusion Query Specification, https://www.khronos.org/registry/OpenGL/extensions/ARB/ARB_occlusion_query.txt

9. Dean Sekulic, 'Efficient Occlusion Culling', https://developer.nvidia.com/gpugems

10. Munshi, A,. Ginsburg, D., Shreiner, D., 'OpenGL ES 2.0 Programming Guide', 2011.

11. Instanced Rendering Specification, https://www.khronos.org/registry/OpenGL/extensions/ARB/ARB_draw_instanced.txt

12. Single Pass Stereo Specification, https://www.khronos.org/registry/OpenGL/extensions/NV/NV_stereo_view_rendering.txt

13. Akenine-Moller, T., Haines, E., Hoffman, N., 'Real-Time Rendering', 2012.

9. AUTHORS BIOGRAPHY

Gordan Šikić holds the current position of Head of Development at USCS. Before joining USCS, he was involved in mathematical modeling, software design and implementation of distributed simulations, real time visual flight simulations, design and implementation of rocket control systems, terrain data visualization manipulation, and vibration calculation and measurement.

WEARABLE APPLICATION IN SHIPBUILDING

G Rogero Fernando, Optio Marine Pvt Ltd, India.

SUMMARY

The paper deals by way of example with some aspects of implementation of wearable computing in shipbuilding. It is a viewpoint and it meant to provoke an integrated vision of the total process from early design to delivery of the finished product. It points out the increasing value and importance of innovative computing tools along the process chain and leads to radically new concepts for the design production integration. The paper indicates a potential for substantial productivity gains resulting from further efforts in research and development.

NOMENCLATURE

Symbols, Abbreviations and Acronyms utilised in the paper ordered alphabetically.

2D	Two dimensional
3D	Three dimensional
AR	Augmented Reality
CAD	Computer aided design
CAM	Computer aided manufacturing
CFD	Computational fluid dynamics
DWT	Dead weight
FEA	Finite element analysis
GPS	Global positioning system
IoT	Internet of Things

1. INTRODUCTION

Ship building has an extremely long tradition and yet only in the last decades it moved from handicraft type heavy industry to industrial processing in the current sense.

Shipbuilding has many appearances depending mainly on the product varying from warfare submarine to luxury mega yachts. This nearly has no similarities with respect to the approach, the technologies and the methods. This paper aims to discuss with scenarios and some future aspects of utilizing present and upcoming computing technologies into the design and ship production work flow.

Shipbuilding market is closely connected to the world trade development and especially seaborne traffic, which in last decade increased from roughly 66 million DWT to 778 million DWT. And increasing demands lead to rapid development in new types of ships, new operation methods and development in production technologies.

Comparing the shipbuilding industry with others shipbuilding has been late acceptors of innovative computing technologies because of the project complexity and uniqueness of every product to market.

2. IT-SOLUTION IN SHIPBUILDING

Whereas in the past the shipbuilding was a sequential chain of individual events this is no longer possible in today's highly competitive market for various reasons:-

- First and most important the clients have no time and want the product fast.
- Second is the shorter the building period the shorter the utilisation of production facility and money.

The answer is to pre plan the production operation using design technologies. With the development of computer graphics and high performing computing devices led to the possibility of CAD. Computer Aided design is the process of preparing the digital mock-up of the real product in a virtual environment. CAD has been helping shipbuilding industry by providing the production requirements and the ideology of the actual product. By this means of engineering and further development in the field like FEA and CFD has helped the industry reduce the shipbuilding time by 1/20th Compare to 1960s.

3. WEARABLE COMPUTING

The computers which once were used by enterprise and scientific institutes were made available to common man by the visionary Steve jobs. From the time computer have been evolving very fast in many forms from the office table to smart phones. Smartphones have already penetrated into day to day life of a human being. Beyond this research and development is slowly foraying into wearable computers.

What are wearable computers? The device which has computing abilities and can be utilised as a part of accessories or clothing's even implanted within human body. These computers are emphasizing computers potential even further utilizing sensors and other nano components.

3.1 WEARABLE HARDWARES

3.1 (a) Smart phones

Over the last decade the immense development in computing hardware has aided the development of smart phones. These are the evolution of traditional communication device the mobile phone which has equal abilities like the personal computers. Smartphones has already been playing a major role in gaming entertainment and banking industry. A small segment of the manufacturing industry also utilises the benefits of this small hand portable devices.

3.1 (b) Smart Glasses

These are small computing devices with a form factor resembling day to day accessories like the specs which are light. These devices come with high computing abilities and graphics rendering capabilities. Their light weight and unique form factors allow it to be worn which provides a different perceptive to human vision.

Figure 1: Representation of Smart Glass.

3.1 (c) Smart Watches

These are small devices with lower computing ability compared to the predecessors mostly have been used as a lifestyle and fashion accessories. Smart watches with proper integration with the above devices can provide many significant benefits mostly in the place of internal communication and alerting systems.

3.1 (d) Embedded computers or IoT

These are small base computers in various forms and sizes. This device also comes in size of a small penny with advance and low energy communication technology. This device comes with the ability to respond to digital and analogy data which can perform real-time arithmetic operations to fit for industrial automation.

4. ASPECTS OF IMPLEMENTATION OF WEARABLE COMPUTING

4.1 AR/MIXED REALITY

Augmented reality brings aspects of the virtual world into the real world. It is closer to the real environment, as opposed to the virtual environments, in the spectrum of reality technologies. This is because in augmented reality the user remains in the real world while experiencing enhanced virtual created visuals, aural and feelings. Augmented reality does this by layering virtual information and graphics in top of a user's view of a real world scene.

Mixed reality is the result of blending the physical world with the digital world. Mixed reality is the next evolution in human, computer, and environment interaction and unlocks possibilities that before now were restricted to our imaginations. It is made possible by advancements in computer vision, graphical processing power, display technology, and input systems.

Devices today can support a specific range within the mixed reality spectrum, and over time new devices should expand that range. In the future, holographic devices will become more immersive, and immersive devices will become more holographic.

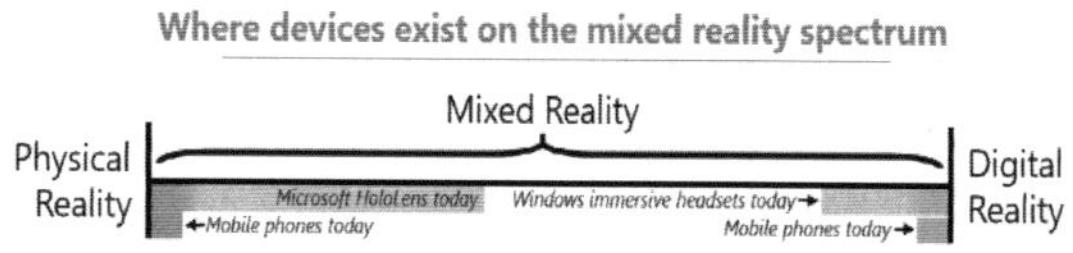

Figure 2: Mixed reality support graph

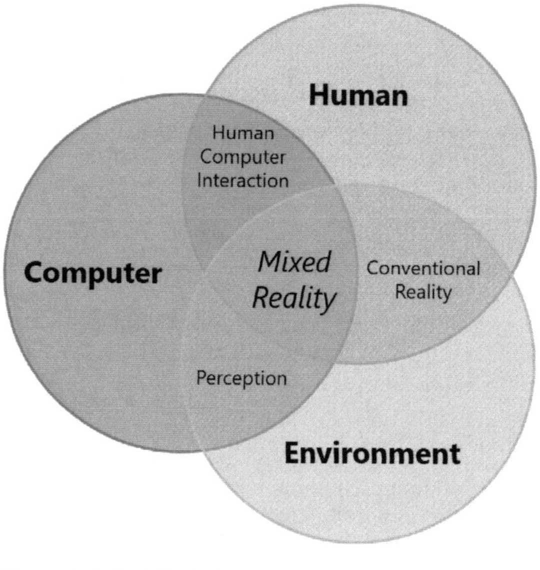

Figure 3: Principles behind AR .

4.1 (a) Aspects of implementation in Design

Over the past decade the computer has been utilised to produce graphical representation of the ship the process is called computer aided designing. This CAD model is utilised for preliminary visualisation and to get an overview of the items required for the fabrication. The present CAD models are utilised to produce numerical coding for cutting of plates, to generate weight report, bill of material and 2D workshop drawings.

The life cycle of the developed model is utilised for only the initial production information identification. Due to the limitation of carrying computers to the shop floors has led to the conversion of 3D models to 2D drawing. The wearable smart glass technology solves this issue by providing the benefit of portability to any location.

4.1(b) 3D enhanced workshop drawings

3D enhanced workshop drawing can be made possible with the marker based augmented reality concept. By providing a unique identification marker with the drawings with integration of the 3D CAD software can provide the ability to overlay the 3 dimensional views of the part or panel or blocks to be fabricated over the drawings. This methodology of visual understanding will help the engineers at the shop floor to understand and to execute their work more effectively.

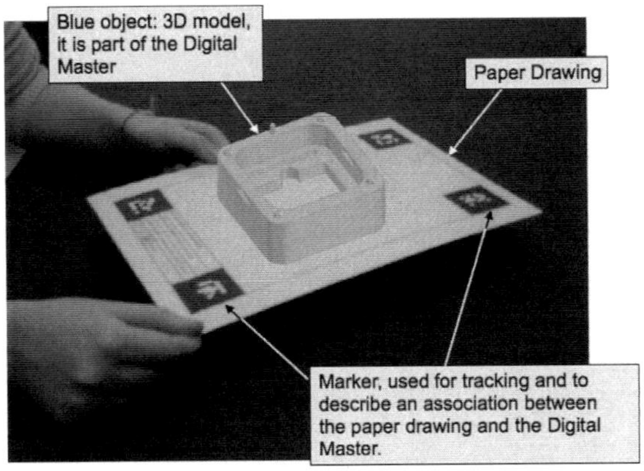

Figure 4: Representation of Enhanced 2D drawing.

4.1(c) Real time job simulation

By utilising the sensors and 3d rendering ability of the wearable computers either in the form of smartphone or smart glass can be utilised to visualise the simulation of lifting operation, transport operation, yard planning and launching study.

This augmented yard planning system can be implemented with the help of the wearable computers systems internal sensors and components as follows:-

Global Positioning Sensor (GPS):- a sensor which keeps track of the device geographical location in terms of latitude and longitude an imaginary grid drawn over the globe by which a user's location information can be retrieved by the software application.

Camera: - a component which consist of a photo light sensor which interprets colours and lights around the environment into digital quantity. In short it works as a human made vision system which gives the device the ability to look around real word environment.

Accelerometer sensor: - a component within the computing hardware which calculate and track the moment of the user with respect to XYZ 3 dimensional coordinate system. By software programs can study the speed and motion of the user.

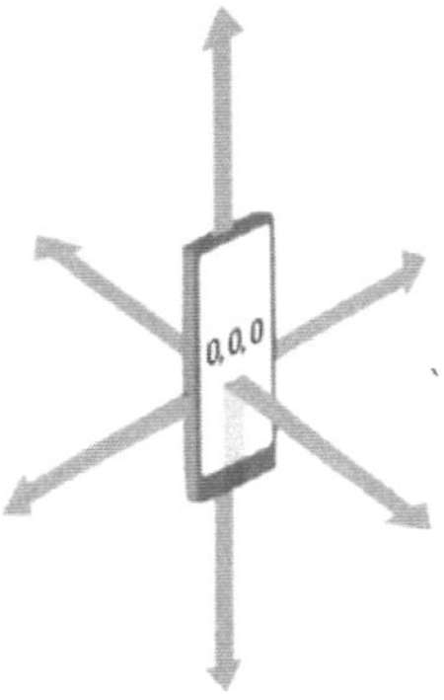

Figure 5: Linear coordinate system.

Gyroscopic sensor: - a component within the computing hardware which calculates and keeps track of the axial rotation of the device with respect to XYZ coordinate system by software programs which can study the angle and direction of the user view.

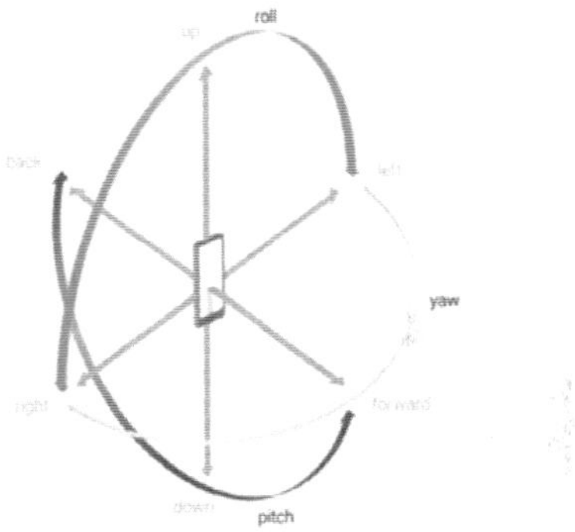

Figure 6: Device six degree of freedom.

The representation below shows a short pictorial representation of the concept behind the mixed reality.

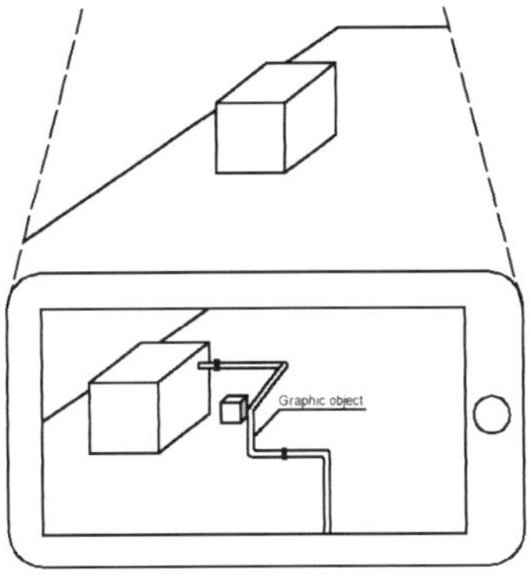

Figure 7: Reorientation of AR rendering.

Consider block lifting operation which is to be studied during initial planning stages. The scenario defined as a super structure block to be lifted and placed over the hull.

Normal practice is to study the life thought CAD simulations. By utilising the wearable holographic devices the 3d simulation can be viewed at the actual environment in real scale. This provides us a better understanding of hindrances in the real environment.

How it works?. The real scale simulation object like the crane is virtually placed over a GPS coordinate which relates to the lifting equipment's real position data in the yard environment.

The user's position, motion and angle of view information are retrieved from the build in sensors of the computer utilized by the user. Based on the data the internal mathematical computation takes place within the device which in turn renders a virtual object overlaid in the visual display of the device. This way of visualization gives the builders a better understanding of lifting operation in turns reduces real operation difficulty.

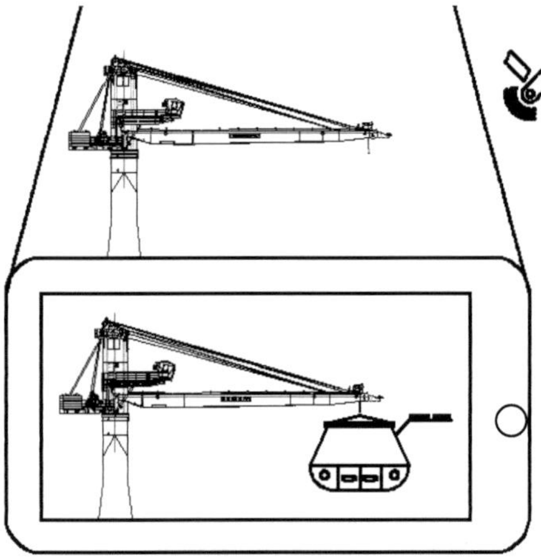

Figure 8: GEO special AR rendering.

4.1(d) Design and engineering

The designer working on the design can utilise the augmented reality to generate a real scale and scaled model at his work place. By experiencing the real environment the designer can understand operational difficulties. This helps in providing better ergonomic design to the operators. With integrated communication programs like Skype can also enhance the experience of coordination between the design and the production team.

Real time access to the production team camera can aid the opportunity of the designer to visualise the yard condition from the design office. This addition provides the designer the freedom of location. A design engineer at India can efficiently coordinate with the fabrication team at Singapore.

4.1(e) Area Learning

The engineering activities at a existing vessel has more difficulties than a new vessel. A small engineering addition requires huge man-hours of design work to replicate the existing environment. Even with the maximum perfect replication also there may be small discrepancy which may lead to big problems.

Presently the new laser scanning technology provides the technology to understand real environment object and can be converted into 3 dimensional computer graphics model. Then these models are imported into CAD software for further addition of piping and machinery.

The latest addition to the wearable computing devices is laser and IR sensors. The combined work of these sensors with the camera provides the ability of area learning. In turn a hand held smart phone with the ability of the laser scanner is obtained. The portability and small form gives easier scanning ability in confined spaces also.

Further development in this space can lead to in site design experience which reflects huge reduction in added engineering works.

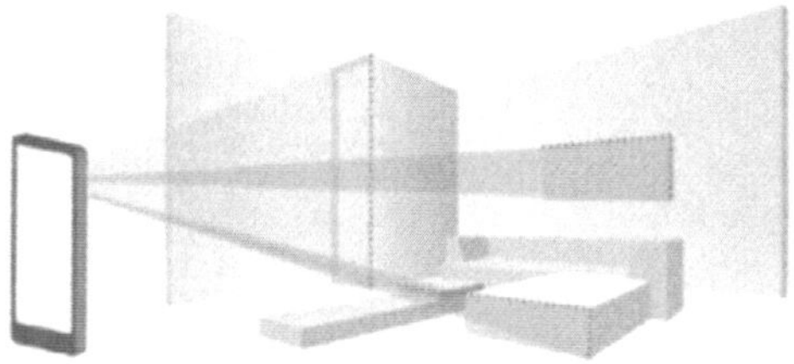

Figure 9: Area Learning.

4.2 THE DIGITAL TWIN

The digital twin is dynamic software model of a physical thing or system. The digital twin model allows a shared conceptualization that can be visualized in exactly the same way by an unlimited amount of individuals and by individuals who do not need to share the same location.

A combination of sensor data from the ship overlaid on the digital twin helps the ship owner and ship management companies to get real-time operational simulation on their work place anytime anywhere.

A digital twin differs from traditional computeraided design (CAD), nor does it serve as merely another sensor-enabled Internet of Things (IoT) solution. It could be much more than either. CAD is completely encapsulated in a computer-simulated environment that has demonstrated moderate success in modeling complex environments and more simple IoT systems measure things such as position and diagnostics for an entire component, but not interactions between components and the full life cycle processes.

Collaboration in the work environment is easily achieved by producing a real time data visualisation over the digital 3d model. A group of people working on management of the offshore platform can view a mixed reality aided visual clone of the physical product which can be utilised in analysing real time situation and forecast data.

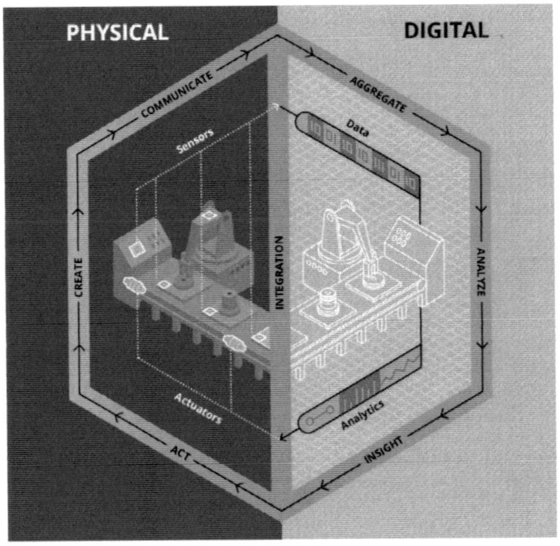

Figure 10: Principles of digital twin.

4.3 NOTIFICATION AND ALERT SYSTEM

Apart from the major wearable devices the smart watches which has a lower computing capabilities can be combined with shipyards existing PLM and ERP. Software can provide a new era of workplace management system. This implementation can increase the efficiency of the work managing system by providing regular notification regarding their jobs and duties at their own place. This bridges the gap between the project management team and team working in various point of shipbuilding. This provides real time information transmission between all parties and emphasis the coordination of the team.

5. INTEGRATION ASPECTS

The new fabrication technologies and methods require considerable efforts regarding the integration of suitable IT and Communication tools with the system. The individual special software solutions need to communicate often down to the shop floor level in a reliable manner and without loss of information. Today's revolutionary communication systems like the wireless communion and internet systems provide long range communication at a lower cost. A central intranet communication facilitated by the wireless communication technology can provide freedom of portability to the individuals working in the shop floor also.

A new task evolves from the necessity to update information along the production lifecycle in order to adjust downstream process data according to built situation. The problem can be addressed by a new form of 3D information transfer topology is to be build. A server in with modular fabrication and assembly data is used to provide the wearable computer with the necessary Product information on each level of fabrication workflow starting from part model , assembly models to the complete ship models. The dàta are retrieved from the central sever on request from the portable computing device.

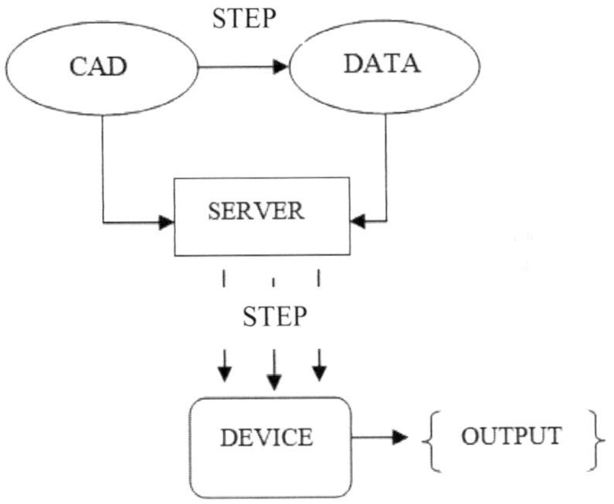

Figure 11: Integration and Data flow.

6. CONCLUSION

The paper has addressed only a few of the many upcoming new aspects of shipbuilding technologies. The underlying vision shows that the shipbuilding is still quite far away from harvesting the potentials which new technologies and methods can offer. Due to the special challenge of the industry new and more efficient solutions cannot be purchased in the market or simply copied from other sectors. The innovation must be developed in house because only then full understanding of the complexity and the interactivity is generated. On the other hand innovation requires intense co-operation with other industry sectors, provided however the benefits are also trans sectorial. The development of FEA and CFD technologies are typical example.

7. REFERENCES

1. RONALD AZUMA, *Recent advances in augmented reality technology*, 2001.

2. R. Azuma, et. al., ìA Motion-Stabilized Outdoor Augmented Reality System,î Proc. IEEE Virtual Reality í99. Houston, TX, 13-17 Mar. 1999.

3. S. Mann, *Wearable computing: a first step toward personal imaging*, IEEE Computer. vol. 30, no. 2, Feb. 1997.

4. I. Kasai, et.al., *A forgettable near eye display, Proc. 4th Int. Symp on Wearable Computers*, (ISWC 2000). Atlanta, 16-17 Oct. 2000.

5. H. Kato, et. al., *Virtual Object Manipulation of a Table-Top AR Environment*, Proc. Intíl Symp. Augmented Reality 2000 (ISAR'00). Munich, 5-6 Oct. 2000.

8. AUTHORS BIOGRAPHY

Rogero Fernando holds the current position of Naval Architect at Optio Marine Pvt Ltd,India. He is responsible for naval architecture and structural engineering for ships and workboats. His previous experience includes junior design engineer at L&T shipbuilding where he involved with naval architecture and structural engineering duties for 98m Indian indigenously developed coast guard offshore patrol vessel. Apart from this experience he also worked as structural engineer at Patel micro data India where he was deployed at drydocks world Dubai to carry out detail and production engineering duties for BORWIN 3 a offshore HVDC platform build to support the German renewable energy vision.

EXPERIENCE CAPTURE IN SHIPBUILDING THROUGH COMPUTER APPLICATIONS AND NEURAL NETWORKS

S. U. Sangeet, K. Sivaprasad, Yashwant R. Kamath, Department of Ship Technology, Cochin University of Science and Technology, Kerala, India

SUMMARY

It has always been a severe loss for any establishment when an experienced hand retires or moves to another firm. The specific details of what his job/position entails will always make the work more efficient. To curtail such losses, it is possible to implement a system that takes input from a new employee regarding the challenges he/she is facing and match it to a previous occurrence where someone else held his/her chair. This system could be made possible with input through the ages from the array of individuals who managed that particular job and processing this data through a neural network that recognizes the pattern. The paper is based on data collected from traditional wooden dhow builders and some of the modern day unconventional shipyards. Since the requirements for successful implementation in such scenarios seems too steep at the moment, an alternate approach has been suggested by implementation through the design processes across multiple shipyards. The process entails the traditional value passed down through generations regarding a particular profession and analysis has been done regarding how this knowledge/experience can be captured and preserved for future generations to work upon. A series of tools including SharePoint, MATLAB, and some similar software working in tandem can be used for the design of the same. This research will provide valuable insight as to how information sharing can be applied through generations for effective application of production capabilities.

NOMENCLATURE

ANN	Artificial Neural Network
EULA	End User License Agreement
GUI	Graphical User Interface
IACS	International Association of Classification Societies
KBMS	Knowledge Base Management System
NPO	Non-profit Organisation

1. INTRODUCTION

Since the advent of ships, there have been accidents at sea. The root causes for all those unfortunate events were lack of attention to detail in both maintenance and newly built ships. Human errors were the main reason for almost 80-85% of maritime accidents [2]. The entities that are responsible to supervise and enforce attention to detail are the classification societies that ensure that shipbuilding conform to certain acceptable standards. It would be a huge leap if we could ensure better quality and increase production efficiency with a single tool. This paper finally provides a proposal that could accomplish just that.

2. METHODOLOGY

Two approaches were considered based on reach and economic viability. Since computational power has been keeping up with the fast advancing requirements for the past decade, this method has acquired relevance only recently. Therefore it is only justified that a case study be conducted to verify the applicability of such a design. Analysis was done through case studies that finally provided an idea as to how it would help or harm the current scenario. Therefore this paper projects the results of a case study.

3. CASE STUDY

3.1 CASE 1- IMPLEMENTATION IN PRODUCTION

3.1 (a) Applicability in a Shipyard

It is possible to implement a database development programme to serve as a base for creating a knowledge system that is preformatted as per the requirements for an artificial neural network that can perform deep learning. Deep Learning recognises intricate patterns in huge databases that has varying levels of hierarchy through a backpropogation algorithm and suggests how a machine should re-compute each layer using different parameters [1]. With this, a probable neural network that applies here can calculate the extent of dependency between databases generated from different hierarchical positions. In the present day scenario, this applies to the different levels of communication between the design, production and outfit departments in a new build shipyard [3]. The most viable way to implement such a database development programme would be in the production department where decisions are made every day as to what the next step should be in assembly, transportation and welding in general. A lot of time can be saved by creating a comprehensive timeline where every department can work in tandem without disrupting another's work. User friendly GUI's can further alleviate the pains associated with training the working professionals to log their work into the database every day. Microsoft SharePoint is a platform where shipyard workers can log their everyday decisions and subsequent effects [4]. So the prime requirement from our solution is anomaly detection in the logs created through Microsoft SharePoint. As shown in Figure 1, there can be different types of anomalies depending on the type of ship being built. Since shipbuilding is one off kind, there can be point anomalies like the one in Figure1-(B) which could simply mean a different type of outfitting as requested by the owner that calls for a specific type of welding in a location or timeframe which is not found among the data from other shipbuilding activities. There can also be collective anomalies like the one shown in the Figure 1-(C) which could be due to an entirely different kind of ship being built which involves entirely different procedures in unit assembly, sub-assembly, welding, outfitting and launching [5]. Together, they account for a collection of anomalies that are grouped together. If we assume that our observations are based on a stochastic model, statistical analysis can be used to forecast further data points with occasional anomalies [5]. Figure 1-(A) represents a typical new build job with no anomalies in any of the different processes involved in the build process.

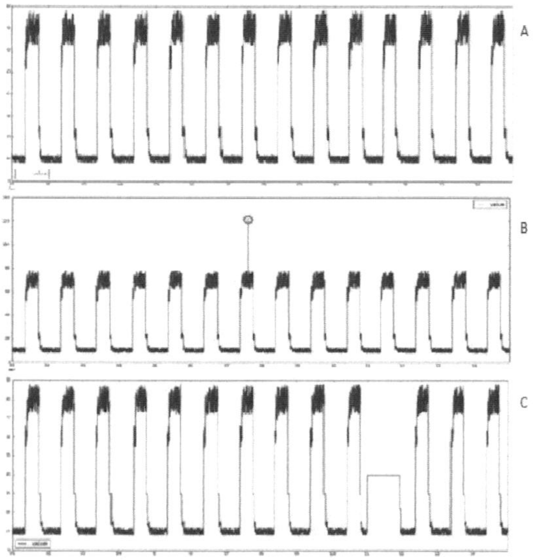

Figure 1- Statistical representation of data points collected. (A): Normal Distribution, (B): Anomaly in single value of data, (C): Collective Anomalies

Simply put, after such a neural network has been subjected to deep learning and is properly trained, the ANN can forecast the entire timeline for the shipbuilding process including many circumstances that may be overlooked generally with solutions to get around those difficulties. This could be a huge impact on the production scenario in shipyards and could easily boost production efficiency drastically. At the same time it could save a lot of time and effort for the planning team. Such a system in its entirety could be attributed the name, KBMS, a knowledge based management system [6]. A series of steps necessary for the implementation of such a system in a shipyard is detailed below in chronological order:

- Training of Personnel
- Infrastructure
- HR allocation
- Database Review Team
- Database Incubation
- Periodical analysis of data points
- Minimum Database Collected
- Integration of Neural Network into the Database System
- Recognising Parameters
- Setting up Input/output
- Defining Neural Network Dynamics
- Training the network
- Pattern Recognition
- Suggestion of solutions (output).

The above process will require at least a decade to properly train the network to better analyse and provide solutions for problems. Since this follows a statistical approach, the only way to improve accuracy is to add more data points provided the network definition is sound. A program with such a long term goal can only be undertaken by a body that has several shipyards working around the world that can provide different databases to efficiently train the network and make it capable of dealing with ships of all types and sizes. Any other shipyard would find it wasteful both economically and in human resources. Even here, classification societies could together (IACS) create a database for the same using data from shipyards all around the world.

3.1 (b) Applicability in Traditional Shipbuilding

Traditional wooden dhow construction carried out in Kerala, India is the perfect example of knowledge capture. The head shipwright also known as 'Moothasari' in local language trains an apprentice as his successor. The apprentice works with the Head shipwright for more than half of his mentor's working tenure until the mentor deems his successor worthy to take over. Therefore this would be the perfect platform to implement the deep learning concept and create an artificial neural network that can later on become the apprentice's personal troubleshooting tool. In the said context, it is still uncertain whether this can be implemented immediately without an overseeing body that can provide technical assistance. Hence this approach requires more resources including financial funding. Therefore this method is not feasible presently but could be very useful in the future provided classification societies can take a more active role in analysing and classifying traditional wooden dhows.

3.1 (c) Applicability in a NPO

A KBMS may contribute differently to different shipyards around the world that predominantly deals in certain types or sizes of ships. Hence it is more sensible to create a global model that can be trained with data from shipyards around the world, capable of dealing with huge amounts of data on all varieties of ocean going vessels. Even though this might require several teraflops or more in computing power, it will certainly yield better efficiency in production and further improve communication of practices among shipyards globally. It is possible to set up a Non-profit organisation (NPO) that is funded and run by classification societies and other institutions like RINA (Royal Institution of Naval Architects). Such an organisation can monitor and provide a common platform for various shipbuilding activities to work in tandem under an array of strict quality control systems that can ensure safety of the vessels at sea while decreasing the production time of every individual ship. This will greatly reduce the time required to train an ANN to comply with all the required norms in shipbuilding. It also adds to the fact that having a great many number of databases to include in the knowledge base will increase the accuracy of all predictions done by the network tremendously. A common data entry form can be provided to all shipyards to log their data and problems faced. Instant troubleshooting will be available to all the employees through real time feedback from the network as well as other working individuals of the same position in a different shipyard. All amendments made by the IMO, IACS and other authorities can be brought to practise with ease and by spending lesser resources through simple parameter manipulation. In an age where macrocosms of computers will become the tools used to govern industries at large, such a tool would be the perfect ground to build upon. It remains to be seen whether a technology will arise that can further simplify the process in the very less timeframe that this approach needs. In event of such a discovery, this method will lose its relevance and applicability.

3.1 (d) Conclusion of Case 1

In all three methods elaborated on above, it was apparent that applying such a tool successfully requires a number of resources both financially and in man power. Hence, we came to the conclusion that it would be easier to implement the tool at the grassroots level. That is in the design department. Already the shipbuilding industry has become very progressive by developing countless comprehensive software that can deal in both design and production. Hence the second case studies the effects and applicability of an ANN in design software.

3.2 CASE 2- IMPLEMENTATION IN DESIGN

With time, ship design has aged from full scale drawings by master loftsmen to complete three dimensional renderings of ship models with computerised structural analysis using advanced finite element analysis.

It is now possible to simulate almost all conditions that the vessel might face during its lifetime on desktop based computers. As shown in Figure 2, all parts including structure, tanks, plumbing, outfitting etc. can be simulated along with the hull for exact real time data replication. Every aspect of the shipbuilding process from the stockyard to the fully outfitted ship at sea can now be designed, planned and constructed using comprehensive software tools that provide naval architects with everything they need to complete the detailed design. Simulations can even tell us what might happen if we start using some new technology, or different equipment mainly focusing on the cost of operation and strict adherence to build timeline [7]. This field is still evolving. All software are now being equipped with user-friendly, interactive, and easy to learn graphical user interfaces that simplify the process for the working personnel in a shipyard's design department. It would be very easy to train an ANN with deep learning using data from widely used ship design software. Since the same software will be used in shipyards all around the world, there will be numerous data points to serve as input while training the neural network. Hence prediction accuracy will be very high. Since classification societies are already taking an active role in promoting ship design software to incorporate eco friendly building techniques as well as the design of hulls, bows and propellers with higher efficiencies, it will not require a great deal of effort to include a knowledge acquiring system in them. Due to the large user base that such software has in the industry, we can train the ANN in a very short span. Qualitative and quantitative data from over a hundred shipyards can easily train the ANN for life. Such a venture can be tailored to perfection by IACS. Similar to the archives of ship data that some of the classification societies now maintain, a knowledge base that supplements the archives with the details of the building process will easily suffice when collected from a large number of shipyards. Once so many data points are amassed, the network will be able to forecast data on a variety of subjects covering all aspects of the design process.

Figure 2- Typical schematic rendered on a design tool [8]

Once it is successfully implemented in the design process, it can be extended to the production stage through class surveyors who supervise the production process. Information can be logged through the small tablet computers that a surveyor takes with him. Such a device can reduce the number of items that he has to carry around as it is an exceptionally task while accessing areas like the double bottoms and wing tanks for survey. If everything starts to be logged electronically starting now, the database we build will be most useful for developing better ways of accomplishing tasks that are otherwise hard to do.

3.2 (a) Sample Case Study in Suggestion Output

Figure 3 shows a sample data entry form that would be used by the designers in the shipyard to train the ANN for future suggestions. In future when the network is capable of extending beyond the design department, it will need both qualitative and quantitative data from more than a hundred shipyards. For example, a welder in his right mind-set can do his work perfectly while a disturbed one cannot. In everyday shipbuilding there are numerous welded joints being created per day. If a welder is careless in his work, the whole region in the newly built ship might become weak and prone to fracture. In a scenario where the ANN extends its reach into production, it can be monitored whether workers are over exerted or if someone is not good at their jobs. This can improve quality assurance at every shipyard. Moreover different problems faced by different shipyards might have a single and easy solution. Such instances can be easily identified and rectified.

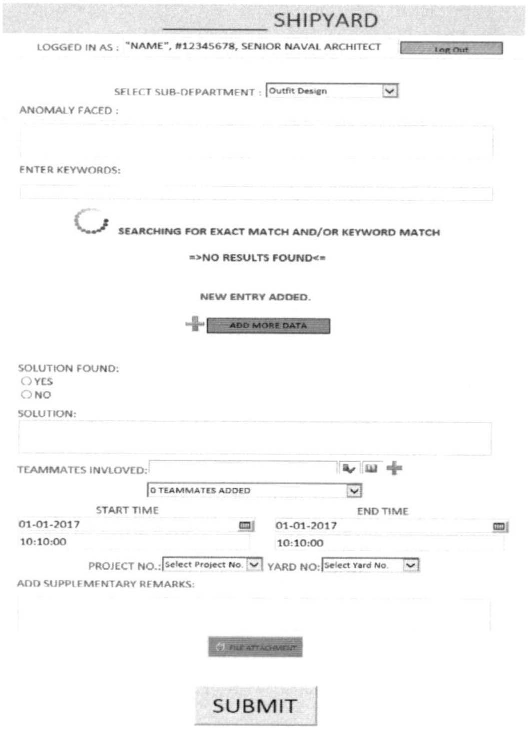

Figure 3- A GUI for Initial Database Creation/ Output

4. CONCLUSION

In short, it can be concluded that the easiest and most economical way to start the process of building a KBMS that employs an ANN capable of deep learning is through the premium design software that are widely used. Once a legal EULA is drawn that safeguards the commercial interests of all the users, data can be used to build a knowledge base that will be useful for future generations to build upon. It is a fact that shipbuilding practises have remained more or less the same save for some advanced tools that simplify the process. There have been no radical technological advancements that questioned the relevance of conventionally modelled ships. Hence it is paramount that a digital database for the same be developed that encases all the core values of traditional and present day shipbuilding.

5. ACKNOWLEDGEMENTS

The work on this paper started amidst a pandemonium of other responsibilities and ended the same way. Therefore, first and foremost I would like to thank my family who provided valuable advice, support and motivation and so became the prime mover in this venture. This paper would have been impossible without the invaluable help and advice provided by my teacher and mentor, Dr. K Sivaprasad.

6. REFERENCES

1. YANN LECUN, YOSHUA BENGIO, GEOFFREY HINTON, 'Deep Learning', Nature 521, 436-444 (28 May 2015).

2. M. ASAMI, F. KANEKO, 'Development of vessel collision model based on Naturalistic Decision Making Model', *Collision and Grounding of Ships and Offshore Structures- Amdahl, Ehlers & Leira (Eds), ISBN 978-1-138-00059-9, 2013*

3. ROBERT BRONSART, STEFFEN GAU, DIANE LUCKAU, WOLFGANG SUCHAROWSKI, 'Enabling Distributed Ship Design and Production Processes by an Information Integration Platform', *Proceedings of ICCAS 2015*

4. YASHWANT R KAMATH, K SIVAPRASAD, 'Experience Capture in Shipbuilding using Microsoft SharePoint', *ICCAS 2017.*

5. ADWAIT BHAVE, 'Druva Blog', *https://www.druva.com/blog/machine-learning-detect-anomalies-application-logs/*

6. ERICH SCHIKUTA, 'Neural Networks and Database Systems', *Austrian Computer Society,* 2008

7. CHARLES MCLEAN, GUODONG SHAO, 'Proceedings of the 33rd conference on Winter Simulation Pages 870-876', ISBN: 0-7803-7309-X, December 2001

8. AVEVA Outfitting Tool, Typical Outfit rendering, *http://www.aveva.com/en/solutions/product_finder/aveva_outfitting/*

7. AUTHORS BIOGRAPHY

K. Sivaprasad currently holds the position of Associate Professor at Department of Ship Technology, Cochin University of Science and Technology, Kerala, India. His area of specialization is in Ship Production and Ship Repair and so handles relevant classes for the B.Tech Program and is also a Research Guide. He has more than 35 published papers to his name and has more than 20 years of experience in teaching. He has also worked as a Senior Naval Architect at Mazagon Docks Ltd., India.

S. U. Sangeet has just completed his bachelor's degree in Naval Architecture and Shipbuilding from Department of Ship Technology, Cochin University of Science and Technology, Kerala, India.

Yashwant R. Kamath currently a student of B.Tech in Naval Architecture and Shipbuilding at Department of Ship Technology, Cochin University of Science and Technology, Kerala, India.

HOLISTIC SHIP DESIGN – HOW TO UTILISE A DIGITAL TWIN IN CONCEPT DESIGN THROUGH BASIC AND DETAILED DESIGN

T-H Stachowski and **H Kjeilen,** Digitread AS, Norway

SUMMARY

People have been designing and building ships successfully for hundreds of years – why change now?

There are many factors influencing the ship designers desire to find new processes and tools to drive their ship design forward. Initiatives like Industry 4.0 and the Digital Thread address the automation and utilization of robots. Other drivers are increased complexity and desire for sustainable solutions. Combined with the demand for higher innovation and focus on productivity and cost cutting, ship designers and yards must find new processes and systems in order to propel into the digital world.

This paper explains the advantages of holistic ship design compared to traditional design approaches and the significant potential of a Digital Twin in such a process.

NOMENCLATURE

[Symbol]	[Definition]
4GD	4th Generation Design
CAD	Computer-Aided Design
CAE	Computer-Aided Engineering
CAM	Computer-Aided Manufacturing
CAx	Computer-Aided technologies
CFD	Computational Fluid Dynamics
ECR	Electronical Change Request
KBE	Knowledge Based Engineering
PLM	Product Lifecycle Management
PMI	Product and Manufacturing Information

1. INTRODUCTION: THE DIGITAL TWIN

The Digital Twin is a virtual image of an asset, maintained throughout the lifecycle and should be easily accessible at any time. It is a central part of our digital asset ecosystem and it will enable a new generation of advanced analytics and understanding of the product. A Digital Twin should be able to integrate data from many different software products and handle them in a managed environment throughout the whole lifecycle. The concept of the Digital Twin will potentially enhance information management, understanding and collaboration, to prevent costly mistakes and rework.
The combination of holistic design and Digital Twin has the potential to revolutionize the ship eco system from early design and planning to design and operation and demobilisation.

"The Digital Twin serves as a virtual replica of what is actually happening on the factory floor in near-real time" [5]. It gives the opportunity to make production related decisions early in the design process and during the full lifecycle of the asset.

Throughout the lifecycle [the asset] is going through different stages of development. It is important to define and maintain those stages to be able to enable good decisions. This traceability also enables better future products due to the captured knowledge in the existing Twins.

A Digital Twin should carry production and manufacturing information (PMI data). Such data is a facilitator for automatization and efficient use of robot and direct linking to different production assets like cutting- and nesting machinery. PMI is both readable by machines and it can be used for 2D prints or 3D visualisation for involved stakeholders.

A Digital Twin is more than just a rich 3D model, it may also include optimisation, simulations and –results and defining documents. Such an approach enables no double data description and provides a single definition source for the asset.

A Digital Twin is suitable to couple with sensors on the physical asset for capturing data that might be used in future designs, fatigue studies and different aspects of maintenance like scheduling replacement of components. Autonomous vessels like the Norwegian "Yara Birkeland" - which will be world's first fully electric and autonomous container ship, with zero emissions – will ultimately require a Digital Twin approach for design and remote operation. [1]

Companies like Wilh. Wilhelmsen Holding ASA together with Ivaldi Group are currently developing methods to utilize 3D printers to manufacture spare parts for their ships. [6] Model information derived from a Digital Twin will take this one step further.

2. THE POTENTIAL OF THE DIGITAL TWIN

As mentioned earlier, many yards and ship designers are now challenging their way of doing ship design and are looking on new process and supporting systems. Legacy processes can involve as many as 15 to 25 different systems across the different design disciplines from early planning to fabrication. Data consistency and changes are hard to drive through and manage when the data definition and information is kept in so many different systems. To avoid failure, many costly efforts must be spent on non-value adding processes. It is also pricey to secure that relevant people can access the correct data in the required time span. In a demanding digital and disruptive world [with focus on cost cutting, improved innovation and fierce competition] yards and ship designers have a must challenge what has been and look for a novel approach for designing vessels for the next decades.

Legacy environments for ship design have been characterized by different specialized systems targeting a limited market. Other industries like Automotive, Aerospace and High-Tech are more or less sharing the same kind of product definition environments. The ongoing consolidation in PLM and CAx has, so far, had less impact to the marine industries. Enrolling such [PLM and CAx] environments to marine industries enables sharing of destiny with other industries, which means the development costs can be distributed on a large customer population.

3. INTRODUCTION HOLISTIC SHIP DESIGN

A holistic approach is dependent on a software platform that can handle the design from cradle to a ship in operation. It is a connected process with a master model guiding and maturing through the total design and life span of the asset. Reproducible scenarios and stages are used for optimization, simulation and understanding of the design.

Holistic design as a concept goes beyond problem solving. It incorporates all aspects of the ecosystem that a product will be subjected to. The concept is mainly focused around aesthetics, sustainability, spirituality and other aspects of "wants" rather than "needs". It is mostly employed in architecture but can be adapted to any form of product or service design. Ship design is especially suited for a holistic approach.

Using a holistic approach, helps to avoid scattered processes [as applied up to now], will result in a limited number of used systems and has a single backbone carrying the information through the various stages and disciplines.

Some key enablers in holistic design are:

- Front Loading
- Backbone
- Knowledge Based Engineering

3.1 FRONT LOADING

Front Loading challenges the traditional way of ship design specifically when it comes to the ability to able to order the steel as early as possible. The legacy processes are primarily 2D based and dependent on skilled workers leaving less opening for automation and robotization.

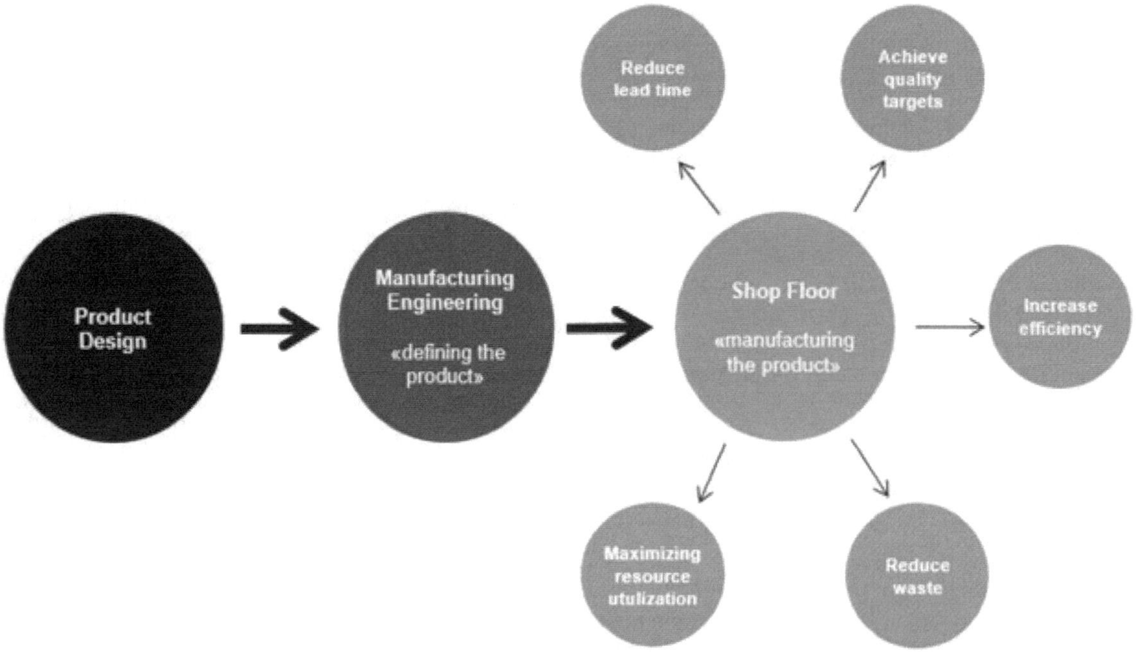

Figure 1: Traditional approach dependent on skilled workers

With Front Loading, more emphasis is put on early design phases by building up a digital representation for decision making, simulation and testing. Typically, models are "matured" and populated with more and more detailed information as the design moves along from concept design, to basic design and detailed design. A rich 3D model with BREP definitions and PMI carries the machinery information to different production equipment like nesting systems, cutting machines and different kind of robots. (Compare figures 1 and 2)

3.2 BACKBONE

A Backbone system is handling all relevant product information through all stages in a managed way. Such systems often called Product Lifecycle Management (PLM) systems. The Backbone is able to handle information from different systems spanning from office programs to CAD, CAM and CAE. It is independent of geography and disciplines and displays all managed information in a user-friendly fashion independent of devices or location.

The Backbone is managing the content, user access, revisions and versions of all data. It should be key to support a Digital Twin approach. (Compare figure 3)

3.3 KNOWLEDGE BASED ENGINEERING (KBE)

In this context, we refer to Knowledge Based Engineering (KBE) as parametric models that incorporate/capsulate knowledge and handles specific tasks and/or processes. Use of KBE is often motivated by a desire to secure a certain process, a way of conducting design, or to speed up the process for cumbersome and time-consuming designs.

Some examples can be KBE parts for railing, ladders, doors, bathrooms, anchor pockets or even complete hull designs. A KBE solution often includes optimization, simulation as well as documentation.

4. DIGITAL TWIN IN CONCEPT PHASE

In this phase the designer/yard is bidding to get the contract, traditionally it has been done in 2D AutoCAD and the handover is a 2D GA drawing. Many ship designers are in addition to this using systems such as Rhinoceros or 3D Studio Max to create a visual presentation of the concept. Simulations and calculations are done by hand or in separate systems for each design discipline.

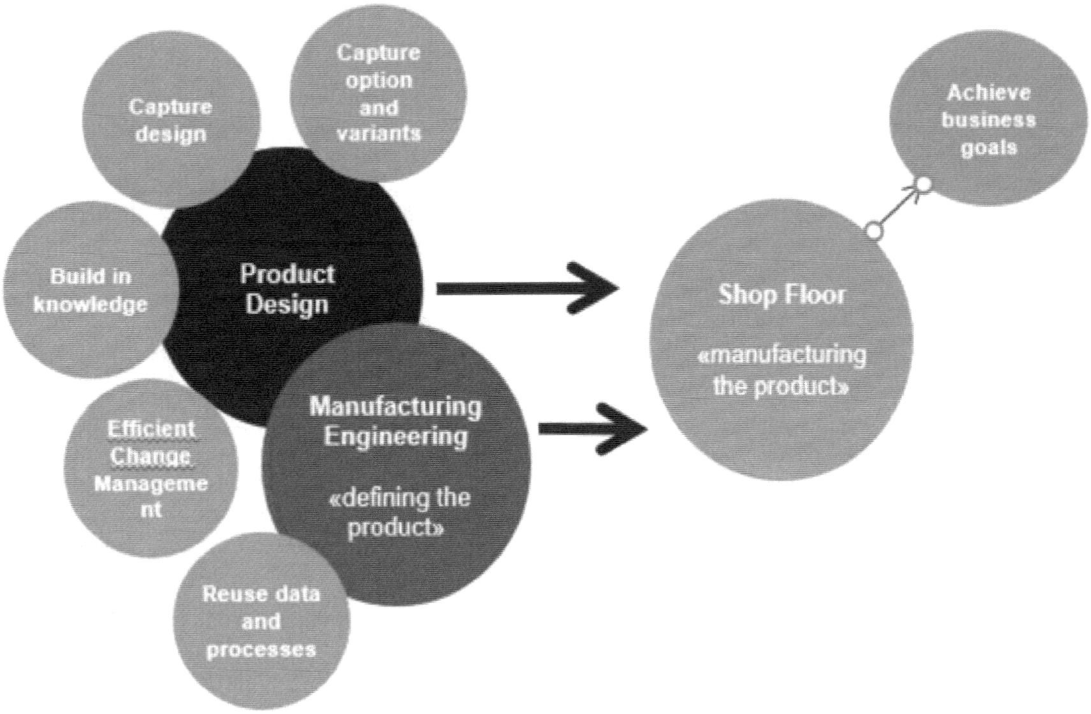

Figure 2: Front loaded process

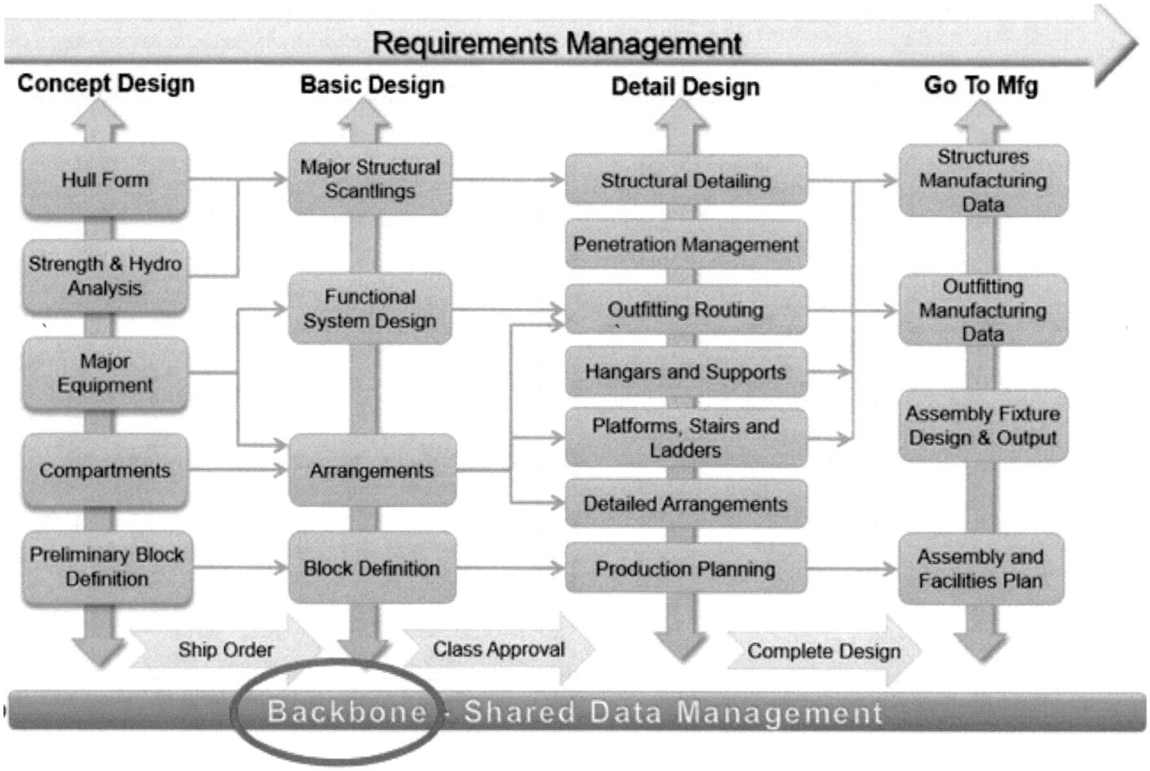

Figure 3: Example of a Process scheme included Backbone for shared data Management

In conventional design methodologies, unconnected data sets are generated, often by copying data from previous projects with little adjustment to meet the requirements of the new project. The experienced Naval Architect manages to find various data from a number of suitable projects and adjust them to the needs of the new project. This process is driven by getting concept design data as fast as possible. It is not very transparent and therefore creates a risk of failure. In addition, unconnected data sets have no proof of compatibility to each other. The design team, collaborating in the concept phase, can work out machinery schematics, general arrangements or hull lines which will not may not fit together. Every involved designer is focusing on the optimal result within his or her own design discipline in the new project, failing to focus on the broader picture.

With a holistic approach, ship designers will utilize a Master Model: building a 3D general arrangement with a linked 2D drawing, aesthetic views and different models suited for analysis and simulation. The Master Model holds both solid and surface representations which are available for different use cases and decision making. In this stage, the number of details are limited to what is necessary for the given purpose.

When defining the objective of the concept design phase, to arrive a signed contract for building the asset, a number of needs are identifiable among the created data. The feasibility, costs and risks are much more predictable when the design is carried out with a high level of detail and precision. However, there is no guarantee that the time invested in creating the concept will pay off, before the contract is signed. This results in the need to reduce the required time for creating sufficient set of data to what is necessary win a contract. As a result, it is vital to reduce the time required to generate sufficient sets of data to win a contract.

In the Digital Twin concept, a team of technical specialists will need to work and collaborate at the same time. Non-technical personnel will need access to information to pursue the project on the administrative- and business side. This requires a platform for collaboration and data access - which state of the art PLM systems can provide. A user-friendly interface with configurable access to data that meets the requirements of both technical and non-technical personnel is essential. Good search capabilities, using indexing technology, ensures that every person can find the data they needs when they needs it.

A Digital Twin in the concept phase, is different from the legacy processes by the fact that all the data is connected. The advantage is that conflicts in compatibility will be identified earlier and can be completely avoided which subsequently will lead to better decision procedures.

To address time efficiency, it is essential to organise the creation of the Digital Twin, as lean and effective as

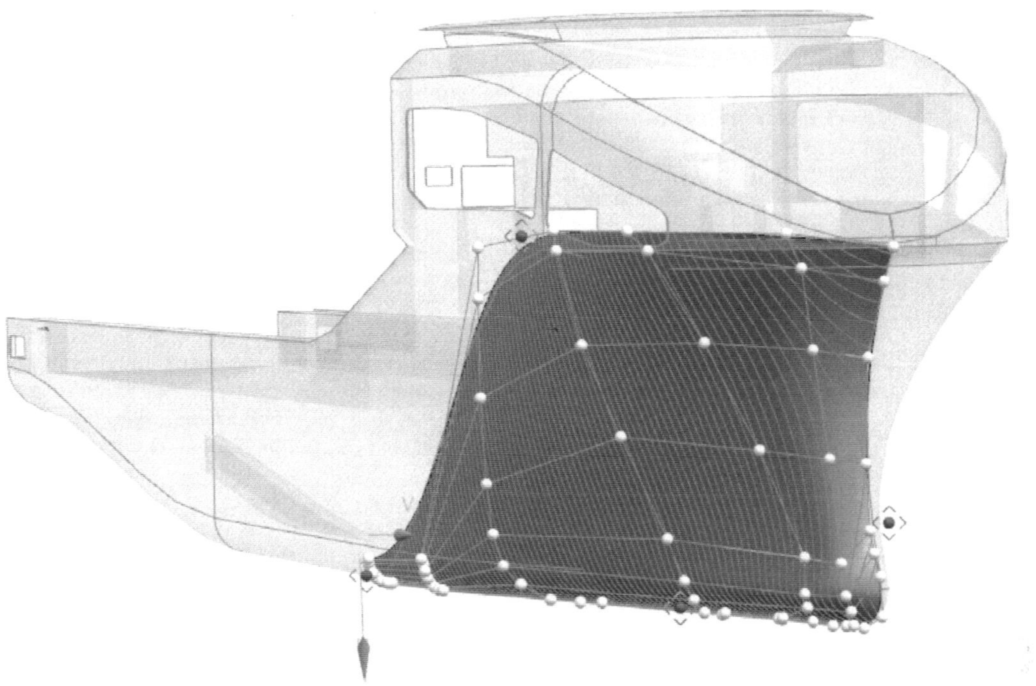

Figure 4: hull fairing on the Digital Twin

possible. Re-Use of existing data for the Digital Twin is key in order to save time. Re-used data might be a library of standard parts or other project neutral components. Those components can come from another Digital Twin. The ability to clone existing Digital Twins and change it to new requirements enables the use of existing project data to be the foundation for a new design. Therefore, it is important to be able to identify the concept phase of a more matured Twin. In a PLM system, this may be addressed by using revision and status management. State of the art product definition platforms support the reuse of data with sophisticated search functionality which is accessing both lifecycle and CAx data from a single and easily accessible user interface.

The Digital Twin in concept design [due to its connected product architecture] is providing a better baseline for CAE analysis. This enables more and better simulations earlier in the design process. Performing simulations and optimization earlier in the product design process will get more important as Industry 4.0 thinking will spread into the maritime industries. [3]

Capturing knowledge [rather than just documenting a product] is the required thinking to generate a Digital Twin, which will helps the concept design team to create a better ship design in the future.

Connected visualisation and Virtual Reality applied to the Digital Twin will give better support to the required sales processes reducing the gap between the different parties involved in winning a new contract.

Hull shape design of the Digital Twin is an integrated process [shown in figure 4], this will enable changes to be made rapidly in the different downstream processes. The defined Hull Shape model usually is the first geometrical definition defined for a Digital Twin. The design impact of the Hull Shape in a ship design usually is considerable in most downstream design processes. At the same time downstream processes defining requirements against the Hull Shape definition. Therefore, Hull Shape definition needs to be centric in the early concept phase of a Digital Twin. Processes analysing the hull shape and providing feedback for changes of the hull shape may be secondary and give change requests towards the hull design.

Using state of the art 3D systems gives the ship designer the opportunity to work on faired Hull Shapes from the very start of the design phase. Such a Hull Shape is parametric and very suitable for CFD calculation early in the design process. These calculations are imperative for achieving more efficient and sustainable hull designs.

While the hull is being analysed and optimised towards stability, seakeeping and resistance the hull is already used in the space subdivision of the vessel for applying functional design and risk analysis. The hull definition has an associative relationship to this subdivision. Any change applied to the hull can immediately be distributed towards the spatial subdivision. This may be managed by automated processes or user interaction.

Knowledge about required outfitting will be collected with traditional schematic sketches. Utilising systems engineering approaches throughout the product definition [the software] will provide part lists and interactive requirements between representations of the early 3D Model, the required functional space and the schematic drawings.

The functional space layout, the hull and interactive part lists [of the schematics] will lead to associative definitions of a three-dimensional General Arrangement model. (As illustrated in figure 5) This creates a comprehensive view of the Digital Twin in the concept phase.

The entire CAD process is supported by a lightweight viewing format, ISO 13406 (JT) which enables all Non-CAD users to display the current status of the geometrical model. The lightweight geometry is embedded in the data Backbone and it is therefore always in sync with the latest version of the Digital Twin. Status and access control gives a whole new level of opportunities for interaction with the classification society, yard and the customer of the ship design team.

5. DIGITAL TWIN IN BASIC DESIGN PHASE

This design phase addresses the need for class information. Classification documents are created together with information for any other authority. This information must be in line with the latest rules and regulations.

Historically customers using drawing automations to generate classification drawings to formally collaborate with the classification society. If those drawings have been generated from a 2.5D or 3D model manual tracking and mapping of changes is required from the communication on the drawing to its model definition.
In the basic design process of the Digital Twin, collaboration with the ship owner and the classification society is enabled and controlled by the Backbone. HTML 5-based technology helps to access Twin's data without complex installation processes. It is platform independent and works on multiple devices and sites. Status and rights management ensures that the related parties will only have access to the relevant data, in its matured state, for their role and project privileges.

During the basic design phase, multiple product views will show the relevant subset of the Digital Twin to the viewing user. [4] The product definition platform is assisting the different groups in providing and generating the required product views dynamically.

The ship structure process is supported by a managed library for standard parts and structural elements. The structural design in basic design is challenged in flexibility, as radical changes may occur. The ship design company can address radical changes quickly based on reasonable technical and economical data and this will give them a competitive edge. To address this a structural systems approach is used. A structural system might be a deck, bulkhead or structural hull definition represented by a sheet body. Seam curves subdivides a system into subsystems. This might be performed based on

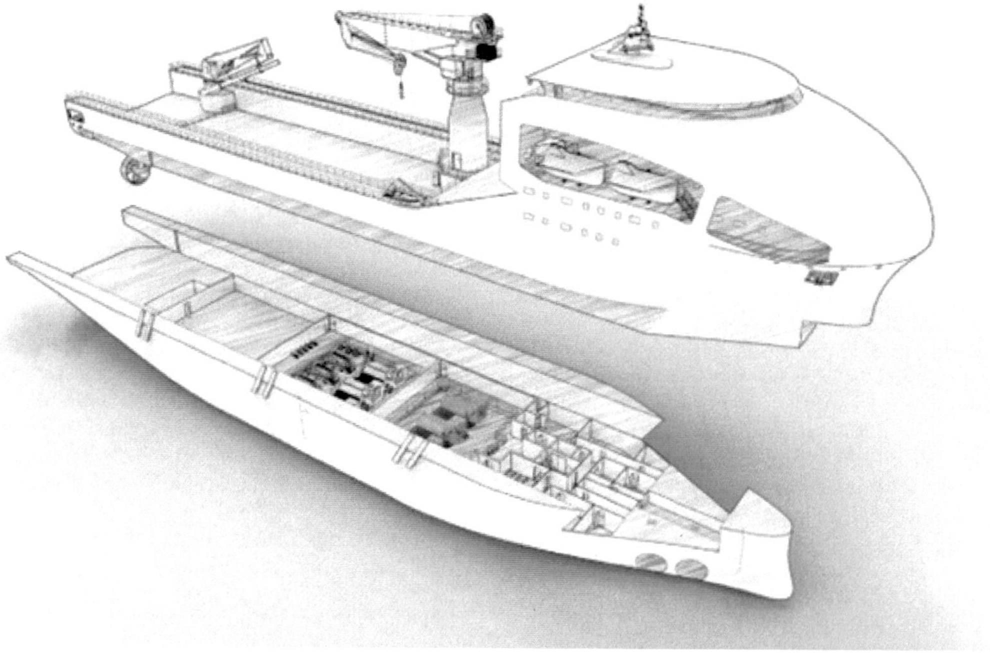

Figure 5: spatial subdivision connected with general arrangement and associative Hull Shape definition

intersections with other systems, different thickness areas, structure block boundaries and/or manufacturable parts. In this phase Stiffener, Edge Reinforcements and Pillars are defined as curves, they can be displayed as solid body for better usability. This topology architecture provides the ship designer the flexibility he/she needs.

The sheet and curve based architecture of the ship structural definition enables an efficient interface for ship structural CAE calculations. The derived CAE model is associatively connected with the ship structural definition, so CAE results are traceable to the structural definition. Changes can be tracked and coordinated transparently across multiple disciplines and sites. Interactive collaboration with the classification society on the Digital Twin [not on documentation of the Digital Twin!] is possible and the way to go forward.

In the future, a model based classification process will improve the collaboration and communication on the relevant Digital Twin data. [2] Certificates and approved structures will be managed and connected directly to the Digital Twins data within the Backbone.

The structure designer is working in context of the 3D conceptual data, such as the General Arrangement assembly. The designer can address required changes while displaying the structural view in context. (Illustrated in figure 6) The ship structure basic design view will be transitioned, with an automated process, into the ship structure detailed design view. This transition is supporting later changes on the basic design view.

6. DIGITAL TWIN IN DETAILED DESIGN PHASE

In the detail design phase, the Digital Twin will be shown in ship structure detail design views in a subdivision of structural blocks. Design features such as end cuts, cutouts, slots, etc. are created. Detail Design can make use of KBE capabilities to implement connections and create design features.

Fourth Generation Design (4GD) in NX/Teamcenter enables the display of the relevant ship structural subsets of the Digital Twin for other design disciplines. [4] Piping and Outfitting, Interior Design and design of specific mission equipment are handled in the context of each other. (see figure 7)

As on the physical building berth the process in a digital yard is applied to a single product – the Digital Twin of the asset. Different people have the opportunity to collaborate their design expertise into a good product. In digital processes around the concept of a Digital Twin it is the Backbone which is supporting and managing this cooperation.

Penetration management is embedded and helps the piping engineer to collaborate with the ship structural engineer. The piping engineer can design his pipes in a ship structural context. He will route his pipes interactively through the defined structural arrangement. This enables him consider i.e. girder or profiles while planning his piping routes. When the engineer intends to have his/her pipe to penetrate a bulkhead or plate, he/she can choose sufficient cutouts from a managed library. Together with remarks of the piping engineer an electronic change request (ECR) will be generated and provided in a task pool. Tasks from this pool are distributed based on organisational rules to the structural engineer in question. The structural engineer can either approve the ECR [which will lead him to the software to create the cutout] or come back with a new suggestion of cutout or position to take the structural design in consideration.

The Backbone with embedded reporting tools enables a light and straight forward communication on the assets definition, without the need [or delay] of an extracted representation such as a drawing. However, if drawings and PDFs are needed they are easily available on demand. They should not be used as information carrier but rather handled as "snap shots".

The ability to have subset views of the Digital Twin for designing in context also connects interior design and further outfitting close to the structural and piping design of the vessel. The rich definition of Digital Twins gives a foundation to utilise capabilities of virtual reality [and augmented reality in later processes]. This creates a new experiences and opportunities for collaboration with both suppliers, clients and internally at the yard for better designs.

In collaboration with our ship building and design customers worldwide, the different requirements of the ship structural definition in the different design processes have been identified. To support the detailed design phases for ship structure design in the best way, a new ship structural breakdown of the Digital Twin will automatically be transitioned by the system. This transition creates a more production- and detailed design focussed view of the ship structural definition. The ship structure detailed design model is built in a component based design architecture.

In the finalised ship structural detailed design definition, the Digital Twin will have a component item for each physical component for optimal support of the Lifecycle Management requirements during detailed planning, production and maintenance of the asset.

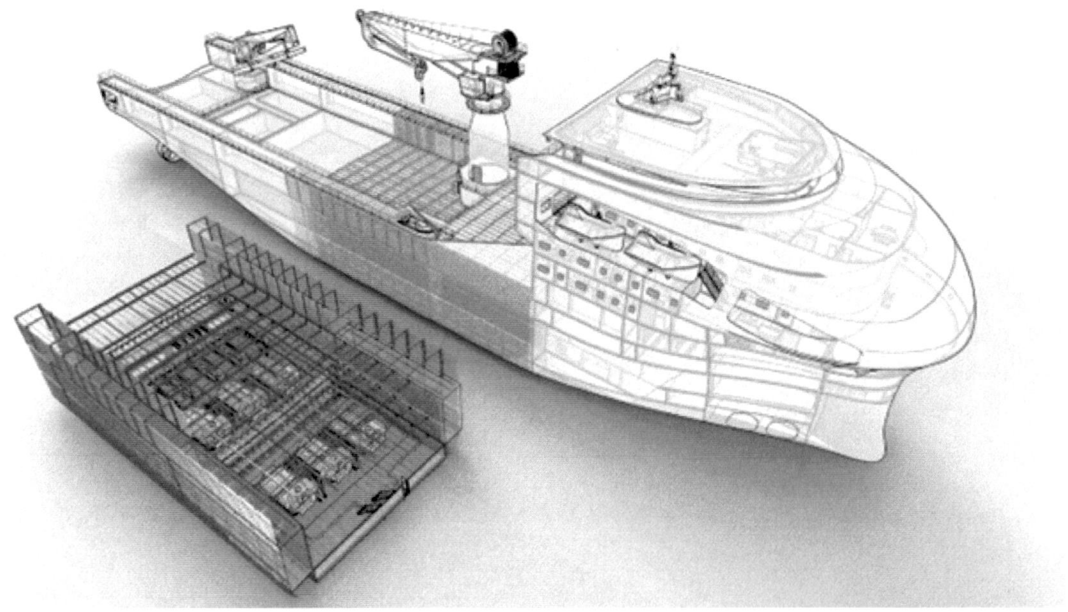

Figure 6: Utilizing concept data of the Digital Twin for the ship structural basic design in context

Figure 7: The Digital Twin enables Piping and Outfitting of an engine room in ship structural and general arrrangement views

7. ENABLING TECHNOLOGY

Our background and experience comes from working with Siemens PLM Software. As a provider, Siemens PLM Software differs from its competition in its openness and a devoted alignment to its customers. Unlike its major competitors, Siemens PLM Software has shared its technology and promoted openness. Good examples are JT that has become an ISO standard (14306) and Parasolid that is used by several hundred competing companies in CAx.

Siemens PLM Software heritage from both CAE and CAM solutions being able to communicate with different systems. This heritage, Siemens builds on in both CAD and PLM.

Major customers such as Samsung, Daimler Benz and Rolls Royce are driving the technology forward together with our 90 000 other customers with more than 11 million seats. The basis of the technology is shared by different industries like Automotive, Defence and Aerospace, Machinery, Marine and High-Tech.

Ship design- and building solutions are built upon a shared technology base with other industries, meaning that Siemens PLM Software can share developing costs on diverse industries and a large number of users and offer the latest state of the art software in the various industries.

Ship design customers in Europe, USA and the Far East have been instrumental in developing the solution for commercial ship design into a leading-edge position. Because we share destiny with more than 11 million seats Siemens PLM Software is able to preserve its leading technology position in a world of accelerating technology and disruptive initiatives.

Ship design- and building is one of the most complex processes among the industries. Legacy vice it has also been a scattered process with many different systems and information silos. Like most other industries the product [the vessel] is becoming more and more complex. Customers have more sophisticated needs and green politics is driving the demands for efficient and innovative solutions. Digitalization will eventually transform the marine industry and create a new generation of Digital Shipyards. To meet these demands ship design customers need a good foundation and partnership [to their software providers] that can meet an ever more complex and compelling environment. Traditional providers of ships design software do not have either the muscles or the technology to fulfil these needs.

Siemens PLM Software with its openness and willingness to share and work together with customers contributes to competitiveness and secures an ongoing improvement. These customers can build upon Digital excellence and leverage the power of emerging technologies across organizations giving them ability to transform data into actionable insight and ultimately gaining business advantages. Those customers will be able to capitalize on digital disruption and make the right decisions going forward in a competitive digital world.

8. CONCLUSION

In a holistic design approach, the Digital Twin is holding all information of the asset. Starting with the definition of a Digital Twin already in concept design is helping the ship designer to have transparent processes and ultimately a better product.

Technology managing the Digital Twin and supporting a digital design processes will help ship design organisations to have a competitive edge in a digital world.

9. REFERENCES

1. **Wilh. Wilhelmsen Holding ASA.** www.wilhelmsen.com. [Online] 26 May 2017. [Cited: 15 August 2017.] http://www.wilhelmsen.com/media-news-and-events/press-releases/2017/from-parts-to-files-wilhelmsen-explores-3d-printing-with-ivaldi-group/.
2. **Astrup, Ole Christian and Cabos, Christian.** *A Model Based Definition for Shipbuilding.* s.l. : DNVGL, 2017.
3. *PERSPECTIVES FOR INTELLIGENT ICT IN SHIPBUILDING.* **Hribernik, Karl.** Bremen : International Conference on Computer Applications in Shipbuilding, 2015.
4. *SHIP WORK BREAKDOWN STRUCTURES THROUGH DIFFERENT SHIP LIFECYCLE.* **Pal, Malay.** Bremen : International Conference on Computer Applications in Shipbuilding, 2015.
5. *Industry 4.0 and the digital twin.* **Parrott, Aaron and Warshaw, Dr. Lane.** s.l. : Deloitte University Press, 2017.
6. **Kongsberg .** www.kongsberg.com. [Online] [Cited: 21 August 2017.] https://www.km.kongsberg.com/ks/web/nokbg0240.nsf/AllWeb/4B8113B707A50A4FC125811D00407045?OpenDocument.
7. *An Object-oriented approach for Virtual prototyping in conceptual ship design.* **Fonseca, Icaro Aragao and Gaspar, Henrique Murilo.** Albena (Varna) : European Conference on Modelling and Simulation, 2015, 2015.
8. Quantifying value robustness of OSV designs taking into consideration medium to long term stakeholders' expectations. **Gaspar, Henrique Murilo, et al.** s.l. : 12th International Marine Design Conference (IMDC 2015), 2015.
9. *Epoch Era Analysis in the Design of the Next Generation Offshore Subsea Construction Vessels.* **Keane, Andre, Gaspar, Henrique Murilo and Brett, Per Olaf.** San Antonio : 10th System of Systems Engineering Conferences (SoSE), 2015.
10. *Data-Driven Documents (D3) Applied to Conceptual Ship Design Knowledge.* **Gaspar, Henrique Murilo, et al.** Redworth : 13th International Conference on Computer and IT Applications in the Maritime Industries (COMPIT), 2014.

10. AUTHORS BIOGRAPHY

Torben-H. Stachowski currently holds the position of Technical Lead Shipdesign in Digitread and, being a Naval Architect, is responsible for activities in the ship building industry.

Since 2010 Torben is dedicated to develop new digital ship design processes. As the link between end user and software development he helps to get Siemens PLM products ready for the commercial ship building market.

Helge Kjeilen is Chairman and owner in Digitread AS distributing and supporting Siemens PLM Software. His career starting in 1981, he founded and co-found several Nordic IT companies. Helge has also worked as Country Manager in Apollo Computer and Prime Computervision. For 25 years Helge has severed as board member in Wittusen & Jensen. Helge Kjeilen has a Master of Business and marketing (MBM), Field of study Business Administration and Royal Norwegian Naval Academy, OKA. Helge has been sensor for Master thesis at Norwegian University of Science and Technology for 7 years.

IMPROVEMENTS IN LEARNING PROCESS AND PRODUCTIVITY. AN ANALYSIS OF THE RETURN OF THE INVESTMENT

R Perez, SENER, Spain

SUMMARY

CAD Systems help engineers and designers in various industries, designing and building 3D models of airplanes, bridges... and of course, ships, submarines and floating structures. Nowadays all the industry, an especially shipyards and design offices, uses CAD in order to design better and cheaper boats and floating artefacts. The use of a CAD System provides not only a cost reduction in design, but also a cost reduction in production, including two major aspects: savings in materials due to a more accurate definition of components and material management, as well as savings in materials due to a significant reduction of design changes, and consequently reduction of rework in production; and savings of production man-hours due to the reduction of changes and rework.

This paper shows an analysis performed with information compiled as a consequence of the author experience in CAD implementation in numerous shipyards around the world during the last years, to provide a clear idea of the Return on Investment of using CAD Systems (in particular, the FORAN® System) in a generic shipyard. Another key aspect of this paper, is the analysis of the learning curve of a new CAD System implemented in a generic shipyard. The data, hours expended designing blocks and units of different weights, collected for this paper during a couple of years from a generic shipyard, have been a fantastic tool for analysing the factors that affect a learning curve process.

NOMENCLATURE

CAD	Computer-Aided Design
CAM	Computer-Aided Manufacturing
CAE	Computer-Aided Engineering
CIM	Computer Integrated Manufacturing
DCF	Cash Flow
DPP	Discounted Payback Period
HVAC	Heat, Ventilation and Air-Conditioning
IRR	Internal Rate of Return
NPV	Net Present Value
PI	Profitability Index
PLM	Product Lifecycle Management
FPLM	FORAN® Product Lifecycle Manag.
QR	Quick Response
ROI	Return On Investment

1. A BIT OF CAD HISTORY

Computer-Aided Design (CAD) Systems help engineers and designers in various industries, designing and building 3D models of airplanes, bridges... and of course, ships, submarines and floating structures. There are other acronyms that are usually accompanying the acronym CAD, such as Computer-Aided Manufacturing (CAM), Computer-Aided Engineering (CAE), and Computer Integrated Manufacturing (CIM), including instructions to Computer Numerical Control (CNC) machines. One could claim that these CAD/CAM/CAE Systems had their origin around the year 350 BC, with the mathematician Euclid of Alexandria. Many of the postulates and axioms used by today's CAD Systems are based on Euclidean geometry. Some 2,300 years later, many see the birth of 3D CAD with the work of Pierre Bezier, a French engineer working at the *Arts et Métiers ParisTech*. After his mathematical work concerning surfaces, he developed UNISURF, between 1966 and 1968, to ease the design of parts and tools for the automotive industry. Then, UNISURF became the working base for the following generations of CAD software. Only one decade later, CAD Systems were introduced in academic courses. The real breakthrough in CAD Systems came, logically, with the development of computers; and during the 1980s the application of CAD Systems matured to a similar stage as known today, Perez and Gonzalez (2015).

CAD tools are usually divided into 2D drawing and 3D modelling programs. The drawing tools are based on 2D vector geometric entities such as points, lines, arcs and polygons. In 3D modellers, solids and surfaces are created. In the early stages of the development of CAD Systems, the software was running on mainframes, which limited use of CAD Systems for manufacturing. With the arrival of workstations and PCs came then widespread use of CAD Systems in engineering on a daily base.

2. STATE OF THE ART IN THE CAD WORLD

There are several aspects where CAD Systems could, and most likely will, improve in the near future. However, in this section the focus is on functionalities that are being improved right now. E.g. in hull forms definition, the global shape modelling or the advance continuity and capping tools could transform complex surfaces with excellent results, less interaction, high accuracy, and full control. These techniques shorten dramatically the design time, from days to few hours while obtaining excellent results, Perez and Toman (2014).

Another area of improvement concerns one of the most time-consuming tasks in outfitting design, the routing of pipes, *Heat, Ventilation and Air-Conditioning* (HVAC) ducts and cable trays. Automatic routing options will minimise the design man-hours, and at the same time will increase the robustness of the design. Automatic routing provides simple solutions, with optimisation of material, and several algorithms exist. But the matter is not only to consider existing elements for future routings; it is also necessary to assign priorities, and eventually handle automatic modifications of existing elements as a consequence of new ones. The complexity of the problem explains that there is not yet fully satisfactory solution for the automatic routing in practice. Current solutions provided by CAD Systems solve partial problems, offering already significant support.

Another area where the CAD companies are active is Virtual Reality, Perez and Alonso (2015). The objective is to create a user-friendly environment in order to review, audit, obtain metrics such as the progress of a project, etc. This type of review process of the model does not need to use tools for designing, just a simplified tool allowing easy access.

Advanced browsers allow incorporating human models in order to study ergonomic aspects, movements of components to perform simulations, etc. Sometimes there is a need to take information from an on-line database and if there is an Ethernet network through the shipyard, it is useful to have viewers that permit connecting to a project. Browsers can connect to the database of a project in order to access information in real time. When there is not an on-line access to the database, viewers should be able to read files with the project information required for 3D modelling of the product and component data with optimal performance. So far it was prevalent to implement viewers on laptops, because laptops are usually equipped with processor and graphics cards that allow navigating through the entire vessel. In recent years there has been a breakthrough in mobile devices like tablets or smartphones. This hardware progressively incorporates new processors that enable enhanced graphics. On the software side, operating Systems have been developed specially adapted for such devices (such as Android or IOS) allowing interfacing naturally by touch gestures, Perez and Gonzalez (2015).

The widespread use of these devices nowadays has precipitated its use by software companies. Software developers have taken their time preparing oriented solutions including those that allow to have project drawings or 3D models on tablets or others electronic displays. In modern projects, there is the need, for supervisors, to use these devices to work better, with a quickly access to the 3D model, with all parts information and construction drawings needed. A Wi-Fi connection would allow accessing to an information server to be updating the information needed, mainly in files, as for example: 3D models, classification or production drawings, between others. Another advantage of mobile devices is that the user interface might interact with the project model or parts using gestures just as everybody does every day with smartphones. A development line of navigator's evolution would incorporate augmented reality technology. It would be helpful for production technicians to scroll through the project and pointing the camera of their mobile device to a particular component to obtain information from it and have the actual image of the same 3D design model displayed. This is possible through the use of markers that help the device to position itself within the project and also for the use of *Quick Response* (QR) codes; Perez and Gonzalez (2015).

The design of a ship needs very different design applications. It seeks to optimize the product from different points of view, which requires design Systems that, to a greater or lesser extent, need to be connected. They must be able to talk to one another, exchange data, models or forms. It is necessary to have data communication agreements, through either APIs or common interfaces. CAD exchange formats have extended their scope. The connection of CAD Systems with the product life cycle management Systems is indispensable to bring value to the product itself. The PLM will not only be part of the design and construction stage but also extend to the operation of the ship. The PLM stands as the controller of the product, the repository of the unique truth, where the information is the only valid one. The different objects and equipment of the ship should not be autonomous entities that decide by themselves, but they should be supervised by a System that takes into account the product as a whole. A ship is a product with a high value by itself, but also has a lot of value for what it contains, so it cannot be left to the edge of technological autonomy. It is necessary to have a higher responsibility and this responsibility must be of a human intelligence, Muñoz and Perez (2017). Fig. 1 shows a simplistic representation how to maintain all the access of the elements of a ship and the ship itself under control.

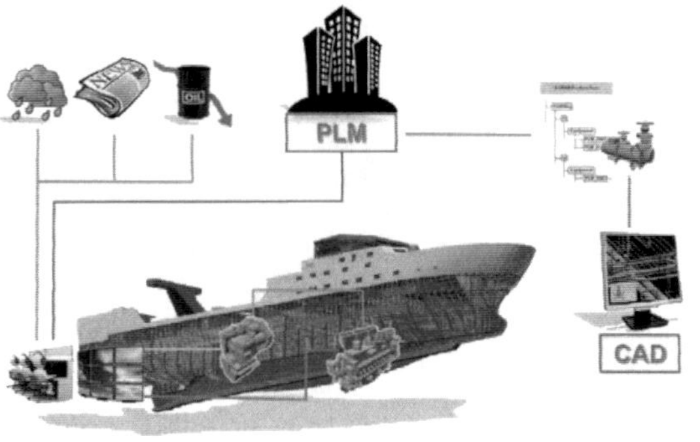

Figure 1: All the ship information under control (Muñoz and Perez, 2017)

Another recent milestone is the integration between different CAD Systems and Product Lifecycle

Management (PLM) tools, e.g. the FORAN® Product Lifecycle Management (FPLM) tool with a neutral architecture. In this case, all the information generated in FORAN® may be transferred to a PLM and may be subject to all processes: control, configuration and releases lifecycle and process management. FPLM consists of a series of tools and features that enable bidirectional integration between different modules of FORAN® and PLM tools, Perez and Penas (2015).

There are many advantages of using CAD in shipbuilding: ease of design, speed of construction, use and reuse of information, etc. It is expected that in the future CAD tools will advance further and allow greater information management and virtual access through smart devices. In general, CAD Systems provide tangible benefits while the process is optimized, reducing design time and production, and therefore costs, Perez et al. (2013).

3. AN ANALYSIS OF THE RETURN ON INVESTMENT OF A CAD IMPLEMENTATION

This paper shows the analyses performed to provide a clear idea of the Return On Investment (ROI) of performing a CAD System implementation (in particular, the FORAN® System) in a generic shipyard.

The analysis on the potential improvements that could be achieved in the different design processes affecting various disciplines in a generic ship design technical office, when working with a specific CAD System (FORAN®), provides the necessary inputs to evaluate the productivity gain that are expected. With those inputs, present paper is intended to provide a financial evaluation of the necessary investment, quantifying the benefits and the costs of FORAN® implementation in a generic shipyard. The quantification of benefits and costs are analysed along a reasonable time period, generating cash inflows and cash outflows.

This financial analysis includes only those aspects that can be estimated or measured by SENER®, company owner of the FORAN® System, including also a list of benefits and costs that, although cannot be directly evaluated by SENER®, should be also taken into consideration. The information included in this paper has been compiled as a consequence of SENER® experience in FORAN® implementation in numerous shipyards around the world between 2007 and 2015.

The following sections show not only the ratios used to evaluate the ROI, but also analyse the benefits of FORAN® implementation, the generated cash flows, the estimated learning process productivity losses, the expected productivity gain, the assumptions taken for the financial analysis, and a sensitivity analysis of the ROI respect to productivity gain and the considered discount rate. The ROI financial analysis included in this paper shows the complete set of cases considered as well as summary tables and charts that helps to the data interpretation.

4. RETURN ON INVESTMENT EVALUATION CRITERIA

It is highly recommended to have some numbers showing the merits of the investment. There are several valuation tools and methodologies used to describe how good an investment such as the implementation of FORAN® is. Those methodologies basically quantify and calculate the different metrics usually used to have a better knowledge of the ROI, because ROI is a standard measure of investment profitability that is based on the calculation of the discounted profits over the life of the project expressed as a percentage of initial investment. Also, ROI is a financial analysis technique used to calculate the benefits of investing. This calculation is based on the following concepts, Perez (2016):

Cost: Total amount of money spent on the new System implementation.

$$\sum_{i=1}^{n} DiscountedCosts_i \tag{1}$$

Benefits: Total amount of money gained from the new System implementation.

$$\sum_{i=1}^{n} DiscountedBenefits_i \tag{2}$$

Net Present Value (NVP): Discounted cash flows.

$$NVP = \sum_{year=1}^{Years} \left(\frac{\sum_{i=1}^{n} Benefits_i - \sum_{i=1}^{n} Costs_i}{(1+DiscountRate)^{year}} \right) - Costs_0 \tag{3}$$

Internal Rate of Return (IRR): Discount rate that makes the Net Present Value zero.

$$IRR = DiscountRate \leftrightarrow \sum_{year=1}^{Years} \left(\frac{\sum_{i=1}^{n} Benefits_i - \sum_{i=1}^{n} Costs_i}{(1+DiscountRate)^{year}} \right) - Costs_0 = 0 \tag{4}$$

Payback Period: Amount of time needed to reach the Break Event Point.

$$PB = year \leftrightarrow \sum_{year=1}^{Years} \left(\frac{\sum_{i=1}^{n} Benefits_i}{(1+DiscountRate)^{year}} \right) = \sum_{year=1}^{Years} \left(\frac{\sum_{i=1}^{n} Costs_i}{(1+DiscountRate)^{year}} \right) \tag{5}$$

Break Event Point (BEP): Value corresponding to the point when benefits meet or exceed cost.

$$BEP = \sum_{year=1}^{Years}\left(\frac{\sum_{i=1}^{n} Benefits_i}{(1+DiscountRate)^{year}}\right) = \sum_{year=1}^{Years}\left(\frac{\sum_{i=1}^{n} Costs_i}{(1+DiscountRate)^{year}}\right) \quad (6)$$

Profitability Index (PI) or Benefit/Cost Ratio: Ration of benefits to costs.

$$PI = \frac{\sum_{i=1}^{n} DiscountedBenefits_i}{\sum_{i=1}^{n} DiscountedCosts_i} \quad (7)$$

ROI: Ratio of adjusted benefits to cost.

$$ROI = \frac{\sum_{i=1}^{n} DiscountedBenefits_i - \sum_{i=1}^{n} DiscountedCosts_i}{\sum_{i=1}^{n} DiscountedCosts_i} \times 100\% \quad (8)$$

The ROI calculation results in a value that represents the benefits received from a project against the total costs of the project, being necessary to define costs and benefits on a timeline when they are expected to occur. Therefore, it is necessary to establish a cash flow (outflows = costs and inflows = benefits) in a conservative and reasonable amount.

5. ECONOMIC STUDY

5.1. BENEFITS IN CAD IMPLEMENTATION

Several shipyards presented its current design process showing a significant complexity, mainly due to the very different data bases and tools used along the process. The use of different data bases drives to a very important effort to maintain their coherence, requiring successive comparison between them (manual and electronic) to maintain data integrity and their consistency. Current modelling process is based in different tools that although stable it is a rather old System with very restrictive development. The implementation of FORAN® in a generic shipyard provides obvious benefits, but not all of them can be directly quantified. The obtained benefits can be tangible or intangible, and they can be estimated or not. Benefits can be classified as

- Tangibles:
  - Those that can be estimated or measured by SENER®.
  - Those that cannot be estimated or measured by SENER®, but that can be estimated or measured by the diverse shipyards.
- Intangibles: they cannot be inherently estimated or measured.

The present financial analysis takes into account only those aspects that can be estimated or measured by SENER®. However, a list of other benefits that cannot be estimated or measured by SENER® is also included, as although not quantified, they could be of remarkable importance. The implementation of a software application as FORAN® should be supported by an analysis of benefits, activity improvement and business added value. Part of this support can be given with a financial analysis of the necessary investment.

5.2. CAD IMPLEMENTATION CASH FLOW

ROI financial analysis is based on the study of the generated cash flows: cash outflows (costs) and cash inflows (gains). FORAN® implementation would have the following tangible cash flows (estimated by SENER®):

- Cash Outflows (cost):
  - Net Licence cost: single initial cost.
  - Software installation in a generic shipyard computers: single initial cost.
  - Training: single initial cost.
  - Yearly maintenance: annual cost, excluding guarantee of six first months.
  - Technical assistance: First year cost.
  - Implementation/adaptation, coordination and management: First 2 years cost.
  - Loss of productivity during training phase: affecting only to the first year.
- Cash Inflows (benefit):
  - Increase of productivity during training phase: affecting only to first year.
  - Increase of productivity once FORAN® is fully implemented: affecting the part of the first year in which the FORAN® is fully implemented, and the subsequent years.
  - Increase of productivity due to an effective concurrent engineering (all disciplines working with a single database).
  - Reduction of production hours due to reduction of changes and reworks.
  - Reduction of material costs (steel, piping, cables, etc.) due to a most precise definition of items, and reduction of changes and rework during production.

There are other concepts that although tangible, they cannot be estimated or measured by SENER®:

- Potential Cash Outflows (cost):
  - Information Technology (IT) infrastructure: PCs and necessary basic software.
  - Effort to adapt shipyard structure to the use of FORAN® in the design processes.
  - Additional migration costs that cannot be specified by SENER®, but that probably can be specified by the shipyard.
- Potential Cash Inflows (benefit):

- Reduction of number of workstations, reduction of maintenance of IT infrastructure.
- Potential internal structure simplification, and therefore structure optimisation.

In addition, there are other intangible benefits, including:

- Use of a design and modelling tool that includes the state of the art technologies used all over the world for ship design and ship construction. FORAN® represents the leading edge CAD/CAE/CAM technology in shipbuilding.
- Use of a tool that is always looking for improvement, developing continuously enhancements, including new functionalities and planning new System improvements.
- Show a better corporate image through the tool used for design and modelling.

5.3. LEARNING CURVE PROCESS

ROI analysis for software implementation must consider the changes in productivity during its implementation period. Those changes in productivity can be generically represented by a transition curve along a training period. After that period, the standard productivity gain is achieved. The transition curve is characterized by a productivity loss during the training courses and some weeks afterwards, characterized by a strong productivity gain progression. After the training courses there is a maturation period when the designer or modeller gets the adequate skills in a more tended slope curve, achieving finally the standard user skills. The productivity loss at beginning of the implementation is part of the so-called migration costs. Migration costs will be smaller if the implementation is performed in the shortest possible period. Fig. 2 shows the generic design productivity evolution of FORAN® implementation for a generic modeller or designer. There is an immediate dip in productivity, as the users get up to speed on the new System. With time, productivity climbs back to where it was with the original System and levels off at a higher point as the new System is fully implemented.

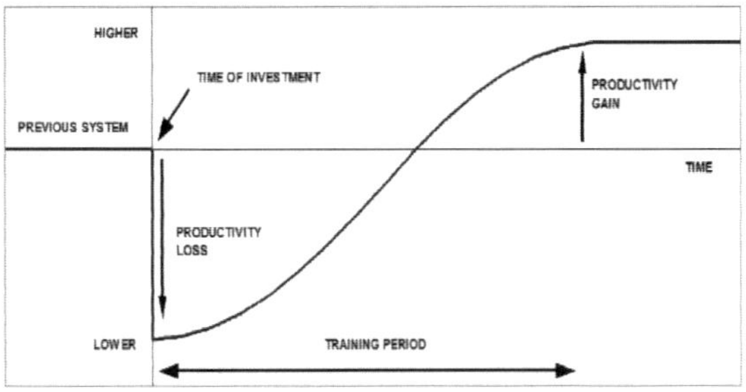

Figure 2: Design productivity evolution of FORAN® implementation for a generic modeller or designer

In the case of the ROI analysis for FORAN® implementation, the transition curve representing the changes in productivity during the implementation period is characterized by a very short period of loss of productivity (slightly longer than the duration of the training courses), and a short to medium time period to achieve the standard common skills, achieving the old productivity level in a very short time after finishing the training courses. After the two or three week training course there is a strong progression in the skills of the users up to four or five weeks, changing to a more tended slope of productivity gain along the following ten to fifteen weeks. Fig. 3 shows the design productivity evolution of a standard skilled designer, as used in present analysis. It is based on a training course of two to three weeks and a post training course period in which the personnel reached the usual skills. We assumed that modellers and designers have a previous standard skill in 3D modelling software packages. Figure 3 includes SENER® experience in the implementation of FORAN® in different shipyards.

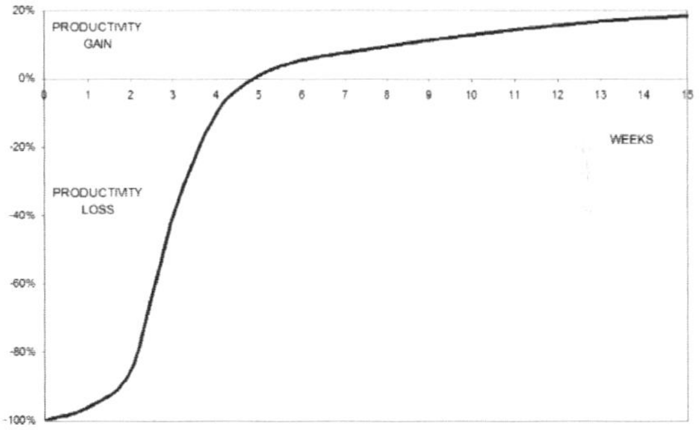

Figure 3: Design productivity evolution of a standard skilled designer, as used in present analysis

After the training courses, there is a very short period necessary to achieve the previous productivity, and after that there is a period of usually eight weeks after training course when the standard skills are achieved reaching the standard gain in productivity. Nevertheless, for this analysis, a longer time period has been considered, assuming twelve weeks instead of eight weeks.

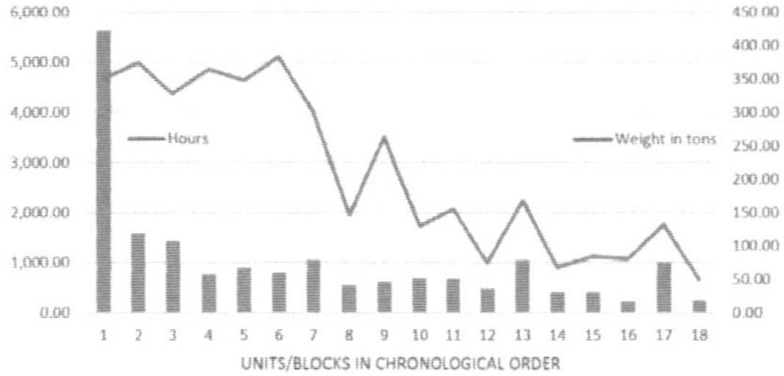

Figure 4: Learning curve with FORAN® for first ship

As the training courses are provided sequentially for several groups of people and for different disciplines along the first year, for present analysis the potential productivity gain in the first year has been reduced to 50%, being that a very conservative assumption. In order to consider a realistic scenario involving all factors, it has been carried out a study about the CAD System learning process in a standard shipyard. Figures. 4 and 5 show the learning curves during the CAD System implementation in the shipyard. During the last two years, data related with hours devoted to design a precise unit (block) with a specific weight were collected in a shipyard. The chosen shipyard was one where FORAN® had been implemented for first time. Figure 4 shows the learning curve during the design of the first ship, figure 5 for the second ship design.

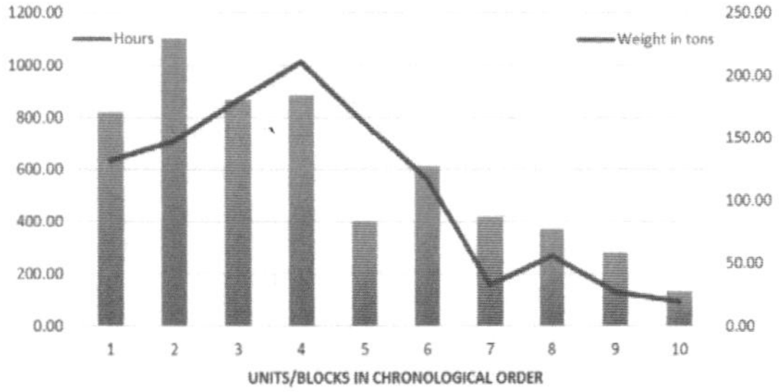

Figure 5: Learning curve for second ship

5.4. AN INCREASE OF PRODUCTIVITY

SENER® experience indicates that an increase of productivity from 10 % to 20 % is very common when implementing FORAN® in shipyards. In present case corresponding to the implementation of FORAN® in a generic shipyard, an estimate has been performed by discipline, including weighting factors to obtain a global productivity increase rate. Table 1 shows disciplines, the weighting factors, and the considered productivity increase rate. These weighting factors, and considered productivity increase rates are only estimates based on impressions obtained from the meetings maintained with different shipyards people. SENER® would recommend a detailed quantitative re-assessment once quantitative data from the shipyard is available. In a generic shipyard, the analysis of the different design processes and disciplines drives us to estimate a design productivity increase of 12 % to 18 %.

Table 1: Disciplines, the weighting factors, and the considered productivity increase rate

Discipline	Weighting Factor	Productivity Gain	
		Relative	Subtotal
Structure	20 %	14 %	2.80 %
Electrical	15 %	18 %	2.70 %
Piping Routed Systems	30 %	21 %	6.30 %
HVAC Routed Systems	15 %	10 %	1.50 %
Design Area Integration	10 %	11 %	1.10 %
Accommodation /Outfit	5 %	11 %	0.55 %
Miscellaneous Applications	5 %	8 %	0.40 %
Overall Process Productivity Gain			15.35 %

However, the ROI financial analysis includes cases with productivity gains of 6%, 9%, 12%, 15%, and 18%.

5.5. RETURN ON INVESTMENT FINANCIAL ANALYSIS STATEMENTS

A ten years' time frame is going to be considered for the FORAN® implementation financial analysis, as FORAN® is a strategic long-term tool. Residual cash flows have not been considered in this analysis. This ROI financial analysis takes into account the time value of money, considering an escalation rate of 3%. The financial analysis includes sensitivity analysis respect to discount rate and to expected increase of productivity rate. For the discount rate the following values are used: 3%, 6%, 9%, 12% and 15%. For the increase of productivity rate the following values are used: 6%, 9%, 12%, 15% and 18% (see expected increase of productivity in previous section). Although taxes and depreciation can also have an impact in the financial analysis, they have not been considered in the present analysis.

5.5 (a) Savings in Design, Materials and Production

Savings in design as a function of productivity gain are estimated as follows:

- Design personnel hourly rate cost (and associated indirect costs) that is going to work with FORAN® is assumed to be 45 €/h. This number may differ depending of region, country, etc.
- Assumed yearly man production hours: 1,760 h/year. As before, this amount could vary.
- Number of designers involved with FORAN® is 100. This figure was chosen as an average.
- Different productivity gains considered are: 6%, 9%, 12%, 15%, and 18%.

The use of FORAN® will provide not only a reduction of changes, but a more accurate definition of the components of the different Systems. It means not only a cost reduction in design, but also a cost reduction in production, including two major aspects:

- Savings in materials due to a more accurate definition of components and material management, as well as savings in materials due to a significant reduction of design changes, and consequently reduction of rework in production.
- Savings of production man-hours due to the reduction of changes and rework.

The estimate of material savings has been done as follows:

- Steel:
  - Steel elaboration: 10,000 t/year. This figure has been chosen as an average.
  - Steel price: 600 €/t. This price could change according to several factors.
  - A steel saving rate of 0.5% is assumed, meaning 30,000 €/year.

- This saving is assumed to be effective from the third and subsequent years.

- Other materials: pipes and cables:
  - The annual cost of such materials is assumed to be about 2 M€/year.
  - Materials saving rate is estimated to be about 1 %, meaning 20,000 €/year.
  - This saving is assumed to be effective from the third and subsequent years.

Savings in production due to reduction of changes and corresponding rework is assumed as follows:

- Production personnel cost (and associated indirect costs) is assumed to be 35 €/h.
- Assumed yearly man production hours: 1,760 h/year.
- Assumed production personnel rate respect to design personnel: 8 (it means 4,000 people in production). Again, this number has been assumed by the author.
- Production hours reduction rate due to reduction of changes and rework is assumed to be about 0.5 %, meaning a production saving of 1,230,000 €/year.
- This saving is assumed to be effective from the third and subsequent years.

5.5 (b) Software License and Services Costs

The costs have not been included in these paragraphs due to it is a confidential information. But although the costs are not included, these have been used for the analysis, and not less important, the concepts of these costs are shown as follow:

- FORAN® Licence cost: single initial cost in Euros.
- Software installation in standard computers (sites): single initial cost in Euros.
- Training cost: single initial cost in Euros. For reducing cost, it could be considered that only receive the trainings a group of super users.
- Annual maintenance cost: excluding first months of guarantee.
- First year technical assistance cost: it depends of how many man-weeks will be necessary.
- First year of implementation/adaptation program, coordination and management: it depends of how many man-weeks will be necessary.

5.6. RETURN ON INVESTMENT FINANCIAL BREAKDOWN

The following financial indexes are used to evaluate the implementation of FORAN®:

- Net Present Value (NPV): represents the difference between investment market value and its costs (sum of present values of net annual benefits minus the initial investment required). The future cash flows that the FORAN® implementation would produce is estimated taking into account outflows (costs, loss of productivity ...) and inflows (cost savings, increase of productivity, other benefits ...). The Discounted Cash Flow (DCF) has been evaluated.
- Internal Rate of Return (IRR): It is the discount rate that makes the NPV of a project equal zero
- Payback Period (discounted): It is the amount of time needed for an investment to generate cash flows (discounted) to recover its initial costs.
- Profitability Index (PI) or Benefit/Cost Ratio: It is the present value of future cash flows divided by the initial investment.
- ROI: It represents the relative value of the project's cumulative net benefits (benefits less costs) over the analysis period, divided by the project's cumulative total costs, expressed as a percentage.

The following tables show the results of the financial analysis including a sensitivity analysis respect to Productivity gain to Discount Rate. Corresponding charts are also included for easier data interpretation. Tables with complete cash flows, discounted cash flows and corresponding financial evaluations are also included for predefined cases (different productivity gains and different discount rates). Table 2 describes the NPV in Euros, which is the difference between the present value of cash inflows and the present value of cash outflows. NPV is used in capital budgeting to analyse the profitability of a projected investment or project. A positive NPV indicates that the projected earnings generated by a project or investment exceed the anticipated costs. This investment has a positive NPV, so it will be profitable.

Table 2: NPV in Euros related to Productivity Gain

Discount Rate	Productivity Gain				
	6 %	9 %	12 %	15 %	18 %
3 %	10,702,675	21,817,329	32,931,983	44,046,636	55,161,290
6 %	7,823,812	17,259,347	26,694,883	36,130,419	45,565,954
9 %	5,575,061	13,668,480	21,761,899	29,855,318	37,948,737
12 %	3,804,320	10,813,417	17,822,513	24,831,609	31,840,705
15 %	2,399,755	8,523,872	14,647,990	20,772,108	26,896,226

Determining the value of a project is a challenging exercise because there are different ways to measure the value of future cash flows. Because of the time value of money, money in the present is worth more than the same amount in the future. This is both because of earnings that could potentially be made using the money during the intervening time and because of inflation. In other words, Euro earned in the future will not be worth as much as one earned in the present.

Table 3 describes the Discounted Payback Period (DPP), which is a capital budgeting procedure used to determine the profitability of a project. In contrast to an NPV analysis, which provides the overall value of a project, a DPP gives the number of years it takes to break even from undertaking the initial expenditure. Future cash flows are considered are discounted to time zero. This procedure is similar to a payback period; however, the payback period only measure how long it take for the initial cash outflow to be paid back, ignoring the time value of money.

Table 3: DPP in years related to Discount Rate

Discount Rate	Productivity Gain				
	6 %	9 %	12 %	15 %	18 %
3 %	5.4	3.8	3.0	2.5	2.2
6 %	5.7	3.9	3.1	2.6	2.2
9 %	6.1	4.1	3.1	2.6	2.3
12 %	6.6	4.2	3.2	2.7	2.3
15 %	7.2	4.4	3.3	2.7	2.3

The payback period is the time required to recover the cost of an investment. The payback period of a given investment or project is an important determinant of whether to undertake the position or project, as longer payback periods are typically not desirable for investment positions. Ship design projects that have a negative net present value will not have a DPP, because the initial outlay will never be fully repaid. This is in contrast to a payback period where the gross inflow of future cash flows could be greater than the initial outflow, but when the inflows are discounted, the NPV is negative.

Table 4: PI versus Productivity Gain and Discount rate

Discount Rate	Productivity Gain				
	6 %	9 %	12 %	15 %	18 %
3 %	2.47	4.00	5.52	7.05	8.57
6 %	2.07	3.36	4.65	5.94	7.23
9 %	1.76	2.86	3.96	5.06	6.16
12 %	1.52	2.47	3.42	4.36	5.31
15 %	1.32	2.15	2.98	3.80	4.63

Table 4 describes the PI, which is an index that attempts to identify the relationship between the costs and benefits of a proposed project through the use of a ratio calculated as the coefficient between the present value of future cash flows and the initial investment. A ratio of 1.0 is logically the lowest acceptable measure on the index. Any value below 1.0 indicates that the project's present value is less than the initial investment.

Table 5 describes the ROI, which is the benefit to the investor resulting from an investment of some resource. A high ROI means the investment gains compare favourably to investment cost. As a performance measure, ROI is used to evaluate the efficiency of an investment or to compare the efficiency of a number of different investments. In purely economic terms, it is one way of considering profits in relation to capital invested. Figure 6 shows the ROI, considering different productivity gains and different discount rates.

Table 5: ROI vs Productivity Gain considering Discount

Discount Rate	Productivity Gain				
	6 %	9 %	12 %	15 %	18 %
3 %	147%	300%	452%	605%	757%
6 %	107%	236%	365%	494%	623%
9 %	76%	186%	296%	406%	516%
12 %	52%	147%	242%	336%	431%
15 %	32%	115%	198%	280%	363%

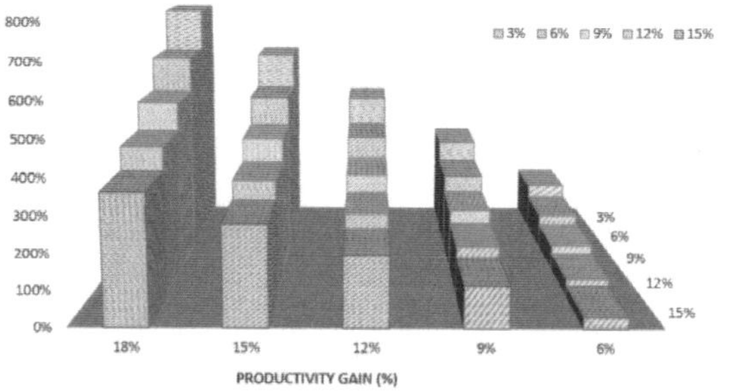

Figure 6: ROI related to Productivity Gain and Discount Rate

Investment is buying an asset with the expectation of capital appreciation, dividends (profit), interest earnings, rents, or some combination of these returns. This may or may not be backed by research and analysis. Most forms of investment involve some form of risk, such as investment in equities, property, and even fixed interest securities which are subject to e.g. inflation risk. It is indispensable for project investors to identify and manage the risks related to the investment. Investment and investing is distinguished from other uses of money (such as saving, speculation, donation, gifting), in that the deployment of money is done for the purposes of obtaining a positive expected return.

6. SHIPBUILDING CASE STUDY

The suitability of one software or another also affects the time to reduce or increase our ROI.

The following example is based on FORAN® CAD/CAM/CAE System, with all the capabilities that make this software one of the most profitable on the market. In FORAN® all the disciplines work in the same working environment with all stages of naval design, from the initial design to the production, to have the part of the structure and armament available in the same database, connecting immediately and without loss of information, two ways connection with the PLM's tools and to be able to manage a series of ships simultaneously were taken into account in the present study.

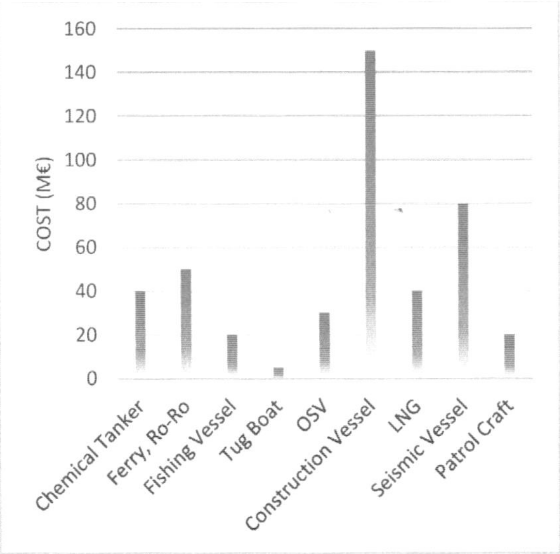

Figure 7: Shipbuilding cost by type of vessel

The use of Natural Gas in this area, and more specifically in Singapore, where more than 90% of Singapore's electricity is generated using imported natural gas, has done that we have focused this study in the construction of LNG vessels that supported in our experience has given us very interesting results from the point of view of the profitability of the software.

We have started with a shipyard that produces three small LNG vessels in a year and taking as reference the price of this type of ships, we can estimate that the hours dedicated to engineering with a naval design software such as FORAN® would be 1%.

Knowing the total number of hours of software engineering of the ships and the number of working hours per year, we get the number of people needed in a year for the project and therefore the number of licenses.

This number of licenses with their corresponding price, plus the associated services of installation, training of staff and maintenance will give us the fundamental data of the investment needed to design this type of ships with FORAN®.

The data on how much money we save in a year, when implementing FORAN®, comes from customer data that estimate that the price of the ship design corresponds to 7% of the total price of the ship. The savings will therefore come from the application of this amount for each of the productivity improvement subtotals in Table 1. Another important saving concept is the one that derives precisely from previous savings, that means, the savings obtained from the design of the first vessel will allow the shipyard to spend more time on the construction of new ones.

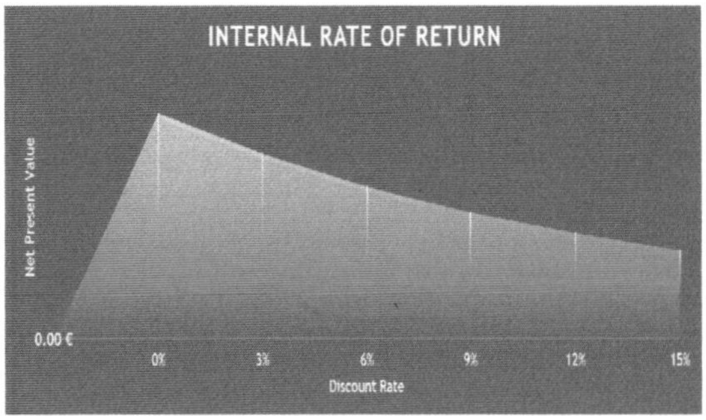

Figure 8: IRR related to NPV and Discount Rate

This saving, increase with the number of ships that the shipyard makes, and will be added to the previous one to obtain the total savings, which in percent and in this case, is equivalent to 26% of the price of the design price of the ship.

Taking into account that the improvement of productivity happens from the second year, applying an annual period with a discount rate of 6%, the result of the Payback period is 2.03, which means, from the second year the investment in software and associated services begins to return.

7. CONCLUSIONS

In the light of the results of the ROI financial analysis and sensitivity analysis, we can state that the implementation of FORAN® in a generic shipyard has a reasonable justification. The expected benefits are considerably much higher that the costs incurred, even with high variability of productivity gain.

As mentioned before, in addition to the factors considered in this ROI analysis, there are others (intangibles or tangibles, but not estimated by SENER®), that could have an important impact in the decision making process. The assumptions considered in this ROI financial analysis are mainly based on qualitative information obtained during the meetings with different shipyards. Therefore, it is highly recommended to recalculate this ROI analysis with more accurate data, if available.

Table 6: IRR in percentage

Productivity Gain				
6 %	9 %	12 %	15 %	18 %
22.5%	40.8%	59.6%	79.5%	100.9%

Table 6 describes the IRR related with the productivity gain. The IRR (a.k.a. discounted cash flow rate of return) is the rate of return used in capital budgeting to measure and compare the profitability of investments. It is important to obtain the IRR of the CAD System implementation versus the productivity gain, so it will provide its degree of lucrativeness.

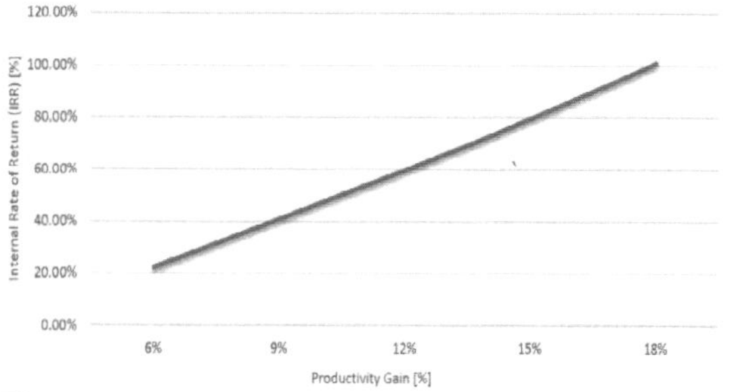

Figure 9: IRR in percentage and related to Productivity Gain

Table 6 illustrates where with an 18% of productivity gain, the IRR will exceed of 100%. IRR for the calculations here has been used to evaluate the desirability of CAD System investments in the shipyards. The higher a CAD System investment's IRR, the more desirable is investment in a new CAD tool. Assuming all CAD System investments require the same up-front investment, the CAD System investment with the highest IRR would be considered the best and undertaken first. Another key aspect of this paper, is the analysis of the learning curve of a new CAD System implemented in a generic shipyard. The data, hours expended designing blocks and units of different weights, collected for this paper during a couple of years from a generic shipyard, have been a fantastic tool for analysing the factors that affect a learning curve process. One of this key factor is the designers' mobility, which explains why the curve has peaks, so a well-trained and experienced draftsman could leave the shipyard, and another less skilled designer needs time to learn the new CAD System.

8. REFERENCES

1. LEE, D.J.; CEBOLLERO, A.; PÉREZ, R. (2013), New approach to design transition in Korean production environment, ICCAS, Busan, pp.109-118

2. MUÑOZ, J.A.; PÉREZ, R. (2017), CAD tools for designing smart ships in the world of the internet of things (IoT), Smart Ship Technology, London, pp.29-35

3. PEREZ, R. (2016), Learning Curve and RoI in the Implementation of a CAD System in a Generic Shipbuilding Environment, COMPIT, Lecce, pp. 19-32

4. PEREZ, R.; GONZALEZ, C. (2015), History and Evolution of Shipbuilding Oriented CAD Tools, 14th COMPIT, Ulrichshusen, pp. 8-21

5. PEREZ, R.; PENAS, R. (2015). Integration between shipbuilding CAD Systems and a generic PLM tool in naval projects. Computer Science and Applications 2/5, pp. 181-191

6. PEREZ, R.; TOMAN, M. (2014), An innovative approach for hull surface fairing stages, 13th COMPIT, Redworth, pp.32-40

7. PEREZ, R.; ALONSO, V. (2015), Virtual Reality in a Shipbuilding Environment, Advances in Engineering Software 81, pp. 30-40

8. PEREZ, R.; ALONSO, V.; VALDERRAMA, A. (2013), The use of a 3D CAD System for early design in shipbuilding, Ship Production Symposium (SPS), Bellevue, pp. 545-553

9. AUTHORS BIOGRAPHY

Rodrigo Perez *PhD* in *Naval Engineering* at the *Universidad Politécnica de Madrid (UPM)*, in Spain. He holds the current position of *Naval Shipbuilding Area Manager* at *SENER®*. He is also professor at the *UPM.*

DRIVING TRANSFORMATION IN THE AGE OF EXPERIENCE

A Tew Kai, Dassault Systèmes, France

SUMMARY

In today's distributed and interconnected Marine and Offshore industry, staying profitable, competitive and relevant means big changes. Stakeholders around the world are forced to rethink the way they design, manufacture, and operate. Extreme cycles in its economics make the Marine and Offshore industry ripe to embrace digital transformation and reap its full benefits to improve performance and achieve new efficiencies. There is a clear need to be smart and connected at all levels of the value creation chain, taking advantage of technology such as Internet of Things, Robotics, Automation, Big Data, or Augmented Reality. The ability to truly transform processes and take on a holistic approach is critical. An integrated business platform that provides end-to-end continuity from ideation to ship in service by integrating design, engineering, simulation, manufacturing, and business processes in a single collaborative environment is an essential enabler in carrying out this profound mutation.

1. INTRODUCTION

Products are no longer enough for today's consumers. True value is linked to the multitude of customized services associated with a product, as well as the experience the user derives from its use. Consumers value experience over all else and demand that their lives be enriched in meaningful ways. Experience Thinking is a framework for innovation, focusing on engaging consumers with breathtaking product experiences that are smarter, intuitive and sustainable. It encourages companies to consider all aspects of the consumer experience journey, driving value for the end user, which is only possible when all the players in the innovation process, from sales, marketing, finance, procurement, management to design, engineering, manufacturing and the supply chain, work collaboratively.

Innovators need to be able to develop a deep understanding of customer's expectations by analyzing social and usage information in a unique data-driven approach. Enterprises can then model, simulate and optimize their offerings to best suit each customer experience. During the operating stream, this model is enriched with valuable insights gained from the usage data, setting the stage for the next cycle of innovation.

Companies must bring their different departments together, in a holistic manner to drive their innovation – in products, in new business models and in customer experience successes. They need to be able to rethink, improve, and connect processes across main functions, feed them with meaningful data analytics, and ensure the automatic propagation of changes across all functions.

2. CHALLENGES

2.1 INCREASING COMPLEXITY

Ships and platforms today are highly sophisticated, large scale systems that are complex to design, build, and maintain throughout their lifecycle. Leading Marine and Offshore players around the world strive to satisfy their clients' requirements for higher value, better performance, and flawless quality. This means embedding new and potentially unproven technology in ships and platforms, which generally implies more complexity and risks for design. Intense competition creates enormous pressure to achieve design excellence and improve productivity. Getting it right the first time is critical. Finding the optimal combination requires the ability to test alternative design scenarios that support innovation and advanced technology while efficiently managing the associated complexity and risk. Shipbuilders expect to be able to engineer complex systems efficiently as well as test and validate design options as early as possible in the cycle to avoid costly rework downstream.

It is fundamental for any shipbuilder to successfully manage the relationships between design and customer requirements. It is also critical to manage the requirement change process properly, that is easily identifying the change, assessing its impact and propagating it adequately across the impacted organizations. All impacted stakeholders have access to requirement information and understand the details and implications. They need a robust integrated requirement management system with full traceability to easily monitor adherence to customer requirements and compliance to classification rules for faster and more effective design validation and certification. Requirement information must be integrated into the design process and progress tracked from the requirement item to the functional, logical, and physical design.

2.2 SHORTER CYCLE TIME AND COST REDUCTION

In a highly competitive landscape where buyers seek the most cost-effective solutions, shipyards and offshore EPCI companies are under constant pressure to be on or ahead of schedule or risk costly penalties. With programs totalling in the millions and sometimes billions of dollars and contracts shifting to a fixed cost approach, it is critical to keep track and contain costs until the ship or platform is delivered. It requires strict process and resource planning across an extensive network of stakeholders and an integrated view of performance combining cost and schedule to help ensure timely and on-budget delivery.

2.3 BREAKING DOWN SILOS

2.3 (a) Silos between stages

Historically, ship design has been developed with little anticipation or thought as to how it would be manufactured, typically causing significant delay and very costly rework. Aware of the significant time and cost savings, shipbuilders have been looking for ways to ensure their designs are well suited to be manufactured. This includes building into the design applications general and shipyard-specific manufacturability rules that may affect how the ship is designed.

2.3 (b) Silos between disciplines

Much of a ship's design is still based on individual engineering disciplines working independently that send information back and forth between departments. The industry is plagued with examples of disjointed multiple design applications across multiple sites that lead to inconsistencies and version management problems. Such a lack of integration between the many applications that are used by various disciplines causes additional coordination delays and costs transferring design information from one team to another. Moreover, information sharing and cross-discipline design review are often inefficient, which, combined with incompatible design data, results in identifying potential issues and design conflicts late in the process which, in turn, leads to significant costs and delays.

2.4 EVOLVING WORKFORCE

2.4 (a) Capture know-how of retiring workers

The traditional employment base in the maritime sector is best described as aging. Today many shipbuilders are faced with a bulk of their manpower and a wealth of shipbuilding knowledge that has been generations in the making sailing toward retirement, and there is no clear line of succession. From the shipbuilder's perspective, it takes many years and hundreds of thousands of dollars to train someone proficient in the shipbuilding craft. Lost with retirement is a collective group of design and manufacturing talent. Shipbuilders that are progressive in optimizing their operations are seeking means to capture existing knowledge in a form and function that can be updated, enhanced, and reused for a generation to come. They are bringing together not only generations of accrued experience, but company process and procedure that can be melded across its entire design and production planning structure, giving a leg-up on competition that opts to 'start from scratch' out the outset of each project.

2.4 (b) Attract and retain new generation of workers

It is easy to argue that 'attracting and retaining a quality workforce' is a top challenge to every CEO in the sector. The on-line, unplugged millennial generation is much more geared toward a life in Silicon Valley than one in a hardhat fitting a pipe or welding steel. Millennials are digital technology natives. They expect the devices and practices they were raised on, namely social media, tablets, virtual and augmented reality.

3. ENABLING CHANGES

3.1 COLLABORATIVE PLATFORM

A collaborative business platform is necessary to eliminate silos within companies, moving from a static, filed-based world to a digitally connected world, where live data drives innovation, processes, and business-decisions.

Unique connectivity technologies and services allow people and communities to connect in a secure and controlled environment. It enables a new era of innovation on extended ecosystems and fosters a truly open platform innovation. It also enables improved project management, conformity to standards and regulations, process certification, and value chain relationships via an interactive 360° program management approach and full traceability.

3.2 A SINGLE 3D MASTER MODEL

The model-based approach is a core enabler of the transformation and is valuable to ensure innovation effectiveness. The innovation process typically calls for multiple views of the master model with varying degrees of fidelity. Early in the design process, a low fidelity view is used to understand the systems interactions and behaviour; while later, as the product definition matures, higher fidelity views are adopted to guide optimization along often-conflicting functional attributes and cost.

This model-based approach serves across the whole lifecycle, from conception to operations. In the design phase for instance, all engineering disciplines whether they're responsible for structure, piping, safety or structural stability can access and work on the same model to do their part. Any potential issue can therefore be immediately detected and corrected. It is important to catch these issues early and not during the build phase when they are more complex and costly to correct. In the manufacturing planning phase, as an example, NAOS Ship and Boat Design [1], an Italian ship design company, took advantage of a single 3D master model to streamline its process. The company had to provide shipyards with high quality manufacturing documentation for hull structure scantling and fabrication, as well as for systems design and installation. This documentation indicated where to cut the plates and instructed on how the ship should be assembled block by block. These manufacturing instructions, comprising 3D assembly documentation and simulations as well as detailed instructions for each task, were automatically generated from the 3D master model. In the past, without the single master model, every design unit had to be created from scratch. Another example is how Bureau Veritas [2] uses a single 3D master model as the basis for its next generation of asset integrity management solution.

3.3 INTEGRATED MULTI-D SIMULATION

Integrating digital simulation to evaluate the performance of ships or platforms before they are built is essential in order to ensure the integrity of the increasingly complex marine and offshore structures involved without resorting to over-specification and the accompanying cost inflation.

3.3 (a) Automation of the simulation process

Using the 3D master model that embeds all of the relevant real-world physical data (material characteristics, forces, pressures, temperatures, restraints, vibrations, etc) as the foundation for all downstream structural, hydrodynamic or electromagnetic simulation of hulls or platforms enables the generation of the finite element mesh exactly matching the design intent. This otherwise complex and time-consuming operation can be performed automatically instead of having to remodel each time the area that needs to be analyzed.

The capacity to run a comprehensive, detailed analysis automatically on design variants has far reaching advantages. For one, designers can explore various design scenarios in far less time than it would have taken to generate a separate model for each thanks to the automatic propagation of design changes to the simulation. But even more importantly, it leads to a progression from verification to optimization and finally to generative design.

At first, digital simulation serves to verify that the design will perform as expected. Next, it becomes a means to optimization, by evaluating several design variants to converge towards the best cost/performance compromise – for example, by minimizing the amount of material used in a hull, while enhancing the experience by reducing the propagation of engine vibrations. And finally, as integration matures and computational power increases, digital simulation helps generate new and better user experiences by automating the exploration of all options in search of those that provide a qualitatively better result.

Virtual models can be correlated with actual behaviour during the operating life of the ship or platform. The knowledge gained from this correlation can be used to enhance the fidelity of the virtual models. Any enhancement required in the operating product is first simulated in the virtual model, fine-tuned and optimized, before incorporating it in the real world. In fact, the accuracy of the simulation can be significantly enhanced by connecting the virtual model to physical systems. When the physical systems are assembled, they are just a twin of the model. The real and virtual worlds reinforce each other – modelling and simulating the real world virtually and enhancing the virtual model with experiences from the real world.

3.3 (b) Dashboarding

Digital simulation generates large amounts of data which must be distilled into relevant Key Performance Indicators for each participant. In the same manner that a single 3D master model is critical to simulation efficiency, a business platform with dedicated and coherent dashboarding capabilities is critical to break down knowledge silos and ensure the sharing of a common view of the program simulation results.

3.4 DATA DRIVEN

Data is at the heart of product innovation and is now the ultimate monetizable asset. However this data sits across many disparate systems today at many companies and is not readily visible or easily accessible. Further, data is not just what's in the enterprise. There's a wealth of relevant data on the internet that can be extracted to improve innovation, products, brands and consumers experiences. It is all about tackling very big data challenges and benefitting from next generation technologies to index, search, sort, filter, navigate, and understand data.

In order to manage and leverage the massive amount of data coming from different sources, search-based applications that combine the sophisticated search and access typically associated with databases with the speed, scalability, and simplicity of the Web are necessary.

In addition to capturing and accessing data through advanced search capabilities, there is also a need to define customized dashboards for data monitoring and data-driven decision making. It is also possible to add dynamical post-treatment in order to define relevant metrics based on the original data for a holistic analysis of a situation.

3.5 END-TO-END DIGITAL CONTINUITY

Digital continuity expresses the idea that the digital world of product engineering and the real world of manufacturing and operations can share a single, digital source for a comprehensive management. All Marine and

Offshore companies can monitor every last detail of their industrial operations for every product and process by adopting a single digital model approach through concept / basic / detail design, production, and operations to optimize overall company performance. This fosters continuous improvement, innovative design and precise traceability. Digital continuity also promotes collaboration, operational excellence and agility.

4. CONCLUSIONS

As an integrated business platform, Dassault Systèmes' **3D**EXPERIENCE® platform is an essential enabler in carrying out digital transformation. Experiential platforms are essential because they bring together talents, ideas, solutions and data. They enable the implementation of innovative supply strategies (interconnected modular manufacturing systems) and demand strategies (customized and connected products and services).

The **3D**EXPERIENCE platform provides a single, real-time source of data/information across the entire ship/platform lifecycle, from conception to design, engineering, manufacturing, operation and maintenance. It uses a single, intelligent 3D master representation for all processes which helps break silos, avoid duplication and inconsistency, reduce costly errors and delays, support decision making, and improve global efficiency/productivity.

Leading companies across a number of industries see the strong value proposition of the **3D**EXPERIENCE business platform to bring their different disciplines together and power processes across different disciplines, from ideation, design, scientific simulation and manufacturing to marketing and sales, using meaningful data analytics, thus enabling end-to-end digital continuity. This results in significant improvement in their innovation capacity and operational efficiency. The move to the **3D**EXPERIENCE platform is accelerated by increasing product complexity, need for connectivity and collaboration, and the necessity to make sense of an exponentially larger amount of data.

The **3D**EXPERIENCE platform provides social and collaborative applications to:

- connect people to imagine sustainable innovations together,
- help all disciplines involved in a marine and offshore project to collaborate more effectively,
- facilitate planning and management of the complete ecosystem including the supply chain

3D Modeling applications to:

- virtually design smart ship or smart platform (high quality, energy efficient, safe, and regulation compliant) with world class 3D modeling capabilities from requirements to manufacturing

Content and simulation applications to:

- virtually experience a ship or a platform before it actually exists with realistic simulation to explore its real-world behaviour
- deliver a unique, stunning immersive buying experience customers can trust
- maximize design and manufacturing efficiency by re-imagining planning, management, and optimization of industrial operations
- ensure compliance with regulations and owners' requirements
- optimize on-board safety, operations, and maintenance

Information Intelligence applications to:

- transform a massive amount of data coming from different sources into tangible insights and business action with real time, dynamic dashboards
- reveal the hidden information intelligence needed to understand, participate, and act
- help track and monitor all project-related information from a single entry point

5. REFERENCES

1. Dassault Systèmes, 'NAOS Ship and Boat Design case study', *3ds.com*, 2017.

2. Bureau Veritas, 'Introducing Veristar AIM – a Bureau Veritas technology solution, powered by Dassault Systèmes', http://www.bureauveritas.com/home/news/business-news/veristar-aim-a-new-bureau-veritas-solution-powered-by-dassault-systemes, 2017

6. AUTHORS BIOGRAPHY

Alexandre Tew Kai holds the current position of Marine & Offshore Industry Senior Business Consultant at Dassault Systèmes. He is responsible for developing consultative approaches to reinforce the business value proposition delivered to Marine & Offshore companies. His previous experience includes working closely with leading Dassault Systèmes' customers including Bureau Veritas and Naval Group.

SHIP CONCEPT DESIGN BASED ON A 3D-CAD-SYSTEM INCLUDING A REQUIREMENT VERIFICATION

H Lindner and **R Bronsart**, University of Rostock, Germany

SUMMARY

A system architecture allowing designers to create a ship concept design in 3D from the first sketches on is described in this paper. The centrepiece of the system architecture is a combination of a CAD system with well-developed modelling capabilities and a product data management system (PDM). This paper will show how the PDM system is extended in order to store and check functional requirements. A modular approach allows the integration of this system in the shipbuilding specific heterogeneous software infrastructure. The example will show how the different tasks in the concept design are supported by the system architecture including basic functionalities for requirement testing.

1. INTRODUCTION

Ship design is almost completely performed with the help of 3D CAD systems. However, the early ship design, especially at the pre-contractual stage, is often still developed and documented based on 2D drawing tools, because modelling and modification of a 3D model is a too complex and time consuming process. Therefore, the need for an efficient-to-use 3D CAD system combined with a PDM system in the concept design phase is apparent.

In order to apply a 3D modelling tool linked with a product data management system within the concept design, a novel system infrastructure consisting of interfaced specialized components has been realised in close cooperation with two shipyards. The system infrastructure is designed to suit the following key requirements:

- 3D-modelling: fast and flexible definition of the ships geometry
- Information-handling: capability of handling all relevant information
- Data-sharing: assuring that all parties involved in highly parallel design tasks work on the same data
- Requirement-tracking: store and associate design requirements with information objects

Requirement management is getting more important with the increasing complexity of the design. Customers more often request documented requirement breakdowns in some industrial sectors. Formalisation of the requirement definition can generate connections between the relationship model of the demands and the ship design model. Based on these links the concept design can be checked against the owners demands and the eligible statutory regulations.

A system infrastructure is presented, in which a general purpose 3D-modelling tool is linked with an external PDM system. The PDM system is the core-component of the infrastructure and acts as information hub for all related tasks. The geometry resides in the CAD system and is linked to the PDM system via a bidirectional interface for an automatic synchronisation of the data.
Based on an example ship it is shown that the implemented system architecture can be applied in the early ship design phase to develop a 3D ship model from the very beginning.

2. SYSTEM ARCHITECTURE FOR THE EARLY SHIP DESIGN STAGES

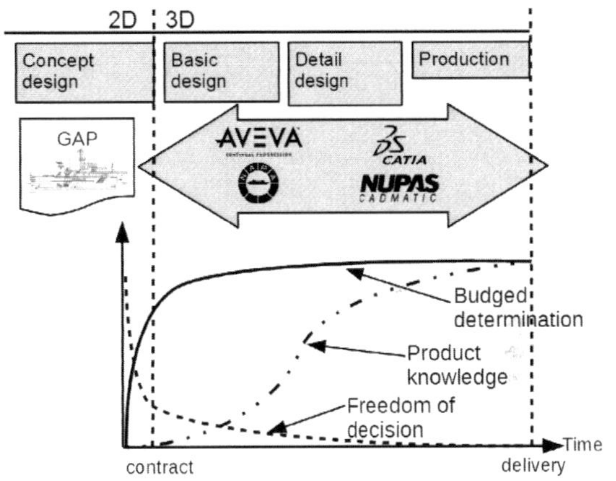

Figure 1: Budget determination, freedom of decision and product knowledge in different design stages. adopted from [1]

Ship design involves typical design stages such as concept design, basic design, detailed design and production. Shipyards usually deal with different important qualitative variables from concept design to production stages which are the following: Budget determination, product knowledge and freedom of decision. These variables alter over time as shown in Fig. 1 (adopted from [1]). In the beginning of the design process the design options are high and the budget constraints are low. Major decisions made in this first phase lead to a highly determined budget. In addition these decisions are made with limited product knowledge. The figure also denotes that most of the budget is determined in the very short time between request and signed contract whilst the product knowledge stays low. This implies a high risk for the shipyard.

Fig. 1 also shows that the main CAD Model in the concept design is the general arrangement plan (GAP) which is generally a 2D drawing. However, many specialised CAD systems (AVEVA, NAPA etc.) are used in the concept design but not as main source of information as the GAP. After completion of the concept design and conclusion of contract, more sophisticated tools are used for different tasks such as the compartmentation or the steel design. These tools are based on 3D models derived from the general arrangement drawing and related documents. With these models and other further details the product knowledge rises, but along with the level of detail increases the freedom of decision decreases.

In summary, Fig. 1 shows that major decisions are made in spite of lack of detailed product knowledge. In order to reduce the technical and financial risk for the shipyard the improvement of the product knowledge especially in the concept design stage is crucial. In this paper a system architecture will be presented to replace the general arrangement drawing with a 3D model developed in a CAD system. Together with a PDM the product knowledge will increase significantly and therefore reduce the risk. To facilitate the process of reviewing the design against requirements the PDM was enhanced to store and check functional requirements.

2.1 GENERAL ARRANGEMENT – MAIN SOURCE OF INFORMATION

The general arrangement model is the main source of information in the concept design. It is common practice to develop and document the general arrangement with 2D drawing tools and associated spreadsheets. In this paper a system architecture is presented which replaces the general arrangement drawings with a 3D CAD model. Every individual task of the concept design process including the related information need to be determined to develop this system.

The concept design is an iterative process to refine the design in each iteration. In Fig. 2 the design spiral of the concept design is presented. This Figure is adopted from Eyres and Bruce [2].

In Fig.2 all tasks which are performed in the CAD system are coloured light grey. One of the first tasks is the subdivision of the ship into compartments and, later on, into tanks and rooms. In the next task the compartments receive attributes such as labelling and purpose. This kind of information is needed to check the design in subsequent tasks. Another major task performed in the CAD system is the arrangement of equipment. The steel design itself is not carried out in the CAD system but all information related to space requirements needs to be transferred to the general arrangement model.

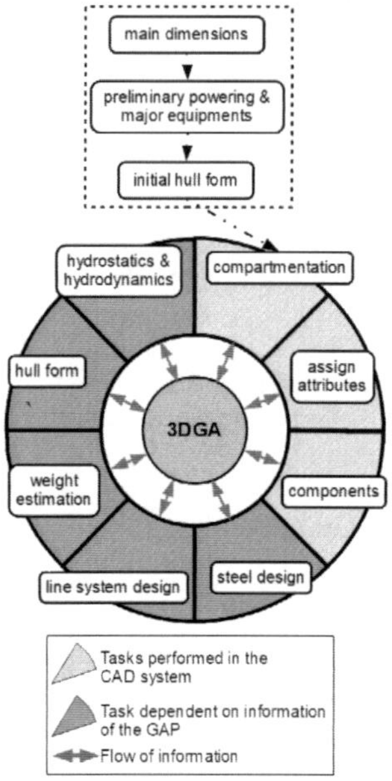

Figure 2: Design tasks in the concept design stage

Tasks relying on the information in the system are shown on the left hand side of Fig. 2. Mostly, these tasks check the design and are performed in specialised software tools that need to be interfaced by the CAD system. The ship should not be completely remodelled in these tools to reduce the time needed for the whole design and the chances of errors. An important task is the development of the hull form including hydrostatic and hydrodynamic calculations. These tasks are performed in specialized software tools like NAPA, AVEVA, etc. Theses tasks rely on information from the 3D general arrangement (3DGA) but also giving back information to the 3DGA like the hull form, hydrostatic and hydrodynamic parameters.

2.2 SYSTEM REQUIREMENTS BASED ON PROCESS ANALYSIS

Based on the concept design process shown in Fig. 2 all tasks are fragmented to utmost simplicity in order to set up a system infrastructure which is tailored to the needs of the concept design. These functions were structured in trees with four main branches shown in Fig. 3. The branches contain functions with the following types:

- 3D-modelling capabilities: The time needed for the concept design should not be extended, therefore the need for fast and flexible creation and modification methods of 3D objects is apparent.
- Information-handling: The product data model should be capable of handling all required information of the

early ship design process. In addition, the PDM must be able to handle the increasing granularity and population of the model information. Especially the information needed to check the technical and economic feasibility should be stored.

- Data-sharing: The coordinated storage of information assures the designers that all parties involved in highly parallel tasks work on the same data. Additionally it is ensured that the information stored in the PDM can be accessed by the numerous specialized design tools via a well-defined interface including an open object oriented application programming interface (API).

- Requirement-tracking: The compliance of requirements needs to be checked. In addition it should be possible to trace the objects which are responsible for the compliance of a requirement. Therefore input and storage of requirements in computer as well as human readable form needs to be possible.

2.3 MODULAR SYSTEM ARCHITECTURE

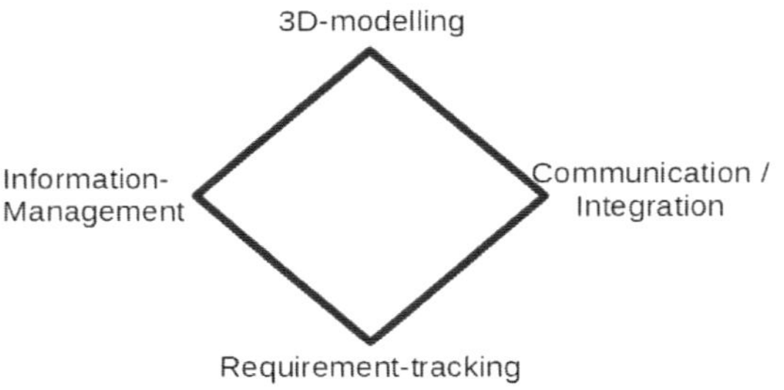

Figure 3: Main functions of the design system

The best system would be optimal for all functions shown in Fig. 3. This kind of system has not been developed yet. There are systems adequate for some functions but containing weaknesses in other aspects. The system architecture presented in this paper is based on a modular approach which allows the usage of tools with the best possible efficiency, whereas the advantages of the individual tools can be combined by connecting them.

The essential requirements in Fig. 3 are "3D-modeling" and "Information-handling". In concept design an efficient performance in the creation and manipulation of geometry is of importance. In order to check the technical and economical feasibility of the design the product data needs to be stored and made accessible in an efficient manner.

Following an intensive market research it became apparent that there is no software system available which is best suited in these two requirement fields. In order to apply the best suited tools a modular interfaced software architecture is introduced.

In this software architecture two core modules shape the backbone of the system. A two step evalution process was carried out to find the best tools for the core modules. The evaluation and assessment is discussed in detail in [3] and [4].

The results of the evaluation process suggest Bentley AECOsim Building Designer as 3D-CAD-module. The modelling functions of this multi-purpose CAD system are well developed and best serve the geometry creation and manipulation tasks in the concept design. This includes the capabilities to derive and configure drawings from the 3D CAD model which are adopted to the specific needs of shipbuilding.

As module for the information handling a PDM system is interfaced. The PDM system uses a database management system to store design information. A single centralised database server avoids redundancy of data and ensures the availability of up-to-date data. In order to link the geometry in the CAD system with the information in the PDM system and make the information available in 3D CAD system a bidirectional interface between these modules is necessary.

The modules associated to the requirement fields of Fig. 3 are defined as follows:

- *3D modelling: Bentley AECOsim Buildingdesigner*
  This module is used for the geometry modelling. Within the CAD system all geometrical information such as distances, position, area or volume are calculated.

- *Information-Management: SQL Database Server (MySQL, ORACLE)*
  The database itself is any SQL server. The configuration and management of the information is done with an object relation mapping (ORM) system. This system also takes care of the mapping between the relational database and the object oriented API of the CAD system.

- *Communication/Integration*
  This module is part of the common API of the CAD system and the ORM system. As it depends on the shipyard's, which specialized tools are used and which needs to be interfaces it has be easy to develop interfaces.

- *Requirement-tracking*
  The requirement-tracking is integrated in the PDM-system. The data model is capable to store requirements as objects. Functions are developed to check these requirements against a design. The requirements will be imported or defined via specialized interfaces.

2.4 SYSTEM ARCHITECTURE

This chapter will briefly explain the system architecture, details can be found in [3].

The 3D CAD system is used as storage for the geometrical data and geometrical model operations, whereas the PDM system stores all additional data. Data redundancy is reduced as scarcely geometrical data is stored in the PDM system. For that the real-world object is split into two virtual objects. One object is the geometrical object in the CAD system. The other is an object in the PDM which holds the identity and any non-geometrical information. This PDM objects contains a unique identifier to the geometrical representation. This implies the prime importancy of robust, reliable and fast identification and synchronisation of objects in both directions between the systems. In addition, the PDM system needs to be seamlessly integrated in the CAD system so that the user is on one hand not hindered in his work-flow by an overcomplicated interface and on the other hand can make efficient use of the information. For that reason a object relational mapping (ORM) system was chosen which is integrated in the API of the CAD system. Bentley AECOsim Building Designer offers an API in .NET with the COM-Interop technology.

As .NET is widely used in multiple industries including web-server applications there are a few ORM systems to configure and access SQL server. In this system architecture Telerik Dataaccess was used, because it allows the usage of different types of SQL server and additionally can evaluate and modify schemes of a SQL server.

In Fig. 4 the developed system architecture is shown. The CAD module is located on the left hand side and the module for the information-handling is located in the centre. On the right hand side there are two possible options shown for interfacing specific software tools in shipbuilding.

The two databases (3D-model and the SQL data storage) combined form the 3D general arrangement. The bidirectional interface between these two components is developed in .NET and integrated in the API of the CAD module. Events automatically detect changes in the geometry objects in the 3D model and update the corresponding data in the PDM.

The shipyard specific data model is defined/modified in the ORM software. This software generates on the one hand the class libraries for the .NET integration. On the other hand the necessary files (DDL scripts) for the setup/modification of the SQL server are provided. Because the data model differs between shipyards and shiptypes an ORM software was chosen, trough which the data model can be adapted to the needs of the shipyard.

The system architecture was set up in that way that only the data model needs to be modified by the shipyard. The SQL server and the modifications in the library are applied automatically. Additionally most of the forms for the GUI integration in the CAD system directly extract their information from the library, and thus adapt to the modified data model automatically.

In Fig. 4 on the right hand side the upper possible interface suggests the usage of the class libraries of the ORM system directly in the API of the shipbuilding specific software tool. This is only possible if the API is based on .NET as well. Nevertheless this is the preferred interface as it allows the shipbuilding specific software to use all information available and ensures up-to-date data.

The lower work flow relies on generated or exported files and command scripts which are then processed in the shipbuilding specific software. This is the most versatile way for an interface because almost all software tools have an API based on batch command scripts. Their disadvantage lies in a relatively high effort to develop export routines in the ORM system.

In Fig. 4 no part is designated to the requirement tracking field. The field of system requirements is part of the information handling and therefore located in the middle section of Fig. 4. The modifications to the data model are discussed in the following chapter.

A system infrastructure based on a centralised database server which is integrated in the client CAD tools reduces the data redundancy significantly. This allows every designer involved to use the latest data as design changes are immediately synchronised to the server. This is also valid for the interface for specialised software tools. The derived models in these tools are representing the actual state of the PDM.

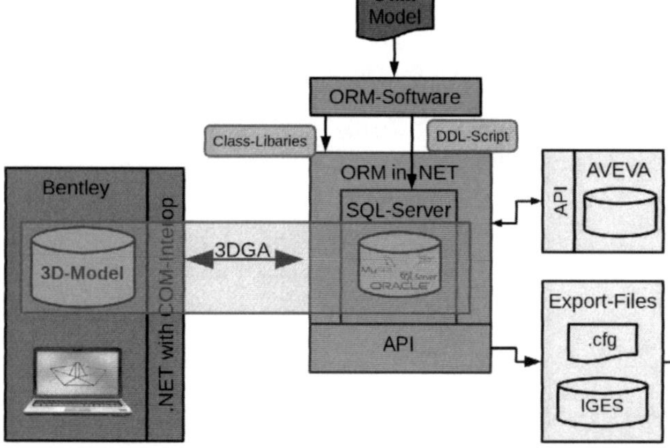

Figure 4: System architecture

3. REQUIREMENT FORMALISATION

The ship concept design is a creative and individual process. A design is developed according to the needs and requirements of the owner and is influenced by the regulations. Any sort of requirement management has always been part of the concept design process but there where no standardised procedures available. The demand for standardised procedures in the requirement management increases also in the commercial shipping industries. This trend is also represented by the process analysis, where a top level field addresses requirement management.

In order to automatically check requirements it is necessary to have a formal description of the requirements implied. Another constraint is that only requirements which test information, which is represented in the design data model can be checked. As the design or rather the data model of the design is based on objects, e.g. components or rooms/tanks, the requirements have to be broken down to this level as well. If a requirement demands for information which is not apparent in the data model, this requirement must either be redrafted to match the data model or the data model has to be adapted to be able to contain that information.

3.1 REQUIREMENT BREAKDOWN

The requirement breakdown is the essential part in the requirement management. This process is described in numerous papers like [5]. The requirement breakdown in the early ship design has been described in detail previously in [6]. This process starts with the owner´s requirements including all derived regulatory requirements and standards. These requirements are written in the voice of the stakeholder (also referred as voice of customer (VoC) in [6]). In the shipbuilding industry this is typically the owner and statutory documents. The shipyard has typically no influence on the wording of the requirements.

The first two phases of the requirement breakdown (requirement elicitation and analysis) are to understand and negotiate the customer´s requirements. The goal of theses steps is to find a common understanding of the requirements between shipyard and owner. The next step is the requirement specification, where the customer requirements are defined and specified by the shipyard in their own voice. In this step the requirements are broken down to functional requirements which should at least have the following quality aspects:

- Unitary
- Complete
- Verifiable
- Unambiguous
- Non-Conjugated

Functional requirements demands for characteristics of certain objects and is in this paper referred to the component level.

3.2 DATA MODEL FOR FORMAL REQUIREMENTS

Requirements exist in different forms and can be of high complexity. Simple requirements guarantee a number of certain well defined objects, such as number of containers, whereas complex requirements can demand the conformity with regulations like SOLAS.

As the requirements to be checked are formulated by the designer the requirement can be assumed unambiguous, testable and solution free. According to [7] a requirement can be interpreted as a demand for an object. This leads to the logical split of the requirement into a descriptive part of the object and a demand to that object as shown in Fig. 5.

In Fig. 5 a simple functional requirement on the component level is shown. The first line shows the requirement text which will be recorded in the documents. In the second line the parts of the sentence are marked which are the description and the demand to the object. Below the description and the demand the necessary properties of them are listed with the values given in the example text.

The description contains at least one object class. The description is refined by adding an arbitrary number of attribute, comparator and value combinations. In this example the tanks, for which the demand is applied, are described in more detail by the purpose type of the tank to be equal to freshwater. This means all objects of the class `Tank` with the attribute `tankpurposetype` equal to `freshwater` are selected for the evaluation of this requirement.

All Tanks with tankpurposetype equals freshwater should have in total a Volume greater than or equals 70m^3.

description		demand	
Objectclass	Tank	Demandtype	PropertyAmount
Attribute	tankpurposetype	Attribute	Volume
Comparator	equals	Comparator	greaterThanOrEquals
Value	freshwater	Value	70m^3

As shown in Fig. 5 the first characteristic of the demand is the demand type. There are various different types of demands. In the system developed so far two demand types are implemented. The first is the object number, where all objects which are described by the requirement are counted and compared to the demanded value (e.g. number of beds). The second demand type is the need for an amount of a certain property of the objects (e.g. volume of a tank). In this example a need to the property `Volume` of the eligible objects is formulated. If this demand type is used a combination of attribute, comparator and value is used as the demand.

In Fig. 5 a simple example is shown, where the description is defined by properties of the object itself. The complexity increases when the description of the objects is not only defined by properties of the object class but by associated classes. For example an element which is described by the room it is in, like "An element of type Bed in a room with type Crewarea_cabin should be installed 20 times.". This is necessary depending on the data model. The data model for storing requirements should be as generic as possible to allow modifications to the product data model at later stages.

This leads to a data model for the requirements as shown in Fig. 6. The class diagram starts at the top with a class called `RequirementSet`. This class groups together some requirements. This class is also used to store an identifier to external software, if the requirement management is organised in special software tools like Rational Doors [8].

The `Requirement`-class in Fig. 6 is representing the actual requirement. This class has an "ExternalID"-property to store an identifier to a requirement in external management software tools. Additionally there is one property to store the requirement text and another one to store the identification in some sort of organisational system identification scheme. In this example a Ship Work Breakdown Structure (SWBS) was used. The `Requirement`-class is a composition of one or more `RequirementSubject`-class entities. One

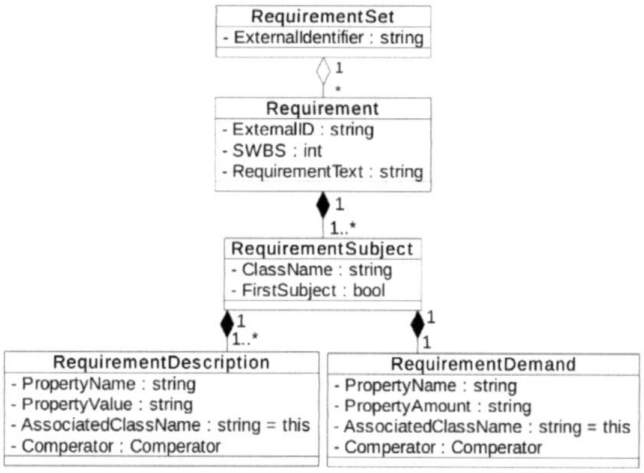

`RequirementSubject`-class entity is representing a single statement like the example requirement given. A requirement can consist of multiple connected statements. Nevertheless the first described object will remain the main subject of the requirement. Most of the requirements have just one `RequirementSubject`. The `RequirementSubject`-class consists of any number of `RequirementDescriptions` and one `RequirementDemand`.

The `RequirementDescription` and `RequirementDemand`-class has the above described properties to store the combination of property, value and comperator as described above.

3.3 REQUIREMENT DEFINITION BASED ON PRODUCT MODEL

As described above a formal way of implementing requirements is shown. In order to be able to automatically check these requirements against the product data model it is required that the class and property names are correctly set in the requirement. This leads to a sensitivity against misspelling due to the "string"-data type of the `Classname` and `Propertyname`-attributes in Fig. 6. To avoid mistakes in spelling by the user and to facilitate the creation of requirements a form was developed to assemble the requirements from prearranged lists. This form is shown in Fig. 7. On the upper left side of this form the description is formulated by multiple `Property`, `Comperator` and `Value` combinations. These are added and removed from the list via the buttons. The demand is set on the upper right hand side. In the middle section ("Related Objects") additional `RequirementSubjects` can be added. In the lower part of the form the requirement text is built together from the data given above. This text can be modified in case the

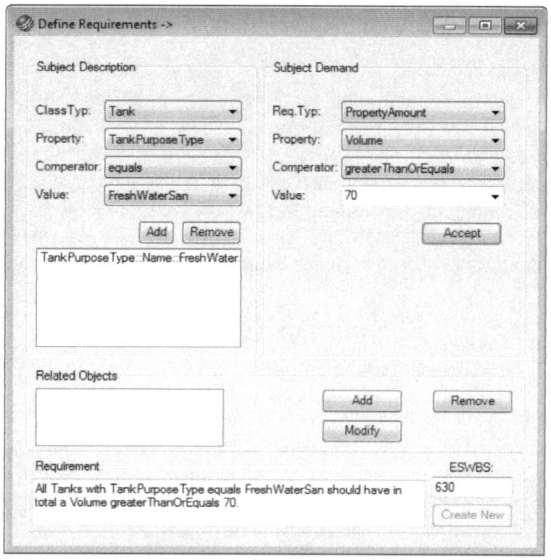

automatic wording is not sufficient. Nonetheless modifications to the text are not modifying the representation of the requirement in the data model.
The form shown in Fig. 7 is mainly based on drop down lists to choose from. On the one hand this speeds up the requirement input process. On the other hand this avoids misspelling of class and property names. The drop down lists are extracted from the product model on every start of the form.

The product data model is changing over time and new properties and types of components are added regularly. As the data model is analysed on every start, the form is adapted to these changes without any modifications. This means that there is no doubled work when introducing new classes or properties, as well as no redundancy of information which is prone to mismatch.

In the first step all classes of the product data model are filtered. According to Fig. 4 any information of interest is inserted trough an interface into the PDM system. The algorithm excludes all non-public classes as they are not of interest. Afterwards classes are excluded which are not subject to requirements, like comments, project management or for the requirement management itself.

In the second step the properties of the filtered classes are analysed in a similar manner. In a first loop the algorithm includes only simple type of attributes, like `double`, `integers`, `strings` and `booleans`. In a second loop it excludes attributes via list to get a short and easy searchable list in the drop down list. In a third loop it adds the properties of associated classes if needed.

In Fig. 8 an excerpt of an exemplary product model is shown. The procedure explained above will find the `Element`-class. This class will be added to the drop down list of classes. Afterwards the drop down list of properties will be filled with the `Element`-properties. Since `AssemblyIdentification` is on the list for including the properties in the `Element`-class it will not be added to the class drop down list. However the properties of the `AssemblyIdentification`-class will be added to the properties drop down list of the `Element`-class. Properties including the substring "`_id`" and "`BentleyID`" are omitted.

The resulting items of the property drop down list of the `Element`-class are:

- `ElementType`
- `Name`
- `Shortname`
- `Annotation`
- `IsDeadweight`
- `COVX`
- `COVY`
- `COVZ`

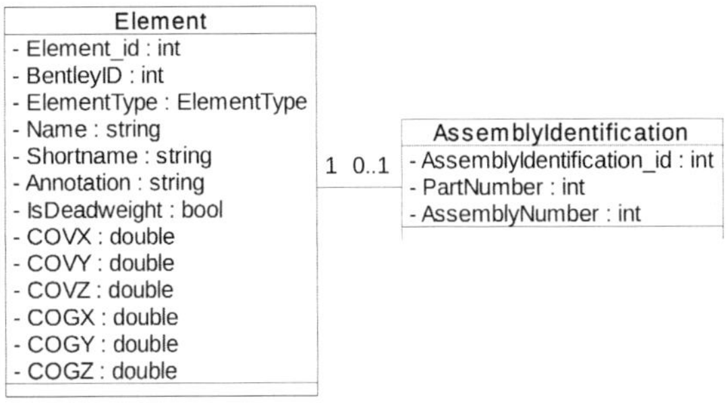

Figure 8: Excerpt of class diagram of Element and AssemblyIdentification

- `COGX`
- `COGY`
- `COGZ`
- `PartNumber`
- `AssemblyNumber`

This information is used to check the requirement against the product data model. The algorithm to check requirements reverses the described procedure. It searches for classes in the product model which have a name equals the class name given. Afterwards every object of this class will be queried. These objects are then filtered by the property, which is defined by the property name. The value of the property is compared to the given value in the description of the requirement.

To check if the requirement meets the demanded value the sum of the described property in the demand is calculated over all found objects. If this accumulated value is in the correct relationship, given by the comparator, to the demand value this requirement will be marked as "comply".

4. CASE STUDY

This section provides case studies of how design tasks are performed based on the system architecture. The tasks are related to Fig. 2. The integration in the GUI of the CAD system is exemplified in Fig. 6. Each object has an update-method implemented which will be executed when the event-handler of the CAD system notices changes to the object.

The first case study will demonstrate the flexibility in creating the compartmentation. This is mainly based on the modelling functionalities which are part of the CAD system. Additionally, the automatic update going along with changing the geometry will be shown. The second case study shows the task "placing components" which is an important step in the concept design. In the third case study an example is given where a requirement is formulated and checked.

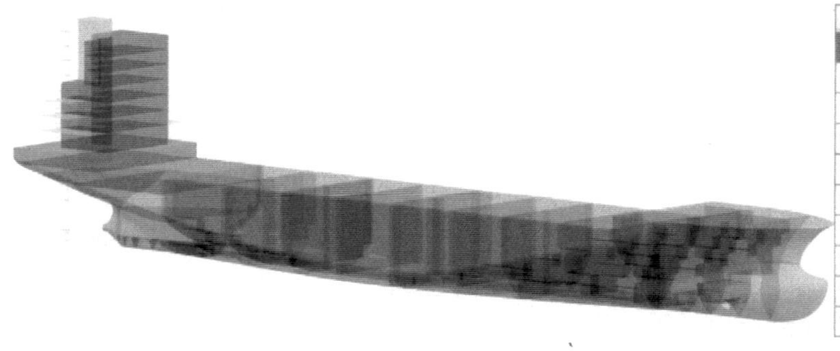

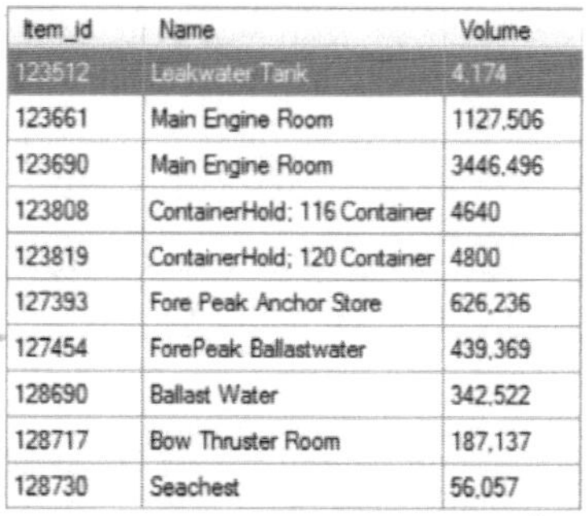

Item_id	Name	Volume
123512	Leakwater Tank	4.174
123661	Main Engine Room	1127.506
123690	Main Engine Room	3446.496
123808	ContainerHold: 116 Container	4640
123819	ContainerHold: 120 Container	4800
127393	Fore Peak Anchor Store	626.236
127454	ForePeak Ballastwater	439.369
128690	Ballast Water	342.522
128717	Bow Thruster Room	187.137
128730	Seachest	56.057

Figure 9: Compartmentation of a 2500TEU container ship design. Engine room and deck house not further subdivided.

4.1 CREATE COMPARTMENTATION

One of the first major steps in design is the subdivision of the whole ship body. This describes the first idea of the different functional areas of the ship. This task is a time-consuming step in the advanced iterations when it comes to e.g. designing complex shaped rooms or adding smaller rooms to the centre of the ship. In Fig. 9 the compartmentation of a small 2500 TEU container ship is shown. This compartmentation was done in about an hour and consists of 85 rooms/tanks. The decks and bulkheads were generated by slicing the ship's volume. The more complicated container holds were modelled outside of the ship, then placed in the ships volume and afterwards subtracted from the ships volume. This procedure speeded up the generation of recesses like the shown monkey seats of the container holds. On the right side of Fig. 9 a list with an excerpt of the information about the rooms/tanks is given. This list represents some data from the PDM and is integrated in the GUI of the CAD System.

4.2 PLACING COMPONENT

Fig. 10 shows the design of a fishery protection vessel. The model is populated with equipment as well as accommodation and sanitary furniture. Each equipment has attributes such as name, weight, type and centre of volume. The form needed to assign these attributes appears on the right side of Fig. 10, where the attributes of the port rudder are shown. The system automatically detects in which compartment the equipment is. As this information is assigned to the equipment, it is easily possible to list equipment in a room or tank.

4.3 AUTOMATED REQUIREMENT CHECKING

In Fig. 7 the methodology to input requirements is shown. This form is either implemented in the specialized requirement management tool or can be used as stand alone. The formulated requirements are then imported into the 3D general arrangement tool to give the designer an overview about the requirements he/she needs to work on. If required the designer is able to trigger the check for a single requirement or the requirements altogether. Some visualisation features have been implemented to easily trace back the objects that fulfill the requirements.

An example is shown in Fig. 11. On the right side the form for managing the requirements is shown. In this list four requirements have been saved. The second requirement of the list is the example given above. In the actual design there is enough technical fresh water tank capacity, but the assigned tank for sanitary fresh water does not have the required volume. The location of these tanks in the ship are shown on the left side of Fig. 11 in blue colours.

5. CONCLUSIONS

In this paper a ship design system based on a 3D-model for the concept design is presented. Due to the modular system approach a 3D CAD system is used to manage the

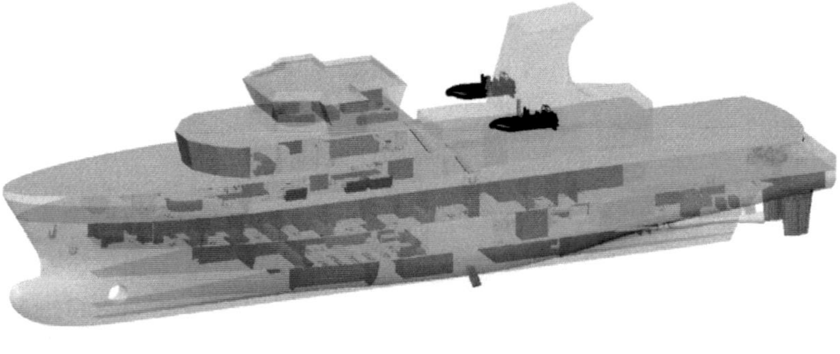

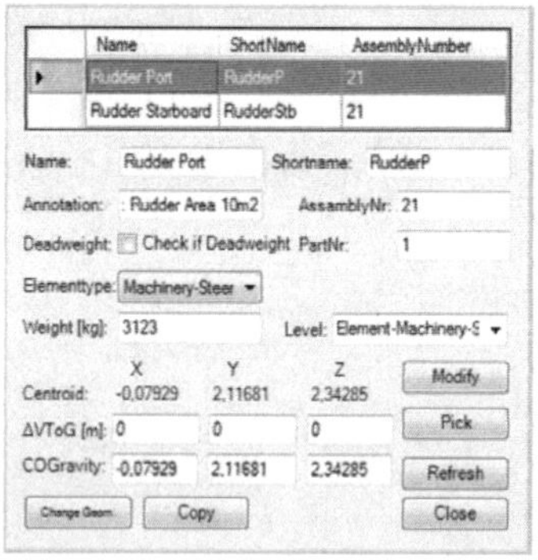

Figure 10: Example for placing equipments. Fishery protection vessel populated with equipments.

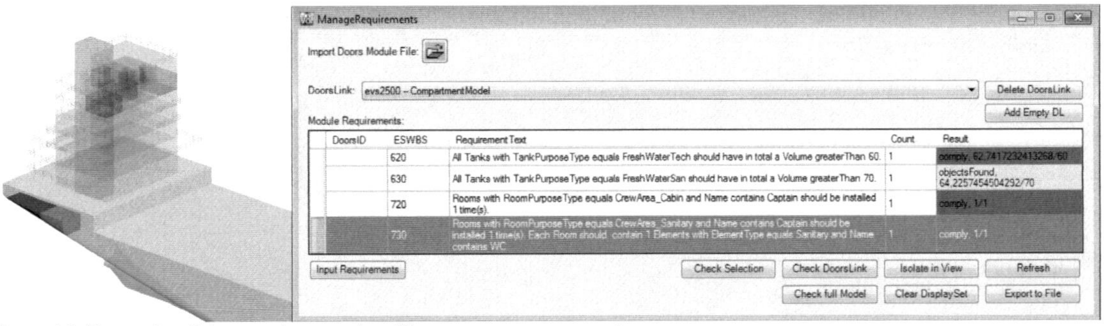

Figure 11: Example of the requirement handling.

geometry. This allows the generation of the 3D general arrangement in the very early design stages without violating the typical time horizon. Accurate geometrical information is calculated and saved in the database. Together with additional data the product knowledge is increased which leads to the reduction of risks in the design.

The presented modular system architecture with the interface options for specialized tools supports the design ultimately. The integration of this software tools avoids out-of-date data and reduces the risks of errors during manual data-transfer and also accelerate the model creation in other tools.

The examination of the design is simplified due to a higher amount of information available in the PDM system. The methodology to check the design with respect to requirements and trace back the associated objects described in this paper builds up confidence in the design. This increases the efficiency in the concept design and therefore excites new developments.

6. ACKNOWLEDGEMENTS

We would like to thank the Flensburger Schiffbau-Gesellschaft and ThyssenKrupp Marine Systems for their valuable support. This work was funded by the German Federal Ministry of Economic Affairs and Energy (BMWi) based on a decision of the German Bundestag, grand No. 03SX336B.

7. REFERENCES

[1] M. Zimmermann, 'Knowledge-Based Design Patterns for Detailed Ship Structural Design', University of Rostock, Rostock, 2010.

[2] D. J. Eyres and G. J. Bruce, *Ship Construction*. Butterworth-Heinemann, 2012.

[3] H. Lindner, S. Schenk, R. Bronsart, B. Ebeling, K. Frömming, and F. Kluwe, 'A Modular System Architecture for the Early Ship Design by Combining a 3D CAD System with a Product Data Management System', in *14th International Conference on Computer and IT Applications in the Maritime Industries*, 2015, pp. 135–146.

[4] R. Bronsart, H. Lindner, and S. Schenk, 'Entwicklung einer 3D-topologischen Produktkoordinierung zur Optimierung des Generalplan-Entwurfs komplexer Schiffstypen - Abschlussbericht'. 2015. (in german)

[5] J. (Roger) Jiao and C.-H. Chen, 'Customer Requirement Management in Product Development: A Review of Research Issues', *Concurr. Eng.*, vol. 14, pp. 173–185, 2006.

[6] S. Schenk, H. Lindner, R. Bronsart, K. Frömming, and B. Ebeling, 'A contribution to the Requirement Formalisation in the Early Stage of Ship Design', in *12th International Marine Design Conference 2015 Proceedings Colume 3*, Tokyo, 2015, vol. 3, pp. 146–157.

[7] L. v. Ruijven and U. Nienhuis, 'Requirement management, traditional and a second generation scenario', in *PROCEEDINGS OF COMPIT '06*, Oegstgeest, The Netherlands, 2006, pp. 286–301.

[8] Ibm.com. (2017). *IBM - Rational DOORS.* [online] Available at: https://www.ibm.com/software/products/en/ratidoor [Accessed 24 Aug. 2017].

8. AUTHORS BIOGRAPHY

Robert Bronsart is holding the Chair of Naval Architecture at the Department of Mechanical Engineering and Marine Technology at the University of Rostock. He has more than 35 years experience in R&D in ship design, production and operation, thereof ten years in industry. In research he is focusing on information and communication systems in collaboration networks, CAE-system integration and on knowledge based methods in ship design and operation.

Hannes Lindner is a research assistant at the Chair of Naval Architecture at the University of Rostock. Prior to that he worked on research projects focused on the early ship design process. He graduated from the University of Rostock with a Diploma in Mechanical Engineering, specialised in Naval Architecture and Ocean Engineering.

International Conference on Computer Applications in Shipbuilding (ICCAS 2017)

Singapore
26 - 28 September 2017

Volume 3 of 3

ISBN: 978-1-5108-8303-1

Printed from e-media with permission by:

Curran Associates, Inc.
57 Morehouse Lane
Red Hook, NY 12571

CURRAN ASSOCIATES INC. proceedings.com

Some format issues inherent in the e-media version may also appear in this print version.

Copyright© (2017) by The Royal Institution of Naval Architects
All rights reserved.

Printed by Curran Associates, Inc. (2019)

For permission requests, please contact The Royal Institution of Naval Architects at the address below.

The Royal Institution of Naval Architects
8-9 Northumberland Street
London, WC2N 5DA
United Kingdom

Phone: 44 207 235 4622
Fax: 44 207 259 5912

publications@rina.org.uk

Additional copies of this publication are available from:

Curran Associates, Inc.
57 Morehouse Lane
Red Hook, NY 12571 USA
Phone: 845-758-0400
Fax: 845-758-2633
Email: curran@proceedings.com
Web: www.proceedings.com

CONTENTS

© 2017: The Royal Institution of Naval Architects

© 2017: The Royal Institution of Naval Architects

© 2017: The Royal Institution of Naval Architects

© 2017: The Royal Institution of Naval Architects

International Conference on Computer Applications in Shipbuilding (ICCAS 2017)

26-28 September 2017, Singapore

Volume 3

SHIPBUILDING
ICCAS
SINGAPORE | 2017

Sponsored by:

Altair AVEVA BETA SIMULATION SOLUTIONS
CADMATIC SOFTWARE SOLUTIONS INTERGRAPH SENER
SIEMENS Ingenuity for life SSI VIRTALIS

THE ROYAL INSTITUTION OF NAVAL ARCHITECTS
8-9 Northumberland Street, London, WC2N 5DA, Telephone: +44 (0)20 7235 4622

A CAD/CAE INTEGRATED SYSTEM OF SHIP STRUCTURE CUSTOM-DEVELOPED BASED ON NX PLATFORM

Chen YouFang, Wang LiRong, Meng FanChong, China Classification Society, China

SUMMARY

With the advent of "intelligent manufacturing", it is becoming main pattern in the ship industry that ship design, inspection, manufacturing and operation are based on 3D digital model. How to build a 3D ship model rapidly, and convert CAD model into CAE model rapidly have become hot topics for ship engineers. This paper introduces a CAD/CAE integrated system for ship structure developed by China Classification Society (CCS) on the top of Siemens NX platform. A parametric and rapid modeling system is developed for 3D ship structure following the modeling habit of ship engineers. The system achieves several key functions, including intelligent creation of compartment, mesh ship structure automatically to finite elements, attach attributes of ship structure to CAE model automatically and update CAE model with the modification of CAD model automatically. The system cannot only save a great deal of time and labor cost, but also shorten design cycle.

NOMENCLATURE

CCS	China Classification Society
FE	Finite Element

1. INTRODUCTION

Following the concept of "Industrial Internet" which was put forward by United States in 2012, the "industry 4.0" was proposed by the German government in 2013. Afterwards in 2015, China officially issued "China Manufacturing 2025". These events indicated that the contemporary era is growing toward the era of fourth industrial revolution. Being the main development tendency of the fourth industrial revolution, Intelligent Manufacturing (IM) is characterized by digitization, networked and intellectualization. Specifically, in ship industry, the concept of IM can be translated into intelligent design, intelligent drawing-recognition, intelligent inspection, intelligent manufacturing, intelligent ship, intelligent management and so on. By constructing a scientific and complete cyber-physical system (CPS) architecture, an intelligent whole-life-cycle information service system integrated with ship designing, inspecting, manufacturing and operating modules can be achieved, laying the foundation for the development of safe, green and personalized intelligent ship industry.

Digital modeling of ship structures is an important issue in the intelligent ship manufacturing. Hence, China Classification Society (CCS) has carried out the research in intelligent drawing recognition and inspection, and developed a new engineering software system by 3D modeling of ship structures (shown in Figure 1). With the concepts of standardization, modularization, integration and intelligence, the system can be used for parametric and rapid ship modeling. Besides, by transforming CAD models to CAE models automatically, rebuilding CAE models with existing CAD models is avoided, which improves modeling efficiency and decreases time cost significantly. The CAD models generated by the system can be directly used for ship stability calculations, ship rule calculations and direct strength assessment.

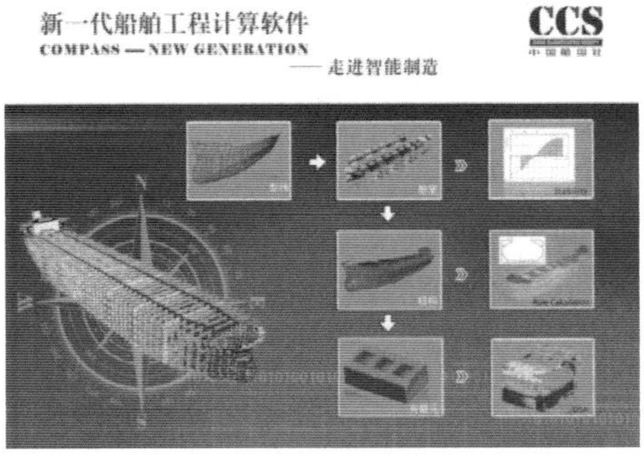

Figure 1: A new generation of ship engineering calculation software system of CCS

2. RAPID MODELLING OF 3D SHIP STRUCTURE (Ship CAD)

A rapid modeling system of 3D ship structure is developed and integrated into Siemens NX software according to the custom designing thought of engineers. Using the modeling system, ship engineers can build a 3D ship model rapidly by merely inputting key parameters. Seven modules included in this system are listed as follows.

2.1 PROFESSIONAL SHELL MODELLING

According to the offset tables input or imported by designers, the system is capable to generate ship hull lines automatically and managed them systematically. With the strong editing functions of curves and surfaces built in the system, hull lines, shell surfaces and deck surfaces can be generated and smoothed rapidly and expediently. Figure 2 shows the effect of shell modelling.

Figure 2: Ship out-shell model with zebra stripes

2.2 PARAMETRIC MODELLING IN BATCHES

Designers are only required to fill the shape parameters and position parameters according to the drawings, with which the system could automatically generate all the same type of ship structure in batches, which can avoid the tedious 3D modeling operations. After ship structures are constructed, the system provides flexible editability for single and multiple ship component models. Parameterized corrugated bulkhead modelling in batches is shown in Figure 3.

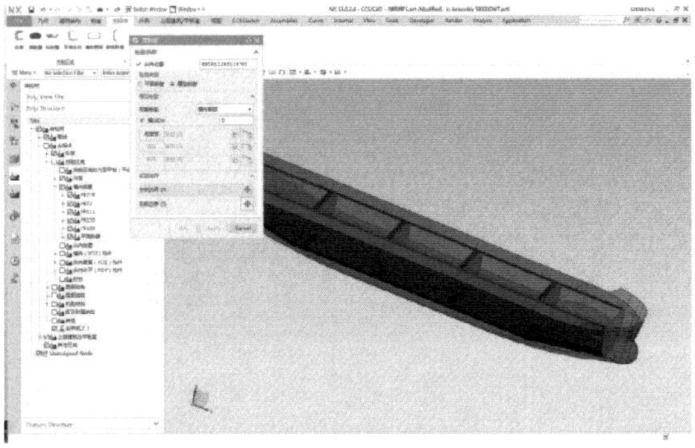

Figure 3: Parameterized modelling in batches for corrugated bulkhead

2.3 INTELLIGENT MANAGEMENT OF STRUCTURE TREE

The system could name the generated structures according to criterions and rules and affiliate the products to a structure tree in accordance with the classification of structure automatically after parametric modeling. There is no need for designers to search for the concerned structure in the whole ship, but directly show or hide the concerned structures through the tree that can lower operating complexity. In addition, designers can also customize specialized structure tree on practical requirements.

2.4 INTELLIGENT CREATION OF COMPARTMENTS

The system could create all compartments rapidly in one touch according to the water-tightness of structure, which boosts efficiency greatly. Additionally, the system could recognize automatically border and internal structures of compartment.

For a VLCC ship, the whole ship has over 260 various sizes of compartments. It will cost an engineer at least two days to build the compartments manually, however, only 30 minutes for the system. It not only saves design time, but also liberates designers from tedious design tasks.

2.5 INTELLIGENT UPDATING STRUCTURE

The associated structures will be updated automatically as one of the structures is modified by designers, and those analysis models which come from the CAD model will be updated automatically and accordingly.

2.6 PRACTICAL STANDARD LIBRARY

A rich and practical standard library of ship structures is established for the system, including the material library, profile library and structure details library, which makes designers easily invoke standards at any time. The system also supports user-defined library (UDL), the UDL will be extended with the use of the system.

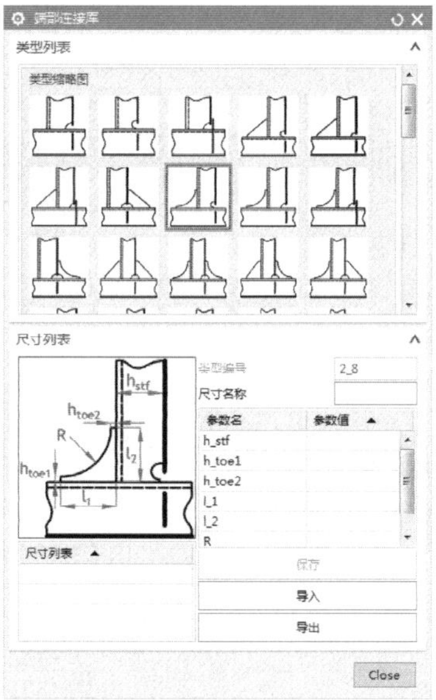

Figure 4: Structure details library

2.7 UNIFORM DATA STANDARDS (UDS) OF SHIP STRUCTURE

The UDS of ship structure includes geometric model data, structural composition, structure identification code, structure name and abbreviation and parameters, which are used to manage ship structure.

3. CAD/CAE INTEGRATION OF SHIP STRUCTURES (Ship CAE)

It is time-consuming and error-prone to establish a finite element model, also called CAE model, of ship structure. The CAE modeling of a whole ship often takes skilled engineers about 3-4 months, leading to high labor cost and error rate. It brings repetitive modeling works for the designers who have already built the CAD model, which results in huge labor waste and probable inconsistency between CAD model and CAE model. Therefore, the automatic transformation of CAD model into CAE model is an expected function in the ship industry. The system proposed has realized the exact function of transforming CAD model into CAE model on NX platform, which satisfies the requirements of ship structure rules (shown in Figure 5).

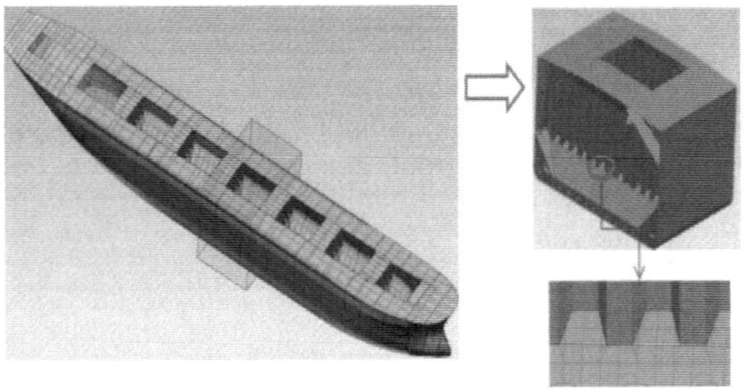

Figure 5:A CAD/CAE integrated system of ship structure

3.1 SPECIAL REQUIREMENTS OF FINITE ELEMENT MODEL OF SHIP STRUCTURE

3.1 (a) Requirements of Calculation Type

The type of finite element analysis includes whole ship FE analysis, cargo hold FE analysis and local detail FE analysis. According to analyzing type, the finite element size is classified into large mesh, coarse mesh, fine mesh and very fine mesh. The modeling requirements are different based on different mesh sizes: for large mesh, an equivalent orthotropic shell element is used; for coarse mesh, 2D shell element and 1D eccentric element are adopted to discretized plates and stiffener respectively; for fine mesh and very fine mesh, 2D shell element for both plate and stiffeners. In this paper, the particularity of finite element of ship structure is illustrated by the example of coarse mesh as follows.

3.1 (b) Size and Shape Requirements of Coarse Mesh

- The special requirements of the general coarse mesh in the finite element calculation of ship structure are as follows[2]: 2D shell elements rather than 3D solid elements for plates, while 1D eccentric element for stiffener.
- The size of the 2D element is in accordance with the distance between stiffeners, with quadrilateral element rather than triangular element adopted as far as possible, and stiffener boundary must be element boundary.
- In the actual hull structure, the seam and stiffener are often not at the same position, and the seam often cannot be used as 2D element boundary for the limitation of element size. For the 1D element discretizing a stiffener, the segmental point of stiffeners often cannot be used as a 1D element node. In Figure 6, the red lines representing intersecting lines of structures and the blue lines representing stiffeners should be the boundary of 2D mesh, but the black lines representing seam lines can't.

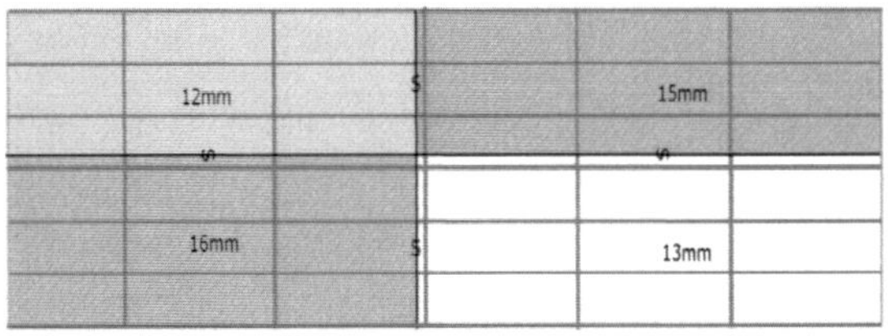

Figure 6: Intersection lines but not seams being edge of 2D element

- In some ship structures, some stiffeners are arranged between the face-plate of other two stiffeners (refer to the bottom tank compartment shown in Figure 7 left). However, in the finite element model, the stiffeners have to be modeled with 1D element between the ends of the connected stiffeners (as shown in Figure 7 right).

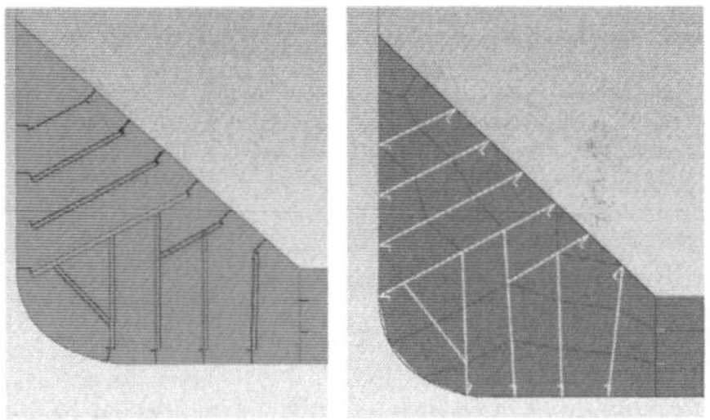

Figure 7: Difference between CAD and CAE model for stiffeners

- Small holes such as water holes, weld holes and some other negligible holes need to be deleted before meshing. For some larger holes, inner edges of the holes should be retained, which are approximate simulated in the finite element model. To ensure meshing quality, a waist hole could be converted to a square hole in CAD model.
- In the hull design, to reduce stress concentration, the toe of the bracket is created as smooth as possible, which, however, tends to produce threadlike toe with its end point being not at the position of the support structure. In order to ensure the connectivity of nodes and avoid the abnormal threadlike elements, the toe of brackets is simplified, i.e. placing the termination point at the

support structure so that the elements in the end satisfy the quality requirement.

With conventional CAE mesh auto-generation, above mentioned requirements are difficult to be satisfied. In other words, to meet the special requirements in ship industry, human intervention is necessary.

3.2 AUTOMATIC PRETREATMENT OF CAD MODEL

In order to satisfy the special requirements of the finite element mesh rules described in the 3.1, the system provides the function of automatic pre-treatment for ship CAD models. The pre-treating contents are as follows:

(1). Panel geometry idealization:
- Delete small panels;
- Search and delete the structure whose opening area accounted for a larger proportion automatically;
- Handle plate seam automatically;
- Break the termination of the bracket toe at the position of nearest support structure; The left picture in Figure 8 shows a toe of bracket in CAD model, and the right picture in Figure 8 shows the termination point of bracket are broken at the position of nearest support structure in FE model.

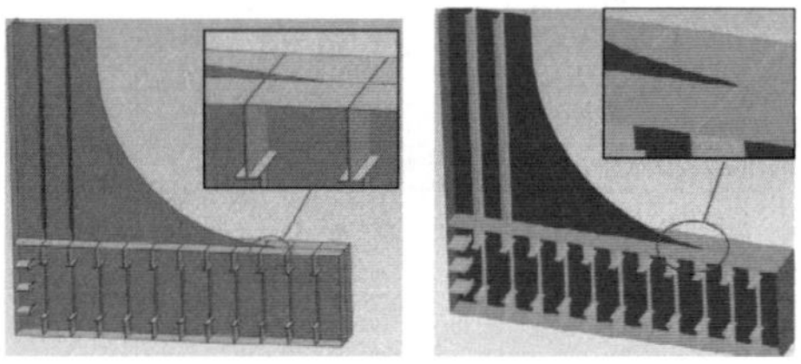

Figure 8: Idealization of the toe of the bracket

(2). Hole idealization:
- Search and delete holes automatically;
- Convert waist round holes to square holes in batches according to the requirements;

(3). Connect the end of the stiffeners automatically as shown in Figure 7;

3.3 AUTOMATIC MESH GENERATION

With the one-click automatic mesh generation, designers only need to define the size of the element. The system automatically discretizes the pre-processed CAD model by finite elements that meet the requirements of ship specifications, which significantly improves the finite element modeling efficiency. The automatic meshing results of cargo hold model and of the connection among corrugate bulkhead, lower stool and shedder plate are shown in Figure 9-a and Figure 9-b respectively. The left picture represents CAD geometry model, while the right picture for FE models.

Figure 9-a: Automatic meshing result of cargo hold model

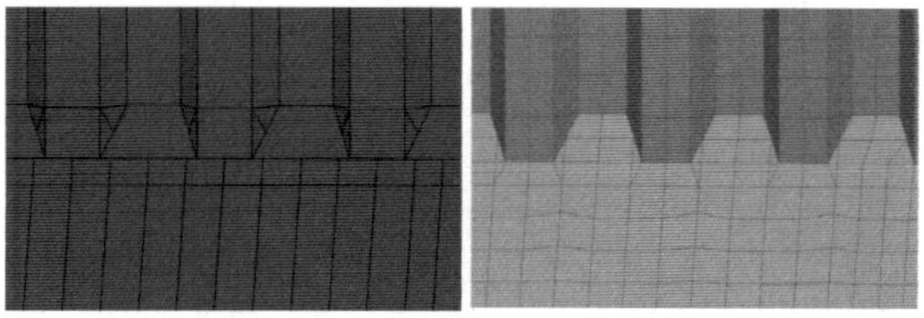

Figure 9-b: Automatic meshing result of the connection of corrugate bulkhead, lower stool and shedder plate

3.4 AUTOMATIC ASSIGN THE SHIP STRUCTURE PROPERITIES TO THE FE MODEL

The system can automatically attach the physical properties of the ship structure in the CAD model to the corresponding finite element. It can not only automatically inherit the material, plate thickness, and profiles of stiffeners, but also can automatically inherit the relationship between the structure and the compartment. Thus, the cumbersome work of the attribute assignment is avoided and human error is also reduced, which decreases the labor and time costs of finite element modeling and improves the FE model quality dramatically.

3.5 INTELLIGENT ASSOCIATION OF CAD/CAE MODEL

The system provides an automatic solution for integrating design and analysis. When the geometric shapes and sizes of the CAD model are modified, the system can update the CAE model automatically. When the attributes of the CAD model (such as thickness, material, stiffener profile, etc.) are modified, the system can quickly update the element attributes of the CAE model.

4. CONCLUSIONS AND APPLICATION

The system realizes the rapid modeling of the 3D ship structure model based on a unified data standard and realizes the function of converting the ship CAD model directly into the ship finite element model. In the system, CAD and CAE models are integrated with various kinds of ship engineering calculation functions, such as ship stability calculation, sea load calculation, ship structure rule calculation and direct strength assessment. The system will provide an integrated ship engineering calculation quick solution for ship design and ship plan approval and realize "All Solutions Based on One Model"(shown in Figure 10), which will be an important phase of ship intelligent design and manufacture.

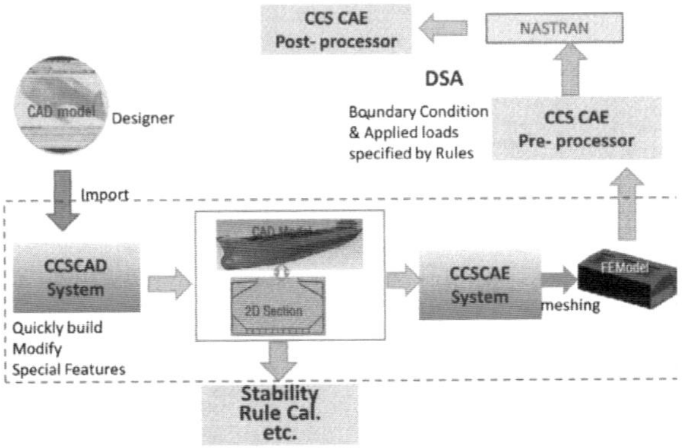

Figure 10 : An integrated solution of ship engineering calculation

5. REFERENCES

1. Chen Ming, Liang Naiming , 'Smart Manufacturing System and practices' , China Machine Press, 2016

2. China Classification Society, Rules for Classification of Sea-going Steel Ships 2015, China Communications Press, 2015

6. AUTHORS BIOGRAPHY

Chen YouFang holds the current position of chief engineer of technology research & development center at China Classification Society. He is responsible for Technical Scheme and Function Specification for the two systems. His previous experience includes developing rapid FE Model system of ship structure, Management System of ship structure, etc.

Wang LiRong holds the current position of software engineer of technology research & development center at China Classification Society. She is responsible for developing CAD/CAE integration system of ship structures. Her previous experience includes developing common structural rule computation software CSR-SDP of CCS.

Meng FanChong holds the current position of software engineer of technology research & development center at China Classification Society. He is responsible for developing 3D geometry modelling system. His previous experience includes developing 3D parametric modelling system of hull structure for CCS.

A STUDY ON DECISION SUPPORT METHODOLOGY FOR EVALUATING IOT TECHNOLOGIES USING SYSTEMS APPROACH

K Hiekata, **T Mitsuyuki**, **R Ueno** and **R Wada,** The University of Tokyo, Japan
B Moser, The University of Tokyo, Japan / Massachusetts Institute of Technology, US

SUMMARY

This paper proposed a methodology for comparing and evaluating the effects of various IoT technologies that can be introduced and examined in the marine industry. The evaluation functions of introducing IoT technologies was selected from system analysis of maritime industry and ship operation. Based on this analysis, ship operation simulator which can evaluate each IoT technology from the viewpoint of QCD of shippers. Developed simulator can simulate one container ship's lifecycle including operation, loading and docking. Proposed methodology was applied to the case studies for evaluating 25 diverse IoT technologies. Results showed that each IoT technology was evaluated relatively and judged from the viewpoint of QCD. In addition, sensitivity analysis in consideration of technology maturity could also discuss the necessity of scrutiny of maturity of IoT technologies and degree of investment in each IoT technology development.

1. INTRODUCTION

With the development of information and communication technology, large amount of data about marine equipment can be transmitted to shipping company during voyage. By analyzing these data, more advanced ship operation and maintenance become possible.

Currently, the entire maritime industry has carried out the activities to take advantage of IoT (Internet of Things) technology. Especially, effective use case scenarios of utilizing IoT technologies have been explored and studied. On the other hand, each IoT technology which has been studied for introducing in the maritime industry has many diverse functions and candidate application sites. Because of this reason, decision-making of introducing one IoT technology has required many types of expertise. In addition, there are a lot of sub-systems affected by introducing one IoT technology. Therefore, it is necessary to consider the whole system of complex ship operation service when considering the decision of introducing one IoT technology.

The objective of this paper is to support the decision-making of introducing many kinds of IoT technologies in the maritime industry. Firstly, we analyze the entire maritime industry by systems approach to define the evaluation criteria and functions of IoT technologies introduction. An IoT technology is evaluated quantitatively by using developed ship operation simulator based on this system analysis.

2. SYSTEM ANALYSIS OF MARITIME INDUSTRY

For the sake of comparison with the effect of introducing IoT technology, it is necessary to determine the evaluation criteria and functions. In this study, the requirements and functions of the maritime industry are analyzed based on previous researches about systems approach [1]. Based on this analysis, this paper defines evaluation criteria and functions of introducing an IoT technology.

First, we select the evaluation criteria of introducing IoT technology based on the request from entire maritime industry. Figure 1 shows the stakeholder value network (SVN) [3] which shows value relationship between stakeholders including maritime industry and its surroundings. A square shows a stakeholder. An arrow shows a flow of value including goods or services between stakeholders. This figure is created based on the interviews with several maritime industry persons.

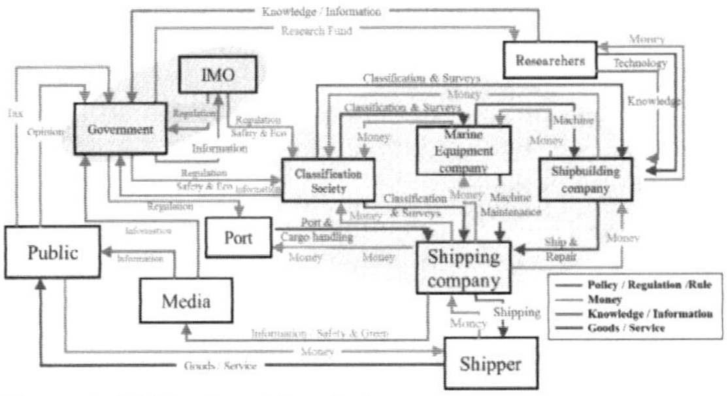

Figure 1: SVN of maritime industry

In this study, we define the motivation of introducing IoT technology as strengthen the competitiveness of the maritime industry. Based on this assumption, this study defines the requirement of maritime industry as improving ship operations services between shipping companies and shippers, which is drawn by thick line in Figure 1.

This paper adopts QCD (Quality, Cost, Delivery) as the evaluation function of ship operation service. The effect of introducing an IoT technology is evaluated based on three criteria; quality transportation, low-cost transportation and delivery reliable transportation.

In order to define the evaluation function of each IoT technology, we analyze the function performed by maritime industry against the requirements which are selected in previous section. This analysis was done based on literature reviews and the interviews with several maritime industry persons.

The objective of the maritime industry functions aimed at quality transportation is to decrease the number of accidents or faults. From the viewpoint of quality transportation, maritime industry has to consider the weather condition, the deterioration or failure of hull and marine equipment, maneuverability, human factor and so on.

In this paper, the following issues are considered for quality transportation.

- Safe operation including heavy weather avoidance, improving maneuverability, improving steering judgement
- Reduction of hull accident rate including improvement of strength and load reduction
- Reduction of marine equipment accident rate

Next, the objective of the maritime industry functions aimed at low cost transportation is to decrease the total transportation cost and downtime loss because of accidents. In this paper, total transportation cost includes ship operation cost, fuel costs, handling cost including freight expenses and port charges, sailors' expenses, construction fee and maintenance fee.

Finally, the objective of the maritime industry functions aimed at delivery reliable transportation is to decrease the delay time. In this paper, handling equipment failure, rough weather and ship failure including hull and maritime equipment accidents are considered for delivery reliable transportation.

Figure 2 shows the results of function analysis in this section. This figure is drawn by Object Process Methodology (OPM) [2][3]. OPM is a conceptual modeling language and methodology for capturing knowledge and designing systems. Each circle shows the function that maritime industry should tackle for improving QCD of ship operation service. IoT technology to be evaluated in this study is the technology for realizing each function which is expressed as circle in Figure 2. Each IoT technology is evaluated by linking these functions.

3. EVALUATION SIMULATOR

Figure 3 shows the necessary requirements of the evaluation simulator based on the analysis of the maritime industry in previous section. In this paper, we develop an evaluation simulator by applying a mathematical expression model to the relation between the evaluation target function surrounded by a square and the three evaluation axes QCD. Regarding cost, simulator evaluates the gross operating profit including downtime loss such as delay with reference to the cost structure of marine industry. In addition, each parameter is assigned to each function to be evaluated surrounded by rectangle in Figure 3. By changing these parameters, this paper can express the change in function due to the introduction of IoT technology.

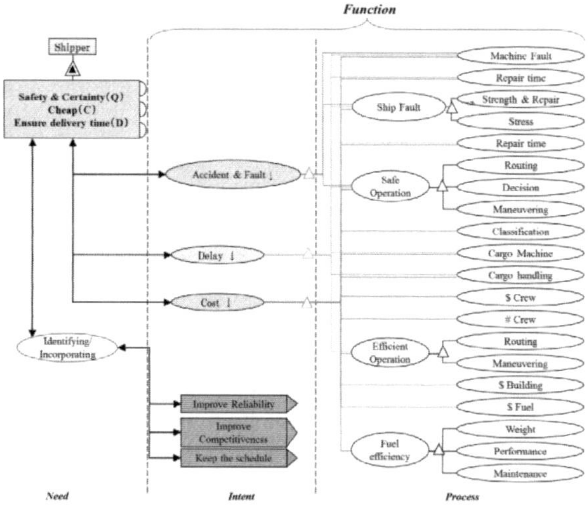

Figure 2: OPM of maritime industry

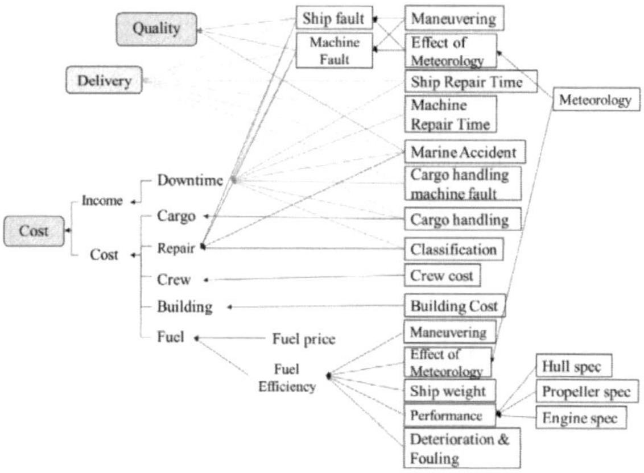

Figure 3: Requirement of evaluation simulator

Figure 4 shows the outline of the evaluation method of introducing IoT technology. According to the requirements described in previous section, this paper construct a ship operation simulator that can evaluate the effect of change in functions on the following three evaluation indicators for one container ship.

- Number of Accidents / Failures (Quality)
- Operating profit (Cost)
- Delay time (Delivery)

This paper assumes that the states of ship in service are "operation", "cargo handling" and "docking". From this assumption, developed simulator utilizes operation model, cargo handling model and docking model for evaluating the functions in Figure 3. The number of accidents / failures, operating profit, flight delay and cargo handling delay are calculated based on these models. In addition, this paper considers only the effects of weather condition as external influences, and do not consider the influence of market fluctuations. The effect of introducing IoT technology is expressed by varying the parameters of simulation models created. This effect is evaluated quantitatively by comparing the value of evaluation index of QCD from simulation results.

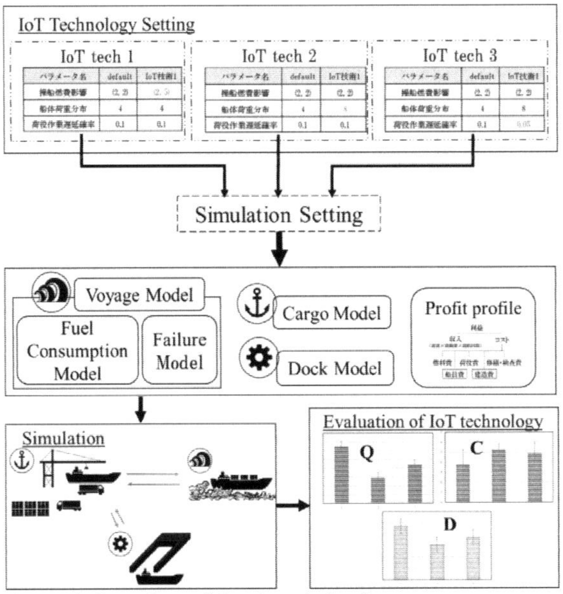

Figure 4: Overview of developed simulator

3.1 SIMULATION MODEL

3.1 (a) Ship operation model

Ship operation model consists of a model for calculating fuel consumption and a model for calculating the number of accidents / faults during operation and the delay time. Fuel consumption is modeled considering following issues:

- Ship performance including the effect of hull form, propeller and engine
- Influence from deterioration or contamination
- Effect of weather condition
- Maneuvering effect
- Hull weight

Ship performance and the influence of deterioration or contamination are modeled by utilizing basic ship theories and previous research [4]. Weather condition is considered by calculating fuel consumption increase rate under constant ship speed conditions with reference to ship speed reduction ratio per each beaufort (BF) scale. The maneuvering effect is expressed by incorporating fuel consumption fluctuation by manuvering $\varepsilon_{maneuvering}$ into the theoretical fuel consumption (FOC theory) as shown in Equation 1. $\varepsilon_{maneuvering}$ is modeled as fluctuating value according to probability density function in the range of 0% to 10% by Equation 2. In this equation, α and β are the parameters of maneuvering effects.

$$FOC = FOC_{theoruy} \cdot (1 + \varepsilon_{maneuvering}) \tag{1}$$

$$f(\varepsilon_{maneuvering}) = \frac{\left(\frac{\varepsilon_{maneuvering}}{0.1}\right)^{\alpha-1}\left(1-\frac{\varepsilon_{maneuvering}}{0.1}\right)^{\beta-1}}{Beta(\alpha,\beta)} \tag{2}$$

Hull weight effect is considered by setting the ship length L [m], the flooded area S [m^2], and the construction cost Cship [$] according to the following equations.

$$L = L_{theoruy} \cdot \sqrt{W} \tag{3}$$

$$S = S_{theoruy} \cdot W \tag{4}$$

$$C_{ship} = C_{ship\,theoruy} \cdot W \tag{5}$$

BF scale of encountered weather condition is generated in accordance with the probability density function in the range of 1 to 7 in the same format as Equation 2.

The accident / fault of voyage is modeled in consideration of ship hull failure, marine equipment failure and marine accident. Regarding the accident / fault of hull and marine equipment, failure rate per voyage is modeled according to the general framework of reliability engineering theory. This failure rate is expressed as R(t) in Equation 6. R(t) is calculated from Weibull distribution by using two parameters (σ, y) and fatigue accumulation value t depending on the operation time.

$$R(t) = 1 - \exp\left[-\left(\frac{t}{\sigma}\right)^{\gamma}\right] \tag{6}$$

In order to consider weather condition and maneuvering effect, fatigue accumulation value t is generated according to the probability density function dependent on BF and the steering fatigue influence parameter in the same format as Equation 2. A certain probability of occurrence [case / voyage] for the maritime accident incidence rate is set in advance. The delay time associated with accidents and failures is set according to the beta distribution (α = 2, β = 2) with these average delay times as parameters.

3.1 (b) Cargo handling model

Cargo handling is modeled taking into consideration the cargo handling equipment fault effect and the cargo handling work delay effect. A certain occurrence probability [case / port] for handling equipment failure and cargo handling delay are set.

The influence of cargo handling equipment failure is set according to the beta distribution ($\alpha = 2$, $\beta = 2$) with the average delay time as parameter. The impact of cargo handling work delay is generated according to average delay time and the probability density function represented by the parameters α and β in the same format as Equation 2.

In this paper, cargo handling delay affects downtime and the cost of cargo handling.

3.1 (c) Docking model

Docking model defines the rule, cost and time of inspection taking account of the effect of classification inspection.

Regarding the inspection rule, the following three rules are defined. At the time of docking, the degree of contamination and fatigue accumulation are updated to 0.

- Docking should be done every 2.5 years
- Docking should be done if the hull or propeller contamination rate exceeds x %.
- Docking should be done if the hull fatigue rate exceeds x %.

3.1 (d) Profit model

In the operation profit model, total operation cost is calculated based on "operation revenue", "fuel cost", "cargo handling cost (cargo cost + port expenses)", "seafarer costs", "construction costs", "repair / inspection fee".

Fuel cost depends on fuel consumption model. Cargo handling cost depends on cargo handling model. Repair / inspection fee depends on accident / failure model and docking model. Total operating revenue also changes depending on delay time. This paper does not take market fluctuation into consideration. Operation revenue and fuel cost per one voyage shall be constant.

3.2 SIMULATION PROCEDURE

After inputting each setting value, developed evaluation simulator repeats the following procedure for each voyage until the end of operation period.

1. Calculating cargo handing work effect: Simulator obtains the handling equipment failure and the presence or absence of cargo handling delay based on the cargo handling model. From these information, simulator calculates the delay time, downtime and cost.
2. Acquiring weather condition: simulator acquires the weather condition that target container ship encounter from the ship operation model.
3. Calculating accident / failure effect: simulator calculates delay time, downtime and cost from the result of hull failure, marine equipment failure and marine accident acquiring from the accident / failure model.
4. Calculating fuel cost: simulator calculates fuel consumption and cost during voyage based on the ship operation model.
5. Recording income and expenditure: simulator calculate delay time, number of accidents / failure, total time and the profit from income and cost based on profit model.
6. Updating aging and fatigue effect: simulator updates ship propulsion performance considering the impact of aging and contamination based on fuel consumption model. In addition, simulator updates fatigue degree of hull and marine equipment based on accident / failure model.
7. Judging whether target ship does docking or not: simulator decide whether target ship does docking or not based on the docking model. If target ship is judged to do docking, simulator sets up to docking cost and docking time, at the same time reset the contamination rate and fatigue degree to 0.

3.3 IOT TECHNOLOGY EVALUATION

The effect of introducing IoT technology is expressed by changing the input parameter value as shown Figure 5. After changing the input parameter value corresponding to the function of the IoT technology, the Monte Carlo simulation of the ship operation is carried out. The evaluation of the introduction of the IoT technology is done by comparing each simulator's output value of the QCD.

name	symbol	unit	default	IoT tech 1	IoT tech 2
Fuel oil consumption rate of maneuvering	α_{maneu}, β_{maneu}	[-]	(2, 2)	(2, 5)	(2, 2)
Fatigue distribution	β_{load}	[-]	4	4	8

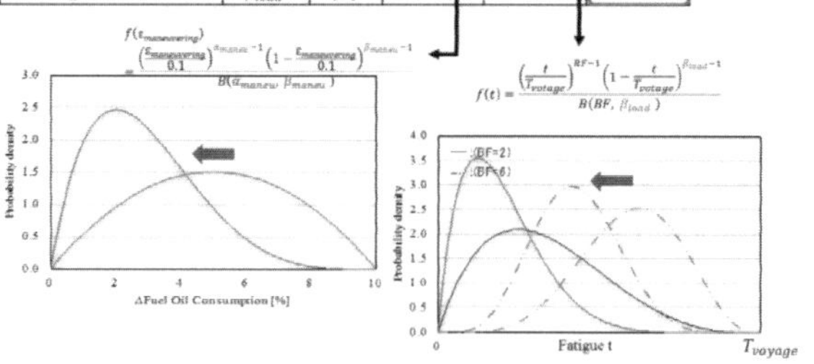

Figure 5: Parameter change by introducing IoT

4. CASE STUDIES

4.1 TARGET IOT TECHNOLOGIES

In this paper, proposed methodology was applied to a case study of evaluating the following IoT technology introduction including currently studied ones for examining the usefulness of this methodology. There are 25 kinds of IoT technologies assumed in this case study as follows:

- Marine equipment monitoring (ID. 1, 2, 3)

By monitoring the engine and other equipment, more appropriate and efficient maintenance will be expected [5]. Abnormal condition of each equipment can be detected by constantly acquiring IoT data of each equipment and analyzing these data from onshore side. Efficient equipment maintenance can be done by carrying out the maintenance only when abnormal condition is detected. In this case study, main engine monitoring (ID. 1), voyage equipment monitoring (ID. 2) and auxiliary power unit monitoring (ID. 3) are the targets IoT technologies.

- Marine equipment remote maintenance (ID. 4, 5, 6)

By sharing information of equipment with marine equipment manufacturers in real time, downtime and cost can be reduced. In addition, it is expected to avoid emergency stops on voyage and to shorten recovery time from emergency stop by realizing remote maintenance from onshore [6]. In this case study, remote maintenance of main engine (ID. 4), voyage equipment (ID. 5) and auxiliary power unit (ID. 6) are the targets IoT technologies.

- Utilization of IoT data for development of high strength ship (ID. 7)

If the measurement and analysis technology develops by hull monitoring services [7], design and construction of vessels with better structural reliability can be expected.

- Hull load control system (ID. 8, 9)

Technologies for controlling the load on the hull using the IoT technology are not put to practical use at the present time. However, in other industry fields, there are cases such as decreasing the fatigue load by using control system. For instance, in the field of wind power generation, the pitch angle of the wing is controlled so as to reduce load fluctuation of the wing.

In marine industry as well, avoidance of excessive load occurring during voyage and reduction of age deterioration of ship hull can be done by realizing the load control function which is done by controlling rudder, thrusters and other additional devices based on IoT data (ID. 8). It can be also assumed that hull weight reduction will be realized with the improvement of hull structure monitoring and load control technology (ID. 9).

- Hull structure monitoring (ID. 10, 11)

Structural reliability of a ship is usually designed to have sufficient yield strength and fatigue strength for the assumed operation service period. However, there are many case that design assumption and voyage condition are different. In addition, a ship sometime encounters unexpected weather condition.

If the damage of ship can be found early during voyage by hull structure monitoring through IoT technology, the maintenance by docking at appropriate time (ID. 10) and the reduction of unexpected downtime or renovation time (ID. 11) can be done.

- Advanced weather routing (ID. 12)

Weather routing has already been in practical use. However, if the accuracy of measuring fuel consumption and weather condition is improved by using IoT technology, optimum fuel efficiency and safe voyage will be realized.

- Safe maneuvering assistance (ID. 13, 14)

Mistakes or misjudgement due to human errors still exist. However, the introduction of IoT technology has led to the reduction of accident rate by accurate ship maneuver decisions (ID. 13), and the reduction of seafarers through automation of their tasks (ID. 14).

- Risk-based inspection (ID. 15)

Ships need to conduct periodic classification inspections during the life cycle period, but inspectors visually or manually confirm numerous checking items. These are time consuming and laborious work. If the constant monitoring during voyage becomes possible with the development of IoT technology, preliminary inspection before docking become possible. As a result, inspection cost and downtime at docking can be decreased.

- Cargo handling equipment monitoring / automation (ID. 16, 17)

By constantly monitoring cargo handling equipment such as cargo handling cranes with IoT technology, maintenance and repair can be carried out in advance before failure of cargo handling equipment (ID. 16). Moreover, with the introduction of IoT technology, efficiency of operation of handling equipment and remote operation can be expected (ID. 17).

- Streamlining port operation (ID. 18)

Logistics planning of all cargos and accuracy of port operation management have large impact of the efficiency of port operation.

By constantly monitoring with IoT technology, it is expected that adequate port operation will be possible.

- Autonomous ship (ID. 19)

The possibility of full remote control from onshore and

autonomous ship are researched. These researches will be expected to proceed more and more. As a result, it is expected to reduce the number of seafarers depending on the progress of these researches.

- Energy saving maneuvering assistance (ID. 20)

As the activities concerning energy saving operation, the optimum trim operation are reported. However, deceleration at heavy weather condition or other decisions are still judged by the tacit knowledge of ship operators. By utilizing IoT technology, fuel efficient operation can be done by accurately grasping the current state of target ship and its surrounding sea condition.

- IoT for constructing ship (ID. 21)

It is assumed that IoT technology will be introduced not only to the shipping industry but also to the shipbuilding industry.

IoT technology will improve the efficiency of constructing, and consequently reduction of construction cost is expected.

- IoT for ship design (ID. 22)

In the process of ship design, a great deal of labor is spent on making drawing and design plan by carrying out tank test and CFD. By introducing IoT technology, information exchange for designing ship will be more efficient.

- Utilization of IoT data for development of energy-saving ship (ID. 23, 24, 25)

By acquiring the operation data in voyage using IoT data and analyzing it, energy-saving ship specialized to scheduled sea area can be designed and developed. In this paper, ship hull (ID. 23), propeller (ID. 24) and main engine (ID. 25) which are specialized to scheduled sea area by utilizing IoT data are the target.

4.2 SIMULATION SETTING

In this case study, this paper assumed a container ship with 6,600 TEU loaded with a shuttle between Los Angeles and Tokyo. The speed of this container ship is 19 knot which is 50% output of main engine. The life cycle period of this ship is 20 years. Comparative evaluation for introducing each IoT technology or not is performed by the average and standard deviation of 1000 times Monte Carlo simulations.

Figure 6 shows the baseline's distribution of the number of accidents / faults, operating profit and delay time. This baseline is assumed the current situation of not introducing IoT technologies.

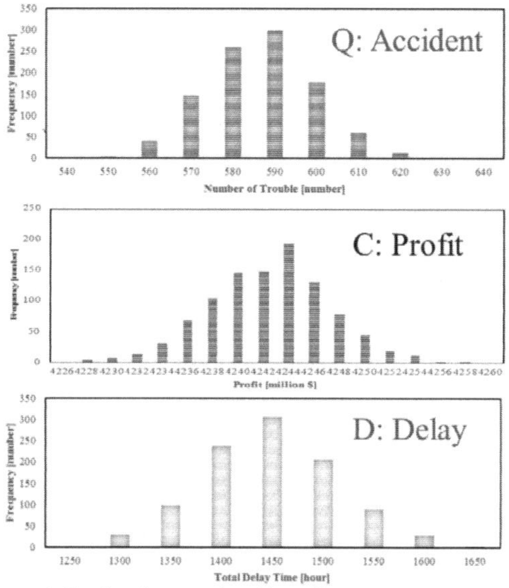

Figure 6: Evaluation result of baseline (no IoT)

Table 1 shows a list of change parameters of IoT technologies to be evaluated. Param 2 in Table 1 is used in case 1. In the sensitivity analysis of case 2, Param 1, 2 and 3 are used. These values are set by the authors from the viewpoint that seem to be achievable within a few years.

4.3 RESULT (CASE 1)

Figure 7 shows the simulation results of the number of accidents, number of failures, operating profit, and delay time by introducing each IoT technology. Each ID on the horizontal axis corresponds to each ID in Table 1. Each error bar indicates the standard deviations. The average value and standard deviation of baseline of not introducing IoT technology are indicated by solid and dotted lines in this figure.

Table .1 : Parameter setting of IoT

ID	Parameter to be changed	Base	Pram1	Pram2	Pram3
1	Main engine error rate[case/h]	0.86 $\times10^{-3}$	0.60 $\times10^{-3}$	0.40 $\times10^{-3}$	0.20 $\times10^{-3}$
2	Navigation equip. error rate[case/h]	1.14 $\times10^{-3}$	0.90 $\times10^{-3}$	0.50 $\times10^{-3}$	0.30 $\times10^{-3}$
3	Auxiliary machinery error rate[case/h]	3.21 $\times10^{-3}$	2.50 $\times10^{-3}$	1.50 $\times10^{-3}$	1.00 $\times10^{-3}$
4	Main engine recovery time[h/case]	2.37	1.80	1.20	0.80
	# of crews[man]	20	18	16	15
5	Navigation equipment recovery time[h/case]	0.93	0.70	0.50	0.30
	# of crews[man]	20	18	16	15
6	Auxiliary machinery recovery time[h/case]	2.15	1.60	1.00	0.70
	# of crews[man]	20	18	16	15
7	Average time between troubles[year]	15	20	25	30
8	Fatigue distribution	Beta (BF,4)	Beta (BF,6)	Beta (BF,8)	Beta (BF,10)
9	Weight ratio of ship	1.00	0.95	0.90	0.85
10	Ship recovery time[h]	24	20	15	10
11	Dock-in timing	2.5 year	1% fatigue	3% fatigue	5% fatigue
12	Weather condition	Beta (2,2)	Beta (2,3)	Beta (2,5)	Beta (2,7)
13	Accident ratio	1.00 $\times10^{-3}$	0.80 $\times10^{-3}$	0.50 $\times10^{-3}$	0.30 $\times10^{-3}$
14	Salary[$/(man*month)]	3,000	2,800	2,500	2,000
15	Inspection cost [$]	1.0 $\times10^{4}$	0.9 $\times10^{4}$	1.0 $\times10^{4}$	0.9 $\times10^{4}$
	Inspection time [h]	96	72	48	48
16	Error rate of handling equipment[case/port]	0.01	0.0008	0.0005	0.0003
	Recovery time[h/case]	24	20	12	12
17	Cargo handling cost[$/port]	50,000	48,000	45,000	40,000
	Delay rate[case/port]	0.10	0.08	0.06	0.04
	Delay time[h]	Beta (2,2)	Beta (2,3)	Beta (2,5)	Beta (2,7)
18	Port work cost[$/port]	50,000	45,000	40,000	37,000
	Delay rate[case/port]	0.10	0.08	0.06	0.04
	Delay time[h]	Beta (2,2)	Beta (2,3)	Beta (2,5)	Beta (2,7)
19	Salary[$/(man*month)]	3,000	3,000	3,000	2,500
	# of crews[man]	20	17	15	15
20	Fuel oil consumption rate of maneuvering	Beta (2,2)	Beta (2,3)	Beta (2,5)	Beta (2,7)
21	Construction cost[$]	200 $\times10^{6}$	190 $\times10^{6}$	185 $\times10^{6}$	180 $\times10^{6}$
22	Construction cost[$]	200 $\times10^{6}$	199 $\times10^{6}$	195 $\times10^{6}$	190 $\times10^{6}$
23	Wave resistance coefficients	As design	-5%	-10%	-15%
24	Propeller efficiency	As design	+3%	+6%	+10%
25	Fuel oil consumption rate of engine	As design	-5%	-10%	-15%

From this figure, it is found that the effect of load control support (ID.8) is high when he main purpose is to reduce the number of accidents / failures. This is because this technology can reduce failures at multiple location by reducing both hull fatigue and marine equipment level. From the viewpoint of improving operating profit, it can be said that load control support technology for reducing hull weight (ID. 9) is most effective. In addition, efficiency of cargo handling work (ID. 17, 18) and engine performance improvement (ID. 25) are also effective.

When the main purpose is to reduce the delay time, the effect of remote maintenance of auxiliary power units (ID. 6) and cargo handling remote control / automatic handling system (ID. 12) are high. This is because that remote maintenance of auxiliary power unit can reduce the marine equipment restoration time and cargo handling remote control / automatic handling system can reduce the delay time of cargo handling directly. In addition, from this figure, it is understood that cargo handling failure detection is less effective than cargo handling work improvement.

As described above, by assigning the influence of various IoT technologies as change parameters, proposed methodology can discuss the effective of introducing IoT technology while comparing each technology from the viewpoint of QCD of shippers.

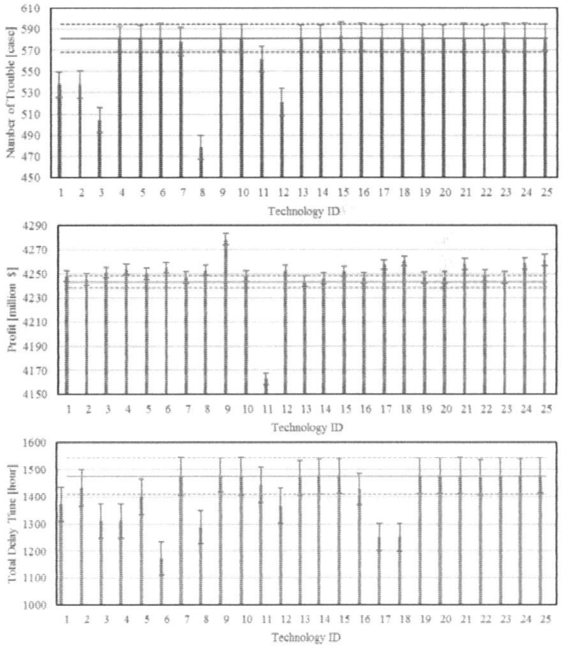

Figure 7: Evaluation result of each IoT technology

4.4 SENSITIVITY ANALYSIS (CASE 2)

In case 2, the sensitivity of parameter values of IoT technology and the effect of improvement of each IoT technology maturity are analyzed. Each stage of IoT technologies maturities are defined as Table 1.

Figure 8 shows the evaluation results in consideration of each IoT technology maturity. i is low in technology maturity, ii is medium in technology maturity and iii is high in technology maturity. In case 1, the value of ii is used.

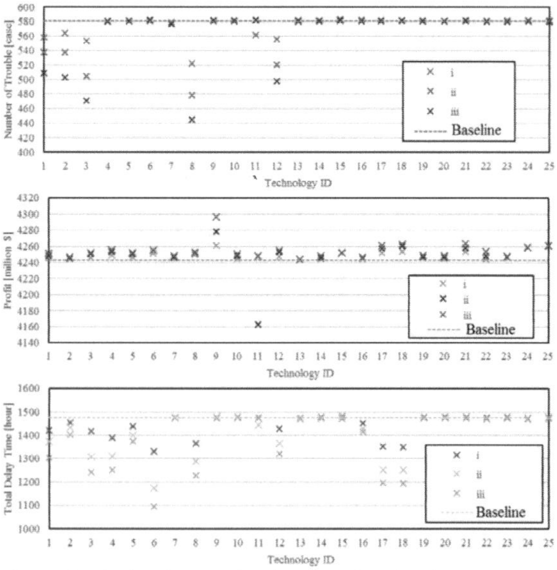

Figure 8: Evaluation result of each IoT technology

From the viewpoint of the number of accidents and failures, the technology maturity of marine equipment monitoring (ID. 1, ID. 2, ID. 3), load control (ID. 8) and weather routing (ID. 12) have high sensitivity. Especially, it can be seen that the technology maturity of weather routing avoidance has high sensitivity comparing to the result in case 1. In addition, these technologies have similar characteristics with respect to delay time.

From the viewpoint of operating profit, it can be seen that the sensitivity of load control support technology (ID. 9) is high. Although the superiority of operation profit is not greatly different from the result of case 1, the flexibility of the docking time due to the hull structure monitoring (ID. 11) was clearly divided high or low efficiency depending on docking rule. Specifically, the operating profit declines sharply when the threshold of hull fatigue for docking is 1% and 5%, and the operating profit slightly increases when the threshold of hull fatigue for docking is 3%.

From the viewpoint of delay time, the technologies related to decrease the number of accidents and failures has high sensitivity. In addition, it can be seen that auxiliary power unit maintenance improvement (ID. 6) and cargo handling related technology (ID. 17, ID. 18) have also high sensitivity.

Through considering technology maturity as this case, we can discuss IoT technologies that need scrutiny of technology maturity and are worth proactive investment. In other words, it is important to carefully examine the technology maturity of each IoT technology in order to make a reasonable evaluation.

5. DISCUSSIONS

This paper proposed a methodology to evaluate the effect of introducing IoT technology in marine industry by selecting QCD of shippers. On the other hand, there are some IoT technologies that cannot be fully evaluated on the evaluation functions defined in this study, such as environmental conservation and CSR. In this case, it is possible to determine the evaluation functions and necessary requirements of evaluating simulator by performing the requirement analysis by the same process as this proposed methodology.

In addition, case studies in this paper do not consider added values, technical difficulty and development cost during the process of introducing IoT technology. When considering these issues, this methodology assumes to change parameters for expressing these factors.

Case studies in this paper showed that it is possible to evaluate superiority / inferiority of IoT technology based on published data. When actually making a decision using this proposed methodology, decision maker have to collect the data and information of technical level of IoT technology assumed by his company. By utilizing these information as input data, quantitative evaluation and decision making can be made according to the situation of decision maker.

The target of evaluation simulator is one container ship. However, there are some IoT technologies that are effective when introduced to a fleet. In addition, the effect of introducing IoT technology varies when the target ship is other types of ship. In this case, we have to create new simulation model based on target situation.

6. CONCLUSIONS

This paper proposed a methodology for comparing and evaluating the effects of various IoT technologies that can be introduced and examined in the marine industry. The evaluation functions of introducing IoT technologies was selected from system analysis of maritime industry and ship operation. Based on this analysis, ship operation simulator which can evaluate each IoT technology from the viewpoint of QCD of shippers. Developed simulator can simulate one container ship's lifecycle including operation, loading and docking.

Proposed methodology was applied to the case studies for evaluating 25 diverse IoT technologies. Results showed that each IoT technology was evaluated relatively and judged from the viewpoint of QCD. In addition, sensitivity analysis in consideration of technology maturity could also discuss the necessity of scrutiny of maturity of IoT technologies and degree of investment in each IoT technology development.

7. ACKNOWLEDGEMENTS

This work was supported by JSPS KAKENHI Grant Number 15H04208.

8. REFERENCES

1. Crawley, E., Cameron, B. and Selva, D., 'System Architecture', PEARSON, 2016.

2. Cameron, B., Crawley, E., Loureiro, G. and Rebentisch, E., 'Value Flow Mapping: Using Networks to Inform Stakeholder Snalysis', Acta Astronautica, Vol. 62, pp. 324-333, 2008.

3. Dori, D., 'Object-Process Methodology: A Holistic Systems Paradigm', Springer, 2002.

4. Hiekata, K., Mitsuyuki, T., Moser, B., Yamato, H., Saito, T. and Wanaka, S., 'A Study on Ship Life Cycle Value Maximization with Flexible Design considering Real Option', Journal of the Japan Society of Naval Architects and Ocean Engineers (in Japanese), Vol.23, pp.231-237, 2016.

5. DNV GL, 'Beyond Condition Monitoring in the Maritime Industry', DNV GL Strategic Research & Innovation Position Paper, 06-2014, 2014.

6. DNV GL, 'Ship Connectivity', DNV GL Strategic Research & Innovation Position Paper, 04-2015, 2015.

7. Yu, H., Choi, J., Park, G., Han, S., Tai, S. and Ha, M., 'Full Scale Measurement of a Large Container Carrier on the Far East - Europe Route', Singapore Maritime Technology Conference, 2008.

9. AUTHORS BIOGRAPHY

Kazuo Hiekata holds the current position of Associate Professor at Graduated School of Frontier Sciences, The University of Tokyo.

Taiga Mitsuyuki holds the current position of Assistant Professor at Graduated School of Engineering, The University of Tokyo.

Bryan Moser holds the current positions of Project Associate Professor at Graduated School of Frontier Sciences, The University of Tokyo, and Lecturer at System Design and Management, Massachusetts of Institute of Technology.

Ryuji Ueno was a master course student at Graduated School of Frontier Sciences, The University of Tokyo. After getting his master degree, he joined NYK Line

Ryota Wada holds the current position of Assistant Professor at Graduated School of Frontier Sciences, The University of Tokyo.

AN ENTERPRISE MODELLING APPROACH FOR THE EARLY SHIP DESIGN

W Jabary Tischreen University, Syria
R Bronsart University of Rostock, Germany

SUMMARY

In order to efficiently supporting the early ship design phase, the Enterprise Modelling Approach is adopted. The main motivation is its principle concept of data integration and interconnected process management being supported by functions like configuration, collaboration and change management, which in turn rely on version and rights management. Through this the state of a design can be monitored continuously in detail which is regarded highly important. Predefined Activities are introduced as a core concept to apply knowledge of the design process itself and at the same time guarantee a high flexibility and design freedom. The fundamental design tasks of the traditional ship design spiral are implemented through input, output, control and tool connectors. However, the underlying simultaneous nature of the design process, particularly in the early stage with frequent changes of any design aspect is supported by an integrated platform which enables a holistic ship design. A highly modular software system is built up in such a way that no restrictions to specific tools are imposed: all partners in a distributed design team will continue to apply any CAE-tool which best meet their requirements to efficiently perform the specific design tasks. The design of a Ro-Ro vessel shows the benefits of the implement CAE-work-bench.

1. INTRODUCTION

Like other design engineering fields, the ship design process has seen significant developments in recent decades as a logical consequence of the continued progress in computational sciences and the ever increasing competition in the global market. This prompted shipyards and design offices to permanently develop new methodologies in order to make this process more effective. Nevertheless, there are some design related issues which can be considered general and fairly constant aspects [1]. One of these aspects is the correct understanding of the owner's requirements. These requirements set the limits of the design process and must be respected by the design team in order to be able to design a ship which is on one hand accepted by the owner and on the other hand satisfies all additional safety and reliability issues, mostly formulated in form of statuary rules. Another important aspect to be considered is the fact that the ship design process is always a team process, the result are achieved through the collaboration between many disciplines like hydrodynamics, machinery, outfitting, etc. and many partners like owner, shipyard, designer, supplier, classification societies, model basin, etc. These partners can be located in geographically dispersed areas which increases the complexity of the design process.

2. SHIP DESIGN PROCESS MODELLING

The ship design process has been represented traditionally by using the design spiral which was introduced by Evan [2] to describe the structural design and then extended to describe the overall ship design process. Many versions of the design spiral exist, some of them represent the general design process and others have been used to describe a specific design stage. Some of these specific versions are shown by D. Andrews et al. [3]. In general, the representation of ship design process by means of the design spiral reflects the following important aspects:

• The iterative nature which describes the complexity of this process.

• The sequential nature through a number of steps followed successively. Each step represents a specific discipline such as (hydrostatics, powering, etc.)

• The complexity nature through the diversity of perspectives of the design participants, for example hydrodynamic-, structure-perspective, ...

• The progressive elaboration nature [4]. Mistree et al. [5] use the "convergent/divergent" approach to describe this property. They point out that as the process is progressed and converged with respect to the general design spiral, the amount of information about the product (ship) is increased reflecting the divergent aspect.

Despite of the advantages, which can be counted to the traditional design spiral representation, it must be clearly stated that this representation is not capable to describe the real nature of the ship design process as of today. The weaknesses of the traditional design spiral representation have been addressed by many authors, and are briefly summarised in the following:

• In practice the design process is not sequential. Gale [1] points out that during the design process the designers may move or jump to perform design tasks, which are not located directly next to each other regarding the order of the design tasks represented in the design spiral. The same aspect is mentioned by Pradillon et al. [6], he states that the ship design process must be regarded as a dynamic process with the consequence that it is common practise to start some independent tasks without considering the sequence giving by the design spiral. Furthermore, Nowacki [7] argued that this representation of the design process in sequential closed loops contrasts with the decision making nature of the design process especially at the early stage. Andrews [8] attempts to overcome this closed nature and considers the interactions between designer and external resources or constraints by adding a new dimension to the design spiral. He represents the design process of the major phases in a three dimensional design spiral as a "tapering corkscrew".

• The parallel or simultaneous nature particularly in the early stage is not considered in the traditional spiral. The model shown in Figure 1, which has been introduced by Papanikolaou et al. [9], represents the integrated nature of the design process. It reflects the necessity and ability to start any design task as soon as the needed information is available.

• The depiction of the design process activities (tasks) in consecutive steps implies many discrepancies with the fact of the interaction and overlapping of activities. Furthermore, it disguises the activities details, for example, it hides the fact that these activities are interdependent or give input or request output from each other.

• Levander [10] points out that the description of the design process by means of the design spiral undermines the innovation and creativity during the design process. The designer is locked to his first assumptions, which has a negative effect regarding the development of innovative design alternatives. Therefore he introduces a "System Based Design" approach with the main objective to reduce the number of loops needed to find a suitable (technical and economical) design solution.

In summary, it can be concluded that the model of the design process as a spiral reflects many relevant aspects but at the same time, it masks a number of important aspects, which have to be fundamentally considered in the ship design process.

2.1 SHIP EARLY DESIGN PHASE

In literature several authors who have addressed the early ship design phase differ in the naming of the phase, such as concept, early, feasibility, initial, etc. However, they unanimously agree that it is demanded to continue the research related to this stage. This is due to the intrinsic importance and the complexity of the tasks to be performed at this stage.
The ship design process however is associated with major challenges which increase its complexity comparing to other design stages [11]. This led Pawling et al. [12] to describe the studies related to the early design stage as a "wicked problem". Some of these challenges are :

• the low amount of available data at this early stage which leads to the reuse of data of the previously built ships,

• rapidly accumulating information have to be taken into consideration which imposes new constraints on the design process. The introduced approach by Gaspar et. al. [30] is based mainly on the "decompositionen-encapsulation" concept by five factors which are: behavioural, structural, contextual, temporal and perceptual.

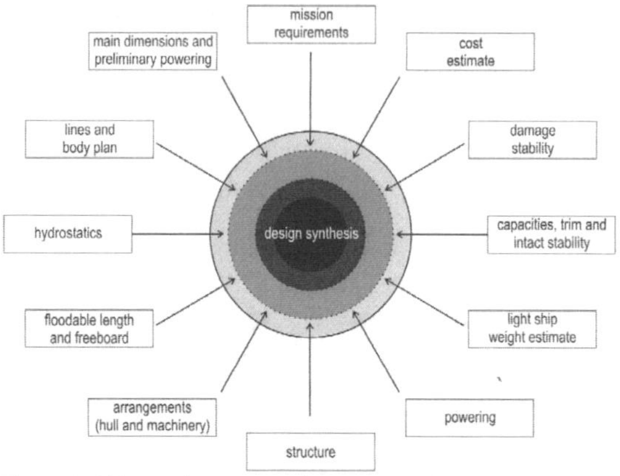

Figure 1: Ship Design Process is not a Spiral [9]

• Designers have just about few weeks to perform all studies before they supply an offer which must be economically safe, achievable and compatible with the design requirements.

• Design aspects are not independent, but highly interrelated. As a result, any change in one of these interdependent design elements will affect many others. Furthermore, these elements can relate to different disciplines which makes this interaction even more difficult to be treated.

• The procedures associated with the design process in the early phase can be classified as design tasks like definition of arrangements (hull form, compartmentation, etc.) and analysis tasks like (damage stability, lightweight estimation, powering, etc.). These procedures are performed sequentially, simultaneously and iteratively.

2.2 ENGINEERING CHANGES

Frequent changes play a dominant role in the early ship design process! An example to illustrate this utmost important aspect are modifications required to update the dimensions of the engine room compartment, which might result from the compartmentation in accordance to the dimensions of the specified main engine (specified depending on the results of the power prediction task in terms of the defined hull form at the required speed range). Required changes of one task might conflict with results or requirements of other tasks due to the interaction between the design elements. Following up the previous example, the modification of the engine room dimensions might affect the damage stability analysis or may affect the dimensions of other compartments such as tanks.

On the other hand, the large flexibility in the early design phase in comparison to the remaining design phases must be considered. However the overall design freedom is rapidly degreasing while the design is proceeding: a change in the advanced design phases can be very expensive to implement especially in cases when it affects design elements of the design solution on which it has been based. The execution of changes leads to new results, which in turn are used by different activities which in turn lead to new results. This can lead to a "disconnect" or "inconsistent" problem.

Summarized it can be stated that the major challenges of the early ship design process are due to the associated data, the data-interaction, activities, activities-dependencies in addition to the interaction between data and activities. Therefore, supporting these aspects can facilitate the early design and can lead to significant advantages regarding the entire design process.

3. EARLY DESIGN CHARACTERISTICS

Taking a high level perspective, the ship design process, like other design processes, is characterized by: *Decision Making*, *Data-Reuse*, *Data Exchange* and *Design Constraints*.

Decision making, alternative and trade-off studies are important but highly interconnected activities designer face in the early design process. Generally there is no unique solution, but rather, many solutions/alternatives are developed for each requirement. Several decisions are made during the design process until a suitable design solution is achieved. These decisions can be characterized as partial decisions. The achieved design alternatives in turn are subjected to selection processes, which are called trade-off studies.

Data-Reuse: Most challenges faced by the design team when starting a new design project are the low level of available data. This fact urged designers (in order to accelerate the early design process) to rely on similar previous works, to re-use their data completely or partially with some modifications regarding the specification of the new design (KPIs). Many design tools have developed functions to support the methodology of reusing design data, but the main focus is in advanced stages of the ship design such as the approach introduced by Malheiros for the reuse of design and production data in operation and maintenance of warships [13].

The reuse of data in the early design phase is still faced with difficulties and obstacles. A major challenge is that data of previous designs are not stored in a way which enables designers to apply them in the early design stage. Often only the data which represent the final results are available, designers are not able to gain an insight into the work-flow of the processes which led to the results. Furthermore the cooperation between different design disciplines such as system engineering, hydrodynamic, etc., which have different perspectives regarding the ship design leads to store data with respect to their specific perspectives.
E. Moredo et al. [14] developed a methodology for reuse data in the tender phase which is based on the combination of a Functional Breakdown Structure (FBS) and a System Breakdown Structure (SBS). FBS helps designers to receive the relevant requirements from the client and save the design data in a structured functional database. SBS is used to enhance the documentation of design data and to improve the data-exchange between shipyards and system-integrators. Another methodology has been developed by M. Toyoda et al. [15], who tried to accelerate the generation process of a 3D-model at the early stage by reusing of ship type specific standard parametric structural components and assemblies stored in libraries.

Data Exchange: The participation of many partners like shipyards, classification society, design offices, suppliers and e.g. model basin raises the need to exchange data in order to efficiently achieve the desired design mission. In general, the difficulties associated with the data exchange are mostly emerged from the variety of data formats, which are available through the tools applied. However, there are some formats like IGES, STL, XML, STEP, etc., which have been introduced and can be considered (to some extent) as general formats to represent design elements such as hull-form, compartments, etc. [16]. Data exchange regarding the early design phase is to be discussed under two perspectives: The exchange of data for the same task between design team members and the exchange between different design tasks. This raises principal questions about the compatibility of data at this stage with the downstream processes and the integration level of tools used for different design tasks.

Design Constraints: Ship design in general is performed with respect to restrictions and limits, to which is referred as design constraints. Requirements formulated by the customer (ship owner) play of course a dominant role like ship mission, service speed, etc., which formulate constraints to be strictly observed for the design. Additional constraints are issued by International Maritime Organization (IMO) like SOLAS, MARPOL, LLC or the IS-Code which are not negotiable. Additionally ships have to satisfy classification rules and regulations issued by the state under which flag it is registered.

The great diversity of ship types impose a significant expansion on the special requirements to be taken into account. However, there are some requirements which can be considered as common requirements relevant to almost all ship types. The correct understanding of the requirements at the beginning of the design process is an essential issue [17]. Design requirements represent the criteria on which the decision-making in the design process will be based. The perception of these requirements is not uniform, C. Koenig et al. [1] classified them in technical like endurance, service speed, manning, accommodations, etc., and commercial such as operating agreements, ship financing, etc. B. Oers [18] introduced another concept, he grouped the requirements into "negotiable" and "non-negotiable". He points out that some requirements must be complied with whereas other requirements can be simply adapted to enhance other requirements. However, requirements are not isolated from each other, they are directly or indirectly interdependent and interrelated and can be contradictory or incompatible.

3.1. DESIGN METHODOLOGIES

Many papers, which address the ship design process and particularly the early design phase have been published. Oers et al. [19] developed an approach to help the naval architect to identify the suitable solutions at the early ship design stage. Their approach is based on two algorithms, a "packing-algorithm" and a "search-algorithm". They automatically generate a large number of feasible ship designs, which support the naval architect to make a decision while selecting the preferable design. Six types of 3D objects (design elements) are used to form the design-models. Spatial and logical rules such as "overlap-rules" are used to pack objects together enabling the designer to control the packing process. Some criteria are used to assess the resulted "packed" objects in order to generate the set of feasible ship designs. The Intelligent Ship Arrangement (ISA) [20] is another approach, which has been developed for the U.S. Navy by the University of Michigan. It has been developed to capture U.S.Navy design rules. ISA aims to support the designer in the early design stage by providing a number of trade-off studies for the ship general arrangement. Many papers discuss the Design Building Block (DBB) approach [21]. This approach depends on a "functional" breakdown of the design into blocks. The designer aggregates blocks together to form "Master Building Blocks" to perform the necessary naval architectural calculations.

A holistic ship design optimization methodology has been introduced by Papanikolaou [22] which is based on the formulation of the ship design problem presented by Nowacki [23]. This methodology is build upon a system approach. In this context, the ship is considered as a complex system composed of subsystems and their components. Papanikolaou argues that the ship design optimization must address the whole ship's life-cycle from the concept/early design to the scrapping/recycling.

3.2 INTEGRATION STRATEGIES

An integration strategy aims mainly to bridge the gap between ship design tasks to be performed. The integration of all phases including the early stage is the main objective of the following design methodologies. An integrated design and multiobjective optimization approach to ship design has been introduced by Papanikolaou et al. and Harries et al. [24]. In addition to an optimization approach, the simultaneous nature of the ship design process has been taken into account. A synthesis model which represents an integrated design software platform has been developed. Therein many tools for the design, calculation, optimization and data-exchange are integrated. Parameters related to the payload, steel weight, strength, oil outflow, stability and hydrodynamics have been simultaneously examined for many variants against determined requirements.

An integrated approach to manage the ship general arrangements has been introduced by A. Cebollero et al. [25] by developing a Foran General Arrangement tool. They tried to overcome the lack of information details at the early design stage by building a product model for the ship general arrangement from the beginning of the design process. This product model has been used as a basis while creating compartments and other general arrangements. Almost the same approach is proposed by Eriksson et al. [26], they also show how to apply an integrated model of a ship from the beginning of the design process. This model is called a "common reference model". Mizutani et al. [27] developed an approach to integrate the design process based on a single NAPA 3D model and their own shipyard's design tools. They state that the usage of a single 3D product model will improve the consistence of the data of various design disciplines, it will support the accuracy of data in the early design stage, and thus will improve the overall design process.

It can be summarised that the common denominator of almost all methodologies, which adopt the integration strategy, is the need to build upon a ship product model.

4. SYSTEM DESIGN ENGINEERING

System engineering is a general framework through which large and complex projects can be addressed. It focuses on the design and management of a system [28]. It guarantees the interaction between technical knowledge and human disciplines of complex projects. However, many aspects like configuration management, work management, etc. should be taken into consideration in the application of the system engineering concepts on a large system. The application of these concepts to the ship design process can lead to advantages and improvements of drawbacks related to the traditional ship design process.

4.1 ENTERPRISE MODELLING

Enterprise modelling is the description of principles, methods and goals of a collection of organizations in coherent models in order to externalize the enterprise knowledge [29]. The main objective of modelling an enterprise is to help persons to communicate, understand and develop efficient solutions of their business. According to ISO 19439:2006, four issues should be taken into consideration by the modelling of an enterprise:

- *information*: the data used and obtained
- *functions*: the operations through the enterprise
- *resources*: the required sources for the operations
- *organization*: the responsibilities authorizations and associations through it

These different concepts can be handled in different abstraction levels regarding the needs of the designer. The modelling approach is based on the integration of two general aspects: data and processes. The interaction of data and processes leads to considerable benefits like the consistency, control capabilities, etc.

4.1 (a) Data Modelling

The Data Model (DM) of an Enterprise, as in information modelling technology, represents the static part of the enterprise model. The DM is a conceptual representation of data elements by means of symbols and text. It describes the addressed data by giving its properties and relationships which help the readers to get a sound understanding of the addressed subject. The DM forms the basis of any database application. The coherent definition of objects, relationships, etc. supports the data-consistency-concept.

4.1 (b) Process Modelling

Enterprise processes are collections of activities within an organization. They are structured and designed in order to achieve the required customer's aims and to fulfil certain goals. The process model represents the dynamic part of the enterprise model. A process model can be divided into several sub-processes complementing each other to accomplish the intended objectives of an organization. The processes are to be defined considering the enterprise goals, translated to software applications, executed and controlled by applying programming techniques.

4.2 ENTERPRISE MANAGEMENT

The organization and coordination procedures applied to the enterprise activities and data like the authority and responsibility restrictions in order to achieve the desired goals and objectives is called enterprise management. This definition implies management disciplines which together incorporate to foster an enterprise model by supporting the efforts of storing, capturing, reporting, processing, visualizing and controlling the available resources. Management disciplines provide technologies and strategies to meet the customer requirements like the assignment of roles, handling of changes, versions and requirements treatment. Principally these functions are not isolated but rather complementing each other. The management should be viewed under two perspectives, the "data view" and the "process view":

The *Enterprise Rights Management* (ERM) is an information protection technique. It is intended to manage rights and usage of data throughout their life-cycle within an organization and among all persons involved. The ERM mainly aims to protect resources by enforcing access and usage restrictions on the data content like, read only, read and copy, modify, etc. In order to efficiently achieve the ERM objectives, an appropriate access control concept is to be carefully defined and executed. A role-based access control can implement a rights management concept within an organization by taking into consideration the functional tasks performed by specific authorized users. Main concepts to be distinguished are: Users, Roles, Permissions, Objects, Operation and Sessions. The management of the security policy may be implemented by a mandatory access control system supplied by the underlying DBMS.

Collaboration Management addresses the interactions of an enterprise. In this context, it reflects the process-interaction aspect of an enterprise model. Collaboration management becomes more and more essential especially in case of large distributed organizations such as shipyards and their numerous partners. Its importance stems from the demands to coordinate and organize the contributions of all participating partners who can be geographically dispersed, are responsible to achieve the activities/tasks associated with the addressed enterprise and consequentially its temporary product "project". The challenge of the collaboration management is to provide all persons at the right time with the right data. Many forms of managing the collaboration relationships exist, which reflect the level of the tasks-interdependencies within an enterprise.

Versions Management can be applied as a stand-alone technique by a Version Control System (VCS). Version management refers to the ability to identify a unique state of a model or any collection of data by using a unique name or number, in addition to the possibility to keep track of the history of these states. Changes are always associated with new states to which is referred to as versions. Therefore change management always requires a suitable versions management. The complexity of versions management is mainly due to issues which have to be addressed such as the capability to access the same object by different users at the same time, merging of versions, the ability to retrieve the difference between two versions of the same object.

Configuration Management (CM) in terms of ISO 10303-44: 2000 meets the static point of view in the data modelling. The main task is to decide products or parts to be specified as configurable objects. Whereas the CM in terms of ISO 10007 as well as of ISO/IEC TR 18018 reflects the functional point of view by describing the activities of the CM-process. Regarding ISO 10007 a CM process involves planning, configuration identification, change control, configuration status accounting and configuration audit activities. Due to the growing demand to manage the integrated activities in the design, CM is becoming increasingly important.

Engineering Change Management (ECM) is the process by which a change within an enterprise is requested, evaluated regarding its impact, implemented, verified and documented. Specially in the early ship design phase changes occur very frequently. This is due to the fact, that the designers' perception regarding the ship is evolved during the time as well as the customer's requirements can be changed. Therefore, an efficient management of changes is an essential issue and can lead to considerable benefits regarding the flexibility within a shipyard.

Work Activity Management (WAM): The product (ship) can be regarded as a "project". It is a temporary activity with an identified start and end. It is an integrated result of all operations taking place through an enterprise. Therefore the more the integrity and coherency of the designed enterprise model, the more the efficiency of its operation results. Moreover the tracking of the tasks performed at any time in order to detect bottlenecks is required. Work Activity Management offers a means to accomplish the objectives by organizing the associated tasks and providing the resources required by the addressed project. Furthermore, it provides a technique to pursue the progress of the activities and the performance of the processes within an organization.

5. EARLY DESIGN INFORMATION MODEL

An information model of the ship considering the early design phase with its related tasks is developed. The information is divided into three major top level categories:

- ship product
- process management
- support

Objects are structured, organized and related to each other reflecting the data interaction concept. Respecting the features and flexibilities provided by the UML method, two kinds of diagrams (class- and package- diagrams) are applied in order to model the information. The ship information model is developed with the main objective of providing a neutral means (independent of any system), on which it can be based to configure an integrated but shareable database. This database will in turn interconnected to the design applications in order to perform the studies related to the early design phase. Furthermore, the existence of a ship information model and accordingly an integrated database will reinforce many important aspects such as the collaborative work and data exchange between different tools applied by different designers.

Considering the complexity of the information in ship design, the modelling of the information objects is performed with several objectives in mind: the grouping of the ship product information objects which relate to each other regarding a specific aspect such as a function in the same unit (represented by a package) simplifies the complexity. An example is the grouping of information objects which represent the general information of a ship in the same package. The modelling of the information objects is achieved in a way that ensures the information consistency which is one of the most important aspects to be considered in order to improve the early design phase. Furthermore, by taking into consideration the potential necessity to expand the information objects to cover more and initially not considered details (related to early or downstream design stages), the grouping approach helps to add more packages or to accelerate the accessibility to the required information. Thus, in spite of the focus on the early design stage, the expandability of the modelled information to support other design stages, is ensured.

Management models such as for change, configuration, etc. are modelled in a way that supports the dynamic nature of the early ship design process. In addition to the ship product- and management-packages, general information objects like status, date and time, etc. is made available in additonal support-packages. Finally the grouping of information objects in packages has benefits regarding their implementation: it facilitates the translating of the product model into the information processing system level.

6. PREDEFINED ACTIVITIES CONCEPT

The process of the ship early design can be regarded as a collection of predefined activities. These predefined activities reflect the tasks which should be achieved in order to complete the studies of the early design phase. The modelling of these activities from the process point of view is accomplished by utilizing two types of methods: the "activity-diagram" (from UML) and the "ICOM"-method. The following three aspects are taken into consideration:

• Each early design task is represented as a predefined activity. It consists of a set of activity partitions, which can be determined from analysing the task respecting the ICOM-concept. The example depicted in Figure 2 shows a Predefined Activity composed of five Activity Partitions. Two of them reflect the input (regarding the ICOM-diagram) while each of the control-, mechanism- and output- aspects is represented by one partition. In

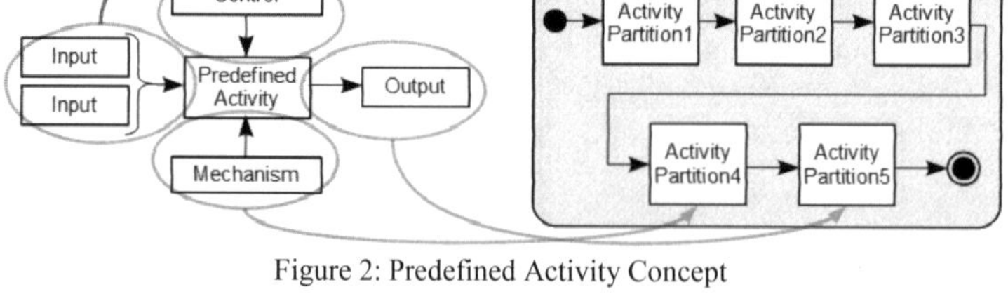

Figure 2: Predefined Activity Concept

this context, the activity partitions represent the consecutive steps which have to be processes in order to accomplish the task.

• The representation of the predefined activities and accordingly their partitions by means of ICOM- as well as UML-diagrams offers the ability to represent the work-flow within the addressed activity. Figure 2 shows that the work-flow of performing the Predefined Activity starting with Activity Partition1 and finishing with Activity Partition 5. It is noteworthy to mention, that the above mentioned utilized diagrams enable the representation of more complex work-flows than a simple sequence.

• Considering the management and reporting necessities in the early ship design process, the predefined activity concept is used to represent the two main activity categories:

1. *Definition/Analysis*: design tasks which have to be performed for definition and analysis purposes, such as Generate Hull Form, Calculate Stability, etc.

2. *Management/Reporting*: activities reflecting management and reporting tasks such as versioning, change management, reporting, etc.

6.1 PREDEFINED ACTIVITIES INTERACTION

Tasks within the ship design process and especially in the early design phase are interdependent to each other. The predefined activity concept offers the ability to analyse task in order to detect their dependencies and status. Early design tasks, which are represented by predefined activities, can be either dependent or independent. Two predefined activities can be considered as "dependent" if at least one of the ship information model objects is interacted with at least one of the activity partitions from any other predefined activity in an input, output or control role. As a result, the same information object can be interacted to represent different Activity Partition Items and accordingly to play different roles in different predefined activity partitions. Independent predefined activities can be performed parallel to each other. Two predefined activities can be considered to be "independent", if the activity partitions of one predefined activity are interacted with information objects completely different from those interacted with the activity partitions of the other one and vice versa.

Triggering any design task available as a formulated Predefined Activity at any time starts with a procedure to check for the availability of the required data. In case all data are available the task is performed with help of the integrated means (=tools). As a final activity a unique ConfigurationItem is defined and stored in the systems database. Depending on the work-flow configuration, users can be informed automatically about the new status of the design objects, which in turn might result in the need to trigger other design activities being dependent on the results of the former one.

6.2 CONFIGURATION CONCEPT

Configurations, realized through the configuration management model, represent an important feature in the enterprise modelling concept. They offer the ability to combine different information objects in order to represent a specific design concept while supporting the data-process as well as the data-data interaction, for an example see Figure 3. Activity Partition Items, which are information objects interacted by the Activity Partitions that are part of a Predefined Activity in order to perform a certain role (input, output, control or mechanism), can be related to each other, combined and attributed by a status. The status in this context can be used to assign a decision-dimension to these combined information objects. Employing the configuration-concept has several advantages regarding the ship early design process. A well structured combination of input, control, mechanism and output supports a rich set of additional functions, among them:

- to navigate dependencies between design tasks,
- to add a design status to a combined set of information objects,
- to provide additional information for data-reuse by providing an insight into the entire design task details and not just the previously achieved task results,
- to set up a change management by providing the possibility to detect the implications of a required change,
- to realize a version management by providing the ability to version the entire design task details.

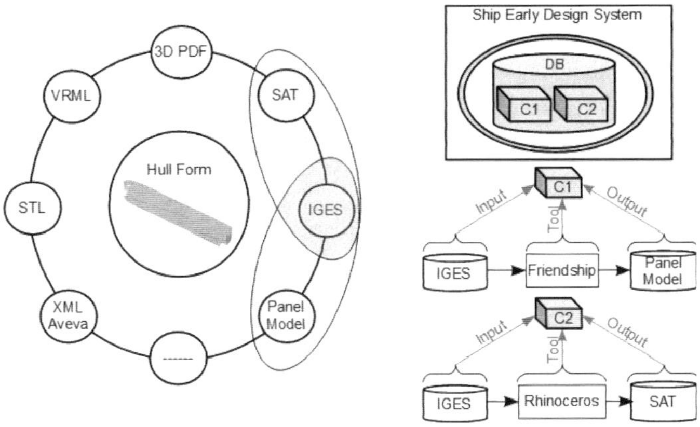

Figure 3: Multiple Hull Form Representations as Configurations

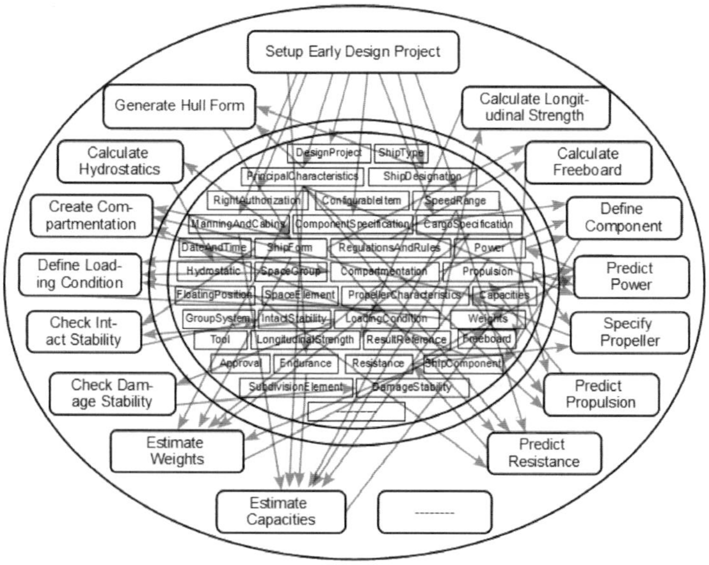

Figure 4: Predefined Activities for Ship Design

Configurations also represent the iterative nature of the ship early design process: multiple ConfigurationItems can be defined for the same design task while the respective design context is made persistent.

6.3 SHIP EARLY DESIGN PREDEFINED ACTIVITIES

Figure 4 shows major predefined activities for the ship design process of category Definition/Analysis. Management/Reporting Predefined Activities, not shown in Figure 4, represented in predefined form and intended to reflect/support management and reporting capabilities in the ship early design process are e.g. Create Design Solution, Update Status, Create Version, Check Change Influences, Request Change, Approve/Reject Requested Change and Create Report.

7. IMPLEMENTATION

In order to proof the enterprise modelling concept for the early ship design phase a prototype system is implemented and integrated with an existing design workbench. Predefined activities are implemented in form of Use-Cases. The ship information model is mapped to the database. This is achieved by using SQLite as a Relational Database Management System (RDBMS), Python as Object Oriented Programming language and SQLAlchemy for the Object Relational Mapping. Each group of information objects (package) is compiled into a Python-module within which the objects with their associations are defined. By mapping the information objects to the system's database, the attributes of these objects are defined as "optional". This offers the ability to exploit these objects with minimum data and complies with the requirement of low level data handling in the early design phase. Once the information model is configured within the database, data can be made persistent.

7.1. PREDEFINED ACTIVITIES AS USE-CASES

Early design tasks represented by predefined activities are implemented in form of Use-Cases which are modelled by means of UML-activity-diagrams. The partitions of these activities, which represent input, control, mechanism and output according to the ICOM-concept are procedures which have to be followed in order to perform the specific design task. In order to realize the mechanism-aspect which represents the applied tool(s) within the Use-cases, interfaces to several CAD and Analysis tools are implemented. The following considerations are taken into account:

- Each Use-case is activated by a specific person.
- The interaction between the design tasks, whether they are dependent or independent, is taken into account. The user is informed automatically in accordance to the available data at the time of activating a specific Use-Case whether the execution is possible or not due to missing data.
- The concurrency-concept within the developed system is limited due to the capabilities provided by the applied DBMS which supports a transaction concept of a single write access but an unlimited number of simultaneous read accesses.
- The results of the executed tasks are available as soon as they are saved within the system database.
- During the execution of the early design tasks, the date/time-data in addition to the user data are associated automatically to the corresponding information objects.

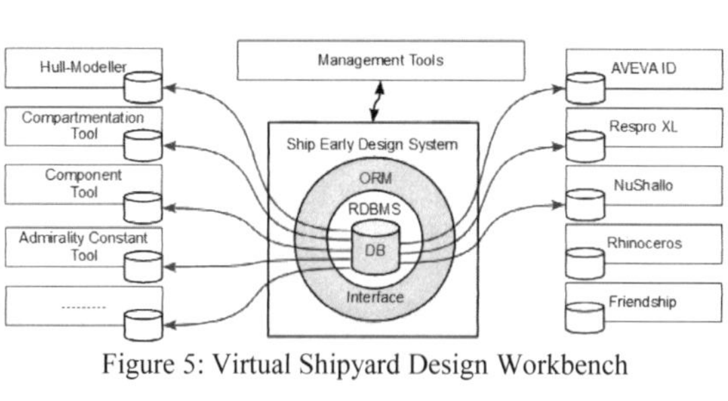

Figure 5: Virtual Shipyard Design Workbench

• The implementation realizes the overall design strategy that each task can be performed at any time. The purpose of the system in this context is to react depending on the availability of data. Moreover, the system is capable to automatically inform the defined responsible user(s) about the existance of requested data as soon as the data are available. Informing the right users substantially increases the efficiency of the design process by avoiding an information overflow.

8. PROOF OF CONCEPT

Some selected early design tasks to be performed for the design of a Ro-Ro vessel serve as an example to proof the above described design concept. For this a virtual shipyard with its available design tools is configured as depicted in Figure 5.

After initiating the system's database with first assumptions of the principal characteristics, a simple, statistics based power assumption is performed through the use-case "Predict-Power". This use-case offers multiple tools (mechanisms) to perform the prediction, ranging from simple algorithms to most advanced CFD systems. The result of this activity is, as for all activities, a configuration which combines the input and control data with the applied tool information (means) yielding to an output which is stored in the systems database. At the same time an URL is stored which points to the data specific to the applied analysis tool, in this case a simple Excel sheet, see Figure 6.

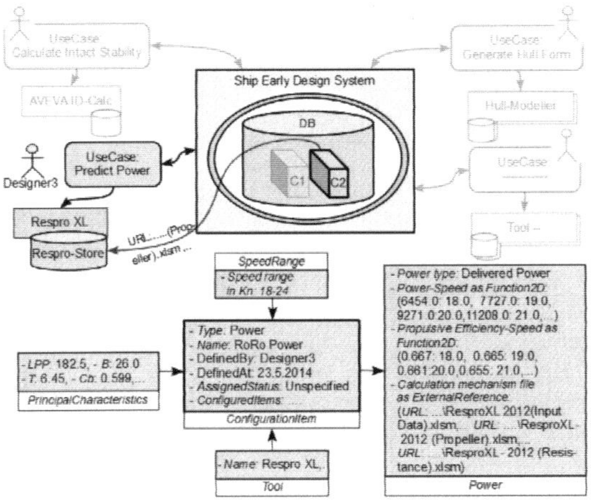

Figure 6: Configuration: Power Prediction based on Statistical Data

In a parallel process, a first hull form is generated by means of the corresponding use-case "Generate Hull Form" as depicted in Figure 7, which in this case is based on a hull form library and a transformation mechanism. Again, a configuration is stored in the systems database by which all information which yield to the specific hull form is made persistent for future use. The status assigned to this configuration representing the first hull form will be certainly "preliminary".

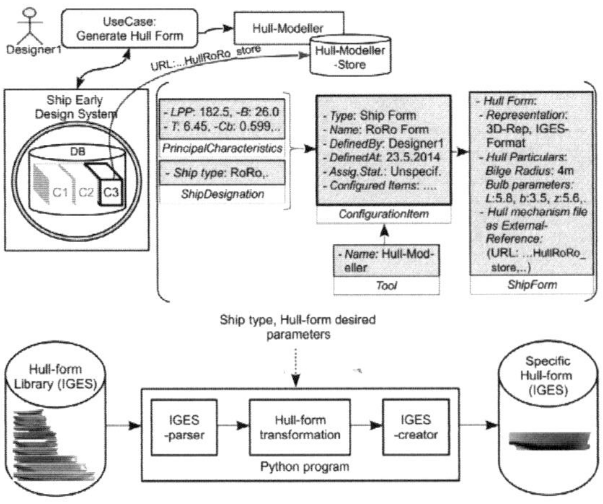

Figure 7: Configuration: Hull Form Modelling

Once a first hull form is available, it might serve as input to several further tasks performed simultaneously to detail the design, through Checking Intact Stability (conformance to IS-Code), Create Compartmentation, more elaborated resistance calculations or e.g. GA development, some results are shown in Figure 8.

As at this stage the hull form, as fundamental input to these tasks, is marked "preliminary", the results achieved, also stored as configurations, can only be of a preliminary nature. Once the hull form changed to a new version or finally frozen through a status set to "final", the system will identify all dependent activities informing defined responsible persons automatically. At any time a reporting activity can be initiated which shows the overall status of the ship design process through navigating all configurations and listing the corresponding status.

To combine a set of achieved design aspects for a specific design project, *Design Solutions* can be defined, which in turn inherit all management functions from the Con-

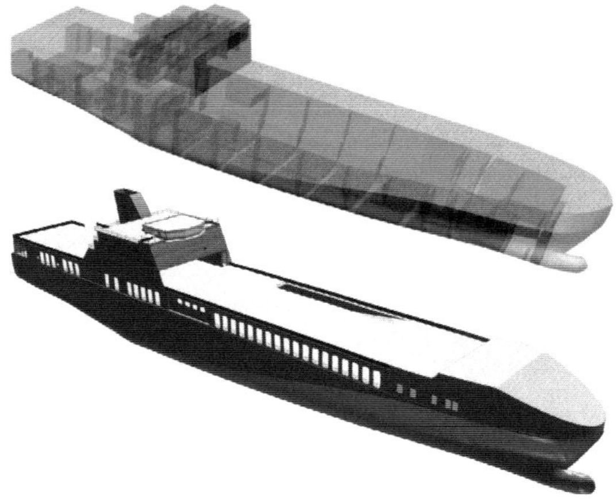

Figure 8: Compartmentation and 3D General Arrangement

figuration concept. Figure 9 depicts an intermediate state made persistent in the systems database in which for a defined hull form a compartmentation is performed and a certain resistance calculation as well conformance calculation with respect to the IS-Code are done. By this feature, a design variant can be retraced.

9. CONCLUSIONS

The efforts of the research presented are subjected to the development of an approach to enhance the early design phase. In order to achieve this goal, an analysis of the early design phase regarding its importance, complexity and its research-needs is discussed. Thereby, the aspects, which can be considered the key development aspects within the early phase such as the decision making, data-reuse, data exchange and design constraints, are introduced. Based on the definition of an enterprise, which is compatible with the domain of system design engineering, several concepts regarding the modelling as well as the management of an enterprise are introduced. An Information Model for the ship in the early design phase is introduced which represents the data perspective of the Ship Early Design Enterprise Model. Predefined Activities are developed to represent the Enterprise Process Model.

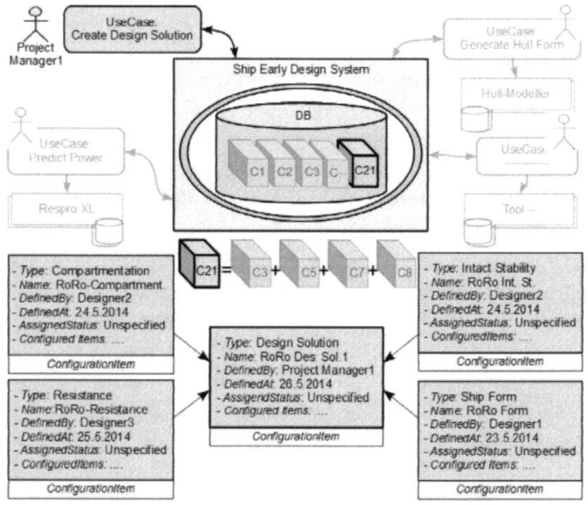

Figure 9: A Design Solution Combining Results from Several Activities

The prototype system supports the dynamic collaborative nature of the ship's early design phase. In order to evaluate its efficiency, it is integrated into a virtual shipyard design workbench. The applied tools are just examples of an assumed shipyard, they can be replaced by any other tools in order to take the diverse software-infrastructures to be found at shipyards into consideration. Moreover, the system accounts for the collaborative nature of the early design phase by the control of the interactions between different design tasks and accordingly between the responsible design team members through a work-flow management component.

10. ACKNOWLEDGEMENTS

The research presented is this paper was partly supported by the German Academic Exchange Service. The authors would like to also express their appreciation to the University of Rostock for the additional funding.

11. REFERENCES

1. T. Lamb et al., "Ship design and construction", Volume 1. *Society of Naval Architects and Marine Engineers*, 2004.
2. J. H. Evans, Basic design concepts. *Journal of the American Society for Naval Engineers*, 71(4): 671 #678, 1959.
3. D. Andrews. The Use Of Simulation In Preliminary Ship Design. *Proceedings of the International Conference on Computer Applications in Shipbuilding, Busan,* Korea, 2005.
4. E. Politecnica et al., Integrating project management, concurrent engineering, and engineering design to improve ship design. *In Third International Conference on Production,* Research, 2006.
5. F. Mistree et al., Decision-based design: a contemporary paradigm for ship design. *Transactions, Society of Naval Architects and Marine Engineers,* 98:565#597, 1990.
6. ISSC, Design Methods, Report of Committee IV.2. *Proceedings of the International Ship and Offshore Structures Congress,* Germany, 2012.
7. H. Nowacki. Five decades of computer-aided ship design. *Computer-Aided Design*, 42(11):956#969, 2010.
8. D. Andrews. Creative ship design. *Transactions RINA*, 123:447#471, 1981
9. A. Papanikolaou et al., Integrated design and multi-objective optimization approach to ship design. *Proceedings of the International Conference on Computer Applications in Shipbuilding*, Italy, 2011.
10. K Levander. System Based Ship Design. Kompendium. *TMR 4110 Marine Design and Engineering, Basic Course*, NTNU. Trondheim, 2006.
11. D. Andrews et al., The true nature of ship concept design and what it means for the future development of CASD. *Proceedings of the International Conference on Computer and IT Applicatios in the Maritime Industries*, Cortona, 2013.
12. R. Pawling et al., Just how valid is the ship design spiral given the existence of cliffs and plateaux?, *Proceedings of the International Marine Design Conference,* Glasgow, volume 1, pages 317#338, 2012.

13. C.V. Malheiros, Ship 3d model and design database reuse for warship operation and maintenance purposes. *Proceedings of the International Conference on Computer and IT Applicatios in the Maritime Industries*, Budapest, 2009.

14. E. Moredo and M. Krikke, Improving the re-use of design data during tender phase. *Proceedings of the International Marine Design Conference*, Glasgow, 2012.

15. M. Toyoda and Y. Nakajima, Competitive conceptual structural design system using 3D-PLM technology. *Proceedings of the International Conference on Computer Applications in Shipbuilding*, South Korea, 2005.

16. A. Cebollero and J. Quiroga, Space management in ship early design. *Proceedings of the International Conference on Computer Applications in Shipbuilding*, Portsmouth, UK, 2007.

17. D. Andrews et al., The true nature of ship concept design and what it means for the future development of CASD, *Proceedings of the International Conference on Computer and IT Applicatios in the Maritime Industries*, Cortona, 2013.

18. B. Oers, *A Packing Approach For The Early Stage Design Of Service Vessels*. PhD thesis, Delft University of Technology, Netherland, 2011.

19. B. van Oers et al., A 3D packing approach for the early stage configuration design of ships. *Proceedings of the 9th International Conference on Computer and IT Applications in the Maritime Industries,* pages 367#381, 2010.

20. M. C. Parker et al., Intelligent ship arrangements: Reseeding scheme development and effectiveness. *Proceedings of the International Conference on Computer Applications in Shipbuilding,* Italy, 2011.

21. D. Andrews and R. Pawling, Research Into The Use Of Computer Aided Graphics In Preliminary Ship Design. *Proceedings of the International Conference on Computer Applications in Shipbuilding*, Portsmouth. UK, 2007.

22. A. Papanikolaou, Holistic ship design optimization, *Computer-Aided Design*, 42(11):1028#1044, 2010.

23. H. Nowacki, Developments in marine design methodology: roots, results and future trends. *Proceedings Tenth International Marine Design Conference*, Volume 1, 2009.

24. S. Harries et al., An integrated approach for simulation in the early ship design of a tanker. *Proceedings of the International Conference on Computer and IT Applicatios in the Maritime Industries,* 2011.

25. A. Cebollero and F. Zurdo, Integrated management of ship compartments. *Proceedings of the International Conference on Computer Applications in Shipbuilding*, China, 2009.

26. U. Eriksson et al., Space Management As An Enabling Technique For The Development Of Ship's Arrangements Based On A Common Reference Model. *Proceedings of the International Conference on Computer Applications in Shipbuilding*, China, 2009.

27. Nakamatsu et al., Total design optimization in initial design stage based on single 3D ship model. *Proceedings of the International Conferernce on Computer Applications in Shipbuilding*, Italy, 2011.

28. Systems Engineering Fundamentals. Supplementary Text Prepared by the Defence Acquisition University Press Fort Belvoir, Virginia, 2001.

29. M. Lankhorst. Enterprise architecture at work: Modelling, communication and analysis. Springer, 2013.

30. M. Gaspar et al., Handling complexity aspects in conceptual ship design. *Proceedings of the International Marine Design Conference*, Glasgow, Volume 1, 2012

11. AUTHORS BIOGRAPHY

Wisam Jabary holds the position of a Lecturer at the Tishreen University in Latakia. He received a PhD degree from the University of Rostock in 2015. His research interest are in ship design optimization including tool integration strategies.

Robert Bronsart is holding the Chair of Naval Architecture at the Department of Mechanical Engineering and Marine Technology at the University of Rostock. He has more than 35 years experience in R&D in ship design, production and operation, thereof ten years in industry. In research he is focusing on information and communication systems in collaboration networks, CAE-system integration and on knowledge based methods in ship design and operation.

ACOUSTIC SIMULATION FOR CABIN ON SHIP BY FINITE ELEMENT METHOD

Y. Wakisaka, Altair Engineering, Inc., Japan
Y. Ohtsuki, Tsuneishi Shipbuilding Co., Ltd., Japan

SUMMARY

This paper introduces new approach of acoustic analysis by Finite Element Method (FEM) [1] [2] for whole ship. Firstly, ship is divided into structural domain and fluid domain. Both domains are consisted of properties only from design phase, like material and geometry data. Structural domain is whole ship structure and structural vibration is propagated through this domain. In addition, main engine and diesel generators are also modelled to apply excitation forces. The excitation forces are estimated from the relation between rotational speed and guide force moment of the engines. On the other hand, fluid domain is air in cabin and structural vibration is propagated as sound pressure through this domain. Secondary, the ship model is used to calculate structural vibration and sound pressure. The sound pressures in cabin is compared with experimental results. In addition, this paper mentions to effect of ballast water for sound pressure result.

1. INTRODUCTION

For acoustic analysis which has ship as target, Statistical Energy Analysis (SEA) is known well as a method to estimate sound pressure [3]. But, SEA requires so many input data from many experiments with complete ship. This means that SEA cannot be applied for ship in design phase, because ship structure comes after design phase.

On the other hand, FEM requires only information from design phase, like material data, geometry data and guide force moment of engine, as input data. Then, this approach is expected as a method to estimate sound pressure for ship.

2. SIMULATION MODEL

2.1 SHIP STRUCTURE

Structural model is whole ship geometry on this simulation. As finite element model, the whole ship consists of 4 edges elements (quadratic elements) and bar elements. Quadratic elements are consisted of ship geometry and bar elements simulate stiffener. Fig. 1 shows the finite element model which is used in this paper. Averaged element size of quadratic element (averaged edge length of an element) is 800 mm. This size is known well as similar size to be consisted of ship model by finite elements. Material constants of all elements are shown in table 1.

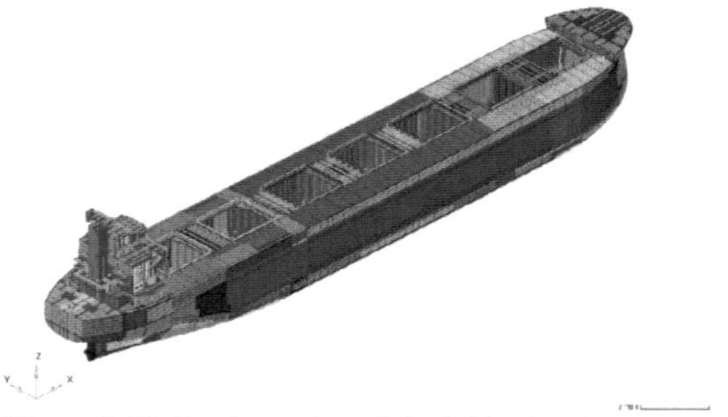

Figure 1: Finite element model of ship structure.

2.2 CABIN MODELING

Fig.2 shows a location of cabin. This simulation calculates sound pressure in cabin (not on structure). Then, air modelling is required by finite elements. Air in cabin consists of hexahedral element of finite elements. In addition, this simulation considers sound-absorbing effect by material of panel wall. Biot material law [4] is applied to panel wall to consider sound-absorbing in this simulation. Whole cabin model is shown in Fig. 3 and all material constants in Table 1.

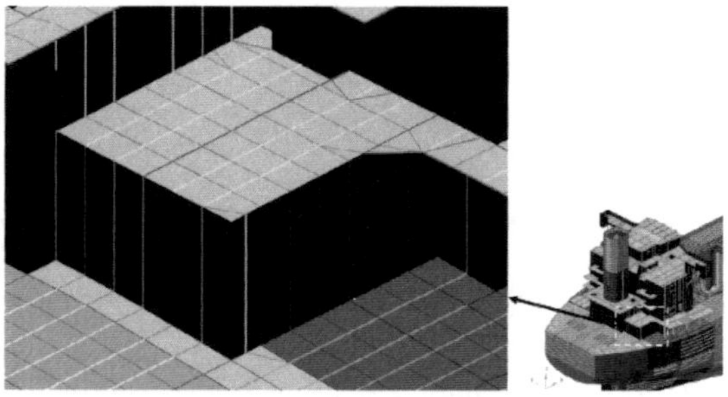

Figure 2: Location of cabin.

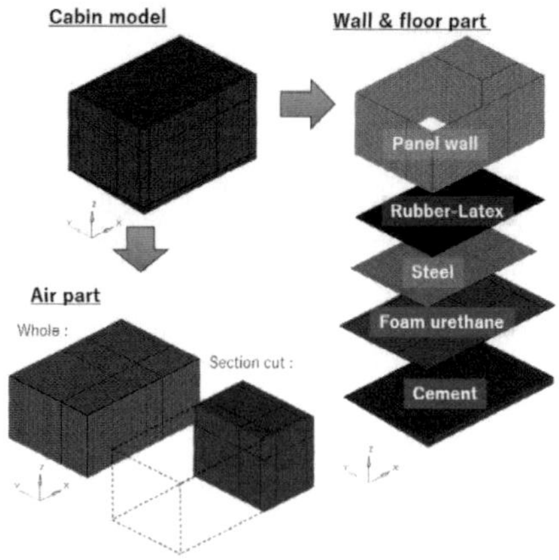

Figure 3: Finite element model of inside of cabin.

Table 1: Material constants for cabin

	Elastic material constants				
	E [MPa]	υ	ρ[ton/mm³]		
Rubber Latex	20000.0	0.3	8.5E-10		
Steel	206000.0	0.3	7.8E-9		
Foam urethane	2000.0	0.2	2.0E-10		
Cement	20000.0	0.2	2.6E-9		
Wall panel	10000.0	0.3	1.0E-9		
	Biot material constants				
Wall panel	Bi	υ	Γ	Pr	ϕ
	1.0	15.12	1.402	0.71	0.1
	τ	σ [Ns/mm⁴]	VL[mm]	TL[mm]	
	1.0	0.5	2.0E-4	2.0E-4	
	Fluid material constants				
	c [mm/s]	ρ [ton/mm³]			
Air	3.4E+5	1.2E-12			

For elastic material
E: Young's modulus
υ: Poisson's ratio
ρ: Density
For Biot material
Bi: Biot factor
υ: Dynamic viscosity
Γ: Fluid ratio of specific heats
Pr: Prandtl number
ϕ: Porosity
τ: Tortuosity
σ: Flow resistivity
VL: Viscous characteristic length
TL: Thermal characteristic length
For Fluid material
c: Sound speed
ρ: Density

2.3 COUPLED PROBLEM

Each part has different governing equations on no coupled problem. For each part, Equation (1) is for structural part and Equation (2) is for fluid part. In these equations, M_S, C_S, K_S and s_S are mass matrix, damping matrix, stiffness matrix and source vector respectively, of the structural part and M_F, C_F, K_F and s_F are mass matrix, damping matrix, stiffness matrix and source vector respectively, of the fluid part.

$$M_S\ddot{u}+C_S\dot{u}+K_Su=s_S \quad (1)$$

$$M_F\ddot{p}+C_F\dot{p}+K_Fp=s_F \quad (2)$$

In this simulation, coupled problem [5] between structural part and fluid part is solved in frequency domain. On this coupled problem, interface element is needed to transfer effects from one part to other part. From fluid part to structural part fluid pressure effect is transferred through interface element. On the other hand, from structural part to fluid part, acceleration effect is transferred through the interface elements. The transfer is written by Equation (3) and (4).

$$M_S\ddot{u}+C_S\dot{u}+K_Su-A_{\mathrm{int}}^T p=s_S \quad (3)$$

$$M_F\ddot{p}+C_F\dot{p}+K_Fp-A_{\mathrm{int}}\ddot{u}=s_F \quad (4)$$

Here, A_{int} represents interface matrix and makes coupling term of acceleration $\ddot{u}$, in fluid part, and of pressure p, in structural part.

3. ASSUMPTION OF EXCITATION

3.1 GUIDE FORCE MOMENT

In this simulation, 2 types of guide force moment (H and X) from main engine and electric generator is used as excitation load. The guide force moment can be gotten from catalogue of the engines. But, it is problems that guide force moments in the catalogue are discrete values at specific frequencies. This means structural and fluid vibration cause only at the specific frequency which guide force moments exist. In addition, guide force moment values exist only at low frequency range (below 19.2Hz). In other word, issues are that guide force moments from catalogue is discrete values and only data in low frequency range. To improve first problem, continuous values for guide force moments are assumed by linear extrapolation. And, it is needed to consider experiment data for main engine to improve second problem. Vibration velocity of the engines from experiment are shown in Fig. 4. The velocity shows inverse proportion. Then, in this paper, guide force moments as input data for simulation is assumed as inverse proportion. These ways are shown as Fig. 5. Assumption with inverse proportion is based on output values (vibration velocity) which follows input values (guide force moments).

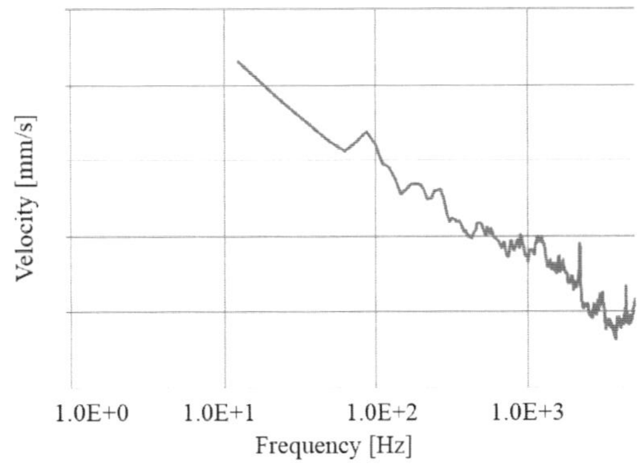

Figure 4: Velocity of main engine from experiment.

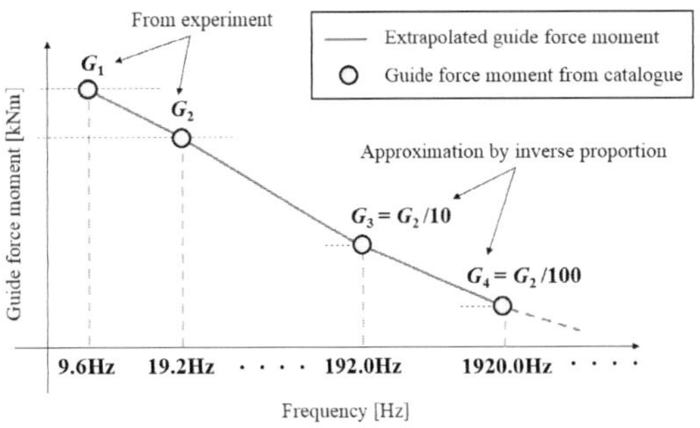

Figure 5: Guide force moment by linear extrapolation.

3.2 ENGINES MODELING

The ship model in this simulation has one main engine and one electric generator. Each engine is simplified by rigid body elements. Master (independent) node plays role of excitation node. Guide force moment is applied to the master node of rigid body element. On the other hand, slave nodes plays role of support of engines. Slave nodes connect ship structure (quadratic elements) directly.

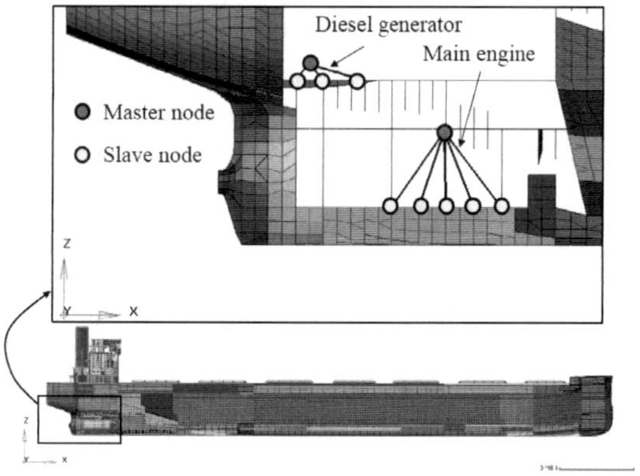

Figure 6: Simplification of engines.

4. ACOUSTIC SIMULATION

4.1 EVALUATION OF RESULT

To compare simulation and experiment, overall values of sound pressure level [dB] is used as metrics in this paper [6]. Sound pressure level and overall value are expressed by Equation (5) and (6) each.The target of frequency range is below 800Hz, because sound pressure over 800Hz has very low contribution for overall value.

$$SPL \text{ [dB]} = 20*\log_{10} (p / p_0) \quad (5)$$

Here, SPL is sound pressure level, p is sound pressure from experiment or simulation, p_0 is reference sound pressure of air and $p_0 = 20\mu\text{Pa}$.

$$O.A \text{ [dB]} = 10*\log_{10}\Sigma\, (10^{(SPL(f)\,/\,10)}) \quad (6)$$

Here, $O.A$ is overall value of sound pressure level, $SPL(f)$ is sound pressure level at frequency f.

Sound pressures p are calculated from simulation and measured in experiment at 2 evaluation points (located at 160cm and 10cm from floor of cabin) in both simulation and experiment each. Each overall values are calculated by Equation (5) and (6) with sound pressure.

4.2 COMPARISON WITH EXPERIMENT

Overall value of sound pressure level from Equation (6) is normalized when comparing simulation result with experiment result. For this normalization of the overall value of sound pressure level, the value at the location of 160cm from experiment is used as reference value. Then, the value at the location of 160cm from experiment becomes 1.0. The comparison summary for normalized overall value of sound pressure level between simulation result and experiment result is shown as Table 2.

Table 2: Overall value from experiment and simulation

	160 cm	10 cm
Experiment	1.000	1.047
Simulation	0.921	1.023

[dB/dB]

5. EFFECT OF BALLAST WATER

5.1 MODELING OF BALLAST WATER

In the experiment, ballast tank of the ship is fulfilled by water. Then, the effect of ballast water should be modelled for more accuracy.

Effect of ballast water can be modelled by various way in simulation. In this paper, ballast water model is consisted of virtual fluid mass. Virtual fluid mass mimics the mass effect of an incompressible inviscid fluid in contact with a structure and does not represent the actual mass of the fluid. This simulation assumes that ballast tank is fulfilled by water.

5.2 COMPARISON WITH EXPERIMENT

Comparison of overall value for sound pressure level between simulation (without ballast water and with ballast water) and experiment is shown as Table 3. The values in Table 3 are normalized by $O.A$ at the location of 160cm from experiment.

Table 3: Overall value from simulation and experiment

	160 cm	10 cm
Experiment	1.000	1.047
Simulation /w.o ballast water	0.921	1.023
Simulation /w ballast water	0.875	0.961

[dB/dB]

6. CONCLUSIONS

In the case of "without" ballast water, normalized overall value on simulation is smaller with 7.9% at the point of 160cm and is smaller with 2.4% at the point of 10cm than each experiment result.

By adding effect of ballast water, comparing with the case of "without" ballast water, 4.6% is decreased at the point of 160cm and 6.2% is decreased at the point of 10cm.

As one of reasons for the difference between experiment and simulation (simulation results are smaller than experiment results), it can be thought to consider effect from air conditioner. In this simulation, sound pressure from air conditioner is not considered. Generally, it is known that air conditioner causes increasing sound pressure level. By considering the effect from air conditioner to, this simulation can predict overall value of sound pressure level with the accuracy to experiment within 5%.

7. ACKNOWLEDGEMENTS

All experiment and simulation data in this work came from Tsuneishi Shipbuilding Co., Ltd and all simulation was solved by Altair OptiStruct.

8. REFERENCES

1. Bathe, K-J., 'Finite Element Procedures', *Prentice Hall*, 1996.

2. MacNeal, R. H. and Harder, R. L., 'A Proposed Standard Set of Problems to Test Finite Element Accuracy', *Finite Elements in Analysis and Design, 1, pp.3-20*, 1985.

3. Tso, Y. K. and Hansen, C. H., 'The Prediction of Structure-borne Noise Transmission in Ships Using Statistical Energy Analysis', *Acoustics Australia, vol. 25, No. 3, pp.97-102*, 1997.

4. Allard, J. F., 'Propagation of Sound in Porous Media', *Elsevier Applied Science*, 1993.

5. Nilsson, A. and Liu, B., 'Vibro-Acoustics, Volume 2', *Science Press Beijing*, 2016.

6. Kinsler, L. E., Frey, A. R., Coppens, A. B., Sanders, J. V., 'Fundamentals of Acoustics', *John Wiley & Sons, Inc.*, 2000.

9. AUTHORS BIOGRAPHY

Yohei Wakisaka holds the current position of technical manager at Altair Engineering, Inc. He engages in technical responsibility for shipbuilding companies

Yasuaki Ohtsuki holds the current position of a director at Tsuneishi Shipbuilding Co., Ltd. He engages in the management of structural design of ship hull.

DEVELOPMENT OF SHIP STRUCTURAL DESIGN AND ASSESSMENT SYSTEM BASED ON HARMONIZED CSR

Ho-Gyun Park, Korean Register, Republic of KOREA

SUMMARY

IACS (International Association of Classification Societies) CSR (Common Structural Rule) for Tankers and Bulk Carriers has been effective on July 1 2015. Harmonized CSR requires much more engineering efforts and time especially than their predecessors, making the user-friendly software be needed for shipbuilders and classification societies. Using CSR-H software, ship designers are able to design ship structure efficiently and accurately.

1. INTRODUCTION

The Korean Register (KR) – an IACS member classification society – has radically overhauled its ground-breaking software package, SeaTrust-HullScan, the ship design support solution that analyzes and appraises a vessel's hull structure. This will deliver efficiencies when applying the incoming Harmonized Common Structural Rules(CSR-H).

CSR-H requires a structural strength appraisal of cargo holds in the foremost and aftermost part of the ship as well as mid-ships. This requires up to five-times as much work. The new SeaTrust-HullScan incorporates additional functionality to support modeling through interfaces with the specialized 3D CAD systems (AVEVA Marine, Smart Marine 3D and others) which are widely used in most international and domestic yards. This radically reduces the time spent applying the requirements of CSR-H.

As shown in Fig. 1, the rule scouring mode for estimating structural members based on classification rules and the DSA mode for direct structural strength evaluation (Pre / Post CAE excluding structural solver) are widely used. Various calculations are made on the compliance of Korean Register with all ship types and compliance with ship design rules based on CSR-H, a common structural rule for oil tankers and bulk carriers. It can be performed based on the 3D geometry model.

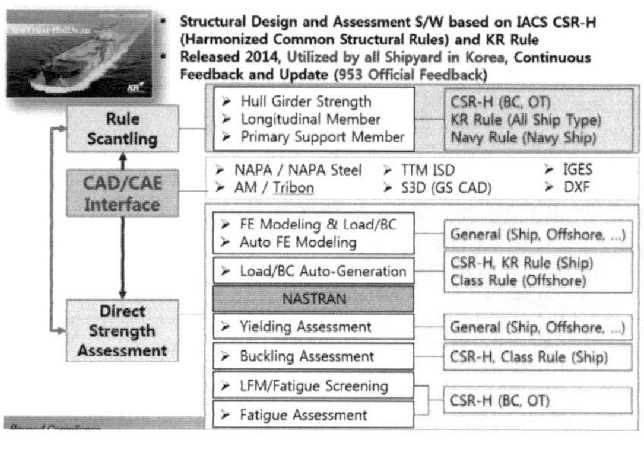

Fig. 1 Configuration of KR Structural Design and Assessement Software for Ship (SeaTrust-HullScan)

Further, Fig. 2, we extend the model paradigm of SeaTrust-HullScan from 2D geometry / 3D finite element model to 3D geometry / 3D finite element model to utilize 3D CAD model that has been created. We have developed the interface with the shipbuilding CAD system used in shipbuilding basic structure design department.

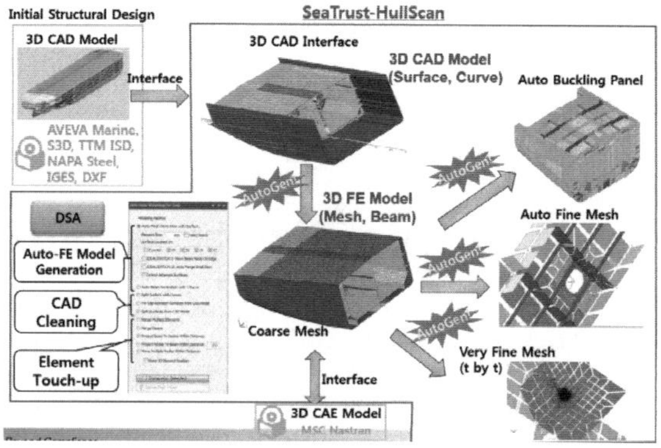

Fig. 2 Interface between CAD Model and FE Model

SeaTrust-HullScan does not need to create a new model, so user can import the 3D CAD model and automatically generate the hull cross-section model for the rule scantling in real time, reducing the number of times buried in modeling in the basic structure design stage with frequent design changes, focusing on the designer's engineering capability. In addition, the 3D CAD model has developed various functions to be automatically modeled as a finite element model. The generated coarse mesh model can be automatically modeled with a fine mesh, a very fine mesh, and the like in accordance with the purpose of design and structural analysis.
In this paper, we will introduce these functions in detail.

2. FEATURE

2.1 POLYGONAL ELEMENT AND CUTTER

There are many limitations in efficiently modeling such complex structures. Structural analysis models generally use quadrangular elements and only use triangular elements if they are unavoidable.

If user makes a complex geometry model as a single plate and then cut it into triangular or quadrangular elements required for structural analysis, user will be able to greatly reduce modeling time and increase its accuracy. To support this modeling method, we developed a polygonal plate element for modeling. Polygonal plate elements can be used to represent most of the complex geometry of ships and offshore plants. In addition, the cutter is provided as a modeling tool as shown in Fig. 3 to cut large elements into small elements

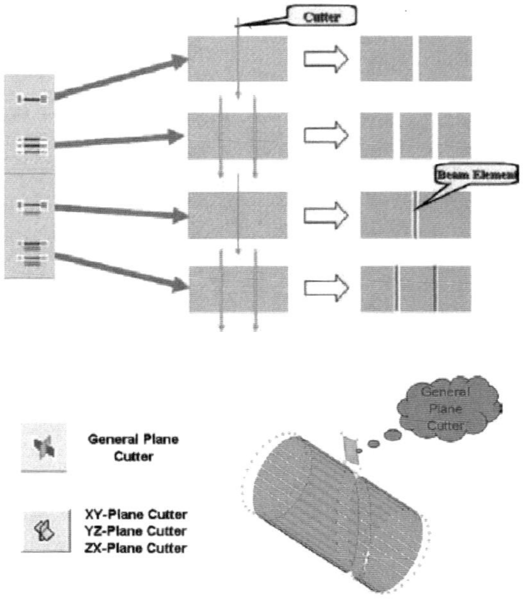

Fig. 3 Type of cutter tools

Using polygonal plate elements and cutter for modeling, user can easily and quickly create 3 or 4 elements needed for structural analysis. A line cutter can be cut a plate or beam element that is on the same plane among elements within a given length, and can automatically generate a beam element along the cutter as needed. The plane cutter can be cut all selected elements into given cutters, which can be used to quickly model corrugated bulkheads of bulk carriers.

2.2 CONVENIENT MODIFICATION

In the modeling for structural analysis, line element has a single node or triangular element has two nodes or quadrangular element has three nodes, which it's called Non-Compatibility Elements. Finding these non-conforming elements in the modeling phase is one of the most difficult tasks for experts, but SeaTrust-HullScan automatically removes the nonconforming elements that are created during modeling, preventing users from creating unintended elements.

If user want to modify the model shown in original model to target model of Fig. 4, user have to click the mouse more than 15 times in case of commercial software, but in case of this software, user can correct by clicking mouse three times.

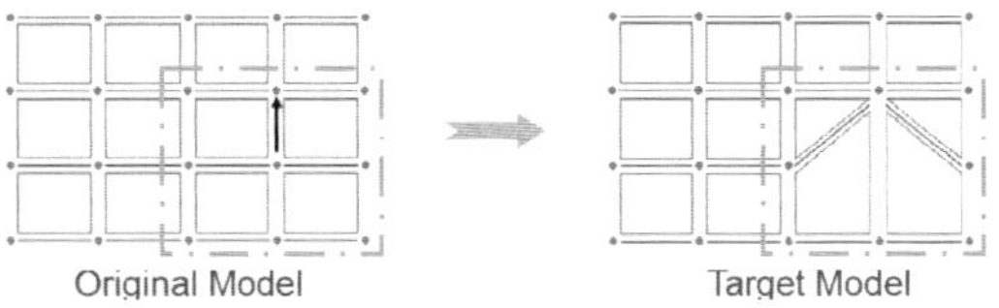

Fig. 4 Modification of Model

2.3 PROVIDE INTUITIVE INFORMATION

The hull structural analysis of bulk carriers and double hull oil tankers is to perform a strength assessment based on net thickness, excluding the corrosion addition thickness. In case of plate element, it is easy to input the dimension based on net properties, but in case of beam element, it is very complicated and time consuming to input the dimension based on net properties. In order to solve this problem, the member size is divided into the total thickness (Gross Thickness) and the corrosion additional thickness, and the data based on net thickness can be generated by using the input separately.

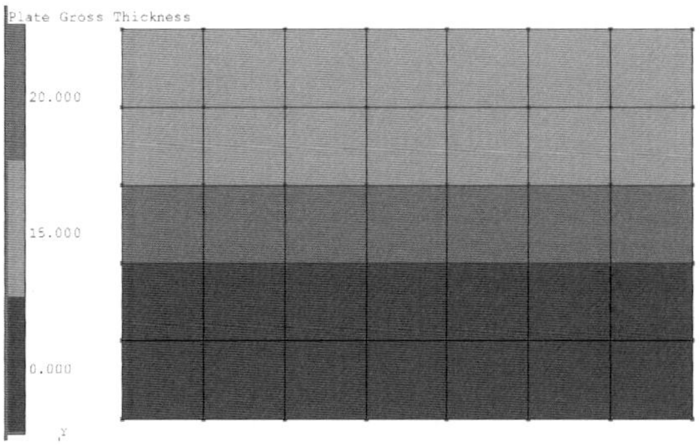

Fig. 5 Representation of plate element thickness

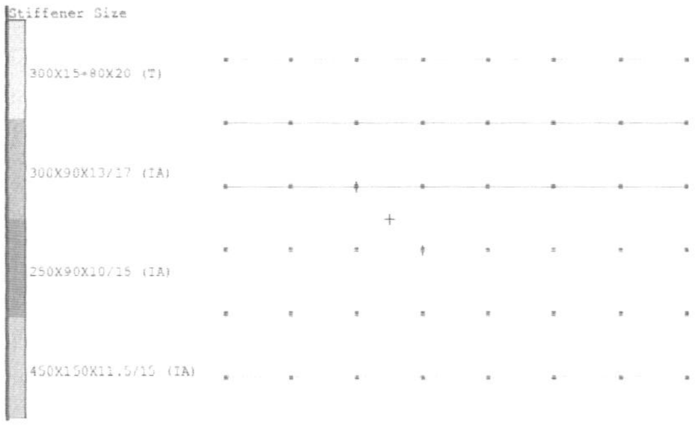

Fig. 6 Representation of stiffener element size

In addition, as shown in Fig. 5, the plate element shows the color that matches the entered thickness, so the user can easily check the input value. As shown in Fig. 6, the beam element supports input and output according to the member dimension notation shown in the drawings.

SeaTrust-HullScan has supported various methods to provide information on modeling and result. Further, it is also possible to use the tool tip function to find the node and element information and result directly without executing the corresponding command, Fig. 7 shows that the tool tip function is used to find the stress

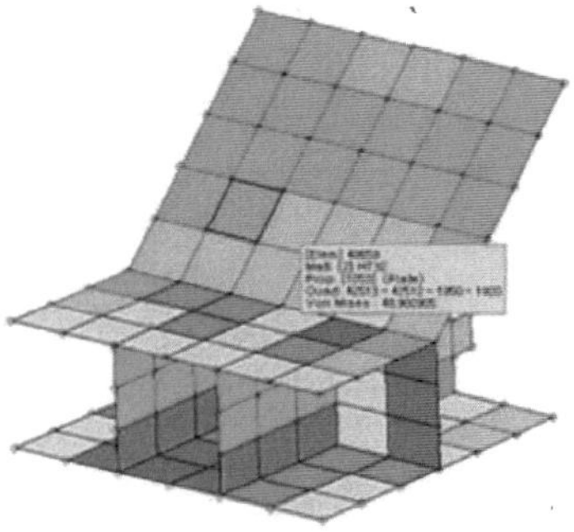

Fig. 7 Result using tooltips

2.4 AUTO MESH AND AUTO FINE MESH

Automatic element generation can be performed on surfaces basically like commercial software. However, it is difficult to show complex hull structural shapes because the surface must be composed of three or four variations. Therefore, in addition to automatic element creation for surface, it also provides automatic element creation function using closed curve chain to easily process the area around the opening. It also provides the ability to create automatic elements directly on the pre-created plate elements. The automatically generated elements are created around the quadrangle elements, and the triangle elements are created only when they are inevitable

After performing whole ship analysis or cargo hold analysis for structural analysis of ships and offshore plants, local structural analysis should be performed by making the fine elements of the weak structure or stressed area. This is referred to as detailed structural analysis, and these areas generally have dozens of locations. In addition, the analysis of fatigue strength shall model the plate element as a square element of plate thickness.

If user can automatically model these areas, user will be able to improve your structural analysis productivity. As shown in Fig. 8, the automatic modeling function of the part has been developed and the parts that require detailed modeling and fatigue strength modeling has been completed, such as the bracket toe, the opening, and the corrugated bulkhead.

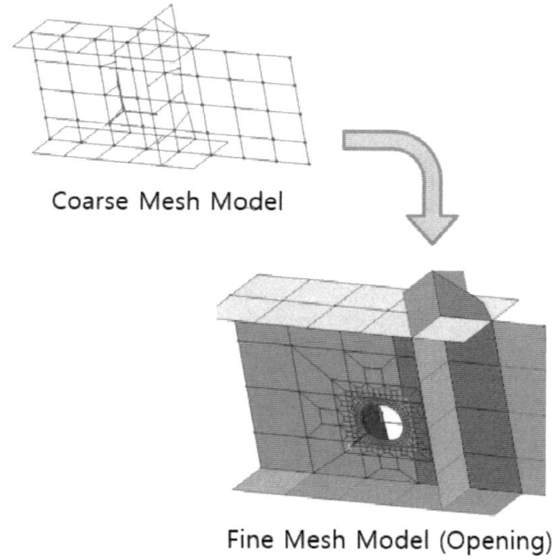

Fig. 8 Auto fine mesh

2.5 FUNCTION OF RESULT COMBINATION

The structural analysis of ships and offshore plants is performed on a number of loading conditions and the results are evaluated. In case of offshore plant, there are cases where there are hundreds of load conditions, and it takes a lot of time to evaluate these after carrying out structural analysis on these conditions. SeaTrust-HullScan provides various features that allow user to combine results to support strength assessment.

■ Maximum result extraction function : extracts only the largest result by element among multiple analysis results
■ Compare two results : ability to select and compare two results, making new results so that you can show the difference in percentage or value for two results
■ Strength of longitudinal strength combination with rules for structural analysis by local load
■ Combination function by customized formula : Using the formula entered by the user as in the below formula.
*Sqrt (R1 ^ 2 + R2 ^ 2) * sin (0.15 + atan2 (R1, R2))*

2.6 ASSESSMENT OF BUCKLING STRENGTH

The functions provided for the buckling strength evaluation include the function of automatically finding the panel for buckling strength review and related property values from the finite element model as shown in Fig. 9

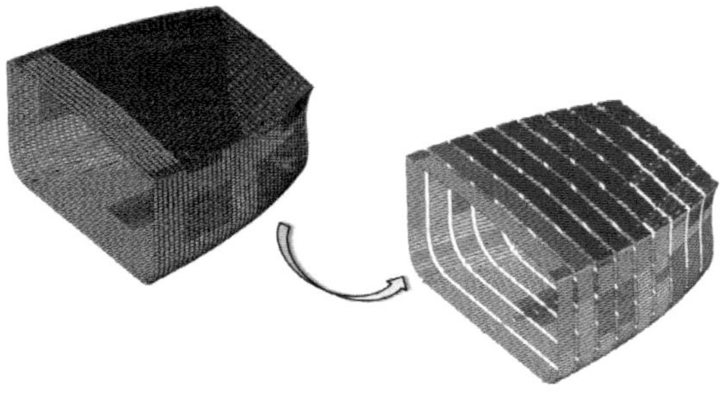

Fig. 9 Transfer FE model to buckling panel model

Using this function, the calculation for the buckling strength evaluation can be performed within one to two minutes on the Pentium IV computer for about 20 results of one cargo hold with about 30,000 elements. SeaTrust-HullScan can output the buckling strength review panel and its results from the finite element model. Due to the complicated buckling evaluation, the user has difficulty in finding the design parameters satisfying the calculation process, the buckling tendency according to design variables, and the requirements. The buckling advanced tool provides a variety of functions to solve this problem as shown in Fig. 10.

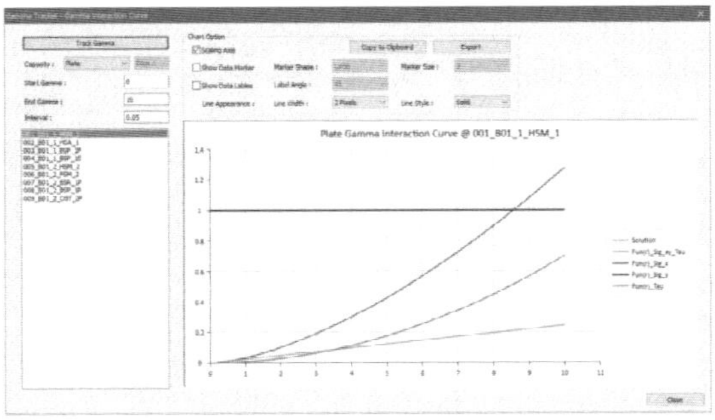

Fig. 10 Buckling Advanced Tools

It is possible to observe the buckling tendency according to the shape of the panel, to perform the simulation according to the change of the design parameter value, and to confirm the buckling change according to these results. It also provides the designer with the main design variables used in the evaluation and automatically finds the design variables corresponding to the required rule values, greatly improving the efficiency and accuracy of the analysis and shortening the design time.

2.7 SUB-MODELING

In CSR-H, the structural analysis time increased as the range of detailed structural analysis increased compared to existing CSR. Also, as the number of detailed structure meshes increases, the total number of meshes increases, which makes the program heavy and difficult to handle. So it takes time for load generation and post processing.

In order to solve these various problems, we developed a sub-modeling function. This function is used to analyze the entire structure with coarse mesh and then import local models to other files to automatically generate the displacement and local pressure from the entire model.

As shown in Fig. 11, the results are almost the same when the model is analyzed in the whole model and local models are analyzed using the sub modeling function. So, using sub modeling function, quick analysis is possible.

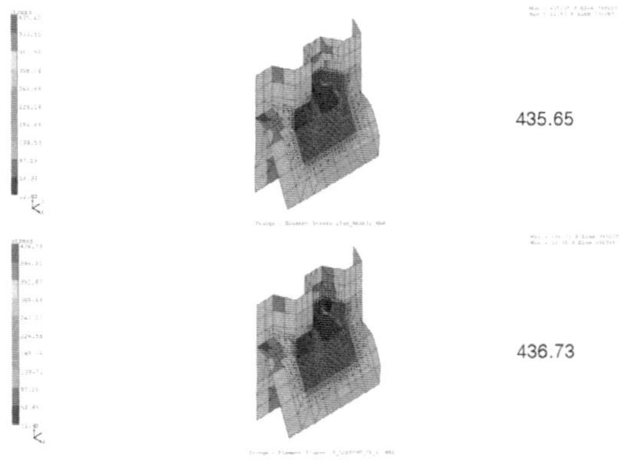

Fig. 11 Comparison of stress result from entire model and local model

2.8 AUTOMATIC REPORTING

It is a very simple and tedious task to perform the structural analysis of ships and offshore plants, then to extract the necessary data and results to report the analysis results and to submit the reports. However, such a structural analysis is generally not terminated at once, but reinterpretation is performed according to structural reinforcement or load change. In case of offshore plants, the structure may change over several tens of times, the structural analysis must be carried out again

This is making engineers very annoyed. Therefore, in order to make it easier to output the result image repeatedly to create a report, we provide the function to automatically generate the Microsoft Word file. It has the ability to generate approximately 1,000 pictures in about 10 minutes.

As shown in Fig 12, user can select the groups and analysis results required for output. If user select the coordinate system and the output item for each selected group, user can easily create the necessary material properties and analysis results.

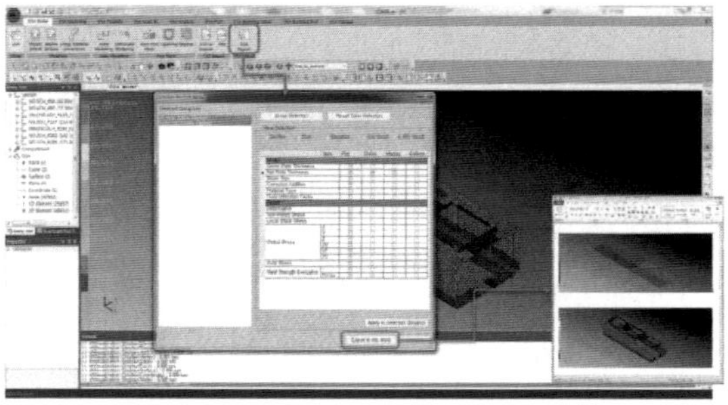

Fig. 12 DSA Reporting

3. CONCLUSIONS

We have briefly introduced the features and functions of SeaTrust-HullScan. This software was developed by Korean Register and supplied to shipbuilding and design company, and it was developed to reflect the improvement function of users who performed structural analysis of ship and offshore plant. It is proving its usefulness through collaboration.

The software is developed exclusively for shipbuilding and offshore plants and can be compared to commercial software for structural analysis modeling and post-processing. It can provide quick feedback of users and develop function of improving productivity.

The SeaTrust-HullScan, a specialization program for shipbuilding design of Korean Register, contributed to the innovative design and construction, the enhancement of user convenience, and quick and accurate results for domestic shipyards and design companies.

4. ACKNOWLEDGEMENTS

This research was performed as a part of the research project below and supported by the organizations indicated. We acknowledge and appreciate the support provided. "Development of a 3D Model-based Consistent System for Ship Design" project funded by Ministry of Trade, Industry and Energy of Korea (No. 10052708).

5. AUTHORS BIOGRAPHY

Ho-Gyun Park

- 1998, Bachelor in Naval Architecture and Ocean Engineering, Ulsan University, Korea.
- 2001, Master of Engineering in Naval Architecture and Ocean Engineering, Ulsan University, Korea.
- 2007 ~ Current, Senior Researcher, ICT Centre, Korean Register of shipping.

A MODEL BASED APPROVAL PROCESS FOR BASIC HULL DESIGN

O C Astrup, DNV GL, Norway and **C Cabos,** DNV GL, Germany

SUMMARY

This paper explores both the barriers to overcome and the requirements needed to establish a fully digital workflow providing a seamless and transparent digital model exchange between stakeholders during the ship newbuilding basic design stage. A 3D digital model established early in the design stage and shared among stakeholders will eliminate the need for producing 2D drawings for the verification of the design by Class. The digital 3D model is the basis for verifying the design against Class Rules and functional goals. A design centric work process involving the various stakeholders is the result of linking the Class rules directly to the Computer Aided Design (CAD) early design tools used by the designer.

1. INTRODUCTION

Traditionally, drawings are used for communicating designs in industry because they are the clearest way to tell a human what to make and how to make it. Technical drawings provide a means to communicate product complexity in a comprehensible and effective manner thanks to visual abstraction. In shipbuilding, it is still an explicit requirement that the designer/yard shall provide the Classification society with Class drawings documenting the design to be approved to the Society's Rules.

At the same time Computer Aided Design (CAD) models are now displacing technical drawings and documentation as the main product definition in several major industries. Within the last ten years or so, the engineering industry in automotive, aerospace and construction has gradually converted to using CAD models directly for communicating designs to manufacturers, builders, maintenance crews and regulators. This switch to creating the engineering record in models, however, presents problems not only for its long-term maintenance and accessibility – due in part to the rapid obsolescence of the hardware, software and file formats involved – but also for recording the evolution of designs, artefacts and products.

One of the major problems facing long-term users of engineering data is a lack of compatibility between software systems, specifically between competing systems and between different generations of the same system. This can partly be explained by market forces, and the consequent tactics used by vendors to encourage customer loyalty, and partly by genuine conceptual differences between systems. In consequence, data created using one software is in danger of becoming inaccessible even to its creators once that piece of software is retired or replaced as part of ongoing modernization.

2. INTEROPERABILITY OF ENGINEERING DATA IN THE SHIPBUILDING INDUSTRY

2.1 USE OF 3D IN THE LIFECYCLE OF A SHIP

Computer-Aided Design (CAD) Systems help engineers and designers in various industries, designing and building 3D models of airplanes, cars, bridges, digital cameras ... and of course, ships, submarines and floating structures.

Many shipyards, designers and at least the larger suppliers already use 3D CAD systems and PDM/PLM including various 3D-based simulation and engineering tools such as FEM or CFD. A large amount of knowledge about the ship is collected in this digital representation: Not only the geometry and details about the production process but also various aspects of the behaviour and the operation procedures. This intensive use of 3D technology in the engineering phase is an important enabler for increasing productivity and managing complexity.

2.2 USE OF 3D IN NEWBUILDING PLAN APPROVAL BY CLASS

The role of the Classification society during the new-building phase is to verify and approve that the design fulfils Class and Statutory requirements [1] (i.e. the Rules). Class carries out the verification through a technical review of the design plans and related documents for a new vessel to verify compliance with the applicable Rules. Current practices in classification is to base the verification job on traditional documents such as 2D General Arrangement drawings and other design information in paper/pdf formats made by the designer/yard. For the hull design verification, 2D drawings have been the single most important design document exchanged between yard/designer and Class.

The yard/design office will prepare 2D drawings as the basis for documenting the design. The society must manually build up their own digital verification model based on these documents. This process is time consuming and error prone. For every design revision, the society may need to do a re-verification adding time to an already inefficient process.

A case study [2] involving the exchange between NAPA Steel and the DNV GL hull calculation tool POSEIDON clearly demonstrated the potential of interoperability. The study concluded:

- Modelling activity for scantling check by the yard can be reduced by more than 50%, saving time for the yard in early design.

- The POSEIDON model can be used for prescriptive Rules Checks and for the complete finite-element buckling analysis based on global FEA results.

- Loads for global FEA can be generated based on input directly transferred from NAPA.

- Calculated scantlings are passed back to NAPA Steel for update by the designer.

From the same study, SHI claimed to achieve the following savings in time [3]:

- Prescriptive rule check calculation 50%
- Global FEM calculation 25%
- FEM calculation for hold analyses 40%
- Local fine mesh FEM check for details 60%

The above study shows the potential of improved interoperability through exchanging design models between yard and Class during the basic design stage.

2.3 THE INTEROPERABILITY CHALLENGE

The variety of engineering tools used to support design, procurement, manufacturing, and support of shipbuilding products has never been greater. From company to company, tools and processes range from manual capture in 2D drawings to sophisticated 3D models that are tightly integrated with other enterprise systems. The challenge is further compounded by the growing need to provide engineering information for support extending beyond the life span of individual applications. This heterogeneity has created both technical and business challenges. Data integrity across the applications and systems that author and consume engineering data is problematic. Point-to-point integration between systems is often so complex and costly that organizations opt for manual data re-entry when faced with program budget and schedule constraints. This leads to high coordination effort between design disciplines and consequently to expensive errors detected only later, e.g. in production planning.

3. CAN A MODEL-BASED DEFINITION REPLACE ENGINEERING DRAWINGS?

3.1 A DIGITAL WORKFLOW

CAD models have traditionally been used in the design, evaluation and manufacturing phases. Up until the turn of the millennium, engineering software was used to support a document-based workflow – CAD packages were used to create virtual models of designs, from which drawings and other design documentation could be produced. The manufacture or construction process was based on the resulting documentation. However, current digital environments allow a model-based flow of information between heterogeneous systems for Computer Aided Design (CAD), Computer Aided Engineering (CAE) and Computer Aided Manufacture (CAM) as well as Enterprise Resource Planning (ERP), Customer Relationship Management (CRM) and Supply Chain Management (SCM). Thus, there is an increasingly greater reliance on CAD models which are now being used as the method for recording product definition.

Traditionally, drawings are used for communication in industry because they are the clearest way to tell human what to make and how to make it. They are considered as a graphic universal language. The fundamental purpose of an engineering drawing is to carry, control and maintain a product's definition in a precise and clear way with no risk of misinterpretation or assumption. Technical drawings provide a means to communicate product complexity in a comprehensible and effective manner to humans thanks to visual abstraction.

Instead of shaping exchanged information towards human perception, digitalization calls for exchanging data sets directly consumable by algorithms, i.e. move from drawings to models with semantical mark-up (as opposed to graphical mark-up). This results in opportunities for automated processing but can lead to challenges for human interpretation.

Both the aviation and automotive industries are moving towards a drawing free product lifecycle. We can see several drivers in this development [4]:

- One master product definition
- Virtual prototyping and simulation
- Computer-aided manufacturing (CAM)
- Assembly automation
- Production of maintenance documentation
- Earlier coordination of design activities

The above list is not exhaustive, but some common denominators are reduced time to market, reduced rework, reduced cost and improved transparency for stakeholders throughout the product lifecycle.

Therefore, it is appropriate to ask the following questions: What are the main barriers that must be overcome to transfer all the different functions of engineering drawings to digital models within the product lifecycle? What are the legal requirements that must be fulfilled? What are the required model characteristics and what specific tools need to be put in place to move to a model based approval?

3.2 WHAT IS A MODEL?

A model is a representation or idealization of the structure, behaviour, operation, or other characteristics of a real-world system. A model is used to convey design information, simulate real world behaviour, or specify a process. ASME [5] and ISO [6] gives the following definition of the term model:

A model is a combination of design model, annotation and attributes that describes a product.

Engineers use models to convey product definition or otherwise define a product's form, fit and function, see Figure 1. According to the Model Based Enterprise [7], models can be applicable to a wide range of domains (systems, software, electronics, mechanics, human behaviour, logistics, and manufacturing). Models can be either computational or descriptive. Computational models are meant for computer interpretation and have a machine-readable format and syntax. Descriptive models are human interpretable and meant for human consumption (symbolic representation and presentation). Core to MBE is the integration of descriptive models with computational models.

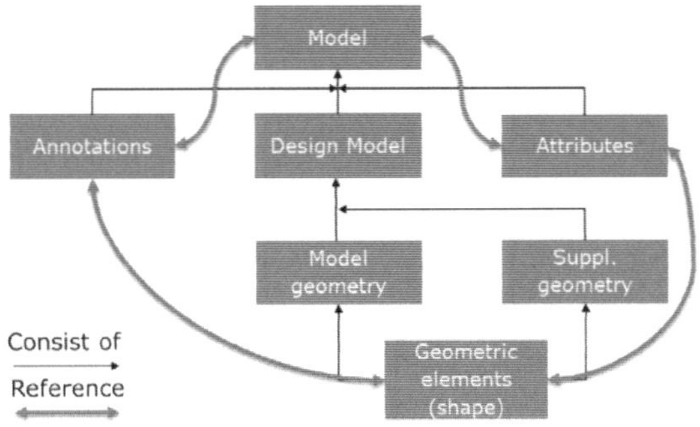

Figure 1: Model data elements, ASME [5]

3.3 MODEL BASED DEFINITION

According to [8] a model-based definition(MBD) is a digital representation (artefact) of an object or system. It is representative of the physical object or system and all its attributes, and is used to communicate information within various MBx activities in a model-based enterprise. The MBD is rich in information – shape, behaviour, and context – and it conveys the information architecture within an enterprise.

Based on a study from the aerospace industry, the following concerns of moving to a full MBD environment were identified, [9] [4]:

- Data accessibility and visualization: Most downstream users (including suppliers and customers) do not have access to CAD software, and therefore a visualization tool that allows them to read and use MBD datasets must be adopted;
- Data content: Downstream users need to be confident that the MBD datasets will carry the drawing's core and the required management elements;
- Data presentation: Engineering drawings follow international standards in terms of how the data is organized and structured and so must MBDs;
- Data management: An appropriate method must be put in place to manage and record revisions of MBD datasets;
- Data security: A mechanism that incorporates security features (confidentiality, authentication, integrity and non-repudiation) when accessing, exchanging and interacting with MBD datasets will be required; and
- Data retention: Today typically only 2D data is stored and used for legal purposes.

4. EXISTING INTEROPERABILITY STANDARDS FOR PRODUCT DATA REPRESENTATION

4.1 ISO AND STEP

ISO and STEP has led to improvements in exchange and sharing of simple CAD information, product models and complete product structures. This is particularly true for application protocols AP203, AP204 and AP214 [10], [11], [12] which are of wide use in the automotive and aviation industries. The recent ISO AP242 [11] consolidates the previous two AP204 and AP214 into one protocol for Managed model-based 3D engineering. Furthermore, STEP has improved communications within the extended enterprise (including suppliers, business partners and customers) and helped to support global collaborations. However, due to the bulk of its documentation and its complexity, there are still some issues that hinder the application of STEP in an industry with many specialized computer codes.

4.2 APPLICATION TO SHIPBUILDING

Considerable effort has been spent in developing standard protocols for product definitions both in the shipbuilding and other industries. Due to its broad scope, the application protocol AP218 is principally capable to support the need for data exchange during the early stages of the design in a consistent manner. Due to the generally limited support by the software vendors, the shipbuilding protocols are not used in actual commercial ship design projects today [13]

The lack of a ship specific standard is of major problem to the shipbuilding industry. This has left a void that needed to be filled and resulted in a continuation of the development of "point-to-point" solutions based on a variety of ad hoc application programming interfaces (API's) and XML-based interfaces. While these have been effective in terms of achieving the result, they have not been cost effective from an industry perspective as the number of interfaces grows at a combinatorial rate as the numbers of applications increase.

The problem of access to data across software and system revisions in the shipbuilding industry is therefore serious, and it is unlikely that a wholly satisfactory solution will be forthcoming soon.

5. A LIGHTWEIGHT FORMAT APPROACH TO MODEL BASED DEFINITION

The initial objective of a product model is to represent the design and engineering aspects of a product during the configuration design, assembly design or detail design stages.

As described in the previous section, interoperability of engineering data is still a major concern to the shipbuilding industry. CAD models are currently experiencing two conflicting demands. On the one hand, they must increasingly support the whole product lifecycle, not just the design stage. On the other hand, there is a need for small files that can easily be shared with partners whilst at the same time withholding Intellectual Property (IP) rights. Compared to full CAD model formats, lightweight geometry representations have some advantages, for example, they are typically generated with approximate geometry using various compression methods, to protect sensitive information. The small file sizes speed the communication of the models. Furthermore, lightweight geometry representations can be viewed without the aid of expensive CAD packages.

Patel [14] proposes the use of annotation to augment the geometric model of the product with additional information supporting the product lifecycle at the same time enabling sharing of 3D models between multiple stakeholders

5.1 MULTI-LAYERED ANNOTATIONS

Annotation can be simply defined as adding any extra information for various purposes, such as dimensioning, feature parameters, further explanations, viewpoint interpretation, extra descriptions of or comments on an existing entity.

Patel [14] identifies two methods of applying annotation or mark-up: 'inline' and 'stand-off'. Inline mark-up involves adding tags directly into the text of a document or representation of a model, whereas the stand-off (external or reference) method allows mark-up information to be stored separately, being linked back to the document or model using references or pointers. The latter is the more appropriate for use with CAD models for several reasons:

- Stand-off annotation allows the 3D geometric representation of a product to be progressively expanded to include additional metadata without changing the representation method used for the geometry of the product.
- Context-specific information can be extracted into several separate files to provide multi-layered mark-up, which can be passed around only as required, reducing the amount of data that need to be transferred and allowing the CAD model to remain smaller in size.
- The same mark-up can be applied to different representations of the same model, granting the mark-up information some independence of the CAD format used.
- Downstream processes (e.g. finite element analysis and manufacturing processes) become capable of independence from the full CAD model.

The ability to transfer information layers between different representations of an object makes this approach particularly valuable. E.g. computational dimensioning results can be transferred back to a CAD model, FE stresses can be used to highlight critical spots in inspection models, etc. [15]

Ding et.al, [16] presents the following advantages of multi-layered annotations:

- Mark-up information is stored outside of CAD models by multiple XML-based documents per different viewpoints and security levels. The advantages of XML include the possibility of manipulation of the mark-up information into different viewpoints, the re-organization of the information for various purposes and applications, and the tailoring of shared information per recipient needs and IP considerations etc.

- It enables partners that only receive lightweight representations of the product to apply mark-up and such mark-up can be linked directly to the original CAD model.

5.2 CHALLENGES USING MULTI-LAYERED ANNOTATIONS

The design process involves complicated interactions among tasks and people. Thus, it is presenting unavoidable inclusion of iterations and rework, which results in the constant insertion, elimination and modification of annotations (e.g. constraints) through the whole process. Such evolutions not only reflect the design process, but also possibly embed the rationale behind the decisions made. Meanwhile, the design process is also accompanied by the evolution of the product model, and therefore there is a high risk of loss of annotations, if the corresponding entity associated with the annotations disappears. To address these issues, a strategy of management and preservation of annotation evolution should be implemented, including:

- Evolution of metadata: The evolution of annotations can be retraced if and only if evolving histories are clearly recorded in an organized way. Normally, the modifications of the associated annotations can happen at two levels: model-level or element-level.

- Reference point: A core limitation identified by previous research [16] is that the annotation may not be retraced when the CAD model changes, e.g. if the entity to which it is anchored is deleted.

While the use of a lightweight geometry model addresses many of the challenges of PLM, there remains a problem in selecting a format which is the most appropriate not only for a use and view of the product, but also for long-term retention.

6. ENABLING A MODEL BASED APPROVAL PROCESS FOR HULL BASIC DESIGN

The key to achieve interoperability across software applications is open standards, i.e., those developed by consensus either within a standards development organization or a consortium of stakeholders. No single software tool can perform all the engineering tasks needed to design and manufacture a product. No single software product can do it all well. Users will mix and match software products matching their business objectives. Standards define an agreed upon syntax and semantics of 3D modelling constructs and annotation so that users can understand one another's models and design intents.

A model based definition provides the core foundation of the product lifecycle. It is one of the first pieces of information that is created during product development. It then grows as the product matures evolving it into the single authoritative source of product definition.

Because of obvious IPR considerations, the full MBD created by the designer will not be shared with other industry stakeholders. This is why the approach of lightweight models with multi-layered annotations is proposed [15]. The information passed to downstream users shall be delivered in a format that is CAD neutral and only contains those elements required for the intended purpose.

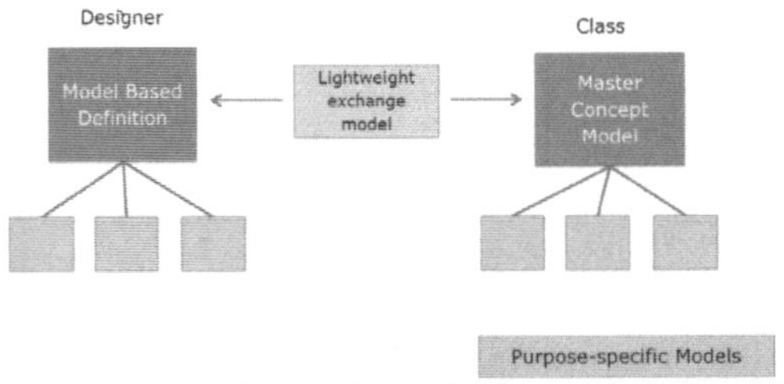

Figure 2: Model Based Definition and lightweight exchange.

The Digital Exchange Specification (DEX) which specifies the lightweight exchange format is instrumental to achieve the ambition of model based approval by the Classification society, see Figure 2.

6.1 THE DIGITAL EXCHANGE SPECIFICATION

The lack of a ship specific standard is of major problem to the shipbuilding industry. The problem of access to data across software and system revisions is therefore serious, and it is unlikely that a wholly satisfactory solution will be forthcoming soon. The authors therefore propose to develop a DEX supporting the necessary exchange capabilities between designer/yard and Class, as there is currently no single working and established standard for the exchange of engineering data within the shipbuilding community for this purpose.

Based on research carried [17] out we want the DEX to have the following capabilities:

- A lightweight and CAD neutral representation of the 3D geometry.
- The ability to carry both precise and approximated geometry representations depending on the intended use.
- The ability to grow the product definition information throughout the design lifecycle.
- The ability to describe function and process related information.
- The ability to reference function and process information to the 3D model using multi-layered annotations.
- A referencing mechanism which is robust to 3D model design changes.

- An easily extendable scheme to cater for new function and process related data.

These design assumptions lead us to logically structure the DEX content into three groups, see Figure 3.

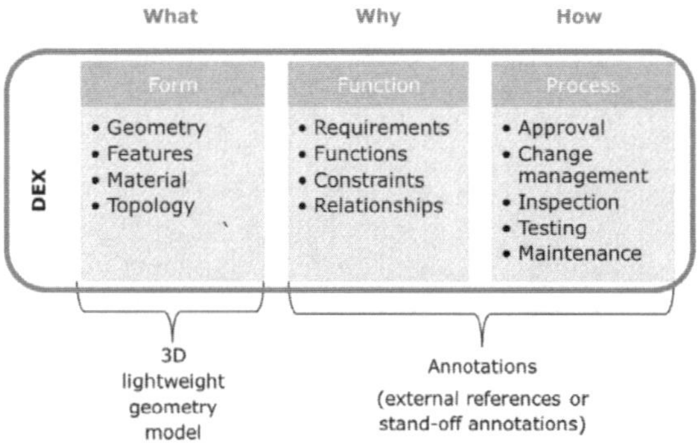

Figure 3: DEX information structure

What: The vessel hull form described by its geometry, features, materials, topology, ...

Why: The functions of the vessel represented by annotation layers describing requirements, constraints, functions or relationships.

How: Process and business related data represented by layered annotations describing process information for e.g. approval, change management, inspection, testing etc.

In this way, it is possible to separate the digital product form definition (the ship itself) from process and life cycle specific data (annotations).

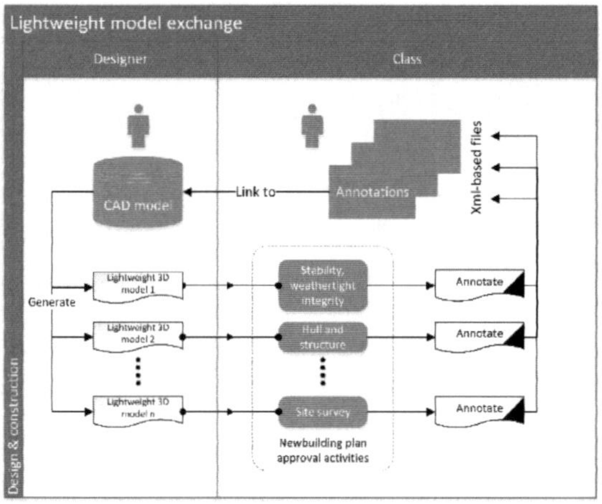

Figure 4: Envisioning model based approval utilising lightweight model exchange and layered annotations.

We envisage that this separation will allow us to establish a "conversation" around a 3D model using layered annotations to resolve any issues during the approval process. Figure 4 above illustrates how a designer and an approval engineer can use this mechanism to resolve comments during the plan approval.

6.2 DEX: ANNOTATION REFERENCE MECHANISM

It will be beneficial to separate function and process specific metadata from the 3D form data. This will be achieved by introducing the concept of the layered annotations described earlier. Each annotation will have a reference to the 3D geometry form. Depending on the annotation type, it can reference the whole geometry or a specific geometry part. As stated earlier, the challenge is to construct a reference mechanism which is robust to design changes.

For annotations referencing a geometry entity in the 3D model, the authors propose to create a reference mechanism which consists of two elements:

- Reference to a unique geometry entity ID in the 3D model
- A reference to an object type and a coordinate position, area or volume in the ship coordinate system

This will introduce a resilient mechanism which can be resolved when the model changes and the entity which is referenced is deleted. The annotation can still be displayed and actions taken to resolve it if necessary. The first mechanism requires authoring systems to keep IDs stable in case of model changes.

As features and topology also need to reference the geometry, they are ideally based on the same referencing mechanisms.

Having a rich toolset for viewing and mark-up of 3D models will be important for preserving efficiency when moving from working with a 2D drawings to a fully 3D based approval process.

6.3 DEX: FORM

There is no single standard covering the complete information scope. ISO and STEP protocols exist for the exchange of 3D data (the 3D Form). These standards can represent both exact and approximate geometries, but have limitations with respect to representing global geometry topologies. Adding to the picture the complexity of a STEP implementation and the fact that current standards do not support the shipbuilding industry, there is a need to define a lightweight schema for the ship form.

AVEVA and Intergraph [18] neutral XML formats meets our requirements stated earlier, and to some extent complement each other. The neutral formats can represent exact B-REP geometries with global topology in native XML, and have mechanisms for referencing external geometry files. With external geometry references several formats can easily be supported (e.g. IGES, JT, STEP, PRC). Through such external referencing of standard geometry formats, the application of commercially available 3D review tools would be facilitated.

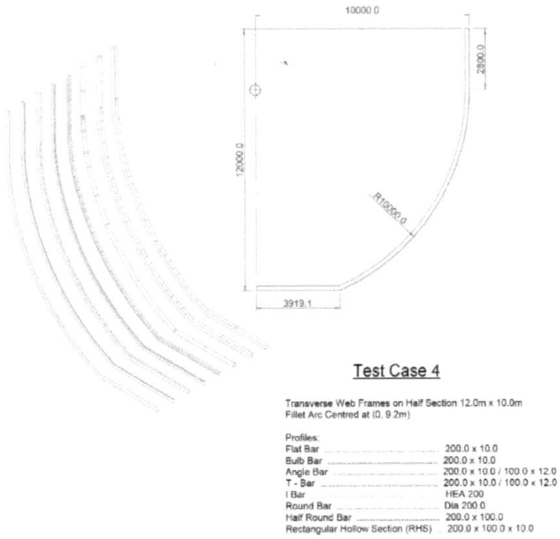

Figure 5: OpenHCM Test Case 4 [19]

Table 1: Comparison of file sizes for test case

STEP	AVEVA	JT	Intergraph
4 344 kB	63 kB	284 kB	120 kB

A simple investigation on file size was carried out using the OpenHCM [19] test case depicted in Figure 5. The investigation demonstrates the compactness of the vendor formats. The results are given in Table 1, and show that both the AVEVA and Intergraph neutral XML files including Siemens JT are compact in size. It is worthwhile to note the large size of the corresponding STEP file, which is almost 100 times the size of the smallest file size.

6.4 DEX: FUNCTION

There is no available standard supporting Class requirements and constraints (the functional part) as per today. There have been attempts to establish such a standard in the past [20], but nothing has materialised.

The functional and regulatory requirements (i.e. the Rules) by Class are today published in paper form, e.g. the DNV GL Rules for Classification: Ships [21]. Such rules must be interpreted by a designer before it can be applied to a new design. Such interpretations may lead to ambiguities which need to be resolved by the designer and Class during the plan approval process, adding time to an already constrained schedule.

What if the Class requirements could be presented in a precise and digital form through a DEX? We believe it is possible to expose Class requirements through a set of standardised Class catalogues. Such catalogues can cover different functional requirements. The following catalogues will be provided in the first version of the DEX:

- Vessel types
- Structure Functions
- End connections
- Cut outs
- Materials catalogue
- Cross section catalogue

Vessel Types: A high level characterization of vessels is fundamental to most technical and commercial topics within the maritime industry. Nevertheless, no universally adopted standard for this exists. Most organisations working in the maritime industry relate to one or more propriety defined lists of 'ship types'. This situation prevents an effective exchange of information. The vessel type catalogue is based on a high-level characterisation of vessels replacing the typical ambiguous vessel type lists currently in use. A unique vessel type is defined through the use of 5 basic variables [22]. The vessel type is the basis for selecting applicable rule sets and Class notations.

Structure Functions: Class structure Rules refer to structural elements of a ship as girders, frames, longitudinal bulkheads, plates and stiffeners. Most ship structural elements have a load bearing function with specific strength rule requirements and acceptance criteria. By classifying the different structure functions and link these to the ship 3D model parts, it will be possible to expose the Class requirements to the designer and automate hull rule calculations.

This is easiest explained by an example:

Minimum plate thickness requirements are defined in the DNV GL Rules for Classification: Ships, Part 3, Chapter 6, Section 3:

$$t = a + b\, L_2 \sqrt{k}$$

where:

a, b are coefficients defined by the plate type
L_2 is he Rule length
k is a material factor based on the material type

From the above, we see that the plate type defines the coefficient a and b. The Rules defines the different applicable structure functions for the plating, see Figure 6.

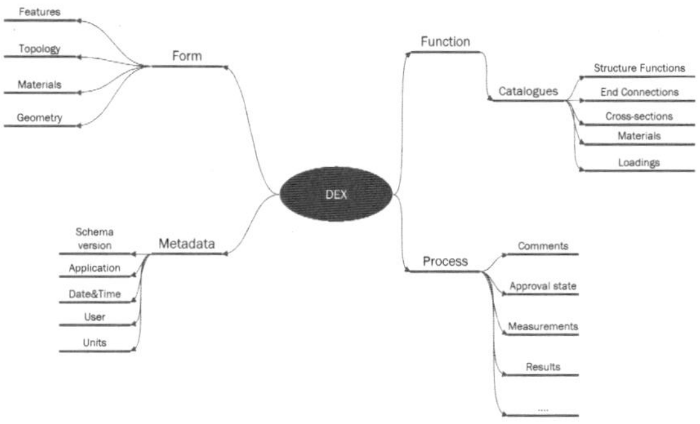

Figure 6: Taxonomy for Structure Function: Plating

The Structure Function catalogue provides the designer with the taxonomy for assigning the structure functions to the 3D design model. We believe that this assignment can be well supported by CAD systems.

When the structure function is assigned to a structure part in the 3D model, most of the minimum requirements can easily be determined. The more complex requirements like cross section ultimate strength (yield and buckling) will need more comprehensive information to be supplied to determine both the loadings and the relevant capacity models.

The same principles can be used to establish structure function catalogues for stiffeners and primary support members.

End connections and cut-outs: The Rules define a set of standardised end connections and cut outs for different ship details. These types will be provided in the form of a catalogue, and the designer must map the relevant standard detail to his own design. For most details, this is a onetime exercise as the designers already are using yard or company specific standard details.

Material catalogue: A catalogue defining the applicable material types defined by the Rules.

Cross Section catalogue: A catalogue defining the standard stiffener cross section types recognised by the society. This can easily be mapped to the yard/designer standards.

6.5 DEX: PROCESS

There are evolving standards targeting the process part, both ISO 10303-239 PLCS (Product Life Cycle Support) [23] and industry standards (e.g. PLM XML [24]). They will be explored to see how the standards, or parts of the standards, can support the plan approval process and change management.

Comments and comments management are central to the plan approval process carried out by Class. For hull plan approval, typically the approval engineer today issues a comment in the form of a mark-up directly on the 2D plans or drawing. In a fully model based approval process, the comments must be issued directly on the 3D model and transferred back to the designer to be resolved, see also Figure 4.

The comment annotations will use the referencing mechanism described previously. In addition, the comments must carry the metadata supporting the comment management process.

One important aspect when attaching a comment to a 3D model, is to also record the camera position and view direction as part of the comment metadata. In this way, it will be possible for each recipient to have the same view as the one that originally issued the comment. It can also be valuable to store the display state of the model as part of the comment annotation meta data, i.e. any model clipping, colour coding, transparencies etc.

Any additional mark-up data (i.e. redlining, measurements etc.) will also be carried on as comment metadata.

6.6 DEX: FORMAT

A specific format (syntax) of a DEX is less important than the actual information content (semantics). We have chosen to start with a native XML format. This has some obvious advantages when there is a need for quick prototyping and testing of new concepts. XML is also human readable and text based, and comes with standard toolsets supporting an implementation.

7. CONCLUSIONS

The authors have described a way forward for establishing a model based approval process for hull plan approval.

The proposed solution has several benefits:

- The industry will benefit from a single neutral exchange format for ship specific 3D shape and topology information avoiding costly point-to-point integrations.
- Engineering models can consistently be derived from one single 3D model source,
- CAD idealizations supporting several engineering analysis purposes can be covered: whole ship models to detailed fatigue models
- Multi-layered annotations are independent of 3D model exchange and CAD formats
- Annotations are loosely coupled to the correct entity in the model and can therefore be transferred between discipline specific model representations

- Annotations can be edited, circulated, and processed independently of the model, whilst the 3D model remains unchanged.
- Multiple independent annotation files can be safely applied to the same 3D model.
- Lightweight 3D shape representation protects owner's IPR
- Annotations are easily extendable and offer support for business specific needs

8. FURTHER WORK

Further research must be carried out to complete a DEX capable of fully supporting a model based approval process. This work needs to solve the following challenges:

Change management: Change management need to address both 3D model revisions and the state of any annotation referencing the 3D model.

Model maturity: Model or design maturity will need to be communicated between stakeholders. We want to explore the layered annotation principle to see if it can be used to express design maturity.

Approval scope: Class must document what has been approved. Today this is documented by the stamping of the 2D drawings. This mechanism needs to be transferred to the 3D model.

Long term storage and retrieval: Class needs to retain the approval documentation for the lifecycle of the vessel, which typically is 25 years. If the approval is based on a 3D model and not drawings, we need to make sure that we can still retrieve this information in a 25-year perspective.

Rule and strength calculations: A natural use of the DEX is as input to rule and strength calculations. This will facilitate recreating models from 2D drawings and fully leverage the potential demonstrated in the use case from section 2.4

The project participants aim to carry out a proof of concept of a fully model based approval when the DEX has been developed.

9. ACKNOWLEDGEMENTS

This work has been supported by the Norwegian Research Council project APPROVED (Project No. 256432) with participations from Rolls Royce Marine, ULSTEIN Design and Solution, STX France, AVEVA, DIGITREAD (Siemens) and Intergraph Corp. (Hexagon).

10. REFERENCES

1. IACS, 'CLASSIFICATION SOCIETIES - their key role', 2011. [Online]. Available: http://www.iacs.org.uk/document/public/explained/CLASS_KEY_ROLE.pdf.

2. C. Cabos, '3D Ship Design from the Start - An Industry Case Study', in *COMPIT*, 2015, pp. 257–268.

3. T. Hulkkonen, B. S. Kang, C. Cabos, and B. Tietgen, 'PROCESS INNOVATION FOR HULL DESIGN AT SAMSUNG HEAVY INDUSTRIES', in *ICCAS*, 2015.

4. V. Quintana, L. Rivest, R. Pellerin, F. Venne, and F. Kheddouci, 'Will Model-based Definition replace engineering drawings throughout the product lifecycle? A global perspective from aerospace industry', *Comput. Ind.*, vol. 61, no. 5, pp. 497–508, 2010.

5. ASME/Y14.41, 'Digital Product Definition Data Practices', ASME International, 2012.

6. ISO, 'ISO 16792:2015 Technical product documentation', International Organization for Standardization, 2015.

7. MBE, 'Model Based Enterprise Overview', 2016. [Online]. Available: http://model-based-enterprise.org/model-based-enterprise-overview.html. [Accessed: 16-Nov-2016].

8. N. Hartman, 'Defining the Minimum Information Model for the Model-based Definition', in *Model-Based Enterprise Summit*, 2017.

9. AIA, 'Aerospace Industry Guidelines for Implementing Interoperability Standards for Engineering Data', 2014.

10. ISO, 'ISO 10303-1:1994 Industrial automation systems and integration -- Product data representation and exchange', 1994.

11. ISO, 'ISO 10303-242:2014 Managed model-based 3D engineering', 2014.

12. ISO, 'ISO 10303-204:2002 Mechanical design using boundary representation', ISO, 2002.

13. R. Bronsart, U. Cantow, W. Grafe, and T. Koch, 'A Collaborative Platform for Ship Design', in *ICCAS*, 2005.

14. M. Patel, A. Ball, and L. Ding, 'Strategies for the Curation of CAD Engineering Models', *Int. J. Digit. Curation*, vol. 4, no. 1, pp. 84–97, Jun. 2009.

15. C. Cabos, 'Towards a DEX Exchange Standard - A Proposed Approach'. DNV GL, Malmø, 2017.

16. L. Ding, D. Davies, and C. McMahon, 'Sharing Information Throughout a Product Lifecycle via Markup of Product Models', pp. 1–9, 2008.

17. O. C. Astrup, 'Report 2017-0413: A Model Based Definition for Shipbuilding', DNV GL, 2017.

18. M. Polini, 'A NEUTRAL XML SCHEMA FOR BASIC DESIGN STAGE INTERFACE TO CLASS SOCIETIES', in *ICCAS*, 2011, vol. 2.

19. OpenHCM, 'OpenHCM Standard Description'. [Online]. Available: https://openhcmstandard.github.io/. [Accessed: 06-Jul-2017].

20. U. Haenisch, J. and Langbecker, 'Exchange of Structural Design Data for Hull Cross Section Approval, EMSA Protocol', 1999.

21. DNV GL, 'Rules for Classification: Ships Part 3 Hull Chapter 1 General principles', July 2016., DNV GL, 2016.

22. V. Vindøy and K. Dalhaug, 'High level vessel characterization', in *Compit' 2016*, 2016, pp. 42–50.

23. ISO, 'ISO 10303-239:2005 Product life cycle support', iso, 2005.

24. Siemens, 'PLM XML: Siemens PLM Software', *White Paper*, 2011. [Online]. Available: https://www.plm.automation.siemens.com/en_us/products/open/plmxml/. [Accessed: 17-Nov-2016].

11. AUTHORS BIOGRAPHY

O.C. Astrup joined DNV GL in 1991 (former DNV) and currently holds the position as Senior Principal Researcher in DNV GL Group Technology and Research. He is currently working on research activities within digitalization of shipping to improve safety and efficiency with focus on dynamic barrier management and model based approval.

C. Cabos joined DNV GL in 1989 (former GL). He is responsible for Information Management Technologies in Maritime Class Development. This includes the development of new computer-based methods in the fields of maintenance, data exchange, PLM and computational mechanics. His professional background is in numerical mathematics and physics.

3D DIGITAL CLASSIFICATION

Lionel LE GUENNIC, Cédric CHEYLAN, Thierry LE GAL, NAVAL Group®, FR
Christophe CHAUVIERE, Olivier DEGRAND, Bureau Veritas®, FR
Alexandre TEW KAI, Glenn DUTRIEUX, Sébastien VERET, Dassault Systèmes®, FR

SUMMARY

The shipbuilding industry is moving towards a new era where 2D drawings will disappear and be replaced by 3D digital mockup. The classification process includes a design approval which, today, is based on drawings review by the classification society in accordance with applicable class rules and standards. Shipyards and class societies will have to modify this process and enable direct 3D digital classification in order to improve the exchange of information between the different stakeholders and ultimately accelerate the classification process. Compared to traditional drawings approval, the advantages include: reducing shipyard workload with fewer drawings to create, improving quality and a common understanding of design and class comments by using the digital mockup directly, optimizing the calculation process by directly interfacing the 3D model with all calculation software such as structural and stability software. A proof of concept has been initiated by Naval Group, Bureau Veritas and Dassault Systèmes to validate that a new 3D-based digital classification process powered by the **3D**EXPERIENCE platform will benefit all the parties involved.

NOMENCLATURE

[Abbreviation]	[Definition]
VPM	VeriSTAR Project Management
BV	Bureau Veritas®
PLM	Product Lifecycle Management
CAD	Computer-Aided Design
R&D	Research and Development
FEM	Finite Element Models
POC	Proof of Concept

1. INTRODUCTION

The shipbuilding industry is moving towards a new era where digital data is present during all process phases (from early design to manufacturing) to reduce the time between the first contract signature and the delivery of the ship, to improve the management and the quality of the data, to increase return on investment.

All shipyards want to remove paper delivery (2D drawings) step by step to directly use digital information based on different part of PLM software. Digitalization will take a major place in the shipbuilding industry in few years and some methodology will be totally modified to analyze, produce and control the quality of the design, production and services.

Naval Group and Bureau Veritas will follow this new trend and they didn't wait the last year to deploy integrated solution based on PLM and CAD software but they want to take another dimension and to become a reference of digital company on their respective fields. This is why they choose the Dassault Systèmes solutions to assist them in this major transformation.

One of the topics analyzed together is to define a new classification process based on 3D Digital mockup to avoid producing 2D drawings.

The classification process includes a design approval which, today, is based on drawings review. Typically, shipyards submit their drawings which are approved by the classification society in accordance with applicable class rules and standards. The final design approved by the classification society is then marked by a set of approved "stamped" drawings.

Shipyards and class societies will have to modify this process and enable direct 3D digital classification in order to improve the exchange of information between the different stakeholders and ultimately accelerate the classification process. Compared to traditional drawings approval, the advantages include: reducing shipyard workload with fewer drawings to create, improving quality and enabling a common understanding of design and class comments by using the digital mockup directly, and optimizing the calculation process by directly interfacing the 3D model with all calculation software such as structural and stability software.

Naval Group, Bureau Veritas and Dassault Systèmes have initiated a proof of concept to validate that a new 3D-based digital classification process powered by Dassault Systèmes' **3D**EXPERIENCE® platform will benefit all the parties involved. In particular the following challenges have been identified: 3D model definition and exchange process, shipyard intellectual property protection, management of model revisions, creation and follow-up of comments in the 3D model, update of class process and access for field surveyors, as well as exchanges with ship owners.

You will find in this document some details about this study, beginning with a short presentation of the companies and the digitalization strategy including some explanations of the contextual use of this new process.

It will then describe the current classification process to allow understanding where we are currently and what are the pains of this methodology.

Then, an important part of this paper is the description of our methodology framework including the objective, the expected gains and how it fits with the digitization strategy of Bureau Veritas and NAVAL Group.

Before the conclusion, we present some illustrations of the theoretical explanations seen previously showing the digital process defined to respond to our problematic. The scope of our next experimentations (Proof of concept) is described clearly and it will give a better understanding about our strategy.

Finally, a conclusion will recap the most important points of this analysis and what we expect for the future.

2. COMPANIES PRESENTATION

2.1. NAVAL GROUP

NAVAL Group (formal DCNS) is the European leader in naval defense and a major player in marine renewable energies. As an international high-tech company, NAVAL Group uses its know-how, unique industrial resources and capacity to arrange innovative strategic partnerships to meet its clients' requirements. The Group designs, produces and supports submarines and surface ships (see Figure1). The Group also provides services for naval shipyards and bases. In addition, the Group offers a wide range of marine renewable energy solutions.

Figure 1: surface vessels designed and built by NAVAL Group

To enforce its leadership, the group decided two years ago to fundamentally transform its process and supported tools. Industry Solution Experiences from Dassault Systèmes were chosen for the future vessels program to replace legacy tools in the PLM and CAD areas and to achieve digital continuity from early designers to controllers, in mobility, in the shipyards (see Figure 2).

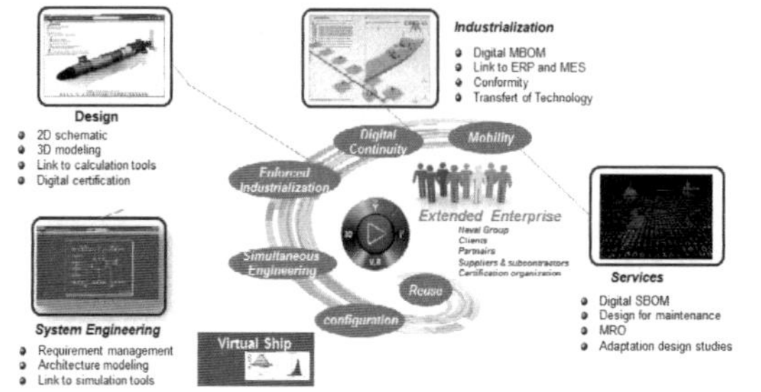

Figure 2: Naval Group **3D**EXPERIENCE® deployment project

2.2. BUREAU VERITAS

A global leader in testing, inspection and certification, Bureau Veritas serves clients' needs in quality, health, safety, environmental protection and social responsibility. For over 180 years, our clients across all industries have looked to us to provide technical support, verify compliance, or obtain certification. Our network of over 1,400 offices and laboratories ensures we meet their needs, wherever they are in the world

Historical partner to the marine industry, Bureau Veritas was founded in 1828 for the initial purpose of collecting, verifying and providing maritime insurance companies with precise and up-to-date information about the condition of ships and their equipment around the world. Our wide range of accreditations and authorizations (over 1,000) across a range of industry sectors prove that we operate professionally and that our reports and certifications are recognized and respected. We hold 150 delegations of authority on behalf of national maritime authorities.

Digital is transforming the marine and offshore industries, bringing opportunities - but also risks. At Bureau Veritas, we have put digital transformation at the heart of our growth strategy. This transformation focuses on two main axes. The first is innovating to develop truly added value tools for our clients that help them achieve their objectives: from higher performance and safety, to optimized costs, throughout the life cycle of their asset. The second is equipping our workforce with the tools they need to become more productive, and deliver better, faster service to our clients.

Today, over 2,600 experts in Marine & Offshore serve our clients' needs worldwide. We are driven by a single commitment: to keep our clients' crew and assets safe, every day.

2.3. DASSAULT SYSTEMES

Dassault Systèmes stands for the **3D**EXPERIENCE® Company and provides business and people with 3DEXPERIENCE universes to imagine sustainable innovations harmonizing product, nature and life.

Founded in 1981, Dassault Systèmes has been developing technologies and solutions that propel digital transformation in twelve industries and is participating in more than 50 global initiatives dedicated to advance world-class production technologies and processes.

Dassault Systèmes has had a long-standing leadership position expanding from 3D Design to Digital Mock-Up, to Product Lifecycle Management and now **3D**EXPERIENCE. Underpinning this market leadership has been the Company's clear and strong commitment to technological innovation.

Dassault Systèmes works closely with customers in different industries in order to develop software solutions that well address its clients' business goals, are highly suited to their industries, and designed to maximize user productivity and experience. The company provides a comprehensive portfolio of dedicated Industry Solution Experiences to assist Marine and Offshore customers to master the complexity of today's ships and a multisite ecosystem, to deliver their products on time and on budget, and to easily design, validate, and ensure manufacturability. This solution is decomposed into several Experiences built to fulfilled specific segment or customer challenges (see figure 3).

- *Optimized Production For Sea* enables shipyards to efficiently define the production plan of their ships to ensure optimized builds to deliver ships with high quality, on time, and on budget
- *Designed For Sea* enables customers to evaluate alternative "What If" design scenarios, simulate, validate, and detail the design
- *On Time To Sea* assists customers to master the complexity of today's marine projects and a multisite ecosystem to deliver their products on time and on budget
- *Winning Bid For Sea* enables customers to imagine, validate, promote, and sell marine/ship concepts in order to win new business

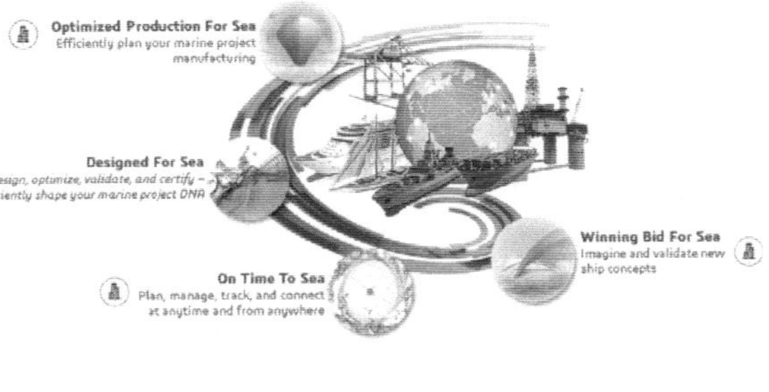

Figure 3: Dassault Systèmes Experience portfolio

3. CURRENT CLASSIFICATION PROCESS

For the follow-up of the full process, BV has developed a web-based collaborative platform called VeriSTAR Project Management (VPM) (see figure 4) where all actors can edit, review and exchange information on a real-time basis.

Figure 4: VeriSTAR Project Management welcome page

At the start of the process shipyards and designers upload the drawings in VPM generally in pdf format. The project manager in Bureau Veritas then assigns the drawings to the relevant design approval surveyor in order to review the drawings in accordance with applicable rules. It is to be noticed that for some disciplines like structure or stability, the BV surveyor has to make an independent calculation and so has to build a 3D model based on the drawings which is obviously the most time-consuming part of the design review process.
When the BV surveyor has finalized his/her review, he/she attaches his/her comments to each drawing in VPM. The shipyards can then access the comments and reply to each of these comments directly in VPM which ensures the traceability of the history for each comment. In addition to the comments, the BV surveyor also stamps the drawings to formalize his/her approval. For a better understanding of this comment, sometimes it is necessary to show the comments directly on the drawings. This is typically the case of structural comments in order to clearly identify the plates or stiffeners for which the comment is attached (see figure 5).

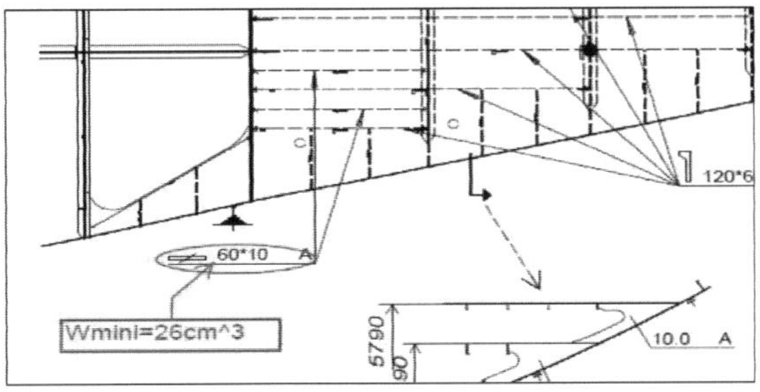

Figure 5: example in red of a structural comment

The shipyards have generally to update their design and then their drawings taking into account these comments and so they upload a new revision of the drawings in VPM and then the process starts again for a new loop.
It is important for BV to keep track of all approved revisions of drawings together with all attached

comments and this history is available in VPM. At the end of the process the final drawings are assumed to be without any more comment and the final design approved by BV is then marked by a set of approved "stamped" drawings. In parallel to this design approval process BV field surveyors who are based in the shipyard follow the construction of the unit and ensure that the construction is made in accordance with approved drawings. In particular they will check that all comments made during design review have indeed been taken into account at the building stage. It is then important that BV field surveyors can get easy access to the comments and this is also done by giving them access to the drawings in VPM.

One of the biggest shortcomings of this process is the fact that shipyards need to generate 2D drawings for classification purpose that are not necessary for their own building process. The generation of some of these drawings can be time consuming, like for example the shell expansion drawing. Another difficulty is to manage revisions of drawings that could have an offset, for example if the "Midship section" is updated to revision B and the "Profiles and decks" are kept at revision A. There could be discrepancies between these drawings that would create confusion during the plan approval. In addition, the understanding of the structure could sometimes be quite difficult in complex areas such as underdeck foundations of main equipment on the mooring area. 3D views would largely facilitate the understanding of both BV surveyors and shipyards when they receive the comments from BV. As previously mentioned, the generation of independent 3D calculation models by BV surveyor is also very time consuming. Indeed today BV surveyor has to create the models from scratch by using the drawings. As an example Finite Element Models (FEM) are necessary to check the structural strength of the large ships (see figure 6). A full ship FEM can take about two man-months to be generated. Exchange of 3D models that would be automatically transferred to calculation tools would largely reduce the time needed for independent calculations.

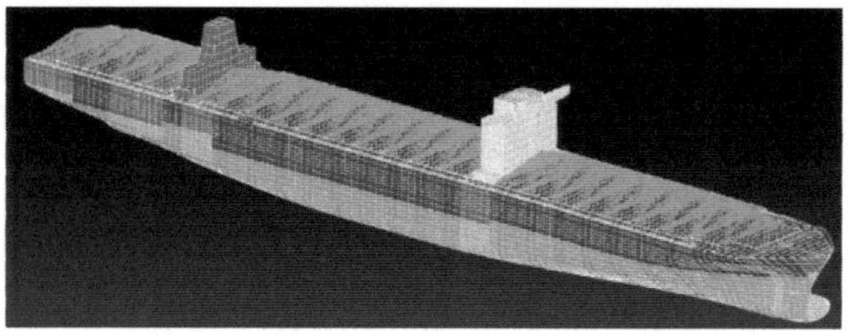

Figure 6: 3D model of a container ship

4. METHODOLOGICAL DIRECTION

Today, in main shipyards, the hull design process is centered on a CAD 3D mockup. This mockup is the design reference for the hull structure. This mockup can be directly linked to FEM software for calculations, it is shared with other disciplines to ensure a perfect consistency of the ship design whatever designs loops and updates are made by any discipline. At the end of the design process the 3D mockup is used to generate the hull production data: material list, production breakdown, drawings, cutting machine files. This 3D approach leads to a more accurate and more efficient design of the hull structure. As mentioned above, issuing 2D plans from this mockup and transmitting them to the classification society has several drawbacks. The main goal of the 3D Digital Classification project is to overcome these drawbacks and be able to exchange 3D mockups directly between shipyards and classification societies to improve the classification process, thanks to the Industry Solution Experiences based on the **3D**EXPERIENCE platform that are used today by both NAVAL Group and Bureau Veritas.

The main goal of this project is to become able to exchange directly 3D mockups between shipyards and classification societies to improve the classification process (see Figure7).

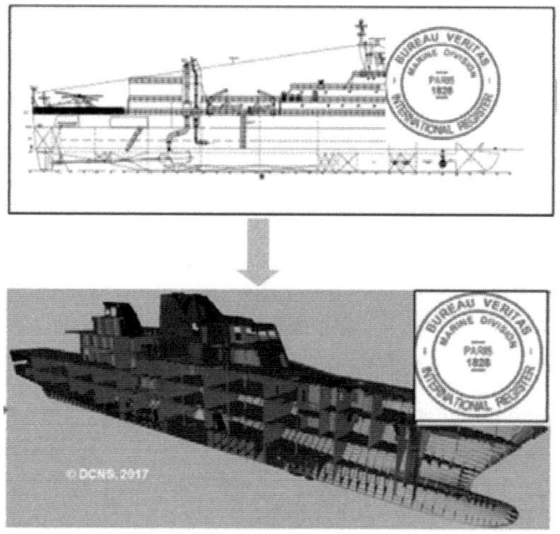

Figure 7: from 2D to 3D Classification

The 3D Digital classification process leads to several improvements for both parties:

- Reducing shipyard workload: the 3D mockup is already built for design purposes. When using 3D Digital classification very few drawings will have to be drawn specifically for the classification process.
- Improving quality: exchanging directly the 3D mockup will reduce the risk of inconsistency between several drawings and the risk of miscopying in the drawing creation process.
- Giving a common and full understanding of the 3D design to the class society instead of re-creating it from the different 2D drawings.
- Easing and speeding-up exchanges between shipyard and class-society by sharing the same understanding of the design.

- Improving comments tracking thanks to their generation and inclusion in the 3D mockup.
- Optimizing the calculation process by directly interfacing the 3D mockup with all class calculation software such as structural and stability software.

The planned project is aimed to illustrate concepts and solve bottle necks that have been raised like:

- Definition of the necessary and optimal mockup for classification exchanges: which type of model should be used, which type of entities may be used from the ` several types existing in the **3D**EXPERIENCE® platform, which level of details is the most useful to have accurate exchanges but not too heavy, which filtering rules may be used to create the exchange mockup, which splitting rules?
- Mock-up sharing: How to preserve confidentiality, how to take into account different confidentiality levels, how to preserve each party's Intellectual property?
- How to manage the different versions of the mock-up, how to highlight and follow modified areas, how to keep history of each version and linked comments?
- Creation and management of comments: How to create comments in the 3D mockup, how to follow them through different revisions, how to specify the comments to the shipyard?
- Change in the Bureau Veritas Process: How can local surveyors have access to comments, how to manage rules changes, how to adapt calculation process?
- Classification: How to formalize design approval. How can the shipyard attest design approval to the ship owner?

5. SOLUTION BACKGROUND

5.1 STARTING ASSUMPTIONS AND EXPECTATIONS

The solution will be first prototyped through a proof of concept that will showcase the full iterative process of approval between the design office and the classification society, replacing the drawings previously used as support of communication with a 3D digital model gathering the design data.

The scope of work is to define and implement the appropriate infrastructure and applications needed to cover the various steps of the process going from the creation of the digital model ready for approval to the exchange of this dataset between stakeholders, the review for classification approval and the communication back to the design office of the result of approval. The defined infrastructure should allow iterative loops of exchanges and enhance the traceability of the approval status. The goal of the solution will be to enhance the quality of the information exchanged thanks to the 3D design objects, propose greater reliability of information transferred to the approval bureau and a better understanding of the approval comments to take the right corrective actions of the design side. To facilitate the adoption of this new way of working the scenario of exchange will not differ much from the existing one and will rely on the existing approval interface proposed by Bureau Veritas.

The main starting infrastructure assumption is that this scenario will be firstly implemented on the basis of the Dassault Systèmes **3D**EXPERIENCE® platform and the Bureau Veritas VeriSTAR Project Management platform. On NAVAL Group side, the **3D**EXPERIENCE® platform will be used to prepare the data and exchange the data for approval, keep track of the approval state of the data, make the link with the design. On Bureau Veritas side, the **3D**EXPERIENCE® platform will be used to receive and approve the data coming from NAVAL Group, create the comments linked to this data, the comment will then be created and stored in the Bureau Veritas VeriSTAR Project Management platform that will still be the place to keep track of comments.

The principle of the study is to keep the classification process as close as possible to the existing one, based on referenced package exchanges and comparisons, but simplify the package preparation and the reviews leveraging 3D: less work for designers, easy understanding for the approver for reviews and for easier interpretation of comments by the designer

5.2 INFRASTRUCTURE

As the solution illustrated during the proof of concept phase will foreshadow the one that could be implemented quickly, it has to be compliant with the existing environments in place on both sides. The **3D**EXPERIENCE® platform will be the business platform deployed at NAVAL Group to design their ship structure on one side and is the one already in use for maintenance activities at Bureau Veritas on the other side.

Moreover, approbation processes relies on rich data, containing enough semantics coming from the design to perform comprehensive analysis, exchange capabilities and also flexible data model to manage approval status as close as possible to the design data.

The heart of Bureau Veritas approbation process is handled by their in-house database named VeriSTAR Project Management. Comments, discussions, external consultations, history of exchanges, drawings archives are all managed by these tools. Moreover this database is today the repository of already classified vessels. For these reasons, it should remain as part of the process with a close integration to the new solution.

6. CHALLENGES

To better understand the choices made for the solution that will be explained later in this document; let's first try to understand what challenges must be overcome to enhance the process of approval.

6.1 LARGE DATA EXCHANGES

One of the main challenges to address is to leverage the existing design data for approval process avoiding alternative data generation and specific preparation work but using comprehensive context and completeness of the 3D. Accordingly and due to the fact that NAVAL Group and Bureau Veritas requested to keep the formal classification process approach, minimizing the change for approvers, the proposal is based on iterative package exchanges between both companies with archives at both sides.

As the approval process can only be performed in the global ship context, almost all the structure data will be involved (only small foundations and features fixed on structure are part of the structure domain but are not taken into account during the approval processes).

The proposal is to extract approval data directly from design, extracting all necessary information for approval as data to be approved but also all other contextual rich semantic data. Consequently, the 3D Digital classification exchange packages contain a large number of features representing the whole structure of the ship instead of a single drawing.

Leveraging a 3D approach leads to involve enough semantic content to be self-explanatory and allow deep analysis like exact analysis (dimensions, measures, thickness, material ...).

Moreover, the package sent back from Bureau Veritas to NAVAL Group should also contain the comprehensive context used during the analysis as it is essential to understand the comment type and status returned. The proposed solution should then offer the ability to send back all measures, viewpoints, sections and so on generated during the approval analysis.

6.2 EXPORT / DATA TYPE

As any process of exchange between two or more companies this scenario raises the question of the type of data to be transmitted. Today there is not one single standard defined for the Marine and Offshore industry to describe the structure of the ship with the appropriate level of semantics so that it can be used directly within the solution. That's why the solution should also be offering a container able to convey the data together with the approval information and the comments.

To guarantee the correct level of exchange the solution will have to carry the information that best describe the design object, its structural characteristics, its state of approval and its link to the rest of the structure. Obviously the 3D visualization of the structural object will help understand its role within the complete structure, but it will not be enough to bring the best analysis capabilities.

Finally, to enhance the reliability of the approved data, the exchange should keep the connection to the design source of information, without changing its nature and its definition. This connection should be kept during all exchange cycles.

6.3 READABILITY OF INFORMATION / SELF EXPLAINING 3D MOCKUP

Today's drawings used for classification approval are made of 2D views of sections or of parts of the ship which are showing in a very explicit way all information for approval. A specific 2D description of the ship structure elements has been developed along the years to show all needed information as symbols, annotation and measures.

The solution will have to replace these well-known 2D drawings by a full 3D model of the ship structure.

The 3D digital model of the structure is a view of the real "as designed" ship and it directly shows how the complete structure will be; a person looking at it can directly appreciate where objects are located and what their shape is. Moreover every single 3D structural object in the **3D**EXPERIENCE® platform contains all the semantics and metadata needed to completely understand it and analyze its function within the global structure. Nevertheless all this information is by default hidden in the 3D visualization to avoid visual clutter. The solution will have to both set the scene to highlight what part of the structure is to be visualized and selectively show detailed information for approval analysis on the 3D view.

6.4 TRACEABILITY (DESIGN/APPROVAL)

This is a common request when trying to set up an exchange procedure. On both side the traceability of the exchanged data set should be provided in order to be able to come back to the state of the design and the status of approval during a past exchange.

In the context of this study we add two major constraints, the first being the legal part of the approval exchange which implies the possibility to come back to the data of the exchange or the final approval data even if the ship is already at sea; and the second being the link to the design data to be completely confident with the fact that the data being approved is really the one figuring the design.

7. SOLUTION PROCESS COVERAGE AND TOOLS

7.1 THE **3D**EXPERIENCE® PLATFORM

Thanks to the common data management architecture for any of the stakeholders, the **3D**EXPERIENCE® platform is, in essence, a collaborative platform leveraging collaboration around a digital 3D model for global benefits. Designers can author and manage design while other profiles manage the exchange and the global consistency.

The platform supports "data-driven" approach meaning that the digital information updated by any authorized people is automatically available and usable by anyone else.

The 3D Digital Classification process will take advantage of a set of standard **3D**EXPERIENCE® platform capabilities. As a matter of facts, Dassault Systèmes has been developing since many years a complete solution to enhance Marine and Offshore design processes. The solution will position at its foundation the capabilities of the Structure Functional Design and the Structure Detail Design applications that provide the Naval Structure Engineer with a detailed data model of structure objects and a complete tool set for design. The Structure Functional Design application is mainly used at early design stage offering a rapid and rich way of designing the complete structure of the ship taking benefit of a functional design approach and a light surfacing model that can easily be used for structural analysis. The Structure Detail Design application is used to enhance the preliminary design with all the detailing features that will complete the structure and prepare it to the manufacturing process definition.

In this solution the Structure Functional Design to Structure Detailed Design synchronisation will be used to establish the reference classification structure composed of exposed features in the platform. That gives the ability to keep classification reference updated by design changes from one side and by classification society's feedbacks by the other side (see figure 8).

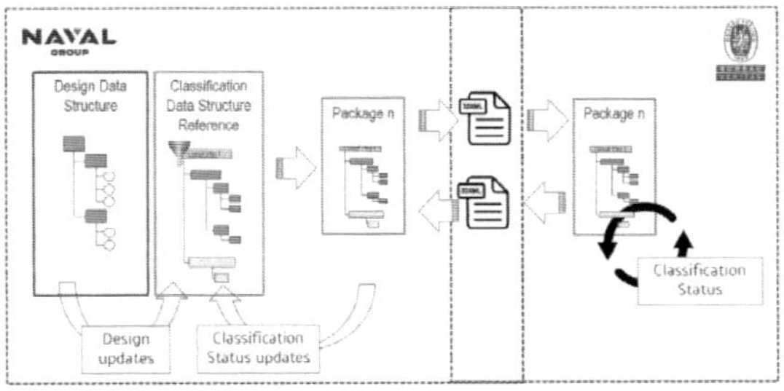

Figure 8: Classification exchange packages principles

7.2 COLLABORATION AND LINK TO THE DESIGN DATA

During the whole design life cycle of the ship structure in the **3D**EXPERIENCE® platform the nature of the data and its granularity may change due to maturity of the design and also design changes.

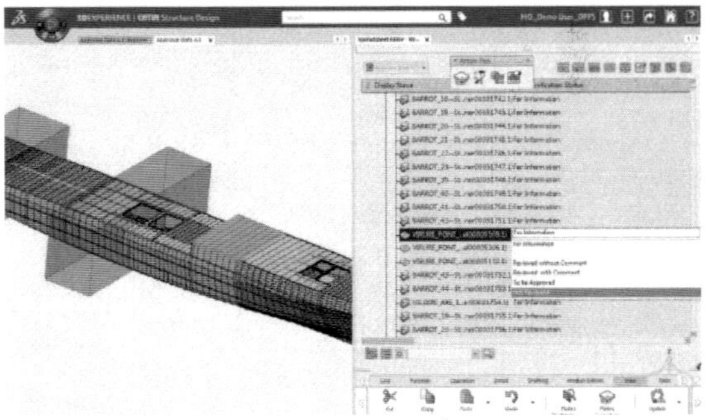

Figure 9: Classification status management

The definition of the Structure Detail Design application will offer the capability to extract the information first from the preliminary structure design, coming from the Structure Functional Design application, thanks to the built-in synchronization functionality. It will also allow a selective filtering of the detail design structure objects by using the filtering functions. Whenever in the design process and regardless of the technique explained just above the generated single structure object will keep a trace of its origin which will closely link it to the design data without creating interferences with the ongoing design of the structure.

This approach will set a stable and unique data model structure for the classification data to be exchanged from the design to the approval office and back.

The data model specialization capabilities give the opportunity to put the classification status as close as possible to the design data; the classification status defined as an attribute of the part allows NAVAL Group to clearly identify the data ready for approval and the data sent for defining the geometrical context: parts ready are specified as "To be approved" and other ones "For Information". The same attribute is used to carry on classification status retuned by Bureau Veritas after the study; Part's classification status could then be moved to "Reviewed without Comment" if the part fulfilled the classification rules, "Reviewed with comment" if the part has been reviewed and has been declared as not compliant or "Not Reviewed" if it has not be analyzed (see figure 9).

Graphical selections tools, as selection by trap (including features that are partly in the trap) combined with mass attributes update capabilities, allows end user to set value on a set of parts in one single step.

Additionally, the B.I. Essentials function, an Information Intelligence application of the **3D**EXPERIENCE® platform providing visual dashboarding of critical information by applying color shadings and labels to the content in the session, allows to reveal key information, and especially in our case, the classification status (see figure 10).

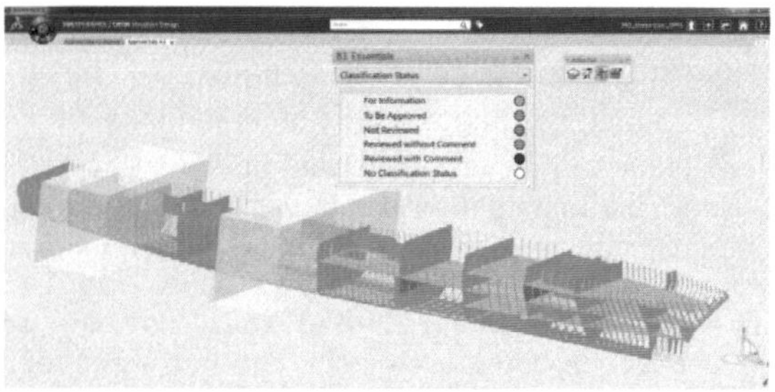

Figure 10: Graphical status analysis

On the one hand, NAVAL Group will then quickly and reliably be able to validate the package's content status before sending it. On the other hand, Bureau Veritas will be able to analyze the approval request, the data sent and will easily identify the work to be done. In the end, parts that have not been approved will be clearly highlighted and the design change process will be initiated quickly.

8. SOLUTION OVERVIEW

8.1 EXTRACTING FROM DESIGN

Earlier in this paper the synchronization of Structure Functional Design Data to Structure Detail Design data has been introduced. Let's see here in little more details the steps of the procedure to extract the information from the design with this approach.

The solution will not force the use of any predefined organization of the design data. The recognition of objects that have to be extracted and exchange for approval will be done by a built-in functionality that allows to automatically detect structural objects in the model. This function will give an overview of the structural objects to be extracted, and allow a first manual filtering of data. In fact through this interface the user will be able to manually select objects that shouldn't be shared or that are not mandatory for approval.

Now that the system knows which data to extract, the synchronization interface will guide the user to create the new data, and during an update will show which data is new or needs to be updated on the approval model.

The dataset generated can now be filtered thanks to the **3D**EXPERIENCE® PLM Filter functionalities. The design office will be able to set a predefined filter removing all structure objects that are not useful for classification or that are not sharable for security reasons from the data.

The PLM Filter can either use manual selection method or base its selection on attributes and types or even on 3D indexing information of the objects.

Finally the solution will offer a predefined organization of the data that will help the understanding of what is to be approved, and which zone of the ship is ready or not to be approved.

8.2 EXPORTING AND SHARING

Filtering functions and data simplification tool allow limiting the weight of the package to send and to archive keeping the requested useful information as the product structure and the attributes.

One of the challenges is to exchange a global comprehensive set of PLM data including the design, the comments, the status and also the links between all of them. The **3D**EXPERIENCE® platform provides the ability to export and import 3D XML files containing all PLM data and the links between them. 3D XML is a universal, lightweight XML-based format that enables users to share live, accurate 3D data quickly and easily. 3D XML is suited for fast and efficient communication, fully based on standard XML.

The data exchanged should not be modified by the other stakeholders and the ownership of the data has to be preserved. "As reference" import enables round trip exchanges allowing data edition only in one location at a time. Exchanges services deliver also the capability to synchronize the modified package updating the classification status on the parts created at the NAVAL Group site and the comments (Design Reviews) created by Bureau Veritas.

8.3 APPROVAL WORK ENVIRONMENT

The main review will be made on the **3D**EXPERIENCE® platform. We can distinguish several needs that the solution will need to fulfil in order to make the dataset comprehensible.

First whether we are talking about extracting the data from the design or analyzing it for classification approval a set of tools need to be put in place to understand how the structure is made. An analysis of the structure data need to be directly embedded in the user working environment to easily highlight:

- Scantling analysis
    - Plate thickness distribution
    - Stiffener section distribution
- Material applied to structural objects

The results of those analyses will be displayed directly on the 3D structural objects as shown in the picture below (Figure 11).

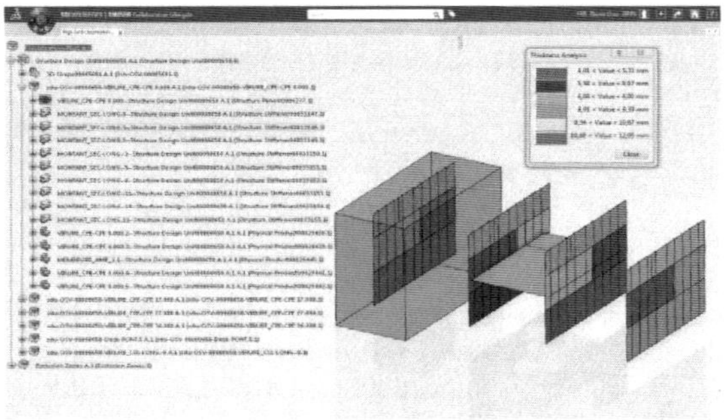

Figure 11: Colouring for thickness analysis

In addition to these analyses, a comparison method and tool need to be provided.

On the one hand being able, on the NAVAL Group side, to compare the package sent for approval with the response package received after classification review. This will compare the status of approval to identify what has been reviewed and will not need any design change and what has been reviewed but need to be modified and implies some engineering tasks.

On the other hand, mainly on Bureau Veritas side the ability to compare the difference between the previously analysed packaged and the newly received package. The comparison should highlight the new created objects, the modified objects and the deleted ones. This is of major importance to Bureau Veritas to decide if previous comments are still valid or if the design has been changed according to these comments and they should therefore be closed.

8.4 APPROVAL COMMENTS ON 3D DATA

For the comments we need to replace the fact that those were mainly annotations and mainly managed on the VeriSTAR Project Management database. The management of the comment itself and its main lifecycle, as explained earlier in this document, will still be done through the Bureau Veritas Project Management database. Nevertheless the comment will have to be created on the 3D Digital Classification Model managed within the **3D**EXPERIENCE® platform, and for this reason all the Design Review capabilities will be used to enhance both the creation and the understanding of the comment.

The **3D**EXPERIENCE® Design Review application used on the 3D Classification model will help the approver to define where the anomaly has been detected in the 3D model by setting up the 3D scene that will clearly show what objects are concerned by the comment and adding appropriate measures and annotations to make the comment more understandable. This application offers all capabilities to introspect and measure the objects visualized in 3D and will be the interface for the approver to analyse the data (see figure 12).

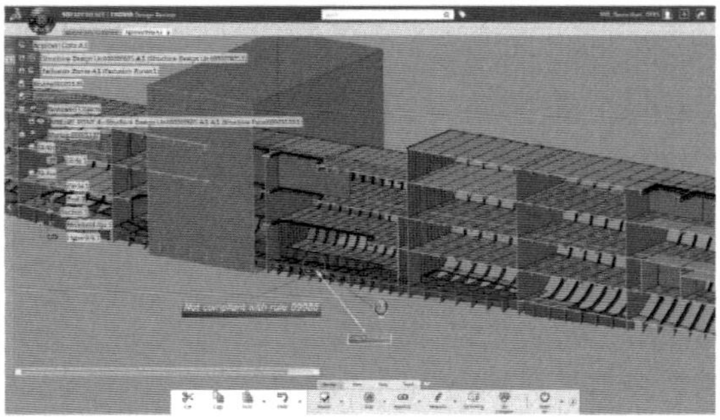

Figure 12: Design Review clipping and annotations

The link created between the object and the review will help both navigating from the comment which is the review itself to the objects and from the objects to the comments. This will define a much direct way for the designer to analyse the approval feedback and for the approver to explain the issue (see figure 13). Moreover the platform will give the possibility to the design office to define and allocate the appropriate task to the engineers to correct this anomaly.

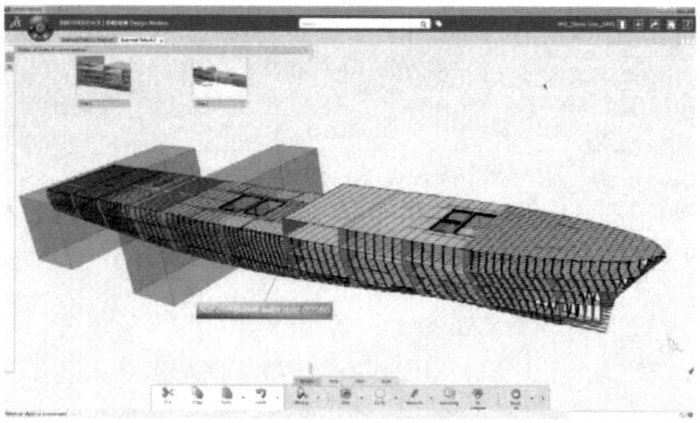

Figure 13: NAVAL Group comment analysis based on sent Design Reviews

Finally this information will be stored in the VeriSTAR Project Management database and a link directly on the 3D object will help navigating to the page of VeriSTAR Project Management where the discussion on the comment will take place.

8.5 INTEGRATION OF APPROVAL COMMENTS ON DESIGN

Into the platform, comments concept will be held by the **3D**EXPERIENCE® Design Review enabling users to conduct collaborative reviews in order to communicate, track and resolve issues in the design. The solution provides a comprehensive set of tools to analyse and communicate issues including slides, 3D markers, measurements, and sectioning to highlight critical issues to investigate which results in more informed decision making. The results are stored in the **3D**EXPERIENCE® platform and could be exported within the 3D XML package.

As previously explained, the comments management still remains so far into the Bureau Veritas in-house solution VPM but in order to propose a deep relationship between the design and the comments, the proposal is to initiate the comment process from the design, into the **3D**EXPERIENCE® platform and then push the creation and the updates to VeriSTAR PM thanks to an integration development.

In one side, during classification studies, Bureau Veritas's approvers need to capture the analysis they performed as measures, sections and view-points. On the other side, as comments revealed issues on design and as those issues should be resolved at the end by NAVAL Group's designers, the comments have to be as clear as possible and sent back to designers; the comments hosted into Bureau Veritas VeriSTAR PM are accessible from external network; this capabilities will be kept and enriched by the **3D**EXPERIENCE® comments sent back in the same way than the classification status.

9. CONCLUSION

The partnership between NAVAL Group, Bureau Veritas and Dassault Systèmes started with a period of analysis to understand today's process of classification and the expectations of both design offices and classification societies. The outcome of this study is the definition of what could be a near future solution to adopt the 3D Digital Classification process and this is what this paper explained. Today the focus will be on defining a first proof of concept solution that will be tested in a real use case scenario to better understand the new procedure and applications to be implemented. The long term goal is to create an industrial solution based on the **3D**EXPERIENCE® platform to answer to a wider scope of classification scenarios not only for the metallic structure of vessels and platforms but also for the approval of global systems.

10. AUTHORS BIOGRAPHY

[Cédric CHEYLAN] Hull Design Office Manager – at NAVAL Group® – Lorient Shipyard. Responsible for both managing the Design Office team and surface ship Hull design internal validation. Previously Manager of the Hull and General Arrangement preliminary design team.

[Lionel LE GUENNIC] PLM Business Analyst – at NAVAL Group® – Lorient Shipyard. Responsible for **3D**EXPERIENCE® deployment for frigate and submarine project on Hull / Structure. Previously Responsible of AVEVA deployment.

[Thierry LE GAL] Enterprise Architect – at NAVAL Group® – Lorient Shipyard. Lead Architect for **3D**EXPERIENCE® deployment for frigate and submarine projects. Previously Lead Architect for CADDS5 & Windchill deployment for previous frigate and submarine projects

[Glenn DUTRIEUX] Industry Solution Experience Manager – at Dassault Systèmes- France. In charge of defining Industry Solution Experiences for the Marine and Offshore industry.

[Sébastien VERET] Business Transformation Solution Consultant– at Dassault Systèmes® – France. In charge of explaining, demonstrating and illustrating the value of the 3DEXPERIENCE solution contributing to the project definition and organization.

[Alexandre TEW KAI] Marine & Offshore Industry Senior Business Consultant - at Dassault Systèmes - France. Responsible for developing consultative approaches to reinforce the business value proposition delivered to Marine & Offshore companies.

[Olivier DEGRAND] Scientific Software Manager - at Bureau Veritas Marine and Offshore - France. Responsible for development and sales of BV Scientific Software.

[Christophe CHAUVIERE] Director of Development Department - at Bureau Veritas Marine and Offshore - France. In charge of rules and software development.

INTEGRATION OF SHIP HULL FORM MODELING BASED ON SUBDVISION SURFACES WITH OTHER SHIP DESIGN TOOLS

SH Greshake, University of Rostock, Germany
R Bronsart, University of Rostock, Germany

SUMMARY

The industry standard for the representation of hull forms are tensor-product B-spline surfaces. However, tensor-product B-splines are limited to four-sided surfaces. Aside of the necessity to compose hull surfaces of several patches, it makes hull form modeling and fairing inefficient. An alternative are generalized B-spline surfaces. They originate from the field of subdivision surfaces, whereas the term subdivision refers to a method that enables B-spline surfaces of arbitrary complexity. Hull form representation based on generalized B-splines improves hull form modeling. To be employed for ship design in practice, it is essential to provide an integration with other design tools. Naturally, this is a matter of data exchange. This article describes a method to convert a generalized B-spline surface to a collection of conventional Bézier patches. These patches may be shared with other design tools using a general purpose file format such as IGES.

1. INTRODUCTION

The industry standard for the representation of hull forms are tensor-product B-spline surfaces. They have a number of characteristics that make them attractive for the representation of free-form surfaces. The most important characteristic is that they are defined with the help of a control mesh which is closely approximated by the surface. This provides a reliable method to define smooth surfaces of any shape. The result is insensitive to changes of the control mesh what simplifies fairing. Unfortunately, tensor-product B-splines are limited to four-sided surfaces and therefore most hull surfaces are composed of several patches, but this considerably increases the number of control points and yields complex dependencies of the individual control meshes. As a consequence control meshes are only rarely used for hull form modeling.

In contrast, hull form design is mostly based on the interpolation of curve networks to define hull surfaces. The complexities of the patch generation and their interdependencies are left to an interpolation algorithm, where most algorithms provide a reliable patch generation, but the result is often not satisfactory in terms of the fairness and sensitive to changes of the curves. Unfortunately, this is an intrinsic property of interpolation. Some algorithms may alleviate this issue of interpolation, but it remains difficult to fair hull surfaces based on curves.

An alternative to tensor-product B-spline surfaces are generalized B-spline surfaces. They originate from the field of subdivision surfaces. The term subdivision refers to the method that enables B-spline surfaces of arbitrary complexity. Thus, the necessity to use several patches for the representation of hull forms is avoided. The most important consequence is, however, not the ability to represent hull forms with a single surface, but the possibility to effectively utilize the control mesh for modeling.

Using the control mesh apparently improves hull design for the reasons given before. Generalized B-splines promote the utilization of control meshes for hull surface modeling, but this worth nothing when the downstream support in the ship design process is missing. This paper addresses this problem and discusses how a generalized B-spline surface could be exchanged with other design tools. Two variants are imaginable: either support for generalized B-splines is added to the relevant design tools or generalized B-splines are converted to tensor-product B-splines which are currently supported by all major design tools and data exchange formats. Although generalized B-splines are compatible to the common architecture of the geometry kernels that are used by such tools, it is unlikely that a majority of the ship design tools will adopt this surface representation in the foreseeable future. For practical purposes, this paper focuses therefore on the second variant, the conversion of a generalized B-spline surface into a collection of tensor-product B-spline patches. The result is finally exported to one of the generic file formats which are widely accepted by other design tools such as IGES.

2. RELATED WORK

The application of control meshes for hull form modeling is currently restricted by the limitation of tensor-product B-splines to regular control meshes and therefore four-sided surfaces. This limitation is addressed by a certain class of subdivision surfaces that generalize B-spline surfaces for control meshes of arbitrary topology. Examples are the generalization of biquadratic B-splines given by Doo and Sabin [1], the generalization of bicubic B-splines given by Catmull and Clark [2], and the arbitrary degree generalizations of Stam [3] and Zorin and Schröder [4].

The subdivision algorithm is the core of any generalized B-spline surface as it defines the surface in the vicinity of irregular control points. It is possible to evaluate surface properties, such as points on the surface or normals, at any parameter values using the method of Stam [5], but the subdivision algorithm has to be stationary. This is the case for the previously mentioned algorithms. However, all algorithms are only generalizations of uniform B-spline surfaces. The generalization on non-uniform B-spline surfaces is covered for the low-degree case by Sederberg et al. [6,7], Müller [8,9], and for arbitrary degrees by Cashman et al. [10], but non-uniform features are only possible when the relevant part of the control mesh is regular or the subdivision algorithm is non-stationary. Therefore non-uniform generalizations of B-splines suffer from either being restrictive in the choice of the control mesh topology or Stam's method for the exact evaluation of surface properties is not applicable at irregular control points. The most important application of non-uniform B-splines is the boundary behavior of the surface. A similar behavior is, however, realizable for uniform B-splines with stationary, but topologically unrestricted, modifications of the subdivision algorithm. Therefore the uniform variants, in particular the variant of Catmull and Clark [2], are primarily used in practice.

Generalized B-splines are only rarely used for hull form modeling. An application of non-uniform bicubic B-splines, which are called T-splines, is shown by Sederberg and Sederberg [11]. Aside of the boundaries, the non-uniform features are employed to define knuckles. The presented examples reflect the complexity of typical yacht hulls and are rather simple compared to the hull forms of other vessel types. The study of Greshake and Bronsart [12] uses uniform bicubic B-splines to represent the hull form of a modern container vessel. It is shown that the fairness is improved compared to the result of conventional methods. The application of control meshes for hull form modeling is discussed by Greshake and Bronsart [13,14] for different types of vessels, likewise based on the application of uniform bicubic B-splines. While the previous work proposes to use the control mesh for hull form modeling, Lee et al. [15] present a method to generate a hull form based on a curve network in the context of generalized B-splines.

The integration of generalized B-splines into an existing CAD system is described by Antonelli et al. [16] for the uniform bicubic variant. The study reviews the typical architecture of CAD systems and shows how generalized B-splines fit into this architecture without the necessity to touch the existing concepts of the system. In contrast, once implemented, the full range of tools of the CAD system is available to modify generalized B-splines.

The conversion of a generalized B-spline surface to a collection of tensor-product patches is described by Peters [17] for the uniform bicubic variant. The algorithm generally produces one bicubic NURBS patch for each quadrilateral face of the control mesh, but next to irregular control points up to 16 patches are produced for each face. The patches join with curvature continuity except for the vicinity of irregular control points where they join only with tangent continuity. Another algorithm is given by Loop and Schaefer [18] which produces one bicubic Bézier patch for each quadrilateral face of the control mesh, even in the presence of irregular control points. Similar to the previous algorithm, the patches generally join with curvature continuity, but as a concession to the lower number of patches and the restriction to relatively simple Bézier surfaces, the patches join only with position continuity in the vicinity of irregular control points. However, it is emphasized that the patches are at least watertight, certainly not a matter of course in hull form modeling, see for example the work of Edessa et al. [19,20].

The conversion of a generalized B-spline surface to a collection of tensor-product B-spline patches represents the original surface exactly wherever the control mesh is regular, but close to irregular control points the initial surface is only approximated. The regular case is a trivial consequence of the definition of generalized B-splines which is summarized in the next section. The irregular case is, however, significantly more difficult. Given the regular patches, the construction of the remaining patches translates to the problem of filling n-sided holes smoothly with tensor-product B-splines. The complexities of this problem and possible algorithms are described for example by Piegl and Tiller [21], Ye et al. [22], Gregory and Zhou [23], or more recently by Fan and Peters [24-26]. These works indicate that constructions with simple bicubic tensor-product patches, such as Bézier patches, do not suffice to obtain a tangent continuous approximation near irregular control points. In contrast, either sophisticated constructions with multiple knots are required, patches of higher degrees have to be used, or each patch incident to the irregular control points has to be subdivided into at least nine smaller patches, see in particular Peters and Fan [25].

3. GENERALIZED B-SPLINE SURFACES

The title of the paper refers to subdivision surfaces whereas the previous text is mostly speaking of generalized B-splines what essentially reflects two complementary points of view on subdivision surfaces. The first point of view considers subdivision as method to refine polygon meshes of all kind. After a few steps of subdivision the result may look like a smooth surface, but it remains a polygon mesh. The linear nature of polygon meshes is generally not qualified for the representation of hull forms in ship design, although they are often used in the context of numerical simulations. The second point of view on subdivision provides a construction of smooth B-spline surfaces in the presence of irregular control meshes. To clarify the difference of both views, the latter is referred to as generalized B-splines. The subsequent material gives a brief introduction to the basic concepts of the generalized B-splines. The focus is laid on the relevant details for the conversion to tensor-product B-splines. For a more general introduction refer to [14] and the references specified therein.

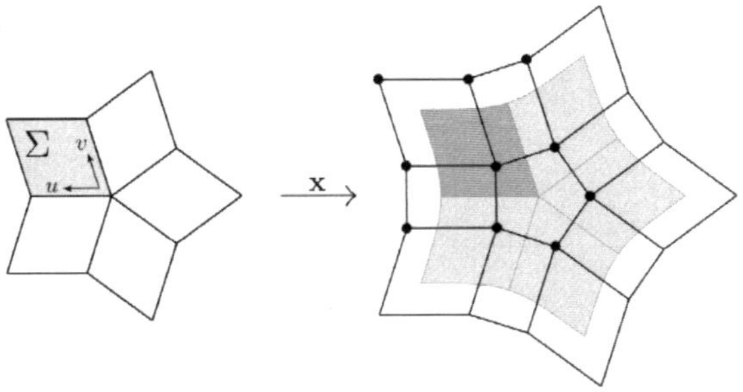

Figure 1: Definition of generalized B-spline surfaces. Left: The domain **S** is given as the composition of unit cells Σ. Right: A schematic view of the surface $\mathbf{x}(u,v,i)$ is shown in gray and the control mesh is shown in black. The surface is composed of patches according to cell layout of the domain. The black dots refer to the control points that influence the shape of the highlighted patch.

3.1 DEFINITION

A generalized B-spline surface is defined over a domain that is composed of unit cells. The restriction of the surface to a certain cell of the domain is called a patch, see Figure 1. This complies with the usual setup in hull form design to represent complex surfaces by a collection of surface patches, but they are now formally treated as a single surface. This may sound trivial at the first glance, but an essential consequence is that the surface is defined on a single control mesh what obviates the need to maintain smooth transitions between the individual patches.

Any point on a B-spline surface is only influenced by a few control points close to it. Therefore each patch is defined only on a subset of the global control mesh, refer to Figure 1 for clarification. In the presence of control meshes of arbitrary topology, this subset is either regular or irregular. Both cases are treated separately.

3.2 REGULAR PATCHES

A regular patch is defined on a regular subset of the global control mesh and can be defined in terms of a tensor-product B-spline. This is characteristic is utilized to convert generalized B-splines to a collection of tensor-product B-splines surfaces when the control mesh is locally regular.

3.3 IRREGULAR PATCHES

An irregular patch is defined on an irregular subset of the global control mesh and cannot be defined in terms of a tensor-product B-spline. The key to define irregular patches is subdivision, the refinement algorithm of meshes which is characteristic for subdivision surfaces. Applied to the control mesh, each step of subdivision introduces new control points, but it does not change the shape of the corresponding B-spline surface. As the control mesh is subdivided again and again, it is possible to define a growing portion of an irregular patch in terms of tensor-product B-splines. Once the part in question is defined in terms of a tensor-product B-spline, surface properties such as points on the surface, normals, or curvature can be computed exactly. Using the method of Stam [5], it is possible to avoid the computational expensive subdivision of the control mesh and to compute these properties directly in terms of the initial mesh.

Unfortunately, neither the recursively definition of an irregular patch, nor the direct evaluation of an irregular patch provides a construction for an equivalent tensor-product B-spline. Hence, the conversion of generalized B-splines to a collection of tensor-product B-splines requires an appropriate approximation where the control mesh is locally irregular.

4. CONVERSION ALGORITHM

The input for the conversion algorithm is a generalized B-spline surface as described in [13]. It is based on a variant of the algorithm of Catmull and Clark [2] and represents a bicubic B-spline surface with creases and corners as additional features. In [12-14] this surface is employed to represent the hull forms of various types of vessels. The conversion algorithm converts this surface to a collection of bicubic Bézier surfaces.

The choice of bicubic surfaces is a natural consequence of the degree of the input surface. Bézier surfaces are the modest variant of non-uniform rational B-spline surfaces (NURBS) which are considered as the industry standard for the representation of hull surfaces.

4.1 ALGORITHM

The conversion algorithm is based on the observation that each patch of a bicubic generalized B-spline surface is associated to a face of the control mesh. The algorithm involves four steps:

Step 1: A bicubic Bézier surface is generated for each regular face of the control mesh. A regular face is four-sided, defined by only regular control points, and all incident faces are four-sided as well. However, the incident faces are allowed to be partially defined by irregular control points. The generated Bézier surfaces are an exact representation of the original surface.

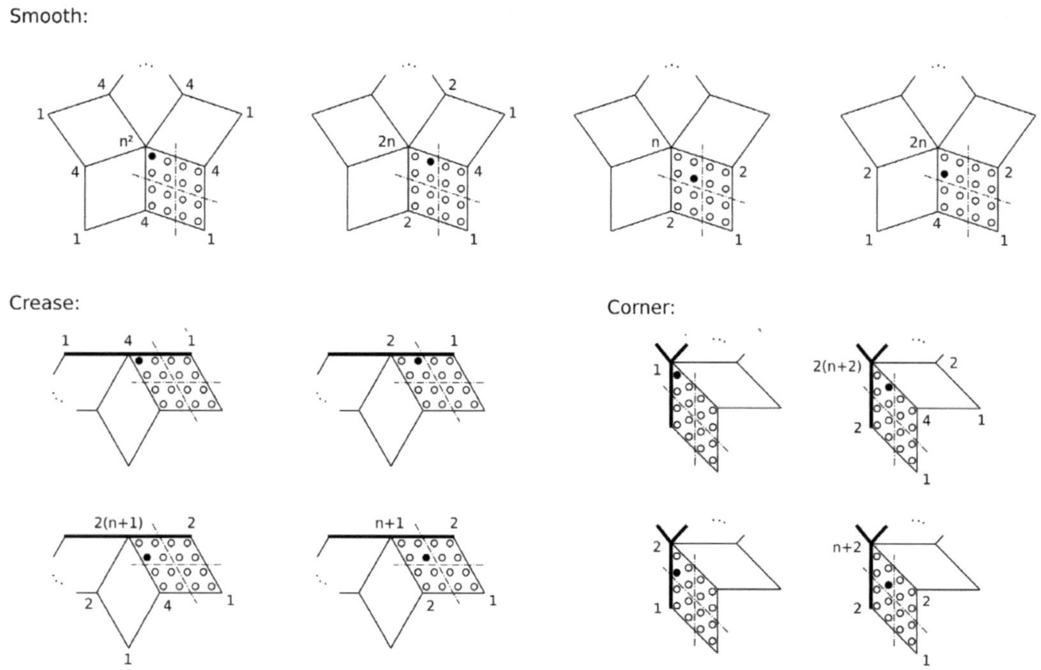

Figure 2: Rules to compute a bicubic Bézier surface which is defined by a grid of 4×4 Bézier points. The rules are specified for a single quadrant of this grid which includes a corner point, two edge points, and one interior point. The other quadrants are given by rotation of the illustrated rules. Each Bézier point is computed by the weighted average of the control points of the control mesh. The appropriate choice of the weights depends on the type of the control point at the quadrant's corner and the type of the incident edges. The illustrated cases are sorted by the control point types: smooth, crease, and corner. Crease eges are illustrated by bold lines, smooth edges are shown with thin lines. The valence of the control point at the quadrant's corner is denoted by n.

Step 2: The control is mesh is once subdivided what splits each n-sided face into n four-sided faces. In the context of this section, the former is referred to as the *parent face*, the latter are called *child faces*. The effect of subdivision is twofold: the control mesh consists only of four-sided faces what is the prerequisite to obtain a collection of watertight Bézier surfaces in the presence of irregular control meshes and irregular control points are isolated because subdivision regularizes the control mesh around them. The subdivision rules for the control mesh are summarized in [13].

Step 3: A bicubic Bézier surface is generated for each child face with an irregular parent face. Irregular parent faces are omitted in the first step of the algorithm. Thus, the third step of the algorithm generates the missing surfaces. Curved surfaces are only an approximation of the original surface, but flat surfaces coincide with the original surface. This is an important characteristic in the context of hull form regions such as the flat side, flat bottom, or flat transom.

Step 4: All generated Bézier surfaces are written to a file for data exchange with other design tools. The authors employ the widely accepted IGES file format, but any other file format which handles B-spline surfaces could be used.

4.2 BÉZIER POINTS

The core of the conversion algorithm is the construction of the Bézier surfaces. A surface is constructed for a face of the control mesh. The surface is defined by a grid of 4×4 Bézier points which are computed based on the control points of the control mesh in the local neighborhood of the face.

The rules proposed by Loop and Schaefer [18] are employed to compute the Bézier points. The rules are reproduced in Figure 2 for a number cases, where each case illustrates the computation of a single point of the grid of 4×4 Bézier points. Only a single quadrant of this grid is covered in the figure because the other quadrants are given by rotation of the illustrated cases. A quadrant includes a corner point, two edge points and one interior point of the grid.

The additional features of the input surface are realized with tags that are applied to the elements of the control mesh. Control points are by default smooth, but can be tagged as creases or corners. Edges are by default smooth or tagged as creases. Each Bézier point is computed as the weighted average of the control points in local neighborhood of the face the Bézier surface is associated to. The appropriate choice of the weights depends on the

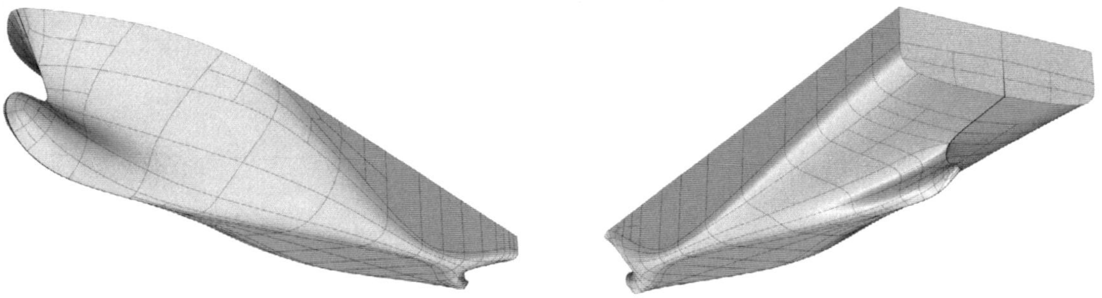

Figure 3: Bézier patches generated for the Generalized B-spline Container Carrier (GCC). In total 189 patches are created to represent the original hull surface. No degenerated patches are generated, what avoids discontinuities in the curved regions of the hull form, but yields a needlessly high number of patches at the forward end of the flat side.

tag of the control point at the quadrant's corner and the tags of the incident edges. The illustrated cases are sorted by the control point type:

- **Smooth:** The corner point depends on the control points of the incident faces. No incident crease edges are allowed for smooth control points.
- Crease: The corner point depends only on the control points along the two incident crease edges. The computation of the edge points differentiates crease edges (top right) and smooth edges (bottom left).
- **Corner:** The corner point interpolates the control point at the quadrant's corner. The computation of the edge points differentiates crease edges (top right) and smooth edges (bottom left).

5. RESULTS

The conversion algorithm is applied to convert the hull form of a container vessel, which is represented by a single generalized B-spline surface, to a collection of Bézier surfaces. As the individual Bézier surfaces are considered as parts of the hull surface, they are referred to as patches in the following text.

5.1 PATCHES

The vessel is called the Generalized B-spline Container Carrier (GCC). It is a modern container vessel and closely imitates the design of the Duisburg Test Case (DTC) [27]. The main characteristics of the hull form are summarized in Table 1.

	GCC	DTC	Δ
L_{PP} [m]	355.0	355.0	
B [m]	51.0	51.0	
T [m]	14.5	14.5	
C_B [-]	0.661	0.658	-0.46%
C_M [-]	0.987	0.989	+0.20%
C_P [-]	0.669	0.665	-0.66%
C_{WP} [-]	0.846	0.845	-0.11%
LCB [%]	-0.967	-1.038	-0.15%

Table 1: Main dimensions of the Generalized B-spline Container Carrier (GCC). The vessel reflects the design of the Duisburg Test Case (DTC).

The generated Bézier patches are shown in Figure 3. In total 189 patches are created. The algorithm does not create any degenerated patches, where one edge of the patch is set to zero length. On the one hand, this is advisable to avoid discontinuities in curved regions of the hull form, but on the other hand flat regions of the hull form may be represented by a needlessly high number of patches. For example seven patches are generated to represent the forward end of the flat side. As all patches are flat, a single degenerated patch at the tip and an ordinary patch next to it would suffice to represent this part of the hull form properly. The hull form of the GCC includes a flat bar to consider production requirements. Upon conversion this yields a layer of Bézier patches with high aspect ratios next to the center plane, see Figure 5 for clarification. Although there is a potential to further minimize the number of generated Bézier patches, the total number agrees with the result of state-of-the-art modeling systems which mostly employ a curve network to construct the surface patches. The hull form of the DTC is for example modeled with this method and the hull surface is composed of even 660 B-spline patches. However, the authors believe that this number could be at least halved without sacrificing the precision of the hull surface definition.

5.2 QUALITY

A previous study [12] shows that using a generalized B-spline surface to represent a hull form improves the quality of the hull representation. The study compares the GCC to the DTC, what brings up the question whether this still applies after the conversion to a collection of Bézier patches using the algorithm given in Section 4.

(a) Generalized B-spline surface (b) Conversion to Bézier patches (c) Duisburg Test Case (DTC)

Figure 4: Quality analysis of the generated Bézier patches. Smooth reflection lines show curvature contiuity, straight reflection lines indicate fairness of the hull surface. Left: GCC represented by a single generalized B-spline surface. Middle: GCC converted to a collection of Bézier patches. The overall quality of the original surface is mainted. Right: The patches of the DTC show a lack of smoothness compared to the two other cases.

(a) Generalized B-spline surface (b) Conversion to Bézier patches (c) Duisburg Test Case (DTC)

Figure 5: Detailed quality analysis of the bulbous bow. Left: GCC represented by a single generalized B-spline surface. Left: The conversion to a collection of Bézier patches introduces continuity errors between a few of the patches. Right: The hull surface of the DTC also also affected by discontinuities. The overall situation is even worse.

The first observation is that the Bézier patches generated for the GCC are watertight as opposed to the patches of the DTC. The patches next the flat of side and flat of bottom of the DTC are not connected to each other, what supposedly results from an inappropriate manual definition of the flat patches after the curved patches are generated from the curve network. The result are gaps and overlaps between adjacent patches. These kind of issues occur frequently in practice and require substantial effort to be handled in downstream analysis methods.

More technically, the former characteristic is referred to as zero-order continuity of the patches. The second component of the quality assessment covers the high-order continuity. The bicubic generalized B-spline surface which is used to represent the GCC is curvature continuous (G^2) everywhere except for a few irregular points where it is only normal continuous (G^1). Across features, such as knuckles, the continuity characteristics of the surface are intentionally different. Ideally, the generated Bézier patches inherit the continuity properties of the original surface. Figure 4a shows the reflection lines of the GCC which are smooth as supposed for curvature continuity, Figure 4b shows the reflection lines for the generated Bézier patches. Figure 4c shows the reflection lines for the DTC as reference of the state-of-the-art. The reflection lines of the DTC are clearly not as smooth as the reflection lines of the GCC. The superior quality of the GCC is mostly maintained by the generated Bézier patches and still better than the DTC. However, the Bézier surfaces are less smooth than the original model as the detailed analysis of the bulbous bow in Figure 5 reveals. Again, the original hull surface of the GCC is compared to the generated Bézier surfaces and the patches of the DTC. In Figure 5b, a mismatch of the reflection lines across some of the patch boundaries indicates discontinuities of the normals. The respective patches form a knuckle along the common boundary which is difficult to spot based on the surface shading, but easily seen by the mismatch of reflection lines. The relevant Bézier surfaces are considered as irregular and the Bézier point computation given in Section 4.2 yields only position continuous surfaces. However, the patches of the DTC suffer from the same discontinuities and the overall situation shown in Figure 5c is even worse.

5.3 ACCURACY

The last analysis covers the accuracy of the generated Bézier patches. The geometrical deviation of the original surface and the generated Bézier patches is illustrated in Figure 6.

As explained in Section 4, the algorithm generates one surface for each face of the control mesh and differentiates between regular and irregular faces. Similarly, the generated Bézier patches may be classified as regular or irregular. Regular Bézier patches are an exact representation of the original surface. As well-behaved control meshes for hull forms are mostly regular for the reasons given in [14], most Bézier patches are regular and consequently large parts of the hull surface of the GCC are exactly represented by the Bézier patches. In contrast, irregular Bézier patches are only an approximation of the

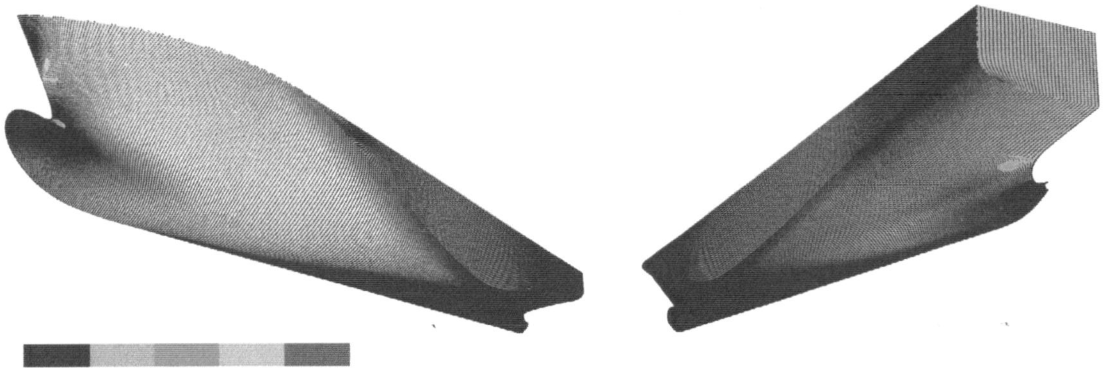

Figure 6: Geometrical deviation of the generated Bézier patches to the original surface. Large parts of the hull surface are exactly represented by the Bézier patches so the deviation is zero. In the vicinity of irregularties the original surface is only approximated by the Bézier patches. In the interior this is neglible, but on the boundary the Bézier patches deviate up to 0.05 m form the original surface (L_{PP} = 355.0 m).

original surface. Irregular interior patches do not show any significant deviation, but the boundary patches deviate up to 0.05m from the original surface.

6. CONCLUSION

Tensor-product B-spline surfaces are the industry standard for the representation of hull forms, but they suffer from their limitation to four-sided surfaces. Generalized B-splines originate from the field of subdivision surfaces and allow to define B-spline surfaces of any complexity.

A practical integration generalized B-splines into the ship design process is the conversion to collection of tensor-product B-splines which can be shared with other design tools that do not support this surface representation. This study presents an algorithm to generate a collection of Bézier surfaces, the simplest variant of tensor-product B-splines. The number of generated Bézier patches is similar to the usual number of patches used to represent hull forms. The generated Bézier patches are always watertight and the superior quality of the generalized B-spline surface is mostly maintained. Only at a few irregular spots the quality is less. The same holds for the accuracy of the Bézier patches as most parts of the original hull surface are exactly represented, but in the vicinity of irregularities the original surface is only approximated.

The approximation error is negligible in the interior of the surface, but may be notable at the boundary and other surface features such as knuckles. Therefore future work should address the surface construction for these cases. In addition, there is a potential identified to further reduce the number of patches generated for flat regions of the hull form.

7. REFERENCES

1. Doo, D., Sabin, M., 'Behaviour of Recursive Division Surfaces near Extraordinary Points', *Computer-Aided Design* 10 (6), 1978.
2. Catmull, E., Clark, J., 'Recursively Generated B-Spline Surfaces on Arbitrary Topological Meshes', *Computer-Aided Design* 10 (6), 1978.
3. Stam, J., 'On Subdivision Schemes Generalizing Uniform B-Spline Surfaces of Arbitrary Degree', *Computer-Aided Geometric Design* 18 (5), 2001.
4. Zorin, D., Schröder, P., 'A Unified Framework for Primal/dual Quadrilateral Subdivision Schemes', *Computer Aided Geometric Design* 18 (5), 2001.
5. Stam, J., 'Exact Evaluation of Catmull-Clark Subdivision Surfaces at Arbitrary Parameter Values', In *Proceedings of the 25th Annual Conference on Computer Graphics and Interactive Techniques,* 1998.
6. Sederberg, T.W., Zheng, Z., Sewell, D., Sabin, M., 'Non-Uniform Recursive Subdivision Surfaces', In *Proceedings of the 25th Annual Conference on Computer Graphics and Interactive Techniques,* 1998.
7. Sederberg, T.W., Zheng, J., Bakenov, A., Nasri, A., 'T-Splines and T-NURCCs', *ACM Trans. Graph.* 22 (3), 2003.
8. Müller, K., Reusche, L., Fellner, D., 'Extended Subdivision Surfaces: Building a Bridge Between NURBS and Catmull-Clark Surfaces', *ACM Trans. Graph.* 25 (2), 2006.

9. Müller, K., Fünfzig, C., Reusche, L., Hansford, D., Farin, G., Hagen, H., 'Dinus: Double Insertion, Nonuniform, Stationary Subdivision Surfaces', *ACM Trans. Graph.* 29 (3), 2010.

10. Cashman, T.J., Augsdörfer, U.H., Dodgson, N.A., Sabin, M.A., 'NURBS with Extraordinary Points: High-Degree, Non-Uniform, Rational Subdivision Schemes', *ACM Trans. Graph.* 28 (3). 2009.

11. Sederberg, M.T., Sederberg, T.W., 'T-Splines: A Technology for Marine Design with Minimal Control Points.', In *2nd Chesapeake Power Boat Symposium 2010*, 2010.

12. Greshake, S.H., Bronsart, R., 'High Quality Ship Hull Form Representation Based on Subdivision Surfaces.', In *RINA, Royal Institution of Naval Architects - International Conference on Computer Applications in Shipbuilding 2015, ICCAS 2015 - Papers*, 2015.

13. Greshake, S.H., Bronsart, R., 'Using Subdivision Surfaces to Address the Limitations of B-Spline Surfaces in Ship Hull Form Modeling', In *Proceedings of the 13th International Symposium on Practical Design of Ships and Other Floating Structures (PRADS 2016)*, 2016.

14. Greshake, S.H., Bronsart, R., 'Application of Subdivision Surfaces in Ship Hull Form Modeling', *Computer-Aided Design*, under review.

15. Lee, K.-Y., Cho, D.-Y., Kim, T.-W., 'Interpolation of the Irregular Curve Network of Ship Hull Form Using Subdivision Surfaces', *Computer-Aided Design and Applications* 1 (1-4), 2004.

16. Antonelli, M., Beccari, C.V., Casciola, G., Ciarloni, R., Morigi. S., 'Subdivision Surfaces Integrated in a CAD System', *Computer-Aided Design* 45 (11), 2013.

17. Peters, J., 'Patching Catmull-Clark Meshes' In *Proceedings of the 27th Annual Conference on Computer Graphics and Interactive Techniques*, 2000.

18. Loop, C., Schaefer, S., 'Approximating Catmull-Clark Subdivision Surfaces with Bicubic Patches', *ACM Trans. Graph.* 27 (1), 2008.

19. Edessa, D.M, Kleinsorge, L., Bronsart, R., 'A Contribution to Automatic Ship Hull Form CAD Data Repairing', *Proceedings of the 12th International Symposium on Practical Design of Ships and Other Floating Structures (PRADS 2013)*, 2013.

20. Edessa, D.M, Kleinsorge, L., Bronsart, R., 'A Tool for Ship Hull Surface Healing and Domain Preparation for Downstream Applications', *International Journal of Computer Aided Engineering and Technology* 9 (1), 2017.

21. Piegl, L.A., Tiller, W., 'Filling N-Sided Regions with NURBS Patches', *The Visual Computer* 15 (2), 1999.

22. Ye, X., Nowacki, H., Patrikalakis, N.M., 'GC1 Multisided Bézier Surfaces', *Engineering with Computers* 13 (4), 1997.

23. Gregory, J.A., Zhou, J., 'Filling Polygonal Holes with Bicubic Patches', *Computer Aided Geometric Design* 11 (4), 1994.

24. Fan, J., Peters, J., 'On Smooth Bicubic Surfaces from Quad Meshes', In *Advances in Visual Computing: 4th International Symposium, ISVC 2008, Las Vegas, NV, USA, December 1-3, 2008. Proceedings, Part I*, 2008.

25. Peters, J., Fan, J., 'On the Complexity of Smooth Spline Surfaces from Quad Meshes', *Computer Aided Geometric Design* 27 (1), 2010.

26. Fan, J., Peters, J.,. 'Smooth Bi-3 Spline Surfaces with Fewest Knots', *Computer-Aided Design* 43 (2), 2011.

27 Moctar, O., Shigunov, V., Zorn, T., 'Duisburg Test Case: Post-Panamax Container Ship for Benchmarking', *Ship Technology Research-Schiffstechnik* 59 (3), 2012.

8. AUTHORS BIOGRAPHY

Sebastian H. Greshake is a research assistant at the Chair of Naval Architecture at the University of Rostock. His research focuses on the application of subdivision surfaces for hull form modeling. He graduated from the University of Rostock with a Master's degree in Naval Architecture and Ocean Engineering. Previously, he completed a cooperative engineering studies program with a Bachelor's degree in Mechanical Engineering.

Robert Bronsart is holding the Chair of Naval Architecture at the Department of Mechanical Engineering and Marine Technology at the University of Rostock. He has more than 35 years experience in R&D in ship design, production and operation, thereof ten years in industry. In research he is focusing on information and communication systems in collaboration networks, CAE-system integration and on knowledge based methods in ship design and operation.

INTELLIGENT PME SOFTWARE FOR PRODUCTIVITY IN DESIGN OFFICES

P S Chopra, MRINA; Manager, Technical & Business Development, SeaTech Solutions International (s) Pte. Ltd.
U R Gadekari, MRINA; Assistant Manager, SeaTech Solutions International (s) Pte. Ltd., Singapore

SUMMARY

The 4th industrial revolution will alter the way we work. Higher productivity growth has become a basic necessity in all spheres of life and commercial enterprise. In a design office, there are multiple design projects running concurrently. These projects are monitored and evaluated on a continuous basis to improve productivity. Ship design is a sequential process with numerous iterations to refine and optimize the design. Each document may undergo multiple revisions. The Project Monitoring System must address specific needs of the ship design process and it requires intelligent Project Monitoring & Evaluation (PME) software that automates the manual process and also provides a live dashboard for the management information system.

As each project progresses, the amount of project data can become very huge. The PME software needs to combine features for project execution, monitoring and evaluation, communication and a project database with ability to handle multiple projects concurrently. This paper presents PME software specially customized for use in a design office and estimates the expected improvement in productivity compared to the current systems.

Keywords: Productivity in Design Office, PME software, Project Execution, Project Monitoring –Dashboard, Project Communication, Project Database Management, Performance Evaluation.

1. INTRODUCTION

The 4th industrial revolution, refers as digital transformation of industry, leveraging third platform technologies, such as Big Data /Analytics and innovation accelerators, such as the Internet of Things; and requiring the convergence of IT (Information Technology) and OT (Operational Technology). New IoT, data analytics technologies are re-shaping the marine industry in ship operations and management at rapid pace. Design offices as frontiers needed to handle large volumes of data, as in creating and managing the ship design data, critical equipment and communication information for each vessel.

In common scenario of working with multiple design projects, the complexity of project management aspects of planning, execution, monitoring, progress, performance evaluation and project communication seldom becomes uncontrollable. As project evolves documents generated undergo various phases of review by client, class society approvals and yard modifications which needs to be controlled and monitored regularly.

The traditional use of spreadsheets or the generic project management softwares proven to be ineffective to meet the design office requirements. These methods observed to create bottle necks in the process flow due to lack of adaptation to document management in iterative process, inefficient response for changes in project structure for various ship types, loss of time on existing data utilization and a non-uniform project management approaches.

Working towards solution for design offices, SeaTech Solutions developed custom made Project Monitoring and Evaluation software for productivity improvement. The PME software ensures design offices to perform efficiently meeting the design deliverables as defined by project deadlines. Software key features are creation of ship design database, hierarchy and controlled work flow with data security, multiple projects handling capability, multiple level user access, resources allocation, e-mail management system, dashboard for projects /departments/employee performances. In the current dynamic market, this software tool encompasses design office productivity and competitiveness meeting demand for quality work in optimum time.

2. PROJECT MANAGEMENT – DESIGN OFFICE – PRODUCTIVITY

The scope of project management in design office is to deliver approved basic ship design drawings to the owner/ shipyard as per owner requirements and contractual technical specifications as shown in figure 1. Further, the project manager establishes required documents to be delivered based on functional and operational requirements considering ship type, class notation and systems onboard. Typically, the planning of resources and scheduling the deliverables carried out in co-ordination with engineering sections. Details of typical project management method are shown figure: 2.

Ship is a high valued asset, which involves collective contribution of owner, designer, shipyard, class society, equipment supplier to realize the final product. The nature of project management in ship design office varies with other industries, as it requires matching the interests of all parties involved. Eventually, design office needs incorporate some modifications and perform iterative design process, result in multiple revisions for each documents generated.

Productivity in design office environment is much focused on efficient utilization of work force to produce quality work and minimizing the time as resource for data preparation and performing work. Decision making & response - a vital aspect of productivity, requires project information and communication to be available readily and precisely. An automated project monitoring plays crucial role in productivity for status reports, weekly progress and setting up action plans.

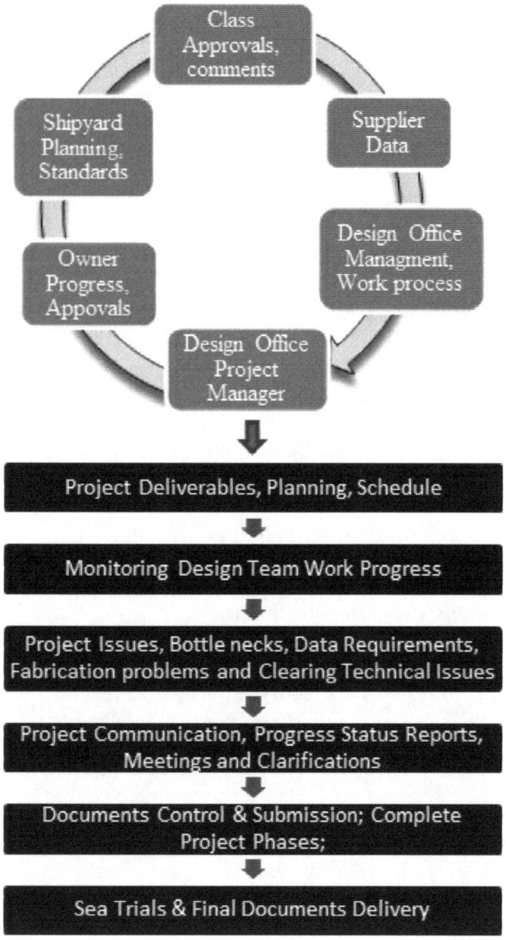

Figure 1: Overall Scope of Project Management in Design Office

In a perspective view, productivity in design office has to be extended to re-usability and best utilization of produced data. Work quality is an important parameter of productivity, which can impact project with reworks at expense of large man-hours and company reputation. For seamless execution and continuous process improvement - feedback, evaluation and knowledge bank of problem – solution needs to be built for future reference of similar project issues.

The objective of this software is to develop a system to handle project management aspects for the design office from the start – enquiry stage - to the end – vessel delivery stage - of the project in standard procedure.

This system will serve to help project managers to consistently manage and monitor the progress of each project deliverables and status. The system developed on web/intranet platform, where application installed on server can be accessed by different sections from local system to fetch project data and update assigned work status. The system should provide role based access to the project to display relevant information and data protection. Database is used to save large volumes of data for the projects.

3. PME SOFTWARE – CONCEPT AND METHODOLOGY

Establishing development scope is essential and foremost step to convert the productivity needs to project goals considering current execution mechanisms and available resources. Operational system of company is divided into organization structure, functional sections and their interaction aspects to get the extent of automation can be carried out and also to establish both issues considered and solutions as goals. Software needs to provide leverage over traditional tools in following areas of design office operations:

- ☐ Database management
- ☐ Progress/Status monitoring
- ☐ Control over project execution
- ☐ Dashboard
- ☐ Data Analytics
- ☐ Feedback and auto alerts
- ☐ Email management

From client enquiry to project closure managing the planning, execution, data and communication effectively with PME Software. It is developed in 4 phases mentioned as below and figure: 3 show the overview of the phases with associated work flow execution.

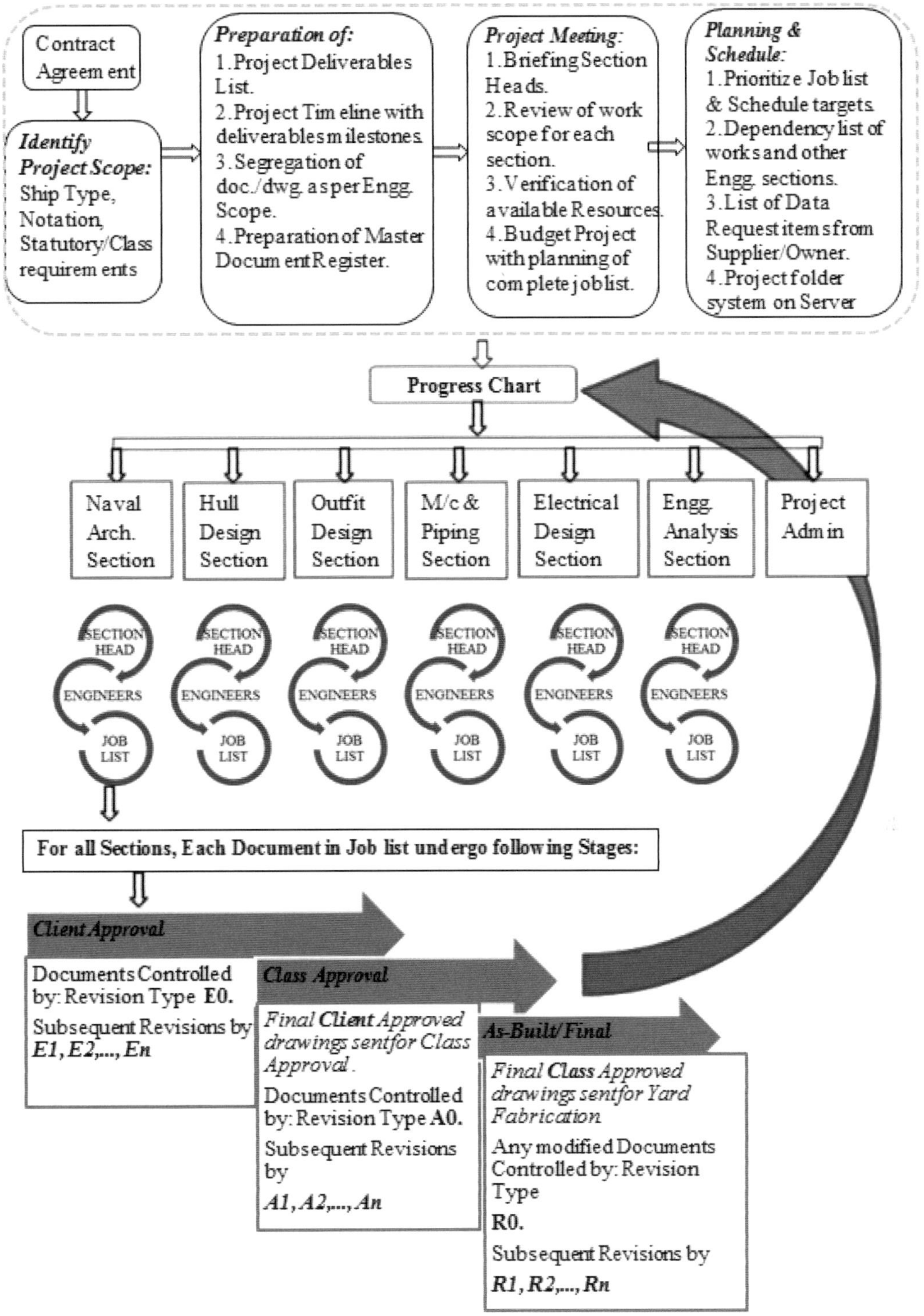

Figure 2: Project Management Method in Design Office

Module 1:	Project Information Management
Module 2:	Project Management and Execution
Module 3:	Dashboard - Project Monitoring and Performance Evaluation
Module 4:	Email Management System & Email Reference Number

3.1 PROJECT INFORMATION MANAGEMENT:

The basic requirement of software is to collect project data through user interface in an organized manner and store in structurally arranged database. This phase enables the design office to generate standard database for organization and enquiry projects, which will become basis for subsequent project development upon confirmation. Database for completed / on-going projects can be prepared in this phase. The framework of this module is as below:

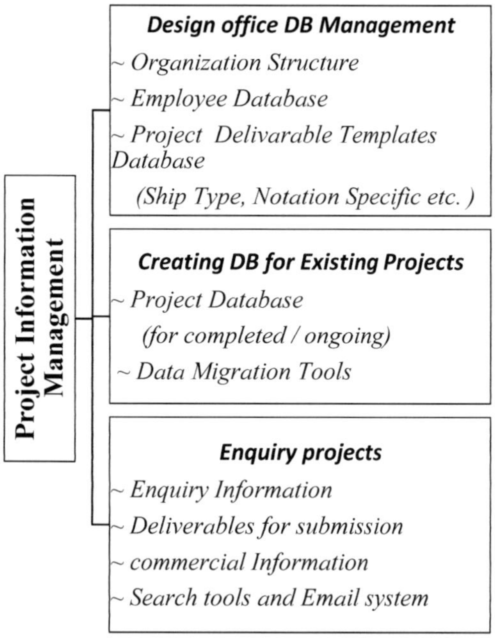

Figure 3: Overview of Project Information Management

In Design office database, organization structure used to create and collect data on hierarchy, designations, departments, office locations and team's categorization. Employee database is resources database connected to organization structure. Project deliverables templates are libraries of master document register prepared categorically based on ship type, Class Notation, Cargo type, Statutory requirements etc.

Creating Database for existing ships is important aspect for design office to preserve the valuable data and future references. The engineering project data is collected related to principle particulars, hull form parameters, tank capacities, accommodation details, lightship data, equipment list etc. Reference documents Technical specifications, GA etc. are made available.

Any project begins as an enquiry, hence enquiries and projects are needs to separate initially and after contract agreement it has to transformed project. Enquiry management is key for design office, where each enquiry is handled with due diligence and quick response to increase success ratio. The potential enquiry is registered as new enquiry and its information, deliverables list, techno- commercial documentation is maintained. Features include communication through Email system and alert system is of great help for design office as large number of enquires needs to be handled.

Apart from preparation of relational database which is essential for resource allocation for multiple projects planning, this concept uses preparation of standard library of deliverables list related to design office based on ship type, notation etc. for all possible cases. Software provided with flexibility to add and modify the library files for any future requirements. These features enable standardization and reusability of data. As creation of this database is one-time activity, it reduces time loss for each project and associated human errors, enhancing the design office productivity. Creation of database for completed projects is referred as digitalization of project data, which is key resource for data analytics and improvement of ship design process for organization.

3.2 PROJECT MANAGEMENT AND EXECUTION

This module collects the real-time project data and stores in database as project progresses. Project Management is the process of breaking the work scope into multi-stage groups and finally identification of activities with defined finishing dates. Further these activities are operated using resources – both man power and time – to obtain data points in the path of planned progress. The data collected is very important source in decision making, setting action plans and processing the data to evaluate performance to improve management processes both in project works and office. This part deals with core functional aspects of project and demands for application of smart technology to add value over all available tools. The software is designed to work on multiple project databases and capable of operated by multi-level users based on their roles and responsibilities at the same time. The framework of this module is as below in figure 4.

Project pre-planning involves automation of manual process of generation of deliverables list, project timeline with man-hours budget and master document register. Also, project manager provided with tools such as check lists, document dependency list and document templates. Project meeting communications and minutes log with work issue to sections in-charge and planning re-schedule. To monitor and resolving comments from client, class and yard manage comments section is used.

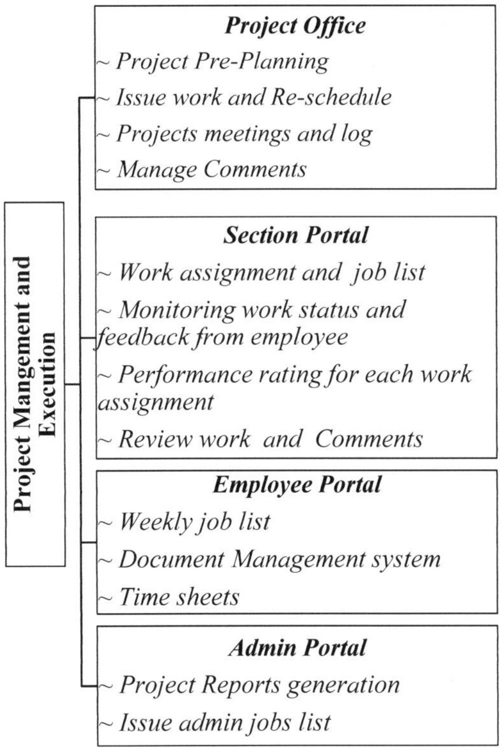

Figure 4: Overview of Project Management and Execution

Section Portal, accessed by designated section in-charges of Naval Arch., Hull, Outfit, M/c & piping, Electrical and Analysis engineering departments. Execution of work carried out by assigning work to engineers after verification of work load on engineers and prioritizing the work based on urgency. The work assignment provided with target dates and standard man-hours to perform. Review of work and performance assessment drawn with provision for noting comments and rating. Work type segregation for reworks, work variations, comments revision is applied for better understanding of project man-hours utilization.

Employee Portal is for engineers to find their work assignments and manage their jobs with standard document system. Engineers prepare database for actual durations for completion of work. Feedback and alert mechanism are used to communicate with section head for requirements of data or issues on progress. Admin portal is to assign admin related works and generate required reports from the database.

Each components of this module is developed as per working nature of design offices, which makes PME software as custom solution for ship design offices. Data protection and security is achieved in two stages, At user end multi- level authentication of user to access data and at server side data transferred by encryption and no credentials given to project team either for accessing and manipulate database. Project activities are maintained by web technologies, where application can be accessed any branch office in the world through internet while all information is available at central server.

3.3 DASHBOARD - PROJECT MONITORING AND PERFORMANCE EVALUATION:

Dashboard is graphical tool to present the real-time project data in a useful productive way through processing the collected project data. The data points of project are essentially the relations of activities (or) man power resources with respect to target dates (or) time periods. The project schedule, work allocation and employee completion dates are collected as project progresses and based on the target date and completion date, the progress and performance charts are generated in real-time for quick review of any project or employee.

- Overall Projects Status monitoring.
- For every project, each department completion status with reference to delivery schedule.
- Project budget time and re-works time monitor.
- Man-hour estimate for Variation Orders
- Employee Work Load Status monitoring.
- Employee Performance Status monitoring.
- Weekly reports generation.

The essence of project management is made readily available using dashboard and gives clear edge over traditional tools. The status data is plotted as percentage compared to planned milestones, which enable user easy to comprehend data. This is powerful tool to determine necessary action plans at right time.

3.4 EMAIL MANAGEMENT SYSTEM & EMAIL REFERENCE NUMBER

Email is common communication mode for projects and project team, where information is transferred between concerned parties. Generally, the mail communication is standardized in design offices, following approaches in usage-

- Direct mailing system to project managers
- Two-tier transfer; where a specific office mail id to receive external mails and then admin to transfer mail to the concerned.

In both cases with its own advantages and disadvantages, requires a better solution to be implemented.

Integration of PME Software with Email management is smart approach for faster communication and security for data transferred. While working in multiple projects, it is also an effective method to segregate data received from the owner, suppliers and class and manage incoming data.

The basic concept is, each project after its status changed from enquiry, automatically a project email id is created by software under project number.
e.g. P2100@seatechsolutions.com

The project mail id is automatically linked to external clientele from database and internal distribution list for project team is made separately. These emails were redirected to by the project manager directly. For sending document attachments, initially files are uploaded through system and an email is sent to the receiving parties to download the file. A notification is sent to office management to inform the successful download of the documents.

System for document submission:

- Email is provided with links to download.
- Attachment files to be uploaded to cloud space before sending download links automatically from office server.
- Passwords for downloading files are sent separately to client.

The automation script is set up on the server to run every 5 min. Following are the features of the automation script:

- Redirecting emails from client to Project Managers on unique project email address.
- Sending received emails to project department members as set in the database for each project.
- Redirecting emails from Project Managers to clients when received at unique project email address.

E-mail Reference Number System:
For outgoing emails, all emails have to be assigned a reference number for record and future tracking.

4. SOFTWARE DEVELOPMENT

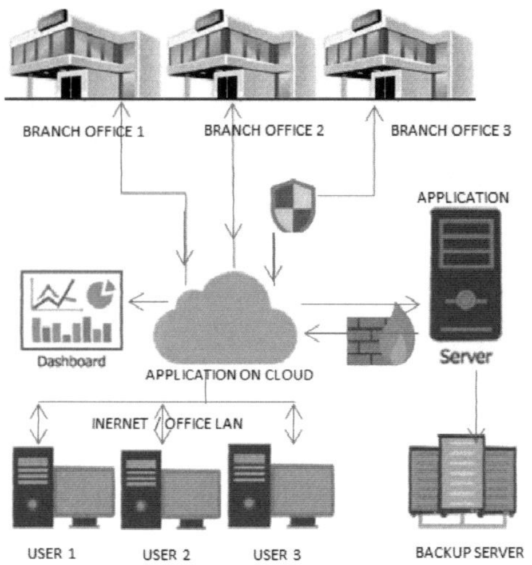

Figure 5: Software Development Concept

PME software is developed using web technology tools PHP, Javascript, JQuery, CSS, HTML5 and Bootstrap as frontend with MSSQL Server as backend. The advantage of these technologies is software can be operated on web/cloud or intranet platform. The project work happening in different countries made simple to manage. Data transfer and communication made secured using data encryption and multi- level user authentication. At server side, project data is controlled and managed using admin protocols to access.

4.1 PROJECT INFORMATION MANAGEMENT

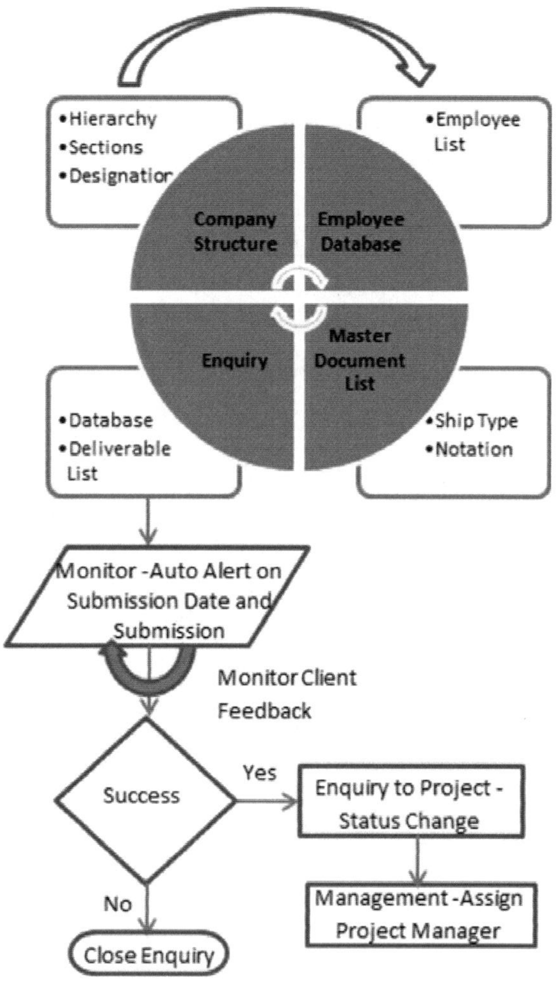

Figure 6: Module 1 Flow chart

Figure 6 shows the flowchart for the Module 1 Project Information Management and the user interface and user inputs are shown in figures 7-10.

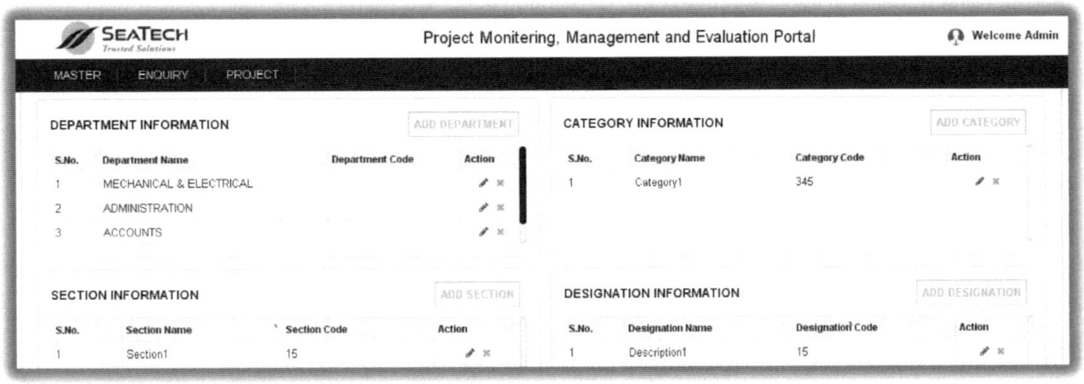

Figure 7: User Interface for Model 1 – Project Information Management – Organization Structure

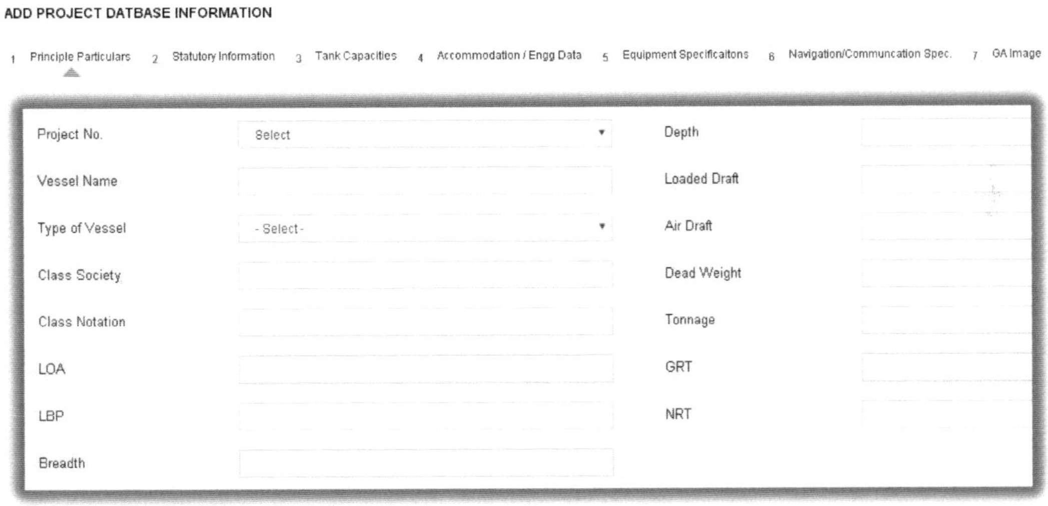

Figure 8: User Interface for Model 1 – Project Information Management – Project Database

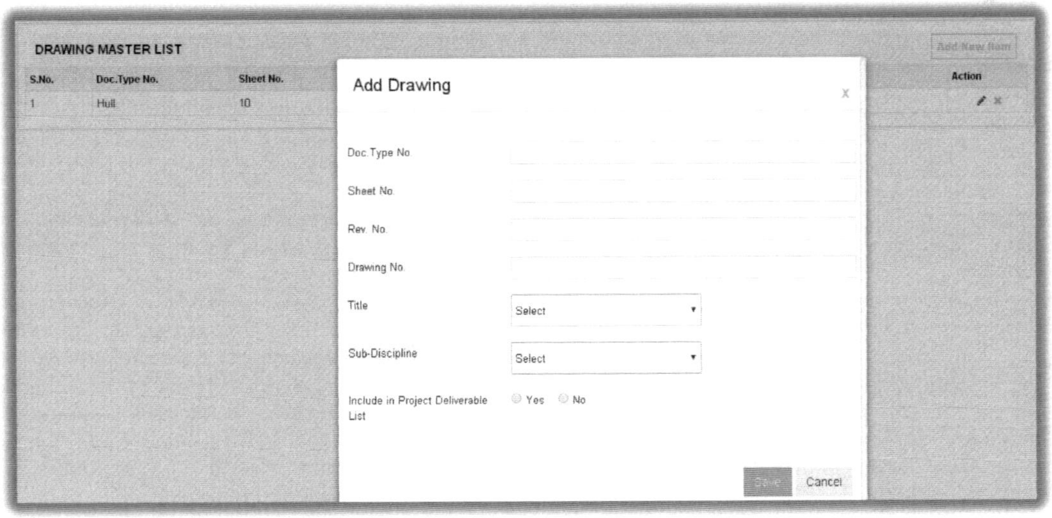

Figure 9: User Interface for Model 1 – Project Information Management –Master Document List Template

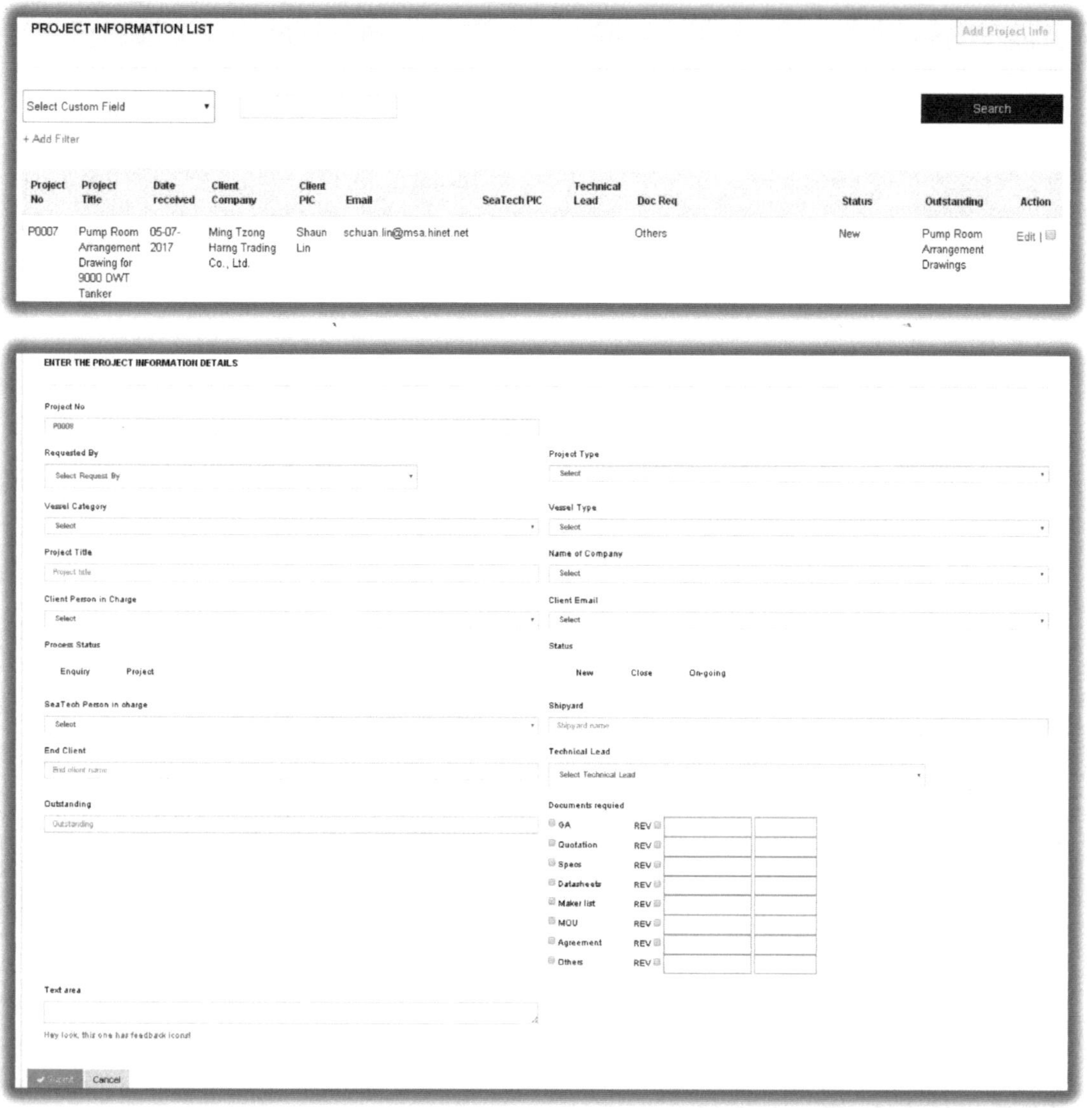

Figure 10: User Interface for Model 1 – Project Information Management – Enquiry Data

<u>Advantages:</u>

- Single platform to manage all project related information.
- Allows users to add and view only their projects and related information.
- With the new system, all information is systematically managed by project numbers on one database system. With a database, the business development department can also track trends of the company and make informed decisions for the company
- Rights of access for each part of the database can be created to ensure timely update of information to the respective departments.
- Information is collected through systematic method into the system. This allows proper search functions and criteria. Engineers can search for what they want quickly and accurately. This helps to reduce time spent from searching all the existing projects and increase accuracy of the information used for future projects or share with clients.
- Eliminate duplicated information.

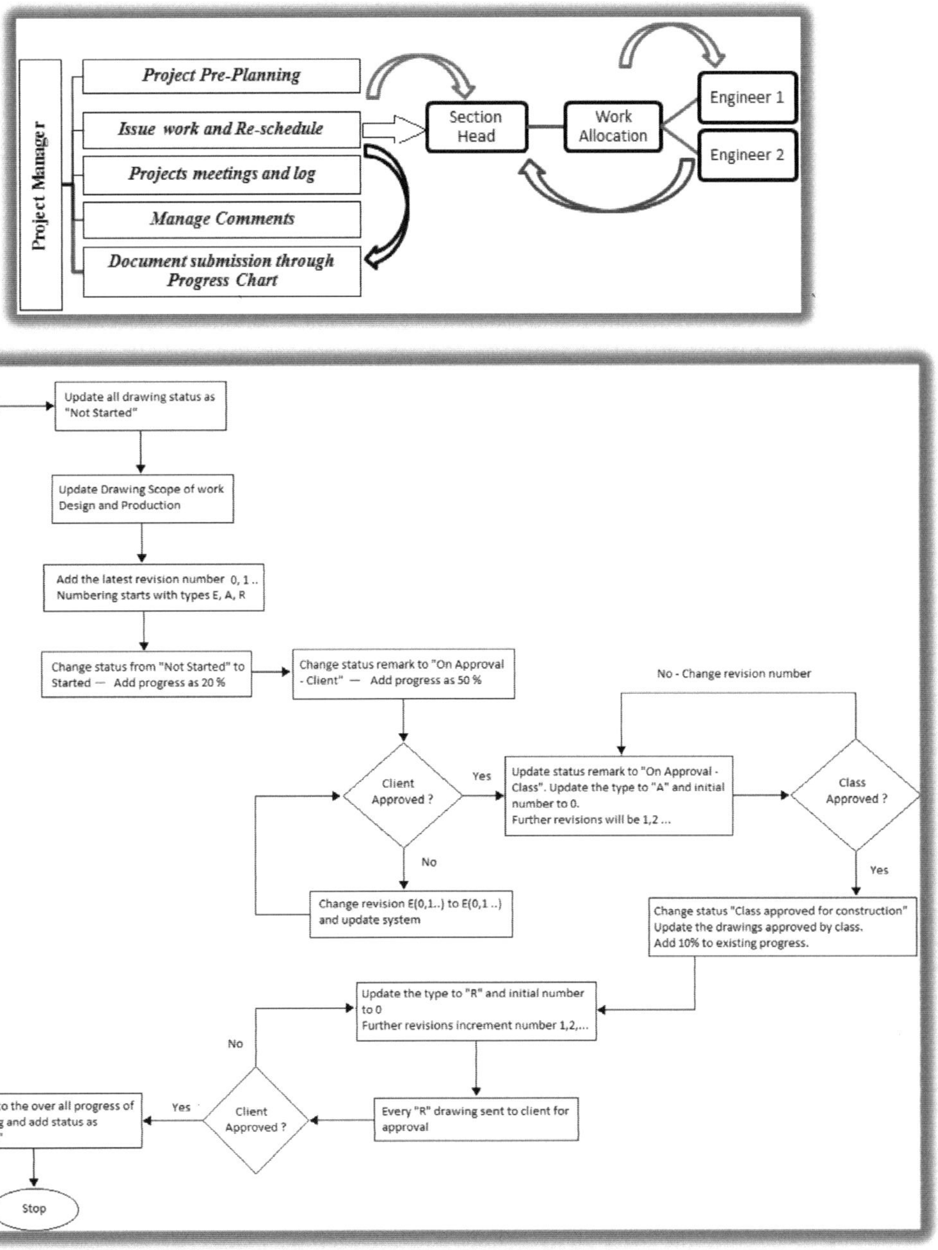

Figure 11: Project Management and Execution of the Individual Deliverables

4.2 PROJECT MANAGEMENT AND EXECUTION:

Figure 11 shows the software workflow control for the project manager activities followed by work assignment to employees. Further after completion of work for each phase, through progress chart email with attachment links is sent to client. The document progress and its % completion status is automatically updated in the database.

Following figures 13-14, typical progress chart with list of drawings and revisions in each phase. By clicking the revision, email will be pop out to send mail with attachment links. Project manager also monitor the meetings, reschedule and comments received as shown in Figure 15-16.

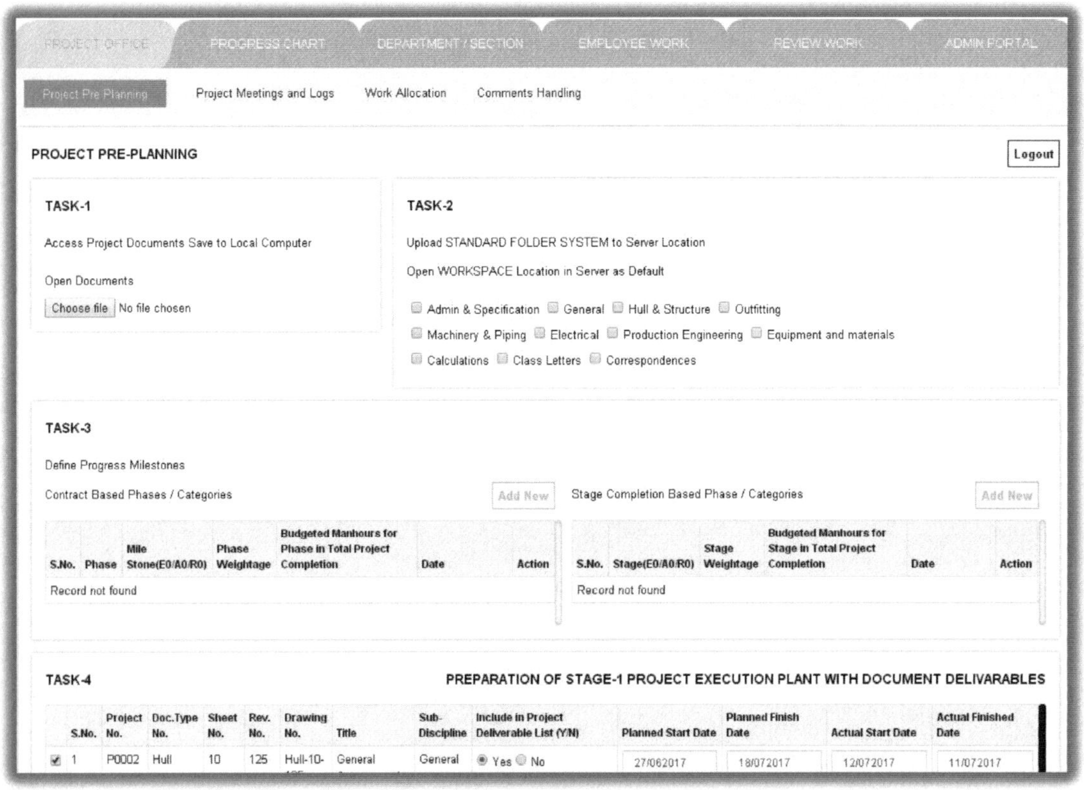

Figure 12: User Interface for Model 2 – Project Management and Execution – Project Office – Pre Planning

Figure 13: User Interface for Model 2 – Project Management and Execution – Project Progress Chart

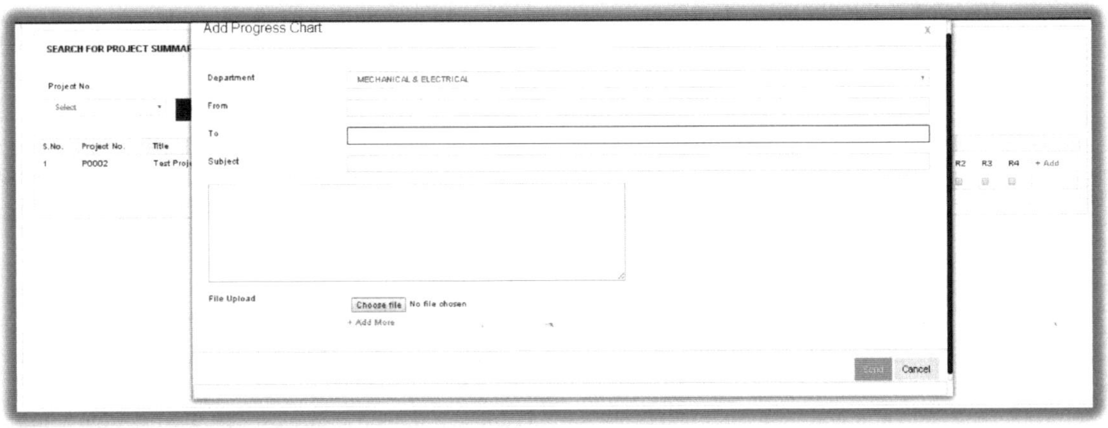

Figure: 14 –User Interface for Model 2 – Project Management and Execution – Project Progress Chart –Update progress & Email to client for submission

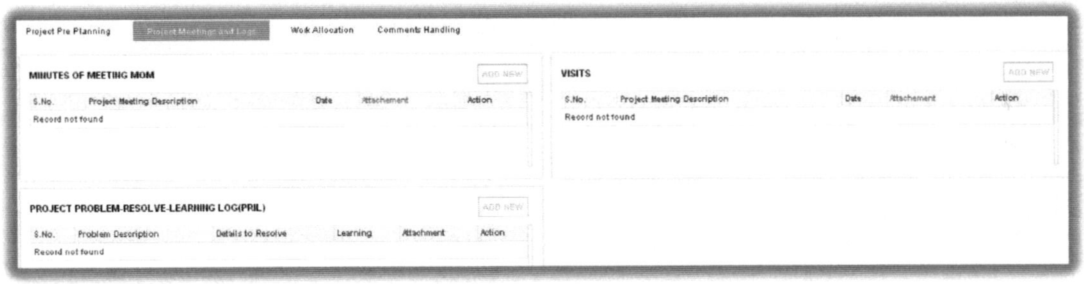

Figure 15: User Interface for Model 2 – Project Management and Execution – Project Office –meetings log

Figure 16: User Interface for Model 2 – Project Management and Execution – Project Office –Manage comments

4.3 DASHBOARD - PROJECT MONITORING AND PERFORMANCE EVALUATION:

Dashboard is the final outcome from the PME Software, which provides real-time status of both projects and evaluation & utility of the resources. Project data collected on work allocation, completion is organized structurally in the database and evaluated against milestones and budgeted project hours using project management principles. The results are plotted to represent the current status as percentage of target work completion.

Dashboard for overview of all active projects shown in figure 17. Where the bar chart shows each project is on X-axis and it's % completion as Y- axis. Data is analyzed and each section data is compiled to calculate the overall % completion. For easy understanding, each project scheduled target completion in Blue bar is plotted along with current status in Orange bar. Similar plotting method is used further on bar charts.

Figure 19 refers to, Dashboard for project performance and sections overview. It is essentially monitoring and tracing of each project's performance and status of work in each section separately. Critical information about man-hours expenses on reworks and variation works are evaluated for management review and subsequent action plans to control.

Figure 20 refers to, Dashboard for Employee performance. From Section heads, data on each job performance rating is collected for every employee in execution module. This data along with actual work completion duration against assigned standard work hours are analyzed to plot performance of employee. Also, based on work allocation, work load on employee is shown in the dashboard.

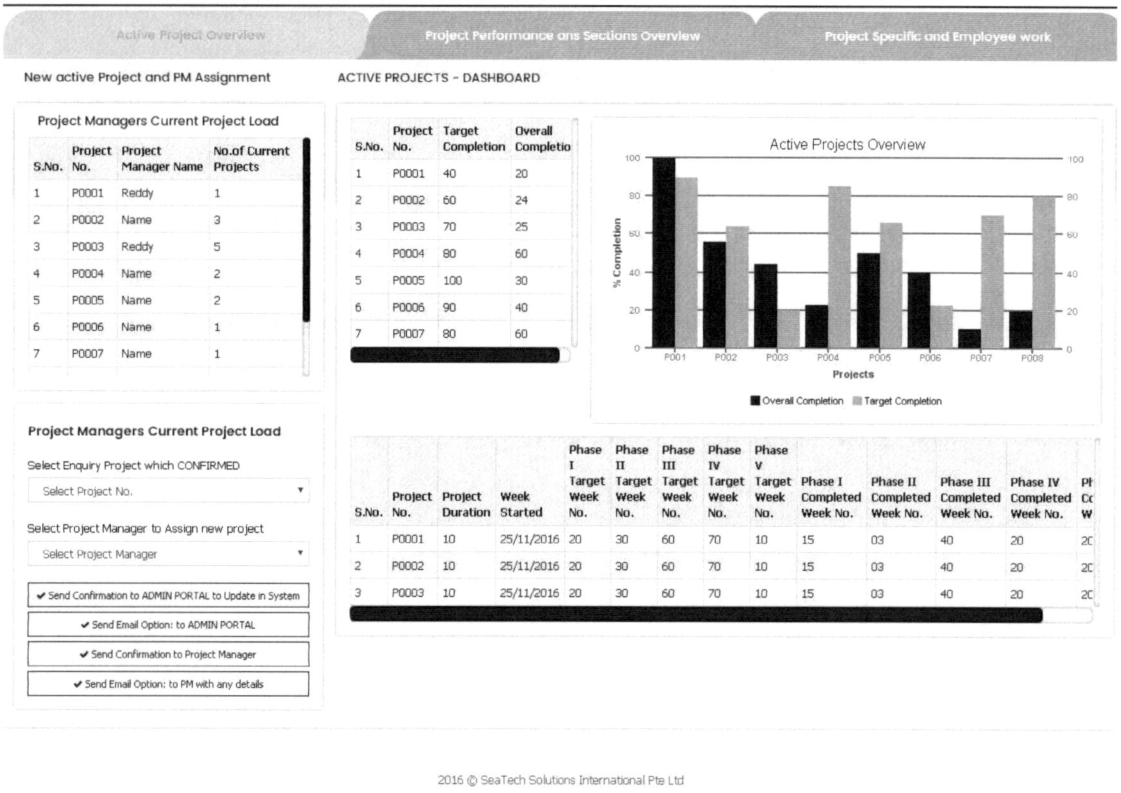

Figure 17: User Interface for Model 3 – Dashboard - Project Monitoring and Performance Evaluation – Active Projects Overview

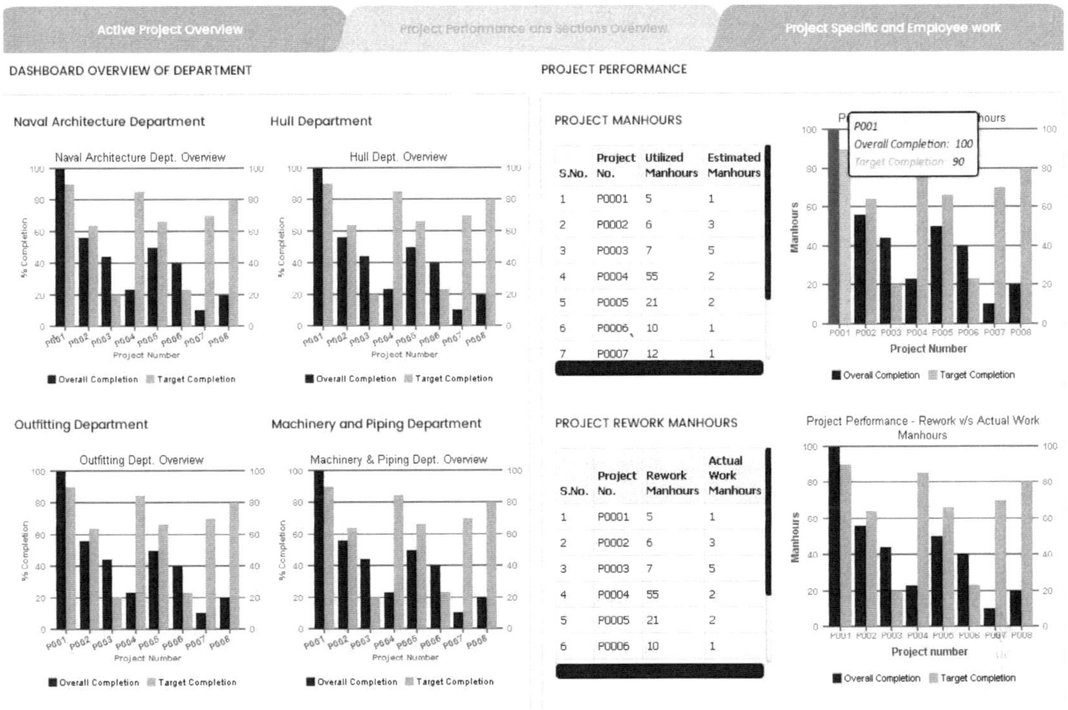

Figure 18: User Interface for Model 3 – Dashboard - Project Monitoring and Performance Evaluation – Project performance and Section Overview

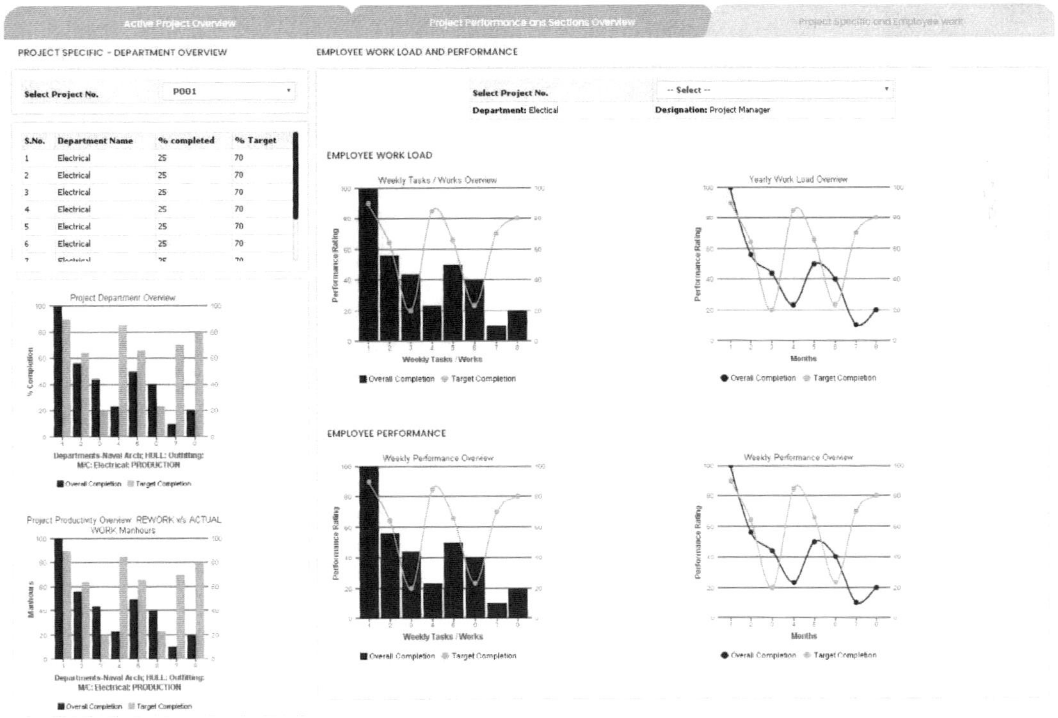

Figure 20: User Interface for Model 3 – Dashboard - Project Monitoring and Performance Evaluation – Employee performance and work load plots

4.4 EMAIL MANAGEMENT SYSTEM & EMAIL REFERENCE NUMBER:

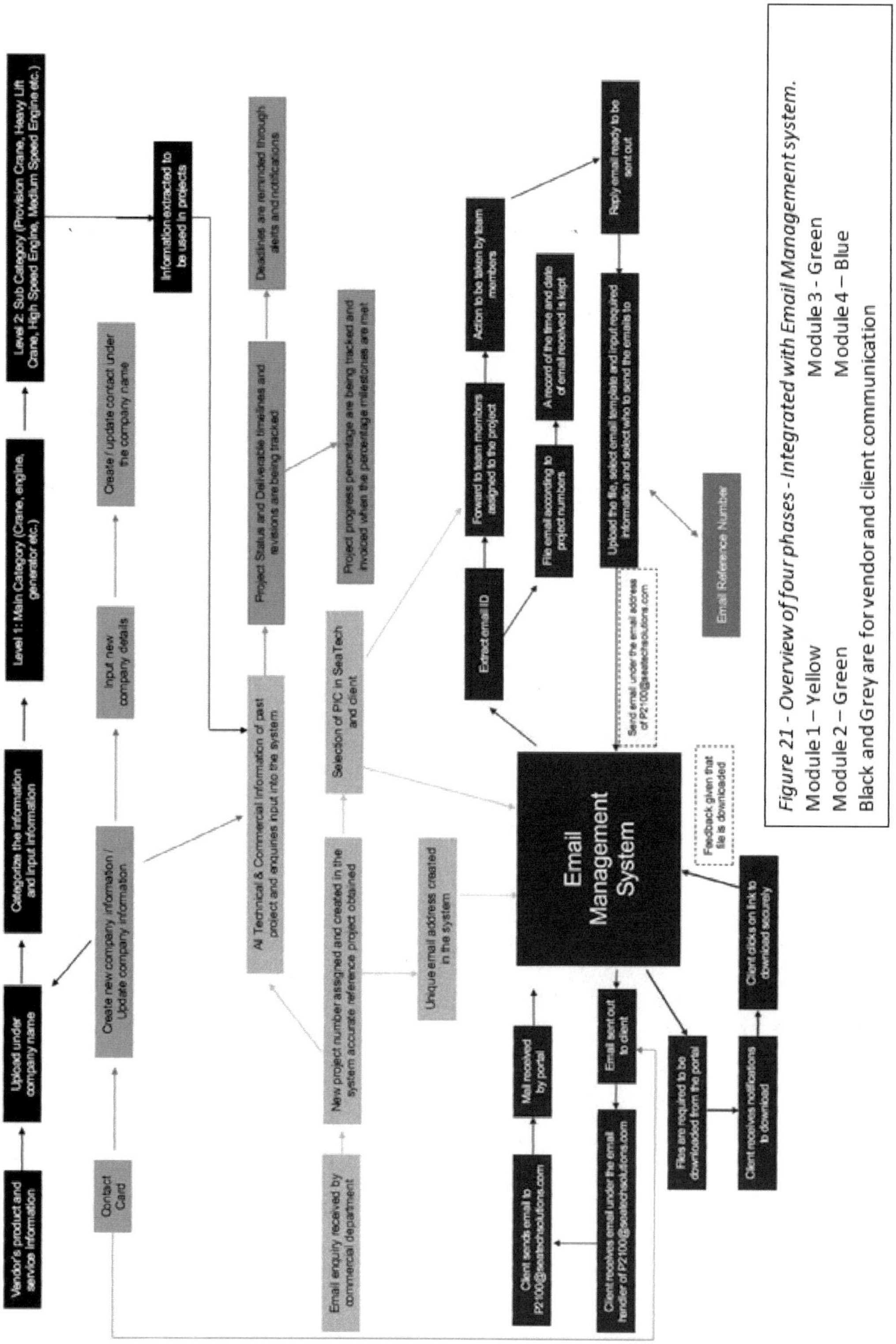

Figure 21: is typical Email communication and management system for each project.

5. PROJECTED BENEFITS OF THE PME SOFTWARE:

S.No.	Process / Actions	Comparison	Time Required	Manpower required	Productivity Savings (Projected)			
					Time (man hours)	No. of Staff	$ Saved	% Savings
1	Retrieval of Data, Duplication of Data, Standardization of Documentation	***Before PME Software***	Average time taken is 12 hours with Average frequency of request: once a week.	1 to 2 staff on ad hoc basis depending on the amount of data per request	Up to 572 man hours per year (approx.) based on an average frequency of once a week	N.A.	$8,580 per year (approx.)	92% (approx.)
		Using PME Software	30 min to 1 hour as all information is available on the database.	No additional staff is required.				
2	Updating of project status, project deliverables, and deadline reminders are done manually via various spreadsheets or email reminders.	***Before PME Software***	6, 000 man hours each year. (approx.)	3 administrative staff	2,000 man hours each year (approx.)	1 staff	$30,000 per year (approx.)	33%
		Using PME Software	4,000 man hours required.	2 administrative staff				
3	Work distribution as subtasks to engineering department, Monitoring and Feedback by Project Manager	***Before PME Software***	2,000 man hours a year (approx.) across 5 engineering departments to be spent.	N.A.	1250 man hours each year (approx.)	N.A.	$50,000 per year (approx.)	75% (approx.)
		Using PME Software	750 man hours a year (approx.) is required.	N.A.				

S.No.	Process / Actions	Comparison	Time Required	Manpower required	Productivity Savings (Projected)			
					Time (man hours)	No. of Staff	$ Saved	% Savings
4	Email Management System - Project Manager Quick access and response to client/yard/ class mail requests	***Before PME Software***	6,000 hours a year (approx.)	3 administrative staff	4,000 man hours a year (approx.)	2 staff	$60,000 per year (approx.)	67%
		Using PME Software	2,000 man hours	1 administrative staff				
5	Email References and tracking history for clarifications	***Before PME Software***	750 man hours a year (approx.)	N.A.	750 man hours a year (approx.)	N.A.	$11,250 per year (approx.)	100%
		Using PME Software	0 man hours	N.A.				
6	Client Relations Management - Database Management. Data is obtained from meetings of Project Managers and Senior Executives	***Before PME Software***	500 man hours a year (approx.)	Up to 9 executives update data received separately	375 man hours a year (approx.)	8 staff	$14, 375 per year (approx.)	100%
		Using PME Software	125 man hours required a year	1 marketing staff given responsibility				
7	Vendor data and publications management.	***Before PME Software***	830 man hours each year (approx.)	N.A.	730 man hours a year (approx.)	N.A.	$29,200 per year (approx.)	87% (approx.)
		Using PME Software	100 man hours required (approx.)	N.A.				

Total Savings per year (approx.)	
Hours saved	9,677 hours
Amount Saved	$177,405

	Current	5 year plan
Estimation of improvement in Project Management in terms of Project to Enquiry success ratio.	Project to Enquiry success ratio is currently approximately 1:20	With the improved efficiency, the success ratio is projected to improve by double to 1:10 gradually over a period of 5 years at 15% growth per annum

5.1 QUALITATIVE BENEFITS OF THE PME SOFTWARE

a) Streamlined operations

- As we try to automate as much as possible and document as many things as far possible with this system, colleagues can focus on their main job scope on hand. Improving Company's overall output as a company.

- Response time with clients will also be improved after we removed the need for administrative department to be the middle person to send out information. The administrative department can now manage the database and ensure that the database is well maintained.

- Monitoring of the deliverable's status through the system will be much easier as compared to existing tools in operation. The system will allow department members to track at every stage which deliverables and drawings are done or pending for action. This will improve the speed of deliverables to our clients.

- With an increased savings and higher revenue, the overall profitability and cash flow will improve too.

b) Better management of human resources.

When we have automated most of the administrative jobs, we can train the administrators to manage the database, tend to the request of data extraction, support or move to other departments like Sales and Marketing to help increase the sales of the company.

- Having employees of better training and higher quality will increase the company's productivity in the long term.

c) Proper data management of the information.

Information of all projects and enquiries are kept and maintained in the system. Such information is essential to us so that we have a better understanding of Company's past and present. Having a comprehensive database also allows us to retrieve information quickly, helping us save time and reduce manpower.

- The system will allow information consistency throughout the project's lifecycle beginning from the enquiry stage till the completion of the project. When the data is saved in the database, redundant and outdated data is eliminated and data recovery is more manageable.
- Information on every stage is entered through the system. Less precious man hours are wasted on repeated searches that are a result of poor data and information management.

d) Better decision making.

- Based on the accurate information that is recorded and obtained in the system, the Senior

Management, Project Managers and Engineers can make more accurate decisions on their subject matter.
- In the short term, data can be used to provide an accurate reference for future projects.
- In the long term, senior management can make use of the information collected over the years to study trends of the market to provide quality information to our clients

e) Improved project success rate.

- Accurate information helps to produce speedy and quality replies to the clients and customers. These will help to build the trust clients and customers have for us; increasing our reliability and improve Company's reputation. This will also improve our enquiry-to-project award conversion rate.

- The improved topline will not only increase our surplus but also provide opportunities for
  - Fuller utilization of existing engineers, and
  - Possible increase in the company manpower leading to stability of resources

6. CONCLUSIONS

In ship design offices, productivity and efficiency can only be achieved by custom made project management system. Project monitoring, dashboard, execution, database management, performance evaluation, alerts and feedback are the key features for the software tool, apart from traditional resource planning and scheduling aspects of PMS. Introduction of web technologies in design offices will lead to effective results in increase in savings of working man-hours and costs for expenses.

Application of project monitoring, management and evaluation software is key to meet design office requirements of handling complex iterative processes and delivering the project under the influence of external sourcesand project team internally. Eliminating human errors and standard work procedures are the essential factors for sustainable company growth under dynamic market conditions.

Real-time assessment of project status, early prediction of bottlenecks in the project progress will enable company management to plan & prioritize the deliverables and resource allocation. Continuous evaluation of personnel performance will help company to enhance suitable skills in project teams.

7. ACKNOWLEDGEMENTS

This work was supported by the SPRING Singapore Capability Development Grant for Productivity Improvement. The authors gratefully acknowledge them for financial support of this work.

The authors deeply thankful to Mr. Govinder Singh Chopra (Managing Director) and Mr. Dilip.G.Sarangdhar (Technical Director) of SeaTech Solutions International Pte Ltd for continuous support in monitoring and review of this work.

We also greatly appreciate the support and participation of our project team.

8. REFERENCES

Seatech Solutions International Pte Ltd,

- Quality Management System documents
- Quality Standards
- Work Instructions & Procedures

9. AUTHORS BIOGRAPHIES

Prabjot Singh Chopra holds the current position of Manager, Technical & Business Development at SeaTech Solutions International (s) Pte Ltd. He helms the Initial Design team at SeaTech Solutions, where he oversees the conceptualization for new designs and customers enquiries. His previous experience includes Structural Analysis of Jack-ups and jackets, Hydrodynamic Analysis on Offshore Fixed and Floating Structures, Earthquake Analysis using USFOS, Dynamic Analysis of Offshore Structures and Numerical Analysis of Free Vibrations including fully coupled interaction between structure and liquid.

Umamaheshwar Reddy Gadekari holds the current position of Assistant Manager at SeaTech Solutions International (s) Pte Ltd. He is responsible for ship design project management, marine software products development, design office project management & engineering applications development and performing design calculations. His previous experience includes managing design projects for tankers, OSV/DSV/AHTS, tugs, heavy lift & pipe lay vessels; software development for Ship Loadicator, Rule design calculations, Expert system for ship design software.

INNOVATIVE PROGRESS MONITORING TOOL FOR ELECTRICAL DETAIL DESIGN

Anish S, Cochin shipyard Ltd, Cochin, India

SUMMARY

This paper focus on methodology which could be adopted for integrating detail design progress monitoring functionality within 3D software platform making use of production information using data analytics. A data exchange mechanism with 3D software presentation layer is developed to automatically capture all type of data modification associated with individual entities in information model and drawings at user end with time stamp. Drawing revision control is also managed automatically. A dashboard capability with customized reporting of various key progress indicators for higher management is also provided. The above methodology allows a ubiquitous arrangement wherein the requisite data for data analysis is obtained from end user automatically. The availability of real time status of detail design updation improve the productivity reducing project timelines and proves to be a handy tool for design and planning departments.

1. INTRODUCTION

Shipboard electrical systems include an electric power plant, lighting, interior communications, radio network, controls, exterior communications, navigation and many other safety and mission related electric/ electronic systems. Shipboard electrical design involves in development of key plans and yard plans associated with the above power generation and distribution, control and monitoring of various ship systems, lighting, navigation and communication systems. A typical electrical design work flow is as detailed in figure 1 below.

Output of electrical design is divided into

- Key plans -Wiring diagrams, connection diagrams and arrangement drawings indicating various electrical components as available in various compartments.
- Yard plans- Fabrication drawings for cable way, cable way supports, equipment foundations and fit up /installation drawing which indicates in detail how to install cable ways, electrical equipments in various areas on-board.

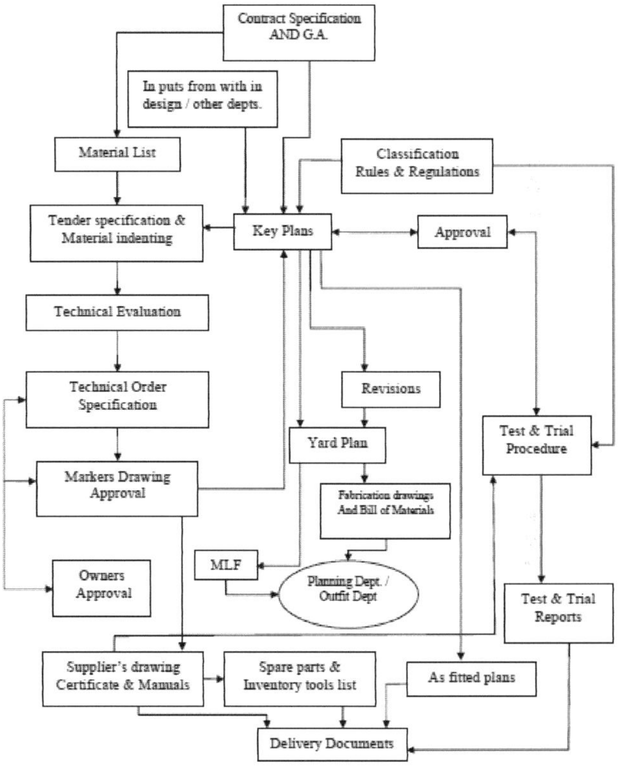

Figure 1: A typical flow diagram for electrical design

2. CAD SOFTWARES – SOFTWARE PERSPECTIVE

Shipyards use a combination of two dimensional (2D) and three dimensional (3D) softwares to develop the key plans (schematics) and yard plans (fit-up /installation drawings) for ship design considering easiness. These software environments are object oriented, modular and support various inter-process communication (IPC) frameworks.

Key plans are usually developed in 2D softwares, which over the period have evolved from simple geometry based systems to very complex object oriented architecture. The new generation 2D software enables data storage of key parameters and properties of individual objects within the drawing to facilitate development of Material list of Fitting (MLF), Bill of Materials (BoM) and other data required for production and procurement. 3D softwares also followed the same path from primitive contours to complex data oriented system which helped in extraction of 2D sectional views, MLFs,BoMs,etc.These softwares gradually evolved from stand alone systems to network centric systems based on client server topology, multi tier architecture and with a relational /proprietary database backend. The concept of one common product information model or "ship database" for each project enabled availability of the complete project information to entire design team. Thus multiple users could work concurrently on the "same model". This resulted in reduction of errors due to proper "collision detection" among various components sharing the same space in ship model.

Normally Computer Aided Design (CAD) environment is not integrated/interfaced to other enterprise level softwares available at organization level like enterprise resource planning environment (ERP) and stay as standalone "islands" with no synchronization.

3. PROGRESS MONITORING – PRESENT SCENARIO

Ship production design/yard plan begins with development of various documents viz master equipment list, drawing list, area definition, block division plan/list, compartment list etc as per build specifications. Electrical detail design involves in development of various yard plans (fabrication drawings for cable way, cable way supports, equipment foundations and fit up /installation drawing). The estimated time lines are developed for each drawing in drawing list considering ship type and specifications. This list goes for various levels of discussion at department level and value additions are incorporated as per experience. Based on the above list, team leader prepares team allocation for detail design.

Based on the details promulgated by yard standards, cable way definition/modelling is done throughout the ship. Foundations and equipment modelling for various electrical control panels are developed based on binding details as made available from system vendors. Equipment binding data availability is subject to the timelines of procurement plan/progress. Final system configuration and disposition onboard shall be subject to key plan approval by owner/classification society and other regulatory bodies. Compartment completion for release production drawings depends on the input data availability as above and also depends on the design progress of other ship design departments including turnkey contracts.

The design progress measurement is usually done by maintaining independent worksheets in which progress made for various design works/drawings are recorded as per interaction with individual team member/team lead. This crude method of manual progress reporting is highly cumbersome and error prone. Here the time needed for reworks, design modifications, version control, delays and other non factored changes in project timelines are rarely recorded/explicitly available for data analysis.

4. METHODOLOGY PROPOSED

Main actors involved in the proposed methodology are as follows and indicated in figure 2 below.

- 3D Software
- Progress monitoring software with following sections
    - Embedded mechanism for detecting and responding to creation, modification and deletion of equipments, foundations and cableways including supports within 3D software
    - Visualization manager for display results
    - Info exchange mechanism framework
    - Front end for displaying analytics
    - Dash boards for higher management
    - Interfaces /Connectors for external enterprise level software including ERP

Initially the method of communication of 3D software and its internal database is analyzed and suitable connectors are developed as per the Application Programmable Interface (API)/interfaces available to communicate with the "ship database model". These connectors are used to query information regarding components used in electrical detail design viz cableways, equipments, equipment foundations associated with project under consideration and parse native/open formats of 3D software information. Suitable mechanisms are developed to detect and extract requisite information on the above components being created, modified and deleted within the application view space. A visualisation manager is also developed using internal APIs to generate views, drawings that were issued to production department. The connectors as above and visualizer shall be made available embedded seamlessly to 3D software at client end.

The info exchange framework collects the results returned by 3D software connectors for doing various analytics, generates various queries for these connectors to fetch details from 3D space and converts the results into neutral formats for future use in the internal database. This framework shall be developed at the server side.

The front end of analytics shall be used to generate various Key performance indicator (KPI) and progress information for project management. Dashboard functionality is also catered for higher management needs. The analytics front end and dashboard are available at client end.
Suitable connectors are developed to undertake requisite data exchange between enterprise level softwares including

enterprise resource planning (ERP) software and document management system (DMS). These connectors are embedded into the analytics front end for manual or automated data exchange.

5. IMPLEMENTATION

The implementation of progress monitoring tool involves development of following sub components using a combination of desktop, web, server/service oriented and mobile technology.

- Embedded scripts for event detection and visualization.
- Info exchange framework
- Analytics front end
- Dashboard

Careful analysis of activities involved in each yard plan was also carried out to ascertain any possible improvement in productivity by usage of automation and customisation of existing process. Data collaboration and traceability was given the topmost priority considering the man-hour savings by avoiding input of similar data on multiple occasions and possible errors during the subject process.

Material control aspects considering high uncertainty of detail design process to extend possible was also considered. Design engineering factors like delivery time, handling, transit route for under deck items, installation time of complex items, dimensionally large and heavy items were also given special attention in process definition and for developing suitable KPIs. Similarly project management aspects like turnkey/sub contracted items, delays in procurement, drawing approvals, risk management using risk index parameter for vital equipments to extend practically feasible is also considered in development of KPIs. Visibility of these KPIs during detail design process shall enable end users to pay more attention to these factors and shall result in efficient and productive designs.

Provision is catered in the environment to enable supervisors to understand the exact load on each team member on a particular time period is also available for efficient load management and to handle future needs with minimum disruption. The data model also helps in restructuring any possible misalignments due to concurrent or sequential tasks in the schedule as the design activity progresses. It is also possible to "hold" design/issue of production drawing for a particular ship area due to nonavailablity data or due to inter departmental process.

The data stored in databases are project specific but are intended to be of use to a wide variety of users. The centralised server configuration of this database enables the progress monitoring software to allow multiple users at the same time. The database schema is developed considering various input attributes and properties viz project info, equipment, equipment foundation, cable ways, drawing list, master equipment list, area code list, compartment list, user info, procurement details and schedule time lines. From these inputs various processed outputs viz BoMs, MLFs, Revision details are also stored for generation of customisable reports as per user needs.

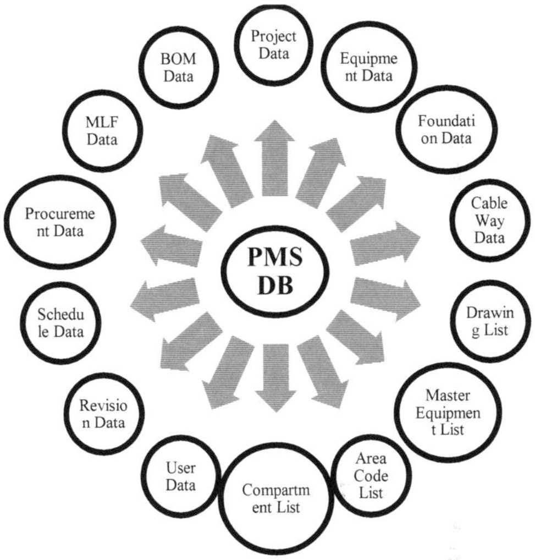

Figure 3: PMS Database outline schema

Modular open source based solutions were preferred in comparison to closed monolithic approach. The selection of technology for implementation was arrived at after careful consideration of various aspects like open source, deployment, maintenance, availability of technical support, obsolescence and interoperability with existing softwares etc.

Based on API/interfaces as made available in 3D software to communicate with the internal "ship database model" in native/open formats, various modules/functions are developed to object attributes and properties of electrical components. These modules are triggered automatically using the infrastructure as made available within the 3D software (eg: using triggers in TRIBON 3D Software) during creation, modification and deletion of individual component/object. Then this data is suitably transmitted automatically to the info exchange framework to be processed and stored in the internal database with proper timestamp and user credentials.

As per the request of end user, MLFs, BoMs and other drawings are generated on the fly using the already available customised scripts or new scripts are developed to automate the above process. The metadata of these drawings are also stored in the database for future use including revision management. Suitable Graphical user Interface (GUI) is developed to enable end user to assign which all components are to be considered in the individual drawings as per drawing lists with minimal manual inputs for the first issue and consequential releases. The changes across various revisions of these individual drawings are automatically analysed and suitable reports are generated internally for dashboards and data analytics front end.

A visualisation manager is also developed using internal APIs as available in 3D software and external API/frameworks to generate various section views and also to display drawings already created. In the visualisation manager it is possible to generate a specific view a ship area with respect to an individual revision of an applicable drawing. A customisable query builder is also being integrated into the data analytics front end and visualisation manager for catering end user needs. The data management scripts and view management scripts are embedded seamlessly to 3D software at client end.

The deployment of Info exchange framework can be done in an application server or as a service. MySQL open-source relational database management system is being considered for database solution. The application server shall also share the hardware resources with the database server considering the data handling requirements. The inputs to the data analytics shall be in the form of text files, spreadsheets, eXtensible Markup Language (XML) as per the integration and external application interfaces. These inputs shall be generated by the users directly or as a result of processing from enterprise softwares like DMS, ERP etc. Data analytics front end can be implemented as a desktop application or a web based application. If developed as a web based application, the same shall be separate from the info exchange framework application but however could coexist in the same physical server. The dashboard application implementation has much wider choice starting from desktop to web or mobile based. Initially a desktop based solution is being considered for prototyping and later a mobile based solution is being planned.

6 CONCLUSIONS

The proposed methodology of progress monitoring tool for electrical design is a paradigm shift in the present manual process of progress and productivity monitoring and reporting. The data analysis is obtained from various details from end user automatically and with minimal interaction with team members, real time status of detail design progress is available to supervisory and higher management level. It is expected that the usage of subject tool shall improve the productivity reducing project timelines and proves to be a handy tool for design and planning departments in shipyards.

7 ACKNOWLEDGEMENTS

The author would like to thank his superiors, colleagues in the Cochin shipyard Limited who had instigated confidence and support during the preparation of subject paper. He would also like to thank his wife, daughter and parents, for their trust and encouragement.

8 REFERENCES

1. Rafael de Góngora, SENER Ingeniería y Sistemas SA, Madrid/Spain, A New Collaborative Engineering Management Tool for Shipbuilding, COMPIT 2004
2. Nick Danese, Nick Danese Applied Research, Antibes/France, Ship CAD Systems - Past, Present and Possible Future,COMPIT 2010
3. Anish.S, Cochin Shipyard Ltd, Economic Ship Building – Design Perspective, 2010
4. Theodoor de Jonge,Nupas-Cadmatic,Groningen /Netherlands,Henk Kramer, Nupas- Cadmatic ,Groningen /Netherlands, Integrating CAD/CAM in a PDM/ERP Environment, COMPIT 2014
5. INCOSE(2015), System engineering handbook

9 AUTHORS BIOGRAPHY

Anish.S holds the current position of Assistant General Manager, Electrical design department at Cochin Shipyard Limited. He is responsible for electrical design of defence and commercial ship building projects and also development of bid designs in defence segment.

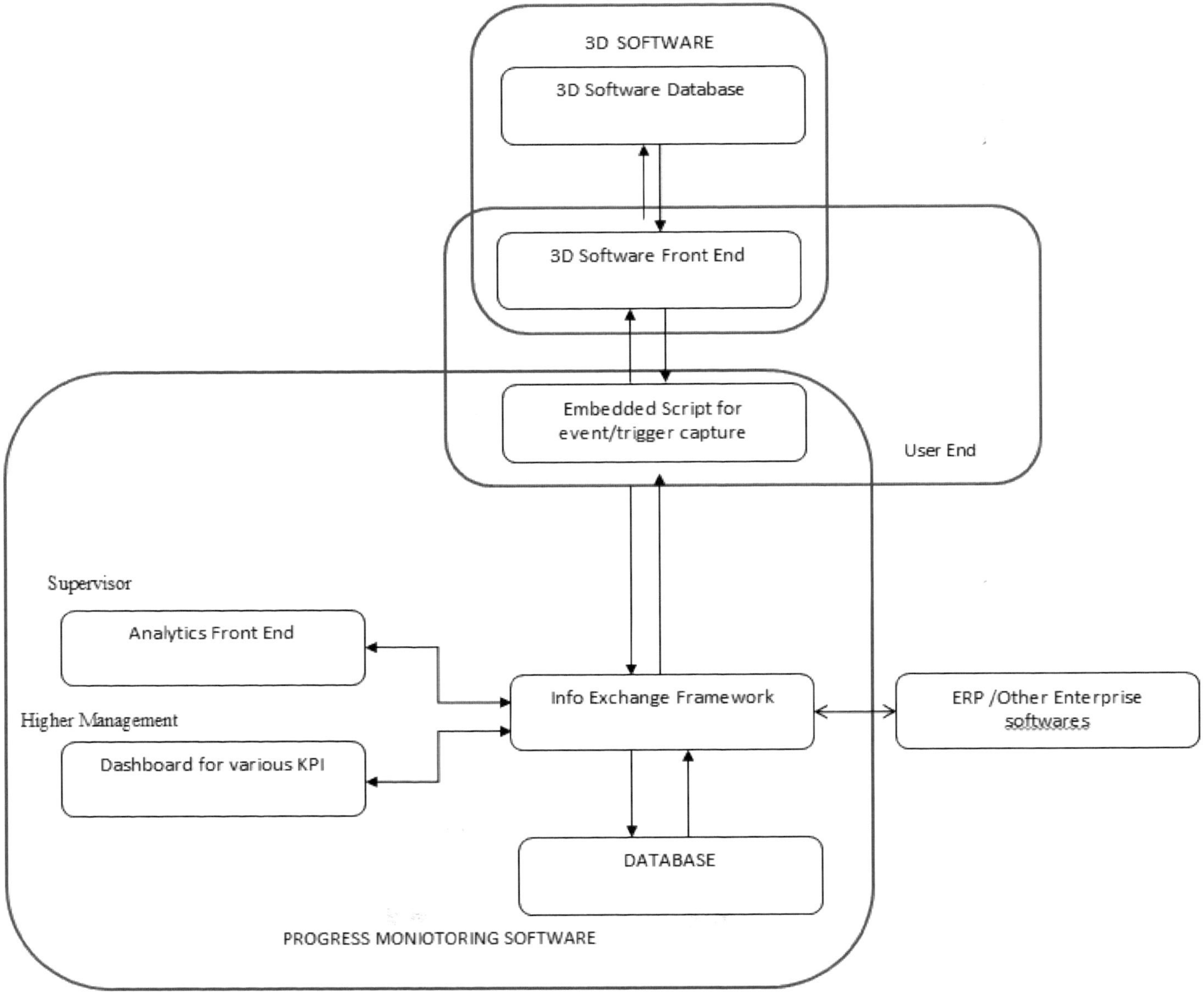

Figure 2: Basic scheme for PMS

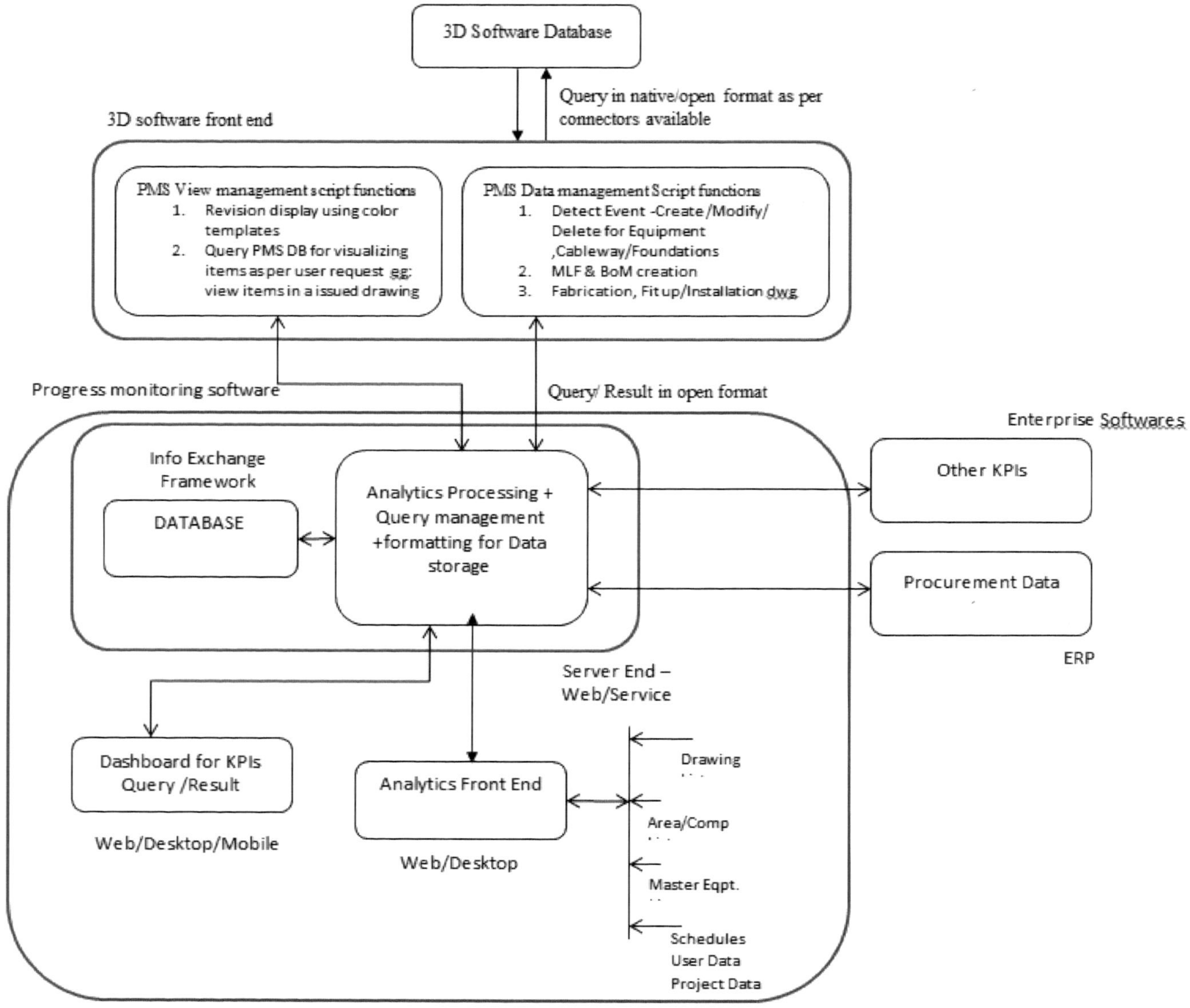

Figure 4: Outline software architecture

AUTOMATED GENERATION OF DETAILED CABLING DOCUMENTATION FOR CRUISE SHIPS', PASSENGER AND COMMERCIAL VESSELS' ACCOMODATION

Dr. A V Lalechos, LePlan, Greece
D Maketas, LePlan, Switzerland
A Pavlou, LePlan, Greece
I Davaris, LePlan, Greece

SUMMARY

This paper presents a novel approach for the complete design to commission lifecycle of electrical & electronic works of modern cruise ships & large passenger vessels accommodation. Smart accommodations are designed and fully documented once technical specifications are defined. Electrical schematics are generated automatically using a set of variable enhanced templates. Templates are selected in an automatic way based on linguistic description of the project. Detailed installation instructions, bill of materials and cable lengths are, among others reports, also generated.

The tool handles among others, low power distribution, lighting, heating/cooling, data networks, voice networks, smart alarm systems, safety systems, access control systems and rack content management. Total processing time is on average less than 22 seconds per cabin.

1. INTRODUCTION

The increasing size and complexity of the accommodation of Cruise ships, passenger and commercial vessels, requires a novel and enhanced way to address all sort of challenges concerning the electrical and electronic aspects of each project from conceptual design phase up to and including commissioning phase. Today tasks such as the conceptual - final drawings design and management, generation of BOMs and instructions to field installers, are commonly handled from different persons or teams, making it challenging to accurately monitor the process and deliverables. Numerous major and minor changes are often hard to communicate to all involved parties timely. Optimising the process is a must in order to reduce design time, cut cost, minimize errors, enhance budget accuracy and compliance, and more importantly generate a complete and easy to follow set of electrical drawings.

Inspired as a 21st century, Particle Accelerators facilities planning tool, LePlan electrical design and documentation tool is an innovative software that converts verbal specifications/concepts using a wording easily understood by naval architects and shipping interior designers into actual electrical plans in an autonomous and automated way. Eventually the tool generates course cable lengths estimates required for budgeting as well as final accurate estimates for procurement purposes.

Unified electrical drawings for all systems, cable installation instructions, potential penetrations sizing, complete vessel's Data-Centre wiring plan, unique cable identification and optimal cable path based on cable interference restrictions, are common reports supplied to the project owner. Finally the resulting drawings are not static and can be utilised by crew during normal operation to timely troubleshoot and repair faults that might occur in the futures

2. CONCEPT

LePlan tool is inspired from modern modular software engineering principles and best practices. Electrical drawings are implemented as a combination of individual single-functions. A subsystem, which is called function, is the combination of several single-functions, which forms a complete system, called group-function is the combination of one or more functions. For simplicity, all levels of functions, irrespectively of their complexity, are commonly called macros.

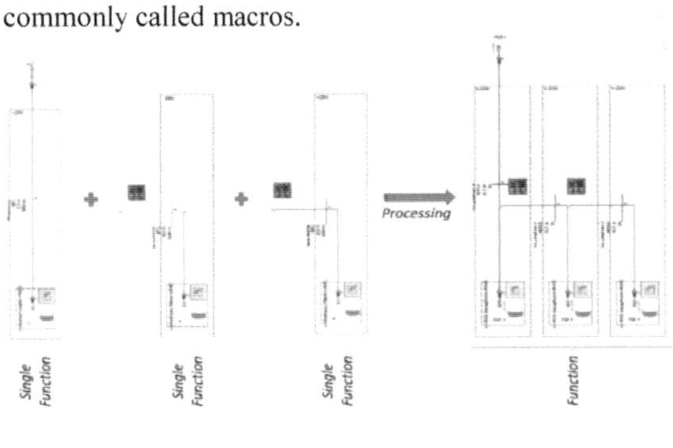

Figure 1: Example of a Function macro built up from three individual single function macros

2.1 MODULAR TEMPLATES/MACROS

The modular design of an electrical system is not something new. Widely available electrical documentation tools have the functionality available for ages; however, in all cases templates remain static.

The term static, refers to templates that are fully predefined. For instance, cables' sizes are pre-fixed and changing the power requirements of a given power line requires the manual alteration in the final drawing, or the adaptation of the template before using it.

LePlan tool goes one-step further and uses a combination of artificial intelligence (AI) scripts that identify which single functions, function or group functions should be used in order to produce the required design and documentation. Therefore, all appropriate and necessary macros are automatically selected, placed on the drawings and required information such as cable's cross-section, cable type, connectors type and size, junction box type, feeding fuses, cables start and end location etc., are automatically populated by the software.

2.2 PROJECT'S SPECIFICATION DATA

Data input for the application originate from a unique source of information, called "Electrical equipment specification form". The data of the form are automatically imported and stored on a local database. Once processed, the data are the only input used by the software in the decision process for sizes, IDs/names, source/destination location etc. The form is mostly drop-down list type to avoid mistype errors and is custom built to fit perfectly each project.

Figure 2: Example of the web-based "Electrical equipment specification form"

An "Electrical equipment specification form" can have any form that can be linked back to a database. For instance, the data can originate from a project's web page, where the client can fill up a dynamic form from a remote location without delaying the specification collection process of the project.

3. DOCUMENTATION GENERATION

Once data collection is verified and initial templates alongside with scripts are designed, the generation of the final documentation is completed within minutes. More importantly, late changes to the data input on the "Electrical equipment specification form" and regeneration of the templates, scripts, and drawings is completed anew within a few minutes.

Making changes can be as simple as reducing low voltage power distribution sockets numbers at a particular location or as substantial as doubling projects size. In a similar way, any closely related project such as a sister vessel can be generated within a matter of hours. Even more significantly, even a non-related project can also be generated within a matter of hours. The need for new single-function templates or not, is the only parameter that can extend the required design time. In most cases no new single functions are needed thus, no new preparatory work is required and once the appropriate form has been submitted, any project can be fully documented within few hours.

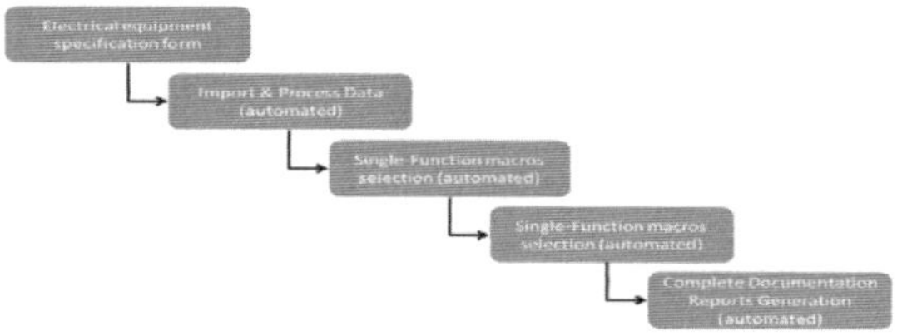

Figure 3: Complete documentation generation flowchart

4. DOCUMENTATION STRUCTURE

The documentation structure follows the ISO/IEC-81346 standard. In principle, a system can be documented using three distinct ways, by answering the simple question "Where I wish to find the documentation," in order to decide how to document a project,

- Location wise:
  When a project is classified by Location, all documents are stored in the source or target location i.e. MSB (main switchboard), EMSB (emergency switchboard), cabinet X etc.

- Product wise
  When a project is classified by product, all documents are typically stored nearby the actual system, i.e. bow-thruster room/cabinet

- Functionally
  In more complex systems, it is common to find documentation is distinct functional folders. A typical example is "Motors List document". Such document holds all the info related to each motor to be found on-board and in most cases includes all/most information related to the automation system handling the operation of each individual motor.

LePlan tool is designed in a way that documentation can be made in all three modes depending on client's needs and preference.

5. NAMING CONVENSIONS

Among the most common problems encountered in most projects, is the use of a non-common naming convention. During the initiation phase of the project, the client and the design team often refer to the same thing with a slightly or even completely different name. Experience and research on these issues, has proved that making changes in later phases of a project involves a huge amount of effort. LePlan tool on the other hand, not only it guides all involved parties into using the same nomenclature but it also allows corrections to be made at any time within a few minutes. All corrections are project-wide and any cross-reference is automatically verified and updated.

6. CABLE ROUTING

In parallel with any other task, the design team analyses 2D and 3D drawings supplied by the naval architect team. The main goal is to identify cable- runways locations and paths as well as penetrations. Once subject information is inserted on the 2D floor plans, they are imported in the tool. This task is time consuming factored to the size of the project and can be as detailed as the cable lengths accuracy needed. When only a course estimate is needed, this task can be typically completed within one or two days in most of the cases.

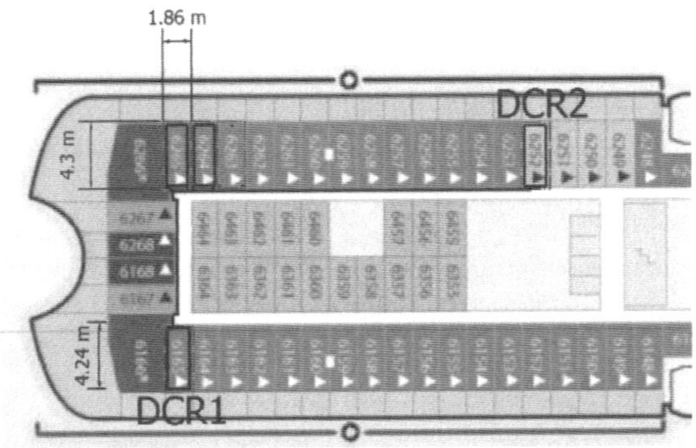

Figure 4: Example of floor plan including potential locations, for two datacentre rooms and one main circular cables runway. Drawing is in 1:1 scale and includes numbered room names.

Once the above step is completed, the manual interaction of the designer and the tool is over. Any cable length can be estimated within few seconds. Changes in the "Electrical equipment specification form" are incorporated in the calculations, almost real-time without any effort from the designer. Just click "update" and wait few seconds. Moving a cabinet to a different deck or room, automatically updates all associated cables. Several reports are available to the design, procurement, and installation team respectively. All reports can be either detailed or generic, with imagination being the only limit.

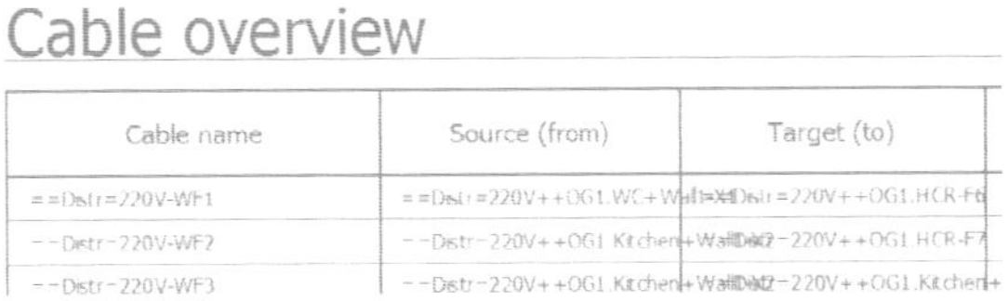

Cable overview

Cable name	Source (from)	Target (to)
==Distr=220V-WF1	==Distr=220V++OG1.WC+Wall	=Distr=220V++OG1.HCR-F6
--Distr-220V-WF2	--Distr-220V++OG1.Kitchen+Wall	-Distr-220V++OG1.HCR-F7
--Distr-220V-WF3	--Distr-220V++OG1.Kitchen+Wall	-Distr-220V++OG1.Kitchen+

Figure 5: Cable Overview list including unique cable ID, source, destination, cable type, etc.

7. DEVICES OVERVIEW

One of the biggest problems when budgeting a new project is over-sizing common items. Especially with IT equipment, it is common practice to plan for size one and procure double size. However, this is no longer required. Taking as an example a simple network patch-panel, the ports usage is monitored in real-time. Adding a new device is as simple and as always automated. A free port is always automatically identified by the software and utilised. Additionally via the overview page, it is always possible to see what devices are connected on subject patch panel.

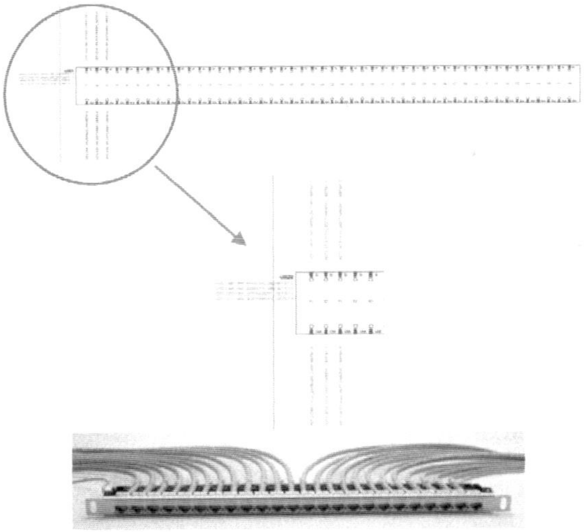

Figure 6: Patch panel ports usage overview

A complete rack view is possible for all common IT equipment or even for custom rack mountable devices. Moving a device from one rack in room X1 deck Z1 to a rack in room X2 on deck Z2 is a trivial task. This gives the opportunity for different vendors to reserve space on racks and reduce the footprint of the equipment used.

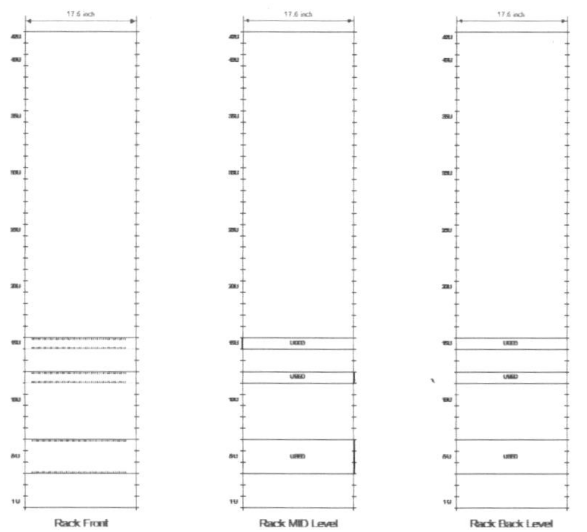

Figure 7: View of a 42U rack including two patch panels and one fully managed network switch[1].

Same principle applies for any component documented using the tool. Among others, these could be a PLC card, an IT device, power and network sockets in a room/area/deck etc. All that is required is to design a function macro for such a device as described in the "concept" section of this paper.

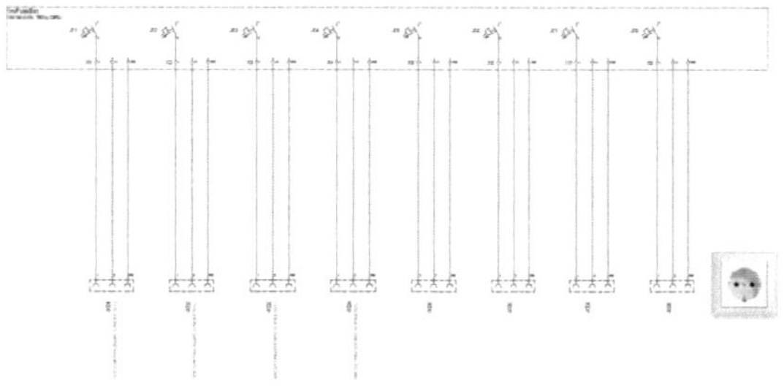

Figure 8: Low voltage power sockets distribution & usage. The green light colour demonstrates that a device is in use and indicates what device is connected on subject socket. The so-called cross-reference appearing in green colour is automatically update once an electrical appliance/device is for example removed (design-wise) or moved to a different connection point / power socket.

8. Processing Time

On average 22 seconds are needed per cabin during the processing time allowing a vessel of 600 cabins to be processed and fully documented within 220 minutes.

The above example includes the complete design for:

- Low power distribution sockets (110V/220V) – 6pcs
- Lighting (controlled via a PLC) – 4pcs
- Data network cabling – 3pcs
- IP-Telephony – 2pcs
- CCTV – 1pc
- En-suite facilities
- Personal & property alarm system – 4 sensors
- Temperature Sensors/controller – 1pcs
- Blinds / shading control – 2pcs

9. CONCLUSIONS

This paper presents a complete unified solution for documenting any project irrespectively of its complexity. Once single function and function macros are designed and implemented, to meet the requirements of a design team or yard, any other project can be documented without extra effort. Repetitive tasks can be incorporated in an effortlessly and quick manner. Detailed or summarised bill of materials are available at any given moment to assist with pre-sales or budget compliance and forecasting. Resulting documentation can be provided to the owner in read only electronic format for easy and accurate troubleshooting in the future.

10. ACKNOWLEDGEMENTS

I would like to express my appreciation to the ePlan Gmbh Company for developing a unique platform serving as the backbone of LePlan tool.

11. REFERENCES

1. Dr. A V Lalechos, '*Design & Documentation at European XFEL Rack Management Concept*', 'KITE DESY Engineering and Innovation Day', 2015.

2. Dr. A V Lalechos, D Maketas, A Pavlou, J Davaris, '*Automatic Generation of Detailed Cabling Documentation for Super & Mega Yachts Accommodation*', RINA Design & Construction of Super & Mega Yachts, Genoa, 2017, pp77-pp80

12. AUTHORS BIOGRAPHY

Dr. Antonios-Vassilios Lalechos, holds PhD from University of Southampton, MBA international from University of Portsmouth, MSc in Control system from London Imperial College and BEng in Electrical & Electronic Engineering from University of Portsmouth. Works as Electrical & Electronic Manager while he also holds the position of Leading R&D Consultant at LePlan supporting businesses working with ePlan P8 and EEC One software. Previously he led the Electrical design works for the Experimental Hall of European XFEL project (1.5 billion Euros project). Among others he developed a tool to unify the documentation of 3.4km tunnels, 6 datacentres and thousands of motors, valves and sensors all part of the same unified design & documentation.

Dimitrios Maketas, holds a MSc in Information Security from Lulea University of Technology in Sweden and a MSc in Software Engineering and BSc (Hons) in Computer Sciences from University of Portsmouth in UK. He has more than 10 years of experience in software and systems engineering with Fortune 500 companies, NASDAQ listed companies and UN agencies. He has co-authored "Formal presentation of fuzzy systems with multiple sensor inputs" with Dr. Alexander Gegov, published in Sensors & Transducers magazine, (S&T e-digest), 08.2005, p. 366-373. He currently works as an IT Analyst for a UN specialised agency.

Antonios Pavlou, holds a MEng in Naval Architecture & Marine Engineering from National Technical University of Athens. Among others, he ranked 3rd at the EU Student Competition 2011, "Visionary Concepts for Ships and Floating Structures – Visions Olympics", organized by WEGEMT and CESA, with the project "E^4-Containership". His dissertation was part of the publication "Energy efficiency parametric design tool in the framework of holistic ship design optimization", Journal of Engineering for the Maritime Environment, August 2011, with authors E.K. Boulougouris, A.D. Papanikolaou and A.C. Pavlou. He currently works as Superintendent Engineer in Shipping Company.

John Davaris, Holds an MSc in Marine Engineering from University of Southampton and a BEng in Mechanical Engineering from Brunel University. Works as a Technical Superintendent in a Shipping Company with extensive experience in Ship repairs and retrofits. Deeply involved in Planned and Condition Based maintenance systems.

INCORPORATING SUBJECTIVE PREFERENCES IN DESIGN SYTHESIS MODELS WITH LIMITED INFORMATION

D C Brefort and **D J Singer,** University of Michigan, USA

SUMMARY

Ship design requires both objective knowledge, found in the form of mathematical models or simulations, and subjective knowledge, found in the form of design requirements or personal experience. This paper presents a method of incorporating human preferences into low fidelity mathematical models. A fuzzy logic system is used to assign preferences to areas of the design space and determine a design's ranking according to multiple subjective preferences. Subjective rankings are overlaid onto the objective results from the mathematical model of the design. This method will help designers incorporate context into low fidelity optimization models and help them differentiate designs along the Pareto front which have seemly equal performance. This paper provides a brief background on fuzzy logic and shows how it can be used to include human preferences into rigid mathematical models to increase one's insight about a design space.

1. INTRODUCTION

In the initial stages of the ship design process, engineers explore the design space to find promising sets of ship principle characteristics from which to start a detailed analysis. To cope with the limited amount of information they have at initial stages of the design, they develop mathematical models which describe the relationship between a vessel's principle characteristics and some objective to be improved upon. These models, which range from low fidelity regressions of historical data to high fidelity simulations, are used to extract knowledge of the design space and understand trade-offs between performance parameters.

Although very useful for determinate problems, model-based design is plagued with inadequacies when solving open-ended complex problems. First, the mathematical formulation of the problem makes it difficult to incorporate qualitative measures of performance or subjective preferences which cannot be described by an equation. Second, it is difficult to include context when evaluating a variable. A design with a large tonnage might be favourable when shipping demand is high, but unfavourable when demand slows. In classical optimization, vessels with a large tonnage are considered more efficient and thus always deemed favourable, independently of the economic context. Additionally, in multi-objective optimization, all designs along the Pareto front are deemed equally good. An improvement in one objective leads to a deterioration of another objective. Thus, two Pareto optimal solutions cannot be ranked objectively. A particular problem occurs in multi-disciplinary optimization, where a top level optimizer (i.e. the ship design manager) manages and integrates objectives for multiple, interdependent, lower level disciplines, in order to identify the best overall solution. The models of the lower level disciplines could be a mix of high and low fidelity. In that case, the disciplines using the low fidelity models might incorrectly drive a solution because their models cannot capture the design environment adequately. Any time and money spent developing high fidelity tools is therefore lost when they are integrated with low fidelity models.

This paper proposes a method which attempts to resolve some of the issues presented above by increasing the modelling flexibility of classical optimization problems. This method uses rule based fuzzy logic systems to include human intelligence and expert opinion into mathematical design synthesis models. Subjective preferences are assigned to variable ranges of the design space. A design's overall subjective ranking is then found by combining these preferences using a rule base. Subjective rankings are overlaid onto the objective results from the mathematical model of the design. It allows designers to use context when evaluating the fitness of a design and can help them compare designs based on both the objective knowledge extracted from a mathematical design synthesis model and a subjective preference given by a designer, providing both a quantitative and qualitative perspective of the solution.

Fuzzy logic is an adequate method to increase the flexibility of classical optimization models. It is a method of representing and operating on data with imprecise boundaries. Originally developed as a method of programming computers using natural language [1], it has since found wide applications in consumer and industrial product control systems, decision support systems [2], market forecasting [3], and even automatic train control systems [4]. Its ease of implementation and ability to "think like a human" are two of its best advantages. Steinberg developed an automatic aircraft carrier landing system for an F/A-18 using fuzzy logic to incorporate "elements of human pilot 'intelligence' with more conventional automatic control laws" [5]. In ship design, fuzzy logic systems have been used for set-based design to assign preferences to areas of the design space [6], for general arrangements optimization [7], and to incorporate human intent into MDO algorithms [8].

This paper is organized as follows. In Section 2, provides an overview of fuzzy logic. Section 3 presents a case study on which subjective preferences are included, and

Section 4 evaluates the resulting effect of these preferences on the ranking of designs.

2. METHODS

Classical set theory describes a binary relationship between an element and a set. An element e is either a member of set A (i.e. $e \in A = 1$), or is not a member of set A (i.e. $e \in A = 0$). Fuzzy set theory expands this concept so that the degree of membership of an element to a set can lie between 0 and 1. For example, when classifying outside temperatures as either hot or cold, we know that 0 Celsius is definitely cold, and 40 Celsius is definitely hot. However, it would be hard to classify 20 Celsius as either hot or cold. Using fuzzy set theory, 20 Celsius could belong to the set cold with degree of membership 0.5 and belong to the set hot with degree of membership 0.5. Fuzzy set theory's ability to manipulate imprecise concepts, much like humans can, is one of its biggest advantages. There is no precise temperature at which hot becomes cold. Instead, the change is gradual.

A rule-based fuzzy logic system (FLS), or fuzzy logic controller, is a method of mapping crisp inputs to a crisp output by using a set of IF-THEN rules (or input-output relationships) that mimic how humans think. It consists of four parts: the fuzzifier, fuzzy rule bank, fuzzy inference engine, and defuzzifer, as shown in Figure 1.

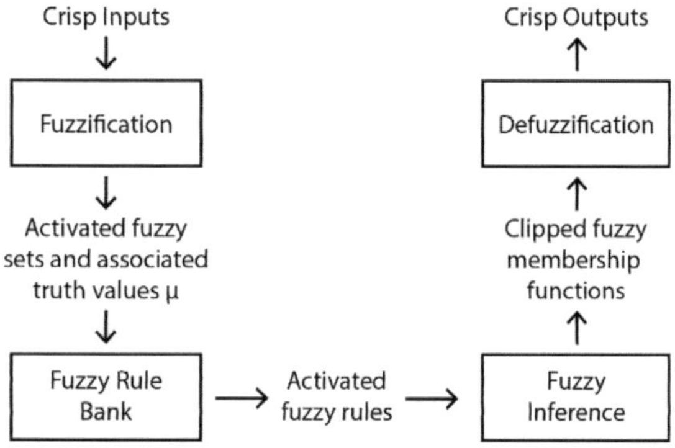

Figure 1: Fuzzy Logic System Process

Fuzzification converts crisp input variables into linguistic input variables using a membership function. Assume we want to determine the reliability of an operating cost model with respect to its fuel price parameter. The time series in Figure 2 shows that fuel prices below \$400/tonne were common in the past, and might temporarily occur again in the future, but that they are unlikely. Thus a fuel price parameter below \$400/tonne would have low preference in an operating cost model. Prices above \$600/tonne were common after 2011 and they are also conservative, thus from an operating cost model reliability point of view, these values have high preference. Values between \$400/tonne and \$600/tonne are transitional, thus they are partially low preference and partially high preference. A fuel price of \$500/tonne could have low preference membership value 0.5 and high preference membership value 0.5.

$$\mu_{\text{low}}(500) = 0.5 \text{ and } \mu_{\text{high}}(500) = 0.5$$

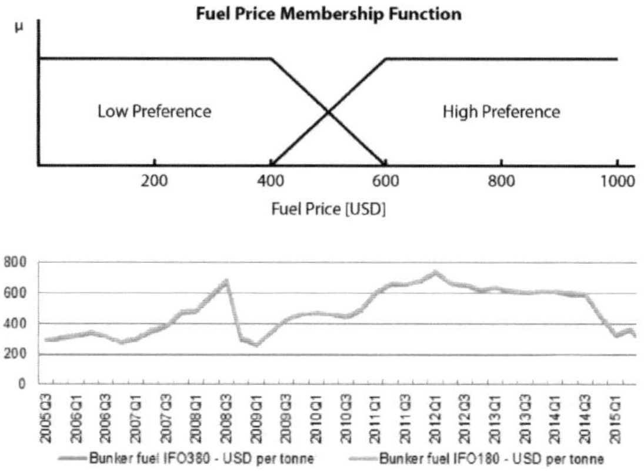

Figure 2: Fuel price membership function in relation to operating cost model reliability

The fuzzy rule bank is at the centre of the FLS. It consists of a set of IF-THEN statements that map linguistic inputs (called antecedents) to linguistic outputs (called consequents). These rules can be populated using expert opinion or extracted from data. In the operating cost model example, assume we only consider fuel price preference and daily consumption preference, where ships with lower consumption have higher preference. A fuzzy rule bank could be given as follows:

IF Fuel Pref. is High *AND* Consumption Pref. is High *THEN* Model Output Pref. is High

IF Fuel Pref. is High *AND* Consumption Pref. is Low *THEN* Model Output Pref. is Medium

IF Fuel Pref. is Low *AND* Consumption Pref. is High *THEN* Model Output Pref. is Low

The first rule states that high fuel price reliability and low fuel consumption have high preference, the second rule states that a high price reliability together with high fuel consumption has a moderate preference. The third rule states that even if a ship has a low fuel consumption, if the fuel price reliability is small that design would still be unfavourable because of the high uncertainty surrounding the fuel price.

The third step of the fuzzy logic system is fuzzy inference, which determines the strength of each rule consequent by mapping the linguistic inputs (antecedents) to linguistic outputs (consequents). In this work we use minimum inference, in which the membership value of the consequent equals the smallest membership value μ, of the rule's antecedents [9].

The last step is defuzzifucation, which combines the consequents from each activated rule into a single crisp output. In this work, we use centroid defuzzification, which calculates the centroid of all calculated rules.

3. CASE STUDY

A case study was conducted to investigate the ability of a fuzzy logic system to incorporate subjective knowledge into low fidelity mathematical models. By superimposing objective and subjective knowledge, this method allows designers to incorporate their experience into models to better understand the design space and evaluate solutions.

For the objective aspect of design knowledge, we use a modified version of the low fidelity preliminary design synthesis model of bulk carriers developed by Yang and Sen [10]. This model takes as inputs the length, beam, depth, draft, block coefficient, and speed of a vessel. These inputs are used to evaluate intermediate functions like round trips per year, steel mass, and fuel cost, among others. Finally, the independent variables of the intermediate functions are used to evaluate three objectives: minimize lightship mass, maximize annual cargo carried, and minimize transportation cost. A detailed formulation of the model can be found in [11].

Parker [12] used network theory to analyse the structure of the Sen bulker model. His goal was to determine which variables and functions have the highest influence on the final objective functions. He built a network from the Sen bulker problem where each variable, parameter, function, and constraint was represented by a node. Each arc between nodes represented the use of a variable or parameter in a function or constraint [13]. The network consists of fifty-nine nodes and ninety-five arcs. After building the network, the influence of variables and functions can be measure by the nodes' Park centrality [14], which ranks nodes according to the difference between its influence and how much it is influenced. Parker found that traditional naval architecture variables like the principle dimensions and stability variables were driving optimization results a lot more than operational variables like fuel cost and cargo deadweight. Fuel cost ranked 44th and cargo deadweight ranked 56th most influential nodes out of a total of 59 nodes. Considering that two of the Sen bulker's objective functions are transportation cost and cargo carrying capacity, subjective preferences were included on fuel cost and annual cargo capacity to understand how an increase in the fidelity of the model with respect to these operational variables would translate to final preferences in design selection. The rest of this section explains how these subjective preferences are modelled.

3.1 FUEL COST PREFERENCE

In the Sen bulker model, the fuel cost per voyage is a function of fuel price, daily consumption, and number of sea days. From an objective point of view, vessels with lower fuel costs are optimal with regard to operating cost. A designer, however, might use different metrics when evaluating a ship based on its fuel cost. Our subjective preference model combines information on fuel price, daily consumption, and sea days to determine the subjective fitness of the design relative to fuel cost. The membership functions, representing a human designer's preference for given variable ranges of a design space, are shown in Figure 3.

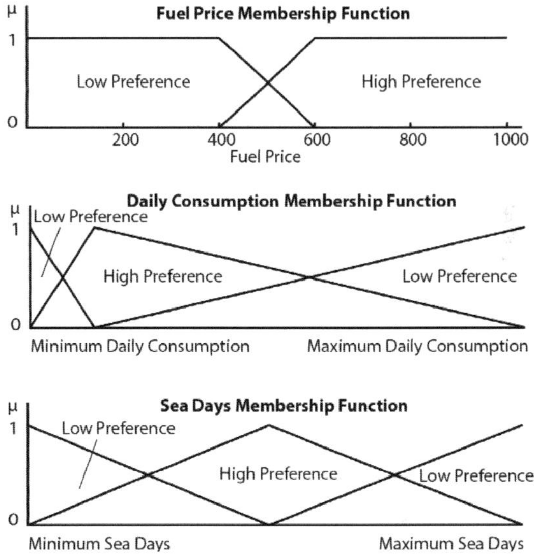

Figure 3: Membership function for fuel cost

The fuel price preference rationale is given in Section 2. For daily consumption, objectively one would likely prefer ships with the smallest consumption. However, a designer's experience and best judgement would tell him or her that the vessels with the very smallest daily consumption do not have enough of a power margin to operate in heavy seas. Thus, designs with the very smallest daily fuel consumption have low preference. From a subjective point of view, the preference membership function for daily fuel consumption is skewed to the left to include designs with low consumption, but avoids designs with the very lowest consumption. The last input variable for fuel cost preferences is the number of days a vessel is at sea during a voyage. In this model, designers have low preference for designs with the smallest number of sea days, to avoid having a high relative downtime waiting to enter ports and to pass through shipping canals. They also have low preference for maximum sea days to avoid being uncompetitive with respect to shipping time.

The fuel cost fuzzy rule bank is given in Table 1. It implies that the designer places higher importance on daily consumption than sea days (a high preference for daily consumption, together with a low preference for sea days, has a high preference output). The preference given by the rule bank is context dependent. For bulkers, which carry low value merchandise, daily fuel consumption is weighted higher than sea days (i.e. transit speed). For vessels carrying higher value merchandise, like container ships, the transit speeds becomes more important, and thus the model would have different rule banks for different vessel types.

Table 1: Fuzzy rule bank for fuel cost preference

Daily Consumption Preferences / Sea Days Preferences	High Daily Consumption Preference	Low Daily Consumption Preference
High Sea Days Preference	High Preference	Medium Preference
Low Sea Days Preference	High Preference	Low Preference

Fuel Cost Preference Rule Bank for High Fuel Price Preference

Daily Consumption Preferences / Sea Days Preferences	High Daily Consumption Preference	Low Daily Consumption Preference
High Sea Days Preference	Medium Preference	Low Preference
Low Sea Days Preference	Medium Preference	Unpreferred

Fuel Cost Preference Rule Bank for Low Fuel Price Preference

3.2 ANNUAL CARGO PREFERENCE

In the Sen bulker model, the annual cargo capacity is a function of the ship's cargo deadweight and the ship's number of round trips per year. From the model's objective point of view, increasing annual cargo capacity is always preferred. From a subjective point of view; however, a shipping overcapacity is detrimental to operators, as experienced in the latest shipping crisis [15]. This aversion to overcapacity is modelled by giving a low preference to designs with the highest cargo deadweight, as shown in Figure 4. The designer still prefers vessels with high cargo deadweight though, since these are more cost effective. The rationale for the round trips per annum (RTPA) membership function is similar to the sea days' membership function rationale. A small number of RTPA is detrimental to customer satisfaction, but a large number of RTPA would increase the vessel's relative downtime, thus the highest preference is given to designs in between the two extremes.

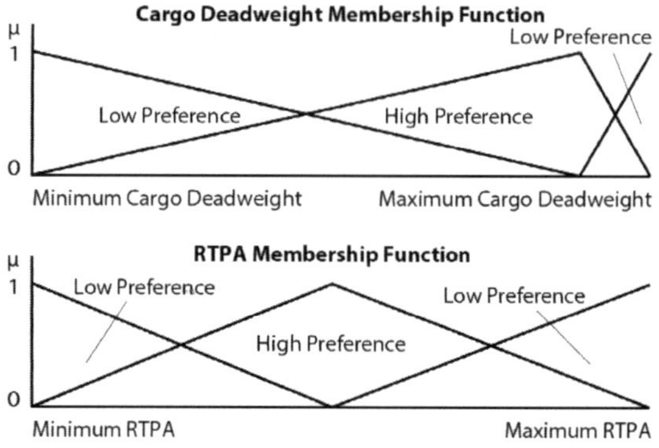

Figure 4: Membership function for annual cargo

The fuel cost fuzzy rule bank is given in Table 2. It implies that the designer places higher importance on cargo capacity than RTPA (a high preference for cargo deadweight, together with a low preference for RTPA, has a high preference output). For a vessel carrying perishable goods, the preference could be given to RTPA to avoid spoilage of goods.

Table 2: Fuzzy rule bank for annual cargo preference

Cargo Deadweight Preferences / RTPA Preferences	High Cargo Deadweight Preference	Low Cargo Deadweight Preference
High RTPA Preference	High Preference	Low Preference
Low RTPA Preference	High Preference	Unpreferred

4. RESULTS

Three sets of results were examined. For each of these result sets, two types of knowledge are present: objective knowledge and subjective knowledge. The objective knowledge is identical in all three sets of results. It is composed of design data from the mathematical Sen bulker model, whose equations and objective functions were derived through regression of historical data. For a given set of input variables (length, beam, depth, draft, block coefficient, speed) the Sen bulker model outputs a set of intermediate variables like Froude number, deadweight, fuel cost, and the three objective functions annual cargo, transport cost, and lightship, which are dependent on the input variables. With respect to objective knowledge, the data points at the same (x, y) coordinate in figure 5, 6, and 7 are identical. They have equal principle dimensions, intermediate function values, and objective function values (i.e. the objective knowledge is represented by a point's location on the figure and the subjective knowledge is represented by the point's colour). In this paper, we only consider transport cost and annual cargo objective functions.

The subjective knowledge differs in all three plots. The first set of results analysed the effects of subjective fuel cost preferences. The second set of results considered the effects of subjective cargo capacity preferences. The last set of results looked at the combined fuel cost and cargo capacity preferences. For all three result sets, each design is given a preference according to the fuzzy membership functions and rule banks presented in Section 3. The crisp output of the fuzzy logic system's defuzzifier combines human preferences given by each membership function and by the rule bank into an overall score, which allows a subjective ranking of each design. Thus, each design has a subjective score output from the fuzzy logic system and an objective function value output from the Sen bulker model. This extra layer of information brings additional context to results from low fidelity mathematical design synthesis model, helping the designer gain additional insight of the design space early in the design process, when this information is most useful.

4.1 FUEL COST PREFERENCE

This section considers subjective fuel cost preferences. The fuzzifier assigns each design a degree of membership into high and low preference sets for fuel price, daily consumption, and number of sea days according to the membership functions (Figure 3) of these three input variables. The fuzzy inference engine and defuzzifier then use these degrees of membership and the rule bank given in Table 1 to assign a crisp fuel cost preference to each design. In Figure 5, the very best design according to this metrics is shown by the star, and the design ranking quartiles are shown in colour, with the first quartile being worst and the fourth quartile being best according to subjective human preference.

A subjective preference for fuel cost translates mostly to transport cost preferences (ie. the quartiles in figure 5 correlate most strongly with transport cost). This is not surprising considering the high correlation between fuel cost and operating cost. Vessels with low transport costs score well according to human preference; however, the low preference for designs with the very smallest daily fuel consumption causes designs with the very smallest transport cost, on the lower portion of the Pareto front, to be unpreferred, directly contradicting the objective knowledge from the mathematical model. The subjective model can thus warn engineers against the negative consequences of overly optimizing a vessel. Finally, for designs with lower annual cargo values (ie. smaller or slower vessels) the separation between the third and fourth quartile is poorly defined. This region of high uncertainty should be avoided by risk averse decision makers.

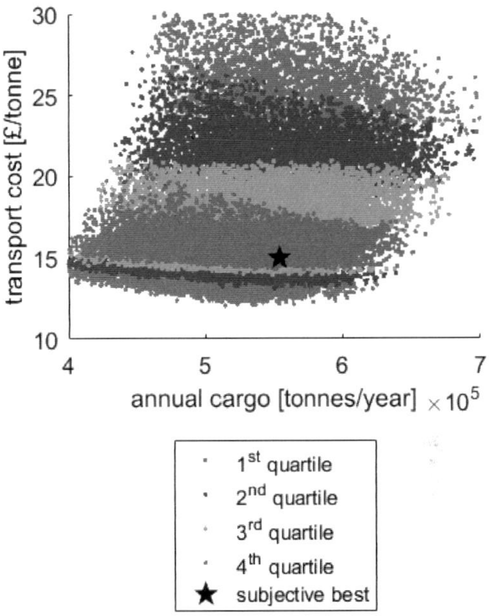

Figure 5: Subjective preferences with respect to fuel cost

4.2 ANNUAL CARGO PREFERENCE

This section presents results for subjective cargo carrying capacity preferences. Here, the fuzzifier assigns each design a degree of membership into high and low preference sets for cargo deadweight and number of round trips per year by using the membership functions given in Figure 4. The inference engine and defuzzifier then provide a crisp preference to each design with respect to annual cargo, using the rule bank given in Table 2. Vessels with large annual cargo capacity score well according to human preference. However, the vessels with the very largest annual cargo capacity, at the very right edge of the Pareto front, are less preferred than vessels with a slightly lower carrying capacity. The designer's reluctance to create overcapacity in the shipping market causes the best subjective design to be shifted inwards slightly as can be seen by the narrow band of quartile one, two, and three designs at the right of Figure 6.

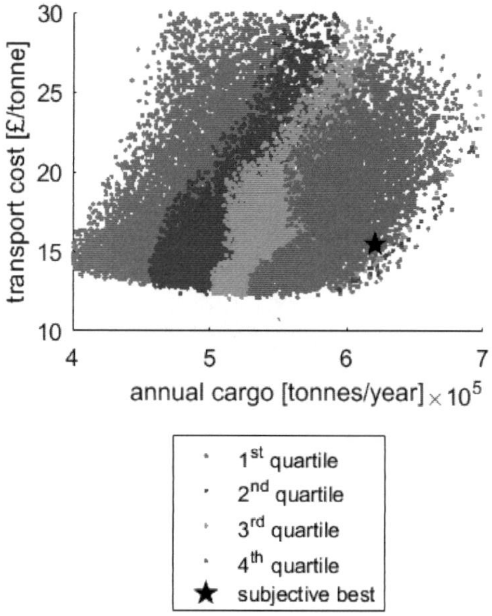

Figure 6: Subjective preferences with respect to cargo capacity

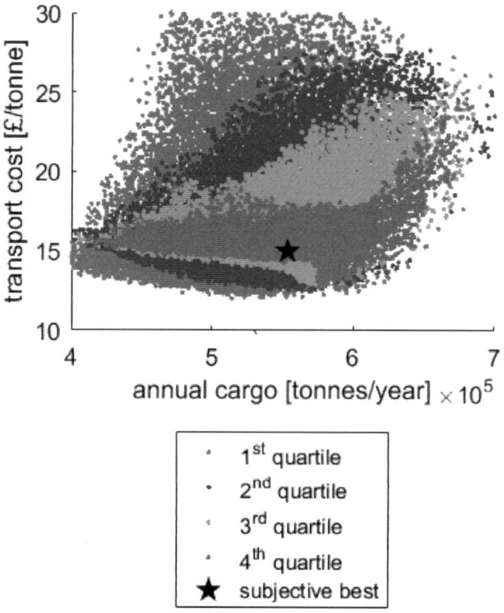

Figure 7: Subjective preferences with respect to fuel cost and cargo capacity

4.3 COMBINED FUEL COST AND CARGO CAPACITY PREFERENCE

The previous two sets of results considered fuel cost and cargo capacity independently. This section considers both variables together, and weighted equally. This is analogous to the decision maker selecting a vessel design based on profit, where both the revenue from the annual cargo and the transport cost must be considered.

In these results, the decision maker's aversion to vessels with the smallest daily consumption and to vessels with the largest cargo deadweight can still be noticed by the lower preferences of designs located at the Pareto front. The fuel cost subjective preference and the annual cargo subjective preference affect the overall subjective ranking differently at different areas of the design space. For smaller transport costs, overall subjective rankings correlate better with fuel cost preferences whereas for higher transport cost the subjective rankings appear to be equally correlated to fuel cost preferences and annual cargo preferences, helping the designer understand how subjective preferences can impact an overall design preference differently for different classes of ships.

5. CONCLUSIONS

This work investigates the ability to incorporate subjective knowledge into low fidelity mathematical models. The introduction of expert opinion and designer experience that cannot be captured by equations increases the amount of information and knowledge that can be leveraged in early stages of design. Fuzzy logic systems are used to model subjective preferences, which are superimposed onto objective knowledge from a mathematical design synthesis model. By superimposing these two types of knowledge, which augments the model's "human intelligence", this method allows designers to better evaluate solutions in relation to objective goals and subjective preferences. Fuzzy logic systems also allow engineers to include qualitative information and context into mathematical models, greatly increasing modelling flexibility. The ability to incorporate personal experience at the front end of ship design, rather than using it only to evaluate completed designs, should greatly increase our ability to make informed decisions when design freedom is still high.

6. ACKNOWLEDGEMENTS

The authors would like to thank Ms. Kelly Cooper and the U.S. Office of Naval Research for their support. This work was funded by the Naval International Cooperative Opportunities in Science and Technology Program (NICOP) contract number N00014-15-1-2752.

7. REFERENCES

1. ZADEH, L.A., 'Fuzzy Sets', *Journal of Information and Control*, 8:338-353, 1965.

2. NAIM, S., HAGRAS, H., 'A Type 2-Hesitation Fuzzy Logic Based Multi-Criteria Group Decision Making System for Intelligent Shared Environments', *Journal of Soft Computing 18:1305-1319*, 2014

3. LI, J., PARSONS, M.G., 'Improved Method for Shipbuilding Market Modeling and Forecasting', *SNAME Transactions*, 1998

4. MENDEL, J., HAGRAS, H., TAN, W.W, MELEK, W.W., YING, H., 'Introduction to Type-2 Fuzzy Logic Control: Theory and Applications', *IEEE Press Series on Computational Intelligence*, 2014

5. STEINBERG, M., 'Development and Simulation of an F/A-18 Fuzzy Logic Automatic Carrier Landing System', *IEEE International Conference on Fuzzy Systems*, 1993

6. SINGER, D.J., 'A Hybrid Agent Approach for Set-Based Conceptual Ship Design Through the Use of a Fuzzy Logic Agent to Facilitate Communications and Negotiation', *Ph.D. Thesis, University of Michigan,* 2003

7. NICK, E., PARSONS, M.G., NEHRLING, B., 'Fuzzy Optimal Allocation of Spaces to Zone-Decks in General Arrangements', *Proceedings of the International Marine Design Conference*, 2006

8. CUNEO, B., 'Development of a Human Centric Multidisciplinary Design Optimization Method Using Fuzzy Logic Systems and Controllers', *Ph.D. Thesis, University of Michigan,* 2013

9. MENDEL, J.M., 'Uncertain Rule-Based Fuzzy Logic Systems: Introduction and New Directions', *Prentice Hall*, 2001

10. YANG, J.B., SEN, P, 'Interactive trade-off analysis and Preference Modelling for Preliminary Multiobjective Ship Design', *Systems Analysis, Modelling and Simulation,* 26:25-55, 1996

11. SEN, P., YANG, J.B, 'Multiple Criteria Decision Support in Engineering Design', *Springer,* pp.113-175, 1998

12. PARKER, M.C., 'A Contextual Multipartite Network Approach to Comprehending the Structure of Naval Design', *Ph.D. Thesis, University of Michigan,* 2014

13. PARKER, M.C., SINGER, D.J., 'The Impact of Design Tools: Looking for Insights with a Network Theoretic Approach', *Proceedings of the International Conference on Computer and Information Technology Applications in the Maritime Industries*, 2013

14. PARK, J., NEWMAN, M. E. J., 'A Network-Based Ranking System for US College Football', *Journal of Statistical Mechanics: Theory and Experiment*, 2005

15. 'Shipping's Blues', *The Economist,* 2017

8. AUTHORS BIOGRAPHY

Dorian C. Brefort is a Graduate Student Research Assistant in the department of Naval Architecture and Marine Engineering at the University Of Michigan College Of Engineering. He earned his B.S.E. and M.S.E. degrees in NA&ME and a M.S.E. in Industrial and Operations Engineering from the University of Michigan. He conducts research in the areas of naval design theory, design optimization, and fuzzy logic.

David J. Singer, Ph.D., is an Associate Professor in the department of Naval Architecture and Marine Engineering at the University of Michigan College of Engineering. He earned his B.S.E., M.Eng., and Ph.D. degrees in NA&ME and a M.S.E. in Industrial and Operations Engineering from the University of Michigan. He conducts research in the areas of naval design theory, design optimization, and ship production.

AUTOMATIC PIPING ARRANGEMENT DESIGN CONSIDERING PIPING SUPPORTS AND CURVED SURFACES OF BUILDING BLOCKS

H Kimura, Kyushu University, JAPAN

SUMMARY

In this paper, a new piping path planning system is proposed in order to automate piping design corresponding to pipe supports and curved hulls. In the proposed system, candidates for positions and directions to which pipes should be passed are given in advance as 'candidate points' from the circumstances of pipe racks and support. Then, the system selects the appropriate candidate points automatically to generate piping paths keeping constraint of many factors, e.g., gravitational flow, or geometrical limitation of the pipe-bending machine, etc. Therefore, it is quite practical. The proposed system is implemented to a computer program, and the performance of the system is demonstrated through several simulations.

1. INTRODUCTION

Pipe arrangement is considered one of the most time-consuming works in the production of vessels, buildings and plants. In recent years the advancement of 3D-CAD systems has helped designers to arrange pipes. Especially, they can grasp the pipe location visually and check the interference between pipes and structures by using the system. Designers have to make right decisions in order to generate optimum piping routes and suitable positions for equipment during the process. However, the most part of the pipe routing problem, which is to design the best routes of the pipeline, demands veteran designer's experiences because the problem involves many regulations or requirements to be concerned. It has become increasingly difficult to secure skilled designers and problems of technology transferring to young engineers are also occurring. Automation of piping design is one promising method for solving these problems.

Many previous works of automatic piping design have been done in many approaches. Some of these studies, such as Ito [1], Park et al.[2], Asmara et al.[3,4], Paulo et al.[5], and Lin et al.[6] applied for Cell Decomposition approach that is composed to divide the design area into meshes and connect them from the start point to the goal point. There are two main advantages in applying this approach. The first is possible to apply maze solving algorithms to find solutions. In the maze algorithms, there exist methods to assure to find optimum solutions such as the Dijkstra's method. The second is possible to set different cost values in each cell. From this feature, the algorithm can draw pipelines near to a ship's hull, while avoiding aisle spaces as possible. in previous works that used Cell Decomposing approach, the mesh size was restricted to be larger than the pipe diameter.

Ando et.al[7, 8] presents a new pipe arrangement algorithm of which mesh sizes are not restricted by the pipe diameter, but it cannot cope with the pipes along the curved structures. In piping design, consideration must be given to the position and direction in which pipes are passed, in order to properly support pipes from pipe racks or structural members with support. In this paper, a new piping path planning system is proposed in order to automate piping design corresponding to pipe supports and curved hulls.

2. ASSUMPRIONS AND PURPOSE

The following assumptions are given in the pipe routing problem:

[Topology of target pipelines]
All pipelines are represented by a tree structure, that is, all pipelines are composed of one root (start) point and multiple leaf (goal) points. If the number of the leaf point is one in a pipeline, no branch exists in it. The piping route in the same pipeline is represented by multiple routes from the common root point to the leaf points. As a result, the piping routes are overlapped from the root point to the branches in the same pipelines. The position and the shape of the branches in the pipeline are given by analyzing the multiple routes.

[Pipes' diameter]
The pipes' diameter is invariable in the same pipeline. You may feel that this assumption is not practical, however, we can easily modify the pipes after the routing.

[Design points]
In advance, we put design points which indicate possible positions and directions of the way of pipes for each pipeline. That is, the design points are candidates of the waypoints of the pipes. In actual piping design, the pipes are put at the position that is certain distance away from structures. Also directions of the pipes on the pipe-supports are along the (curved) structures, especially in ships. For the automatic pipe routing, putting the design points is a promising approach to cope with these situations.

[Obstacles]
Structures or equipments are regarded as obstacles. The geometric information of obstacles is represented by several primitive shapes, that is, triangles, boxes, spheres

or cylinders. It is not allowed that arranged pipes interfere with these obstacles.

The algorithm searches pipe routes with following design objectives:

1) To connect the pipes from start points to goal points without interference between obstacles or the other pipelines.
2) To minimize the total length of pipes,
3) To minimize the number of elbows and bends,
4) To maximize the length of the overlapping pipes in the same pipelines. The overlapping pipes in the same pipelines contribute to cut material costs.

In order to regard the pipe arrangement as a single purpose optimization, a routing cost which is proportional to the total length of routes is provided. Costs elbows and bends are given in advance. The proposed algorithm tries to find optimal routes with minimized sum of these costs.

3. PIPING PATH SEARCH ALGORITHM

3.1 PREPARATION FOR PIPING

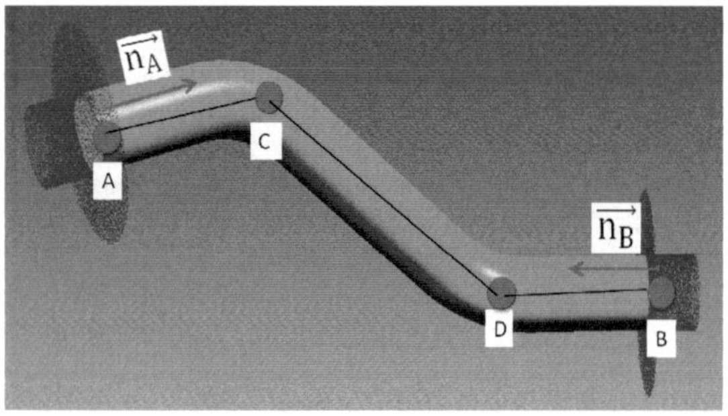

Figure 1: An example of a pipe unit connecting two design points (A and B) with two bends. The centre lines of the straight pipes are put along the AC, CD and DP.

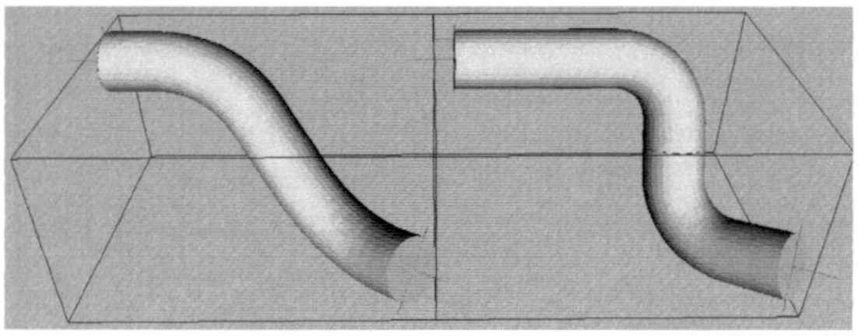

Figure 2: Left side: A piping example between two waypoints with two bends. Right side: A piping example in the same waypoints with two 90-degree elbows.

A pipe in Figure 1 shows a sample that connects two design points which have arbitrary positions and directions. Notice that geometrically, all the pipes with two bends can connect arbitrary two design points which have any positions and directions. Two pipes shown in Figure2 are examples which have the same start and goal positions and directions. In the left pipe in Figure 2, the bends have certain angles, but the position of the bend is given in advance such as AC:CD:DP=1:2:1. In the right pipe in Figure 2, bends are constrained to use only 90 degree elbows, and the position of the bends are derived from mathematical calculations.

Shapes of real pipes are constrained by the limitations of the pipe bending machines. Pipes are bent with arbitrary angles, but there exists a limit of the maximum angle. The radius of the bend is constrained by the pipe bender's moulds. When pipes are bent, straight parts between the bend are required to grasp the pipes.

3.2 PATH PLANNING FOR SINGLE PIPE

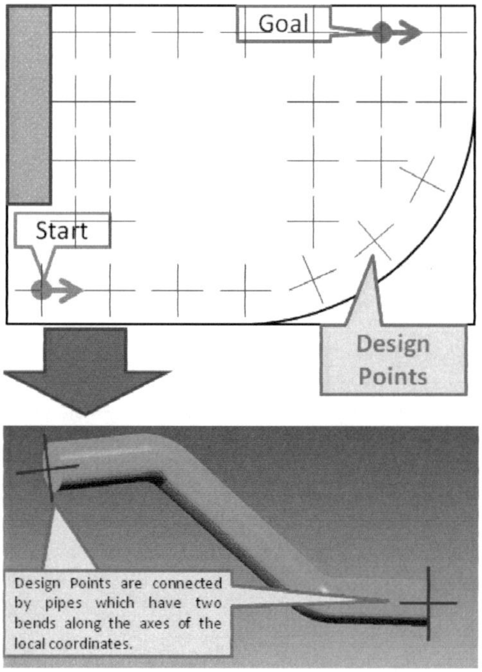

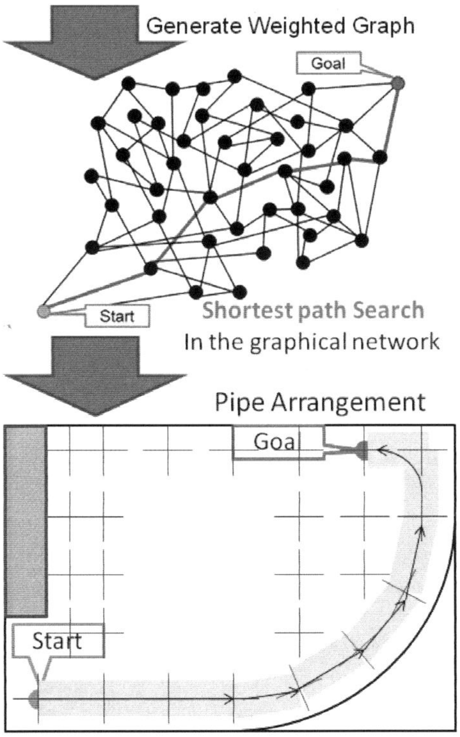

Figure 3: Process of the routing algorithm for single pipe.

Figure 3 shows an overview of the proposed algorithm for single pipe arrangement. First, we put design points which indicate possible positions and directions of the way of pipes for each pipeline considering pipe supports, pipe racks or structures. The design points are given manually in advance, however, it is easy to generate automatically. In the next step, the system tries to generate weighted and directed graph by judging that arbitrary two design points (and directions) can be physically connectable or not. Conditions of the construction constraints or gravitational flow or pockets of the pipe is also considered in the judgement of the edges. When the weighted graph is generated, the system applies a shortest path search algorithm, e.g., Dijkstra's method or A* search method. Note that both the graph generation and the path search can be executed concurrently. The calculation result in the directed graph is converted to the path of the single pipe.

3.3 STANBLE BLOCKS IN PATH PLANNING FOR MULTIPLE PIPELINES

Usually there exist multiple pipelines in practical pipe arrangement problems. However, when we apply the path planning algorithm presented in the previous section to the pipes one by one, we would get largely different piping arrangement design depending on the following two reasons.

3.3 (a) Influence on the Arrangement By Search Order

In practice of piping design, a route search is generally performed in order of expensive pipes. When the material is equivalent, it is arranged from a pipe with a large diameter. However, when there are multiple pipes of the same diameter, arbitrariness arises in the order of processing, and the final route plan changes dramatically depending on the order.

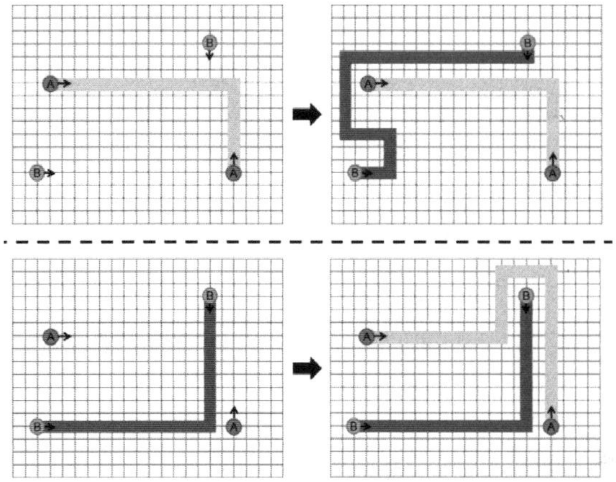

Figure 3: Influence of routing search order.

Figure 3 shows an example in which the piping route acquired by the search order is different whereas the route of two pipelines (pipe A, pipe B) have the same diameter. For the sake of clarity, the route search within a two-dimensional plane is performed. In this case, since the diameters of the two pipes are equal, there are two possible search order, one from the search from pipe A as shown in the upper part of Figure 3, and the other from pipe B. The resulted route plan is completely different depending on the pipe search order.

3.3 (a) Influence on the Arrangement by Selection from Multiple Optimum Solutions

In one piping route search, there are cases where there are multiple solutions as optimal solutions of routes. In this case, depending on which route is selected from among the optimum route candidates, the arrangement finally obtained may be greatly changed in some cases.

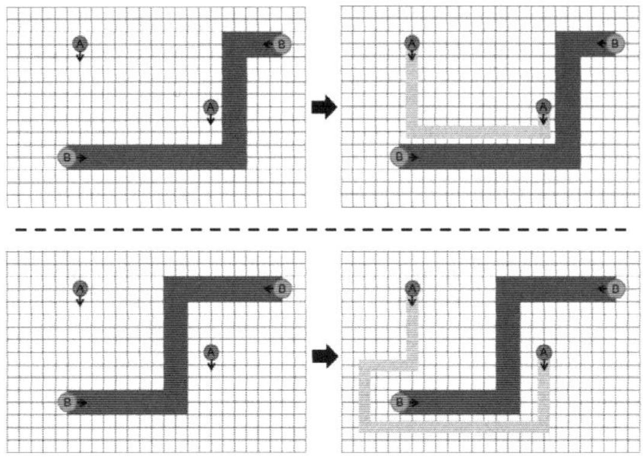

Figure 4: Influence of selection from optimum routes.

Figure 4 shows the case of searching the route of the large diameter pipe B and the small diameter pipe A on the 2-dimensional plane. Here, as the optimum path candidate for pipe B, the upper route in the figure and the lower route are considered. In the figure below, due to the influence of the route of the pipe B, the pipe A greatly bypasses. Since there are few obstacles in the design target space at the beginning of the route search, there are more optimum route proposals. However, depending on which one of these route proposals is chosen, it can be seen that the result of the subsequent pipe arrangement change drastically.

3.4 PATH PLANNING FOR MULTIPLE PIPELINES: TOUCH AND CROSS METHOD

Touch and cross method is known as a heuristics for wire-routing in electric circuits. The basic idea of this algorithm is that the arrangement of shortest paths without considering interferences has important information to avoid the other pipelines (or to overlap the same pipelines). Here shows the process of the Touch and Cross method:

1) Find optimum paths ignoring interference between the pipelines.

2) Redraw all pipes one by one, adding penalties of interference between the other pipelines to the costs. The penalties are proportional to the interfered volumes. Also in the cost calculation of the piping, when the target pipe is overlapped with pipes which are the same pipelines, the cost in the overlapped section is discounted as $Cost \times 1/\sqrt{N_{dup}}$, where N_{dup} is the number of overlapped pipes.

3) If all the pipelines do not interfere with each other, then finish. Else, increase the rate of the penalties of interference and go to the step 2).

The proposed path-planning method for single pipe shown in the Section 3.2 is applied to search piping paths in the step 1) and 2). Figure 5 and 6 shows examples of the piping path searching process using the Touch and Cross method on the two dimensional space. In the early stage of the process in Figure 5, all the pipes are interfered, but it is gradually solved by redrawing the pipes in order of A, C and B. In contrast, the length of the overlapped section is gradually increased by redrawing the pipes in order A, B, (C), A and B in the same pipeline as shown in Figure 6.

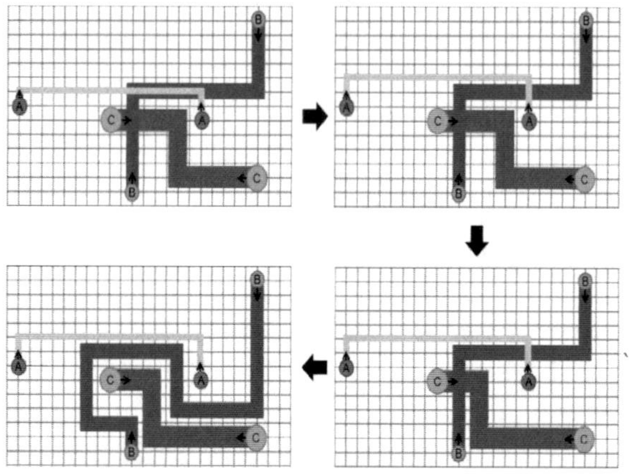

Figure 5: Process of Touch and Cross method to avoid interfering pipes.

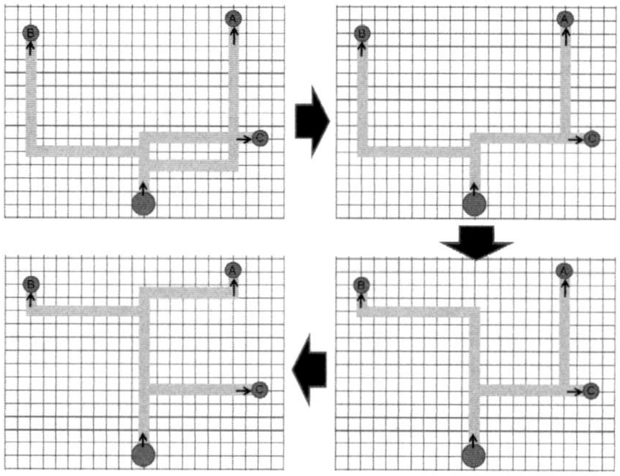

Figure 6: Process of Touch and Cross method to arrange low-cost branched paths in the same pipeline.

4. EXPERIMENTS

4.1 PIPING ALONG CURVED STRUCTURE

Using the proposed algorithms, pipes of ϕ 100 (orange), ϕ 400 (yellow), ϕ 600 (green) are placed along a wall-like structure consisting of a half cylinder with a radius of 3000 and a flat part with a length of 2000. Arrangement of the design points are: at 400 intervals in the height direction, and in the direction of radius of the cylinder, away 200 from the wall, the points are put at 400 intervals, and in the circumferential direction, divided 180 degrees into 20 (at 9 degrees intervals), then the total number of the design points are 870. The pipes have constraints of the bends that radius of the bend is 2400 or 1200 at the ϕ 600 pipe, 1600 or 1200 at the ϕ 400 pipe and 300 or 150 at the ϕ 100 pipe. Note that the angle of the bend is only 90 degrees in the ϕ 100 pipe, and the others are arbitrary degrees of pipes. The minimum length of the straight pipe section between the bends (or elbows) is 200 in the ϕ 600 or ϕ 400 pipes and 100 in the ϕ 100 pipe. Figure 7 shows a result by a standard PC with 64bit Microsoft Windows7, IntelCOREi7-2.7GHz processor, 16GB memories and the proposed system is developed on Java-1.8.0 environment. It takes about 1 or 2 minutes for the calculation, but it was sometimes failed to obtain piping paths without interfering pipes depending on the initial paths of the Touch-and-Cross method. Notice that the orange pipe is arranged along the curved wall despite using only the 90 degrees' elbows.

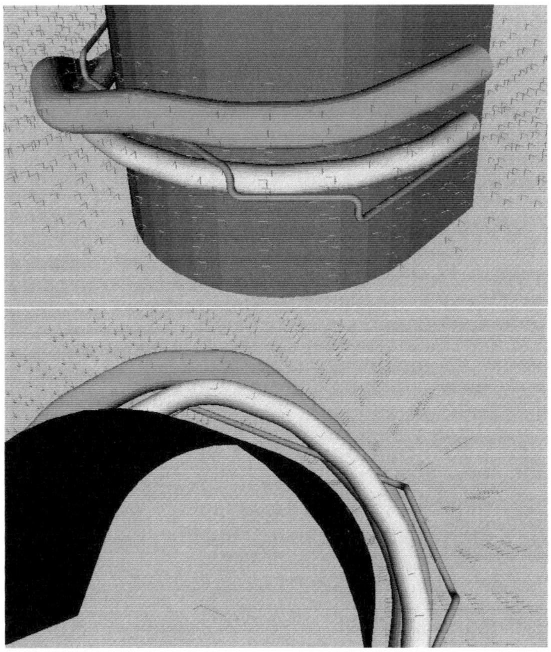

Figure 7: A result of the path-planning along curved structure.

4.2 PIPING CONSIDERING SUPPORTS

Two pipes of ϕ 200, one pipe of ϕ 150 and two pipes of ϕ 100 are arranged in a cage lining side by side along a wall on the side inside a passage shaped space of width 1500 depth 4000 height 2000. Assuming that pipe supports of two positions in the horizontal position of 100 and 400 from the ceiling and pipe supports of two stages in the vertical direction of 100 and 400 from one wall side are installed at 500 intervals in the depth direction, the design points are placed in advance assuming use of these pipe supports. A rectangular parallelepiped of width 800, height 1300 from the floor and depth 4000 was set as aisle space opposite the wall with supports, and invasion of pipes is prohibited there.

Figure 8: Candidates of design points for the φ200 pipes.

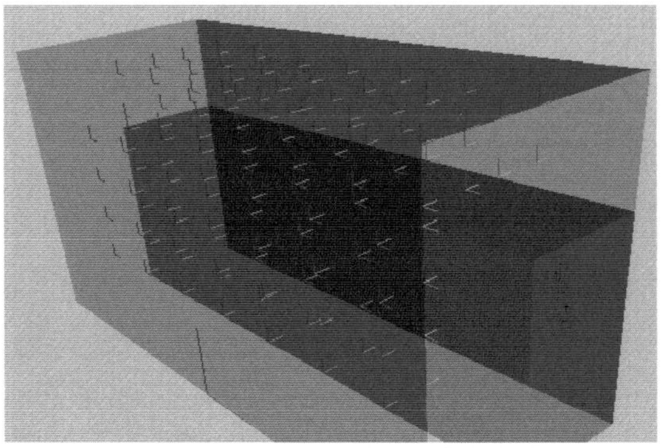

Figure 9: Candidates of design points for the φ150 pipe.

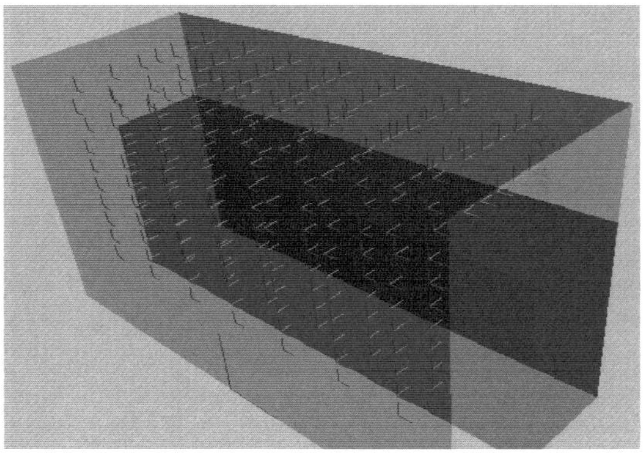

Figure 10: Candidates of design points for the φ100 pipes.

All pipes' pieces are constrained to connect two design points placed at a distance of 2000 or less because of construction limitations. Figure 8, 9 and 10 shows the arrangements of design points for ϕ 200, 150 and 100 respectively. The number of the design points are 112 for ϕ 200 and ϕ 150, 210 for ϕ 100. The transparent dark-red box represents the aisle space. The pipes have constraints of the bends that radius of the bend is 400 or 300 at the φ200 pipes, 300 or 150 at the φ150 pipe, and 200 or 100 at the φ100 pipes. The minimum length of the straight pipe section between the bends (or elbows) is 100 in all the pipes.

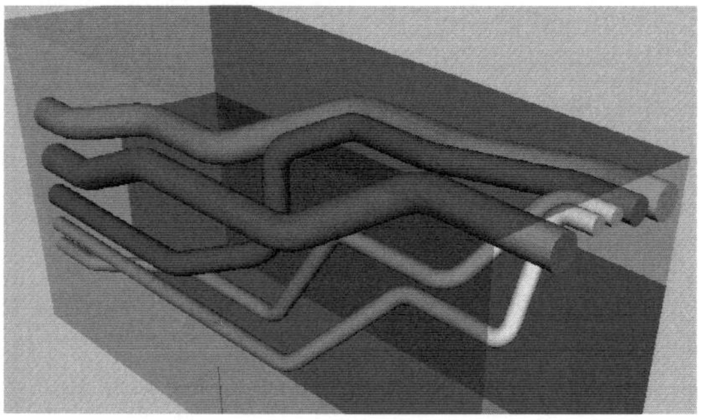

Figure 11: A result using free angle bends.

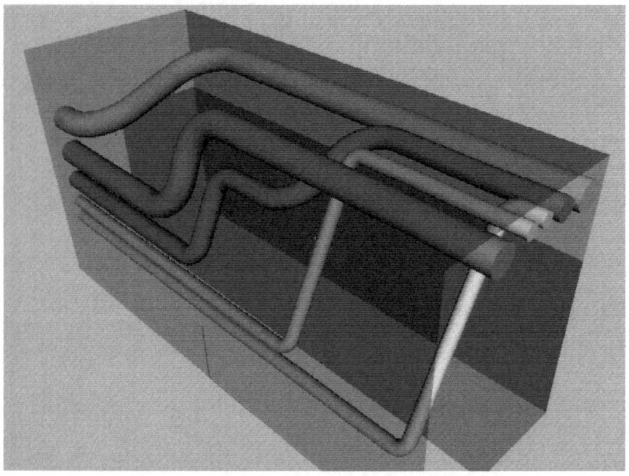

Figure 12: A result using only 90 degrees' elbows.

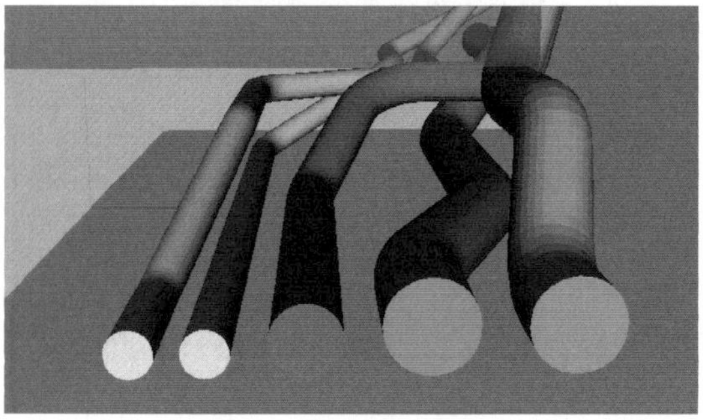

Figure 13: Another view of the solution in Fig.11.

Figure 11 shows a successful calculation result under the condition that all the pipes use free angle of bends. However, 18 out of 20 trials failed to explore. Figure 12 shows a result using only 90 degrees' elbows. Notice that although the bends are 90 degrees, some pipes are arranged diagonally. Thanks to the provision of different design points for each pipe diameter, pipes with different diameters can be fixed with the same pipe supports as shown in Figure 13. It takes about 3-5 minutes for the calculation under the same computational condition as the Section 4.1.

4.3 PIPING WITH BRANCHES

Two pipelines (green and yellow) which are composed of one root point and three leaf points are arranged behind the ceiling of a building as shown in Figure 14. The number of the design points is 1288, the number of the primitives of the obstacles is 28, and all pipes are restricted to use only 90 degrees elbows. It takes about 18 minutes for the calculation under the same computational conditions as the previous sections. The red sections in each pipeline represent branch pipe-parts which are detected by the path planning system. In this experiment, all ten trials resulted in quite similar piping paths.

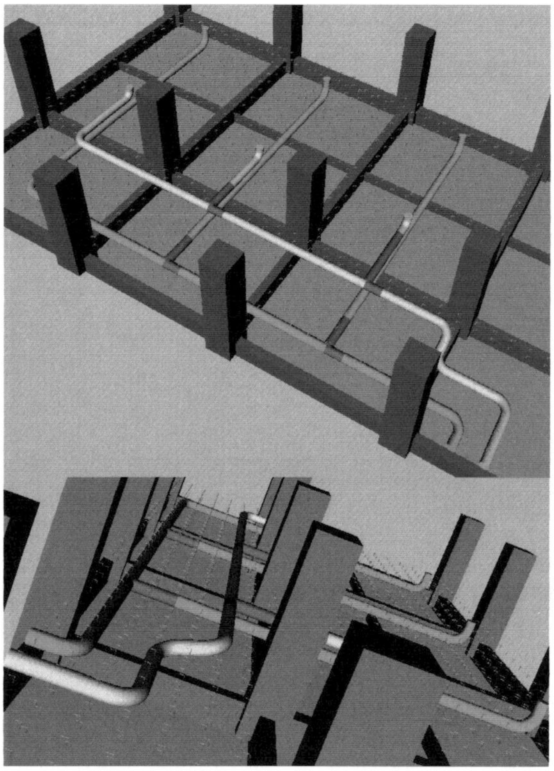

Figure 14: A simulation result for two branched pipelines.

The proposed system is also applied to a real practical problem that consists three pipelines with six leaf points (total 18 pipes), 108 primitives of the obstacles and 4002 design points for each pipeline, however, it takes 8 days for the calculation under the same computational environment as the previous sections.

4.4 PATH PLANNING OF MULTIPLE DRONES

The proposed system is applied to a path planning problem of multiple drones. Figure 15 and 16 represent a calculation result of the path planning of three drones. The number of the design points is 2000, and it takes about 25 minutes for the calculation. In order to generate short-cutting paths, the system makes a directed graph using long edges that are generated by connecting two distant design points. For this reason, the directed graph grows very large, and it takes a lot of time to solve it.

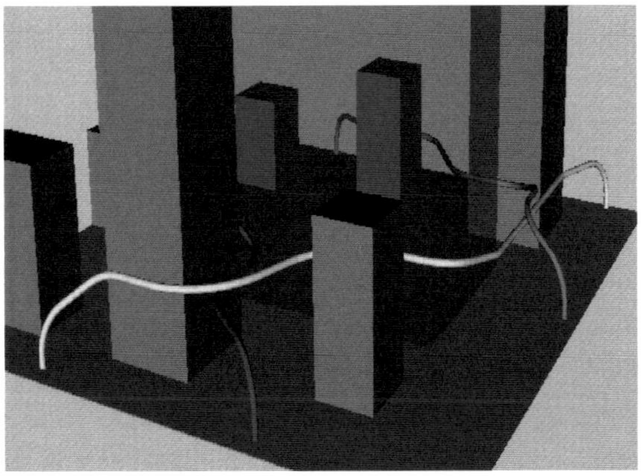

Figure 15: A path planning result of 3 drones.

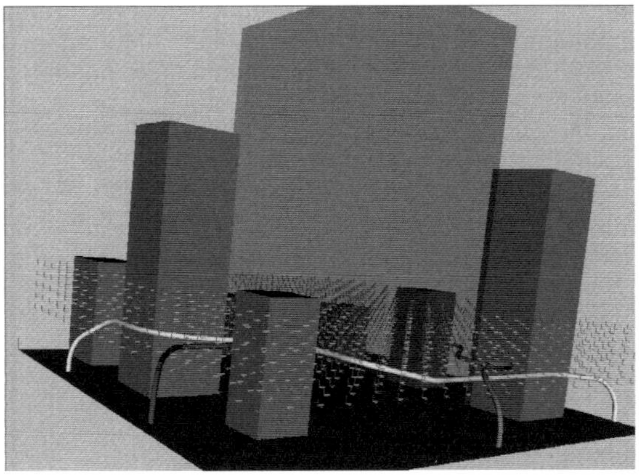

Figure 16: The obtained drones' paths with design points.

5. DISCUSSIONS

5.1 ARRANGEMENT OF THE DESIGN POINTS

In this system, design points are assumed to be arranged appropriately in advance, and the quality of the piping routes obtained by the system and the calculation time are largely affected by it. If it is too many, the computation time and required memory dramatically increase, and if it is too few, the quality of the solution goes down, or in the worst case the route cannot be found. In piping design where gravity flow is considered often, it is better to narrow the vertical direction of the pitch between the design points than the pitch of the horizontal direction

It is troublesome to input if manually assigning via design points, and if it is arranged uniformly in the design target space, it becomes a piping route passing through a place where it is difficult to install support. We are currently developing a system that automatically generates via design points to positions where the installation of pipes becomes easier from the position and shape of structural members and supports.

5.2 CHOICE OF THE INITIAL SOLUTION IN THE TOUCH AND CROSS METHOD

In the first step of the touch and cross method, all pipes are arranged with the shortest (i.e., the lowest cost) paths ignoring interfering each other. In this time, when a pipe has several shortest paths, the system must choose one from the paths. The choice of the initial paths would affect the quality of the final solution of the touch and cross method.

5.3 REDRAWING PIPE ORDER IN THE TOUCH AND CROSS METHOD

In the touch and cross method, redrawing order of the pipes also largely affects the quality of the final solution. In our experiences, following the order of the pipes' length is a good strategy, because the longer piping path tends to have the more number of shortest paths which would avoid the other piping paths. However, when this strategy fails, repetition of the same strategy no longer passes. The proposed system takes mixture strategy of the pipes' length order and a random order. Research of the appropriate strategy is an important future work.

5.4 DEALING WITH BRANCHES IN THE TOUCH AND CROSS METHOD

The touch and cross method for branching pipes seems working well in the experiments, however, it would be biased easily to wrong solutions depending on the initial paths of the touch and cross method.

In practical piping design, there exist unfavourable shapes of the branch-pipe parts, e.g., the branch forces the main stream to strike on the wall, etc., however, the proposed system does not consider it. Also in the branch, the diameters of the pipes at the entrance and the exit of the branch are different, but currently the proposed system cannot deal with it because considering it in the automatic piping arrangement is too complicated. Modification of the pipes' diameters in post-processing is a promising way.

6. CONCLUSIONS

On the routing problem for one pipe, a new algorithm to deal with supports and arrangement along curved structure is proposed. It is assumed that design points that represent directions and positions for the candidates of the piping waypoints of the pipe are arranged in advance, and the system generates weighted graph checking whether two design points can connect by a pipe which satisfies the constraint of the pipe bender. After that, the system generates piping path from a shortest path search in the weighted graph. For arrangements of multiple pipelines with branches, a new approach that is composed of the proposed routing algorithm and the touch and cross method is also presented. Experimental results show the proposed system can withstand practical use.

7. ACKNOWLEDGEMENTS

The author thanks SUGA CO,LTD and SHINRYO corporation for valuable comments on the piping arrangement algorithms. This research is partially supported by the Matching Planner Program MP28116808445 from Japan Science and Technology Agency (JST).

8. REFERENCES

1. ITO T. and FUKUDA S., 'A genetic algorithm approach to piping route path planning', Journal of Intelligent Manufacturing, 1999.

2. PAEK J. H. and STORCH R. L., 'Pipe-routing Algorithm Development: Case Study of a Ship Engine Room Design', Journal of Expert Systems with Application, 2002.

3. ASMARA, A. and U. NIENHUIS 'Automatic Piping System in Ship', International Conference on Computer and IT Application (COMPIT), 2006.

4. ASMARA, A. and U. NIENHUIS 'Automatic piping system Implementation', International Conference on Computer and IT Application (COMPIT), 2007.

5. PAULO T. M. and LOBO V. J. A. S., 'A Tool for Automatic Routing of Auxiliary Circuits in Ships', Portuguese Conference on Artificial Intelligence (EPIA), 2009.

6 LIN C., WANG L., PENG J., "Engine Room Pipeline Arrangement with Consideration of Equipments Operability Space by Genetic Algorithm", International Conference on Computer Application in Shipbuilding (ICCAS), 2011

7. Y. ANDO and H. KIMURA 'An Automatic Piping Algorithm including Elbows and Bends', International Conference on Computer Application (ICCAS), 2011.

8. Y. ANDO and H. KIMURA 'An Automatic Pipe Arrangement Algorithm Considering Elbows and Bends', International Marine Design Conference (IMDC), 2012.

9. AUTHOR BIOGRAPHY

Hajime Kimura holds the current position of Associate Professor at the Department of Maritime Engineering, Kyushu University. His previous experience includes optimization in shipbuilding.

SPATIAL AUGMENTED REALITY FOR MANUFACTURING INFORMATION OF CURVED SHELL PLATES

K Hiekata, T Mitsuyuki, M Enomoto and **K Okada**, The University of Tokyo, Japan
Y Furukawa, National Institute of Advanced Industrial Science and Technology, Japan

SUMMARY

This study designs a system to introduce the Spatial Augmented Reality technology to curved shell plates' manufacturing. Firstly, the current process was analysed and the need and *ilities* of the system, which are characteristics required for the system, were defined. Subsequently, Morphological Matrix (MM) to express design candidates of system exhaustively was created, and some Pareto optimal solutions from all combinations were derived by analyzing MM. One solution was selected as a design proposal and a prototype was constructed. As a case study, a projecting experiment using a plate model was conducted, and the result showed that the proposal could work as expected.

1. INTRODUCTION

Manufacturing process for ship curved shell plates is not standardized because it is heavily depending on implicit knowledge of workers. As a solution of that problem, some IT systems using laser scanner were proposed. For example, an accuracy evaluation system was developed [1]. The system calculates displacement errors comparing design shape and measured shape, and visualizes the result as 3D computer graphics on a digital display. However, it is often difficult for workers to recognize the intended spots correctly, because the plates have unique and big surface.
In this paper, the introduction of Spatial Augmented Reality (SAR) for manufacturing of curved shell plates is discussed with the aim of developing an efficient interface between workers and IT systems which calculate the manufacturing information. SAR is a Human-Computer Interaction technique introduced by Raskar [2], and it is also called as projection mapping. It can provide realistic information for users by projecting a variety of contents onto a real object using projector, and already introduced in some kinds of fields including entertainment, promotion and product design evaluation [3]. SAR system for manufacturing of curved shell plates was previously proposed [4], but system requirement was not considered sufficiently to realize.
The objective of this paper is to design a spatial augmented reality system to project manufacturing information of curved shell plates. Based on the design proposal, a prototype is constructed and a projecting experiment using a plate model is conducted to confirm its behavior.

2. SYSTEM DESIGN

2.1 CURRENT PROCESS ANALYSIS

The goal of this paper is to use SAR technology as an interface between workers and existing IT systems for curved shell plates, projecting manufacturing information onto the real plates and support workers to recognize smoothly and effectively. Projected information generated by IT systems are as follows.

- Accuracy information of plate

This information is calculated using accuracy evaluation system developed by Hiekata [1]. Point cloud data of a curved shell plate measured by a laser scanner and the design data are registered and its displacement errors that refer to the deviation distances based on the design surface are evaluated. Each evaluated point is colored according to its error, and it enables workers to see output visualized as color map and grasp the overall shape and the local unevenness of the target plate.

- Curvature error information of frame lines

This information is calculated using one engine of Sun's system, "heating factor proposing engine" [5]. Curvature error of each frame line between measured point cloud data and design data is calculated and visualized as a color map. Workers can understand the curvature difference where they should check and heat for forming.

These two system are used according to the process shown as Figure 1. Firstly, a laser scanner is located aside the target plate and measure it to obtain point cloud data, and systems start to calculate and analyze. After calculation, workers check the output, such as color map, on the digital display and compare the real plate and display to recognize the intended spots. This recognition process is conducted based on workers' visual confirmation or using tapelines and chalks to mark. Finally, workers evaluate the spots using wooden template and make heating plans, and do thermal forming using gas burners.
The problem of its operation lies in the recognizing process. For example, recognizing the wrong spots sometimes happens. And, working efficiency is decreased when they try to recognize correct spots using some measurement tools such as tapelines.

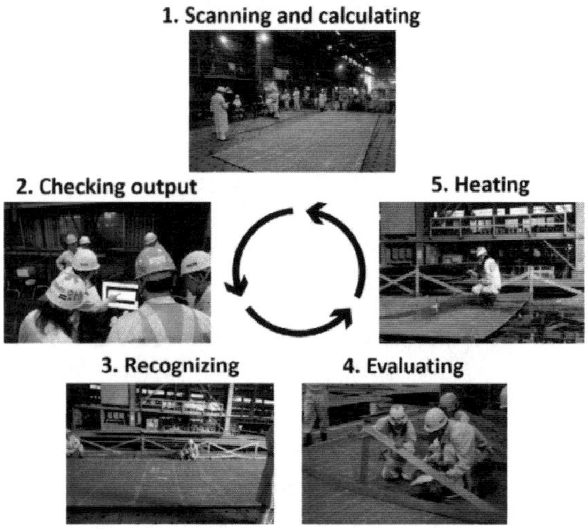

Figure 1: Current working flow

2.2 REQUIREMENT DEFINITION

This section explains the way how to define the requirement for the system using SAR.

2.2 (a) Need

Firstly, the need of the system should be clarified. The goal of the manufacturing is to finish forming plates comparing design shape, and workers use some tools to get some manufacturing information for their works. Therefore, in this research, the need which should be achieved by system is "providing the manufacturing information for workers while working". The information is what workers refer to for forming plates, such as local unevenness, overall shape (like "twist") and curvature of the plates.

2.2 (b) Ilities

Next, "*ilities*" of the system are considered. The *ilities* are desired properties of systems, such as flexibility or maintainability, that are not necessarily part of the fundamental set of functions or constraints and sometimes not in the requirements [6].
Typical *ilities* are known as quality, safety, operability and reliability. However, their definitions are various and perspectives are different depending on situation. Thus the four top *ilities* considering research objective are selected and defined as follows. System's performance can be evaluated by the four ones.

- *Quality*; ability of accuracy to provide manufacturing information for workers. It is composed of both the accuracy of information itself and accuracy of workers' recognition of information.
- *Operability*; cost required for workers when the system works. In this research, operability can be expressed by only time cost, and it has "high" operability when less time is needed for its working.
- *Feasibility*; likelihood of system's realization considering the requirement and constraint of applied environment.
- *Technology maturity*; readiness of technology for realizing the system. If existing method is not sufficient and new technological development is required, it can be said that it has "low" technology maturity.

2.3 ARCHITECHTING

This section explains the way how to make a decision of system's architecting. A design framework which utilizes Morphological Matrix was referred [7]. There is an existing research to propose design method using Morphological Matrix [8].

2.3 (a) Morphological matrix creation

Morphological Matrix (MM) is a table in which individual explanatory variables constituting the system are arranged in each row and those choices are arranged in each corresponding column.
The top seven decisions which mainly architect the system are selected and corresponding choices are arranged as shown in Table 1. The top three decisions show combinations of the way how to estimate and recognize manufacturing information. Scanning time means required time using laser scanner to measure a plate's shape. Relative position between a projector and a scanner, and also device mobility are decisions about hardware of the system. The last one, projector calibration method, is for achievement of SAR. To project a specific image onto the intended position, projector calibration is needed for calculating the perspective projection matrix of a projector, which enables us to conduct rendering process for making a projected image using 3D point cloud data. Various calibration methods are proposed and each method has different characteristics against *ilities*.

2.3 (b) Combinations evaluation

Using MM created in previous section, characteristic values are set into all choices from view point of each *ility*. The values are set based on current status and knowledge according to three levels, "0.5 - 0.7 - 1". It is a relative evaluation comparing among each choice. Table 2 shows an example of characteristic values of the first decision; how to grasp local unevenness of plates. For example, wooden template is not better at detecting the local unevenness of a plate than IT system, so the value of *Quality (information)* is smaller.
Each combination of all choices, 4608 combinations in total, can express one design, and evaluation values are calculated. There are 2 types of calculation ways, "sum"

Table 1: Morphological Matrix; seven decisions and corresponding choices

Decision	alt A	alt B	alt C	alt D
How to grasp local uneveness (estimation/recognition)	wooden template visual confirmation	IT system visual confirmation	IT system measurement tools	IT system SAR
How to grasp overall shape (estimation/recognition)	wooden template visual confirmation	IT system visual confirmation	IT system measurement tools	IT system SAR
How to grasp curvature error (estimation/recognition)	wooden template visual confirmation	IT system visual confirmation	IT system measurement tools	IT system SAR
Scanning time	short	medium	long	
Relative position between projector and scanner	unfixed	fixed physically	fixed automatically	
Device mobility	movable	unmovable		
Projector calibration method	none	manual totally	manual partially	automatically

Table 2: An example of set characteristic values of each choice

Decision	alt A	alt B	alt C	alt D
How to grasp local uneveness (estimation/recognition)	wooden template visual confirmation	IT system visual confirmation	IT system measurement tools	IT system SAR
***Ilities* and corresponding characteristic value**				
Quality (information)	0.7	1	1	1
Quality (recognition)	1	0.5	1	1
Operability	0.5	0.7	0.5	1
Feasibility	1	1	1	0.7
Technology maturity	1	1	1	1

and "multiply". An evaluation value of *Operability* is calculated as sum of all characteristic values of the combination, and the other evaluation values of *Quality*, *Feasibility* and *Technology maturity* are calculated as product of all characteristic values of the combination. Though there are some impossible combinations realistically, they are excluded after evaluation.

Finally, Pareto optimal solutions are derived from all combinations. The Pareto optimal solution is a solution in which other characteristics value are worsened when it tries to improve any of the characteristic values.

In this case, 14 Pareto optimal solutions are derived and one example is shown in Table 3.

Table 3: One example of Pareto optimal solutions

Decision	selected choice
How to grasp local uneveness (estimation/recognition)	IT system SAR
How to grasp overall shape (estimation/recognition)	IT system SAR
How to grasp curvature error (estimation/recognition)	IT system SAR
Scanning time	medium
Relative position between projector and scanner	fixed physically
Device mobility	movable
Projector calibration method	manual partially

2.4 DESIGN PROPOSAL

One Pareto optimal solution expressed in Table 3 is adopted as a design proposal. This section explains it in detail.

2.4 (a) Overview

Figure 2 shows the proposal overview and its working flow. The system consists of a laser scanner, a projector and a server. The scanner and the projector are combined as one movable device and calibrated beforehand by partially manual using markers. As the relative position is fixed physically, there is no need to conduct calibration after it is done at once. Manufacturing information using measured shape of a target plate is generated by IT systems mentioned in 2.1, and both can be projected onto plate from projector.

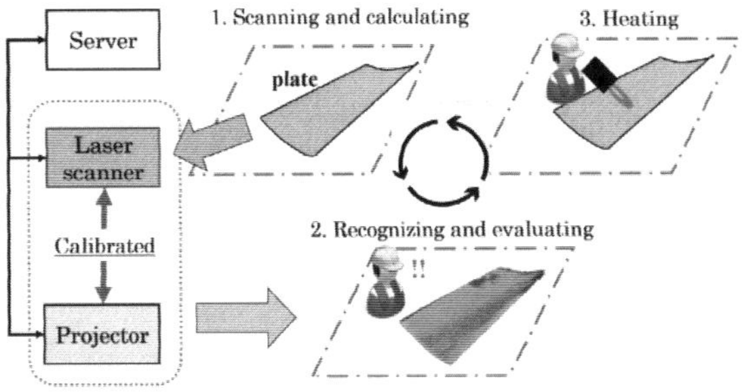

Figure 2: Overview of design proposal and working flow

In the first stage, workers bring the device to their working place and start the program, measuring the shape of the target plate as point cloud data by laser scanner. The second phase is to analyze the data and generate manufacturing information using existing systems. The third phase is to generate a projected image and project it onto the real plate from projector. Workers can select the required information that they want to know as appropriate, and they can recognize it by seeing the projected image. Then they can discuss and make some decision for next working plan. Finally, they move aside the device and restart their work such as thermal forming based on their decision.

2.4 (b) Function

System has two main functions as follows.

- POINT CLOUD ANALYSIS FUNCTION

This is to generate manufacturing information. Two type of IT systems developed by Hiekata and Sun are used, as mentioned in 2.1. Each calculated data is expressed with point cloud data having color information (RGB).

- Displacement error of surface
- Curvature error of frame lines

- PROJECTING INFORMATION FUNCTION

Using SAR technology, this is to generate a projected image and enable workers to recognize information by projecting it onto the plate according to process shown in Figure 3. A projector is calibrated using Raskar's method [3] and projector's parameter is computed. Using the parameter and point cloud data which have manufacturing information, a projected image is rendered, and projected from a projector.

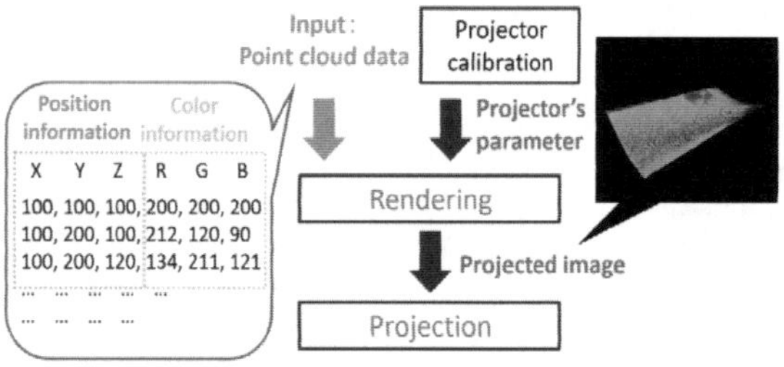

Figure 3: Image generating process

3. PROTOTYPING SYSTEM

A physical prototype shown as Figure 4 was constructed according to the design proposal. CASIO XJ-A245V projector displaying at 1280x800 resolution and Microsoft Kinect for Windows v2 as a scanner instead of a laser scanner were used. Also, Microsoft Surface Pro 4, with the configuration of Intel® Core™ i5-6300U CPU @2.40GHz and 8GB memory, was used as a server. The Open GL was used as the rendering program for generating a projected image. The Kinect can measure the shape of a target as point cloud data like a laser scanner, even though the precision falls. The projector and Kinect are attached to top of a device which is composed of stainless steel pipes, and their orientation are vertically downward. The device has 4 posts and each has a caster with fastener at the bottom, and users can move and fix it easily. The device's height is about 230 cm and width is about 200 cm.

Figure 4: Snapshot of prototype system

4. CASE STUDY

4.1 PREPARATION

An experiment to apply the constructed prototype system to a curved shell plate model shown as Figure 5 was conducted. Model has 6 frames made with wood, and they are covered with a smooth cloth, to make a surface like a curved shell plate. It is about 120 cm in width and 80 cm in length. The projector was calibrated beforehand at once using Nagaoka's method [9], which utilizes Raskar's method for Kinect using markers.

Figure 5: Curved shell plate model

4.2 EXPERIMENTAL FLOW

Firstly, the device was moved over the model. When the system starting, the measurement, analysis of point cloud data, and generating a projected image was executed automatically. In this case study, the manufacturing information was displacement error of surface visualized as color map. After that, the generated image was projected onto the model from the projector.

Figure 6 and 7 show the results, and the total time from moving device to projection was almost 1 minute and projecting error was around 4 cm at the edge of the model. The projected image could be perceived visually and it was confirmed to support users to recognize the information more intuitively compared to only seeing results visualized on the digital display.

Figure 6: Snapshot of projected image

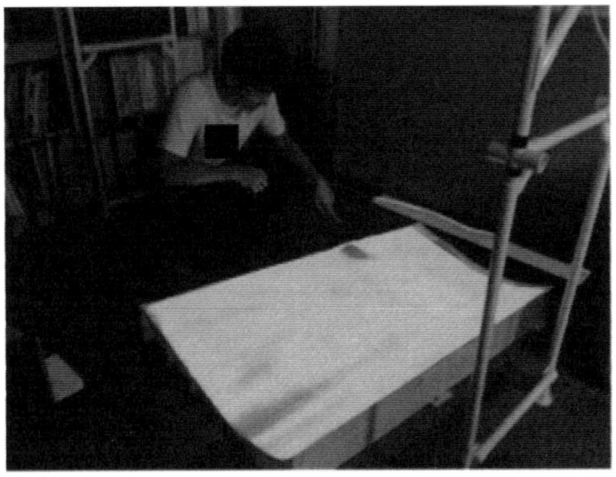

Figure 7: Snapshot that shows a user recognizes the manufacturing information by seeing the projected image

5. DISCUSSIONS

5.1 EVALUATION OF DESIGN PROPOSAL

In the case study, a prototype system combining existing technologies was constructed according to Table 3, based on the set characteristic values. It is considered that the total time required for workers from moving device to projection is not particularly long, and it can be said that it is sufficient for *Operability*, *Feasibility*, and *Technology maturity*. On the other hand, *Quality* was only evaluated qualitatively.

When using SAR, the accuracy of workers' recognition can be affected mainly by projection accuracy, but its tolerance should differ depending on the purpose of the information and also the information accuracy itself calculated by IT systems. Thus when more detailed requirement is needed, a tolerance of information that is finally provided for workers should be decided according to manufacturing information, and the improvement of IT systems or SAR are considered independently, if necessary.

5.2 POSSIBILITY OF OTHER DESIGNS

In the design method based on MM used in this research, the Pareto optimal solutions can differ according to the characteristic values depending on the requirements and constraints of applied environment. For example, if it is possible to develop an instant calibration method of a projector or an automatic device control system, workers do not need to move the device like the design proposal. Therefore, even if in a different environment, it is possible to derive different solutions by setting corresponding characteristic values considering the environment, and select an appropriate design.

6. CONCLUSIONS

In this research, a system to introduce the Spatial Augmented Reality (SAR) technology to curved shell plates' manufacturing was designed.

Firstly, the current process was analyzed and the need and *ilities* of the system, which are characteristics required for the system, were defined. Subsequently, Morphological Matrix (MM) to express design candidates of system exhaustively was created, and some Pareto optimal solutions from all combinations were derived by analyzing MM. One solution was selected as a design proposal and a prototype was constructed. As a case study, a projecting experiment using a plate model was conducted, and the result showed that the proposal could work as expected.

7. AKNOWLEDGEMENTS

Photos were provided by Imabari Shipbuilding Co., LTD.

8. REFERENCES

1. Hiekata, K., Yamato, H., Oida, Y., Furukawa, Y., Makino, Y. and Sugihiro, T., 'Development and case studies of accuracy evaluation system for curved shell plates by laser scanner', Journal of Ship Production and Design, Vol. 27, No. 2, pp.84-90, 2011.

2. Raskar, R. and Oliver, B., 'Spatial augmented reality: merging real and virtual worlds', CRC press, 2005.
3. Park, M., Lim, K., Seo, M., Jung, S. and Lee, K., 'Spatial augmented reality for product appearance design evaluation', Journal of Computational Design and Engineering, Vol.2, pp.38-46, 2015.

4. Hiekata, K., Mitsuyuki, T., Enomoto, M., Okada, K. and Furukawa, Y., 'Introduction of Spatial Augmented Reality Technology to Curved Shell Plates' Manufacturing Process', Proc. of the Conference of the Japan Society of Naval Architects and Ocean Engineers, Vol.24, pp.191-192, 2017 [published in Japanese].

5. Sun, J., Hiekata, K., Yamato, H., Nakagaki, N. and Sugawara, A., 'Virtualization and automation for knowledge extraction in curved shell plates', Int. J. Agile Systems and Management, Vol.7, pp.282-303, 2014.

6. L. de Weck, O., Roos, D. and L.Magee, C., 'Engineering Systems', THE MIT PRESS, 2011.

7. Simmons, W.L., 'A framework for decision support in systems architecting', Doctoral thesis, Department of Aeronautics and Astronautics, Massachusetts Institute of Technology, 2008.

8. Hiekata, K., Mitsuyuki, T. and Ishihara, S., 'Design method of remote monitoring service for elderly considering community characteristics', Proc. of the 24th ISPE International Conference on Transdisciplinary Engineering (TE2017), Vol.5, pp.713-720, 2017.

9. Nagaoka, A. and Hashimoto, N., 'Easy projector calibration with depth sensor', Technical report of the Institute of Image Information and Television Engineers, Vol.37, No.7, pp.3-6, 2013 [published in Japanese].

9. AUTHORS BIOGRAPHY

Kazuo Hiekata holds the current position of Associate Professor at Graduated School of Frontier Sciences, The University of Tokyo.

Taiga Mitsuyuki holds the current position of Assistant Professor at Graduated School of Engineering, The University of Tokyo.

Masakazu Enomoto holds the current position of Senior Technical Specialist at Graduated School of Engineering, The University of Tokyo.

Kota Okada is 2nd year master course student at Graduated School of Frontier Sciences, The University of Tokyo.

Yoshiyuki Furukawa holds the current position of Senior Researcher at Advanced Manufacturing Research Institute, National Institute of Advanced Industrial Science and Technology.

PRE-PROCESSING MODULE FOR THE WELDING DISTORTION ANALYSIS ACCORDING TO THE DIFFERENT ASSEMBLY SEQUENCES

M S Kang, **H Chung**, Korea Advanced Institute of Science and Technology, Republic of Korea

SUMMARY

In order to reduce the pre-processing cost for simulating multiple assembly sequences, an input file generation module is proposed for the welding distortion simulation by ABAQUS. This module is based on the input file information of an assembly sequence and considers contact relation between parts and weld lines in the assembly. The proposed module is applied to assembly structures and the results show that it is useful to reduce the pre-processing cost for the welding distortion simulation with regard to the different assembly sequences.

1. INTRODUCTION

Ship and offshore structures are manufactured by joining dozens of blocks and each block is composed of tens of hundreds of plates. Most of joining processes are carried out by welding. Even though the welding process provides many advantages in terms of manufacturing cost and productivity, it also requires a lot of efforts to remain the quality from the welding distortion.

Welding distortion is basically due to the heating and cooling process in the local region. It is gradually accumulated during the joining process and may require rework to meet the allowable tolerance of an assembly. Therefore the control of welding distortion is closely related to the manufacturing quality and productivity.

In order to evaluate the effects of welding distortion during the assembly process in advance, experiments and welding distortion simulation based on finite element method (FEM) have been performed. Experiments offer relatively accurate data for the analysis. However, its application range is highly limited. So welding distortion analysis for the block unit assembly is mostly analysed by FEM based 3D thermal elasto-plastic simulation. From the simulation results, it is possible to analyse dominant regions and quantitative amount of welding distortion during the assembly processes. However, as the size and complexity of the assembly model increase, the computational cost of the analysis is exponentially increased [1, 2]. In addition, a considerable amount of time is required to create an analysis input file as a pre-processing step of the analysis. Particularly, in the case of analysing the influence of the welding distortion due to the different assembly sequences, even if there is an analysis input file made for an assembly sequence, a user has to create the analysis input file for the different assembly sequence again. It is considerably inefficient and time-consuming process.

In this paper, a pre-processing module is developed to create an analysis input file for different assembly sequences when ABAQUS software is used for welding distortion simulations. Basic principle of the developed module is to use the information of existing analysis input file for an assembly sequence. The proposed module considers contact relations between parts and weld lines in the assembly model. It is applied to determine the welding sequence from the assembly sequence information. This module was applied to the two different assembly models and results show that it is useful to reduce the pre-processing cost for the welding distortion simulation with regard to the different assembly sequences.

2. WELDING DISTORTION SIMULATION USING ABAQUS

Welding distortion simulation using ABAQUS is usually performed by sequential procedure of heat transfer analysis and structural analysis. It is usually called thermal elasto-plastic analysis.

The heat transfer analysis is to simulate time varied temperature variations in the local area near the weld line. It is due to the effects of heat flux in the weld line. For the heat transfer analysis using ABAQUS, there are different element types for the analysis. In the solid element based analysis, DC3D10 or DC3D8 is used as an element type. On the other hand, the element type of DS3 or DS4 is used for the heat transfer analysis using shell elements [3]. Surface film conditions that take into account the convection and radiation effects are also applied to the heat transfer analysis. As material properties, specific heat, density, and thermal conductivity are considered.

Structural of the assembly model is performed based on the time varied temperature history obtained from the results of heat transfer analysis. There are also specific element types for the analysis. In the solid element based analysis, C3D8R is used as an element type. On the other hand, the element type of S4R is used for the structural analysis using shell element [3]. Thermal deformation of the assembly model is caused by the effects of thermal expansion coefficient and temperature variation. Constraint condition of degree of the specific nodes is applied as a boundary condition. Also, elasto-plasticity and thermal expansion coefficient are taken into account as material properties.

For the convenience of the analysis of welding distortion simulation, keywords of `*Nset`, `*Surface` and

`*Elset` are used for grouping nodes, surfaces and elements respectively.

3. PRE-PROCESSING MODULE

The developed pre-processing module is available to generate an input file for simulating heat transfer or structural analysis about an arbitrary assembly sequence by ABAQUS software. The process of the module is based on the information of an existing input file for an assembly sequence. Fig. 1 shows the flow diagram for the generation of an input file.

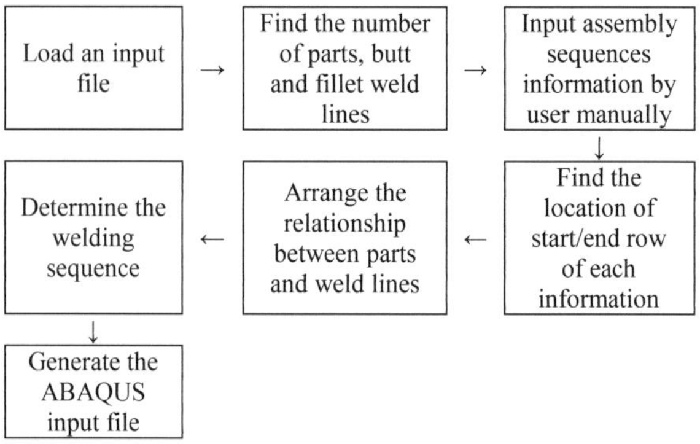

Figure 1: Flow diagram of an input file generation.

The whole procedure is carried out by three steps of analysis of reference input file, determination of welding sequence, and rearrangement of the information.

3.1 ANALYSIS OF REFERENCE INPUT FILE

As a first step, an existing ABAQUS input file about an assembly sequence is analysed as reference information. In order for an ABAQUS input file to be used as reference file, following information and format are required.

- Each part in the assembly model is named as P1, P2, ...
- Butt and fillet beads are named as B1, B2, ... and F1, F2, ... respectively.
- Reference input file contains node information of parts and beads in the assembly model and the name of each node set include _node after the part or bead name like B1_NODE, F1_NODE, P1_NODE, ...

From the reference input file, following information is extracted.

- The number of parts, butt and fillet beads
- Location of start/end row of some information
- Contact relation between parts and beads

The information of the number of parts, butt and fillet weld lines is counted and used to define the contact relation matrix and determine the welding sequence.

In order to generate ABAQUS input file of an arbitrary assembly sequence by rearranging the information of the reference input file, locations of start/end row of some information are extracted. The extracted information contains node sets of each part and bead in the assembly model, step related information. It can be used to extract a specific group of information.

In order to determine the welding sequence from the information of an arbitrary assembly sequence, contact relation between parts and beads is expressed as follows [4,5].

$$\begin{bmatrix} C_{11} & \cdots & C_{1j} \\ \vdots & \ddots & \vdots \\ C_{i1} & \cdots & C_{ij} \end{bmatrix}$$

where C_{ij} represents a contact relation between part j and weld bead i. C_{ij} has the value of 0 or 1 depending on whether contact relation exists or not. Existence of contact relations are evaluated by comparing node set information. It means that a part and a weld line are under contact if they share common nodes. However, there is an exceptional case of contact relation between a butt weld line and a stiffener. Even though a butt weld line and a stiffener perpendicular to this butt weld line share some nodes in the small region, it is regarded as no contact relation because these are not directly related to the joining process by welding.

3.2 DETERMINATION OF WELDING SEQUENCE

In a previous step, the assembly sequence information was obtained by a user manually. Once the assembly sequence is determined, the corresponding welding sequence should be determined. In the pre-processing module, the contact relation matrix is used to determine the welding sequence. When a part is joined to the sub-assembly, a weld line which has contact relation with both a part in the sub-assembly and the joined part becomes a welding sequence between them.

3.3 GENERATION OF ANALYSIS INPUT FILE

The final step of pre-processing module is to generate an input file considering an arbitrary assembly sequence. It is basically a rearrangement process of total information in the reference input file. Especially, `*Step` procedure which constructs the whole scenario of the expected assembly process is rearranged. This procedure contains element add/remove, simulation time, applying heat flux to the weld line elements and so on. So, contents of `*Step` is changed according to the difference assembly sequence except the first step and the final step. The first `*Step` is considered to set the conditions to be applied over all steps continuously. So the first `*Step` includes element removal, interaction condition like convection

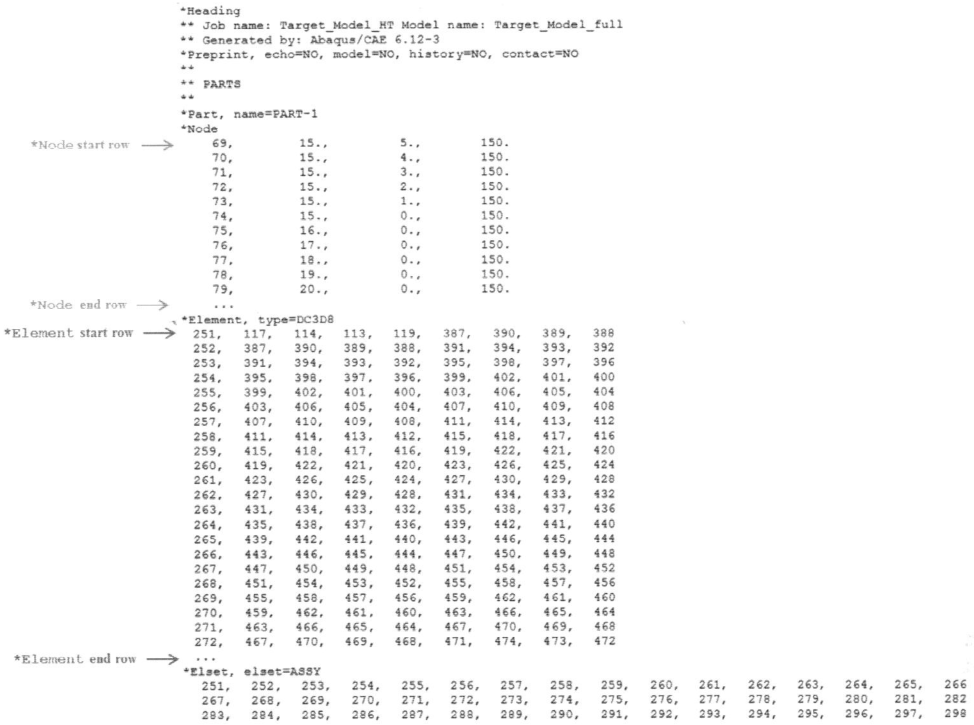

```
                         *Heading
                         ** Job name: Target_Model_HT Model name: Target_Model_full
                         ** Generated by: Abaqus/CAE 6.12-3
                         *Preprint, echo=NO, model=NO, history=NO, contact=NO
                         **
                         ** PARTS
                         **
                         *Part, name=PART-1
                         *Node
*Node start row    -->       69,      15.,      5.,      150.
                             70,      15.,      4.,      150.
                             71,      15.,      3.,      150.
                             72,      15.,      2.,      150.
                             73,      15.,      1.,      150.
                             74,      15.,      0.,      150.
                             75,      16.,      0.,      150.
                             76,      17.,      0.,      150.
                             77,      18.,      0.,      150.
                             78,      19.,      0.,      150.
                             79,      20.,      0.,      150.
*Node end row      -->       ...
                         *Element, type=DC3D8
*Element start row -->     251,  117,  114,  113,  119,  387,  390,  389,  388
                           252,  387,  390,  389,  388,  391,  394,  393,  392
                           253,  391,  394,  393,  392,  395,  398,  397,  396
                           254,  395,  398,  397,  396,  399,  402,  401,  400
                           255,  399,  402,  401,  400,  403,  406,  405,  404
                           256,  403,  406,  405,  404,  407,  410,  409,  408
                           257,  407,  410,  409,  408,  411,  414,  413,  412
                           258,  411,  414,  413,  412,  415,  418,  417,  416
                           259,  415,  418,  417,  416,  419,  422,  421,  420
                           260,  419,  422,  421,  420,  423,  426,  425,  424
                           261,  423,  426,  425,  424,  427,  430,  429,  428
                           262,  427,  430,  429,  428,  431,  434,  433,  432
                           263,  431,  434,  433,  432,  435,  438,  437,  436
                           264,  435,  438,  437,  436,  439,  442,  441,  440
                           265,  439,  442,  441,  440,  443,  446,  445,  444
                           266,  443,  446,  445,  444,  447,  450,  449,  448
                           267,  447,  450,  449,  448,  451,  454,  453,  452
                           268,  451,  454,  453,  452,  455,  458,  457,  456
                           269,  455,  458,  457,  456,  459,  462,  461,  460
                           270,  459,  462,  461,  460,  463,  466,  465,  464
                           271,  463,  466,  465,  464,  467,  470,  469,  468
                           272,  467,  470,  469,  468,  471,  474,  473,  472
*Element end row   -->     ...
                         *Elset, elset=ASSY
                           251,  252,  253,  254,  255,  256,  257,  258,  259,  260,  261,  262,  263,  264,  265,  266
                           267,  268,  269,  270,  271,  272,  273,  274,  275,  276,  277,  278,  279,  280,  281,  282
                           283,  284,  285,  286,  287,  288,  289,  290,  291,  292,  293,  294,  295,  296,  297,  298
```

Figure 2: Example of extracting start and end row location

and radiation, boundary condition and so on. The final `*Step` is considered as the cooling stage after assembly process by welding. Even though an assembly sequence is changed, this `*Step` remains the same. Contents of other `*Step` are changed according to the difference assembly sequence. It is carried out by using the start and end row locations of the specific information. That is, specific information is extracted by using the location information and it is rearranged to another position corresponding to an arbitrary assembly sequence. Particularly, `*Model Change` in the heat transfer and structural analysis is the one of the most complex information for the rearrangement. For example, in the second `*Step`, at least two parts should be added for one weld line. As another example, in the case of two-sided fillet welding, when a part is added for fillet welding to one side, there is no addition of any part for fillet welding to the remaining side. That is, even though add/remove of weld lines at each step can be easily determined by considering the welding sequence information, add/remove of parts at each step should be determined based on the existing information. A solution to determine the add/remove parts at each step is to use a matrix. The first-row values of this matrix are zero and the contact relation between parts and weld lines is appended to the second row. When there is a weld line to be added in a `*Step`, the information of the row corresponding to the weld line in the matrix becomes reference row. Based on reference row information, a new matrix is defined and it is composed of the row information of $(row_{refer}, :) + (1,:)$, $(row_{refer},:) + (2,:)$, …, $(row_{refer},:) + (row_{refer} - 1, :)$. From the new matrix, the sum of each column values are compared. If the minimum value of a column is 1, then a part corresponding the column information becomes added part. By repeating the following method until the last welding sequence, add/remove parts in each step can be determined. Finally, the input file for an arbitrary assembly sequence is generated.

3.4 PRE-PROCESS MODULE GUI

Fig. 3 shows graphic user inter face (GUI) of the developed pre-processing module.

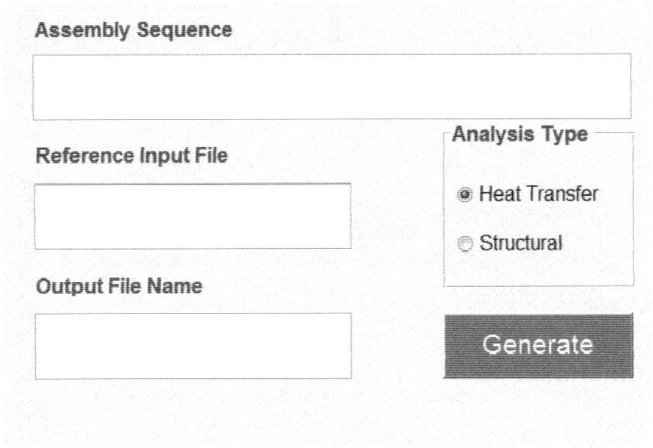

Figure 3: Pre-processing module GUI

It is composed of five windows. The meaning and usage of each window are as follows

Assembly Sequence User enters numbers corresponding to the assembly sequence. For example, there is an assembly model which is composed of four parts of P1, P2, P3 and P4. If you want to specify the assembly sequence as P1 → P3 → P2 → P4, you just enter 1324 in the Assembly Sequence window.

Reference Input File User enters the name of the completed analysis input file for an assembly sequence. The prerequisite is that the input file should be located in the same folder with the module file location.

Output File Name User enters the name of file to be saved when an analysis input file is generated by the pre-processing module. The file extension of generated file is .inp.

Analysis Type According to the analysis type of the reference input file, user selects appropriate analysis mode. Even though whole procedures to create an input file about an arbitrary assembly sequence is similar, the operation mode in the module is distinguished due to some structure differences between heat transfer and structural analysis.

Generate When the input of necessary information and the selection of the analysis mode are completed, the analysis input file about the user-specified assembly sequence can be created by just clicking this window.

4. CASE STUDY

4.1 TEST MODEL INFORMATION

The performance of the proposed model was tested using the following two assembly models. Both models are consisted of four parts and five weld lines. P3 and P4 in the first model are parallel to the butt weld line. On the other hand, P3 and P4 in the second model are perpendicular to the butt weld line.

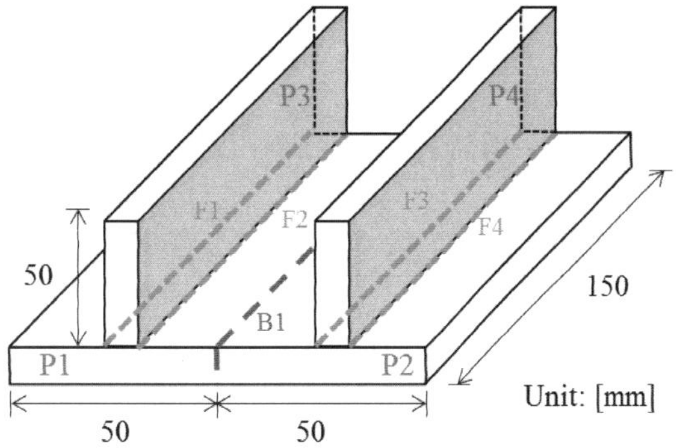

Figure 4: Test model 1 with stiffeners parallel to the butt weld line

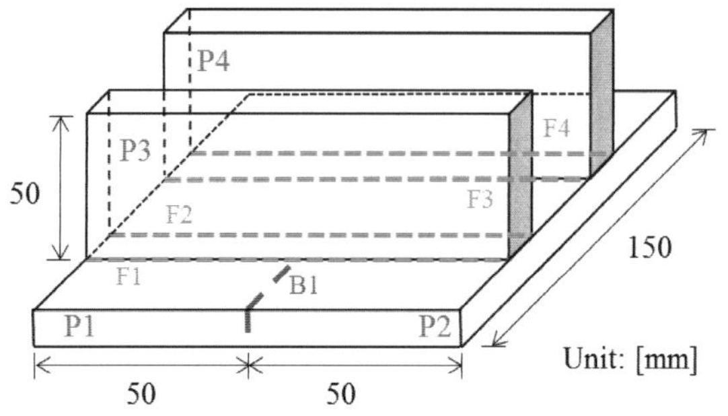

Figure 5: Test model 2 with stiffeners perpendicular to the butt weld line

4.2 PRE-PROCESSING MODULE APPLICATION

The developed module has been made using MATLAB program and its performance was tested using Intel i7-4770 PC.

At each model, two assembly sequences of P1 → P3 → P2 → P4 and P2 → P4 → P1 → P3 were tested in the developed pre-processing module. The reference file of each test model is about the assembly sequence of P1 → P2 → P3 → P4.

$$\begin{bmatrix} 1 & 1 & 0 & 0 \\ 1 & 0 & 1 & 0 \\ 1 & 0 & 1 & 0 \\ 0 & 1 & 0 & 1 \\ 0 & 1 & 0 & 1 \end{bmatrix} \qquad \begin{bmatrix} 1 & 1 & 0 & 0 \\ 1 & 1 & 1 & 0 \\ 1 & 1 & 1 & 0 \\ 1 & 1 & 0 & 1 \\ 1 & 1 & 0 & 1 \end{bmatrix}$$

Model 1 **Model 2**

Figure 6: Contact relation matrix

In the first test model, P1 has contact relation with welding lines of B1, F1 and F2. In the second model, however, P1 has contact relation with all welding lines. In case of P3 in the second model, there is no contact relation with the welding line B1 even though they are sharing nodes. The pre-processing module makes the contact relation matrix during the process of creating an input file about different assembly sequence. This matrix is used to determine the welding sequence corresponding to the specified assembly sequence.

Table 1 shows the processing time of the pre-processing module. The processing time of an analysis input file generation about two models was similar and generation of structural analysis input file required less time than heat transfer analysis input file. Even though generation time will be increased if the number of parts and weld lines is increase, the developed module still can be useful tool to reduce pre-processing cost in terms of preparing the analysis input file.

Table 1: Comparison of processing time

Analysis	Test Model 1		Test Model 2	
	Heat Transfer	Structural	Heat Transfer	Structural
Time [second]	11.751	9.256	10.155	8.867

Fig. 6 and Fig. 7 show the simulation results of welding distortion analysis about the assembly sequence of P1 → P2 → P3 → P4.

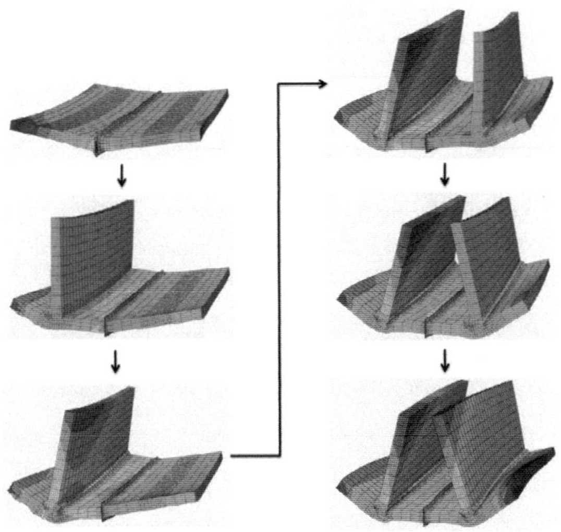

Figure 6: Welding distortion simulation of test model 1 by ABAQUS software

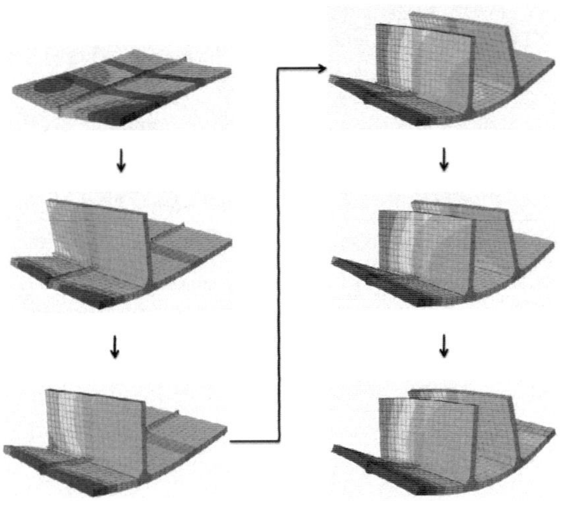

Figure 7: Welding distortion simulation of test model 2 by ABAQUS software

In this simulation, heat transfer and structural analysis were carried out sequentially. All weld lines were single pass welds and uniform body heat flux were applied. In addition, the element add/remove technique was applied about the parts and weld lines in the assembly model.

According to the different assembly sequence, the results of welding simulation have variations about the primary points of the assembly model. The comparison of the results can provide guidance on the assembly sequence to minimize the welding distortion. For these simulations, the developed module can reduce the pre-processing cost in terms of preparing analysis input files.

5. CONCLUSIONS

This paper represented an analysis input file generation module to simulate multiple assembly sequences using ABAQUS. This module is based on the reference input file and offers creating the analysis input file about an arbitrary assembly sequence automatically. Generating an input file using the proposed module just takes a few seconds. In addition, it is available to create an input file about multiple assembly sequences even if the assembly model is complex by considering contact relation and arrangement direction of parts and weld lines in the target assembly model. In conclusion, the proposed module is expected to reduce pre-processing cost for simulating multiple assembly sequences.

6. ACKNOWLEDGEMENTS

This material is based upon work supported by the Ministry of Trade, Industry & Energy (MOTIE, Korea) under Industrial Technology Innovation Program No. 10067156, 'Development of heavy structure dimensional accuracy measurement and management software system for improving productivity in shipbuilding company'

7. REFERENCES

1. HA, Y. S. 'Analytical methodology obtaining an optimal welding sequence for least distortion of welded structure', *Journal of Welding and Joining*, 2013

2. YOO, J. S., HA, Y. S., 'Pre-pre-processing system for converting shell to solid at selected weldment in shell FE model', *Journal of Welding and Joining*, 2016

3. Dassault Systems Simulia Corp., 'ABAQUS analysis User's guide', Version 6.12'

4. SEO, J. Y., 'Assembly sequence planning for ship and offshore structures considering welding distortion', *Master Thesis in KAIST*, 2015

5. Shipeng, Q., Zuhua, J., Ningrong, T., 'An integrated method for block assembly sequence planning in shipbuilding', *International Journal of Advanced Manufacturing Technology*, 2013

8. AUTHORS BIOGRAPHY

Minseok Kang is a Ph.D candidate in Mechanical Engineering department in Korea Advanced Institute of Science and Technology (KAIST). His research interests include stability and quality of GMAW welding process. Particularly, his research focus is modelling of metal transfer, determination of assembly sequence considering welding distortion.

Hyun Chung holds the current position of assistant professor at Korea Advanced Institute of Science and Technology. He is responsible for Marine Design and Production Systems Engineering laboratory, and has worked in the field of ship production systems engineering, especially for welding, welding distortion, compliant assembly system, as well as accuracy control and tolerance analysis and optimization.

INFLUENCE OF COMPUTER APPLICATIONS IN RESISTANCE PREDICTION OF MARINE VESSELS

Sri Harish Kalidass, Blue Bear Systems and Research, UK and University of Plymouth, UK

SUMMARY

The efficient and convenient way of predicting resistance and other performance parameters of marine vessels are necessary for preliminary design phase for a better design and hassle free deadline approach. It's therefore been the motivation of this research work for standardising and amending the existing method of predicting the resistance of planing and displacement type vessels. This report predominantly focuses on predicting and comparing the resistance of marine vessels by traditional method (currently followed) and the computer model (Simulink model created using MATLAB software) based method. It also details the features, advantages and future scopes that are associated with the developed computer based resistance prediction model. This report is focused on the experiment objective, methods and results. It also includes an appraisal and conclusion to establish the test efficacy and outcomes of the study.

NOMENCLATURE

[Symbol]	[Parameter]	[Units]
τ	Trim Angle	Degrees
Cos(τ)	Cos(Trim angle)	Radians
Sin(τ)	Sin(Trim angle)	Radians
Tan(τ)	Tan(Trim angle)	Radians
Cv	Speed Coefficient	No units
Δ	Displacement of vessel	Kgs
b	Beam length	M
d	Draft length	M
L	Overall Length of vessel	M
(VCG)	vertical centre of gravity	M
(LCG)	Longitudinal centre of gravity	M
β	Deadrise Angle	Radians
ρw	Water density	Tonnes/m^3
ρa	Air density	Tonnes/m^3
s	Speed	Knots
V	Velocity	m/s
g	Gravity	m/s^2
CLβ	Lift Coefficient at deadrise	no units
CL0	Lift Coefficient at zero deadrise	no units
λ	Mean wetted length-beam ratio	no units
λb	Wetted length	M
ϒ	Kinematic viscosity	m/s^2
Vm	Mean velocity	m/s
Re	Reynolds number	no units
CF	Friction drag coefficient	no units
ΔCF	Δ Friction drag coefficient	no units
CF_T	Total Friction drag coefficient	no units
DF	resistance component due to pressure	KN
CP	Distance of centre of pressure measured along keel forward to transom	no units
LP	Distance from transom to point of hydrodynamic pressure	no units
C	Distance between component of resistance normal to bottom and gravity	M
a	Distance between frictional drag force and gravity	M
f	Distance between propeller thrust and gravity	M
ε	Inclination of thrust line relative to keel line	Degrees
M	Moment	Nm
M	moment (Absolute value)	Nm
IF	interpolation factor	no units
D	Drag force(Resistance)	KN
P	Power	KW

1. INTRODUCTION

The understanding and prediction of resistance of a ship is more essential because of its necessity in evaluation of propulsive power requirement at given speed. There is various ways for calculating the resistance. Most importantly, the resistance prophecy depends on the type of hull considered.

The planing hulls and displacement hulls are the two main types of hulls for marine vessels. The displacement hulls are those which are supported by the pressure of the water that it displaces. These types of hulls are preferred more for efficient average speed applications and they don't essentially have a lift. The planing hull balances the vessels and operates it by the hydrodynamic lift.[1]*(Mark Kennett boat designs ,2009)*.The planing hulls comparatively require more power for the same speed than displacement hulls. The planing hulls are usually preferred for their speed, because the displacement hulls cannot operate at high speeds. [2] *Kalkomey (2016)*

A SIMPLE SCHEMATIC STRUCTURE OF COMPONENTS OF RESISTANCE

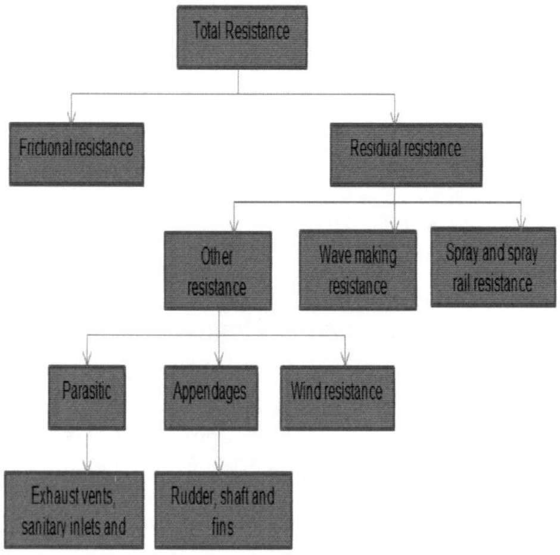

The frictional resistance is proportional to the wetted area and is concerning 70-90% of the ship total resistance for low-speed ships and sometimes less than 40% for high-speed ships. The residual resistance usually represents 10-25% of the overall resistance for low-speed ships and up to 40-60% for high-speed ships. Air resistance usually accounts to 2% of the overall resistance. However, for loaded container ships in head wind, the air resistance is the maximum amount as 10%.During the voyage of ship, the paint film on the hull breaks down and erosion starts, This facilitates the growth and accumulation of marine plants and barnacles, etc. on the surface of the hull. Additionally, the propeller surface will become rough and fouled. The overall resistance caused by fouling may increase by 25-50% throughout the period of a ship.[3]

2. RESISTANCE PREDICTION FOR DISPLACEMENT HULL TYPE VESSELS

The Displacement hulls are generally efficient when operated at certain speed range although they are limited to low speeds. A good hull design and more efficient engine facilitates to less overall total resistance. The ITTC(International Towing Test Conference) 1957 method for calculating the frictional co efficient and the ITTC 1978 method which is more accurate and includes air resistance and other affecting factors is one of the preferred method for determining the resistance of displacement hull type vessels.

To make it less complicated, cheap and less time consumption the total resistance is calculated for the model scale of the actual ship and that gives the resistance of actual ship (full scale).A picture of displacement hull type vessel is furnished below in figure 1.1.

3. RESISTANCE PREDICTION FOR PLANNING HULL TYPE VESSELS

The planing hulls comparatively require more power for the same speed than displacement hulls. The planing hulls are usually preferred for their speed, because the displacement hulls cannot operate at high speeds. The Savitsky's method (1964) is one of the most adopted methods used for the planning hull resistance analysis. A picture of planing hull type vessel is furnished below in

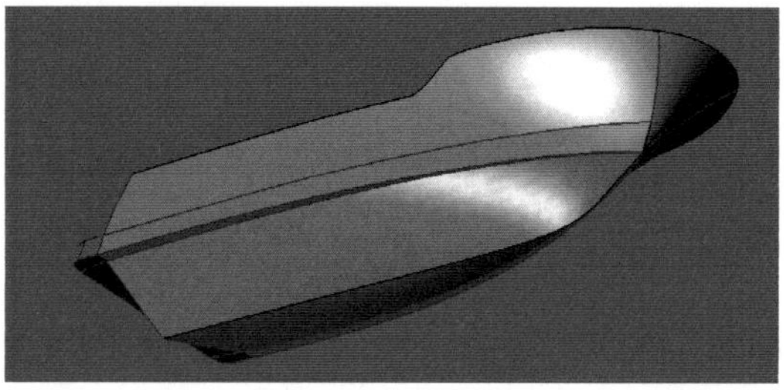

Figure 3.1.

3.1. AN INTRODUCTION ON PLANING HULLS-SAVITSKY METHOD (1964)

The Savitsky's method is used to predict the performance of planing hulls. The hydrodynamics of planing hulls was introduced by Daniel Savitsky in 1964 [4], after this most of the vessel designers consider this method to predict the performance of planing vessels. Even though the planing hulls are designed to operate at high speeds, the Savitsky's method of planing hulls are generally used for planing hulls which operates at lower speeds and therefore the buoyancy is taken into account. Savitsky also believed that a hull is planing hull if it satisfies the condition that Cv/√Lambda >1.This method is used to determine the optimum trim angle, moment and resistance for the given planing hull, where the LCG is fixed. Also the forces and moments acting on the vessel are considered to be in equilibrium.

4. PROCEDURE

The hydrodynamic parameters are and geometric data of a ship is required to compare the traditional method, the

Simulink based computer model method and the . This requisite was fulfilled by retrieving the data form a hull design, which was designed using commercial designing software called MAXSURF. The table of parameters that was used for testing this Simulink model is furnished below.

Table 4.1

	Item	Value	Units
1	LWL	22.523	m
2	Beam	6.931	m
3	Draft	1.955	m
4	Displaced volume	157.145	m^3
5	Wetted area	179.267	m^2
6	Prismatic coeff. (Cp)	0.77	
7	Waterpl. area coeff. (Cwp)	0.766	
8	1/2 angle of entrance	22.7	deg.
9	LCG from midships(+ve for'	-2.449	m
10	Transom area	0.006	m^2
11	Transom wl beam	0	m
12	Transom draft	0.771	m
13	Max sectional area	9.06	m^2
14	Bulb transverse area	0.496	m^2
15	Bulb height from keel	1.618	m
16	Draft at FP	1.955	m
17	Deadrise at 50% LWL	23.8	deg.
18	Hard chine or Round bilge	Round bilge	
19			
20	Frontal Area	0	m^2
21	Headwind	0	kn
22	Drag Coefficient	0	
23	Air density	0.001	tonne/m^3
24	Appendage Area	0	m^2
25	Nominal App. length	0	m
26	Appendage Factor	1	
27			
28	Correlation allow.	0.0004	
29	Kinematic viscosity	0.0000011883	m^2/s
30	Water Density	1.026	tonne/m^3

The resistance prediction in general can be narrowed down to resistance prediction of displacement type vessels and resistance prediction of planning hull type vessels. The Simulink model was developed for both type hulls and the images of the model are furnished in appropriate areas of this report. For simplicity the resistance prediction of planning hull was done using the above ship data using the traditional method, MAXSURF resistance analysis software and Simulink model based method in MATLAB software and the comparison was carried. Also, a particular sensible trim angle was considered at each speed to make it less complicated.

The project was carried to A) enhance the understanding of design and working of different types of hulls, B) determine the optimum trim angle, moment, power and drag force (resistance) theoretically by using empirical equations at different speeds with constant LCG and to perform the resistance analysis for the same hull using MAXSURF software and Simulink model based method created using MATLAB software and to compare the results. C) Understand the effects of relative hydrodynamic parameters on the performance of planing hulls with respect to speed, by way of evaluating a practical example.

4.1. TRADITIONAL METHOD OF RESISTANCE PREDICTION OF PLANING VESSELS

Savitsky method of planning hull procedure

- Trim angle guess (preferably somewhere between 1 degree-10degrees)
- Calculation of wetted length
- Average or mean velocity estimation
- Frictional coefficient calculation using ITTC 1957 formula
- Calculate the centre of pressure in the hull of the vessel
- Force and power calculations

Table 4.2

Parameter	Units	lower limit	upper limit	Best Trim angle
Trim Angle	degrees	9.1	9.2	9.184737334
Trim Angle^1.1	degrees	11.34868474	11.48594177	11.46498299
Cos(Trim angle)	Radians	0.987413807	0.987136265	0.98717882
Sin(Trim angle)	Radians	0.158158067	0.159881188	0.159618225
Tan(Trim angle)	Radians	0.160174049	0.161964658	0.161691298
CV	no units	1.247986291	1.247986291	1.247986291
Displacement	Kgs	161230.77	161230.77	161230.77
Beam	m	6.931	6.931	6.931
Draft	m	1.995	1.995	1.995
Length	m	22.523	22.523	22.523
VCG	m	0.74	0.74	0.74
LCG	m	8.8125	8.8125	8.8125
Deadrise Angle	Degrees	23.8	23.8	23.8
Deadrise Angle	Radians	0.415388362	0.415388362	0.415388362
Water density	Tonnes/m^3	1.026	1.026	1.026
Air density	Tonnes/m^3	0.001	0.001	0.001
Speed	knots	20	20	20
Velocity	m/s	10.28888	10.28888	10.28888
Gravity	m/s^2	9.80665	9.80665	9.80665
CL Beta	no units	0.606070627	0.606070627	0.606070627
CL Beta (goal seek for CL0)	no units	0.606964669	0.606964669	0.606959292
CL0	no units	0.735638087	0.735638087	0.73563208
CL0 (goal seek value for lambda)	no units	0.735586233	0.735597254	0.735589703
Lambda	no units	2.764314105	2.746730907	2.749386548
Wetted length	m	19.15946106	19.03759191	19.05599816
Kinamatic viscosity	m/s^2	0.0000011	0.0000011	0.0000011
Vmean	m/s	9.976994223	9.971605086	9.972429442
Reynolds number	no units	173776211.2	172577589.4	172758724.7
CF	no units	0.001926165	0.001928022	0.00192774
ΔCF	no units	0.0004	0.0004	0.0004
CF total	no units	0.002326165	0.002328022	0.00232774
DF	KN	17.23989093	17.12539183	17.14270952
CP	no units	0.460304255	0.461444384	0.461271363
LP	no units	8.819181452	8.784789867	8.789986243
C	m	-0.006681452	0.027710133	0.022513757
a	m	-0.024233809	-0.024233809	-0.024233809
b/4 tan beta	no units	0.764233809	0.764233809	0.764233809
Sin(Trim angle+E)	no units	0.526955795	0.528438335	0.528212164
1-sin(Trim angle)*Sin(trim angle+E)	no units	0.91665769	0.915512651	0.915687712
1-sin(Trim angle)*Sin(trim angle+E) - C/Cos(trim angle)	no units	0.923424308	0.887441417	0.892881553
f	m	0.13	0.13	0.13
e	degrees	4	4	4
fsin(trim angle)	no units	0.020560549	0.020784554	0.020750869
moment	Nm	-42318.89044	7768.655636	207.5686593
moment (Absolute value)	Nm	42318.89044	7768.655636	207.5686593
interpolation factor	no units	0.844898458		
Best Trim Angle	degrees	9.184489846		
Drag force (Resistance)	KN	270.7162327	273.4363432	273.0209194
Power	KW	2700.934289	2726.599231	2722.681855

The above table 4.2 with hydrodynamic parameters were calculated using Savitsky method and the method is detailed below with the equations and theories. This table is framed for resistance and power at a particular speed. The different speed range prediction tables are nor furnished due to simplicity reasons and thereby abiding the conference paper regulations on limitations of tables/graphs and Images per paper.

The trim angle is the difference of the drafts between forward and aft. Since the planning vessels have a lift while on motion, there will be a difference in draft at aft and forwards. It's easier for further calculations if the Trim angle is known, if not there should be a sensible guess made ranging between 1 degree and 10 Degrees. It should have an upper limit and lower limit. For example, an Upper limit of 8.5degrees and lower limit of 6.5 degrees.
Wetted length is to be calculated using the formula below:

Wetted length (λb) = (Mean wetted length to beam ratio) × (Beam length) 4.1.1

The calculations for Lift coefficient at zero deadrise (CL0) and Mean wetted length to beam ratio (λ) are done using the goal seek option in the Excel spread-sheet.

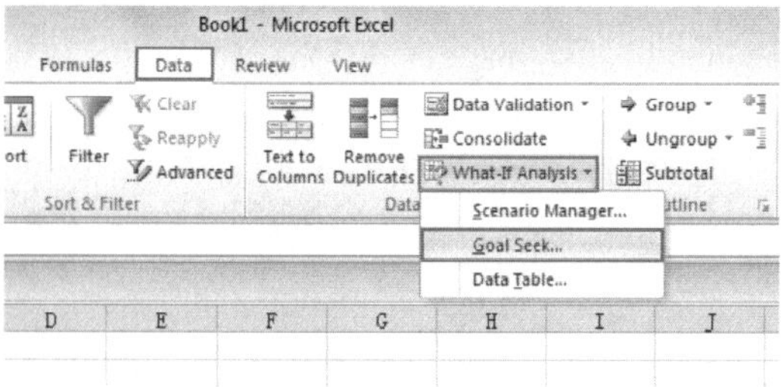

Figure 4.1

As shown in the figure 4.1, the goal seek option is selected from the "data tab" available on the top of the excel window and then the "what if analysis" is selected which gives a drop down menu where the goal seek option can be found.

$$C_{L\beta} = \frac{\Delta g}{\frac{1}{2}\rho V^2 b^2}$$

where

$C_{L\beta}$ = lift coeff for deadrise surface
b = beam of planing surface
V = forward velocity

4.1.2

$$C_{L\beta} = C_{L0} - 0.0065\beta C_{L0}^{0.6}$$

where

C_{L0} = lift coeff for zero deadrise
β = deadrise angle (degrees)

4.1.3

$$C_{L0} = \tau^{1.1}\left(0.0120\lambda^{1/2} + \frac{0.0055\lambda^{5/2}}{C_V^2}\right)$$

4.1.4

where

τ = trim angle (degrees)

λ = mean wetted length beam ratio $= \frac{L_C + L_K}{2b}$

C_V = Froude number (based on beam) $= \frac{V}{\sqrt{gb}}$

The following procedure is to be followed if the Mean wetted length to beam ratio (λ) value is not provided in order to calculate the wetted length of the vessel.
For example to find CL0, the CLβ is evaluated from the equation 4.1.2 and a new parameter is created as "CLβ1 [goal seek for CL0]".Now a value is assumed for CL0 (say 1) based on this value the CLBeta1 [goal seek for CL0] is calculated using equation 4.1.3. By replacing the CLβ1 [goal seek for CL0] value by the previously found value CLβ using 4.1.2, the CL0 cell in excel automatically changes to the actual desired value with respect to the CLβ value.

To find Mean wetted length to beam ratio (λ), Now a value is assumed for Mean wetted length to beam ratio (λ) (say 1) based on this value CL01 [goal seek for λ] is calculated using equation 4.1.4. By replacing the CL01 [goal seek for λ] value by the previously found value CL0 using goal seek method, the Mean wetted length to beam ratio (λ) cell in excel automatically changes to the actual desired value with respect to the CL0 value.

To find Wetted length (λb) , Now a value is assumed for Wetted length (λb) (say 1) based on this value Mean wetted length to beam ratio1 (λ) [goal seek for λb] is calculated using equation 4.1.1. By replacing the Mean wetted length to beam ratio1 (λ) [goal seek for λb] value by the previously found value Mean wetted length to beam ratio (λ) using goal seek method, the Wetted length (λb) cell in excel automatically changes to the actual desired value with respect to the Mean wetted length to beam ratio (λ) value.

Average velocity (or) mean velocity calculation

$$V_{mean} = V\left(1 - \frac{0.0120\tau^{1.1}\left(1-0.0065\beta\left(0.0120\lambda^{1/2}\tau^{1.1}\right)^{-0.4}\right)}{\lambda^{1/2}\cos\tau}\right)^{1/2}$$

4.1.5

Where,

V – Velocity
λ - Mean wetted length to beam ratio
β – Dead rise angle
τ – Trim angle

The frictional coefficient CF is found by using ITTC 1957 frictional coefficient formula:

$$C_F = \frac{0.075}{(\log_{10}(Re) - 2)^2}$$

4.1.6

Where,
Re- Reynolds number of the ship at given speed

Position of Centre of pressure (hydrodynamic pressure measured from keel)

$$C_p = 0.75 - \frac{1}{5.21\frac{C_v^2}{\lambda^2} + 2.39}$$

4.1.7

λ - Mean wetted length to beam ratio
Cv – $V/\sqrt{g * b}$
V – Velocity
g- Gravity
b- Beam

The resistance or Drag force for planning hull type vessel is found using the formula below

$$\frac{D}{\Delta g} = \tan\tau + \frac{\left(\frac{V_1}{V}\right)^2 C_f \lambda}{C_{L\beta} \cos\tau \cos\beta}$$

4.1.8

where
D = total drag (N)
Δ = displacement (kg)
V_1 = mean vel over planing surface = V_{mean}
C_f = total coeff of friction

The moment of the vessel at a given trim angle is can be estimated using the equation 4.1.9:

$$\Delta g\left(\frac{(1-\sin\tau\ \sin(\tau+\varepsilon))c}{\cos\tau} - f\sin\tau\right) + D_f(a-f) = 0$$

4.1.9

Df – Frictional Drag force
Lp - hydrodynamic pressure measured from transom along the keel
C – Distance between hydrodynamic pressure (Lp) and longitudinal centre of gravity (LCG)
A – Distance between Frictional Drag force (Df) and Vertical centre of gravity (VCG)
f – Distance between propeller thrust and gravity
ε – Inclination of thrust line relative to keel line
g- Gravity
τ – Trim angle
Δ - Displacement of the vessel

Frictional Drag force (Df) of the vessel calculation
Df = (Cft × 0.5 ×ρ× [λb × b/cosβ] ×Vmean^2)

4.1.10

Where,
Cft = Cf + ΔCf
ΔCf = 0.0004
Hydrodynamic pressure (Lp) measured from transom along the keel

Lp = (Cp × λb)

4.1.11

Distance between hydrodynamic pressure (Lp) and longitudinal centre of gravity (LCG)

C= (LCG - Lp)

4.1.12

Distance between Frictional Drag force (Df) and Vertical centre of gravity (VCG)

a = (VCG – (b/4 × Tanβ))

4.1.13

The above procedure from a guessing a trim angle to moment evaluation is repeated for both upper limit and lower limit trim angle. Then the optimum trim angle is found by estimating the interpolation factor.

Interpolation factor (I.F)

$$\text{Interpolation factor (I.F)} = \frac{\text{Moment 1}}{\text{Moment 1+Moment 2}}$$

4.1.14

Moment 1 is Moment at assumed upper trim angle limit
Moment 2 is Moment at assumed lower trim angle limit

Optimum trim angle (Oτ)

(Oτ) = lower trim angle + Interpolation factor × (upper trim angle-lower trim angle)

4.1.15

4.2. MAXSURF RESISTANCE ANALYSIS SOFTWARE METHOD

MAXSURF includes tools for hull modelling, stability, motions and resistance prediction, structural modelling, structural analysis, and export to vessel detailing.

MAXSURF Resistance estimates the resistance and power requirements for any MAXSURF design using industry standard prediction techniques. This resistance analysis tool was used for resistance prediction of an example Hull design which was used to experiment the new concept and for comparison of different evaluation methods. (The hull parameters are displayed in the figure 4.1).

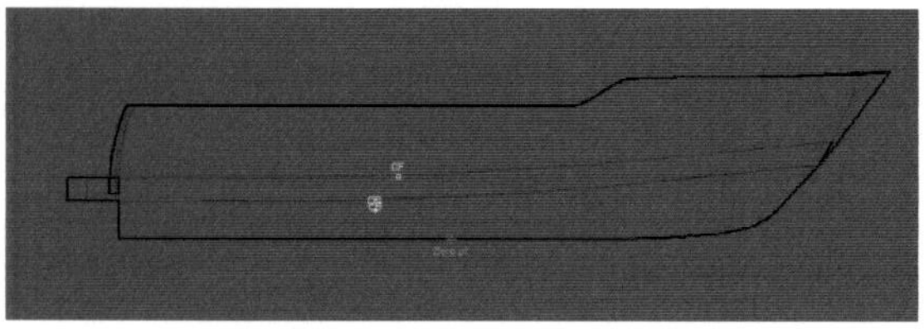

Figure 4.2: Profile view

This given hull design was uniform throughout this work even for different resistance methods and the results were extracted and tabulated for comparison studies.

4.3. SIMULINK MODEL – MATLAB SOFTWARE METHOD

The extended and one of the most useful capabilities of MATLAB is SIMULINK environment.

Simulink, developed by MathWorks, is a graphical programming environment for modelling, simulating and analysing multi domain dynamic systems. Its primary interface is a graphical block schematization tool and a customizable set of block libraries
Using the Simulink's simulation environment and the MATLAB tool the resistance prediction of the given planning hull was carried.

The empirical equations, Unit conversions, formulas, sea states, sea spectra, encounter frequency, angle of heel while turning, rate of turn and few other associated parameters were modelled as blocks in Simulink

Procedure:

- Load the input variables and their respective values in MATLAB.
- Run the simulation model by switching to the required speed and trim angle (only for planning vessel).
- Perform Goal seek for the parameter 1. Lift coefficient at zero dead rise (CL0) and 2. Mean wetted length to beam ratio (Lambda- λ) by performing what If Analysis in Excel spread sheet.
- Import those two goal seek values to Simulink and run the simulation.

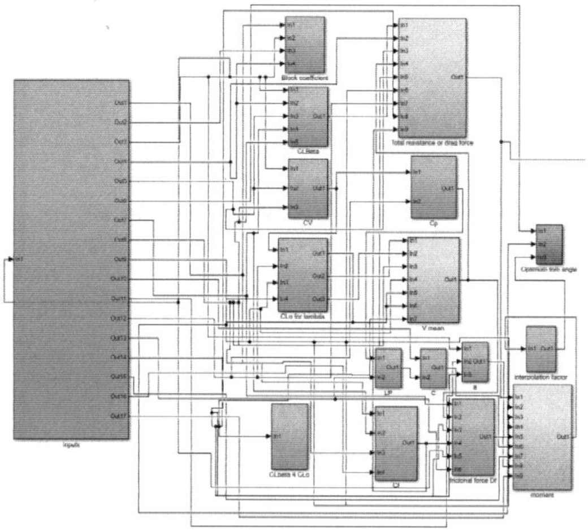

Figure 4.3: A view of planning vessel resistance pred. model

5. DISCUSSION AND OBSERVATIONS

The performance parameters of the planing hull obtained from theoretical calculations, Simulink model and the values obtained from the MAXSUF software were compared. The key parameters like, drag force (KN), Power (Kw), speed and LCG are considered to have major impact on the performance of the planing hull. The trim in the vessel is due to the flow separation which influences the pressure along the vessel creates the lift. [5].

From the graph 5.1, The drag force or the resistance varied with the trim angle and speed. The drag force had ups and downs at varying speeds. Initially the drag force increased with increase in speed and after crossing 30 knots the drag force decreased gradually with further increase in speed. The power consumed will also be affected with the drag force because this drag acts on the opposite direction of the motion of vessel, so this negative force must be avoided as much as possible for more efficient operation of the vessel. The drag force values obtained from Simulink model was similar to the theoretical evaluation and the MAXSURF software values. There were some differences for same given speed.

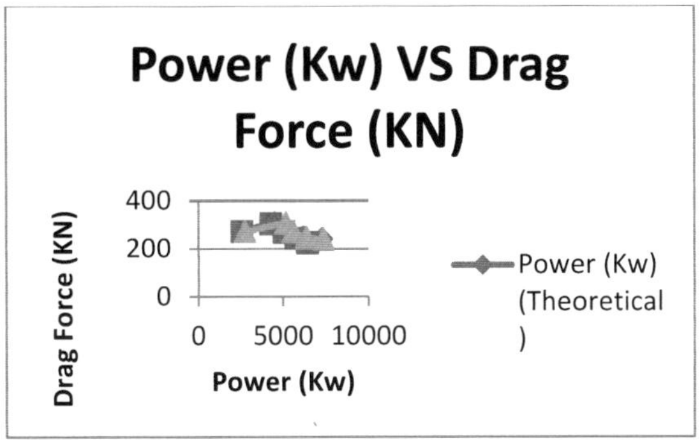

Graph 5.1

From the graph 5.2, The power (Kw) consumed by the vessel varied with the speed and drag force. Initially for 20 knots the power consumed was relatively low and increased gradually with a growth in speed. Also the drag force increase caused increase in power consumption. On comparing the values obtained from all three methods, it was found that there were some deviations in values for same speed. The Simulink model was lying below the theoretical prediction values and the Maxsurf was lying above the theoretical and Simulink based model.

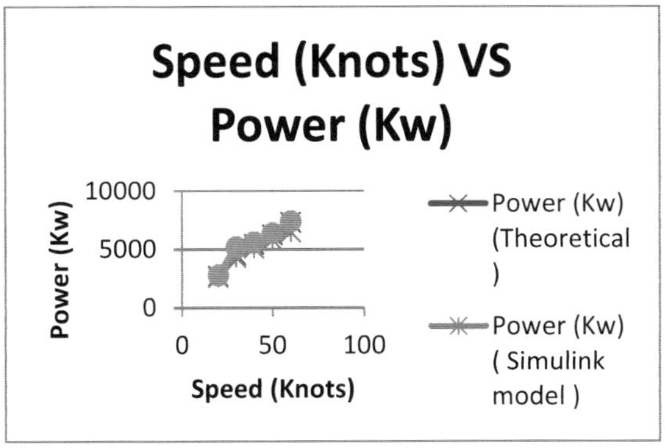

Graph 5.2

The differences and mismatching of values between the Savitsky's method of planing hulls, Simulink based computer model and MAXSURF software for given same speed is discussed below. The optimum trim angle, drag force and power were considered to be three main parameters which describe how good a hull is designed. These key parameters are related to almost all relative parameters mentioned in the nomenclature.

From the graph 5.1,5.2 & 5.3 there were some inaccuracies spotted and the inaccuracies in these values were due to the A] rounding of values during calculations for simplicity. B] Constants - some of the constants value assumed during the theoretical calculation like gravity was 9.80665), but the MAXSURF software might be programmed with 9.81, which causes difference.

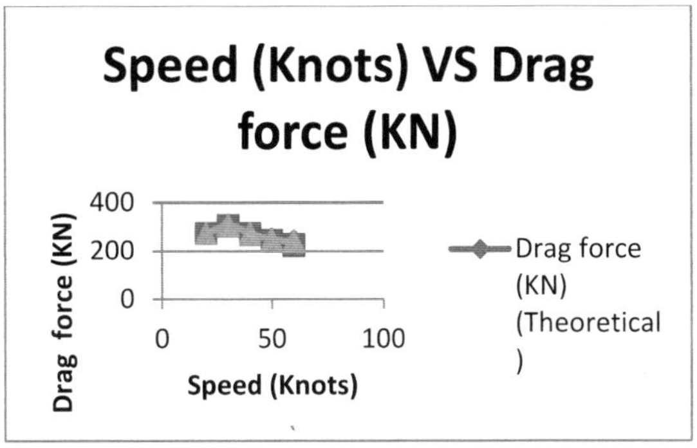

Graph 5.3

C] The software issue-the values were calculated in excel-spread sheet which might have some limitations in the maximum decimal digit length which is considered to vary with Simulink based computer model and the MAXSURF software's decimal digit length. D] The goal seek option – The goal seek option- It was used to calculate the lift coefficient at zero dead-rise(CL0) and the mean wetted length and beam ratio(Lambda), it was noticed that the goal seek did not exactly match the expected value. For example ,for a speed of 20 knots, The expected lift coefficient at zero dead-rise(CL0) value was 0.735638087 but the goal seek option in EXCEL generated 0.735586233 which affected the accuracy of values. E] The imperial units and covert ions- The Savitsky's research paper on hydrodynamic and planing hulls had imperial units like pounds and feet for his calculation but the method carried by MAXSURF software is unknown and the excel spread sheet calculation had SI units. The Simulink model has been modelled to take both the types of units and for this resistance case study S.I unit has been used to compare the outputs. So, this might have impact on the results. F] The human error, assumptions, theoretical non considerations of external effects for simplicity and uncertainty errors are believed to have affected the accuracy of values carried during the comparison study. The mismatching and differences cannot be avoided, but can be minimized by careful evaluation and avoiding the assumptions, rounding of figures and neglecting of external effects as much as possible.

The variation of other associated hydrodynamic parameters with speed and trim angle are discussed below:

Cv- This parameter increased with increase in speed but had no effect on change in trim angle.

CL beta- This value had a gradual increase with increase in speed of the vessel, but no change with trim angle.

CL0 – This value had no effect with trim angle, but dropped with climbing speed.

Lambda- The lambda was relatively had low value for higher speeds and higher value for lower speed.

Reynolds number- This constant had increase in value with increase in speed and no effect on change of trim angle at different speeds.

Cp- This parameter increased with increase in speed.

6. RESULTS

THE SAVITSKY'S RESISTANCE PREDICTION METHOD

S.No	Speed (Knots)	Speed (m/sec)	Trim angle (deg) (Theoretical)	Drag force (KN) (Theoretical)	Power (Kw) (Theoretical)
1	20	10.28888	9.1	270.7162327	2700.934289
2	30	15.43332	9.9	305.6478594	4539.221711
3	40	20.57776	8.1	271.9708061	5427.664244
4	50	25.7222	6.4	247.0493106	6212.904762
5	60	30.86664	5.1	239.9905412	7286.832646

Table 6.1

THE MAXSURF RESISTANCE ANALYSIS SOFTWARE METHOD

S.No	Speed (Knots)	Speed (m/sec)	Trim angle (deg) (Maxsurf)	Drag force (KN) (Maxsurf)	Power (Kw) (Maxsurf)
1	20	10.28888	9.109726	270.0356289	2781.66582
2	30	15.43332	9.940493	305.802871	5182.932852
3	40	20.57776	8.136487	271.978274	5596.708473
4	50	25.7222	6.422927	246.422471	6381.44871
5	60	30.86664	5.17086	238.84636	7372.393317

Table 6.2

THE SIMULINK BASED COMPUTER MODEL METHOD

S.No	Speed (Knots)	Speed (m/sec)	Trim angle (deg) (Simulink model)	Drag force (KN) (Simulink model)	Power (Kw) (Simulink model)
1	20	10.28888	9.1	269.738655779854	2590.839788765498
2	30	15.43332	9.9	303.3183674060531	4290.13252229208
3	40	20.57776	8.1	268.1176235178090	5085.246516132269
4	50	25.7222	6.4	241.3956389559529	5783.400173918185
5	60	30.86664	5.1	223.5175802758035	6460.776023272045

Table 6.3

7. CONCLUSION

More than 90% of the world's trade is being transported in seaways [6]. The implementation of efficient systems and engineering on board, on site and preliminary research phase will have a direct influence on improving functionality of the industry/system and increase effectiveness of the existing system. Thereby, the selection of the smarter, efficient and cheaper way of resistance Scrutiny is the key in Marine vessel designing and building for the vessel to perfectly fit the design requirement and to perform the task perfectly for which the system was actually designed. So by comparing the results, way of performing the resistance examination, time and other associated factors, we propose that Simulink based computer model of predicting resistance is the enhanced method. This proposal is based on the outcomes of the resistance prediction study using three different approaches.

8. REFERENCES:

1. Mark Kennett boat designs (2009) ,Displacement motor boats versus planing motor boats.

2. Kalkomey (2016).Hulls types and how they operate. Available at: https://www.boat-ed.com/pennsylvania/studyGuide/Hull-Types-and-How-They-Operate/101039_101039011/

3. Jan Babicz.2008, ship resistance chapter, Wartsila Encyclopedia of Ship

4.Daniel Savitsky, "Hydrodynamic Design of Planing Hulls," Marine Technology, October 1964 Issue, SNAME, Paramus, NJ

5. Odd M. Faltinsen. (2006).Hydrodynamics of High-Speed Marine Vehicles. page number 345-347.Cambridge University Press,uk.

6. IMO (International Maritime Organization),(2017), IMO profile. Available at: https://business.un.org/en/entities/13

9. ACKNOWLEDGEMENT

I would like to express my deep gratitude to 1.Blue Bear systems and research limited, UK. 2. University of Plymouth, UK and 3.Professor Jahir Rizvi, (Department of Marine Science and engineering University of Plymouth, UK) for their patient guidance, advice and assistance in keeping my progress on schedule. My grateful thanks are also extended to the Editor of this Journal and RINA (Royal institute of Naval Architects).

This chapter includes use of software like MATLAB (Mathworks), Simulink (Mathworks), Microsoft excel and MAXSURF software. This does not anyway support or promote the use of the above mentioned software/(s).

10. AUTHORS BIOGRAPHY

Sri Harish Kalidass holds the current position of Student Design Engineer at Blue Bear Systems Research Ltd and a continuing student at University of Plymouth pursuing Final year of studies in BENG (HONS) Marine Technology. He is also a recent undergraduate in B.E. Mechanical Engineering from Saveetha University, Chennai, Tamil Nadu, India. He is skilled in Marine engineering and technology, Image processing, computer vision, robotics and unmanned systems & engineering. Mechanical engineering, Sustainable Business, and Urban Infrastructure & development,.

SINGLE SOURCE DATA – HELP or HINDRANCE?

J L Martin, SAMOSC Ltd., UK

SUMMARY

There is a continuing drive towards single large volume database systems to integrate all data created and used in a shipyard. "Single Source of Data" is one of the common marketing expressions used to describe a shared data capture and management system for use across all departments and all aspects of ship design, manufacture, build and in-service support. The single source data concept removes the need for numerous interfaces, and the complexities of data quality and integrity management. The comprehensive shared data environment is presented as the ultimate in efficient data capture, storage and management, with considerable benefits and advantages from using inclusive database systems technology in a shipyard environment. However, it is often assumed that ALL data captured during the lifecycle of a ship must be included in the single source database irrespective of its use, with superficial regard for logical categorisation or data modelling. The result can be a single source of data unnecessarily very large, convoluted, difficult to control, clogged up with unnecessary data, and not easy to interpret. For an effective and efficient database, careful consideration must be given to the format content, quality, and use of all data items.

This paper will describe how data modelling may be used to ensure that all data in a shared environment single source database is necessary and sufficient, and how shipbuilding data is logically categorised by its function and purpose.

1. INTRODUCTION

1.1 OVERVIEW

Shipyards use computing technology to achieve productivity and efficiency in the design and build of a ship. They must also provide sufficient information to the customer to enable the delivered ship to be maintained during operational life, often for twenty years or more.

Shipyards traditionally use comprehensive and concise data structures to define the ship and how to design and build it. Utilising these structures as the basis of a Single Source Database is an effective and efficient approach to data and information management.

1.2 HISTORIC DATA CAPTURE

Drawings were introduced to capture the detail of a ship, initially providing the principle characteristics and general arrangement layout as a guide to the concept design, evolving to providing considerable detail of the vessel. Drawings are produced in the specialist disciplines of structural, mechanical and electrical. Drawings include the location, orientation and connectivity of the structure, routed systems and fitted items, providing dimensional details sufficient to resolve uncertainties, with 'fit at ship' instruction for selected components difficult to define accurately on a drawing. Drawings also include supplementary information, in the form of a Parts List, to enable necessary material and manufactured or procured items to be identified and specified. Scantling, Schematic, and System Diagram drawings include functional design information such as classification rules for structural design, and cable, valve, and instrument and control lists for routed systems.

Information used in a shipyard involves the creator of the data presenting it to several 'follow-on' departments with the receiving departments using and/or enhancing the data prior to passing in on to other departments and back to the originator. Data is passed back and forth between departments frequently as they contribute to the progressively evolving vessel design and build. In many shipyards drawings remain the primary information provided from design to manufacturing, production and associated departments.

Drawings with their associated data are historically the 'Master Information' for the design.

Of the thousands of drawings produced in a shipyard, four categories of drawings, relatively few in number, are relevant for use on the ship in-service, they are:

- General Arrangement Drawings – the layout of the vessel showing compartments with the location of all equipment and fittings.
- Scantling Drawings – detail of the steelwork structure based on classification rules.
- Schematic Drawings – detail of the electrical systems, instrument and control systems with cable and cable tray lists.
- Systems Diagram Drawings – detail of the pipe and HVAC mechanical systems, connectivity with valve/equipment/component and instrument/control lists.

All other drawings are exclusively to do with the shipyard, often produced in accordance with shipyard build strategies and are archived after use.

© 2017: The Royal Institution of Naval Architects

1.3 DATA IN A SHIPYARD

The shipyard process creates data that is high in volume and complex in content for very small build quantities, often only one, such as:

- a feasible concept (requirements)
- a functional design (the design intent PBS)
- a physical solution for the design (CAD)
- procurement and materials data (BoM)
- manufacturing information (WBS)
- installation, build and test data (WBS)
- project management and control (Planning)
- finance (cost) data
- in-Service data (included in PBS)

Computing technologies enable this data to be captured, stored and managed electronically and selectively extracted. The data is often captured and used in a variety of independent systems, with local storage. Interfaces are developed for copying/converting data between systems for inter-departmental use.

To address this, system vendors have progressively developed data management systems and seek to provide a single source of data that can be used by all functional departments. To be effective and maintain integrity, the Single Source Database must be the 'Master' information. It must not be a copy of data held in the system used to capture it, and it must be able to provide the data in a usable format to all systems needing access for reading, processing or enhancement.

1.4 EVOLUTION OF COMPUTING SYSTEMS

Shipbuilding is a leading industry in the use of computing systems, for design, materials management, planning, manufacturing, project and financial control. From the early days of evolution of Computer Aided Design, CAD systems were implemented in shipyards. It was quickly apparent that CAD generated 2D drawings or 3D models needed a capability to add data to drawing or model geometry. Attribute capture, a means to add text data to CAD geometry was introduced but very soon became incapable of the ever increasing demand for more and more data to be associated with CAD models.

Engineering Data Management (EDM) databases were developed as an enhancement to CAD systems. Advancements in database technology enabled EDM to include non-engineering data and become Product Data Management (PDM) enabling multi-department use of shared data.

The latest contemporary development is a single underlying database to be interfaced to well proven and established systems. The result is database solutions such as Product Lifecycle Management (PLM) and Enterprise Resource Planning (ERP) being offered as centralised software, interfacing and managing all data across an organisation. Both are based on the concept of 'Single Source Data', PLM to define a ship through all aspects of its life from concept design to in-service operation, ERM to manage the all company activities, and are promoted by vendors as the optimum database solution.

2 WHY SINGLE SOURCE DATA?

2.1 PERCEPTION OF PROBLEM

A common approach to defining the problem of data management in a shipyard is based on functional departments, as indicated in Fig 1.

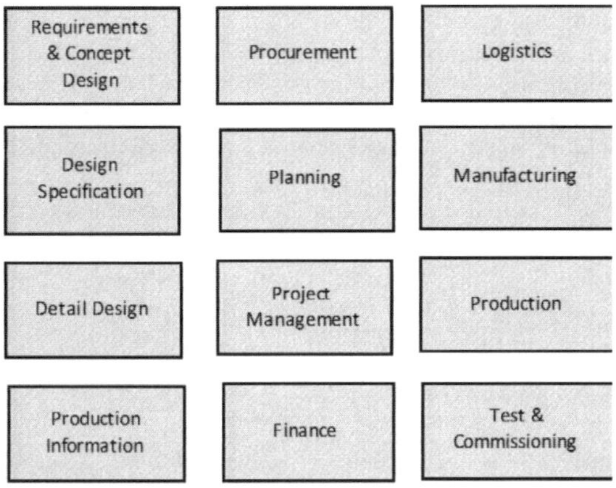

Fig. 1 Functional departmens

The problem is classically illustrated by showing the spider web of routes of data transferred between the departments, as illustrated in Fig 2. Although not exhaustive, this form of chart clearly shows the extent of data movement and use across shipyard departments. Often with each department running disparate computing systems designed for a particular role. Data conversion interfaces are necessary to enable both the source and receiving systems to use the data. Retaining data integrity is not trivial and involves identifying the 'master' data and updating all copies when there is a change.

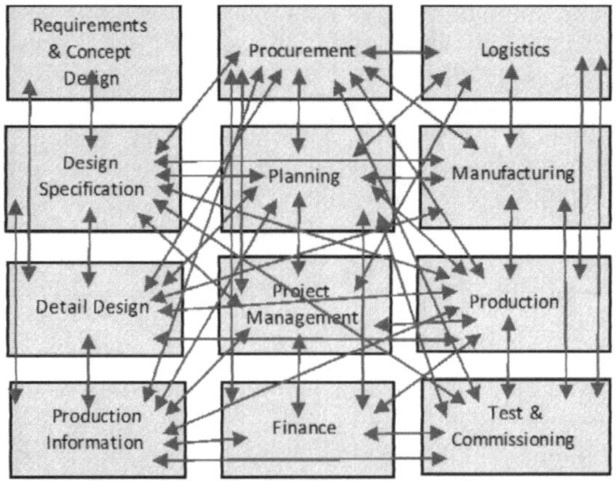

Fig. 2 Departmental data transfer

© 2017: The Royal Institution of Naval Architects

The increase in use computing technology and particularly vendor developed systems has substantially improved department productivity, and improved the quality and content of data captured by each department. Increasing functionality enables computing systems to be used by several departments that have a close common data interest. A single computing system can be used effectively in such departments, minimising the need for data transfer/conversion. CAD, for example, is now used for design specification, detail design, production information, planning and manufacturing. Material Requirements Planning (MRP) systems are used by procurement, planning, logistics, and manufacturing.

2.2 THE ASSUMED SOLUTION

Computing systems used in shipyards may be loosely grouped into distinct categories such as:

- Engineering Design and calculation
- Computer Aided Design (with engineering database)
- Project management and Planning
- Material Requirement Planning (MRP)
- Computer Aided Manufacturing (with NC tools and production processes)
- Finance

Data created in each of these systems is stored in an associated database such as EDM associated with CAD, or materials management associated with MRP. They all involve multi department access but still need to transfer data between them. A particular issue is that several of these systems store the same data. EDM will store a Bill of Materials as it is being created by the CAD engineers. The MRP System will store a Bill of Materials for procurement purposes. The Planning system and Manufacturing system also need the Bill of Materials to build the vessel.

To resolve this, computer system vendors offer the option of a single source of data to be used and accessed by all systems used in the shipyard processes. See Fig 3. This is assumed to be a solution to the problems of data synchronisation and integrity with data transfer and conversion, done from the 'master data' held in the single source.

Each individual computing system interfaced to the Single Source Database will access data from it, process the data and return data to it. In some cases, this will still need some form of data conversion. Difficulties arise with some vendor products (CAD for example) where data is integral to the system. In such a case, it may be feasible to hold a reference link in the database with tools to access the 'master' data via the link. To have a true Single Source Database, all data stored in it must be exclusive and not a copy of data from interfaced systems.

A Single Source Database for shipbuilding is a significant improvement in the capture and management of data.

*It is claimed that this is a major **HELP** in shipbuilding data management!*

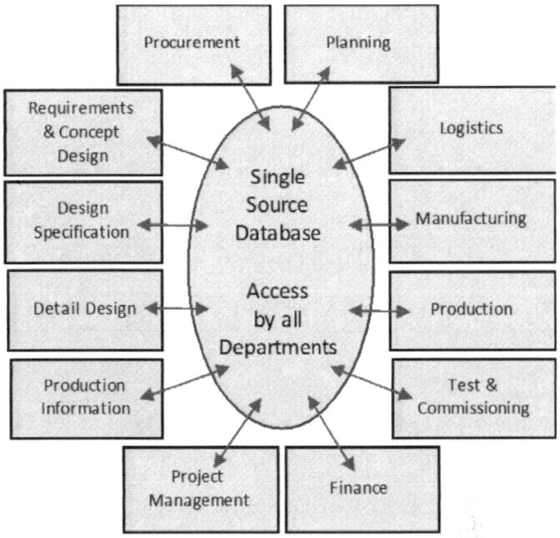

Fig. 3 Single source solution

2.3 PROBLEM WITH ASSUMED SOLUTION

The concept of a Single Source Database as the master information for a ship design, build and operation is sound. It ensures there is only one occurrence of each data field and that this data is up to date. It also ensures all users are accessing the correct data and that data fields created by other departments will be accessible along with the data created by the using department.

The approach to developing such a database is considerably different between vendors, with each promoting an extension of a previously proven database. Engineering database systems developed in support of CAD models are well used and understood with vendors claiming that this form of database can also manage data from other categories of system, such as MRP. Vendors providing generic databases supporting procurement and materials management systems also claim their well proven database can be enhanced to include engineering data, including CAD.

In many cases the vendors have limited understanding of the data and information requirements and practices for all categories of shipbuilding activity and although 'expert' in their particular field need extensive and expensive time consuming study and analysis to address the complex multi-functional shipbuilding data needs. As a result, the methodology for the Single Source Database is often based on the functional department model illustrated in Fig 3 to capture, store, access and manage data and information, and fine-tuned to suit departmental practices, tradition and processes.

© 2017: The Royal Institution of Naval Architects

Departmental disciplines have varied requirements and accountabilities for data creation or use and resolving the Single Source Database to address these multi-department issues involves a data model that is less than optimal. A Single Source Database needs well-structured data to address the complexity in extracting information accurately in the format necessary to be used by all functional departments throughout the total ship lifecycle.

*Ship data is **NOT** based on functional department, and such a data structure can become a **HINDERANCE** to the Single Source Database*

3. DIVERSITY OF SHIPBUILDING DATA

3.1 SHIPYARD DATA TYPES

Shipbuilding data is not based on department or department activities and can be divided into distinct groupings each providing data for different purposes.

Requirements, concept design, and function design data is used to define the ship and its operational capabilities to suit the needs of the owner. Requirements elucidation together with concept design studies and selection of a feasible ship [1] provide the characteristics, principle dimensions, general layout and operability of the ship. Requirements information is ship-owner data shared with the shipyard when defining the vessel and its operational performance.

A Product Breakdown Structure (PBS) provides a hierarchical relationship of the ship structure, equipment, systems and components. A PBS is a well-structured unambiguous description of the ship. The PBS is applicable to the ship through design, manufacture, procurement, build, testing, and in-service support. Every item of data in the product breakdown has relevance to the ship, rather than the shipyard, and each item is uniquely identified with a systems based instance or occurrence number.

Physical design data defines the location, orientation, and connectivity of each item in the product breakdown in the form of a general arrangement often a CAD 3D geometric model. The 3D model may be considered shipyard data, although it can be useful, although not essential, information for in-service support. Once the ship is built, the product breakdown will reference systems diagrams identifying systems and their components, including compartment location, sufficient to locate each item on the ship for in service maintenance. The 3D model can graphically show the item location and if available would complement the systems diagram functional information, rather than replace it.

A Bill of Materials (BoM) will contain the data necessary to procure each item in the PBS and associated items defined in the specification of the item such as bolts, gaskets etc. It will also contain a list of 'non-ship' items to be procured specifically for the contract in support of design, manufacture, build and testing of the vessel. The Bill of Materials is a shipyard dataset used in the manufacturing and procurement of items and materials for the ship.

A Work Breakdown Structure (WBS) provides all of the tasks and activities necessary to design, manufacture, build, launch, outfit, test and commission, and sea trial the ship. This data is exclusively for the shipyard and is of no value for long term use such as in-service support. Even a similar ship in a class will have a work breakdown with some differences in activities due to availability of resources and shipyard loading. A work breakdown includes non-productive work such as fire patrols, surveying, and formal events such as keel laying and naming ceremonies. It also includes work on non-ship items such as staging, temporary lighting, ladders, walkways, temporary ventilation etc. which as well as being defined as a work effort must also include non-ship material items in the Bill of Materials for procurement. Work breakdown data can be deleted or archived for reference on completion of the vessel.

Production Planning data is exclusively for the shipyard and has no relevance to the ship after build and delivery and may be deleted or archived after use for future reference.

Project cost data may be derived from the cost of all materials in the product breakdown and its derived bill of materials plus the cost of work in the work breakdown, and adjusted to include overhead costs, such as plant, energy, and general consumables. as per the shipyard process for recording and recovering such costs. Project cost data is exclusively for the shipyard and may be deleted or archived on completion of the ship.

3.2 NUMBERING

Numbering is extensively used in shipbuilding with many and varying numbering systems developed by the shipyard in accordance with practices, processes and systems being used. Numbers are used for part identification, planning, finance recording, systems, material identification, work packages, documents, drawings, and many other uses during the design and build of the vessel. Two numbers define what the item is, the system it belongs to and where it is on the ship

- An Instance Number (also called Occurrence Number) to identify each item on a ship uniquely.
- A part number is used to identify equipment, components or material for each instance. A part number may be associated with several instances and used many times on the ship.

© 2017: The Royal Institution of Naval Architects

The Instance Number is defined during functional design, used during physical design, planning, installation and in-service maintenance support. The number used on diagrams and schematics and on the vessel to uniquely identify an item such as component, valve, cable, equipment etc. This number is unique and only occurs once on the ship.

A Part Number will be generated during physical design when a part is selected for procurement, in accordance with specifications and allocated to one or more particular 'Instances' on the ship.

For efficient operation of the ship, instance numbers often include a global standard systems reference for the item, such as FW for fresh water system, LO for lubricating oil system, etc.

4. PRIMARY DATA STRUCTURES

4.1 PRODUCT BREAKDOWN

Product Breakdown Structures (PBS) are well understood and have been used for many years in shipbuilding. There are several worldwide standards such as those based on weight groupings using a five-digit code, with the first three digits describing the system, the other two digits describing component parts and equipment for that system. The structure logically categorises every item in the ship in a hierarchical relationship. Although such breakdown structure standards differ in detail, they are all based on a similar numbering system. A typical example is:

1 Hull
2 Propulsion
3 Electrical
4 Navigation & Communications
or Command & Surveillance (Naval)
5 Auxiliary Systems
6 Outfitting & Furnishings
7 Armaments (Naval)

The next two digits define in increasing detail the system, for example:

1 Hull Structure
  100 Hull Structure General
    120 Hull Structure Bulkheads
      121 Longitudinal Bulkheads
      122 Transverse Bulkheads
      123 Trunks and Enclosures

5 Auxiliary Systems
  500 Auxiliary Systems General
    530 Fresh Water Systems (main system)
      531 Distilling Plant (sub-system)
      532 Cooling Water (sub-system)
      533 Potable Water (sub-system)

Each item installed on a ship can be associated with a line in the product breakdown and include information about the item such as:

PBS number for system
PBS number for System entity
Unique Instance Number (with System Identifier)
Description
Drawings (numbered to suit Document Control)
Documents – digital and hard copy (such as valve, instrument, and cable lists)
Specifications
Operational Criteria (Temperatures, pressures etc.)
Part or Material used
Part Number (with reference to manufacturers part number as appropriate)
Weight
Location (3D model)
Connectivity (Structure, pipes and/or cables connected to the item)
Pipe Code
Maintenance Criteria (frequency, method, spares etc.)
Training Instructions (if applicable)

The above list is not exhaustive and may also include other data about the item as necessary.

An illustrative example is as follows:

533 Potable Water
  5331 Galley

Instance No.	FW 6750 (1st item in galley FW system)
Description	Valve
Drawing No.	12-4596-65 03 (Systems Diagram for Potable water system)
Document No.	12-5645-100 01 (Potable Water requirement, specification, and rules).
Specifications	Stainless, ball, 80mm. PN16, etc.
Operation	-20° – 150° @ 14 bar
Part No.	5642
Manufacturer:	Johnson Valves
Manf. Part No.	BV926
Manf. Serial No.	56BV756879
Weight:	18.9Kg
Location	*reference to GA 3D model or drawing*
Connectivity	FW13654, FW13655 (attached pipe numbers)
ISO Pipe Code	Blue-White-Blue stripes (Fresh Water)
Maintenance	Gasket replace, lubrication instructions.

Instance No.	FW 6751(2nd item in galley FW system)
Description	Valve
Drawing No.	12-4596-65 03 (Systems Diagram for Potable water system)

© 2017: The Royal Institution of Naval Architects

Document No. 12-5645-100 01 (Potable Water requirement and rules).
Specifications Stainless, ball, 100mm. PN16, etc.

Etc. repeated for all items in Fresh Water Potable galley system.

This information defines the items in the galley potable water system, each item in the system uniquely identified by the instance number.

In a Single Source Database, all of the data and information about the ship should be captured in the form of a PBS. After build, the PBS data can be extracted from the database to become the principal information for the life of the vessel in operation, providing the comprehensive information required to maintain the ship efficiently whilst in service. Modifications to the ship during life can easily be recorded in the PBS, including a new instance number if required, to ensure integrity and accuracy of information.

4.2 WORK BREAKDOWN

Work breakdown Structures (WBS) are often presented as a method of defining the ship, with a hierarchical format including data on how the ship is built. The philosophy is that all items on the ship need work to be designed, manufactured (or procured) and installed, and thus need to be included in the work breakdown structure.

However, the traditional problem in shipbuilding is that ships are designed by system and built by block. As a routed system may span several ship blocks, it will be split into work packages that suit the block build sequence in the WBS, rather than defined as a complete system with all components and equipment as in the PBS. A work breakdown based on block build methodology does not efficiently define the ship, and it can be extremely difficult to extract ship information by system, as necessary for in-service support, from such a WBS.

The work required to design and build a ship can be summarised as:

PRODUCT WORK
- Work for Component development
  - Specification writing
  - Research
  - Evaluation
  - Assessment/Approval
- Work to Design (by system or area)
  - Functional design
  - Structural Design
  - Physical Design
  - Integration
  - Outputs for Manufacturing
  - etc.
- Work to Manage
  - Planning
  - Monitoring
  - Reporting
  - Project Control
  - etc.
- Work to Build
  - Manufacturing
  - Compartment build/outfit
  - Block build/outfit
  - Assembly
  - Installation
  - Test and Commissioning
  - etc.
- Work for Logistics
  - Material handling
  - Spares Provision
  - Maintenance
  - etc.

NON PRODUCT WORK
- Work for Support
  - Production Engineering
  - Weight Control
  - Standards
  - Process Development /Implementation
  - Procurement
- Work in Technical Departments
  - Quality Management & inspection
  - Technical Publications
  - Health and Safety
  - Estimating
  - Resource Planning
  - Sea Trials
  - Rework/defect repair
  - etc.
- Sundry Work
  - Radiography/Non-destructive testing
  - Temporary services
  - Physical weighing
  - Launch
  - Training
  - Temporary protection of equipment
  - Tools and Accessories
  - Cleaning
  - Maintenance of stores and fitted equipment
  - Ceremonials
  - etc.

4.3 RELATIONSHIP BETWEEN PBS AND WBS

In the product lifecycle, work is carried out at each stage and the deliverable from that work is the creation or use of a PBS item, such as steelwork component, equipment, pipe/cable/HVAC routed system, component, or outfit item, as illustrated in Fig 4.

© 2017: The Royal Institution of Naval Architects

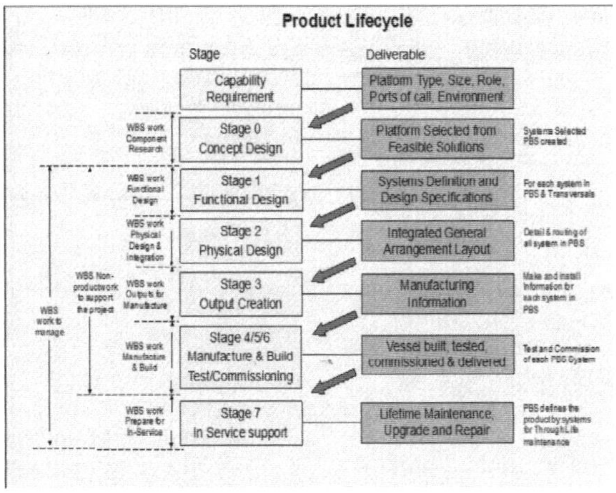

Fig. 4 Lifecycle Relationship between PBS and WBS

It is clear from above that the work necessary to design and build a ship is considerably different from the product structure that defines the ship. A WBS must not be confused with a PBS, and needs a separate format, content and structure to be effective.

Of course, all of the items in the PBS will need work done and must be included in the WBS. However, in the PBS they are defined by system in a logical hierarchy of the completed vessel. In the WBS a steelwork assembly or routed system will be often subdivided into sections to suit design area, compartment, unit or block build and outfit –or perhaps part of a sub-assembly such as a module - with little relation to the Product Breakdown.

4.4 BILL OF MATERIAL (BoM)

The Bill of Material is related to, and a key component of, both the PBS and the WBS. Each item in the PBS will be included in the BoM for procurement. In many cases a BoM item, frequently called a 'part', will be used many times in the PBS. As the part or material will be used on the ship, it will need work and be included in at least one activity in the work breakdown.

The BoM will have three component parts:

- Engineering BoM (EBoM)
- Manufacturing BoM (MBoM)
- Integrated Logistics Support BoM (ILSBoM)

The engineering bill of material will consist of all items, including materials, identified in the PBS. The manufacturing bill of materials will show the materials/components to be used in work packages to manufacture an item (including steelwork assemblies, pipes, HVAC vents, cables, and other shipyard produced items). It will be used to muster these items in the appropriate workshop, unit, block or ship when the work is scheduled to be started. The ILS BoM will include PBS information on maintainable items installed on the ship with the identification and number of spares, such as gaskets, filters, etc. to be supplied for routine maintenance at sea.

The Engineering and ILS BoM content will be referenced in the PBS as they form part of the definition of the vessel. The Manufacturing BoM will be temporary information used only during the building of the ship and may be archived or deleted once complete.

5. SINGLE SOURCE DATA MODELLING

5.1 PBS/BoM/WBS DATA SOLUTION

From above we see that data and information used in a shipyard may be logically categorised and includes three key primary and distinct datasets, the PBS defining the functional design of the ship, the WBS identifying all of the work needed to design and build the ship and the BoM listing all material to be procured. Modelling these primary datasets such that the PBS is the 'driver' supported by the WBS and the BoM, both of which reference data from the PBS, offers the basis of an efficient Single Source Database.

Although the PBS/WBS/BoM approach to database design is completely different from the departmental approach previously discussed, the database can still offer the necessary information to the users across all departments. A user interface simplifies the capture, storage and access of data and is based on a unique key to reference the correct data field. This applies equally to whatever format of underlying data modelling, or data schema, is implemented.

5.2 'BIG DATA'

There is considerable development in the idea and implementation of 'Big Data' solutions for product support. At sea, the capture and analysis of data recorded from instrumentation and control systems on board a ship is providing exceptional knowledge and understanding of ship performance and ensures timely maintenance and optimum operational costs.

To establish a 'Big Data' environment, a suitable underlying database structure has to be defined and implemented in order that the very high volumes of data collected is stored efficiently and the resultant information managed effectively. A Product Breakdown Structure comprehensively and unambiguously defines the ship and offers the ideal base environment for a successful big data technology application. A PBS based dataset, therefore, will be a ***HELP*** for Big Data applications.

If a Single Source Database with content arranged to suit departmental processes is used, this would be a major ***HINDERANCE*** to potential Big Data applications. This is due to the shipyard data co-existing with the ship data making the ship data difficult to extract, or in some cases

© 2017: The Royal Institution of Naval Architects

to identify, from the complex format and content of such data structures.

5.3 CROSS REFERENCE CONCERNS.

The relationship between the PBS, WBS and supplementary data structures is critical and, as we have seen, can be identified uniquely using two types of identifier:

- An Instance Number.
- A Part Number

Once an instance number is allocated, it is used whenever data is created, accessed, updated or used for **all** shipyard and ship operation activities during the life of the vessel.

A part number, allocated to an instance, will define the item to be used at the location of that instance, and a common part may be used over many instances. If the item needs maintenance, particularly when based on running hours or some other operational criteria, and several of these items are installed on-board the ship, then a further number – the manufacturers serial number – is used to uniquely identify a particular part to an instance.

Together, the instance and part offer a reference across all stages of the product life, as illustrated in Fig. 5.

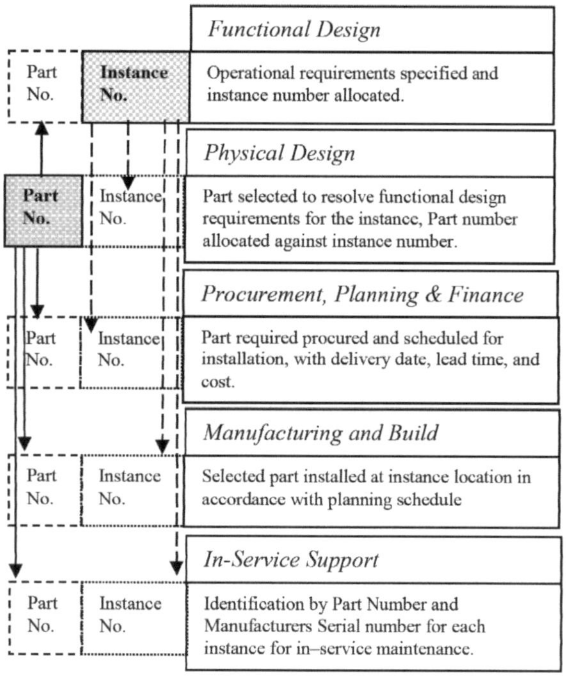

Fig. 5 Cross reference numbering

The Instance Number/Part Number duo are the critical key to access database fields in the Single Source Database. They are directly used in the PBS, WBS and BoM as natural identification and search keys, and can be extended as the reference when using supplementary database structures such as planning, project management and cost control for example

6. PRINCIPLES FOR AN EFFECTIVE DATABASE

6.1 SHIP DATA v SHIPYARD DATA

The volume of data generated in a shipyard is primarily and understandably biased towards the shipyard processes, procedures, and practices. This data defines what must be done to design and build the vessel accurately, on schedule and cost effectively. With data quality and improved productivity being major shipyard drivers, PLM/ERP and other forms of single source database are developed and implemented to improve efficiency, usability, and accuracy of data capture, storage, and integrity, often without regard for identification and separation of the ship data from the shipyard data.

As only a small subset of the data (the PBS) defines the ship, a database can be structured to store this definition concisely without negative effect on the productivity improvements attained when using the database. Furthermore, data structured in such a manner can be readily extracted and offered to the ship-owner as the basis of in-service support of the vessel, eliminating the frequent struggle by owners to extract relevant ship data from a database in which ship and shipyard data co-exist without distinction.

Users don't have any knowledge or understanding of database technologies or underlying schemas for structuring data, only a desire to access their data efficiently and simply when required. The database designer thus has an opportunity to logically model product and shipyard data as separate database entities, and to further divide the shipyard data into functional subsets without detrimental effect on the user interface or user means of data access.

6.2 PBS – THE DATABASE MASTER

A database in which the PBS is the principal dataset, supported by the WBS and BOM cross referenced using instance numbers and part numbers is an elegant solution to single source data, as illustrated in Fig 6. It ensures that product information is well structured and efficiently accessed as well as providing the search key for the supporting datasets. Data, such as WBS items, that have a temporary life in the shipyard process can be deleted, or archived, after use keeping the database 'clean' without risk of losing product related data. Around these three principle datasets all other shipyard information can be arranged including:

© 2017: The Royal Institution of Naval Architects

REQUIREMENTS

Requirements are ship data and are complimentary to the PBS

CAD

The CAD model is a general arrangement based on PBS content. CAD compliments the PBS and can be a useful dataset for in-service support if the ship-owner has the capability and procedures to maintain the CAD model.

PROJECT MANAGEMENT & PLANNING

Project Management is based on the Work defined in the WBS on items defined in the PBS and may use CAD as a monitoring tool.

Planning uses the PBS and CAD to determine build sequences, and the BoM for materials requirement dates, and identifies work to be added to the WBS.

COST/FINANCE

The material cost of the ship is determined by the sum of the items from the PBS. The work cost of the ship is determined by the sum of items in the WBS, including non-product tasks. Cost is adjusted to suit overhead rules applied by the company, and project related consumables in BoM.

Note: The CAD model is *useful* but *not essential* for in-service support of the ship at sea, and ship owners do not need CAD expertise or to use CAD tools to maintain the ship in-service. Often ship-owners struggle to implement a CAD system to use data supplied by the shipyard and don't have the bespoke developments the shipyard have used to enhance functionality for creating the CAD model. It is very difficult, complex and expensive keep a CAD system maintained during the ship's life, and the benefits are minimal.

A generic visualisation tool able to view the CAD data may be more appropriate and cost effective if seeing and 'flying through' the 3D model is considered important.

A PBS database, on the other hand, is text based and relatively easy to convert to current state of the market database technologies over the long term of a ship's operational life.

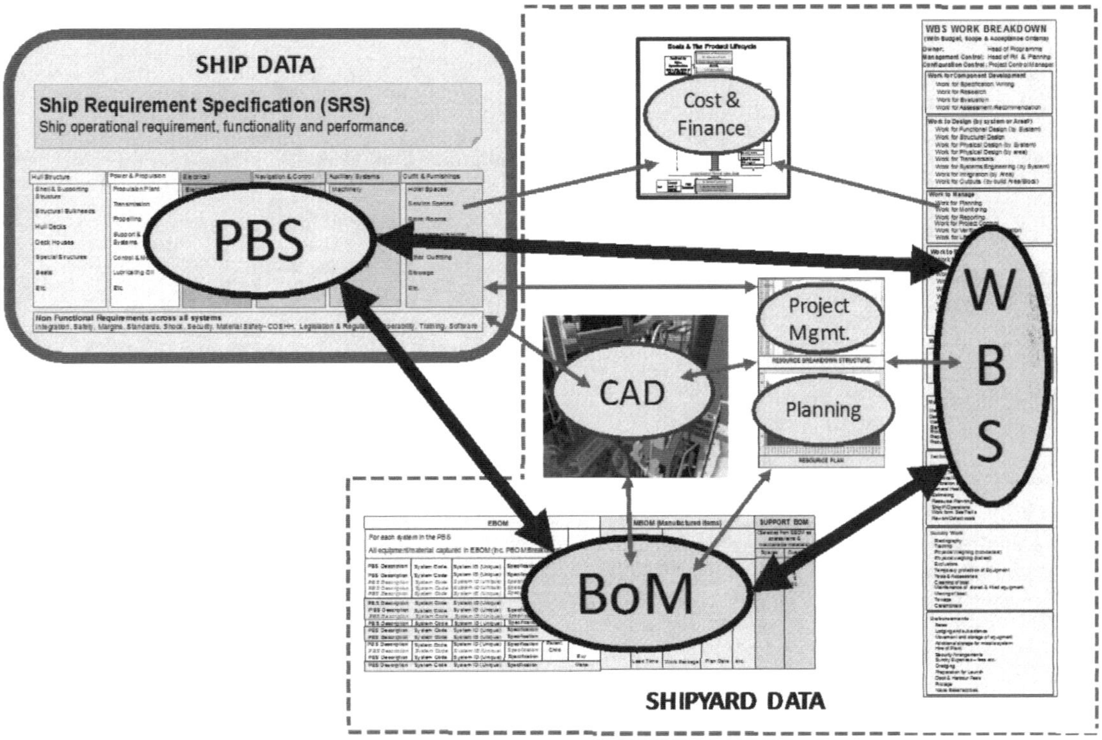

Fig. 6 PBS-WBS-BoM Based Data Structure

© 2017: The Royal Institution of Naval Architects

7. CONCLUSIONS

As an evolution from the success of EDM, MRP and other effective multi-department computing systems, Single Source Database systems addressing all aspects of shipyard activity are increasingly presented by vendors as a major contributor to productivity and efficiency. They are developed with analysis of the requirement of user groups and company working practices. Often the focus is on departmental processes and managing the exchange of data between departments. As a result, data structures and database schemas are based on perceived understanding and knowledge of the collective data created and used by departments with data modelling based on this philosophy without concern of separating ship and shipyard data.

Successful application of such single source databases is limited, due to complexity, but if achieved, they can be a considerable ***HELP*** to shipyard productivity and data management.

If, however, the shipyard wishes to maintain a full and unambiguous definition of the ship design for future reference, and to present the ship-owner with a comprehensive and concise data set on completion and delivery of the vessel, a department based database as described above will be a considerable ***HINDERANCE*** in achieving these aims.
Product Breakdown (particularly weight breakdown), Work Breakdown, and Bill of Materials datasets are ***ALWAYS*** created and used in a shipyard, and are critical in the process of designing and building a ship. In some cases, when not readily available, they are created and managed locally by appropriate departments by extracting a subset of data from the database systems in use.

By using the PBS, WBS and BoM data as the principle structure of a Single Source Database, these important datasets will be readily available and well defined, enabling high quality data management. With the PBS definition of the ship at the core, selected WBS and BoM data items may often be deleted after use, to avoid superfluous storage of unnecessary data. On completion of the ship, the Product Breakdown Structure, with its associated and referenced material is a ready-made 'Datum Pack' of information to provide to the ship-owner and to the ship crew for long term support of the ship. This is particularly useful if a 'big data' approach is to be adopted on board as all data collected from sensors and instruments can be associated with PBS entities to give a well-structured data management environment.

Database technologies are evolving from long term well proven technologies such as Index Linked, Relational, SQL, and Object Oriented to web-based technologies, such as the cloud. The PBS/WBS/BoM concept presented above is simple in definition, based on a hierarchical arrangement of text data with reference links to supporting external material, such as documents and drawings. It is thus relatively simple in data modelling terms and applicable to **ALL** underlying database technologies for success in Single Source Database implementation.

Many of the difficulties in implementing Single Source Database solutions, with their limited success in shipbuilding organisations, are based on lack of appreciation of shipbuilding data and its underlying well understood and well used data structures such as a PBS based on internationally recognised weight groupings codes. This is compounded by the shipyard defining requirements based on functional department working practices and how data is created, used and passed between each department, with the vendor developing the database to suit these requirements.

A PBS driven approach as presented in this paper offers an alternative simple and effective solution for success in implementing a Single Source Data system.

In such a case, the Single Source Database will be a ***CONSIDERABLE HELP*** to both the shipyard and the ship-owner

8. REFERENCES

1. Andrews: 'Marine Requirements Elucidation and the Nature of Preliminary Ship Design', IJME Vol 153 2011. UK.

9. AUTHORS BIOGRAPHY

John Martin. BA C.Eng MRINA has 50 years of practical experience in international industries including over 40 years in shipbuilding. He has been involved in the selection, implementation, and application of computing systems since 1977 with multiple proven success.

Prior to retirement from BAE Systems in 2011 he was employed as a consultant engineer, specialising in the practical application of computing technologies and in 1997-8 was particularly noted for the successful introduction of world leading advanced visualisation systems for design and manufacturing of ships and submarines. He was awarded the status of BAE Systems Engineering Fellow for his contribution to excellence, one of 12 Fellows selected from a population of 18,000 engineers worldwide.

Semi-retired, he is the founder and Director of SAMOSC Ltd. providing 'specialist expertise' for organisations, projects, forums, and conferences, based on his extensive and wide-ranging knowledge, experience, and ability.

© 2017: The Royal Institution of Naval Architects

He has the following papers published in ICCAS proceedings:

- "Virtual Reality Techniques for Ship and Submarine Design", *Yokohama, Japan, 1997*
- "Geometric Modelling for Simulation Based Ship Design", *Cambridge, USA 1999*
- "The Actual Reality of Virtual Reality", *Malmö, Sweden, 2002*
- "Visualisation Technologies in Ship Design and Build" *Bussan, South Korea, 2005*
- "Processes – a Basis for capturing Knowledge", *Portsmouth, UK 2007*
- "Data Modelling – its use and Contribution to Ship Design, Manufacture and Build", *Shanghai, China, 2009*
- "A Cost Effective Process for Simulation Based `Design in Shipbuilding", *Trieste, Italy, 2011.*
- "How Classic Systems Engineering process must change to be effective for use in ship design", *Bussan, South Korea, 2013*
- **"Accessible Immersive Visualisation for Shipbuilding", *Bremen, Germany, 2015***

10. LIABILITY

Whilst every effort has been made to ensure the accuracy of the contents of this contribution, SAMOSC Ltd. accepts no liability whatsoever for any errors or omissions therein nor for the use made of any said contents. All implied warranties and conditions are excluded, to the maximum extent permitted by the law.

SAMOSC LTD. 2017

© 2017: The Royal Institution of Naval Architects

© 2017: The Royal Institution of Naval Architects

CAD/CAM Integration with PLM in Naval Shipbuilding

D Morais, M Waldie, SSI, Canada **B Benevelo,** ARCOS Innovations, Canada

SUMMARY

This paper highlights a case study of integrating a CAD/CAM system with PLM at Ingalls Shipyard in a naval shipbuilding context. Naval shipbuilding requires powerful tools to manage large and complex projects. Regulations are stringent, extensive documentation is necessary, and frequent change is a constant component of the process. To manage this complexity, multiple programs must constantly leverage up-to-date and accurate engineering data from a CAD/CAM product model. Faced with the requirement to dramatically reduce costs, Ingalls identified that implementing a PLM system was the key to improvement. SSI worked with Ingalls to implement the system using Agile methodologies. While working on the project, Ingalls embraced the idea that it was undergoing a business transformation, not simply digitizing old processes. In less than a year, Ingalls was already seeing benefits, and due to the software's resiliency and flexibility, continues to move forward with innovative initiatives.

1. INTRODUCTION

In the wake of the ongoing challenges in much of the shipbuilding market, one of the bright lights is the construction of naval warships. Yet, even in that sector, there is now intense cost-cutting pressure. The United States Government plans to significantly increase the size of its navy but administration officials have notified companies both publicly and privately that the cost per ship must dramatically be reduced. To accomplish this, the United States Government is moving towards fixed price contracts rather than cost-plus. In response, naval shipyards have been driven to seek sizeable efficiencies.

As a major builder of Navy and Coast Guard vessels, Ingalls Shipbuilding, is emblematic of this trend and the story of this company's journey towards increased efficiency is a lesson for shipbuilders in all sectors of marine construction. With 11,500+ employees and a shipyard encompassing 800 acres, coordinating activities is complex. Thus, Ingalls realized that to improve, it had to undergo a business transformation. A key element in this transformation was to integrate the wealth of data found in its CAD/CAM software with an information platform. This paper shows how that was accomplished.

2. COMPLEXITY AND CONFUSION

Ingalls faced a challenge regarding managing information. It had thousands of databases, Excel files, and custom scripts that were created by various departments to solve particular problems. Unfortunately, the information produced was copies of other data or even copies of copies. It was difficult to ensure that everyone was working from the latest information and a large support staff was required. This situation was no surprise to us. In fact, based on site visits to shipyards around the globe, we can say with confidence that the standard way that information flows within most shipbuilding organizations is a labour-intensive, complicated, error-prone, highly manual process. (Figure 1 on the next page shows a typical scenario.)

3. DIGITAL TRANSFORMATION

But before talking about exactly what SSI and Ingalls did on this project to solve the problem, let us step back and discuss the overall philosophy we had which guided our thinking. Anyone reading this paper and/or attending ICCAS is most likely interested in improving how their business operates. There is a desire to reduce redundancies and enhance communication. Beyond that, attendees are interested in applying innovative technologies and techniques to improve long-term effectiveness and efficiency. These include items such as smart manufacturing, integrating design data with machines, artificial intelligence, machine learning, big data, the internet of things, 3D printing, and harvesting the benefits of model based definitions and being a model based enterprise. In other words, the subjects people believe will help their business address challenges typically are subjects involving a digital transformation.

4. BUSINESS TRANSFORMATION

However, "going digital" is not enough. A successful *digital* transformation requires a *business* transformation to really take advantage of the latest technology. To illustrate this point, consider how one might go about digitizing a map. You could just take a picture of a fold-out paper map and make a pdf. It would now be digital and there would be some benefits but it would be nothing compared to using a GPS system that gives you turn by turn directions as you drive.

The point we are making is that what businesses must do is to think through what they are really trying to accomplish, look at all the tools now available with modern technology and explore how to maximize what is possible. But just as important as analysing new tools is the point that a business must consider what transformations they need to undergo regarding developing processes that might be totally different. They also must consider what kind of people they will need and the skills those people should have.

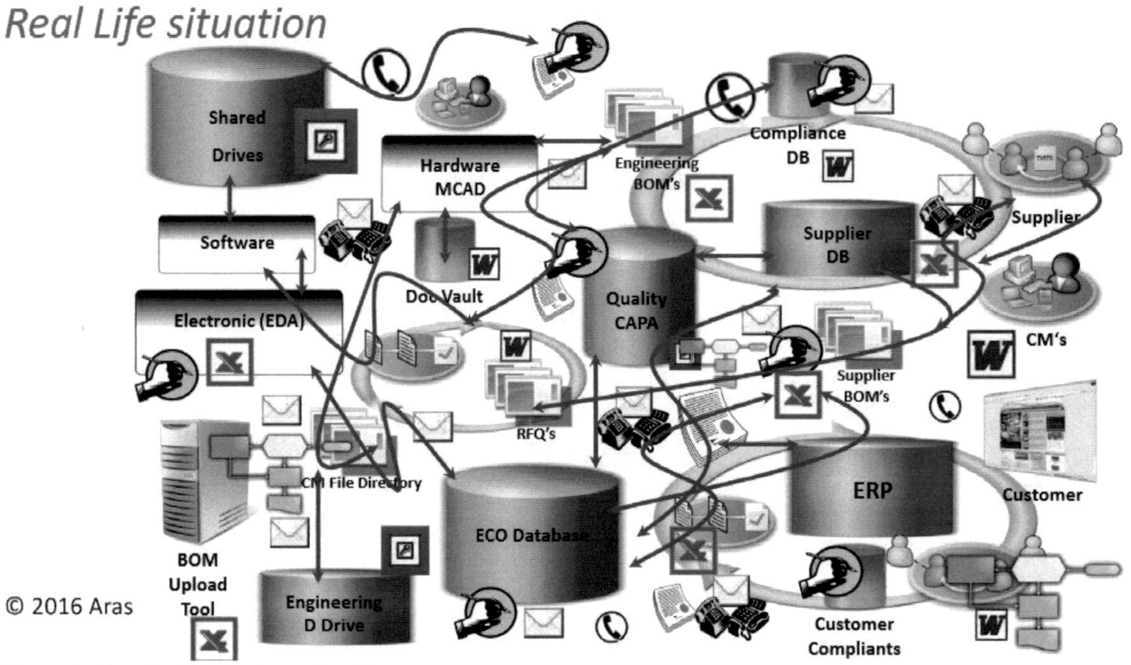

Figure 1: Complexity of Information Flows

There are multiple things to consider but when you get things right, there can be enormous improvements. A notable example of this (albeit in a non-shipbuilding context) is the tremendous advancements that have occurred over the years in auto racing. If you happen to see old footage of a pit stop in the 1950s, it is almost painful to watch a crew take over a minute to awkwardly change four tires. Now, Formula One teams routinely change tires in less than three seconds. Watching this is incredible. It is a perfect mix of skilled people, a precision process, and smart applications of technology. That is the kind of transformation you must be aiming for in every implementation or integration project.

5. WHY AN INFORMATION PLATFORM

With that goal in mind, there are several things that can be done to transform your business for the better. In a previous paper, we argued that because the engineering information found in a CAD system's virtual product model is at the heart of so much of the rest of shipbuilding, organizations should find ways to leverage that rich trove of data. [1] We have since followed up by giving examples of how to do this in presentations and papers [2] featuring a product called SSI EnterprisePlatform which makes CAD data available in the correct format and context to various programs, processes and production machines. Examples range from easily creating interactive 3D visualizations of CAD data with attribute information, to more exotic applications such as directly driving welding robots without programming, as well as supporting Virtual Reality applications.

In this paper, we want to build on the point of leveraging engineering data but wish to highlight in-depth a more comprehensive, enterprise-wide solution, the kind of solution that has been successfully implemented at Ingalls Shipyard. The solution we are talking about involves an information platform.

Terminology is often inconsistent, but generally speaking, what we call an information platform is currently commonly referred to as a PLM system. There are many reasons why a system such as this is crucial for solving the chaos that characterizes information management in most shipyards.

First of all, an information platform simplifies and streamlines the typical messy, labour-intensive, error-prone flow of information by reducing the number of disconnected systems. People and programs receive the information they need seemingly invisibly; it is all transparent to the end-user.

Secondly, an information platform synchronizes all stakeholders so that all departments in the shipyard can move forward smoothly together. It does this by connecting different software and hardware systems and then providing the information each system and person needs in the correct context and format required for efficient consumption.

A third major benefit is better change management. An information management platform allows you to respond to change in a structured way and not just react in a panic. A system like this makes it easier to plan and see the impact of changes in advance. You can find things earlier, calmly have various stakeholders assess options, and have more opportunity for developing different solutions. This reduces cost and risk.

Finally, having an information platform enables you to support new ways of doing business and increases your capabilities as an organization. Indeed, implementing an information platform / PLM system is a key component for adopting many new digital technologies and undergoing a significant business transformation.

6. CHALLENGES FOR SHIPBUILDING

However, despite all the benefits of adoption, before the PLM system was implemented at Ingalls Shipyard, we identified three overarching challenges that we think all shipbuilders in similar situations should be aware of.

Firstly, there are no PLM tools that have a shipbuilding DNA. PLM systems are designed for discreet manufacturing processes and generally force shipbuilders to work as if they are building a series. However, shipbuilding is *not* discrete manufacturing. In shipbuilding, you have to design, procure parts and construct the ship simultaneously. This means that to use PLM in a shipbuilding context, you are doing something that the program was not meant to accommodate.

This leads to a second challenge. The PLM systems that most shipyards typically down-select are typically costly to implement in a way that fits with a shipyard's processes and there are many integrations necessary to achieve any benefit. This means that it takes longer to achieve a return on investment.

Another cost-related challenge that is typically faced is the challenge of upgrading to newer versions. PLM systems are massive programs that are usually difficult to upgrade so all too often, this is not done. That means that an organization ends up using obsolescent technology.

Addressing all these challenges was paramount in the mind of the team at Ingalls when down selecting programs to implement.

7. HOW WE WENT FORWARD

Realizing that we were involved in a complete business transformation project, we came up with a plan that addressed all the critical elements involving people, processes and tools.

7.1 VISION AND GOALS

Before we did anything, we made sure that all the key stakeholders agreed on the purposes of the project. We created a clear and concise vision and several short, attainable goals that were supportive of the vision. This really helped bring focus to the project and gave us a clear way of measuring where we were at.

7.2 AGILE CULTURE

To help us achieve each goal, we embraced an Agile strategy and encouraged everyone to embrace an Agile culture. For those unfamiliar with what "Agile" means in this context, it refers to completing work in small increments but with each iteration adding value in and of itself. This is different from the traditional waterfall style of project management which involves massive amounts of planning upfront in the naïve hope that you can create a perfect plan for a complex project from which you will never have reason to deviate. In complex situations, this rarely works.

By contrast, an Agile approach dramatically increases the likelihood of having success—multiple successes, in fact. To use a simplified example, an Agile way of gradually working up to constructing a car would be to first build a skateboard then add a handle to make it a scooter. Then, once you had mastered building those, you would progress to building a bicycle, then adding an engine to make a complete car. This may seem inefficient but in reality, in any scenario where you lack almost perfect knowledge, an Agile approach is better. With each small project completed, you deliver something of value and if there is a failure, you just learn from it, pivot and continue. By working in an Agile way, the project at Ingalls was set on the path to success.

Figure 2: Non-Agile (top) vs. Agile (bottom)

7.3 STEERING COMITTEE

To keep us on the right path and ensure that we maintained our focus, we assembled a steering committee that was made up of representatives from various teams who all had knowledge of different disciplines. It was truly a diverse team that was ship-class agnostic. This ensured that we were implementing a solution that would work for everyone and ensured that the entire shipyard had buy-in on this project. Having a great steering committee of knowledgeable people who were committed to the vision and who insisted on keeping us focussed was a key to making the project a success. It cannot be stressed enough, having the right people guiding you makes an enormous difference.

7.4 TOOL SELECTION

Thus, once the group agreed on what we wanted to achieve and how we were going to go about doing it, Ingalls had to make the actual selection of tools to use. Here is what was selected and why:

7.4 (a) ShipConstructor for CAD + EnterprisePlatform

Ingalls Shipbuilding was already using ShipConstructor CAD software to model ships such as the LHA Amphibious Assault Ships, DDG 51 Arleigh Burke Destroyers and the Legend-Class National Security Cutters. Therefore, selecting ShipConstructor as the company' CAD/CAM tool was natural. However, there were several additional reasons for choosing ShipConstructor to integrate with a PLM system.

ShipConstructor software in and of itself is a powerful tool that combines the familiarity of the world's most popular drafting program (AutoCAD) with shipbuilding-specific tools and menus covering all the engineering disciplines. The program is intuitive to use because it is designed to follow common shipbuilding workflows.

The power of this software is enhanced by the fact that ShipConstructor incorporates a Marine Information Model (MIM) which stores a virtual product model of the entire vessel including the complete 3D model, attribute information for each part and assembly, the assembly sequence for the project, the industry-specific rules, the standards used in the product model's composition and the associative relationships between all standards, geometry, parts and assemblies within the model.

All this information is easily available for connecting to another system due to ShipConstructor's open architecture which is not common for a shipbuilding CAD system. More importantly, the SSI EnterprisePlatform makes it possible, through automated processes, to gather, format and provide, meaningful, actionable information from the MIM for distribution and consumption in a wide variety of formats. Since the information is provided in the correct representation and context, it allows all stakeholders throughout a shipyard to make use of the information effectively, especially when making decisions. Therefore, ShipConstructor was the ideal foundation for a CAD/PLM integration because it had the ability to easily share its enormous wealth of information with a PLM system.

7.4 (b) Aras PLM: Flexible & Adaptable PLM

The PLM system that Ingalls selected was Aras. One of the biggest reasons for choosing Aras was that its data model architecture was flexible enough so that it could support standard shipbuilding processes. As explained earlier, most PLM systems that are typically chosen cannot easily accommodate frequent change and concurrent workflow. However, after analysing their options, Ingalls Shipyard determined that Aras would be able to support a shipbuilding environment and this has been proven by subsequent experience.

As for the challenge of having to implement too much at once, fortunately, like ShipConstructor, the Aras PLM system has an open architecture which allows an Agile, incremental implementation rather than a "big bang". We saw that this staggered approach would allow Ingalls to replace legacy systems and processes without disrupting its current business which was important in implementing the system cost-effectively.

Also of importance was the fact that upgrading would not be a problem. The Aras PLM system is designed in such a way that the company guarantees that upgrading to newer versions will retain all data, workflows and even customizations. Meanwhile, all this flexibility and resiliency does not mean a trade-off in terms of functionality. Aras PLM is full-featured and is used in a variety of industries by major corporations such as GE, Airbus, Microsoft and GM.

7.5 SYSTEM ARCHITECTURE

Figure 3 (next page) shows the system architecture integrating ShipConstructor, Aras PLM and also an MRP system at Ingalls. Catalogs containing supplier preferences and materials information etc. are fed into the ShipConstructor CAD program from the MRP system and ShipConstructor creates and stores information about Engineering Work in Progress (WIP). Production documentation etc. flows out of ShipConstructor to the Aras PLM system. There, after review, information can be released as MBOMs to MRP, or as schedules, engineering change orders or as various user defined attributes for parts etc. back to the CAD system.

7.6 IMPLEMENTED RAT AND MVP

As the team worked on the project, we used standard Agile methodologies in that we were always seeking to fail fast and get to value quickly. The Agile methodologies used go by the names RAT (Riskiest Assumption Test) and MVP (Minimum Viable Product). To test out how the entire system would work when integrated, we first focussed on doing 3D planning, in other words, using interactive 3D visualizations for planning purposes (more on this later). This was a completely new way of doing things for Ingalls and is innovative for shipbuilding in general since the industry has always planned using 2D drawings and Excel. But this innovation fit with our firm belief that Ingalls had to be looking at a business transformation, not just a technology implementation.

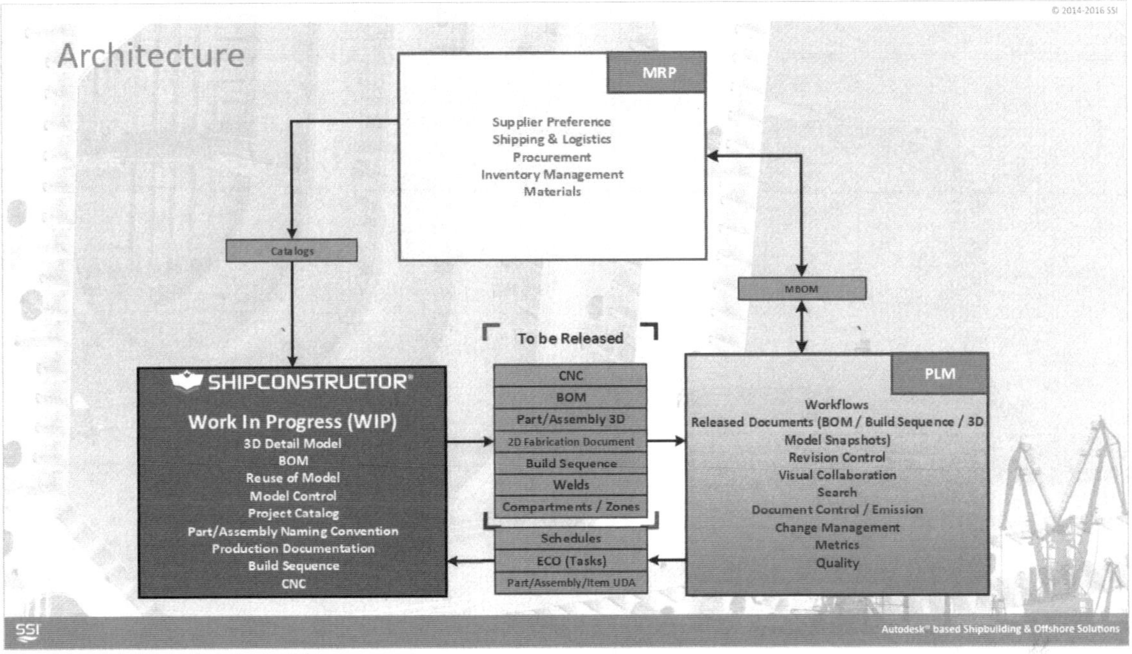

Figure 3: PLM/CAD/MRP Architecture

If 3D planning could be implemented, we would learn a lot about implementing everything else—this was the Riskiest Assumption Test. Furthermore, all by itself, 3D planning was a valuable product—our MVP. We had an idea on how to proceed and within one month we had a proof of concept that we could demonstrate, learn from, and then pivot off of towards the solution in use today.

8. WHAT WAS ACHIEVED IN < 1 YEAR

Below is only a subset of what was accomplished with SSI working with Ingalls on this project.

8.1 COMPREHENSIVENESS

To clean up the confusion and complexity identified at the beginning of the paper, we wanted to gather a comprehensive amount of engineering, planning and production information in the PLM system. We did that. All engineering, detailed planning, and production data with relationships is now published to the Aras PLM from ShipConstructor. This happens via a task management system in ShipConstructor which automatically publishes when a user specifies that their modelling task is complete. A slick feature of this is that the user does not need to track or remember what they created, modified, or deleted in a project. It is all tracked for them in the background.

The types of information stored includes parts' BOMs, model drawings, production drawings, isometrics, nest tapes, profile cutting machine exchange info, the 3D viewable and a model snapshot.

Plus, a key point to note is that it is not just the raw information that is stored, it is the relationships between the information. These are critical to understanding the context and working with other systems.

Of course, the ShipConstructor CAD system's Marine Information Model already stored massive amounts of useful information but the advantage of having the information in a PLM system is configuration management.

In other words, because information is configuration managed, it is version controlled. This gives Ingalls so much more control of its processes. It can now more effectively investigate future issues, respond to clients' requests, support its future digital thread initiative and provide metrics for analysing progress and change. The key was getting a comprehensive amount of engineering into the Aras PLM system.

8.2 CENTRALIZED INFO PLATFORM

Once the information was all in a centralized location, we now had a centralized information platform and this offered so many benefits. Instead of having to gather information from multiple places, e.g. Excel files, Access databases, etc.), it was all in one place and this really helped eliminate confusion and complexity.

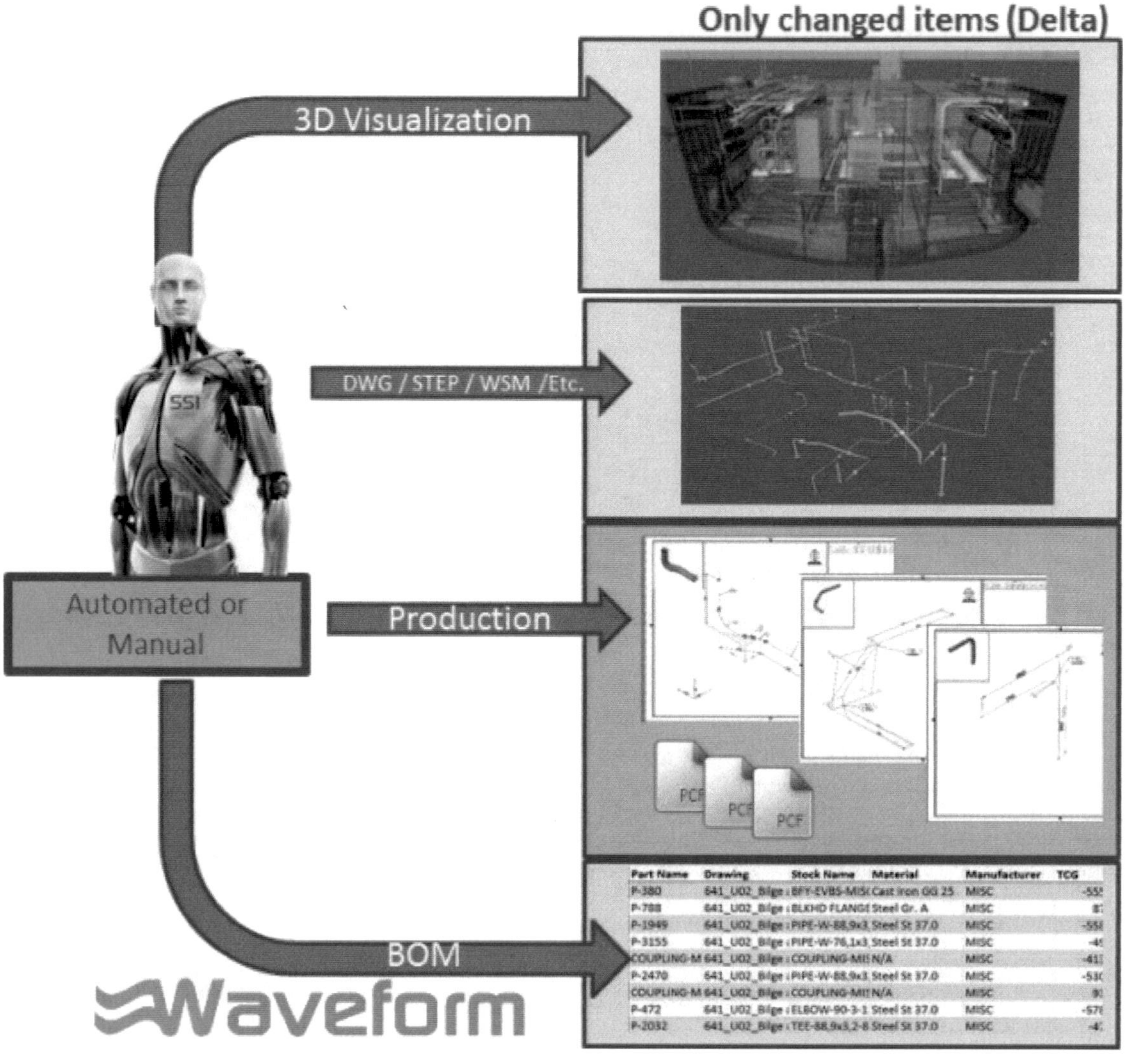

Figure 4: Communicating Change

The biggest benefit of PLM is that it interconnects all the departments and disciplines in the shipyard via a single hub. With a centralized platform, now everyone can easily access the most up-to-date and accurate information in the context they need.

This also allows true concurrent engineering because with configuration management and the high level of control a centralized information platform enables, planners can more easily coordinate and give feedback to designers and modelers working in multiple connected disciplines. It is all tied to the centralization of information.

And note, we are not merely referring to information flows between the PLM system and people in different departments, we are also talking about production machines using accurate CAD information without the need for programming. As previously mentioned, we at SSI have written about connecting CAD data with machines before but what PLM adds to the mix is enhanced coordination. For instance, ShipConstructor CAD data can be used to drive HGG CNC profiling machines via the SSI EnterprisePlatform. However, that alone, can still lead to problems because the plate nest information still has to be synchronized with the CNC cutting information. This is what PLM is ideal for, coordinating the flow of information.

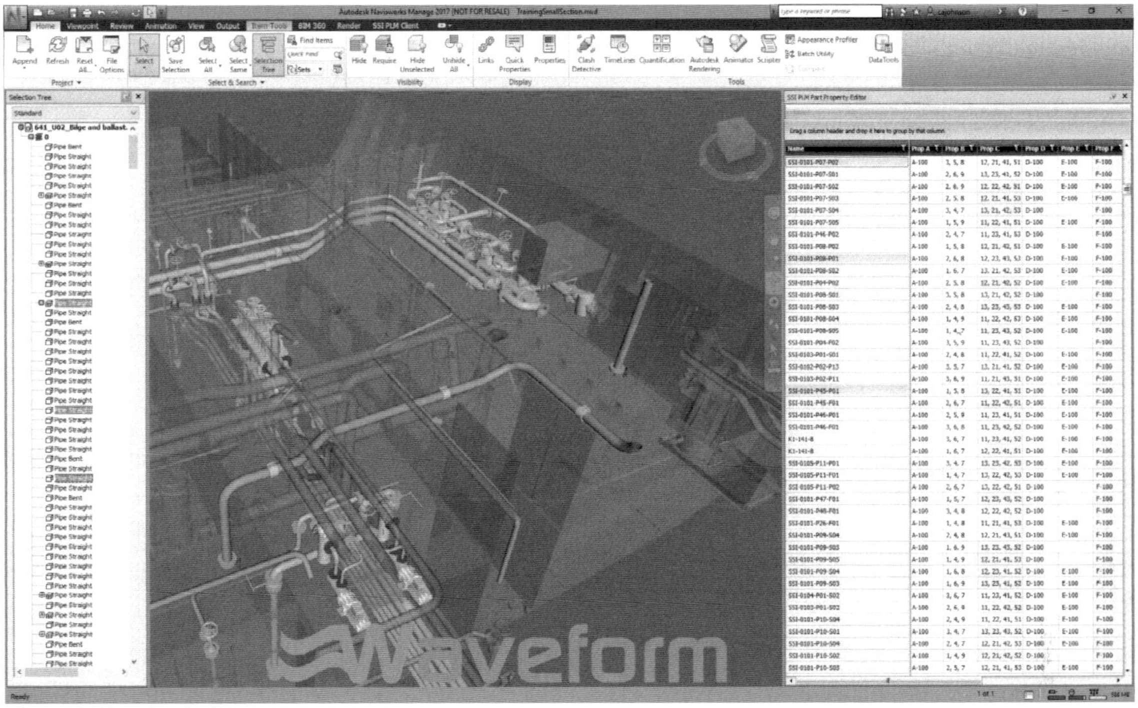

Figure 5: Planning in 3D

Another way PLM enables information coordination is that it provides information in the correct context and amount—not too much, not too little.

Consider how "green" on a plate (additional stock) is handled in different departments. Purchasing still needs to take this into account when calculating length and quantity but green must be disregarded for most weight calculations. Therefore, the PLM provides the green information to the purchasing department and not to engineers calculating weight. In summary, now at Ingalls with PLM integrated with CAD, the right information (up to date and accurate info) is provided to the right people at the right time.

And as far as most users are concerned, it all happens in the background, automatically, invisibly. PLM now has allowed Ingalls to set up workflows and notifications so that workers are notified of additions or modifications automatically. There is no requirement for engineers to initiate actions to communicate information about changes because this is all taken care of for them.

Automated communication of change improves efficiency because it allows highly skilled engineers to focus on tasks of maximum value instead generating information in different formats and having to worry about who to communicate that information to. This is a great benefit that was only made possible through having a centralized information platform.

8.3 NEW INNOVATIVE WAYS OF WORKING

Beyond the benefits of the centralization was the fact that Ingalls truly embraced SSI's philosophy of recognizing that they had to undergo a business transformation to maximize the use of their PLM/CAD implementation. Ingalls realized that in less than a year, they could implement completely new and improved ways of working so that is what we helped them put in place.

8.3 (a) Change Management

For instance, as mentioned earlier, now that Ingalls had a powerful PLM system, it was able to manage changes in a much more sophisticated manner. This was because the right people are now notified of what has changed in the best context they can consume. For example, when there is a change in the design, the purchasing department can get a list of all the parts created, deleted and modified with corresponding attributes such as the catalog items. The weight engineers can get a list of parts per SWBS that were created, deleted and modified with the corresponding weight information. The production floor may want to see the changes to the shop floor documents and corresponding CNC files. Other departments may want to see a 3D visual of the parts change to understand the impact. Ingalls now has the capability to set up these kinds of notifications and has been doing so.

8.3 (b) Comparing with Previous Versions

Another new capability that Ingalls has adopted is the ability to analyse different versions of a design. The system allows anyone to have a visual display showing changes in any item that is configuration managed.

8.3 (c) 3D Planning

The system is highly visual, rather than text-based. It lets Ingalls visualize ShipConstructor data in 3D using Autodesk Navisworks. Within the same interface, Ingalls can edit and plan with the Aras Innovator PLM system. Ingalls can also configure the system to show visually how the design has changed over time. Planners thus have exactly the information they need to plan accordingly.

8.3 (d) Metrics

Another new method that Ingalls has adopted is looking at how well they are designing over time. Because they now have the ability to trace the history and impact of changes, they can identify what has happened and its impacts on labour and material.

9. CONTINUOUS IMPROVEMENT

What SSI accomplished with the team at Ingalls in less than a year is something to be proud of but it is only the beginning. The philosophy regarding business transformation implemented via Agile techniques has taken root and the company is excited about further improving its business.

With the PLM system as implemented, Ingalls has a firm foundation upon which to build that does not limit it. Because of the underlying architecture of ShipConstructor and the Aras PLM system, Ingalls can adopt new programs, technology and techniques without having to worry that this will wreck the PLM implementation. This means that Ingalls can now make business decisions without having to worry about options being closed due to its PLM and CAD tool selection.

Ingalls continues to have a good team focussed on continuous improvement and the Agile culture is strong. It is a very forward-thinking company that is planning to make advancements on other initiatives such as Industry 4.0, increased use of Mobile technology, paperless production, MBD and MBE, artificial intelligence, machine learning, and rule based design. It is exciting.

10. CONCLUSIONS

In summary, as a US Naval shipbuilder, Ingalls has been forced to dramatically improve efficiencies. It wisely noted that better managing information was the key to improvement. The company got a diverse team together of good people who had developed a vision and then identified goals they wished to achieve. SSI worked with Ingalls to choose CAD and PLM software tools and then implement the system in an incremental fashion. In a short timeframe, the company immediately started seeing benefits, especially since it bought into the idea of business transformation; Ingalls now does things differently with PLM, it did not just digitize old processes. As Ingalls moves forward, it will continue to improve its business because it has the right, people, process and tools in place.

11. REFERENCES

1. MORAIS, D.; WALDIE, M.; LARKINS, D. (2013), 'Leveraging Engineering Information', *Conference on Computer Applications in Shipbuilding*, 2014

2. WALDIE, M.; MORAIS, D.; DANESE, N. (2017), '2D, 3D and Virtual Reality: Creating the Natural Way to Work', *Go-3D*, 2017

12. AUTHORS BIOGRAPHY

Denis Morais is CTO of SSI where he is responsible for developing and directing the strategic product vision for SSI's software and services. This involves analysing industry needs and technological trends to prioritize SSI product creation and enhancement. Denis oversees SSI's development team and consults with clients directly to implement optimized solutions that address their unique constraints and business requirements. Denis is a world authority regarding the future of shipbuilding and technology. He is an award-winning author, writes a popular blog, and is a sought out presenter at international conferences. He is an active participant in several shipbuilding and technology organizations such as RINA, SNAME, the Autodesk Developer Network, VIATeC and Autodesk University.

Mark Waldie is the PR Writer for SSI where he is responsible for promoting SSI's innovative software solutions and success stories worldwide via various forms of media. He in an award-wining author of various technical papers and has more than 17 years' experience marketing advanced technology for use by the world's leading naval forces.

Bruno Benevolo has taken a lead role in several significant CAD/CAM, EDMS/PDM/PLM projects in the shipbuilding, offshore EPC, industrial and nuclear plant industries. As founder of Sincronia, based in Rio de Janeiro, he was the initial software architect of Sincronia IFM, one of the most successful EDMS/PDM applications developed for the Brazilian shipbuilding, offshore, industrial and nuclear plant industries

BIG DATA PLATFORM FOR PLM (PRODUCT LIFECYCLE MANAGEMENT) SYSTEMS IN SHIPBUILDING AND OFFSHORE INDUSTRY

S H Kim, M I Roh, M J Oh and S W Park, Seoul National University, Republic of Korea
N K Ku, Dong-eui University, Republic of Korea
S H Myung, Youngsan University, Republic of Korea

SUMMARY

In shipyards, PLM (Product Lifecycle Management) systems are being used to create, store, and manage all the information for design and production of ships and offshore structures. As the ships and offshore structures are getting complicated, the PLM systems are required to handle an enormous amount of data called big data. However, it's hard to handle efficiently such big data with the existing method for data processing. Therefore, the need to apply big data technologies to the PLM systems has emerged these days. In this study, the big data technology for the PLM systems was implemented. It was applied to solving a problem of material purchase process in the actual shipyard, to evaluate the applicability of the big data technology. The result shows that the big data technology can be effectively linked with the PLM systems in shipbuilding and offshore industry.

1. INTRODUCTION

1.1 BACKGROUND

Big data, literally a large amount of data, becomes a concept or philosophy which supports to collect, handle, categorize a large amount of data. From this next generation technology and architecture, data can be easily gathered and analyzed. Characteristic of big data could be represented by 3V, each of V is volume, velocity, and variety. The 'volume' refers to a big volume of data which exceed 100 TB (Tera Byte). The 'velocity refers to how fast data is processed. In usual, it should be fast enough to get a result in real time without any delay. The 'variety' means that there are many types of processable data even it is unstructured.
Shipbuilding and offshore industry, especially for a shipyard, produce a lot of data. As a shipyard constructs many ships or offshore structures, data becomes large enough to call it big data. Therefore, this study analyses application for PLM data using big data technology.

1.2 RELATED WORKS

Research of big data and methodology to analyse big data is progressing in various industries. Media and social network service are one of the leading industry in this area. Besides, the importance of big data in shipbuilding and offshore industry is increasing, but related research seems still staying early stage. Therefore, we have broadened the scope and analysed researches related to big data in related industries, as well as the shipbuilding and offshore industry. Some of the analysis results are summarized below.

There are several types of research for application of big data in PLM system. Nedelcu (Nedelcu, 2013) proposed big data application in the manufacturing industry. Also, it can be found that some attempt to apply big data in the field of shipbuilding and offshore industry. Kim (Kim et al. 2013) suggest an application of big data to build a methodology of ship engineering automation. Recently, Perera (Perera, 2017) proposed a big data handling framework for harnessing ship performance and navigation monitoring data.

However, it is hardly found that research of big data application which is focused on shipbuilding and offshore industry. In this paper, big data technology for shipbuilding and offshore industry will be introduced and applied to an example.

2. BIG DATA TECHNOLOGY

Each big data technology which applied to the big data analysis will be described in this chapter.

2.1 HADOOP

Hadoop is an open source framework which provides distributed processing and data storage system. It is composed by HDFS and MapReduce. HDFS (Hadoop Distributed File System) is a data storage technology that can store data distributed. MapReduce provides fast data processing function using distributed computing.

2.2 SPARK

Spark, an open source distributed processing system which is usual for big data, processes data set at once. Compared with Spark, MapReduce processes data set step by step and spark show ten times faster performance than MapReduce. Spark also has various libraries which can be used easily and supports various programming language. Spark is replacing MapReduce due to these advantages and used in the application proposed by this study.

3. APPLICATION

In this chapter, an example is applied to confirm the utility of big data technology. The example is from material procurement process of offshore structure material in a shipyard.

3.1 MATERIAL PURCHASE PROCESS OF OFFSHORE INSTALLATION

The problem is seen from material purchase process. Predicted consumption plan can be inaccurate because any margin during construction is not included.
In this study, to solve the problem, regression analysis has applied using big data technology.

3.2 ESTIMATION OF MATERIAL CONSUMPTION BY SCHEDULE

Raw data from records from 10 offshore structures saved in HDFS has been data-mined by Spark for predicting consumption of material per period which takes to construct one module of the structure. Applying regression analysis, it can be confirmed that predicted consumption of material is quite similar to the record of material consumption. It seems valuable to extend this application to other materials if data is increased and if variables of regression can be diversified.

4. CONCLUSION

In this paper, the big data technology was applied to material procurement process of offshore structure to confirm the utility of it. From the result of the analysis, it has been found that there is a potential for this application.

5. ACKNOWLEDGEMENTS

This work was partially supported by:

(a) MSIP (Ministry of Science, ICT and Future Planning), Korea, under "Development of Ship Design Standard PLM Platform based on Big Data" (NIPA-2016-S1106-16-1025) supervised by the NIPA (National IT Industry Promotion Agency),

(b) BK21 Plus Program (Education and Research Centre for Creative Offshore Plant Engineers), funded by the Ministry of Education, Korea.

(c) Research Institute of Marine Systems Engineering of Seoul National University.

(d) Program of University Specialized for Offshore Plant Engineering, funded by the Ministry of Trade, Industry, and Energy, Republic of Korea.

6. REFERENCES

1. Nedelcu. B, 'About Big Data and its Challenges and Benefits in Manufacturing', Database System Journal, 4(3), pp. 10-19 2013.
2. Kim, Y.J., Park, J.G., Lee, J.H., Yang, H.Y., Jeong, M.A., 'A Study on the Big data technology and analysis technique for vessel design automation', Korean Institute of Communication and Information Sciences, 2013.
3. Perera, LP., 'Handling Big Data in Ship Performance & Navigation Monitoring', Smart Ship Technology, 2017.

7. AUTHORS BIOGRAPHY

Seong-Hoon Kim is M.Sc. candidate naval architecture and ocean engineering from Seoul National University, Korea. His main area of research interests includes big data analysis, deep learning, optimization, and so on.

Myung-Il Roh is a professor of naval architecture and ocean engineering at Seoul National University, Korea. He holds a B.S., M.S., and Ph.D. in naval architecture and ocean engineering from Seoul National University. His main area of teaching and research interests includes ship and offshore plant design, simulation-based design and production, optimization, CAD, CAM, CAE, CAPP, and so on.

Min-Jae Oh is a post doctor of Research Institute of Marine Systems Engineering at Seoul National University, Korea. He holds a B.S., M.S., and Ph.D. in naval architecture and ocean engineering from Seoul National University. He worked at American Bureau of Shipping for 4 years. His main area of research interests includes CAD, CAM, Big Data, Software Quality Assurance, and so on.

Sung-Woo Park is M.Sc. candidate interdisciplinary program in offshore plant engineering from Seoul National University, Korea. His main area of research interests includes pipe routing automation of offshore structure and optimum design and so on.

Namkug Ku is a professor of naval architecture and ocean engineering at Dong-Eui University, Korea. He holds a Ph.D. in naval architecture and ocean engineering from Seoul National University. His research interests are in shipbuilding, multi-body dynamics, and modelling and simulation, and so on.

Sehyun Myung is a professor of automotive engineering at Youngsan University, Korea. He holds an M.S. and Ph.D. in mechanical engineering from Korea Advanced Institute of Science and Technology. His main area of teaching and research interests includes automotive design, product lifecycle management, digital manufacturing and so on.

A STUDY ON THE EFFECTIVE COLLABORATION CONCEPT BETWEEN DIFFERENT SEVERAL DESIGN STAGES

T Kunisada, Sanoyas Shipbuilding Corporation, Japan
T Murakami, NAPA Japan Ltd., Japan
S Kasai, SENER Ingeniería y Sistemas

SUMMARY

The authors have been studying the methods to achieve efficiency and costs minimization for design and production stage by collaboration between several 3D design systems. This concept is based on "CIMS", which was suggested in 1980s to acquire international competitiveness of Japanese shipbuilding industry by fully integrated software. However, practical system full integration have not realized yet and many shipyards are struggling to establish individually optimized design systems for each design stage. According to recent tough market situation, the efficiency and costs minimization is pressuring issue and total optimized system is really needed. The authors suggest practical total optimization; 1) Using intermediate format to collaborate several 3D design systems. 2) Developing converter of intermediate format for collaboration between individual optimized tools. In this paper, feasibility and benefit of our proposal are discussed.

1. INTRODUCTION

Sanoyas Shipbuilding Corporation has been building various types of merchant ships, bulk carriers, tankers, wood chip carriers, and so on since the establishment in 1911. In the design department, 3D design systems have been fully utilized to enhance work efficiency, minimize design cost, and increase product quality. However we have to take more improvement in 3D system utilization due to getting stricter regulations and shorter design period due to tough market competition, nowadays it is industry-wide concern.

3D design system has been widely used and developed in ship design process because of the complicated curved hull shape, it's very difficult to grasp overall condition by 2D based system. The ship design process is divided into several stages and utilizing individual 3D design system, so-called "Individual optimization", is achieved in each design process, but the strong demand of "Total optimization" is rising due to recent market situation. Regulation changes and fiercer competition decline series vessel building business model lead to frequent design improvement and shorting lead-time, so that total optimization of design process is urgent industrial issue.
One of the biggest preventing factors for total optimization is the coexistence of different 3D design systems in each design stage. It causes difficulties of design information handover and many duplicated work is required. Sanoyas uses 3D design system, NAPA and NAPA Steel (NAPA Ltd.) in Initial and Basic design stages and FORAN (SENER) in Detail/Production design stage.

In this paper we focus on the 3D system for structural design and collaboration between existing different systems, we are planning to develop data linkage system between different design stages and 3D design systems as a practical and feasible solution. The specification of data linkage system is discussed based on the interviewing session with designers, and we estimated the positive effect quantitatively. Our future plan is also described in the last part as a conclusion.

2. CURRENT STATE ANALYSIS

2.1 DESIGN STAGES

Figure 1 shows the schematic diagram of overall ship design process based on the previous research [1]. Generally, ship design process is divided into 3 stages; Initial design, Basic design, and Detail/Production design splitting at the timing of shipbuilding contract and class approval. In Figure 1, design work proceeds from top to bottom and area of each stage implies the amount of work. The transition of the design stage is one-way design take over, the design process does not return to the prior stage. The reason of one-way process is that design spiral in each design stage is an internal process and designers can improve the design only with internal discussion, on the other hand, making a contract and getting approval require the external communication, and it is usually difficult to make a great design change across these timings.

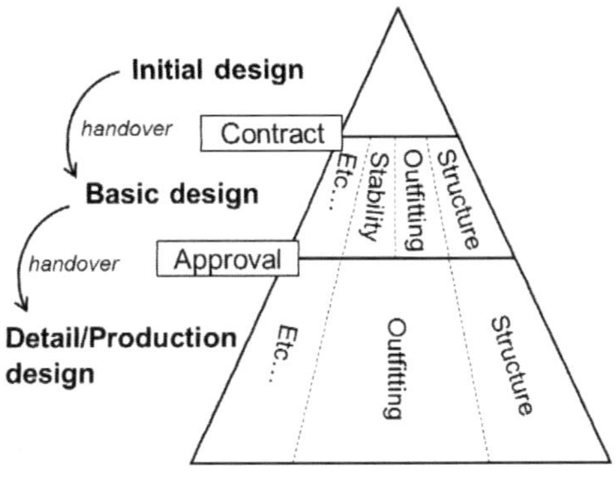

Figure 1: Overall process of ship design [1]

2.2 REQUIREMENTS FOR 3D DESIGN SYSTEMS IN EACH STRUCTURAL DESIGN STAGE

Structural design process belongs to two stages, prior is Basic design stage and posterior is Detail/Production design stage. Basic design stage handles strength assessment and drawings of ship structures referring the hull shape, compartment and specifications decided in the prior Initial design stage, and main output is approval drawings for classification society and customer. Detail/Production design stage takes over the structural drawings from the Basic design stage and prepare the all production information like working plans, parts number, detailed drawings, steel plate ordering data, and so on. Reflecting the differences of goal in each design stage, the requirements for 3D design system is also different (see Table 1).

In the Basic design stage, designers have to conduct strength assessment using FE analysis software and classification society's software, mold line (surface) based structural model is needed for the calculation. To avoid duplicated modeling work, data linkage function with these softwares is required. Additionally, structural arrangement varies frequently for design study by trial and error, so flexibility for model change is also required. Therefore, many 3D design systems for this stage adopt surface geometry like NAPA Steel, 3D model is created along mold line (surface), and the plate thickness is recorded as numerical property instead of geometrical plate thickness. This is also suitable for generating 2D key/yard drawing from the 3D model. In short, the somewhat simplified 3D model is preferable for Basic design stage.

On the contrary, in Detail/Production design stage, designers have to prepare all part information for production stage and business transaction by complementing the information taken over from the Basic design stage. Production information like part number and welding groove is also to be added in this stage. This work is like a building virtual ship on the computer system so that high accuracy and completeness are required for the 3D model. And almost all 3D design system for Detail/Production design stage adopts solid geometry like FORAN, it's an essential feature to design plate clearance and alignment. In exchange for accuracy, much larger data volume and high computer power are required, so the 3D model is usually made and handled per construction block.

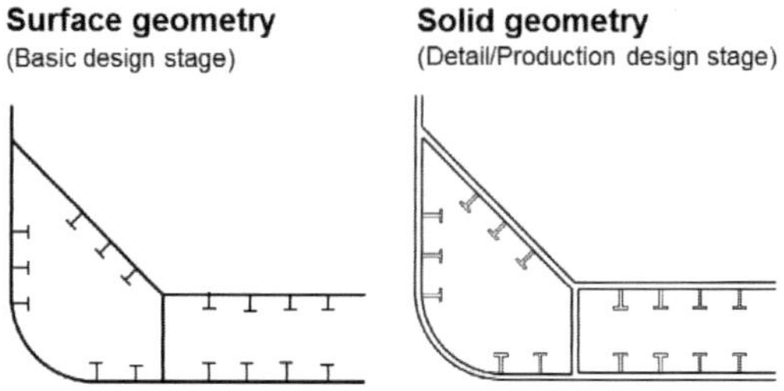

Figure 2: Schematic diagram of surface and solid geometry in 3D design system

Table 1: Comparison between structural design stages

Basic design stage

Goal	• Approval drawing
Customer	• Classification society • Ship owner • Detail/Production design stage
Need for 3D model	• Linkage with calculation software • Flexibility of model change
Common geometry	• Surface

Detail/Production design stage

Goal	• Production information • Purchase information
Customer	• Field workers • Factory machines
Need for 3D model	• High accuracy and completeness
Common geometry	• Solid

Above geometrical difference between 3D design systems in Basic and Detail/Production design stage is described in Figure 2. These are the background why it is difficult to handle overall structural design process by single 3D design system.

2.3 DIFFICULTIES AND CHALLENGES

Many shipyards are struggling with total optimization of the design process. However, as we discussed in Section 2.2, Basic and Detail/Production design stage has individual goals and requirements. Especially the difference of appropriate geometry disturbs to establish seamless design process, and most of the existing achievements are individual optimization in each design stage. Individual optimization has a limitation in improving working efficiency, for example, designers are modeling same object on a different system in the different design stage, and it is obviously duplicated work. Thus, total optimization is required for establishing more efficient design process by avoiding duplicated work and enhancing collaboration between different design stages.

The possible means of system collaboration are categorized into three types, and pros and cons are summarized in Table 2.

Table 2: Comparison of collaboration means between different design systems

Collaboration means	Pros	Cons
Totally comprehensive system	• Most preferable and effective	• Practical solution is not in the market • Function is fully rely on the software vendor (**uncontrollable**)
Data linkage via API	• Detailed information can be exchanged	• Development is fully rely on the vendor (**uncontrollable**)
Data linkage via intermediate file	• Shipyard can handle the development (**controllable**) • Minimum future risk	• Combination of 3D system is shipyard specific, so shipyard has to develop the I/F.

It's widely discussed that totally comprehensive, end to end, single 3D design system is ideal for ship design, and such a concept has been discussed from 1980s and called as CIMS, a Computer Integrated Manufacturing System. The achievements are summarized in several reports e.g. [2, 3, and 4]. But authors recognize that there is no practical solution and existing success case of in real design project. Such an ideal solution probably appear in the future, but improving the effectiveness of design process is pressing issue so that we have to solve the issue by readily available technology.

There are two possibilities of data linkage, utilizing API or intermediate file. Generally, API can realize much more functionality compared with utilizing intermediate file, but the implementation and API specifications fully rely on the software vendor, so some uncertainty will remain considering the long-term operation. On the other hand, data linkage using intermediate file has some limitation due to the format of the intermediate file, but shipyards can control overall operation if we choose public data format. Moreover, we can keep same operational procedure even though we change the 3D design system in the future.

Through above discussion, we decided to develop data collaboration feature between NAPA Steel and FORAN utilizing intermediate file.

3. DATA COLLABORATION PLANNING

3.1 OVERALL PLAN

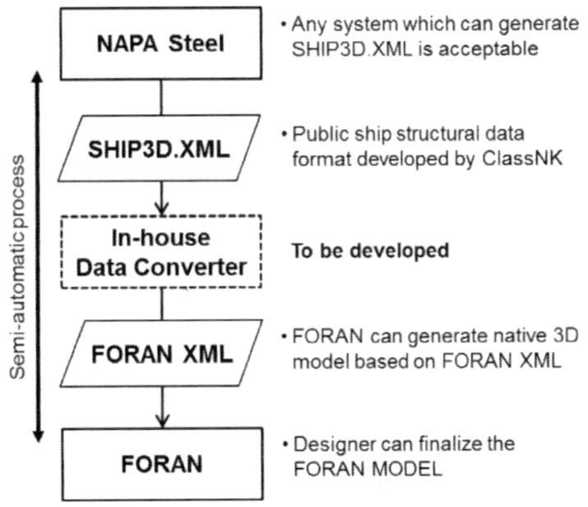

Figure 3: Data collaboration plan

In this study, we are planning to develop data linkage function between NAPA Steel and FORAN, utilizing SHIP3D.XML as an intermediate format, convert it into FORAN XML format, and the last step is to import FORAN XML into FORAN system (see Figure 3).

The main study and development part is data converter, and note that it is not only digital data format conversion but also Surface to Solid geometry conversion and some additional information should be added considering the difference of design stages. Moreover, converted data has to comply with Sanoyas in-house standard, e.g. direction of plate clearance and the naming rule of parts.

3.2 SHIP3D.XML

We selected the SHIP3D.XML format as an intermediate file by comparison of the existing formats. There are many 3D CAD formats like IGES, STEP, PARASOLID, JT, and so on, but these formats express 3D shape with a little property like object name. In the Basic design stage, many 3D design systems handle plate thicknesses, stiffeners, and seams are treated as the property of objects, so these 3D formats are not suitable for data linkage between Basic and Detail/Production design stage.

SHIP3D.XML was developed as "Research and Development for the common Hull CAD data format and interface program" by ClassNK as part of the ClassNK Joint R&D for Industry Program [5]. ClassNK developed SHIP3D.XML format for data linkage between 3D design system and their rule calculation software, PrimeShip-HULL (Rules). SHIP3D.XML designed only for ship structural study, it contains surface geometry, plate thicknesses, stiffeners, and seams as property, and hull form is expressed precisely by NURBS surface, so that good compatibility with NAPA Steel model.

3.3 FORAN XML FORMAT

FORAN has own XML format called "FORAN XML", it contains structural geometry and properties, and FORAN can import the XML file. FORAN XML is not compatible with SHIP3D.XML, but most of the required data for generating FORAN XML is included in SHIP3D.XML, so we can utilize SHIP3D.XML to create FORAN model if we prepare the data converter.

The advantage of this procedure is FORAN can generate the "native model", same model as a manually created model. For example, if we use a common 3D formats like IGES or STEP, FORAN can generate precise shape, however in many cases, we cannot edit the imported object freely, and it is difficult to finalize the model based on it. Each 3D design system has its own modeling procedure, so key success factor of data linkage between different 3D systems is whether we can create a native model or not. Utilizing FORAN XML, we can overcome this difficulties and improvement of working efficiency is expected.

4. BENEFIT FOR SOFTWARE COLLABORATION CONCEPT

4.1 THE PRESENT SITUATION

About the approach of the model collaboration concept between the different 3D design systems, aimed to optimization of the whole design process, the possibility is considered as high from a technical point of view, as mentioned above. On the other hand, for developing the effective and useful function, opinions suggested from the real user (NAPA Steel and FORAN) is necessary. Sanoyas carried out the interview and research. In addition, we calculated the inspection and a quantitative effect for this approach.

Generally, 3D design system needs man-hours to construct the model. As mentioned in Introduction, under the present conditions for applying several 3D design systems on each design stage, the model construction in each design stage requires a huge man-hours.

In addition, the hull basic design information, which is input by 3D design system in Basic design stage, can't be converted to Detail/Production Design stages. This fact causes "Non-collaboration with each design stage", and it is occurred by applying different 3D design systems on each design stage. Each design stage does quite similar work, it is so-called "double work", because of "Non-collaboration". This situation obstructs the reduction of the huge design man-hours and the efficiency of the design process.

4.2 THE INTERVIEW AND RESEARCH

The interview for actual NAPA Steel & FORAN users was carried out, it was paid to inputting information into a model, precision of the information, a man-hour of the model construction. The target items are given to Table 3.

Table 3: Interview for End users

NAPA Steel Definition	FORAN Definition
The accuracy ✓ Plate ✓ Stiffener ✓ Bracket ✓ Opening ✓ Cut out	Man-hours ✓ Definition Section ✓ Plate definition ✓ Stiffener definition ✓ Bracket definition ✓ Opening definition ✓ Collar & Slot definition
Defined Attributes ✓ Scantlings ✓ Thickness Orientation	

Sanoyas analyses what information, which NAPA Steel model includes, will be converted to FORAN model via SHIP3D.XML by the result of interview from NAPA Steel users. And Sanoyas also calculates reducing man-hours, due to converting information from NAPA Steel to FORAN, by the result of interview from FORAN users. Depends on our analysis and calculation depends on the result of hearing, Sanoyas estimates the reduced ratio of the man-hours from whole FORAN modelling process.

4.3 THE RESULTS

As a result of user interview, the definition of plates, stiffeners and openings can be converted from NAPA Steel to FORAN via SHIP3D.XML, but a few attributes of plates and stiffeners, brackets, slots and collar plates will be difficult to convert to FORAN. Table 4 is shown the result of interview.

Table 4: Convertible items and reasons

Necessary information for defining FORAN model		Convertible From NAPA	Reasons
Plate	Plate Boundary	→	
	Plate Thickness	→	
	Thickness Orientation	NG	Not defined
	Assigned Block	NG	Not defined
	Symmetry Parts	→	
Stiffener	Stiffener Scantlings	→	
	Web Orientation	→	
	Face Orientation	NG	Based on Web
	Attached Direction	→	
	Attached Angle	→	
	Start/End Position	→	
	Assigned Block	NG	Not defined
	Symmetry Parts	→	

Bracket	Place	NG	Not definer
	Size / Thickness	NG	Not defined
	Type	NG	Not defined
Opening	Opening Shape	→	
	Size Width/Height	→	
	Centre Coordinate	→	
Slot/Collar	Slot Shape Tightness	NG	Not defined
	Attached direction	NG	Not defined
	Collar Thickness	NG	Not defined

Under these conditions, mentioned to the below, and the result of interview, converting ratio was estimated. It is necessary to calculate the reducing value of FORAN modelling by Interface.

- ✓ Items, undefined in NAPA Steel, is not converted by Interface. It should be defined by FORAN same as the past.
- ✓ Items, FORAN can ' t reproduce NAPA Steel definition, should be defined by FORAN same as the past.

The man-hours of FORAN modelling, with applying Interface, was estimated by considering these results and present man-hours. Figure 4 is shown the estimated man-hours compared with the present. This graph is shown that the man-hours of brackets, slots and collar plate definition are same, but for plates, stiffeners and openings definition are possible to be reduced as about 50% from present man-hours.

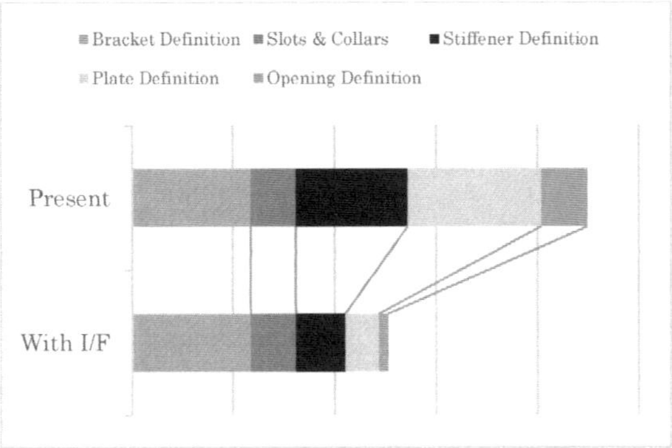

Figure 4: Compare the man-hours – With / Without I/F

This result is a proof that Interface can reduce FORAN modelling man-hours by converting NAPA Steel information. In addition, the basic information, decided the beginning of design stage, can be passed to Detail design stage and Product design stage. It will be possible to improve the model quality and improve the model value.

In the future, to increase input information for NAPA Steel (It is called "front-loading") will be reduced more man-hours of FORAN modelling, because more information will be converted via SHIP3D.XML.

5. ACHIEVEMENT AND DEVELOPMENT

5.1 RECENT ACHIEVEMENT

In Sanoyas, several efficiency and optimization approaches were studied in the past [1] [6]. Now the following approaches, practical 3D model collaboration between 3D Viewer and FORAN with using general 3D intermediate format file, have been already started.

- ✓ The approach for work shop efficiency and saving costs, by the collaboration between VridgeR (Product made in DIGITAL PROCESS LTD.) and FORAN via intermediate format.
- ✓ The approach for calculating welding length and painting area, by the collaboration between Beagle (Product made in NTT DATA ENGINEERING SYSTEMS Corporation) and FORAN via intermediate format.

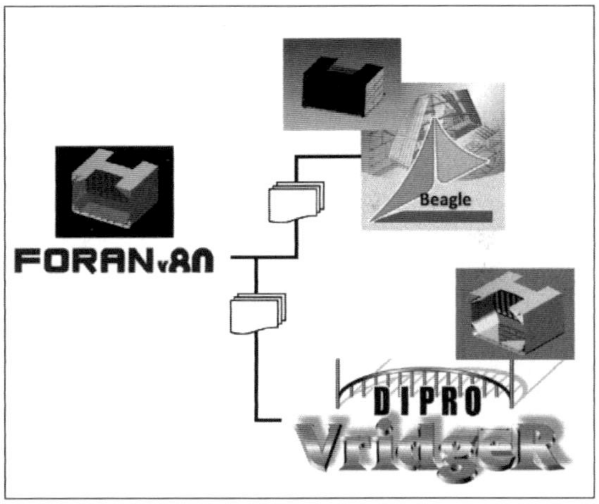

Figure.5: The dataflow between 3D Viewer & FORAN

Figure 5 is shown the workflow for relation between several 3D tools. These approaches achieved the several 3D tool collaboration. And this collaboration has implemented the multi 3D model usage in the Production design stage and Workshop.

These approaches contributes to give the value to 3D model, and also to improve the effectiveness of model. It means "Improvement the value to 3D model".

About these approaches, Sanoyas aims to establish further collaboration strength between the several 3D tools, and aim to efficiency, reducing man-hours and quality improvement.

5.2 FUTURE DEVELOPMENT PLAN

This study leads "Improvement the value to 3D model". It will be provided many aspects in Sanoyas, mentioned in the previous sections. In addition, the practical benefit, depends on the following characteristics, will be given.

- ✓ The achievement of the individual optimization, by optimized 3D design system.
- ✓ The further strength of model collaboration, by including enough hull attribute information format "SHIP3D.XML".
- ✓ Possible to modify, by importing FORAN via FORAN XML format.
- ✓ The achievement of the seamless design collaboration between Detail/Production Design, by developed Interface which is considered the difference of 3D model depends on each design stages requirement.

6. CONCLUSIONS

A concept of "CIMS" was not only aimed to ship design optimization, but it also aimed to contribute the whole shipbuilding procedure. This study is an approach to realize the concept of former "CIMS" by application of several 3D design systems and 3D model collaboration by the SHIP3D.XML, and it seems the similar form with original "CIMS" concept was realized. The establishment of this concept will be possible to achieve "The efficiency and the advancement by the single software" and "To strengthen the international competitiveness by the productivity improvement". Our final target for collaboration is shown in Figure 6.

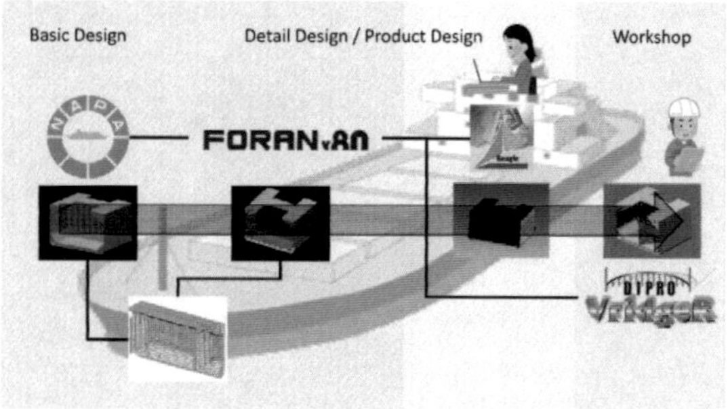

Figure 6: Final target for model collaboration

Now the concept of former "CIMS" was not realize by this study completely, therefore this approach should be go forward to improve, mentioned to the below.

- ✓ The concrete study, for the collaboration of the model information between NAPA Steel & FORAN, should be carried out. In addition, the problem during development the Interface system and the practice for converting 3D model should be cleared and solved.
- ✓ The Individual optimization for each design stage should be achieved completely, by the improvement each workability and the function of applied 3D design system at the each design stage.
- ✓ Further "Seamless" 3D model collaboration will be constructed, with the function extensions for SHIP3D.XML.
- ✓ Due to inputting information at the early design stage and the expansion of Interface system for converting information, the approach for reducing man-hour for construction 3D model should be continued.

The interview for users has proved that the suggested concept, to collaborate between several 3D design systems, can provide a huge benefit for whole ship design and ship building. The benefit will be bigger, result of having the function accorded with the required definition and development by users. Sanoyas will continue the suggested approach in future, in order to contribute the strength of the international competitiveness and further improvement of the design quality.

7. ACKNOWLEDGEMENTS

The authors would like to thank the valuable contributions from DIGITAL PROCESS LTD., NTT DATA ENGINEERING SYSTEMS Corporation and colleagues in Sanoyas Shipbuilding Corporation. SHIP3D.XML was developed with the support of ClassNK as part of the ClassNK Joint R&D for Industry Program.

8. REFERENCES

1. Nakamatsu, T., Mizutani, K., Mizutani, N., 'Total Design Optimization in Initial Design Stage based on Single 3D Ship Model', *Proceedings of ICCAS2011, Trieste.*

2. Technical memorandums of Shipbuilding Research Association of Japan SR210, 1-263, 1989-03

3. Aya, H., Study on product modelling and system architecture for integrated CAD/CAM of shipbuilding, Journal of the Society of Naval Architecture of Japan, No.153, 1985.

4. Amemiya, T., Process Planning System for CIMS in Shipbuilding, Journal of the Society of Naval Architecture of Japan, No.168, 1990.

5. ClassNK, 'Research and development on common format and interface program of CAD data for hull structures'(in Japanese), http://report.classnkrx.com/, 2014.

6. Kasai S, Kunisada T, Ohtsuki Y, Ikeda R, Tamura Y, Sasaki Y 'FE Analysis and Lead Time Reduction by Combined Use of CAD and CAE', *Proceedings of ICCAS2013, Busan.*

9. AUTHORS BIOGRAPHY

Taisuke Kunisada holds the current position of Manager at Sanoyas Shipbuilding Corporation. He is responsible for Introduction and customization 3D Design CAD. His previous experience includes hull basic design and using FORAN & NAPA system.

Takashi Murakami holds the current position of Business Developer at NAPA Japan Ltd. He is responsible for business development of NAPA Shipping Solutions. He holds M.Sc. in Environmental Science & Technology from the Okayama University and has five years of structural design experience in a Japanese Shipyard. Professional Engineer, Jp (Marine & Ocean).

Shigeru Kasai holds the current position of Project Manager at SENER Ingeniería y Sistemas. His previous experience includes ship design work in shipbuilder.

TOWARDS A MARINE DIGITAL TWIN, PREDICTIVE ENGINEERING ANALYTICS FOR SHIP DESIGN

S W Ferguson, Siemens PLM Software, UK

SUMMARY

For over 130 years the shipbuilding industry has relied on model scale testing in order to predict the performance of new ship designs. Although these methods have served the industry well, they are no longer capable of delivering the level of performance improvement required to meet stringent IMO guidelines on emissions and fuel efficiency. In this paper we propose combining a number of computer aided engineering simulation tools into a single model in order to form a fully functioning "Performance Digital Twin" of a proposed vessel that can be used early in the design process to predict the actual performance of vessel under a range of operating conditions.

1. INTRODUCTION

Since the first commercial ship basin was commissioned in 1883, towing tanks have provided naval architects with a reliable method of predicting the performance of a ship at sea. Towing tanks are used for both resistance and propulsion tests, with towed and self-propelled ship models used to determine how much power the engine will have to provide to achieve the speed specified in the contract between shipyard and ship owner.

The performance of a vessel depends on the hydrodynamic interaction between the hull, its propulsion system and its rudder, which all combine to interact with the environmental conditions. The flow past the hull influences the flow past the rudder, which affects the quality of flow "seen" by the propeller and its ability to generate thrust. This in turn determines the loading on the engine

This is a very complex engineering problem that involves the tightly coupled dynamic interaction between many disparate physical phenomena, spread across a wide range of engineering disciplines. As is usually the case with challenges of this nature, engineers have historically tackled the problem by splitting it into a number of separate "systems", which are only loosely coupled together.

The cost and effort of producing a model and testing it, means that towing tanks are usually deployed relatively late in the design cycle, verifying and fine tuning an established design, rather than providing engineering data that could be used to drive the design into different, better, directions. In addition, any novel solution tested at model scale has increased uncertainty of actual performance at ship scale due to deficiencies of the scaling process.

While it is certainly possible to obtain useful design information from experiments (or simulations) that investigate these systems individually, in order to predict the at-sea performance of a vessel with a high degree of accuracy, it is necessary to include all systems in a single model. This is particularly important with the current demand for energy efficient "green ships" which is driven by a combination of legislation (via the IMO and measured by the EEDI) and economic necessity. Energy savings of a few percent can significantly influence the operational viability of a vessel.

In this paper we propose combining a number of computer aided engineering simulation tools into a single model, to form a fully functioning "Performance Digital Twin" of a proposed vessel, which can be used early in the design process to predict the actual performance of a vessel under a range of operating conditions.

2. THE DIGITAL TWIN

Digital Twins are digital replicas of physical assets; in this instance we are using the term to refer to a digital replica of a marine vessel. In many ways the Digital Twin concept is a realisation of the "Product Lifecycle Management" concept that purports to manage the entire lifecycle of a product from inception, through engineering design and manufacture, to service and disposal. Although there are different types of Digital Twins, one of the most prominent is the "Performance Digital Twin" in which a collection of high-fidelity simulation tools are combined to provide a complete picture of product performance from early in the design process. In this section we discuss how state-of-the-art simulation tools can be combined to construct a Performance Digital Twin of a vessel design that can be used to overcome many of the shortcomings of the traditional design process.

In the discussion of this concept, we will reference simulation tools that form part of the Siemens Simcenter Portfolio of leading computer aided engineering tools.

2.1 HYDRODYNAMIC PERFORMANCE

Computational Fluid Dynamics (or CFD) has long been offered as a credible alternative to experimental testing, providing a "numerical" towing tank that could, at least in principle, be deployed much earlier in the design process, providing naval architects with a stream of engineering data that could be used to influence and improve the design. CFD simulation carries the significant advantage of result accuracy independent of the scale at which the simulations are performed. In practice this means that CFD calculations can predict the performance of the actual vessel at full scale, without having to employ the "scaling corrections" that are used to extrapolate full-scale performance from model scale tests in a towing tank.

CFD simulations solve the fundamental equations of fluid dynamics, through a process known as "discretization" in which a volume occupied by the fluids (both water and air) surrounding the vessel is subdivided into a mesh of much smaller control volumes (known as computational cells). In practical terms, a "fine mesh" that is constructed from a large number of small computational cells provides a more accurate prediction than a "coarse mesh" of larger cells. However, a greater number of cells results in a larger computational cost, requiring more computer resources and longer simulation times compared with a coarser mesh. In the past constructing a computational mesh was at difficult task that involved a degree of manual intervention from the engineer performing the simulation, a balancing act between capturing the geometry of the vessel in adequate detail and resolving important flow details, while ensuring that the final cell count is practical enough to allow the simulations to actually be performed with the computing resources available.

Thankfully these concerns are largely in the past as modern CFD tools such as Simcenter STAR-CCM+ take the guesswork out of meshing by providing tools that both automate the mesh generation process and apply "best practice" through templates that ensure that the resolution is adequate in critical regions.

Figure 1: Automatic templated mesh generation using Simcenter STAR-CCM+

Unlike the simulation of an aircraft or road-vehicle, which in ideal circumstances moves forward in a single direction, the forward progress of a ship is heavily influenced by the surrounding sea-state. Even in still water, establishing the dynamic position of the ship in relation to the sea surface ("sink and trim") is critical to providing accurate resistance predictions. In rough seas, the full motion of the vessel in six-degrees-of-freedom must be correctly accounted for, as the vessel pitches, rolls and heaves in response to oncoming waves.

Simcenter STAR-CCM+ accounts for 6DOF vessel motion in an automatic manner. The "Dynamic Fluid-Body Interaction" model integrates the forces acting on the vessel at every time step, and adjusts its position (in all-six-degrees-of-freedom) accordingly.

"Adjusts its position" means moving the computational mesh, which historically has been a difficult proposition, and various methods have been used to account for this motion. For relatively small movements, the vertices of cells in the mesh can be adjusted on a step-by-step basis. However, for large movements, this becomes impractical as individual cells become highly distorted, leading to inaccuracies in, or failure of, the simulation. Simcenter STAR-CCM+, uniquely among commercial CFD codes, solves this problem using "overset" or "chimera" meshes, in which the mesh around the vessel is independent of the mesh used to represent the sea. This allows the simulated ship to move as much as necessary. Furthermore, it can be used to model the interaction between multiple vessels or objects, such as one ship moving independently in the wake of another, or the collision of two vessels. Also, with overset mesh, the rotation of the propeller and rudder motion, in addition to propeller pitching, can all be modeled in relation to the ship motion, leading to robust, accurate self-propulsion and maneuvering analysis. Additionally, Simcenter STAR-CCM+ also includes an extensively validated cavitation model that can be used to predict and manage the phase changes caused by the propeller.

Figure 2: Six-degree-of-freedom simulation of an FPSO in heavy seas performed with Simcenter STAR-CCM+

This leads us to our final consideration, in order to be useful a CFD simulation tool needs to be able to apply a wide range of sea states to the design vessel, from the relatively calm conditions in which most sea trials are conducted, to rough sea states that will likely provided the stiffest test of the robustness of a given vessel design. Simcenter STAR-CCM+ also provides a range of built-in higher-order wave models that can be used to test the vessel under realistic sea states.

2.2 POWERTRAIN PERFORMANCE

An important part of the ship design and certification process is selecting an appropriate propulsion system for the vessel. The minimum requirement is that vessels achieve a certain speed during sea trials, which are normally performed in calm seas and favourable weather conditions. Traditionally this selection process has been based on information gathered from model scale measurements in a towing facility, including so called "self-propulsion" tests in which the model is equipped with an electric motor that powers its propeller.

Although this process is useful in meeting these minimum requirements, it is much less effective in choosing a propulsion system that is optimized across all of the potential operating conditions that it will face during its working life. For example, the loading on the propeller and engine is greatly increased in heavy sea states, or during maneuvering when the flow is not aligned with the propeller.

Aside from meeting the regulatory requirements imposed by the IMO, and the contractual obligations between the shipyard and the shipowner, significant savings can be made in operating costs (both in improved fuel efficiency and reduced engine maintenance) by designing the propulsion system that properly represents all of the operating modes of the vessel. In practical terms all that prevents this is limitations in the traditional experimental toolset.

Finnish engine supplier Wärtsilä, who provide propulsion systems to one third of the vessels currently in operation, recently proposed [1] an innovative solution to this problem. Extending the approach described above, Wärtsilä coupled a 1D simulation of the ship powertrain (constructed using Simcenter Amesim) with a full scale CFD simulation of the vessel (constructed using Simcenter STAR-CCM+). In comparison with the traditional methods, this is akin to replacing the electric motor in a model scale self-propulsion test, with a fully functioning engine (although obviously at full scale).

The methodology in principle allows Wärtsilä to simulate the full performance of the actual operating conditions, accounting for all aspects of hydrodynamic (and aerodynamic) performance using the Simcenter STAR-CCM+ and all aspects of the powertrain performance and other onboard systems using Simcenter Amesim. Since, in performance terms, this is an almost complete replica of an actual vessel (albeit one that does not actually exist yet) it can be considered as the first real world example of a Performance Digital Twin of a ship. Since all engine design extensively relies upon semi-empirical 1D simulation (using tools such as Simcenter Amesim), many of these models are readily available for off the shelf powertrain solutions.

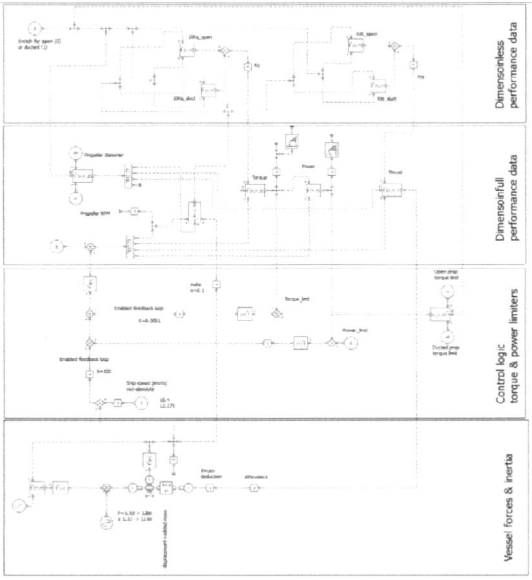

Figure 3: Wärtsilä 1D model of ship powertrain constructed in Simcenter Amesim

Although this Digital Twin approach has only been used as proof of concept thus far, it has many potential uses above and beyond optimising the powertrain performance of a vessel. Using process automation tools such as Simcenter HEEDS, the Digital Twin could be driven across the full operating range of the vessel. This opens the door to both "automated design exploration," where the proposed vessel is subjected to a wide range of potential operating scenarios, and "optimization," where the design of the vessel is automatically adjusted to account for deficiencies in the performance identified in previous simulations. Information gathered from these simulation model can be used as a basis for other numerical simulations including acoustic and structural simulations provided by tools such as Simcenter 3D.

Furthermore, as the ship design matures so will the Performance Digital Twin. As the ship enters service, actual operational data can be gathered and compared to the predictions made during the design stage. Not only will this provide validation for the simulation tools, but it will also provide a more complete understanding of real life usage scenarios that can be used to inform and improve the simulation process for future generations of the ship design.

3. VALIDATION

Although it is too soon to publish any meaningful data on the Performance Digital Twin described above, the "full scale CFD for ship design" approach is very well validated, and is beginning to be applied across the shipbuilding industry. The American Bureau of Shipping (ABS) recently published [2] the results of a study that demonstrated the superiority of full-scale simulations. They used Simcenter STAR-CCM+ to run simulations on multiple hull variants, based on the KCS 3600 TEU container vessel geometry. ABS ran the same cases at both model scale and full scale, and then compared the results, along with extrapolated model scale to full-scale results (using ITTC 1978). The results gave two conclusions. For situations where viscosity is not dominant, such as bulbous bow investigations, all three methods ranked the different hull variants in the same order of efficiency. In cases where hull changes mostly affect the viscous part of the overall resistance, such as aft-body optimization, each of the three methods gave different results (though the relative differences were small in most cases).

Lloyd's Register's recently conducted [3] full-scale hydrodynamics workshop showed a remarkable alignment of CFD results from multiple participants with full-scale sea trial data on a 16.9k DWT general cargo vessel. CFD results were provided blind by a wide range of participants, with almost all participants predicting the actual performance of the vessel within a few percentage points. The vast majority of participants used Simcenter STAR-CCM+ for their simulation. The most accurate predictions came from IBMV [4], who predicted resistance and propulsion to within the accuracy of the measurements, across all three trial conditions.

4. CONCLUSIONS

Having addressed the main challenges to replicating the performance tests, CFD is now able to provide a useful tool to augment, if not replace, towing tank testing. Furthermore, CFD simulations also have the advantage that they can easily be deployed at full-scale if desired, reducing the uncertainty inherent in model scaling. With the recent addition of 1D simulation tools to model the performance of the powertrain, it is evident that numerical simulation is moving beyond simply being an alternative to towing tanks, by providing high quality design information that could never be derived from an experiment.

Although it is unlikely that any large vessel will be designed in the foreseeable future without some aid from towing tanks, ship builders and naval architects are now routinely using numerical simulation as part of the design process across the world. Used effectively, simulation can be used to reduce the amount and cost of physical towing tank tests by providing a more refined and optimized design that requires fewer modifications in order to meet contractual obligations.

What of the future? Unlike towing tanks, once you have developed a robust process for simulating the performance of a vessel, it is relatively easy to automate it. Widespread adoption of this approach will not only lead to more innovative and efficient ship designs (which can be developed at lower cost), but also more robust vessels that have been numerically tested against a much wider range of real-world operating conditions than could ever be considered using a towing tank alone.

5. REFERENCES

1. Bulten and Stoltenkamp, 'Improved Sustainable Speed Due to Thrusters with Ducted Propellers', *Proceedings of the ASME 2017 36th International Conference on Ocean, Offshore and Artic Engineering*, June 2017.

2. Cairoli, 'Using cutting-edge CFD for production type ship design evaluation and optimization', *Webinar Presentation,* April 2017.

3. Ponkratov, 'Proceedings', *2016 Workshop on Ship Scale Hydrodynamic Computer Simulation,* 2016.

4. Guiard, 'Full-Scale Self-Propulsion Calculations Including Roughness Effects and Validation with Sea Trial Data', *STAR Global Conference 2017,* April 2017.

6. AUTHOR BIOGRAPHY

Stephen Ferguson holds the current position of Marketing Director at Siemens Industry Software. He is a 25-year veteran of the CFD industry, and has held engineering positions at WS Atkins, BMW, and CD-adapco.

IMPROVING LAUNCH AND RECOVERY OPERATIONS THROUGH QUIESCENT PERIOD PREDICTION FROM RADAR

M. Al-Ani, **J. Christmas** and **M.R. Belmont**, University of Exeter, UK.
J M Duncan and **J Duncan**, Ministry of Defence Equipment and Support HQ, UK.
B Ferrier, Hoffman Engineering Corp., USA.

SUMMARY

Of major international maritime interest is how to improve the safety of a wide range of launch and recovery operations, or to increase the sea state under which they can be safely undertaken. The key wave-height-critical sub-tasks that limit the sea state under which they can be carried out is short, typically less than 1 minute. Prediction of quiescent periods will enable these key sub-tasks to be scheduled to be safely undertaken in sea states that would normally prohibit them. This requires the prediction of the actual shape of the sea surface and its temporal evolution, not the traditional discipline of wave forecasting through predicting the statistical properties of the sea. In this paper, we describe a method for short-term deterministic sea wave prediction based on radar observations of sea surface profiles, and present results obtained from sea trials undertaken off the western coast of Scotland.

1. INTRODUCTION

Of major international maritime interest is how to improve the safety of a wide range of launch and recovery operations, or to increase the sea state under which they can be safely undertaken [1-4]. Typical such operations include the launch and recovery of fixed and rotary wing aircraft, or small boats, from ships. These operations share two common elements. Firstly, while the overall execution of such tasks may take a significant amount of time, the key wave-height-critical sub-tasks that actually limit the sea state under which they can be carried out is short, typically less than one minute. Secondly, a fundamental property of most sea conditions that are of relevance to launch and recovery is that sets of large waves alternate with sets of smaller waves, with the smaller waves being of lower height than the standard sea statistics for the prevailing conditions (see Figure 1). Consequently the intervals during which these lower amplitude waves occur are referred to as quiescent periods. If these quiescent periods can be predicted, then the safety-critical sub-tasks can be scheduled such that the overall maritime operation could be safely undertaken under conditions whose overall sea state statistics would normally prohibit its execution. Note that this involves the prediction of the actual shape of the sea surface and its temporal evolution, not the traditional discipline of wave forecasting through predicting the statistical properties of the sea.

2. QUIESCENT PERIOD PREDICTION (QPP)

In order to predict, in real time, quiescent periods, a QPP system must make observations of the sea's surface profile around the site of the maritime operation of interest. From these observations a deterministic model is constructed, from which predictions of the surface motion for a short period into the future at a specified position can be made. These predictions, combined with a ship's frequency response function, may then be used to predict the vessel motion.

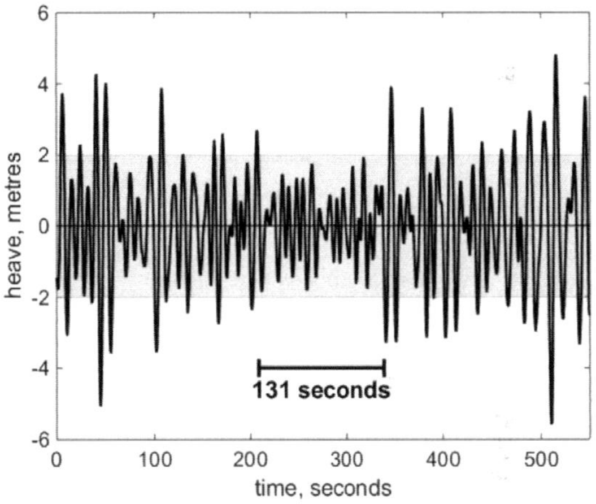

Figure 1: In real sea, with excursions of up to 6m, a quiescent period of more than 2 minutes is marked during which time the excursions do not exceed 2m.

2.1 SEA SURFACE OBSERVATIONS

The traditional method of observing sea wave profile is through moored buoys. These only provide observations at particular points in the sea, and are clearly not feasible in the context of maritime operations where the vessel is under way.

In [5] we explored the feasibility of using a wave profiling radar, WaMoS II [6], to make the observations. This provides a number of advantages: (i) the system receives and processes signals from an X-band radar antenna (the standard navigation radar for ships), (ii) the observations cover a large two-dimensional patch of sea, and (iii) the patch surrounds the site of the proposed operation. Figure 2 shows an example surface profile measured by this system during the sea trial; the shading indicates the surface elevation, in metres, as given by the colourbar on the right.

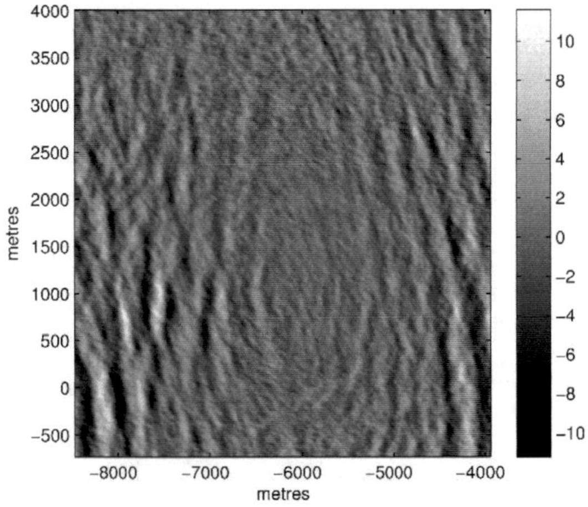

Figure 2: Example wave profile measurement made by WaMoS II. The shading represents the sea surface elevation, according to the metre scale shown on the right.

Each observation represents the surface recorded by a single sweep of the radar antenna. If we consider each such "image" to have been recorded instantaneously in time, then we may use a two-dimensional Fourier transform against each observation to model the waves and predict their future motion. However, the data in each of these observations were not recorded instantaneously in time, but over the period of a single sweep of the radar antenna, i.e. 2.52 sec. In the next subsection, we describe the developed prediction technique for these radar measurements.

2.2 SEA SURFACE PREDICTION TECHNIQUE

The fundamental linear oceanographic wave model of the sea surface, $h(x,y,t)$, at the coordinates x, y, and time t is given by

$$h(x,y,t) = \sum_{r=1}^{R}\sum_{n=1}^{N} c_{n,r}\cos(k_n x\cos(\theta_r) + k_n y\sin(\theta_r) - \omega_n t + \varphi_{n,r}) \quad (1)$$

where $c_{n,r}$ is the directional magnitude spectral coefficient at the wave number k_n and its angular frequency ω_n, $\varphi_{n,r}$ is its initial phase, N is the number of spectral components in the propagation direction θ_r of the wave system, and R is the number of wave systems. The directions of the wave systems $\{\theta_r\}_{r=1}^{R}$are set to span $[0,2\pi)$. The wave numbers $\{k_n\}_{n=1}^{N}$ are set to uniformly range over the longest and shortest waves to represent. As we mentioned above, the wave profiling radar provides consecutive images of the sea surface. The images are captured using a rotating radar with period of a few seconds. Accordingly, the time stamp associated with the pixels in the images varies, and so the classical Fourier transform cannot be employed. Instead a least-squares fitting must be applied to solve this Vandermonde system. However, this solution is infeasible considering the size of the data and poor condition of the matrix that is required to be inverted. Hence, we adopt the sequential least-squares fitting based on the well-known work of Lomb and Scargle [7, 8] in literature regarding uneven data spectral analysis. The assumption is that the correlation between the sinusoidal components of different frequencies is negligible. This assumption is sensible given the theory of Gaussian processes of sea waves generation [9] that $c_{n,r}$ are independent random variables of Rayleigh distribution and $\varphi_{n,r}$ are independent random variables with uniform distribution on $[0,2\pi)$. We start by letting

$$a_{n,r} \triangleq c_{n,r}\cos(\varphi_{n,r}), \quad (2)$$

$$b_{n,r} \triangleq -c_{n,r}\sin(\varphi_{n,r}). \quad (3)$$

Accordingly, (1) can be written as,

$$h(x,y,t) = \sum_{r=1}^{R}\sum_{n=1}^{N}[a_{n,r}\alpha_{n,r}(x,y,t) + b_{n,r}\beta_{n,r}(x,y,t)]. \quad (4)$$

where

$$\alpha_{n,r}(x,y,t) \triangleq \cos(k_n x\cos(\theta_r) + k_n y\sin(\theta_r) - \omega_n t), \quad (5)$$

$$\beta_{n,r}(x,y,t) \triangleq \sin(k_n x\cos(\theta_r) + k_n y\sin(\theta_r) - \omega_n t) \quad (6)$$

For compact representation, we denote (x,y,t) by (s) and $\{h_d(s_m)\}_{m=1}^{M}$ are the radar data of the sea surface. The coefficients $a_{n,r}$ and $b_{n,r}$ can be determined by

$$\begin{bmatrix}\hat{a}_{n,r}\\ \hat{b}_{n,r}\end{bmatrix} = \mathbf{H}^{-1}\mathbf{p}, \quad (7)$$

where

$$\mathbf{H} = \sum_{m=1}^{M}\begin{bmatrix}\alpha_{n,r}(s_m)\\ \beta_{n,r}(s_m)\end{bmatrix}[\alpha_{n,r}(s_m) \quad \beta_{n,r}(s_m)], \quad (8)$$

$$\mathbf{p} = \sum_{m=1}^{M}\begin{bmatrix}\alpha_{n,r}(s_m)\\ \beta_{n,r}(s_m)\end{bmatrix} h_d(s_m). \quad (9)$$

The spectral coefficients obtained from a single scan using the fitting of (7) or standard Fourier transform are Hermitian symmetrical around the origin. That is, there is 180° ambiguity in the interpretation of the spectral directions of each coefficient. A direct approach is to run the fitting for several scans to introduce a new coordinate in the spectral coefficients across ω_n which identifies the true directions. This is unsuitable for our purpose and we alternatively adopt a simpler approach. We run the fitting for two consecutive scans only. Then, at a wave propagation angle θ_r the sum of its spectral coefficients $\sum_{n=1}^{N}(\hat{a}_{n,r}^2 + \hat{b}_{n,r}^2)^{0.5}$ is compared with the sum of the coefficients at $\theta_r - \pi$, and the direction with the higher value would represent the right direction and marked with 1 whereas the other will be marked with 0. Accordingly, a "map" of 1's and 0's can be generated that identifies the directions of the wave systems. After the map of the correct spectral directions is obtained, the necessary spectral coefficients can be calculated and correctly masked from a *single* image. Updating the map could be a secondary task that runs in the background every a few minutes.

3. EXPERIMENTAL SETUP

The data was recorded during the Golden Arrow sea trial, which took place in November 2014 off the west coast of Scotland. During the period of this trail, the sea consisted of large swell waves coming from the west and south-west. A WaMoS II wave profiling system was integrated into the ship's own X-band navigation radar. Between 13:00 and 14:00, WaMoS II provided sea surface images of the area around the ship every approximately 2.52 sec.

One radar image at time t_s was used to build the prediction model that predicts the sea surface motion for the region near the ship for a period of $(t_s, t_s + T_p)$ sec in the future, where T_p is the maximum prediction time. Figure 3 shows an example prediction at the ship location made from a single radar scan. In assessing the predictive quality of the prediction model, we have used the linear correlation between the prediction and the truth at every spatial location. This would give us a value in the range of -1 to 1 that measures the correctness of the shape of the wave profile, where a value of 1 means that the prediction is exactly the same shape as the truth, 0 means that there is no correlation between the shapes of the two, and -1 means that the predictions are exactly out of phase with the truth. The "future" radar measurements available to us are used as the truth in this comparison. For every 10 sec in the period of 13:01–13:58, the prediction model is built and generates its predicted sea surface for T_p sec ahead. The correlation is evaluated for every spatial location generating a frame of [-1,1] ranged values every 10 sec. First to show all the results in one plot, we average the correlation values for every spatial location across time. The results are shown in Figure 4 in the region that expands to 1200 m from the ship in the direction of the significant waves and stretches for 2400 m across. The maximum prediction time T_p is set to 60 sec and 120 sec.

In Figure 5, we show the correlation across time for selected locations: the centre of the ship, 60 m north of the ship, and 90 m east of the ship. We can see that the correlations are consistently high throughout the hour for all the locations, especially for $T_p = 60$ sec. Figure 6 illustrates the normalised error against the prediction time instants. The error is normalised over the local dynamic range of the sea elevation at the prediction location over the $[0,120]$ sec prediction period. We can see that the deviation of the error monotonically increases with prediction time. Thus, the maximum prediction time could be decided based on the sought prediction accuracy and reliability.

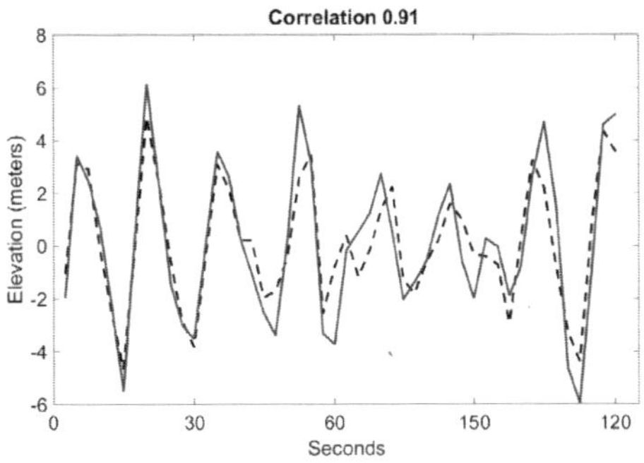

Figure 3: An example prediction (solid) for 2 min of sea surface motion made from a single radar observation, plotted against the true motion (dashed) at the ship.

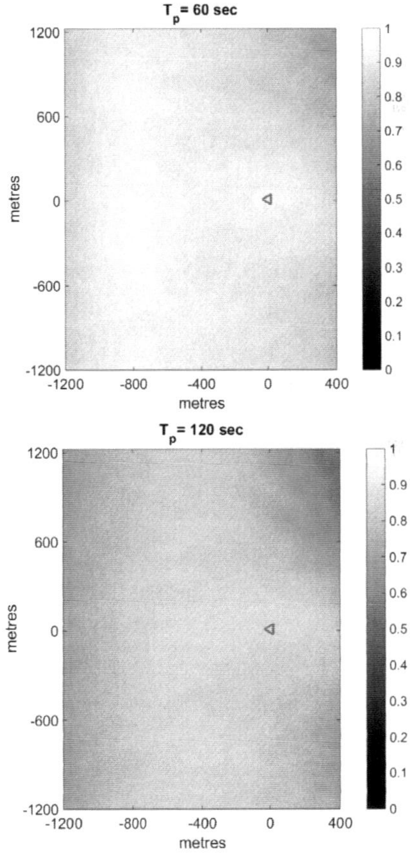

Figure 4: The mean correlations coefficients of the prediction area over the 13:01–13:58 period for a maximum prediction time T_p of 60 sec and 120 sec. The ship's location is marked by a triangle.

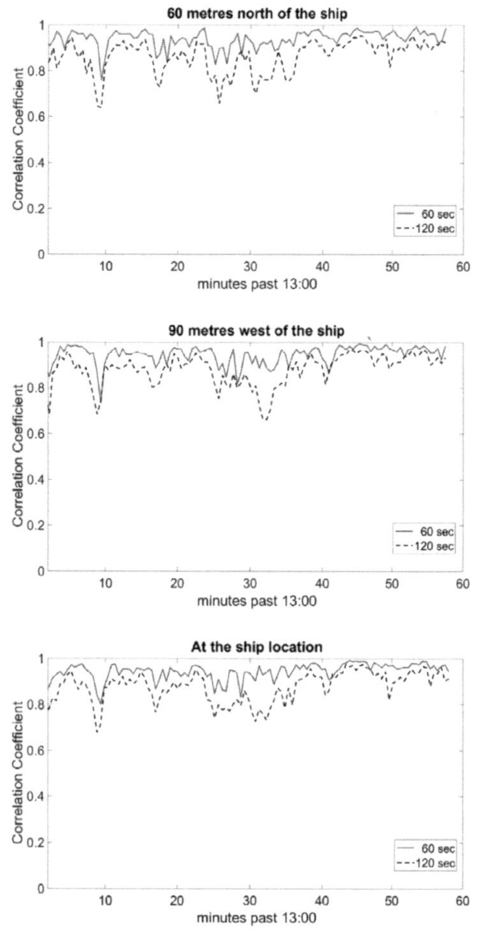

Figure 5: The correlation coefficients for T_p of 60 sec and 120 sec over the 13:01–13:58 period at the labelled locations.

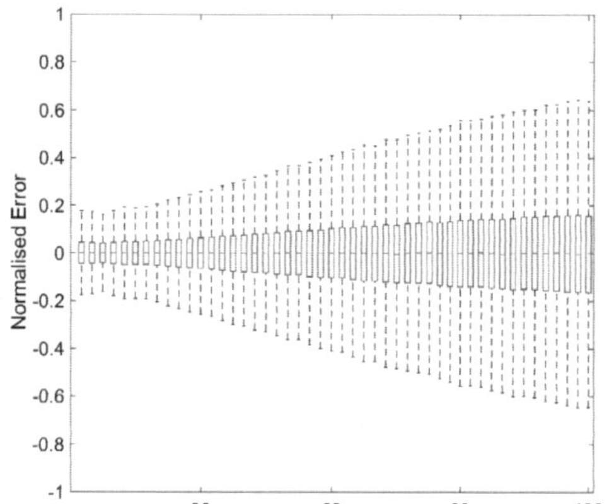

Figure 6: The normalised prediction error versus the prediction time instants. Each box marks the upper and lower quartiles, with the centre being the median; the black dashed lines indicate the full range of values

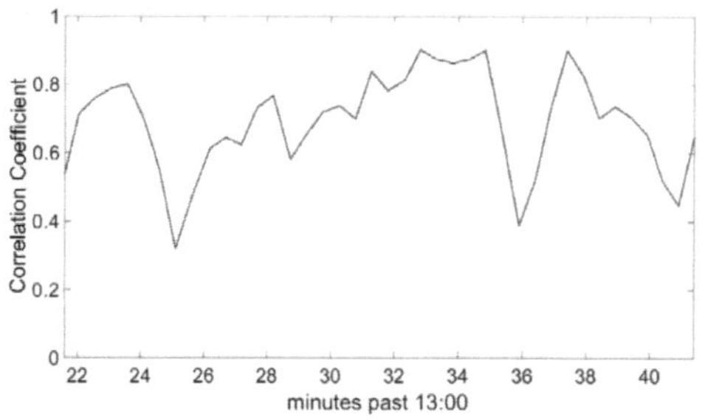

Figure 7: The correlation between the WaMoS II based prediction and the ship motion sensor data.

A high-quality motion sensor system was installed approximately 14 m forward of and 2 m above the ship's centre of mass. The motion sensor recorded the ship's heave at 5 Hz for the time span 13:21–13:44. The ship motion model is unfortunately not available to us, and hence a ship heave predictor cannot be built at the moment. However, under such sea state, the ship's heave will more or less follow the shape of the wave profile. That is, the typical period of these waves was such that the ship acted as a cork, so that we may ignore the impulse response of the vessel and treat sea surface predictions as ship motion predictions. Thus, we plot the correlation between the motion sensor data and the WaMoS II predicted sea profile directly in Figure 7. We can see a relatively high correlation overall. The low correlation values can be linked to the time periods when the ship was manoeuvring sharply.

4. CONCLUSIONS

A new technique of QQP that explicitly incorporates the time variation across the radar sweep has been developed. It was shown to make a good-quality prediction of the motion of the sea's surface at particular locations up to 2 minutes ahead, and that the prediction quality can be maintained over the period of an hour. Clearly, because of the nature of sea waves, this quality deteriorates as the prediction window is increased; we have shown that this relationship is approximately linear.

5. ACKNOWLEDGEMENTS

The authors wish to express their thanks to the UK Ministry of Defence for their support of the work presented here.

6. REFERENCES

1. M. Pourzanjani, M.R. Belmont, H. Zienkiewicz and M. Morris, 'Applications of a sea-surface estimator in predictive ship control,' *Proc. of the 2nd. Int. Conf. on Manoeuvring and Control of Marine Craft*, Southampton, UK, pp 581–591, 1992.

2. B. Ferrier, A. Baitis and A. Manning, 'Evolution of the landing period designator (LPD) for shipboard air operations', *Naval Engineers Journal*, vol 112, no 4, pp 297–316, 2000.

3. P. Crossland, J.M. Duncan, and B. Ferrier, 'The feasibility of developing a quiescent period prediction system in a simulation environment', *Proc. of the Int. Conf. on Computer Applications in Shipbuilding (ICCAS)*, Shanghai, China, vol 2, 2009.

4. B. Ferrier, J.M. Duncan, M. Belmont, A. Curnow and P. Crossland, 'Using simulation to justify and develop quiescent period prediction systems for the Royal Navy', *Proc. of the Int. Conf. on Computer Applications in Shipbuilding (ICCAS)*, Busan, South Korea, vol 2, pp 99–107, 2013.

5. M.R. Belmont, J. Christmas, J. Dannenberg, T. Hilmer, J.M. Duncan, J. Duncan and B. Ferrier, 'An examination of the feasibility of linear deterministic sea wave prediction in multi-directional seas using wave profiling radar: theory, simulation and sea trials', *Journal of Atmospheric and Oceanic Technology*, vol 31, no 7, pp 1601–1614, 2014.

6. OceanWaveS GmbH, 'WaMoS II - Wave and Current Monitoring System', accessed 27 July 2017. [Available online at http://www.rutter.ca/files/download/eb860547e4f9c2b.]

7. N. R. Lomb, 'Least-squares frequency analysis of unequally spaced data', *Astrophysics and Space Science*, vol 39, no 2, pp 447–462, 1976.

8. J. D. Scargle, 'Studies in astronomical time series analysis. II - Statistical aspects of spectral analysis of unevenly spaced data', *Astrophysical Journal,* pp 263 – 835, 1982.

9. L. H. Koopmans, 'The Spectral Analysis of Time Series', *Academic Press*, 1974

7. AUTHORS' BIOGRAPHIES

M Al-Ani holds the current position of Research Fellow in Mathematics at the University of Exeter. He is responsible for developing and testing the theoretical tools in sea waves prediction. His previous experience includes image processing, spectral analysis and nonuniform data processing.

J Christmas holds the current position of Lecturer in Computer Science at the University of Exeter where she works on maritime systems research and statistical machine learning. She has been working on the QPP system for a number of years.

M R Belmont holds the current position of Professor of Marine Dynamics at the University of Exeter where he is responsible for maritime systems research. The main application of his present research is in enhancing the wave limits under which marine operations can be safely conducted. Underpinning this work are the new subjects of Deterministic Sea Wave Prediction and the intimately linked field of Quiescent Period Prediction, both of which he pioneered.

JM Duncan is Group Leader for Simulation Based Acquisition and Technical Computing for Maritime Combat Systems, UK Defence Equipment & Support (DE&S). The Group's focus is on whole ship and submarine safety and technical services including new modelling techniques for complex dynamic interfaces such as for manned and unmanned air and sea vehicles. Duncan is the NATO Principal Representative for the Naval Armaments Group on Virtual Ships. He was previously chairman of NATO Subgroup 61 and the Specialist Team on Simulation Based Design and Virtual Prototyping for ship acquisition. He led the development of Allied Naval Engineering Publications '61' for ship virtual prototyping and '84' for Virtual Ship simulation standards.

J Duncan has served for 6 years as the manager of sea trials data collection and coordination on behalf of MOD. He has supervised 10 sea trials data collection activities on behalf of the NATO Submarine Rescue System team including substantial planning, post-trials analysis and reporting. He has been involved on various simulation projects for MOD supporting Type 45, Merlin, QEC, Ship Air Interface Framework and Quiescent Period Prediction. He has worked on the development of NATO Virtual Ships documents STANAG 4684 and ANEP 84.

B Ferrier holds the current position of Systems Engineering and Technology Associate (SETA) in the offices of NAVAIR PMA 266. He is responsible for assisting NAVAIR in integrating Fire Scout Endurance upgrade technologies, such as the ONR Landing Period Designator, into the Fire Scout Launch and Recovery System. He is also the Engineering Head of the Hoffman Engineering's Dynamic Interface Lab in Arlington, VA and holds an honourary chair at the University of Exeter.

ALL WEATHER SHIP OPERATIONAL PREDICTION USING SIMULATION – TECHNOLOGY DEVELOPMENTS AND RESULTS FROM A DEDICATED ROYAL NAVY AND RELATED SEA TRIALS

Prof. B Ferrier, Hoffman Engineering, Dynamic Interface Laboratory, USA
Prof. J Duncan, Defence Equipment & Support, MOD, UK
Prof. M. R. Belmont, Exeter University, UK
Dr. J. T. Christmas, Exeter University, UK
Mr. J. Duncan, Defence Equipment & Support, MOD, UK

SUMMARY

The operational capability of a ship in all its major roles is limited owing to the effects of the maritime environment. Excessive motions and motion related phenomena in bad weather increases the length of time to undertake a particular task or indeed prevent the task from being undertaken at all. The ever increasing pressure to extract extra value from existing naval expenditure makes it clear that technologies capable of extending the operational envelope of air/sea operations is potentially very valuable. Two primary environmental factors limiting such operations are wave induced deck motions and airflow over the deck. A new Integrated Marine Environment approach which combines the recent advances in using Doppler LIDAR for airflow with the new ability to predict the quiescent periods that are known that to exist in high sea states offers the promise of just such operational enhancement. Typical applications are: aircraft launch and recovery, cargo transfer, off-loading from small vessels and small boat launch and recovery.

Under the Framework of the NATO Virtual Ships MOU, there is a desire within a Project Arrangement (PA) to develop, in a simulation environment, an onboard Quiescent Period Prediction and Airwake Profile system. The aim of the project is to build a prototype system by developing a simulation to demonstrate the capability. The simulation consists of modular "plug and play" components contributed by the project participants assembled in a High Level Architecture (HLA) with federates reflecting the ship environment, the long and short term prediction methodologies, mapping shedding air vortices and the ship's wave driven motions. This allows the comparison of slight modifications to a ship's design and methods of operation. The result of the simulation are compared with actual at sea measurements and confirm the tendency of certain parameters to be well represented by simulation. Examples case studies using air vehicle launch and recovery activities are presented forming primary focus of this article.

1. INTRODUCTION

The Simulation Based Acquisition (SBA) Group is the technical authority across the UK Ministry of Defence's Defence Equipment & Support Organisation, Maritime Combat Systems for Maritime Simulation development. The Simulation Based Acquisition Programme on behalf of the RN is an initiative set up to support MOD use of simulation in its role as intelligent customer. To deliver its remit, the programme relies on a corporate strategy of engineering simulation technology ownership underpinned by external sourcing of world class subject matter expertise. An overarching technical infrastructure and discrete technology components are readily available for RN projects to use; the lead contractor acting as a co-ordinating prime and the primary representative for other SME supplier organisations.

SBA has been used effectively and extensively by MOD since 2000 to support the management and reduction of key risks across a wide range of projects. MOD policy places clear emphasis on managing and reducing project risks particularly during the earlier stages to minimise the likelihood of cost and time overruns. Consideration of the use of simulation is often the preferred alternative to traditional approaches, such as sea trials or scale model testing, to predict and verify the safety and performance of new systems. Simulation technology is being used to contribute to safety case assessment, performance prediction and project assurance during the major acquisition decision milestones.

Since 2000, SBA has been responsible for providing the UK technical support package to NATO on Virtual Ships.

An International Replenishment at Sea project known as IRAS formed under the Virtual Ships MOU involving Canada, France, Germany, Italy and UK is verifying NATO STANAG 4684 and ANEP 84 simulation standards by delivering a set of fully-tested, validated and interoperable simulation components that can be used to construct a federation that can be used internationally to simulate ship-ship highline transfer of liquids and solids for any pair of ships in any sea state using any type of transfer rig and any size of load or any type of liquid and flow rate.

Future simulation project arrangements under consideration by various groups of nations that are able to leverage the outputs above include:-

- Extending High Sea State Operations using Quiescent Period Prediction;
- Launching and Recovering Small Boats and UAVs in Higher Sea States,
-

Both scenarios are discussed later in this article.

2. SIMULATION IN ACQUISITION

NATO nations have been pursuing new initiatives and processes reducing the risk, cost and time involved in developing, building and operating new naval systems. Advances in computer science, telecommunications and naval engineering analysis, have enabled the 'virtual prototyping' of warship systems and processes with high-fidelity performance prediction, visualization, interference detection, automated generation of production data and virtual manufacturing.

Evolving simulation techniques, begun during the NATO/PfP Interoperability and Re-Use Study (NIREUS), are offering a substantial contribution to meeting desired risk, cost and time reductions within ship programs **[1]**. SBA processes and technologies are further improving engineering integrity, whole system integration and super-system optimization. Simulation is permitting testing and evaluation of designs in advance of the production of the real system by integrating:

- 3D walk-through;
- Equipment interference checking and operating limits;
- Human factors aspects such as fields of view and task performance;
- Digital demonstrators and mock-ups;
- Assessment of design alternatives;
- Crew traffic flow studies;
- Deployment & recovery of equipment;
- Manoeuvring, sea keeping and hydrodynamics;
- Performance and interoperability of sensors, weapons and command & control systems
- Survivability (susceptibility, vulnerability and recoverability);
- Simulation of the performance of mechanical systems.

For use in design validation and assessing safety, typical maritime simulations may require the modelling of:

- ship motion;
- hydrodynamic and aerodynamic effects;
- system dynamic interfaces;
- propulsion and helm control systems;
- mechanical systems used in vehicle deployment and recovery, ship-to-ship transfer of solids and liquids, etc;
- sensors such as optical, radar, sonar.

When to Use Simulation

The situations in which a simulation approach might provide an appropriate solution are:

- Where the range of variables to be studied are large and/or where the relationships between the variables are complex;
- Where the emergent properties of components exercised in combination need to be explored;
- Where many options need to be assessed, and need to be kept open as long as possible;
- When more than one operational scenario needs to be considered;
- When time-scale, cost or practicality precludes more direct methods of assessment;
- Where the impact of matters such as future technologies not currently developed need to be anticipated and evaluated;
- When there is no other means of predicting and assessing an attribute (e.g. through life cost of ownership);
- When system complexity demands early, pre-hardware, assessment;
- When multiple system dependencies and relationships need to be examined and traded-off;
- When safety or security considerations preclude practical testing, or where confidence needs to be developed before proceeding to a live trial;
- When different teams working on a project need structured access to consistent or common information;
- When seeking to demonstrate that the envelope of performance can be/has been met without having to physically trial each and every scenario.

Clearly, making the decision as to whether or not a simulation solution is viable will require subject matter experts to consider the 'art of the possible'.

3 – HELICOPTER – SHIP INTEFACE

Experimental / Simulated Dynamic Interface

It is important to develop options to reduce the cost and cycle time associated with the test and evaluation process, to enhance the productivity of flight test team members, and to improve the safety of flight test operations, especially in adverse environments. Flight testing is required in both land based and sea based environments with a variety of test aircraft and ships. Simulation is often listed as one option with the potential to help reduce the cost associated with flight testing **[2]**. The simulation of helicopter operations from naval vessels provides a unique set of challenges.

As described in reference **[3]**, simulated dynamic interface strategies have been developed over a number of simulation programs. The earliest High Level Architecture (HLA) simulation was called NIREUS (NATO Interoperability and RE-Use Study) which was followed by the SAIF (Ship/Air Interface Framework) programs. The purpose of these complex programs is to use the HLA standards to integrate air vehicle simulations, ship simulations and environment models to aid assessment of the dynamic interface for a range of helicopter / ship and UAV / ship combinations. The initial phase of the SAIF program focused -on SHOL prediction where operations may involve recovery in other environmental factors including, high levels of turbulence to new naval vessels. The primary future SAIF objective is to use the simulation to minimize the time and cost required for first of class sea trials.

The purpose of conducting these tests independently is two-fold. First, dynamic interface activities are defined as they apply to SHOL tests and aircraft/ship dynamic interface expertise and analysis. Second, to provide a platform to test devices like the LPD software to demonstrate the aid; for example, to signal the initiation of helicopter launch and recovery. The objective is to recover the aircraft aboard a moving vessel within reasonable safety margins regardless of the seaway.

HELICOPTER-SHIP LAUNCH AND RECOVERY DYNAMIC INTERFACE

A summary from earlier articles is presented below **[4]**. Dynamic Interface is defined as the study of the relationship between an air vehicle and a moving platform. It is performed to reduce risks and maximize operational flexibility. Globally, DI is concerned with the effects that one free body has in respect to another. Historically, this means the effects that a ship may have on a recovering or launching air vehicle. However, recent studies have concluded that the same principles apply to other motion related activities, such as, the boarding of Landing Craft vessels or LCACs into the wells of Amphibious Warfare Ships, the docking of submarines or the launching of unsophisticated missiles.

Dynamic Interface is divided into two broad categories: experimental or at-sea measurement and analysis, and analytical which is concerned with mathematical analysis and solution. The methods are not mutually exclusive. Neither method alone can produce a comprehensive and timely solution of the DI problem.

The traditional approach is experimental DI. It investigates operational launch and recovery of vehicles, engaging and disengaging of rotors, vertical replenishment and helicopter in-flight refuelling envelopes. "Shipboard suitability testing" assesses the adequacy, effectiveness, and safety of shipboard aviation. Testing methodologies and procedures have been standardized by laboratories, such as NAWCAD (Patuxent River) assisted by NSWC (Carderock), and QinetiQ (Boscombe Down). While experimental testing has numerous objectives, the primary focus is on launch and recovery envelope development and expansion.

DI testing of unmanned vehicles is not straight forward. The methodology of replacing piloted evaluations with operator estimates and the corresponding test criteria, is established prior to actual testing. However, during the testing, it is often found that assumptions applied early in the test may need to be adapted to circumstance found during the test.

NATOPS or SHOL limits are computed from wind and ship motion data as assembled by conducting various approaches and launch and recovery events. Central in the SHOL determination is the pilot evaluation. The unmanned configuration normally is instrumented to provide, amongst other parameters, oleo compression and deflection, torque monitor along with indications of precise weight on wheels. The number of landings per condition becomes almost academic since there is no piloted variation or technique the piloting is always the same as originally programmed. The settled approach is to assess system performance using multiple launch and recovery cycles but only one recovery is required to justify an envelope expansion.

In the course of conducting the unmanned sea tests it became evident that greater emphasis is needed on the simulation side to test for contingencies normally diagnosed by a pilot to deal with an unplanned event. In addition, a greater understanding of the control logic is needed representing a pilot decision making process. Several metrics are used to develop the relationship between the simulated and real world.

HELO-SHIP DYNAMIC INTERFACE SIMULATION

Whether a simulation represents an unmanned or manned system, the system must be capable of accurately responding in a variety of environmental conditions. The easiest is to evaluate the device in a closed and controlled environment. The primary difference between a manned and unmanned system revolves around pilot driven commands and controls. The pilot is represented in a UAV system by a series of flight laws and mission commands. The primary elements of the VTUAV system can be generalized: Unmanned rotary wing air vehicle, Ground Control Station (GCS) composed of a tactical common data link (TCDL) and tactical control station (TCS), along with a navigational aid and tracking system (UAV Common Automatic Recovery System version 2 or UCARS-V2), and normally a traverser system with landing grid. Each of these systems are federates along with the simulated environment which were also composed of federates.

A typical HLA design defines 6 separate federates (Figure 1), connected via the HLA Run-time Infrastructure (RTI) software. The structure is applicable to either a manned or unmanned scenario **[5]**.

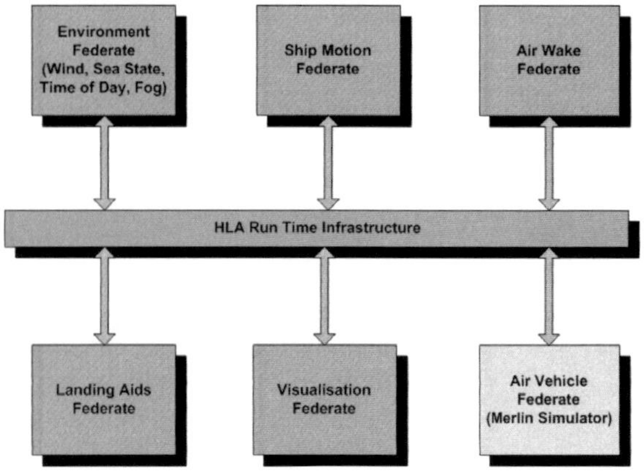

Figure 1: Federation Architecture

The initial software package was designed to use a full motion simulator (in the manned case) and a TCS (in the UAV case), to estimate system effectiveness as a function of simulated ship motion, visual environment and synthetic operational systems, and to compare the results to related analytic data.

By discipline the Federation is reduced to figure 2.

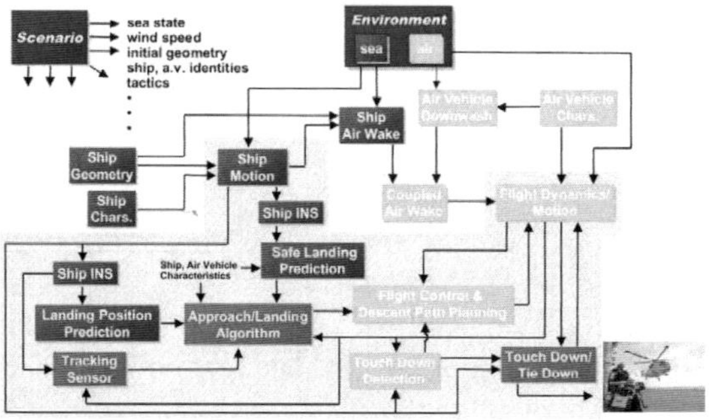

Figure 2: HLA Federation by Discipline

SHIP MOTION FEDERATE

The SMS Model is derived from the relationship between the wave and ship motion spectrum **[6]**. It incorporates seakeeping philosophy and applies various definitions of seaway spectral formulation, such as, Bretschneider **[7]**. SMS defines a seaway, computes the hydrodynamic and hydrostatic forces imposed on a ship (defined as the product of its transfer function and the seaway) and calculates a resulting ship time history. The simulation is an extensive treatment of a floating object's response to the dynamic loads on its structure. The simulation has been validated periodically after every major program improvement **[8]**. Figure 3 displays the simulation flow diagram along with the corresponding reports. The simulation strategy is divided into eight (8) subject routines identified as NAV and then the series number. The simulation establishes in the frequency domain the forcing function (seaway), selects an appropriate definition of the ship transfer function or response amplitude operator (RAO), producing the ship response definition as a series of harmonic components. The transfer from the frequency domain to the time domain is next made by summing over time the series of harmonic components which produces ship motion time histories. The time histories are used to calculate aircraft deck limits, encountered deck forces and air vehicle motion limits. The transfer to the real-time is made by running the ship motion time histories through LPDLOOP which calculates deck availability or clearance.

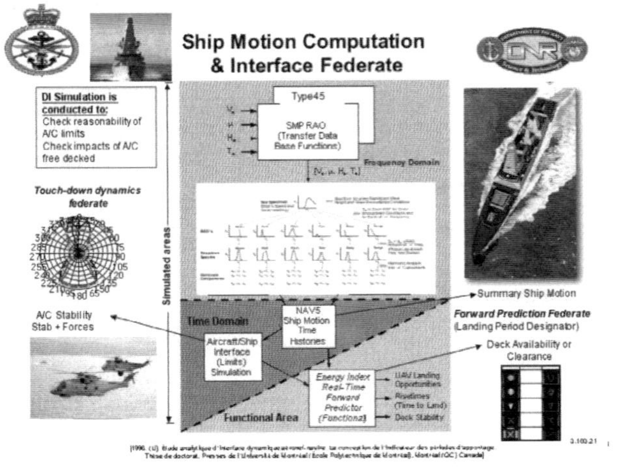

Figure 3: DI Simulation Calculation Methodology

The flight deck movement is defined by time history (deterministic) motion derived stochastically from a probabilistic spectrum. Ship speed, relative wave heading, significant wave height and modal period are the primary ship motion parameters. The relative motions are calculated at the point of interest (bullseye or landing point).

The program methodology uses essentially stochastic spectral probabilities to produce deterministic synthetic time histories **[9]**.

The ship motion program is divided into two basic themes, spectral analysis and the calculation motion histories in the time domain. The SMS fundamental relationship is:

$$Sr = S_W(w) \cdot RAO \cdot \boldsymbol{f}(V, m) \tag{1}$$

where:
- Sr: Ship response spectrum
- $S_W(w)$: Seaway spectrum
- RAO: Ship transfer functions
- f(V,m): Frequency mapping
- V: Velocity
- m: Relative wave angle

The seaway is defined by a forcing function called the Bretschneider, which is given by:

$$S_W(w) = \frac{483.5}{w^5 T_\emptyset^4} H_S^2 \, \mathbf{e}^{\left(\frac{-1994.5}{w^4 T_\emptyset^4}\right)} \tag{2}$$

where:
- $T_\emptyset$: period (sec)
- w : wave frequency (rad/sec)
- $S_W(w)$ seaway spectrum (m^2-sec)
- H_S : significant wave height (m)

The spectral characteristic of a vessel is defined by experimental or computational developed transfer functions termed Response Amplitude Operators (RAO). The response amplitude operators define the dynamic ship responses for a specified load/operating condition **[10]**. Referring to figure 4, the SMS portion of the program is divided into two basic themes, spectral analysis and the calculation of motion histories in the time domain.

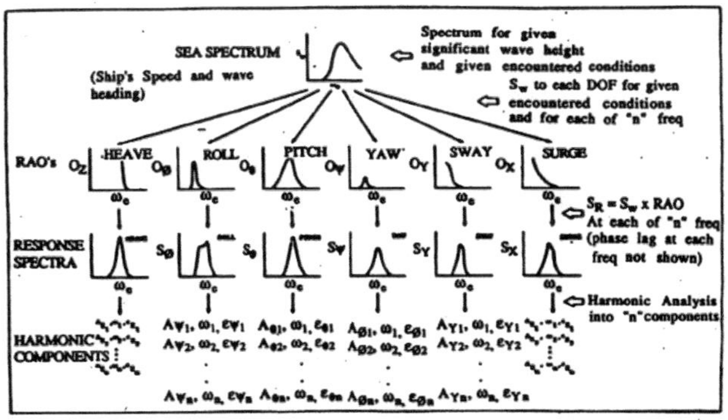

Figure 4: SMS Calculation Methodology

The ship response spectrum is created as the product of the RAO and the driving sea spectrum over the entire range of frequencies. The response spectrum is reduced to sets of harmonic components for each degree-of freedom. Synthetic time histories are created stochastically by summing the harmonic components over a given time period. A typical time history equation is given by:

$$A_Z = \sum_{n=1}^{k} \left(A_{Z_n} \cos\left(w_n - e_{Z_n}\right)\right) \tag{3}$$

where

- A_Z: DOF amplitude
- w: a circular frequency
- e: phase angle

Time histories are produced by the sum of 48 synthetic functions (k=48). In summary, the Ship Motion Simulation creates deterministic measures of ship motion from a probabilistic spectrum.

AIR VEHICLE ON DECK FEDERATE

Motion of an aircraft on the flight deck is calculated in terms of ship motion as a function of the aircraft model. The aircraft model is considered an extension of the ship. The model is defined by: aircraft landing gear footprint; on-deck location and orientation; aircraft weight, inertias, centre of gravity, lateral drag area and centre of pressure. The aircraft experiences ship transferred forces and moments that create rectilinear and angular accelerations on the air vehicle. The accelerations can be numerically integrated to determine the position and attitude of the helicopter relative to the ship as function of time, for various ship motions **[11]**. In essence, the aircraft is displaced as the sum of all forces to which it is exposed. Figure 5 displays a resumé of the moments acting on a skid-based aircraft at sea.

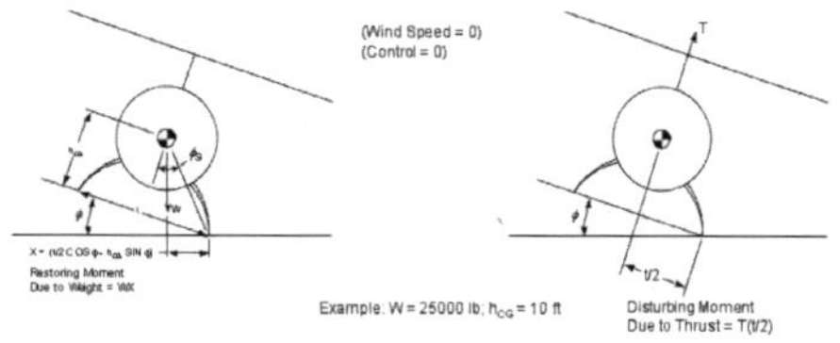

Figure 5- Moments Acting on a Skidded Helicopter at Sea

The inertial loads at the helicopter center of gravity induced by ship motion are given by;

$$F_{i_x} = W * AXcg$$
$$F_{i_y} = W * AYcg \qquad (4)$$
$$F_{i_z} = W * AZcg$$

where,

$\begin{pmatrix} F_{ix} \\ F_{iy} \\ F_{iz} \end{pmatrix}$ = inertial forces due to ship motion

W = aircraft weight

$\begin{pmatrix} A_{xcg} \\ A_{ycg} \\ A_{zcg} \end{pmatrix}$ = acceleration

In the longitudinal, lateral and vertical directions, these inertial loads become:

$$\begin{pmatrix} X \\ Y \\ Z \end{pmatrix} = \begin{pmatrix} T_{11}T_{12}T_{13} \\ T_{21}T_{22}T_{23} \\ T_{31}T_{32}T_{33} \end{pmatrix} \begin{pmatrix} F_{ix} \\ F_{iy} \\ F_{iz}+W \end{pmatrix} \qquad (5)$$

where: Tij = T (ϕ, q, y) (transformation matrix from ship's axis system to horizontal level/vertical axis system).
and

ϕ = roll
Θ = pitch
Ψ = yaw

A wind force is added to the ship motion induced forces. In SMS, a unidirectional continuous wind model (simplistic model), with its vector in the same direction as the seaway, is applied. Figure 6 displays the inertial components associated with the introduction of wind on the computation of the sum of forces at the centre of gravity of the air vehicle. The wind vector is defined by its magnitude (Vwod) and its direction (□□ wod). To compute the lateral force applied at the aircraft Centre of Pressure due to the wind, the Vwod is resolved along the normal to the aircraft centre line ($V_{W_{long}}$ and $V_{W_{lat}}$). The lateral component is used to compute the lateral force, as follows:

$$F_{W_y} = 3.5\ A_y \left(\frac{V_{W_{lat}}}{100}\right)^2 \qquad (6)$$

where:

Ay=Aircraft projected area normal to the $V_{W_{lat}}$ component

F_{W_y} =Lateral force applied to aircraft centre of pressure due to wind

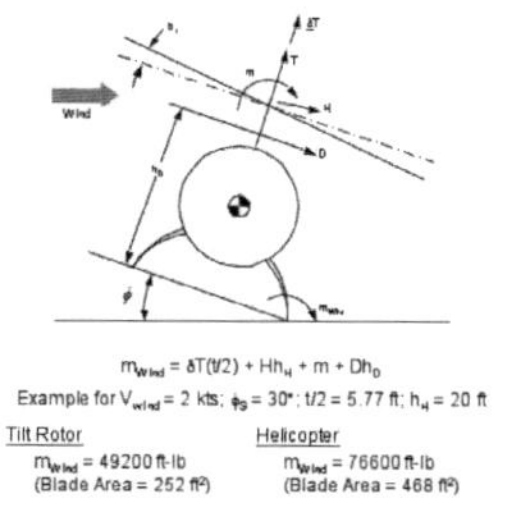

Figure 6- Wind Acting on a Skid-based Air Vehicle on the Deck

The standard tricycle coordinate system used to develop the stability of air vehicle on the deck of a moving vessel is displayed in Figure 7. In this study, the helicopter handling system probe replaces the nose wheel and the skids are reduced to point references replacing the main gear.

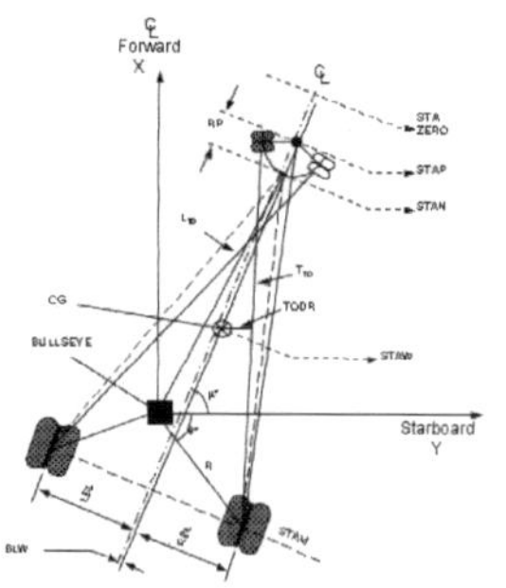

Figure 7: Typical Air Vehicle Freebody Diagram

The axial forces on the main landing gear due to the wind force F_{W_y} is given by:

$$F_{RMG_{WIND}} = (F_{W_y}) \frac{(WL_{CP} - WL_G)}{(L_{BL} - R_{BL})} \quad (7)$$

where:

$F_{RMG_{WIND}}$ = Main Gear (right) axial force
F_{W_y} = Wind lateral force component
WL_{CP} = Centre of pressure waterline
WL_G = Ground waterline
L_{BL} = Left wheel butteline
R_{BL} = Right wheel butteline

The incremental aircraft roll due to the wind is given by:

$$D_{f(WIND)} = \tan^{-1}\left(\frac{F_{RMG_{WIND}}}{(K)(L_{BL})}\right) \quad (8)$$

where: K = spring constant

Axial forces on the main landing gear due to aircraft inertial forces in the plane of the main gear, is given by:

$$F_{RMG(Inertia)} = Y\left(\frac{WL_{CG} - WL_G}{L_{BL} + R_{BL}}\right) \quad (9)$$

where:
WL_G =Centre of gravity waterline
$F_{RMG(Inertia)}$ =Right main gear axial force due to the lateral inertial force Y defined in equation (5).

Assuming perfect rocking, the axial force on the left main gear is opposite to the force vector acting on the right main gear:

$$F_{LMG(Inertia)} = -F_{RMG(Inertia)} \quad (10)$$

where: $F_{LMG(Inertia)}$ = Left main gear axial force

The incremental aircraft roll due to inertial loads is determined by:

$$D_{f(Inertia)} = \tan^{-1}\left(\frac{F_{RMG(Inertie)}}{(K)(L_{BL})}\right) \quad (11)$$

The simulation model assumes a constant wind, therefore, $D_{f(wind)}$ is constant throughout the simulation run. However, $D_{f\,(Inertia)}$ is continuously changing with ship motion. The total incremental change in the aircraft roll with respect to the ship is given by:

$$D_{f(total)} = D_{f(wind)} + D_{f(Inertia)} \quad (12)$$

Turnover incidents are static or dynamic in character. Static turnover is the same as on shore. The resolved weight vector migrates beyond either the friction forces causing the aircraft to displace or the reaction forces causing the aircraft to turnover. Figure 8 develops dynamic turnover caused by the rotor disk (uneven loading of the rotor) or by ship motion, the same phenomena occur.

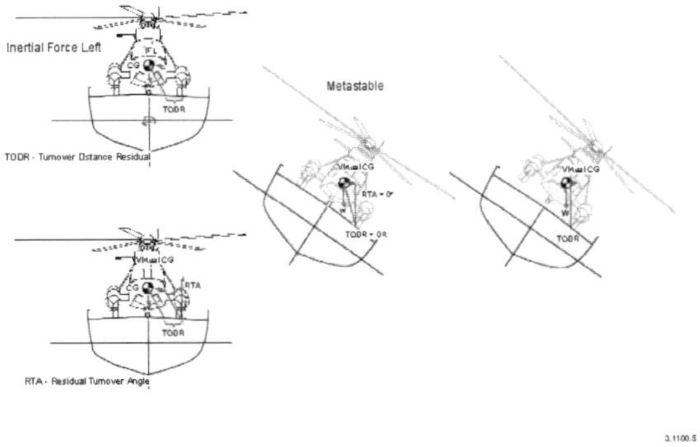

Figure 8- On-deck Dynamic Turnover

The aircraft centre of gravity is in motion. In the sum of forces, the weight vector is continually modified in response to inertial forces applied by either the rotor disk or ship motion or both. The distances TODR, TODL, and PBD essentially reflect system stability. At the point where a distance becomes negative, the system is unstable and will seek to find a more stable, but usually undesirable geometric solution. In similar fashion, when the landing gear friction values are exceeded by the combination of aircraft apparent weight and induced inertial forces, slippage will occur. Aircraft slide will continue until the aircraft frictional forces are greater than the disturbing inertial forces. Finally, when the vertical inertial force equals and opposes the aircraft weight, the deck friction goes to zero and an unintentional liftoff is indicated. The sum of these incidents trace aircraft-ship operational envelopes.

DECK INTERFACE (ENERGY INDEX) FEDERATE

In order for the helicopter to operate to the deck without a deck officer, it needs to know what the ship is doing now and what it will be doing during the final descent to touchdown. The Landing Period Designator (LPD) Energy Index (EI) algorithm is designed to identify quiescent periods of ship motion suitable for the recovery of the helicopter.

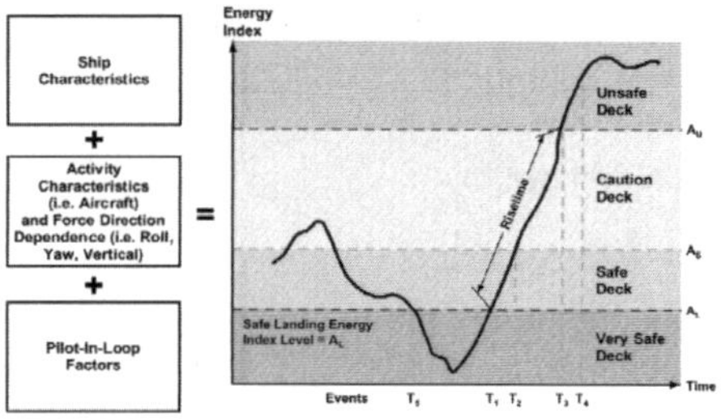

Figure 9: Deck Status and Rise time

The time required for the deck motion to rise from minimal motion (or very safe deck) to unacceptable motion is called the rise time. In terms of the EI scale, the rise time is defined as the period of time that is measured from the end of a green signal to the positive side of the red line. This is given as ($T_3 - T_1$) as shown on Figure 9. The rise time is a thumb print characteristic of the ship's response and remains fairly constant for each ship class.

The very safe deck is a special condition in which there is insufficient energy in the aircraft-ship system to raise the deck out of limit for some defined time period or rise time, and it is this concept that was used in the simulators to indicate that the deck was safe to initiate landing.

Assuming the minimum average rise time for a given vessel remains constant, the aircraft descends within the rise time value and the aircraft is assured a recovery within the deck motion limits. Essentially, the methodology summarized above is a formulation to quantify operating beyond the static deck limits as defined in NATOPS or SHOL. Figure 10 displays graphically beyond static limit operations. The base envelope is taken from 10 knots of ship speed. As before, any points within the hour-glass structure are conditions within limits and contain no appreciable probability of out-of-limit deck motions. Outside of the structure contains motions which are considered by static reference as out-of-limit **[12]**.

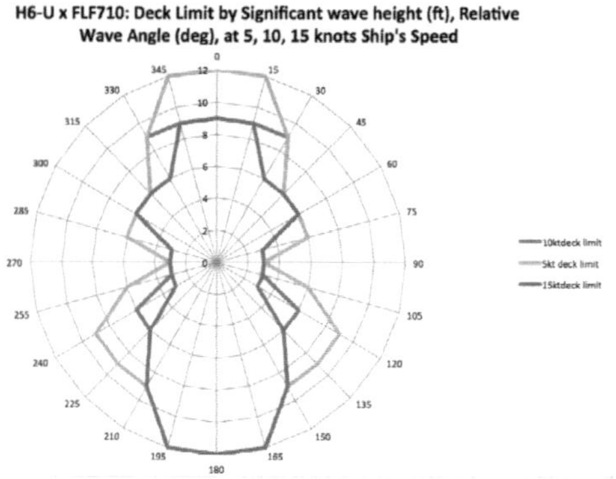

Figure 10: Safe Motion Operations

This is particularly important for vehicles without pilots seated in them (Unmanned Air Vehicles or UAV). A schematic representation of how the LPD is used for UAV autorecovery to signal the onset of the descent (and of deck quiescence) is displayed in Figure 11. Throughout the approach and initial hover, LPD is monitoring the deck. It is only at the low hover that the UAV autoland system would accept a Green Deck signal from the LPD. When the EI signals the onset of quiescence and through the uplink sets the air vehicle on its descent to the deck.. Figure 11 displays current VTUAV Flight Patterns.

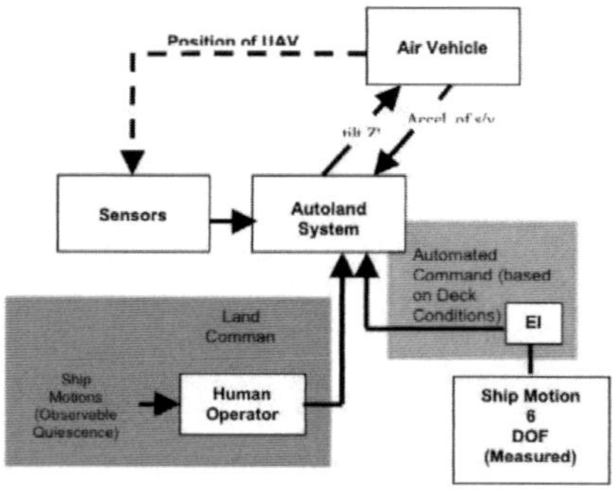

Figure 11: Notional Autoland System

4. IMPACTS OF DECK MOTION ON DECK LIMITS

The simulation of ship platforms and systems to assess their interfaces to other platforms and systems need to be made more available. Early use of trusted simulation can allow major ship projects to be de-risked during design to a greater degree than has been possible in the past. During ship design phases, the performance of platform operations and platform interactions such as helicopter recovery, replenishment at sea, vehicle docking and vehicle deployment and recovery can now be assessed in all required sea and weather conditions **[13]**. Establishing the accuracy of simulations developed for supporting design is vitally important perhaps more so than for simulations that have been developed specifically for training. Numerical inaccuracies in training systems can normally be identified early by experienced operators (not necessarily my experience, particularly wrt to ship air wake).

The calculated system limitations provide experimental DI with the necessary data to more effectively set testing strategy to probe the limiting conditions **[14].** Continuing the theme of ship's motion, and having created a large library of data on the target ship, it is possible to derive tendencies of ship's characteristics to the most basic of parameters, such as, ship's speed. Figure 12 displays the effect that ship's speed has on roll motion (here assessed at 10 feet significant wave height).

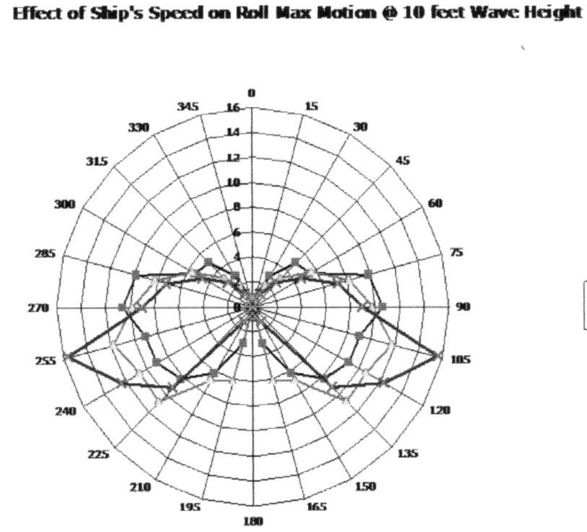

Figure 12: Effect Ship's speed on Roll Motion

Roll appears to improve slightly with increasing ship's speed in ahead seas through 90° relative wave angles. From mid ship's to 135° relative wave angles, roll angle becomes larger with increasing ship's speed. It then falls away at angles greater than 150° to 180° relative wave angles. Studying pitch (see figure 13), the tendencies are opposite those of roll.

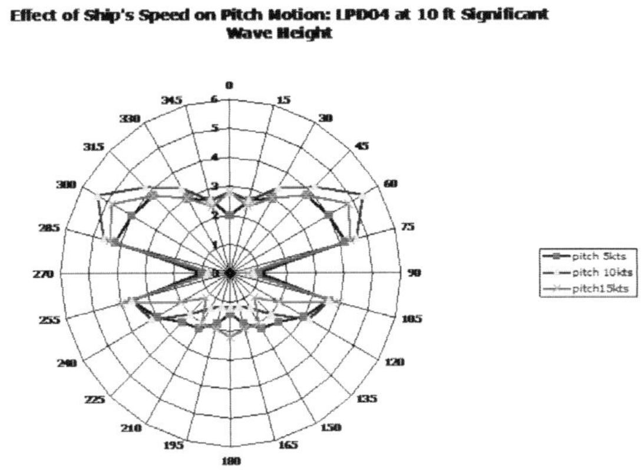

Figure 13: Effects of Ship's Speed on Pitch Motion

Much as expected, in the ahead seas condition pitch motion increases with increasing ship's speed. The following seas cases are more complicated, but overall follow the same theme.

Assessment was made concerning landing gear responses to encountered deck loads from several ship models. Identification of incidents of gear slide (coefficient of friction = 0) and gear dampener stroke are produced. This is done for each motion scenario. A sample analysis is given in Figure 14. The example was randomly taken from data file 10301511, which translated means, 10 knots ship's velocity, and 30 degrees relative wave angle 15 feet significant wave height and 11 seconds wave period. Calculated encountered axial forces are shown over the course of the time history. Fz (vertical forces) is the largest contributor to the resultant force vector (Fr). For this scenario, Fx (longitudinal force vector) is statistically slightly greater than those of Fy (lateral force vector). At the point where the coefficient of friction measures zero (or the index of slide is greater than one), an aircraft displacement is indicated. The sum of the force distribution studies produces a helicopter encountered force deck limit diagram (rondelle, speed polar or O'Reilly plot).

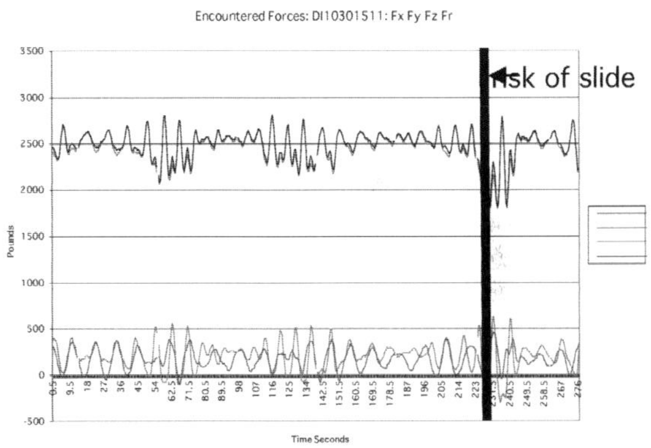

Figure 14- Landing Gear and Probe Encountered Deck Forces, File 10301511

As developed earlier for vehicle interface simulation, Figure 15 represents deck limits as a function of deck encountered forces, in this case onboard the LPD-4 running at 5 to 15 knots of ship's speed. Reading the diagram is similar to that for aircraft stability. Anything inside the hourglass figure is within limits where there no measurable risk of aircraft incident recorded. Anywhere outside the hourglass, an incident of landing gear faults was recorded and is, therefore, out of limit. As an example, at 60 degrees relative wave angle, seaway significant wave heights lower than 5 feet are within limits. In interpreting the diagram, it should be apparent that the large size of the Austin Class ship has a great deal to do with deck stability. Ahead seas shows good stability between 330 – 360/0 – 30 degrees relative wave angles and in following seas 150 – 180 – 210 degrees of relative wave angles. Helicopter displacement risk appears to be focused at the beams. Areas without risk on the beams occur with less than 3 feet significant wave height (or sea state three). The most heavily constrained areas are formed from 60 – 120 degrees of relative wave angles and from 240 – 300 degrees of relative wave angles.

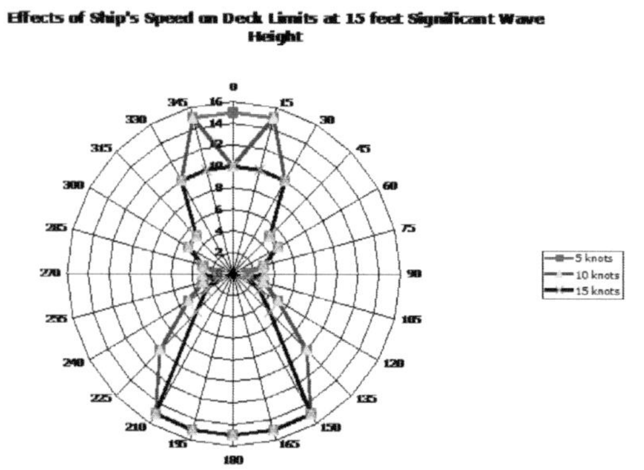

Figure 15- Encountered Deck Forces of air vehicle x LPD-4

The effects of ship's speed on the deck limits are displayed in Figure 15. The interface between the LPD-4 response characteristics combined with the Fire Scout aircraft stability parameters, appears to be particularly sensitive, with minor adjustments in following seas, to ahead seas at higher ship's speeds. From the diagram, ahead seas (330 – 30 degrees relative wave angle) is most stable at 9 feet significant wave height. The stable condition quickly decays at relative wave angles greater than 030 – 135 and 225 – 330 degrees of relative wave angles.

IMPACT DECK MOTIONS ON AIR VEHICLE LIMITS

Figure 16 is composed of the launch and recovery events, energy index, ship's roll, pitch and yaw traces along with the deck energy levels. From the time history, with LPD off, a normal evolution had taken place with both the launch and recovery events occurring on a green-amber or safe deck, but not quiescent or green deck. The other channels analysed, corresponding translational traces, engine torque and landing gear compression all appeared to have shown normal operation.

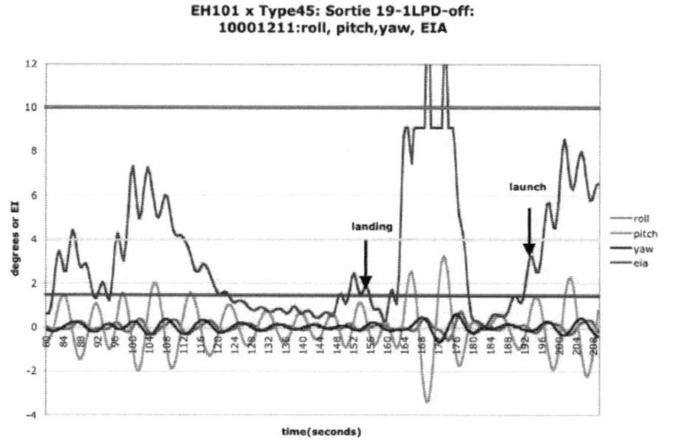

Figure 16: Evolution 19-1(Day) with Launch and Recovery Points

From the time history in Figure 17 , with LPD on, a normal evolution had taken place with both the launch and recovery events occurring on a green or quiescent deck.

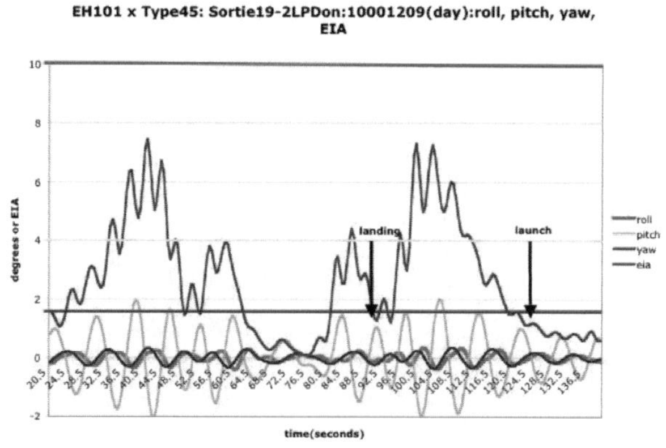

Figure 17 - Evolution 19-2(Day) with Launch and Recovery Points

The other channels analysed, corresponding translational traces, and engine torque appeared to have shown normal operation. Oleo compression trace appeared to show a peak compression on the nose gear (Figure 18).

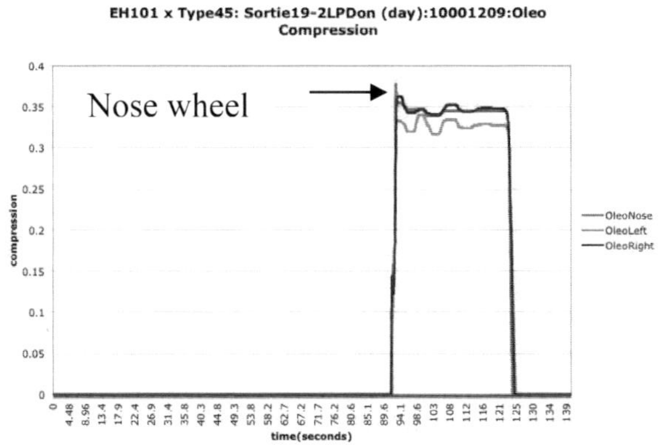

Figure 18: Sortie 19-2 Oleo Compression Trace

Figure 19 time history, with LPD off, the recovery event occurred on a quiescent deck, but the launch happened from a high amber or caution deck. Indeed, the deck was very nearly out-of-limit. The corresponding translational traces showed similar large displacements at launch.

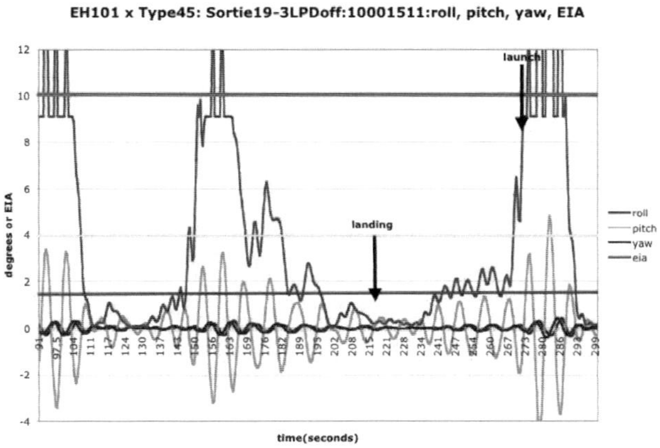

Figure 19 - Evolution 19-3(Day) with Launch and Recovery Points

Oleo compression appeared normal, but the, engine torque was measured greater than 100% at launch (figure 20).

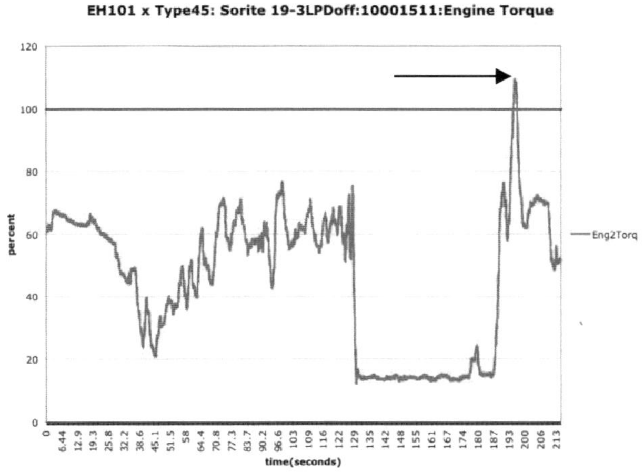

Figure 20: Sortie 19-3 Engine Torque

5 AT- SEA VERIFICATION OF SIMULATION

Dynamic Interface– establish requirements

In the course of analysing the ship's motion, other characteristics which may impact the launch, recovery and deck handling, were identified. The time necessary to raise the deck from a stable to an unavailable condition can be derived experimentally from the calculation of the maximum EImax. As described earlier, the time required to raise the deck from minimal motion to unacceptable motion is called the risetime (written as one or two words). The risetime is a thumb print characteristic of the ship's response and rarely changes. In terms of the EI scale, it is defined as the period of time that is measured from the end of a green signal to the positive side of the red line. The risetime is mirrored by a droptime, which is the time period measured from the negative side of the red line to the negative side of the green line [8]. A typical trace as recently recorded, is displayed in Figure 21.

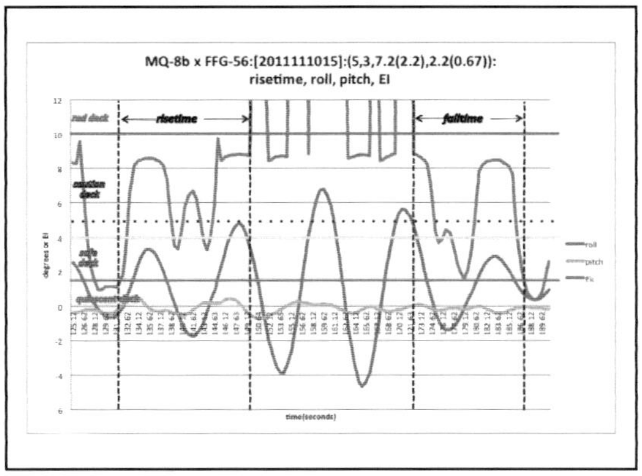

Figure 21: Rise time events in file "2011111015"

The example shows a roll axis divergence eventually exceeding the MQ-8B deck roll limit of 5.0 degrees. In this case the time it took for the deck to rise from quiescence was 18 seconds. There appears to be a 1° roll list to port whilst pitch appears to be essentially in trim. The impulse converges in a fall time measuring 15 seconds. Interestingly, the index exceeded 10.0 prior to the roll axis exceeding the maximum deck motion limit. Figure 22 displays the deck motion rates for the same time period.

The Y' axis diverged like roll, but is 1/4 cycle out of phase with roll. There was sufficient energy on the second limit crossing to cause a roll strong enough to exceed deck limits. This is another example of the coupling between Y' and roll. During this test, the ship managed to record at various points 18.9° roll, 4.7° pitch, 2.3 m/s Z' vertical rate, and 2.7 m/s lateral rate maximum deck displacement.

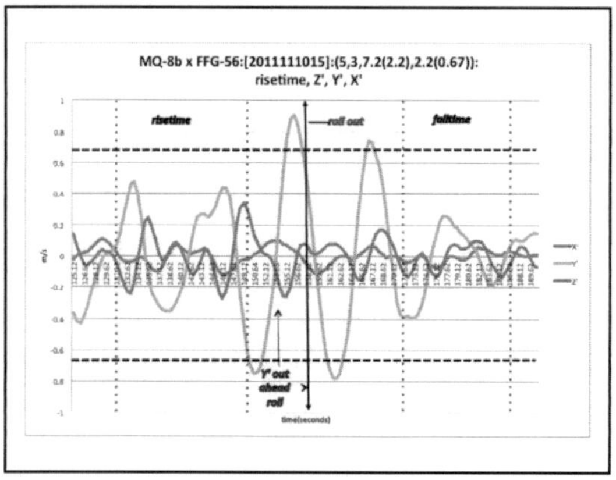

Figure 22: Risetime events (rates) in 2011111015

Dynamic Interface – All Weather Operations

The majority of the motions recorded alongside the encountered deck accelerations or forces, appeared to equate to "textbook" motion traces for given study platforms. Recorded shipboard data are often compared with related information from other sources such as the National Data Buoy Center run by National Oceanic and Atmospheric Administration (NOAA).

Figure 23 displays a NOAA wave height forecast chart similar to those received during DI Fire Scout trials. It is often compared with anchored NOAA wave buoys mainly as a reference for the shipboard data recordings.

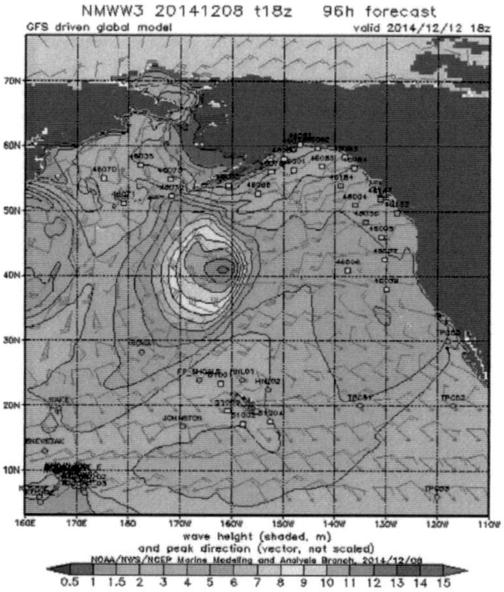

Figure 23: NOAA Wave Ht Chart

Events during the LCS class DI gave reason for pause. On what appeared to have been a low sea state day, unexpected motion events occurred exceeding air vehicle deck limits uniquely in roll. Figure 24 displays the hour file recorded in which the excessive motions were recorded. Of particular note (green) were the extent of roll cycles. Figure 17 displays a typical Fire Scout (MQ-8B) recovery [9].

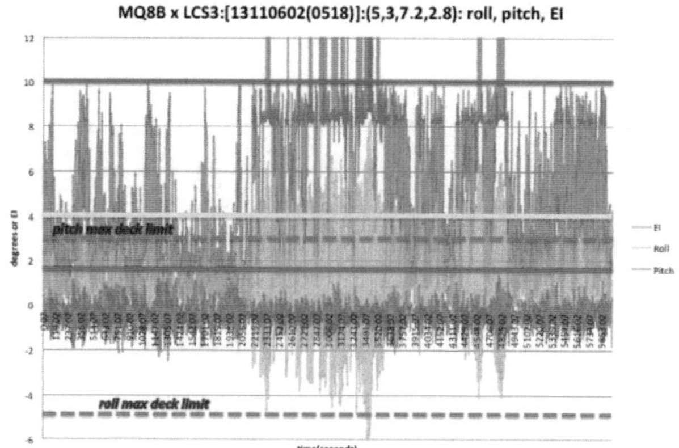

Figure 24: Typical Time History Trace

Figure 25: MQ-8B on Final

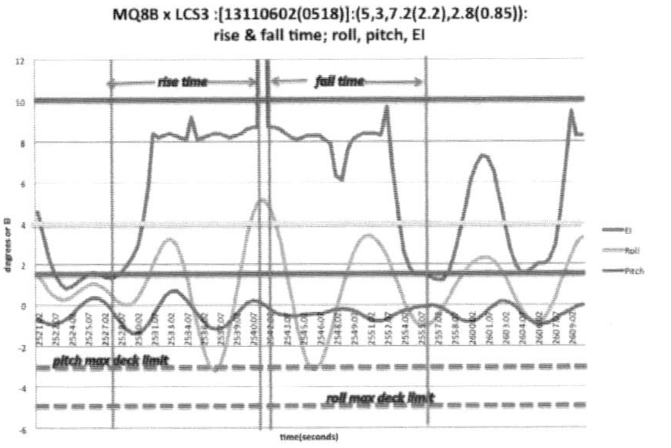

Figure 26: Rise / Fall time events recorded

Examining the related traces (figures 24, 26, 27), and lining up the graphs (figure 28), it would appear a rare event occurred in which all the degrees of freedom appear to be in phase, but only roll contains sufficient energy to move the ship beyond the MQ-8 landing limits.

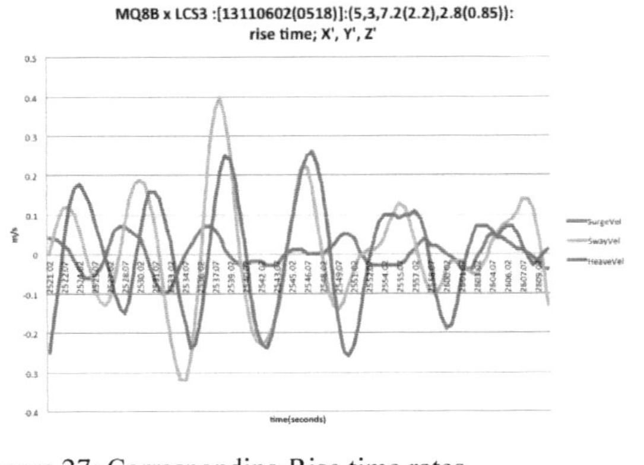

Figure 27: Corresponding Rise time rates

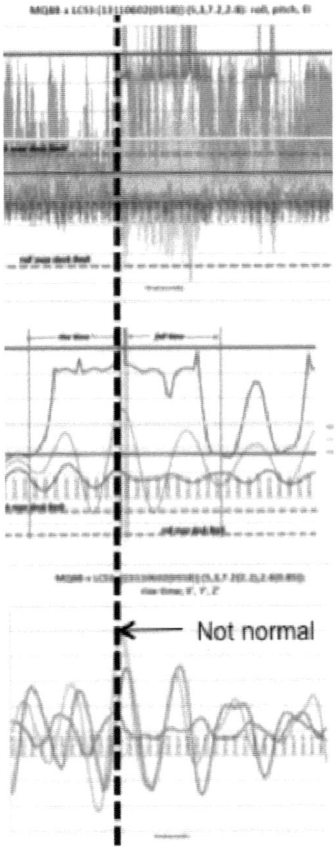

Figure 28: All DOF axes in-phase in Rise time Event

On closer inspection, it was determined this event was not an isolated case. Figure 29 displays other examples of rise time events composed of in-phase axial motions even those that are diametrically opposed.

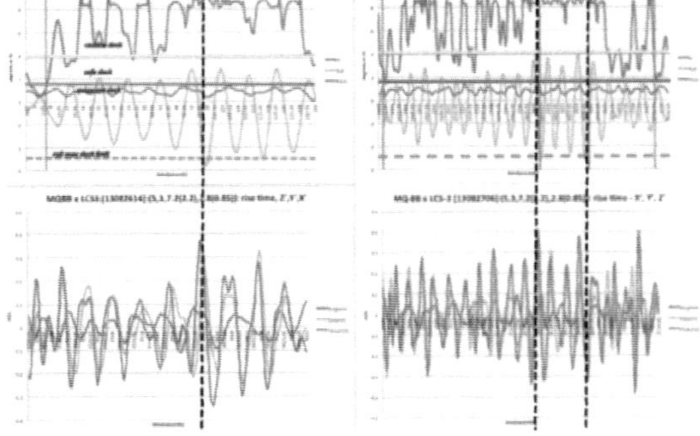

Figure 29: Parametric-like roll events

Table 1 lists the motion statistics for one of the LCS DI trials.

Table 1: DI Trial Motion Statistics

- **Total hours recorded at sea: 240**
  - ***Sea State: 0 – 3*** Ship handling: 0 – 5 Ship Speeds: 2 – 15 knots
  - Max motions: roll – 9.9° pitch – 2.5° Z' – 0.7 m/s Z" – 1.0 m/s² Y' – 0.6 m/s Y" – 0.6 m/s² X'- .28 m/s X" - .29 m/s²
  - MQ-8 deck limits: roll – 5°, pitch – 3°, Z'– 2.2 m/s(7.2'/s), Y'– 0.86m/s(2.8'/s)
- **The number of rise time events: 252 (non slam)**
  - Average minimum rise time recorded: 10 seconds
  - Computational minimum rise time: 6.3 seconds
  - No of Slam events: 0 (non aviation);
  - ***Excessive ship motion axis causing rise time events***:
    - ***Roll : 100.0%***
    - Y, Z' : 0.0%
    - Pitch 0.0%
- **The number of launch and recovery events recorded = 177 (86 launch; 86 recovery; 5 wave-off): 64 piloted serials; 22 autonomous serials;**
  - Launch: 55 green, 20 gr-amber, 25 amber, 0 red events
  - Recovery: 60 green, 15 gr-amber, 25 amber, 0 red events
- ***Other: ship's power failure events (some occurring during flight ops). System robustness was a requirement for the success of this test.***

Red flags

This went unnoticed until mugs fell over; raised eyebrows. These events did not interfere generally with the DI.

According to the NOAA wave charts based on local wave buoy data, the test box off of California was experiencing sea states 0 – low sea state 3. Yet the LCS platform recorded a 10° roll. There was motion along the other degrees of freedom but nothing to support a 10 degree roll. Lateral and vertical rates never exceeded the respective deck limits of the MQ-8. Based on the statistics table, only roll exceeded deck limits of the MQ-8. This is very unusual. Referring to figure 30, a video was taken of the deck whilst it was in a roll cycle.

Figure 30: FireScout on Platform in roll cycle

These resonance cycles of ship motions appears to have occurred while the ship was moving relatively slowly through the sea to an overnight track.

DEFINING MOTION CHARACTERISTICS

The impacts on motion sensitive tasks by ship platform traits as those described in this article, could be significant. Throughout this article, the apparent motions described have been classified as parametric-like roll. While the ship's 6 degrees-of-freedom motions are all in phase, only elevated roll motions are evident. However, there are many related functional variables that have not been verified. Careful analysis of the seaway to include currents, swells, direction, speeds along with significant wave heights and their periods, is required to gain the complete picture or definition. To this point, much of the analysis has been using motions in the ship coordinate system without attentive documentation of the ship's heading and speed relative to any existing seaway.

Logbook climatic observations indicated most of the motion events occurred in calm sea conditions with the ship navigating at under 10 knots of ship's speed. These conditions are contrary to the conditions reported in much of the literature concerning parametric roll. The vast majority of the publications on the subject describes an elevated sea condition equating to storm waves .

Future work

Envelope expansion dynamic interface trials are expected to be conducted with an MQ-8C on board one of the LCS class vessels later this year. A key factor related to increased operational capability in landing UASs onboard ship is the ability to repeatedly L&R safely from a ship moving in response to the seaway. The operational benefits of an unmanned air vehicle are increased if it can be safely captured autonomously, i.e. without the aid of an experienced pilot and Landing Signal Officer (LSO) to a quiescent ship. The test instrumentation for this task during DI trails includes ship motion recording devices. The recordings from these trials will join the library of ship motions data already in existence. Coupled with these recordings, careful attention will be paid to maintaining an eye on the ships speeds and headings. This is combined with external seaway recordings both from "nearby" anchored wave buoys managed by NOAA and close up wave buoy information recorded by University of California San Diego's (UCSD) Scripps Institution of Oceanography.

In the case of the MQ-8C, some of the possible impact on the aircraft will be mitigated by the system's LPD unit identifying the quiescent deck period.

6. SUMMARY AND CONCLUSIONS

Parts of the MQ-8 future improvements involve the enhancement of the associated system components, such as that being done for the L&R segment. The primary goal for conducting DI analysis is to expand existing operating envelopes and increase AV's availability thereby improving overall naval effectiveness. The objective of DI is to determine the maximum safe air vehicle/ship platform operational limitations. Given an air/ship system and inherent operational limitations, DI strives to increase tactical flexibility for any set of environmental conditions. Modelling and Simulation complements DI testing to delineate system limitations. The calculated system limitations provide experimental DI with the necessary data to more effectively set testing strategy to probe the limiting conditions.

This paper provides a discussion of the improvements associated with the MQ-8 system focusing on the L&R process and equipment. These improvements are driven by a need for L&R systems requiring few or no people to operate; systems requiring reduced time to execute high tempo L&R on the deck; systems that provides improved reliability whilst being simple to maintain. One potential problem that could seriously affect MQ-8's L&R on both of the LCS class ships is the large magnitude of roll experienced in seemingly benign sea conditions. This phenomenon should be further investigated to determine its cause, whether by a swell or parametric conditions, so that it possibly can be avoided.

There has been a great deal of interest in this project from its inception. The team looks forward to providing a status update at the conclusion of the development and start of deployments.

7. ACKNOWLEDGEMENTS

The authors wish to express their thanks to the UK Ministry of Defence for their support of the work presented here. This paper is dedicated to the loving memory of Dr. Peter J. F. O'Reilly, a founding father of the DI discipline, and one of the first people to propose the use of the EI to automatically recover Unmanned Air Vehicles.

8. ABBREVIATIONS

AO FNC	Autonomous Operations Future Naval Capabilities
ASIST	Aircraft/Ship Integrated Secure and Traverse System also RSD
CBT	Computer Based Training
CD	Clear Deck
CDS	Cockpit Dynamic Simulator
DI	Dynamic Interface Study
DOF	Degree-of-freedom
EI	Energy Index (calculated by LPD)
FD	Free Deck Recovery (RAST trap used only)
HLA	High Level Architecture
LPD	Landing Period Designator
LPDLOOP	Landing Period Designator software
MFS	Manned Flight Simulator
MRU	Motion Reference Unit
NATOPS	Naval Air Training and Operating Procedures Standardization
NIREUS	NATO Interoperability and RE-Use Study
ONR	Office of Naval Research
RA	Recovery Assist
RAO	Response Amplitude Operator
RAST	Recovery, Assist, Securing Traversing
RN	Royal Navy (UK)
RNAS	Royal Navy Air Station (UK)
SAIF	Ship/Air Interface Framework
SHOL	Ship-Helicopter Operational Limit
SMP	Ship Motion Program
SMS	Ship Motion Simulation composed of routines identified as NAV
TCS	Tactical Control Station
UAV	Unmanned Aerial Vehicle
VTOL	Vertical Takeoff and Landing
VTUAV	Vertical takeoff and Landing Unmanned Air Vehicle
WOD	Wind-over-deck

9. REFERENCES

1. White, S., Reading, R. (2001). NATO/PfP HLA Federation of VTOL Operations Supporting Simulation Based Acquisition. 01E-SIW-032. Brussels

2. Applebee, TR., Baitis, AE., Meyers, W.G. (1981) "User's Manual for the Standard Ship Motion Program, SMP" David W. Taylor Naval Ship R&D Centre Report DTNSRDC/SPD-0936-01.

3. Blackwell, J and Feik, R. A (1988). "A Mathematical Model of the On-Deck Helicopter/Ship Dynamic Interface (U)". Aerodynamics Technical Memorandum 405. Aeronautical Research Laboratory. Melbourne.

4. Cox, I., Turner, G., Duncan, J.(2005). Applying a Networked Architecture to the Merlin Helicopter Simulator. Royal Aeronautical Society Conference. London.

5. Crossland, P., Ferrier, B., Applebee, TR. (2004). Providing Decision Making Support to Reduce Opeator Workload for Shipboard Air Operations. Royal Institute of Naval Architects. London.

6. Ferrier, B. & Carico, D.[2007], Testing and Evaluation of Landing Aids to Improve Helicopter/Ship Operational Limits. Proceedings of the AHS. Washington.

7. Ferrier, B., Applebee, T., Manning, A., James, CDR D.[2000], Landing Period Designator Visual Helicopter

Recovery Aide; Theory and Real-Time Application. Proceedings of the AHS. Washington.

8.Nelson, J. & Danielson, J.(2013). MQ-8B Fire Scout Initial Ship Suitability Tests Aboard USS McInerney (FFG 8). NAWCADPAX/RTR-2009/118. Lexington Park.

9.France, W. N., et al. (2001). "An Investigation of Head-Sea Parametric Rolling and its Influence on Container Lashing Systems". Proceedings of the SNAME. New York.

8. AUTHORS' BIOGRAPHIES

M R Belmont holds the current position of Professor of Marine Dynamics at the University of Exeter where he is responsible for maritime systems research. His previous experience includes modelling and land and sea trials of Doppler LiDAR systems.

J Christmas holds the current position of Lecturer in Computer Science at the University of Exeter where she works on maritime systems research and statistical machine learning. Her previous experience includes land and sea trials of Doppler LiDAR systems.

JM Duncan is Group Leader for Simulation Based Acquisition and Technical Computing for Maritime Combat Systems, UK Defence Equipment & Support (DE&S). The Group's focus is on whole ship and submarine safety and technical services including new modelling techniques for complex dynamic interfaces such as for manned and unmanned air and sea vehicles. Duncan is the NATO Principal Representative for the Naval Armaments Group on Virtual Ships. He was previously chairman of NATO Subgroup 61 and the Specialist Team on Simulation Based Design and Virtual Prototyping for ship acquisition. He led the development of Allied Naval Engineering Publications '61' for ship virtual prototyping and '84' for Virtual Ship simulation standards.

J Duncan has served for 6 years as the manager of sea trials data collection and coordination on behalf of MOD. He has supervised 10 sea trials data collection activities on behalf of the NATO Submarine Rescue System team including substantial planning, post-trials analysis and reporting. He has been involved on various simulation projects for MOD supporting Type 45, Merlin, QEC, Ship Air Interface Framework and Quiescent Period Prediction. He has worked on the development of NATO Virtual Ships documents STANAG 4684 and ANEP 84.

B Ferrier holds the current position of Systems Engineering and Technology Associate (SETA) in the offices of NAVAIR PMA 266. He is responsible for assisting NAVAIR in integrating Fire Scout Endurance upgrade technologies, such as the ONR Landing Period Designator, into the Fire Scout Launch and Recovery System. He is also the Engineering Head of the Hoffman Engineering's Dynamic Interface Lab in Arlington, VA and holds an honorary chair at the University of Exeter.

REAL-TIME SHIP AIR-WAKE AND FREE STREAM MEASUREMENTS USING DOPPLER LIDAR

M R Belmont and **J Christmas**, University of Exeter, UK.
J M Duncan and **J Duncan**, Ministry of Defence Equipment and Support HQ, UK.
B Ferrier, Hoffman Engineering Corp., USA.
R Potts, BAE Systems, UK.

SUMMARY

Understanding the air-wake around a ship is of vital importance for safe air operations. While CFD may be used during ship design, its computational complexity means that only a limited number of conditions can be simulated; and any subsequent changes to the design will result in changes to the air-wake pattern. The traditional measurement technology is anemometers, with one or more devices being permanently installed to measure the free stream and additional devices installed temporarily over the deck during sea trials to measure the air-wake structure. Physical limitations mean that these can only measure a limited range of distances directly above the deck.

We describe the outcome of recent land-based trials of Doppler LIDAR for measuring both the envelope and the structure of a ship's air-wake, above the deck and laterally away from it, and results from research into improving the resolution of the devices from 40m down to 5m.

1. INTRODUCTION

Understanding the air-wake around a ship, and the location and form of the free stream outside the wake, is of vital importance for safe air operations, particularly in the launch and recovery of fixed and rotary wing aircraft. While CFD modelling may be used during the design phase of a ship, the computational complexity of CFD means that only a limited number of conditions can be simulated, and any subsequent changes to the ship's superstructure will, in any case, result in changes to the pattern of air-wake.

The tradition approach to measuring the airflow over a vessel is via an array of vector anemometers of some kind (e.g. [9]). Usually one or more reference anemometers are permanently installed, usually above the bridge, to measure the free stream. However, it is well known that these reference anemometers are often installed within the air-wake, giving erroneous information for air operations. During sea trials, additional anemometers are temporarily installed at various positions on a ship's decks in order to measure the characteristics of the air-wakes themselves. When synchronous data is required at a number of locations simultaneously this can be an extremely time consuming, costly and cumbersome process. Furthermore, data is not available at any significant distance above or laterally off the deck.

A potentially attractive alternative is Laser Doppler based airflow measurement which is a mature technique in meteorology with several types of commercial equipment being available (e.g. [7]). Such instruments can scan over a full upper-hemisphere at several Hertz out to ranges of many kilometres if required. Clearly this would make it possible, for the first time, to undertake measurement campaigns that include the very significant down-flow region behind an aircraft carrier. Each Doppler LIDAR returns the line of sight (along the beam direction) component of the air-flow velocity so three separate instruments are needed to obtain fully three-dimensional vector flow data.

A constraint imposed by the generalised uncertainty principle means that in order to achieve sufficient wave-number resolution it is typically necessary for Doppler LIDAR to collect scattered light over distances typically in the range 20m to 40m, which inherently spatially averages the line of sight component of the air velocity. This is generally acceptable in atmospheric meteorological applications, but it means that the spatial resolution of Doppler LIDAR is considerably coarser than has been deemed useful in the measurement of airflows both over and off the deck of vessels. The goal is to improve this figure down to approximately 5m. The proposed approach is to model the Doppler LIDAR measurement process and then to employ deconvolution to mitigate the effects of the inherently low-pass spatial filtering process. This technique has attracted very modest attention in the scientific literature [1-5]. Most of the attention has been paid (mainly by a Bulgarian group) to the effects of laser pulse profile [1-4], and there has also been some more recent interest in using wavelet shrink techniques for noise reduction [5].

The present work is primarily concerned with applying aspects of the Wiener deconvolution approach to the results of a high resolution land based field trial designed specifically to assess the viability of employing Doppler LIDAR resolution enhancement methods to ship-borne applications.

2. FUNDAMENTAL LIMITATIONS OF DECONVOLUTION

Given that the Doppler LIDAR measurement process is a linear function of the *true* line of sight velocity component, $v(x,t)$, at distance x from the LIDAR and time t, then, in the ideal noise-free case of an instantaneous snapshot measurement over the whole spatial domain, the Doppler LIDAR estimate of that velocity, $v_e(x,t)$, can be described by the spatial convolution

$$v_e(x,t) = \int_{-\infty}^{\infty} v(\beta,t)h(x-\beta)\,\mathrm{d}\beta \qquad (1)$$

where the function $h(x)$ is the time-independent impulse response of the overall measurement and velocity estimation process. Exploiting the convolution theorem of Fourier Transforms, equation (1) can be recast in the wave-number (k) domain into the form

$$V_e(k,t) = V(k,t)H(k) \qquad (2)$$

where the upper-case symbols are the Fourier Transforms of the corresponding lower-case symbols.

For the noise-free case, equation (2) offers the prospect of simply dividing $V_e(k,t)$ by $H(k)$ to recover the Fourier Transform, $V(k,t)$, of the original data. All real-world physical processes have a finite bandwidth, i.e. $|H(k)| \to 0$ above some range of k; however, another fundamental property of the Fourier Transforms of *causal realisable systems* is that this decay to zero is continuous, i.e. $H(k)$ does not abruptly drop to zero above some $|k_{\max}|$. Physically this means that such systems will continue to exhibit some very small response to stimuli even at very large wave-numbers, well beyond the point where they have large overall responses. The consequence of this is that dividing $V_e(k,t)$ by $H(k)$ allows, at least in principle, the recovery of information about $V(k,t)$ well beyond what might be expected from purely examining the raw Doppler LIDAR output.

The above process is termed deconvolution and can either be performed in the wave-number domain, as described above, or, alternatively, the process can be thought of as saying that, given a system with a wave-number response function, $G(k,t)$, where

$$G(k,t) = \frac{1}{H(k)}$$

then a filter can be produced with a spatial impulse response function, $g(x,t)$, that processes the raw Doppler LIDAR data, $v_e(x,t)$, to give, the true data, $v(x,t)$, as described in equation (3):

$$v(x,t) = \int_{-\infty}^{\infty} v_e(\alpha,t)g(x-\alpha)\,\mathrm{d}\alpha \qquad (3)$$

where, $g(x)$, is the inverse Fourier Transform of, $G(k)$.

2.1 WIENER DECONVOLUTION

The deconvolution approach appears to be obtaining something for nothing and one would intuitively expect that there is a limit to how high in wave-number this process can be pushed in practice. Such intuition is correct, and this limit is set by the inevitable uncertainties, treated as noise, so that in the real world equation (1) has the form

$$v_e(x,t) = \int_{-\infty}^{\infty} v(\beta,t)h(x-\beta)\,\mathrm{d}\beta + n(x,t) \qquad (4)$$

where $n(x,t)$ represents the unknown (noise) contribution to the velocity estimates. Clearly at wave-numbers where the contributions to $v_e(\beta,t)$ become comparable with that of the noise it gets increasingly less meaningful to extract information about $V(k,t)$, and hence $v_e(x,t)$, by the simple deconvolution division by $H(k)$ process. Norbert Wiener looked at this issue [6] and showed that, given equation (3), the best estimate of $v_e(x,t)$ (in a least squares error sense) could be obtained by filtering the raw data with a filter whose wave-number response function is given by

$$G(k) = \frac{H^*(k)V_p(k,t)}{H(k)H^*(k)V_p(k,t)+N_p(k,t)} \qquad (5)$$

where the quantities $V_p(k,t)$ and $N_p(k,t)$ are the Power Density Spectra respectively of the true input velocity data, $v(x,t)$, and of the noise, $n(x,t)$. The interpretation of equation (5) is that over wave-number ranges where the output in the noise-free case would be large compared to the noise (as present in the real case), the simple $V_e(k,t)/H(k)$ process is used, but over ranges of wave-number where this condition no longer holds then the contribution from the corresponding wave-number range is reduced to

$$A(k)\frac{V(k,t)}{H(k)}$$

where the factor $A(k)$ is given by

$$A(k) = \frac{H(k)H^*(k)}{H(k)H^*(k)+\frac{H(k)H^*(k)}{N_p(k,t)}} \qquad (6)$$

which is, by definition, less than unity.

The practical issue with using Wiener deconvolution is that it requires statistical knowledge (via the Power Density Spectrum) of both the unknown velocity and the noise. However, sensible estimates of these are often available and in any case a reasonable approximation to Wiener deconvolution is simply to terminate contributions from the deconvolution process from wave-numbers where the noise becomes comparable with that of $v_e(x,t)$.

2.2 LIMIT TO RESOLUTION ENHANCEMENT

From the above discussion it is clear that, in practice, it is only viable to compensate for the affects on the true data of the Doppler LIDAR measurement process over regions where the spectrum of the signal is comparable or larger than that of the noise. So recovering data from regions where the signal spectrum is very small, by simply multiplying by a large value of $1/H(k)$, is not possible. Thus the benefit that can be had is as follows. Over regions of the wave-number spectrum where: (i) $H(k)$ falls off steadily (rather than abruptly) to zero, as is the case in Doppler LIDAR, and (ii) the signal-to-noise ratio is "reasonable", then the effects of the low-pass filtering, which is inherent in pulsed Doppler LIDAR measurements, can be compensated for. This is equivalent to increasing the spatial resolution. The enhanced spatial resolution, $l_{\min}$, is relatively easy to estimate, assuming that deconvolution is terminated at the point where the signal power density spectrum is equal to that of the noise. This termination occurs at a wave-number, k_m, given by

$$\frac{H(k_m)H^*(k_m)}{N(k_m,t)} \approx 1 \tag{6}$$

This value of k_m corresponds to the maximum value that can be included in the spectrum of the enhanced data. Now, taking into account the requirements of the Spatial Sampling Theorem, this value of k_m corresponds to a length scale, $l_{\min}$, given by

$$l_{\min} \approx \frac{\pi}{k_m} \tag{7}$$

2.3 RESOLUTION *VS* TIME RESPONSE

It is the signal-to-noise ratio that controls the level of resolution enhancement which can be achieved. It is well-known that the signal-to-noise ratio can be enhanced by averaging, because if R different measurements are taken of $v_e(x,t)$, i.e. incorporating R different realisations of the noise $n(x,t_r)$ at R different times, then provided that $n(x,t_r)$ is statistically stationary over the total time involved, then $N_p(k,t)$ will reduce with increasing R. In particular, as R increases, the Central Limit Theorem states that, irrespective of the probability distribution of the noise, the reduction in $N_p(k,t)$ will tend to become directly proportion to R. This allows one to trade spatial resolution enhancement for temporal bandwidth, because if the values of t_r are a fixed time interval apart, then the temporal bandwidth reduces in proportion to R. In other words, the product of *spatial resolution* x *temporal bandwidth* is constant and so signal-to-noise ratio (and hence resolution) can be traded for bandwidth, i.e. for time response.

2.4 INCREASED HIGH WAVE-NUMBER NOISE

Irrespective of the details, any form of resolution enhancement will involve compensating for the reduction in size of the higher wave-number components present in the true data which is introduced by inherent spatial filtering present in the Doppler LIDAR measurement system. This compensation is generically termed high-pass filtering, and it is very easy to see that it inevitably worsens the signal to noise ratio at higher wave-numbers and hence increases the scatter in the data at those higher wave-numbers.

This occurs because while the spectrum of the true data inevitably falls off at large wave-numbers, in contrast the spectrum of the noise typically does not (this is apart from data at very low wave-numbers where so-called pink noise occurs). In fact spatial white noise has a flat spectrum. Consequently, given that $H(k)$ also falls off with k, dividing by $H(k)$ involves multiplying the relatively constant noise spectrum by an increasingly large number. Hence the overall signal-to-noise ratio is inevitably worsened at higher wave-numbers.

Again this is an entirely general result and hence it has to be accepted that increasing the spatial resolution by processing the raw data is inevitably at the cost of increasing the scatter in the enhanced data at high wave-numbers.

3. EXPERIMENTAL SETUP

In order to assess the effectiveness of the proposed resolution enhancement techniques, a land-based trial was established in which the beam from a pulsed Doppler LIDAR system (the Halo Photonics "Stream Line" system [7]) was aligned a few centimetres above the top of an array of twelve METEK sonic vector anemometers [8]. The sonic anemometers recorded at 20Hz with a velocity resolution of $0.5\text{cm}\,\text{s}^{-1}$. The Doppler LIDAR was capable of measuring wind variations at up to 8.3Hz, at a velocity resolution of $2\text{cm}\,\text{s}^{-1}$, with each velocity estimate representing an average over 1,500 measurements. The pulse length of the Doppler LIDAR was set to its minimum viable value of 18.5m. The data from the LIDAR system and the array of anemometers were logged time-synchronously.

Figure 1: The linear sonic anemometers. The LIDAR beam was targetted along and a few centimetres above the top of the sensor heads.

The trial arrangement is shown in figure 1, with the anemometers being spaced 3m apart in order to assess the target of 5m for the resolution enhanced Doppler LIDAR results. Clearly a uniform airflow would not allow assessment of the resolution enhancement; hence it was necessary to ensure that some spatial variation existed in the air stream along the sensor array. To this end, as shown in figure 1, the array was sited next to a building with the intention of inducing some curvature into the flow stream under the prevailing wind conditions. The general features of the resulting flow field were assessed on-site using a bubble stream generator.

The raw Doppler LIDAR line-of-sight velocity estimates were processed using the variant of Wiener deconvolution discussed in section 2. The cut-off point that arises naturally in the numerical execution of the deconvolution enhancement process suggested an expected resolution enhancement by a factor of approximately 3.5.

4. RESULTS

Figure 2 shows time series results obtained at the mid position anemometer (number 6) using the Doppler LIDAR in an unfocussed mode. Figure 3 shows spatial results where the absolute root mean square error (with respect to the anemometers) is plotted against anemometer number in the array.

5. CONCLUSIONS

The spatial and temporal results obtained show that the variant of the Wiener deconvolution process employed allows the velocity estimates to track the point anemometer data considerably closer than the raw Doppler LIDAR values. Furthermore, the cut-off point achievable in the Wiener deconvolution indicated an improvement in spatial resolution by approximately a factor of 3.5. Given these findings, together with the 3m spacing of the anemometer array employed, it is reasonable to claim that the resolution enhancement process is producing reliable velocity estimators at close to the 5m requirement.

6. ACKNOWLEDGEMENTS

The authors wish to express their thanks to the UK Ministry of Defence for their support of the work presented here.

7. REFERENCES

1. D.V. Stoyanov, L.L. Gurdev and T.N. Dreischuh. Reduction in the pulse spike-cut error in Fourier-deconvolved lidar profiles. Applied Optics, vol 35, no 24, pp 4798-4802, 1996.

2. L.L. Gurdev, T.N. Dreischuh and D.V. Styanov. Deconvolution algorithms for improving the resolution of exponential-shaped pulse lidars. Proc. of the Ninth International School on Quantum Electronics: Lasers-Physics and Applications. SPIE vol 3052, pp 310-315, 1996.

3. T.N. Dreischuh, L.L. Gurdev and D.V. Stoyanov. Effect of pulse-shape on the accuracy of deconvolved lidar profiles. Journal of the Optical Society of America A, vol 12, no 2, pp 301-306, 1995.

4. J. Bühl, R. Engelmann and A. Ansmann. Removing the Laser-Chirp from Coherent Doppler Lidar Datasets by Two-Dimensional Deconvolution. Journal of Atmosospheric and Oceanic Technology, vol 29, no 8, pp 1042-1051, 2012.

5. A.R. Bahrampour and A.A. Askari. Fourier-wavelet regularized deconvolution (ForWaRD) for lidar systems based upon TEA-CO2 laser. Optics Communications, vol 257, no 1, pp 97-111, 2006.

6. R.C. Gonzalez, R.E. Woods and S.L. Eddins. *Digital Image Processing Using Matlab*, Prentice Hall, 2003.

7. Halo Photonics, http://halo-photonics.com/Halo%20Photonics%20Stream%20Line%20LiDAR%20system.htm, last accessed 11 July 2017.

8. METEK GmbH, http://metek.de/product/usonic-3-omni/, last accessed 11 July 2017.

9. C.J. Brownell, L. Luznik, M.R. Snyder and H.S. Kang. In Situ Velocity Measurements in the Near-Wake of a Ship Superstructure, Journal of Aircraft, vol 49, no 5, pp 1440-1450, 2012.

8. AUTHORS' BIOGRAPHIES

M R Belmont holds the current position of Professor of Marine Dynamics at the University of Exeter where he is responsible for maritime systems research. His previous experience includes modelling and land and sea trials of Doppler LiDAR systems.

J Christmas holds the current position of Lecturer in Computer Science at the University of Exeter where she works on maritime systems research and statistical machine learning. Her previous experience includes land and sea trials of Doppler LiDAR systems.

JM Duncan is Group Leader for Simulation Based Acquisition and Technical Computing for Maritime Combat Systems, UK Defence Equipment & Support (DE&S). The Group's focus is on whole ship and submarine safety and technical services including new modelling techniques for complex dynamic interfaces such as for manned and unmanned air and sea vehicles. Duncan is the NATO Principal Representative for the Naval Armaments Group on Virtual Ships. He was previously chairman of NATO Subgroup 61 and the Specialist Team on Simulation Based Design and Virtual Prototyping for ship acquisition. He led the development of Allied Naval Engineering Publications '61' for ship virtual prototyping and '84' for Virtual Ship simulation standards.

J Duncan has served for 6 years as the manager of sea trials data collection and coordination on behalf of MOD. He has supervised 10 sea trials data collection activities on behalf of the NATO Submarine Rescue System team including substantial planning, post-trials analysis and reporting. He has been involved on various simulation projects for MOD supporting Type 45, Merlin, QEC, Ship Air Interface Framework and Quiescent Period Prediction. He has worked on the development of NATO Virtual Ships documents STANAG 4684 and ANEP 84.

B Ferrier holds the current position of Systems Engineering and Technology Associate (SETA) in the offices of NAVAIR PMA 266. He is responsible for assisting NAVAIR in integrating Fire Scout Endurance upgrade technologies, such as the ONR Landing Period Designator, into the Fire Scout Launch and Recovery System. He is also the Engineering Head of the Hoffman Engineering's Dynamic Interface Lab in Arlington, VA and holds an honourary chair at the University of Exeter.

R Potts holds the current position of Senior Aerodynamicist at BAE Systems. He is responsible for Air-Vehicle Ship Integration. His previous experience includes land and ship based development test of aircraft flying qualities as well as land and sea based LiDAR trials.

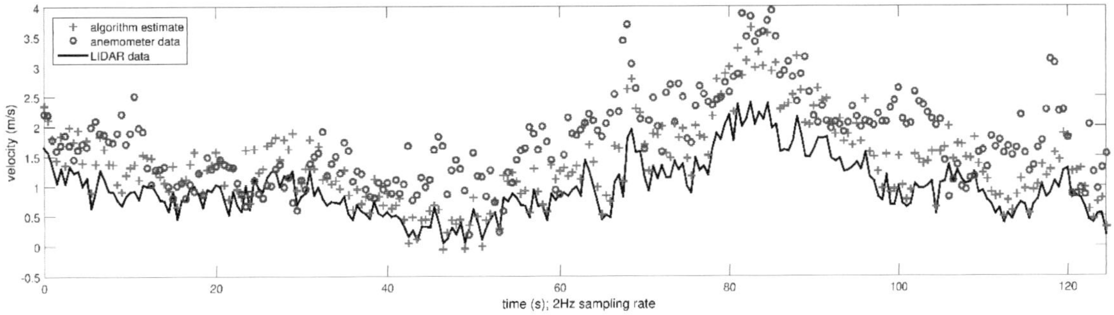

Figure 2: A comparison of the raw and processed Doppler LIDAR estimates against the anemometer data at the mid-point of the anemometer array.

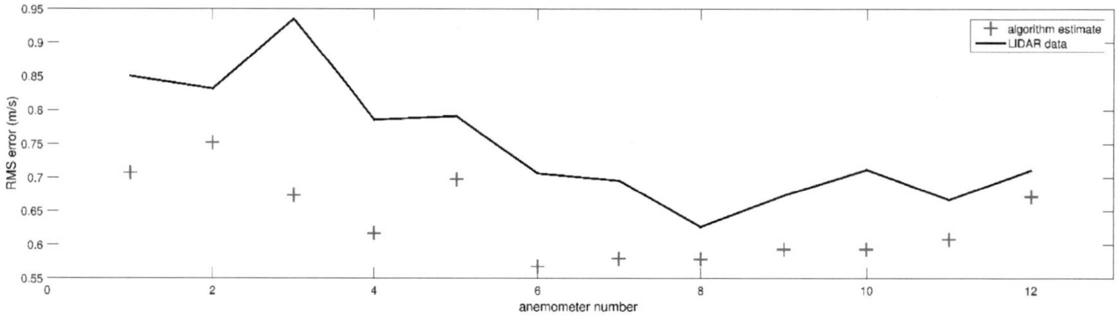

Figure 3: Absolute root mean square error for the raw and processed Doppler LIDAR.

ENSURING SOFTWARE RELIABILITY, SAFETY, AND SECURITY: LESSONS LEARNED

D.N. Card, DNV GL, USA
L.G. Chua, DNV GL, Singapore

SUMMARY

This article describes the motivation, approach, and results of an effort to introduce modern software quality assurance practices into the construction of mobile offshore drilling units (MODUs). This approach is applicable to the construction of other complex vessels such as cruise ships and LNG carriers where reliability, safety, and security are principal concerns. Experience with high specification drilling vessels indicates that undependable software contributes significantly to operational down-time and loss of revenue for operators. DNV GL published the optional class notation OS D-203, Integrated Software Dependent Systems (ISDS), in 2010, in an effort to address these concerns. ISDS defines a set of best practices for software engineering and system integration, based on generic international standards. DNV GL's software verification approach focuses on ensuring that software arrives at the shipyard thoroughly tested and ready to be integrated. ISDS has been applied to eight Mobile Offshore Drilling Units built in Korean shipyards.

NOMENCLATURE

ISDS	Integrated Software Dependent Systems
LNG	Liquefied Natural Gas
MODU	Mobile Offshore Drilling Unit
SQA	Software Quality Assurance

1. INTRODUCTION

This article explores the status of software quality standards development and application related to the offshore industry. It summarizes the standards currently in use, as well as trends. This article elaborates on the framework for software quality introduced in [1]. That framework includes process, product, operational, and customer perspectives. As will be shown, security, safety, and reliability are aspects of software quality. This article summarizes the quality standards currently in use, as well as ongoing trends.

Recent experience with high specification drilling vessels indicates that undependable software contributes significantly to operational down-time and loss of revenue for operators [2]. These software and integration challenges are magnified in large-scale projects with many software-dependent systems. Roy Robinson [3], Global Contracts Manager for Repsol, identifies 14 points of failure for E & P megaprojects. One of them is integration of software and control systems.

In order to set the stage for a discussion of software quality standards for the offshore industry, it is necessary to discuss the nature of software quality and software quality assurance in order to understand what can be standardized. The first question we have to answer is "What is software quality?" Quality may be viewed from many perspectives. Figure 1 illustrates some of them.

The **software engineering process** may be more or less capable of producing a good product. Process quality may be defined in terms of process standards or process reference models based on "best" (or at least "good") practices. Of course, in order to provide benefit, these practices must be implemented by the developers.

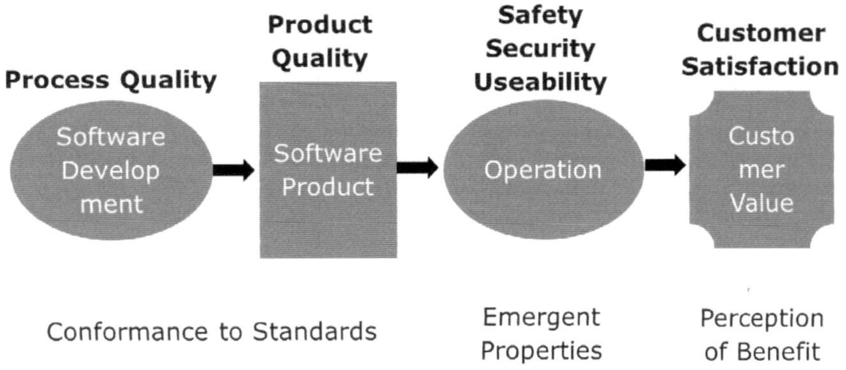

Figure 1. Software Quality Perspectives

The **software product**, itself, may be evaluated in terms of conformance to appropriate product standards. The process perspective has proven to be more important for software quality than for typical manufacturing industries, because software product standards have proven harder to agree upon, and logically, different standards are needed for different products. Software product standards typically define characteristics that are associated with successful operational use and maintenance of the software. Software product standards commonly focus on user interface and networking issues. Data may also be considered a software product, that can be standardized for specific applications such as maps

The **operating software** may be evaluated in terms of the quality of service provided, such as security, safety and usability. As emergent properties, these qualities are not amenable to standards. They must be evaluated in the overall system context. For example, the safety of a software component depends on how the user interacts with it as well as the mechanical equipment attached to it. Conformance to product standards (such as "no single point of failure") cannot ensure safety, but can make it easier to achieve. While Safety is a special concern for offshore vessels, most safety analyses assume that software is 100% reliable, but reliability also depends on the context of use. Security has largely been ignored in this industry. For example, many suppliers deliver software with a hard-coded password. Nevertheless, quality (as freedom from defects and nonconformances to requirements), is a pre-requisite for both safety and security. Research shows that 50% of all security vulnerabilities are due to software defects [4]. Clearly, software quality has an immense effect on the customer's perception of the operational product, even if it is not the only factor.

The **customer's satisfaction** is based on his experience with the product as compared to his expectations of it. Satisfaction is a state of mind, not a physical state. Managing customer expectations is as important in achieving customer satisfaction as managing operational quality. Establishing standards for customer satisfaction is very difficult.

As indicated in Figure 1, the software process and software product are the views of software quality that are most amenable to standards. How do standards fit into software quality assurance? Standards provide expectations of activities to be performed or outcomes to be achieved. Satisfaction of these expectations can be confirmed by direct observation (verification) or quality assurance.

2. SOFTWARE QUALITY ASSURANCE

Software quality assurance is often equated with software testing. That is wrong. The following definition from Wikipedia helps to clarify the current consensus:

Software quality assurance (SQA) consists of a means of monitoring the software engineering processes and methods used to ensure quality. SQA encompasses the entire software development process, which includes processes such as requirements definition, software design, coding, source code control, code reviews, software configuration management, testing, release management, and product integration.

The first activity recognizable as software quality assurance was implemented in the Atlas missile and Mercury spacecraft programs during the 1960s. The term "Software Quality Assurance" was introduced during the same 1968 NATO conference [5] at which the term "software engineering" was coined. Unfortunately, many vendors in the maritime and offshore market rely only on testing to assure the quality and dependability of software.

The traditional V-model (Figure 2, [6]) can be recognized as the basis of most software development and quality standards. Generally these standards area not written at a level that is directly executable. A more detailed level of operating instructions must be developed for an organization to implement the standards.

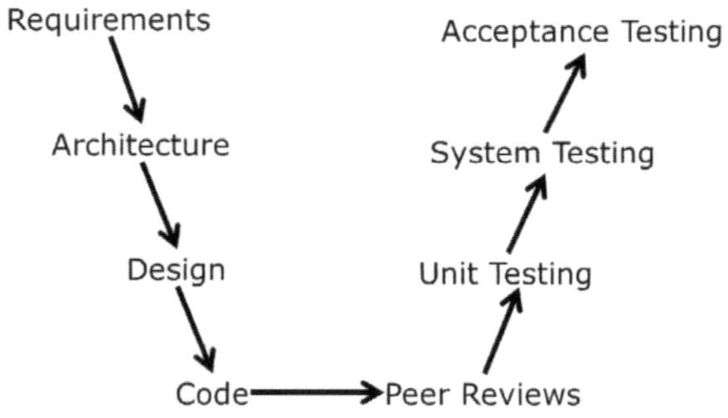

Figure 2. V-model of Software Development

Consequently, these standards are often referred to as process refence models. A process reference model is a standard that provides a high-level description of a process. An organization implements a process reference model by developing more detailed guidance (e.g., standards, procedures, training, tools). Some widely accepted process reference models are as follows:

- SEI Capability Maturity Model – Integrated [7]
- ISO/IEC/IEEE 15288: Systems Engineering [8]
- ISO/IEC/IEEE 12207: Software Engineering [9]
- ISO/IEC 90003: Software Quality Systems [10]

Independent verification is especially important to purchasers of software-intensive systems, because it provides objective information about the quality of the supplier's process. Baker, et al. [11], explain the need for independent verification.

Even if quality is everyone's job (as widely proclaimed), there must be a coordinating function to ensure that all necessary actions get carried out. More detailed guidance for the implementation of a *software quality assurance function* internal to an organization can be found in IEEE 730:2014 – IEEE Standard for Software Quality Assurance Processes. This is a widely accepted, non-industry-specific standard. It provides requirements for initiating, planning, controlling, and executing the Software Quality Assurance processes of a software development or maintenance project for any application. This standard is harmonized with the software life cycle process of ISO/IEC/IEEE 12207:2008 (a *comprehensive process reference model* for software engineering organizations [9]).

The general software engineering and standards communities have provided a lot of material for the maritime and offshore industry to work with, but what has been done with it?

4. SQA IN MARITIME AND OFFHSORE

There is no specific national or international regulation addressing Software Quality Assurance (SQA) in the Maritime and Offshore industry. Some recommendations and guidelines are emerging. IMO Circular 645/1994 requires that for DP systems: "The software *should* be produced in accordance with an appropriate international quality standard recognized by the Administration." (Note that the clause does not say *shall.*) The international quality standards, ISO 90003 [10] and ISO/IEC 15288 [8], have appeared in Technical Building Specifications for vessels in Korea. A software quality assurance guideline for the IMO e-Navigation initiative is under development (NCSR 2/6-28 March 2014). The IMO is encouraging class societies to develop more detailed standards to implement the SQA guideline. However, adoption of these standards is still voluntary.

Individual system-specific standards also have appeared in the maritime and offshore industry, for example: IEC 60945:2002 – Maritime Navigation and Radiocommunication Equipment and Systems – General Requirements – Methods of Testing and Test Results. While this is a welcome contribution, it also raises the question of whether some more generally applicable rules might be easier to adopt and provide greater value to the industry than a plethora of equipment-specific standards.

Three classification societies (DNV GL, ABS, and LR) have published software process reference models (standards) for offshore vessels. The DNV GL and ABS standards already have been applied to vessels constructed in Korea. These standards are based on widely accepted ISO/IEC standards that have been endorsed by national standards bodies. The LR standard is a direct implementation of ISO/IEC 15288 [8]. Suppliers working on projects requiring one of the class rules for software quality have formalized and improved their software quality assurance programs in response. However, the number of such projects is still small, so progress across the industry has been slow.

5. THE ISDS EXAMPLE

Integrated Software Dependent Systems (ISDS) [12] is a software process reference model customized to the maritime and offshore industry. It provides more detail than the ISO/IEC standards on which it is based, but it still is not intended to be executable. That is, the implementing organization is expected to develop more detailed standards and procedures for staff to work with.

ISDS is an optional rule. The owner of a new-build vessel must request the application of ISDS, and pay extra for suppliers and the shipyards to implement it. DNV GL serves as the independent verifier to ensure that the ISDS requirements are implemented effectively. The ISDS standard is undergoing its third revision to incorporate lessons learned from experience to date.

Figure 3 provides an overview of ISDS. This standard includes 119 activities, each of which is mapped to a confidence level, role, and life-cycle phase. Confidence levels are cumulative, so that it is necessary to implement the Level 1 and 2 activities in order to achieve Confidence Level 3. Thus, the choice of a confidence level is a choice about the amount of investment in software quality assurance activities. ISDS also attempts to account for the fact that vessel systems are developed following a component-based approach, rather than town-down [13].

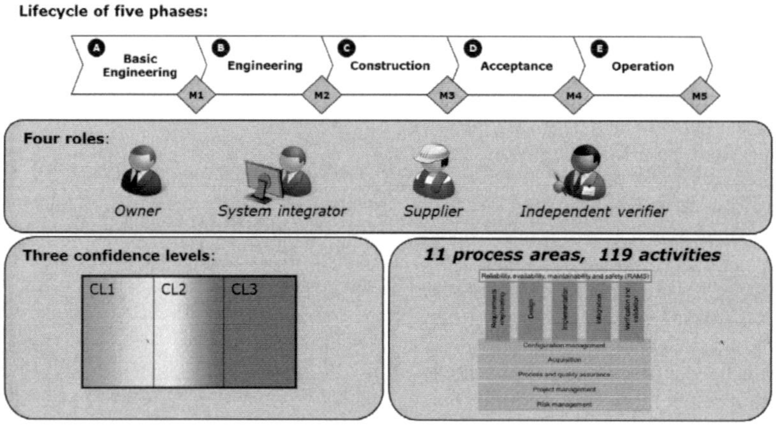

Figure 3. Structure of ISDS rule.

A comparison of ISDS with ISO/IEC 90003 (available from the author) shows that ISDS covers all ISO/IEC 90003 requirements except those related to training, organization, environment, and measuring devices.

As cybersecurity has become an increasing significant concern, ISDS has been extended to include basic security practices, such as development of a security policy and performance of audits of conformance to it. DNV GL has also developed a recommended practice for maritime cyber security [14]. Most recently, DNV GL has partnered with insurers to offer a cybersecurity audit as part of the underwriting process (Cyber Risk 360).

These audits consider cybersecurity risks in four dimensions:

- Threats from potentially malicious actors
- Intentions specifically targeting an asset or company
- Vulnerabilities due to good and bad cybersecurity practices implemented
- Potential Impact of cyber attack on the asset

An owner whose assets score well may get a better rate on insurance covering cyber attacks. Alternatively, coverage of cyber attacks may be excluded from the policy.

Of course, these audits provide a snapshot at a point in time. The cybersecurity threat is continually evolving, so an effective cybersecurity program must be able to adapt to the changing threats.

6. LESSONS FROM ISDS APPLICATION

To date, ISDS has been applied to 8 mobile offshore drilling units constructed in Korea and the early stages on one built in Singapore. Five of these vessels have gone into operation without experiencing significant problems with the control systems. Three are in the final stages of construction. One vessel upgrade has also been completed according to ISDS in Norway. In addition to classification of complete new build offshore projects, aspects of ISDS have also been used in verification of control systems for cruise ships as well as technology qualification of an autonomous underwater vehicle.

ISDS cannot ensure a successful outcome, by itself. ISDS provides information about status and quality that helps make better decisions during the management of the construction process. Someone must act on that information. One of the key lessons of ISDS is the importance of the on-site owner project team. They are in the best position to make use of ISDS data and take appropriate action to affect the outcome of a newbuild project.

Related to this is the need for good communication within the project team. The natural tendency of shipbuilders and vendors seems to be to restrict the flow of information to the shipbuilder and owners. This is partially motivated by a concern for the loss of proprietary information, but also, at times, due to a reluctance to admit to delays and mistakes. Good communication requires a conscious effort. ISDS requires milestone meetings (involving all stakeholders) at least at the end of each phase. Some projects scheduled additional meetings because they were so helpful.

Another key consideration is that ISDS should be implemented as part of the normal engineering process, not as a separate activity. Design reviews, FMECAs, etc., should be extended to encompass software issues, rather than organizing an expensive parallel system of activities focused only on software. Of course, this means personnel have to be trained in handling software tasks.

During our verification of software installations for ISDS projects, we discovered many cases where software and/or firmware was out of date. In some cases, the vendor had installed a version with known security vulnerabilities even though a more recent, more secure version was available. Owners of all vessels should pay attention to this, whether or not ISDS or a similar standard is applied.

7. CONCLUSIONS

Process and product standards are recognized in most industries as essential to consistently delivering good quality products. Conformance to such standards is a condition of being allowed to do business in industries such as medical devices, aerospace, and automobiles. Many studies have confirmed that following such standards leads to improved quality and reduces the probability of delays, e.g. [15].

Regulations and liability (insurance) concerns will eventually force the maritime and offshore industry to accept greater requirements for software quality assurance, as has happened in other major industries. *For now, if a customer wants quality, he/she must ask for it. These new standards make that easier.*

8. ACKNOWLEDGEMENTS

The authors would like to acknowledge the contributions of owner site teams and shipbuilders' ISDS teams to the evolution and success of ISDS.

9. REFERENCES

1. CARD, D., 'The Role of Software Quality in Maritime Safety and Security', *Proceedings: ASME Conference on Ocean, Offshore, and Arctic Engineering*, 2016.

2. CARD, D., T. COQ, and R.B. JOHANSEN, 'Reducing Commissioning Delays and Operational Downtime for High Specification Drilling Units', *Proceedings: Rio Oil & Gas Conference*, 2012.

3. ROBINSON, R., 'Offshore Megaprojects – Why we fail and how to fix it' *Proceedings: Rio Oil & Gas Conference*, 2012.

4. WOODY, C., *Predicting Software Assurance Using Quality and Reliability Measures*, Software Engineering Institute, 2014

5. NATO, *Software Engineering, Report on a conference sponsored by the NATO Science Committee*, October 7-11, 1968

6. German Federal Armed Forces, Das V Modell, as per *German Directive 250, Software Development Standard for the German Federal Armed Forces, V-Model, Software Lifecycle Process Model*, August 1992

7. CHRISSIS, M., M. PAULK, and S. SHRUM, *Capability Maturity Model - Integrated*, Addison Wesley, 2003

8. ISO/IEC 90003:2004 Software Engineering – Guidelines for Application of ISO 9001:2000 to Computer Software.

9. ISO/IEC/IEEE 12207:2008 Systems and Software Engineering – Software Life Cycle Processes

10. ISO/IEC/IEEE 15288:2008 Systems Engineering – System Life Cycle Processes

11. BAKER, E., M. FISHER, and W. GOETHERT, *Basic Principles and Concepts for Achieving Quality*, Software Engineering Institute, December 2007

12. Det Norske Veritas (DNV), OS D-203, *Integrated Software Dependent Systems (ISDS)*, December 2012

13. CARD, D., A Software Integration and Process Model for Offshore Vessels, *Proceedings: Asia-Pacific Software Engineering Conference*, 2013

14. DNV GL AS, DNVGL-RP-0496, *Cyber Security Resilience Management for Ships and Mobile Offshore Units in Operation*, 2016

15. GOLDENSEN, D., and D. GIBSON, *Demonstrating the Impact and Benefits of CMMI®*, Software Engineering Institute, 2003

10. AUTHORS BIOGRAPHIES

David N. Card is a Senior Principal Specialist in software quality and reliability working in the Systems and Components Approval Department of DNV GL, USA. He spent five years in Korea leading software quality assurance projects for newbuild Mobile Offshore Drilling Units.

Lian-Gee Chua holds the current position of Business Development Manager, Principal Engineer at DNV GL Singapore Pte Ltd. He is responsible for classification business development of the Tanker and FSRU segments in SE Asia. His previous experience includes ISDS verification activities for MOU newbuild and equipment suppliers.

PROTECTING INTELLECTUAL PROPERTY RIGHTS IN DISTRIBUTED CAD ENVIRONMENTS

L. Seppälä, CADMATIC, Finland

SUMMARY

Mega trends such as globalization and digitalization shape the future of the marine industry and shipbuilding. They have impacted significantly on the way design processes are handled and business processes conducted in the whole life cycles of marine industry. As projects become more complex and the demands for digital 3D models and production data increase, the most valuable intellectual know-how of organizations will increasingly be stored in digital project environments.

This paper discusses the evolution of business processes related to Intellectual Property Rights (IPR) in distributed marine projects, future intellectual property issues in design tooling, and the role that modern CAD tools can play in it with respect to extended use of digital assets.

1. INTRODUCTION

Through the centuries of its history, shipbuilding always was considered as a science-intensive industry. Shipbuilding requires knowledge, expertise, and skills in a rather narrow field of science and with constantly changing technology. Historically, ship design was one of the first industries to harness the power of computer technology in the design process. Already at the earliest stages of computer technology development, there were use cases for its computational power being utilized in ship design. Shape forms and stability calculations were performed using computers back in the era of punched cards. At that time, only a few people could forecast what later became the everyday life of ship designers: the use of complex, interactive multi-user CAD systems with embedded best practices of design and possibility to output numerical data directly to CNC machines, warehouse storage systems, and welding robots for production. Further advancements include the utilization of digital, data-rich 3D models in the life cycle of vessels and constructions, and the possibility for remote manipulation of the operations.

Analyzing current trends that shape the marine industry at present provides a basis for understanding the future. This overview aims to map current trends in digitalization and to look into the relevance of IPR in the design process and what role CAD tools play in it.

2. DESIGN AND BUSINESS PROCESS

The design process in shipbuilding and marine design has changed significantly over the last decades, mostly due to the rapid development of IT infrastructures and CAD software.

2.1 SHIFTS TOWARDS CAD AS ESSENTIAL STEP

The process evolved from using computers as complex calculus machines to being used as visualization engines and data storage facilities. In the early days of CAD, the 3D model was considered as a complimentary, additional entity. Today, the 3D model is a central piece of the design.

Originally, the goal was to use the 3D model for facilitating arrangements and emulating different possibilities quickly and without costly workshop hours. Automatic or semi-automatic extraction of production data was an immediate gain in productivity. Instead of manual drafting which often caused mistakes and inaccuracies, it became possible to check the 3D model for consistency, often automatically.

Knowledge that used to be spread between the design and drafting processes is now focused in 3D modelling. The simplicity and intuitiveness of 3D allowed people who might have no previous shipbuilding experience to access the design.

Intangible expertise and innovations are increasingly stored in the CAD environments. These include components and architecture, layouts, new technologies and other means. Having all that data inside the CAD environment means easy access to it for the company's own designers and possibility to ease knowledge transfer within the company.

During that time several major players emerged to the CAD market, each with their own philosophy behind the product. Some were pure IT companies, facilitating 3D and 3D CAD/CAM engineering needs, others were born out of engineering companies' own resources to facilitate and automate the tasks of their designers.

2.2 MULTI USER AND MULTI SITE DISTRIBUTED CAD ENVIRONMENTS

As usage of CAD environments grew and spread quickly, it started to shape the business processes around it. The natural need to involve many professionals in the design process resulted in the creation of multi-user and multi-site CAD environments. Simultaneous access to the 3D model and the possibility to share designs through viewer files are the most important features of a modern CAD system. The need for this functionality originated in the business process – there was a need to outsource parts of the design work and involve narrowly specialized professionals.

From the knowledge perspective it means an extensive sharing and exchange process. Data rich 3D models can be seen as a storage of the most valuable knowledge. Through sharing of 3D models and opening access to the data, companies provide access to their most valuable assets. However, at the same time, they gain the advantage of receiving external knowledge and expertise. This process is normally backed up by various legal agreements and contracts, often including a Non-Disclosure Agreement (NDA). But, one can ask whether these measures are sufficient, and what role can CAD itself take in this process.

2.3 FROM DATA MODELS OR PRODUCTION TOWARDS DIGITAL ASSETS

Historically, the final goal of the shipbuilding process was to build a vessel: from the initial design phase to a sea trial. Automating 2D documentation extraction and workshop production output were the topmost concerns of the design process. In our days, there's a clear understanding that a digital asset such as a data-rich 3D model has value far beyond the hand-over event. This is where the process goes to the boundary area of the business processes – to operation and maintenance and fleet operations.

3. MEGA TRENDS SHAPING THE FUTURE OF MARINE INDUSTRY

Shipbuilding and marine industries are an integral part of industrial development and, in a wider context, of the society as a whole. Therefore, any trends shaping the future will unavoidably affect these industries. At the global level, there are several mega trends that prevail in future development. These are presented in figure 1 below and discussed in more details in this section.

The presented overview and analysis mostly target the exploration of alternative futures – for knowing what needs to be done in the present, not for knowing what will happen in the future [1]. Contemporary futures studies focus on analyzing existing trends and signals in order to draw a complex picture of the possible, probable and preferable futures. Identifying preferable scenarios helps to develop a strategy in the present. In the context of this paper, a few main mega trends were collected to illustrate the current changing environment for CAD with respect to data handling. Understanding the dynamism and rapid changes in a world that is becoming ever more complex is a primary goal to succeed in the future [2].

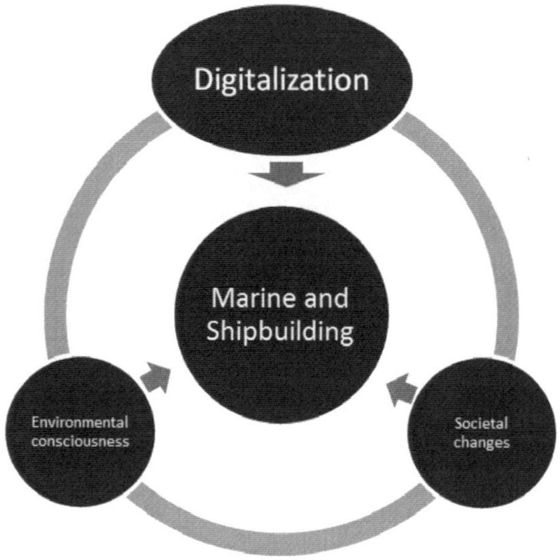

Figure 1: Global mega trends shaping the futures of marine design and shipbuilding

3.1 DIGITALIZATION

In the shipbuilding industry, the process of digitalization started a long time ago and it remains as a prevailing megatrend that is shaping the industry. From facilitating calculations to 2D drafting, 3D models, automatic extraction of production data to using 3D model as a central hub for collecting information about the project, digital handouts, connection of design and building models to maintenance and operational purposes and using AR and VR for autonomous ships and digital fleet management, and as for example in the latest Rolls Royce vision [3].

This significant step forward in the marine domain was not possible without mature IT technology and hardware. Despite numerous predictions and visions from the past, only now we're reaching the needed level of technology to enable such innovations as autonomous sailing and optimized automatic handling of logistics in the transportation.

Digitalization as a global trend includes various aspects of the influence of digital innovations, such as Big Data, AI, IoT, SaaS, cloud based technologies, robotics, cyber security, blockchain technology and others. The process of digitalization started decades ago reaches the level of highest impact to overall shipbuilding and marine industries. It impacts organizations, ways of working, means of everyday work and business processes.

Expectations for this domain are significant. There are strong legal and ethical concerns and need for research and development to make this technology into use in commercial projects, and implementation takes its time.

3.2 SOCIETAL CHANGES: Globalization versus Nationalization, Millennials and Gen Y, and knowledge based business.

Globalization prevailed in the previous decades; it affected the openness of markets and trade, free sharing of knowledge and accessibility of global resources. EU, WTO and other big steps in the direction of global environment made most of the companies to target global operations. Long ongoing discussion raises questions of the pros and cons of this process, as for example in Forbes [4]. In recent years, the rise of a nationalization trend has made people consider the interests of their nation over the global market, as for example was demonstrated in Brexit and in the elections in the USA. The conflict of the two trends poses a threat to democracy, as it affects all areas of the societies [5].

This conflict seems at first glance to be far away from the marine domain. However, the marine industry gained a significant advantage from the global approach of the last decades; from the possibility to involve specialists from many countries and continents to implement the best available practices, to access the cheaper labor market for bulk work and meet tight budgets and ever shortening time schedules of the projects. Now, this relatively well established process faces national boundaries and the need for IPR protection.

Shifts in generation's values also increasingly affect the industry. Societal changes are not explicitly evident, but make a great impact on the industry. Younger generations of engineers burst with creativity and desire to make a change, and at the same time do not accept difficult processes or interfaces; they rely on technology to handle routine or background tasks up to a great extent. Knowledge definition shifts from passive storage of various data and practices towards the ability to apply these already stored standard solutions for innovation. This expectation means that younger engineers instinctively presume software to serve as a knowledge bank and provide them with best available practice, while their role is to select the most suitable solutions, and to combine and improve the existing ones.

3.3 ENVIRONMENTAL CONSCIOUSNESS, Resource Scarcity and Sustainable Development.

Increasing awareness of our environment and the impact of human activities on it creates an additional force that shapes the marine industry. Tight regulations on ballast water, CO, NOX, etc. aim to minimize the damage caused by ship operations. Energy efficiency drives innovations in propulsion and power generation mechanisms. This overall shift is a needed change of paradigm in post-industrial society [2].

In the case of marine industry, most of these aspects are included in the digitalization movement. Reducing the impact on the environment and ensuring sustainability by means of manipulating virtual 3D models before construction begins was the greatest change that CAD brought into shipbuilding decades ago. Further digitalization process is expected to have an even bigger impact.

4. EVOLUTION OF DESIGN AND BUSINESS PROCESSES

It is easy to trace the influence of the mega trends mentioned above in the evolution of the design processes in shipbuilding.

4.1 TECHNOLOGY

Technology advancements enable effective data creation, storage, and usage. Advanced tools for data creation within CAD environments facilitate design process, in-built knowledge help to design more complex constructions with fewer errors, based on predefined rules, specifications, and best-in–the-industry knowledge.

Vessels are created by several parties (owner, design company, architects, shipyard, subcontractors). At the same time the vessel or construction itself is not a final product anymore. This is reflected in a shift in demand from ship owners and fleet operators. These bodies need complete digital models of a vessel for operation and maintenance purposes, and aim for further digitalization that automated production can provide.

This process changes business approach from a product offering (a ship) towards co-created value of service – involving all players into the process and ensuring the chain of information flow from one stage of the life cycle to another.

4.2 GLOBALIZATION

For shipbuilders, globalization opened access to the world of diversity of resources. Diversification of the marine business area means that often specialists from other fields are involved. Shipbuilding is no longer a closed world of like-minded and similarly educated professionals. Often workforce from subcontractors involve dealing with larger amounts of bulk engineering work or professionals from other fields participate in the process. For example for interior design, in cruise liners often hotel interior design companies are involved to make the best possible customer experience on board. Growing complexity of the vessels and platforms demand narrowly specialized professionals in the field. Often, such professionals are not available inside a single design company or shipyard, but they need to access the whole project data to be effectively involved.

4.3 INNOVATIONS IN MARINE INDUSTRY

After being a rather closed industry for a long time, the marine industry is under increasing pressure for more openness and integration into the global values chains. More players enter the field of providers and suppliers for marine projects, and the use of floating constructions expands from traditional vessels to floating houses and power stations. All processes mentioned above impact the design companies, their needs and how software providers can serve them.

Over the decades, shipbuilding was a separate field from airplane building or automobile industries, for example. Partly this can be explained by the complexity and traditions of the field, but at most the reason is in the different nature of the design. The automobile industry focuses on mass production and operates with small-sized models where curved geometry plays a significant role. Airplanes involve heavy calculations, while production can be effectively reproduced once the prototype is tested. In shipbuilding, there are mixed demands on extensive knowledge and 3D arrangements with constantly changing types of equipment used. No sister ship is exactly the same, and in this regard shipbuilding is more similar to houses or architectural buildings.

5. ROLES OF CAD

One of the greatest impacts of digitalization is the value of the co-creation process. It includes all participants of the design, building and operation phases.

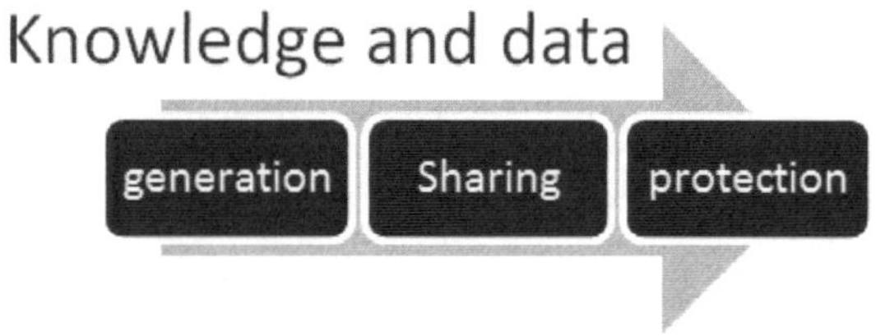

Figure 2: CAD roles in knowledge and data

Software providers are not traditional suppliers of a tool itself anymore. They become deeply involved in the value created during the life cycle of a design project. This way software is not only a tool, but it is also a hub of knowledge. It stores libraries of parts and fittings according to various standards, in-built design rules and specifications, predefined modules and parts of the vessels. In the future, this might be one possible scenario of the role CAD plays in the Marine area – as a keeper of knowledge and provider of the platform.

Presently roles in shipbuilding and marine design process are already divided into smaller fields. It becomes seldom that you meet a specialist who would be equally knowledgeable in hull structure, outfitting, automation and electric fields. The amount of knowledge needed for all these fields exceeds the capacity possibilities of the human mind. However, CAD environments store even more expertise. Highly sophisticated programs cannot substitute engineers, but can simplify work and help greatly. Predefined rules, class society regulations, libraries, specifications, compatibility and consistency checks eliminate the need for the engineer to remember or check these details every time.

This process of changing the role that CAD plays goes on slowly but steadily. The main outline, as shown in Figure 2, originates from knowledge facilitation or simple modelling to knowledge sharing and goes further in need of knowledge protection.

6. IPR IN CAD DISTRIBUTED PROJECTS

Conflicting demands for facilitating knowledge creation and sharing and need for IPR protection at the same time set the requirements for the digital facilitator of the marine design – CAD environment. For example, existing EU regulations for IPR in industries [6] or more specific for shipbuilding industry [7] do not set any particular regulations towards data sharing within distributed CAD environments.

This leaves CAD providers to decide the level and the means of the software to ensure company's IPR are controlled without damage to the fragile value co-creation process within the design and in the stages that come after it.

6.1 EXPORTS OF DATA

One possible source of information leak is various export interfaces. This was seen as a dangerous point in the digital data historically. Often the unwillingness of a CAD provider to create interfaces and disclose their database structures is related to the fear of allowing others to access their core data. This was described well in details in several publications, for example in [8], and causes artificial borders in the interoperability of CAD software.

Lately, with the development of neutral data formats such as IFC and JT or STEP, this problem fades away. Often these interfaces are separately licensed and can be easily controlled with simple administration settings.

A similar process with mechanical CAD was significantly simpler as data models are lighter and do not contain similar topological data. Due to these reasons mechanical CAD is easier to integrate with PDM/PLM. However, these deeper concepts target mostly building process and facilitation of business processes around it and leave without much attention further use of digital assets. An example of such process was described in details in [8].

6.2 CORE DATABASE ACCESS

Most common mechanisms provide restriction of access to certain areas or parts of the databases. So called filtered replication provides control of the access to connected sites.

This way the main design company can control what parts of the project will be visible for other participants. Often this process is complex and needs additional definitions and setup. However, it could be approached from the designer's perspective: using the 3D spaces as a definition of area. This is often a way to separate design team's work and make sure the integrity of the design is met by placing the responsibility for each area inside the team and controlling only overlapping areas with all teams involved.

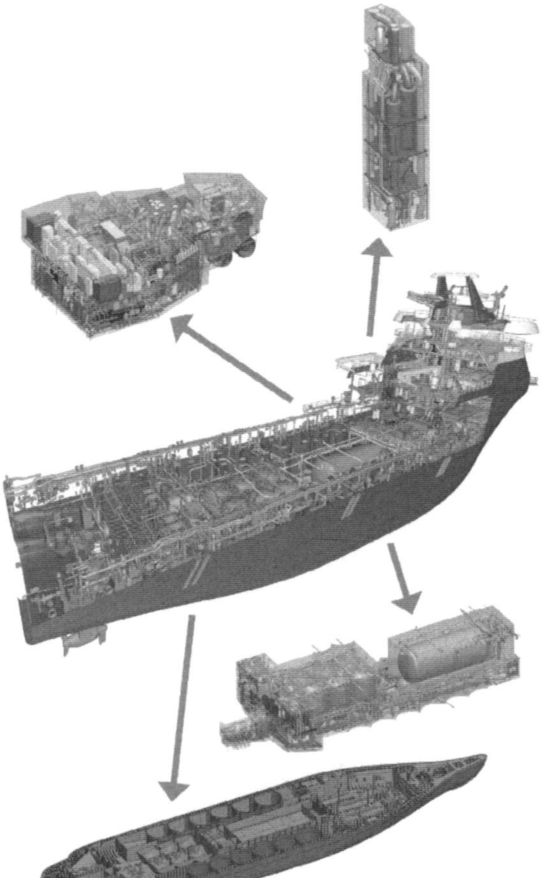

Figure 3: Illustration of the distributed design based on the area: different involved parties access only designated area of the whole project. Project model courtesy of Wärtsilä Ship Design Norway AS.

6.3 VIEWER FILES

Besides the data-rich native CAD database, traditionally for a business manager and other participants who are less involved in the core design process, extracts from the database visualized through 3D viewers are used to facilitate access to the design. These files contain a limited amount of data and do not present an imminent risk of losing the IPR. However, design solutions used in particular areas can be considered as a valuable asset to the design company and therefore should be protected. With a filtered replication process as described above, viewer files available on each site can not include areas that are not meant for the site.

An additional layer of security is a set validity date of the viewer files. Providing access to see and walk around the 3D during the design and building phases might be essential, while leaving the indefinite 3D model that potentially can be used as a sketch for next projects is often seen as a needed limitation by the owners of the design.

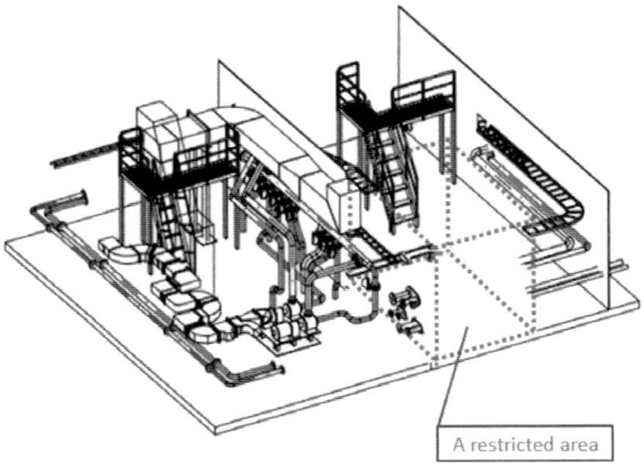

Figure 4: Illustration of the distributed design 3D model with restricted access area

7. CADMATIC APPROACH TO IPR IN DISTRIBUTED CAD ENVIRONMENT

CADMATIC has the roots of its history in the era of plastic 3D models of engine rooms created by engineering companies to facilitate its demanding tasks. Originating from two engineering companies (Elomatic, Finland and CIG, The Netherlands) it has deep connections to the marine industry. The product quickly gained popularity in the engineering world, under the former brand name Nupas-Cadmatic, pioneering in CAD usage with main goals of using the power of the 3D arrangements and assistance in 2D automatic production documentation for shipyards needs. At CADMATIC, most of the staff are naval architects and mechanical engineers with hands-on experience of work in the industry. This approach provides software solutions based on the knowledge of the industry and needs of designers.

CADMATIC offers an extremely robust yet resource-friendly project distribution system. It enables users to work simultaneously on the same project. This distribution system suits both large multinational operators and smaller companies and is called the CADMATIC Co-Designer. It was first offered to the users over 12 years ago, and since then it has been proved to be very effective in even the biggest cruise liner projects with over 20 participating design companies and over 200 simultaneous designers working across the globe.

CADMATIC Distribution design solution is based on a smart database-centric client server system that efficiently stores 3D ship models, documents and component libraries in master and replica databases hosted by a database server system. Distributing ship design work globally or adding new design teams to a project can be accomplished with a few clicks.

Ensuring the integrity and security of project data at all times receives the utmost priority in the CADMATIC Distributed Design system. In globally distributed projects, data is updated at set intervals between remote design sites via an online network such as the internet, or by simply exchanging the file in an email attachment.

CADMATIC provides filtered replication to protect intellectual property in design projects. If a company prefers a subcontractor to only have access to their parts of the project, it is possible to achieve without any extra licensing or tools. Project replication setup could include a filter for 3D objects via 3D box based selection. The remote design team will get only the part of the 3D model they need for their tasks. With filtered replication, the subcontractor has access only to designated areas of the project, and can't, for example, publish a viewer model of the whole project. When drawings are made it is possible to use drawing mask boxes to restrict visibility in drawing views. Objects that are completely inside a mask box are not shown in drawing views. Objects that are partially inside a mask box are clipped in visualization, and the part of the object that is inside the box is not shown.

Expiration dates for viewer files provide the possibility to control access to the design solutions after the completions of the project, ensuring that outdated files are not used.

Our approach provides all users with an easy-to-set-up replication mechanism without any extra modules or fees involved, and effective tools to protect the IPR of the 3D designs.

At CADMATIC we aim to continuously develop our products, not only serving the needs of the customer, but also innovating and forecasting the needs of the futures changes in the marine domain.

8. CONCLUSIONS

The overview of trends influencing the marine design and CAD tools functionality shows that increasing the complexity of the projects and raised expectations of the digital era defines the course of CAD development. The ovverall tendency of data creation and sharing facilitated by CAD reaches the level when data protection starts playing an essential role. Filtered replication and in-built protection of viewer level exports, along with export files containing different levels of exported data from 3D models are the main directions of the development.

The explicit need of using 3D models as digital assets to the purposes beyond production opens a wide range of possibilities to make further integrations of CAD towards information management and a broader range of IT systems, such as marine traffic control and others.

Providing significant in-built knowledge within CAD and enabling multi-user and multi-site access to intelligent design environment is not a sufficient level. Providing efficient mechanisms to protect IPR within distributed environments and ensuring data filtering according to the access rights becomes a must-have for modern CAD.

The role CAD environments play in the future of the marine industry depends largely on willingness of CAD solution providers to be open and seize the opportunities digitalization unlocks. Besides obvious globalization possibilities when CAD would be natively integrated into larger systems of the same provider, there are possibilities to retain top level core functionality and be able to integrate with other IT systems that specialize in border fields.

9. REFERENCES

1. Dufva, M., Könnölä, T., Koivisto, R., Multi-layered foresight, *Elsevier Futures Vol. 73*, 2015.

2. Wilenius, M. Patterns of the Future. Understanding the next 40 Years of Global Change. *World Scientific*, 2017

3. Rolls Royce https://www.rolls-royce.com/~/media/Files/R/Rolls-Royce/documents/customers/marine/ship-intel-160316.pdf, 2017

4. Forbes, https://www.forbes.com/sites/mikecollins/2015/05/06/the-pros-and-cons-of-globalization/#4752282accce, 2015

5. Anttila J. The revolution of global democracy. *Sitra*, 2017

6. Intellectual, Industrial and Commercial Property, *Fact Sheets on the European Union*, 2017

7. LeaderSHIP 2015, *Commission of the European Communities*, 2003

8. Grau M.,Liese H., Stjepandic J., Intellectual property protection in the Maritime industry – State-of-the-art and solution approaches, *RINA,* 2013

10. AUTHOR BIOGRAPHY

Ludmila Seppälä (MSc Technology in Shipbuilding and Ocean engineering, MBA) holds the position of Product Marketing and Information Manager at CADMATIC Oy. She is responsible for information about the products and positioning in the market. Her previous experience includes marine engineering, project management, international business development and marketing, and futures research.

EXPERIENCE CAPTURE IN SHIPBUILDING USING MICROSOFT SHAREPOINT

Yashwant R. Kamath and **K. Sivaprasad,** Department of Ship Technology, Cochin University of Science and Technology, Kerala, India.

SUMMARY

Shipbuilding being heavy, one off kind, and made-to-order industry has many characteristics that add to the complexity of the production process. Ships are generally multifunctional and should have multiuser capabilities while in operation. Requirements to fulfil contradicting rules and regulations are a headache for the shipbuilders. Capture of shipbuilding experience and preservation of it would definitely help to improve problem-solving capacity in a shipyard production centre. Systematic collection, storage, compilation and categorisation of shipbuilding information broadly explain the term 'Experience Capture'. Recording of experiences in a traditional shipyard context is highly optional and therefore the function has limited scope for integration within the shipyard and with other life cycle activities of ships. Microsoft Share Point, an online content management system, which is an efficient data handler for problem solving provides ample scope to be implemented in shipbuilding context. The features of Microsoft SharePoint would make the Experience Capture implementation more productive in a shipyard. Systematic implementation of Microsoft Share Point as a Data handler tool for Experience Capture in shipbuilding is proposed in this paper. Microsoft SharePoint being a powerful and robust software with limitless capabilities which includes document management, team collaboration, intranet and extranet can be implemented very effectively in the shipbuilding industry. With recent advancements in cloud integration, authorised personnel can access the database globally. Moreover the Share Point has highly secure database with exceptional indexing and searching capabilities, by which data can be retrieved easily.

1. INTRODUCTION

The history of ship and boat building dates back to time immemorial. Water borne vessels were in operation as a means of passenger and cargo transport even before the invention of the wheel. Ships and boats have been designed and built well before the advent of modern naval architecture and shipbuilding processes. In this context, it is surprising to note that the age-old traditional ship and boat building are still seen in some locations of the world. Ocean going *Oorus* of the Malabar region of South India is one of the examples, which points to support this argument. Vijith et al. [1] states that the master craftsmen of *Ooru* making boat yards build seaworthy ocean going wooden vessels without any documented naval architecture calculations and formal shipbuilding drawings.

The craftsmen capture, preserve and reuse the experience of shipbuilding in traditional ways, which are not usually neither documented nor disclosed to others. Though this type of Experience Capture is very unique, traditional boat and shipbuilding offers a little variation in the product mix i.e. the products the industry manufacture are typical vessels few categories. Modern shipbuilding has grown massively and the shipbuilding has attained status of knowledge intensive industry [2]. Wang et al. [3] has categorically stated that knowledge base is one of the strategic components that decide the performance of modern shipbuilding industry. Six major characteristics of the shipbuilding industry suggested by Sivaprasad et al. [4] point to the fact that shipbuilding industry is a platform where management of engineering knowledge plays an important role in successful implementation of the projects. There exist several information technology solutions to the issue of knowledge management in the manufacturing industry. The Experience Capture, which is practiced in engineering industry, varies from individual record keeping [5] to complicated Knowledge Association Engine [6], which is supported by high end computing machines like Apache Tomcat Server v6.0.

The Experience Capture mechanism in a shipyard should be implemented in all levels of work centers starting from skilled laborers who are employed in shop floor to managers sitting in executive cabins. In order to find an effective and efficient use of Experience Capture tool, the implementation proposal should focus on easy and user-friendly information technology solutions. In this context, SharePoint, One of the IT products from Microsoft offers suitable solution. The main feature of Microsoft SharePoint is that it is less complex in usage and offers immense scope for knowledge capture, database buildup for multi level sharing and ability to adopt intelligent case based reasoning.

2. STATE OF THE ART

The concept of Experience Capture is mentioned in many different ways in the already published literature. The equivalent terminologies used in the literature are Knowledge Capture, Experience Reuse, Knowledge Reuse and Knowledge Transfer. Llamas et al. [7] have stressed the need of having processes based on capitalization and the use of experiences and knowledge to improve the agility in knowledge driven industries, which include the manufacturing industries such as ship and boat building.

Traditional Information sharing methods lack procedures to capture, transfer and reuse of knowledge and this is one of the major omissions in knowledge management systems in modern manufacturing industries [8]. The authors have also pointed out in same paper that the knowledge base what so ever captured has no unified structure to connect in between. G. Vianello [9] has gone further deep into the implementation of Experience Capture by stating that the knowledge management should cover the entire life cycle of engineering product and capture of in-house knowledge. Lianying Z. et al. [10] has argued that knowledge capture and reuse can act as an effective tool in engineering industries to reduce cost, improve quality and reduce lead time for project completion thereby providing better performance.

Hao Q et al. [11] has stressed the need for a web based knowledge capture mechanism for working efficiently in a collaborative working environment. Konstantianos et al. [6] have pointed out the need for a knowledge repository. The proposed knowledge repository by the authors is one of the vital platforms for the development of knowledge Association Engine, which is main engine of Virtual Factory Framework. Dixon N. M. [12] has mentioned that effective knowledge sharing or reuse requires a complete and user-friendly solution. Orlikowski [5] has suggested that personal documentation are of great use for solving future problems and the only problem with storage is regarding the lack of systematic approach which are followed in Experience Captured for reuse. According to D Mourtiz et al. [13], capturing tacit knowledge regarding the experiences that encountered by the engineers is a complex activity and separate knowledge capture and reuse models have to be prepared to suit the type and characteristics of the industries.

3. ISSUES RELATED TO EXPERIENCE CAPTURE & KNOWLEDGE BASED APPROACHES IN SHIPBUILDING

There has been step-by-step advancement in development of ship building technology. Some of the critical areas of shipbuilding still employ traditional methods of technology. Unlike the automobile or habitat engineering industry the advent of automation and mechanization in shipbuilding has been very limited. Ship has a unique product status as an engineering entity i.e. she has a very unique product development structure. These unique characteristics of shipbuilding have been presented by Sivaprasad [14] and the salient features presented were:

- One off Kind and made to order product
- Multi user and multi functional design, and operational attributes.
- Ever increasing extensive and contradicting rules and regulations (within the seaworthiness requirements)
- Eventful and live life cycle

These unique features along with the massiveness of the product and labor-intensive life cycle activities make ship very complex engineering product. Though shipbuilding is a heavy industry with inherent product development complication, the parent ship concept, which is followed during the design and the sister ships procedures adopted in series shipbuilding projects are two other important features related to the product development.

During the various product development stages, a huge amount of data is created and exchanged in each phase and also across the phases. The generated and exchanged data are of great use in case of sister ship construction or series ships are built. These can become more useful where series ships are attempted after a time break. Simple recording of very huge data during the product development would be of much use in the later stages due to various changes in technology, methodology and human resources, which occurs with time. Also a big cloud of data with no specific tag and address may act as a burden to the production engineers. This paper addresses the question of capturing, storing, arranging, analyzing and reusing using a simple and available knowledge management tool, i.e. Microsoft SharePoint for the first time in shipbuilding experience capture context.

4. NEED FOR EXPERIENCE CAPTURE IN SHIPBUILDING

Ships have usually a long life cycle with multiple stages. Eight distinct phases have been identified in the entire lifecycle of an ocean going vessel [15].

In all these life cycle phases, the product experiences number of engineering issues. All these issues make the life cycle phases live and eventful. With respect to recording, compiling and analyzing of engineering experience, the engineers and skilled workers, at various levels of production encounter problems. Management of engineering experience database becomes a real challenge as systematic knowledge management offer a potential tool for efficient and profitable shipbuilding.

Shipyards have more number of production shops compared to any other engineering and manufacturing industry. These shops employ number of engineering trades and skills. Thus in a shipyard, shipbuilding knowledge can be found widely scattered across different work platforms.

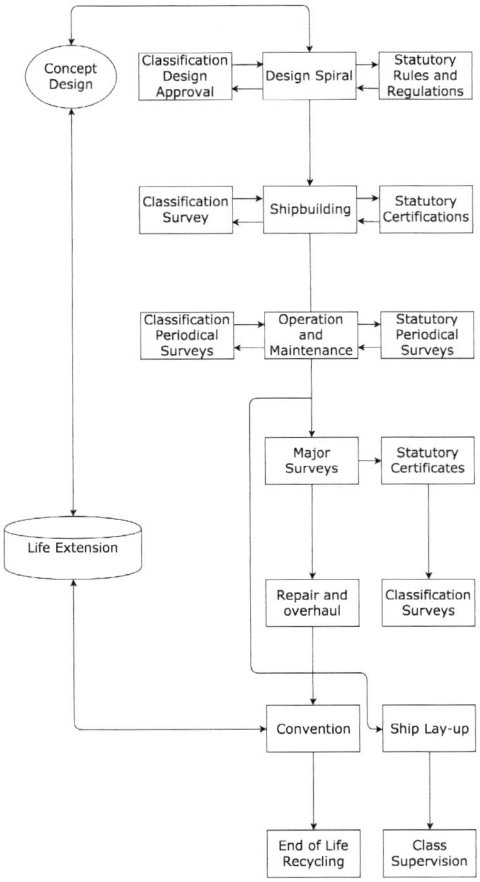

Figure 1: Different Phases in Ship's Lifecycle

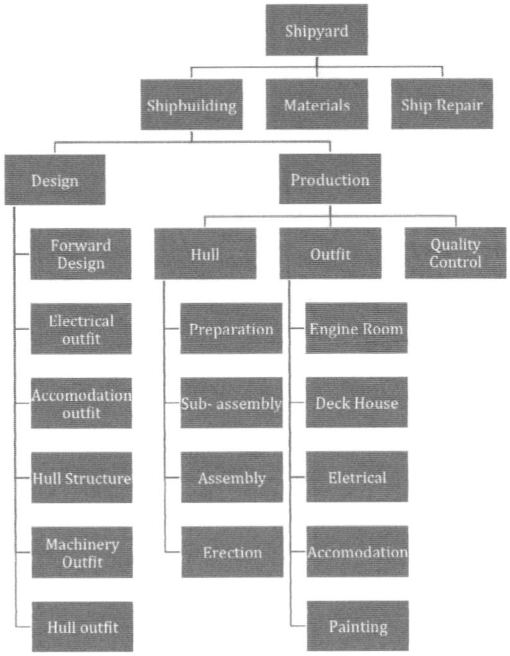

Figure 2: Various Departments in a shipyard [16]

Figure 2 shows the production centers in a typical shipyard, which builds ocean-going ships.

Shipbuilding departments in a shipyard generate a knowledge base, which are partly rulebook based and partly subjective. Subjective Experience Capture is difficult but offer more scope in development of a Knowledge Management System in shipbuilding. A simple and easy mechanism to capture the subjective, thereby tacit knowledge base is discussed in the following sections.

5. SHAREPOINT AND KNOWLEDGE MANAGEMENT

5.1 WHAT IS SHAREPOINT

SharePoint is a web-based collaboration tool from Microsoft Corporation. It is generally used as simple collaboration platform a document management and storage system, albeit some organizations configure it extensively suiting its needs and also into a public facing website. With the latest version, it is integrated with Microsoft Office, so that it simplifies integration of documents onto website. [17] [18]

SharePoint is mated to a Microsoft SQL Server as a Data server and Microsoft Internet Information Services (IIS) as web server, both of which are backed by a Windows Server. The installation of SharePoint may be 'On-Premises' or 'Online' depending on the location of this above said Windows server. 'SharePoint Online' is inherently cloud based, so it does not require additional cloud computing, while, 'SharePoint On-Premises' need to be mated to an Internet Backbone to ensure a completely mobile access.

This paper exploits SharePoint in the following domains:

- SharePoint as a collaboration tool: To store and share documents like drawings, Key Plans, Reports etc.
- Information Portal: To store the information or data, which in this case is "Experience".
- Enterprise content management: To utilize the record Management capability of Microsoft SharePoint, to organize the Documents, above said information, according to the metadata and assigning appropriate usage access.
- Internal Website: To use the SharePoint website for Intranet services.

In a shipyard, the Experience Capture or Knowledge Capture platform can be enforced in four different types of knowledge reuse situations as described by Lynn [19]

- Shared Work Producers: In this situation people work as a team homogenously and cross-functional, constantly acquiring knowledge for their own and their successors' reuse. E.g.: The shipyard a whole.

- Shared Work Practitioners: In this situation people do similar work in different situations, acquiring knowledge for each other's reuse. E.g.: various sub-departments under a department, say Design.

- Expert Seeking Novices: Here, people need occasional information or knowledge and guidance from experts. E.g.: Beginners

- Secondary Knowledge Miners: These are people who look into answering questions and seek for developing new knowledge by analyzing the records available.

5.2 ADVANTAGES OF SHAREPOINT

Advantages of using SharePoint as an Experience Capture tool are as follows.

5.2 (a) Enterprise Search

Since 2013, SharePoint features Enterprise search capabilities, which indexes all the data (Experience) in the entire corporate network (of shipyard), which may be highly heterogeneous and contrasting in nature. It ensures periodic reorganization of the data, thereby creating a summarized transcripts and purging all the redundant data.

5.2 (b) Maintaining context of data

The database can be so arranged and linked such that each record can be linked to all documents in the context of that. It also helps in documenting the related rulebook knowledge.

5.2 (c) Mobile Access

The SharePoint website is actually a web-app, therefore it ensures seamless access across all the devices irrespective of the operating system or screen size. Moreover, It can be more easily accessed on-the-go with SharePoint mobile app available currently on Android, iOS and Windows Phone devices, thereby an individual can feed the data or request for or browse the data instantly and irrespective of his/her workplace.

5.2 (d) Customizability and Ease of Setup

The SharePoint website can be configured as easily as on configures his/her social network page, thereby reducing the company's investment in a core software team. In case of choosing 'SharePoint Online' there are no procedures for installation of SharePoint and zero investment in hardware for SharePoint hosting. Each SharePoint sub site can be customized and configured to suit each and every department's needs.

5.2 (e) Usage Rights

The SharePoint Administrator can provide usage rights to each and every user in the organization to restrict his/her usage of data in any horizontal or vertical hierarchy, thereby ensuring security to sensitive data, which will be prevalent in shipyards.

5.2 (f) Data Entry and Data Extraction

Data Entry to the database can be made through customized forms, which makes it highly user-friendly. The forms ensure the data entry in all the mandatory fields. Similarly submitting data extraction requests through forms can facilitate data extraction more easily.

5.2 (g) Monitoring The Users

The SharePoint Administrator can monitor and track each and every user, their activity and data.

6. SHAREPOINT IN SHIPBUILDING

6.1 PROBLEMS IN SHIPYARD

The entire shipbuilding, ship repair and ship operation industry is valued on the basis if time. Therefore the deadlines and schedules fixed for a designer in shipbuilding industry or an engineer in ship repair industry etc. will be so tight, such that, even if a unique problem arises, the individuals won't be free to discuss this among the entire team. Upon solving the problem, the knowledge or 'experience' thus generated fails to be captured in midst of the tight schedules which results in extra work re-solving when the same problem is faced again.

A very similar situation set in due to presence of multiple works at the same time and also due to varying requirements to be incorporated or carried out in the product.

6.2 PRESENT DAY SOLUTION

Currently knowledge is transferred among individuals and various projects by following mean as mentioned by HS Sæther et al. [20]

- Personal meetings
- Project evaluation meeting
- Technical inspections
- Informal inspections
- Owners meeting
- Job rotation
- Common project meetings

Some engineers keep a record of problems faced by themselves during their period of service in a book, handwritten. Maintaining a spreadsheet is also a valid option when in small scale or for a single project.

6.3 NEED FOR IMPROVEMENT

The current solutions do not provide a permanent solution to problem of knowledge lapse or loss of experience.
As mentioned by P Chimm et al. [8], firstly, the above-mentioned solutions are completely made with natural language and will be plagued with lot of syntactical and semantic errors. Therefore searching for a solution from a large data is pretty tiresome process. Secondly, the documents and consequently the knowledge or experienced captured in them exist independently and is not bound by any unified structure again making it difficult to search for a solution. It is obvious that these solutions are not mobile in nature.

All these and the advantages of using SharePoint support our argument that SharePoint is excellent content management software, which in this case is exploited as an Experience Capture Platform.

6.4 THE EXPERIENCE CAPTURE PLATFORM

The basic building blocks of SharePoint Platform are SharePoint Site, Site Collections and Sub Sites. In the Platform we propose, a SharePoint Site will be declared for a shipyard, under which various Site Collections will be declared corresponding to each of the departments in that shipyard. Under each of the Site Collections there will be a number of Sub Sites. This is represented in Figure 3.

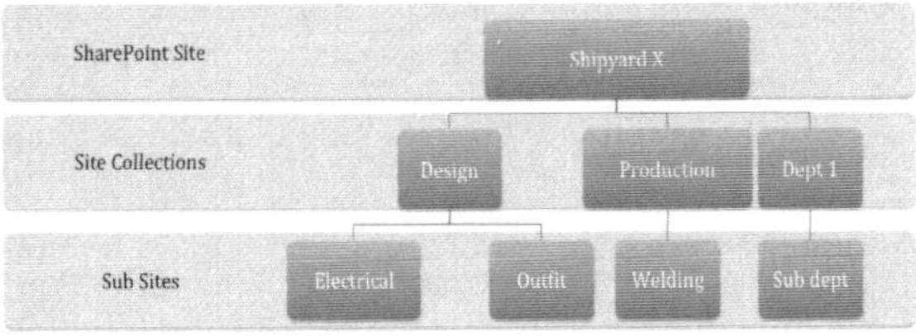

Figure 3: Part of SharePoint Platform Organisation

The SharePoint site will be having the SharePoint Administrator who is in charge of the entire Platform, the one who can monitor all the information. The Administrator can create the Site Collection for each department and assign administrator privilages limited to each Site Collection, to the individuals in charge of each department.

Each department will be having its own Sub Sites based on various factors:

- Sub Sites may be corresponding to sub-departments working under the Department. E.g. 'Electrical Design', 'Hull Structure Design' Sub Sites may be created under the Site Collection 'Design'

- Sub Sites may be also configured to cater for problems in a regular work under a department. E.g. 'Procurement' Sub Site in 'Materials' Site Collection.

The usage access can be set for each element of the SharePoint Platform, from public anonymous access to secured single-handed access. The Site Collection owner i.e., the administrator will have administration permissions limited to that particular SharePoint Site Collection.

The Data Entry will be made through Data Entry Forms. Data extraction is also achieved through a form interface. The SharePoint Administrator can customize the form interface according to the need. A process flow of data entry is shown in Figure 4.

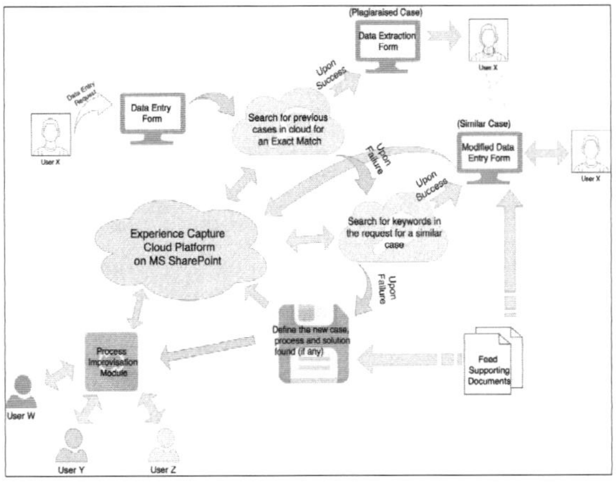

Figure 4: Data Entry Flowchart

When a User starts the Data extraction Request, the entry is searched for an exact match in the database. Upon a successful match, i.e. a condition of plagiarized case, the result is displayed to user in the Data Output Form. If there is no exact match, a search for main keywords is made on the database. Upon successful match with one or more keywords, a part of context will be adopted from the record with which match was found. The User will feed additional related documents to be stored along the 'Experience' he is feeding to the cloud, through the modified Data Entry Form. This is the condition when problems of similar type or context are registered.

If no match is found, then it is the condition of unique and first-time experience for an individual in that shipyard. In such a case, all the details along with related documents will be fed to the Experience Capture Platform. After this, during course of time, it will be enhanced by continuous entry and extraction from different users in the shipyard.

7. A SAMPLE CASE STUDY IN DATA EXTRACTION

The entered data can be accessed through a data request form and output is displayed in a Data Output Form.
A sample Data Output form is shown in Figure 5.

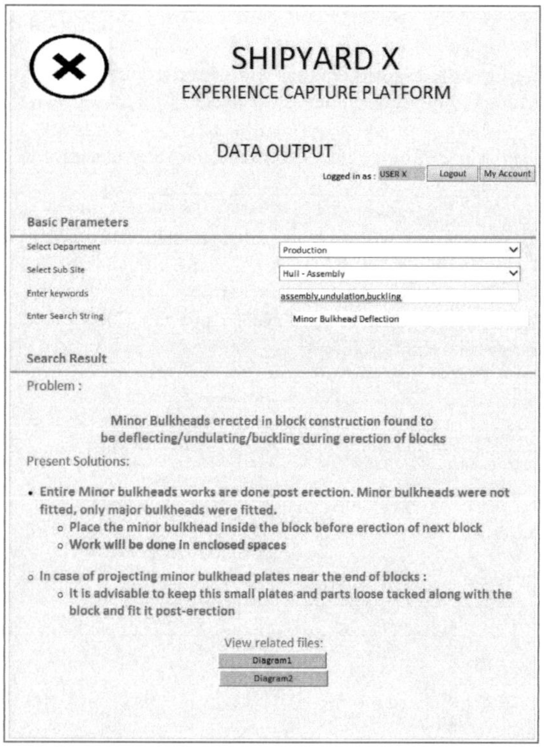

Figure 5: A sample Data Output form

8. CONCLUSION

Experience Capture and data reuse in shipbuilding have been effectively addressed in this paper and a user friendly & easy solution have been proposed. Use of commonly available software i.e. Microsoft SharePoint has been proposed after making suitable modification in its implementation in shipbuilding context. A sample data extraction form has been generated as a case study to show the practical use of SharePoint in experience capture and data reuse. However, scope for addition of more and more features are felt. The limitations and future scope are discussed below.

8.1 LIMITATIONS

Current limitations include the extra work an individual should do in entering the data. Also, the platform is effective only after it is fed with sufficient data, more the data stored in system, more effective it is.

8.2 SCOPE FOR IMPROVEMENT

When the platform has years of existence, i.e. it has abundant data stored in it over the years; the platform can be made to evolve into a more intelligent platform by following means:

8.2 (a) Incorporating DFMEA And PFMEA

Design Failure Mode and Effect Analysis (DFMEA) or Process Failure Mode and Effect Analysis (PFMEA) is an analytical tool maintained by organization to analyze the design or process and analyze effect of potential failures. This tool can be effectively coupled with Microsoft SharePoint for tapping the reuse of experience.

8.2 (b) Case Based Reasoning

With large amount of data available, the Experience Capture Platform will become an Expert System by incorporating a Case Based Reasoning methodology.
The Expert System thus evolved ill be able to suggest or specify the probability of a solution to problems faced by users.

8.2 (c) Incorporating Neural Networks

The system can be made more intelligent or human-like by exporting these data to an Experience Capture Platform using Neural Networks, or incorporating the same
in the SharePoint Platform.

9. REFERENCES

1. VIJITH, V., SANTHANU BOSE, 'An Overview of Traditional boat building in Kerala', *Shipstechnic Vol. 8, Society of Naval Architecture Students, Department of Ship Technology, Cochin University of Science and Technology*, 1991

2. SÆTHER, H. S., 'Knowledge transfer in shipbuilding projects: A study of facilitating mechanisms', *International Journal of Project Organization and management Vol. 4 No. 3,* 2012.

3. WANG, J., GWEBU, K., SHANKER, M., TROUTT, M. D., 'Application of agent-based simulation to knowledge sharing', *Decision Support Systems, Vol. 46 No. 2*, 2008

4. SIVAPRASAD, K., RAMESH, R. S., 'Quality Function Deployment in Shipbuilding', *Shipstechnic Society of Naval Architecture Students, Department of Ship Technology, Cochin University of Science and Technology*, 1997

5. ORLIKOWSKI, W. J., 'Learning from notes: organizational change around groupware development.' *The Information Society Vol. 9 No. 3,* 1993

6. KONSTANTINOS, E., KONSTANTINOS, S., DIMITRIS. M., GEORGE. C., 'On knowledge reuse for manufacturing systems design and planning: A semantic technology approach', *CIRP Journal of Manufacturing Science and Technology, Volume 8*, 2015,

7. LLAMAS, VALENTINA AND COUDERT, THIERRY AND GENESTE, LAURENT AND ROMERO BEJARANO, JUAN CAMILO AND DE VALROGER, AYMERIC 'Experience reuse to improve agility in knowledge-driven industrial processes', *'IEEE International Conference on Industrial Engineering and Engineering Management'*, 2016

8. P., CHIMM, R. B., CHINNAM, N., SADAWI, 'Product design and manufacturing process based ontology for manufacturing knowledge reuse', N. J Intell Manuf, 2017.

9. G. VIANELLO, S. AHMED, 'Transfer of knowledge from service phase: a case study from oil industry', *Res Eng Design,* 2012

10. LIANYING, Z., XIAONAN, L., 'How to reduce the negative impacts of knowledge heterogeneity in engineering design team: Exploring the role of knowledge reuse', *'International Journal of Project Management, Volume 34, Issue 7'*, 2016.

11. HAO, Q., HONGWEI, W., AYLMER, L. J., 'A RFBSE model for capturing engineers' useful knowledge and experience during the design process', *'Robotics and Computer-Integrated Manufacturing, Volume 44'*, 2017.

12. DIXON, N.M., 'Common Knowledge: How Companies Thrive by Sharing What They Know', *'Boston: Harvard Business School Press',* 2000

13. D. MOURTZIS, M. DOUKAS, 'Knowledge Capturing and Reuse to Support Manufacturing of Customized Products: A Case Study from the Mould Making Industry', *'Procedia CIRP Volume 21'*, 2014

14. SIVAPRASAD, K., NANDAKUMAR, C. G., 'Design for Ship Recycling', *Ships and Offshore Structures, Vol. 8, No. 2.,* 2013

15. SIVAPRASAD, K., 'Development of Best Practices in Ship Recycling Processes', *Cochin University of Science and Technology, 2010*

16. YASHWANT R. K., *'Summer Internship Report', Department of Ship Technology, Cochin University of Science and Technology* 2017

17. *'SharePoint 2016 for dummies', John Wiley and sons,* 2016

18. MICROSOFT, *https://products.office.com/en-in/sharepoint/sharepoint-server*

19. LYNNE, M. M., 'Toward a Theory of Knowledge Reuse: Types of Knowledge Reuse Situations and Factors in Reuse Success, *'Journal of Management Information Systems'*, 2001

10. AUTHOR'S BIOGRAPHY

Sivaprasad K. currently holds the position of Associate Professor at Department of Ship Technology, Cochin University of Science and Technology, Kerala, India. His area of specialization is Ship Production and Ship Repair and handles relevant classes for B.Tech Program and also Research Guide. He has published more than 35 papers to his name and has more than 20 years of teaching experience and worked as Senior Naval Architect at Mazagon Docks Ltd., India.

Yashwant R. Kamath currently a student of B.Tech in Naval Architecture and Shipbuilding at Department of Ship Technology, Cochin University of Science and Technology, Kerala, India.

MODELLING AND SIMULATION OF ORGANIC RANKINE CYCLE FOR WASTE HEAT RECOVERY ON AN OFFSHORE SERVICE VESSEL

C W Ng, I C K Tam and D Wu, Newcastle University, NE1 7RU, UK

SUMMARY

Organic Rankine cycle (ORC) have been used widely in the geothermal and solar industry to generate electrical power from thermal heat sources. Recently, there had been more research on applying this thermodynamic cycle for waste heat recovery from internal combustion engine. Most simulation studies forecast ORC efficiencies between 15-20% while realistic expectations lie between 7-10% which make fuel savings of around 10% feasible [1].

In choosing the optimum ORC system to recover waste heat from a thermal plant, there will be various options to examine, for example, selecting the working organic fluid from among the hundreds available, identifying the waste heat source temperature and quantity which will in turn affect the choice of the cycle components such as evaporator, expander, condenser and feed pump.

To overcome this challenge and provide flexibility in detailed thermal analyses, a commercial multi-disciplinary 1D software called Siemens LMS Imagine.Lab Amesim which is based on model-based systems engineering is used to derive the optimised system design for the ORC system. Using this software and base on operational data from a seagoing offshore service vessel, it is found that annual fuel savings of 9% is possible through the use of ORC to recover waste heat from diesel engine exhaust and high temperature cooling water. When the cost of installing such system is estimated, payback period is about 8 years and can be interesting to shipowners with investment horizons of 10 years.

1. INTRODUCTION

There is little doubt that regulations will get more stringent on carbon emissions with the Paris Agreement entered into force on 4 Nov 2016. In the maritime world, the Energy Efficiency Design Index (EEDI) implemented under MARPOL Annex VI by International Maritime Organisation (IMO) is one such measure that would encourage shipowners to adopt energy efficient systems onboard their ships.

The EEDI calculated for a ship measures the environmental cost in terms of grams of CO_2 divided by the transport work in ton-miles and take into account the amount of carbon dioxide generated by the diesel engines for propulsion and power generation. A technical file will need to be prepared for each ship detailing the EEDI calculations and it is to be shown that the calculated EEDI is less than the reference EEDI mandated by IMO for each ship type. IMO has also dictated that the reference EEDI be lowered gradually to challenge the industry to provide even more energy efficient designs.

Needless to say, the use of fuel efficient diesel engines will have a direct impact on lowering the EEDI but the regulations also allow the use of energy efficient technologies like waste heat recovery systems, wind propulsion, air lubrication, solar panels, etc to reduce the EEDI.

Waste heat recovery entails reusing thermal energy from exhaust gases of combustion engines, and mostly onboard seagoing ships, this will be the diesel engines.

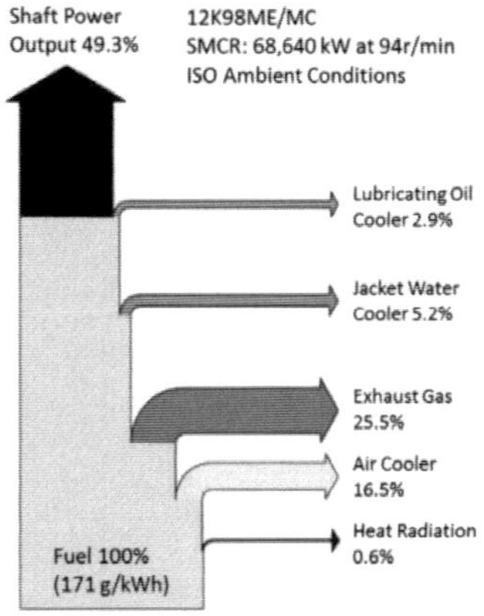

Figure 1: Heat balance diagram for MAN 12K98ME/MC [2]

As reviewed by Singh and Pedersen [2], about 50% of the total fuel heat energy onboard a ship is rejected to the surroundings via different streams without doing any useful work (Figure 1).

Waste heat sources can be categorised by their temperature range: low (<230°C), medium (230 to 650°C) and high (>650°C). While steam Rankine cycle (SRC) are very economical to operate at high power output (>1MW) and medium to high temperature heat source. Other thermal cycles like ORC and Kalina cycles have advantages in the smaller low to medium temperature range with the former being preferred due to a more simple setup [3].

ORC involves using an organic fluid like hydrocarbon or refrigerant as the working fluid to generate electrical power from waste heat energy from diesel engines. Compared to water, the organic fluid has lower boiling point with lower specific heat of vaporisation, meaning that it is able to exploit waste heat of lower temperatures sources like HT cooling water and lubrication oil thereby improving the overall energy efficiency by utilising waste heat from even low to medium temperature sources which would otherwise be lost to the surrounding.

In designing such an ORC system, one will encounter a wide range of possible solutions in that for a given waste heat source temperature and quantity, wide ranging options can be found in selection of different cycle configurations, working fluids, component type of evaporators, expanders, condensers and feed pumps.

To arrive at the optimum solution, several permutations would have to be studied and compared in detail. Using traditional method of thermal analysis of academic software like Matlab can be tedious and time consuming as models need to be built.

A more modern approach will be to use commercial off the shelf software like Siemens LMS Imagine.Lab Amesim [4] which is based on model based system engineering techniques. The use of system engineering software in maritime industry had been discussed in recent research as given below:

Table 1: List of recent research in system engineering software for energy systems

Authors	Framework /library	Software	Systems modelled
Dimopoulos et al. [5]	COSSMOS	gProms	Combined cycle on bulk carrier, dynamic simulation of electric propulsion system
Ginnetti et al. [6]	Ship Energy Systems	SimulationX	Waste heat recovery system onboard a large containership
Lepisto et al. [7]	NA	Apros	Dynamic simulation of energy system of cruise ferry
Mondejar et al. [8]	NA	IPSEpro/Matlab	ORC onboard passenger vessel

Each of this research demonstrate how a commercial system engineering software can be used to analyse the energy systems onboard ships both under steady state and dynamic conditions. Similar to softwares above, the software to be used in this paper is Siemens LMS Imagine.Lab Amesim which is a multi-disciplinary 1-dimensional software which means that it is able to analyse across physical domains of thermal, fluid, mechanical, electromechanical, control, etc which is an advantage for energy systems that often straddle between these disciplines. In Amesim, there are more than 38 validated libraries with more than 5000 components that can be used directly without creating complicated codes that is usually required in scientific software like Matlab.

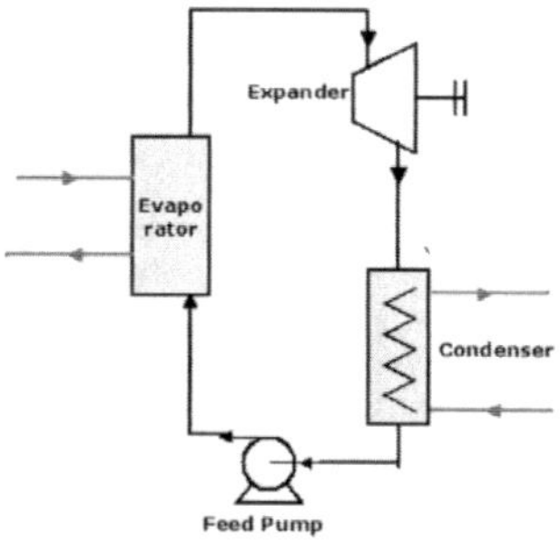

Figure 2 Schematic diagram of simple ORC

Among the listed research above, Mondejar et al. [8] presented how her team had modelled the ORC for a passenger vessel which covers a daily trip between Stockholm and Finland and concluded from the simulation that ORC net power production of approximately 22% of the total power consumption onboard. Several other researches on ORC onboard ships had been done on other ship types like LNG carrier, cruise vessel, air craft carrier and container ship. To the authors' knowledge, none had been done for offshore service vessel which unlike the former are not primarily cargo/passenger carrying ships but supports offshore rig operations and have a variable load profile. This paper will demonstrate how a commercial system engineering software, Siemens LMS Imagine.Lab Amesim can be applied to modelling and simulation of an ORC onboard an offshore service vessel.

2. THE CASE SHIP

Offshore service vessels (OSVs) refers to a class of shiptype that serves the offshore oil and gas industry and can include diverse operations like platform supply, anchor handling, construction, fire-fighting, stand-by and rescue, diving support, accommodation etc. Due to different operations, their load patterns can also differ widely.

As a case study to examine the thermodynamic performance of ORC, we have selected as our case ship to be a 88.8m 5200 DWT multipurpose platform supply vessel (MPSV) which is designed to operate in offshore Malaysia with its main operations of carrying supplies like fuel oil, drilling material and chemicals to offshore rigs in the region. The general specifications of the case ship are as below:

Table 2: General specifications of case ship

Length overall:	88.8m
Breadth:	20.0m
Depth:	8.4m
Deadweight:	5200t
Class notation:	American Bureau of Shipping, +A1, (E) Offshore Support Vessel, Supply-HNLS, FFV 1, OSR-S1, +DPS-2, SPS, +AMS,
Area of operations:	Offshore Malaysia (seawater temperature 32°C)

The MPSV is powered by 4 nos. of Wartsila main diesel generators with Wartsila electric propulsion & azimuthing stern thrusters to perform dynamic positioning operations (DPS-2) in offshore locations in Malaysia. Other main specifications are as below:

Table 3: Technical specifications of diesel generators

Engine make/model:	4 x Wartsila 6L26
Rated power:	1950kWm @ 900rpm
No. of cylinders:	6
Fuel consumption:	184 g/kWh
Generator make/model:	4 x AVK DSG 00 L1/8 W

2.1 ENGINE LOAD

The operational profile of the MPSV will dictate the power consumption which will affect the diesel generator load and hence the temperature and quantity of the exhaust gas and cooling water.

The typical operation profile of a MPSV includes the following:

- loading of supplies from supply base at harbour mode,
- transit to offshore location at steaming mode
- unloading of supplies on DPS-2 mode
- loading of waste from rig on DPS-2 mode
- transit back to supply base at steaming mode
- unloading of waste to shore at harbour mode
- standby mode

The engine load at each operational mode and number of operating hours assuming 350 days of operating days is listed as below.

Table 4: Percentage of time at each mode vs engine load

Operation Modes	Units	Steaming 13.5 knots	Steaming 12 knots	Transit low, 7-8 knots	Dyn Pos Heavy	Dyn Pos Light	Standby	Harbour
% of operating hours	%	20%	15%	5%	10%	30%	10%	10%
Operating hours	h	1680	1260	420	840	2520	840	840
Propulsion load	kW	5000	2400	1400	3600	2000	300	0
Manoevering load	kW	0	0	0	1200	800	400	0
Other ship load	kW	300	300	300	500	500	200	300
Total load	kW	5300	2700	1700	5300	3300	900	300
Rated power of each engine	kW	1950	1950	1950	1950	1950	1950	1950
No. of running engines		4	2	2	4	2	1	1
Engine load		68%	69%	44%	68%	85%	46%	15%

2.2 ENGINE HEAT SOURCES

The possible heat sources from the diesel generators include exhaust gas, lubricating water, high temperature (HT) and low temperature (LT) cooling water etc. The heat source temperature range as well as the flow rates is listed in Table 3.

As the MPSVs are still under construction at the shipyard with sea trials only in October 2017, only engine test results from the Factory Acceptance Test (FAT) is available. This will act as the main input data for the model to find out how much heat can be recovered at different engine load i.e. 25/50/75/100%.

Table 5: Heat source temperature and flow rate

Heat source	Temperature range category	Temperature (°C)	Flow rate (m^3/h)
Exhaust gas	Medium	353-392	2.08-3.82 kg/s
Lub oil	Low	66-67	60 m^3/h
HT cooling water	Low	84-89	35 m^3/h
LT cooling water	Low	42-57	42 m^3/h

3. MODELLING & SIMULATION

3.1 WORKING FLUID SELECTION

The selection of a suitable working fluid is one of the most important task in the design of the ORC. There are hundreds of possible options of organic fluids and can be categorised into seven main classes [9]:

- Hydrocarbons
- Perfluorocarbons
- Siloxanes
- Partial fluro-substituted straight chain hydrocarbons
- Ethers and fluorinated ether
- Alcohols
- Inorganics

Bao [9] compiled a list of recommended fluids for different heat source temperature range and for low temperature range the popular fluids include refrigerants like R123, R245fa and R134a and for medium temperature range, hydrocarbons like n-Octane and n-Pentane are preferred. For comparison, the Temperature-Entropy diagram is plotted on the same chart for these fluids against that of water on Figure 3. This makes it easy to see that SRC is the preferred cycle for high temperature and high power output heat sources.

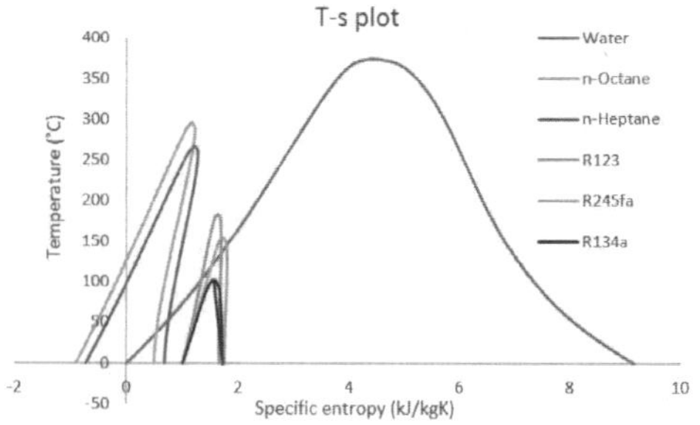

Figure 3: Temperature - Specific Entropy for selected organic fluids

3.2 SELECTION OF LIMITATIONS FOR WORKING PRESSURES AND TEMPERATURES AT HEAT SOURCE/SINK AND WORKING FLUID

At the heat source, the exhaust gas will be supplied at temperature of more than 300°C, as sulphur oxides present in exhaust react with steam to form sulphuric acids that has corrosive effects on the exhaust pipings, a minimum temperature of about 100°C is required before it leaves the evaporator [10].

For the heat sink using seawater cooled condenser, the inlet temperature is 32°C and outlet temperature is set at 45°C. Temperature higher than this can lead to increased corrosion on the seawater pipings as well as per normal marine knowledge.

When the pressure ratio between evaporation and condensation, $\frac{p_e}{p_c}$ is increased, this will lead to higher efficiency but there is a limit to this. For marine applications, the p_e will be limited to about 30 barg before it becomes expensive to install, a condensing pressure lower than 1 bara is selected to maximise the pressure ratio. The implication of this is that atmospheric air can sucked into the system that can reduce efficiency.

Organic fluids also exhibit chemical instability when heated to higher temperatures and hence it is necessary to limit the maximum process temperature after heating at evaporator to 327°C [11]. Considering the engine exhaust gas is at 300°C, the thermal instability temperature of the organic fluids will not be reached.

For the safety aspects, working temperatures of organic fluids should not exceed its autoignition temperature which is the temperature at which the organic fluids will combust with contact with hot surfaces. Flash point of organic fluids should be kept in mind as the maritime regulatory regime will require additional safety features when handling fluids with low flash points. These limitations are especially relevant when hydrocarbon fluids are selected as working fluids.

3.3 INITIAL SCREENING OF ORGANIC FLUIDS

To be suitable for installation onboard a ship, the working fluids need to meet the following requirements:

- Good performance given the heat sources from diesel engines and heat sink using seawater.
- Thermal stability at high temperature
- Safe to be used onboard, working fluid shall have low flammability and toxicity.
- Little impact on environment in terms of ozone depleting potential (ODP) and greenhouse warming potential (GWP).
- Availability and low cost.

With respect to the above criteria, Table 6 is compiled for the 6 selected organic fluids and water.

Table 6: Main properties of the selected organic fluids, partly from [12]

Fluid	Formula		$T_{boiling}$ (°C)	T_{crit} (°C)	P_{crit} (bar)	GWP	ODP	Flashpoint (°C)	Autoignition temperature (°C)
Water	H_2O		99.97	373.95	220.64	Low	0	None	None
n-Octane	C_8H_{18}	HC	125.62	296.17	24.97	Low	0	12.9	205.9
n-Heptane	C_7H_{16}	HC	98.38	266.98	27.36	Low	0	-4.1	203.9
R123	$CHCl_2CF_3$	HCF(	27.82	183.68	36.72	120	0.01	None	770
	CHF_2CH_2								
R245fa	CF_3	HFC	15.04	153.86	36.51	1300	0	None	412
R134a	CF_3CH_2F	HFC	-26.07	101.06	40.59	950	0	None	770

Thermodynamic data of the organic fluids is derived from an open source software Coolprop which is a library similar to NIST's REFPROP that contains equation of state and transport properties for 122 pure and pseudo-pure fluids [13]. Its data is able to be interfaced with commonly used programming languages like Python, Modelica, Matlab, Microsoft Excel etc.

With the proposed ORC system layout from Figure 4, to keep the cycle sub-critical i.e. evaporation pressure is less than the fluid's critical pressure, only. Considering thermal stability, refrigerants starts exhibiting thermal instability at lower temperatures than hydrocarbons.

Hence, for this study, only the two hydrocarbons, n-Octane and n-Heptane will be modelled and analysed.

3.4 ORC SYSTEM DESIGN

ORC can be categorised based on the number of pressure levels the working fluid is subjected to that includes single pressure level cycles and multiple pressure level cycles. For simplicity and economy, single pressure cycle is often selected for most applications. If there is a need for higher cycle efficiency, for example in geothermal applications, multiple pressure level ORC is employed which can justify for its higher capex.

For this study, a single pressure ORC will be modelled as the case ship is a comparatively small-sized ship and as she is already under construction and a more simple design will be more beneficial for future retrofits considering that space is a premium onboard a small vessel like PSV.

From earlier section on heat sources available from diesel generator, only the engine exhaust and HT cooling water streams will be selected as the heat sources due to its higher temperatures and flow rates.

For the two selected hydrocarbons, it can be observed that their autoignition temperature is lower than the engine exhaust temperature, this means that an intermediate heating circuit will be needed to prevent spontaneous combustion of organic fluids. A thermal oil suitable for high temperature usage is selected for this.

The thermal oil is first heated in a economiser by exhaust gas and is circulated to transfer heat to the ORC circuit via another heat exchanger. The working fluid is then expanded in the positive displacement expander and the torque produced is converted into electrical energy in the generator.

The working fluid leaves the turbine and enters the seawater cooled plate type condenser to be cooled to condensation temperature. Another positive displacement feed pump then raises the pressure of working fluid to evaporation pressure. To maximise the recovery of heat energy, the working fluid passes through a plate type preheater heated by the HT cooling water.

The proposed cycle is as below:

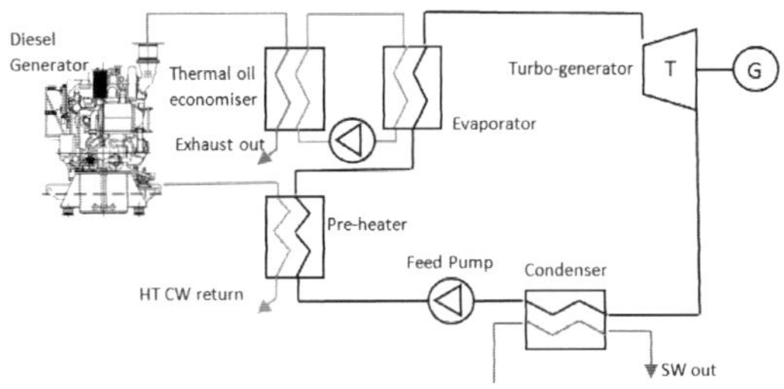

Figure 4: Proposed ORC system layout

3.5 MODELLING IN AMESIM SOFTWARE OF THE ORC

The ORC is modelled in Amesim using submodels from the Pneumatic, Thermal Hydraulic and Two-Phase Flow libraries. In Amesim, components are known as submodels. The organic fluid circuit is modelled using the Two-Phase Flow library as there is a change in phase along the cycle, the exhaust, HT cooling, SW cooling and thermal oil lines are modelled using the Thermal Hydraulic library. Figure 5 shows the simplified model of the ORC system.

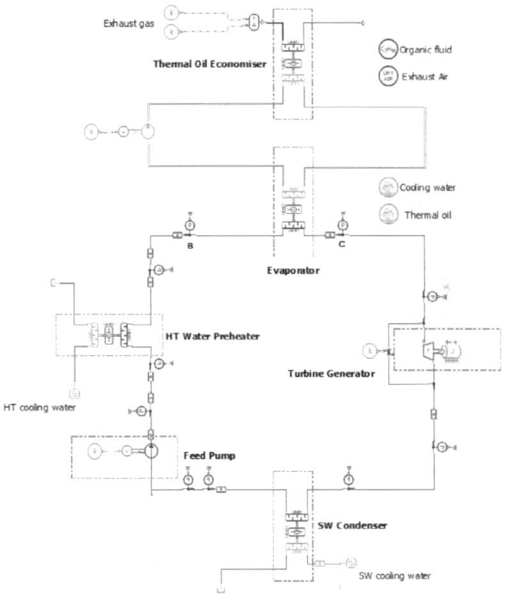

Figure 5: Model of ORC in Amesim

3.5.1 Modelling of heat sources

Heat sources are typically modelled by a component called Modulated Flow Source with 2 inputs of mass flow rate and temperature, each of this is derived from an input signal which is given by the engine heat sources from earlier section.

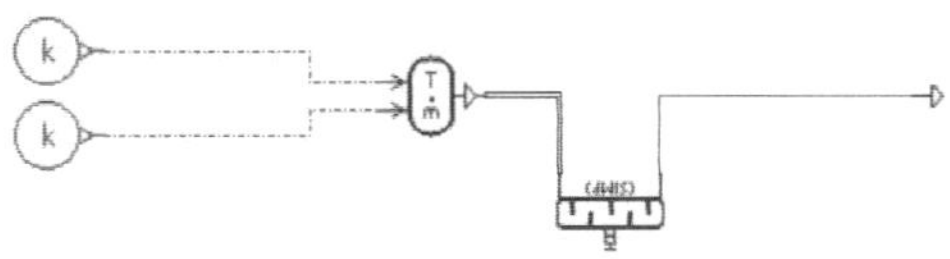

Figure 3: Model of heat source

The output is passed on to the next component called the Half Heat Exchanger which is used to globally model a heat exchanger relying on few parameters and is based on the efficiency-NTU method. There is an external port that is connected to a heat flux component to account for heat flow to the other fluid.

The last component is the atmospheric pressure source to model the exhaust into atmosphere.

3.5.2 Modelling of heat exchangers

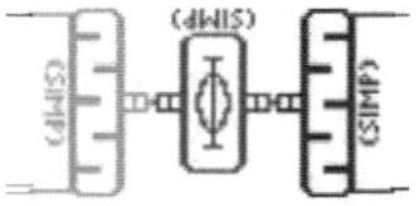

Figure 4: Model of heat exchangers

Heat exchangers are modelled with two half heat exchangers connected via a heat flux component relying on experimental thermal efficiency and is geometry independent. Calculations are set on efficiency-NTU method and mass of heat exchanger can be set as well.

3.5.3 Modelling of expander

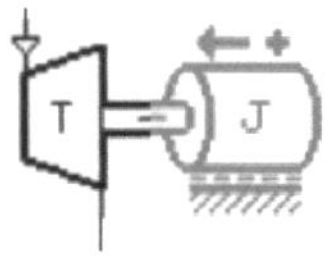

Figure 5: Model of expander

The expander is modelled with two components, the Two-Phase Turbine and the Rotary Load. The Two-Phase Turbine takes its input of pressure and density from the working fluid circuit and provides the torque in Nm to drive the rotary load. With the moment of inertia known, the rpm can be derived the power output established. The volumetric, isentropic and mechanical efficiencies are set at 0.9, 0.8, 0.9 respectively.

3.5.4 Modelling of feedpump

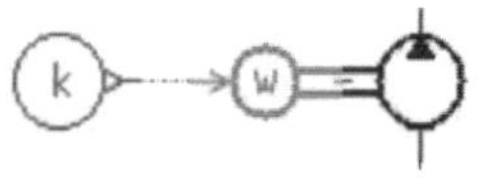

Figure 6: Model of feedpump

The feedpump is modelled by a Two-Phase Fixed Positive Displacement Pump, a converter that takes a signal input and outputs the RPM to the pump. The volumetric, isentropic and mechanical efficiencies are set at 0.9, 0.8, 0.9 respectively.

To account for electrical efficiencies in both feed pump and expander, an electrical efficiency of 0.9 is assumed.

4 RESULTS & DISCUSSIONS

4.1 CYCLE EFFICIENCY AND NET WORK OUTPUT

Referring to Table 2, as the data is not available for all engine load from the FAT, the next higher engine load is selected and the flow rate and temperature of exhaust and HT cooling water compiled as follow.

Operation Modes	Units	13.5 knots	12 knots	7-8 knots	DP Heavy	DP Light	Stby	Hbr
No. of running engines		4	2	2	4	2	1	1
Engine Load		75%	75%	50%	75%	85%	50%	25%
Exhaust flow rate	kg/s	2.98	2.98	2.08	2.98	3.32	2.08	1.22
Temperature	°C	372	372	392	372	365	392	353
HT water flow	m3/h	35	35	35	35	35	35	35
Temperature	°C	87	87	85	87	87	85	84

Figure 10: Flow rate and temperature of heat source at different engine load

Simulation is performed for each of the two organic fluids, n-Heptane, n-Octane at four different engine loads. The cycle efficiency and net power output is plotted against time. It can be observed that steady state is reached after 20 seconds.

Cycle Efficiency and Net Work Output are derived from the following relations:

$$\eta_{cycle} = \frac{W_{turbine} - W_{pumps}}{Q_{in}}$$

$$W_{net} = W_{turbine} - W_{pumps}$$

where

η_{cycle} = Cycle efficiency (%)
$W_{turbine}$ = Work output by turbine (kW)
W_{pumps} = Work input by pumps (kW)
Q_{in} = Heat input (kW)
W_{net} = Net work output (kW)

The following graphs are obtained for n-Heptane.

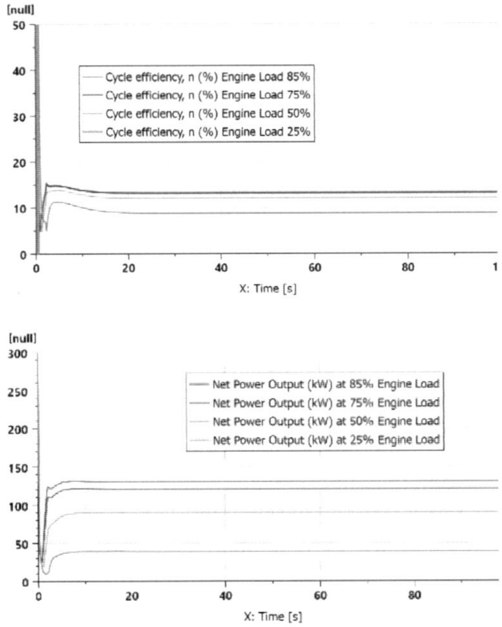

Figure 11: Cycle efficiency and net power output for n-Heptane

Highest efficiency is obtained when engine is operating at 85% engine load at 13.4%. However, efficiency drops to 9.2% at 25% engine load.

For the net work output, maximum is 130.9 kW also at 85% engine load. When operating at 25% engine load, it drops rapidly to 40kW.

Highest efficiency is obtained when engine is operating at 85% load at 12.6% and drops to 9.2% when operating at 25% engine load.

The following graphs are obtained for n-Octane.

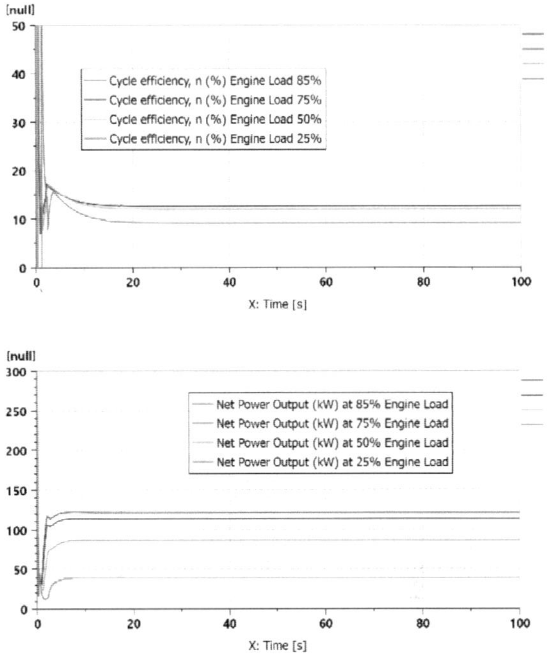

Figure 12: Cycle efficiency and Net work output for n-Octane

For turbine power output, maximum is 114 kW at 85% engine load and drops to 40kW when operating at lower engine load of 25%.

From these results, it can be seen that n-Heptane offers higher efficiency and net work output than n-Octane. Also, the engine load has an influence on cycle efficiency and especially net power output. This means that it will be suited to ships with constant high engine load.

4.2 DERIVING THE FUEL SAVINGS AND PAYBACK TIME WHEN ORC IS INSTALLED ON MPSV

In earlier discussions,we can obtain higher cycle efficiency and net work output when using n-Heptane as working fluid.

Hence, it will be used as the working fluid to derive the fuel savings by correlating the net power output to the specific fuel consumption of the diesel engines.

$$Fuel\ saving\ per\ year$$
$$= W_{net}\ x\ n_{engines}\ x\ t_{year} x\ SFC\ x\ Price_{fuel}$$

where

$n_{engines}$ = No. of running engines
t_{year} = Operating hours per year (h)
SFC = Specific fuel consumption (kg/kWh)
$Price_{fuel}$ = Fuel price (USD$/ton)

Table 7: Fuel Savings per year from ORC

Operation Modes	Units	13.5 knots	12 knots	7-8 knots	DP Heavy	DP Light	Stby	Hbr
Operating hours	h	1680	1260	420	840	2520	840	840
No. of running engines		4	2	2	4	2	1	1
Net Work Output	kW	121.4	121.4	90.1	121.4	130.9	90.1	39.5
Cycle efficiency	%	13.4	13.4	12.5	13.4	13.6	12.5	9.2
Energy	kWh	815808	305928	75684	4E+05	659736	75684	33180
MDO saved	kg	150108.7	56291	13926	75054	121391	13926	6105.1

A total of 437 ton of fuel can be saved in one year which will cost about USD$218,401. As the typical fuel cost for this vessel is about USD$2.5 mil, the potential fuel savings is about 9%.

Song [10] managed to compile a list of ORC module costs against Net Power Output collected from non-exhaustive set of ORC manufacturers and from published scientific publications. From the graph, for a ORC with a Net Power Output of 120kW, the unit cost is expected to be about €3000/kW. This will lead to CAPEX of about USD $1.7mil for four units of ORC. Hence, a simple payback time of 8 years is expected for the installation of ORC on MPSV. This could be interesting for shipowners with investment horizon of 10 years and above.

5. CONCLUSIONS

This paper discusses the possibility of installing an Organic Rankine Cycle (ORC) system onboard a Multi-Purpose Service Vessel (MPSV).

Design and test data from the MPSV which will operate in Malaysian waters serving the offshore rigs there is used as input data to the ORC system to provide a more realistic result.

Two organic fluids, n-Heptane and n-Octane are used as working fluids for the ORC and is modelled and simulated in a commercial off the shelf 1D system simulation software known as Siemens LMS Imagine.Lab Amesim which saves time building model from scratch and is able to perform transient simulations to see how the systems react to changing inputs.

With varying engine load based on actual design operational profile of the MPSV, it is found that n-Heptane offers higher cycle efficiency and net work output than n-Octane and hence N-Heptane is selected for further analysis.

Using n-Heptane as working fluid, it is found that fuel savings of USD$218,401 is possible and this is about 9% of the annual fuel cost for the MPSV.

When the estimated cost of installing the ORC system onboard is considered, payback time of about 8 years is possible and can be interesting for shipowners with investment horizon of 10 years.

In conclusion, ORC systems can be a viable means of achieving higher energy efficiency onboard ships to meet tightening environmental regulations and also reduce carbon emissions.

6. ACKNOWLEDGEMENTS

The authors would like to thank Messrs. Javier Lopez and Sylvain Pluchart from Siemens Industry Software and Messrs. Kwan Seng Fatt and Lu Zhi Jun from Nam Cheong Limited for their valuable advice and guidance on this project.

7. REFERENCES

[1] C. Sprouse and C. Depcik, “Review of organic Rankine cycles for internal combustion engine exhaust waste heat recovery,” *Appl. Therm. Eng.*, vol. 51, no. 1–2, pp. 711–722, 2013.

[2] D. V. Singh and E. Pedersen, “A review of waste heat recovery technologies for maritime applications,” *Energy Convers. Manag.*, vol. 111, no. X, pp. 315–328, 2016.

[3] B. F. Tchanche, G. Lambrinos, A. Frangoudakis, and G. Papadakis, “Low-grade heat conversion into power using organic Rankine cycles - A review of various applications,” *Renew. Sustain. Energy Rev.*, vol. 15, no. 8, pp. 3963–3979, 2011.

[4] Siemens Industry Software, “LMS Imagine.Lab Amesim.” Roanne, France, 2016.

[5] G. G. Dimopoulos, C. A. Georgopoulou, I. C. Stefanatos, A. S. Zymaris, and N. M. P. Kakalis, “A general-purpose process modelling framework for marine energy systems,” *Energy Convers. Manag.*, vol. 86, pp. 325–339, 2014.

[6] A. Ginnetti, J. Lampe, and A. Magdanz, “Energy simulation for waste heat recovery systems,” in *International Conference on Computer Applications in Shipbuilding 2015*, 2015.
[7] V. Lepisto, J. Lappalainen, K. Sillanpaa, and P. Ahtila, “Dynamic process simulation promotes energy efficient ship design,” *Ocean Eng.*, vol. 111, pp. 43–55, 2016.
[8] M. E. Mondejar, F. Ahlgren, M. Thern, and M. Genrup, “Quasi-steady state simulation of an organic Rankine cycle for waste heat recovery in a passenger vessel,” *Appl. Energy*, vol. 185, pp. 1324–1335, 2017.
[9] J. Bao and L. Zhao, “A review of working fluid and expander selections for organic Rankine cycle,” *Renew. Sustain. Energy Rev.*, vol. 24, pp. 325–342, 2013.
[10] J. Song, Y. Song, and C. Gu, “Thermodynamic analysis and performance optimization of an Organic Rankine Cycle (ORC) waste heat recovery system for marine diesel engines,” *Energy*, vol. 82, pp. 976–985, Mar. 2015.
[11] U. Drescher and D. Bruggemann, “Fluid selection for the Organic Rankine Cycle (ORC) in biomass power and heat plants,” *Appl. Therm. Eng.*, vol. 27, no. 1, pp. 223–228, 2007.
[12] C. L. Yaws, Yaws’ handbook of thermodynamic and physical properties of chemical compounds physical, thermodynamic and transport properties for 5,000 organic chemical compounds. Norwich, N.Y.: Norwich, N.Y. : Knovel, 2003.
[13] I. H. Bell, J. Wronski, S. Quoilin, and V. Lemort, “Pure and pseudo-pure fluid thermophysical property evaluation and the open-source thermophysical property library coolprop,” *Ind. Eng. Chem. Res.*, vol. 53, no. 6, pp. 2498–2508, Feb. 2014.

8. AUTHORS BIOGRAPHY

ChunWee Ng is a part-time PhD student at Newcastle University Singapore. His research interest covers energy efficiency and waste heat recovery onboard ships and how system modelling and simulation methods can be applied to further enhance their performances. He is working as a senior engineer in the DNV GL Singapore and is responsible for classification plan approval of ships and offshore units.

Ivan C K Tam is an Associate Professor in Marine Engineering Design & Technology at Newcastle University Singapore. He has strong interest in combustion process, exhaust emission control as well as energy management in renewable energy technology. His recent research interest is the application of cryogenic technology in the reduction of ship hull resistance as well as ballast water and sewage water treatment technologies.

Dawei Wu is a lecturer in Marine Engineering at Newcastle University. His research interests include waste heat recovery and trigeneration (combined cooling, heating and power), new heat engine cycles, bio-energy and other renewable energy for power generation and refrigeration and energy storage.

EMD-BASED NATURAL EXCITATION TECHNIQUE FOR MODAL PARAMETER IDENTIFICATION OF SHIP STRUCTURES

Hongyu Cui*, Xin Xu, Zhonghua Zhou, Weiqiang Peng and **Ming Hong,** State Key Laboratory of Structural Analysis for Industrial Equipment, School of Naval Architecture & Ocean Engineering, Dalian University of Technology, China

SUMMARY

Modal parameters, such as the natural frequency, damping ratio and mode shapes, play a vital role in the structural modification, optimization and health monitoring of ships and offshore drilling platforms. However, identifying the modal parameters for such structures using traditional modal parameter identification methods is difficult due to the large scale and complex characteristics of the structures; this challenge has led to increasing interest in the development of modal parameter identification methods based on response signals. Thus, this paper defines a novel modal parameter identification method based on the empirical mode decomposition (EMD) method, natural excitation technique (NExT), and Hilbert transform (HT). An experiment using a ship model was conducted to demonstrate the accuracy of this method in identifying modal parameters. Using this method, the original signal is first decomposed into a series of intrinsic mode functions (IMFs) by EMD. Then, the NExT is applied to obtain the impulse response function. Finally, the modal parameters are identified by the HT. The results demonstrate that the proposed method is effective, robust and promising for modal parameter identification of ship structures.

Keywords—modal parameter identification, ship structure, empirical mode decomposition, Hilbert transform, ambient excitation

1. INTRODUCTION

The main task of modal parameter identification is extracting the structural modal parameters from dynamic test data. Traditional modal parameter identification methods require both input and output data. However, for large structures, such as ships and offshore platforms, the excitation is expensive and can also easily cause structural damage due to excessive excitation force. Therefore, many modal parameter identification methods based solely on the output response have been developed and applied in the past ten years, such as peak picking [1], the natural excitation technique (NExT) [2], stochastic subspace methods [3], and frequency domain decomposition [4]. These methods assume that the excitation is white noise and that the structural response is a steady-state response. However, the ambient excitation is typically not stationary random excitation. Thus, modal parameter identification under non-stationary excitation has become the focus of considerable attention [5,6]. Huang et al. [7] proposed the empirical mode decomposition (EMD) method, under which signals are decomposed into a series of approximate stationary intrinsic mode functions (IMFs) according to the local-scale characteristics. Non-stationary and nonlinear signals can also be handled well. Scholars have used the EMD method to identify the modal parameters of practical engineering structures, and satisfactory results have been obtained [8,9].

This paper presents a new method combining the EMD, NExT, and Hilbert transform (HT). First, the response signal is decomposed into a series of IMFs using the EMD technique. Second, the impulse response function is obtained using the NExT. Third, the modal parameters are identified by the HT. A modal parameter identification experiment of a ship model is performed, and the results obtained using the proposed method are compared with those obtained using the eigensystem realization algorithm (ERA). The results of this study demonstrate that the proposed method can identify the natural frequency and damping ratio of the structure precisely and can also provide the corresponding modal shapes.

2. THE NEXT FOR MODAL PARAMETER IDENTIFICATION BASED ON EMD

2.1 EMPIRICAL MODE DECOMPOSITION

The hypotheses of EMD, an efficient time-frequency analysis method for nonlinear and non-stationary signals, are as follows: (1) the signal has at least two extrema, a maximum point and a minimum point; (2) the characteristic time scale is defined by the time delay between the extrema; and (3) the extrema can be obtained by differentiating once or more if the data have no extrema.

For an arbitrary nonlinear and non-stationary signal, $X(t)$ can be decomposed into n-order IMF components and a residue:

$$X(t)=\sum_{j=1}^{n}C_j(t)+r_n(t) \tag{1}$$

where $C_j(t)$ is the IMF, and $r_n(t)$ is a residue. Frequency aliasing occurs in the EMD method and is restrained by bandpass filtering in this paper.

© 2017: The Royal Institution of Naval Architects

2.2 NATURAL EXCITATION TECHNIQUE

The basic concept of the NExT is that when a linear system is excited by stationary white noise, the cross-correlation function of the response signals has a similar expression as the impulse response function. Therefore, the impulse response function can be replaced by the correlation function. Then, the time-domain method can be used for identification.

For a linear multi-degree-of-freedom system, the impulse response function between input k and output i can be expressed as

$$x_{ik}(t) = \sum_{r=1}^{n} \frac{\phi_{ir}\phi_{kr}}{m_r \omega_{dr}} \exp(-\zeta_r \omega_{nr} t) \sin(\omega_{dr} t) \tag{2}$$

where ϕ_{ir} is the ith component of the rth mode shape, and ω_{dr}, ω_{nr}, m_r and ζ_r are the rth damped modal frequency, rth modal frequency, rth modal mass and rth damping ratio, respectively.

Under the excitation of white noise, the cross-correlation function between responses i and j can be can be written as

$$R_{ij}(t) = \sum_{r=1}^{n} \frac{\phi_{ir} A_{jr}}{m_r \omega_{dr}} \exp(-\varsigma_r \omega_{nr} t) \sin(\omega_{dr} t + \theta_r) \tag{3}$$

where A_{jr} is a constant, and θ_r is a phase angle. Eqs. (2) and (3) have similar mathematical expressions. Therefore, the cross-correlation function can be used instead of the impulse response function for modal parameter identification.

2.3 HILBERT TRANSFORM

According to the mode superposition method, the decoupled dynamics equations can be obtained by decoupling the multi-degree-of-freedom system:

$$\ddot{q}_j(t) + 2\zeta_j \omega_j \dot{q}_j(t) + \omega_j^2 q_j(t) = \Phi_j(t) F(t) / m_j \tag{4}$$

where ζ_j, ω_j, and m_j are the jth damping ratio, jth modal frequency and jth modal mass, respectively, and Φ_j is the jth modal matrix. Suppose that the system is excited by a pulse at point k: $f_k(t) = F_0(t)\delta(t)$. For arbitrary $j \neq k$ and $f_j(t) = 0$, the jth response in modal coordinates can be obtained as

$$\ddot{q}_j(t) = \frac{F_0 \phi_{kj} \omega_j}{m_j \sqrt{1-\zeta_j^2}} e^{-\zeta_j \omega_j t} \cos\left(\omega_{dj} t + \varphi_j + \frac{\pi}{2}\right) \tag{5}$$

where ϕ_{kj} is the kth component of Φ_j, ω_{dj} is the jth damped modal frequency, and $\varphi_j = \arctan\left[2\zeta_j\left(1-\zeta_j^2\right)/\left(1-2\zeta_j^2\right)\right]$ is the phase difference. Therefore, the pulse response of the structure at the point p can be written as

$$\ddot{x}_p(t) = \sum_{j=1}^{n} \phi_{pj} \ddot{q}_j(t) = \sum_{j=1}^{n} \ddot{x}_{pj}(t) \tag{6}$$

$$\ddot{x}_{pj}(t) = \phi_{pj} \ddot{q}_j(t) = B_{pj,k} e^{-\zeta_j \omega_j t} \cos\left(\omega_{dj} + \varphi_j + \frac{\pi}{2} + \varphi_{pj,k}\right) \tag{7}$$

$$B_{pj,k} = \frac{F_0 |\phi_{pj}| |\phi_{kj}| \omega_j}{m_j \sqrt{1-\zeta_j^2}} \tag{8}$$

where $\ddot{x}_{pj}(t)$ is the jth modal response at point p, and $\phi_{pj,k}$ is the phase difference between point k and point p. Vibration signals contain not only valuable features but also considerable noise; thus, the signal measured can be expressed as

$$\ddot{Z}(t) = \ddot{X}(t) + V(t) \tag{9}$$

where $V(t) = \left[v_1(t) \quad v_2(t) \quad \cdots \quad v_3(t)\right]^T$ is the noise signal. The acceleration signals at point p can be written as

$$\ddot{z}_p(t) = \ddot{x}_p(t) + v_p(t) = \sum_{j=1}^{n} \ddot{x}_{pj}(t) + v_p(t) \tag{10}$$

Considering a special case of light damping and high frequency, the Hilbert transform of $\ddot{x}_{pj}(t)$ can be defined as

$$\tilde{\ddot{x}}_{pj}(t) = B_{pj,k} e^{-\xi_j \omega_j t} \sin\left(\omega_{dj} t + \phi_j + \frac{\pi}{2} + \phi_{pj,k}\right) \tag{11}$$

Define the signal as

$$Y_{pj}(t) = \ddot{x}_{pj}(t) + i\tilde{\ddot{x}}_{pj}(t) = A_{pj}(t) e^{i\theta_{pj}(t)} \tag{12}$$

The instantaneous amplitude $A_{pj}(t)$ and instantaneous phase $\theta_{pj}(t)$ can be defined as

$$A_{pj}(t) = B_{pj,k} e^{-\xi_j \omega_j t} \tag{13}$$

$$\theta_{pj}(t) = \omega_{dj} t + \phi_j + \frac{\pi}{2} + \phi_{pj,k} \tag{14}$$

© 2017: The Royal Institution of Naval Architects

Using logarithmic and differential operators, the following expressions are obtained:

$$\ln A_{pj}(t) = -\xi_j \omega_j t + \ln B_{pj,k} \tag{15}$$

$$\omega_j(t) = d\theta_{pj}(t)/dt = \omega_{dj} \tag{16}$$

The damped modal frequency can be directly obtained by the slope of Eq. (16). Moreover, the modal frequency and damping ratio can be obtained based on the relationship between the modal frequency and damped modal frequency. With regard to the mode shapes, a reference point q should be selected first, and the amplitude ratio and phase difference between the remaining points and the reference point can be written as

$$|\phi_{pj}|/|\phi_{qj}| = \exp\left[A'_{pj}(t_0) - A'_{qj}(t_0)\right] \tag{17}$$

$$\varphi_{pj,q} = \theta'_{pj}(t_0) - \theta'_{qj}(t_0) \tag{18}$$

where A'_{pj} and $\theta'_{pj}(t_0)$ are the instantaneous amplitude and instantaneous phase after the fitting process, respectively.

3. EXPERIMENTAL STUDY ON A SHIP MODEL

Fig. 1: Ship model.

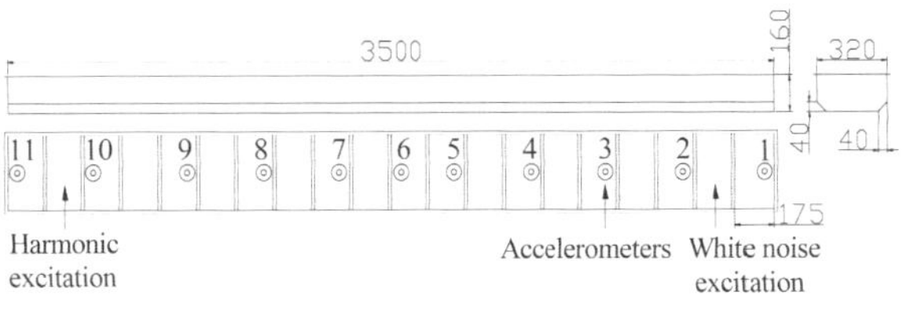

Fig. 2: Experimental arrangement.

As shown in Fig. 1, a ship model, which has similar structural features as a practical ship, was created to demonstrate the validity of the proposed method. Eleven accelerometers were installed along the bottom of ship model (Fig. 2). The accelerometers were located near the bulkhead to avoid the influence of local modes. The sampling frequency was set to 512 Hz. In the experiments, white noise and harmonic excitation (15 Hz) were set jointly to simulate the operating conditions of ships.

The time-domain and frequency-domain data of the acceleration response at accelerometer 4 are plotted in Figs. 3 and 4, respectively. As shown in Fig. 4, the first modal frequency is in the range of 30-35 Hz; the second modal frequency is in the range of 75-80 Hz; and the third modal frequency is in the range of 100-104 Hz. Moreover, the first three orders of IMFs can be obtained by bandpass filtering combined with the EMD method (Fig. 5).

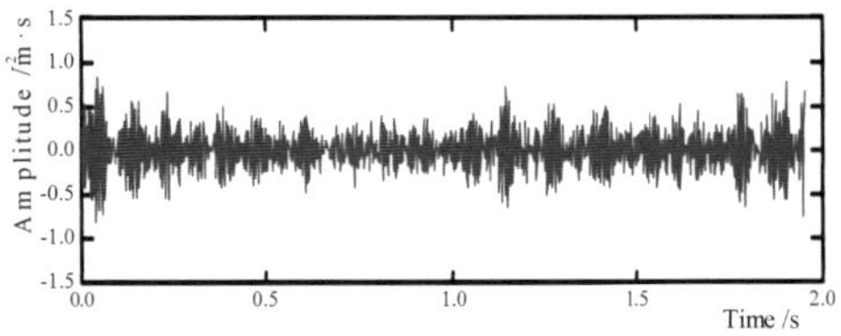

Fig. 3: Time-domain data of the acceleration response (accelerometer 4).

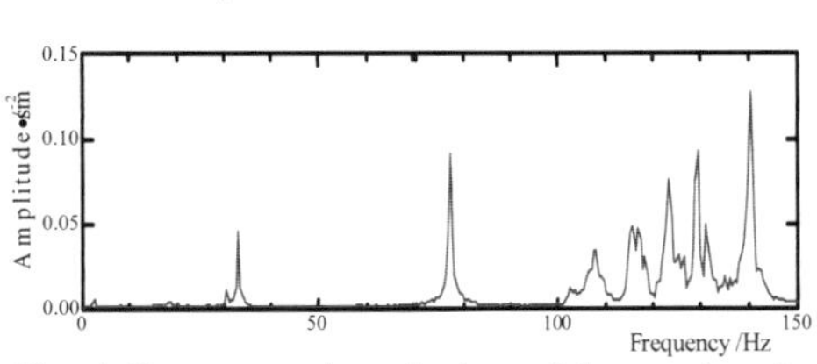

Fig. 4: Frequency-domain data of the acceleration response (accelerometer 4).

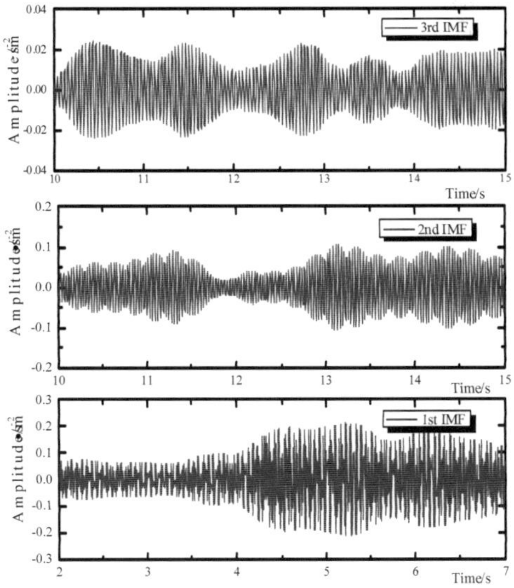

Fig. 5: IMFs.

The first three orders of the impulse response function can be obtained by the NExT based on the first three orders of IMFs, respectively (Fig. 6). The instantaneous

© 2017: The Royal Institution of Naval Architects

Table 1: Results of the modal test

Mode	Modal frequencies (Hz)			Damping ratio (%)		
	Hammering method	EMD-NExT	ERA	Hammering method	EMD-NExT	ERA
1st	32.98	33.02	33.04	0.45	0.46	0.41
2nd	77.37	77.08	77.08	0.32	0.30	0.26
3rd	103.78	102.42	102.34	0.20	0.18	0.45

amplitude and instantaneous phase can be obtained by the HT of the impulse response function. The logarithmic amplitude spectrum and instantaneous frequency spectrum can be obtained according to Eqs. (13) and (14). Moreover, $\xi_j \omega_j$ and ω_{dj} can be acquired by least squares linear fitting, and then the modal frequency and damping ratio can be obtained.

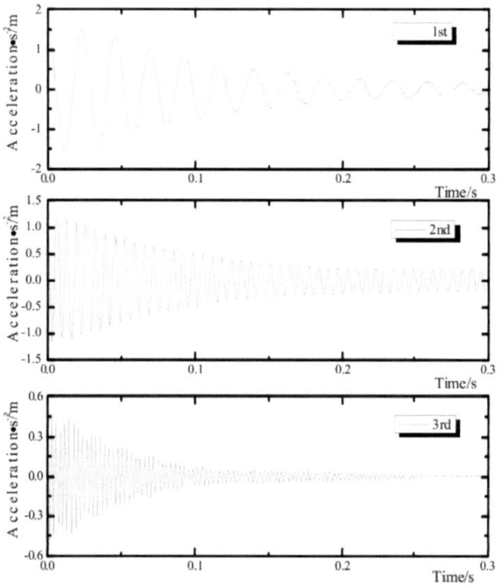

Fig. 6: First three orders of the impulse response function.

The results of the proposed method and ERA are shown in Table 1 [10]. The modal frequencies of the two methods are similar. However, the ERA is based on the state space equation and has the problem of system order determination, which can lead to false modes. The proposed method can effectively avoid noise pollution and false modes. Therefore, the proposed method significantly improves the accuracy of the damping ratio identification. The mode shapes are illustrated in Figs. 7-9.

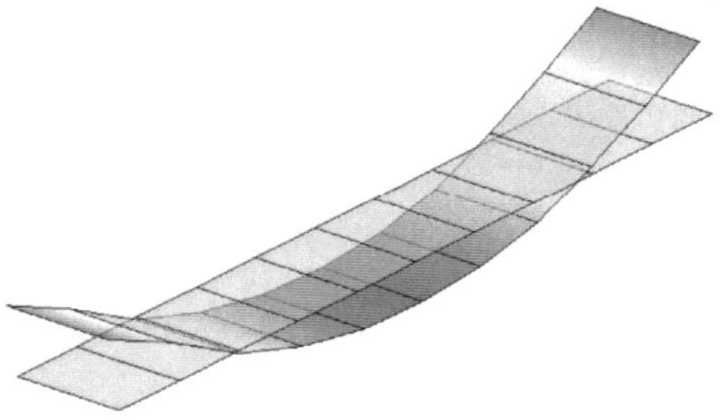

Fig. 7: First mode

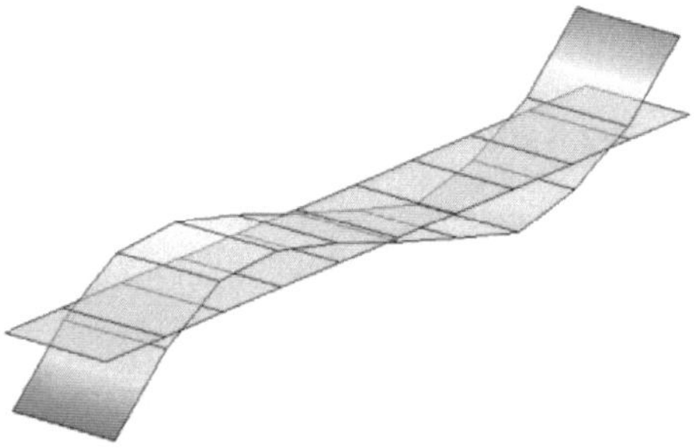

Fig. 8: Second mode

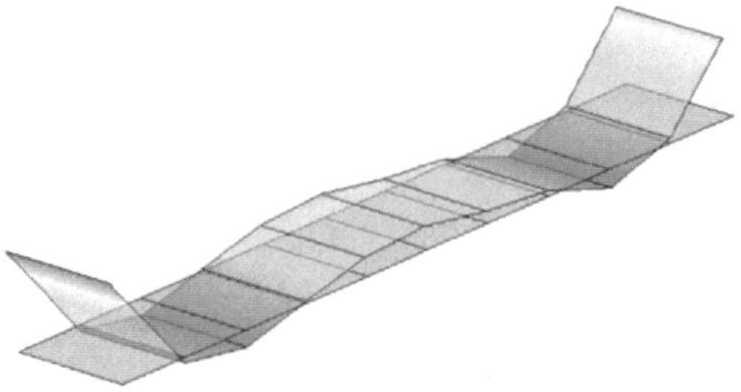

Fig. 9: Third mode

© 2017: The Royal Institution of Naval Architects

4. CONCLUSION

This paper proposes a new method for identifying the modal parameters of ship structures by combining the EMD, NExT and HT. The experimental results demonstrate that the proposed method can accurately identify the modal parameters. Moreover, the results obtained using the proposed method demonstrate that the proposed method is superior to the ERA method in many respects:

1. The proposed method can identify the modal parameters based solely on the output response of the structure, which is of great practical value.
2. The IMFs are obtained using the proposed method, and the false modes are avoided during the modal parameter identification.
3. The actual excitation state of ships was simulated in this paper, and the modal parameters were successfully obtained. Therefore, the proposed method is suitable for the modal parameter identification and health monitoring of ship Structures.

5. REFERENCES

1. Guido, De. Roeck., Bart, Peeters., Wei, Xin, Ren., 'Study on System Identification through Ambient Vibration Measurements', *18th IMAC*, 2000.

2. Caicedo, J. M., Dyke, S. J., Johnson, E. A., 'Natural Excitation Technique and Eigensystem Realization Algorithm for Phase I of the IASC-ASCE Benchmark Problem: Simulated Data', *Journal of Engineering Mechanics*, 2004.

3. B. Peeters., G. De. Roeck., T. Pollet., L. Schueremans., 'Stochastic Subspace Techniques Applied to Parameter Identification of Civil Engineering Structures', *Proceeding of the New Advances in Modal Synthesis of large structures: Non-Linear, Damped and Non-deterministic case, Lycn*, 1995.

4. R. Brincker., L. Zhang., P. Andersen., 'Modal Identification from Ambient Responses using Frequency Domain Decomposition', *18th IMAC*, 2000.

5. Yan, B. F., Miyamotoa, A., Brühwilerb, E., 'Wavelet transform-based modal parameter identification considering uncertainty', *Journal Sound and Vibration*, 2006.

6. Min, Z. H., Sun, L. M., 'Wavelet-based structural modal parameter identification', *Structural Control & Health Monitoring*, 2013.

7. Huang, NE., Zheng, S., Steven, RL., 'The empirical mode decomposition and the Hilbert spectrum for nonlinear random-stationary time series analysis', Proceedings Mathematical Physical & Engineering Sciences, 1998.

8. C. Geng., F. Wang., J. Zhang., Z. Jin., 'Modal parameters identification of power transformer winding based on improved Empirical Mode Decomposition method', *Electric Power Systems Research*, 2014.

9. Yan, Zhang., Ji-jian, Lian., Fang, Liu., 'Modal parameter identification for a roof overflow powerhouse under ambient excitation', *Water Science and Engineering*, 2016.

10. Junchen, Xu., Ming, Hong., Hongyu, Cui., 'The Contrast Experimental Study on Operational Modal Analysis of Ship Structural Model', *Applied Mechanics and Materials*, 2012.

6. AUTHORS BIOGRAPHY

Hongyu Cui received a Ph.D from Dalian University of Technology, Dalian, China, in 2009. Now he is a lecturer at the School of Naval Architecture Engineering, Dalian University of Technology. His research interests include fault diagnosis, pattern recognition and active control of structure vibration.

Xin Xu received his B.Sc. degree from Jiangsu University of Science and Technology, Zhenjiang, China, in 2015. Now he is a Master Candidate in the School of Naval Architecture Engineering, Dalian University of Technology. His research interests include signal processing, fault diagnosis, and machine learning.

Ming Hong received a Ph.D from Dalian University of Technology, Dalian, China, in 2003. Now he is a professor at the School of Naval Architecture Engineering, Dalian University of Technology. His research interests include ship vibration analysis, control, and experimentation as well as acoustic transmission in multi-media.

©2017: The Royal Institution of Naval Architects

© 2017: The Royal Institution of Naval Architects

METHODOLOGY FOR EFFICIENT APPLICATION OF 3D SHIP MODELLING SOFTWARE

D Kolich, University of Rijeka, Croatia
A Becirevic University of Rijeka, Croatia
R L Storch, University of Washington, USA

SUMMARY

Most of today's Naval Architecture and Ocean Engineering University Departments prepare students with much theoretical knowledge. However, students will need to develop more practical skills in 3D modelling than what is presently provided. Since many shipyards are subcontracting the design documentation out of yard to exclusively design companies, more and more students upon graduation are finding jobs in small and medium enterprise companies (SMEs) that prepare detailed production documentation using advanced 3D modelling software. Thus, it has become normal for many design engineers to work locally without moving to a distant country since the internet has enabled efficient communication between shipyards and design houses. However, the need for engineers with practical 3D modelling skills is growing, and Universities need to meet this ever growing demand. Therefore, the development of a methodology whereupon students can receive the practical skills in 3D modelling during University studies through cooperation with SMEs would be useful. In this paper, an innovative approach to bridging the 3D modelling gap is explained. A case study at an actual University Naval Architecture and Ocean Engineering Department in cooperation with an SME describes how students are receiving hands on experience. Likewise, a practical demonstration in designing a foundations for ship equipment is described in a succinct and clear manner. The theoretical knowledge of understanding how to read classification drawings is explained as is the practical 3D modelling application. Nupas Cadmatic 3D modelling software is used in the case study.

NOMENCLATURE

SME	Small and medium enterprise
3D	Three dimensional
SFI	Ship classification standards

1. INTRODUCTION

The importance of Naval Architecture/Marine Engineering University programs cannot be understated. Whereas it is possible for students from other Engineering programs to eventually gain experience necessary to work on ship design and in shipyards, Naval Architecture/Marine Engineering programs still produce most of the naval architects and marine engineers in the worldwide industry today.

The case is that even though the shipbuilding industry in Europe has declined during the past few decades, the Naval Architecture/Marine Engineering programs continue to attract and produce engineers employable by both local and international naval and maritime companies. The reason why is due to the advent and increase in computer 3D modelling of ship structures, and equipment. Up to 50 naval architecture/marine engineering small and medium enterprises (SMEs) exist in the city of Rijeka as a result of meeting the need for international maritime customers. These include shipyards such as Meyer Werft, Neptun Werft and Meyer Turku, as well as many others.

The example given in this paper includes explaining how students can be influenced to learn what the maritime customer wants, whether that be a shipyard or a large international vessel design office. The need for this proactive approach arose as a result of the need to answer the needs of the local industry.

Whereas the University program teaches students scientific and engineering theory during the extensive course load, there has been a lack of formal and state of the art ship 3D modelling training. Once students graduate from either 3 year or 5 year programs, they realize that companies are using specific software with which they had little to no experience. The reason why is because of the budget limitations to purchasing the plethora of software available on the market.

At the same time these SMEs approach the students as well as 3D modelling companies such as Nupas-Cadmatic [4] and FORAN. Whereas, it is impossible to spend extensive time in training students to use all of the 3D ship modelling software on the market, it is possible to immerse students to use software as it is made available due to partnership agreements between SMEs, the University and software companies.

In this paper an example of achievements in adapting avante garde 3D modelling through cooperation with industry in a novel way is the new paradigm in Naval Architecture/Marine Engineering education.

2. SHIPYARD DOCUMENTATION

The shipyard documentation and drawings are broken down into three basic groups. The Evans spiral is one typical illustration of this. [1], [2]. Likewise there are offshoots of this spiral as the one in Figure 1. The ship has more parts and requires the use of all engineering fields.

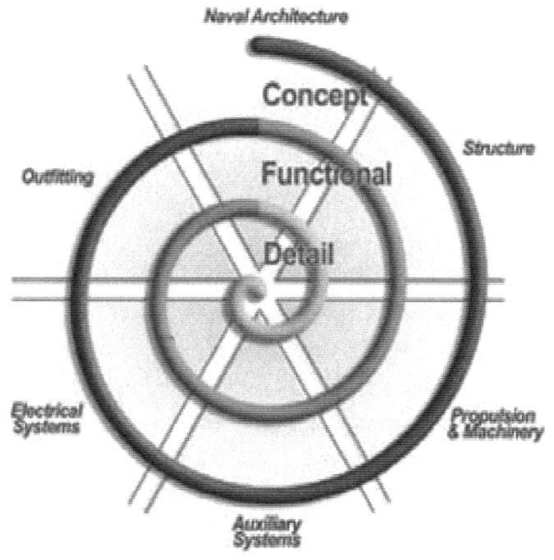

Figure 1: Ship design spiral [2]

The entire ship design can be broken up into three major milestones.

1) Contract design drawings
2) Classification Society design drawings
3) Production design drawings

The contract design documentation includes the following:

1) General Arrangement plan
2) Midship section drawing,
3) Capacity plan,
4) Preliminary trim and stability calculations,
5) Damaged stability calculations,
6) Heat and energy balance
7) Damaged stability

The vessel displacement, speed and general dimensions are defined prior to contract drawing as is the midship section. However, the plethora of other details which need to be approved by a classification society leads to the development of Classification society drawings which follows the Norwegian Ship Research Institute, formerly known as Skipsteknisk forskningsinstitutt or the popular abbreviation SFI system of classification [3]. All drawings are labelled in a uniform way. This means that there are 10 main group breakdowns along with the sub breakdowns. The ten main groups are:

1) Main group
2) Hull
3) Cargo handling equipment
4) Ships equipment
5) Equipment for crew and passengers
6) Main engine and related parts
7) Systems for parts of the main engine
8) Ship systems

Finally, the third and final set of documentation are the drawings used by shipyard production personnel. It is important to understand that the ship design process requires much time and resources, as well as engineers of all trades, not just naval architects and marine engineers, but also mechanical engineers, electrical/electronic engineers and even civil engineers, not to mention computer engineers.

All three design stages are necessary for a ship to be built. However, the production drawings define exactly how the interim products are to be made and outfitted. Likewise production drawings need to comply with the technological capabilities of each individual shipyard. For instance, whereas the contract design and classification society design for a specific ship design can be the same for different shipyards, the production design will always require some degree of changes to the drawings. This is due to the fact that the interim products differ from shipyard to shipyard and there are no two shipyards that are exactly the same [5]. This is due to differences in the shipyard layout and the production facilities. For instance, while the shipyards 3.Maj in Rijeka and Uljanik in Pula are managed under the Uljanik group, the production drawings are different in order to maximize the production facilities of each shipyard. One good example are the erection blocks. Since, 3.Maj shipyard in Rijeka has a larger gantry crane over the slipway (300 tons capacity), the erection blocks are much bigger than that in Pula. Therefore, the drawings need to reflect this.

3. STUDENT INDEPENDENT PROJECTS USING LATEST 3D MODELLING SOFTWARE

Students are required to undergo a few independent projects during the duration of their education. This is one opportunity for proactive students to opt to do a project related to 3D modelling in shipbuilding using some of the most modern ship design software on the market. The case that will be reviewed in this case study is an example of work performed by a student at a local Naval architecture/Marine engineering SME. The rest of this section are instructions on how to use Nupas Cadmatic software for defining equipment foundation modelling for ships.

Whereas, contract documentation and drawings do not require the development of a 3D model, the production drawings require the creation of very detailed and accurate 3D models. With the advent and improvement of 3D models, shipyard production is receiving better documentation with isometric as well as two-dimensional views in the various assembly phases. This enables more accurate planning of shipyard jobs. Likewise, the shipyard workers are more readily able to read the drawings without having to learn the archaic symbols that are used on traditional two-dimensional drawings.

3.1 FOUNDATIONS OF SHIP EQUIPMENT

The fifth year student in the Masters program had done an internship at a local SME. The specific design involved designing and modelling foundations for ship equipment. The prerequisites for designing any foundation are that the ship structure must be defined and modelled as well as the actual equipment. Only then is it possible to determine the exact characteristics of the foundation. Likewise there are certain rules and constraints which the designer must abide by. For instance, it is forbidden to weld anything to the edges of structural stiffeners, girders, and openings. The foundations are always manually welded to the hull structure where permitted according to the rules. The constraints that designers must pay attention to are:

1) The dimensions and mass of the equipment
2) The exact position
3) Whether the equipment will touch other equipment or pipes
4) Choosing the proper dimensions of stiffeners or steel plates
5) How to make the connection to the hull structure

In order to make the above listed design decisions, it is necessary to constantly check drawings and specifications. For instance drawings of the equipment, and the preliminary drawings of equipment locations.

3.2 CASE STUDY OF FOUNDATION DESIGN MODELLING

The foundation that the student needed to design was for a starting air tank in the engine room. The starting air tank is used to initially get the pistons of the main diesel engine moving in order to then achieve the necessary compression which is necessary to create internal combustion which then moves the pistons (See Figure 1)

Figure 1: Starting air tank

First, the designer checks the dimensions and shape of the starting air tank, and then the mass. The greater the mass of the equipment, the greater the required strength of the stiffeners as well. Sometimes the mass of the equipment is not known, and it is necessary to estimate it based on the dimensions and the specific gravity of steel. Since the starting air tank is made of steel with considerable dimensions, it can be assumed that the mass is considerable as well (over 100 kg). Therefore the stiffeners will also have to be of greater dimension than usual. Likewise, it is necessary to choose stiffeners which are readily available at the shipyard.

The bulb stiffeners are inserted by using the "Insert" option of the Structural menu in Nupas-Cadmatic (See Figure 2).

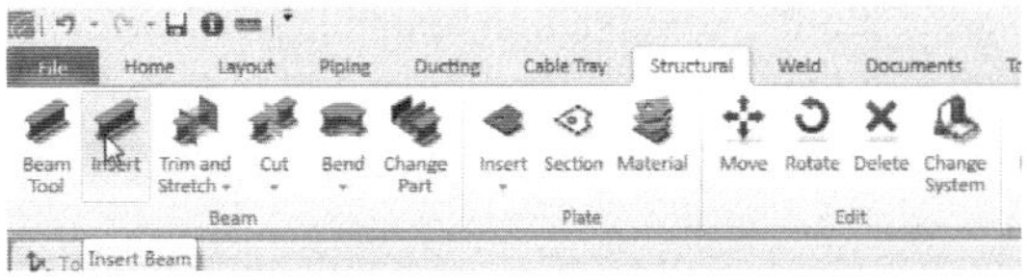

Figure 2: Inserting stiffeners from the Structural menu

Once the profiles are placed in position, then it is necessary to define the exact type from a menu of choices as in Figure 3 below.

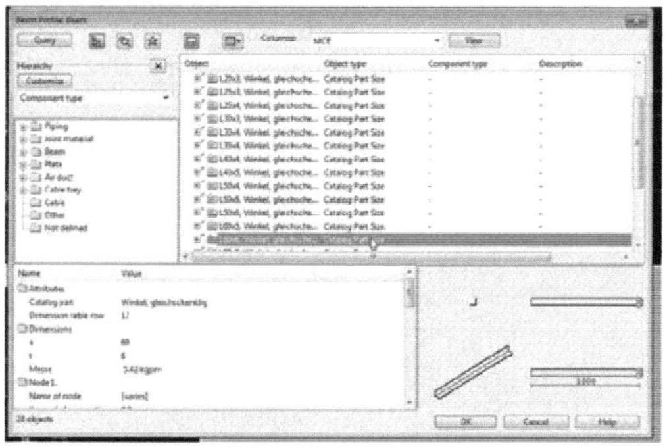

Figure 3: Menu of stiffener choices

After the profile type is chosen, then the stiffener is positioned by choosing the start and end positions as in Figure 4 below.

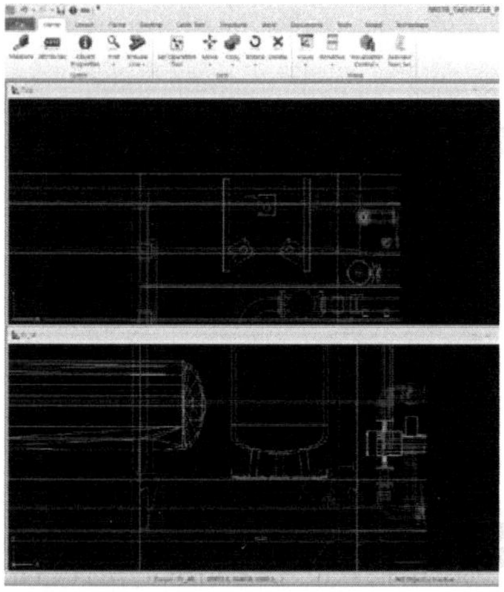

Figure 4: Position of the bulb stiffeners in three views (plan, side, and cross-section views)

When the coordinates are known for the position of the profiles, then they are manually entered by pressing the letter "C" as in Figure 5 below. A different method is by using the snap command ("E") key. The condition is to find the exact same point in at least two views.

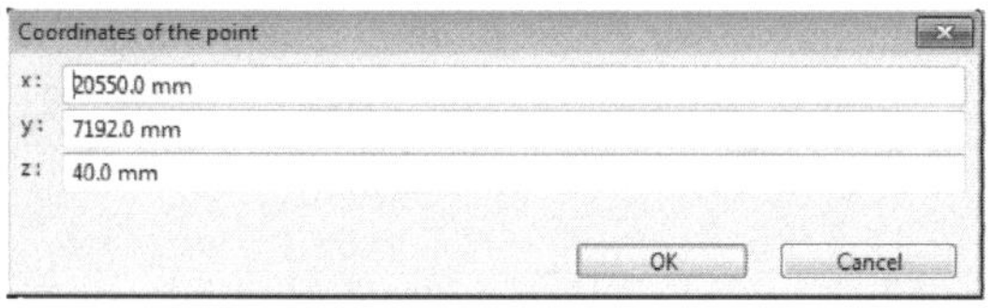

Figure 5: Coordinates

Once the final point is chosen, the direction of the profile stiffener is oriented by pressing the "Enter" key. A new profile stiffener is shown in the model (Figure 6).

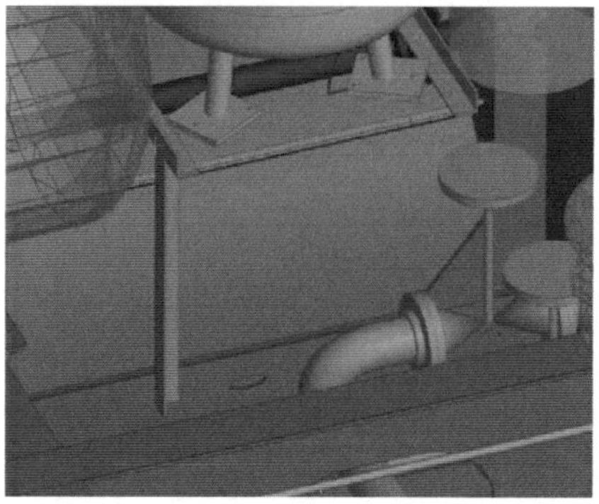

Figure 6: Second placed stiffeners

The next step is to add a stiffener to the opposite side, which is necessary to act as a foundation. Since the profile stiffeners have identical dimensions, it is quicker to just copy and use the mirror command as can be seen in Figure 7 below.

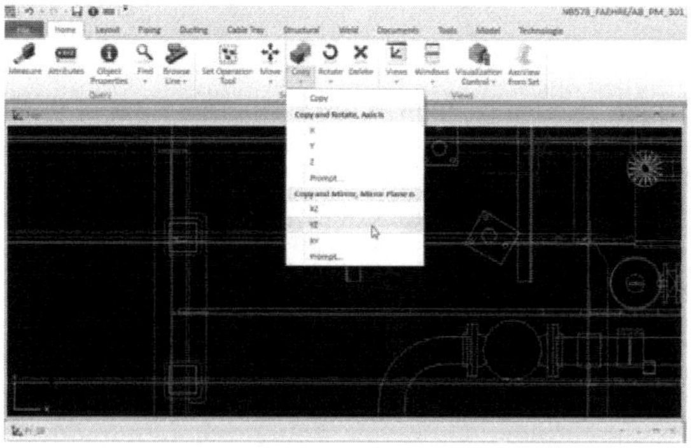

Figure 7: Profile stiffener copy and mirror procedure

While using the command, it is necessary to choose the proper surface the element will be mirrored upon. Using the "O" button from the keyboard to choose the profile stiffener. Afterwards the "Enter" key is pressed. Then it is necessary to choose the point where the mirroring will take place. It is possible to press the snap key "E" or the midpoint by pressing the "G" key. The midpoint is used and the situation is as in Figure 8. After the midpoint appears, the "Enter" key is pressed and the cursor positions itself in the middle as in Figure 9. Then, by clicking the mouse button the actual mirroring of the profile stiffener is executed and completed (Figure 10).

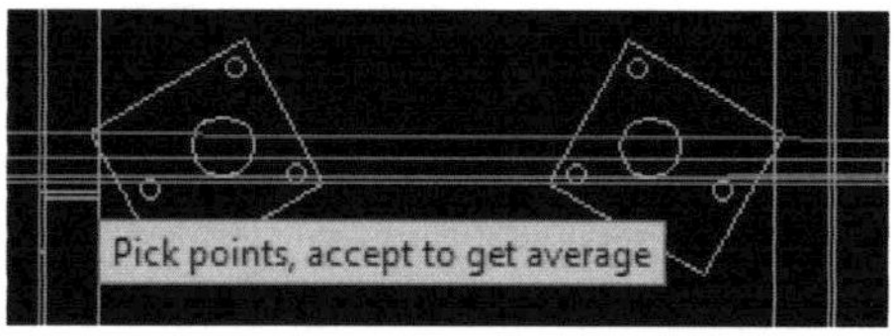

Figure 8: Pressing the "G" key yields a choice of the midpoint

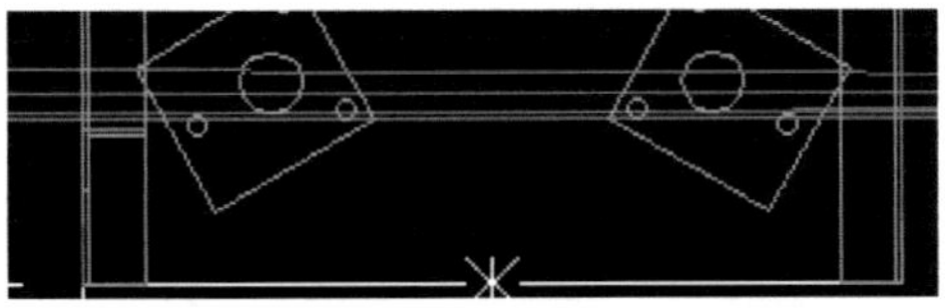

Figure 9: Midpoint between two profile stiffeners

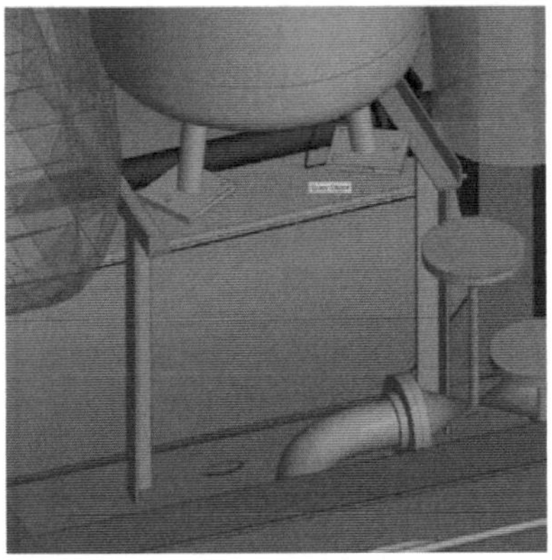

Figure 10: Result of the mirroring

A small space is left between the profile stiffeners of the foundation and the profile stiffeners attached to the hull structure. The reason is to allow space for a steel plate known as a doubler which allow a connection between profile stiffeners and also does not directly touch the hull structure.

These steel plates are characteristically thicker than the profile stiffeners by 2mm. Therefore, it is necessary to use an 8mm thick doubler. The steel plates are inserted in a similar manner as profile stiffeners (See Figure 11) through choosing the type of steel plate and thickness (Figure 12).

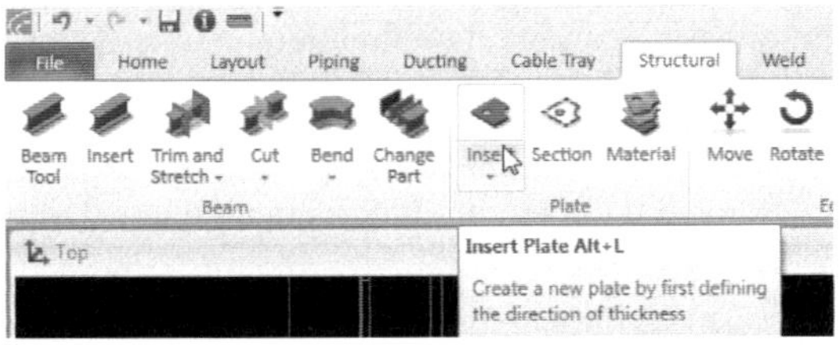

Figure 11: Insert command for placing steel plates

Inserting steel plates also requires the choosing of a reference point which will be the origin of the steel plate. Only one point is chosen for modelling a steel plate. Then the menu for the shape of the steel plate is chosen as in Figure 12 below.

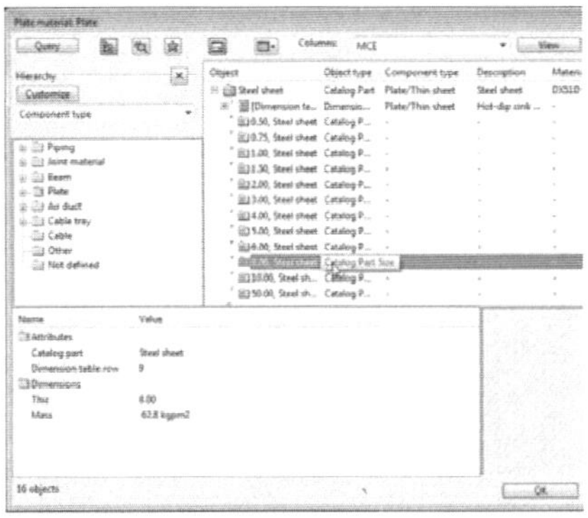

Figure 12: Choosing steel plate type and thickness

There are various shape types as in Figure 13. However, for ships equipment it is normal to use a rounded rectangle. After clicking rounded rectangle then another menu opens up, where it is necessary to define the size of the steel plate dimensions (Figure 14)

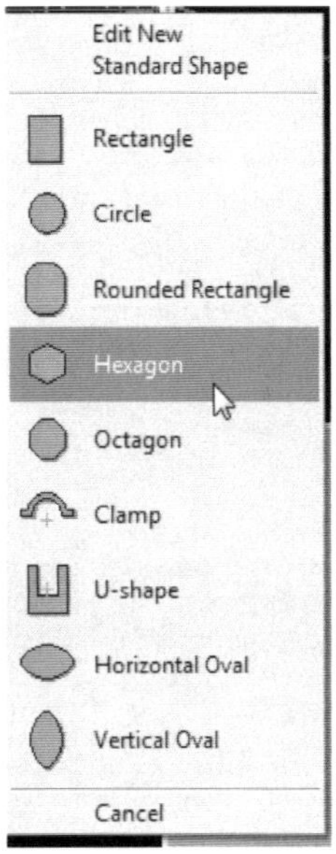

Figure 13: Choosing steel plate shape

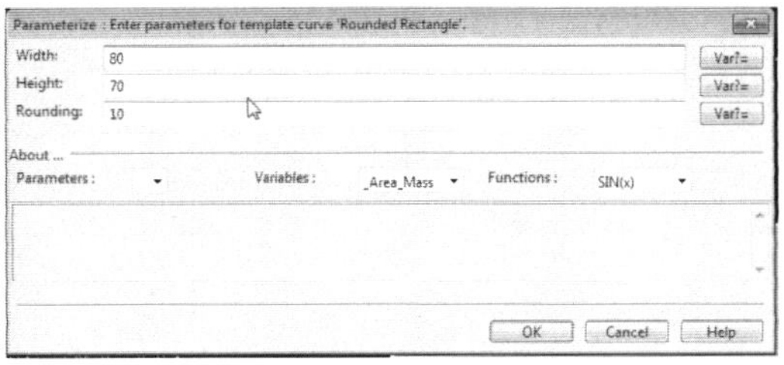

Figure 14: Menu with steel plate parameters

Upon clicking "OK" shown in Figure 14 above, the steel plate gets placed in position. The steel plate must be dimensioned so that it is large enough to satisfy a sound connection between the hull structural stiffener and the profile stiffener of the tank (See Figure 15).

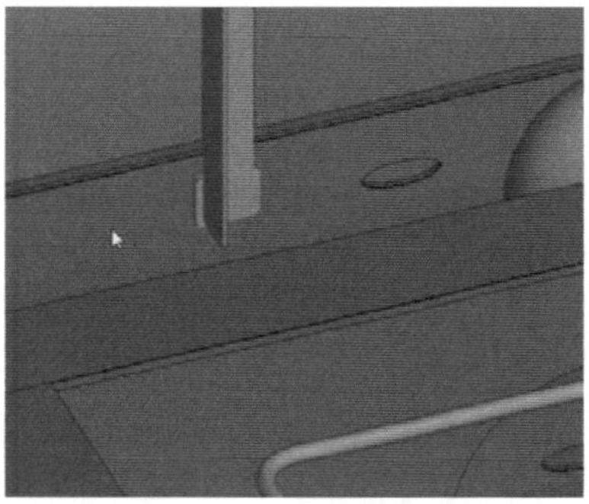

Figure 15: Doubler placed between hull stiffener and tank support stiffener

It is necessary to repeat this for all four vertical tank support profile stiffeners. Since all four stiffener profiles are identical, it is possible to use the "Copy" command (Figure 16). The doubler to be copied is chosen by pressing the "O" key (Figure 17). Afterwards, an arrow appears and it is necessary to choose the direction of the copied doubler (Figure 18), upon which a new menu appears (Figure 19) where the number of elements to be copied are chosen as well as the distance. With a click of "OK", the new doubler is successfully copied to the prescribed position (See Figure 20).

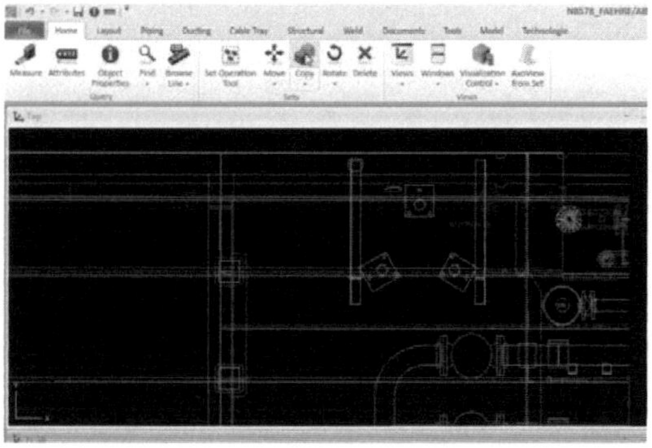

Figure 16: "Copy" command with the number of copies and the distance

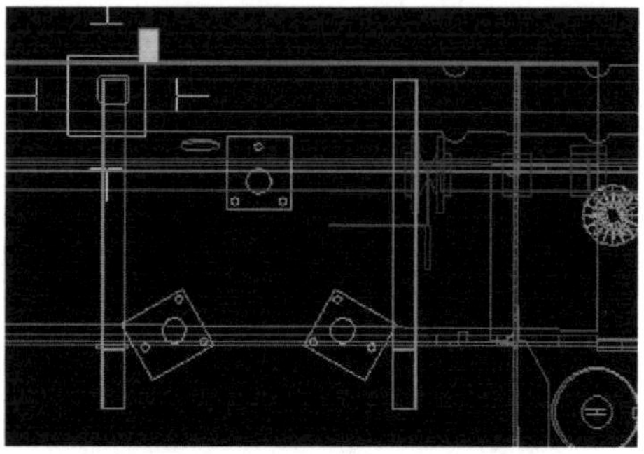

Figure 17: Pressing the "O" key copies the doubler

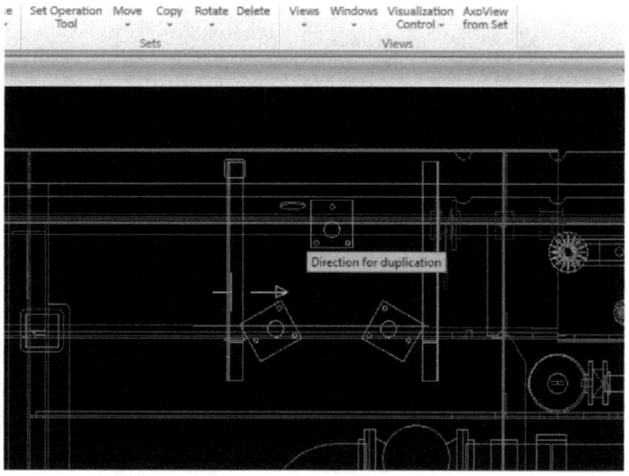

Figure 18: Direction for duplication

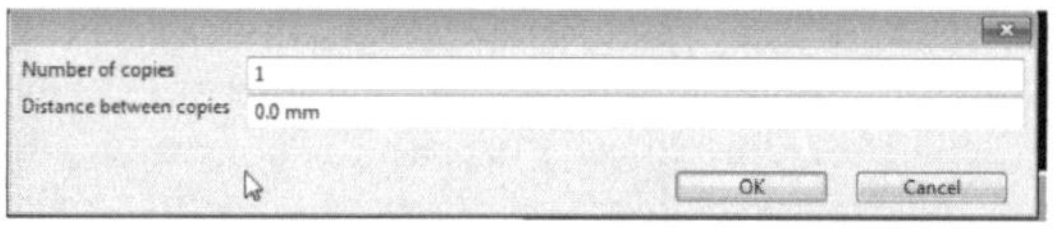

Figure 19: "Copy" command with the number of copies and the distance

Figure 20: Doubler copied to the right side

Next it is necessary to choose the tank from the database by selecting "File" and "Check Out" as in Figure 20 below. After the correct tank from the database is chosen, it is possible to use the "Move" in order to position the tank to the correct position.

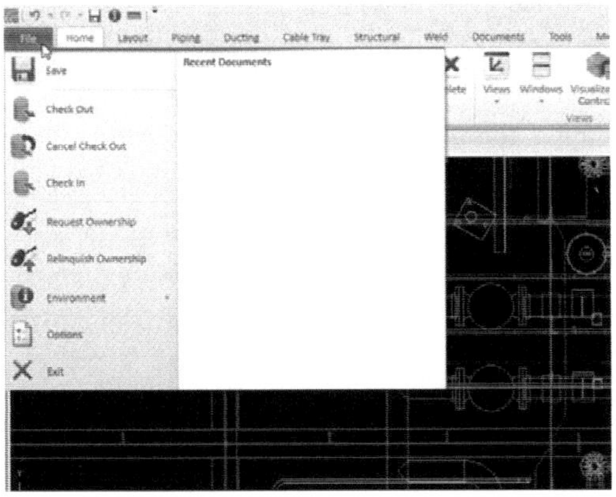

Figure 21: "File" and then "Check Out" commands

The next command that is necessary to use is the "Move" command in order to move the tank to the optimal position in order not to change the positions of any pipes which need to be connected to the tank (Figure 22).

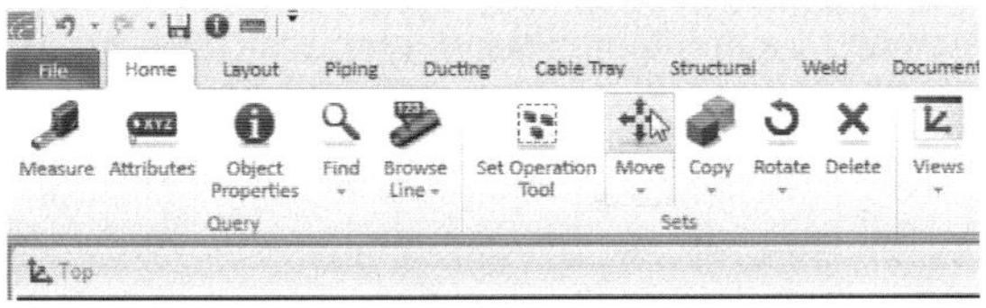

Figure 22: "Move" command

It is next necessary to attach horizontal profile stiffeners to the vertical profile stiffeners (Figure 23), upon which a horizontal steel plate is added on top by using the insert plate command (Figure 24). This is necessary in order to properly assemble the cylindrical tank to the foundation. The strength of the resulting foundation structure will satisfy classification society requirements.

Figure 23: Placement of horizontal profile stiffeners

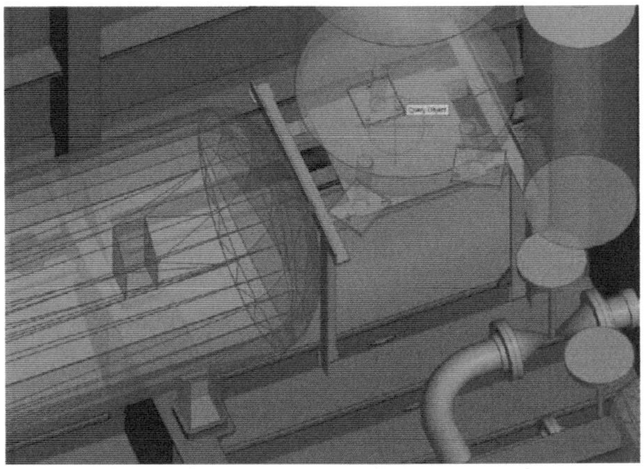

Figure 24: Placement of horizontal steel plate on horizontal profile stiffeners

Since the steel plate is not properly dimensioned, it is necessary to press the "Section" tab from the top menu and then update its exact size (Figure 25).

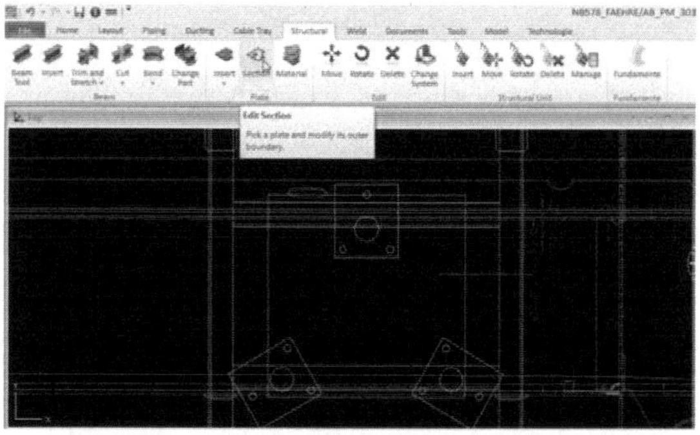

Figure 25: Clicking on Section to make dimension adjustments

Using the right click of the mouse makes it possible to choose from an entire list of commands (Figure 26).

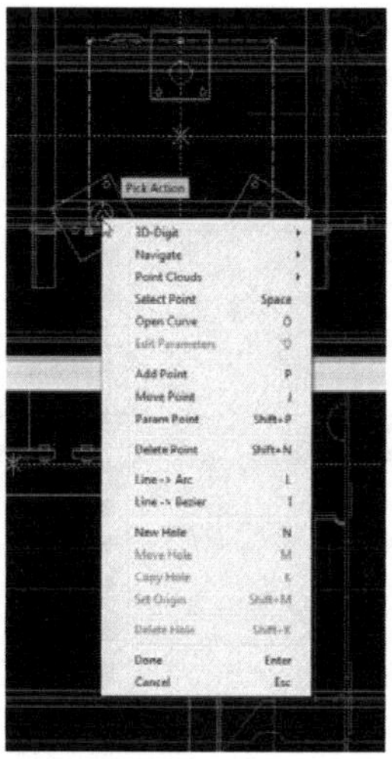

Figure 26: Menu of Section options

It is necessary to make sure there is enough space between the profile stiffeners and the steel plate in order to allow for accessible welding. Finally the completed foundation structure appears as in Figure 27 below.

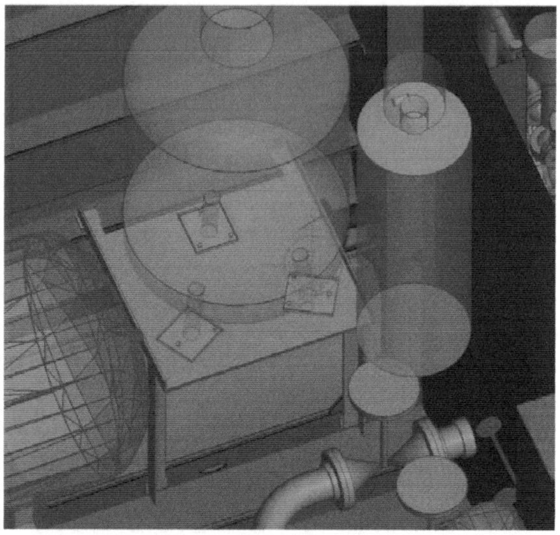

Figure 27: Completed foundation

4. CONCLUSIONS

In conclusion, it is very useful for students to be exposed to the latest ship design tools used by industry. Since University Naval Architecture/Marine Engineering programs have a core structure which cannot cover all the detailed software solutions due to constraints in time and resources, it is very practical to allow for student exchanges at the design houses where the advanced year students are able to get hands on experience while learning the avante garde tools of the trade. In this way, as soon as the students graduate, they have the necessary theoretical as well as the corresponding complementary practical skills to immediately get to working on real shipbuilding design projects.

5. REFERENCES

1. STORCH, R.L., HAMMON, C.P., BUNCH, H.M. and MOORE, R.C., 'Ship Production', SNAME, New Jersey, 1995.

2. http://www.neely-chaulk.com/narciki/Design_spiral32-36.

3. SFI Group System Book, Ship Research Institute of Norway, 3. Maj Shipyard, 1974.

4. Nupas Cadmatic Strucutral Design User Training Tutorial, Numeriek Centrum Groningen B.V., 2015.

5. Kolich, D., Storch R.L., Fafandjel N, 'Lean Manufacturing in Shipbuilding with Monte Carlo Simulation', *Proceedings of the International Conference on Computer Applications in Shipbuilding*, Trieste, Italy, pp. 159 -167, 2011.

6. AUTHORS BIOGRAPHY

Damir Kolich is an Assistant Professor (PhD) in the Naval Architecture and Ocean Engineering Department at the University of Rijeka, Croatia. He is a member of RINA and a Chartered Engineer.

Armin Becirevic is a Masters degree student in the Naval Architecture and Ocean Engineering Department at the University of Rijeka, Croatia.

Richard Lee Storch is a Professor Emeritus (PhD) and former Chair of the Industrial and Systems Engineering Department at the University of Washington, Seattle, WA, USA.